What is Materials Science and Engineering?

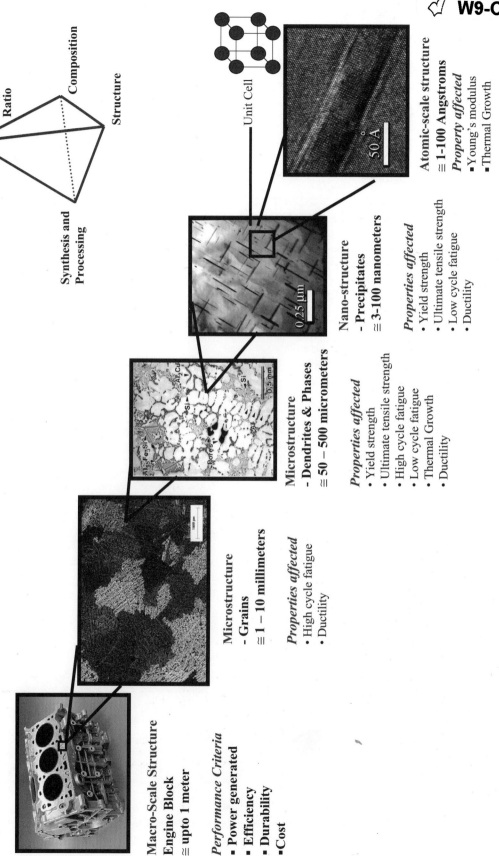

Performance or Properties to Cost Ratio

Composition

Structure

Synthesis and Processing

Unit Cell

50 Å

Atomic-scale structure
≅ 1-100 Angstroms
Property affected
- Young's modulus
- Thermal Growth

0.25 μm

Nano-structure
- Precipitates
≅ 3-100 nanometers

Properties affected
- Yield strength
- Ultimate tensile strength
- Low cycle fatigue
- Ductility

Al₅Fe₃Si — Si — Al₂Cu — Si — pores — 0.5 mm

Microstructure
- Dendrites & Phases
≅ 50 – 500 micrometers

Properties affected
- Yield strength
- Ultimate tensile strength
- High cycle fatigue
- Low cycle fatigue
- Thermal Growth
- Ductility

1000 μm

Microstructure
- Grains
≅ 1 – 10 millimeters

Properties affected
- High cycle fatigue
- Ductility

Macro-Scale Structure
Engine Block
≅ upto 1 meter

Performance Criteria
- Power generated
- Efficiency
- Durability
- Cost

A real-world example of important microstructural features at different length-scales, resulting from the sophisticated synthesis and processing used, and the properties they influence. The atomic, nano, micro, and macro-scale structures of cast aluminum alloys (for engine blocks) in relation to the properties affected and performance are shown. The materials science and engineering (MSE) tetrahedron that represents this approach is shown in the upper right corner.

(Illustrations Courtesy of John Allison and William Donlon, Ford Motor Company).

The
Science
and Engineering
of Materials

FOURTH EDITION

The Science and Engineering of Materials

FOURTH EDITION

Donald R. Askeland
University of Missouri—Rolla, Emeritus

Pradeep P. Phulé
University of Pittsburgh

THOMSON

BROOKS/COLE

Australia • Canada • Mexico • Singapore • Spain
United Kingdom • United States

THOMSON

BROOKS/COLE

Publisher: *Bill Stenquist*
Development Editor: *Rose Kernan*
Editorial Coordinator: *Valerie Boyiajian*
Technology Project Manager: *Burke Taft*
Executive Marketing Manager: *Tom Ziolkowski*
Project Manager, Editorial Production: *Mary Vezilich*
Print/Media Buyer: *Karen Hunt*
Permissions Editor: *Sue Ewing*
Production Service: *RPK Editorial Services*

Text Designer: *Terri Wright*
Photo Researcher: *RPK Editorial Services*
Illustrator: *J.B. Woolsey Associates*
Cover Designer: *Terri Wright*
Cover Image: *PhotoDisc/Nick Rowe*
Cover Printer: *Lehigh Press*
Compositor: *Asco Typesetters*
Printer: *Quebecor World–Taunton*

Printed in the United States of America

1 2 3 4 5 6 7 06 05 04 03 02

> For more information about our products, contact us at:
> **Thomson Learning Academic Resource Center**
> **1-800-423-0563**
> For permission to use material from this text, contact us
> by:
> **Phone:** 1-800-730-2214   **Fax:** 1-800-730-2215
> **Web:** http://www.thomsonrights.com

Library of Congress Control Number: 2002107447

ISBN 0-534-95373-5

Brooks/Cole–Thomson Learning
511 Forest Lodge Road
Pacific Grove, CA 93950
USA

Asia
Thomson Learning
5 Shenton Way #01-01
UIC Building
Singapore 068808

Australia/New Zealand
Thomson Learning
102 Dodds Street
Southbank, Victoria
Australia 3006

Canada
Nelson
1120 Birchmount Road
Toronto, Ontario M1K 5G4
Canada

Europe/Middle East/Africa
Thomson Learning
High Holborn House
50/51 Bedford Row
London WC1R 4LR
United Kingdom

Latin America
Thomson Learning
Seneca, 53
Colonia Polanco
11560 Mexico D.F.
Mexico

Spain
Paraninfo
Calle/Magallanes, 25
28015 Madrid, Spain

To Mary Sue and Tyler
— *Donald R. Askeland*

To Janakibai Fulay, and
Prabhakar and Pratibha Fulay
— *Pradeep P. Phulé*

Brief Contents

Contents

Chapter 4   Imperfections in the Atomic and Ionic Arrangements   129

Chapter 5   Atomic and Ionic Movements in Materials   173

PART 2 Controlling the Microstructure and Mechanical Properties of Materials 229

Chapter 6 Mechanical Properties and Behavior 231

Chapter 7 Strain Hardening and Annealing 315

Chapter 8 Principles of Solidification 357

Chapter 9 Solid Solutions and Phase Equilibrium 407

Chapter 10   Dispersion Strengthening and Eutectic Phase Diagram   451

Chapter 11   Dispersion Strengthening by Phase Transformations and Heat Treatment   497

PART 3   Engineering Materials   541

Chapter 12   Ferrous Alloys   543

Chapter 13   Nonferrous Alloys   591

Chapter 14   Ceramic Materials   625

Chapter 15   Polymers   669

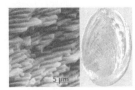

Chapter 16   Composites: Teamwork and Synergy in Materials   721

Chapter 17   Construction Materials   765

PART 4   Physical Properties of Engineering Materials   787

Chapter 18   Electronic Materials   789

Chapter 19   Magnetic Materials   849

PART 5   Protection Against Deterioration and Failure of Materials   941

Preface

The Fourth Edition of *The Science and Engineering of Materials* has two primary objectives. The first continues the theme of earlier editions by providing a clear presentation of the relationship between the structure of a material, its processing, and its properties (see Chapter 1). The principal goals of a materials scientist and engineer are to make existing materials better and to invent or discover phenomena that will allow development of new materials, devices, and applications. All engineers, in all specialties, can benefit from a thorough knowledge of materials science. Materials science underlies many technological advances related to energy, information, and the environment. Having a basic understanding of materials and their application not only makes for a better engineer, but also aids in the design process **for any device or application under consideration**. It is important for today's engineers to understand the constraints of material behavior in order to select the best material for a given application. We believe that engineers who have a thorough understanding of the basic concepts of atomic and molecular structure, and who know which material properties lend themselves to the most efficient and cost-effective processes, will make design decisions to develop the best products and applications.

The second objective of this text is to motivate engineering students to want to study and understand materials science using exciting and current developments in the field. Considerable effort has been spent to bring interesting and modern applications into discussions about basic theoretical presentations in order to tie the theory to actual applications. The diversity of applications and the unique uses of materials are presented in the Example sections of each chapter in order to help illustrate why a good engineer needs to thoroughly understand and know how to apply the principles of materials science and engineering. Considerable advances have occurred in the areas of nano-technology information technology, energy technology and biomedical engineering. In this edition you will see many examples that directly relate to these areas of technology.

Audience and Prerequisites

This text is intended for engineering students who have completed courses in general chemistry, physics, engineering and calculus. Most students will take this course when they are sophomores or juniors and the presentation has been carefully geared towards this audience. To facilitate understanding, a brief description of key ideas from earlier prerequisite courses is provided to help students refresh their memories. The text does not presume that students have had any of the prerequisite engineering courses in statics, dynamics, or mechanics of materials.

We feel that while reading and using this book, students will really find materials science and engineering very interesting, and they will clearly see the relevance of what they are learning. We have presented many examples of modern applications of materials science and engineering that impact students' lives. Our feeling is that, if students recognize that many of today's technological marvels depend on the availability of engineering materials, they will be more motivated and remain interested in learning about materials science and materials engineering.

Changes to the Fourth Edition

The Fourth Edition of *The Science and Engineering of Materials* has been thoroughly revised to include discussions of recent breakthroughs in materials science such as nano-materials and micro-electro mechanical systems (Chapter 1), flexible micro-electric circuits comprised of polymers (Chapter 18), smart materials (Chapters 11 and 18), fiber optics (Chapters 2 and 20), carbon nanotubes (Chapter 18), polymers as semiconductors (Chapter 15), and advanced electronic, magnetic and photonic materials, just to name a few. Every chapter presents examples of the latest exciting real-world applications.

Our text continues to be in five parts, as in previous editions, and we have combined our discussions of failure with our discussions of properties and behavior to create a revised Chapter 6. The Third Edition Chapter 23 on failure has been eliminated. Based on user feedback, we decided the discussion of failure is better suited to appear with the discussion of material properties and testing. These topics have been integrated with Chapters 6 and 22. We have expanded and thoroughly updated our discussions of ceramics (Chapter 14), polymers (Chapter 15), composites (Chapter 16), electronic materials (Chapter 18), magnetic materials (Chapter 19), and photonic materials (Chapter 20) to reflect recent developments and applications. We have also made a deliberate effort to integrate the applications of fundamental concepts, as opposed to strictly compartmentalizing them as ceramics, metals and polymers.

New Features

We have added many new and unique features to the Fourth Edition:

CD-ROM Included with the Text   The accompanying CD-ROM contains many helpful software tools:

- One of the most useful tools is a student version of CaRIne Crystallography™. This is computational materials science software designed to aid students in the visualization of the basic concepts of atomic and molecular structure and material properties. We feel that the inclusion of CaRIne will be invaluable in those learning concepts related to crystal structures.

- A collection of QuickTime™ animated visualizations development by Dr. John Russ (North Carolina State University) covering such important material science concepts as crystal deformation, phase diagrams, shearing, and molecular bonding.

Chapter Opening Photos and Captions Each chapter begins with a photo and caption illustrating an interesting application. This provides an extra tidbit of information designed to pique the reader's curiosity about the material presented in that chapter.

Have You Ever Wondered? Questions The opening chapter photo is followed by a section entitled "*Have You Ever Wondered?*" These questions are designed to arouse the reader's interest, put things in perspective, and form the framework for what the reader will learn in that chapter.

Chapter Introductions The chapter introductions provide a breakdown of how the material will be presented in the chapters, as well as providing important information for the student to consider while reading the chapter. This section is not gratuitous, as the chapter material is developed from these introductory discussions.

Examples As mentioned before, many real-world Examples have been integrated to accompany the chapter discussions. These Examples specifically cover design considerations, such as operating temperature, presence of corrosive material, economic considerations, recyclability, and environmental constraints. The examples also apply to theoretical material and numeric calculations to further reinforce the presentation.

Chapter Summaries The Summaries at the end of the chapters have been expanded and are more thorough than those in previous editions.

Glossary The Glossary terms at the end of the chapters have been expanded. All of the Glossary terms that appear in the chapter are set in boldface type the first time they appear within the text. This provides an easy reference to the definitions provided in the end of each chapter Glossary.

Additional Information At the end of each chapter, we have provided a list of references for the chapter. These references can be used, by students or instructors, for further reading on chapter subjects. We have refrained from making specific references to web sites, as these often change and it is quite easy for most users to find the information using the Internet.

End-of-Chapter Problems More than 150 new end-of-chapter problems have been added to give the student practice in applying the principles presented in each chapter. We have added more Design Problems, as well as Computer Problems. The instructor can choose to assign these Design and Computer Problems as requirements or as extra credit material. Enhanced computer skills and familiarity with the use of softwares and databases are considered as highly desirable by most employers and we feel that the computer problems can help students develop these skills. The use of computers can be an important consideration in the ABET accreditation process. We feel that the problems suggested here would be useful in this regard as well.

Art and Photo Program The art program has been completely redrafted and over 50 new photographs and micrographs have been added. Many new photographs illustrate fascinating real-world applications. Many new illustrations communicate technical concepts in a clearer fashion.

Four-Color Insert   The full-color insert shows photographically the relationship between the structure of a material, the processing of that material, and how the properties of the material can then be used in a final product.

Periodic Table   We have provided the Periodic Table as a two-page spread in Chapter 2 and tinted the pages with blue to the edge to provide a quick and easy reference to this important item. This Periodic Table includes helpful information such as the atomic number, electron configurations, melting point, boiling point, atomic mass, oxidation states, density, and electronegativity.

Answers to Selected Problems   The answers to the selected problems are provided at the end of the text to help the student work through the end-of-chapter problems.

Appendices and Endpapers   Appendix A provides a listing of selected physical properties of metals, Appendix B presents the atomic and ionic radii of selected elements, and Appendix C provides the electronic configuration for each of the elements. The Endpapers include SI Conversion tables and Selected Physical Properties of elements. One of the endpapers contains a useful illustration that helps answer the question most often posed by students—"What is materials science and engineering?"

Chapter Outline

Part I introduces the student to Atomic Structure, Atomic Arrangements, and Atomic Movements:

Chapter 1—Introduction to Material Science and Engineering: This chapter presents an introduction to the study of material science and engineering using the MSE (Material Science and Engineering) tetrahedron to explain the interconnections between composition, structure, synthesis, processing, and performance. Chapter 1 goes on to discuss the classification of materials (functional and structure), as well as material design and selection. These principles of materials science and engineering are illustrated through several real-world applications,

Chapter 2—Atomic Structure: This chapter introduces the student to the basics of a material's atomic structure and bonding mechanisms. Many examples of how bonding affects the properties of materials are discussed.

Chapter 3—Atomic And Ionic Arrangements: This chapter discusses atomic and ionic arrangements, including short-range versus long-range order; lattices, unit cells, crystal structures, points, directions, and planes. This chapter also covers amorphous materials, allotropic and polymorphic transformations, and crystal structures of ionic- and covalent-bonded structures.

Chapter 4—Imperfections in the Atomic and Ionic Arrangements: This chapter continues the discussion of atomic and ionic arrangements with discussions of imperfections such as point defects, dislocations, crystal structures, and surface defects.

Chapter 5—Atom and Ion Movements in Materials: This chapter covers the basic concepts of diffusion—the movement of atoms and ions. The discussion of diffusion includes applications, mechanisms, activation energy, rate, and factors affecting diffusion. Discussion of Fick's First and Second Laws and the permeability of polymers is also included.

Part II introduces the student to the concepts of Mechanical Properties of Materials and Controlling the Microstructure:

Chapter 6—Mechanical Properties and Behavior: This chapter introduces the student to the properties and behavior of various materials through the examination of testing methods and the reasons for failure. Discussions include the terminology of mechanical properties, the tensile test, the bend test, true stress-true strain, hardness, impact, fracture mechanics, Weibull statistics, fatigue, creep, stress rupture, and stress corrosion.

Chapter 7—Strain Hardening and Annealing: This chapter introduces the student to the concepts of strain hardening and annealing through discussions of cold working, the stress-strain curve, strain-hardening mechanisms, texture strengthening, the three stages of annealing, control of annealing, hot working, and superplastic forming.

Chapter 8—Principles of Solidification: This chapter introduces the student to the concept of solidification. Discussions include nucleation, growth mechanisms, time and dendrite size, cooling curves, cast structure, defects, casting processes, continuous and ingot casting, directional solidification, single crystal growth, epitaxial growth, solidification of polymers and glasses, and joining of metallic materials.

Chapter 9—Solid Solutions and Phase Equilibrium: This chapter introduces the student to the important concepts of phase diagrams and solid solutions. Discussions include phases, unary phase diagrams, solubility and solid solutions, conditions for unlimited solid solubility, solid solution strengthening, isomorphous phase diagrams, properties and phase diagrams, solidification of solid solution alloys, non-equilibrium solidification and segregation. Examples include discussions on phase diagrams of metallic, ceramic and polymeric systems.

Chapter 10—Dispersion Strengthening and Eutectic Phase Diagram: This chapter continues the discussion of phase diagrams to introduce and discuss dispersion strengthening, intermetallic compounds, three-phase reactions, the eutectic phase diagram, eutectic alloys, eutectics and materials processing, nonequilibrium freezing, and ternary phase diagrams.

Chapter 11—Dispersion Strengthening by Phase Transformations and Heat Treatment: This chapter continues the discussion of dispersion strengthening by phase transformation, but also includes a discussion of heat treatment. Discussions include nucleation and growth in solid-state reactions, age and precipitation hardening, effects of aging temperature and time, the eutectoid reactions, the martensitic reaction, and shape-memory alloys.

Part III introduces the student to Engineered Materials:

Chapter 12—Ferrous Alloys: This chapter introduces the designations, classifications, and heat treatments used for steels, cast irons, and metallic alloys. The synthesis and processing of these materials are also introduced.

Chapter 13—Nonferrous Alloys: This chapter introduces the designations, classifications, and treatments of nonferrous alloys.

Chapter 14—Ceramic Materials: This chapter introduces the student to the applications, properties, synthesis and processing of advanced ceramic materials. Concepts related to powder synthesis and processing are also introduced.

Chapter 15—Polymers: This chapter introduces the student to the classifications, structure, mechanical properties, processing techniques, and recyclability of polymers. Several examples illustrate applications of engineered plastics.

Chapter 16—Composites: Teamwork and Synergy in Materials: This chapter introduces the concepts of dispersion-strengthened, particulate, fiber-reinforced, and laminar composites.

Chapter 17—Construction Materials: This chapter introduces the basic material properties and behavior of construction materials such as wood, concrete, and asphalt.

Part IV introduces the student to the Physical Properties of Engineering Materials:

Chapter 18—Electronic Materials: This chapter introduces the student to electrical behavior and properties and how they can be applied to materials science. Discussions include Ohm's Law and electrical conductivity; band structures of solids; conductivity of metals and alloys; superconductivity; conductivity in other materials such as plastics, semiconductors and insulators; dielectric properties; electrostriction; piezoelectricity; ferroelectricity and pyroelectricity.

Chapter 19—Magnetic Materials: This chapter introduces the concepts of magnetism within materials and their classifications. Discussions include magnetic dipoles and moments, magnetization, permeability, diamagnetic, paramagnetic, ferrimagnetic, ferromagnetic and superparamagnetic materials, domain structure and the hysteresis loop, the Curie temperature, and metallic and ceramic magnetic materials. Applications of magnetically hard and soft materials are also discussed.

Chapter 20—Photonic Materials: This chapter introduces the student to behavior and properties of photonic materials and how they are used in industry, including fiber optic communication systems.

Chapter 21—Thermal Properties of Materials: This chapter discusses the thermal properties of materials including heat capacity, specific heat, thermal expansion, thermal conductivity, and thermal shock.

Part V includes a discussion on the Protection Against Deterioration and Failure of Materials:

Chapter 22—Corrosion and Wear: This chapter covers the topics of corrosion and wear. Discussions include chemical corrosion, electrochemical corrosion, microbial degradation, biodegradable polymers, oxidation, gas reactions, wear, and erosion. Strategies for prevention of corrosion are also discussed.

Strategies for Teaching from the Book

All of the material presented here admittedly cannot or should not be covered in a typical one-semester course. By selecting the appropriate topics, however, the instructor can emphasize the desired materials (i.e., metals, alloys, ceramics, polymers, composites, etc.), provide an overview of materials, concentrate on behavior, or focus on physical properties. In addition, the text provides the student with a useful reference for subsequent courses in manufacturing, design, and materials selection. For students specializing in materials science and engineering, or closely related disciplines, sections related to synthesis and processing could be discussed in greater detail.

Supplements

Supplements for the student include:

- As noted before, one of the most useful additions to the CD-ROM is the CaRIne software. This will really enable students to better understand crystal structures and related concepts.
- A student website at: www.brookscole.com/engineering_d/.

Supplements for the instructor include:

- The Instructor's Solutions Manual that provides complete, worked-out solutions to selected text problems and additional text items.
- PDF files of all figures from the textbook in a multimedia presentation format.

Acknowledgments

It takes a team of many people and a lot of hard work to create a quality textbook. We are indebted to all of the people who provided the assistance, encouragement, and constructive criticism leading to the preparation of this Fourth Edition.

First, we wish to acknowledge the many instructors who have read and used the text and provided helpful feedback to our initial survey:

C. Maurice Balik, North Carolina State University

Brian Cousins, University of Tasmania

Arthur F. Diaz, San Jose State University

Richard S. Harmer, University of Dayton

Prashant N. Kumta, Carnegie Mellon University

Rafael Manory, Royal Melbourne Institute of Technology

Sharon Nightingale, University of Wollongong, Australia

Christopher K. Ober, Cornell University

David Poirier, University of Arizona

Ramurthy Prabhakaran, Old Dominion University

Lew Rabenberg, The Unviersity of Texas at Austin

Wayne Reitz, North Dakota State University

John Schlup, Kansas State University

Robert L. Snyder, Rochester Institute of Technology

The following reviewed individual chapters and provided specific direction for us: Harvey Abramowitz, Purdue University–Calumet; Arthur F. Diaz, San Jose State University; Hassan M. Rejali, California State Polytechnic University, Pomona; Judy Schneider, Mississippi State University; and Supapan Seraphin, Unversity of Arizona.

We are particularly indebted to our two reviewers who have been with us through the entire manuscript and galley stages—Professor Susan James of Colorado State University and Professor Norman E. Dowling of Virginia Tech. Their hard work, dedication, and insight helped us create this quality text. We also wish to acknowledge

Richard McAfee and Dr. Ian Nettleship of the University of Pittsburgh. Thanks are also due to Dr. Cyrille Boudias and Dr. Daniel Monceau, authors of the CaRIne software.

Thanks most certainly to everyone at Brooks/Cole Publishing Company for their encouragement, knowledge, and patience in seeing this text to fruition: Valerie Boyajian, Mary Vezilich, Tom Ziolkowksi, and Vernon Boes.

We wish to thank three people, in particular, for their diligent efforts: Many thanks to Bill Stenquist, our publisher, who set the tone for excellence and who provided the vision, expertise, and leadership to create such a quality product; to Rose Kernan, our developmental and production editor, who worked long hours to improve our prose and produce this quality text from the first pages of manuscript to the final, bound product; and to Dr. Deepa Godbole, of the University of Pittsburgh, whose valiant efforts, hard work, and dedication can never be repaid.

Pradeep Phulé would like specifically to thank his wife, Dr. Jyotsna Phulé and children, Aarohee and Suyash, for their patience, understanding, and encouragement. Thanks are also due to Professor S.H. Risbud, University of California–Davis, for his advice and encouragement and to all of our colleagues who provided many useful illustrations.

Donald R. Askeland
University of Missouri–Rolla, Emeritus

Pradeep P. Phulé
University of Pittsburgh

About the Authors

Donald R. Askeland is a Distinguished Teaching Professor Emeritus of Metallurgical Engineering at the University of Missouri–Rolla. He received his degrees from the Thayer School of Engineering at Dartmouth College and the University of Michigan prior to joining the faculty at the University of Missouri–Rolla in 1970. Dr. Askeland taught a number of courses in materials and manufacturing engineering to students in a variety of engineering and science curricula. He received a number of awards for excellence in teaching and advising at UMR. He served as a Key Professor for the Foundry Educational Foundation and received several awards for his service to that organization. His teaching and research were directed primarily to metals casting and joining, in particular lost foam casting, and resulted in over 50 publications and a number of awards for service and best papers from the American Foundry Society.

Pradeep P. Phulé is a Professor of Materials Science and Engineering at the University of Pittsburgh. He joined the University of Pittsburgh in 1989, was promoted to Associate Professor in 1994, and then to full professor in 1999. Dr. Phulé received a Ph.D. in Materials Science and Engineering from the University of Arizona (1989) and a B. Tech (1983) and M. Tech (1984) in Metallurgical Engineering from the Indian Institute of Technology Bombay (Mumbai) India.

He has authored close to 60 publications and has two U.S. patents issued. He has received the Alcoa Foundation and Ford Foundation research awards.

He has been an outstanding teacher and educator and was recently listed on the Faculty Honor Roll at the University of Pittsburgh (2001) for outstanding services and assistance. From 1992–1999, he was the William Kepler Whiteford Faculty Fellow at the University of Pittsburgh. From August to December 2002, Dr. Phulé was a visiting scientist at the Ford Scientific Research Laboratory in Dearborn, MI.

Dr. Phulé's primary research areas are chemical synthesis and processing of ceramics, electronic ceramics and magnetic materials, development of smart materials and systems. Part of the MR fluids technology Dr. Phulé has developed is being transferred to industry. He was the Vice President of Ceramic Educational Council (2001–2002) and a Member of the Program Committee for the Electronics Division of the American ceramic society since 1996.

He has also served as an Associate Editor for the *Journal of the American Ceramic Society* (1994–2000). He has been the lead organizer for symposia on ceramics for sol-gel processing, wireless communications, and smart structures and sensors. In 2002, Dr. Phulé was elected as a Fellow of the American Ceramic Society. Dr. Phulé's research has been supported by National Science Foundation (NSF) and many other organizations.

The
Science
and Engineering
of Materials

FOURTH EDITION

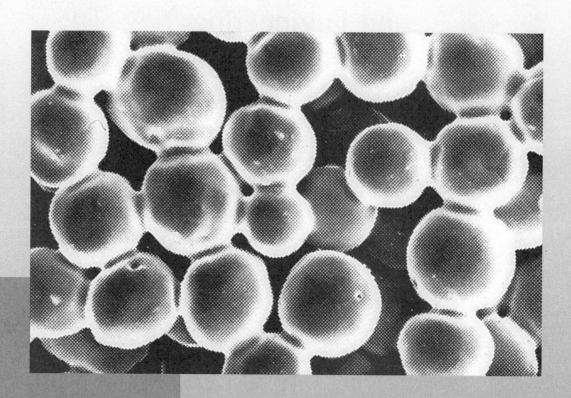

Powder consolidation of particles is a common method for manufacturing metal, ceramic, and composite materials. Diffusion of atoms to points of contact between the particles—in this case, spherical copper powders—during sintering causes the particles to become bonded. Continued sintering and diffusion eventually cause the pores between the particles to disappear. (*From* Metals Handbook, *Vol. 9, Metallography and Microstructure (1985), ASM International, Materials Park, OH 44073.*)

PART 1

Atomic Structure, Arrangement, and Movement

We classify materials into several major groups: metals, ceramics, polymers, semiconductors, and composites. The behavior of the materials in each of these groups is determined by their structure. The electronic structure of an atom determines the nature of atomic bonding, which helps govern the mechanical and physical properties of a given material.

The arrangement of atoms into a crystalline or amorphous structure also influences the properties of materials. Imperfections in atomic arrangement play a critical role in our understanding of deformation, mechanical and other properties.

Finally, the movement of atoms, known as diffusion, is important for many heat treatments and manufacturing processes, as well as for both physical and mechanical properties of materials.

In the following chapters, we introduce the structure-property-processing concept for controlling the behavior of materials and examine the roles of atomic structure, atomic arrangement, imperfections, and atom movement. This examination lays the groundwork needed to understand the structure and behavior of engineered materials discussed later on.

Chapter

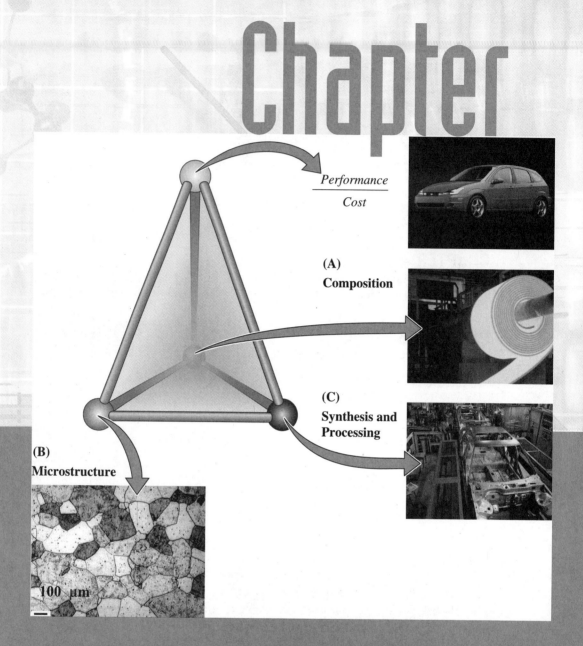

$$\frac{Performance}{Cost}$$

(A) Composition

(B) Microstructure

100 µm

(C) Synthesis and Processing

The principal goals of a materials scientist and engineer are to: (1) make existing material better, and (2) invent or discover new phenomena, materials, devices, and applications. Breakthroughs in the materials science and engineering field are applied to many other fields of study such as biomedical engineering, physics, chemistry, environmental engineering, and information technology. The materials science and engineering tetrahedron shown here represents the heart and soul of this field.[1] As shown in this diagram, a materials scientist and engineer's main objective is to develop materials or devices that have the best performance for a particular application. In most cases, the performance-to-cost ratio, as opposed to the performance alone, is of utmost importance. This concept is shown as the apex of the tetrahedron and the three corners are representative of A—the composition, B—the microstructure, and C—the synthesis and processing of materials. These are all interconnected and ultimately affect the performance-to-cost ratio of a material or a device. The accompanying micrograph shows the microstructure of stainless steel.

For materials scientists and engineers, materials are like a palette of colors to an artist. Just as an artist can create different paintings using different colors, materials scientists create and improve upon different materials using different elements of the periodic table, and different synthesis and processing routes. (*Car image courtesy of Ford Motor Company. Steel manufacturing image courtesy of Donovan Reese/PhotoDisc. Car chassis image courtesy of Digital Vision/Getty Images. Micrograph courtesy of Dr. A.J. Deardo, Dr. M. Hua, and Dr. J. Garcia.*)

1

Introduction to Materials Science and Engineering

Have You Ever Wondered?

- What do materials scientists and engineers study?

- How sheet steel can be processed to produce a high-strength, lightweight, energy absorbing, malleable material used in the manufacture of car chassis?

- Can we make flexible and lightweight electronic circuits using plastics?

- What is a "smart material?"

In this chapter, we will first introduce you to the field of materials science and engineering (MSE) using different real-world examples. We will then provide an introduction to the classification of materials. Although most engineering programs require students to take a materials science course, you should approach your study of mate-rials science as more than a mere requirement. A thorough knowledge of materials science and engineering will make you a better engineer and designer. Materials science underlies all techno-logical advances and an understanding of the basics of materials and their applications will not only make you a better engineer, but will help you during the design process. In order to be a good designer, you must learn what materials will be appropriate to use in different applications. The most important aspect of materials is that they are enabling; materials make things happen. For example, in the history of civilization, materials such as stone, iron, and bronze played a key role in mankind's development.[2] In today's fast-paced world, the discovery of silicon single crystals and an understanding of their properties have enabled the information age.

In this chapter and throughout the book, we will provide compelling examples of real-world applications of engineered materials. The diversity of applications and the unique uses of materials illustrate why a good engineer needs to thoroughly understand and know how to apply the principles of materials science and engineering. In each chapter, we begin with a section entitled *Have You Ever Wondered?* These questions are designed to pique your curiosity, put things in perspective, and form a framework for what you will learn in that chapter.

1-1   What is Materials Science and Engineering?

Materials science and engineering (MSE) is an interdisciplinary field concerned with inventing new materials and improving previously known materials by developing a deeper understanding of the microstructure-composition-synthesis-processing relationships. The term **composition** means the chemical make-up of a material. The term **structure** means a description of the arrangement of atoms, as seen at different levels of detail. Materials scientists and engineers not only deal with the development of materials, but also with the **synthesis** and **processing** of materials and manufacturing processes related to the production of components. The term "synthesis" refers to how materials are made from naturally occurring or man-made chemicals. The term "processing" means how materials are shaped into useful components to cause changes in the properties of different materials. One of the most important functions of materials scientists and engineers is to establish the relationships between a material or a device's properties and performance and the microstructure of that material, its composition, and the way the material or the device was synthesized and processed. In **materials science**, the emphasis is on the underlying relationships between the synthesis and processing, structure and properties of materials. In **materials engineering**, the focus is on how to translate or transform materials into a useful device or structure.

One of the most fascinating aspects of materials science involves the investigation of a material's structure. The structure of materials has a profound influence on many properties of materials, even if the overall composition does not change! For example, if you take a pure copper wire and bend it repeatedly, the wire not only becomes harder but also becomes increasingly brittle! Eventually, the pure copper wire becomes so hard and brittle that it will break! The electrical resistivity of wire will also increase as we bend it repeatedly. In this simple example, take note that we did not change the material's composition (i.e., its chemical make up). The changes in the material's properties are due to a change in its internal structure. If you look at the wire after bending, it will

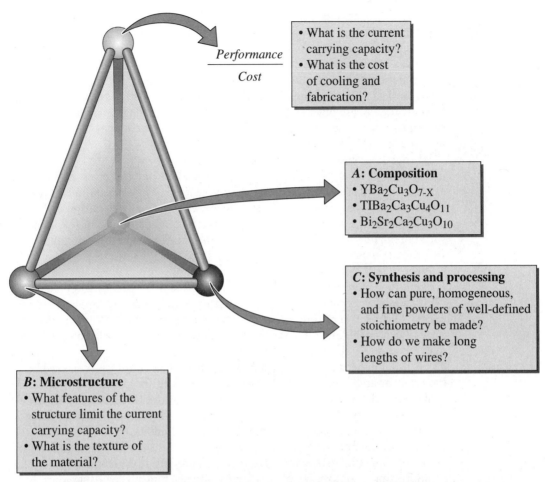

Performance / Cost

• What is the current carrying capacity?
• What is the cost of cooling and fabrication?

A: **Composition**
• $YBa_2Cu_3O_{7-X}$
• $TlBa_2Ca_3Cu_4O_{11}$
• $Bi_2Sr_2Ca_2Cu_3O_{10}$

C: **Synthesis and processing**
• How can pure, homogeneous, and fine powders of well-defined stoichiometry be made?
• How do we make long lengths of wires?

B: **Microstructure**
• What features of the structure limit the current carrying capacity?
• What is the texture of the material?

Figure 1-1 Application of the tetrahedron of materials science and engineering to ceramic superconductors. Note that the microstructure-synthesis and processing-composition are all interconnected and affect the performance-to-cost ratio.

look the same as before; however, its structure has been changed at a very small or microscopic scale. The structure at this microscopic scale is known as **microstructure**. If we can understand what has changed microscopically, we can begin to discover ways to control the material's properties.

Let's put the materials science and engineering tetrahedron presented on the chapter opening page in even better perspective by examining a sample product–ceramic superconductors invented in 1986 (Figure 1-1). You may be aware that ceramic materials usually do not conduct electricity. Scientists found serendipitously that certain ceramic compounds based on yttrium barium copper oxides (known as *YBCO*) can actually carry electrical current without any resistance under certain conditions. Based on what was known then about metallic superconductors and the electrical properties of ceramics, superconducting behavior in ceramics was not considered as a strong possibility. Thus, the first step in this case was the *discovery* of superconducting behavior in ceramic materials. One limitation that was discovered is that these materials can superconduct only at low temperatures (<150 K).

The next step was to determine how to make these materials better. By "better" we mean, how can we retain superconducting behavior in these materials at higher and higher temperatures or how can we transport a large amount of current over a long distance. This involves materials processing and careful structure-property studies. Materials scientists wanted to know how the composition and microstructure (arrangement of the atomic structure, etc.) affect the superconducting behavior. They also want to know if there are other compounds that exhibited similar superconductivity. Through experimentation, the scientists developed controlled *synthesis* of ultrafine powders that are used to create a superconducting ceramic material.

From a *materials engineering* perspective, let's say we want to know if we can make motors, very powerful magnets (that could used for magnetic resonance imaging, for example), wires used in electronic applications or tiny microelectronic devices using ceramic superconductors. Let's say that, in the electrical power transmission application, we ultimately want to know if we can make reliable and reproducible long lengths of superconducting wires that are superior to the current copper and aluminum wires. Can we produce such wires in a cost-effective way?

The next challenge was to make long lengths of ceramic superconductor wires. Ceramic superconductors are brittle, so making long lengths of wires was a challenge. Thus, *materials processing* techniques had to be developed to create these wires. Successful solutions were found through trial and error. One successful way of creating these superconducting wires was to fill hollow silver tubes with powders of superconductor ceramic and then draw wires.

Although the discovery of ceramic superconductors did cause a lot of excitement, the path toward translating that discovery into useful products has been met by many challenges related to the synthesis and processing of these materials.[3–6] Recently, superconductivity has been observed in a compound known as magnesium diboride (MgB_2). This compound appears not to have the limitations of some of the other, previously used ceramic superconductors, but further tests and development will be needed before this material can be manufactured in a cost-effective manner.[7]

Sometimes, discoveries of new materials, phenomena, or devices are heralded as *revolutionary*. Today, as we look back, the discovery of the silicon-based transistor used in computer chips is considered revolutionary. On the other hand, materials that have evolved over a period of time can be just as important. These materials are known as *evolutionary*. Many alloys based on iron, copper, and the like are examples of evolutionary materials. Of course, it is important to recognize that what are considered as evolutionary materials now, did create revolutionary advances many years back. It is not uncommon for materials or phenomena to be discovered first and then many years go by before commercial products or processes appear in the marketplace. The transition from the development of novel materials or processes to useful commercial or industrial applications can be slow and difficult.[8]

Let's examine another example using the materials science and engineering tetrahedron presented on the chapter opening page. Let's look at "sheet steels" used in the manufacture of car chassis (Figure 1-2). Steels, as you may know, have been used in manufacturing for more than a hundred years, but they probably existed in a crude form during the Iron Age, thousands of years ago.[2,3] In the manufacture of automobile chassis, a material is needed that possesses extremely high strength but is easily formed into aerodynamic contours. Another consideration is fuel-efficiency, so the sheet steel must also be thin and lightweight. The sheet steels should also be able to absorb significant amounts of energy in the event of a crash, thereby increasing vehicle safety. These are somewhat contradictory requirements.

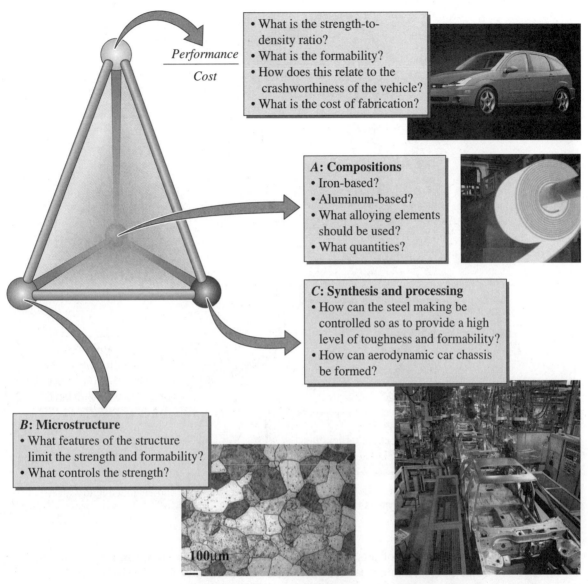

Performance
───────
Cost

- What is the strength-to-density ratio?
- What is the formability?
- How does this relate to the crashworthiness of the vehicle?
- What is the cost of fabrication?

A: **Compositions**
- Iron-based?
- Aluminum-based?
- What alloying elements should be used?
- What quantities?

C: **Synthesis and processing**
- How can the steel making be controlled so as to provide a high level of toughness and formability?
- How can aerodynamic car chassis be formed?

B: **Microstructure**
- What features of the structure limit the strength and formability?
- What controls the strength?

100μm

Figure 1-2 Application of the tetrahedron of materials science and engineering to sheet steels for automotive chassis. Note that the microstructure-synthesis and processing-composition are all interconnected and affect the performance-to-cost ratio.

Thus, in this case, materials scientists are concerned with the sheet steel's

- composition;
- strength;
- weight;
- energy absorption properties; and
- malleability (formability).

Materials scientists would examine steel at a microscopic level to determine if its properties can be altered to meet all of these requirements. They also would have to

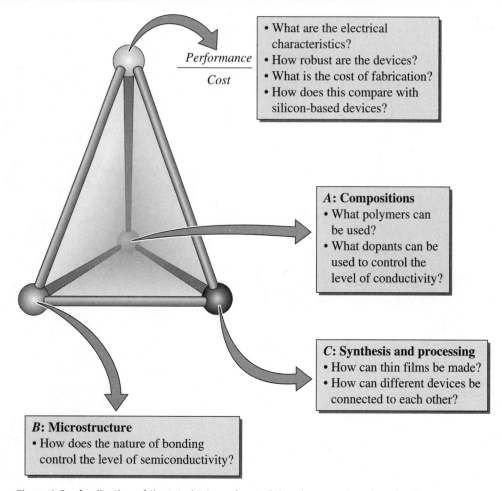

Figure 1-3   Application of the tetrahedron of materials science and engineering to semiconducting polymers for microelectronics.

consider the cost of processing this steel along with other considerations. How can we shape such steel into a car chassis in a cost-effective way? Will the shaping process itself affect the mechanical properties of the steel? What kind of coatings can be developed to make the steel corrosion resistant? In some applications, we need to know if these steels could be welded easily. From this discussion, you can see that many issues need to be considered during the design and materials selection for any product.

Let's look at one more example of a relatively new class of materials known as semiconducting polymers (Figure 1-3). Alan J. Heeger, Alan G. MacDiarmid, and Hideki Shirakawa won the Nobel Prize in Chemistry in 2000 for their discovery of electrically conducting polymers. Many semiconducting polymers have now been processed into light emitting diodes (LEDs). Most of you have seen LEDs in alarm clocks, watches, and other displays. These displays often use inorganic compounds based on gallium arsenide (GaAs) and other materials. The advantage of using plastics for microelectronics is that they are lightweight and flexible. The questions materials scientists and engineers must answer with applications of semiconducting polymers are:

- What are the relationships between the structure of polymers and their electrical properties?
- How can devices be made using these plastics?
- Will these devices be compatible with existing silicon chip technology or will they create a technology of their own?
- How robust (strong) are these devices?
- How will the performance and cost of these devices compare with traditional devices?

These are just a few of the factors that engineers and scientists must consider during the development, design, and manufacture of semiconducting polymer devices.

1-2 Classification of Materials

There are different ways of classifying materials. One way is to describe five groups (Table 1-1):

1. **metals** and **alloys**;
2. ceramics, **glasses**, and **glass-ceramics**;
3. polymers (plastics);
4. semiconductors; and
5. composite materials.

Materials in each of these groups possess different structures and properties. The differences in strength, which are compared in Figure 1-4, illustrate the wide range of properties engineers can select from. Since metallic materials are extensively used for load-bearing applications, their mechanical properties are of great practical interest. We briefly introduce these here. The term "stress" refers to load or force per unit area. "Strain" refers to elongation or change in dimension divided by original dimension. Application of "stress" causes "strain." If the strain goes away after the load or applied stress is removed, the strain is said to be "elastic." If the strain remains after the stress is removed, the strain is said to be "plastic." When the deformation is elastic and stress and strain are linearly related, the slope of the stress-strain diagram is known as the elastic or <u>Young's modulus</u>. A level of stress needed to initiate plastic deformation is known as "yield strength." The maximum percent deformation we can get is a measure of the ductility of a metallic material. These concepts are discussed further in Chapter 6.

Metals and Alloys These include steels, aluminum, magnesium, zinc, cast iron, titanium, copper, and nickel. In general, metals have good electrical and thermal conductivity. Metals and alloys have relatively high strength, high stiffness, ductility or formability, and shock resistance. They are particularly useful for structural or load-bearing applications. Although pure metals are occasionally used, combinations of metals called alloys provide improvement in a particular desirable property or permit better combinations of properties. The cross section of a jet engine shown in Figure 1-5 illustrates the use of metallic materials for a number of critical applications.

TABLE 1-1 ■ *Representative examples, applications, and properties for each category of materials*

	Examples of Applications	Properties
Metals and Alloys		
Copper	Electrical conductor wire	High electrical conductivity, good formability
Gray cast iron	Automobile engine blocks	Castable, machinable, vibration-damping
Alloy steels	Wrenches, automobile chassis	Significantly strengthened by heat treatment
Ceramics and Glasses		
SiO_2-Na_2O-CaO	Window glass	Optically transparent, thermally insulating
Al_2O_3, MgO, SiO_2	Refractories (i.e., heat-resistant lining of furnaces) for containing molten metal	Thermally insulating, withstand high temperatures, relatively inert to molten metal
Barium titanate	Capacitors for microelectronics	High ability to store charge
Silica	Optical fibers for information technology	Refractive index, low optical losses
Polymers		
Polyethylene	Food packaging	Easily formed into thin, flexible, airtight film
Epoxy	Encapsulation of integrated circuits	Electrically insulating and moisture-resistant
Phenolics	Adhesives for joining plies in plywood	Strong, moisture resistant
Semiconductors		
Silicon	Transistors and integrated circuits	Unique electrical behavior
GaAs	Optoelectronic systems	Converts electrical signals to light, lasers, laser diodes, etc.
Composites		
Graphite-epoxy	Aircraft components	High strength-to-weight ratio
Tungsten carbide-cobalt (WC-Co)	Carbide cutting tools for machining	High hardness, yet good shock resistance
Titanium-clad steel	Reactor vessels	Low cost and high strength of steel, with the corrosion resistance of titanium

Ceramics, Glasses, and Glass-Ceramics   Ceramics can be defined as inorganic crystalline materials. Ceramics are probably the most "natural" materials. Beach sand and rocks are examples of naturally occurring ceramics. Advanced ceramics are materials made by refining naturally occurring ceramics and other special processes. Advanced ceramics are used in substrates that house computer chips, sensors and actuators, capacitors, wireless communications, spark plugs, inductors, and electrical insulation. Some ceramics are used as barrier coatings to protect metallic substrates in turbine engines. Ceramics are also used in such consumer products as paints, plastics, and tires, and for industrial applications such as the tiles for the space shuttle, a catalyst support, and the oxygen sensors used in cars. Traditional ceramics are used to make bricks, tableware, sanitaryware, refractories (heat-resistant material), and abrasives. In general,

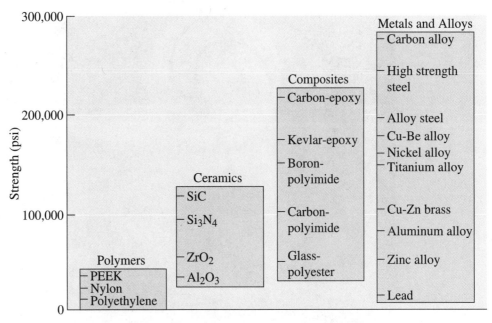

Figure 1-4   Representative strengths of various categories of materials.

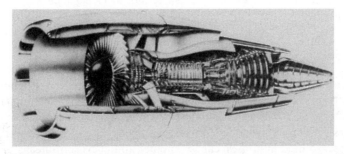

Figure 1-5   A section through a jet engine. The forward compression section operates at low to medium temperatures, and titanium parts are often used. The rear combustion section operates at high temperatures and nickel-based superalloys are required. The outside shell experiences low temperatures, and aluminum and composites are satisfactory. (*Courtesy of GE Aircraft Engines.*)

due to the presence of porosity (small holes), ceramics do not conduct heat well and must be heated to very high temperatures before melting. Ceramics are strong and hard, but also very brittle. We normally prepare fine powders of ceramics and convert these into different shapes. New processing techniques make ceramics sufficiently resistant to fracture that they can be used in load-bearing applications, such as impellers in turbine engines (Figure 1-6). Ceramics have exceptional strength under compression. Can you believe that an entire fire truck can be supported using four ceramic coffee cups?

Figure 1-6 A variety of complex ceramic components, including impellers and blades, which allow turbine engines to operate more efficiently at higher temperatures. (*Courtesy of Certech, Inc.*)

Glass is an amorphous material, often, but not always, derived from molten silica. The term "amorphous" refers to materials that do not have a regular, periodic arrangement of atoms. Amorphous materials will be discussed in detail in Chapter 3. The fiber optics industry is founded on optical fibers based on high-purity silica glass. Glasses are also used in houses, cars, computer and television screens, and hundreds of other applications. Glasses can be thermally treated (tempered) to make them stronger. Forming glasses and nucleating (forming) small crystals within them by a special thermal process creates materials that are known as glass-ceramics. Zerodur™ is an example of a glass-ceramic material that is used to make the mirror substrates for large telescopes (e.g., the Chandra and Hubble telescopes). Glasses and glass-ceramics are usually processed by melting and casting.

Polymers Polymers are typically organic materials. They are produced using a process known as **polymerization**. Polymeric materials include rubber (elastomers) and many types of adhesives. Many polymers have very good electrical resistivity. They can also provide good thermal insulation. Although they have lower strength, polymers have a very good **strength-to-weight ratio**. They are typically not suitable for use at high temperatures. Many polymers have very good resistance to corrosive chemicals. Polymers have thousands of applications ranging from bulletproof vests, compact disks (CDs), ropes, and liquid crystal displays (LCDs) to clothes and coffee cups. **Thermoplastic** polymers, in which the long molecular chains are not rigidly connected, have good

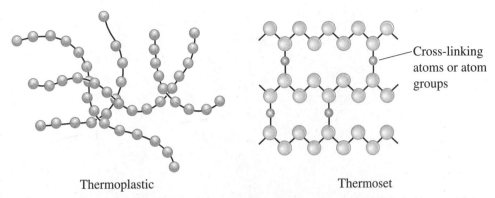

Thermoplastic   Thermoset

Figure 1-7 Polymerization occurs when small molecules, represented by the circles, combine to produce larger molecules, or polymers. The polymer molecules can have a structure that consists of many chains that are entangled but not connected (thermoplastics) or can form three-dimensional networks in which chains are cross-linked (thermosets).

ductility and formability; **thermosetting** polymers are stronger but more brittle because the molecular chains are tightly linked (Figure 1-7). Polymers are used in many applications, including electronic devices (Figure 1-8). Thermoplastics are made by shaping their molten form. Thermosets are typically cast into molds. **Plastics** contain additives.

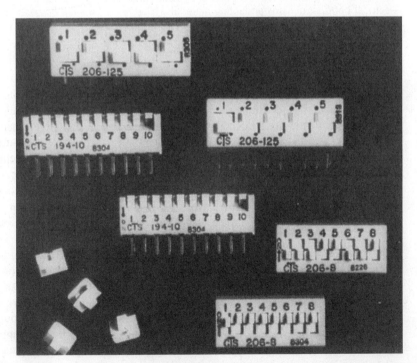

Figure 1-8 Polymers are used in a variety of electronic devices, including these computer dip switches, where moisture resistance and low conductivity are required. (*Courtesy of CTS Corporation.*)

Figure 1-9 Integrated circuits for computers and other electronic devices rely on the unique electrical behavior of semiconducting materials. (*Courtesy of Rogers Corporation.*)

Semiconductors Silicon-, germanium-, and gallium arsenide-based semiconductors such as those used in computers and electronics are part of a broader class of materials known as electronic materials. The electrical conductivity of semiconducting materials is between that of ceramic insulators and metallic conductors. **Semiconductors** have enabled the information age. In some semiconductors, the level of conductivity can be controlled to enable their use in electronic devices such as transistors, diodes, and integrated circuits (Figure 1-9). In many applications, we need large single crystals of semiconductors. These are grown from molten materials. Often, thin films of semiconducting materials are also made using specialized processes.

Composite Materials The main idea in developing **composites** is to blend the properties of different materials. These are formed from two or more materials, producing properties not found in any single material. Concrete, plywood, and fiberglass are examples of composite materials. Fiberglass is made by dispersing glass fibers in a polymer matrix. The glass fibers make the polymer stiffer, without significantly increasing its density. With composites we can produce lightweight, strong, ductile, high temperature-resistant materials or we can produce hard, yet shock-resistant, cutting tools that would otherwise shatter. Advanced aircraft and aerospace vehicles rely heavily on composites such as carbon-fiber-reinforced polymers (Figure 1-10). Sports equipment such as bicycles, golf clubs, tennis rackets, and the like also make use of different kinds of composite materials that are light and stiff.[9,10]

Figure 1-10 The X-wing for advanced helicopters relies on a material composed of a carbon-fiber-reinforced polymer. (*Courtesy of Sikorsky Aircraft Division—United Technologies Corporation.*)

1-3 Functional Classification of Materials

A functional classification of materials can be useful. We can classify materials based on whether the most important function they perform is mechanical (structural), biological, electrical, magnetic, or optical. This classification of materials is shown in Figure 1-11. Some examples of each category are shown. These categories can be broken down further into subcategories.

Aerospace Light materials such as wood and an aluminum alloy (that accidentally strengthened the engine even more by picking up copper from the mold used for casting) were used in the Wright brothers historic flight.[11] Today, NASA's space shuttle makes use of aluminum powder for booster rockets. Aluminum alloys, plastics, silica for space shuttle tiles, and many other materials belong to this category.

Biomedical Our bones and teeth are made, in part, from a naturally formed ceramic known as hydroxyapatite. A number of artificial organs, bone replacement parts, cardiovascular stents, orthodontic braces, and other components are made using different plastics, titanium alloys, and nonmagnetic stainless steels. Ultrasonic imaging systems make use of ceramics known as *PZT* (lead zirconium titanate). Magnets used for magnetic resonance imaging make use of metallic niobium tin-based superconductors.

Electronic Materials As mentioned before, semiconductors, such as those made from silicon, are used to make integrated circuits for computer chips. Barium titanate ($BaTiO_3$), tantalum oxide (Ta_2O_5), and many other dielectric materials are used to make ceramic capacitors and other devices. Superconductors are used in making powerful magnets. Copper, aluminum, and other metals are used as conductors in power transmission and in microelectronics.

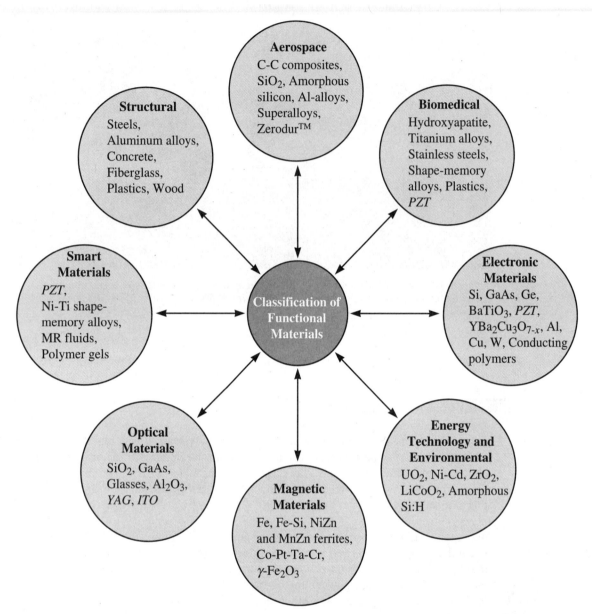

Figure 1-11   Functional classification of materials. Notice that metals, plastics, and ceramics occur in different categories. A limited number of examples in each category are provided.

Energy Technology and Environmental Technology   The nuclear industry uses materials such as uranium dioxide and plutonium as fuel. Numerous other materials, such as glasses and stainless steels, are used in handling nuclear materials and managing radioactive waste. New technologies related to batteries and fuel cells make use of many ceramic materials such as zirconia (ZrO_2) and polymers. The battery technology has gained significant importance owing to the need for many electronic devices that require longer lasting and portable power. Fuel cells will also be used in electric cars.[12] The oil and petroleum industry widely uses zeolites, alumina, and other materials as catalyst substrates. They use Pt, Pt/Rh and many other metals as catalysts. Many membrane technologies for purification of liquids and gases make use of ceramics and plastics. Solar power is generated using materials such as amorphous silicon (a:Si:H).

Magnetic Materials   Computer hard disks and audio and video cassettes make use of many ceramic, metallic, and polymeric materials. For example, particles of a special form of iron oxide, known as gamma iron oxide (γ-Fe_2O_3) are deposited on a polymer substrate to make audio cassettes. High-purity iron particles are used for making videotapes. Computer hard disks are made using alloys based on cobalt-platinum-tantalum-chromium (Co-Pt-Ta-Cr) alloys. Many magnetic ferrites are used to make inductors and components for wireless communications. Steels based on iron and silicon are used to make transformer cores.

Photonic or Optical Materials   Silica is used widely for making optical fibers. Almost ten million kilometers of optical fiber has been installed around the world.[13] Optical materials are used for making semiconductor detectors and lasers used in fiber optic communications systems and other applications. Similarly, alumina (Al_2O_3) and yttrium aluminum garnets (*YAG*) are used for making lasers. Amorphous silicon is used to make solar cells and photovoltaic modules. Polymers are used to make liquid crystal displays (*LCDs*).

Smart Materials   A **smart material** can sense and respond to an external stimulus such as a change in temperature, the application of a stress, or a change in humidity or chemical environment. Usually a smart-material-based system consists of sensors and actuators that read changes and initiate an action. An example of a passively smart material is lead zirconium titanate (*PZT*) and shape-memory alloys. When properly processed, *PZT* can be subjected to a stress and a voltage is generated. This effect is used to make such devices as spark generators for gas grills and sensors that can detect underwater objects such as fish and submarines. Other examples of smart materials include magnetorheological or MR fluids.[14] These are magnetic paints that respond to magnetic fields. These materials are being used in suspension systems of automobiles, such as the 2003 Cadillac^{TM}. Still other examples of smart materials and systems are photochromic glasses and automatic dimming mirrors.

Structural Materials   These materials are designed for carrying some type of stress. Steels, concrete, and composites are used to make buildings and bridges. Steels, glasses, plastics, and composites are also used widely to make automotives. Often in these applications, combinations of strength, stiffness, and toughness are needed under different conditions of temperature and loading.

1-4   Classification of Materials Based on Structure

As mentioned before, the term "structure" means the arrangement of a material's atoms; the structure at a microscopic scale is known as "microstructure." We can view these arrangements at different scales, ranging from a few angstrom units to a millimeter. We will learn in Chapter 3 that some materials may be **crystalline** (where the material's atoms are arranged in a periodic fashion) or they may be amorphous (where the material's atoms do not have a long-range order). Some crystalline materials may be in the form of one crystal and are known as **single crystals**. Others consist of many crystals or **grains** and are known as **polycrystalline**. The characteristics of crystals or grains (size, shape, etc.) and that of the regions between them, known as the **grain boundaries**, also affect the properties of materials. We will further discuss these concepts in later chapters. A micrograph of stainless steel showing grains and grain boundaries is shown in the chapter opening image for this chapter.

1-5   Environmental and Other Effects

The structure-property relationships in materials fabricated into components are often influenced by the surroundings to which the material is subjected during use. This can include exposure to high or low temperatures, cyclical stresses, sudden impact, corrosion or oxidation. These effects must be accounted for in design to ensure that components do not fail unexpectedly.

Temperature   Changes in temperature dramatically alter the properties of materials (Figure 1-12). Metals and alloys that have been strengthened by certain heat treatments or forming techniques may suddenly lose their strength when heated. A tragic reminder of this is the collapse of the World Trade Center towers on September 11, 2001.

High temperatures change the structure of ceramics and cause polymers to melt or char. Very low temperatures, at the other extreme, may cause a metal or polymer to fail in a brittle manner, even though the applied loads are low. This low temperature embrittlement was a factor that caused the *Titanic* to fracture and sink.[15,16] Similarly, the 1986 *Challenger* accident, in part, was due to embrittlement of rubber O-rings.[17] The reasons why some polymers and metallic material become brittle are different. We will discuss these concepts in later chapters.

The design of materials with improved resistance to temperature extremes is essential in many technologies, as illustrated by the increase in operating temperatures of aircraft and aerospace vehicles (Figure 1-13). As faster speeds are attained, more heating of the vehicle skin occurs because of friction with the air. At the same time, engines operate more efficiently at higher temperatures. So, in order to achieve higher speed and better fuel economy, new materials have gradually increased allowable skin and engine temperatures. But materials engineers are continually faced with new challenges. The *X-33* and *Venturestar* are examples of advanced reusable vehicles intended to carry passengers into space using a single stage of rocket engines. Figure 1-14 shows a schematic of the *X-33* prototype.[18] The development of even more exotic materials and processing techniques is necessary in order to tolerate the high temperatures that will be encountered.

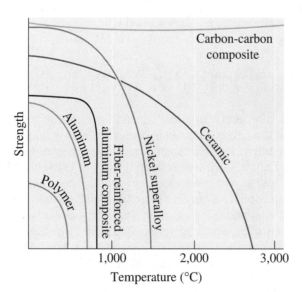

Figure 1-12
Increasing temperature normally reduces the strength of a material. Polymers are suitable only at low temperatures. Some composites, such as carbon-carbon composites, special alloys, and ceramics, have excellent properties at high temperatures.

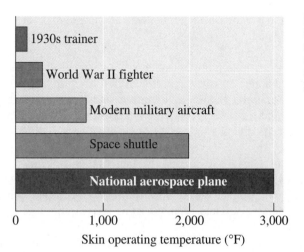

Figure 1-13
Skin operating temperatures for aircraft have increased with the development of improved materials. (*After M. Steinberg*, Scientific American, *October 1986.*)

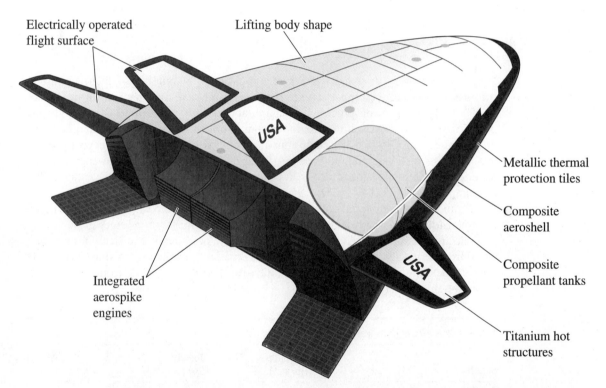

Figure 1-14   Schematic of a *X-33* plane prototype. Notice the use of different materials for different parts. This type of vehicle will test several components for the *Venturestar* (*From "A Simpler Ride into Space," by T.K. Mattingly, October, 1997,* Scientific American, *p. 125. Copyright © 1997 Slim Films.*)

Corrosion   Most metals and polymers react with oxygen or other gases, particularly at elevated temperatures. Metals and ceramics may disintegrate and polymers and non-oxide ceramics may oxidize. Materials are also attacked by corrosive liquids, leading to premature failure. The engineer faces the challenge of selecting materials or coatings that prevent these reactions and permit operation in extreme environments. In space applications, we may have to consider the effect of the presence of radiation and the presence of atomic oxygen.

Fatigue   In many applications, components must be designed such that the load on the material may not be enough to cause permanent deformation. However, when we do load and unload the material thousands of times, small cracks may begin to develop and materials fail as these cracks grow. This is known as **fatigue failure**. In designing load-bearing components, the possibility of fatigue must be accounted for.

Strain Rate   Many of you are aware of the fact that Silly Putty®, a silicone- (not silicon-) based plastic, can be stretched significantly if we pull it slowly (small rate of strain). If you pull it fast (higher rate of strain) it snaps. A similar behavior can occur with many metallic materials. Thus, in many applications, the level and rate of strain have to be considered.

In many cases, the effects of temperature, fatigue, stress, and corrosion may be interrelated, and other outside effects could affect the material's performance.

1-6   Materials Design and Selection

When a material is designed for a given application, a number of factors must be considered. The material must acquire the desired physical and mechanical properties, must be capable of being processed or manufactured into the desired shape, and must provide an economical solution to the design problem. Satisfying these requirements in a manner that protects the environment—perhaps by encouraging recycling of the materials—is also essential. In meeting these design requirements, the engineer may have to make a number of tradeoffs in order to produce a serviceable, yet marketable, product.

As an example, material cost is normally calculated on a cost-per-pound basis. We must consider the **density** of the material, or its weight-per-unit volume, in our design and selection (Table 1-2). Aluminum may cost more per pound than steel, but it is only one-third the weight of steel. Although parts made from aluminum may have to be thicker, the aluminum part may be less expensive than the one made from steel because of the weight difference.

TABLE 1-2 ■ *Strength-to-weight ratios of various materials*

Material	Strength (lb/in.^{2})	Density (lb/in.^{3})	Strength-to-weight ratio (in.)
Polyethylene	1,000	0.030	0.03×10^6
Pure aluminum	6,500	0.098	0.07×10^6
Al_2O_3	30,000	0.114	0.26×10^6
Epoxy	15,000	0.050	0.30×10^6
Heat-treated alloy steel	240,000	0.280	0.86×10^6
Heat-treated aluminum alloy	86,000	0.098	0.88×10^6
Carbon-carbon composite	60,000	0.065	0.92×10^6
Heat-treated titanium alloy	170,000	0.160	1.06×10^6
Kevlar-epoxy composite	65,000	0.050	1.30×10^6
Carbon-epoxy composite	80,000	0.050	1.60×10^6

In some instances, particularly in aerospace applications, weight is critical, since additional vehicle weight increases fuel consumption and reduces range. By using materials that are lightweight but very strong, aerospace or automobile vehicles can be designed to improve fuel utilization. Many advanced aerospace vehicles use composite materials instead of aluminum. These composites, such as carbon-epoxy, are more expensive than the traditional aluminum alloys; however, the fuel savings yielded by the higher strength-to-weight ratio of the composite (Table 1-2) may offset the higher initial cost of the aircraft. There are literally thousands of applications in which similar considerations apply. Usually the selection of materials involves trade-offs between many properties.

By this point of our discussion, we hope that you can appreciate that the properties of materials depend not only on composition, but also how the materials are made (synthesis and processing) and, most importantly, their internal structure. This is why it is not a good idea for an engineer to refer to a handbook and select a material for a given application. The handbooks may be a good starting point. A good engineer will consider: the effects of how the material was made, what exactly is the composition of the candidate material for the application being considered, any processing that may have to be done for shaping the material or fabricating a component, the structure of the material after processing into a component or device, the environment in which the material will be used, and the cost-to-performance ratio. Earlier in this chapter, we had discussed the need for you to know the principles of materials science and engineering. If you are an engineer and you need to decide which materials you will choose to fabricate a component, the knowledge of principles of materials science and engineering will empower you with the fundamental concepts. These will allow you to make technically sound decisions in designing with engineered materials.

SUMMARY

◈ Materials Science and Engineering (MSE) is an interdisciplinary field concerned with inventing new materials and devices and improving previously known materials by developing a deeper understanding of the microstructure-composition-synthesis-processing relationships.

◈ Engineered materials are materials designed and fabricated considering MSE principles.

◈ The properties of engineered materials depend upon their composition, structure, synthesis, and processing. An important performance index for materials or devices is their cost-to-performance ratio.

◈ The structure of a material refers to the arrangement of atoms or ions in the material.

◈ The structure at a microscopic level is known as the microstructure.

◈ Many properties of materials depend strongly on the structure, even if the composition of the material remains the same. This is why the structure-property or microstructure-property relationships in materials are extremely important.

◈ Materials are classified as metals and alloys, ceramics, glasses, and glass ceramics, composites, polymers, and semiconductors.

◈ Metals and alloys have good strength, good ductility, and good formability. Metals have good electrical and thermal conductivity. Metals and alloys play an indispensable role in many applications such as automotives, buildings, bridges, aerospace, and the like.

◈ Ceramics are inorganic crystalline materials. They are strong, serve as good electrical and thermal insulators, are often resistant to damage by high temperatures and corrosive environments, but are mechanically brittle. Modern ceramics form the underpinnings of many of the microelectronic and photonic technologies.

◈ Polymers have relatively low strength; however, the strength-to-weight ratio is very favorable. Polymers are not suitable for use at high temperatures. They have very good corrosion resistance, and—like ceramics—provide good electrical and thermal insulation. Polymers may be either ductile or brittle, depending on structure, temperature, and strain rate.

◈ Semiconductors possess unique electrical and optical properties that make them essential for manufacturing components in electronic and communication devices.

◈ Composites are made from different types of materials. They provide unique combinations of mechanical and physical properties that cannot be found in any single material.

◈ Functional classification of materials includes aerospace, biomedical, electronic, energy and environmental, magnetic, optical (photonic), and structural materials.

◈ Materials can also be classified as crystalline or amorphous. Crystalline materials may be single crystal or polycrystalline.

◈ Properties of materials can depend upon the temperature, level and type of stress applied, strain rate, oxidation and corrosion, and other environmental factors.

◈ Selection of a material having the needed properties and the potential to be manufactured economically and safely into a useful product is a complicated process requiring the knowledge of the structure-property-processing-composition relationships.

GLOSSARY

Alloy A metallic material that is obtained by chemical combinations of different elements (e.g., steel is made from iron and carbon). Typically, alloys have better mechanical properties than pure metals.

Ceramics A group of crystalline inorganic materials characterized by good strength, especially in compression, and high melting temperatures. Many ceramics have very good electrical and thermal insulation behavior.

Composition The chemical make-up of a material.

Composites A group of materials formed from mixtures of metals, ceramics, or polymers in such a manner that unusual combinations of properties are obtained (e.g., fiberglass).

Crystal structure The arrangement of the atoms in a crystalline material.

Crystalline material A material comprised of one or many crystals. In each crystal atoms or ions show a long-range periodic arrangement.

Density Mass per unit volume of a material, usually expressed in units of g/cm^3 or $lb/in.^3$

Fatigue failure Failure of a material due to repeated loading and unloading.

Glass   An amorphous material derived from the molten state, typically, but not always, based on silica.

Glass-ceramics   A special class of materials obtained by forming a glass and then heat treating it to form small crystals.

Grains   Crystals in a polycrystalline material.

Grain boundaries   Regions between grains of a polycrystalline material.

Materials engineering   An engineering oriented field that focuses on how to translate or transform materials into a useful device or structure.

Materials science and engineering (MSE)   An interdisciplinary field concerned with inventing new materials and improving previously known materials by developing a deeper understanding of the microstructure-composition-synthesis-processing relationships between different materials.

Materials science   A field of science that emphasizes studies of relationships between the internal or microstructure, synthesis and processing and the properties of materials.

Materials science and engineering tetrahedron   A tetrahedron diagram showing how the performance-to-cost ratio of materials depends upon the composition, microstructure, synthesis, and processing.

Mechanical properties   Properties of a material, such as strength, that describe how well a material withstands applied forces, including tensile or compressive forces, impact forces, cyclical or fatigue forces, or forces at high temperatures.

Metal   An element that has metallic bonding and generally good ductility, strength, and electrical conductivity.

Physical properties   Describe characteristics such as color, elasticity, electrical or thermal conductivity, magnetism, and optical behavior that generally are not significantly influenced by forces acting on a material.

Polycrystalline material   A material comprised of many crystals (as opposed to a single-crystal material that has only one crystal).

Polymerization   The process by which organic molecules are joined into giant molecules, or polymers.

Polymers   A group of materials normally obtained by joining organic molecules into giant molecular chains or networks. Polymers are characterized by low strengths, low melting temperatures, and poor electrical conductivity.

Plastics   These are polymeric materials consisting of other additives.

Processing   Different ways for shaping materials into useful components or changing their properties.

Semiconductors   A group of materials having electrical conductivity between metals and typical ceramics (e.g., Si, GaAs).

Single crystal   A crystalline material that is made of only one crystal (there are no grain boundaries).

Smart material   A material that can sense and respond to an external stimulus such as change in temperature, application of a stress, or change in humidity or chemical environment.

Strength-to-weight ratio   The strength of a material divided by its density; materials with a high strength-to-weight ratio are strong but lightweight.

Structure   Description of the arrangements of atoms or ions in a material. The structure of materials has a profound influence on many properties of materials, even if the overall composition does not change!

Synthesis   The process by which materials are made from naturally occurring or other chemicals.

Thermoplastics   A special group of polymers in which molecular chains are entangled but not interconnected. They can be easily melted and formed into useful shapes. Normally, these polymers have a chainlike structure (e.g., polyethylene).

Thermosets   A special group of polymers that decompose rather than melting upon heating. They are normally quite brittle due to a relatively rigid, three-dimensional network structure (e.g., polyurethane).

ADDITIONAL INFORMATION

1. CAHN, R., *The Coming of Materials Science*. 2001: Pergamon.
2. BRONOWSKI, J., *The Ascent of Man*. 1973: Back Bay Books. 124–131.
3. FORESTER, T., ed. *The Materials Revolution*. 1988, MIT Press.
4. LEHNDORFF, B.R., "High T_c Superconductors for Magnet and Energy Technology," *Springer Tracts in Modern Physics*. 2001: Springer.
5. LIVINGSTON, J.D., *Driving Force*. 1996: Harvard Press.
6. MANNHART, J. and P. CHAUDHARI, "High-T_c Bicrystal Grain Boundaries," *Physics Today*, 2001 (November): p. 48–53.
7. COOLEY, L.D., et al., "Potential Applications of Magnesium Diboride for Accelerator Magnet Applications," *Proc. of the 2001 Particle Accelerator Conf.* 2001. Chicago: IEEE.
8. *Materials Science and Engineering: Forging Stronger Links to Users*. 1999: National Academy Press.
9. FROES, F.H., "Materials for Sports," *MRS Bulletin*, 1998. 23(3).
10. MUSCAT, A.J., et al., "Interdisciplinary Teaching and Learning in a Semiconductor Processing Course," *J. Engineering Education*, 1998. 87: p. 413.
11. GAYLE, F.W. and M. GOODWAY, "Precipitation Hardening in the First Aerospace Aluminum Alloy: The Wright Flyer Crankcase," *Science*, 1994. 266(5187): p. 1015–1017.
12. WOUK, V., "Hybrid Electric Vehicles," *Scientific American*, 1997. October: p. 70–74.
13. PALAIS, J.C., *Fiber Optic Communications*. 4th ed. 1998.
14. PHULÉ, P.P. AND GINDER, J.M., *MRS Bulletin*, August 2001.
15. GANNON, R., "What Really Sank the Titanic," *Popular Science*, 1995. 246(2): p. 49–55.
16. HILL, S., "The Mystery of the Titanic: A Case of Brittle Fracture?" *Materials World*, 1996. 4(6): p. 334–335.
17. VAUGHAN, D., *The Challenger Launch Decision: Risky Technology, Culture, and Deviance at NASA*. 1996: Chicago Press.
18. MATTINGLY, T.K., "A Simpler Ride into Space," *Scientific American*, 1997. October: p. 121–125.

✓ PROBLEMS

1-1 What is Materials Science and Engineering?

1-1 Define Material Science and Engineering (MSE)?

1-2 Define the following terms: **(a)** composition, **(b)** structure, **(c)** synthesis, **(d)** processing, and **(e)** microstructure.

1-3 Explain the difference between the terms materials science and materials engineering.

1-4 Name one revolutionary discovery. Name one evolutionary discovery.

1-2 Classification of Materials

1-3 Functional Classification of Materials

1-4 Classification of Materials Based on Structure

1-5 Environmental and Other Effects

1-5 Iron is often coated with a thin layer of zinc if it is to be used outside. What characteristics do you think the zinc provides to this coated, or galvanized, steel? What precautions should be considered in producing this product? How will the recyclability of the product be affected?

1-6 We would like to produce a transparent canopy for an aircraft. If we were to use a ceramic (that is, traditional window glass) canopy, rocks or birds might cause it to shatter. Design a material that would minimize damage or at least keep the canopy from breaking into pieces.

1-7 Coiled springs ought to be very strong and stiff. Si_3N_4 is a strong, stiff material. Would you select this material for a spring? Explain.

1-8 Temperature indicators are sometimes produced from a coiled metal strip that uncoils a specific amount when the temperature increases. How does this work; from what kind of material would the indicator be made; and what are the important properties that the material in the indicator must possess?

1-9 You would like to design an aircraft that can be flown by human power nonstop for a distance of 30 km. What types of material properties would you recommend? What materials might be appropriate?

1-10 You would like to place a three-foot diameter microsatellite into orbit. The satellite will contain delicate electronic equipment that will send and receive radio signals from earth. Design the outer shell within which the electronic equipment is contained. What properties will be required, and what kind of materials might be considered?

1-11 What properties should the head of a carpenter's hammer possess? How would you manufacture a hammer head?

1-12 The hull of the space shuttle consists of ceramic tiles bonded to an aluminum skin. Discuss the design requirements of the shuttle hull that led to the use of this combination of materials. What problems in producing the hull might the designers and manufacturers have faced?

1-13 You would like to select a material for the electrical contacts in an electrical switching device which opens and closes frequently and forcefully. What properties should the contact material possess? What type of material might you recommend? Would Al_2O_3 be a good choice? Explain.

1-14 Aluminum has a density of 2.7 g/cm^3. Suppose you would like to produce a composite material based on aluminum having a density of 1.5 g/cm^3. Design a material that would have this density. Would introducing beads of polyethylene, with a density of 0.95 g/cm^3, into the aluminum be a likely possibility? Explain.

1-15 You would like to be able to identify different materials without resorting to chemical analysis or lengthy testing procedures. Describe some possible testing and sorting techniques you might be able to use based on the physical properties of materials.

1-16 You would like to be able to physically separate different materials in a scrap recycling plant. Describe some possible methods that might be used to separate materials such as polymers, aluminum alloys, and steels from one another.

1-17 Some pistons for automobile engines might be produced from a composite material containing small, hard silicon carbide particles in an aluminum alloy matrix. Explain what benefits each material in the composite may provide to the overall part. What problems might the different properties of the two materials cause in producing the part?

Chapter

Optical fibers used in the communications industry are a form of silica. Silica is the main chemical found in beach sand. The silica used to make optical fibers is amorphous and of extremely high purity. (*Courtesy of Corning Incorporated.*)

2

Atomic Structure

Have You Ever Wondered?

- *What is nanotechnology?*

- *Why carbon, in the form of diamond, is one of the hardest materials known, but as graphite is very soft and can be used as a solid lubricant?*

- *How silica, which forms the main chemical in beach sand, is used in an ultrapure form to make optical fibers?*

The goal of this chapter is to describe the underlying physical concepts related to the structure of matter. We will examine atomic structure in order to lay a foundation for understanding how it affects the properties, behaviors, and resulting applications of engineering materials. You will learn that the structure of atoms affects the types of bonds that hold materials together. These differing types of bonds directly affect suitability of materials for real-world engineering applications.

Both the **composition** and the **structure** of a material have a profound influence on its properties and behavior. Engineers and scientists who study and develop materials must understand their atomic structure. We will learn that the properties of materials are controllable and can actually be tailored to the needs of a given application by controlling their structure and composition.

We can examine and describe the structure of materials at five different levels:

1. macrostructure;
2. microstructure;
3. nanostructure.
4. short- and long-range atomic arrangements; and
5. atomic structure.

Engineers and scientists concerned with development and practical applications of advanced materials need to understand the **microstructure** and **macrostructure** of various materials and the ways of controlling them. Microstructure is the structure of material at a **length-scale** of ~10 to 1000 nm. Length-scale is a characteristic length or range of dimensions over which we are describing the properties of a material or the phenomena occurring in materials. Microstructure typically includes such features as average grain size, grain size distribution, grain orientation, and other features related to defects in materials. (A grain is a portion of the material within which the arrangement of atoms is nearly identical.) Macrostructure is the structure of a material at a macroscopic level where the length-scale is ~>1000 nm. Features that constitute macrostructure include porosity, surface coatings, and such features as internal or external micro-cracks.

It is also important to understand **atomic structure** and how the atomic bonds lead to different atomic or ionic arrangements in materials. The atomic structure includes all atoms and their arrangements, which constitute the building blocks of matter. It is from these building blocks that all the nano, micro, and macrolevels of structures emerge. The insights gained by understanding atomic structure and bonding configurations of atoms and molecules are essential for the proper selection of engineering materials, as well as for developing new advanced materials.

A close examination of atomic arrangement allows us to distinguish between materials that are **amorphous** (those that lack a long-range ordering of atoms or ions) or **crystalline** (those that exhibit periodic geometrical arrangements of atoms or ions). Amorphous materials have only **short-range atomic arrangements** while crystalline materials have short- and **long-range arrangements**. In short-range atomic arrangements the atoms or ions show a particular order only over relatively short distances. For crystalline materials, the long-range atomic order is in the form of atoms or ions arranged in a three dimensional pattern that repeats over much larger distances (from ~>100 nm to few cm).

2-1   The Structure of Materials: Technological Relevance

In today's world, information technology (IT), biotechnology, energy technology, environmental technology, and many other areas require smaller, lighter, faster, portable, more efficient, reliable, durable, and inexpensive devices. We want batteries that are smaller, lighter, and last longer. We need cars that are relatively affordable, lightweight, safe, highly fuel efficient, and "loaded" with many advanced features, ranging from global positioning systems (GPS) to sophisticated sensors for airbag deployment.

Some of these needs have generated considerable interest in **nanotechnology** and **micro-electro-mechanical systems** (MEMS).[1,2] A real-world example of the MEMS technology, Figure 2-1 shows a small accelerometer sensor obtained by the micromachining of silicon (Si). This sensor is used to measure acceleration in automobiles. The information is processed to a central computer and then used for controlling airbag

(a)   (b)

Figure 2-1 Micro-machined silicon sensors used in automotives to control airbag deployment. (*Courtesy of Dr. John Yasaitis, Analog Devices, Inc.*)

deployment. Properties and behavior of materials at these "micro" levels can vary greatly when compared to those in their "macro" or bulk state. As a result, understanding the structure at nano-scale or **nanostructure** (i.e., the structure and properties of materials at a **nano-scale** or ~length-scale 1–100 nm) and microstructure are areas that have received considerable attention. The term "nanotechnology" is used to describe a set of technologies that are based on physical, chemical, and biological phenomena occuring at nano-scale.

The applications shown in Table 2-1 and accompanying figures (Figures 2-2 through 2-7) illustrate how important the different levels of structure are to the material behavior. The applications illustrated are broken out by their levels of structure and their length-scales (the approximate characteristic length that is important for a given application). Examples of how such an application would be used within industry, as well as an illustration, are also provided.

We now turn our attention to the details concerning the structure of atoms, the bonding between atoms, and how these form a foundation for the properties of materials. Atomic structure influences how atoms are bonded together. An understanding of this helps categorize material as metals, semiconductors, ceramics, or polymers. It also permits us to draw some general conclusions concerning the mechanical properties and physical behaviors of these four classes of materials.

2-2 The Structure of the Atom

An atom is composed of a nucleus surrounded by electrons. The nucleus contains neutrons and positively charged protons and carries a net positive charge. The negatively charged electrons are held to the nucleus by an electrostatic attraction. The electrical charge q carried by each electron and proton is 1.60×10^{-19} coulomb (C). Because the numbers of electrons and protons in the atom are equal, the atom as a whole is electrically neutral.

The **atomic number** of an element is equal to the number of electrons or protons in each atom. Thus, an iron atom, which contains 26 electrons and 26 protons, has an atomic number of 26.

Most of the mass of the atom is contained within the nucleus. The mass of each proton and neutron is 1.67×10^{-24} g, but the mass of each electron is only 9.11×10^{-28} g. The **atomic mass** M, which is equal to the average number of protons and neutrons in the atom, is also the mass in grams of the **Avogadro number** N_A of atoms. The quantity $N_A = 6.02 \times 10^{23}$ atoms/mol is the number of atoms or molecules in a mole. Therefore, the atomic mass has units of g/mol. An alternate unit for atomic mass

TABLE 2-1 ■ *Levels of structures*

Level of Structure	Example of Technologies
Atomic Structure	*Diamond*: Diamond is based on carbon-carbon (C-C) covalent bonds. Materials with this type of bonding are expected to be relatively hard. Thin films of diamonds are used for providing a wear-resistant edge in cutting tools.
Atomic Arrangements: Long-Range Order (LRO)	*Lead-zirconium-titanate [Pb(Zr$_x$Ti$_{1-x}$)] or PZT:* When ions in this material are arranged such that they exhibit tetragonal and/or rhombohedral crystal structures, the material is piezoelectric (i.e., it develops a voltage when subjected to pressure or stress). PZT ceramics are used widely for many applications including gas igniters, ultrasound generation, and vibration control.
Atomic Arrangements: Short-Range Order (SRO)	*Ions in silica (SiO$_2$) glass* exhibit only a short-range order in which Si^{+4} and O^{-2} ions are arranged in a particular way (each Si^{+4} is bonded with 4 O^{-2} ions in a tetrahedral coordination). This order, however, is not maintained over long distances, thus making silica glass amorphous. Amorphous silica glasses based on silica and certain other oxides form the basis for the entire fiber optical communications industry.

Approximate Length-Scale

$\sim$Up to 10^{-10} m or 1 Å

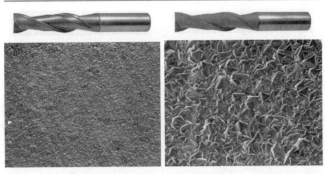

Figure 2-2   Diamond-coated cutting tools. (*Courtesy of OSG Tap & Die, Inc.*)

$\sim 10^{-10}$ to 10^{-9} m (1–10 Å), ordering can exist up to a few cm in larger crystals

Figure 2-3   Piezoelectric PZT-based gas igniters. When the piezoelectric material is stressed (by applying a pressure) a voltage develops and a spark is created between the electrodes. (*Courtesy of Morgan Electro Ceramics, Inc.*)

$\sim 10^{-10}$ to 10^{-9} m (1 to 10 Å)

Figure 2-4   Optical fibers based on a form of silica that is amorphous. (*Courtesy of Corning Incorporated.*)

TABLE 2-1 (continued)

Level of Structure	Example of Technologies
Nanostructure	Nano-sized particles (~5–10 nm) of iron oxide are used in ferrofluids or liquid magnets. These nano-sized iron oxide particles are dispersed in liquids and commercially used as ferrofluids. An application of these liquid magnets is as a cooling (heat transfer) medium for loudspeakers.
Microstructure	The mechanical strength of many metals and alloys depends very strongly on the grain size. The grains and grain boundaries in this accompanying micrograph of steel are part of the microstructural features of this crystalline material. In general, at room temperature a finer grain size leads to higher strength. Many important properties of materials are sensitive to the microstructure.
Macrostructure	Relatively thick coatings, such as paints on automobiles and other applications, are used not only for aesthetics, but to provide corrosion resistance.

Approximate Length-Scale

$\sim 10^{-9}$ to 10^{-7} m (1 to 100 nm)

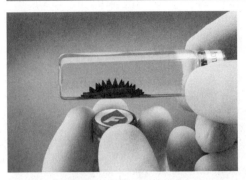

Figure 2-5   Ferrofluid. (*Courtesy of Ferro Tec, Inc.*)

$\sim > 10^{-8}$ to 10^{-6} m (10 nm to 1000 nm)

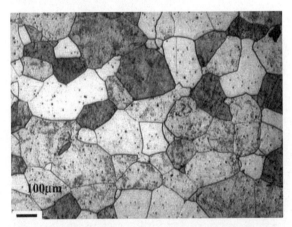

Figure 2-6   Micrograph of stainless steel showing grains and grain boundaries. (*Courtesy Dr. Hua and Dr. Deardo—University of Pittsburgh.*)

$\sim > 10^{-4}$ m (1000 nm)

Figure 2-7   A number of organic and inorganic coatings protect the steel in the car from corrosion and provide a pleasing appearance. (*Courtesy of Ford Motor Company.*)

is the **atomic mass unit**, or amu, which is 1/12 the mass of carbon 12 (i.e., carbon atom with 12 protons). As an example, one mole of iron contains 6.02×10^{23} atoms and has a mass of 55.847 g, or 55.847 amu. Calculations including a material's atomic mass and Avogadro's number are helpful to understanding more about the structure of a material. Example 2-1 illustrates how to calculate the number of atoms for silver, a metal and a good electrical conductor. Examples 2-2 and 2-3 illustrate applications to magnetic and semiconducting materials, respectively.

EXAMPLE 2-1   *Calculating the Number of Atoms in Silver*

Calculate the number of atoms in 100 g of silver (Ag).

SOLUTION

The number of atoms can be calculated from the atomic mass and Avogadro's number. From Appendix A, the atomic mass, or weight, of silver is 107.868 g/mol. The number of atoms is:

$$\text{Number of Ag atoms} = \frac{(100 \text{ g})(6.023 \times 10^{23} \text{ atoms/mol})}{107.868 \text{ g/mol}}$$

$$= 5.58 \times 10^{23}$$

EXAMPLE 2-2   *Nano-Sized Iron-Platinum Particles For Information Storage*

Scientists are considering using nano-particles of such magnetic materials as iron-platinum (Fe-Pt) as a medium for ultrahigh density data storage. Arrays of such particles potentially can lead to storage of trillions of bits of data per square inch—a capacity that will be 10 to 100 times higher than any other devices such as computer hard disks. If these scientists considered iron (Fe) particles that are 3 nm in diameter, what will be the number of atoms in one such particle?

SOLUTION

You will learn in a later chapter on magnetic materials that such particles used in recording media tend to be acicular (needle like). For now, let us assume the magnetic particles are spherical in shape.

The radius of a particle is 1.5 nm.

$$\text{Volume of each iron magnetic nano-particle} = (4/3)\pi(1.5 \times 10^{-7} \text{ cm})^3$$

$$= 1.4137 \times 10^{-20} \text{ cm}^3$$

Density of iron = 7.8 g/cm^3. Atomic mass of iron is 56 g/mol.

$$\text{Mass of each iron nano-particle} = 7.8 \text{ g/cm}^3 \times 1.4137 \times 10^{-20} \text{ cm}^3$$

$$= 1.102 \times 10^{-19} \text{ g.}$$

One mole or 56 g of Fe contains 6.023×10^{23} atoms, therefore, the number of atoms in one Fe nano-particle will be 1186. This is a very small number of atoms. Compare this with the number of atoms in an iron particle that is 10 micrometers in diameter. Such larger iron particles often are used in breakfast cereals, vitamin tablets, and other applications.

EXAMPLE 2-3 *Dopant Concentration In Silicon Crystals*

Silicon single crystals are used extensively to make computer chips. Calculate the *concentration* of silicon atoms in silicon, or the number of silicon atoms per unit volume of silicon. During the growth of silicon single crystals it is often desirable to deliberately introduce atoms of other elements (known as **dopants**) to control and change the electrical conductivity and other electrical properties of silicon. Phosphorus (P) is one such dopant that is added to make silicon crystals *n*-type semiconductors. Assume that the concentration of P atoms required in a silicon crystal is 10^{17} atoms/cm^3. Compare the concentrations of atoms in silicon and the concentration of P atoms. What is the significance of these numbers from a technological viewpoint? Assume that density of silicon is 2.33 g/cm^3.

SOLUTION

First, we want to calculate the number of silicon atoms per unit volume. From the periodic table of elements (Figure 2-11), we know that the atomic mass of silicon is 28.09 g/mol. That is 28.09 g of silicon contain 6.023×10^{23} atoms. Therefore, 2.33 g of silicon will contain $(2.33 \times 6.023 \times 10^{23}/28.09)$ atoms = 4.99×10^{22} atoms. We also know the mass of one cm^3 of Si is 2.33 g. Therefore, the concentration of silicon atoms in pure silicon is ~5×10^{22} atoms/cm^3.

Significance of comparing dopant and Si atom concentrations: If we were to add phosphorus (P) into this crystal, such that the concentration of P is 10^{17} atoms/cm^3, the ratio of concentration of atoms in silicon to that of P will be $5 \times 10^{22}/10^{17} = 5 \times 10^5$. This says that only 1 out of 500,000 atoms of the doped crystal will be that of phosphorus (P)! This is equivalent to one apple in 500,000 oranges! This explains why the single crystals of silicon must have exceptional purity and at the same time very small and uniform levels of dopants. Materials scientists and engineers have developed the technology to grow large, high-purity crystals of silicon and other materials.

2-3 The Electronic Structure of the Atom

Electrons occupy discrete energy levels within the atom. Each electron possesses a particular energy, with no more than two electrons in each atom having the same energy. This also implies that there is a discrete energy difference between any two energy levels.

Quantum Numbers The energy level to which each electron belongs is determined by four **quantum numbers**. Quantum numbers are the numbers in an atom that assign electrons to discrete energy levels. The four quantum numbers are the principal quantum number n, the Azimuthal quantum number l, the magnetic quantum number m_l, and the spin quantum number m_s. Azimuthal quantum numbers describe the energy levels in each quantum shell. The **spin quantum number** (m_s) is assigned values $+1/2$ and $-1/2$ and reflects the different electronic spins.

The number of possible energy levels is determined by the first three quantum numbers.

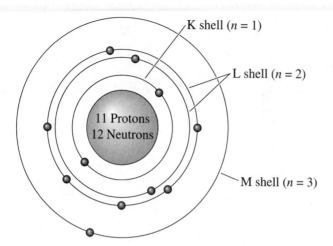

K shell ($n = 1$)

L shell ($n = 2$)

11 Protons
12 Neutrons

M shell ($n = 3$)

Figure 2-8
The atomic structure of
sodium, atomic number 11,
showing the electrons in the K,
L, and M quantum shells.

1. The principal quantum number n is assigned integral values 1, 2, 3, 4, 5, ... that refer to the quantum shell to which the electron belongs (Figure 2-8). A **quantum shell** is a set of fixed energy levels to which electrons belong. Each electron in the shell is designated by four quantum numbers.

   Quantum shells are also assigned a letter; the shell for $n = 1$ is designated K, for $n = 2$ is L, for $n = 3$ is M, and so on.

2. The number of energy levels in each quantum shell is determined by the **azimuthal quantum number** l and the **magnetic quantum number** m_l. The magnetic quantum number describes the number of energy levels for each azimuthal quantum number. The azimuthal quantum numbers are also assigned numbers: $l = 0, 1, 2, \ldots, n - 1$. If $n = 2$, then there are also two azimuthal quantum numbers, $l = 0$ and $l = 1$. The azimuthal quantum numbers are designated by lowercase letters:

$$s \text{ for } l = 0 \qquad d \text{ for } l = 2$$
$$p \text{ for } l = 1 \qquad f \text{ for } l = 3$$

3. The magnetic quantum number m_l gives the number of energy levels, or orbitals, for each azimuthal quantum number. The total number of magnetic quantum numbers for each l is $2l + 1$. The values for m_l are given by whole numbers between $-l$ and $+l$. For $l = 2$, there are $2(2) + 1 = 5$ magnetic quantum numbers, with values $-2, -1, 0, +1,$ and $+2$.

4. The **Pauli exclusion principle** specifies that no more than two electrons, with opposing electronic spins, may be present in each orbital. Figure 2-9 shows the quantum numbers and energy levels for each electron in a sodium atom.

The shorthand notation frequently used to denote the electronic structure of an atom combines the numerical value of the principal quantum number, the lowercase letter notation for the azimuthal quantum number, and a superscript showing the number of electrons in each orbital. The shorthand notation for germanium, which has an atomic number of 32, is:

$$1s^2 2s^2 2p^6 3s^2 3p^6 3d^{10} 4s^2 4p^2$$

The electronic configurations for the elements are summarized in Appendix C; the energy levels are summarized in Table 2-2.

$3s^1$ _____

electron 11 $n = 3,\ l = 0,\ m_l = 0,\quad m_s = +\frac{1}{2}$ or $-\frac{1}{2}$

$2p^6$ _____

$\begin{cases} \text{electron 10} \quad n = 2,\ l = 1,\ m_l = +1,\ m_s = -\frac{1}{2} \\ \text{electron 9} \quad n = 2,\ l = 1,\ m_l = +1,\ m_s = +\frac{1}{2} \end{cases}$

$\begin{cases} \text{electron 8} \quad n = 2,\ l = 1,\ m_l = 0,\ m_s = -\frac{1}{2} \\ \text{electron 7} \quad n = 2,\ l = 1,\ m_l = 0,\ m_s = +\frac{1}{2} \end{cases}$

$\begin{cases} \text{electron 6} \quad n = 2,\ l = 1,\ m_l = -1,\ m_s = -\frac{1}{2} \\ \text{electron 5} \quad n = 2,\ l = 1,\ m_l = -1,\ m_s = +\frac{1}{2} \end{cases}$

$2s^2$ _____

$\begin{cases} \text{electron 4} \quad n = 2,\ l = 0,\ m_l = 0,\ m_s = -\frac{1}{2} \\ \text{electron 3} \quad n = 2,\ l = 0,\ m_l = 0,\ m_s = +\frac{1}{2} \end{cases}$

$1s^2$ _____

$\begin{cases} \text{electron 2} \quad n = 1,\ l = 0,\ m_l = 0,\ m_s = -\frac{1}{2} \\ \text{electron 1} \quad n = 1,\ l = 0,\ m_l = 0,\ m_s = +\frac{1}{2} \end{cases}$

Figure 2-9
The complete set of quantum numbers for each of the 11 electrons in sodium.

TABLE 2-2 ■ *The pattern used to assign electrons to energy levels*

	$l = 0$ (s)	$l = 1$ (p)	$l = 2$ (d)	$l = 3$ (f)	$l = 4$ (g)	$l = 5$ (h)
$n = 1$ (K)	2					
$n = 2$ (L)	2	6				
$n = 3$ (M)	2	6	10			
$n = 4$ (N)	2	6	10	14		
$n = 5$ (O)	2	6	10	14	18	
$n = 6$ (P)	2	6	10	14	18	22

Note: 2, 6, 10, 14, ... refer to the number of electrons in the energy level.

Deviations from Expected Electronic Structures   The orderly building up of the electronic structure is not always followed, particularly when the atomic number is large and the d and f levels begin to fill. For example, we would expect the electronic structure of iron, atomic number 26, to be:

$$1s^2 2s^2 2p^6 3s^2 3p^6 \boxed{3d^8}$$

The actual structure, however, is:

$$1s^2 2s^2 2p^6 3s^2 3p^6 \boxed{3d^6 4s^2}$$

The unfilled $3d$ level causes the magnetic behavior of iron, as shown in Chapter 19.

Valence   The **valence** of an atom is the number of electrons in an atom that participate in bonding or chemical reactions. Usually, the valence is the number of electrons in the outer s and p energy levels. The valence of an atom is related to the ability of the atom to enter into chemical combination with other elements. Examples of the valence are:

Mg: $1s^2 2s^2 2p^6$ $\boxed{3s^2}$ valence = 2

Al: $1s^2 2s^2 2p^6$ $\boxed{3s^2 3p^1}$ valence = 3

Ge: $1s^2 2s^2 2p^6 3s^2 3p^6 3d^{10}$ $\boxed{4s^2 4p^2}$ valence = 4

Valence also depends on the immediate environment surrounding the atom or the neighboring atoms available for bonding. Phosphorus has a valence of five when it combines with oxygen. But the valence of phosphorus is only three—the electrons in the $3p$ level—when it reacts with hydrogen. Manganese may have a valence of 2, 3, 4, 6, or 7!

Atomic Stability and Electronegativity   If an atom has a valence of zero, the element is inert (non-reactive). An example is argon, which has the electronic structure:

$$1s^2 2s^2 2p^6 \quad \boxed{3s^2 3p^6}$$

Other atoms prefer to behave as if their outer s and p levels are either completely full, with eight electrons, or completely empty. Aluminum has three electrons in its outer s and p levels. An aluminum atom readily gives up its outer three electrons to empty the $3s$ and $3p$ levels. The atomic bonding and the chemical behavior of aluminum are determined by the mechanism through which these three electrons interact with surrounding atoms.

On the other hand, chlorine contains seven electrons in the outer $3s$ and $3p$ levels. The reactivity of chlorine is caused by its desire to fill its outer energy level by accepting an electron.

Electronegativity describes the tendency of an atom to gain an electron. Atoms with almost completely filled outer energy levels—such as chlorine—are strongly electronegative and readily accept electrons. However, atoms with nearly empty outer levels—such as sodium—readily give up electrons and have low electronegativity. High atomic number elements also have low electronegativity because the outer electrons are at a greater distance from the positive nucleus, so that they are not as strongly attracted to the atom. Electronegativities for some elements are shown in Figure 2-10. Example 2-4 illustrates a comparison of electronic structures using what we just learned about electronegativity. Elements with low electronegativity (i.e., <2.0) are sometimes described as electropositive.

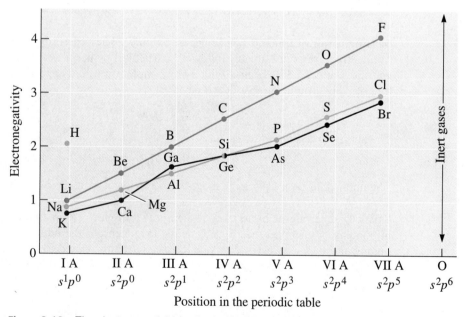

Figure 2-10   The electronegativities of selected elements relative to the position of the elements in the periodic table.

EXAMPLE 2-4   *Comparing Electronegativities*

Using the electronic structures, compare the electronegativities of calcium and bromine.

SOLUTION

The electronic structures, obtained from Appendix C, are:

$$\text{Ca: } 1s^2 2s^2 2p^6 3s^2 3p^6 \; \boxed{4s^2}$$

$$\text{Br: } 1s^2 2s^2 2p^6 3s^2 3p^6 3d^{10} \; \boxed{4s^2 4p^5}$$

Calcium has two electrons in its outer $4s$ orbital and bromine has seven electrons in its outer $4s4p$ orbital. Calcium, with an electronegativity of 1.0, tends to give up electrons and has low electronegativity, but bromine, with an electronegativity of 2.8, tends to accept electrons and is strongly electronegative. This difference in electronegativity values suggests that these elements may react readily to form a compound.

2-4   The Periodic Table

The periodic table contains valuable information about specific elements, and can also help identify trends in atomic size, melting point, chemical reactivity, and other properties. The familiar periodic table (Figure 2-11) is constructed in accordance with the electronic structure of the elements. Not all elements in the periodic table are naturally occurring. Rows in the periodic table correspond to quantum shells, or principal quantum numbers. Columns typically refer to the number of electrons in the outermost s and p energy levels and correspond to the most common valence. In engineering, we are mostly concerned with:

(a) polymers (plastics) (primarily based on carbon, which appears in group 4B);

(b) ceramics (typically based on combinations of many elements appearing in Groups 1 through 5B, and such elements as oxygen, carbon, and nitrogen); and

(c) metallic materials (typically based on elements in Groups 1, 2 and transition metal elements).

Many technologically important semiconductors appear in group 4B (e.g., silicon (Si), diamond (C), germanium (Ge)). Semiconductors also can be combinations of elements from groups 2B and 6B (e.g., cadmium selenide (CdSe), based on cadmium (Cd) from group 2 and seleneium (Se) based on Group 6). These are known as **II–VI** (two-six) **semiconductors**. Similarly, gallium arsenide (GaAs) is a **III–V** (three-five) **semiconductor** based on gallium (Ga) from group 3B and arsenic (As) from group 5B. Many **transition elements** (e.g., titanium (Ti), vanadium (V), iron (Fe), nickel (Ni), cobalt (Co), etc.) are particularly useful for magnetic and optical materials due to their electronic configuration that allows multiple valencies.

Trends in Properties   The periodic table contains a wealth of useful information (e.g., atomic mass, atomic number of different elements, etc.). It also points to trends in

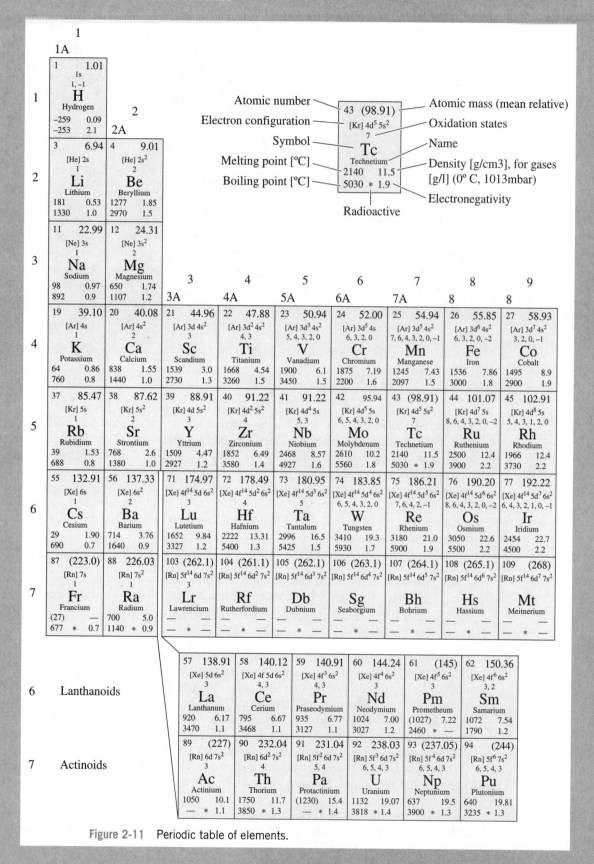

Figure 2-11 Periodic table of elements.

18

0

2	4.00
$1s^2$	
He	
Helium	
—	0.18
−269	—

13 **14** **15** **16** **17**

3B **4B** **5B** **6B** **7B**

5	10.81
[He] $2s^2\,2p$	
3	
B	
Boron	
(2030)	2.35
2550	2.0

6	12.01
[He] $2s^2\,2p^2$	
4, 2, −4	
C	
Carbon	
(3550)	2.2
4830	2.5

7	14.01
[He] $2s^2\,2p^3$	
5, 4, 3, 2, −3	
N	
Nitrogen	
−210	1.25
−196	3.0

8	16.00
[He] $2s^2\,2p^4$	
−2, −1	
O	
Oxygen	
−219	1.43
−183	3.5

9	19.00
[He] $2s^2\,2p^5$	
−1	
F	
Fluorine	
−220	1.7
−188	4.0

10	20.18
[He] $2s^2\,3p^5$	
Ne	
Neon	
−249	0.9
−246	—

13	26.98
[Ne] $3s^2\,3p$	
3	
Al	
Aluminum	
660	2.70
2450	1.5

14	28.09
[Ne] $3s^2\,3p^2$	
4, −4	
Si	
Silicon	
1410	2.33
2680	1.8

15	30.97
[Ne] $3s^2\,3p^3$	
5, 3, −3	
P	
Phosphorus	
44	1.82
280	2.1

16	32.06
[Ne] $3s^2\,3p^4$	
6, 4, 2, −2	
S	
Sulfur	
119	2.07
445	2.5

17	35.45
[Ne] $3s^2\,3p^5$	
7, 5, 3, 1, −1	
Cl	
Chlorine	
−101	3.2
−35	3.0

18	39.95
[Ne] $3s^2\,3p^6$	
Ar	
Argon	
−189	1.78
−183	—

10 **11** **12**

8 **1B** **2B**

28	58.70
[Ar] $3d^8\,4s^2$	
3, 2, 0	
Ni	
Nickel	
1453	8.9
2730	1.9

29	63.55
[Ar] $3d^{10}\,4s$	
2, 1	
Cu	
Copper	
1453	8.9
2730	1.9

30	65.38
[Ar] $3d^{10}\,4s^2$	
2	
Zn	
Zinc	
420	7.13
906	1.6

31	69.72
[Ar] $3d^{10}\,4s^2\,4p$	
3	
Ga	
Gallium	
30	5.91
2237	1.6

32	72.59
[Ar] $3d^{10}\,4s^2\,4p^2$	
4	
Ge	
Germanium	
937	5.32
2830	1.8

33	74.91
[Ar] $3d^{10}\,4s^2\,4p^3$	
5, 3, −3	
As	
Arsenic	
Subl.	5.72
	2.0

34	74.91
[Ar] $3d^{10}\,4s^2\,4p^4$	
6, 4, 2	
Se	
Selenium	
217	4.79
685	2.4

35	79.90
[Ar] $3d^{10}\,4s^2\,4p^5$	
7, 5, 3, 1, −1	
Br	
Bromine	
−7	3.12
58	2.8

36	83.80
[Ar] $3d^{10}\,4s^2\,4p^6$	
Kr	
Krypton	
−157	3.7
−152	—

46	106.42
[Kr] $4d^{10}$	
4, 2, 0	
Pd	
Palladium	
1552	12.0
3140	2.2

47	107.87
[Kr] $4d^{10}\,5s$	
2, 1	
Ag	
Silver	
961	10.5
2210	1.9

48	112.41
[Kr] $4d^{10}\,5s^2$	
2	
Cd	
Cadmium	
321	8.65
765	1.7

49	114.82
[Kr] $4d^{10}\,5s^2\,5p$	
3	
In	
Indium	
156	7.31
2080	1.7

50	118.69
[Kr] $4d^{10}\,5s^2\,5p^2$	
4, 2	
Sn	
Tin	
232	7.30
2270	1.8

51	121.75
[Kr] $4d^{10}\,5s^2\,5p^3$	
5, 3, −3	
Sb	
Antimony	
631	6.69
1380	1.9

52	127.60
[Kr] $3d^{10}\,5s^2\,5p^4$	
6, 4, −2	
Te	
Tellurium	
450	6.24
990	2.1

53	126.90
[Kr] $4d^{10}\,5s^2\,5p^5$	
7, 5, 1, −1	
I	
Iodine	
114	4.94
183	2.5

54	131.29
[Kr] $4d^{10}\,5s^2\,5p^6$	
Xe	
Xenon	
−112	5.89
−108	—

78	195.08
[Xe] $4f^{14}\,5d^9\,6s^2$	
4, 2, 0	
Pt	
Platinum	
1769	21.4
3830	2.2

79	196.97
[Xe] $4f^{14}\,5d^{10}\,6s$	
3, 1	
Au	
Gold	
1063	19.3
2970	2.4

80	200.59
[Xe] $4f^{14}\,5d^{10}\,6s^2$	
2, 1	
Hg	
Mercury	
−38	13.6
357	1.9

81	204.38
[Xe] $4f^{14}\,5d^{10}\,6s^2\,6p$	
3, 1	
Tl	
Thallium	
303	11.85
1457	1.8

82	207.20
[Xe] $4f^{14}\,5d^{10}\,6s^2\,6p^2$	
4, 2	
Pb	
Lead	
327	11.4
1725	1.9

83	208.98
[Xe] $4f^{14}\,5d^{10}\,6s^2\,6p^3$	
5, 3	
Bi	
Bismuth	
271	9.8
1560	1.9

84	(209)
[Xe] $4f^{14}\,5d^{10}\,6s^2\,6p^4$	
6, 4, 2	
Po	
Polonium	
254	9.3
962	2.0

85	(210)
[Xe] $4f^{14}\,5d^{10}\,6s^2\,6p^5$	
7, 5, 3, 1, −1	
At	
Astatine	
(302)	
337	2.2

86	(222)
[Xe] $4f^{14}\,5d^{10}\,6s^2\,6p^6$	
Rn	
Radon	
(−71)	9.73
−62	—

110	(269)
[Rn] $5f^{14}\,6d^8\,7s^2$	
Uun	
—	—
—	* —

111	(272)
[Rn] $5f^{14}\,6d^{10}\,7s^2$	
Uuu	
—	* —

112	(277)
Uub	
—	* —

114	(289)
Uuq	
—	* —

116	(289)
Uuh	
—	* —

118	(293)
Uuo	
—	* —

63	151.96
[Xe] $4f^7\,6s^2$	
3, 2	
Eu	
Europium	
828	5.26
1439	—

64	157.25
[Xe] $4f^7\,5d\,6s^2$	
3	
Gd	
Gadolinium	
1312	7.89
3000	1.1

65	158.93
[Xe] $4f^9\,6s^2$	
4, 3	
Tb	
Terbium	
1356	8.27
2800	1.2

66	162.50
[Xe] $4f^{10}\,6s^2$	
3	
Dy	
Dysprosium	
1407	8.54
2600	—

67	164.93
[Xe] $4f^{11}\,6s^2$	
3	
Ho	
Holmium	
1461	8.80
2600	1.2

68	167.26
[Xe] $4f^{12}\,6s^2$	
3	
Er	
Erbium	
1497	9.05
2900	1.2

69	168.93
[Xe] $4f^{13}\,6s^2$	
3, 2	
Tm	
Thulium	
1545	9.33
1727	1.2

70	173.04
[Xe] $4f^{14}\,6s^2$	
3, 2	
Yb	
Ytterbium	
824	6.98
1196	1.1

95	(243)
[Rn] $5f^7\,7s^2$	
6, 5, 4, 3	
Am	
Americium	
994	13.7
—	* 1.3

96	(247)
[Rn] $5f^7\,6d\,7s^2$	
4, 3	
Cm	
Curium	
(1340)	13.51
3100	* —

97	(247)
[Rn] $5f^9\,7s^2$	
4, 3	
Bk	
Berkelium	
—	* —

98	(251)
[Rn] $5f^{10}\,7s^2$	
4, 3	
Cf	
Californium	
—	* —

99	(254)
[Rn] $5f^{11}\,7s^2$	
Es	
Einsteinium	
—	* —

100	(257)
[Rn] $5f^{12}\,7s^2$	
Fm	
Fermium	
—	* —

101	(258)
[Rn] $5f^{13}\,7s^2$	
Md	
Mendelevium	
—	* —

102	(259)
[Rn] $5f^{14}\,7s^2$	
3, 2	
No	
Nobelium	
—	* —

atomic size, melting points, and chemical reactivity. For example, carbon (in its diamond form) has the highest melting point (3550°C). Melting points of the elements below carbon decrease (i.e., silicon (Si) (1410°C), germanium (Ge) (937°C), tin (Sn) (232°C), and lead (Pb) (327°C). Note that the melting temperature of Pb is higher than that of Sn. What we can conclude are trends and not exact variations in properties.

We also can discern trends in other properties from the periodic table. Diamond (carbon), a group 4B element, is a material with a very large bandgap (i.e., it is not a very effective conductor of electricity). This is consistent with the fact that it has the highest melting point among group 4 elements, which suggests the interatomic forces are strong (see Section 2-6). As we move down the column, the bandgap decreases (the bandgaps of Si and Ge are 1.11 and 0.67 eV, respectively). Moving further down column 4, one form of tin is a semiconductor. Another form of tin is metallic. If we look at group 1A, we see that lithium is highly **electropositive** (i.e., an element whose atoms want to participate in chemical interactions by donating electrons and are therefore highly reactive). Likewise, if we move down column 1A, we can see that the chemical reactivity of elements decreases.

Thus, the periodic table gives us useful information about formulas, atomic numbers, and atomic masses of elements. It also helps us in predicting or rationalizing trends in properties of elements and compounds. This is why the periodic table is very useful to both scientists and engineers.

2-5 Atomic Bonding

There are four important mechanisms by which atoms are bonded in engineered materials. These are:

1. **metallic bond**;
2. **covalent bond**;
3. **ionic bond**; and
4. **van der Waals bond**.

In the first three of these mechanisms, bonding is achieved when the atoms fill their outer *s* and *p* levels. These bonds are relatively strong and are known as **primary bonds** (relatively strong bonds between adjacent atoms resulting from the transfer or sharing of outer orbital electrons). The van der Waals bonds are secondary bonds and originate from a different mechanism and are relatively weaker. Let's look at each of these types of bonds.

The Metallic Bond   The metallic elements have more electropositive atoms that donate their valence electrons to form a "sea" of electrons surrounding the atoms (Figure 2-12). Aluminum, for example, gives up its three valence electrons, leaving behind a core consisting of the nucleus and inner electrons. Since three negatively charged electrons are missing from this core, it has a positive charge of three. The valence electrons move freely within the electron sea and become associated with several atom cores. The positively charged ion cores are held together by mutual attraction to the electron, thus producing a strong metallic bond.

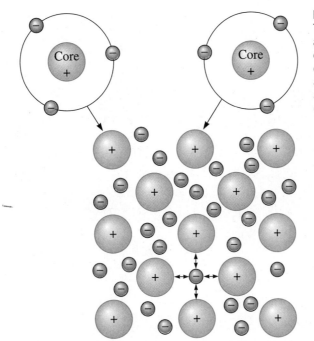

Figure 2-12
The metallic bond forms when atoms give up their valence electrons, which then form an electron sea. The positively charged atom cores are bonded by mutual attraction to the negatively charged electrons.

Because their valence electrons are not fixed in any one position, most pure metals are good electrical conductors of electricity at relatively low temperatures ($\sim T <$ 300 K). Under the influence of an applied voltage, the valence electrons move (Figure 2-13), causing a current to flow if the circuit is complete. Example 2-5 explores the conductivity and valence electrons of silver.

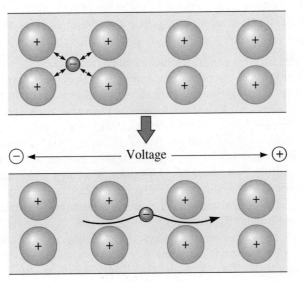

Figure 2-13
When voltage is applied to a metal, the electrons in the electron sea can easily move and carry a current.

EXAMPLE 2-5   Calculating the Conductivity of Silver

Calculate the number of electrons capable of conducting an electrical charge in ten cubic centimeters of silver.

SOLUTION

The valence of silver is one, and only the valence electrons are expected to conduct the electrical charge. From Appendix A, we find that the density of silver is 10.49 g/cm^3. The atomic mass of silver is 107.868 g/mol.

$$\text{Mass of 10 cm}^3 = (10 \text{ cm}^3)(10.49 \text{ g/cm}^3) = 104.9 \text{ g}$$

$$\text{Atoms} = \frac{(104.9 \text{ g})(6.023 \times 10^{23} \text{ atoms/mol})}{107.868 \text{ g/mol}} = 5.85 \times 10^{23}$$

$$\text{Electrons} = (5.85 \times 10^{23} \text{ atoms})(1 \text{ valence electron/atom})$$

$$= 5.85 \times 10^{23} \text{ valence electron/atom per 10 cm}^3$$

Materials with metallic bonding exhibit relatively high Young's modulus since the bonds are strong. Metals also show good ductility since the metallic bonds are non-directional. There are other important reasons related to microstructure that can explain why metals actually exhibit *lower strengths* and *higher ductility* than what we may anticipate from their bonding. **Ductility** refers to the ability of materials to be stretched or bent without breaking. We will discuss these concepts in greater detail in Chapter 6. In general, the melting points of metals are relatively high. From an optical properties viewpoint, metals make good reflectors of visible radiation. Owing to their electropositive character, many metals such as iron tend to undergo corrosion or oxidation. Many pure metals are good conductors of heat and are effectively used in many heat transfer applications. We emphasize that metallic bonding is *one of the factors* in our efforts to rationalize the trends in observed properties of metallic materials. As we will see in some of the following chapters, there are other factors related to microstructure that also play a crucial role in determining the properties of metallic materials.

The Covalent Bond   Materials with **covalent bonding** are characterized by bonds that are formed by sharing of valence electrons among two or more atoms. For example, a silicon atom, which has a valence of four, obtains eight electrons in its outer energy shell by sharing its electrons with four surrounding silicon atoms (Figure 2-14). Each instance of sharing represents one covalent bond; thus, each silicon atom is bonded to four neighboring atoms by four covalent bonds. In order for the covalent bonds to be formed, the silicon atoms must be arranged so the bonds have a fixed **directional relationship** with one another. A directional relationship is formed when the bonds between atoms in a covalently bonded material form specific angles, depending on the material. In the case of silicon, this arrangement produces a tetrahedron, with angles of 109.5° between the covalent bonds (Figure 2-15).

Covalent bonds are very strong. As a result, covalently bonded materials are very strong and hard. For example, diamond (C), silicon carbide (SiC), silicon nitride (Si_3N_4), and boron nitride (BN) all exhibit covalency. These materials also exhibit very high melting points, which means they could be useful for high-temperature applica-

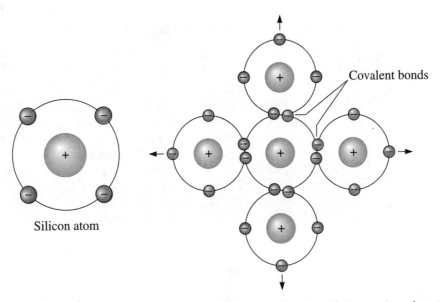

Covalent bonds

Figure 2-14 Covalent bonding requires that electrons be shared between atoms in such a way that each atom has its outer *sp* orbital filled. In silicon, with a valence of four, four covalent bonds must be formed.

tions. On the other hand, the temperature resistance of these materials present challenges in their processing. The materials bonded in this manner typically have limited ductility because the bonds tend to be directional. The electrical conductivity of many covalently bonded materials (i.e., silicon, diamond, and many ceramics) is not high since the valence electrons are locked in bonds between atoms and are not readily available for conduction. With some of these materials such as Si, we can get useful and controlled levels of electrical conductivity by deliberately introducing small levels of other elements known as dopants. Conductive polymers are also a good example of covalently bonded materials that can be turned into semiconducting materials.[3] The development of conducting polymers that are lightweight has captured the attention of many scientists and engineers for developing flexible electronic components.

We cannot simply predict whether or not a material will be high or low strength, ductile or brittle, simply based on the nature of bonding! We need additional information on the atomic, microstructure, and macrostructure of the material. However, the nature of bonding does point to a trend for materials with certain types of bonding and chemical compositions. Example 2-6 explores how one such bond of oxygen and silicon join to form silica.

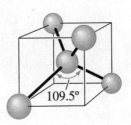

Figure 2-15
Covalent bonds are directional. In silicon, a tetrahedral structure is formed, with angles of 109.5° required between each covalent bond.

EXAMPLE 2-6   *How Do Oxygen and Silicon Atoms Join to Form Silica?*

Assuming that silica (SiO_2) has 100% covalent bonding, describe how oxygen and silicon atoms in silica (SiO_2) are joined.

SOLUTION

Silicon has a valence of four and shares electrons with four oxygen atoms, thus giving a total of eight electrons for each silicon atom. However, oxygen has a valence of six and shares electrons with two silicon atoms, giving oxygen a total of eight electrons. Figure 2-16 illustrates one of the possible structures. Similar to silicon (Si), a tetrahedral structure also is produced. We will discuss later in this chapter how to account for the ionic and covalent nature of bonding in silica.

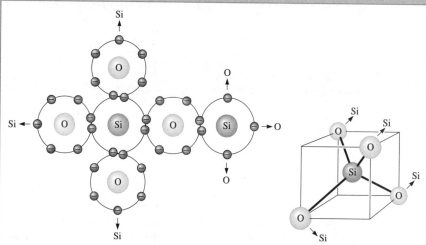

Figure 2-16   The tetrahedral structure of silica (SiO_2), which contains covalent bonds between silicon and oxygen atoms (for Example 2-6).

EXAMPLE 2-7   *Design of a Thermistor*

A **thermistor** is a device used to measure temperature by taking advantage of the change in electrical conductivity when the temperature changes. Select a material that might serve as a thermistor in the 500 to 1000°C temperature range (see Figure 2-17).

SOLUTION

The resistance of a thermistor can be made to increase or decrease with increasing temperature. These are known as positive temperature coefficient of resistance (*PTCR*) or negative temperature coefficient of resistance (*NTCR*) thermistors, respectively. The fact that a thermistor changes its resistance in response to a temperature change is used to control temperature or switch (turn "on" and "off") the operation of an electrical circuit when a particular device (i.e., a refrigerator, hairdryer, furnace, oven, or reactor) reaches a certain temperature.

Packaging Configurations

PTC Applications

Figure 2-17
Photograph of a commercially available thermistor (for Example 2-7). (*Courtesy of Vishay Intertechnology, Inc.*)

Two design requirements must be satisfied. First, a material with a high melting point must be selected. Second, the electrical conductivity of the material must show a systematic and reproducible change as a function of temperature. Covalently bonded materials might be suitable. They often have high melting temperatures, and, as more covalent bonds are broken when the temperature increases, increasing numbers of electrons become available to transfer electrical charge.

The semiconductor silicon is one choice: Silicon melts at 1410°C and is covalently bonded. A number of ceramic materials also have high melting points and behave as semiconducting materials. Silicon will have to be protected against oxidation. We will have to make sure the changes in conductivity in the temperature range are actually acceptable. Some thermistors that show a predictable decrease in the resistance with increasing temperature are made from semiconducting materials.

Polymers would *not* be suitable, even though the major bonding is covalent, because of their relatively low melting, or decomposition, temperatures. Many thermistors that can be used for switching applications make use of barium titanate ($BaTiO_3$) based formulations. Many useful *NTCR* materials are based on Fe_3O_4-$ZnCr_2O_4$, Fe_3O_4-$MgCr_2O_4$, or Mn_3O_4, doped with Ni, Co, or Cu.[4]

In almost any design situation, once the technical performance criteria are met we should always pay attention to and take into account the cost of raw materials, manufacturing costs, and other important factors such as the durability. In some applications, we also need to pay closer attention to the environmental impact including the ability to recycle materials.

The Ionic Bond   When more than one type of atom is present in a material, one atom may donate its valence electrons to a different atom, filling the outer energy shell of the second atom. Both atoms now have filled (or emptied) outer energy levels, but both have acquired an electrical charge and behave as ions. The atom that contributes the electrons is left with a net positive charge and is called a **cation**, while the atom that accepts the electrons acquires a net negative charge and is called an **anion**. The oppositely charged ions are then attracted to one another and produce the **ionic bond**. For

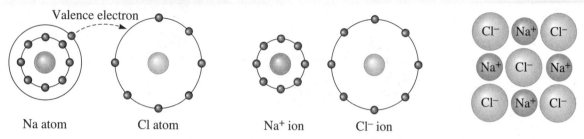

Valence electron

Na atom   Cl atom   Na⁺ ion   Cl⁻ ion

Figure 2-18 An ionic bond is created between two unlike atoms with different electronegativities. When sodium donates its valence electron to chlorine, each becomes an ion; attraction occurs, and the ionic bond is formed.

example, the attraction between sodium and chloride ions (Figure 2-18) produces sodium chloride (NaCl), or table salt.

EXAMPLE 2-8   *Describing the Ionic Bond Between Magnesium and Chlorine*

Describe the ionic bonding between magnesium and chlorine.

SOLUTION

The electronic structures and valences are:

$$Mg: 1s^2 2s^2 2p^6 \quad \boxed{3s^2} \qquad \text{valence} = 2$$

$$Cl: 1s^2 2s^2 2p^6 \quad \boxed{3s^2 3p^5} \qquad \text{valence} = 7$$

Each magnesium atom gives up its two valence electrons, becoming a Mg^{2+} ion. Each chlorine atom accepts one electron, becoming a Cl^- ion. To satisfy the ionic bonding, there must be twice as many chloride ions as magnesium ions present, and a compound, $MgCl_2$, is formed.

Solids that exhibit considerable ionic bonding are also often mechanically strong because of the strength of the bonds. Electrical conductivity of ionically bonded solids is very limited. A large fraction of the electrical current is transferred via the movement of ions (Figure 2-19). Owing to their size, ions typically do not move as easily as electrons. However, in many technological applications we make use of the electrical conduction that can occur via movement of ions as a result of increased temperature, chemical potential gradient, or an electrochemical driving force. Examples of these include, lithium ion batteries that make use of lithium cobalt oxide, conductive indium tin oxide coatings on glass for touch sensitive screens for displays, and solid oxide fuel cells based on compositions based on zirconia (ZrO_2).

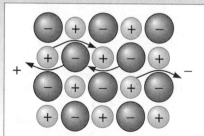

Figure 2-19
When voltage is applied to an ionic material, entire ions must move to cause a current to flow. Ion movement is slow and the electrical conductivity is poor (for Example 2-8).

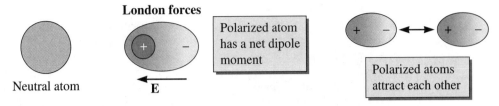

London forces

Neutral atom

E

Polarized atom has a net dipole moment

Polarized atoms attract each other

Figure 2-20   Illustration of London forces, a type of a van der Waals force, between atoms.

Van der Waals Bonding   The origin of van der Waals forces between atoms and molecules is quantum mechanical in nature and a meaningful discussion is beyond the scope of this book. We present here a simplified picture. If two electrical charges $+q$ and $-q$ are separated by a distance d, the dipole moment is defined as $q \times d$. Atoms are electrically neutral. Also, the centers of the positive charge (nucleus) and negative charge (electron cloud) coincide. Therefore, a neutral atom has no dipole moment. When a neutral atom is exposed to an internal or external electric field the atom gets polarized (i.e., the centers of positive and negative charges separate). This creates or induces a dipole moment (Figure 2-20). In some molecules, the dipole moment does not have to be induced—it exists by virtue of the direction of bonds and the nature of atoms. These molecules are known as **polar molecules**. An example of such a molecule that has a permanently built-in dipole moment is water (Figure 2-21).

Molecules or atoms in which there is either an induced or permanent dipole moment attract each other. The resulting force is known as the van der Waals force. Van der Waals forces between atoms and molecules have their origin in interactions between dipoles that are induced or in some cases interactions between permanent dipoles that are present in certain polar molecules. What is unique about these forces is they are present in every material.

There are three types of **van der Waals** interactions, namely London forces, Keesom forces, and Debye forces. If the interactions are between two dipoles that are induced in atoms or molecules, we refer to them as **London forces** (e.g., carbon tetrachloride) (Figure 2-20). When an induced dipole (that is, a dipole that is induced in what is otherwise a non-polar atom or molecule) interacts with a molecule that has a permanent dipole moment, we refer to this interaction as a **Debye interaction**. An example of Debye interaction would be forces between water molecules and those of carbon tetrachloride.

If the interactions are between molecules that are permanently polarized (e.g., water molecules attracting other water molecules or other polar molecules), we refer to these as **Keesom interactions**. The attraction between the positively charged regions of one water molecule and the negatively charged regions of a second water molecule provides an attractive bond between the two water molecules (Figure 2-21).

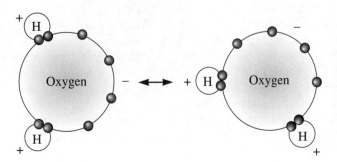

Figure 2-21
The Keesom interactions are formed as a result of polarization of molecules or groups of atoms. In water, electrons in the oxygen tend to concentrate away from the hydrogen. The resulting charge difference permits the molecule to be weakly bonded to other water molecules.

The bonding between molecules that have a permanent dipole moment, known as Keesom force, is often referred to as **hydrogen bond**, where hydrogen atoms represent one of the polarized regions. Thus, hydrogen bonding is essentially a Keesom force and is a type of van der Waals force. The relatively strong Keesom force between water molecules is the reason why surface tension (72 mJ/m^2 or dyne/cm at room temperature) and the boiling point of water ($100°C$) are much higher than those of many organic liquids of comparable molecular weight (surface tension ~ 20–25 dyne/cm, boiling points up to $80°C$).

Note that van der Waals bonds are **secondary bonds**, but the atoms within the molecule or group of atoms are joined by strong covalent or ionic bonds. Heating water to the boiling point breaks the van der Waals bonds and changes water to steam, but much higher temperatures are required to break the covalent bonds joining oxygen and hydrogen atoms.

Although termed "secondary", based on the bond energies, van der Waals forces play a very important role in many areas of engineering. Van der Waals forces between atoms and molecules play a vital role in determining the surface tension and boiling points of liquids. In materials science and engineering, the surface tension of liquids and the surface energy of solids come into play in different situations. For example, when we want to process ceramic or metal powders into dense solid parts, the powders often have to be dispersed in water or organic liquids. Whether we can achieve this dispersion effectively depends upon the surface tension of the liquid and the surface energy of the solid material. Surface tension of liquids also assumes importance when we are dealing with processing of molten metals and alloys (e.g., casting) and glasses.

Surface tension considerations also come into play in the welding of metals and alloys. We note that the surface tension of molten glasses and molten metals is considerably higher (~ 1000–1500 mJ/m^2 or dyne/cm) than that of water and organic liquids. This is *not* because of van der Waals bonding. The high values of surface tension and surface energy in metals and glasses are primarily due to the strength of the ionic, metallic, or covalent bonds between the constituent atoms.

Van der Waals bonds can change dramatically the properties of certain materials. For example, graphite and diamond have very different mechanical properties. In many plastic materials, molecules contain polar parts or side groups (e.g., cotton or cellulose, PVC, Teflon). Van der Waals forces provide an extra binding force between the chains of these polymers (Figure 2-22).

Polymers in which van der Waals forces are present tend to be relatively stiffer and exhibit relatively higher glass temperatures (T_g). **Glass temperature** is a temperature below which polymers tend to behave as brittle materials (i.e., they show poor ductility). As a result, polymers with van der Waals bonding (in addition to the covalent bonds in the chains and side groups) are relatively brittle at room temperature (e.g., PVC). In processing such polymers, they need to be "plasticized" by adding other smaller polar molecules that interact with the polar parts of the long polymer molecule (chains), thereby lowering the T_g and enhancing flexibility.

What is also fascinating about van der Waals forces is that they begin as interactions between atoms and molecules that ultimately lead to considerable forces between fine particles of any material. These forces are almost always attractive and play a key role in the processing of paints, magnetic inks for recording media, powders, and dispersions. We make use of metal and ceramic powders (dry or slurries) in processing many ceramic and metallic materials through routes collectively known as **powder processing** (processing technique for metals involving the solid-state bonding of a fine grained powder into a fully dense product). When we want to prepare stable suspen-

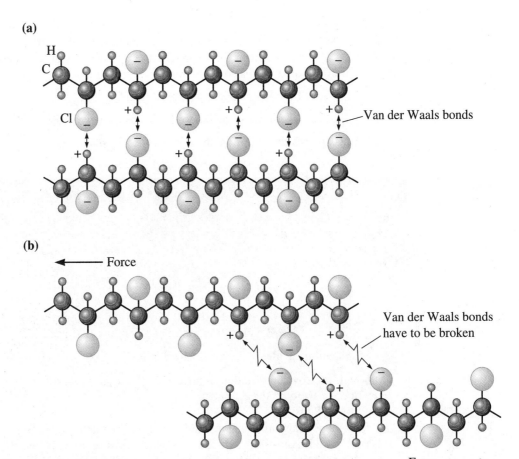

(a)

(b)

← Force

Force →

Van der Waals bonds

Van der Waals bonds
have to be broken

Figure 2-22   (a) In polyvinyl chloride (PVC), the chlorine atoms attached to the polymer chain have a negative charge and the hydrogen atoms are positively charged. The chains are weakly bonded by van der Waals bonds. This additional bonding makes PVC stiffer. (b) When a force is applied to the polymer, the van der Waals bonds are broken and the chains slide past one another.

sions of ceramic or metal powders we have to overcome van der Waals and other attractive forces between the particles. We can accomplish this by introducing like charges on the particles (electrostatic stabilization) or adsorbing soap like surfactant molecules on the surfaces of particles (steric stabilization).

Mixed Bonding In most materials, bonding between atoms is a mixture of two or more types. Iron, for example, is bonded by a combination of metallic and covalent bonding that prevents atoms from packing as efficiently as we might expect.

Compounds formed from two or more metals (**intermetallic compounds**) may be bonded by a mixture of metallic and ionic bonds, particularly when there is a large difference in electronegativity between the elements. Because lithium has an electronegativity of 1.0 and aluminum has an electronegativity of 1.5, we would expect AlLi to have a combination of metallic and ionic bonding. On the other hand, because both aluminum and vanadium have electronegativities of 1.5, we would expect Al_3V to be bonded primarily by metallic bonds.

Many ceramic and semiconducting compounds, which are combinations of metallic and nonmetallic elements, have a mixture of covalent and ionic bonding. As the electronegativity difference between the atoms increases, the bonding becomes more ionic. The fraction of bonding that is covalent can be estimated from the following equation:

$$\text{Fraction covalent} = \exp(-0.25\Delta E^2), \tag{2-1}$$

where ΔE is the difference in electronegativities.

Example 2-9 explores the nature of the bonds found in silica.

EXAMPLE 2-9   *Determine if Silica is Ionically or Covalently Bonded*

In a previous example, we used silica (SiO_2) as an example of a covalently bonded material. In reality, silica exhibits ionic and covalent bonding. What fraction of the bonding is covalent? Give examples of applications in which silica is used.

SOLUTION

From Figure 2-10, we estimate the electronegativity of silicon to be 1.8 and that of oxygen to be 3.5. The fraction of the bonding that is covalent is:

$$\text{Fraction covalent} = \exp[-0.25(3.5 - 1.8)^2] = \exp(-0.72) = 0.486$$

Although the covalent bonding represents only about half of the bonding, the directional nature of these bonds still plays an important role in the eventual structure of SiO_2.

Silica has many applications. Silica is used for making glasses and optical fibers. We add nano-sized particles of silica to tires to enhance the stiffness of the rubber. High-purity silicon (Si) crystals are made by reducing silica to silicon.

EXAMPLE 2-10   *Design Strategies for Silica Optical Fibers*

Silica is used for making long lengths of optical fibers (Figure 2-4). Being a covalently and ionically bonded material, the strength of Si-O bonds is expected to be high. Other factors such as susceptibility of silica surfaces to react with water vapor in atmosphere have a deleterious effect on the strength of silica fibers. Given this, what design strategies can you think of such that silica fibers could still be bent to a considerable degree without breaking?

SOLUTION

Based on the mixed ionic and covalent bonding in silica we know that the Si-O bonds are very strong. We also know that covalent bonds will be directional and hence we can anticipate silica to exhibit limited ductility. Therefore, our choices to enhance ductility of optical fibers are rather limited since the composition is essentially fixed. Most other glasses are also brittle. We can make an argument that silica fibers will exhibit better ductility at higher temperatures. However, we have to use them for making long lengths of optical fibers (most of which are to be buried underground or under the sea) and hence keeping them at an elevated temperature is not a practical option.

Therefore, we need to understand, beyond what the nature of bonding consideration can offer us, why glass fibers exhibit limited ductility. Is this a property that is *intrinsic* to the glass or are there *external variables* that are causing a change in the chemistry and structure of the glass? Materials scientists and engineers have recognized that the lack of ductility in optical glass fibers is linked to the ability of the silica surface to react with water vapor in the atmosphere. They have found that water vapor in the atmosphere reacts with the surface of silica leading to micro-cracks on the surface. When subjected to stress these cracks grow rapidly and the fibers break quite easily! They have also tested silica fibers in a vacuum and found that the levels to which one can bend fibers are much higher. So what about protecting the surface of silica fibers, just like we paint cars and bridges to prevent them from rusting? This is what is done by manufacturers of optical fibers such as Corning and Lucent. When the optical fibers are manufactured, they are immediately coated with a polymeric film. Later, bundles of such fibers are encased in metallic cables and used in the fiber optics network.

2-6 Binding Energy and Interatomic Spacing

Interatomic Spacing The equilibrium distance between atoms is caused by a balance between repulsive and attractive forces. In the metallic bond, for example, the attraction between the electrons and the ion cores is balanced by the repulsion between ion cores. Equilibrium separation occurs when the total inter-atomic energy (IAE) of the pair of atoms is at a minimum, or when no net force is acting to either attract or repel the atoms (Figure 2-23).

The **interatomic spacing** in a solid metal is *approximately* equal to the atomic diameter, or twice the atomic radius r. We cannot use this approach for ionically bonded materials, however, since the spacing is the sum of the two different ionic radii. Atomic and ionic radii for the elements are listed in Appendix B and will be used in the next chapter.

The minimum energy in Figure 2-23 is the **binding energy**, or the energy required to create or break the bond. Consequently, materials having a high binding energy also have a high strength and a high melting temperature. Ionically bonded materials have a particularly large binding energy (Table 2-3) because of the large difference in electronegativities between the ions. Metals have lower binding energies because the electronegativities of the atoms are similar.

TABLE 2-3 ■ *Binding energies for the four bonding mechanisms*

Bond	Binding Energy (kcal/mol)
Ionic	150–370
Covalent	125–300
Metallic	25–200
Van der Waals	<10

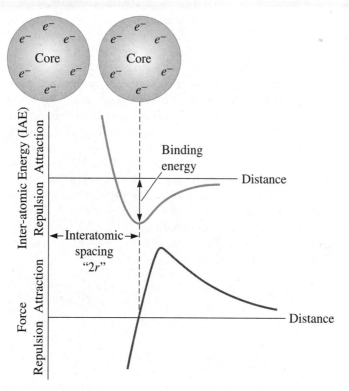

Figure 2-23
Atoms or ions are separated by an equilibrium spacing that corresponds to the minimum inter-atomic energy for a pair of atoms or ions (or when zero force is acting to repel or attract the atoms or ions).

Other properties can be related to the force-distance and energy-distance expressions in Figure 2-24. For example, the **modulus of elasticity** of a material (the slope of the stress-strain curve in the elastic region (E), also known as Young's modulus) is related to the slope of the force-distance curve (Figure 2-24). A steep slope, which correlates with a higher binding energy and a higher melting point, means that a greater force is required to stretch the bond; thus, the material has a high modulus of elasticity.

An interesting point that needs to be made is that not all properties of engineered materials are microstructure sensitive. Modulus of elasticity is one such property. If we have two aluminum samples that have essentially the same chemical composition but different grain size, we can expect that the modulus of elasticity of these samples will be about the same. However, the **yield strength**, a level of stress at which the material begins to deform easily with increasing stress, of these samples will be quite different. The

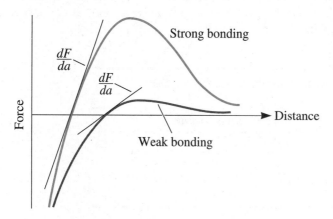

Figure 2-24
The force-distance curve for two materials, showing the relationship between atomic bonding and the modulus of elasticity. A steep *dF/da* slope gives a high modulus.

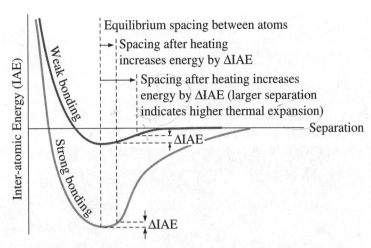

Figure 2-25 The inter-atomic energy (IAE)–separation curve for two atoms. Materials that display a steep curve with a deep trough have low linear coefficients of thermal expansion.

yield strength, therefore, is a microstructure sensitive property. We will learn in subsequent chapters that, compared to other mechanical properties such as yield strength and tensile strength, the modulus of elasticity does not depend strongly on the microstructure. The modulus of elasticity can be linked directly to the strength of bonds between atoms. Thus, the modulus of elasticity depends primarily on the atoms that make up the material.

Another property that can be linked to the binding energy or interatomic force-distance curves is the **coefficient of thermal expansion** (CTE). CTE is defined as $\alpha = (1/L)(dL/dT)$, where the overall dimensions of the material in a given direction, L, will increase with increasing temperature, T. The CTE describes how much a material expands or contracts when its temperature is changed. It is also related to the strength of the atomic bonds. In order for the atoms to move from their equilibrium separation, energy must be supplied to the material. If a very deep inter-atomic energy (IAE) trough caused by strong atomic bonding is characteristic of the material (Figure 2-25), the atoms separate to a lesser degree and have a low, linear coefficient of thermal expansion. Materials with a low coefficient of thermal expansion maintain their dimensions more accurately when the temperature changes. It is important to note that there are microstructural features (e.g., anistropy, or varying properties, in thermal expansion along with different crystallographic directions and other effects) that also have a significant effect on the overall thermal expansion coefficient of an engineered material. Indeed, materials scientists and engineers have developed materials that show very low, zero or negative coefficients of thermal expansion (Chapter 21).[5]

Materials that have very low expansion are useful in many applications where the components are expected to repeatedly undergo relatively rapid heating and cooling. For example, cordierite ceramics (used as catalyst support in catalytic converters in cars), ultra-low expansion (ULE) glasses, Visionware™, and other glass-ceramics developed by Corning, have very low thermal expansion coefficients. In the case of thin films or coatings on substrates, we are not only concerned about the actual values of thermal expansion coefficients but also the difference between thermal expansion coefficients between the substrate and the film or coating. Too much difference between these causes development of stresses that can lead to delamination or warping of the film or coating (Chapter 21).

EXAMPLE 2-10   *Design of a Space Shuttle Arm*

NASA's space shuttles have a long manipulator robot arm, also known as the Shuttle Remote Manipulator System or SRMS (Figure 2-26), that permits astronauts to launch and retrieve satellites. It is also used to view and monitor the outside of the space shuttle using a mounted video camera. Select a suitable material for this device.

Figure 2-26   NASA's Shuttle Remote Manipulator System: SRMS (for Example 2-10). (*Courtesy of Getty Images.*)

SOLUTION

Let's look at two of the many material choices. First, the material should be stiff so that little bending occurs when a load is applied; this feature helps the operator maneuver the manipulator arm precisely. Generally, materials with strong bonding and high melting points also have a high modulus of elasticity, or stiffness. Second, the material should be light in weight to permit maximum payloads to be carried into orbit; a low density is thus desired. It is estimated that it costs about US $100,000 to take the weight of a beverage can into space! Thus, the density must be as low as possible.

Good stiffness is obtained from high-melting-point metals (such as beryllium and tungsten), from ceramics, and from certain fibers (such as carbon). Tungsten, however, has a very high density, while ceramics are very brittle. Beryllium, which has a modulus of elasticity that is greater than that of steel and a density that is less than that of aluminum, might be an excellent candidate. However, toxicity of Be and its compounds must be considered. The preferred material is a composite consisting of carbon fibers embedded in an epoxy matrix. The carbon fibers have an exceptionally high modulus of elasticity, while the combination of carbon and epoxy provides a very low-density material. Other factors such as exposure to low and high temperatures in space and on earth must also be considered. The current shuttle robot arm is about 45 feet long, 15 inches in diameter and weighs about 900 pounds. When in space it can manipulate weights up to 260 tons.

SUMMARY

◈ Similar to composition, structure of a material has a profound influence on the properties of a material.

◈ Structure of materials can be understood at various levels: atomic structure, long- and short-range atomic arrangements, nanostructure, microstructure, and macrostructure. Engineers concerned with practical applications need to understand the structure at both the micro and macro levels. Given that atoms and atomic arrangements constitute the building blocks of advanced materials, we need to understand the structure at an atomic level. There are many emerging novel devices centered on micro-electro-mechanical systems (MEMS) and nanotechnology. As a result, understanding the structure of materials at nano-scale is also very important for some applications.

◈ The electronic structure of the atom, which is described by a set of four quantum numbers, helps determine the nature of atomic bonding, and, hence, the physical and mechanical properties of materials.

◈ Atomic bonding is determined partly by how the valence electrons associated with each atom interact. Types of bonds include metallic, covalent, ionic, and van der Waals. Most engineered materials exhibit mixed bonding.

◈ A metallic bond is formed as a result of atoms of low electronegativity elements donating their valence electrons and leading to the formation of a "sea" of electrons. Metallic bonds are non-directional and relatively strong. As a result, most pure metals show a high Young's modulus and ductility. They are good conductors of heat and electricity and reflect visible light.

◈ A covalent bond is formed between two atoms when each atom donates an electron that is needed in bond formation. Covalent bonds are found in many polymeric and ceramic materials. These bonds are strong, and most inorganic materials with covalent bonds exhibit high levels of strength, hardness, and limited ductility. Most plastic materials based on carbon-carbon (C-C) and carbon-hydrogen (C-H) bonds show relatively lower strengths and good levels of ductility. Most covalently bonded materials tend to be relatively good electrical insulators. Some materials such as Si and Ge are semiconductors.

◈ The ionic bonding found in many ceramics is produced when an electron is "donated" from one electropositive atom to an electronegative atom, creating positively charged cations and negatively charged anions. As in covalently bonded materials, these materials tend to be mechanically strong and hard, but brittle. Melting points of ionically bonded materials are relatively high. These materials are typically electrical insulators. In some cases, though, the microstructure of these materials can be tailored so that significant ionic conductivity is obtained.

◆ The van der Waals bonds are formed when atoms or groups of atoms have a non-symmetrical electrical charge, permitting bonding by an electrostatic attraction. The asymmetry in the charge is as a result of dipoles that are induced or dipoles that are permanent. London forces are a result of interactions between induced dipoles. Interactions between permanent dipoles are known as Keesom interactions. Interactions between permanent dipoles and induced dipoles are known as Debye interactions. Although their magnitude is relatively small, van der Waals forces play a crucial role in determination of surface tension and boiling point of liquids, especially those that are polar. Van der Waals forces between particles of ceramic and other materials play an important role in factors that affect the stability of slurries, paints, dispersions, etc. Van der Waals forces also play an important role in the mechanical behavior of plastics containing polar groups (e.g., PVC).

◆ The binding energy is related to the strength of the bonds and is particularly high in ionically and covalently bonded materials. Materials with a high binding energy often have a high-melting temperature, a high modulus of elasticity, and a low coefficient of thermal expansion.

◆ Not all properties of materials are microstructure sensitive, and modulus of elasticity is one such property.

◆ In designing components with materials, we need to pay attention to the base composition of the material. We also need to understand the bonding in the material and make efforts to tailor it so that certain performance requirements are met. Finally, the cost of raw materials, manufacturing costs, environmental impact, and factors affecting durability also must be considered.

GLOSSARY

Amorphous material   A material that does not have a long-range order of atoms (e.g., silica glass).

Anion   A negatively charged ion produced when an atom, usually of a nonmetal, accepts one or more electrons.

Atomic mass   The mass of the Avogadro number of atoms, g/mol. Normally, this is the average number of protons and neutrons in the atom. Also called the atomic weight.

Atomic mass unit   The mass of an atom expressed as 1/12 the mass of a carbon atom.

Atomic number   The number of protons or electrons in an atom.

Atomic structure   All atoms and their arrangements that constitute the building blocks of matter.

Avogadro number   The number of atoms or molecules in a mole. The Avogadro number is 6.02×10^{23} per mole.

Azimuthal quantum number   A quantum number that designates different energy levels in principal shells.

Binding energy   The energy required to separate two atoms from their equilibrium spacing to an infinite distance apart. Alternately, the binding energy is the strength of the bond between two atoms.

Cation   A positively charged ion produced when an atom, usually of a metal, gives up its valence electrons.

Coefficient of thermal expansion (CTE)   The amount by which a material changes its dimensions when the temperature changes. A material with a low coefficient of thermal expansion tends to retain its dimensions when the temperature changes.

Composition   The chemical make-up of a material.

Covalent bond   The bond formed between two atoms when the atoms share their valence electrons.

Crystalline materials   Materials in which atoms are arranged in a periodic fashion exhibiting a long-range order.

Debye interactions   Van der Waals forces that occur between two molecules, with only one of them with a permanent dipole moment.

Directional relationship   The bonds between atoms in covalently bonded materials form specific angles, depending on the material.

Dopant   An element deliberately added into a semiconductor (e.g., P in Si).

Ductility   The ability of materials to be stretched or bent without breaking.

Electronegativity   The relative tendency of an atom to accept an electron and become an anion. Strongly electronegative atoms readily accept electrons.

Glass temperature   A temperature above which many polymers and inorganic glasses no longer behave as brittle materials. They gain a considerable amount of ductility above the glass temperature.

Hydrogen bond   A Keesom interaction (a type of van der Waals bond) between molecules in which a hydrogen atom is involved (e.g., bonds *between* water molecules).

Interatomic spacing   The equilibrium spacing between the centers of two atoms. In solid elements, the interatomic spacing equals the apparent diameter of the atom.

Intermetallic compound   A compound such as Al_3V formed by two or more metallic atoms; bonding is typically a combination of metallic and ionic bonds.

Ionic bond   The bond formed between two different atom species when one atom (the cation) donates its valence electrons to the second atom (the anion). An electrostatic attraction binds the ions together.

Keesom interactions   Van der Waals forces that occur between molecules that have a permanent dipole moment.

Length-scale   A relative distance or range of distances used to describe materials-related structure, properties or phenomena.

London forces   Van der Waals forces that occur between molecules that do not have a permanent dipole moment.

Long-range atomic arrangements   Repetitive three-dimensional patterns with which atoms or ions are arranged in crystalline materials.

Magnetic quantum number   A quantum number that describes energy levels for each azimuthal quantum number.

Macrostructure   Structure of a material at a macroscopic level. The length-scale is $\sim > 100$ nm. Typical features include porosity, surface coatings, and internal or external micro-cracks.

Metallic bond   The electrostatic attraction between the valence electrons and the positively charged ion cores.

Micro-electro-mechanical systems (MEMS)   These consist of miniaturized devices typically prepared by micromachining.

Microstructure   Structure of a material at a length-scale of ~ 10 to 1000 nm. This typically includes such features as average grain size, grain size distribution, grain orientation and those related to defects in materials.

Modulus of elasticity   The slope of the stress-strain curve in the elastic region (E). Also known as Young's modulus.

Nano-scale   A length scale of 1–100 nm.

Nanostructure   Structure of a material at a nano-scale ($\sim$length-scale 1–100 nm).

Nanotechnology   An emerging set of technologies based on nano-scale devices, phenomena, and materials.

Pauli exclusion principle   No more than two electrons in a material can have the same energy. The two electrons have opposite magnetic spins.

Polarized molecules   Molecules that have developed a dipole moment by virtue of an internal or external electric field.

Powder processing   Processing technique for metals involving the solid-state bonding of a fine-grained powder into a polycrystalline product.

Primary bonds   Strong bonds between adjacent atoms resulting from the transfer or sharing of outer orbital electrons.

Quantum numbers   The numbers that assign electrons in an atom to discrete energy levels. The four quantum numbers are the principal quantum number n, the azimuthal quantum number l, the magnetic quantum number m_l, and the spin quantum number m_s.

Quantum shell   A set of fixed energy levels to which electrons belong. Each electron in the shell is designated by four quantum numbers.

Secondary bond   Weak bonds, such as van der Waals bonds, that typically join molecules to one another.

Short-range atomic arrangements   Atomic arrangements up to a distance of few nm.

Spin quantum number   A quantum number that indicates spin of an electron.

Structure   Description of spatial arrangements of atoms or ions in a material.

Thermistor   A device used to measure temperature by taking advantage of the change in electrical conductivity as the temperature changes.

Transition elements   A set of elements whose electronic configurations are such that their inner d and f levels begin to fill up. These elements usually exhibit multiple valence and are useful for electronic, magnetic and optical applications.

III-V semiconductor   A semiconductor that is based on group 3A and 5B elements (e.g. GaAs).

II-VI semiconductor   A semiconductor that is based on group 2B and 6B elements (e.g. CdSe).

Valence The number of electrons in an atom that participate in bonding or chemical reactions. Usually, the valence is the number of electrons in the outer s and p energy levels.

Van der Waals bond A secondary bond developed between atoms and molecules as a result of interactions between dipoles that are induced or permanent.

Yield strength The level of stress above which a material begins to show permanent deformation.

ADDITIONAL INFORMATION

1. "Special Issue: Nanotech—The Science of the Small Gets Down to Business," *Scientific American*, 2001 (September).

2. BISHOP, D., P.G. GAMMEL, and C. RANDY, "Micromachines on Rise," *Physics Today*, 2001 (October): p. 38–44.

3. KULKARNI, V.G., "Polyanilines: Progress in Processing and Applications," *ACS Symp. Series.* 1999: American Chemical Society.

4. HERBERT, J.M. and A.J. MOULSON, *Electroceramics—Materials, Properies, and Applications.* 1990.

5. ROY, R., D.K. AGRWAL, J. ALAMA, and R.A. ROY, "A New Structural Family of Near Zero Expansion Ceramics", *Materials Research Bulletin*, Vol. 19, 1984: p. 471–477.

✓ PROBLEMS

Section 2-1 The Structure of Materials— An Introduction

2-1 What is meant by the term *composition* of a material?

2-2 What is meant by the term *structure* of a material?

2-3 What are the different levels of structure of a material?

2-4 Why is it important to consider the structure of a material while designing and fabricating engineering components?

2-5 What is the difference between the microstructure and the macrostructure of a material?

Section 2-2 The Structure of the Atom

2-6 (a) Aluminum foil used for storing food weighs about 0.3 g per square inch. How many atoms of aluminum are contained in one square inch of foil?

(b) Using the densities and atomic weights given in Appendix A, calculate and compare the number of atoms per cubic centimeter in (i) lead and (ii) lithium.

2.7 (a) Using data in Appendix A, calculate the number of iron atoms in one ton (2000 pounds).

(b) Using data in Appendix A, calculate the volume in cubic centimeters occupied by one mole of boron.

2-8 In order to plate a steel part having a surface area of 200 in.^{2} with a 0.002 in. thick layer of nickel: (a) How many atoms of nickel are required? (b) How many moles of nickel are required?

Section 2-3 The Electronic Structure of the Atom

2-9 Suppose an element has a valence of 2 and an atomic number of 27. Based only on the quantum numbers, how many electrons must be present in the $3d$ energy level?

2-10 Indium, which has an atomic number of 49, contains no electrons in its $4f$ energy levels. Based only on this information, what must be the valence of indium?

2-11 Without consulting Appendix C, describe the quantum numbers for each of the 18 electrons in the M shell of copper, using a format similar to that in Figure 2-9 (see next page).

2-12 Electrical charge is transferred in metals by movement of valence electrons. How many potential charge carriers are there in aluminum wire 1 mm in diameter and 100 m in length?

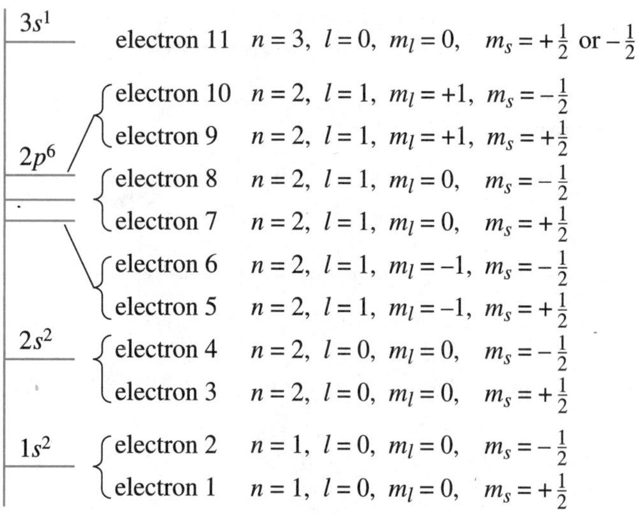

$3s^1$ — electron 11 $n = 3$, $l = 0$, $m_l = 0$, $m_s = +\frac{1}{2}$ or $-\frac{1}{2}$

$2p^6$
- electron 10 $n = 2$, $l = 1$, $m_l = +1$, $m_s = -\frac{1}{2}$
- electron 9 $n = 2$, $l = 1$, $m_l = +1$, $m_s = +\frac{1}{2}$
- electron 8 $n = 2$, $l = 1$, $m_l = 0$, $m_s = -\frac{1}{2}$
- electron 7 $n = 2$, $l = 1$, $m_l = 0$, $m_s = +\frac{1}{2}$
- electron 6 $n = 2$, $l = 1$, $m_l = -1$, $m_s = -\frac{1}{2}$
- electron 5 $n = 2$, $l = 1$, $m_l = -1$, $m_s = +\frac{1}{2}$

$2s^2$
- electron 4 $n = 2$, $l = 0$, $m_l = 0$, $m_s = -\frac{1}{2}$
- electron 3 $n = 2$, $l = 0$, $m_l = 0$, $m_s = +\frac{1}{2}$

$1s^2$
- electron 2 $n = 1$, $l = 0$, $m_l = 0$, $m_s = -\frac{1}{2}$
- electron 1 $n = 1$, $l = 0$, $m_l = 0$, $m_s = +\frac{1}{2}$

Figure 2-9   (Repeated for Problem 2-11)
The complete set of quantum numbers for each of the 11 electrons in sodium.

Section 2-4 The Periodic Table

2-13   The periodic table of elements can help us better rationalize trends in properties of elements and compounds based on elements from different groups. Search the literature and obtain the co-efficients of thermal expansions of elements from group 4B. Establish a trend and see if it correlates with the melting temperatures and other properties (e.g., bandgap) of these elements.

2-14   Bonding in the intermetallic compound Ni_3Al is predominantly metallic. Explain why there will be little, if any, ionic bonding component. The electronegativity of nickel is about 1.8.

2-15   Plot the melting temperatures of elements in the 4A to 8–10 columns of the periodic table versus atomic number (i.e., plot melting temperatures of Ti through Ni, Zr through Pd, and Hf through Pt). Discuss these relationships, based on atomic bonding and binding energies: (a) as the atomic number increases in each row of the periodic table and (b) as the atomic number increases in each column of the periodic table.

2-16   Plot the melting temperature of the elements in the 1A column of the periodic table versus atomic number (i.e., plot melting temperatures of Li through Cs). Discuss this relationship, based on atomic bonding and binding energy.

Section 2-5 Atomic Bonding

2-17   Increasing the temperature of a semiconductor breaks covalent bonds. For each broken bond, two electrons become free to move and transfer electrical charge. (a) What fraction of the total valence electrons are free to move? (b) What fraction of the covalent bonds must be broken in order that 5×10^{15} electrons conduct electrical charge in 50 g of silicon? (c) What fraction of the total silicon atoms must be replaced by arsenic atoms to obtain one million electrons that are free to move in one pound of silicon?

2-18   Methane (CH_4) has a tetrahedral structure similar to that of SiO_2 with a carbon atom of radius 0.77×10^{-8} cm at the center and hydrogen atoms of radius 0.46×10^{-8} cm at four of the eight corners. Calculate the size of the tetrahedral cube for methane.

2-19   The compound aluminum phosphide (AlP) is a compound semiconductor having mixed ionic and covalent bonding. Calculate the fraction of the bonding that is ionic.

2-20   Calculate the fraction of bonding of MgO that is ionic.

2-21   What is the type of bonding in diamond? Are the properties of diamond commensurate with the nature of bonding?

2-22   What are some of the industrial applications of diamond?

2-23   Such materials as silicon carbide (SiC) and Si_3N_4 are used for grinding and polishing applications, rationalize the choice of these materials for this application.

2-24   What type of van der Waals forces acts between argon gas atoms?

2-25   What type of van der Waals forces acts between water molecules?

2-26   Explain why surface tension of water is higher than comparable non-polar organic liquids.

2-27 Explain the role of van der Waals forces in PVC plastic.

2-28 Why are van der Waals forces important in preparation of ceramic slurries and other dispersions?

Section 2-6 Binding Energy and Interatomic Spacing

2-29 Beryllium and magnesium, both in the 2A column of the periodic table, are lightweight metals. Which would you expect to have the higher modulus of elasticity? Explain, considering binding energy and atomic radii and using appropriate sketches of force versus interatomic spacing.

2-30 Boron has a much lower coefficient of thermal expansion than aluminum, even though both are in the 3B column of the periodic table. Explain, based on binding energy, atomic size, and the energy well, why this difference is expected.

2-31 Would you expect MgO or magnesium to have the higher modulus of elasticity? Explain.

2-32 Would you expect Al_2O_3 or aluminum to have the higher coefficient of thermal expansion? Explain.

2-33 Aluminum and silicon are side-by-side in the periodic table. Which would you expect to have the higher modulus of elasticity (E)? Explain.

2-34 Explain why the modulus of elasticity of simple thermoplastic polymers, such as polyethylene and polystyrene, is expected to be very low compared with that of metals and ceramics.

2-35 Steel is coated with a thin layer of ceramic to help protect against corrosion. What do you expect to happen to the coating when the temperature of the steel is increased significantly? Explain.

2-36 Why is the modulus of elasticity considered a structure insensitive property?

Design Problems

2-37 You wish to introduce ceramic fibers into a metal matrix to produce a composite material, which is subjected to high forces and large temperature changes. What design parameters might you consider to ensure that the fibers will remain intact and provide strength to the matrix? What problems might occur?

2-38 Turbine blades used in jet engines can be made from such materials as nickel-based superalloys. We can, in principle, even use ceramic materials such as zirconia or other alloys based on steels. In some cases, the blades also may have to be coated with a thermal barrier coating (TBC) to minimize exposure of the blade material to high temperatures. What design parameters would you consider in selecting a material for the turbine blade and for the coating that would work successfully in a turbine engine. Note that different parts of the engine are exposed to different temperatures, and not all blades are exposed to relatively high operating temperatures. What problems might occur? Consider the factors such as temperature and humidity in the environment that the turbine blades must function.

2-39 An extrinsic *n*-type semiconductor can be produced by introducing dopants into pure silicon. By doing so, additional electrons beyond those needed to participate in the bonding mechanism become part of the structure and can move. Design an alloy system that will cause this extrinsic semiconductivity in silicon.

Chapter

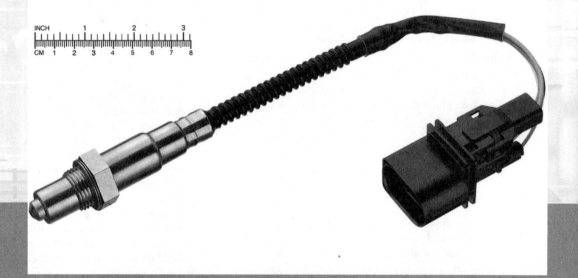

Many materials exhibit a periodic arrangement of atoms or ions. The photograph here shows an oxygen sensor made from Zirconia (ZrO_2). This material shows different crystal structures. It is capable of conducting electricity through movement of ions in the crystal structure. Zirconia is also used to make fuel cells. In this chapter, we will examine crystal structures of different engineered materials. (*Image courtesy of Bosch © Robert Bosch GmbH.*)

3

Atomic and Ionic Arrangements

Have You Ever Wondered?

- *What is amorphous silicon and how is it different from the silicon used to make computer chips?*

- *How do automobile oxygen sensors work?*

- *What are liquid crystals?*

- *If you were to pack a cubical box with uniform-sized spheres, what is the maximum packing possible?*

- *How can we calculate the density of different materials?*

Arrangements of atoms and ions play an important role in determining the microstructure and properties of a material. The main objectives of this chapter are to:

(a) learn classification of materials based on atomic/ionic arrangements; and

(b) describe the arrangements in crystalline solids based on **lattice**, **basis**, and **crystal structure**.

For crystalline solids, we will illustrate the concepts of Bravais lattices, unit cells, crystallographic directions, and planes by examining the arrangements of atoms or ions in many technologically important materials. These include metals (e.g., Cu, Al, Fe, W, Mg, etc.), semiconductors (e.g., Si, Ge, GaAs, etc.), advanced ceramics (e.g., ZrO_2, Al_2O_3, $BaTiO_3$, etc.), ceramic superconductors, diamond, and other materials. We will develop the necessary nomenclature used to characterize atomic or ionic arrangements in crystalline materials. We will examine the use of **x-ray diffraction** (XRD), **transmission electron microscopy** (TEM), and **electron diffraction**. These techniques allow us to probe the arrangements of atoms/ions in different materials. We will present an overview of different types of **amorphous materials** such as amorphous silicon, metallic glasses, polymers, and inorganic glasses.

Chapter 2 highlighted how interatomic bonding influences certain properties of materials. This chapter will underscore the influence of atomic and ionic arrangements on the properties of engineered materials. In particular, we will concentrate on arrangements of atoms or ions in what are crystalline solids with "perfect" arrangements of atoms or ions.

The concepts discussed in this chapter will prepare us for understanding how *deviations* from these perfect arrangements in crystalline materials create what are described as **atomic level defects**. The term **defect** in this context refers to a lack of perfection in atomic or ionic order of crystalline material, and not to any flaw or quality of an engineered material. In Chapter 4, we will describe how these atomic level defects actually enable the development of formable, yet stronger, steels used in cars and buildings, aluminum alloys for aircrafts, solar cells and photovoltaic modules for satellites, and semiconductors and many other technologies.

3-1   Short-Range Order versus Long-Range Order

In different states of matter, we can find four types of atomic or ionic arrangements (Figure 3-1).

No Order   In monoatomic gases, such as argon (Ar) or plasma created in a fluorescent tubelight, atoms or ions have no orderly arrangement. These materials randomly fill up whatever space is available to them.

Short-Range Order (SRO)   A material displays **short-range order (SRO)** if the special arrangement of the atoms extends only to the atom's nearest neighbors. Each water molecule in steam has a short-range order due to the covalent bonds between the hydrogen and oxygen atoms; that is, each oxygen atom is joined to two hydrogen atoms, forming an angle of 104.5° between the bonds. However, the water molecules in steam have no special arrangement with respect to each other's position.

A similar situation exists in materials known as inorganic glasses. In Chapter 2, we described the **tetrahedral structure** in silica that satisfies the requirement that four oxygen ions be bonded to each silicon ion (Figure 3-2). However, as will be discussed later, in a glass individual tetrahedral units are joined together in a random manner. These tetrahedra may share corners, edges, or faces. Thus, beyond the basic unit of a $(SiO_4)^{4-}$ tetrahedron, there is no periodicity in their arrangement. In contrast, in quartz or other forms of crystalline silica, the $(SiO_4)^{4-}$ tetrahedra are indeed connected in different periodic arrangements.

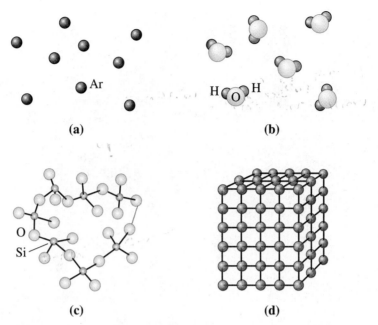

(a)   **(b)**

(c)   **(d)**

Figure 3-1 Levels of atomic arrangements in materials: (a) Inert monoatomic gases have no regular ordering of atoms. (b,c) Some materials, including water vapor, nitrogen gas, amorphous silicon and silicate glass have short-range order. (d) Metals, alloys, many ceramics and some polymers have regular ordering of atoms/ions that extends through the material.

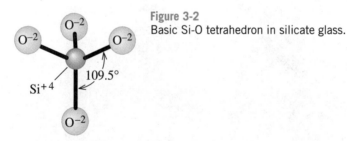

Figure 3-2
Basic Si-O tetrahedron in silicate glass.

Many polymers also display short-range atomic arrangements that closely resemble the silicate glass structure. Polyethylene is composed of chains of carbon atoms, with two hydrogen atoms attached to each carbon. Because carbon has a valence of four and the carbon and hydrogen atoms are bonded covalently, a tetrahedral structure is again produced (Figure 3-3). Tetrahedral units can be joined in a random manner to produce polymer chains.

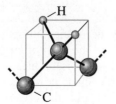

Figure 3-3
Tetrahedral arrangement of C-H bonds in polyethylene.

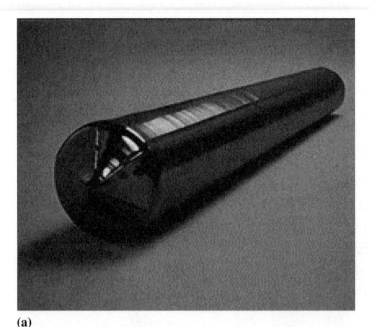

(a)

100μm

(b)

Figure 3-4
(a) Photograph of a silicon single crystal. (b) Micrograph of a polycrystalline stainless steel showing grains and grain boundaries (*Courtesy Dr. M. Hua, Dr. I. Garcia, and Dr. A.J. Deardo.*)

Long-Range Order (LRO) Most metals and alloys, semiconductors, ceramics, and some polymers have a crystalline structure in which the atoms or ions display **long-range order (LRO)**; the special atomic arrangement extends over much larger length scales $\sim >100$ nm. The atoms or ions in these materials form a regular repetitive, grid-like pattern, in three dimensions. We refer to these materials as **crystalline materials**. If a crystalline material consists of only one large crystal, we refer to it as a **single crystal material**. Single crystal materials are useful in many electronic and optical applications. For example, computer chips are made from silicon in the form of large (up to 12-inch diameter) single crystals [Figure 3-4(a)]. Similarly, many useful electro-optical devices are made from crystals of lithium niobate ($LiNbO_3$). Single crystals can also be made as thin films and used for many electronic and other applications. Certain types of tur-

Figure 3-5
Liquid crystal display. These materials are amorphous in one state and undergo localized crystallization in response to an external electric field and are widely used in liquid crystal displays. (*Courtesy of Nick Koudis/PhotoDisc/Getty Images.*)

bine blades may also be made from single crystals of nickel-based super alloys. A **polycrystalline material** is comprised of many small crystals with varying orientations in space. These smaller crystals are known as **grains**. A polycrystalline material is similar to a collage of several tiny single crystals. The borders between tiny crystals, where the crystals are in misalignment, are known as **grain boundaries**. Figure 3-4(b) shows the microstructure of a polycrystalline stainless steel material. Many crystalline materials we deal with in engineering applications are polycrystalline (e.g., steels used in construction, aluminum alloys for aircrafts, etc.). We will learn in later chapters that many properties of polycrystalline materials depend upon the physical and chemical characteristics of both grains and grain boundaries. The properties of single crystal materials depend upon the chemical composition and specific directions within the crystal (known as the crystallographic directions). Long-range order in crystalline materials can be detected and measured using techniques such as **x-ray diffraction** or **electron diffraction** (Section 3-9).

Liquid crystals (LCs) are polymeric materials that have a special type of order. Liquid crystal polymers behave as amorphous materials (liquid-like) in one state. However, when an external stimulus (such as an electric field or a temperature change) is provided some polymer molecules undergo alignment and form small regions that are crystalline, hence the name "liquid crystals." These materials have many commercial applications in liquid crystal display (LCD) technology (Figure 3-5).[1]

The Nobel Prize in Physics for 2001 went to Eric A. Cornell, Wolfgang Ketterle, and Carl E. Wieman. These scientists have verified a new state of matter known as the **Bose-Einstein condensate** (BEC). The existence of this state of matter had been predicted by Dr. Satyendra Nath Bose and Albert Einstein back in 1924 (Figure 3-6).[2,3] In this highly unusual state of matter, a group of atoms cooled to very low temperatures (just above 0 Kelvin), using lasers and magnetic traps, has the same quantum ground state. Although the BECs do not have any engineering applications today, they could be useful in studies related to development of nano-devices such as atom lasers. Figure 3-7 shows a summary of classification of materials based on the type of atomic order.

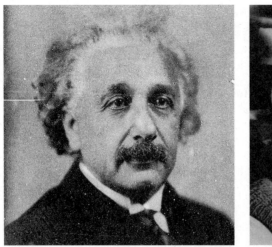 

Figure 3-6   Albert Einstein and Satyendranath Bose predicted the existence of BEC in 1924. (*Einstein image courtesy of the University of Pennsylvania Library; Bose image courtesy of Indian National Council of Science Museums, Emilio Segre Visual Archives.*)

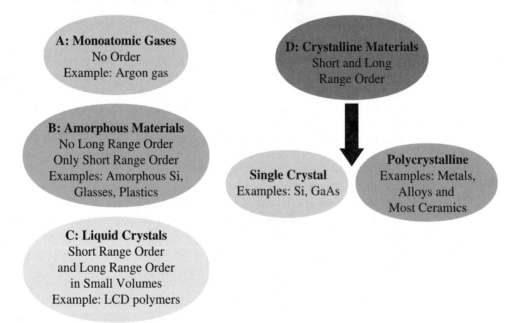

A: Monoatomic Gases
No Order
Example: Argon gas

B: Amorphous Materials
No Long Range Order
Only Short Range Order
Examples: Amorphous Si,
Glasses, Plastics

C: Liquid Crystals
Short Range Order
and Long Range Order
in Small Volumes
Example: LCD polymers

D: Crystalline Materials
Short and Long
Range Order

Single Crystal
Examples: Si, GaAs

Polycrystalline
Examples: Metals,
Alloys and
Most Ceramics

Figure 3-7   Classification of materials based on the type of atomic order.

3-2   Amorphous Materials: Principles and Technological Applications

Any material that exhibits only a short-range order of atoms or ions is an **amorphous material**; that is, a noncrystalline one. In general, most materials want to form periodic arrangements since this configuration maximizes the thermodynamic stability of the material. Amorphous materials tend to form when, for one reason or other, the kinetics of the process by which the material was made did not allow for the formation of periodic arrangements. **Glasses**, which typically form in ceramic and polymer systems, are good examples of amorphous materials. Similarly, certain types of polymeric or colloidal gels, or gel-like materials, are also considered amorphous. Amorphous materials often offer a unique and unusual blend of properties since the atoms or ions are not assembled into their "regular" and periodic arrangements. Note that often many engineered materials labeled as "amorphous" may contain a fraction that is crystalline. Such techniques as electron diffraction and x-ray diffraction (Section 3-9) cannot be used to characterize the short-range order in amorphous materials. Scientists use neutron scattering and other methods to investigate the short-range order in amorphous materials.

A common example of an amorphous material is a silicate glass. Different ingredients, such as silica (SiO_2), alumina (Al_2O_3), sodium oxide (Na_2O), calcium oxide (CaO), etc., are melted to form glasses used in automobile windshields of cars or windowpanes by using the **float-glass process**. In this process, other oxides are added to silica to lower the melting temperature because, even at temperatures of $\sim 1400°C$, a pure silica glass melt is too viscous for the process to work. The molten glass is floated on a bath of molten tin and hence the name float glass. As the glass solidifies, the $(SiO_4)^{-4}$ tetrahedra (Figure 3-2), that are present even in the molten glass, do not get a chance to form a regular periodic arrangement. If we deliberately raise the temperature of a silicate glass and hold it at a high temperature (e.g., 1000°C) for a long period of time, small portions of this originally amorphous glass will begin to "crystallize" and the $(SiO_4)^{4-}$ tetrahedra will begin to organize themselves into tiny crystals. As tiny crystals grow they begin to scatter light and the glass will begin to lose clarity. If the formation of crystals is not controlled, a stress will develop causing the glass to break.

On the other hand, **crystallization** of glasses can be controlled. Materials scientists and engineers, such as Donald Stookey [Figure 3-8(a)], have developed ways of deliberately nucleating ultrafine crystals into amorphous glasses. The resultant materials, known as **glass-ceramics** [Figure 3-8(b)], can be made up to $\sim 99.9\%$ crystalline and are quite strong. Some glass-ceramics can be made optically transparent by keeping the size of the crystals extremely small ($\sim <100$ nm). The major advantage of glass-ceramics is that they are shaped using glass-forming techniques, yet they are ultimately transformed into crystalline materials that do not shatter like glass.[4] We will consider this topic in greater detail in Chapter 8.

Similar to inorganic glasses, many plastics are also amorphous. They do contain small portions of material that is crystalline. During processing, relatively large chains of polymer molecules get entangled with each other, like spaghetti. Entangled polymer molecules do not organize themselves into crystalline materials. During polymer processing, mechanical stress is applied to the preform of the bottle (e.g., the manufacturing of a standard 2-liter soft drink bottle using polyethylene terephthalate (PET plastic)). This process is known as **blow-stretch forming** (Figure 3-9). The radial (blowing) and longitudinal (stretching) stresses during bottle formation actually untangles some of the polymer chains, causing **stress-induced crystallization**. The formation of crystals adds to the strength of the PET bottles. Similar to the situation of glass-ceramics, the crystallization process has to be controlled so that the crystals actually make the PET plastic stronger, and the crystals are not big enough to scatter light and compromise the optical clarity.

Compared to plastics and inorganic glasses, metals and alloys tend to form crystalline materials rather easily. As a result, special efforts must be made to quench the metals and alloys quickly; a cooling rate of $>10^6°C/s$ is required to form **metallic glasses**. This technique of cooling metals and alloys very fast is known as rapid solidification. Many metallic glasses have both useful and unusual properties. Some metallic glasses, such as Metglas^{TM}, have commercial applications as more efficient magnetic

(a)

(b)

Figure 3-8   (a) Donald Stookey, credited with pioneering research on glass-ceramics. (b) Derived from amorphous glasses, glass-ceramics represent a useful family of engineered materials. (*Courtesy of Corning Corporation.*)

materials for transformers and other applications. The topic of solidification is discussed further in Chapter 8.

Amorphous silicon, denoted a:Si-H, is another important example of a material that has the basic short-range order of crystalline silicon (Figure 3-10).[5] The H in the symbol tells us that this material also contains some hydrogen. In amorphous silicon, the silicon tetrahedra are not connected to each other in the periodic arrangement seen in crystalline silicon. Also, some bonds are incomplete or "dangling." This lack of pe-

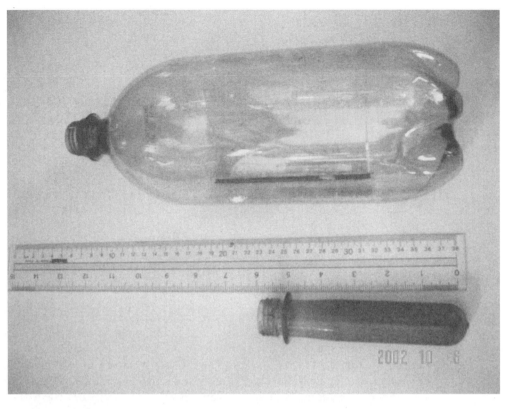

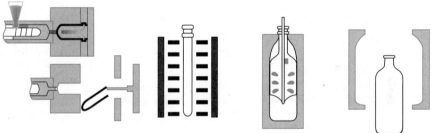

Figure 3-9   The top figure shows a standard 2-liter bottle and a preform from which this sized bottle is made. The bottom figure shows a schematic of the blow-stretch process used for fabrication of a standard two-liter PET (polyethylene terephthalate) bottle from a preform. The stress induced crystallization leads to formation of small crystals that help reinforce the remaining amorphous matrix. (*Source: Bottle making process from paysmart.com/usedcentral.*)

riodic arrangement decreases the mobility of carriers (i.e., the speed with which charge carriers, such as electrons, move under an electric field) in amorphous silicon when compared to crystalline silicon. However, amorphous silicon can be made on larger area substrates than can crystalline silicon; this is crucial for electronic applications. Amorphous silicon is made by a **chemical-vapor deposition** (CVD) process that involves decomposing silane (SiH_4) gas. This process represents an example of producing amorphous materials without melting. During the CVD process, silicon atoms from the silane precursor deposit on a substrate and do not get a chance to form crystalline silicon. Thin films of amorphous silicon are used to make transistors for active matrix displays in computers. Amorphous silicon is also widely used for such applications as solar cells and solar panels.

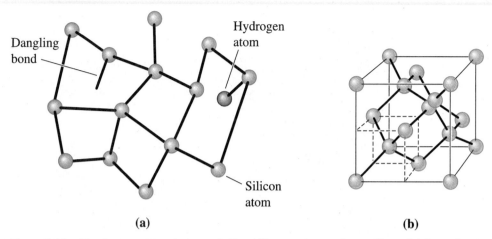

Figure 3-10   Atomic arrangements in crystalline silicon and amorphous silicon. (a) Amorphous silicon. (b) Crystalline silicon. Note the variation in the inter-atomic distance for amorphous silicon.

Gels such as those of colloidal silica are also examples of amorphous materials. These gel-like materials are useful in the formulations of cosmetics, to make molds for precision casting of metals and alloys, paints, and pharmaceuticals.

To summarize, amorphous materials can be made by restricting the atoms/ions from assuming their "regular" periodic positions. This means that amorphous materials do not have a long-range order. This allows us to form materials with many different and unusual properties. Many materials labeled as "amorphous" can contain some level of crystallinity. Since atoms are assembled into nonequilibrium positions, the natural tendency of an amorphous material is to crystallize (i.e., since this leads to a thermodynamically more stable material). This can be done by providing a proper thermal (e.g., a silicate glass), thermal and mechanical (e.g., PET polymer), or electrical (e.g., liquid crystal polymer) driving force. Amorphous materials, and the crystalline materials derived using them, have many useful technological applications.

3-3   Lattice, Unit Cells, Basis, and Crystal Structures

A **lattice** is a collection of points, called **lattice points**, which are arranged in a periodic pattern so that the surroundings of each point in the lattice are identical. A lattice may be one, two, or three dimensional. In materials science and engineering, we use the concept of "lattice" to describe arrangements of atoms or ions. A group of one or more atoms, located in a particular way with respect to each other and associated with each lattice point, is known as the **motif** or **basis**. We obtain a **crystal structure** by adding the lattice and basis (i.e., crystal structure = lattice + basis).

The **unit cell** is the subdivision of a lattice that still retains the overall characteristics of the entire lattice. A unit cell is shown in Figure 3-11. By stacking identical unit cells, the entire lattice can be constructed. There are seven unique arrangements, known as **crystal systems**, which fill in a three-dimensional space. These are cubic, tetragonal, orthorhombic, rhombohedral (also known as trigonal), hexagonal, monoclinic, and triclinic. Although there are seven crystal systems, we have a total of 14 distinct arrangements of lattice points. These unique arrangements of lattice points are known as the **Bravais lattices**, named after Auguste Bravais (1811–1863), an early French

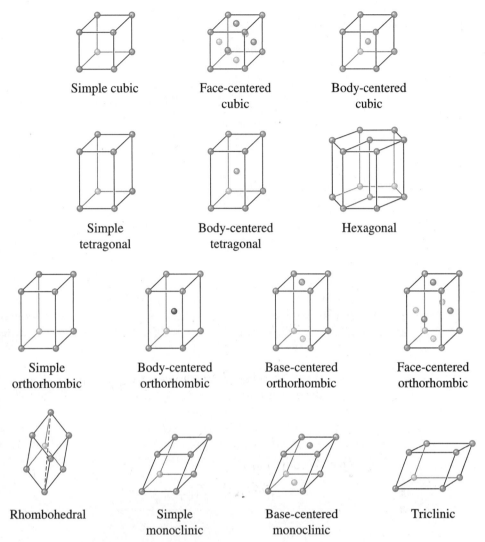

Simple cubic   Face-centered cubic   Body-centered cubic

Simple tetragonal   Body-centered tetragonal   Hexagonal

Simple orthorhombic   Body-centered orthorhombic   Base-centered orthorhombic   Face-centered orthorhombic

Rhombohedral   Simple monoclinic   Base-centered monoclinic   Triclinic

Figure 3-11 The fourteen types of Bravais lattices grouped in seven crystal systems. The actual unit cell for a hexagonal system is shown in Figures 3-12 and 3-16.

crystallographer. We identify 14 types of Bravais lattices, grouped in seven crystal systems (Figure 3-11 and Table 3-1). Lattice points are located at the corners of the unit cells and, in some cases, at either faces or the center of the unit cell. Note that for the cubic crystal system we have simple cubic (SC), face-centered cubic (FCC), and body-centered cubic (BCC) Bravais lattices. Similarly, for the tetragonal crystal system, we have simple tetragonal and body centered tetragonal Bravais lattices. Any other arrangement of atoms can be expressed using these 14 Bravais lattices. Note that the concept of a lattice is mathematical and does not mention atoms, ions or molecules. It is only when we take a Bravais lattice and begin to define the basis (i.e., one or more atoms associated with each lattice point) that we can describe a crystal structure. For example, if we take the face-centered cubic lattice and assume that at each lattice point we have one atom, then we get a face-centered cubic crystal structure.

Note that although we have only 14 Bravais lattices, we can have many more bases. Since crystal structure is derived by adding lattice and basis, we have hundreds of different crystal structures. Many different materials can have the same crystal structure.

TABLE 3-1 ■ *Characteristics of the seven crystal systems*

Structure	Axes	Angles between Axes	Volume of the Unit Cell
Cubic	$a = b = c$	All angles equal 90°	a^3
Tetragonal	$a = b \neq c$	All angles equal 90°	a^2c
Orthorhombic	$a \neq b \neq c$	All angles equal 90°	abc
Hexagonal	$a = b \neq c$	Two angles equal 90°. One angle equals 120°.	$0.866a^2c$
Rhombohedral or trigonal	$a = b = c$	All angles are equal and none equals 90°	$a^3\sqrt{1 - 3\cos^2\alpha + 2\cos^3\alpha}$
Monoclinic	$a \neq b \neq c$	Two angles equal 90°. One angle (β) is not equal to 90°	$abc \sin\beta$
Triclinic	$a \neq b \neq c$	All angles are different and none equals 90°	$abc\sqrt{1 - \cos^2\alpha - \cos^2\beta - \cos^2\gamma + 2\cos\alpha\cos\beta\cos\gamma}$

For example, copper and nickel have the face-centered cubic crystal structure. In this book, for the sake of simplicity, we will assume that each lattice point has only one atom (i.e., the basis is one), unless otherwise stated. This assumption allows us to refer to the lattice and the crystal structure interchangeably. Let's look at some of the characteristics of a lattice or unit cell.

Lattice Parameter   The **lattice parameters**, which describe the size and shape of the unit cell, include the dimensions of the sides of the unit cell and the angles between the sides (Figure 3-12). In a cubic crystal system, only the length of one of the sides of the cube is necessary to completely describe the cell (angles of 90° are assumed unless otherwise specified). This length is the lattice parameter a (some times designated as a_0). The length is often given in nanometers (nm) or Angstrom (Å) units, where:

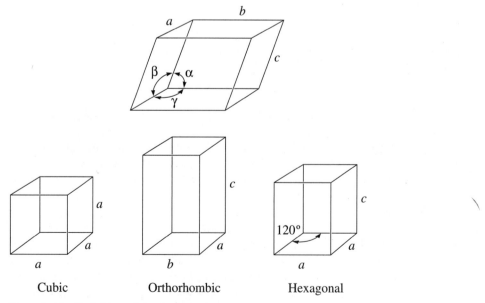

Cubic   Orthorhombic   Hexagonal

Figure 3-12   Definition of the lattice parameters and their use in cubic, orthorhombic, and hexagonal crystal systems.

$$1 \text{ nanometer (nm)} = 10^{-9} \text{ m} = 10^{-7} \text{ cm} = 10 \text{ Å}$$

$$1 \text{ angstrom (Å)} = 0.1 \text{ nm} = 10^{-10} \text{ m} = 10^{-8} \text{ cm}$$

Several lattice parameters are required to define the size and shape of complex unit cells. For an orthorhombic unit cell, we must specify the dimensions of all three sides of the cell: a_0, b_0, and c_0. Hexagonal unit cells require two dimensions, a_0 and c_0, and the angle of 120° between the a_0 axes. The most complicated cell, the triclinic cell, is described by three lengths and three angles.

Number of Atoms per Unit Cell A specific number of lattice points define each of the unit cells. For example, the corners of the cells are easily identified, as are the body-centered (center of the cell) and face-centered (centers of the six sides of the cell) positions (Figure 3-11). When counting the number of lattice points belonging to each unit cell, we must recognize that lattice points may be shared by more than one unit cell. A lattice point at a corner of one unit cell is shared by seven adjacent unit cells (thus a total of eight cells); only one-eighth of each corner belongs to one particular cell. Thus, the number of lattice points from all of the corner positions in one unit cell is:

$$\left(\frac{1}{8} \frac{\text{lattice point}}{\text{corner}} \right) \left(8 \frac{\text{corners}}{\text{cell}} \right) = 1 \frac{\text{lattice point}}{\text{unit cell}}$$

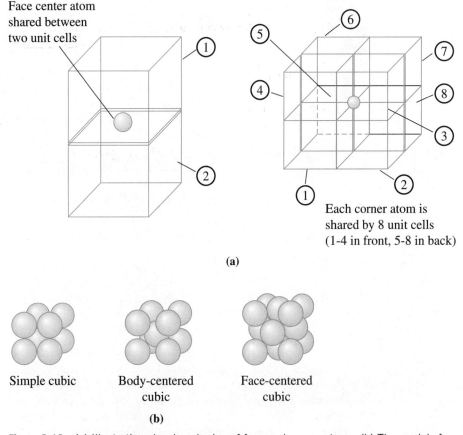

(a)

Simple cubic   Body-centered   Face-centered
                          cubic   cubic

(b)

Figure 3-13 (a) Illustration showing sharing of face and corner atoms. (b) The models for simple cubic (SC), body-centered cubic (BCC), and face-centered cubic (FCC) unit cells, assuming only one atom per lattice point.

Note that mathematically a point cannot be divided. What we mean here is the atom located at a lattice point can be thought to be shared between unit cells. Corners contribute 1/8 of a point, faces contribute 1/2, and body-centered positions contribute a whole point [Figure 3-13(a)].

The number of atoms per unit cell is the product of the number of atoms per lattice point and the number of lattice points per unit cell. In most metals, one atom is located at each lattice point. The structures of simple cubic (SC), body-centered cubic (BCC), and face-centered cubic (FCC) unit cells, with one atom located at each lattice point, are shown in Figure 3-13(b). In more complicated structures, particularly polymer, ceramic, and biological materials, several atoms may be associated with each lattice point (i.e., the basis is greater than one), forming very complex unit cells. Example 3-1 illustrates how to determine the number of lattice points in cubic crystal systems. To develop a better understanding of these and other crystal structures, please make use of the CaRIne™ software provided on the CD-ROM that accompanies this book.

EXAMPLE 3-1   *Determining the Number of Lattice Points in Cubic Crystal Systems*

Determine the number of lattice points per cell in the cubic crystal systems. If there is only one atom located at each lattice point, calculate the number of atoms per unit cell.

SOLUTION

In the SC unit cell, lattice points are located only at the corners of the cube:

$$\frac{\text{lattice point}}{\text{unit cell}} = (8 \text{ corners})\left(\frac{1}{8}\right) = 1$$

In BCC unit cells, lattice points are located at the corners and the center of the cube:

$$\frac{\text{lattice point}}{\text{unit cell}} = (8 \text{ corners})\left(\frac{1}{8}\right) + (1 \text{ center})(1) = 2$$

In FCC unit cells, lattice points are located at the corners and faces of the cube:

$$\frac{\text{lattice point}}{\text{unit cell}} = (8 \text{ corners})\left(\frac{1}{8}\right) + (6 \text{ faces})\left(\frac{1}{2}\right) = 4$$

Since we are assuming there is only one atom located at each lattice point, the number of atoms per unit cell would be 1, 2, and 4, for the simple cubic, body-centered cubic, and face-centered cubic, unit cells, respectively.

Atomic Radius versus Lattice Parameter Directions in the unit cell along which atoms are in continuous contact are **close-packed directions**. In simple structures, particularly those with only one atom per lattice point, we use these directions to calculate the relationship between the apparent size of the atom and the size of the unit cell. By geometrically determining the length of the direction relative to the lattice parameters, and then adding the number of **atomic radii** along this direction, we can determine the desired relationship. Example 3-2 illustrates how the relationships between lattice parameters and atomic radius are determined.

EXAMPLE 3-2   *Determining the Relationship between Atomic Radius and Lattice Parameters*

Determine the relationship between the atomic radius and the lattice parameter in SC, BCC, and FCC structures when one atom is located at each lattice point.

SOLUTION

If we refer to Figure 3-14, we find that atoms touch along the edge of the cube in SC structures. The corner atoms are centered on the corners of the cube, so:

$$a_0 = 2r \tag{3-1}$$

In BCC structures, atoms touch along the body diagonal, which is $\sqrt{3}a_0$ in length. There are two atomic radii from the center atom and one atomic radius from each of the corner atoms on the body diagonal, so

$$a_0 = \frac{4r}{\sqrt{3}} \tag{3-2}$$

In FCC structures, atoms touch along the face diagonal of the cube, which is $\sqrt{2}a_0$ in length. There are four atomic radii along this length—two radii from the face-centered atom and one radius from each corner, so:

$$a_0 = \frac{4r}{\sqrt{2}} \tag{3-3}$$

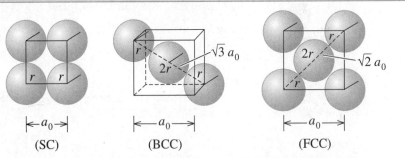

Figure 3-14   The relationships between the atomic radius and the lattice parameter in cubic systems (for Example 3-2).

Coordination Number   The **coordination number** is the number of atoms touching a particular atom, or the number of nearest neighbors for that particular atom. This is one indication of how tightly and efficiently atoms are packed together. For ionic solids, the coordination number of cations is defined as the number of nearest anions. The coordination number of anions is the number of nearest cations. We will discuss the crystal structures of different ionic solids and other materials in Section 3-7.

In cubic structures containing only one atom per lattice point, atoms have a coordination number related to the lattice structure. By inspecting the unit cells in Figure 3-15, we see that each atom in the SC structure has a coordination number of six, while each atom in the BCC structure has eight nearest neighbors. In Section 3-5, we will show that each atom in the FCC structure has a coordination number of 12, which is the maximum.

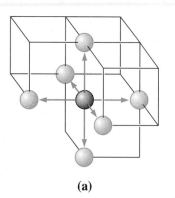
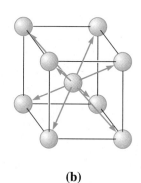

(a)   (b)

Figure 3-15
Illustration of coordinations in (a) SC and (b) BCC unit cells. Six atoms touch each atom in SC, while the eight atoms touch each atom in the BCC unit cell.

Packing Factor The **packing factor** is the fraction of space occupied by atoms, assuming that atoms are hard spheres sized so that they touch their closest neighbor. The general expression for the packing factor is:

$$\text{Packing factor} = \frac{(\text{number of atoms/cell})(\text{volume of each atom})}{\text{volume of unit cell}} \qquad (3\text{-}4)$$

Example 3-3 illustrates how to calculate the packing factor for a FCC cell.

EXAMPLE 3-3 *Calculating the Packing Factor*

Calculate the packing factor for the FCC cell.

SOLUTION

In a FCC cell, there are four lattice points per cell; if there is one atom per lattice point, there are also four atoms per cell. The volume of one atom is $4\pi r^3/3$ and the volume of the unit cell is a_0^3.

$$\text{Packing factor} = \frac{(4 \text{ atoms/cell})\left(\frac{4}{3}\pi r^3\right)}{a_0^3}$$

Since, for FCC unit cells, $a_0 = 4r/\sqrt{2}$:

$$\text{Packing factor} = \frac{(4)\left(\frac{4}{3}\pi r^3\right)}{(4r/\sqrt{2})^3} = \frac{\pi}{\sqrt{18}} \cong 0.74$$

The packing factor of $\pi/\sqrt{18} \cong 0.74$ in the FCC unit cell is the most efficient packing possible. BCC cells have a packing factor of 0.68 and SC cells have a packing factor of 0.52. Notice that the packing factor is independent of the radius of atoms, as long as we assume that all atoms have a fixed radius. What this means is that it does not matter whether we are packing atoms in unit cells or packing basketballs or table tennis balls in a cubical box. The maximum packing factor you can get is $\pi/\sqrt{18}$! This discrete geometry concept is known as **Kepler's conjecture**. Johannes Kepler proposed this conjecture back in the year 1611 and it remained an unproven conjecture until 1998 when Thomas C. Hales actually proved this to be true.[6]

The FCC arrangement represents a **close-packed structure** (CP) (i.e., the packing fraction is the highest possible with atoms of one size). The SC and BCC structures are relatively open. We will see in the next section that it is possible to have a hexagonal structure that has the same packing efficiency as the FCC structure. This structure is known as the **hexagonal close-packed structure** (HCP). Metals with only a metallic bonding are packed as efficiently as possible. Metals with mixed bonding, such as iron, may have unit cells with less than the maximum packing factor. No commonly encountered engineering metals or alloys have the SC structure, although this structure is found in ceramic materials.

Density   The theoretical **density** of a material can be calculated using the properties of the crystal structure. The general formula is:

$$\text{Density } \rho = \frac{(\text{number of atoms/cell})(\text{atomic mass})}{(\text{volume of unit cell})(\text{Avogadro's number})} \tag{3-5}$$

If a material is ionic and consists of different types of atoms or ions, this formula will have to be modified to reflect these differences. Example 3-4 illustrates how to determine the density of BCC iron.

EXAMPLE 3-4   *Determining the Density of BCC Iron*

Determine the density of BCC iron, which has a lattice parameter of 0.2866 nm.

SOLUTION

For a BCC cell,

$$\text{Atoms/cell} = 2$$

$$a_0 = 0.2866 \text{ nm} = 2.866 \times 10^{-8} \text{ cm}$$

$$\text{Atomic mass} = 55.847 \text{ g/mol}$$

$$\text{Volume of unit cell} = a_0^3 = (2.866 \times 10^{-8} \text{ cm})^3 = 23.54 \times 10^{-24} \text{ cm}^3/\text{cell}$$

$$\text{Avogadro's number } N_A = 6.02 \times 10^{23} \text{ atoms/mol}$$

$$\text{Density } \rho = \frac{(\text{number of atoms/cell})(\text{atomic mass of iron})}{(\text{volume of unit cell})(\text{Avogadro's number})}$$

$$\rho = \frac{(2)(55.847)}{(23.54 \times 10^{-24})(6.02 \times 10^{23})} = 7.882 \text{ g/cm}^3$$

The measured density is 7.870 g/cm³. The slight discrepancy between the theoretical and measured densities is a consequence of defects in the material. As mentioned before, the term "defect" in this context means imperfections with regard to the atomic arrangement.

The Hexagonal Close-Packed Structure   A special form of the hexagonal structure, the hexagonal close-packed structure (HCP), is shown in Figure 3-16. The unit cell is the skewed prism, shown separately. The HCP structure has one lattice point per cell—one from each of the eight corners of the prism—but two atoms are associated with each lattice point. One atom is located at a corner, while the second is located within the unit cell. Thus, the basis is 2.

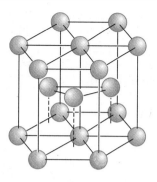

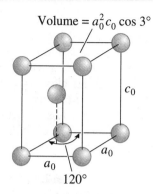

Volume = $a_0^2 c_0 \cos 3°$

Figure 3-16
The hexagonal close-packed (HCP) structure (left) and its unit cell.

TABLE 3-2 ■ Crystal structure characteristics of some metals

Structure	a_0 versus r	Atoms per Cell	Coordination Number	Packing Factor	Examples
Simple cubic (SC)	$a_0 = 2r$	1	6	0.52	Polonium (Po), α-Mn
Body-centered cubic	$a_0 = 4r/\sqrt{3}$	2	8	0.68	Fe, Ti, W, Mo, Nb, Ta, K, Na, V, Zr, Cr
Face-centered cubic	$a_0 = 4r/\sqrt{2}$	4	12	0.74	Fe, Cu, Au, Pt, Ag, Pb, Ni
Hexagonal close-packed	$a_0 = 2r$ $c_0 \approx 1.633 a_0$	2	12	0.74	Ti, Mg, Zn, Be, Co, Zr, Cd

In metals with an ideal HCP structure, the a_0 and c_0 axes are related by the ratio $c_0/a_0 = 1.633$. Most HCP metals, however, have c_0/a_0 ratios that differ slightly from the ideal value because of mixed bonding. Because the HCP structure, like the FCC structure, has the most efficient packing factor of 0.74 and a coordination number of 12, a number of metals possess this structure. Table 3-2 summarizes the characteristics of crystal structures of some metals.

Structures of ionically bonded materials can be viewed as formed by the packing (cubic or hexagonal) of anions. Cations enter into the interstitial sites or holes that remain after the packing of anions. Section 3-7 discusses this in greater detail.

3-4   Allotropic or Polymorphic Transformations

Materials that can have more than one crystal structure are called allotropic or polymorphic. The term **allotropy** is normally reserved for this behavior in pure elements, while the term **polymorphism** is used for compounds. You may have noticed in Table 3-2 that some metals, such as iron and titanium, have more than one crystal structure. At low temperatures, iron has the BCC structure, but at higher temperatures, iron transforms to an FCC structure. These transformations result in changes in properties of materials and form the basis for the heat treatment of steels and many other alloys.

Many ceramic materials, such as silica (SiO_2) and zirconia (ZrO_2), also are polymorphic. A volume change may accompany the transformation during heating or

Figure 3-17
Oxygen gas sensors used in cars and other applications are based on stabilized zirconia compositions. (*Image courtesy of Bosch © Robert Bosch GmbH.*)

cooling; if not properly controlled, this volume change causes the brittle ceramic material to crack and fail. For zirconia (ZrO_2), for instance, the stable form at room temperature ($\sim 25°C$) is monoclinic. As we increase the temperature, more symmetric crystal structures become stable. At 1170°C, the monoclinic zirconia transforms into a tetragonal structure. The tetragonal form is stable up to 2370°C. At that temperature, zirconia transforms into a cubic form. The cubic form remains stable from 2370°C to a melting temperature of 2680°C.[7] Zirconia also can have orthorhombic form, when high pressures are applied.

Ceramics components made from pure zirconia typically will fracture as the temperature is lowered and as zirconia transforms from tetragonal to monoclinic form because of volume expansion (the cubic to tetragonal phase change does not cause much change in volume). As a result, pure monoclinic or tetragonal polymorphs of zirconia are not used. Instead, materials scientists and engineers have found that adding dopants such as yttria (Y_2O_3) make it possible to stabilize the cubic phase of zirconia, even at room temperature. This yttria stabilized zirconia (*YSZ*) contains up to 8 mol. % Y_2O_3. A similar stabilization effect is achieved using CaO and the resultant zirconia is known as calcia stabilized zirconia (*CSZ*).[8] Fully or partially stabilized zirconia formulations are used in many applications, including thermal barrier coatings (TBCs) for turbine blades and electrolytes for oxygen sensors and solid oxide fuel cells. Virtually every car made today uses an oxygen sensor that is made using stabilized zirconia compositions (Figure 3-17). Zirconia conducts electricity through the motion of oxygen ions. In an oxygen sensor, a voltage signal is produced across zirconia using partial pressures of oxygen—one standard oxygen partial pressure, and a variable partial pressure of oxygen in the air–fuel mixture. The voltage produced indicates the leanness or the richness of the air–fuel mixture. This allows for the most efficient use of fuel. The zirconia sensor is only one example of how materials scientists and engineers develop materials and devices that benefit society, while simultaneously optimizing the use of natural resources and limiting environmental pollution. It is also quite possible that some of the cars coming out soon will have zirconia-based fuel cell systems.

The transformation of zirconia from tetragonal to monoclinic is made to work to our advantage in enhancing the toughness of ceramic materials. Toughness means the ability of a material to absorb considerable amount of energy before fracturing. In these materials, the expansion associated with the tetragonal to monoclinic zirconia helps generate a compressive stress in front of a crack tip. This helps increase the toughness of a ceramic material containing zirconia particles/grains of appropriate size.

Polymorphism is also of central importance to several other applications. The properties of some materials can depend quite strongly on the type of polymorph. For example, the dielectric properties of such materials as *PZT* and $BaTiO_3$ depend upon the particular polymorphic form. Example 3-5 illustrates how to calculate volume changes in polymorphs of zirconia. Example 3-6 discusses the extent of such volume changes that can occur during such transformations.

EXAMPLE 3-5   *Calculating Volume Changes in Polymorphs of Zirconia*

Calculate the percent volume change as zirconia transforms from a tetragonal to monoclinic structure.[9] The lattice constants for the monoclinic unit cells are: $a = 5.156$, $b = 5.191$, and $c = 5.304$ Å, respectively. The angle β for the monoclinic unit cell is 98.9°. The lattice constants for the tetragonal unit cell are $a = 5.094$ and $c = 5.304$ Å, respectively.[10] Does the zirconia expand or contract during this transformation? What is the implication of this transformation on the mechanical properties of zirconia ceramics?

SOLUTION

The volume of a tetragonal unit cell is given by $V = a^2c = (5.094)^2(5.304) = 134.33$ Å^3.

The volume of a monoclinic unit cell is given by $V = abc \sin \beta = (5.156)(5.191)(5.304) \sin(98.9) = 140.25$ Å^3.

Thus, there is an expansion of the unit cell as ZrO_2 transforms from a tetragonal to monoclinic form.

The percent change in volume = (final volume − initial volume)/(initial volume) $* 100 = (140.25 - 134.33$ Å$^3)/140.25$ Å$^3 * 100 = 4.21\%$.

Most ceramics are very brittle and cannot withstand more than a 0.1% change in volume. (We will discuss mechanical behavior of materials in Chapters 6 and 7.) The conclusion here is that ZrO_2 ceramics cannot be used in their monoclinic form since, when zirconia does transform to the tetragonal form, it will most likely fracture. Therefore, ZrO_2 is often stabilized in a cubic form using different additives such as CaO, MgO, and Y_2O_3.

EXAMPLE 3-6   *Designing a Sensor to Measure Volume Change*

To study how iron behaves at elevated temperatures, we would like to design an instrument that can detect (with a 1% accuracy) the change in volume of a 1-cm^3 iron cube when the iron is heated through its polymorphic transformation temperature. At 911°C, iron is BCC, with a lattice parameter of 0.2863 nm. At 913°C, iron is FCC, with a lattice parameter of 0.3591 nm. Determine the accuracy required of the measuring instrument.

SOLUTION

The volume change during the transformation can be calculated from crystallographic data. The volume of a unit cell of BCC iron before transforming is:

$$V_{BCC} = a_0^3 = (0.2863 \text{ nm})^3 = 0.023467 \text{ nm}^3$$

This is the volume occupied by two iron atoms, since there are two atoms per unit cell in the BCC crystal structure.

The volume of the unit cell in FCC iron is:

$$V_{FCC} = a_0^3 = (0.3591 \text{ nm})^3 = 0.046307 \text{ nm}^3$$

But this is the volume occupied by *four* iron atoms, as there are four atoms per FCC unit cell. Therefore, we must compare two BCC cells (with a volume of $2(0.023467) = 0.046934$ nm^3) with each FCC cell. The percent volume change during transformation is:

$$\text{Volume change} = \frac{(0.046307 - 0.046934)}{0.046934} \times 100 = -1.34\%$$

This indicates that the iron contracts upon heating during the temperature range that causes the BCC to FCC change.

The 1-cm^3 cube of iron contracts to $1 - 0.0134 = 0.9866$ cm^3 after transforming; therefore, to assure 1% accuracy, the instrument must detect a change of:

$$\Delta V = (0.01)(0.0134) = 0.000134 \text{ cm}^3$$

3-5   Points, Directions, and Planes in the Unit Cell

Coordinates of Points We can locate certain points, such as atom positions, in the lattice or unit cell by constructing the right-handed coordinate system in Figure 3-18. Distance is measured in terms of the number of lattice parameters we must move in each of the x, y, and z coordinates to get from the origin to the point in question. The coordinates are written as the three distances, with commas separating the numbers.

Directions in the Unit Cell Certain directions in the unit cell are of particular importance. **Miller indices** for directions are the shorthand notation used to describe these directions. The procedure for finding the Miller indices for directions is as follows:

1. Using a right-handed coordinate system, determine the coordinates of two points that lie on the direction.

2. Subtract the coordinates of the "tail" point from the coordinates of the "head" point to obtain the number of lattice parameters traveled in the direction of each axis of the coordinate system.

3. Clear fractions and/or reduce the results obtained from the subtraction to lowest integers.

4. Enclose the numbers in square brackets [ ]. If a negative sign is produced, represent the negative sign with a bar over the number.

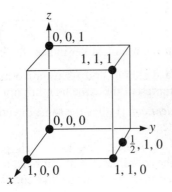

Figure 3-18
Coordinates of selected points in the unit cell. The number refers to the distance from the origin in terms of lattice parameters.

Example 3-7 illustrates a way of determining the Miller indices of direction.

EXAMPLE 3-7   *Determining Miller Indices of Directions*

Determine the Miller indices of directions A, B, and C in Figure 3-19.

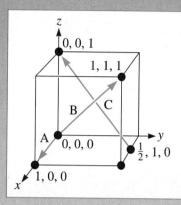

Figure 3-19
Crystallographic directions and coordinates (for Example 3-7).

SOLUTION

Direction A

1. Two points are 1, 0, 0, and 0, 0, 0
2. 1, 0, 0 −0, 0, 0 = 1, 0, 0
3. No fractions to clear or integers to reduce
4. [100]

Direction B

1. Two points are 1, 1, 1 and 0, 0, 0
2. 1, 1, 1 −0, 0, 0 = 1, 1, 1
3. No fractions to clear or integers to reduce
4. [111]

Direction C

1. Two points are 0, 0, 1 and $\frac{1}{2}$, 1, 0
2. 0, 0, 1 $-\frac{1}{2}$, 1, 0 $= -\frac{1}{2}, -1, 1$
3. $2\left(-\frac{1}{2}, -1, 1\right) = -1, -2, 2$
4. $[\bar{1}\bar{2}2]$

Several points should be noted about the use of Miller indices for directions:

1. Because directions are vectors, a direction and its negative are not identical; [100] is not equal to [$\bar{1}$00]. They represent the same line, but opposite directions.

2. A direction and its multiple are *identical*; [100] is the same direction as [200]. We just forgot to reduce to lowest integers.

3. Certain groups of directions are *equivalent*; they have their particular indices because of the way we construct the coordinates. For example, in a cubic system, a [100] direction is a [010] direction if we redefine the coordinate system as

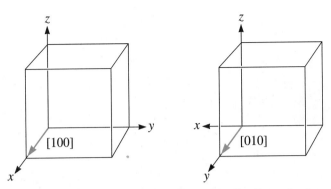

Figure 3-20 Equivalency of crystallographic directions of a form in cubic systems.

TABLE 3-3 ■ *Directions of the form ⟨110⟩ in cubic systems*

$$\langle 110 \rangle = \begin{cases} [110] \ [\bar{1}\bar{1}0] \\ [101] \ [\bar{1}0\bar{1}] \\ [011] \ [0\bar{1}\bar{1}] \\ [1\bar{1}0] \ [\bar{1}10] \\ [10\bar{1}] \ [\bar{1}01] \\ [01\bar{1}] \ [0\bar{1}1] \end{cases}$$

shown in Figure 3-20. We may refer to groups of equivalent directions as **directions of a form**. The special brackets ⟨ ⟩ are used to indicate this collection of directions. All of the directions of the form ⟨110⟩ are shown in Table 3-3. We would expect a material to have the same properties in each of these 12 directions of the form ⟨110⟩.

Significance of Crystallographic Directions Crystallographic directions are used to indicate a particular orientation of a single crystal or of an oriented polycrystalline material. Knowing how to describe these can be useful in many applications. Metals deform more easily, for example, in directions along which atoms are in closest contact. Another real-world example is the dependence of the magnetic properties of iron and other magnetic materials on the crystallographic directions. It is much easier to magnetize iron in the [100] direction compared to [111] or [110] directions. This is why the grains in Fe-Si steels used in magnetic applications (e.g., transformer cores) are oriented in the [100] or equivalent directions. In the case of magnetic materials used for recording media, we have to make sure the grains are aligned in a particular crystallographic direction such that the stored information is not erased easily. Similarly, crystals used for making turbine blades are aligned along certain directions for better mechanical properties.

Repeat Distance, Linear Density, and Packing Fraction Another way of characterizing directions is by the **repeat distance** or the distance between lattice points along the direction. For example, we could examine the [1̄10] direction in an FCC unit cell (Figure 3-21); if we start at the 0, 0, 0 location, the next lattice point is at the center of a face, or a 1/2, 1/2, 0 site. The distance between lattice points is therefore one-half of the face diagonal, or $\frac{1}{2}\sqrt{2}a_0$. In copper, which has a lattice parameter of 0.36151 nm, the repeat distance is 0.2556 nm.

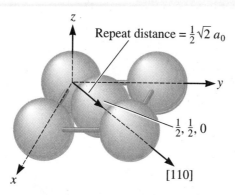

Figure 3-21
Determining the repeat distance, linear density, and packing fraction for a [110] direction in FCC copper.

The **linear density** is the number of lattice points per unit length along the direction. In copper, there are two repeat distances along the [110] direction in each unit cell; since this distance is $\sqrt{2}a_0 = 0.51125$ nm, then:

$$\text{Linear density} = \frac{2 \text{ repeat distances}}{0.51125 \text{ nm}} = 3.91 \text{ lattice points/nm}$$

Note that the linear density is also the reciprocal of the repeat distance.

Finally, we could compute the **packing fraction** of a particular direction, or the fraction actually covered by atoms. For copper, in which one atom is located at each lattice point, this fraction is equal to the product of the linear density and twice the atomic radius. For the [110] direction in FCC copper, the atomic radius $r = \sqrt{2}a_0/4 = 0.12781$ nm. Therefore, the packing fraction is:

$$\text{Packing fraction} = (\text{linear density})(2r)$$
$$= (3.91)(2)(0.12781)$$
$$= (1.0)$$

Atoms touch along the [110] direction, since the [110] direction is close-packed in FCC metals.

Planes in the Unit Cell   Certain planes of atoms in a crystal also carry particular significance. For example, metals deform along planes of atoms that are most tightly packed together. The surface energy of different faces of a crystal depends upon the particular crystallographic planes. This becomes important in crystal growth. In thin film growth of certain electronic materials (e.g., Si or GaAs), we need to be sure the substrate is oriented in such a way that the thin film can grow on a particular crystallographic plane.

Miller indices are used as a shorthand notation to identify these important planes, as described in the following procedure

1. Identify the points at which the plane intercepts the x, y, and z coordinates in terms of the number of lattice parameters. If the plane passes through the origin, the origin of the coordinate system must be moved!

2. Take reciprocals of these intercepts.

3. Clear fractions but do *not* reduce to lowest integers.

4. Enclose the resulting numbers in parentheses ( ). Again, negative numbers should be written with a bar over the number.

The following example shows how Miller indices of planes can be obtained.

EXAMPLE 3-8   *Determining Miller Indices of Planes*

Determine the Miller indices of planes *A*, *B*, and *C* in Figure 3-22.

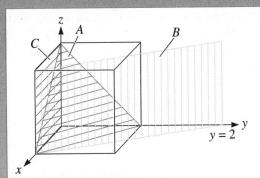

Figure 3-22
Crystallographic planes and intercepts (for Example 3-8).

SOLUTION

Plane *A*

1. $x = 1$, $y = 1$, $z = 1$

2. $\frac{1}{x} = 1, \frac{1}{y} = 1, \frac{1}{z} = 1$

3. No fractions to clear

4. (111)

Plane *B*

1. The plane never intercepts the *z* axis, so $x = 1$, $y = 2$, and $z = \infty$

2. $\frac{1}{x} = 1, \frac{1}{y} = \frac{1}{2}, \frac{1}{z} = 0$

3. Clear fractions: $\frac{1}{x} = 2, \frac{1}{y} = 1, \frac{1}{z} = 0$

4. (210)

Plane *C*

1. We must move the origin, since the plane passes through 0, 0, 0. Let's move the origin one lattice parmaeter in the *y*-direction. Then, $x = \infty$, $y = -1$, and $z = \infty$

2. $\frac{1}{x} = 0, \frac{1}{y} = -1, \frac{1}{z} = 0$

3. No fractions to clear.

4. $(0\bar{1}0)$

Several important aspects of the Miller indices for planes should be noted:

1. Planes and their negatives are identical (this was not the case for directions). Therefore, $(020) = (0\bar{2}0)$.

2. Planes and their multiples are not identical (again, this is the opposite of what we found for directions). We can show this by defining planar densities and planar packing fractions. The **planar density** is the number of atoms per unit area whose centers lie on the plane; the packing fraction is the fraction of the area of that plane actually covered by these atoms. Example 3-9 shows how these can be calculated.

3. In each unit cell, **planes of a form** represent groups of equivalent planes that have their particular indices because of the orientation of the coordinates. We represent these groups of similar planes with the notation { }. The planes of the form {110} in cubic systems are shown in Table 3-4.

4. In cubic systems, a direction that has the same indices as a plane is perpendicular to that plane.

TABLE 3-4 ■ *Planes of the form {110} in cubic systems*

$$
\{110\}
\begin{cases}
(110) \\
(101) \\
(011) \\
(1\bar{1}0) \\
(10\bar{1}) \\
(01\bar{1})
\end{cases}
$$

Note: The negatives of the planes are not unique planes.

EXAMPLE 3-9   *Calculating the Planar Density and Packing Fraction*

Calculate the planar density and planar packing fraction for the (010) and (020) planes in simple cubic polonium, which has a lattice parameter of 0.334 nm.

SOLUTION

The two planes are drawn in Figure 3-23. On the (010) plane, the atoms are centered at each corner of the cube face, with 1/4 of each atom actually in the face of the unit cell. Thus, the total atoms on each face is one. The planar density is:

$$
\text{Planar density (010)} = \frac{\text{atom per face}}{\text{area of face}} = \frac{1\ \text{atom per face}}{(0.334)^2}
$$

$$
= 8.96\ \text{atoms/nm}^2 = 8.96 \times 10^{14}\ \text{atoms/cm}^2
$$

The planar packing fraction is given by:

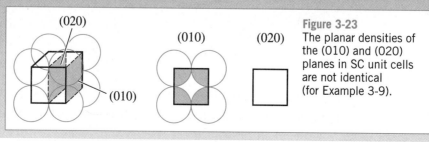

(020)

(010)   (020)

Figure 3-23
The planar densities of the (010) and (020) planes in SC unit cells are not identical (for Example 3-9).

$$\text{Packing fraction (010)} = \frac{\text{area of atoms per face}}{\text{area of face}} = \frac{(1 \text{ atom})(\pi r^2)}{(a_0)^2}$$

$$= \frac{\pi r^2}{(2r)^2} = 0.79$$

However, no atoms are centered on the (020) planes. Therefore, the planar density and the planar packing fraction are both zero. The (010) and (020) planes are not equivalent!

Construction of Directions and Planes To construct a direction or plane in the unit cell, we simply work backwards. Example 3-10 shows how we might do this.

EXAMPLE 3-10   *Drawing Direction and Plane*

Draw (a) the $[1\bar{2}1]$ direction and (b) the $(\bar{2}10)$ plane in a cubic unit cell.

SOLUTION

a. Because we know that we will need to move in the negative y-direction, let's locate the origin at 0, +1, 0. The "tail" of the direction will be located at this new origin. A second point on the direction can be determined by moving +1 in the x-direction, -2 in the y-direction, and +1 in the z-direction [Figure 3-24(a)].

b. To draw in the $(\bar{2}10)$ plane, first take reciprocals of the indices to obtain the intercepts, that is:

$$x = \frac{1}{-2} = -\frac{1}{2} \quad y = \frac{1}{1} = 1 \quad z = \frac{1}{0} = \infty$$

Since the x-intercept is in a negative direction, and we wish to draw the plane within the unit cell, let's move the origin +1 in the x-direction to 1, 0, 0. Then we can locate the x-intercept at $-1/2$ and the y-intercept at +1. The plane will be parallel to the z-axis [Figure 3-24(b)].

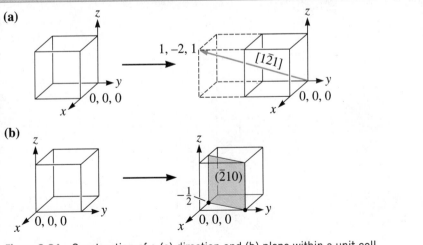

Figure 3-24 Construction of a (a) direction and (b) plane within a unit cell (for Example 3-10).

Miller Indices for Hexagonal Unit Cells A special set of **Miller-Bravais indices** has been devised for hexagonal unit cells because of the unique symmetry of the system (Figure 3-25). The coordinate system uses four axes instead of three, with the a_3 axis being redundant. The procedure for finding the indices of planes is exactly the same as before, but four intercepts are required, giving indices of the form $(hkil)$. Because of the redundancy of the a_3 axis and the special geometry of the system, the first three integers in the designation, corresponding to the a_1, a_2, and a_3 intercepts, are related by $h + k = -i$.

Directions in HCP cells are denoted with either the three-axis or four-axis system. With the three-axis system, the procedure is the same as for conventional Miller indices; examples of this procedure are shown in Example 3-11. A more complicated procedure, by which the direction is broken up into four vectors, is needed for the four-axis system. We determine the number of lattice parameters we must move in each direction to get from the "tail" to the "head" of the direction, while for consistency still making sure that $h + k = -i$. This is illustrated in Figure 3-26, showing that the [010] direction is the same as the $[\bar{1}2\bar{1}0]$ direction.

We can also convert the three-axis notation to the four-axis notation for directions by the following relationships, where h', k', and l' are the indices in the three-axis system:

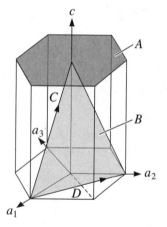

Figure 3-25
Miller-Bravais indices are obtained for crystallographic planes in HCP unit cells by using a four-axis coordinate system. The planes labeled A and B and the directions labeled C and D are those discussed in Example 3-11.

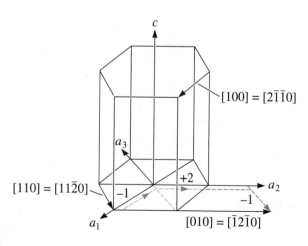

Figure 3-26
Typical directions in the HCP unit cell, using both three- and four-axis systems. The dashed lines show that the $[\bar{1}2\bar{1}0]$ direction is equivalent to a [010] direction.

$$h = \frac{1}{3}(2h' - k')$$

$$k = \frac{1}{3}(2k' - h')$$

$$i = -\frac{1}{3}(h' + k')$$

$$l = l'$$

(3-6)

After conversion, the values of h, k, i, and l may require clearing of fractions or reducing to lowest integers.

EXAMPLE 3-11   *Determining the Miller-Bravais Indices for Planes and Directions*

Determine the Miller-Bravais indices for planes A and B and directions C and D in Figure 3-25.

SOLUTION

Plane A

1. $a_1 = a_2 = a_3 = \infty$, $c = 1$
2. $\frac{1}{a_1} = \frac{1}{a_2} = \frac{1}{a_3} = 0$, $\frac{1}{c} = 1$
3. No fractions to clear
4. (0001)

Plane B

1. $a_1 = 1$, $a_2 = 1$, $a_3 = -\frac{1}{2}$, $c = 1$
2. $\frac{1}{a_1} = 1$, $\frac{1}{a_2} = 1$, $\frac{1}{a_3} = -2$, $\frac{1}{c} = 1$
3. No fractions to clear.
4. $(11\bar{2}1)$

Direction C

1. Two points are 0, 0, 1 and 1, 0, 0.
2. 0, 0, 1 −1, 0, 0 = −1, 0, 1
3. No fractions to clear or integers to reduce.
4. $[\bar{1}01]$ or $[\bar{2}113]$

Direction D

1. Two points are 0, 1, 0 and 1, 0, 0.
2. 0, 1, 0 −1, 0, 0 = −1, 1, 0
3. No fractions to clear or integers to reduce.
4. $[\bar{1}10]$ or $[\bar{1}100]$

Close-Packed Planes and Directions In examining the relationship between atomic radius and lattice parameter, we looked for close-packed directions, where atoms are in continuous contact. We can now assign Miller indices to these close-packed directions, as shown in Table 3-5.

We can also examine FCC and HCP unit cells more closely and discover that there is at least one set of close-packed planes in each. Close-packed planes are shown in Figure 3-27. Notice that a hexagonal arrangement of atoms is produced in two dimensions. The close-packed planes are easy to find in the HCP unit cell; they are the (0001) and (0002) planes of the HCP structure and are given the special name **basal planes**. In fact, we can build up an HCP unit cell by stacking together close-packed planes in an ... *ABABAB* ... **stacking sequence** (Figure 3-27). Atoms on plane *B*, the (0002) plane, fit into the valleys between atoms on plane *A*, the bottom (0001) plane. If another plane identical in orientation to plane *A* is placed in the valleys of plane *B*, the HCP structure is created. Notice that all of the possible close-packed planes are parallel to one another. Only the basal planes—(0001) and (0002)—are close-packed.

From Figure 3-27, we find the coordination number of the atoms in the HCP structure. The center atom in a basal plane is touched by six other atoms in the same plane. Three atoms in a lower plane and three atoms in an upper plane also touch the same atom. The coordination number is 12.

In the FCC structure, close-packed planes are of the form {111} (Figure 3-28). When parallel (111) planes are stacked, atoms in plane *B* fit over valleys in plane *A* and atoms in plane *C* fit over valleys in both planes *A* and *B*. The fourth plane fits directly over atoms in plane *A*. Consequently, a stacking sequence ... *ABCABCABC* ... is produced using the (111) plane. Again, we find that each atom has a coordination number of 12.

Unlike the HCP unit cell, there are four sets of nonparallel close-packed planes— (111), (11$\bar{1}$), (1$\bar{1}$1), and ($\bar{1}$11)—in the FCC cell. This difference between the FCC and

TABLE 3-5 ■ *Close-packed planes and directions*

Structure	Directions	Planes
SC	$\langle 100 \rangle$	None
BCC	$\langle 111 \rangle$	None
FCC	$\langle 110 \rangle$	{111}
HCP	$\langle 100 \rangle$, $\langle 110 \rangle$ or $\langle 11\bar{2}0 \rangle$	(0001), (0002)

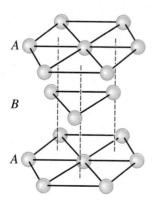

Figure 3-27
The *ABABAB* stacking sequence of close-packed planes produces the HCP structure.

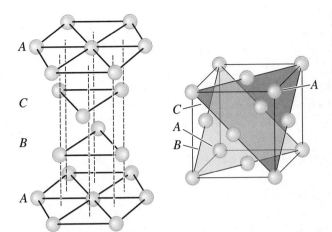

Figure 3-28
The *ABCABCABC* stacking sequence of close-packed planes produces the FCC structure.

HCP unit cells—the presence or absence of intersecting close-packed planes—affects the behavior of metals with these structures.

Isotropic and Anisotropic Behavior   Because of differences in atomic arrangement in the planes and directions within a crystal, some properties also vary with direction. A material is crystallographically **anisotropic** if its properties depend on the crystallographic direction along which the property is measured. For example, the modulus of elasticity of aluminum is 75.9 GPa (11×10^6 psi) in $\langle 111 \rangle$ directions, but only 63.4 GPa (9.2×10^6 psi) in $\langle 100 \rangle$ directions. If the properties are identical in all directions, the material is crystallographically **isotropic**. Note that a material such as aluminum, which is crystallographically anisotropic, may behave as an isotropic material if it is in a polycrystalline form. This is because the random orientations of different crystals in a polycrystalline material will mostly cancel out any effect of the anisotropy as a result of crystal structure. In general, most polycrystalline materials will exhibit isotropic properties. Materials that are single crystals or in which many grains are oriented along certain directions (natural or deliberately obtained by processing) will typically have anisotropic mechanical, optical, magnetic, and dielectric properties.

Interplanar Spacing   The distance between two adjacent parallel planes of atoms with the same Miller indices is called the **interplanar spacing** (d_{hkl}). The interplanar spacing in *cubic* materials is given by the general equation

$$d_{hkl} = \frac{a_0}{\sqrt{h^2 + k^2 + l^2}} \tag{3-7}$$

where a_0 is the lattice parameter and h, k, and l represent the Miller indices of the adjacent planes being considered. The interplanar spacings for non-cubic materials are given by more complex expressions.[11]

3-6   Interstitial Sites

In any of the crystal structures that have been described, there are small holes between the usual atoms into which smaller atoms may be placed. These locations are called **interstitial sites**.

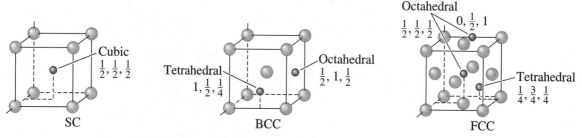

Figure 3-29 The location of the interstitial sites in cubic unit cells. Only representative sites are shown.

An atom, when placed into an interstitial site, touches two or more atoms in the lattice. This interstitial atom has a coordination number equal to the number of atoms it touches. Figure 3-29 shows interstitial locations in the SC, BCC, and FCC structures. The **cubic site**, with a coordination number of eight, occurs in the SC structure. **Octahedral sites** give a coordination number of six (not eight). They are known as octahedral sites because the atoms contacting the interstitial atom form an octahedron with the larger atoms occupying the regular lattice points. **Tetrahedral sites** give a coordination number of four. As an example, the octahedral sites in BCC unit cells are located at faces of the cube; a small atom placed in the octahedral site touches the four atoms at the corners of the face, the atom in the center of the unit cell, plus another atom at the center of the adjacent unit cell, giving a coordination number of six. In FCC unit cells, octahedral sites occur at the center of each edge of the cube, as well as in the center of the unit cell.

EXAMPLE 3-12   *Calculating Octahedral Sites*

Calculate the number of octahedral sites that *uniquely* belong to one FCC unit cell.

SOLUTION

The octahedral sites include the 12 edges of the unit cell, with the coordinates

$$\tfrac{1}{2},0,0 \quad \tfrac{1}{2},1,0 \quad \tfrac{1}{2},0,1 \quad \tfrac{1}{2},1,1$$

$$0,\tfrac{1}{2},0 \quad 1,\tfrac{1}{2},0 \quad 1,\tfrac{1}{2},1 \quad 0,\tfrac{1}{2},1$$

$$0,0,\tfrac{1}{2} \quad 1,0,\tfrac{1}{2} \quad 1,1,\tfrac{1}{2} \quad 0,1,\tfrac{1}{2}$$

plus the center position, 1/2, 1/2, 1/2. Each of the sites on the edge of the unit cell is shared between four unit cells, so only 1/4 of each site belongs uniquely to each unit cell. Therefore, the number of sites belonging uniquely to each cell is:

$$(12 \text{ edges})\left(\tfrac{1}{4} \text{ per cell}\right) + 1 \text{ center location} = 4 \text{ octahedral sites}$$

Interstitial atoms or ions whose radii are slightly larger than the radius of the interstitial site may enter that site, pushing the surrounding atoms slightly apart. However, atoms whose radii are smaller than the radius of the hole are not allowed to fit into the interstitial site, because the ion would "rattle" around in the site. If the interstitial

TABLE 3-6 ■ *The coordination number and the radius ratio*

Coordination Location of Number	Interstitial	Radius Ratio	Representation
2	Linear	0–0.155	
3	Center of triangle	0.155–0.225	
4	Center of tetrahedron	0.225–0.414	
6	Center of octahedron	0.414–0.732	
8	Center of cube	0.732–1.000	

atom becomes too large, it prefers to enter a site having a larger coordination number (Table 3-6). Therefore, an atom whose radius ratio is between 0.225 and 0.414 enters a tetrahedral site; if its radius is somewhat larger than 0.414, it enters an octahedral site instead. When atoms have the same size, as in pure metals, the radius ratio is one and the coordination number is 12, which is the case for metals with the FCC and HCP structures.

Many ionic crystals (Section 3-7) can be viewed as being generated by close packing of larger anions. Cations then can be viewed as smaller ions that fit into the interstitial sites of the close packed anions. Thus, the radius ratios described in Table 3-6 also apply to the ratios of radius of the cation to that of the anion. The packing in ionic crystals is not as tight as that in FCC or HCP metals.

EXAMPLE 3-13   *Design of a Radiation-Absorbing Wall*

We wish to produce a radiation-absorbing wall composed of 10,000 lead balls, each 3 cm in diameter, in a face-centered cubic arrangement. We decide that improved absorption will occur if we fill interstitial sites between the 3-cm balls with smaller balls. Design the size of the smaller lead balls and determine how many are needed.

SOLUTION

We can apply our knowledge of crystal structures to this design. For instance, we may decide to introduce small lead balls that just fit into all of the octahedral sites between the 3-cm balls. First, we can calculate the diameter of the octahedral sites located between the 3-cm diameter balls. Figure 3-30 shows the arrangement of the balls on a plane containing an octahedral site.

$$\text{Length } AB = 2R + 2r = 2R\sqrt{2}$$

$$r = \sqrt{2}R - R = (\sqrt{2} - 1)R$$

$$r/R = 0.414$$

This is consistent with Table 3-6. Since $r/R = 0.414$, the radius of the small lead balls is

$$r = 0.414 * R = (0.414)(3 \text{ cm}/2) = 0.621 \text{ cm}.$$

From Example 3-12, we find that there are four octahedral sites in the FCC arrangement, which also has four lattice points. Therefore, we need the same number of small lead balls as large lead balls, or 10,000 small balls. (As an exercise, you may wish to determine the change in packing factor due to the smaller balls; the reader may also compare tetrahedral sites to octahedral sites.)

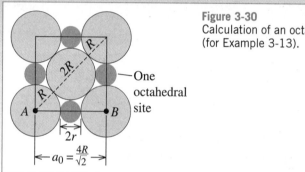

Figure 3-30
Calculation of an octahedral interstitial site (for Example 3-13).

3-7   Crystal Structures of Ionic Materials

Many ceramic materials contain considerable fraction of ionic bonds between the anions and cations. These ionic materials must have crystal structures that assure electrical neutrality, yet permit ions of different sizes to be packed efficiently. As mentioned before, ionic crystal structures can be viewed as close-packed structures of anions. Anions form tetrahedra or octahedra, allowing the cations to fit into their appropriate interstitial sites. In some cases, it may be easier to visualize coordination polyhedra of cations with anions going to the interstial sites. Recall, from Chapter 2, that very often in real materials with engineering applications the bonding is never 100% ionic. We still use this description of the crystal structure, though, to discuss the crystal structure of most ceramic materials. The following factors need to be considered in order to understand crystal structures of ionically bonded solids.

Ionic Radii   The crystal structures of the ionically bonded compounds often can be described by placing the anions at the normal lattice points of a unit cell, with the cations then located at one or more of the interstitial sites described in Section 3-6 (or vice-versa). The ratio of the sizes of the ionic radii of anions and cations influences both the manner of packing and the coordination number (Table 3-6). Note that the radii of atoms and ions are different. For example, the radius of an oxygen atom is 0.6 Å, however the radius of an oxygen anion (O^{2-}) is 1.32 Å. This is because an oxygen anion has acquired two additional electrons and has become larger. As a general rule, anions are larger than cations. Cations, having acquired a positive charge by losing electrons, are expected to be smaller. Strictly speaking, the radii of cations and anions also depend upon the coordination number.[12,13] For example, the radius of Al^{+3} ion is 0.39 Å when the coordination number is four (tetrahedral coordination). However, the radius of Al^{+3} is 0.53 Å when the coordination number is 6 (octahedral coordination). Also, note that the coordination number for cations is the number of nearest anions and vice-versa. This is also true for atoms. For example, the radius of an iron atom in the FCC and BCC polymorphs is different! This tells us that atoms and ions are not "hard spheres", but they can be deformed. Appendix B in this book contains the atomic and ionic radii for different elements.

Electrical Neutrality   The overall material has to be electrically neutral. If the charges on the anion and the cation are identical, and the coordination number for each ion is identical to assure a proper balance of charge, then the compound will have a formula AX (A: cation, X: anion). As an example, each cation may be surrounded by six anions, while each anion is, in turn, surrounded by six cations. However, if the valence of the cation is +2 and that of the anion is −1, then twice as many anions must be present, and the formula is AX_2. The structure of the AX_2 compound must assure that the coordination number of the cation is twice the coordination number of the anion. For example, each cation may have 8 anion nearest neighbors, while only four cations touch each anion.

Connection between Anion Polyhedra   As a rule, the coordination polyhedra (formed by the close packing of anions) will share corners, as opposed to faces or edges. This is because in corner sharing polyhedra electrostatic repulsion between cations is reduced considerably and this leads to the formation of a more stable crystal structure. A number of common structures in ceramic materials are described in the following discussions. Compared to metals, ceramic structures are more complex. You will also note that the lattice constants of ceramic materials tend to be larger than those for metallic materials because electrostatic repulsion between ions prevents close packing of both anions and cations (Figure 3-31).

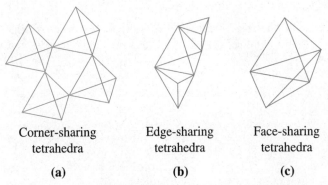

Corner-sharing tetrahedra   Edge-sharing tetrahedra   Face-sharing tetrahedra

(a)   **(b)**   **(c)**

Figure 3-31   Connection between anion polyhedra. Different possible connections include sharing of corners, edges, or faces. In this figure, examples of connections between tetrahedra are shown.

Visualization of Crystal Structures Using Computers Before, we begin to describe different crystal structures, it is important to note that there are many new software programs and tools that have become available recently. These are quite effective in better understanding the crystal structure concepts as well as other concepts discussed in Chapter 4. One example of a useful program is the CaRIne™ software. (See the CD-ROM supplied with this text.) Many of the commercial software packages such as a professional version of CaRIne™ allow the user to "build" a material on the computer. Some also have features that allow calculation of electron and x-ray diffraction patterns.

Cesium Chloride Structure Cesium chloride (CsCl) is simple cubic, with the "cubic" interstitial site filled by the Cl anion [Figure 3-32(a)]. The radius ratio, r_{Cs}^+/r_{Cl}^{-1} = 0.167 nm/0.181 nm = 0.92, dictates that cesium chloride has a coordination number of eight. We can characterize the structure as a simple cubic structure with two ions— one Cs^+ and one Cl^{-1}—associated with each lattice point (or a basis of 2). This structure is possible when the anion and the cation have the same valence.

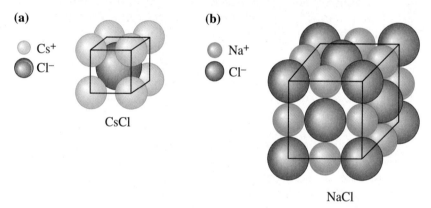

(a) Cs⁺ Cl⁻ CsCl

(b) Na⁺ Cl⁻ NaCl

Figure 3-32 (a) The cesium chloride structure, a SC unit cell with two ions (Cs^+ and Cl^-) per lattice point. (b) The sodium chloride structure, a FCC unit cell with two ions ($Na^+ + Cl^-$) per lattice point. *Note*: Ion sizes not to scale.

EXAMPLE 3-14 *Radius Ratio for KCl*

For potassium chloride (KCl), (a) verify that the compound has the cesium chloride structure and (b) calculate the packing factor for the compound.

SOLUTION

a. From Appendix B, r_{K^+} = 0.133 nm and r_{Cl^-} = 0.181 nm, so:

$$\frac{r_{K^+}}{r_{Cl^-}} = \frac{0.133}{0.181} = 0.735$$

Since 0.732 < 0.735 < 1.000, the coordination number for each type of ion is eight and the CsCl structure is likely.

b. The ions touch along the body diagonal of the unit cell, so:

$$\sqrt{3}a_0 = 2r_{K^+} + 2r_{Cl^-} = 2(0.133) + 2(0.181) = 0.628 \text{ nm}$$

$$a_0 = 0.363 \text{ nm}$$

$$\text{Packing factor} = \frac{\frac{4}{3}\pi r_{K^+}^3(1 \text{ K ion}) + \frac{4}{3}\pi r_{Cl^-}^3(1 \text{ Cl ion})}{a_0^3}$$

$$= \frac{\frac{4}{3}\pi(0.133)^3 + \frac{4}{3}\pi(0.181)^3}{(0.363)^3} = 0.725$$

This structure has been shown in Figure 3-32(a).

Sodium Chloride Structure   The radius ratio for sodium and chloride ions is $r_{Na^+}/r_{Cl^-} = 0.097 \text{ nm}/0.181 \text{ nm} = 0.536$; the sodium ion has a charge of +1; the chloride ion has a charge of −1. Therefore, based on the charge balance and radius ratio, each anion and cation must have a coordination number of six. The FCC structure, with Cl^{-1} ions at FCC positions and Na^+ at the four octahedral sites, satisfies these requirements [Figure 3-32(b)]. We can also consider this structure to be FCC with two ions—one Na^{+1} and one Cl^{-1}—associated with each lattice point. Many ceramics, including magnesium oxide (MgO), calcium oxide (CaO), and iron oxide (FeO) have this structure.

EXAMPLE 3-15   *Illustrating a Crystal Structure and Calculating Density*

Show that MgO has the sodium chloride crystal structure and calculate the density of MgO.

SOLUTION

From Appendix B, $r_{Mg^{+2}} = 0.066$ nm and $r_{O^{-2}} = 0.132$ nm, so:

$$\frac{r_{Mg^{+2}}}{r_{O^{-2}}} = \frac{0.066}{0.132} = 0.50$$

Since $0.414 < 0.50 < 0.732$, the coordination number for each ion is six, and the sodium chloride structure is possible.

The atomic masses are 24.312 and 16 g/mol for magnesium and oxygen, respectively. The ions touch along the edge of the cube, so:

$$a_0 = 2r_{Mg^{+2}} + 2r_{O^{-2}} = 2(0.066) + 2(0.132) = 0.396 \text{ nm} = 3.96 \times 10^{-8} \text{ cm}$$

$$\rho = \frac{(4Mg^{+2})(24.312) + (4O^{-2})(16)}{(3.96 \times 10^{-8} \text{ cm})^3(6.02 \times 10^{23})} = 4.31 \text{ g/cm}^3$$

Zinc Blende Structure   Although the Zn ions have a charge of +2 and S ions have a charge of −2, zinc blende (ZnS) cannot have the sodium chloride structure because $\frac{r_{Zn^{+2}}}{r_{S^{-2}}} = 0.074 \text{ nm}/0.184 \text{ nm} = 0.402$. This radius ratio demands a coordination number of four, which in turn means that the sulfide ions enter tetrahedral sites in a unit cell, as indicated by the small "cubelet" in the unit cell (Figure 3-33). The FCC structure, with Zn cations at the normal lattice points and S anions at half of the tetrahedral sites, can accommodate the restrictions of both charge balance and coordination number. A

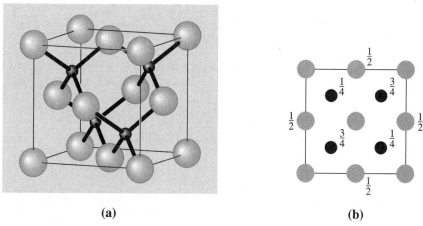

Figure 3-33   (a) The zinc blende unit cell, (b) plan view.

variety of materials, including the semiconductor GaAs and many other III–V semi-conductors (Chapter 2) have this structure.

EXAMPLE 3-16   *Calculating the Theoretical Density of GaAs*

The lattice constant of gallium arsenide (GaAs) is 5.65 Å. Show that the theoretical density of GaAs is 5.33 g/cm^3.

SOLUTION

For the "zinc blende" GaAs unit cell, there are four Ga and four As atoms per unit cell.

From the periodic table (Chapter 2):

Each mole (6.023×10^{23} atoms) of Ga has a mass of 69.7 g. Therefore, the mass of four Ga atoms will be ($4 * 69.7/6.023 \times 10^{23}$) g.

Each mole (6.023×10^{23} atoms) of As has a mass of 74.9 g. Therefore, the mass of four As atoms will be ($4 * 74.9/6.023 \times 10^{23}$) g.

These atoms occupy a volume of $(5.65 \times 10^{-8})^3$ cm^3.

$$\text{density} = \frac{\text{mass}}{\text{volume}} = \frac{4(69.7 + 74.9)/6.023 \times 10^{23}}{(5.65 \times 10^{-8}\ \text{cm})^3} = 5.33\ \text{g/cm}^3$$

Therefore, the theoretical density of GaAs will be 5.33 g/cm^3.

Fluorite Structure   The fluorite structure is FCC, with anions located at all eight of the tetrahedral positions (Figure 3-34). Thus, there are four cations and eight anions per cell and the ceramic compound must have the formula AX_2, as in calcium fluorite, or CaF_2. In the designation AX_2, A is the cation and X is the anion. The coordination number of the calcium ions is eight, but that of the fluoride ions is four, therefore assuring a balance of charge. One of the polymorphs of ZrO_2 known as cubic zirconia exhibits this crystal structure. Other compounds that exhibit this structure include UO_2, ThO_2, and CeO_2.

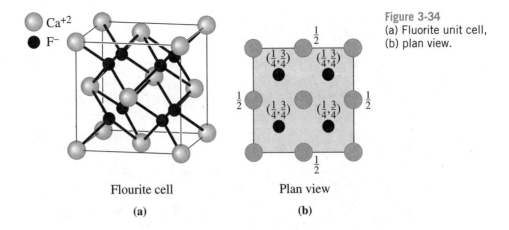

Ca^{+2}
F$^-$

Flourite cell

(a)

Plan view

(b)

Figure 3-34
(a) Fluorite unit cell,
(b) plan view.

Perovskite Structure   This is the crystal structure of a mineral named perovskite (CaTiO$_3$). In general, these compounds could be described as ABO$_3$ (Figure 3-35). In CaTiO$_3$, oxygen anions occupy the face centers of the perovskite unit cell, the corners or the A-sites are occupied by the Ca^{+2} ions and the octahedral B-site at the cube center is occupied by the Ti^{+4} ions. Billions of capacitors for electronic applications are made using formulations based on a material known as barium titanate (BaTiO$_3$). This ceramic material exhibits the perovskite crystal structure in one of its forms. Lead zirconate titanate (*PZT*), also in one of its polymorphs, exhibits this crystal structure. Many new ceramic superconductors, such as yttrium barium copper oxide (YBa$_2$Cu$_3$O$_{7-x}$) and others have structures that are derived from the perovskite structure. An example of a new ceramic superconductor based on yttrium barium copper oxide is shown in Figure 3-36. Again, using the CaRIne^{TM} software provided on the CD-ROM will be very valuable in learning these crystal structures.

Corundum Structure   This is one of the crystal structures of alumina known as alpha alumina (α-Al$_2$O$_3$). In alumina, the oxygen anions pack in a hexagonal arrangement and the aluminum cations occupy some of the available octahedral positions (Figure

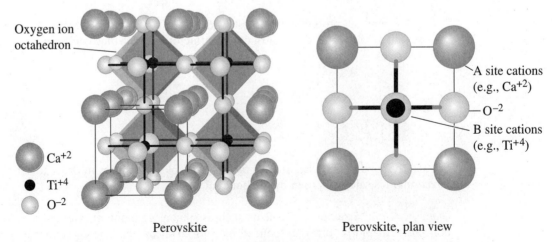

Oxygen ion octahedron

A site cations (e.g., Ca^{+2})
O^{-2}
B site cations (e.g., Ti^{+4})

Ca^{+2}
Ti^{+4}
O^{-2}

Perovskite

Perovskite, plan view

Figure 3-35   The perovskite unit cell showing the A and B site cations and oxygen ions occupying the face-center positions of the unit cell. *Note*: Ions are not shown to scale.

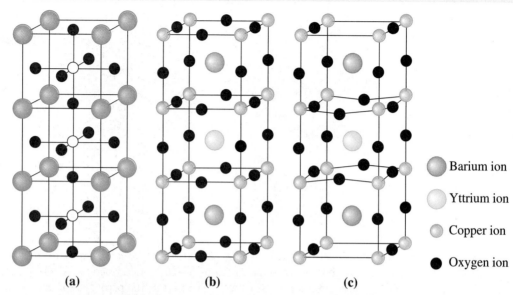

(a)   **(b)**   **(c)**

Barium ion

Yttrium ion

Copper ion

Oxygen ion

Figure 3-36   Crystal structure of a new high T_c ceramic superconductor based on a yttrium barium copper oxide. These materials are unusual in that they are ceramics, yet at low temperatures their electrical resistance vanishes. (*Source: ill.fr/dif/3D-crystals/ superconductor.html; © M. Hewat 1998.*)

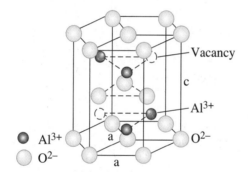

Vacancy

c

Al^{3+}

O^{2-}

Al^{3+}

O^{2-}

a

a

Figure 3-37
Corundum structure of alpha-alumina (α-Al_2O_3).

3-37). Alumina is probably the most widely used ceramic material. Applications include, but are not limited to, spark plugs, refractories, electronic packaging substrates, and abrasives.

3-8   Covalent Structures

Covalently bonded materials frequently have complex structures in order to satisfy the directional restraints imposed by the bonding.

Diamond Cubic Structure   Elements such as silicon, germanium (Ge), α-Sn, and carbon (in its diamond form) are bonded by four covalent bonds and produce a **tetrahedron** [Figure 3-38(a)]. The coordination number for each silicon atom is only four, because of the nature of the covalent bonding.

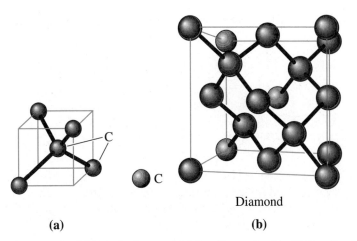

(a)

○ C

Diamond

(b)

Figure 3-38 (a) Tetrahedron and (b) the diamond cubic (DC) unit cell. This open structure is produced because of the requirements of covalent bonding.

As these tetrahedral groups are combined, a large cube can be constructed [Figure 3-38(b)]. This large cube contains eight smaller cubes that are the size of the tetrahedral cube; however, only four of the cubes contain tetrahedra. The large cube is the **diamond cubic** (DC) unit cell. The atoms on the corners of the tetrahedral cubes provide atoms at the regular FCC lattice points. However, four additional atoms are present within the DC unit cell from the atoms in the center of the tetrahedral cubes. We can describe the DC crystal structure as an FCC lattice with two atoms associated with each lattice point (or a basis of 2). Therefore, there must be eight atoms per unit cell.

EXAMPLE 3-17   *Determining the Packing Factor for Diamond Cubic Silicon*

Determine the packing factor for diamond cubic silicon.

SOLUTION

We find that atoms touch along the body diagonal of the cell (Figure 3-39). Although atoms are not present at all locations along the body diagonal, there

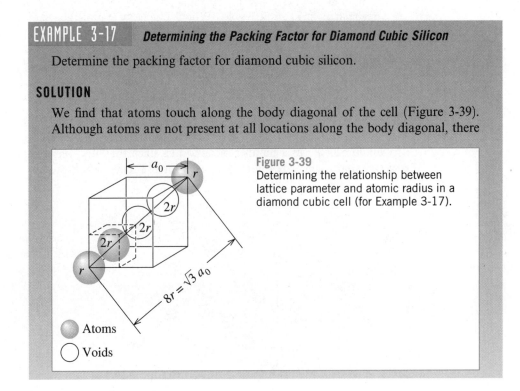

Figure 3-39
Determining the relationship between lattice parameter and atomic radius in a diamond cubic cell (for Example 3-17).

○ Atoms

○ Voids

are voids that have the same diameter as atoms. Consequently:

$$\sqrt{3}a_0 = 8r$$

$$\text{Packing factor} = \frac{(8 \text{ atoms/cell})\left(\frac{4}{3}\pi r^3\right)}{a_0^3}$$

$$= \frac{(8)\left(\frac{4}{3}\pi r^3\right)}{(8r/\sqrt{3})^3}$$

$$= 0.34$$

Compared to close packed structures this is a relatively open structure. In Chapter 5, we will learn that openness of the structure is one of the factors that affects the rate at which different atoms can diffuse inside a given material.

EXAMPLE 3-18   *Calculating the Radius, Density, and Atomic Mass of Silicon*

The lattice constant of Si is 5.43 Å. What will be the radius of a silicon atom? Calculate the theoretical density of silicon. The atomic mass of Si is 28.1 gm/mol.

SOLUTION

Silicon has a diamond cubic structure. As shown in Example 3-17, for the diamond cubic structure.

$$\sqrt{3}a = 8r$$

Therefore, substituting $a = 5.43$ Å, the radius of silicon atom = 1.176 Å. This is the same radius listed in Appendix B. For the density, we use the same approach as in Example 3-16. Recognize that there are eight Si atoms per unit cell.

$$\text{density} = \frac{\text{mass}}{\text{volume}} = \frac{8(28.1)/6.023 \times 10^{23}}{(5.43 \times 10^{-8} \text{ cm})^3} = 2.33 \text{ g/cm}^3$$

This is the same density value listed in Appendix A.

Crystalline Silica   In a number of its forms, silica (or SiO_2) has a crystalline ceramic structure that is partly covalent and partly ionic. Figure 3-40 shows the crystal structure

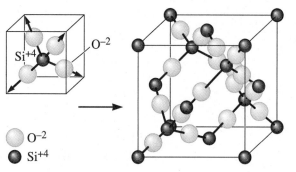

Figure 3-40
The silicon-oxygen tetrahedron and the resultant β-cristobalite form of silica.

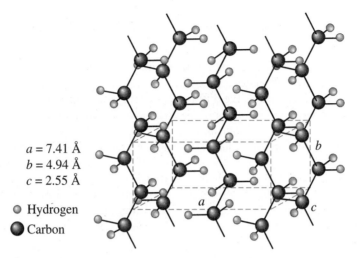

$a = 7.41$ Å
$b = 4.94$ Å
$c = 2.55$ Å

○ Hydrogen
● Carbon

Figure 3-41   The unit cell of crystalline polyethylene.

of one of the forms of silica, β-cristobalite, which is a complicated FCC structure basis. The ionic radii of silicon and oxygen are 0.042 nm and 0.132 nm, respectively, so the radius ratio is $r_{Si^{+4}}/r_{O^{-2}} = 0.318$ and the coordination number is four.

Crystalline Polymers   A number of polymers may form a crystalline structure. The dashed lines in Figure 3-41 outline the unit cell for the lattice of polyethylene. Polyethylene is obtained by joining C_2H_4 molecules to produce long polymer chains that form an orthorhombic unit cell. Some polymers, including nylon, can have several polymorphic forms. Most engineered plastics are partly amorphous and may develop crystallinity during processing. It is also possible to grow single crystals of polymers.

EXAMPLE 3-19   *Calculating the Number of Carbon and Hydrogen Atoms in Crystalline Polyethylene*

How many carbon and hydrogen atoms are in each unit cell of crystalline polyethylene? There are twice as many hydrogen atoms as carbon atoms in the chain. The density of polyethylene is about 0.9972 g/cm^3.

SOLUTION

If we let x be the number of carbon atoms, then $2x$ is the number of hydrogen atoms. From the lattice parameters shown in Figure 3-41:

$$\rho = \frac{(x)(12 \text{ g/mol}) + (2x)(1 \text{ g/mol})}{(7.41 \times 10^{-8} \text{ cm})(4.94 \times 10^{-8} \text{ cm})(2.55 \times 10^{-8} \text{ cm})(6.02 \times 10^{23})}$$

$$0.9972 = \frac{14x}{56.2}$$

$$x = 4 \text{ carbon atoms per cell}$$

$$2x = 8 \text{ hydrogen atoms per cell}$$

3-9 Diffraction Techniques for Crystal Structure Analysis

A crystal structure of a crystalline material can be analyzed using x-ray diffraction (XRD) or electron diffraction. Max von Laue (1879–1960) won the Nobel Prize in 1912 for his discovery related to the diffraction of x-rays by a crystal (Figure 3-42). William Henry Bragg (1862–1942) and his son William Lawrence Bragg (1890–1971) won the 1915 Nobel Prize for their contributions to XRD (Figure 3-42).

When a beam of x-rays having a single wavelength on the same order of magnitude as the atomic spacing in the material strikes that material, x-rays are scattered in all directions. Most of the radiation scattered from one atom cancels out radiation scattered from other atoms. However, x-rays that strike certain crystallographic planes at specific angles are reinforced rather than annihilated. This phenomenon is called **diffraction**. The x-rays are diffracted, or the beam is reinforced, when conditions satisfy **Bragg's law**,

$$\sin \theta = \frac{\lambda}{2d_{hkl}} \qquad (3-8)$$

where the angle θ is half the angle between the diffracted beam and the original beam direction, λ is the wavelength of the x-rays, and d_{hkl} is the interplanar spacing between the planes that cause constructive reinforcement of the beam (see Figure 3-43).

When the material is prepared in the form of a fine powder, there are always at least some powder particles (tiny crystals or aggregates of tiny crystals) whose planes (hkl) are oriented at the proper θ angle to satisfy Bragg's law. Therefore, a diffracted beam, making an angle of 2θ with the incident beam, is produced. In a **diffractometer** [Figures 3-44 and 3-45(a)] a moving x-ray detector records the 2θ angles at which the beam is diffracted, giving a characteristic diffraction pattern [see Figure 3-45(b)]. If we know the wavelength of the x-rays, we can determine the interplanar spacings and, eventually, the identity of the planes that cause the diffraction. In a XRD instrument, x-rays are produced by bombarding a metal target with a beam of high-energy electrons. Typically, x-rays emitted from copper have a wavelength $\lambda \cong 1.54060$ Å (K-α_1 line) and are used.

In the Laue method, which was the first diffraction method ever used, the specimen is in the form of a single crystal. A beam of "white radiation" consisting of x-rays of different wavelengths is used. Each diffracted beam has a different wavelength. In the transmission Laue method, a photographic film is placed behind the crystal. In the back-reflection Laue method, the beams that are back diffracted are recorded on a film located between the source and sample. From the diffraction patterns recorded, the orientation and quality of the single crystal can be determined. It is also possible to determine crystal structure using a rotating crystal and a fixed wavelength x-ray source.[11,14]

Typically, XRD analysis can be conducted relatively rapidly ($\sim$30 minutes to 1 hour per sample), on bulk or powdered samples and without detailed sample preparation. This technique can also be used to determine whether the material consists of many grains oriented in a particular crystallographic direction (texture) in bulk materials and thin films. Typically, a well-trained technician can conduct the analysis as well as interpret the powder diffraction data rather easily. As a result, XRD is used in many industries as one tool for product quality control purposes. Analysis of single crystals and materials containing several phases can be more involved and time consuming.

Figure 3-42   Photographs of William Henry Bragg (top picture, right), Lawrence Bragg (top picture, left) and Max von Laue (below). (*Images courtesy of the University of Pennsylvania Library.*)

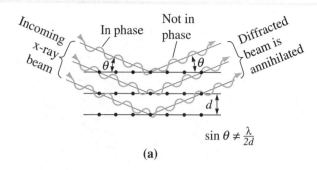

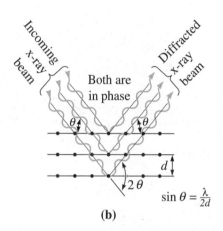

$$\sin \theta \neq \frac{\lambda}{2d}$$

(a)

$$\sin \theta = \frac{\lambda}{2d}$$

(b)

Figure 3-43
(a) Destructive and (b) reinforcing interactions between x-rays and the crystalline material. Reinforcement occurs at angles that satisfy Bragg's law.

Figure 3-44
Photograph of a XRD diffractometer. (*Courtesy of H&M Analytical Services.*)

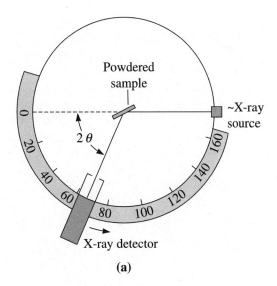

(a)

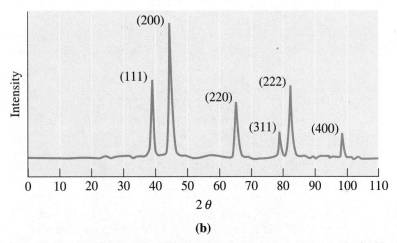

(b)

Figure 3-45   (a) Diagram of a diffractometer, showing powder sample, incident and diffracted beams. (b) The diffraction pattern obtained from a sample of gold powder.

To identify the crystal structure of a cubic material, we note the pattern of the diffracted lines—typically by creating a table of $\sin^2 \theta$ values. By combining Equation 3-7 with Equation 3-8 for the interplanar spacing, we find that:

$$\sin^2 \theta = \frac{\lambda^2}{4a_0^2}(h^2 + k^2 + l^2)$$

In simple cubic metals, all possible planes will diffract, giving an $h^2 + k^2 + l^2$ pattern of 1, 2, 3, 4, 5, 6, 8, ... In body-centered cubic metals, diffraction occurs only from planes having an even $h^2 + k^2 + l^2$ sum of 2, 4, 6, 8, 10, 12, 14, 16, ... For face-centered cubic metals, more destructive interference occurs, and planes having $h^2 + k^2 + l^2$ sums of 3, 4, 8, 11, 12, 16, . . . will diffract. By calculating the values of $\sin^2 \theta$ and then finding the appropriate pattern, the crystal structure can be determined for metals having one of these simple structures, as illustrated in Example 3-20.

EXAMPLE 3-20   *Examining X-ray Diffraction*

The results of a x-ray diffraction experiment using x-rays with $\lambda = 0.7107$ Å (a radiation obtained from molybdenum (Mo) target) show that diffracted peaks occur at the following 2θ angles

Peak	2θ	Peak	2θ
1	20.20	5	46.19
2	28.72	6	50.90
3	35.36	7	55.28
4	41.07	8	59.42

Determine the crystal structure, the indices of the plane producing each peak, and the lattice parameter of the material.

SOLUTION

We can first determine the $\sin^2 \theta$ value for each peak, then divide through by the lowest denominator, 0.0308.

Peak	2θ	$\sin^2 \theta$	$\sin^2 \theta/0.0308$	$h^2 + k^2 + l^2$	(hkl)
1	20.20	0.0308	1	2	(110)
2	28.72	0.0615	2	4	(200)
3	35.36	0.0922	3	6	(211)
4	41.07	0.1230	4	8	(220)
5	46.19	0.1539	5	10	(310)
6	50.90	0.1847	6	12	(222)
7	55.28	0.2152	7	14	(321)
8	59.42	0.2456	8	16	(400)

When we do this, we find a pattern of $\sin^2 \theta/0.0308$ values of 1, 2, 3, 4, 5, 6, 7, and 8. If the material were simple cubic, the 7 would not be present, because no planes have an $h^2 + k^2 + l^2$ value of 7. Therefore, the pattern must really be 2, 4, 6, 8, 10, 12, 14, 16,... and the material must be body-centered cubic. The (hkl) values listed give these required $h^2 + k^2 + l^2$ values.

We could then use 2θ values for any of the peaks to calculate the interplanar spacing and thus the lattice parameter. Picking peak 8:

$$2\theta = 59.42 \quad \text{or} \quad \theta = 29.71$$

$$d_{400} = \frac{\lambda}{2 \sin \theta} = \frac{0.7107}{2 \sin(29.71)} = 0.71699 \text{ Å}$$

$$a_0 = d_{400} \sqrt{h^2 + k^2 + l^2} = (0.71699)(4) = 2.868 \text{ Å}$$

This is the lattice parameter for body-centered cubic iron.

Electron Diffraction and Microscopy Louis de Broglie had theorized that electrons behave like waves. In electron diffraction we make use of high-energy ($\sim$100,000 to 400,000 eV) electrons. These electrons are diffracted from electron transparent samples of materials. The electron beam that exits from the sample is also used to form an image of the sample. Thus, transmission electron microscopy and electron diffraction are used for imaging microstructural features and determining crystal structures.

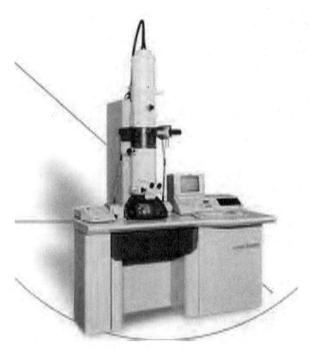

Figure 3-46
Photograph of a transmission electron microscope (TEM) used for analysis of the microstructure of materials. (*Courtesy of JEOL USA, Inc.*)

A 100,000 eV electron has a wavelength of about 0.004 nm![15] This ultra-small wavelength of high-energy electrons allows a transmission electron microscope (TEM) to simultaneously image the microstructure at a very fine scale. If the sample is too thick, electrons cannot be transmitted through the sample and an image or a diffraction pattern will not be observed. Therefore, in transmission electron microscopy and electron diffraction the sample has to be made such that portions of it are electron transparent. A transmission electron microscope is the instrument used for this purpose (Figure 3-46). Figure 3-47 shows a TEM image of and an electron diffraction pattern from an area of the sample. The large bright spots correspond to the grains of the matrix. The smaller spots originate from small crystals of another please.

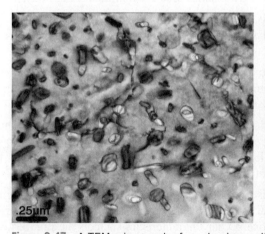
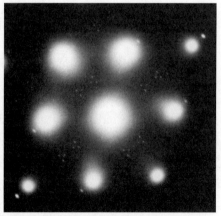

Figure 3-47 A TEM micrograph of an aluminum alloy (Al-7055) sample. The diffraction pattern at the right shows large bright spots that represent diffraction from the main aluminum matrix grains. The smaller spots originate from the nano-scale crystals of another compound that is present in the aluminum alloy. (*Courtesy of Dr. Jörg M.K. Wiezorek, University of Pittsburgh.*)

Another advantage to using a TEM is the high spatial resolution. Using TEM, it is possible to determine differences between different crystalline regions and between amorphous and crystalline regions at very small length-scales ($\sim$1–10 nm). This analytical technique and its variations (e.g., high-resolution electron microscopy (HREM), scanning transmission electron microscopy (STEM), etc.) are also used to determine the orientation of different grains and other microstructural features discussed in later chapters. Advanced and specialized features associated with TEM also allow chemical mapping of elements in a given material. Some of the disadvantages associated with TEM include:

(a) the time-consuming preparation of samples that are almost transparent to the electron beam;

(b) considerable amount of time and skill are required for analysis of the data from a thin, three-dimensional sample, yet represented in a two-dimensional image and diffraction pattern;

(c) only a very small volume of the sample is examined; and

(d) the equipment is relatively expensive and requires great care in use.

In general, TEM has become a widely used and accepted research method for analysis of microstructural features at micro and nano-length scales.

SUMMARY

◇ Atoms or ions may be arranged in solid materials with either a short-range or long-range order.

◇ Amorphous materials, such as silicate glasses, metallic glasses, amorphous silicon and many polymers, have only a short-range order. Amorphous materials form whenever the kinetics of a process involved in the fabrication of a material do not allow the atoms or ions to assume the equilibrium positions. These materials often offer very novel and unusual properties. Many amorphous materials can be crystallized in a controlled fashion. This is the basis for the formation of glass-ceramics and strengthening of PET plastics used for manufacturing bottles.

◇ Crystalline materials, including metals and many ceramics, have both long- and short-range order. The long-range periodicity in these materials is described by the crystal structure.

◇ The atomic or ionic arrangements of crystalline materials are described by seven general crystal systems, which include 14 specific Bravais lattices. Examples include simple cubic, body-centered cubic, face-centered cubic, and hexagonal lattices.

◇ A lattice is a collection of points organized in a unique manner. The basis or motif refers to one or more atoms associated with each lattice point. Crystal structure is derived by adding lattice and basis. Although there are only 14 Bravais lattices there are hundreds of crystal structures.

◇ A crystal structure is characterized by the lattice parameters of the unit cell, which is the smallest subdivision of the crystal structure that still describes the overall structure of the lattice. Other characteristics include the number of lattice points and atoms per unit cell, the coordination number (or number of nearest neighbors) of the atoms in the unit cell, and the packing factor of the atoms in the unit cell. Crystal structure = lattice + basis.

◆ Allotropic, or polymorphic, materials have more than one possible crystal structure. The properties of materials can depend strongly on the type of particular polymorph or allotrope. For example, barium titanate has very different dielectric properties in its cubic and tetragonal polymorphs.

◆ The atoms of metals having the face-centered cubic and hexagonal close-packed crystal structures are closely packed; atoms are arranged in a manner that occupies the greatest fraction of space. The FCC and HCP structures achieve the closest packing by different stacking sequences of close-packed planes of atoms.

◆ The closest packing fraction with spheres of one size is 0.74 and is independent of the radius of the spheres (i.e., atoms and basketballs pack with the same efficiency as long as we deal with a constant radius of an atom and a fixed size basketball).

◆ Points, directions, and planes within the crystal structure can be identified in a formal manner by the assignment of coordinates and Miller indices.

◆ Mechanical, magnetic, optical, and dielectric properties may differ when measured along different directions or planes within a crystal; in this case, the crystal is said to be anisotropic. If the properties are identical in all directions, the crystal is isotropic. The effect of crystallographic anisotropy may be masked in a polycrystalline material because of the random orientation of grains.

◆ Interstitial sites, or holes between the normal atoms in a crystal structure, can be filled by other atoms or ions. The crystal structure of many ceramic materials can be understood by considering how these sites are occupied. Atoms or ions located in interstitial sites play an important role in strengthening materials, influencing the physical properties of materials, and controlling the processing of materials.

◆ Crystal structures of many ionic materials form by the packing of anions (e.g., oxygen ions (O^{-2})). Cations fit into coordination polyhedra formed by anions. These polyhedra typically share corners and lead to crystal structures. The conditions of charge neutrality and stoichiometry have to be balanced. Crystal structures of many ceramic materials (e.g., Al_2O_3, ZrO_2, $YBa_2Cu_3O_{7-x}$) can be rationalized from these considerations.

◆ Crystal structures of covalently bonded materials tend to be open. Examples include diamond cubic (e.g., Si, Ge).

◆ Although most engineered plastics tend to be amorphous, it is possible to have significant crystallinity in polymers, and it is also possible to grow single crystals of certain polymers.

◆ XRD and electron diffraction are used for the determination of the crystal structure of crystalline materials. Transmission electron microscopy can also be used for imaging of microstructural features in materials at smaller length scales.

GLOSSARY

Allotropy   The characteristic of an element being able to exist in more than one crystal structure, depending on temperature and pressure.

Amorphous materials   Materials, including glasses, that have no long-range order, or crystal structure.

Anisotropic   Having different properties in different directions.

Atomic level defects Defects such as vacancies, dislocations etc. occurring over a length-scale comparable to few interatomic distances.

Atomic radius The apparent radius of an atom, typically calculated from the dimensions of the unit cell, using close-packed directions (depends upon coordination number).

Basal plane The special name given to the close-packed plane in hexagonal close-packed unit cells.

Basis A group of atoms associated with a lattice point (same as motif).

Blow-stretch forming A process used to form plastic bottles.

Bose-Einstein condensate (BEC) A newly experimentally verified state of a matter in which a group of atoms occupy the same quantum ground state.

Bragg's law The relationship describing the angle at which a beam of x-rays of a particular wavelength diffracts from crystallographic planes of a given interplanar spacing.

Bravais lattices The fourteen possible lattices that can be created using lattice points.

Chemical vapor deposition (CVD) A process used for production of inorganic thin films or powders involving decomposition of precursors followed by chemical reaction in the vapor phase.

Close-packed directions Directions in a crystal along which atoms are in contact.

Close-packed structure Structures showing a packing fraction of 0.74 (FCC and HCP).

Coordination number The number of nearest neighbors to an atom in its atomic arrangement.

Crystal structure The arrangement of the atoms in a material into a regular repeatable lattice.

Crystal systems Cubic, tetragonal, orthorhombic, hexagonal, monoclinic, rhombohedral and triclinic arrangements of points in space that lead to 14 Bravais lattices and hundreds of crystal structures.

Crystalline materials Materials comprised of one or many small crystals or grains.

Crystallization The process responsible for the formation of crystals, typically in an amorphous material.

Cubic site An interstitial position that has a coordination number of eight. An atom or ion in the cubic site touches eight other atoms or ions.

Defect A microstructural feature representing a disruption in the perfect periodic arrangement of atoms/ions in a crystalline material. This term is not used to convey the presence of a flaw in the material.

Density Mass per unit volume of a material, usually in units of g/cm^3.

Diamond cubic (DC) A special type of face-centered cubic crystal structure found in carbon, silicon, and other covalently bonded materials.

Diffraction The constructive interference, or reinforcement, of a beam of x-rays or electrons interacting with a material. The diffracted beam provides useful information concerning the structure of the material.

Directions of a form Crystallographic directions that all have the same characteristics, although their "sense" is different. Denoted by $\langle\ \rangle$ brackets.

Electron diffraction A method to determine the level of crystallinity at relatively smaller length scale. Based on the diffraction of electrons typically involving use of a transmission electron microscope.

Float-glass process A process to manufacture large pieces of flat glass in which molten glass is floated on a bath of tin.

Glass-ceramics A family of materials typically derived from molten inorganic glasses and processed into crystalline materials with very fine grain size and improved mechanical properties.

Glasses Solid, non-crystalline materials (typically derived from the molten state) that have only short-range atomic order.

Grain A small crystal in a polycrystalline material.

Grain boundaries Regions between grains of a polycrystalline material.

Interplanar spacing Distance between two adjacent parallel planes with the same Miller indices.

Interstitial sites Locations between the "normal" atoms or ions in a crystal into which another—usually different—atom or ion is placed. Typically, the size of this interstitial location is smaller than the atom or ion that is to be introduced.

Isotropic Having the same properties in all directions.

Kepler's conjecture A conjecture made by Johannes Kepler in 1611 that stated that the maximum packing fraction with spheres of uniform size could not exceed $\pi/\sqrt{18}$. In 1998 Thomas Hales proved this to be true.

Lattice A collection of points that divide space into smaller equally sized segments.

Lattice parameters The lengths of the sides of the unit cell and the angles between those sides. The lattice parameters describe the size and shape of the unit cell.

Lattice points Points that make up the lattice. The surroundings of each lattice point are identical anywhere in the material.

Linear density The number of lattice points per unit length along a direction.

Liquid crystals Polymeric materials that are typically amorphous but can become partially crystalline when an external electric field is applied. The effect of the electric field is reversible. Such materials are used in liquid crystal displays.

Long-range order (LRO) A regular repetitive arrangement of atoms in a solid which extends over a very large distance.

Metallic glass Amorphous metals or alloys obtained using rapid solidification.

Miller-Bravais indices A special shorthand notation to describe the crystallographic planes in hexagonal close-packed unit cells.

Miller indices A shorthand notation to describe certain crystallographic directions and planes in a material. Denoted by [ ] brackers. A negative number is represented by a bar over the number.

Motif A group of atoms affiliated with a lattice points (same as basis).

Octahedral site An interstitial position that has a coordination number of six. An atom or ion in the octahedral site touches six other atoms or ions.

Packing factor The fraction of space in a unit cell occupied by atoms.

Packing fraction The fraction of a direction (linear-packing fraction) or a plane (planar-packing factor) that is actually covered by atoms or ions. When one atom is located at each lattice point, the linear packing fraction along a direction is the product of the linear density and twice the atomic radius.

Planar density The number of atoms per unit area whose centers lie on the plane.

Planes of a form   Crystallographic planes that all have the same characteristics, although their orientations are different. Denoted by { } braces.

Polycrystalline material   A material comprised of many grains.

Polymorphism   Compounds exhibiting more than one type of crystal structure.

Rapid solidification   A technique used to cool metals and alloys very quickly.

Repeat distance   The distance from one lattice point to the adjacent lattice point along a direction.

Short-range order   The regular and predictable arrangement of the atoms over a short distance—usually one or two atom spacings.

Stacking sequence   The sequence in which close-packed planes are stacked. If the sequence is *ABABAB*, a hexagonal close-packed unit cell is produced; if the sequence is *ABCABCABC*, a face-centered cubic structure is produced.

Stress-induced crystallization   The process of forming crystals by the application of an external stress. Typically, a significant fraction of many amorphous plastics can be crystallized in this fashion, making them stronger.

Transmission electron microscopy (TEM)   A technique for imaging and analysis of microstructures using a high-energy electron beam.

Tetrahedral site   An interstitial position that has a coordination number of four. An atom or ion in the tetrahedral site touches four other atoms or ions.

Tetrahedron   The structure produced when atoms are packed together with a four-fold coordination.

Unit cell   A subdivision of the lattice that still retains the overall characteristics of the entire lattice.

X-ray diffraction (XRD)   A technique for analysis of crystalline materials using a beam of x-rays.

ADDITIONAL INFORMATION

1. COLLINGS, P.J. and M. HIRD, *Introduction to Liquid Crystals: Chemistry and Physics.* 1997: Taylor and Francis.

2. BOSE, S., *Z. Phys.*, 1924(26): p. 178.

3. EINSTEIN, A., *Sitzber. Kgl. Preuss. Akad. Wiss*, 1924: p. 261.

4. STOOKEY, DONALD, "Profiles in Ceramics," *American Ceramic Society Bulletin*, 2000.

5. SINKE, W.C., "The Photovoltaic Challenge," *MRS Bulletin*, 1993, Vol. 18, No. 10, p. 18.

6. HALES, T.C., *The Kepler Conjecture.* http://xxx.lanl.gov/abs/math.MG/9811078, 1998.

7. MOULSON, A.J. and J.M. HERBERT, *Electroceramics-Materials-Properties-Applications*, 1990: Chapman and Hall.

8. SMART, L. and E. MOORE, *Solid State Chemistry, An Introduction.* 1992: Chapman and Hall.

9. NETTLESHIP, I. *Personal communication*, 2001.

10. STEVENS, R., *Zirconia and Zirconia Ceramics*, 2nd Ed. 1986: Magnesium Elektron Ltd.

11. CULLITY, B.D., *Elements of X-ray Diffraction*, 2nd Ed. 1978: Allison Wesley.

12. CHIANG, Y.M., D. BIRNIE, and W.D. KINGERY, *Physical Ceramics: Principles for Ceramic Science and Engineering*, 1997: Wiley.

13. KINGERY, W.D., H.K. BOWEN, and D.R. UHLMANN, *Introduction to Ceramics*, 2nd Ed 1976: Wiley.

14. JENKINS, R. and R.L. SNYDER, *Introduction to X-ray Powder Diffractometry* 1996: Wiley Interscience.

15. WILLIAMS, D.B. and C.B. CARTER, *Transmission Electron Microscopy Basics I*. 1996: Plenum Press.

✓ PROBLEMS

Section 3-1 Short-Range Order versus Long-Range Order

3-1 What is a "crystalline" material?

3-2 What is a single crystal material?

3-3 State any two applications where single crystal materials are used.

3-4 What is a polycrystalline material?

3-5 What is a liquid crystal material?

3-6 What is an amorphous material?

3-7 Why do some materials assume an amorphous structure?

3-8 State any two applications of amorphous silicate glasses.

3-9 Give an example of an amorphous material that is not derived by melting.

3-10 Compare and contrast the atomic arrangements in amorphous silicon and crystalline silicon.

Section 3-2 Amorphous Materials: Principles and Technological Applications

3-11 Large area films of amorphous silicon are used to make photovoltaic modules. How is amorphous silicon made?

3-12 State any two applications of amorphous silicon.

3-13 Compare and contrast the atomic arrangements in a silicate glass and crystalline silica.

3-14 What is meant by the term glass-ceramic?

3-15 Is glass ceramic an amorphous or a crystalline material?

3-16 Briefly compare the mechanical properties of glasses and glass-ceramics.

3-17 State any two applications of glass-ceramics.

3-18 Sketch the schematic of the blow-stretch process for making PET bottles.

3-19 Explain how PET plastic actually gets stronger during the process of converting a preform into a PET bottle.

3-20 What is stress-induced crystallization?

3-21 What is rapid solidification? What is the main application of this technique?

3-22 Why are gels considered amorphous?

Section 3-3 Lattice, Unit Cells, Basis, and Crystal Structures

3-23 Define the terms lattice, unit cell, basis, and crystal structure.

3-24 Explain why there is no face-centered tetragonal Bravais lattice.

3-25 Calculate the atomic radius in cm for the following:
(a) BCC metal with $a_0 = 0.3294$ nm and one atom per lattice point; and
(b) FCC metal with $a_0 = 4.0862$ Å and one atom per lattice point.

3-26 Determine the crystal structure for the following:
(a) a metal with $a_0 = 4.9489$ Å, $r = 1.75$ Å, and one atom per lattice point; and
(b) a metal with $a_0 = 0.42906$ nm, $r = 0.1858$ nm, and one atom per lattice point.

3-27 The density of potassium, which has the BCC structure and one atom per lattice point, is 0.855 g/cm^3. The atomic weight of potassium is 39.09 g/mol. Calculate
(a) the lattice parameter; and
(b) the atomic radius of potassium.

3-28 The density of thorium, which has the FCC structure and one atom per lattice point, is 11.72 g/cm^3. The atomic weight of thorium is 232 g/mol. Calculate

(a) the lattice parameter; and

(b) the atomic radius of thorium.

3-29 A metal having a cubic structure has a density of 2.6 g/cm^3, an atomic weight of 87.62 g/mol, and a lattice parameter of 6.0849 Å. One atom is associated with each lattice point. Determine the crystal structure of the metal.

3-30 A metal having a cubic structure has a density of 1.892 g/cm^3, an atomic weight of 132.91 g/mol, and a lattice parameter of 6.13 Å. One atom is associated with each lattice point. Determine the crystal structure of the metal.

3-31 Indium has a tetragonal structure, with $a_0 = 0.32517$ nm and $c_0 = 0.49459$ nm. The density is 7.286 g/cm^3 and the atomic weight is 114.82 g/mol. Does indium have the simple tetragonal or body-centered tetragonal structure?

3-32 Bismuth has a hexagonal structure, with $a_0 = 0.4546$ nm and $c_0 = 1.186$ nm. The density is 9.808 g/cm^3 and the atomic weight is 208.98 g/mol. Determine

(a) the volume of the unit cell; and

(b) the number of atoms in each unit cell.

3-33 Gallium has an orthorhombic structure, with $a_0 = 0.45258$ nm, $b_0 = 0.45186$ nm, and $c_0 = 0.76570$ nm. The atomic radius is 0.1218 nm. The density is 5.904 g/cm^3 and the atomic weight is 69.72 g/mol. Determine

(a) the number of atoms in each unit cell; and

(b) the packing factor in the unit cell.

3-34 Beryllium has a hexagonal crystal structure, with $a_0 = 0.22858$ nm and $c_0 = 0.35842$ nm. The atomic radius is 0.1143 nm, the density is 1.848 g/cm^3, and the atomic weight is 9.01 g/mol. Determine

(a) the number of atoms in each unit cell; and

(b) the packing factor in the unit cell.

3-35 A typical paper clip weighs 0.59 g and consists of BCC iron. Calculate

(a) the number of unit cells; and

(b) the number of iron atoms in the paper clip. (See Appendix A for required data.)

3-36 Aluminum foil used to package food is approximately 0.001 inch thick. Assume that all of the unit cells of the aluminum are arranged so that a_0 is perpendicular to the foil surface. For a 4 in. × 4 in. square of the foil, determine

(a) the total number of unit cells in the foil; and

(b) the thickness of the foil in number of unit cells. (See Appendix A.)

Section 3-4 Allotropic or Polymorphic Transformations

3-37 What is the difference between an allotrope and a polymorph?

3-38 What are the different polymorphs of zirconia?

3-39 Above 882°C, titanium has a BCC crystal structure, with $a = 0.332$ nm. Below this temperature, titanium has a HCP structure with $a = 0.2978$ nm and $c = 0.4735$ nm. Determine the percent volume change when BCC titanium transforms to HCP titanium. Is this a contraction or expansion?

3-40 α-Mn has a cubic structure with $a_0 = 0.8931$ nm and a density of 7.47 g/cm^3. β-Mn has a different cubic structure with $a_0 = 0.6326$ nm and a density of 7.26 g/cm^3. The atomic weight of manganese is 54.938 g/mol and the atomic radius is 0.112 nm. Determine the percent volume change that would occur if α-Mn transforms to β-Mn.

3-41 Calculate the theoretical density of the three polymorphs of zirconia. The lattice constants for the monoclinic form are $a = 5.156$, $b = 5.191$, and $c = 5.304$ Å, respectively. The angle β for the monoclinic unit cell is 98.9°. The lattice constants for the tetragonal unit cell are $a = 5.094$ and $c = 5.304$ Å, respectively. The cubic zirconia has a lattice constant of 5.124 Å.

3-42 From the information in this chapter, calculate the volume change that will occur when the cubic form of zirconia transforms into a tetragonal form.

3-43 Monoclinic zirconia cannot be used effectively for manufacturing oxygen sensors or other devices. Explain.

3-44 What is meant by the term stabilized zirconia?

3-45 State any two applications of stabilized zirconia ceramics.

3-46 What is meant by the term transformation toughened ceramics?

3-47 What are the two allotropes of iron?

Section 3-5 Points, Directions, and Planes in the Unit Cell

3-48 Explain the significance of crystallographic directions using an example of an application.

3-49 Why are Fe-Si alloys used in magnetic applications "grain oriented?"

3-50 How is the influence of crystallographic direction on magnetic properties used in magnetic materials for recording media applications?

3-51 Determine the Miller indices for the directions in the cubic unit cell shown in Figure 3-48.

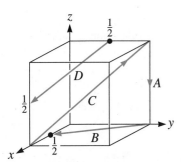

Figure 3-48 Directions in a cubic unit cell for Problem 3-51.

3-52 Determine the indices for the directions in the cubic unit cell shown in Figure 3-49.

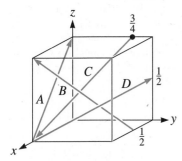

Figure 3-49 Directions in a cubic unit cell for Problem 3-52.

3-53 Determine the indices for the planes in the cubic unit cell shown in Figure 3-50.

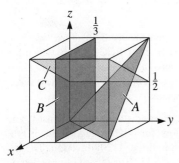

Figure 3-50 Planes in a cubic unit cell for Problem 3-53.

3-54 Determine the indices for the planes in the cubic unit cell shown in Figure 3-51.

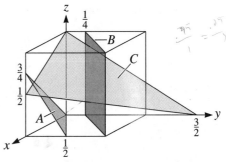

Figure 3-51 Planes in a cubic unit cell for Problem 3-54.

3-55 Determine the indices for the directions in the hexagonal lattice shown in Figure 3-52, using both the three-digit and four-digit systems.

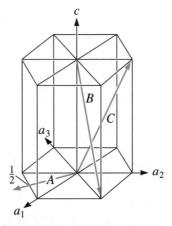

Figure 3-52 Directions in a hexagonal lattice for Problem 3-55.

3-56 Determine the indices for the directions in the hexagonal lattice shown in Figure 3-53, using both the three-digit and four-digit systems.

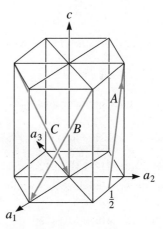

Figure 3-53 Directions in a hexagonal lattice for Problem 3-56.

3-57 Determine the indices for the planes in the hexagonal lattice shown in Figure 3-54.

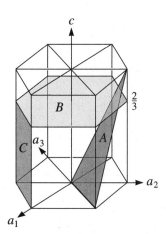

Figure 3-54   Planes in a hexagonal lattice for Problem 3-57.

3-58 Determine the indices for the planes in the hexagonal lattice shown in Figure 3-55.

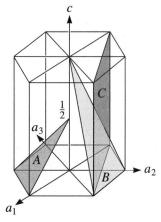

Figure 3-55   Planes in a hexagonal lattice for Problem 3-58.

3-59 Sketch the following planes and directions within a cubic unit cell:

(a) [101] (b) [0$\bar{1}$0] (c) [12$\bar{2}$] (d) [301]
(e) [$\bar{2}$01] (f) [2$\bar{1}$3] (g) (0$\bar{1}\bar{1}$) (h) (102)
(i) (002) (j) (1$\bar{3}$0) (k) ($\bar{2}$12) (l) (3$\bar{1}\bar{2}$)

3-60 Sketch the following planes and directions within a cubic unit cell:

(a) [1$\bar{1}$0] (b) [$\bar{2}\bar{2}$1] (c) [410] (d) [0$\bar{1}$2]
(e) [$\bar{3}$2$\bar{1}$] (f) [1$\bar{1}$1] (g) (11$\bar{1}$) (h) (01$\bar{1}$)
(i) (030) (j) ($\bar{1}$21) (k) (11$\bar{3}$) (l) (0$\bar{4}$1)

3-61 Sketch the following planes and directions within a hexagonal unit cell:

(a) [01$\bar{1}$0] (b) [11$\bar{2}$0] (c) [$\bar{1}$011]
(d) (0003) (e) ($\bar{1}$010) (f) (01$\bar{1}$1)

3-62 Sketch the following planes and directions within a hexagonal unit cell:

(a) [$\bar{2}$110] (b) [11$\bar{2}$1] (c) [10$\bar{1}$0] (d) (1$\bar{2}$10)
(e) ($\bar{1}\bar{1}$22) (f) (12$\bar{3}$0)

3-63 What are the indices of the six directions of the form ⟨110⟩ that lie in the (11$\bar{1}$) plane of a cubic cell?

3-64 What are the indices of the four directions of the form ⟨111⟩ that lie in the ($\bar{1}$01) plane of a cubic cell?

3-65 Determine the number of directions of the form ⟨110⟩ in a tetragonal unit cell and compare to the number of directions of the form ⟨110⟩ in an orthorhombic unit cell.

3-66 Determine the angle between the [110] direction and the (110) plane in a tetragonal unit cell; then determine the angle between the [011] direction and the (011) plane in a tetragonal cell. The lattice parameters are $a_0 = 4.0$ Å and $c_0 = 5.0$ Å. What is responsible for the difference?

3-67 Determine the Miller indices of the plane that passes through three points having the following coordinates:

(a) 0, 0, 1; 1, 0, 0; and 1/2, 1/2, 0
(b) 1/2, 0, 1; 1/2, 0, 0; and 0, 1, 0
(c) 1, 0, 0; 0, 1, 1/2; and 1, 1/2, 1/4
(d) 1, 0, 0; 0, 0, 1/4; and 1/2, 1, 0

3-68 Determine the repeat distance, linear density, and packing fraction for FCC nickel, which has a lattice parameter of 0.35167 nm, in the [100], [110], and [111] directions. Which of these directions is close packed?

3-69 Determine the repeat distance, linear density, and packing fraction for BCC lithium, which has a lattice parameter of 0.35089 nm, in the [100], [110], and [111] directions. Which of these directions is close packed?

3-70 Determine the repeat distance, linear density, and packing fraction for HCP magnesium in the [$\bar{2}$110] direction and the [11$\bar{2}$0] direction. The lattice parameters for HCP magnesium are given in Appendix A.

3-71 Determine the planar density and packing fraction for FCC nickel in the (100), (110), and (111) planes. Which, if any, of these planes is close-packed?

3-72 Determine the planar density and packing fraction for BCC lithium in the (100), (110), and (111) planes. Which, if any, of these planes is close packed?

3-73 Suppose that FCC rhodium is produced as a 1-mm thick sheet, with the (111) plane parallel to the surface of the sheet. How many (111) interplanar spacings d_{111} thick is the sheet? See Appendix A for necessary data.

3-74 In a FCC unit cell, how many d_{111} are present between the 0, 0, 0 point and the 1, 1, 1 point?

3-75 What are the stacking sequences in FCC and HCP structures?

3-76 What is meant by the term anisotropic material?

3-77 The dielectric constant of a barium titanate crystal is significantly different along the a and c directions of the crystal structure. Is $BaTiO_3$ crystal to be considered an isotropic or anisotropic material?

3-78 Explain why most often polycrystalline materials exhibit isotropic properties.

Section 3-6 Interstitial Sites

3-79 Determine the minimum radius of an atom that will just fit into:

   (a) the tetrahedral interstitial site in FCC nickel; and
   (b) the octahedral interstitial site in BCC lithium.

3-80 What are the coordination numbers for octahedral and tetrahedral sites?

Section 3-7 Crystal Structures of Ionic Materials

3-81 What is meant by coordination polyhedra?

3-82 Is the radius of an atom or ion constant? Explain.

3-83 Explain why we consider anions to form the close-packed structures and cations enter the interstitial sites?

3-84 What is the coordination number for the titanium ion in the perovskite crystal structure?

3-85 What is the coordination number for Ca^{+2} in $CaTiO_3$ structure?

3-86 What is the radius of an atom that will just fit into the octahedral site in FCC copper without disturbing the crystal structure?

3-87 Using the ionic radii given in Appendix B, determine the coordination number expected for the following compounds:

   (a) Y_2O_3 (b) UO_2 (c) BaO (d) Si_3N_4
   (e) GeO_2 (f) MnO (g) MgS (h) KBr

3-88 Would you expect NiO to have the cesium chloride, sodium chloride, or zinc blende structure?

Based on your answer, determine

   (a) the lattice parameter;
   (b) the density; and
   (c) the packing factor.

3-89 Would you expect UO_2 to have the sodium chloride, zinc blende, or fluorite structure? Based on your answer, determine

   (a) the lattice parameter;
   (b) the density; and
   (c) the packing factor.

3-90 Would you expect BeO to have the sodium chloride, zinc blende, or fluorite structure? Based on your answer, determine

   (a) the lattice parameter;
   (b) the density; and
   (c) the packing factor.

3-91 Would you expect CsBr to have the sodium chloride, zinc blende, fluorite, or cesium chloride structure? Based on your answer, determine

   (a) the lattice parameter;
   (b) the density; and
   (d) the packing factor.

3-92 Sketch the ion arrangement of the (110) plane of ZnS (with the zinc blende structure) and compare this arrangement to that on the (110) plane of CaF_2 (with the fluorite structure). Compare the planar packing fraction on the (110) planes for these two materials.

3-93 MgO, which has the sodium chloride structure, has a lattice parameter of 0.396 nm. Determine the planar density and the planar packing fraction for the (111) and (222) planes of MgO. What ions are present on each plane?

3-94 Sketch the perovskite crystal structure of $(Ba_xSr_{1-x})TiO_3$ (x: mole fraction of Ba^{+2}). *Note*: The crystal structure is the same as that for $CaTiO_3$, however, there are two types of A-site cations. Assume that these randomly occupy the A-site positions.

3-95 Draw the crystal structure of the perovskite polymorph of *PZT* ($Pb(Zr_xTi_{1-x})O_3$) (x: mole fraction of Zr^{+4}). Assume the two B-site cations occupy random B-site positions.

Section 3-8 Covalent Structures

3-96 Calculate the theoretical density of α-Sn. Assume diamond cubic structure and obtain the radius information from Appendix B.

3-97 Calculate the theoretical density of Ge. Assume diamond cubic structure and obtain the radius information from Appendix B.

3-98 What are the different polymorphs of carbon?

3-99 Why are covalent structures more open than metallic or ionic crystal structures?

3-100 Polypropylene forms an orthorhombic unit cell with lattice parameters of $a_0 = 1.450$ nm, $b_0 = 0.569$ nm, and $c_0 = 0.740$ nm. The chemical formula for the propylene molecule, from which the polymer is produced, is C_3H_6. The density of the polymer is about 0.90 g/cm^3. Determine the number of propylene molecules, the number of carbon atoms, and the number of hydrogen atoms in each unit cell.

3-101 The density of cristobalite is about 1.538 g/cm^3, and it has a lattice parameter of 0.8037 nm. Calculate the number of SiO_2 ions, the number of silicon ions, and the number of oxygen ions in each unit cell.

Section 3-9 Diffraction Techniques for Crystal Structure Analysis

3-102 Explain the principle of XRD.

3-103 Explain why the x-ray radiation has to be monochromatic in the powder diffraction method.

3-104 Explain the basic principle by which the Laue method works.

3-105 A diffracted x-ray beam is observed from the (220) planes of iron at a 2θ angle of 99.1° when x-rays of 0.15418 nm wavelength are used. Calculate the lattice parameter of the iron.

3-106 A diffracted x-ray beam is observed from the (311) planes of aluminum at a 2θ angle of 78.3° when x-rays of 0.15418 nm wavelength are used. Calculate the lattice parameter of the aluminum.

3-107 Figure 3-56 shows the results of an x-ray diffraction experiment in the form of the intensity

of the diffracted peak versus the 2θ diffraction angle. If x-rays with a wavelength of 0.15418 nm are used, determine:

(a) the crystal structure of the metal;
(b) the indices of the planes that produce each of the peaks; and
(c) the lattice parameter of the metal.

3-108 Figure 3-57 shows the results of an x-ray diffraction experiment in the form of the intensity of the diffracted peak versus the 2θ diffraction angle. If x-rays with a wavelength of 0.07107 nm are used, determine:

(a) the crystal structure of the metal;
(b) the indices of the planes that produce each of the peaks; and
(c) the lattice parameter of the metal.

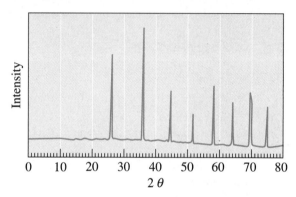

Figure 3-57   XRD pattern for Problem 3-108.

3-109 A sample of zirconia contains cubic and monoclinic polymorphs. What will be a good analytical technique to detect the presence of these two different polymorphs?

3-110 A sample of a polycrystalline ceramic material is suspected of having some minor impurity phases. A XRD analysis was conducted and did not reveal any such impurities. A valuable customer dealing with the ceramics is convinced that there is definitely some impurity phase(s) present. How can you analyze this problem and identify the impurities?

3-111 Can a glass or a glass-ceramic sample be analyzed using XRD? Explain.

3-112 A light emitting diode (LED) is made by depositing an AlGaAs thin film on a GaAs single crystal. How will you analyze the orientation of the thin film? How will you analyze the orientation of the single crystal substrate?

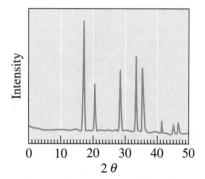

Figure 3-56   XRD pattern for Problem 3-107.

✳ Design Problems

3-113 An oxygen sensor is to be made to measure dissolved oxygen in a large vessel containing molten steel. What kind of material would you choose for this application? Explain.

3-114 You would like to sort iron specimens, some of which are FCC and others BCC. Design an x-ray diffraction method by which this can be accomplished.

3-115 You want to design a material for making kitchen utensils for cooking. The material should be transparent and withstand repeated heating and cooling. What kind of materials could be used to design such transparent and durable kitchenware?

3-116 The float-glass process is used for making windshields and windowpanes. Sometimes NiS impurities appear in these glasses and such particles are known as "stones." These NiS stones can undergo a phase transformation at room temperature. What do you think will happen to glass as a result of a phase change in NiS stones in glass? (*Hint*: Many phase transformations are accompanied by volume changes and glasses are brittle.)

3-117 Many magnetic materials have strongly anisotropic behavior i.e. the magnetic properties depend strongly on crystallographic directions. This is known as magneto-crystalline anisotropy. Although for some applications this may be advantageous, there are applications in which this strong magneto-crystalline anisotropy is not desirable. How will you go about designing a process that will lead to metallic alloys of magnetic materials so that there will be virtually no magneto-crystalline anisotropy?

▢ Computer Problems

Note: You should consult your instructor on the use of computer language. In principle, it does not matter what computer language is used. Some suggestions are using C/C++, Fortran or Java. If these are unavailable you can also solve most of these problems using spreadsheet softwares.

3-118 Table 3-1 contains formulas for the volume of different types of unit cells. Write a computer program to calculate the unit cell volume in the units of $Å^3$ and nm^3. Your program should prompt the user to input the **(a)** type of unit cell, **(b)** necessary lattice constants, and **(c)** angles. The program then should recognize the inputs made and use the appropriate formula for the calculation of unit cell volume.

3-119 Write a computer program that will ask the user to input the atomic mass, atomic radius, and cubic crystal structure for an element. The program output should be the packing fraction and the theoretical density.

3-120 Write a computer program that will ask the user to input the radius of an interstitial atom and the radius of the host atom. The program should then calculate the radius ratio and predict the appropriate coordination number expected. Modify the program so that it will work for ionic solids as well (i.e., the inputs should be the radius of cations and anions and the program should calculate the expected cation coordination number).

Chapter

What makes a ruby red? The addition of about 1% chromium oxide in alumina creates defects. A transition between these defect levels makes a chromium oxide containing alumina the red ruby crystal. Similarly, incorporation of Fe^{+2} and Ti^{+4} makes the blue sapphire. (*Courtesy of Lawrence Lawry/PhotoDisc/Getty Images.*)

4

Imperfections in the Atomic and Ionic Arrangements

Have You Ever Wondered?

- Why silicon crystals used in the manufacture of semiconductor wafers contain trace amounts of dopants, such as phosphorous or boron?

- What makes steel considerably harder and stronger than pure iron?

- What limits the current carrying capacity of a ceramic superconductor?

- What makes a ruby crystal red and a sapphire crystal blue?

- Why do we use very high-purity copper as a conductor in electrical applications?

- Why FCC metals (such as copper and aluminum) tend to be more ductile than BCC and HCP metals?

The arrangement of the atoms or ions in engineered materials contains imperfections or defects. These defects often have a profound effect on the properties of materials. In this chapter, we introduce the three basic types of imperfections: point defects, line defects (or dislocations), and surface defects. These imperfections only represent defects in or deviations from the perfect or ideal atomic or ionic arrangements expected in a given crystal structure. The material is not considered defective from an application viewpoint. In many applications, the presence of such defects is useful. There are a few applications, though, where we will strive to minimize a particular type of defect. For example, defects known as dislocations are useful in increasing the strength of metals and alloys. However, in single crystal silicon, used for manufacturing computer chips, the presence of dislocations is undesirable. Often the "defects" may be created intentionally to produce a desired set of electronic, magnetic, optical, and mechanical properties. For example, pure iron is relatively soft, yet, when we add a small amount of carbon, we create defects in the crystalline arrangement of iron and turn it into a plain carbon steel that exhibits considerably higher strength. Similarly, a crystal of pure alumina is transparent and colorless, but, when we add a small amount of chromium, it creates a special defect, resulting in a beautiful red ruby crystal. In the processing of Si crystals for microelectronics, we add very small concentrations of P or B atoms to Si. These additions create defects in the arrangement of atoms in silicon that impart special electrical properties to different parts of the silicon crystal. This, in turn, allows us to make useful devices such as transistors—the basic building blocks that enabled computers and the information technology revolution. Small concentrations of elements in an otherwise pure metal almost always lowers its electrical conductivity. When we want to use copper for microelectronics as a conductor, we use the highest purity available. This is because even small levels of impurities will cause orders of magnitude increase in the electrical resistivity of copper! The effect of point defects is, therefore, not always desirable.

Grain boundaries, regions between different grains of a polycrystalline material, represent one type of defect. The new ceramic superconductors, under certain conditions, can conduct electricity without any electrical resistance. Materials scientists and engineers have made long wires or tapes of such materials. They have also discovered that, although the current flows quite well within the grains of a polycrystalline superconductor, there is considerable resistance to the flow of current from one grain onto another—across the grain boundary.[1] On the other hand, the presence of grain boundaries actually helps strengthen metallic materials. In later chapters, we will show how we can control the concentrations of these defects through tailoring of composition or processing techniques. In this chapter, we explore the nature and effects of different types of defects.

4-1   Point Defects

Point defects are localized disruptions in otherwise perfect atomic or ionic arrangements in a crystal structure. Even though we call them point defects, the disruption affects a region involving several atoms or ions. These imperfections, shown in Figure 4-1, may be introduced by movement of the atoms or ions when they gain energy by heating, during processing of the material, by introduction of impurities, or doping. The distinction between an impurity and a dopant is as follows: Typically, **impurities** are elements or compounds that are present from raw materials or processing. For example, silicon crystals grown in quartz crucibles contain oxygen as an impurity. **Dopants**, on the other hand, are elements or compounds that are deliberately added, in known concentrations, at specific locations in the microstructure, with an intended beneficial effect on properties or processing. In general, the effect of impurities is deleterious, whereas the effect of dopants on the properties of materials is useful. Phosphorus (P) and boron (B) are examples of dopants that are added to silicon crystals to improve the electrical properties of pure silicon (Si).

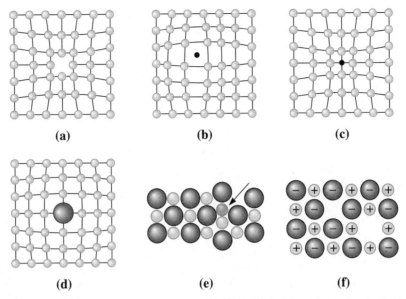

Figure 4-1   Point defects: (a) vacancy, (b) interstitial atom, (c) small substitutional atom, (d) large substitutional atom, (e) Frenkel defect, and (f) Schottky defect. All of these defects disrupt the perfect arrangement of the surrounding atoms.

A point defect typically involves one atom or ion, or a pair of atoms or ions, and thus is different from **extended defects,** such as dislocations, grain boundaries, etc. An important "point" about point defects is that although the defect occurs at one or two sites, their presence is "felt" over much larger distances in the crystalline material.

Vacancies   A **vacancy** is produced when an atom or an ion is missing from its normal site in the crystal structure as in Figure 4-1(a). When atoms or ions are missing (i.e., when vacancies are present), the overall randomness or entropy of the material increases, which increases the thermodynamic stability of a crystalline material. All crystalline materials have vacancy defects. Vacancies are introduced into metals and alloys during solidification, at high temperatures, or as a consequence of radiation damage. Vacancies play an important role in determining the rate at which atoms or ions can move around, or diffuse in a solid material, especially in pure metals. We will see this effect in greater detail in Chapter 5.

In ceramic materials, vacancies can appear during processing as well. For example, barium titanate ($BaTiO_3$) ceramics are used to make ceramic capacitors. The process used to make capacitors involves a heat treatment known as **sintering** ($\sim 1300°C$). Sintering is a process for forming a dense mass by heating compacted powdered materials. The process may involve formation of a liquid. Normally, for $BaTiO_3$ this heat treatment is conducted in air. If we conduct this sintering treatment in a reducing or inert atmosphere (e.g., hydrogen or nitrogen), some of the oxygen ions from $BaTiO_3$ will leave the material in the form of oxygen gas. This will lead to the formation of oxygen ion vacancies. This formation of oxygen ion vacancies will also cause other changes in $BaTiO_3$, for example, turning it into a semiconductor. This is not useful if we are making capacitors to store electric charge! In some cases, we can introduce vacancies at specific ion sites so as to enhance the movement of ions in ceramic materials, which can help sinter ceramics at lower temperatures. In some other applications, we make use of the vacancies created in a ceramic material to tune its electrical properties. This includes many ceramics that are used as conductive and transparent oxides such as indium tin oxide (*ITO*) and zirconia oxygen sensors.

At room temperature ($\sim$298 K), the concentration of vacancies is small, but the concentration of vacancies increases exponentially as we increase the temperature, as shown by the following Arrhenius type behavior:

$$n_v = n \exp\left(\frac{-Q_v}{RT}\right) \tag{4-1}$$

where:

n_v is the number of vacancies per cm^3;

n is the number of atoms per cm^3;

Q_v is the energy required to produce one mole of vacancies, in cal/mol or Joules/mol;

R is the gas constant, $1.987\,\dfrac{\text{cal}}{\text{mol} - \text{K}}$ or $8.31\,\dfrac{\text{Joules}}{\text{mol} - \text{K}}$; and

T is the temperature in degrees Kelvin.

Due to the large thermal energy near the melting temperature, there may be as many as one vacancy per 1000 atoms. Note that this equation provides for equilibrium concentration of vacancies at a given temperature. It is also possible to retain the concentration of vacancies produced at a high temperature by quenching the material rapidly. Thus, in many situations the concentration of vacancies observed at room temperature is not the equilibrium concentration predicted by Equation 4-1.

EXAMPLE 4-1   *The Effect of Temperature on Vacancy Concentrations*

Calculate the concentration of vacancies in copper at room temperature (25°C). What temperature will be needed to heat treat copper such that the concentration of vacancies produced will be 1000 times more than the equilibrium concentration of vacancies at room temperature? Assume that 20,000 cal are required to produce a mole of vacancies in copper.

SOLUTION

The lattice parameter of FCC copper is 0.36151 nm. The basis is 1, therefore, the number of copper atoms, or lattice points, per cm^3 is:

$$n = \frac{4 \text{ atoms/cell}}{(3.6151 \times 10^{-8}\text{ cm})^3} = 8.47 \times 10^{22} \text{ copper atoms/cm}^3$$

At room temperature, $T = 25 + 273 = 298$ K:

$$n_v = n \exp\left(\frac{-Q_v}{RT}\right)$$

$$= \left(8.47 \times 10^{22}\,\frac{\text{atoms}}{\text{cm}^3}\right) \cdot \exp\left(\frac{-20{,}000\,\dfrac{\text{cal}}{\text{mol}}}{1.987\,\dfrac{\text{cal}}{\text{mol} - \text{K}} \times 298 \text{ K}}\right)$$

$$= 1.815 \times 10^{8} \text{ vacancies/cm}^3$$

We wish to find a heat treatment temperature that will lead to a concentration of vacancies which is 1000 times higher than this number, or $n_v = 1.815 \times 10^{11}$ vacancies/cm^3.

We could do this by heating the copper to a temperature at which this number of vacancies forms:

$$n_v = 1.815 \times 10^{11} = n \exp\left(\frac{-Q_v}{RT}\right)$$

$$= (8.47 \times 10^{22}) \exp(-20{,}000/(1.987 \times T))$$

$$\exp\left(\frac{-20{,}000}{1.987 \times T}\right) = \frac{1.815 \times 10^{11}}{8.47 \times 10^{22}} = 0.214 \times 10^{-11}$$

$$\frac{-20{,}000}{1.987T} = \ln(0.214 \times 10^{-11}) = -26.87$$

$$T = \frac{20{,}000}{(1.987)(26.87)} = 375 \text{ K} = 102°\text{C}$$

By heating the copper slightly above 100°C, until equilibrium is reached, and then rapidly cooling the copper back to room temperature, the number of vacancies trapped in the structure may be one thousand times greater than the equilibrium number of vacancies at room temperature. Thus, vacancy concentrations encountered in materials are often dictated by both the thermodynamic and kinetic factors.

EXAMPLE 4-2   *Vacancy Concentrations in Iron*

Determine the number of vacancies needed for a BCC iron crystal to have a density of 7.87 g/cm^3. The lattice parameter of the iron is 2.866×10^{-8} cm.

SOLUTION

The expected theoretical density of iron can be calculated from the lattice parameter and the atomic mass. Since the iron is BCC, two iron atoms are present in each unit cell.

$$\rho = \frac{(2 \text{ atoms/cell})(55.847 \text{ g/mol})}{(2.866 \times 10^{-8} \text{ cm})^3 (6.02 \times 10^{23} \text{ atoms/mol})} = 7.8814 \text{ g/cm}^3$$

We would like to produce iron with a lower density. We could do this by intentionally introducing vacancies into the crystal. Let's calculate the number of iron atoms and vacancies that would be present in each unit cell for the required density of 7.87 g/cm^3:

$$\rho = \frac{(X \text{ atoms/cell})(55.847 \text{ g/mol})}{(2.866 \times 10^{-8} \text{ cm})^3 (6.02 \times 10^{23} \text{ atoms/mol})} = 7.87 \text{ g/cm}^3$$

$$X \text{ atoms/cell} = \frac{(7.87)(2.866 \times 10^{-8})^3 (6.02 \times 10^{23})}{55.847} = 1.9971$$

Or, there should be $2.00 - 1.9971 = 0.0029$ vacancies per unit cell. The number of vacancies per cm^3 is:

$$\text{Vacancies/cm}^3 = \frac{0.0029 \text{ vacancies/cell}}{(2.866 \times 10^{-8} \text{ cm})^3} = 1.23 \times 10^{20}$$

If additional information, such as the energy required to produce a vacancy in iron, was known, we might be able to design a heat treatment (as we did in Example 4-1) to produce this concentration of vacancies.

Interstitial Defects   An **interstitial defect** is formed when an extra atom or ion is inserted into the crystal structure at a normally unoccupied position, as in Figure 4-1(b). The interstitial sites were illustrated in Table 3-6. Interstitial atoms or ions, although much smaller than the atoms or ions located at the lattice points, are still larger than the interstitial sites that they occupy; consequently, the surrounding crystal region is compressed and distorted. Interstitial atoms such as hydrogen are often present as impurities; whereas carbon atoms are intentionally added to iron to produce steel. For small concentrations, carbon atoms occupy interstitial sites in the iron crystal structure, introducing a stress in the localized region of the crystal in their vicinity. If there are dislocations in the crystals trying to move around these types of defects, they face a resistance to their motion, making it difficult to create permanent deformation in metals and alloys. This is one important way of increasing the strength of metallic materials. Unlike vacancies, once introduced, the number of interstitial atoms or ions in the structure remains nearly constant, even when the temperature is changed.

EXAMPLE 4-3   *Sites for Carbon in Iron*

In FCC iron, carbon atoms are located at *octahedral* sites at the center of each edge of the unit cell (1/2, 0, 0) and at the center of the unit cell (1/2, 1/2, 1/2). In BCC iron, carbon atoms enter *tetrahedral* sites, such as 1/4, 1/2, 0. The lattice parameter is 0.3571 nm for FCC iron and 0.2866 nm for BCC iron. Assume that carbon atoms have a radius of 0.071 nm. (1) Would we expect a greater distortion of the crystal by an interstitial carbon atom in FCC or BCC iron? (2) What would be the atomic percentage of carbon in each type of iron if all the interstitial sites were filled?

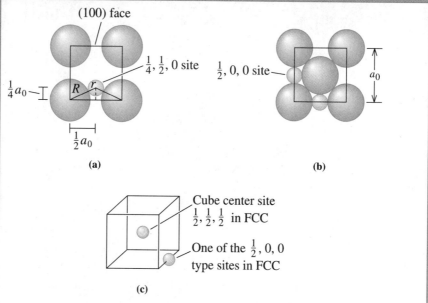

Figure 4-2   (a) The location of the $\frac{1}{4}, \frac{1}{2}$, 0 interstitial site in BCC metals, showing the arrangement of the normal atoms and the interstitial atom (b) $\frac{1}{2}$, 0, 0 site in FCC metals, (for Example 4-3). (c) Edge centers and cube centers are some of the interstitial sites in the FCC structure (Example 4-3).

SOLUTION

1. We could calculate the size of the interstitial site at the 1/4, 1/2, 0 location with the help of Figure 4-2(a). The radius R_{BCC} of the iron atom is:

$$R_{BCC} = \frac{\sqrt{3}a_0}{4} = \frac{(\sqrt{3})(0.2866)}{4} = 0.1241 \text{ nm}$$

From Figure 4-2(a), we find that:

$$\left(\frac{1}{2}a_0\right)^2 + \left(\frac{1}{4}a_0\right)^2 = (r_{interstitial} + R_{BCC})^2$$

$$(r_{interstitial} + R_{BCC})^2 = 0.3125a_0^2 = (0.3125)(0.2866 \text{ nm})^2 = 0.02567$$

$$r_{interstitial} = \sqrt{0.02567} - 0.1241 = 0.0361 \text{ nm}$$

For FCC iron, the interstitial site such as the 1/2, 0, 0 lies along $\langle 100 \rangle$ directions. Thus, the radius of the iron atom and the radius of the interstitial site are [Figure 4-2(b)]:

$$R_{FCC} = \frac{\sqrt{2}a_0}{4} = \frac{(\sqrt{2})(0.3571)}{4} = 0.1263 \text{ nm}$$

$$2r_{interstitial} + 2R_{FCC} = a_0$$

$$r_{interstitial} = \frac{0.3571 - (2)(0.1263)}{2} = 0.0522 \text{ nm}$$

The interstitial site in the BCC iron is smaller than the interstitial site in the FCC iron. Although both are smaller than the carbon atom, carbon distorts the BCC crystal structure more than the FCC crystal. As a result, fewer carbon atoms are expected to enter interstitial positions in BCC iron than in FCC iron.

2. In BCC iron, two iron atoms are expected in each unit cell. We can find a total of 24 interstitial sites of the type 1/4, 1/2, 0; however, since each site is located at a face of the unit cell, only half of each site belongs uniquely to a single cell. Thus:

$$(24 \text{ sites})\left(\frac{1}{2}\right) = 12 \text{ interstitial sites per unit cell}$$

If all of the interstitial sites were filled, the atomic percentage of carbon contained in the iron would be:

$$\text{at } \% \text{ C} = \frac{12 \text{ C atoms}}{12 \text{ C atoms} + 2 \text{ Fe atoms}} \times 100 = 86\%$$

In FCC iron, four iron atoms are expected in each unit cell, and the number of octahedral interstitial sites is:

$$(12 \text{ edges})\left(\frac{1}{4}\right) + 1 \text{ center} = 4 \text{ interstitial sites per unit cell [Figure 4-2(c)]}$$

Again, if all the octahedral interstitial sites were filled, the atomic percentage of carbon in the FCC iron would be:

$$\text{at } \% \text{ C} = \frac{4 \text{ C atoms}}{4 \text{ C atoms} + 4 \text{ Fe atoms}} \times 100 = 50\%$$

As we will see in a later chapter, the maximum atomic percentage of carbon present in the two forms of iron under equilibrium conditions is:

BCC: 1.0%

FCC: 8.9%

Because of the strain imposed on the iron crystal structure by the interstitial atoms—particularly in the BCC iron—the fraction of the interstitial sites that can be occupied is quite small.

Substitutional Defects   A **substitutional defect** is introduced when one atom or ion is replaced by a different type of atom or ion as in Figure 4-1(c) and (d). The substitutional atoms or ions occupy the normal lattice site. Substitutional atoms or ions may either be larger than the normal atoms or ions in the crystal structure, in which case the surrounding inter-atomic spacings are reduced, or smaller causing the surrounding atoms to have larger inter-atomic spacings. In either case, the substitutional defects disturb the surrounding crystal. Again, the substitutional defect can be introduced either as an impurity or as a deliberate alloying addition, and, once introduced, the number of defects is relatively independent of temperature.

Examples of substitutional defects include incorporation of dopants such as phosphorus (P) or boron (B) into Si. Similarly, if we added copper to nickel, copper atoms will occupy crystallographic sites where nickel atoms would normally be present. The substitution atoms will often increase the strength of the metallic material. Substitutional defects also appear in ceramic materials. For example, if we add MgO to NiO, Mg^{+2} ions occupy Ni^{+2} sites and O^{-2} ions from MgO occupy O^{-2} sites of NiO. Whether atoms or ions added go into interstitial or substitutional sites depends upon the size and valence of guest atoms or ions compared to the size and valence of host ions. The size of the available sites also plays a role in this.

Let's look at the example of P or B atoms in a little more detail. We saw in Chapter 3 that silicon has a diamond cubic structure. Semiconductors which do not contain appreciable levels of dopants or impurities are known as **intrinsic semiconductors.** When we add small concentrations of P or B to silicon crystal, these atoms occupy crystallographic sites where silicon atoms usually reside. Recall that Si has a valence of 4. Normally, P has a valence of 5. When P atoms occupy Si sites, they bond to four other neighboring Si atoms. However, P atoms cannot find one more Si atom to form a bond with. The result is each P atom added has one "extra" electron that cannot pair with a Si atom to form a fifth Si-P bond. At low temperatures ($\sim$<50 K) this "extra" electron remains bonded to the P atom via electrostatic attraction. As the temperature increases, this electron breaks away from the P atom and becomes available for conduction. We refer to P as a donor dopant for Si. Each P atom added brings in one electron that can be made available for conduction. This is how P in Si imparts **n-type semiconductivity** to Si. A n-type semiconductor is a semiconductor containing elements that donate electron (e.g., P in Si). Similarly, boron, which has a valence of 3 and thus needs an electron to form four bonds with Si atoms, provides p-type semiconductivity to Si. A **p-type semiconductor** is a semiconductor containing elements that accept electrons (e.g., B in Si). **Extrinsic semiconductors** are semiconductors that contain dopants or impurities. Since we can control how much of P or B is added, the level of conductivity in Si can be controlled precisely. We make useful devices such as transistors by selectively doping different small regions of silicon crystal using dopants such as P, arsenic (As), or B. These devices, of course, enabled the computer and information technology revolution!

EXAMPLE 4-4   *Dopants in Germanium Semiconductor*

Three separate samples of germanium (Ge) crystals contain small concentrations of either silicon (Si), arsenic (As), or boron (B) as dopants. Based on the valence of these elements, what type of semiconductivity is expected from these materials? Assume that these elements will occupy Ge sites.

SOLUTION

Germanium similar to silicon, is a semiconductor that has a diamond cubic crystal structure. Each Ge atom is bonded to four other Ge atoms. When Si is added to Ge, silicon atoms can form four bonds with neighboring Ge atoms. As a result, there is no need to donate or accept an electron. The resultant material then does not show either "n-type" or "p-type" conductivity. It behaves as an intrinsic semiconductor. Recently, alloys based on Si and Ge have been shown to be useful for ultrafast computer chips.

When we add As, we expect n-type conductivity since each As atom brings in five valence electrons. However, it can bond only with four Ge atoms and hence one electron can become available for conduction. On the contrary, when we add small concentrations of B to Ge we expect p-type conductivity for the resultant material, since B has a valence of 3.

4-2   Other Point Defects

An **interstitialcy** is created when an atom identical to those at the normal lattice points is located in an interstitial position. These defects are most likely to be found in crystal structures having a low packing factor.

A **Frenkel defect** is a vacancy-interstitial pair formed when an ion jumps from a normal lattice point to an interstitial site, as in Figure 4-1(e) leaving behind a vacancy. Although, this is described for an ionic material a Frenkel defect can occur in metals and covalently bonded materials. A **Schottky defect**, Figure 4-1(f), is unique to ionic materials and is commonly found in many ceramic materials. In this defect vacancies occur in an ionically bonded material; where a stoichiometric number of anions and cations must be missing from the crystal if electrical neutrality is to be preserved in the crystal. For example, one Mg^{+2} and one O^{-2} missing in MgO constitute a Schottky pair. In ZrO_2, for one missing Zr^{+4} ion there will be 2 O^{-2} ions missing.

An important substitutional point defect occurs when an ion of one charge replaces an ion of a different charge. This might be the case when an ion with a valence of +2 replaces an ion with a valence of +1 (Figure 4-3). In this case, an extra positive charge

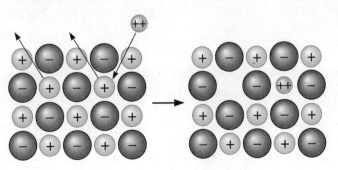

Figure 4-3
When a divalent cation replaces a monovalent cation, a second monovalent cation must also be removed, creating a vacancy.

is introduced into the structure. To maintain a charge balance, a vacancy might be created where a +1 cation normally would be located. Again, this imperfection is observed in materials that have pronounced ionic bonding.

Thus, in ionic solids, when point defects are introduced the following rules have to be observed:

(a) a charge balance must be maintained so that the crystalline material as a whole is electrically neutral;

(b) a mass balance must be maintained; and

(c) the number of crystallographic sites must be conserved.

For example, in nickel oxide (NiO) if one oxygen ion is missing, it creates an oxygen ion vacancy (designated as $V_O^{..}$). Each dot (.) on the subscript indicates an *effective* positive charge of one. To maintain stoichiometry, mass balance and charge balance we must also create a vacancy of nickel ion (designated as V_{Ni}''). Each accent (') in the superscript indicates an *effective* charge of negative 1.

We use the **Kröger-Vink notation** to write the defect chemistry equations. The main letter in this notation describes a vacancy or the name of the element. The superscript indicates the effective charge on the defect and the subscript describes the location of the defect. A dot (.) indicates an effective positive charge of 1 and an accent (') represents an effective charge of -1. Sometimes x is used to indicate no net charge. Any free electrons or holes are indicated as e and h, respectively. Clusters of defects or defects that have association are shown in parentheses. Associated defects, which can affect mass transport in materials, are sometimes neutral and hard to detect experimentally. Concentrations of defects are shown in square brackets.

The following example illustrates the use of the Kröger-Vink notation for writing **defect chemical reactions**. Sometimes, it is possible to write down different defect chemistry reactions that are valid to describe the possible defect chemistry. In such cases, it is necessary to take into account the energy that is needed to create different defects and an experimental verification is necessary. This notation is useful in describing defect chemistry in semiconductors and many ceramic materials used as sensors, dielectrics, and in other applications.

EXAMPLE 4-5   *Application of the Kröger-Vink Notation*

Write the appropriate defect reactions for (1) incorporation of magnesium oxide (MgO) in nickel oxide (NiO), and (2) formation of a Schottky defect in alumina (Al_2O_3).

SOLUTION

1. MgO is the guest and NiO is the host material. We will assume that Mg^{+2} ions will occupy Ni^{+2} sites and oxygen anions from MgO will occupy O^{-2} sites of NiO.

$$MgO \xrightarrow{\text{NiO}} Mg_{Ni}^x + O_O^x$$

We need to ensure that the equation has charge, mass, and site balance. On the left-hand side, we have one Mg, one oxygen, and no net charge. The same is true on the right-hand side. The site balance can be a little tricky—one Mg^{+2} occupies one Ni^{+2} site. Since we are introducing MgO in NiO, we use one Ni^{+2} site and therefore we must use 1 O^{-2} site. We can see that this is true by examining the right-hand side of this equation.

2. A Schottky defect in alumina will involve two aluminum ions and three oxygen ions missing and leaving the crystal. When one aluminum ion is missing, there is a vacancy at the aluminum site and since a positive 3 charge is missing the site has an effective negative charge of -3. Thus V_{Al}''' describes one vacancy of an Al^{+3}. Similarly, $V_O^{\cdot\cdot}$ represents an oxygen ion vacancy. For site balance in alumina, we need to ensure that for every 2 aluminum ion sites used we use three oxygen ion sites. Since we have vacancies, the mass on the right-hand side is zero, and so we write the left-hand side as null. Therefore, the defect reaction will be:

$$\text{null} \xrightarrow{Al_2O_3} 2V_{Al}''' + 3V_O^{\cdot\cdot}$$

EXAMPLE 4-6   *Point Defects in Stabilized Zirconia for Solid Electrolytes*

Write the appropriate defect reactions for the incorporation of calcium oxide (CaO) in zirconia (ZrO_2) using the Kröger-Vink notation.

SOLUTION

We will assume that Ca^{+2} will occupy Zr^{+4} sites. If we send one Ca^{+2} to Zr^{+4}, the site will have an effective negative charge of -2 (instead of having a charge of $+4$ we have a charge of $+2$). We have used one Zr^{+4} site and site balance would require us to utilize *two oxygen sites*. We can send one O^{-2} from CaO to one of the O^{-2} sites in ZrO_2. The other oxygen site must be used and since mass balance must also be maintained we will have to keep this site vacant (i.e., an oxygen ion vacancy will have to be created).

$$CaO \xrightarrow{ZrO_2} Ca_{Zr}'' + O_O^x + V_O^{\cdot\cdot}$$

The concentration of oxygen vacancies in ZrO_2 (i.e., $[V_O^{\cdot\cdot}]$) will increase with increasing CaO concentration. These oxygen ion vacancies make CaO stabilized ZrO_2 an ionic conductor. This provides the use of this type of ZrO_2 in oxygen sensors used in automotives and solid oxide fuel cells, as discussed in Chapter 3.

4-3   Dislocations

Dislocations are line imperfections in an otherwise perfect crystal. They are introduced typically into the crystal during solidification of the material or when the material is deformed permanently. Although dislocations are present in all materials, including ceramics and polymers, *they are particularly useful in explaining deformation and strengthening in metallic materials*. We can identify three types of dislocations: the screw dislocation, the edge dislocation, and the mixed dislocation.

Screw Dislocations   The **screw dislocation** (Figure 4-4) can be illustrated by cutting partway through a perfect crystal, then skewing the crystal one atom spacing. If we follow a crystallographic plane one revolution around the axis on which the crystal was skewed, starting at point x and traveling equal atom spacings in each direction, we finish one atom spacing below our starting point (point y). The vector required to complete the loop and return us to our starting point is the **Burgers vector b**.[2] If we continued our rotation, we would trace out a spiral path. The axis, or line around which

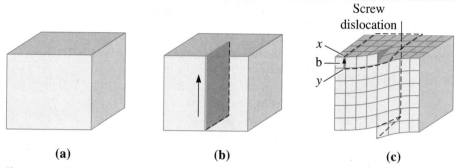

(a) **(b)** **(c)**

Figure 4-4 The perfect crystal (a) is cut and sheared one atom spacing, (b) and (c). The line along which shearing occurs is a screw dislocation. A Burgers vector **b** is required to close a loop of equal atom spacings around the screw dislocation.

we trace out this path, is the screw dislocation. The Burgers vector is parallel to the screw dislocation.

Edge Dislocations   An edge dislocation (Figure 4-5) can be illustrated by slicing partway through a perfect crystal, spreading the crystal apart, and partly filling the cut with an extra plane of atoms. The bottom edge of this inserted plane represents the edge dislocation. If we describe a clockwise loop around the edge dislocation, starting at point x and going an equal number of atoms spacings in each direction, we finish, at point y, one atom spacing from starting point. The vector required to complete the loop is, again, the Burgers vector. In this case, the Burgers vector is perpendicular to the dislocation. By introducing the dislocation, the atoms above the dislocation line are squeezed too closely together, while the atoms below the dislocation are stretched too far apart. The surrounding region of the crystal has been disturbed by the presence of the dislocation. [This is illustrated later on in Figure 4-8(b).] Unlike an **edge dislocation**, a screw dislocation cannot be visualized as an extra half plane of atoms.[3]

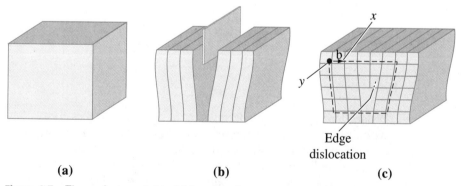

(a) **(b)** **(c)**

Figure 4-5 The perfect crystal in (a) is cut and an extra plane of atoms is inserted (b). The bottom edge of the extra plane is an edge dislocation (c). A Burgers vector **b** is required to close a loop of equal atom spacings around the edge dislocation. (*Adapted from J.D. Verhoeven*, Fundamentals of Physical Metallurgy, *Wiley, 1975.*)

Mixed Dislocations   As shown in Figure 4-6, **mixed dislocations** have both edge and screw components, with a transition region between them. The Burgers vector, however, remains the same for all portions of the mixed dislocation.

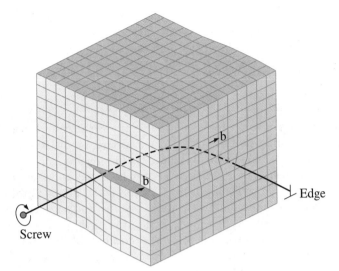

Figure 4-6
A mixed dislocation. The screw dislocation at the front face of the crystal gradually changes to an edge dislocation at the side of the crystal. (*Adapted from W.T. Read*, Dislocations in Crystals. *McGraw-Hill, 1953.*)

A schematic for slip line, slip plane, and slip vector (Burgers vector) for an edge and a screw dislocation are shown in Figure 4-7. The Burgers vector and the plane are helpful in explaining how materials deform.

When a shear force acting in the direction of the Burgers vector is applied to a crystal containing a dislocation, the dislocation can move by breaking the bonds between the atoms in one plane. The cut plane is shifted slightly to establish bonds with the original partial plane of atoms. This shift causes the dislocation to move one atom spacing to the side, as shown in Figure 4-8(a). If this process continues, the dislocation moves through the crystal until a step is produced on the exterior of the crystal; the crystal has then been deformed. Another analogy is the motion by which a caterpillar moves [Figure 4-8(d)]. A caterpillar will lift some of its legs at any given time and use that motion to move from one place to another rather than lifting all the legs at one time. A major difference between the motion of a caterpillar and a dislocation is in the speed with which they move! The speed with which dislocations move in materials is close to or greater than the speed of sound![4] Another way to visualize this is to think

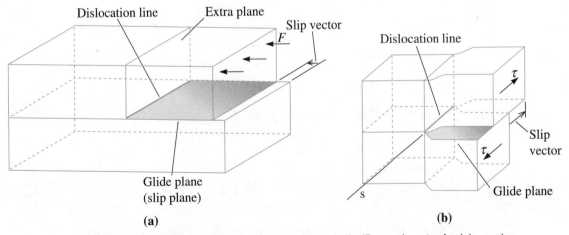

Figure 4-7 Schematic of slip line, slip plane, and slip (Burgers) vector for (a) an edge dislocation and (b) for a screw dislocation. (*Adapted from J.D. Verhoeven*, Fundamentals of Physical Metallurgy, *Wiley, 1975.*)

about how a fold or crease in a carpet would move if we were trying to remove it by pushing it across rather than by lifting the carpet. If dislocations could be introduced continually into one side of the crystal and moved along the same path through the crystal, the crystal would eventually be cut in half.

Slip   The process by which a dislocation moves and causes a metallic material to deform is called **slip**. The direction in which the dislocation moves, the **slip direction**, is the direction of the Burgers vector for edge dislocations as shown in Figure 4-8(b). During slip, the edge dislocation sweeps out the plane formed by the Burgers vector and the dislocation. This plane is called the **slip plane**. The combination of slip direction and slip plane is the **slip system**. A screw dislocation produces the same result; the dislocation moves in a direction perpendicular to the Burgers vector, although the crystal deforms in a direction parallel to the Burgers vector. Since the Burgers vector of a screw dislocation is parallel to the dislocation line, specification of Burgers vector and dislocation line does not define a slip plane for a screw dislocation. As mentioned in Chapter 3, there are new software packages that have been developed and use of these can be very effective in visualizing some of these concepts.

During slip, a dislocation moves from one set of surroundings to an identical set of surroundings. The **Peierls-Nabarro stress** (Equation 4-2) is required to move the dislocation from one equilibrium location to another,

$$\tau = c \, \exp(-kd/b), \tag{4-2}$$

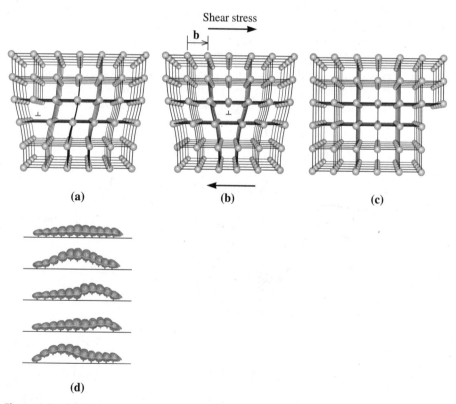

Figure 4-8   (a) When a shear stress is applied to the dislocation in (a), the atoms are displaced, causing the dislocation to move one Burgers vector in the slip direction (b). Continued movement of the dislocation eventually creates a step (c), and the crystal is deformed. (*Adapted from A.G. Guy*, Essentials of Materials Science, *McGraw-Hill, 1976.*) (b) Motion of caterpillar is analogous to the motion of a dislocation.

TABLE 4-1 ■ *Slip planes and directions in metallic structures*

Crystal Structure	Slip Plane	Slip Direction
BCC metals	{110}	$\langle 111 \rangle$
	{112}	
	{123}	
FCC metals	{111}	$\langle 110 \rangle$
HCP metals	{0001}	$\langle 100 \rangle$
	{11$\bar{2}$0} ⎫ See	$\langle 110 \rangle$
	{10$\bar{1}$0} ⎬ Note	or $\langle 11\bar{2}0 \rangle$
	{10$\bar{1}$1} ⎭	
MgO, NaCl (ionic)	{110}	$\langle 110 \rangle$
Silicon (covalent)	{111}	$\langle 110 \rangle$

Note: These planes are active in some metals and alloys or at elevated temperatures.

where τ is the shear stress required to move the dislocation, d is the interplanar spacing between adjacent slip planes, b is magnitude of the Burgers vector, and both c and k are constants for the material. The dislocation moves in a slip system that requires the least expenditure of energy. Several important factors determine the most likely slip systems that will be active:

1. The stress required to cause the dislocation to move increases exponentially with the length of the Burgers vector. Thus, the slip direction should have a small repeat distance or high linear density. The close-packed directions in metals and alloys satisfy this criterion and are the usual slip directions.

2. The stress required to cause the dislocation to move decreases exponentially with the interplanar spacing of the slip planes. Slip occurs most easily between planes of atoms that are smooth (so there are smaller "hills and valleys" on the surface) and between planes that are far apart (or have a relatively large interplanar spacing). Planes with a high planar density fulfill this requirement. Therefore, the slip planes are typically close-packed planes or those as closely packed as possible. Common slip systems in several materials are summarized in Table 4-1.

3. Dislocations do not move easily in materials such as silicon or polymers, which have covalent bonds. Because of the strength and directionality of the bonds, the materials typically fail in a brittle manner before the force becomes high enough to cause appreciable slip. We must keep in mind that in many engineering polymers dislocations play a relatively minor role in deformation. (As discussed in Chapter 3, most polymers contain substantial volume fractions that are amorphous. Significant deformation without fracture is possible in many polymers since when we apply a stress polymer chains become entangled.)

4. Materials with ionic bonding, including many ceramics such as MgO, also are resistant to slip. Movement of a dislocation disrupts the charge balance around the anions and cations, requiring that bonds between anions and cations be broken. During slip, ions with a like charge must also pass close together, causing repulsion. Finally, the repeat distance along the slip direction, or the Burgers vector, is larger than that in metals and alloys. Again, brittle failure of ceramic material typically occurs owing to the presence of flaws such as small holes (pores) before the applied level of stress is sufficient to cause dislocations to move. Ductility in ceramic materials can be obtained by:

(a) phase transformations (known as transformation plasticity, an example would be fully stabilized zirconia);

(b) mechanical twinning;

(c) dislocation motion; and

(d) grain boundary sliding.[5]

Typically, higher temperatures and compressive stresses lead to higher ductility. Recently, it has been shown that certain ceramics such as strontium titanate ($SrTiO_3$) can exhibit considerable ductility.[6] We will also learn later that under certain conditions ceramics can exhibit very large deformations under certain conditions. This behavior is known as superplasticity and is discussed in later chapters.

EXAMPLE 4-7 *Dislocations in Ceramic Materials*

A sketch of a dislocation in magnesium oxide (MgO), which has the sodium chloride crystal structure and a lattice parameter of 0.396 nm, is shown in Figure 4-9. Determine the length of the Burgers vector.

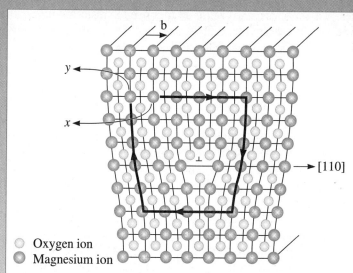

Figure 4-9 An edge dislocation in MgO showing the slip direction and Burgers vector (for Example 4-7). (*Adapted from W.D. Kingery, H.K. Bowen, and D.R. Uhlmann*, Introduction to Ceramics, *John Wiley, 1976.*)

SOLUTION

In Figure 4-9, we begin a clockwise loop around the dislocation at point x, then move equal atom spacings to finish at point y. The vector **b** is the Burgers vector. Because **b** is a [110] direction, it must be perpendicular to {110} planes. The length of **b** is the distance between two adjacent (110) planes. From Equation 3-7,

$$d_{110} = \frac{a_0}{\sqrt{h^2 + k^2 + l^2}} = \frac{0.396}{\sqrt{1^2 + 1^2 + 0^2}} = 0.280 \text{ nm}$$

The Burgers vector is a $\langle 110 \rangle$ direction that is 0.280 nm in length. Note, however, that two extra half-planes of atoms make up the dislocation-one composed of oxygen ions and one of magnesium ions (Figure 4-9). Note that this formula for calculating the magnitude of the Burgers vector will not work for non-cubic systems. It is better to consider the magnitude of the Burgers vector as equal to the repeat distance in the slip direction.

EXAMPLE 4-8   *Burgers Vector Calculation*

Calculate the length of the Burgers vector in copper.

SOLUTION

Copper has an FCC crystal structure. The lattice parameter of copper (Cu) is 0.36151 nm. The close-packed directions, or the directions of the Burgers vector, are of the form $\langle 110 \rangle$. The repeat distance along the $\langle 110 \rangle$ directions is one-half the face diagonal, since lattice points are located at corners and centers of faces [Figure 4-10(a)].

$$\text{Face diagonal} = \sqrt{2}a_0 = (\sqrt{2})(0.36151) = 0.51125 \text{ nm}$$

The length of the Burgers vector, or the repeat distance, is:

$$b = \tfrac{1}{2}(0.51125 \text{ nm}) = 0.25563 \text{ nm}$$

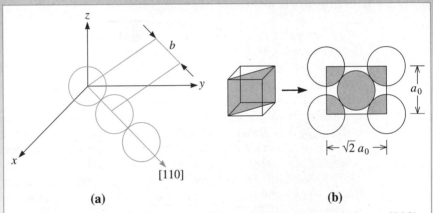

Figure 4-10   (a) Burgers vector for FCC copper. (b) The atom locations on a (110) plane in a BCC unit cell (for Example 4-8 and 4-9, respectively).

EXAMPLE 4-9   *Identification of Preferred Slip Planes*

The planar density of the (112) plane in BCC iron is 9.94×10^{14} atoms/cm². Calculate (1) the planar density of the (110) plane and (2) the interplanar spacings for both the (112) and (110) planes. On which plane would slip normally occur?

SOLUTION

The lattice parameter of BCC iron is 0.2866 nm or 2.866×10^{-8} cm. The (110) plane is shown in Figure 4-10(b), with the portion of the atoms lying within the unit cell being shaded. Note that one-fourth of the four corner atoms plus the center atom lie within an area of a_0 times $\sqrt{2}a_0$.

1. The planar density is:

$$\text{Planar density (110)} = \frac{\text{atoms}}{\text{area}} = \frac{2}{(\sqrt{2})(2.866 \times 10^{-8} \text{ cm})^2}$$

$$= 1.72 \times 10^{15} \text{ atoms/cm}^2$$

Planar density (112) = 0.994×10^{15} atoms/cm^2 (from problem statement)

2. The interplanar spacings are:

$$d_{110} = \frac{2.866 \times 10^{-8}}{\sqrt{1^2 + 1^2 + 0}} = 2.0266 \times 10^{-8} \text{ cm}$$

$$d_{112} = \frac{2.866 \times 10^{-8}}{\sqrt{1^2 + 1^2 + 2^2}} = 1.17 \times 10^{-8} \text{ cm}$$

The planar density and interplanar spacing of the (110) plane are larger than those for the (112) plane; therefore, the (110) plane would be the preferred slip plane.

4-4   Observing Dislocations

The existence of dislocations was postulated about 25 years before the first reports of their observation.[7,8] The places where dislocations intersect the surface of a crystal are relatively high-energy sites. When a metallic material is "etched" (a chemical reaction treatment that involves exposure to an acid or a base), the areas where dislocations intersect the surface of the crystal react more readily than the surrounding parts. These reacted regions show up in the microstructure as **etch pits**. Figure 4-11 shows a schematic of the slip planes, dislocations, and etch pits. Figure 4-12 shows the etch pit distribution on a surface of a silicon carbide (SiC) crystal. A transmission electron microscope (TEM), described in Chapter 3, is also used to observe dislocations. In a typical TEM image, dislocations appear as dark lines at very high magnifications as shown in Figures 4-13(a). When thousands of dislocations move on the surface of a crystal they produce a visible lines known as a **slip lines** [Figure 4-13(b)]. A group of slip lines is known as a **slip band** [Figure 4-13(c)].

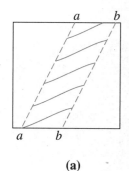

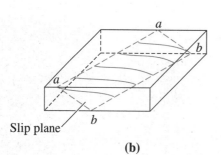

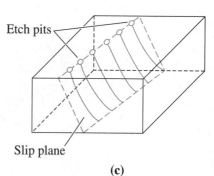

(a)   (b)   (c)

Figure 4-11   A sketch illustrating dislocations, slip planes, and etch pit locations. (*Source: Adapted from* Physical Metallurgy Principles, *Third Edition, by R.E. Reed-Hill and R. Abbaschian, p. 92, Figs. 4-7 and 4-8. Copyright © 1992 Brooks/Cole Thomson Learning. Adapted with permission.*)

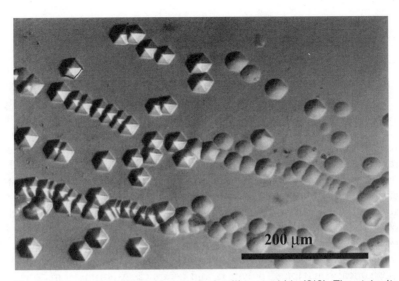

Figure 4-12   Optical image of etch pits in silicon carbide (SiC). The etch pits correspond to intersection points of pure edge dislocations with Burgers vector $\frac{a}{3}\langle 1\bar{1}20 \rangle$ and the dislocation line direction along [0001] (perpendicular to the etched surface). Lines of etch pits represent low angle grain boundaries (*Courtesy of Dr. Marek Skowronski, Carnegie Mellon University.*)

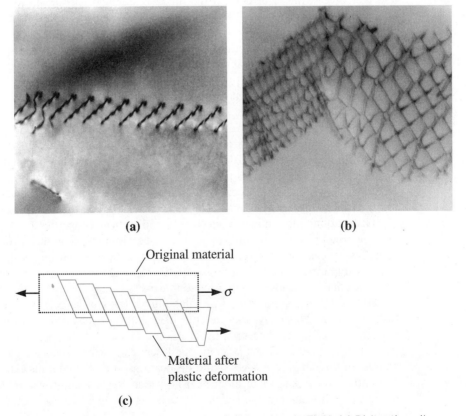

Figure 4-13   Electron photomicrographs of dislocations in Ti₃Al: (a) Dislocation pileups (×36,500). (b) Micrograph at ×100 showing slip lines and grain boundaries in Al. (c) Schematic of slip bands development.

4-5   Significance of Dislocations

First, dislocations are most significant in metals and alloys since they provide a mechanism for plastic deformation, which is the cumulative effect of slip of numerous dislocations. **Plastic deformation** refers to irreversible deformation or change in shape that occurs when the force or stress that caused it is removed. This is because the applied stress causes dislocation motion that in turn causes permanent deformation. When we use the term "plastic deformation" the implication is that it is caused by dislocaiton motion. There are, however, other mechanisms that cause permanent deformation. We will see these in later chapters. The plastic deformation is to be distinguished from **elastic deformation**, which is a temporary change in shape that occurs while a force or stress remains applied to a material. In elastic deformation, the shape change is a result of stretching of interatomic bonds, however, no dislocation motion occurs. Slip can occur in some ceramics and polymers. However, other factors (e.g., porosity in ceramics, entanglement of chains in polymers, etc.) dominate the near room-temperature mechanical behavior of polymers and ceramics. Amorphous materials such as silicate glasses do not have a periodic arrangement of ions and hence do not contain dislocations. *The slip process, therefore, is particularly important to us in understanding the mechanical behavior of metals.* First, slip explains why the strength of metals is much lower than the value predicted from the metallic bond. If slip occurs, only a tiny fraction of all of the metallic bonds across the interface need to be broken at any one time, and the force required to deform the metal is small. It can be shown that the actual strength of metals is 10^3 to 10^4 times *smaller* than that expected from the strength of metallic bonds.[10]

Second, slip provides ductility in metals. If no dislocations were present, an iron bar would be brittle and the metal could not be shaped by metalworking processes, such as forging, into useful shapes.

Third, we control the mechanical properties of a metal or alloy by interfering with the movement of dislocations. An obstacle introduced into the crystal prevents a dislocation from slipping unless we apply higher forces. Thus, the presence of dislocations helps strengthen metallic materials.

Enormous numbers of dislocations are found in materials. The **dislocation density**, or total length of dislocations per unit volume, is usually used to represent the amount of dislocations present. Dislocation densities of 10^6 cm/cm^3 are typical of the softest metals, while densities up to 10^{12} cm/cm^3 can be achieved by deforming the material.

Dislocations also influence electronic and optical properties of materials.[9] For example, the resistance of pure copper increases with increasing dislocation density. We mentioned previously that the resistivity of pure copper also depends strongly on small levels of impurities. Similarly, the speed with which charge carriers such as electrons can move in silicon depends on the dislocation density. We prefer to use silicon crystals that are essentially dislocation free since this allows the charge carriers such as electrons to move faster. Normally, the presence of dislocations has a deleterious effect on the performance of photo detectors, light emitting diodes, lasers, and solar cells.[11] These devices are often made from compound semiconductors such as gallium arsenide-aluminum arsenide (GaAs-AlAs) and dislocations in these materials can originate from concentration inequalities in the melt from which crystals are grown or stresses induced because of thermal gradients that the crystals are exposed to during cooling from growth temperature.[12]

4-6   Schmid's Law

We can understand the differences in behavior of metals that have different crystal structures by examining the force required to initiate the slip process. Suppose we apply a unidirectional force F to a cylinder of metal that is a single crystal (Figure 4-14). We can orient the slip plane and slip direction to the applied force by defining the angles λ, and ϕ. The angle between the slip direction and the applied force is λ , and ϕ is the angle between the normal to the slip plane and the applied force. Note that the sum of angles ϕ and λ can be but does not have to be 90°.

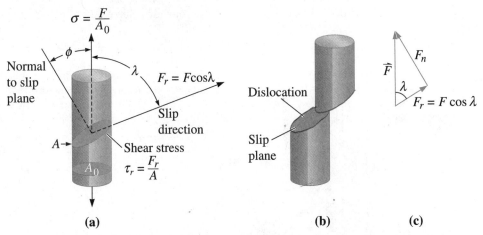

(a)   **(b)**   **(c)**

Figure 4-14 (a) A resolved shear stress τ is produced on a slip system. (Note: $(\phi + \lambda)$ does not have to be 90°.) (b) Movement of dislocations on the slip system deforms the material. (c) Resolving the force.

In order for the dislocation to move in its slip system, a shear force acting in the slip direction must be produced by the applied force. This resolved shear force F_r is given by:

$$F_r = F \cos \lambda$$

If we divide the equation by the area of the slip plane, $A = A_0/\cos \phi$, we obtain the following equation known as **Schmid's law**,

$$\boxed{\tau_r = \sigma \cos \phi \cos \lambda,} \tag{4-3}$$

where:

$$\tau_r = \frac{F_r}{A} = \text{resolved shear } \textit{stress} \text{ in the slip direction}$$

$$\sigma = \frac{F}{A_0} = \text{unidirectional } \textit{stress} \text{ applied to the cylinder}$$

EXAMPLE 4-10   *Calculation of Resolved Shear Stress*

Apply the Schmid's law for a situation in which the single crystal is at an orientation so that the slip plane is perpendicular to the applied tensile stress.

SOLUTION

Suppose the slip plane is perpendicular to the applied stress σ, as in Figure 4-15. Then, $\phi = 0°$, $\lambda = 90°$, $\cos \lambda = 0$, and therefore $\tau_r = 0$. As noted before, the angles ϕ and λ can but do not always add up to 90°. Even if the applied stress σ is enormous, no resolved shear stress develops along the slip direction and the dislocation cannot move. (You could perform a simple experiment to demonstrate this with a deck of cards. If you push on the deck at an angle, the cards slide over one another, as in the slip process. If you push perpendicular to the deck, however, the cards do not slide.) Slip cannot occur if the slip system is oriented so that either λ or ϕ is 90°.

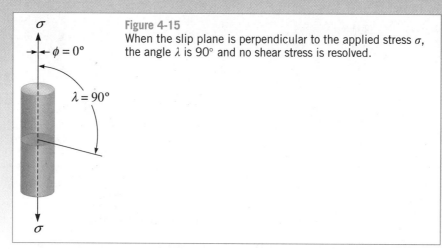

Figure 4-15
When the slip plane is perpendicular to the applied stress σ, the angle λ is 90° and no shear stress is resolved.

The **critical resolved shear stress** τ_{crss} is the shear stress required to break enough metallic bonds in order for slip to occur. Thus slip occurs, causing the metal to plastically deform, when the *applied* stress (σ) produces a *resolved* shear stress (τ_r) that equals the *critical* resolved shear stress.

$$\tau_r = \tau_{crss} \tag{4-4}$$

EXAMPLE 4-11   *Design of a Single Crystal Casting Process*

We wish to produce a rod composed of a single crystal of pure aluminum, which has a critical resolved shear stress of 148 psi. We would like to orient the rod in such a manner that, when an axial stress of 500 psi is applied, the rod deforms by slip in a 45° direction to the axis of the rod and actuates a sensor that detects the overload. Design the rod and a method by which it might be produced.

SOLUTION

Dislocations begin to move when the resolved shear stress τ_r equals the critical resolved shear stress, 148 psi. From Schmid's law:

$$\tau_r = \sigma \cos \lambda \cos \phi, \quad \text{or}$$

$$148 \text{ psi} = (500 \text{ psi}) \cos \lambda \cos \phi$$

Because we wish slip to occur at a 45° angle to the axis of the rod, $\lambda = 45°$, and:

$$\cos \phi = \frac{148}{500 \cos 45} = \frac{148}{(500)(0.707)} = 0.419$$

$$\phi = 65.2°$$

Therefore, we must produce a rod that is oriented such that $\lambda = 45°$ and $\phi = 65.2°$. Note that ϕ and λ do not add to 90°.

We might do this by a solidification process. We would orient a seed crystal of solid aluminum at the bottom of a mold. Liquid aluminum could be introduced into the mold. The liquid begins to solidify from the starting crystal and a single crystal rod of the proper orientation is produced.

4-7 Influence of Crystal Structure

We can use Schmid's law to compare the properties of metals having BCC, FCC, and HCP crystal structures. Table 4-2 lists three important factors that we can examine. We must be careful to note, however, that this discussion describes the behavior of nearly perfect single crystals. Real engineering materials are seldom single crystals and always contain large numbers of defects. They also tend to be polycrystalline. Since different crystals or grains are oriented in different random directions, we can not apply Schmid's law to predict the mechanical behavior of polycrystalline materials.

TABLE 4-2 ■ *Summary of factors affecting slip in metallic structures*

Factor	FCC	BCC	HCP $\left(\frac{c}{a} > 1.633\right)$
Critical resolved shear stress (psi)	50–100	5,000–10,000	50–100[a]
Number of slip systems	12	48	3[b]
Cross-slip	Can occur	Can occur	Cannot occur[b]
Summary of properties	Ductile	Strong	Relatively brittle

[a] For slip on basal planes.
[b] By alloying or heating to elevated temperatures, additional slip systems are active in HCP metals, permitting cross-slip to occur and thereby improving ductility.

Critical Resolved Shear Stress If the critical resolved shear stress in a metal is very high, the applied stress σ must also be high in order for τ_r to equal τ_{crss}. A higher τ_{crss} implies a higher stress is necessary to deform a metal, which in turn indicates the metal has a high strength! In FCC metals, which have close-packed {111} planes, the critical resolved shear stress is low—about 50 to 100 psi in a perfect crystal; FCC metals tend to have low strengths. On the other hand, BCC crystal structures contain no close-packed planes and we must exceed a higher critical resolved shear stress—on the order of 10,000 psi in perfect crystals—before slip occurs; therefore, BCC metals tend to have high strengths and lower ductilities.

We would expect the HCP metals, because they contain close-packed basal planes, to have low critical resolved shear stresses. In fact, in HCP metals such as zinc that have a c/a ratio greater than or equal to the theoretical ratio of 1.633, the critical resolved shear stress is less than 100 psi, just as in FCC metals. In HCP titanium, however, the c/a ratio is less than 1.633; the close-packed planes are spaced too closely together. Slip now occurs on planes such as $(10\bar{1}0)$, the "prism" planes or faces of the hexagon, and the critical resolved shear stress is then as great as or greater than in BCC metals.

Number of Slip Systems   If at least one slip system is oriented to give the angles λ and ϕ near 45°, then τ_r equals τ_{crss} at low applied stresses. Ideal HCP metals possess only one set of parallel close-packed planes, the (0001) planes, and three close-packed directions, giving three slip systems. Consequently, the probability of the close-packed planes and directions being oriented with λ and ϕ near 45° is very low. The HCP crystal may fail in a brittle manner without a significant amount of slip. However, in HCP metals with a low c/a ratio, or when HCP metals are properly alloyed, or when the temperature is increased, other slip systems become active, making these metals less brittle than expected.

On the other hand, FCC metals contain four nonparallel close-packed planes of the form {111} and three close-packed directions of the form ⟨110⟩ within each plane, giving a total of 12 slip systems. At least one slip system is favorably oriented for slip to occur at low applied stresses, permitting FCC metals to have high ductilities.

Finally, BCC metals have as many as 48 slip systems that are nearly close-packed. Several slip systems are always properly oriented for slip to occur, allowing BCC metals to also have ductility.

Cross-Slip   Consider a screw dislocation moving on one slip plane that encounters an obstacle and is blocked from further movement. This dislocation can shift to a second intersecting slip system, also properly oriented, and continue to move. This is called **cross-slip**. In many HCP metals, no cross-slip can occur because the slip planes are parallel, i.e., not intersecting. Therefore, polycrystalline HCP metals tend to be brittle. Fortunately, additional slip systems become active when HCP metals are alloyed or heated, thus improving ductility. Cross-slip is possible in both FCC and BCC metals because a number of intersecting slip systems are present. Consequently, cross-slip helps maintain ductility in these metals.

EXAMPLE 4-12   *Ductility of HCP Metal Single Crystals and Polycrystalline Materials*

A single crystal of magnesium (Mg), which has a HCP crystal structure, can be stretched into a ribbon-like shape four to six times its original length. However, *polycrystalline* Mg and other metals with a HCP structure show limited ductilities. Use the values of critical resolved shear stress for metals with different crystal structures and the nature of deformation in polycrystalline materials to explain this observation.

SOLUTION

From Table 4-2, we note that for HCP metals such as Mg, the critical resolved shear stress is low (50–100 psi). We also note that slip in HCP metals will occur readily on the basal plane—the primary slip plane. When a single crystal is deformed, assuming the basal plane is suitably oriented with applied stress, a very large deformation can occur. This explains why single crystal Mg can be stretched into a ribbon four to six times the original size. When we have a polycrystalline Mg, the deformation is not as simple. Each crystal must deform such that the strain developed in any one crystal is accommodated by its neighbors. In HCP metals, there are no intersecting slip systems, thus dislocations cannot glide over from one slip plane in one crystal (grain) onto another slip plane in a neighboring crystal. As a result, polycrystalline HCP metals such as Mg show limited ductility.

4-8   Surface Defects

Surface defects are the boundaries, or planes, that separate a material into regions, each region having the same crystal structure but different orientations.

Material Surface   The exterior dimensions of the material represent surfaces at which the crystal abruptly ends. Each atom at the surface no longer has the proper coordination number and atomic bonding is disrupted. This is very often an important factor in making Si based microelectronic devices. The exterior surface may also be very rough, may contain tiny notches, and may be much more reactive than the bulk of the material.

In nano-structured materials, the ratio of the number of atoms or ions at the surface to that in the bulk is very high. As a result these materials have a large surface area per unit mass. In petroleum refining and many other areas of technology we make use of high surface area catalysts for enhancing the kinetics of chemical reactions. Similar to nano-scale materials, porous materials, gels, and ultra fine powders are examples of materials where the surface area is very high. You will learn later that reduction in surface area is the thermodynamic driving force for sintering of ceramics and metal powders.

Grain Boundaries   The microstructure of many engineered ceramic and metallic materials consists of many grains. A **grain** is a portion of the material within which the arrangement of the atoms is nearly identical. However, the orientation of the atom arrangement, or crystal structure, is different for each adjoining grain. Three grains are shown schematically in Figure 4-16(a); the arrangement of atoms in each grain is identical but the grains are oriented differently. A **grain boundary**, the surface that separates the individual grains, is a narrow zone in which the atoms are not properly spaced. That is to say, the atoms are so close together at some locations in the grain boundary that they cause a region of compression, and in other areas they are so far apart that they cause a region of tension. Figure 4-16(b) shows a micrograph of a stainless steel sample.

One method of controlling the properties of a material is by controlling the grain size. By reducing the grain size, we increase the number of grains and, hence, increase the amount of grain boundary area. Any dislocation moves only a short distance before encountering a grain boundary and being stopped, and the strength of the metallic material is increased. The **Hall-Petch equation** relates the grain size to the **yield strength**,

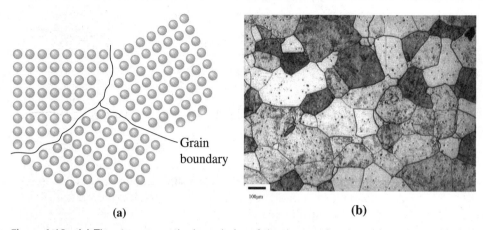

(a) **(b)**

Figure 4-16 (a) The atoms near the boundaries of the three grains do not have an equilibrium spacing or arrangement. (b) Grains and grain boundaries in a stainless steel sample. (*Courtesy Dr. A. Deardo.*)

$$\sigma_y = \sigma_0 + Kd^{-1/2}, \tag{4-5}$$

where σ_y is the yield strength (the level of stress necessary to cause a certain amount of permanent deformation), d is the average diameter of the grains, and σ_0 and K are constants for the metal. Recall from Chapter 1 that yield strength of a metallic material is the minimum level of stress that is needed to initiate plastic (permanent) deformation. Figure 4-17 shows this relationship in steel. The Hall-Petch equation is not valid for materials with unusually large or ultrafine grains. In the chapters that follow, we will describe how the grain size of metals and alloys can be controlled through solidification, alloying, and heat treatment.

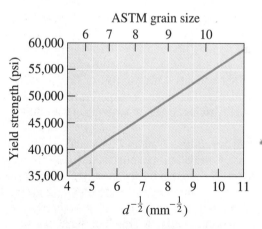

Figure 4-17
The effect of grain size on the yield strength of steel at room temperature.

EXAMPLE 4-13 **Design of a Mild Steel**

The yield strength of mild steel with an average grain size of 0.05 mm is 20,000 psi. The yield stress of the same steel with a grain size of 0.007 mm is 40,000 psi. What will be the average grain size of the same steel with a yield stress of 30,000 psi? Assume the Hall-Petch equation is valid and that changes in the observed yield stress are due to changes in grain size.

SOLUTION

$$\sigma_y = \sigma_0 + Kd^{-1/2},$$

Thus, for a grain size of 0.05 mm the yield stress is

$$20 \times 6.895 \text{ MPa} = 137.9 \text{ MPa}.$$

(Note: 1,000 psi = 6.895 MPa). Using the Hall-Petch equation

$$137.9 = \sigma_0 + \frac{K}{\sqrt{0.05}}$$

For the grain size of 0.007 mm, the yield stress is 40×6.895 MPa = 275.8 MPa. Therefore, again using the Hall-Petch equation:

$$275.8 = \sigma_0 + \frac{K}{\sqrt{0.007}}$$

Solving these two equations $K = 18.43$ MPa-mm$^{1/2}$, and $\sigma_0 = 55.5$ MPa. Now we have the Hall-Petch equation as

$$\sigma_y = 55.5 + 18.43 \times d^{-1/2}$$

If we want a yield stress of 30,000 psi or $30 \times 6.895 = 206.9$ MPa, the grain size will be 0.0148 mm.

Optical microscopy is one technique that is used to reveal microstructural features such as grain boundaries that require less than about 2000 magnification. The process of preparing a metallic sample and observing or recording its microstructure is called **metallography**. A sample of the material is sanded and polished to a mirror-like finish. The surface is then exposed to chemical attack, or *etching*, with grain boundaries being attacked more aggressively than the remainder of the grain. Light from an optical microscope is reflected or scattered from the sample surface, depending on how the surface is etched. When more light is scattered from deeply etched features such as the grain boundaries, these features appear dark (Figure 4-18). In ceramic samples, a technique

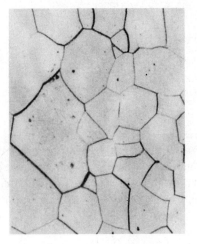

Figure 4-18
Microstructure of palladium (×100). (*From* ASM Handbook, *Vol. 9, Metallography and Microstructure (1985), ASM International, Materials Park, OH 44073.*)

known as **thermal grooving** is often used to observe grain boundaries. It involves polishing and heating, for a short time, a ceramic sample to temperatures below the sintering temperature.

One manner by which grain size is specified is the **ASTM grain size number** (ASTM is the American Society for Testing and Materials). The number of grains per square inch is determined from a photograph of the metal taken at magnification ×100. The number of grains per square inch N is entered into Equation 4-6 and the ASTM grain size number n is calculated:

$$N = 2^{n-1}$$ (4-6)

A large ASTM number indicates many grains, or a fine grain size, and correlates with high strengths for metals.

When describing a microstructure, whenever possible, it is preferable to use a micrometer marker or some other scale on the micrograph, instead of stating the magnification as X, as in Figure 4-25. That way, if the micrograph is enlarged or reduced, the micrometer marker scales with it and we do not have to worry about the changes in the magnification of the original micrograph. A number of sophisticated **image analysis** programs are also available through the Internet. Using such programs, it is possible not only to obtain information on the ASTM grain size number but also quantitative information on average grain size, grain size distribution, porosity, second phases (Chapter 9), etc. A number of optical and scanning electron microscopes can be purchased with image analysis capabilities or the images can be processed using image analysis softwares. The following example illustrates the calculation of the ASTM grain size number.

EXAMPLE 4-14   *Calculation of ASTM Grain Size Number*

Suppose we count 16 grains per square inch in a photomicrograph taken at magnification ×250. What is the ASTM grain size number?

SOLUTION

If we count 16 grains per square inch at magnification ×250, then at magnification ×100 we must have:

$$N = \left(\frac{250}{100}\right)^2 (16) = 100 \text{ grains/in.}^2 = 2^{n-1}$$

$$\log 100 = (n-1) \log 2$$

$$2 = (n-1)(0.301)$$

$$n = 7.64$$

Small Angle Grain Boundaries   A **small angle grain boundary** is an array of dislocations that produces a small misorientation between the adjoining crystals (Figure 4-19). Because the energy of the surface is less than that of a regular grain boundary, the small angle grain boundaries are not as effective in blocking slip. Small angle boundaries

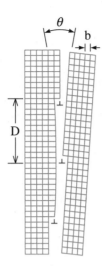

Figure 4-19
The small angle grain boundary is produced by an array of dislocations, causing an angular mismatch θ between the lattices on either side of the boundary.

formed by edge dislocations are called **tilt boundaries**, and those caused by screw dislocations are called **twist boundaries**.

Stacking Faults   **Stacking faults**, which occur in FCC metals, represent an error in the stacking sequence of close-packed planes. Normally, a stacking sequence of *ABC ABC ABC* is produced in a perfect FCC crystal. But, suppose the following sequence is produced:

$$ABC\ ABABC\ ABC$$

In the portion of the sequence indicated, a type *A* plane replaces where a type *C* plane would normally be located. This small region, which has a HCP stacking sequence instead of the FCC stacking sequence, represents a stacking fault. Stacking faults interfere with the slip process.

Twin Boundaries   A **twin boundary** is a plane across which there is a special mirror image misorientation of the crystal structure (Figure 4-20). Twins can be produced when a shear force, acting along the twin boundary, causes the atoms to shift out of position. Twinning occurs during deformation or heat treatment of certain metals. The twin boundaries interfere with the slip process and increase the strength of the metal. Movement of twin boundaries can also cause a metal to deform. Figure 4-20 shows that the formation of a twin has changed the shape of the metal. Twinning also occurs in some ceramic materials such as monoclinic zirconia and dicalcium silicate.[12,13]
     The effectiveness of the surface defects in interfering with the slip process can be judged from the surface energies (Table 4-3). The high-energy grain boundaries are much more effective in blocking dislocations than either stacking faults or twin boundaries.

Domain Boundaries   **Ferroelectrics** are materials that develop spontaneous and reversible dielectric polarization (e.g., PZT or $BaTiO_3$) (Chapter 18). Materials that develop a magnetization in a similar fashion are **ferromagnetic** (e.g., Fe, Co, Ni) or **ferrimagnetic**

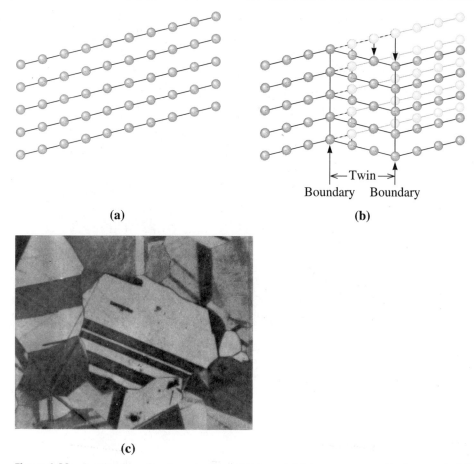

(a)   (b)

(c)

Figure 4-20   Application of a stress to the perfect crystal (a) may cause a displacement of the atoms, (b) causing the formation of a twin. Note that the crystal has deformed as a result of twinning. (c) A micrograph of twins within a grain of brass (×250).

TABLE 4-3 ■ *Energies of surface imperfections in selected metals*

Surface Imperfection (energy/cm²)	Al	Cu	Pt	Fe
Stacking fault	200	75	95	—
Twin boundary	120	45	195	190
Grain boundary	625	645	1000	780

(e.g., Fe_3O_4 and other magnetic ferrites) (Chapter 19). These electronic and magnetic materials contain domains. A **domain** is a small region of the material in which the direction of magnetization or dielectric polarization remains the same. In these materials, many small domains form so as to minimize the total free energy of the material. Figure 4–21 shows an example of domains in tetragonal ferroelectric barium titanate. The presence of domains influences the dielectric and magnetic properties of many electronic and magnetic materials. We will discuss these materials in later chapters.

Figure 4-21 Domains in ferroelectric barium titanate. (*Courtesy of Dr. Rodney Roseman, University of Cincinnati.*) Similar domain structures occur in ferromagnetic and ferrimagnetic materials.

4-9 Importance of Defects

Extended and point defects play a major role in influencing mechanical, electrical, optical and magnetic properties of engineered materials. In this section, we recapitulate the importance of defects on properties of materials. We emphasize that the effect of dislocations is most important in metallic materials only.

Effect on Mechanical Properties via Control of the Slip Process Any imperfection in the crystal raises the internal energy at the location of the imperfection. The local energy is increased because, near the imperfection, the atoms either are squeezed too closely together (compression) or are forced too far apart (tension).

One dislocation in an otherwise perfect metallic crystal can move easily through the crystal if the resolved shear stress equals the critical resolved shear stress. However, if the dislocation encounters a region where the atoms are displaced from their usual positions, a higher stress is required to force the dislocation past the region of high local energy; thus, the material is stronger. *Defects in materials, such as dislocations, point defects, and grain boundaries, serve as "stop signs" for dislocations.* We can control the strength of a metallic material by controlling the number and type of imperfections. Three common strengthening mechanisms are based on the three categories of defects in crystals. Since dislocation motion is relatively easier in metals and alloys, these mechanisms typically work best for metallic materials. We need to keep in mind that very often the strength of ceramics in tension and at low temperatures is dictated by the level of porosity (presence of small holes). Polymers are often amorphous and hence dislocations play very little role in their mechanical behavior, as discussed in a later chapter. The strength of inorganic glasses (e.g., silicate float glass) depends on the distribution of tiny flaws on the surface.

Strain Hardening Dislocations disrupt the perfection of the crystal structure. In Figure 4-22, the atoms below the dislocation line at point *B* are compressed, while the atoms above dislocation *B* are too far apart. If dislocation *A* moves to the right and passes near dislocation *B*, dislocation *A* encounters a region where the atoms are not properly arranged. Higher stresses are required to keep the second dislocation moving; consequently, the metal must be stronger. Increasing the number of dislocations further increases the strength of the material since increasing the dislocation density causes more stop signs for dislocation motion. The dislocation density can be shown to increase markedly as we strain or deform a material. This mechanism of increasing the strength of a material by deformation is known as **strain hardening**, which is discussed formally in Chapter 7. We can also show that dislocation densities can be reduced substantially by heating a metallic material to a relatively high temperature and holding it there for a long period of time. This heat treatment is known as **annealing** and is used to impart ductility to metallic materials. Thus, controlling the dislocation density is an important way of controlling the strength and ductility of metals and alloys.

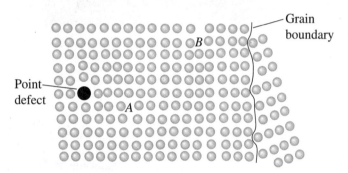

Figure 4-22 If the dislocation at point *A* moves to the left, it is blocked by the point defect. If the dislocation moves to the right, it interacts with the disturbed lattice near the second dislocation at point *B*. If the dislocation moves farther to the right, it is blocked by a grain boundary.

Solid-Solution Strengthening Any of the point defects also disrupt the perfection of the crystal structure. A solid solution is formed when atoms or ions of a guest element or compound are assimilated completely into the crystal structure of the host material. This is similar to the way salt or sugar in small concentrations dissolve in water. If dislocation *A* moves to the left (Figure 4-22), it encounters a disturbed crystal caused by the point defect; higher stresses are needed to continue slip of the dislocation. By intentionally introducing substitutional or interstitial atoms, we cause *solid-solution strengthening*, which is discussed in Chapter 9. This mechanism explains why plain carbon steel is stronger than pure Fe or why alloys of copper containing small concentrations of Be are much stronger than pure Cu. Pure gold or silver, both FCC metals with many active slip systems, are mechanically too soft.

Grain-Size Strengthening Surface imperfections such as grain boundaries disturb the arrangement of atoms in crystalline materials. If dislocation *B* moves to the right (Figure 4-22), it encounters a grain boundary and is blocked. By increasing the number of grains or reducing the grain size, **grain-size strengthening** is achieved in metallic materials. Control of grain size will be discussed in a number of later chapters.

There are two more mechanisms for strengthening of metals and alloys. These are known as **second-phase strengthening** and **precipitation strengthening**. We will discuss these in later chapters.

EXAMPLE 4-15   *Design/Materials Selection for a Stable Structure*

We would like to produce a bracket to hold ceramic bricks in place in a heat-treating furnace. The bracket should be strong, should possess some ductility so that it bends rather than fractures if overloaded, and should maintain most of its strength up to 600°C. Design the material for this bracket, considering the various crystal imperfections as the strengthening mechanism.

SOLUTION

In order to serve up to 600°C, the bracket should *not* be produced from a polymer material. Instead, a metal or ceramic would be considered.

In order to have some ductility, dislocations must move and cause slip. Because slip in ceramics is difficult, the bracket should be produced from a metallic material. The metal should have a melting point well above 600°C; aluminum, with a melting point of 660°C, would not be suitable; iron, however, would be a reasonable choice.

We can introduce point, line, and surface imperfections into the iron to help produce strength, but we wish the imperfections to be stable as the service temperature increases. As shown in Chapter 5, grains can grow at elevated temperatures, reducing the number of grain boundaries and causing a decrease in strength. As indicated in Chapter 7, dislocations may be annihilated at elevated temperatures—again, reducing strength. The number of vacancies depends on temperature, so controlling these crystal defects may not produce stable properties.

The number of interstitial or substitutional atoms in the crystal does not, however, change with temperature. We might add carbon to the iron as interstitial atoms or substitute vanadium atoms for iron atoms at normal lattice points. These point defects continue to interfere with dislocation movement and help to keep the strength stable.

Of course, other design requirements may be important as well. For example, the steel bracket may deteriorate by oxidation or may react with the ceramic brick.

Effects on Electrical, Optical, and Magnetic Properties   In previous sections, we stated how profound is the effect of point defects on the electrical properties of semiconductors. The entire microelectronics industry critically depends upon the successful incorporation of substitutional dopants such as P, As, B, and Al in Si and other semiconductors. These dopant atoms allow us to have a significant control of the electrical properties of the semiconductors. Devices made from Si, GaAs, amorphous silicon (a:Si:H), etc. critically depend on the presence of dopant atoms. We can make n-type Si by introducing P atoms in Si. We can make p-type Si using B atoms. Similarly, a number of otherwise unsatisfied bonds in amorphous silicon are completed by incorporating H atoms.

The effect of such defects as dislocations on the properties of semiconductors is usually deleterious. Dislocations and other defects (including other point defects) can interfere with motion of charge carriers in semiconductors. This is why we make sure that the dislocation densities in single crystal silicon and other materials, used in optical, and electrical applications, are very small. Point defects also cause increased resistivity levels in metals.

In some cases, defects can enhance certain properties. For example, incorporation of CaO in ZrO_2 causes an increase in the concentration of oxygen ion vacancies. This has a beneficial effect on the conductivity of zirconia and allows us to use such compositions for oxygen gas sensors and solid oxide fuel cells. Defects can convert many otherwise insulating dielectric materials into useful semiconductors! These are then used for many sensor applications (e.g., temperature, humidity, and gas sensors, etc.).

Addition of about 1% chromium oxide in alumina creates defects that lead to a defect that makes a chromium oxide containing alumina ruby red. Similarly, incorporation of Fe^{+2} and Ti^{+4} makes the blue sapphire. Nanocrystals of materials such as cadmium sulfide (CdS) in inorganic glasses provide us glasses that have a brilliant color. Nanocrystals of silver halide and other crystals also allow formation of photochromic and photosensitive glasses.

Many magnetic materials can be processed such that grain boundaries and other defects can make it harder to reverse the magnetization in these materials. The magnetic properties of many commercial ferrites, used in magnets for loudspeakers and devices in wireless communication networks, depend critically on the distribution of different ions on different crystallographic sites in the crystal structure. As mentioned before, the presence of domains affects the properties of ferroelectric, ferromagnetic, and ferrimagnetic materials (Chapters 18 and 19).

SUMMARY

◈ Imperfections, or defects, in a crystalline material are of three general types: point defects, line defects or dislocations, and surface defects.

◈ The term "defect" is somewhat misleading. There are countless number of technologies and applications that rely on the usefulness of defects.

◈ The number of vacancies depends on the temperature of the material; interstitial atoms (located at interstitial sites between the normal atoms) and substitutional atoms (which replace the host atoms at lattice points) are often deliberately introduced and are typically unaffected by changes in temperature.

◈ Frenkel and Schottky defects are commonly seen in ionic materials.

◈ The Kröger-Vink notation is used to describe defect chemistry in materials. These reactions must be balanced in charge, mass, and crystallographic sites used.

◈ Dislocations are line defects which, when a force is applied to a metallic material, move and cause a metallic material to deform.

◈ The critical resolved shear stress is the stress required to move the dislocation.

◈ The dislocation moves in a slip system, composed of a slip plane and a slip direction. The slip direction, or Burgers vector, is typically a close-packed direction. The slip plane is also normally close packed or nearly close packed.

◈ In metallic crystals, the number and type of slip directions and slip planes influence the properties of the metal. In FCC metals, the critical resolved shear stress is low and an optimum number of slip planes are available; consequently, FCC metals tend to be ductile. In BCC metals, no close-packed planes are available and the critical resolved shear stress is high; thus, the BCC metals tend to be strong. The number of slip systems in HCP metals is limited, causing these metals to behave in a brittle manner.

◇ Point defects, which include vacancies, interstitial atoms, and substitutional atoms, introduce compressive or tensile strain fields that disturb the atomic arrangements in the surrounding crystal. As a result, dislocations cannot easily slip in the vicinity of point defects and the strength of the metallic material is increased.

◇ Surface defects include grain boundaries. Producing a very small grain size increases the amount of grain boundary area; because dislocations cannot easily pass through a grain boundary, the material is strengthened (Hall-Petch equation).

◇ The number and type of crystal defects control the ease of movement of dislocations and, therefore, directly influence the mechanical properties of the material.

◇ Strain hardening is obtained by increasing the number of dislocations; solid-solution strengthening involves the introduction of point defects; and grain-size strengthening is obtained by producing a small grain size.

◇ Annealing is a heat treatment used to undo the effects of strain hardening by reducing the dislocation density. This leads to increased ductility in metallic materials.

◇ Defects in materials have a significant influence on their electrical, optical, and magnetic properties.

GLOSSARY

ASTM   American Society for Testing and Materials.

ASTM grain size number (*n*)   A measure of the size of the grains in a crystalline material obtained by counting the number of grains per square inch at magnification ×100.

Annealing   A heat treatment that typically involves heating a metallic material to a high temperature for an extended period of time, conducted with a view to lower the dislocation density and hence impart ductility.

Burgers vector   The direction and distance that a dislocation moves in each step, also known as slip vector.

Dopants   Elements or compounds typically added, in known concentrations and appearing at specific places within the microstructure, to enhance the properties or processing of a material.

Critical resolved shear stress   The shear stress required to cause a dislocation to move and cause slip.

Cross-slip   A change in the slip system of a dislocation.

Defect chemical reactions   Reactions written using the Kröger-Vink notation to describe defect chemistry. The reactions must be written in such a way that mass and electrical charges are balanced and stoichiometry of sites is maintained. The existence of defects predicted by such reactions needs to be verified experimentally.

Dislocation   A line imperfection in a crystalline material. Movement of dislocations helps explain how metallic materials deform. Interference with the movement of dislocations helps explain how metallic materials are strengthened.

Dislocation density   The total length of dislocation line per cubic centimeter in a material.

Dispersion strengthening   A mechanism by which grains of an additional compound or phase are introduced in the grains of a polycrystalline metallic material. These second phase crystals serve as stop signs for dislocations thereby causing an increase in the strength of a metallic material.

Domain   A small region of a ferroelectric, ferromagnetic or ferrimagnetic materials in which the direction of dielectric polarization (for ferroelectric) or magnetization (for ferromagnetic or ferrimagnetic) remains the same.

Domain boundaries   Region between domains in a material.

Edge dislocation   A dislocation introduced into the crystal by adding an "extra half plane" of atoms.

Elastic deformation   Deformation that is fully recovered when the stress causing it is removed.

Etch pits   Tiny holes created at areas where dislocations meet the surface. These are used to examine the presence and number density of dislocations.

Extended defects   Defects that involve several atoms/ions and thus occur over a finite volume of the crystalline material (e.g., dislocations, stacking faults, etc.).

Extrinsic semiconductor   Semiconductors containing dopants or impurities.

Ferrimagnetic   A material that also develops a spontaneous and reversible magnetization (e.g., Fe_3O_4). It is different from a ferromagnetic material in that the magnetic moments of different ions are not all in the same direction.

Ferroelectric   A dielectric material that develops a spontaneous and reversible electric polarization (e.g., *PZT*, $BaTiO_3$).

Ferromagnetic   A material that develops a spontaneous and reversible magnetization (e.g., Fe, Co, Ni).

Frenkel defect   A pair of point defects produced when an ion moves to create an interstitial site, leaving behind a vacancy.

Grain   One of the crystals present in a polycrystalline material.

Grain boundary   A surface defect representing the boundary between two grains. The crystal has a different orientation on either side of the grain boundary.

Hall-Petch equation   The relationship between yield strength and grain size in a metallic material—that is, $\sigma_y = \sigma_0 + Kd^{-1/2}$.

Image analysis   A technique that is used to analyze images of microstructures to obtain quantitative information on grain size, shape, grain size distribution, etc.

Impurities   Elements or compounds that find their way into a material, often originating from processing or raw materials and typically having a deleterious effect on the properties or processing of a material.

Interstitial defect   A point defect produced when an atom is placed into the crystal at a site that is normally not a lattice point.

Interstitialcy   A point defect caused when a "normal" atom occupies an interstitial site in the crystal.

Intrinsic semiconductor   A semiconductor that contains no dopants or impurities.

Kröger-Vink notation   A system used to indicate point defects in materials. The main body of the notation indicates the type of defect or the element involved. The subscript indicates the location of the point defect and the superscript indicates the effective positive (.) or negative (') charge.

Line defects   Defects such as dislocations in which atoms or ions are missing in a row.

Metallography   Preparation of a metallic sample of a material by polishing and etching so that the structure can be examined using a microscope.

Mixed dislocation   A dislocation that contains partly edge components and partly screw components.

n-type semiconductor   A semiconductor that is doped with elements that donate electrons (e.g., P-doped Si).

p-type semiconductor   A semiconductor that is doped with elements that accept electrons (e.g., B-doped Si).

Peierls-Nabarro stress   The shear stress, which depends on the Burgers vector and the interplanar spacing, required to cause a dislocation to move—that is, $\tau = c \exp(-kd/b)$.

Point defects   Imperfections, such as vacancies, that are located typically at one (in some cases a few) sites in the crystal.

Precipitation strengthening   Strengthening of metals and alloys by formation of tiny precipitates inside the grains. The small precipitates help resist dislocation motion.

Schmid's law   The relationship between shear stress, the applied stress, and the orientation of the slip system—that is, $\tau = \sigma \cos \lambda \cos \phi$.

Schottky defect   A point defect in ionically bonded materials. In order to maintain a neutral charge, a stoichiometric number of cation and anion vacancies must form.

Screw dislocation   A dislocation produced by skewing a crystal so that one atomic plane produces a spiral ramp about the dislocation.

Second-phase strengthening   A mechanism by which grains of an additional compound or phase are introduced in the grains of a polycrystalline material. These second phase crystals serve as stop signs for dislocations thereby causing an increase in the strength of a metallic material.

Sintering   A process for forming a dense mass by heating compacted powders.

Slip   Deformation of a metallic material by the movement of dislocations through the crystal.

Slip band   Collection of many slip lines, often easily visible.

Slip direction   The direction in the crystal in which the dislocation moves. The slip direction is the same as the direction of the Burgers vector.

Slip line   A visible line produced at the surface of a metallic material by the presence of several thousand dislocations.

Slip plane   The plane swept out by the dislocation line during slip. Normally, the slip plane is a close-packed plane, if one exists in the crystal structure.

Slip system   The combination of the slip plane and the slip direction.

Small angle grain boundary   An array of dislocations causing a small misorientation of the crystal across the surface of the imperfection.

Stacking fault   A surface defect in FCC metals caused by the improper stacking sequence of close-packed planes.

Substitutional defect   A point defect produced when an atom is removed from a regular lattice point and replaced with a different atom, usually of a different size.

Surface defects   Imperfections, such as grain boundaries, that form a two-dimensional plane within the crystal.

Thermal grooving   A technique used for observing microstructures in ceramic materials, involves heating, for a short time, a polished sample to a temperature slightly below the sintering temperature.

Tilt boundary   A small angle grain boundary composed of an array of edge dislocations.

Transmission electron microscope (TEM)   An instrument that, by passing an electron beam through a material, can detect microscopic structural features.

Twin boundary   A surface defect across which there is a mirror image misorientation of the crystal structure. Twin boundaries can also move and cause deformation of the material.

Twist boundary   A small angle grain boundary composed of an array of screw dislocations.

Vacancy   An atom or an ion missing from its regular crystallographic site.

ADDITIONAL INFORMATION

1. MANNHART, J. and P. CHAUDHARI, "High-T_c Bicrystal Grain Boundaries," *Physics Today*, 2001 (November): p. 48–53.

2. BURGERS, J.M., *Proc. Kon. Ned. Akad. Wet.*, 1948: 42(293): p. 378.

3. VERHOEVEN, J.D., *Fundamentals of Physical Metallurgy*, 1975: Wiley.

4. GUMBSCH, P. and H. GAO, "Dislocations Faster than the Speed of Sound," *Science*, 1999. 283: p. 965–968.

5. CASTAING, J. and A. DOMINGUEZ-RODRIGUEZ, "Ductility of Ceramics," *The Encyclopedia of Advanced Materials*, D. Bloor, et al., Editors. 1994. p. 659–664.

6. BRUNNER, D., et al., "Surprising Results of a Study on the Plasticity in Strontium Titanate," *J. Am. Ceram. Soc*, 2001. 84(5): p. 1161–1163.

7. OROWAN, E., *Z. Phys.*, 1934. 89: p. 634.

8. POLYANI, Z., *Z. Phys.*, 1934. 550.

9. MAHAJAN, S., "Dislocations in Compound Semiconductor Crystals and Improvement by Heavy Doping," in *The Encyclopedia of Advanced Materials*, D. BLOOR, M. FLEMINGS, and S. MAHAJAN, Editors. 1994: Permagon. p. 649–654.

10. REED-HILL, R.E. and R. ABBASCHIAN, *Physical Metallurgy Principles, 3rd Edition* 1991: PWS.

11. MAHAJAN, S. and K.S. SREE HARSHA, *Principles of Growth and Processing of Semiconductors.* 1999: McGraw Hill.

12. NETTLESHIP, I. and R. STEVENS, Tetragonal Zirconia Polycrystal (TZP)—A Review," *Int. J. High Technol. Ceram*, 1987. 3(1): p. 1–32.

13. KIM, Y.J., I. NETTLESHIP, and W.M. KRIVEN, "Phase Transformations in Dicalcium Silicate: II, TEM studies of Crystallography, Microstructure, and Mechanisms," *J. Am. Ceram. Soc*, 1992. 75(9): p. 2407–19.

PROBLEMS

Section 4-1 Point Defects

4-1 Calculate the number of vacancies per cm^3 expected in copper at 1080°C (just below the melting temperature). The energy for vacancy formation is 20,000 cal/mol.

4-2 The fraction of lattice points occupied by vacancies in solid aluminum at 660°C is 10^{-3}. What is the energy required to create vacancies in aluminum?

4-3 The density of a sample of FCC palladium is 11.98 g/cm^3 and its lattice parameter is 3.8902 Å Calculate

   (a) the fraction of the lattice points that contain vacancies; and

   (b) the total number of vacancies in a cubic centimeter of Pd.

4-4 The density of a sample of HCP beryllium is 1.844 g/cm^3 and the lattice parameters are $a_0 = 0.22858$ nm and $c_0 = 0.35842$ nm. Calculate

   (a) the fraction of the lattice points that contain vacancies; and

   (b) the total number of vacancies in a cubic centimeter.

4-5 BCC lithium has a lattice parameter of 3.5089×10^{-8} cm and contains one vacancy per 200 unit cells. Calculate

   (a) the number of vacancies per cubic centimeter; and

   (b) the density of Li.

4-6 FCC lead (Pb) has a lattice parameter of 0.4949 nm and contains one vacancy per 500 Pb atoms. Calculate

   (a) the density; and

   (b) the number of vacancies per gram of Pb.

4-7 A niobium alloy is produced by introducing tungsten substitutional atoms in the BCC structure; eventually an alloy is produced that has a lattice parameter of 0.32554 nm and a density of 11.95 g/cm^3. Calculate the fraction of the atoms in the alloy that are tungsten.

4-8 Tin atoms are introduced into a FCC copper crystal, producing an alloy with a lattice parameter of 3.7589×10^{-8} cm and a density of 8.772 g/cm^3. Calculate the atomic percentage of tin present in the alloy.

4-9 We replace 7.5 atomic percent of the chromium atoms in its BCC crystal with tantalum. X-ray diffraction shows that the lattice parameter is 0.29158 nm. Calculate the density of the alloy.

4-10 Suppose we introduce one carbon atom for every 100 iron atoms in an interstitial position in BCC iron, giving a lattice parameter of 0.2867 nm. For the Fe-C alloy, find the density and the packing factor.

4-11 The density of BCC iron is 7.882 g/cm^3 and the lattice parameter is 0.2866 nm when hydrogen atoms are introduced at interstitial positions. Calculate

   (a) the atomic fraction of hydrogen atoms; and

   (b) number of unit cells on average that contain hydrogen atoms.

Section 4-2 Other Point Defects

4-12 Suppose one Schottky defect is present in every tenth unit cell of MgO. MgO has the sodium chloride crystal structure and a lattice parameter of 0.396 nm. Calculate

   (a) the number of anion vacancies per cm^3; and

   (b) the density of the ceramic.

4-13 ZnS has the zinc blende structure. If the density is 3.02 g/cm^3 and the lattice parameter is 0.59583 nm, determine the number of Schottky defects

   (a) per unit cell; and

   (b) per cubic centimeter.

4-14 Suppose we introduce the following point defects.

   (a) Mg^{2+} ions substitute for yttrium ions in Y_2O_3;

   (b) Fe^{3+} ions substitute for magnesium ions in MgO;

   (c) Li^{1+} ions substitute for magnesium ions in MgO; and

   (d) Fe^{2+} ions replace sodium ions in NaCl.

What other changes in each structure might be necessary to maintain a charge balance? Explain.

4-15 Write down the defect chemistry equation for introduction of $SrTiO_3$ in $BaTiO_3$ using the Kröger-Vink notation.

4-16 Write down the defect chemistry equation for introduction of $PbZrO_3$ in $PbTiO_3$ using the Kröger-Vink notation.

4-17 Write down the defect chemistry equation for introduction of Nb_2O_5 in $BaTiO_3$ using the Kröger-Vink notation.

4-18 Write down the defect chemistry equation for introduction of Si as a dopant in GaAs using the Kröger-Vink notation.

4-19 Write down the defect chemistry equation for creation of a Schottky defect in Cr_2O_3 using the Kröger-Vink notation.

4-20 Write down different possible equations that can express dissolution of MgO in alumina. Use the Kröger-Vink notation and be sure to balance the number of sites, electrical charge and mass.

4-21 Explain why addition of small amounts of CaO to zirconia increases its conductivity. Where is this type of material used?

Section 4-3   Dislocations

4-22 What are the Miller indices of the slip directions:

(a) on the (111) plane in an FCC unit cell?
(b) on the (011) plane in a BCC unit cell?

4-23 What are the Miller indices of the slip planes in FCC unit cells that include the [101] slip direction?

4-24 What are the Miller indices of the {110} slip planes in BCC unit cells that include the [111] slip direction?

4-25 Calculate the length of the Burgers vector in the following materials:

(a) BCC niobium;
(b) FCC silver; and
(c) diamond cubic silicon.

4-26 Determine the interplanar spacing and the length of the Burgers vector for slip on the expected slip systems in FCC aluminum. Repeat, assuming that the slip system is a (110) plane and a $[1\bar{1}1]$ direction. What is the ratio between the shear stresses required for slip for the two systems? Assume that k = 2 in Equation 4-2.

4-27 Determine the interplanar spacing and the length of the Burgers vector for slip on the $(110)/[1\bar{1}1]$ slip system in BCC tantalum. Repeat, assuming that the slip system is a $(111)/[1\bar{1}0]$ system. What is the ratio between the shear stresses required for slip for the two systems? Assume that k = 2 in Equation 4-2.

4-28 Can ceramic and polymeric materials contain dislocations?

4-29 Why is it that ceramic materials are brittle?

4-30 What are the different ways by which ductility can be observed in ceramics?

Section 4-4   Observing Dislocations

4-31 What is an "etch pit?"

4-32 What techniques are used to observe dislocations?

4-33 What is a slip line and what is a slip band?

Section 4-5   Significance of Dislocations

4-34 What is meant by the terms plastic and elastic deformation?

4-35 A slab of steel is transformed into a car chassis. Does this process involve plastic or elastic deformation? Explain.

4-36 Why is the theoretical strength of metals much higher than that observed experimentally?

4-37 How many grams of aluminum, with a dislocation density of 10^{10} cm/cm^3, are required to give a total dislocation length that would stretch from New York City to Los Angeles (3000 miles)?

4-38 The distance from Earth to the Moon is 240,000 miles. If this were the total length of dislocation in a cubic centimeter of material, what would be the dislocation density?

4-39 Why would metals behave as brittle materials without dislocations?

4-40 Why is it that dislocations play an important role in controlling the mechanical properties of metallic materials, however, they do not play a role in determining the mechanical properties of glasses?

4-41 Suppose you would like to introduce an interstitial or large substitutional atom into the crystal near a dislocation. Would the atom fit more easily above or below the dislocation line shown in Figure 4-8(b)? Explain.

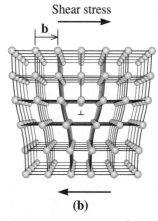

Shear stress

(b)

Figure 4-8(b)   (Repeated for Problem 4-41) Continued movement of the dislocation eventually creates a step.

4-42 Compare the c/a ratios for the following HCP metals, determine the likely slip processes in each, and estimate the approximate critical resolved shear stress. Explain. (See data in Appendix A.)

(a) zinc   (b) magnesium   (c) titanium
(d) zirconium   (e) rhenium   (f) beryllium

Section 4-6 Schmid's Law

4-43 A single crystal of an FCC metal is oriented so that the [001] direction is parallel to an applied stress of 5000 psi. Calculate the resolved shear stress acting on the (111) slip plane in the [$\bar{1}$10], [0$\bar{1}$1], and [10$\bar{1}$] slip directions. Which slip system(s) will become active first?

4-44 A single crystal of a BCC metal is oriented so that the [001] direction is parallel to the applied stress. If the critical resolved shear stress required for slip is 12,000 psi, calculate the magnitude of the applied stress required to cause slip to begin in the [1$\bar{1}$1] direction on the (110), (011), and (10$\bar{1}$) slip planes.

4-45 Our discussion of Schmid's law dealt with single crystals of a metal. Discuss slip and Schmid's law in a polycrystalline material. What might happen as the grain size gets smaller and smaller?

Section 4-7 Influence of Crystal Structure

4-46 Why is it that single crystal and polycrystalline copper are both ductile, however, single crystal, but not polycrystalline, zinc can exhibit considerable ductility?

4-47 Why is it that cross slip in BCC and FCC metals is easier than that in HCP metals? How does this influence the ductility of BCC, FCC, and HCP metals?

4-48 Arrange the following metals in the expected order of increasing ductility. Cu, Ti, and Fe.

Section 4-8 Surface Defects

4-49 The strength of titanium is found to be 65,000 psi when the grain size is 17×10^{-6} m and 82,000 psi when the grain size is 0.8×10^{-6} m. Determine

(a) the constants in the Hall-Petch equation; and
(b) the strength of the titanium when the grain size is reduced to 0.2×10^{-6} m.

4-50 A copper-zinc alloy has the following properties

Grain Diameter (mm)	Strength (MPa)
0.015	170 MPa
0.025	158 MPa
0.035	151 MPa
0.050	145 MPa

Determine

(a) the constants in the Hall-Petch equation; and
(b) the grain size required to obtain a strength of 200 MPa.

4-51 For an ASTM grain size number of 8, calculate the number of grains per square inch

(a) at a magnification of 100 and
(b) with no magnification.

4-52 Determine the ASTM grain size number if 20 grains/square inch are observed at a magnification of 400.

4-53 Determine the ASTM grain size number if 25 grains/square inch are observed at a magnification of 50.

4-54 Determine the ASTM grain size number for the materials in: Figure 4-18 and Figure 4-23.

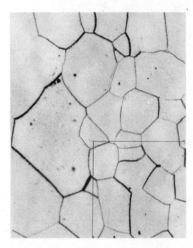

Figure 4-18 (Repeated for Problem 4-54) Microstructure of palladium (×100). (*From* ASM Handbook, *Vol. 9, Metallography and Microstructure (1985), ASM International, Materials Park, OH 44073.*)

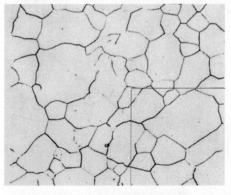

Figure 4-23 Microstructure of iron, for Problem 4-54 (×500). (*From* ASM Handbook, *Vol. 9, Metallography and Microstructure (1985), ASM International, Materials Park, OH 44073.*)

4-55 Certain ceramics with special dielectric properties are used in wireless communication systems. Barium magnesium tantalate (BMT) and barium zinc tantalate (BZT) are examples of such materials. Determine the ASTM grain size number for a barium magnesium tantalate (BMT) ceramic microstructure shown in Figure 4-24.

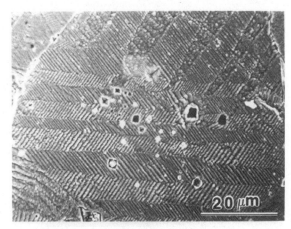

Figure 4-24   Microstructure of a barium magnesium tantalate (BMT) ceramic.

4-56 Alumina is the most widely used ceramic material. Determine the ASTM grain size number for the polycrystalline alumina sample shown in Figure 4-25.

Figure 4-25   Microstructure of an alumina ceramic. (*Courtesy of Richard McAfee and Dr. Ian Nettleship.*)

4-57 What is image analysis? How is it useful for microstructure determination?

4-58 The angle θ of a tilt boundary is given by $\sin(\theta/2) = b/2D$ (See Figure 4-19.) Verify the correctness of this equation.

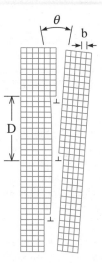

Figure 4-19   (Repeated for Problems 4-58, 4-59, and 4-60) The small angle grain boundary is produced by an array of dislocations, causing an angular mismatch θ between the lattices on either side of the boundary.

4-59 Calculate the angle θ of a small-angle grain boundary in FCC aluminum when the dislocations are 5000 Å apart. (See Figure 4-19 and equation in Problem 4-58.)

4-60 For BCC iron, calculate the average distance between dislocations in a small-angle grain boundary tilted 0.50°. (See Figure 4-19.)

4-61 Why do we use ultrafine particles of metals and alloys for catalysis reactions used in petroleum refining?

4-62 What is added to an alumina crystal to turn it into a ruby crystal?

4-63 Why is it that a single crystal of a ceramic superconductor is capable of carrying much more current per unit area than a polycrystalline ceramic superconductor of the same composition?

4-64 What is a domain? What types of materials contain domains? Give examples.

4-65 Give any one application of a ferromagnetic, ferrimagnetic, and ferroelectric material.

Section 4-9 Importance of Defects

4-66 What makes plain carbon steel harder than pure iron?

4-67 Why is jewelry made out of gold or silver alloyed with copper?

4-68 Why do we prefer to use semiconductor crystals that contain as small a number of dislocations as possible?

4-69 In structural applications (e.g., steel for bridges and buildings or aluminum alloys for aircraft), why do we use alloys rather than pure metals?

4-70 How is the conductivity of silicon altered using P or B as dopants?

4-71 Do dislocations control the strength of a silicate glass? Explain.

4-72 What is meant by the term strain hardening?

4-73 Which mechanism of strengthening is the Hall-Petch equation related to?

4-74 Pure copper is strengthened by addition of small concentration of Be. What mechanism of strengthening is this related to?

✳ Design Problems

4-75 The density of pure aluminum calculated from crystallographic data is expected to be 2.69955 g/cm^3.

   (a) Design an aluminum alloy that has a density of 2.6450 g/cm^3.
   (b) Design an aluminum alloy that has a density of 2.7450 g/cm^3.

4-76 You would like a metal plate with good weldability. During the welding process, the metal next to the weld is heated almost to the melting temperature and, depending on the welding parameters, may remain hot for some period of time. Design an alloy that will minimize the loss of strength in this "heat-affected zone" during the welding process.

4-77 We need a material that is optically transparent but electrically conductive. Such materials are used for touch screen displays. What kind of materials can be used for this application? (*Hint*: Think about coatings of materials that can provide electronic or ionic conductivity: the substrate has to be transparent for this application.)

▣ Computer Problems

4-78 *Temperature dependence of vacancy concentrations.* Write a computer program that will provide a user equilibrium concentration of vacancies in a metallic element as a function of temperature. The user should specify a meaningful and valid range of temperatures (e.g., 100 to 1200 K for copper). Assume that the crystal structure originally specified is valid for this range of temperature. Ask the user to input the activation energy for formation of one mole of vacancies (Q_v). The program then should ask the user to input the density of the element, and crystal structure (FCC, BCC, etc.). You can use character variables to detect the type of crystal structures (e.g., "F" or "f" for FCC, "B" or "b" for BCC, etc.). Be sure to pay attention to use the correct units for temperature, density etc. The program should ask the user if the temperature range that has been provided is in °C, °F, or K and convert the temperatures properly into K before any calculations are performed. The program should then use this information to establish the number of atoms per unit volume and provide an output for this value. The program then should calculate the equilibrium concentration of vacancies at different temperatures. The first temperature will be the minimum temperature specified and then temperatures should be increased by 100 K or another convenient increment. You can then make use of any graphical software to plot the data showing equilibrium concentration of vacancies as a function of temperature. Think about what scales will be used to best display the results.

4-79 *Hall-Petch equation.* Write a computer program that will ask the user to enter two sets of values of σ_y and grain size (d) for a metallic material. The program should then utilize the data to calculate and print out the Hall-Petch equation. The program then should prompt the user to input another value of grain size and calculate the yield stress or vice-versa.

4-80 *ASTM grain size number calculator.* Write a computer program that will ask the user to input the magnification of a micrograph of the sample whose ASTM number is being calculated. The program should then ask the user for the number of grains counted and the area (in square inches) from which these grains were counted. The program then should calculate the ASTM number, taking into consideration the fact that the micrograph magnification is not 100 and the area may not have been one square inch.

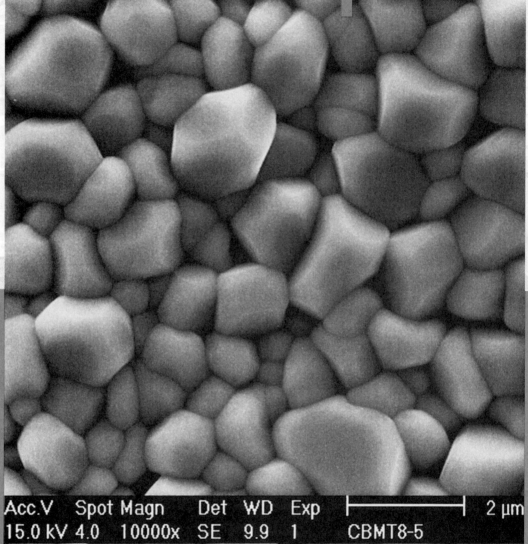

Acc.V Spot Magn   Det WD Exp   2 µm
15.0 kV 4.0 10000x SE 9.9 1   CBMT8-5

Sintered barium magnesium tantalate (BMT) ceramic microstructure. This ceramic material is useful in making electronic components used in wireless communications. The process of sintering is driven by the diffusion of atoms and ions.

5

Atom and Ion Movements in Materials

Have You Ever Wondered?

- ■ *Aluminum oxidizes more easily than iron, so why do we say aluminum normally does not "rust?"*

- ■ *What kind of plastic is used to make carbonated beverage bottles?*

- ■ *How are the surfaces of certain steels hardened?*

- ■ *Why do we encase optical fibers using a polymeric coating?*

- ■ *Who invented the first contact lens?*

- ■ *What is galvanized steel?*

- ■ *How does a tungsten filament in a light bulb fail?*

In Chapter 4, we learned that the atomic and ionic arrangements in materials are never perfect. We also saw that most materials are not pure elements; they are alloys or blends of different elements or compounds. Different types of atoms or ions typically "diffuse", or move within the material, so the differences in their concentration are minimized. Diffusion refers to an observable net flux of atoms or other species. It depends upon the initial concentration gradient and temperature. Just as water flows from a mountain towards the sea to minimize its gravitational potential energy, atoms and ions have a tendency to move in a predictable fashion to eliminate concentration differences and produce homogeneous, uniform compositions that make the material thermodynamically more stable.

In this chapter, we will learn that temperature influences the kinetics of diffusion and that the concentration difference contributes to the overall net flux of diffusing species. The goal of this chapter is to examine the principles and applications of diffusion in materials. We'll illustrate the concept of diffusion through examples of several real-world technologies dependent on the diffusion of atoms, ions, or molecules.

We will present an overview of Fick's laws that describe the diffusion process quantitatively. We will also see how the relative openness of different crystal structures and the size of atoms or ions, temperature, and concentration of diffusing species affect the rate at which diffusion occurs. We will discuss specific examples of how diffusion is used in the synthesis and processing of advanced materials as well as manufacturing of components using advanced materials.

5-1   Applications of Diffusion

Diffusion refers to the net flux of any species, such as ions, atoms, electrons, holes, and molecules. The magnitude of this flux depends upon the initial concentration gradient and temperature. The process of diffusion is central to a large number of today's important technologies. In materials processing technologies, control over the diffusion of atoms, ions, molecules, or other species is key. There are hundreds of applications and technologies that depend on either enhancing or limiting diffusion. The following are just a few examples.

Carburization for Surface Hardening of Steels   Let's say we want a surface, such as the teeth of a gear, to be hard. However, we do not want the entire gear to be hard. Carburization processes are used to increase surface hardness. In carburization, a source of carbon, such as a graphite powder or gaseous phase containing carbon, is diffused into steel components such as gears (Figure 5-1). In later chapters, you will learn how increased carbon concentration on the surface of the steel increases the steel's hardness. Similar to the introduction of carbon, we can also use a process known as **nitriding**, in which nitrogen is introduced into the surface of a metallic material. Diffusion also plays a central role in the control of the phase transformations needed for the heat treatment of metals and alloys, the processing of ceramics, and the solidification and joining of materials (Section 5-9).

Dopant Diffusion for Semiconductor Devices   The entire microelectronics industry, as we know it today, would not exist if we did not have a very good understanding of the diffusion of different atoms into silicon or other semiconductors. In Chapters 3 and 4, we have seen what is meant by n- and p-type dopants in a semiconductor. The creation of the **p-n junction** involves diffusing dopant atoms, such as phosphorous (P), arsenic (As), antimony (Sb), boron (B), aluminum (Al), etc., into selected ultra-small regions of silicon wafers (see Figure 5-2).[1] A p-n junction is a region of the semiconductor, one side of which is doped with *n-type* dopants (e.g., As in Si) and the other side is doped with *p-type* dopants (e.g., B in Si). In this example, not only do different regions of silicon contain a specified concentration of dopants, but these dopants actually occupy the

Figure 5-1 Furnace for heat treating steel using the carburization process. (*Courtesy of Cincinnati Steel Treating.*)

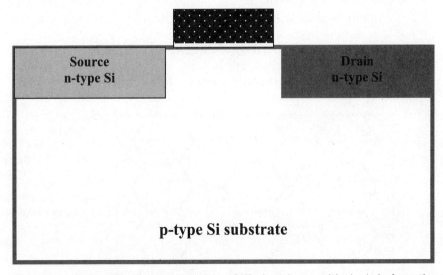

Figure 5-2 Schematic of a n-p-n transistor. Diffusion plays a critical role in formation of the different regions created in the semiconductor substrates. The creation of millions of such transistors is at the heart of microelectronics technology.

silicon sites. When we study the electrical properties of materials in Chapter 18, you will learn that the diffusion of electrons and holes plays an important role in the operation of semiconductor devices such as transistors.

Conductive Ceramics In general, polycrystalline ceramics tend to be good insulators of electricity. Strong covalent and ionic bonds along with microstructural features contribute to the relatively poor electrical conductivity exhibited by many ceramics. However, many ceramics can conduct electricity and, in fact, some are superconductors (Chapter 1). Diffusion of ions, electrons, or holes also plays an important role in the electrical conductivity of many **conductive ceramics**, such as partially or fully stabilized zirconia (ZrO_2) or indium tin oxide (also commonly known as *ITO*). Lithium cobalt oxide ($LiCoO_2$) is an example of an ionically conductive material that is used in lithium ion batteries. These ionically conductive materials are used for such products as oxygen sensors in cars, touch-screen displays, fuel cells, and batteries. The ability of ions to diffuse and provide a pathway for electrical conduction plays an important role in enabling these applications.

Magnetic Materials for Hard Drives Cobalt alloys are used in the manufacture of magnetic hard disks for personal computers. Thin layers of these alloys can be produced using a process called **sputtering**. In the sputtering process, the targets are bombarded with ion beams. The targets are made from elements or compounds containing the necessary constituents from which the thin films are made. The atoms "sputter off" from the targets and deposit on to a substrate, producing a thin film. With the sputtering of cobalt alloy thin films, chromium is often added to enhance their magnetic properties. It has been found that the sputtered films of chromium (Cr) atoms are found primarily within the magnetic grains of the cobalt alloy. For chromium to have a useful magnetic effect, we need a heat treatment that diffuses the chromium atoms to the grain boundaries of the magnetic material. We achieve this using an **annealing** heat treatment that involves heating the sputtered film to a high temperature for a reasonably long period of time in order to allow diffusion of chromium atoms, from the grain interior to grain boundaries, to occur. Figure 5-3 shows a schematic of an annealed Co-Pt-Ta-Cr sputtered thin film. The chromium diffuses to the grain boundaries after annealing. This helps improve the magnetic properties of hard disks that are made using these films.[2]

Creation of Plastic Beverage Bottles/Mylar™ Balloons The occurrence of diffusion may not always be beneficial. In some applications, we may want to limit the occurrence of diffusion for certain species. For example, in the creation of certain plastic

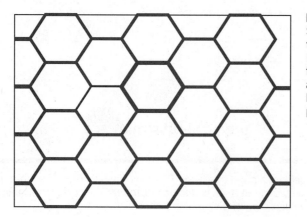

Figure 5-3
Schematic of the microstructure of the Co-Pt-Ta-Cr film after annealing. Most of the chromium diffuses from the grains to the grain boundaries after the annealing process. This helps improve the magnetic properties of the hard disk.

bottles, the diffusion of carbon dioxide (CO_2) must be minimized. This is one of the major reasons why we use polyethylene terephthalate (PET) to make bottles which ensure that the carbonated beverages they contain will not loose their fizz for a reasonable period of time! Another example of an application where diffusion should be limited would be that of Mylar™ balloons. Helium balloons made out of Mylar™ (aluminum-coated PET) retain helium much longer since the thin aluminum coating helps minimize the occurrence of out-diffusion of helium.

Oxidation of Aluminum   You may have heard or know that aluminum does not "rust." In reality, aluminum oxidizes (rusts) more easily than iron. However, the aluminum oxide (Al_2O_3) forms a very protective but thin coating on the aluminum's surface preventing any further diffusion of oxygen hindering further oxidation of the underlying aluminum. The oxide coating does not have a color and is thin and, hence, invisible. This is why we think aluminum does not rust. In Chapter 22, you will also see some examples of aluminum alloys that undergo corrosion.

Coatings and Thin Films   Coatings and thin films are often used in the manufacturing process to limit the diffusion of water vapor, oxygen, or other chemicals. For example,

Figure 5-4   Hot dip galvanized parts and structures prevent corrosion. (*Courtesy of Casey Young and Barry Dugar of the Zinc Corporation of America.*)

we apply paint to steel structures to prevent or minimize rusting. When a corrosive environment is more aggressive, we can apply a zinc (Zn) coating onto steel structures, such as car chassis, chemical refinery structures, food processing equipment, guard rails etc., to reduce rusting of the underlying steel (Figure 5-4).[3] This process of applying a zinc coating by immersing the components in a large bath of molten zinc (melting temperature ∼ 420°C) is known as **hot dip galvanizing**. The chemical nature of zinc, and not just their barrier for diffusion, plays a major role in protecting the underlying steel.

Thermal Barrier Coatings for Turbine Blades   In an aircraft engine, some of the nickel superalloy-based turbine blades are coated with ceramic oxides such as yttria stabilized zirconia (YSZ). These ceramic coatings protect the underlying alloy from high temperatures; hence the name **thermal barrier coatings** (TBCs) (Figure 5-5).[4] The diffusion of oxygen through these ceramic coatings and the subsequent oxidation of the underlying alloy play a major role in determining the lifetime and durability of the turbine blades. In Figure 5-5, EBPVD means electron beam physical vapor deposition. The bond coat is either a platinum or molybdenum-based alloy. It provides adhesion between TBC and the substrate.

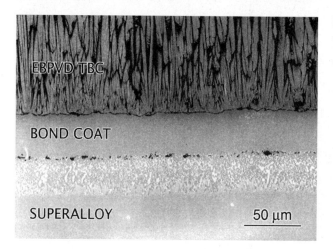

Figure 5-5
A thermal barrier coating on nickel-based superalloy. (*Courtesy of Dr. F.S. Pettit and Dr. G.H. Meier, University of Pittsburgh.*)

Optical Fibers and Microelectronic Components   Another example of diffusion is that of the coatings around optical fibers. Optical fibers made from silica (SiO_2) are coated with polymeric materials to prevent diffusion of water molecules. Water vapor has two related but separate effects on the performance of glass optical fibers. First, incorporation of water in the form of hydroxyl groups (OH^-) increases optical losses (i.e., the signal can not travel far enough without being attenuated significantly). Second, water vapor attacks the glass surface, causing microcracks that lower the strength of glass (Chapter 6). We use plastic coatings to encase microelectronic circuitry components, such as capacitors and other components, to prevent diffusion of water vapor or other chemicals that can deteriorate their performance.

Water and Air Treatment   In environmental engineering applications, such as the treatment of water or air, we are often interested in how certain ions or molecules diffuse through a filter or an ion exchange resin. In certain drug delivery applications, doctors and other health care professionals are interested in the diffusion of drugs or other biochemical molecules through small capillaries carrying blood.

Drift and Diffusion In some engineering applications, we are concerned with the movement of small particles as a result of forces other than concentration gradient and temperature. We distinguish between the movement of atoms, molecules, ions, electrons, holes, etc. as a result of concentration gradient and temperature (diffusion) or some other driving force, such as gradients in density, electric field, or magnetic field gradients. The movement of particles, atoms, ions, electrons, holes, etc. under driving forces other than diffusion is called **drift**. For example, if a bottle of perfume is opened in one end of the room, we can smell the perfume at the other end. This transport of molecules contained in the perfume is mainly due to convective mass transfer. Diffusion will occur and play a role. However, diffusion over large distances would take a much longer time and, therefore, is not a major factor in this situation. Another example where the process of drift is important is **electrophoresis**, or the movement of charged fine particles (or large molecules) under an electric field. This process is used for applying paints to cars. We also use electrophoresis to understand the surface chemistry of ceramic and other fine particles. Electrophoresis, similarly, is also used in certain bioengineering applications concerned with separation of DNA molecules of different molecular weights. The movement of charged ions in a solution (often water based) under an electric field produces thin layers of metals and alloys. This process is known as **electroplating**. The basic principle of this process is also used in extraction and refining of such metals as copper (Cu) and aluminum (Al). In Examples 5-1 and 5-2, we discuss the difference between drift and the diffusion of carriers in a semiconductor.

EXAMPLE 5-1 *Diffusion of Ar/He and Cu/Ni*

Consider a box containing an impermeable partition that divides the box into equal volumes (Figure 5-6). On one side, we have pure argon (Ar) gas; on the other side, we have pure helium (He) gas. Explain what will happen when the partition is opened? What will happen if we replace the Ar side with a Cu single crystal and the He side with a Ni single crystal?

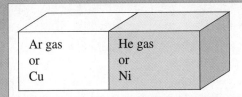

Figure 5-6
Illustration for Diffusion of Ar/He and Cu/Ni (for Example 5-1).

SOLUTION

Before the partition is opened, one compartment has no argon and the other has no helium (i.e., there is a concentration gradient of Ar and He). When the partition is opened, Ar atoms will diffuse toward the He side, and vice versa. This diffusion will continue until the entire box has a uniform concentration of both gases. There may be some density gradient driven convective flows as well. If we took random samples of different regions in this box after a few hours, we would get statistically uniform concentration of Ar and He. Note that Ar and He atoms will continue to move around in the box as a result of thermal; however, there will be no concentration gradients.

If we open the hypothetical partition between the Ni and Cu single crystals at room temperature, we would find that, similar to the Ar/He situation, the concentration gradients exist but the temperature is too low to see any significant diffusion of Cu atoms into Ni single crystal and vice-versa. This is an example of a situation in which there exists a concentration gradient diffusion, however because of the lower temperature the kinetics for diffusion are not favorable. Certainly, if we increase the temperature (say to 600°C) and waited for a longer period of time (e.g., ~24 hours), we would see diffusion of copper atoms into Ni single crystal and vice versa. After a very long time (say a few months), the entire solid will have a uniform concentration of Ni and Cu atoms. The new solid that forms consists of Cu and Ni atoms completely dissolved in each other and the resultant material is termed a "solid solution", a concept we will study in greater detail in Chapter 9.

This example also states something many of you may know by intuition. The diffusion of atoms and molecules occurs faster in gases and liquids than in solids. As we will see in Chapter 8 and other chapters, diffusion has a significant effect on the evolution of microstructure during the solidification of alloys, the heat treatment of metals and alloys, and the processing of ceramic materials.

EXAMPLE 5-2   *Diffusion and Drift of Charge Carriers in a Semiconductor*

The p-n junction is the basis for all transistors (Figure 5-2) and other devices.[1] A p-n junction is formed in single crystal silicon (Si) by **doping** the n-side with phosphorous (P) atoms and the p-side with boron (B) atoms (Figure 5-7). The doping can be achieved by diffusing atoms from a liquid, solid, or gaseous source of dopant atoms known as a **precursor**. Sometimes, the **ion implantation** process, in which dopant atoms are incorporated using high-energy ion beams, is also used instead of thermally diffusing dopant atoms. As discussed in Chapter 3, each phosphorus (P) atom makes available an extra electron, and each boron (B) atom added on the p-side has a deficit of one electron. We call this missing electron a **hole** and treat it as a particle having a positive charge. The magnitude of the charge is the same as that of an electron (1.6×10^{-19} C). Consider that diffusion of a species is initiated by temperature and concentration gradients and that external electric and magnetic fields can initiate the drift of carriers, then:

(a) Show schematically which way the electrons and holes will diffuse when a p-n junction is formed.

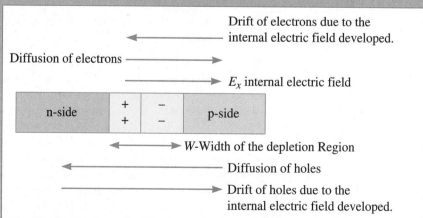

Figure 5-7 Directions for diffusion and drift of charge carriers in a semiconductor (for Example 5-2).

(b) Compare this situation with the diffusion of Cu and Ni atoms in the previous example.

(c) Based on this comment on the electric field driven drift of electrons and holes in the p-n junction. Assume the temperature is 300 K.

SOLUTION

(a) We will designate the concentration of holes and electrons by "p" and "n", respectively. We will designate the side of the p-n junction by subscripts p and n. On the n-side, we have P atoms that provide extra electrons. As a result, the concentration of electrons on n-side (n_n) is greater than that on the p-side (n_p). On the p-side, the concentration of holes (p_p) is higher. We do have a small concentration of electrons (n_p) on the p-side, because bonds between silicon atoms break due to thermal energy resulting in creation of electron–hole pairs. As a result we would expect movement of electrons from the n-side to the p-side. Since this movement is induced by concentration gradient, we say that electrons will diffuse from the n-side to the p-side. Similarly, since the concentration of holes on the p-side (p_p) is greater than that on the n-side (p_n), the holes will diffuse from the p-side to the n-side (Figure 5-7). Notice that across the p-n junction there is a concentration gradient of P and B atoms as well. However, these atoms, which have now actually become ions after donating or accepting an electron, do not diffuse since the temperature of 300 K is too low!

(b) As electrons diffuse from the n-side to the p-side, a portion of the n-side crystal becomes positively charged since each P atom has given up one electron and is now a positively charged P ion. Similarly, each B atom is converted into a negatively charged B ion. When some number of electrons has diffused, they leave behind positively charged P ions on a portion of the crystal on the n-side. When holes have diffused from the p- to the n-side, they leave behind negatively charged B ions. The diffusion of electrons and holes, however, does not continue until the concentrations of electrons and holes become equal. As more electrons try to diffuse from the n-side to the p-side, they "see" or encounter the negatively charged B ions. Similarly, as more and more holes try to diffuse from the p-side to the n-side, they encounter the positively charged P ions and their motion is resisted. Thus, unlike the situation of Ar and He atoms (Example 5-1) the concentration of electrons and holes does not become equal. An internal electric field (E_x) sets up over a width W and prevents equalization of electron and hole concentrations. The resultant internal voltage is known as the contact potential V_0.

(c) The internal electric field E_x is such that it causes the electrons from the p-side to drift to the n-side. Also, the same internal electric field drifts the holes from the n-side to the p-side. Thus, the flow of electrons due to diffusion and drift are in opposite directions. Similarly, the motion of holes due to diffusion and drift are also in opposite directions. It can be shown that the electrical currents due to diffusion and drift of each type of carrier cancel out. As a result a p-n junction under equilibrium does not carry a net current. You will learn later that by applying an external voltage bias (V), the resistance of and, hence, the current flow in a p-n junction can be tuned, thus enabling useful devices such as transistors (Chapter 18).

5-2   Stability of Atoms and Ions

In Chapter 4, we showed that imperfections are often present and also can be deliberately introduced into a material. However, these imperfections and, indeed, even atoms or ions in their normal positions in the crystal structures are not stable or at rest. Instead, the atoms or ions possess thermal energy and they will move. For instance, an atom may move from a normal crystal structure location to occupy a nearby vacancy. An atom may also move from one interstitial site to another. Atoms or ions may jump across a grain boundary, causing the grain boundary to move.

The ability of atoms and ions to diffuse increases as the temperature, or thermal energy, possessed by the atoms and ions increases. The rate of atom or ion movement is related to temperature or thermal energy by the *Arrhenius equation:*

$$\text{Rate} = c_0 \exp\left(\frac{-Q}{RT}\right) \tag{5-1}$$

where c_0 is a constant, R is the gas constant $\left(1.987 \frac{\text{cal}}{\text{mol} \cdot \text{K}}\right)$, T is the absolute temperature (K), and Q is the **activation energy** (cal/mol) required to cause an Avogadro's number of atoms or ions to move. This equation is derived from a statistical analysis of the probability that the atoms will have the extra energy Q needed to cause movement. The rate is related to the number of atoms that move.

We can rewrite the equation by taking natural logarithms of both sides:

$$\ln(\text{rate}) = \ln(c_0) - \frac{Q}{RT} \tag{5-2}$$

If we plot ln(rate) of some reaction versus $1/T$ (Figure 5-8), the slope of the curve will be $-Q/R$ and, consequently, Q can be calculated. The constant c_0 corresponds to the intercept at $\ln c_0$ when $1/T$ is zero.

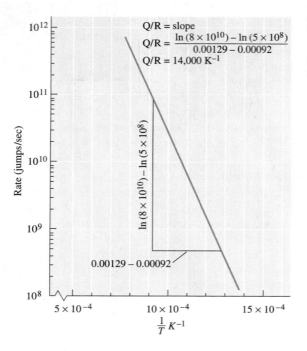

$Q/R = \text{slope}$

$Q/R = \dfrac{\ln(8 \times 10^{10}) - \ln(5 \times 10^8)}{0.00129 - 0.00092}$

$Q/R = 14,000 \text{ K}^{-1}$

Figure 5-8
The Arrhenius plot of ln (rate) versus $1/T$ can be used to determine the activation energy required for a reaction.

Figure 5-9
Svante August Arrhenius (1859–1927) did pioneering work in the field of kinetics of chemical reactions and introduced the notion of activation energy. (*Courtesy University of Pennsylvania Library.*)

Svante August Arrhenius (1859–1927) (Figure 5-9), a Swedish chemist who won the Nobel Prize in Chemistry in 1903 for his research on electrolytic theory of dissociation, had applied this idea to the rates of chemical reactions in aqueous solutions. His basic idea of activation energy and rates of chemical reactions as functions of temperature has since been applied to many applications, including diffusion. The activation energy is the amount of energy needed to cause the rate process to occur, as defined in more detail later.

EXAMPLE 5-3   *Activation Energy for Interstitial Atoms*

Suppose that interstitial atoms are found to move from one site to another at the rates of 5×10^8 jumps/s at 500°C and 8×10^{10} jumps/s at 800°C. Calculate the activation energy Q for the process.

SOLUTION

Figure 5-8 represents the data on a ln(rate) versus $1/T$ plot; the slope of this line, as calculated in the figure, gives $Q/R = 14{,}000$ K^{-1}, or $Q = 27{,}880$ cal/mol. Alternately, we could write two simultaneous equations:

$$\text{Rate}\left(\frac{\text{jumps}}{\text{s}}\right) = c_0 \exp\left(\frac{-Q}{RT}\right)$$

$$5 \times 10^8 \left(\frac{\text{jumps}}{\text{s}}\right) = c_0 \left(\frac{\text{jumps}}{\text{s}}\right) \exp\left[\frac{-Q\left(\frac{\text{cal}}{\text{mol}}\right)}{\left[1.987\left(\frac{\text{cal}}{\text{mol}\cdot\text{K}}\right)\right](500 + 273)\text{K}}\right]$$

$$= c_0 \exp(-0.000651Q)$$

$$8 \times 10^{10} \left(\frac{\text{jumps}}{\text{s}}\right) = c_0 \left(\frac{\text{jumps}}{\text{s}}\right) \exp\left[\frac{-Q\left(\frac{\text{cal}}{\text{mol}}\right)}{\left[1.987\left(\frac{\text{cal}}{\text{mol}\cdot\text{K}}\right)\right](800 + 273)\text{K}}\right]$$

$$= c_0 \exp(-0.000469Q)$$

Note the temperatures were converted into K.
    Since

$$c_0 = \frac{5 \times 10^8}{\exp(-0.000651Q)} \left(\frac{\text{jumps}}{\text{s}}\right),$$

then

$$8 \times 10^{10} = \frac{(5 \times 10^8) \exp(-0.000469Q)}{\exp(-0.000651Q)}$$

$$160 = \exp[(0.000651 - 0.000469)Q] = \exp(0.000182Q)$$

$$\ln(160) = 5.075 = 0.000182Q$$

$$Q = \frac{5.075}{0.000182} = 27,880 \text{ cal/mol}$$

5-3   Mechanisms for Diffusion

As we saw in Chapter 4, defects known as vacancies exist in materials. The disorder these vacancies create (i.e., increased entropy) helps minimize the free energy and, therefore, the thermodynamic stability of a crystalline material. Crystalline materials also contain other types of defects. In materials containing vacancies, atoms move or "jump" from one lattice position to another. This process, known as **self-diffusion**, can be detected by using radioactive tracers. As an example, suppose we were to introduce a

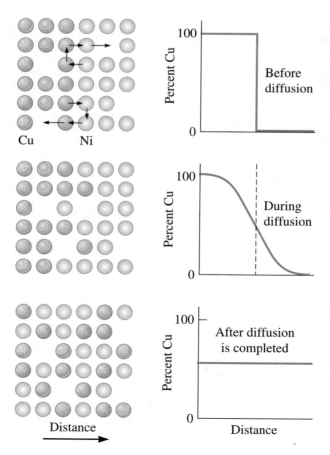

Figure 5-10
Diffusion of copper atoms into nickel. Eventually, the copper atoms are randomly distributed throughout the nickel.

Motion of atom ⟶

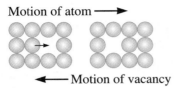

⟵ Motion of vacancy

(a) Vacancy mechanism

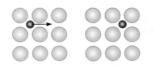

(b) Interstitial mechanism

Figure 5-11
Diffusion mechanisms in materials: (a) vacancy or substitutional atom diffusion and (b) interstitial diffusion.

radioactive isotope of gold (Au^{198}) onto the surface of normal gold (Au^{197}). After a period of time, the radioactive atoms would move into the normal gold. Eventually, the radioactive atoms would be uniformly distributed throughout the entire regular gold sample. Although self-diffusion occurs continually in all materials, its effect on the material's behavior is generally not significant.

Diffusion of unlike atoms in materials also occurs (Figure 5-10). Consider a nickel sheet bonded to a copper sheet. At high temperatures, nickel atoms gradually diffuse into the copper and copper atoms migrate into the nickel. Again, the nickel and copper atoms eventually are uniformly distributed. There are two important mechanisms by which atoms or ions can diffuse (Figure 5-11).

Vacancy Diffusion In self-diffusion and diffusion involving substitutional atoms, an atom leaves its lattice site to fill a nearby vacancy (thus creating a new vacancy at the original lattice site). As diffusion continues, we have countercurrent flows of atoms and vacancies, called **vacancy diffusion**. The number of vacancies, which increases as the temperature increases, helps determine the extent of both self-diffusion and diffusion of substitutional atoms.

Interstitial Diffusion When a small interstitial atom or ion is present in the crystal structure, the atom or ion moves from one interstitial site to another. No vacancies are required for this mechanism. Partly because there are many more interstitial sites than vacancies, interstitial diffusion occurs more easily than vacancy diffusion. Interstitial atoms that are relatively smaller can diffuse faster. In Chapter 3, we have seen that many ceramics with ionic bonding the structure can be considered as close packing of anions with cations in the interstitial sites. In these materials, smaller cations often diffuse faster than larger anions.

5-4 Activation Energy for Diffusion

A diffusing atom must squeeze past the surrounding atoms to reach its new site. In order for this to happen, energy must be supplied to force the atom to its new position, as is shown schematically for vacancy and interstitial diffusion in Figure 5-12. The atom is originally in a low-energy, relatively stable location. In order to move to a new location, the atom must overcome an energy barrier. The energy barrier is the activation energy Q. The thermal energy supplies atoms or ions with the energy needed to exceed this barrier. Note that the symbol Q is often used for activation energies for different processes (rate at which atoms jump, a chemical reaction, energy needed to produce vacancies, etc.) and we should be careful in understanding the specific process or phenomenon to which the general term for activation energy Q is being applied, as the value of Q depends on the particular phenomenon.

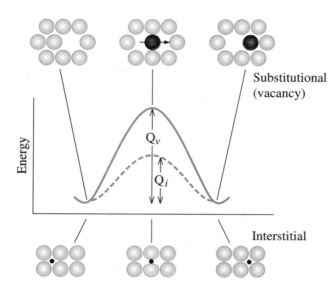

Substitutional
(vacancy)

Interstitial

Normally, less energy is required to squeeze an interstitial atom past the surrounding atoms; consequently, activation energies are lower for interstitial diffusion than for vacancy diffusion. A good analogy for this is as follows. In basketball, relatively smaller and shorter players can quickly "diffuse by" taller and bigger players and score baskets (Figure 5-13). Typical values for activation energies for diffusion of different atoms in different host materials are shown in Table 5-1. We use the term **diffusion couple** to indicate a combination of an atom of a given element (e.g., carbon) diffusing in a

Figure 5-13   In basketball, smaller players can "diffuse" through the open spaces between bigger players. (*Courtesy of Getty Images.*)

TABLE 5-1 ■ *Diffusion data for selected materials*

Diffusion Couple	Q *activation energy* (cal/mol)	D_0 (cm²/s)
Interstitial diffusion:		
C in FCC iron	32,900	0.23
C in BCC iron	20,900	0.011
N in FCC iron	34,600	0.0034
N in BCC iron	18,300	0.0047
H in FCC iron	10,300	0.0063
H in BCC iron	3,600	0.0012
Self-diffusion (vacancy diffusion):		
Pb in FCC Pb	25,900	1.27
Al in FCC Al	32,200	0.10
Cu in FCC Cu	49,300	0.36
Fe in FCC Fe	66,700	0.65
Zn in HCP Zn	21,800	0.1
Mg in HCP Mg	32,200	1.0
Fe in BCC Fe	58,900	4.1
W in BCC W	143,300	1.88
Si in Si (covalent)	110,000	1800.0
C in C (covalent)	163,000	5.0
Heterogeneous diffusion (vacancy diffusion):		
Ni in Cu	57,900	2.3
Cu in Ni	61,500	0.65
Zn in Cu	43,900	0.78
Ni in FCC iron	64,000	4.1
Au in Ag	45,500	0.26
Ag in Au	40,200	0.072
Al in Cu	39,500	0.045
Al in Al_2O_3	114,000	28.0
O in Al_2O_3	152,000	1900.0
Mg in MgO	79,000	0.249
O in MgO	82,100	0.000043

From several sources, including Adda, Y. and Philibert, J., La Diffusion dans les Solides, Vol. 2, 1966.

host material (e.g., BCC Fe). A low-activation energy indicates easy diffusion. In self-diffusion, the activation energy is equal to the energy needed to create a vacancy and to cause the movement of the atom. Table 5-1 also shows values of D_0, the pre-exponent term, and the constant c_0 from Equation 5-1 where the rate process is diffusion. We will see later that D_0 is the diffusion coefficient when $1/T = 0$.

5-5   Rate of Diffusion [Fick's First Law]

Adolf Eugen Fick (1829–1901) was the first scientist to provide a quantitative description of the diffusion process. Interestingly, Fick, was also the first to experiment with contact lenses in animals and the first to implant a contact lens in human eyes in 1887–1888!

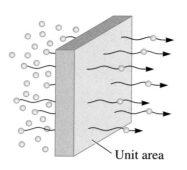

Figure 5-14
The flux during diffusion is defined as the number of atoms passing through a plane of unit area per unit time.

Unit area

The rate at which atoms, ions, particles or other species diffuse in a material can be measured by the **flux** J. Here we are mainly concerned with diffusion of ions or atoms. The flux J is defined as the number of atoms passing through a plane of unit area per unit time (Figure 5-14). **Fick's first law** explains the net flux of atoms:

$$J = -D\frac{dc}{dx} \tag{5-3}$$

where J is the flux, D is the **diffusivity** or **diffusion coefficient** $\left(\frac{cm^2}{s}\right)$, and dc/dx is the **concentration gradient** $\left(\frac{atoms}{cm^3 \cdot cm}\right)$. Depending upon the situation, concentration may be expressed as atom percent (at%), weight percent (wt%), mole percent (mol%), atom fraction, or mole fraction. The units of concentration gradient and flux will also change respectively.

Several factors affect the flux of atoms during diffusion. If we are dealing with diffusion of ions, electrons, holes, etc., the units of J, D, and $\frac{dc}{dx}$ will reflect the appropriate species that are being considered. The negative sign in Equation 5-3 tells us that the flux of diffusing species is from higher to lower concentrations, making the $\frac{dc}{dx}$ term negative, and hence J will be positive. Thermal energy associated with atoms, ions etc. causes the random movement of atoms. At a microscopic scale the thermodynamic driving force for diffusion is concentration gradient. A net or an observable flux is created depending upon temperature and concentration gradient.

Concentration Gradient   The concentration gradient shows how the composition of the material varies with distance: Δc is the difference in concentration over the distance Δx (Figure 5-15). The concentration gradient may be created when two materials of different composition are placed in contact, when a gas or liquid is in contact with a solid material, when nonequilibrium structures are produced in a material due to processing, and from a host of other sources.

The flux at a particular temperature is constant only if the concentration gradient is also constant—that is, the compositions on each side of the plane in Figure 5-14 remain unchanged. In many practical cases, however, these compositions vary as atoms are redistributed, and thus the flux also changes. Often, we find that the flux is initially high and then gradually decreases as the concentration gradient is reduced by diffusion. The examples that follow illustrate calculations of flux and concentration gradients for diffusion of dopants in semiconductors and ceramics, but only for the case of constant concentration gradient. Later in this chapter, we will consider non-steady state diffusion with the aid of Fick's second law.

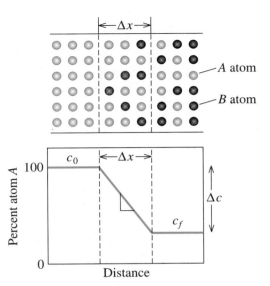

Figure 5-15
Illustration of the concentration gradient.

EXAMPLE 5-4   *Semiconductor Doping*

One way to manufacture transistors, which amplify electrical signals, is to diffuse impurity atoms into a semiconductor material such as silicon (Si). Suppose a silicon wafer 0.1 cm thick, which originally contains one phosphorus atom for every 10 million Si atoms, is treated so that there are 400 phosphorous (P) atoms for every 10 million Si atoms at the surface (Figure 5-16). Calculate the concentration gradient (a) in atomic percent/cm and (b) in $\frac{\text{atoms}}{\text{cm}^3 \cdot \text{cm}}$. The lattice parameter of silicon is 5.4307 Å.

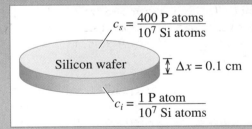

Figure 5-16
Silicon wafer showing variation in concentration of P atoms (for Example 5-4).

SOLUTION

First, let's calculate the initial and surface compositions in atomic percent:

$$c_i = \frac{1 \text{ P atom}}{10^7 \text{ atoms}} \times 100 = 0.00001 \text{ at% P}$$

$$c_s = \frac{400 \text{ P atoms}}{10^7 \text{ atoms}} \times 100 = 0.004 \text{ at% P}$$

$$\frac{\Delta c}{\Delta x} = \frac{0.00001 - 0.004 \text{ at% P}}{0.1 \text{ cm}} = -0.0399 \frac{\text{at% P}}{\text{cm}}$$

To find the gradient in terms of $\frac{\text{atoms}}{\text{cm}^3 \cdot \text{cm}}$, we must find the volume of the unit cell:

$$V_{\text{cell}} = (5.4307 \times 10^{-8} \text{ cm})^3 = 1.6 \times 10^{-22} \frac{\text{cm}^3}{\text{cell}}.$$

The volume occupied by 10^7 Si atoms, which are arranged in a diamond cubic (DC) structure with 8 atoms/cell, is:

$$V = \left[\frac{10^7 \text{ atoms}}{8 \frac{\text{atoms}}{\text{cell}}} \right] \left[1.6 \times 10^{-22} \left(\frac{\text{cm}^3}{\text{cell}} \right) \right] = 2 \times 10^{-16} \text{ cm}^3$$

The compositions in atoms/cm^3 are:

$$c_i = \frac{1 \text{ P atom}}{2 \times 10^{-16} \text{ cm}^3} = 0.005 \times 10^{18} \text{ P} \left(\frac{\text{atoms}}{\text{cm}^3} \right)$$

$$c_s = \frac{400 \text{ P atoms}}{2 \times 10^{-16} \text{ cm}^3} = 2 \times 10^{18} \text{ P} \left(\frac{\text{atoms}}{\text{cm}^3} \right)$$

$$\frac{\Delta c}{\Delta x} = \frac{0.005 \times 10^{18} - 2 \times 10^{18} \text{ P} \left(\frac{\text{atoms}}{\text{cm}^3} \right)}{0.1 \text{ cm}}$$

$$= -1.995 \times 10^{19} \text{ P} \frac{\text{atoms}}{\text{cm}^3 \cdot \text{cm}}$$

EXAMPLE 5-5   *Diffusion of Nickel in Magnesium Oxide (MgO)*

A 0.05 cm layer of magnesium oxide (MgO) is deposited between layers of nickel (Ni) and tantalum (Ta) to provide a diffusion barrier that prevents reactions between the two metals (Figure 5-17). At 1400°C, nickel ions are created and diffuse through the MgO ceramic to the tantalum. Determine the number of nickel ions that pass through the MgO per second. The diffusion coefficient of nickel ions in MgO is 9×10^{-12} cm^2/s, and the lattice parameter of nickel at 1400°C is 3.6×10^{-8} cm.

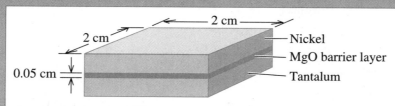

Figure 5-17 Diffusion couple (for Example 5-5).

SOLUTION

The composition of nickel at the Ni/MgO interface is 100% Ni, or

$$c_{\text{Ni/MgO}} = \frac{4 \text{ Ni} \frac{\text{atoms}}{\text{unit cell}}}{(3.6 \times 10^{-8} \text{ cm})^3} = 8.57 \times 10^{22} \frac{\text{atoms}}{\text{cm}^3}$$

The composition of nickel at the Ta/MgO interface is 0% Ni. Thus, the concentration gradient is:

$$\frac{\Delta c}{\Delta x} = \frac{0 - 8.57 \times 10^{22} \frac{\text{atoms}}{\text{cm}^3}}{0.05 \text{ cm}} = -1.71 \times 10^{24} \frac{\text{atoms}}{\text{cm}^3 \cdot \text{cm}}$$

The flux of nickel atoms through the MgO layer is:

$$J = -D\frac{\Delta c}{\Delta x} = (9 \times 10^{-12} \text{ cm}^2/\text{s})\left(-1.71 \times 10^{24} \frac{\text{atoms}}{\text{cm}^3 \cdot \text{cm}}\right)$$

$$J = 1.54 \times 10^{13} \frac{\text{Ni atoms}}{\text{cm}^2 \cdot \text{s}}$$

The total number of nickel atoms crossing the 2 cm × 2 cm interface per second is:

$$\text{Total Ni atoms per second} = (J)(\text{Area}) = \left(1.54 \times 10^{13} \frac{\text{atoms}}{\text{cm}^2 \cdot \text{s}}\right)(2 \text{ cm})(2 \text{ cm})$$

$$= 6.16 \times 10^{13} \text{ Ni atoms/s}$$

Although this appears to be very rapid, we would find that in one second, the volume of nickel atoms removed from the Ni/MgO interface is:

$$\frac{6.16 \times 10^{13} \frac{\text{Ni atoms}}{\text{cm}^3}}{8.57 \times 10^{22} \frac{\text{Ni atoms}}{\text{cm}^3}} = 0.72 \times 10^{-9} \frac{\text{cm}^3}{\text{s}}$$

Or, the thickness by which the nickel layer is reduced each second is:

$$\frac{0.72 \times 10^{-9} \frac{\text{cm}^3}{\text{s}}}{4 \text{ cm}^2} = 1.8 \times 10^{-10} \frac{\text{cm}}{\text{s}}$$

For one micrometer (10^{-4} cm) of nickel to be removed, the treatment requires

$$\frac{10^{-4} \text{ cm}}{1.8 \times 10^{-10} \frac{\text{cm}}{\text{s}}} = 556{,}000 \text{ s} = 154 \text{ h}$$

5-6 Factors Affecting Diffusion

Temperature and the Diffusion Coefficient   The kinetics of process of diffusion are strongly dependent on temperature. The diffusion coefficient D is related to temperature by an Arrhenius-type equation, *don't need to know*

$$D = D_0 \exp\left(\frac{-Q}{RT}\right) \tag{5-4}$$

where Q is the activation energy (in units of cal/mol) for diffusion of species under consideration (e.g., Al in Si), R is the gas constant $\left(1.987\frac{\text{cal}}{\text{mol} \cdot \text{K}}\right)$, and T is the absolute temperature (K). D_0 is the pre-exponential term, similar to c_0 in Equation 5-1.

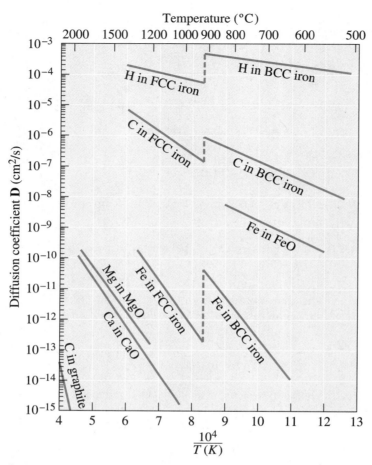

Figure 5-18 The diffusion coefficient D as a function of reciprocal temperature for some metals and ceramics. In this Arrhenius plot, D represents the rate of the diffusion process. A steep slope denotes a high activation energy.

It is a constant for a given diffusion system and is equal to the value of the diffusion coefficient at $1/T = 0$ or $T = \infty$. Typical values for D_0 are given in Table 5-1, while the temperature dependence of D is shown in Figure 5-18 for some metals and ceramics. Covalently bonded materials, such as carbon and silicon (Table 5-1), have unusually high activation energies, consistent with the high strength of their atomic bonds. Figure 5-19 shows the diffusion coefficients for different dopants in silicon.[5]

In ionic materials, such as some of the oxide ceramics, a diffusing ion only enters a site having the same charge. In order to reach that site, the ion must physically squeeze past adjoining ions, pass by a region of opposite charge, and move a relatively long distance (Figure 5-20). Consequently, the activation energies are high and the rates of diffusion are lower for ionic materials than those for metals (Figure 5-21).[6] We take advantage of this in many situations. For example, in processing of silicon (Si), we create a thin layer of silica (SiO$_2$) on top of a silicon wafer (Chapter 18). We then create a window by removing part of the silica layer. This window allows selective diffusion of dopant atoms such as phosphorus (P) and boron (B). However, the silica layer is essentially impervious to the dopant atoms. Slower diffusion in most oxides and other ceramics usually is disadvantageous in that it leads to higher temperatures for their

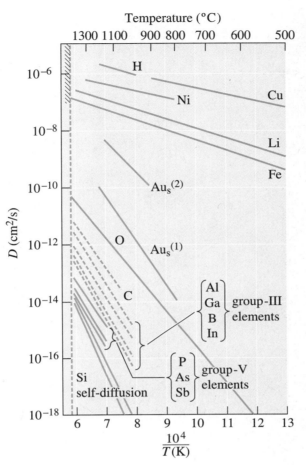

Figure 5-19
Diffusion coefficients for different dopants in silicon. (*Source: From "Diffusion and Diffusion Induced Defects in Silicon," by U. Gösele. In R. Bloor, M. Flemings, and S. Mahajan (Eds.), Encyclopedia of Advanced Materials, Vol. 1, 1994, p. 631, Fig. 2. Copyright © 1994 Pergamon Press. Reprinted with permission of the authors.*)

processing. However, this is an advantage in applications in which components are required to withstand high temperatures.

When the temperature of a material increases, the diffusion coefficient D increases (according to Equation 5-4) and, therefore, the flux of atoms increases as well. At higher temperatures, the thermal energy supplied to the diffusing atoms permits the atoms to overcome the activation energy barrier and more easily move to new sites in the atomic arrangements. At low temperatures—often below about 0.4 times the absolute melting temperature of the material—diffusion is very slow and may not be significant. For this reason, the heat treatment of metals and the processing of ceramics are done at high temperatures, where atoms move rapidly to complete reactions or to reach equilibrium

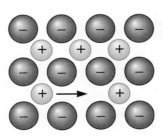

Figure 5-20
Diffusion in ionic compounds. Anions can only enter other anion sites. Smaller cations tend to diffuse faster.

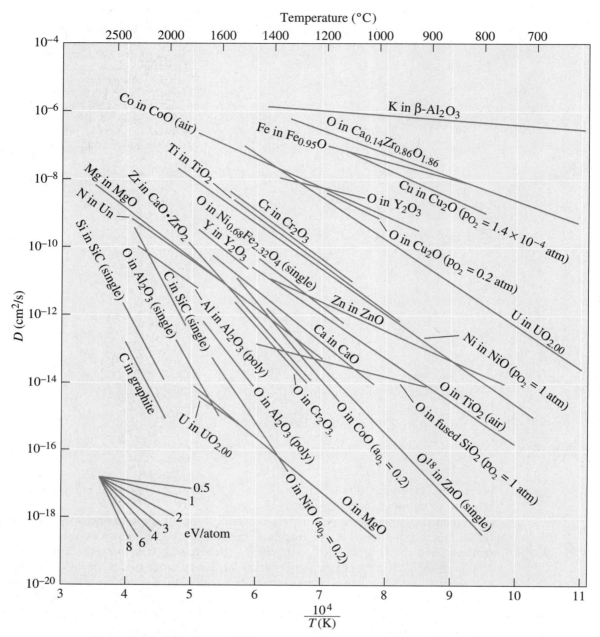

Figure 5-21   Diffusion coefficients of ions in different oxides. (*Source: Adapted from* Physical Ceramics: Principles for Ceramic Science and Engineering, *by Y.M. Chiang, D. Birnie, and W.D. Kingery, Fig. 3-1. Copyright © 1997 John Wiley & Sons. Adapted with permission.*)

conditions. Because less thermal energy is required to overcome the smaller activation energy barrier, a small activation energy Q increases the diffusion coefficient and flux. The following example illustrates how Fick's first law and concepts related to temperature dependence of D can be applied to design an iron membrane.

EXAMPLE 5-6   *Design of an Iron Membrane*

An impermeable cylinder 3 cm in diameter and 10 cm long contains a gas that includes 0.5×10^{20} N atoms per cm^3 and 0.5×10^{20} H atoms per cm^3 on one side of an iron membrane (Figure 5-22). Gas is continuously introduced to the pipe to assure a constant concentration of nitrogen and hydrogen. The gas on the other side of the membrane includes a constant 1×10^{18} N atoms per cm^3 and 1×10^{18} H atoms per cm^3. The entire system is to operate at 700°C, where the iron has the BCC structure. Design an iron membrane that will allow no more than 1% of the nitrogen to be lost through the membrane each hour, while allowing 90% of the hydrogen to pass through the membrane per hour.

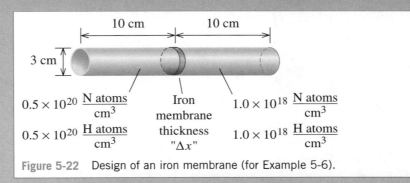

Figure 5-22   Design of an iron membrane (for Example 5-6).

SOLUTION

The total number of nitrogen atoms in the container is:

$$(0.5 \times 10^{20} \text{ N/cm}^3)(\pi/4)(3 \text{ cm})^2(10 \text{ cm}) = 35.343 \times 10^{20} \text{ N atoms}$$

The maximum number of atoms to be lost is 1% of this total, or:

$$\text{N atom loss per h} = (0.01)(35.34 \times 10^{20}) = 35.343 \times 10^{18} \text{ N atoms/h}$$

$$\text{N atom loss per s} = (35.343 \times 10^{18} \text{ N atoms/h})/(3600 \text{ s/h})$$

$$= 0.0098 \times 10^{18} \text{ N atoms/s}$$

The flux is then:

$$J = \frac{(0.0098 \times 10^{18}(\text{N atoms/s})}{\left(\dfrac{\pi}{4}\right)(3 \text{ cm})^2}$$

$$= 0.00139 \times 10^{18} \text{ N} \frac{\text{atoms}}{\text{cm}^2 \cdot \text{s}}$$

The diffusion coefficient of nitrogen in BCC iron at 700°C = 973 K is:

From Equation 5-4 and Table 5-1:

$$D = D_0 \exp\left(\frac{-Q}{RT}\right)$$

$$D_N = 0.0047 \frac{cm^2}{s} \exp\left[\frac{-18,300 \frac{cal}{mol}}{1.987 \frac{cal}{K \cdot mol}(973)K}\right]$$

$$= 0.0047 \exp(-9.4654) = (0.0047)(7.749 \times 10^{-5}) = 3.64 \times 10^{-7} \frac{cm^2}{s}$$

From Equation 5-3:

$$J = -D\left(\frac{\Delta c}{\Delta x}\right) = 0.00139 \times 10^{18} \frac{N \text{ atoms}}{cm^2 \cdot s}$$

$$\Delta x = -D\Delta c/J = \frac{\left[(-3.64 \times 10^{-7} cm^2/s)\left(1 \times 10^{18} - 50 \times 10^{18} \frac{N \text{ atoms}}{cm^3}\right)\right]}{0.00139 \times 10^{18} \frac{N \text{ atoms}}{cm^2 \cdot s}}$$

$$\Delta x = 0.0128 \text{ cm} = \text{minimum thickness of the membrane}$$

In a similar manner, the maximum thickness of the membrane that will permit 90% of the hydrogen to pass can be calculated:

$$\text{H atom loss per h} = (0.90)(35.343 \times 10^{20}) = 31.80 \times 10^{20}$$

$$\text{H atom loss per s} = 0.0088 \times 10^{20}$$

$$J = 0.125 \times 10^{18} \frac{H \text{ atoms}}{cm^2 \cdot s}$$

From Equation 5-4,

$$D_H = 0.004 \frac{cm^2}{s} \exp\left[\frac{-18,300 \frac{cal}{mol}}{1.987 \frac{cal}{K \cdot mol}(973)K}\right] = 1.86 \times 10^{-4} cm^2/s$$

Since

$$\Delta x = -D \, \Delta c/J$$

$$\Delta x = \frac{\left(1.86 \times 10^{-4} \frac{cm^2}{s}\right)\left(49 \times 10^{18} \frac{H \text{ atoms}}{cm^3}\right)}{0.125 \times 10^{18} \frac{H \text{ atoms}}{cm^2 \cdot s}}$$

$$= 0.0729 \text{ cm} = \text{maximum thickness}$$

An iron membrane with a thickness between 0.0128 and 0.0729 cm will be satisfactory.

TABLE 5-2 ■ *The effect of the type of diffusion for thorium in tungsten and for self-diffusion in silver**

	Diffusion Coefficient (*D*)			
	Thorium in Tungsten		Silver in Silver	
Diffusion Type	D_0 cm²/s	*Q* cal/mole	D_0 cm²/s	*Q* cal/mole
Surface	0.47	66,400	0.068	8,900
Grain boundary	0.74	90,000	0.24	22,750
Volume	1.00	120,000	0.99	45,700

Given by parameters for Equation 5-4.

[handwritten: Totally surrounded]

Types of Diffusion In **volume diffusion**, the atoms move through the crystal from one regular or interstitial site to another. Because of the surrounding atoms, the activation energy is large and the rate of diffusion is relatively slow.

However, atoms can also diffuse along boundaries, interfaces, and surfaces in the material. Atoms diffuse easily by **grain boundary diffusion** because the atom packing is poor in the grain boundaries. Because atoms can more easily squeeze their way through the disordered grain boundary, the activation energy is low (Table 5-2). **Surface diffusion** is easier still because there is even less constraint on the diffusing atoms at the surface.

[handwritten: e⁻ are more vulnerable]

Time Diffusion requires time; the units for flux are $\frac{\text{atoms}}{\text{cm}^2 \cdot \text{s}}$! If a large number of atoms must diffuse to produce a uniform structure, long times may be required, even at high temperatures. Times for heat treatments may be reduced by using higher temperatures or by making the **diffusion distances** (related to Δx) as small as possible.

We find that some rather remarkable structures and properties are obtained if we prevent diffusion. Steels quenched rapidly from high temperatures to prevent diffusion form nonequilibrium structures and provide the basis for sophisticated heat treatments. Similarly, in forming metallic glasses (Chapter 4) we have to quench liquid metals at a very high cooling rate ($\sim10^{6\circ}$C/second). This is to avoid diffusion of atoms by taking away their thermal energy and encouraging them to assemble into nonequilibrium amorphous arrangements. Melts of silicates glasses, on the other hand, are viscous and diffusion of ions through these is slow. As a result, we do not have to cool these melts very rapidly. There is a myth that many old buildings contain windowpanes that are thicker at the bottom than on the top because the glass has flowed down over the years. Based on kinetics of diffusion it can be shown that even several hundred or thousand years will not be sufficient to cause such flow of glasses at near-room temperature. In certain thin film deposition processes, such as sputtering, we sometimes obtain amorphous or microcrystalline thin films if the atoms or ions are quenched rapidly after they land on the substrate. If these films are subsequently heated (after deposition) to sufficiently high temperatures, diffusion will occur and the amorphous thin films will eventually crystallize. We previously have seen (Figure 5-3) how sputtered thin films for magnetic hard drives are annealed so as to cause diffusion of chromium atoms to the grain boundaries. In the following example, we examine different mechanisms for diffusion.

EXAMPLE 5-7   *Tungsten Thorium Diffusion Couple*

Consider a diffusion couple setup between pure tungsten and a tungsten alloy containing 1 at.% thorium. After several minutes of exposure at 2000°C, a transition zone of 0.01 cm thickness is established. What is the flux of thorium atoms at this time if diffusion is due to (a) volume diffusion, (b) grain boundary diffusion, and (c) surface diffusion? (See Table 5-2.)

SOLUTION

The lattice parameter of BCC tungsten is about 3.165 Å. Thus, the number of tungsten atoms/cm^3 is:

$$\frac{\text{W atoms}}{\text{cm}^3} = \frac{2 \text{ atoms/cell}}{(3.165 \times 10^{-8})^3 \text{ cm}^3/\text{cell}} = 6.3 \times 10^{22}$$

In the tungsten-1 at.% thorium alloy, the number of thorium atoms is:

$$c_{\text{Th}} = (0.01)(6.3 \times 10^{22}) = 6.3 \times 10^{20} \text{ Th atoms/cm}^3$$

In the pure tungsten, the number of thorium atoms is zero. Thus, the concentration gradient is:

$$\frac{\Delta c}{\Delta x} = \frac{0 - 6.3 \times 10^{20} \frac{\text{atoms}}{\text{cm}^3}}{0.01 \text{ cm}} = -6.3 \times 10^{22} \text{ Th} \frac{\text{atoms}}{\text{cm}^3 \cdot \text{cm}}$$

1. Volume diffusion

$$D = 1.0 \frac{\text{cm}^2}{\text{s}} \exp\left(\frac{-120,000 \frac{\text{cal}}{\text{mol}}}{\left(1.987 \frac{\text{cal}}{\text{deg} \cdot \text{mol}}\right)(2273 \text{ K})}\right) = 2.89 \times 10^{-12} \text{ cm}^2/\text{s}$$

$$J = -D \frac{\Delta c}{\Delta x} = -\left(2.89 \times 10^{-12} \frac{\text{cm}^2}{\text{s}}\right)\left(-6.3 \times 10^{22} \frac{\text{atoms}}{\text{cm}^3 \cdot \text{cm}}\right)$$

$$= 18.2 \times 10^{10} \frac{\text{Th atoms}}{\text{cm}^2 \cdot \text{s}}$$

2. Grain boundary diffusion

$$D = 0.74 \frac{\text{cm}^2}{\text{s}} \exp\left(\frac{-90,000 \frac{\text{cal}}{\text{mol}}}{\left(1.987 \frac{\text{cal}}{\text{deg} \cdot \text{mol}}\right)(2273 \text{ K})}\right) = 1.64 \times 10^{-9} \text{ cm}^2/\text{s}$$

$$J = -\left(1.64 \times 10^{-9} \frac{\text{cm}^2}{\text{s}}\right)\left(-6.3 \times 10^{22} \frac{\text{atoms}}{\text{cm}^3 \cdot \text{cm}}\right) = 10.3 \times 10^{13} \frac{\text{Th atoms}}{\text{cm}^2 \cdot \text{s}}$$

3. Surface diffusion

$$D = 0.47 \frac{\text{cm}^2}{\text{s}} \exp\left(\frac{-66,400 \frac{\text{cal}}{\text{mol}}}{\left(1.987 \frac{\text{cal}}{\text{mol} \cdot \text{deg}}\right)(2273 \text{ K})}\right) = 1.94 \times 10^{-7} \text{ cm}^2/\text{s}$$

$$J = -\left(1.94 \times 10^{-7} \frac{\text{cm}^2}{\text{s}}\right)\left(-6.3 \times 10^{22} \frac{\text{atoms}}{\text{cm}^3 \cdot \text{cm}}\right) = 12.2 \times 10^{15} \frac{\text{Th atoms}}{\text{cm}^2 \cdot \text{s}}$$

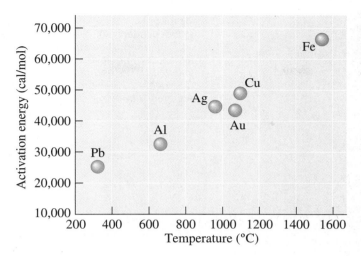

Figure 5-23
The activation energy for self-diffusion increases as the melting point of the metal increases.

Dependence on Bonding and Crystal Structure A number of factors influence the activation energy for diffusion and, hence, the rate of diffusion. Interstitial diffusion, with a low-activation energy, usually occurs much faster than vacancy, or substitutional diffusion. Activation energies are usually lower for atoms diffusing through open crystal structures than for close-packed crystal structures. Because the activation energy depends on the strength of atomic bonding, it is higher for diffusion of atoms in materials with a high melting temperature (Figure 5-23).

We also find that, due to their smaller size, cations (with a positive charge) often have higher diffusion coefficients than those for anions (with a negative charge). In sodium chloride, for instance, the activation energy for diffusion of chloride ions (Cl^-) is about twice that for diffusion of sodium ions (Na^+).

Diffusion of ions also provides a transfer of electrical charge; in fact, the electrical conductivity of ionically bonded ceramic materials is related to temperature by an Arrhenius equation. As the temperature increases, the ions diffuse more rapidly, electrical charge is transferred more quickly, and the electrical conductivity is increased. As mentioned before, these are examples of ceramic materials that are good conductors of electricity.

Dependence on Concentration of Diffusing Species and Composition of Matrix The diffusion coefficient (D) depends not only on temperature, as given by Equation 5-4, but also on the concentration of diffusing species and composition of the matrix. These effects have not been included in our discussion so far. In many situations, the dependence of D on concentration of diffusing species can be ignored, for example, if the concentration of dopants is small. Thus, in the data shown in Table 5-1 and Figures 5-18 and 5-19, we are effectively ignoring the dependence of D on the concentration of diffusing species. In gaseous mixtures, it is observed experimentally that D does depend on concentration; however, the D value (for any given temperature) quickly approaches a relatively constant value even as the concentration of gaseous species changes.[7] In solids, the dependence of D on concentration can be very strong. For example, in the gold-nickel system, the diffusion coefficient of gold (Au) in nickel (Ni) increases 10 times as the concentration of gold in the Au-Ni alloy changes from 20 to 80 atomic percent gold. Figure 5-24 shows the dependence of diffusion coefficient of Au in different alloys.

In some cases, the diffusion appears to occur "uphill" (i.e., from lower concentration to higher concentration). This is similar to the optical illusion where water seems to

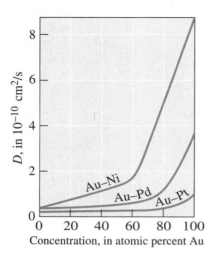

Figure 5-24
The dependence of diffusion coefficient of Au on concentration. (*Source: Adapted from* Physical Metallurgy Principles, *Third Edition, by R.E. Reed-Hill and R. Abbaschian, p. 363, Fig. 12-3. Copyright © 1991 Brooks/Cole Thomson Learning. Adapted with permission.*)

flow up an incline plane and defy gravity! This process, known as **uphill diffusion**, can be shown to be in accordance with the Fick's law, once we understand that we have to consider the *chemical potential* (i.e., the effective and not nominal *concentration* of the diffusing species).[8]

In certain ceramic materials, the value of D can depend strongly upon the overall chemical composition of the material. The changes in diffusivities in the ceramic materials are usually related to the type of point defects introduced as a result of dopant addition or non-stoichiometry. The results are not always intuitive, as the following example illustrates.

EXAMPLE 5-8   *Diffusion in Ionic Conductors*

Consider two compositions of yttria (yttrium oxide, Y_2O_3)-stabilized zirconia (ZrO_2). The first sample contains 6 mole percent yttria (Y_2O_3).[9] Since each mole of yttria contains two moles of yttrium, the mole fraction of element yttrium (Y) in the first sample would be 0.12. The second sample of yttria-stabilized zirconia contains 15 mole percent yttria (Y_2O_3). Therefore, in the second sample, the mole fraction of element yttrium (Y) is 0.30. The introduction of yttria (Y_2O_3) creates oxygen vacancies and defects into which the yttrium ions go on the Zr^{+4} sites. Write down the defect chemistry equation using the Kröger-Vink notation. Show that the concentration of oxygen ion vacancies would be approximately one-half the concentration of yttrium oxide (Y_2O_3). Given this, predict which composition of yttria (Y_2O_3) will likely exhibit higher diffusivity of oxygen ions. Compare your prediction with the data shown in Figure 5-25.[6,10]

SOLUTION

First, let us write down the defect chemistry equation for the introduction of yttria in zirconia using the Kröger-Vink notation (Chapter 4). We need to ensure that the conditions for electrical neutrality, mass balance, and site balance are satisfied. We write an equation to express the addition of yttrium oxide

$$Y_2O_3 \xrightarrow{ZrO_2} 2Y'_{Zr} + 3O^x_o + V^{\bullet\bullet}_o$$

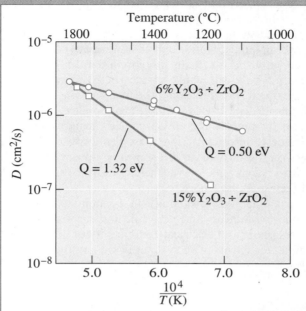

Figure 5-25 Diffusivity of oxygen ions in yttria stabilized zirconia ceramics (for Example 5-8). (*Source: Adapted from* Physical Ceramics: Principles for Ceramic Science and Engineering, *by Y.M. Chiang, D. Birnie, and W.D. Kingery, Fig. 3-14. Copyright © 1997 John Wiley & Sons, Inc. Based on* Transport in Nonstoichiometric Compounds, *by G. Simkovich and U.S. Stubican (Eds.), p. 188–202, Plenum Press. Adapted with permission.*)

In writing this equation, we are assuming that the oxygen available from the surrounding atmosphere is not playing any role in the determination of the defect chemistry. Two yttrium ions go to two zirconium ion sites. Since we used two zirconium sites, we must use four oxygen sites in zirconia. Three oxygen ions from yttria go to three oxygen ion sites of zirconia and one oxygen ion site in zirconia remains empty. Thus, for every $2Y'_{Zr}$ defects, there is one oxygen ion vacancy $V_o^{\cdot\cdot}$.

Therefore,

$$[Y'_{Zr}] = 2[V_o^{\cdot\cdot}]$$

Recall that square brackets here indicate concentration of defects.

For the 6% Y_2O_3 sample, we will have 12 mol% yttrium ions (Y^{+3}). These will go to the zirconium sites and therefore

$$[Y'_{Zr}] = 12 \text{ mol\%}$$

Therefore, the concentration of oxygen ion vacancies would be given by the following equations:

$$[Y'_{Zr}] = 12\% = 2[V_o^{\cdot\cdot}]$$

$$\therefore [V_o^{\cdot\cdot}] = 6\%$$

Similarly, for the second sample containing 15% yttrium oxide, we will have 30% yttrium ions (Y^{+3}) and, therefore, oxygen ion vacancies concentration will be 15%.

We would expect that the diffusivity of oxygen ions would be higher in 15% yttria composition since it has more oxygen ion vacancies (i.e., $[V_o^{..}] = 15\%$).

Figure 5-25 shows experimental data on measured oxygen ion diffusivities in these two materials. Notice that the diffusivity of oxygen in the sample with 6 mole percent yttria is actually *more* than that for the 15 mole percent yttria!

This is not what we would have expected based on our knowledge of the diffusion process and intuition. The reasons why the oxygen ions are less mobile in the 15% mole yttria sample are complex. The decreased diffusivity of oxygen ions in samples containing higher vacancy concentrations may be related to association or clustering of oxygen ion vacancies and yttrium ions on zirconium ion sites. These defects somehow appear to be tying down a considerable portion of oxygen ion vacancies and hence suppress diffusion of oxygen ions and their mass transport.

This example illustrates how, in ceramic oxides, the defect chemistry changes due to additives and non-stoichiometry can have a major impact on the mass transport and electrical conductivity of materials. As is illustrated in this example, the effects of such changes are not always intuitively obvious.

5-7   Permeability of Polymers

In polymers, we are most concerned with the diffusion of atoms or small molecules between the long polymer chains. As engineers, we often cite the permeability of polymers and other materials, instead of the diffusion coefficients. The permeability is expressed in terms of the volume of gas or vapor that can permeate per unit area, per unit time, or per unit thickness at a specified temperature and relative humidity. Polymers that have a polar group (e.g., ethylene vinyl alcohol) have higher permeability for water vapor than that for oxygen gas. Polyethylene, on the other hand, has much higher permeability for oxygen than for water vapor.[11] In general, the more compact the structure of polymers, the lesser the permeability. For example, low-density polyethylene has a higher permeability than high-density polyethylene. Polymers used for food and other applications need to have the appropriate barrier properties. For example, polymer films are typically used as packaging to store food. If air diffuses through the film, the food may spoil. Similarly, care has to be exercised in the storage of ceramic or metal powders that are sensitive to atmospheric water vapor, nitrogen, oxygen, or carbon dioxide. For example, zinc oxide powders used in rubbers, paints, and ceramics must be stored in polyethylene bags to avoid reactions with atmospheric water vapor. If air diffuses through the rubber inner tube of an automobile tire, the tire will deflate.

Diffusion of some molecules into a polymer can cause swelling problems. For example, in automotive applications, polymers used to make O-rings can absorb considerable amounts of oil, causing them to swell. On the other hand, diffusion is required to enable dyes to uniformly enter many of the synthetic polymer fabrics. Selective diffusion through polymer membranes is used to cause desalinization of water. Water molecules pass through the polymer membrane and the ions in the salt are trapped. As mentioned before, Mylar^{TM} helium balloons are either PET (or polyethylene and nylon) coated with a metal film. In each of these examples, the diffusing atoms, ions, or molecules penetrate between the polymer chains rather than moving from one location to another within the chain structure. Diffusion will be more rapid through this structure when the diffusing species is smaller or when larger voids are present between the chains. Diffusion through crystalline polymers, for instance, is slower than that through amorphous polymers, which have no long-range order and, consequently, have a lower density. The following example illustrates how permeability and other factors need to be accounted for in selection of polymers for beverage bottles.

EXAMPLE 5-9 *Design of Carbonated Beverage Bottles*

You want to select a polymer for making plastic bottles that can be used for storing carbonated beverages. What factors would you consider in choosing a polymer for this application?

SOLUTION

For selecting a plastic for manufacturing beverage bottles, several technical and non-technical factors need to be considered. First, since the bottles are to be used for storing carbonated beverages, a plastic material with a small diffusivity for carbon dioxide gas should be chosen. The carbon dioxide dissolved in the beverage is under pressure and hence any values of diffusivities that are to be compared should account for this. In addition, the raw materials and processing of polymers chosen must be relatively cheap. Preferably, we should be able to produce bottles that have different colors, if required, and the bottles should retain optical clarity. The polymers should be non-toxic, non-reactive, and recyclable. The bottles should have enough strength so that they can survive a fall of about six feet. This is often tested using a "drop test." The surface of the polymer should also be amenable to printing of labels or other product information. Materials and processing costs should be considered as well. Based on these initial considerations, we would choose polyethylene terephthalate (PET) for this application.

The effect of processing on the resultant microstructure of polymers must also be considered. For example, PET made by fast cooling from processing temperature of about 95°C (glass temperature is about 76°C) will be optically clear but quite permeable to gases! The blow-stretch process used to make bottles (Chapters 3 and 4) also affects the strength of the bottles and their optical clarity. Hence, the processing of PET should also be optimized. As discussed in previous chapters, a blow-stretch process is suitable for making bottles.

5-8 Composition Profile (Fick's Second Law)

Fick's second law, which describes the dynamic, or nonsteady state, diffusion of atoms, is the differential equation

$$\frac{\partial c}{\partial t} = \frac{\partial}{\partial x}\left(D\frac{\partial c}{\partial x}\right) \tag{5-5}$$

If we assume that the diffusion coefficient D is not a function of location x and the concentration (c) of diffusing species, we can write a simplified version of Fick's second law as follows

$$\frac{\partial c}{\partial t} = D\left(\frac{\partial^2 c}{\partial x^2}\right) \tag{5-6}$$

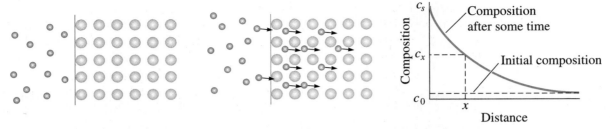

Figure 5-26   Diffusion of atoms into the surface of a material illustrating the use of Fick's second law.

The solution to this equation depends on the boundary conditions for a particular situation. One solution is

$$\frac{c_s - c_x}{c_s - c_0} = \text{erf}\left(\frac{x}{2\sqrt{Dt}}\right) \tag{5-7}$$

where c_s is a constant concentration of the diffusing atoms at the surface of the material, c_0 is the initial uniform concentration of the diffusing atoms in the material, and c_x is the concentration of the diffusing atom at location x below the surface after time t. These concentrations are illustrated in Figure 5-26. In these equations we have assumed basically a one-dimensional model (i.e., we assume that atoms or other diffusing species are moving only in the direction x). The function "erf" is the error function and can be evaluated from Table 5-3 or Figure 5-27. Note that most standard spreadsheet and

TABLE 5-3 ■ Error function values for Fick's second law

Argument of the error function $\dfrac{x}{2\sqrt{Dt}}$	Value of the error function $\text{erf}\,\dfrac{x}{2\sqrt{Dt}}$
0	0
0.10	0.1125
0.20	0.2227
0.30	0.3286
0.40	0.4284
0.50	0.5205
0.60	0.6039
0.70	0.6778
0.80	0.7421
0.90	0.7970
1.00	0.8427
1.50	0.9661
2.00	0.9953

Note that error function values are available on many software packages found on personal computers.

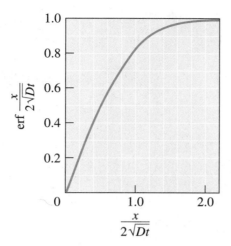

Figure 5-27
Graph showing the argument and value of error function encountered in Fick's second law.

other software programs available on a personal computer (e.g., Excel™) also provide error function values.

The mathematical definition of the error function is as follows:

$$\mathrm{erf}(x) = \frac{2}{\sqrt{\pi}} \int_0^x \exp(-y^2)\, dy \qquad (5\text{-}8)$$

In Equation 5-8, y is known as the argument of the error function. We also define a complementary error function as follows:

$$\mathrm{erfc}(x) = 1 - \mathrm{erf}(x) \qquad (5\text{-}9)$$

This function is used in certain solution forms of Fick's second law.

As mentioned previously, depending upon the boundary conditions, different solutions (i.e. different equations describe the solutions to Fick's second law).[6,8,12] These solutions to Fick's second law permit us to calculate the concentration of one diffusing species as a function of time (t) and location (x). Equation 5-7 is *a possible solution* to Fick's law that describes the variation in concentration of different species near the surface of the material as a function of time and distance, provided that the diffusion coefficient D remains constant and the concentrations of the diffusing atoms at the surface (c_s) and at large distance (x) within the material (c_0) remain unchanged. Fick's second law can also assist us in designing a variety of materials processing techniques, including the steel carburizing heat treatment and dopant diffusion in semiconductors as described in the following examples.

EXAMPLE 5-10   *Design of a Carburizing Treatment*

The surface of a 0.1% C steel gears is to be hardened by carburizing. In gas carburizing, the steel gears are placed in an atmosphere that provides 1.2% C at the surface of the steel at a high temperature (Figure 5-1). Carbon then diffuses from the surface into the steel. For optimum properties, the steel must contain 0.45% C at a depth of 0.2 cm below the surface. Design a carburizing heat treatment that will produce these optimum properties. Assume that the temperature is high enough (at least 900°C) so that the iron has the FCC structure.

SOLUTION

Since the boundary conditions for which Equation 5-7 was derived are assumed to be valid we can use this equation.

$$\frac{c_s - c_x}{c_s - c_0} = \text{erf}\left(\frac{x}{2\sqrt{Dt}}\right)$$

We know that $c_s = 1.2\%$ C, $c_0 = 0.1\%$ C, $c_x = 0.45\%$ C, and $x = 0.2$ cm. From Fick's second law:

$$\frac{c_s - c_x}{c_s - c_0} = \frac{1.2\%C - 0.45\%C}{1.2\%C - 0.1\%C} = 0.68 = \text{erf}\left(\frac{0.2 \text{ cm}}{2\sqrt{Dt}}\right) = \text{erf}\left(\frac{0.1 \text{ cm}}{\sqrt{Dt}}\right)$$

From Table 5-3, we find that:

$$\frac{0.1 \text{ cm}}{\sqrt{Dt}} = 0.71 \quad \text{or} \quad Dt = \left(\frac{0.1}{0.71}\right)^2 = 0.0198 \text{ cm}^2$$

Any combination of D and t whose product is 0.0198 cm² will work. For carbon diffusing in FCC iron, the diffusion coefficient is related to temperature by Equation 5-4:

$$D = D_0 \exp\left(\frac{-Q}{RT}\right)$$

From Table 5-1:

$$D = 0.23 \exp\left(\frac{-32,900 \text{ cal/mol}}{1.987 \frac{\text{cal}}{\text{mol-K}} T(\text{K})}\right) = 0.23 \exp\left(\frac{-16,558}{T}\right)$$

Therefore, the temperature and time of the heat treatment are related by:

$$t = \frac{0.0198 \text{ cm}^2}{D \frac{\text{cm}^2}{\text{s}}} = \frac{0.0198 \text{ cm}^2}{0.23 \exp(-16,558/T) \frac{\text{cm}^2}{\text{s}}} = \frac{0.0861}{\exp(-16,558/T)}$$

Some typical combinations of temperatures and times are:

If $T = 900°C = 1173$ K, then $t = 116,174$ s $= 32.3$ h

If $T = 1000°C = 1273$ K, then $t = 36,360$ s $= 10.7$ h

If $T = 1100°C = 1373$ K, then $t = 14,880$ s $= 4.13$ h

If $T = 1200°C = 1473$ K, then $t = 6,560$ s $= 1.82$ h

The exact combination of temperature and time will depend on the maximum temperature that the heat treating furnace can reach, the rate at which parts must be produced, and the economics of the tradeoffs between higher temperatures versus longer times. Another factor which is important is to consider changes in microstructure that occur in the rest of the material. For example, while carbon is diffusing into the surface, the rest of the microstructure can begin to experience grain growth or other changes.

Example 5-11 shows that one of the consequences of Fick's second law is that the same concentration profile can be obtained for different processing conditions, so long as the term Dt is constant. This permits us to determine the effect of temperature on the time required for a particular heat treatment to be accomplished.

EXAMPLE 5-11   *Design of a More Economical Heat Treatment*

We find that 10 h are required to successfully carburize a batch of 500 steel gears at 900°C, where the iron has the FCC structure. We find that it costs $1000 per hour to operate the carburizing furnace at 900°C and $1500 per hour to operate the furnace at 1000°C. Is it economical to increase the carburizing temperature to 1000°C? What other factors must be considered?

SOLUTION

Again assuming, we can use the solution to Ficks's second law given by Equation 5-7

$$\frac{c_s - c_x}{c_s - c_0} = \text{erf}\left(\frac{x}{2\sqrt{Dt}}\right)$$

Note that, since we are dealing with only changes in heat treatment time and temperature, the term Dt must be constant.

To achieve the same carburizing treatment at 1000°C as at 900°C:

$$D_{1273}t_{1273} = D_{1173}t_{1173}$$

The temperatures of interest are 900°C = 1173 K and 1000°C = 1273 K. For carbon diffusing in FCC iron, the activation energy is 32,900 cal/mol. Since we are dealing with the ratios of times, it does not matter whether we substitute for the time in hours or seconds. It is, however, always a good idea to use units that balance out. Therefore, we will show time in seconds. Note that temperatures must be converted into Kelvin.

$$D_{1273}t_{1273} = D_{1173}t_{1173}$$

$$D = D_0 \exp(-Q/RT)$$

$$t_{1273} = \frac{D_{1173}t_{1173}}{D_{1273}}$$

$$= \frac{D_0 \exp\left[-\dfrac{32{,}900 \frac{\text{cal}}{\text{mol}}}{1.987 \frac{\text{cal}}{\text{K} \cdot \text{mol}} 1173 \text{ K}}\right] (10 \text{ hours})(3600 \text{ sec/hour})}{D_0 \exp\left[-\dfrac{32{,}900 \frac{\text{cal}}{\text{mol}}}{1.987 \frac{\text{cal}}{\text{K} \cdot \text{mol}} 1273 \text{ K}}\right]}$$

$$t_{1273} = \frac{\exp(-14.11562)(10)(3600)}{\exp(-13.00677)} = (10) \exp(-1.10885)$$

$$= (10)(0.3299)(3600) \text{ s}$$

$$t_{1273} = 3.299 \text{ h} = 3 \text{ h and } 18 \text{ min}$$

Notice, we really did not need the value of the pre-exponential term D_0, since it canceled out.

At 900°C, the cost per part is ($1000/h) (10 h)/500 parts = $20/part

At 1000°C, the cost per part is ($1500/h) (3.299 h)/500 parts = $9.90/part

Considering only the cost of operating the furnace, increasing the temperature reduces the heat-treating cost of the gears and increases the production rate. Another factor to consider is if the heat treatment at 1000°C could cause some other microstructural or other changes? For example, would increased temperature cause grains to grow significantly? If this is the case, we will be weakening the bulk of the material. How does the increased temperature affect the life of the other equipment such as the furnace itself and any accessories? How long would the cooling take? Will cooling from a higher temperature cause residual stresses? Would the product still meet all other specifications? These and other questions should be considered. The point is, as engineers, we need to ensure that the solution we propose is not only technically sound and economically sensible, it should recognize and make sense for the system as a whole (i.e., bigger picture). A good solution is often simple, solves problems for the system, and does not create new problems.

EXAMPLE 5-12 *Silicon Device Fabrication*

Devices such as transistors (Figure 5-2) are made by doping semiconductors with different dopants to generate regions that have p- or n-type semiconductivity.[1] The diffusion coefficient of phosphorus (P) in Si is $D = 6.5 \times 10^{-13}$ cm^2/s at a temperature of 1100°C.[13] Assume the source provides a surface concentration of 10^{20} atoms/cm^3 and the diffusion time is one hour. Assume that the silicon wafer contains no P to begin with.

(a) Calculate the depth at which the concentration of P will be 10^{18} atoms/cm^3. State any assumptions you have made while solving this problem.

(b) What will happen to the concentration profile as we cool the Si wafer containing P?

(c) What will happen if now the wafer has to be heated again for boron (B) diffusion for creating a p-type region?

SOLUTION

(a) We assume that we can use one of the solutions to Fick's second law (i.e., Equation 5-7):

$$\frac{c_s - c_x}{c_s - c_0} = \mathrm{erf}\left(\frac{x}{2\sqrt{Dt}}\right)$$

We will use concentrations in atoms/cm^3, time in seconds, and D in $\frac{cm^2}{s}$. Notice that the left-hand side is dimensionless. Therefore, as long as we use concentrations in the same units for c_s, c_x, and c_0 it does not matter what those units are.

$$\frac{c_s - c_x}{c_s - c_0} = \frac{10^{20}\,\frac{atoms}{cm^3} - 10^{18}\,\frac{atoms}{cm^3}}{10^{20}\,\frac{atoms}{cm^3} - 0\,\frac{atoms}{cm^3}} = 0.99$$

$$= \mathrm{erf}\left(\frac{x}{2\sqrt{\left(6.5 \times 10^{-13} \frac{\mathrm{cm}^2}{\mathrm{s}}\right)(3600 \text{ s})}} \right)$$

$$= \mathrm{erf}\left(\frac{x}{9.67 \times 10^{-5}} \right)$$

From the error function values in Table 5-3 (or from your calculator/ computer),

If $\mathrm{erf}(z) = 0.99$, $z = 1.82$, therefore,

$$1.82 = \frac{x}{9.67 \times 10^{-5}}$$

or

$$x = 1.76 \times 10^{-4} \text{ cm}$$

or

$$x = (1.76 \times 10^{-4} \text{ cm})\left(\frac{10^4 \ \mu\mathrm{m}}{\mathrm{cm}} \right)$$

$$x = 1.76 \ \mu\mathrm{m}$$

Note that we have expressed the final answer in micrometers since this is the length scale that is appropriate for this application. The main assumptions we made are (i) the D value does not change while phosphorus (P) gets incorporated in the silicon wafer and (ii) the diffusion of P is only in one dimension (i.e., we ignore any lateral diffusion).

(b) When the wafer is cooled down to room temperature, the profile will essentially stay the same. Thus, at room temperature we will have essentially the same distribution of P as we had at 1100°C. The temperature has a strong effect on the diffusion process and, as we cool the wafer down, the diffusion rate drops substantially and hence the profile remains constant.

(c) Temperature and time have not been specified for diffusion of boron (B). These will have to be accounted for. Typical temperatures for boron (B) diffusion are between 900–1200°C and we can expect that at these temperatures P diffusion will also be significant. When we heat the wafer again for boron diffusion, we will have to make sure P does not begin to substantially diffuse into parts of the wafer (dopant redistribution). We also have to ensure that boron does not get diffused into regions containing P and if it does we have to account for the effects on processing and electrical properties of the resultant device.

Limitations to Applying the Error-Function Solution Given by Equation 5-7 Note that in the equation describing Fick's second law (Equation 5-7):

(a) It is assumed that D is independent of concentration of diffusing species;

(b) the surface concentration of diffusing species (c_s) is always constant.

There are situations under which these conditions may not be met and hence the concentration profile evolution will not be predicted by the error-function solution shown in Equation 5-7. If the boundary conditions are different from the ones we assumed, different solutions to Fick's second law must be used.

Equation 5-7 requires that there be a constant composition c_0 at the interface; this is the case in a process such as carburizing of steel (Examples 5-10 and 5-11), in which carbon is continuously supplied to the steel surface. In many cases, however, the surface concentration gradually changes during the process. In these cases, **interdiffusion** of atoms occurs, as shown in Figure 5-10, and Equation 5-7 is no longer valid. The term interdiffusion means diffusion of one species in one direction (e.g., carbon diffusing inwards from the surface) and diffusion of another species in the opposite direction (e.g., Fe or other atoms diffusing outward toward the surface).

Sometimes interdiffusion can cause difficulties. For example, when aluminum is bonded to gold at an elevated temperature, the aluminum atoms diffuse faster into the gold than gold atoms diffuse into the aluminum. Consequently, more total atoms eventually are on the original gold side of the interface than on the original aluminum side. This causes the physical location of the original interface to move towards the aluminum side of the diffusion couple. Any foreign particles originally trapped at the interface also move with the interface. This movement of the interface due to unequal diffusion rates is called the **Kirkendall effect**. In certain cases, voids form at the interface as a result of the Kirkendall effect. In tiny integrated circuits, gold wire is welded to aluminum to provide an external lead for the circuit. During operation of the circuit, voids may form by coalescence of vacancies involved in the diffusion process; as the voids grow, the Au-Al connection is weakened and eventually may fail. Because the area around the connection discolors, this premature failure is called the **purple plague**. One technique for preventing this problem is to expose the welded joint to hydrogen. The hydrogen dissolves in the aluminum, fills the vacancies, and prevents self-diffusion of the aluminum atoms. This keeps the aluminum atoms from diffusing into the gold and embrittling the weld.

In other cases, even if boundary conditions are the same as those used to derive Equation 5-7 or can be handled using alternate equations, the diffusion coefficient D may show concentration dependence.[6,12,13] Also note that since D has a strong temperature dependence, assuming a constant D assumes isothermal processing situations. In reality, the temperatures in processing of materials are rarely constant in the process. Sometimes complex defect chemistries can also play a major role in causing diffusion profiles to be substantially different from those predicted by Equation 5-7. As a result, it is not surprising that often we observe concentration profiles that are different from those predicted by the various solutions to Fick's second law. For example, diffusion profiles of silicon (Si) and some other dopants in GaAs cannot be described by the error function.[14] Complex defect chemistries influence the dopant diffusion, thereby causing D to be dependent upon the concentration of diffusing species.

5-9 Diffusion and Materials Processing

We briefly discussed applications of diffusion in processing materials in Section 5-1. Many important examples related to solidification, phase transformations, heat treatments, etc. will be discussed in later chapters. In this section, we provide more information to highlight the importance of diffusion in the processing of engineered materials. Diffusional processes become very important when materials are used or processed at elevated temperatures.

Melting and Casting   One of the most widely used methods to process metals, alloys, many plastics, and glasses involves melting and casting of materials into a desired shape. Diffusion plays a particularly important role in solidification of metals and alloys. During the growth of single crystals of semiconductors, for example, we must ensure that the differences in the diffusion of dopants in both the molten and solid forms are accounted for. This also applies for the diffusion of elements during the casting of alloys. Similarly, diffusion plays a critical role in the processing of glasses. In inorganic glasses, for instance, we rely on the fact that diffusion is slow and inorganic glasses do not crystallize easily. We will examine this topic further in Chapter 8.

Sintering   Although casting and melting methods are very popular for many manufactured materials, the melting points of many ceramic and some metallic materials are too high for processing by melting and casting. These relatively refractory materials are manufactured into useful shapes by a process that requires the consolidation of small particles of a powder into a solid mass (Chapter 14). **Sintering** is the high-temperature treatment that causes particles to join, gradually reducing the volume of pore space between them. Sintering is a frequent step in the manufacture of ceramic components (e.g., alumina, barium titanate etc.) as well as in the production of metallic parts by **powder metallurgy**—a processing route by which metal powders are pressed and sintered into dense, monolithic components. A variety of composite materials such as tungsten carbide-cobalt based cutting tools, superalloys, etc., are produced using this technique. With finer particles, many atoms or ions are at the surface for which the atomic or ionic bonds are not satisfied. As a result, a collection of fine particles of a certain mass has higher energy than that for a solid cohesive material of the same mass. Therefore, the driving force for solid state sintering of powdered metals and ceramics is the *reduction in the total surface area* of powder particles. When a powdered material is compacted into a shape, the powder particles are in contact with one another at numerous sites, with a significant amount of pore space between them. In order to reduce the total energy of the material, atoms diffuse to the points of contact, permitting the particles to be bonded together and eventually causing the pores to shrink.

Lattice diffusion from the bulk of the particles into the neck region causes densification. Surface diffusion, gas or vapor phase diffusion, and lattice diffusion from curved surfaces into the neck area between particles do not lead to densification (Chapter 14). If sintering is carried out over a long period of time, the pores may be eliminated and the material becomes dense (Figure 5-28). In Figure 5-29, particles of a powder of a ceramic material known as barium magnesium tantalate ($Ba(Mg_{1/3}Ta_{2/3})O_3$ or BMT) are shown. This ceramic material is useful in making electronic components known as

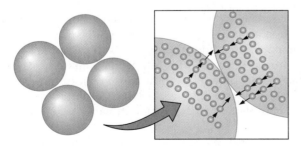
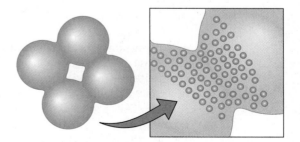

Compacted product   Partly sintered product

Figure 5-28   Diffusion processes during sintering and powder metallurgy. Atoms diffuse to points of contact, creating bridges and reducing the pore size.

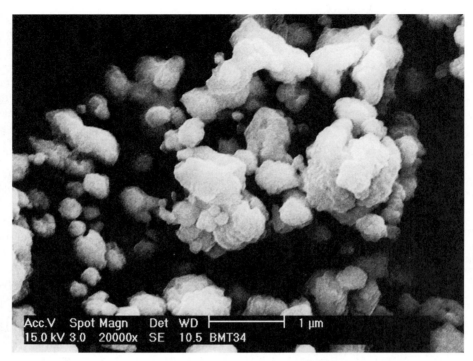

Figure 5-29   Particles of barium magnesium tantalate (BMT) (Ba(Mg$_{1/3}$ Ta$_{2/3}$)O$_3$) powder are shown. This ceramic material is useful in making electronic components known as dielectric resonators that are used for wireless communications. (*Courtesy of H. Shirey.*)

dielectric resonators used in wireless communication systems. The microstructure of BMT ceramics is shown in Figure 5-30. These ceramics were produced by compacting the powders in a press and sintering the compact at a high temperature (∼1500°C).

The extent and rate of sintering depends on (a) the initial density of the compacts, (b) temperature, (c) time, (d) mechanism of sintering, (e) the average particle size, and (f) the size distribution of the powder particles. In some situations, a liquid phase forms in localized regions of the material while sintering is in process. Since diffusion of species, such as atoms and ions, is faster in liquids than in the solid state, the presence of a liquid phase can provide a convenient way for accelerating the sintering of many refractory metal and ceramic formulations. The process in which a small amount of liquid forms and assists densification is known as **liquid phase sintering**. For the liquid phase to be effective in enhancing sintering it is important to have a liquid that can "wet" the grains, similar to how water wets a glass surface. If the liquid is non-wetting, similar to how mercury does not wet glass, then the liquid phase will not be helpful for enhancing sintering. In some cases, compounds are added to materials to cause the liquid phase to form at sintering temperatures. In some other situations, impurities can react with the material and cause formation of a liquid phase. In most applications, it is desirable if the liquid phase is transient or converted into a crystalline material during cooling. This way a glassy and brittle amorphous phase does not remain at the grain boundaries.

When exceptionally high densities are needed, pressure (either uniaxial or isostatic) is applied while the material is being sintered. These techniques are known as **hot pressing**, when the pressure is unidirectional, or **hot isostatic pressing** (HIP), when the pressure is isostatic (i.e., applied in all directions). Many superalloys and ceramics such

Figure 5-30 The microstructure of BMT ceramics obtained by compaction and sintering of BMT powders. (*Courtesy of H. Shirey.*)

as lead lanthanum zirconium titanate (*PLZT*) are processed using these techniques. Hot isostatic pressing leads to high density materials with isotropic properties (Chapter 14).

Grain Growth   A polycrystalline material contains a large number of grain boundaries, which represent a high-energy area because of the inefficient packing of the atoms. A lower overall energy is obtained in the material if the amount of grain boundary area is reduced by grain growth. **Grain growth** involves the movement of grain boundaries, permitting larger grains to grow at the expense of smaller grains (Figure 5-31). If you have watched froth, you have probably seen the principle of grain growth! Grain growth; is similar to the way smaller bubbles in the froth disappear at the expense of bigger bubbles. Another analogy is big fish getting bigger by eating small fish! For grain growth in materials, diffusion of atoms across the grain boundary is required, and,

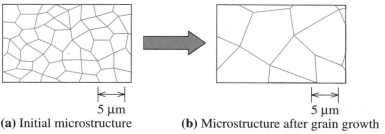

| $\leftrightarrow$ | | $\leftrightarrow$ |
| 5 μm | | 5 μm |
(a) Initial microstructure   **(b)** Microstructure after grain growth

Figure 5-31 Grain growth occurs as atoms diffuse across the grain boundary from one grain to another.

(a)   **(b)**

Figure 5-32   Grain growth in alumina ceramics can be seen from the SEM micrographs of alumina ceramics. (a) The left micrograph shows the microstructure of an alumina ceramic sintered at 1350°C for 15 hours. (b) The right micrograph shows a sample sintered at 1350°C for 300 hours. (*Courtesy of I. Nettleship and R. McAfee.*)

consequently, the growth of the grains is related to the activation energy needed for an atom to jump across the boundary. The increase in grain size can be seen from the sequence of micrographs for alumina ceramics shown in Figure 5-32.[15] Another example where grain growth plays a role is in the tungsten (W) filament in a light bulb. As the tungsten filament gets hotter, the grains grow causing it to get weaker. This grain growth, vaporization of tungsten, and oxidation via reaction with remnant oxygen contribute to the failure of tungsten filaments in a light bulb.

The **driving force** for grain growth is reduction in grain boundary area. Grain boundaries are defects and their presence causes the free energy of the material to increase. Thus, the thermodynamic tendency of polycrystalline materials is to transform into materials that have a larger average grain size. High temperatures or low activation energies increase the size of the grains. Many heat treatments of metals, which include holding the metal at high temperatures, must be carefully controlled to *avoid* excessive grain growth. This is because as the average grain size grows the grain boundary area decreases and there is now less resistance to motion of dislocations. As a result, the strength of a metallic material will decrease with increasing grain size. We have seen this concept before in the form of the Hall-Petch equation (Chapter 4). In **normal grain growth** the average grain size increases steadily and the width of the grain size distribution is not affected severely. In **abnormal grain growth**, the grain size distribution tends to become bi-modal (i.e., we get a few grains that are very large and then we have few grains that remain relatively small). Certain electrical, magnetic, and optical properties of materials also depend upon the grain size of materials. As a result, in the processing of these materials attention has to be paid to factors that affect diffusion rates and grain growth.

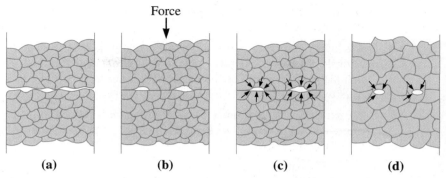

Figure 5-33 The steps in diffusion bonding: (a) Initially the contact area is small; (b) application of pressure deforms the surface, increasing the bonded area; (c) grain boundary diffusion permits voids to shrink; and (d) final elimination of the voids requires volume diffusion.

Diffusion Bonding A method used to join materials, **diffusion bonding** occurs in three steps (Figure 5-33). The first step forces the two surfaces together at a high temperature and pressure, flattening the surface, fragmenting impurities, and producing a high atom-to-atom contact area. As the surfaces remain pressed together at high temperatures, atoms diffuse along grain boundaries to the remaining voids; the atoms condense and reduce the size of any voids in the interface. Because grain boundary diffusion is rapid, this second step may occur very quickly. Eventually, however, grain growth isolates the remaining voids from the grain boundaries. For the third step—final elimination of the voids—volume diffusion, which is comparatively slow, must occur. The diffusion bonding process is often used for joining reactive metals such as titanium, for joining dissimilar metals and materials, and for joining ceramics.

SUMMARY

◆ The net flux of atoms, ions, etc. resulting from diffusion depends upon the initial concentration gradient.

◆ The kinetics of diffusion depend strongly on temperature. In general, diffusion is a thermally activated process and the dependence of the diffusion coefficient on temperature is given by the Arrhenius equation.

◆ The extent of diffusion depends on temperature, time, nature and concentration of diffusing species, crystal structure, and composition of the matrix, stoichiometry, and point defects.

◆ Encouraging or limiting the diffusion process forms the underpinning of many important technologies. Examples include the processing of semiconductors, heat treatments of metallic materials, sintering of ceramics and powdered metals, formation of amorphous materials, solidification of molten materials during a casting process, diffusion bonding, barrier plastics, films, and coatings.

◆ Magnetic, electric, convective, or electrochemical forces can also lead to the movement of particles or other species such as electrons or holes. We refer to these movements as drift. Battery technology, electrophoresis, electroplating, and the operation of microelectronic devices such as transistors make use of the drift of different species.

◆ Fick's laws describe the diffusion process quantitatively. Fick's first law defines the relationship between chemical potential gradient and the flux of diffusing species. Fick's second law describes the variation of concentration of diffusing species under nonsteady state diffusion conditions.

◆ For a particular system, the amount of diffusion is related to the term Dt. This term permits us to determine the effect of a change in temperature on the time required for a diffusion-controlled process.

◆ The two important mechanisms for atomic movement in crystalline materials are vacancy diffusion and interstitial diffusion. Substitutional atoms in the crystalline materials move by the vacancy mechanism.

◆ The rate of diffusion is governed by the Arrhenius relationship—that is, the rate increases exponentially with temperature. Diffusion is particularly significant at temperatures above about 0.4 times the melting temperature (in Kelvin) of the material.

◆ The activation energy Q describes the ease with which atoms diffuse, with rapid diffusion occurring for a low activation energy. A low activation energy and rapid diffusion rate are obtained for (1) interstitial diffusion compared with vacancy diffusion, (2) crystal structures with a smaller packing factor, (3) materials with a low melting temperature or weak atomic bonding, and (4) diffusion along grain boundaries or surfaces.

◆ The total movement of atoms, or flux, increases when the concentration gradient and temperature increase.

◆ Diffusion of ions in ceramics is usually slower than that of atoms in metallic materials. Diffusion in ceramics is also affected significantly by non-stoichiometry, dopants, and the possible presence of liquid phases during sintering.

◆ Diffusion in polymers can be measured through measurement of their permeability. Polymers are used as barrier layers in many applications such as the food industry, optical fibers, and microelectronic components.

◆ Atom diffusion is of paramount importance because many of the materials processing techniques, such as sintering, powder metallurgy, and diffusion bonding, require diffusion. Furthermore, many of the heat treatments and strengthening mechanisms used to control structures and properties in materials are diffusion-controlled processes. The stability of the structure and the properties of materials during use at high temperatures depend on diffusion.

GLOSSARY

Abnormal grain growth   A type of grain growth observed in metals and ceramics. In this mode of grain growth, a bimodal grain size distribution usually emerges as some grains become very large at the expense of smaller grains. See "Grain growth" and "Normal grain growth."

Activation energy   The energy required to cause a particular reaction to occur. In diffusion, the activation energy is related to the energy required to move an atom from one lattice site to another.

Annealing A heat treatment in which a material is held at a relatively high temperature for a long period of time and then cooled very slowly. Annealing is used in processing of metals, ceramic, and glasses. The purpose varies with application.

Carburization A heat treatment for steels to harden the surface using a gaseous or solid source of carbon. The carbon diffusing into the surface makes the surface harder and more abrasion resistant.

Concentration gradient The rate of change of composition with distance in a nonuniform material, typically expressed as $\frac{atoms}{cm^3 \cdot cm}$ or $\frac{at\%}{cm}$.

Conductive ceramics Ceramic materials that are good conductors of electricity as a result of their ionic and electronic charge carriers (electrons, holes, or ions). Examples of such materials are stabilized zirconia, indium tin oxide, etc.

Dielectric resonators Hockey puck-like pieces of ceramics such as barium magnesium tantalate (BMT) or barium zinc tantalate (BZN). These are used in wireless communication systems as low loss filters.

Diffusion The net flux of atoms, ions, or other species within a material caused by temperature and concentration gradient.

Diffusion bonding A joining technique in which two surfaces are pressed together at high pressures and temperatures. Diffusion of atoms to the interface fills in voids and produces a strong bond between the surfaces.

Diffusion coefficient (*D*) A temperature-dependent coefficient related to the rate at which atoms, ions, or other species diffuse. The diffusion coefficient depends on temperature, the composition and microstructure of the host material and also concentration of diffusing species.

Diffusion couple A combination of elements involved in diffusion studies (e.g., if we are considering diffusion of Al in Si, then Al-Si is a diffusion couple).

Diffusion distance The maximum or desired distance that atoms must diffuse; often, the distance between the locations of the maximum and minimum concentrations of the diffusing atom.

Diffusivity Another term for diffusion coefficient (*D*).

Dopant An element or a compound added deliberately, usually in known concentrations and known locations, so as to have a positive effect on the properties of a material. Examples are phosphorus (P) or boron (B) in silicon (Si), or CaO or Y_2O_3 in zirconia (ZrO_2).

Doping A process used for introducing dopant atoms (e.g., P, B) into semiconductors such as silicon. Solid, liquid or gaseous sources can be used as precursors for diffusing dopant atoms. Ion implantation is another technique used to introduce dopants in semiconductors.

Drift Movement of electrons, holes, ions or particles as a result of a gradient in temperature, electric, or magnetic field.

Driving force A cause that induces an effect. For example, an increased gradient in chemical potential enhances diffusion, similarly reduction in surface area of powder particles is the driving force for sintering.

Electrophoresis Movement of small particles (or relatively large molecules) under an electric field. The particles carry a net electrical charge.

Electroplating A process for reducing metal ions in solutions to produce metallic coatings on substrates.

Fick's first law The equation relating the flux of atoms by diffusion to the diffusion coefficient and the concentration gradient.

Fick's second law The partial differential equation that describes the rate at which atoms are redistributed in a material by diffusion. Many solutions exist to Fick's second law; Equation 5-7 is one possible solution.

Flux The number of atoms or other diffusing species passing through a plane of unit area per unit time. This is related to the rate at which mass is transported by diffusion in a solid.

Grain boundary diffusion Diffusion of atoms along grain boundaries. This is faster than volume diffusion, because the atoms are less closely packed in grain boundaries.

Grain growth Movement of grain boundaries by diffusion in order to reduce the amount of grain boundary area. As a result, small grains shrink and disappear and other grains become larger, similar to how large bubbles in soap froth become larger at the expense of smaller bubbles. In many situations, grain growth is not desirable.

Hole An imaginary particle in a semiconductor that represents a missing electron. The hole has a positive charge of 1.6×10^{-19} C.

Hot dip galvanizing A process in which a zinc coating is applied by dipping the components in a molten zinc bath. The process is used mainly for steel structures, car chassis, and other components to protect them from corrosion.

Hot isostatic pressing A sintering process in which an isostatic pressure is applied during sintering. This process is used for obtaining very high densities and isotropic properties.

Hot pressing A sintering process conducted under uniaxial pressure, used for achieving higher densities.

Ion implantation A technique that uses high energy ion beams for introducing dopant atoms in semiconductors.

Interdiffusion Diffusion of different atoms in opposite directions. Interdiffusion may eventually produce an equilibrium concentration of atoms within the material.

Interstitial diffusion Diffusion of small atoms from one interstitial position to another in the crystal structure.

Kirkendall effect Physical movement of an interface due to unequal rates of diffusion of the atoms within the material.

Liquid phase sintering A sintering process in which a liquid phase forms. Since diffusion is faster in liquids, if the liquid can wet the grains, it can accelerate the sintering process.

Nitriding A process in which nitrogen is diffused into the surface of a material, such as a steel, leading to increased hardness and wear resistance.

Normal grain growth Grain growth that occurs in an effort to reduce grain boundary area. This type of grain growth is to be distinguished from abnormal grain growth in that the grain size distribution remains unimodal but the average grain size increases steadily.

Permeability A relative measure of the diffusion rate in materials, often applied to plastics and coatings. It is often used as an engineering design parameter that describes the effectiveness of a particular material to serve as a barrier against diffusion.

p-n junction A region of a semiconductor material one side of which is doped with p-type dopants (e.g., B in Si) and the other side is doped with n-type dopants (e.g., P in Si). The p-n junction forms the building block for microelectronic devices such as transistors.

Powder metallurgy   A method for producing monolithic metallic parts; metal powders are compacted into a desired shape, which is then heated to allow diffusion and sintering to join the powders into a solid mass.

Precursor   A chemical used in a thin film or other deposition process for introducing atoms in a material.

Purple plague   Formation of voids in gold-aluminum welds due to unequal rates of diffusion of the two atoms; eventually failure of the weld can occur.

Self-diffusion   The random movement of atoms within an essentially pure material. No net change in composition results.

Sintering   A high-temperature treatment used to join small particles. Diffusion of atoms to points of contact causes bridges to form between the particles. Further diffusion eventually fills in any remaining voids. The driving force for sintering is a reduction in total surface area of the powder particles.

Sputtering   A deposition process for making thin films. Targets of elements or compounds whose films are to be made are bombarded with ions. The ions that sputter off from these targets are deposited onto a substrate.

Surface diffusion   Diffusion of atoms along surfaces, such as cracks or particle surfaces.

Thermal barrier coatings (TBC)   Coatings used to protect a component from heat. For example, some of the turbine blades in an aircraft engine are made from nickel-based superalloys and are coated with ytrria stabilized zirconia (YSZ).

Uphill diffusion   A diffusion process in which species move from regions of lower concentration to higher concentration. The process is still consistent with Fick's law (i.e., species diffuse from higher chemical potential to lower chemical potential). The diffusion is, however, "uphill" if we compare nominal concentrations and not chemical potentials (i.e., diffusion in this case does occur such that species diffuse from lower concentration to higher concentration).

Vacancy diffusion   Diffusion of atoms when an atom leaves a regular lattice position to fill a vacancy in the crystal. This process creates a new vacancy and the process continues.

Volume diffusion   Diffusion of atoms through the interior of grains.

ADDITIONAL INFORMATION

1. PACKAN, P., "Scaling Transistors into the Deep-Submicron Regime," *MRS Bulletin*, 2000. 25(6): p. 18–21.

2. KLEMMER, T., Seagate, *Technical discussions and assistance is acknowledged*, 2002.

3. YOUNG, C. and B. DUGAN, Zinc Corporation of America, *Technical discussions and assistance is acknowledged*, 2002.

4. PETTIT, F. and G. MEIER, *Technical assistance and discussions are acknowledged*, 2002.

5. GÖSELE, U., "Diffusion and Diffusion-Induced Defects in Silicon," *Encyclopedia of Advanced Materials*, R.J.B.D. BLOOR, M.C. FLEMINGS, and S. MAHAJAN, Editors. 1994: Pergamon Press. p. 629–635.

6. CHIANG, Y.M., D. BIRNIE, and W.D. KINGERY, *Physical Ceramics: Principles for Ceramic Science and Engineering*, 1997: Wiley.

7. REED-HILL, R.E. and R. ABBASCHIAN, *Physical Metallurgy Principles*, Third Edition 1991: PWS.

8. WILKINSON, D.S., *Mass Transport in Solids and Fluids*, 2000: Cambridge University Press.

9. EROR, N.G., *Technical discussions are acknowledged*, 2002.

10. OISHI, Y. and K. ANDO, in *Transport in Nonstoichiometric Compounds*, G. SIMKOVICH and U.S. STUBICAN, Editors: Plenum, p. 188–202.

11. ROSATO, D., D. DIMATTIA, and D. ROSATO, *Designing with Plastics and Composites*, 1991: New York, NY, Van Nostrand Reinhold.

12. CRANK, J., *The Mathematics of Diffusion*, Second Edition, 1980: Oxford, Clarendon Press.

13. MIDDLEMAN, S. and A.K. HOCHBERG, *Process Engineering Analysis in Semiconductor Device Fabrication*. 1993: McGraw Hill.

14. YAN, T.Y., "Diffusion in Gallium Arsenide and Aluminum Arsenide-Gallium Arsenide Materials," in *Encyclopedia of Advanced Materials*, R.J.B.D. BLOOR, M.C. FLEMINGS, and S. MAHAJAN, Editor, 1994: Pergmaon Press. p. 635–641.

15. MCAFEE, R. and I. NETTLESHIP, *Technical discussions are acknowledged*, 2002.

16. DARKEN, L., *Trans. AIME*, 1949. 180: p. 430.

17. KANO, K., *Semiconductor Devices*, 1998: Prentice Hall. p. 168–170.

 PROBLEMS

Section 5-1 Applications of Diffusion

5-1 What is the driving force for diffusion?

5-2 In the carburization treatment of steels, what are the diffusing species?

5-3 Why do we use PET plastic to make carbonated beverage bottles?

5-4 What is hot dip galvanizing? What are the applications of this process?

5-5 Why is that aluminum metal oxidizes more readily than iron but aluminum is considered to be a metal that usually does not "rust?"

5-6 What is electrophoresis? What is the driving force for this phenomenon? What are the applications?

5-7 What is a thermal barrier coating? Where are such coatings used?

5-8 What is the difference between diffusion and drift of charge carriers in a semiconductor? What is the driving force for each?

5-9 Draw a sketch showing which way the electrons and holes diffuse and drift when p- and n-type semiconductors form a p-n junction.

Section 5-2 Stability of Atoms and Ions

5-10 For what did Svante Arrhenius win the Nobel prize?

5-11 Write down the Arrhenius equation and explain the different terms.

5-12 Atoms are found to move from one lattice position to another at the rate of $5 \times 10^5 \frac{\text{jumps}}{\text{s}}$ at 400°C when the activation energy for their movement is 30,000 cal/mol. Calculate the jump rate at 750°C.

5-13 The number of vacancies in a material is related to temperature by an Arrhenius equation. If the fraction of lattice points containing vacancies is 8×10^{-5} at 600°C, determine the fraction of lattice points at 1000°C.

5-14 The Arrhenius equation was originally developed for comparing rates of chemical reactions. Compare the rates of a chemical reaction at 20 and 100°C by calculating the ratio of the chemical reaction rates. Assume that activation energy for liquids in which the chemical reaction is conducted is 10 kJ/mol and that the reaction is limited by diffusion.

Section 5-3 Mechanisms for Diffusion

5-15 What are the different mechanisms for diffusion?

5-16 Why is it that the activation energy for diffusion via the interstitial mechanism is smaller than those for other mechanisms?

5-17 How is self-diffusion of atoms in metals verified experimentally?

5-18 Examine Figure 5-19. Based on this it can be stated that the diffusion of B, P, As, and Sb is always faster than Si self diffusion. What does this say about the possible mechanisms of diffusion of these elements?

5-19 From Figure 5-19, what is the most likely mechanism of diffusion for Ni, Cu, and H in silicon?

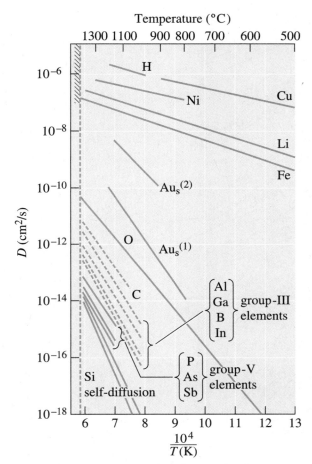

Figure 5-19 (Repeated for Problems 5-18, 5-19, 5-21, 5-22, and 5-23) Diffusion coefficients for different dopants in silicon. (*Source: From "Diffusion and Diffusion Induced Defects in Silicon," by U. Gösele. In R. Bloor, M. Flemings, and S. Mahajan (Eds.),* Encyclopedia of Advanced Materials, *Vol. 1, 1994, p. 631, Fig. 2. Copyright © 1994 Pergamon Press. Reprinted with permission of the authors.*)

Section 5-4 Activation Energy for Diffusion

5-20 Activation energy is sometimes expressed as (eV/atom). For example, see Figure 5-21 illustrating the diffusion coefficients of ions in different oxides. Convert eV/atom into Joules/mole.

5-21 Calculate the activation energy of diffusion of P in Si (Figure 5-19).

5-22 Calculate the activation energy of diffusion of B in Si (Figure 5-19).

5-23 Calculate the activation energy of diffusion of H in Si (Figure 5-19).

5-24 The diffusion coefficient for Cr^{+3} in Cr_2O_3 is 6×10^{-15} cm^2/s at 727°C and is 1×10^{-9} cm^2/s at 1400°C. Calculate

(a) the activation energy; and
(b) the constant D_0.

5-25 The diffusion coefficient for O^{-2} in Cr_2O_3 is 4×10^{-15} cm^2/s at 1150°C, and 6×10^{-11} cm^2/s at 1715°C. Calculate

(a) the activation energy; and
(b) the constant D_0.

5-26 Without referring to the actual data, can you predict whether the activation energy for diffusion of carbon in FCC iron will be higher or lower than that in BCC iron? Explain.

5-27 Using data on diffusion coefficients of different ions in oxides (Figure 5-21), calculate the activation energy of diffusion of O^{-2} in calcia (CaO) stabilized zirconia. Assume that the CaO is 14 mole percent.

Section 5-5 Rate of Diffusion (Fick's First Law)

5-28 Who was the person to have experimented with implanting contact lenses in human eyes?

5-29 Write down Fick's first law of diffusion. Clearly explain what each term means.

5-30 What is the difference between diffusivity and diffusion coefficient?

Section 5-6 Factors Affecting Diffusion

5-31 Write down the equation that describes the dependence of D on temperature.

5-32 Explain briefly the dependence of D on the concentration of diffusing species.

5-33 What does the term "uphill diffusion" mean?

5-34 Darken experimented with a diffusion couple of two alloys. The first alloy had 0.44% C and balance was iron.[8,16]. The other alloy had 0.48% C, 3.8% silicon and the balance was iron. After the diffusion couple was heated to 1050°C for several days, he found that carbon actually had diffused from lower carbon concentration to higher carbon concentration sample. Explain how carbon

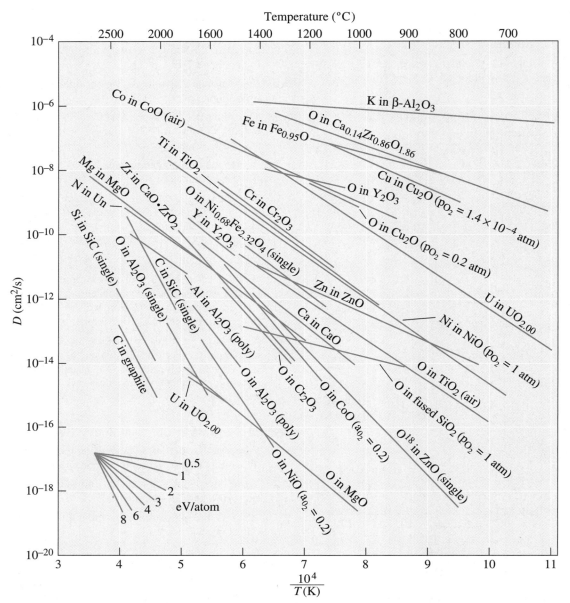

Figure 5-21 (Repeated for Problems 5-20, 5-27, 5-38, and 5-39) Diffusion coefficients of ions in different oxides. (*Source: Adapted from* Physical Ceramics: Principles for Ceramic Science and Engineering, *by Y.M. Chiang, D. Birnie, and W.D. Kingery, Fig. 3-1. Copyright © 1997 John Wiley & Sons. Adapted with permission.*)

can diffuse from lower concentrations to higher concentrations.

5-35 Why is it that the diffusion coefficients expected from defect chemistry equations for yttria stabilized zirconia do not match well with the experimental data?

5-36 Explain how a stabilized zirconia composition works as an oxygen sensor used in cars.

5-37 Write down the defect chemistry equation for incorporation of CaO into ZrO_2.

5-38 Using data on diffusion coefficients of different ions in oxides (Figure 5-21), calculate the activation energy of diffusion of K ions in beta alumina. Materials such as beta alumina show rapid transport of ions and have been considered for batteries based on sodium and sulfur.

5-39 Compare the diffusion coefficients of oxygen ions in single crystal and polycrystalline alumina (Figure 5-21). What can be said about the mechanism of diffusion of ions in these materials based on the difference in these values?

5-40 In solids the process of diffusion of atoms and ions takes time. Explain how this is used to our advantage while forming metallic glasses.

5-41 Why is it that inorganic glasses form upon relatively slow cooling of melts, while rapid solidification is necessary to form metallic glasses?

5-42 A 0.2-mm thick wafer of silicon is treated so that a uniform concentration gradient of antimony is produced. One surface contains 1 Sb atom per 10^8 Si atoms and the other surface contains 500 Sb atoms per 10^8 Si atoms. The lattice parameter for Si is 5.407 Å (Appendix A). Calculate the concentration gradient in

(a) atomic percent Sb per cm; and
(b) Sb $\frac{atoms}{cm^3 \cdot cm}$

5-43 When a Cu-Zn alloy solidifies, one portion of the structure contains 25 atomic percent zinc and another portion 0.025 mm away contains 20 atomic percent zinc. The lattice parameter for the FCC alloy is about 3.63×10^{-8} cm. Determine the concentration gradient in

(a) atomic percent Zn per cm;
(b) weight percent Zn per cm; and
(c) Zn $\frac{atoms}{cm^3 \cdot cm}$

5-44 A 0.001-in. BCC iron foil is used to separate a high hydrogen gas from a low hydrogen gas at 650°C. 5×10^8 H atoms/cm³ are in equilibrium on one side of the foil, and 2×10^3 H atoms/cm³ are in equilibrium with the other side. Determine

(a) the concentration gradient of hydrogen; and
(b) the flux of hydrogen through the foil.

5-45 A 1-mm sheet of FCC iron is used to contain nitrogen in a heat exchanger at 1200°C. The concentration of N at one surface is 0.04 atomic percent and the concentration at the second surface is 0.005 atomic percent. Determine the flux of nitrogen through the foil in N atoms/cm² · s.

5-46 A 4-cm-diameter, 0.5-mm-thick spherical container made of BCC iron holds nitrogen at 700°C. The concentration at the inner surface is 0.05 atomic percent and at the outer surface is 0.002 atomic percent. Calculate the number of

grams of nitrogen that are lost from the container per hour.

5-47 A BCC iron structure is to be manufactured that will allow no more than 50 g of hydrogen to be lost per year through each square centimeter of the iron at 400°C. If the concentration of hydrogen at one surface is 0.05 H atom per unit cell and is 0.001 H atom per unit cell at the second surface, determine the minimum thickness of the iron.

5-48 Determine the maximum allowable temperature that will produce a flux of less than 2000 H atoms/cm² · s through a BCC iron foil when the concentration gradient is $-5 \times 10^{16} \frac{atoms}{cm^3 \cdot cm}$. (Note the negative sign for the flux.)

Section 5-7 Permeability of Polymers

5-49 What are barrier polymers?

5-50 What parameter defines the barrier properties of polymers?

5-51 What factors, other than permeability, are important in selecting a polymer for making plastic bottles?

5-52 Amorphous PET is more permeable to CO_2 than PET that contains micro-crystallites. Explain why.

5-53 Explain why a rubber balloon filled with helium gas deflates over time.

5-54 Compared to rubber balloons, why do Mylar™ balloons deflate more slowly?

Section 5-8 Composition Profile (Fick's Second Law)

5-55 Consider a 2-mm-thick silicon (Si) wafer to be doped using antimony (Sb). Assume that the dopant source (gas mixture of antimony chloride and other gases) provides a constant concentration of 10^{22} atoms/m³. If we need a dopant profile such that the concentration of Sb at a depth of 1 micrometer is 5×10^{21} atoms/m³. What will be the time for the diffusion heat treatment? Assume that the silicon wafer to begin with contains no impurities or dopants. Assume the activation energy for diffusion of Sb in silicon is 380 kJ/mole and D_0 for Sb diffusion in Si is 1.3×10^{-3} m²/s.

5-56 Consider doping of gallium (Ga) in Si. Assume that the diffusion coefficient of gallium (Ga) in Si at 1100°C is 7×10^{-13} cm²/s. Calculate the concentration of Ga at a depth of 2.0 micrometer, if the surface concentration of Ga is 10^{23} atoms/cm^3. The diffusion times are 1, 2, and 3 hours.

5-57 In semiconductor processing we want to selectively dope different regions of the semiconductor substrate. When we diffuse P into silicon, we want to introduce P in certain region. However, we do not want to diffuse P into certain other regions. Explain how this can be achieved. (*Hint*: Diffusion of dopant ions is much slower in silicon oxide than in silicon.)

5-58 The diffusion coefficient of phosphorous (P) in Si is 6.5×10^{-13} cm^2/s at a temperature of 1100°C. Assume the source provides a surface concentration of 10^{21} atoms/cm^3 and the diffusion time is two hours. Assume that the silicon wafer contains no P to begin with. Calculate the concentrations of P at depths of 0.5, 1.0, and 2.0 micrometers.

5-59 The electrical conductivity of Mn_3O_4 is 8×10^{-18} $ohm^{-1}\text{-}cm^{-1}$ at 140°C and is 1×10^{-7} $ohm^{-1}\text{-}cm^{-1}$ at 400°C. Determine the activation energy that controls the temperature dependence of conductivity. Explain the process by which the temperature controls conductivity.

5-60 Compare the rate at which oxygen ions diffuse in alumina (Al_2O_3) with the rate at which aluminum ions diffuse in Al_2O_3 at 1500°C. Explain the difference.

5-61 Compare the diffusion coefficients of carbon in BCC and FCC iron at the allotropic transformation temperature of 912°C and explain the difference.

5-62 Compare the diffusion coefficients for hydrogen and nitrogen in FCC iron at 1000°C and explain the difference in their values.

5-63 What is carburizing? Explain why this process is expected to cause an increase in the hardness of the surface of plain carbon steels?

5-64 What is nitriding heat treatment?

5-65 A certain mechanical component is being heat treated using carburization. A common engineering problem encountered is that we need to machine a certain part of the component and this part of the surface should not be hardened. Explain how we can achieve this objective.

5-66 A carburizing process is carried out on a 0.10% C steel by introducing 1.0% C at the surface at 980°C, where the iron is FCC. Calculate the carbon content at 0.01 cm, 0.05 cm, and 0.10 cm beneath the surface after 1 h.

5-67 Iron containing 0.05% C is heated to 912°C in an atmosphere that produces 1.20% C at the surface and is held for 24 h. Calculate the carbon content at 0.05 cm beneath the surface if

(a) the iron is BCC; and
(b) the iron is FCC. Explain the difference.

5-68 What temperature is required to obtain 0.50% C at a distance of 0.5 mm beneath the surface of a 0.20% C steel in 2 h, when 1.10% C is present at the surface? Assume that the iron is FCC.

5-69 A 0.15% C steel is to be carburized at 1100°C, giving 0.35% C at a distance of 1 mm beneath the surface. If the surface composition is maintained at 0.90% C, what time is required?

5-70 A 0.02% C steel is to be carburized at 1200°C in 4 h, with a point 0.6 mm beneath the surface reaching 0.45% C. Calculate the carbon content required at the surface of the steel.

5-71 A 1.2% C tool steel held at 1150°C is exposed to oxygen for 48 h. The carbon content at the steel surface is zero. To what depth will the steel be decarburized to less than 0.20% C?

5-72 A 0.80% C steel must operate at 950°C in an oxidizing environment where the carbon content at the steel surface is zero. Only the outermost 0.02 cm of the steel part can fall below 0.75% C. What is the maximum time that the steel part can operate?

5-73 A steel with BCC crystal structure containing 0.001% N is nitrided at 550°C for 5 h. If the nitrogen content at the steel surface is 0.08%, determine the nitrogen content at 0.25 mm from the surface.

5-74 What time is required to nitride a 0.002 N steel to obtain 0.12% N at a distance of 0.002 in. beneath the surface at 625°C? The nitrogen content at the surface is 0.15%.

5-75 We can successfully perform a carburizing heat treatment at 1200°C in 1 h. In an effort to reduce the cost of the brick lining in our furnace, we propose to reduce the carburizing temperature to 950°C. What time will be required to give us a similar carburizing treatment?

5-76 What is meant by the term "Kirkendall effect?"

5-77 What does the term "purple plague" mean?

Section 5-9 Diffusion and Materials Processing

5-78 Arrange the following materials in increasing order of self-diffusion coefficient: Ar gas, water, single crystal aluminum, and liquid aluminum at 700°C.

5-79 Most metals and alloys can be processed using the melting and casting route, but we do not prefer to apply this method for the processing of specific metals (e.g., W) and most ceramics. Explain.

5-80 What is sintering? What is the driving force for sintering?

5-81 What does the term powder metallurgy mean?

5-82 What is meant by liquid phase sintering? How does the liquid phase form?

5-83 What happens to the liquid phase after sintering is completed?

5-84 Why does grain growth occur? What is meant by the terms normal and abnormal grain growth?

5-85 Why is the strength of many metallic materials expected to go down with increasing grain size?

5-86 During freezing of a Cu-Zn alloy, we find that the composition is nonuniform. By heating the alloy to 600°C for 3 hours, diffusion of zinc helps to make the composition more uniform. What temperature would be required if we wished to perform this homogenization treatment in 30 minutes?

5-87 A ceramic part made of MgO is sintered successfully at 1700°C in 90 minutes. To minimize thermal stresses during the process, we plan to reduce the temperature to 1500°C. Which will limit the rate at which sintering can be done: diffusion of magnesium ions or diffusion of oxygen ions? What time will be required at the lower temperature?

5-88 A Cu-Zn alloy has an initial grain diameter of 0.01 mm. The alloy is then heated to various temperatures, permitting grain growth to occur. The times required for the grains to grow to a diameter of 0.30 mm are:

Temperature (°C)	Time (minutes)
500	80,000
600	3,000
700	120
800	10
850	3

Determine the activation energy for grain growth. Does this correlate with the diffusion of zinc in copper? (*Hint*: Note that rate is the reciprocal of time.)

5-89 What are the advantages of using hot pressing and hot isostatic pressing, compared to using normal sintering?

5-90 How does a tungsten filament in an incandescent light bulb fail?

5-91 A sheet of gold is diffusion-bonded to a sheet of silver in 1 h at 700°C. At 500°C, 440 h are required to obtain the same degree of bonding, and at 300°C, bonding requires 1530 years. What is the activation energy for the diffusion bonding process? Does it appear that diffusion of gold or diffusion of silver controls the bonding rate? (*Hint*: Note that rate is the reciprocal of time.)

✳ Design Problems

5-92 Design a spherical tank, with a wall thickness of 2 cm that will assure that no more than 50 kg of hydrogen will be lost per year. The tank, which will operate at 500°C, can be made of nickel, aluminum, copper, or iron. The diffusion coefficient of hydrogen and the cost per pound for each available material is listed below:

Diffusion Data

Material	D_0 (cm²/s)	Q cal/mol	Cost ($/lb)
Nickel	0.0055	8,900	4.10
Aluminum	0.16	10,340	0.60
Copper	0.011	9,380	1.10
Iron (BCC)	0.0012	3,600	0.15

5-93 A steel gear initially containing 0.10% C is to be carburized so that the carbon content at a depth of 0.05 in. is 0.50% C. We can generate a carburizing gas at the surface that contains anywhere from 0.95% C to 1.15% C. Design an appropriate carburizing heat treatment.

5-94 When a valve casting containing copper and nickel solidifies under nonequilibrium conditions, we find that the composition of the alloy varies substantially over a distance of 0.005 cm. Usually we are able to eliminate this concentration difference by heating the alloy for 8 h at 1200°C; however, sometimes this treatment causes the alloy to begin to melt, destroying the part. Design a heat treatment that will permit elimination of the nonuniformity without danger of melting. Assume that the cost of operating the furnace per hour doubles for each 100°C increase in temperature.

5-95 Assume that the surface concentration of phosphorus (P) being diffused in Si is 10^{21} atoms/cm^3. We need to design a process, known as the pre-deposition step, such that the concentration of P (c_1 for step 1) at a depth of 0.25 micrometer is 10^{13} atoms/cm^3. Assume that this is conducted at a temperature of 1000°C and the diffusion coefficient of P in Si at this temperature is 2.5×10^{-14} cm^2/s. Assume that the process is carried out for a total time of 8 minutes. Calculate the concentration profile (i.e., C as a function of depth, which in this case is given by the following equation). Notice the use of the complementary error function.[17]

$$c_1(x, t_1) = c_s \left[1 - \text{erf}\left(\frac{x}{4Dt} \right) \right]$$

Use different values of x to generate and plot this profile of P during the pre-deposition step.

█ Computer Problems

5-96 *Calculation of Diffusion Coefficients.* Write a computer program that will ask the user to provide the data for activation energy Q and the value of D_0. The program should then ask the user to input a valid range of temperatures. The program, when executed, provide the values of D as a function of temperature in increments chosen by the user. The program should convert the units of Q, D_0 and temperature in proper units. For example, if the user provides temperatures in °F the program should check and recognize that and convert the temperature into K. The program should also carry a cautionary statement about the standard assumptions made. For example, the program should caution the user that effects of any possible changes in crystal structures in the temperature range specified are not accounted for. Check the validity of your programs using examples in the book and also other problems that you may have solved using a calculator.

5-97 *Comparison of Reaction Rates.* Write a computer program that will ask the user to input the activation energy for a chemical reaction. The program should then ask the user to provide two temperatures for which the reaction rates need to be compared. Using the value of the gas constant and activation energy the program should then provide a ratio of the reaction rates. The program should take into account different units for activation energy.

5-98 *Carburization Heat Treatment.* The program should ask the user to provide an input for the carbon concentration at the surface (c_s), and the concentration of carbon in the bulk (c_0). The program should also ask the user to provide temperature and value for D (or values for D_0 and Q, that will allow for D to be calculated). The program should then provide an output of the concentration profile in a tabular form. The distances at which concentrations of carbon are to be provided can be provided by the user or defined by the person writing the program. The program should also be able to handle calculation of heat treatment times if the user provides a level of concentration that is needed at a specified depth. This program will require calculation of the error function. The programming language or spreadsheet you use may have an inbuilt function that calculates the error function. If that is the case, use that function. You can also calculate the error function as expansion of the following series:

$$\text{erf}(z) = 1 - \left[\frac{1}{\sqrt{\pi}} e^{-z^2} \left(\frac{1}{z} - \frac{1}{2z^2} + \frac{1 \times 3}{2^2} \frac{1}{z^5} - \cdots \right) \right]$$

or use an approximation

$$\text{erf}(z) = 1 - \left(\left[\frac{1}{\sqrt{\pi}} \right] \frac{e^{-z^2}}{z} \right)$$

In these equations z is the argument of the error function. Also, under certain situations you will know the value of the error function (if all concentrations are provided) and you will have to figure out the argument of the error function. This can be handled by having part of the program try and compare different values for the argument of the error function and by minimizing the difference between the value of the error function you require and the value of error function that was approximated.

5-99 *Diffusion of Dopants in Semiconductors.* Develop a computer program that will ask the user to provide surface concentration for a dopant. The program should then ask the user to provide the diffusion coefficient for the dopant and a temperature value. Then, similar to the previous problem, the program should generate a depth of profile for the concentration of the diffused dopant for a given time. Alternatively, the program should also be able to calculate how much time would be needed to accomplish a certain concentration of dopant at a given depth.

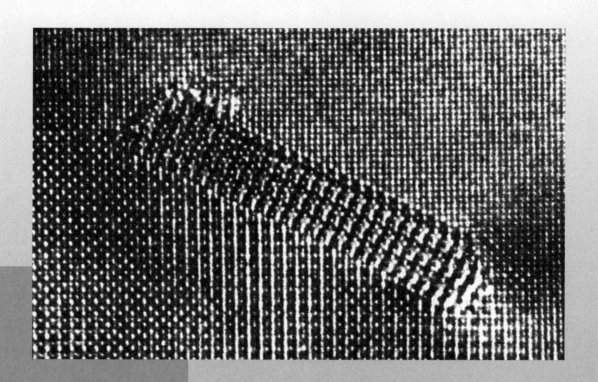

An Al_2MgCu precipitate is shown at the interface of an aluminum matrix (upper left) and an Al_3Li phase (lower right). The individual atoms in each phase are visible using atomic resolution microscopy. Precipitates such as these can significantly increase the strength of many alloys without affecting their density. (*Courtesy of V. Radmilovic and G.J. Shiflet, University of Virginia.*)

PART 2

Controlling the Microstructure and Mechanical Properties of Materials

In this part, we examine several methods used to control the structure and mechanical properties of materials. Three of these processes—grain-size strengthening, solid-solution strengthening, and strain hardening—rely on introducing and controlling the imperfections in crystal structures discussed (Chapter 4).

We also discuss strengthening by creating multiple-phase materials, where each phase has a different composition or crystal structure. The interface between the phases provides strengthening by interfering with the dislocation motion in metallic materials. Dispersion strengthening, age hardening, and a variety of phase transformations—which often rely on allotropic transformations—enable us to control the size, shape, and distribution of the phases in a metallic material. The processing of the material, such as casting, deformation processing, and heat treatment, is also employed to control microstructure and properties.

Before examining the strengthening mechanisms, we first briefly examine the mechanical properties and testing of materials.

Chapter

Some materials can become brittle when temperatures are low and/or strain rates are high. The special chemistry of the steel used on the *Titanic* and the stresses associated in the fabrication and embitterment of this steel when subjected to lower temperatures have been identified as factors contributing to the failure of the ship's hull. (*Courtesy of Getty Images.*)

6

Mechanical Properties and Behavior

Have You Ever Wondered?

- *Why Silly Putty® can be stretched a considerable amount when pulled slowly, but snaps when pulled fast?*

- *Why is it that glass fibers of different lengths have different strengths?*

- *Why we can load the weight of a fire truck on four ceramic coffee cups, however, ceramic cups tend to break easily when we drop them on the floor?*

- *What materials related factors played an important role in the sinking of the* Titanic*?*

- *What factors played a major role in the 1986* Challenger *space shuttle accident?*

- *Why does latex paint feel thinner when you stir it?*

- *Why do some metals and plastics become brittle at low temperatures?*

- *Why do aircrafts have a finite service life?*

The mechanical properties of materials depend on their composition and microstructure. In Chapters 2, 3, and 4, we learned that a material's composition, nature of bonding, crystal structure, and defects such as dislocations, grain size, etc., have a profound influence on the strength and ductility of metallic materials. In this chapter, we will begin to evaluate other factors that affect the mechanical properties of materials, such as how lower temperatures can cause many metals and plastics to become brittle. Lower temperatures contributed to the brittleness of the plastic used for the O-rings, causing the 1986 *Challenger* accident.[1] Similarly, the special chemistry of the steel used on the *Titanic* and the stresses associated in the fabrication and embrittlement of this steel when subjected to lower temperatures have been identified as factors contributing to the failure of the ship's hull.[2,3] Some researchers have shown that weaker rivets and design flaws also contributed to the failure.

The main goal of this chapter is to introduce the basic concepts associated with mechanical properties. We will learn basic terms such as hardness, stress, strain, elastic and plastic deformation, viscoelasticity, strain rate, fracture toughness, fatigue, creep, etc. We will also review some of the basic testing procedures that engineers use to evaluate many of these properties. These concepts will be discussed using illustrations from real-world applications.

6-1 Technological Significance

With many of today's emerging technologies, the primary emphasis is on the mechanical properties of the materials used. For example, in aircraft manufacturing, aluminum alloys or carbon-reinforced composites used for aircraft components must be lightweight, strong, and able to withstand cyclic mechanical loading for a long and predictable period of time (Figure 6-1). Steels used in the construction of structures such as buildings and bridges must have adequate strength so that these structures can be built without compromising safety. The plastics used for manufacturing pipes, valves, floor-

Figure 6-1 Aircraft, such as the one shown here, makes use of aluminum alloys and carbon-fiber-reinforced composites. (*Courtesy of Getty Images.*)

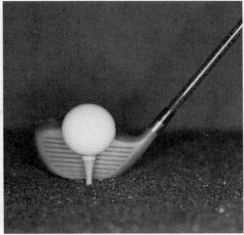

Figure 6-2 The materials used in sports equipment must be lightweight, stiff, tough, and impact resistant. (*Courtesy of Getty Images.*)

ing, and the like also must have adequate mechanical strength. Materials such as pyrolytic graphite or cobalt chromium tungsten alloys, used for prosthetic heart valves, must not fail.[4] Similarly, the performance of baseballs, cricket bats, tennis rackets, golf clubs, skis, and other sport equipment depends not only on the strength and weight of the materials used, but also on their ability to perform under an "impact" loading (Figure 6-2). The importance of mechanical properties is easy to appreciate in many of these "load-bearing" applications.

In Chapters 1 and 2, we learned that advanced engineered materials are used for technologies that capitalize on the electronic, magnetic, optical, biological, and other properties. In most of these applications, the mechanical properties of the material also play an important role. For example, an optical fiber must have a certain level of strength to withstand the stresses encountered in its application. A biocompatible titanium alloy used for a bone implant must have enough strength and toughness to survive in the human body for many years without failure. A scratch-resistant coating on optical lenses must resist mechanical abrasion. An aluminum alloy or a glass-ceramic substrate used as a base for building magnetic hard drives must have sufficient mechanical strength so that it will not break or crack during operation that requires rotation at high speeds. Similarly, electronic packages used to house semiconductor chips and the thin-film structures created on the semiconductor chip must be able to withstand stresses encountered in various applications, as well as those encountered during the heating and cooling of electronic devices. The mechanical robustness of small devices prepared using micro-electro mechanical systems (*MEMS*) and nano-technology is also important. Float glass used in automotive and building applications must have sufficient strength and shatter resistance. Many components designed from plastics, metals, and ceramics must not only have adequate toughness and strength at room temperature but also at relatively high and low temperatures. The point we want to make is very simple. The mechanical properties of a material and that of a component are critical in many applications in which the functional emphasis may be on the electronic, optical, magnetic, biological or some other properties.

For load-bearing applications, engineered materials are selected by matching their mechanical properties to the design specifications and service conditions required of the component. The first step in the selection process requires an analysis of the material's application to determine its most important characteristics. Should it be strong, stiff, or ductile? Will it be subjected to an application involving high stress or sudden intense force, high stress at elevated temperature, cyclic stresses, corrosive or abrasive conditions? Once we know the required properties, we can make a preliminary selection of the appropriate material using various databases. We must, however, know how the properties listed in the handbook are obtained, know what the properties mean, and realize that the properties listed are obtained from idealized tests that may not apply exactly to real-life engineering applications. Materials with the same nominal chemical composition and other properties can show significantly different mechanical properties as dictated by microstructure. Furthermore, changes in temperature; the cyclical nature of stresses applied; the chemical changes due to oxidation, corrosion, or erosion; microstructural changes due to temperature; the effect of possible defects introduced during machining operations (e.g., grinding, welding, cutting, etc.); or other factors can also have a major effect on the mechanical behavior of materials. A competent professional engineer will be aware of these possibilities and be able to take these into account, along with safety, cost, environmental impact, and other requirements when designing and fabricating different components.

The mechanical properties of materials must also be understood so that we can process materials into useful shapes using materials processing techniques. Materials processing such as the use of steels and plastics to fabricate car bodies, requires a detailed understanding of the mechanical properties of materials at different temperatures and conditions of loading, for example, the mechanical behavior of steels and plastics used to fabricate such items as aerodynamic car bodies. One of the reasons we draw, roll, forge, extrude, and stamp steels and many other alloys by first heating them to high temperatures is that they become ductile at high temperatures. Similarly, we make use of favorable changes in properties of plastics and glasses during their processing into different shapes (e.g., optical fibers).

In the sections that follow, we discuss mechanical properties of materials. We will define and discuss different terms that are used to describe the mechanical properties of engineered materials. Different tests used to determine mechanical properties of materials are discussed.

6-2   Terminology for Mechanical Properties

There are different types of forces or "stresses" that are encountered in dealing with mechanical properties of materials. In general, we define **stress** as force acting on the unit area over which the force is applied. Tensile, compressive, shear, and bending stresses are illustrated in Figure 6-3(a). **Strain** is defined as the change in dimension per unit length. Stress is typically expressed in psi (pounds per square inch) or Pa (Pascals). Strain has no dimensions and is often expressed as in./in. or cm/cm.

When discussing stress and strain, it may be useful to think about stress as the *cause* and strain as the *effect*. Typically, tensile and shear stresses are designated by the symbols σ and τ, respectively. Tensile and shear strains are represented by the symbols ε and γ, respectively. Many load-bearing applications involve tensile or compressive stresses. Shear stresses are often encountered in the processing of materials using such techniques as polymer extrusion. Shear stresses are also found in structural applications. Note that even a simple tensile stress applied along one direction will cause a shear stress to components in other directions (similar to the situation discussed in Schmid's law, Chapter 4).

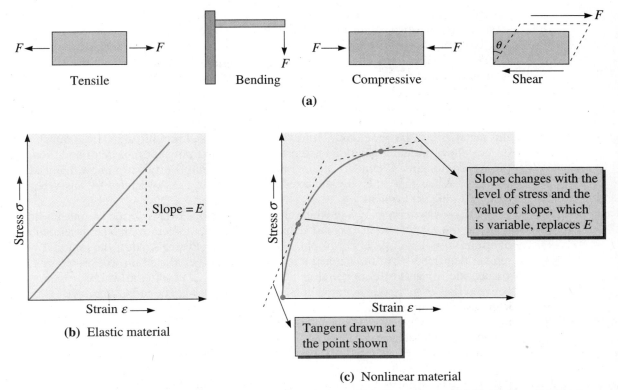

Figure 6-3 (a) Tensile, compressive, shear and bending stresses. (b) Illustration showing how Young's modulus is defined for an elastic material. (c) For nonlinear materials, we use the slope of a tangent as a variable quantity that replaces the Young's modulus constant.

Elastic strain is defined as fully recoverable strain resulting from an applied stress. The strain is "elastic" if it develops instantaneously (i.e., the strain occurs as soon as the force is applied), remains as long as the stress is applied, and is removed as soon as the force is withdrawn. A material subjected to an elastic strain does not show any permanent deformation (i.e., it returns to its original shape after the force or stress is removed). Consider stretching a stiff metal spring by a small amount and letting go. If the spring goes back quickly to its original dimensions, the strain developed in the spring was elastic.

In many materials, elastic stress and elastic strain are linearly related. The slope of a tensile stress-strain curve in the linear regime defines the **Young's modulus** or **modulus of elasticity** (E) of a material [Figure 6-3(b)]. The units of E are measured in pounds per square inch (psi) or Pascals (Pa) (same as those of stress). Large elastic deformations are observed in **elastomers** (e.g., natural rubber, silicones), where the relationship between elastic strain and stress is non-linear. In elastomers, the large elastic strain is related to the coiling and uncoiling of spring-like molecules (Chapter 15). In dealing with such materials, we use the slope of the tangent at any given value of stress or strain and consider that as a variable quantity that replaces the Young's modulus [Figure 6-3(b)]. The inverse of Young's modulus is known as the **compliance** of the material. Similarly, we define **shear modulus** (G) as the slope of the linear part of the shear stress-shear strain curve.

Permanent or **plastic deformation** in a material is known as the **plastic strain**. In this case, when the stress is removed, the material does *not* go back to its original shape.

A dent in a car is plastic deformation! Note that the word "plastic" here does not refer to strain in a plastic (polymeric) material, but rather to a type of strain in any material.

The rate at which strain develops in a material is defined as **strain rate** ($\dot{\varepsilon}$ or $\dot{\gamma}$ for tensile and shear strain rates, respectively). Units of strain rate are s^{-1}. You will learn later in this chapter that the rate at which a material is deformed is important from a mechanical properties perspective. Many materials considered to be ductile, behave as brittle solids when the strain rates are high. Silly Putty® (a silicone polymer) is an example of such a material. When stretched slowly (smaller rate of strain), we can stretch this material by a large amount. However, when stretched rapidly (high strain rates), we do not allow the untangling and extension of the large polymer molecules and, hence, the material snaps. When the strain rates are low, Silly Putty® can show significant ductility. When materials are subjected to high strain rates we refer to this type of loading as **impact loading**.

A **viscous material** is one in which the strain develops over a period of time and the material does not go to its original shape after the stress is removed. The development of strain takes time and is not in phase with the applied stress. Also, the material will remain deformed when the applied stress is removed (i.e., the strain will be plastic). A **viscoelastic** (or **anelastic**) material can be thought of as a material whose response is between that of a viscous material and an elastic material. The term "anelastic" is typically used for metals, while the term "viscoelastic" is usually associated with polymeric materials. Many plastics (solids and molten) are viscoelastic. A common example of a viscoelastic material is Silly Putty®.

In a viscoelastic material, the development of a permanent strain is similar to that in a viscous material. However, unlike a viscous material, when the applied stress is removed, part of the strain will recover over a period of time. Recovery of strain refers to a change in shape of a material after the stress causing deformation is removed. A qualitative description of development of strain as a function of time in relation to an applied force in elastic, viscous, and viscoelastic materials is shown in Figure 6-4. In viscoelastic materials held under constant strain, if we wait, the level of stress decreases over a period of time. This is known as **stress relaxation**. Recovery of strain and stress relaxation are different terms and should not be confused. A common example of stress relaxation is the nylon strings strung in a tennis racket. We know that the level of stress, or the "tension", as the tennis players call it, decreases with time.

While dealing with molten materials, liquids, and dispersions, such as paints or gels, a description of the resistance to flow under an applied stress is required. If the relationship between the applied stress and **shear strain rate** ($\dot{\gamma}$) is linear, we refer to that material as **Newtonian**. The slope of the shear stress versus the steady-state shear strain rate curve is defined as the **viscosity** (η) of the material. Water is an example of a Newtonian material. The following relationship defines viscosity:

$$\tau = \eta\dot{\gamma} \tag{6-1a}$$

The units of η are Pa-s (in the SI system) or Poise (P) or $\dfrac{g}{cm - s}$ in the cgs system. Sometimes the term centipoise (cP) is used, 1 cP $= 10^{-2}$ P.

Conversion between these units is given by 1 Pa-s $= 10$ P $= 1000$ cP.

The **kinematic viscosity** (v) is defined as:

$$v = \eta/\rho \tag{6-1b}$$

where viscosity (η) is in Poise and density (ρ) is in g/cm^3. The kinematic viscosity unit is in Stokes (St). In this, St is cm^2/s. Sometimes the unit of centiStokes (cSt) is used, 1 cSt $= 10^{-2}$ St.

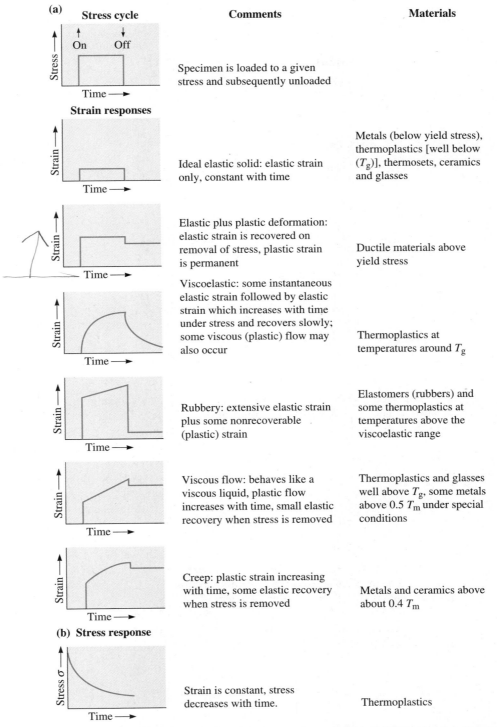

Figure 6-4 (a) Various types of strain response to an imposed stress. (*Source: Reprinted from Materials Principles and Practice, by C. Newey and G. Weaver (Eds.), 1991 p. 300, Fig. 6-9. Copyright © 1991 Butterworth-Heinemann. Reprinted with permission from Elsevier Science.*) (b) Stress relaxation in a viscoelastic material. *Note the y-axis is stress.* Strain is constant.

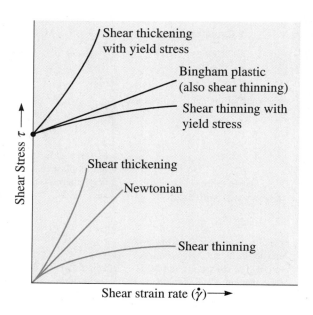

Figure 6-5
Shear stress-shear strain rate
relationships for Newtonian and non-
Newtonian materials.

For many materials the relationship between shear stress and shear strain rate is nonlinear. These materials are **non-Newtonian**. The stress versus steady state shear strain rate relationship in these materials can be described as:

$$\tau = \eta \dot{\gamma}^m \tag{6-2}$$

where the exponent m is not equal to 1.

Non-Newtonian materials are classified as **shear thinning** (or pseudo-plastic) or **shear thickening** (or dilatant). The relationships between the shear stress and shear strain rate for different types of materials are shown in Figure 6-5. As shown in Figure 6-6(a), the **apparent viscosity** (η_{app}) of the material decreases with increasing steady-state shear strain rate. If we take the slope of the line obtained by joining the origin to any point on the curve what we determine is the apparent viscosity. The apparent viscosity of a Newtonian material will remain constant with changing shear strain rate. In shear thinning materials, the apparent viscosity decreases with increasing shear strain rate. In shear thickening materials the apparent viscosity increases with increasing shear strain rate. If you have a can of paint sitting in storage, for example, the shear strain rate that the paint is subjected to is very small and the paint behaves as if it is very viscous. When you take a brush and paint, the paint is subjected to high shear strain rate. The paint now behaves as if it is quite thin or less viscous (i.e., it exhibits a small apparent viscosity). This is the shear thinning behavior.

Some materials have "ideal plastic" behavior. For an ideal plastic material the shear stress does not change with shear strain rate. Many useful materials can be modeled as **Bingham plastics** and are defined by the following equations:

$$\tau = G\gamma \quad \text{(when } \tau \text{ is less than } \tau_y) \tag{6-3a}$$

$$\tau = \tau_y + \eta\dot{\gamma} \quad \text{(when } \tau \geq \tau_y) \tag{6-3b}$$

This is illustrated in Figure 6-6(b) and 6-6(c).

In these equations, τ_y is the apparent **yield strength** obtained by intrapolating the shear stress-shear strain rate data to zero shear strain rate. We define yield strength as the stress level that has to be exceeded so that the material begins to deform plastically. The existence of a true yield strength (sometimes also known as yield stress) has not

(a)

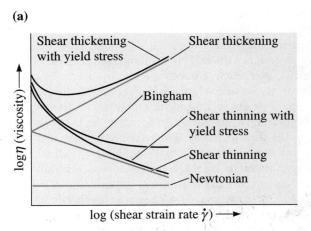

Figure 6-6
(a) Apparent viscosity as a function
of shear log (shear rate $\dot{\gamma}$) strain rate.
(b) and (c) illustration of a Bingham
plastic (Equations 6-3a & b). Note
the x-axis on (b) is shear strain.

(b)

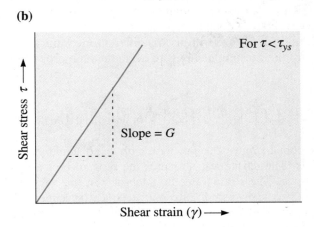

(c)

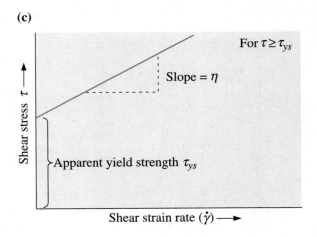

been proven unambiguously for many plastics and dispersions such as paints. To prove
the existence of yield stress, separate measurements of stress versus strain are needed.
For these materials, a critical yielding strain may be a better way to describe the mechanical behavior.[5] Many ceramic slurries (dispersions such as those used in ceramic
processing), polymer melts (used in polymer processing), paints and gels, and food

products (yogurt, mayonnaise, ketchups, etc.) exhibit Bingham plastic-like behavior. Note that Bingham plastics exhibit shear thinning behavior (i.e., the apparent viscosity decreases with increasing shear rate).

Shear thinning materials also exhibit a **thixotropic behavior** (e.g., paints, ceramic slurries, polymer melts, gels, etc.). Thixotropic materials usually contain some type of network of particles or molecules. When a sufficiently large shearing strain (i.e., greater than the critical yielding strain) is applied, the thixotropic network or structure breaks and the materials begin to flow. As the shearing stops, the network begins to form again and the resistance to the flow increases. The particle or molecular arrangements in the newly formed network are different from those in the original network. Thus, the behavior of thixotropic materials is said to be time and deformation history dependent. Some materials show an increase in the apparent viscosity as a function of time and at a constant shearing rate. These materials are known as **rheopectic**.

The rheological properties of materials are determined using instruments known as a viscometer or a **rheometer**. In these instruments a constant stress or constant strain rate is applied to the material being evaluated. Different geometric arrangements (e.g., cone and plate, parallel plate, Couette, etc.) are used.

In the sections that follow, we will discuss different mechanical properties of solid materials and some of their testing methods to evaluate these properties.

6-3   The Tensile Test: Use of the Stress-Strain Diagram

The tensile test is popular since the properties obtained could be applied to design different components. The tensile test measures the resistance of a material to a static or slowly applied force. The strain rates in a tensile test are very small ($\dot{\varepsilon} = 10^{-4}$ to $10^{-2}s^{-1}$). A test setup is shown in Figure 6-7; a typical specimen has a diameter of 0.505 in. and a gage length of 2 in. The specimen is placed in the testing machine and a force F, called the **load**, is applied. A universal testing machine on which tensile and compression can be performed is shown in Figure 6-8. A **strain gage** or **extensometer** is used to measure the amount that the specimen stretches between the gage marks when the force is applied. Thus, what is measured is the change in length of the specimen (Δl)

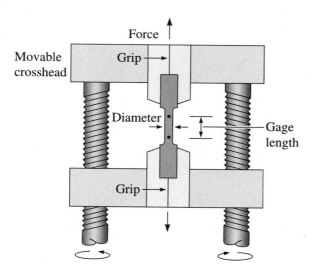

Figure 6-7
A unidirectional force is applied to a specimen in the tensile test by means of the moveable crosshead. The cross-head movement can be performed using screws or a hydraulic mechanism.

Figure 6-8   Students conducting a tensile test on an automated tensile testing machine. (*Courtesy of Pradeep Phulé.*)

over a particular original length (l_0). Information concerning the strength, Young's modulus, and ductility of a material can be obtained from such a tensile test. Typically, a tensile test is conducted on metals, alloys, and plastics. Tensile tests can be used for ceramics, however, these are not very popular because the sample may fracture while it is being aligned. The following discussion mainly applies to the tensile testing of metals and alloys. We will briefly discuss the stress-strain behavior of polymers as well.

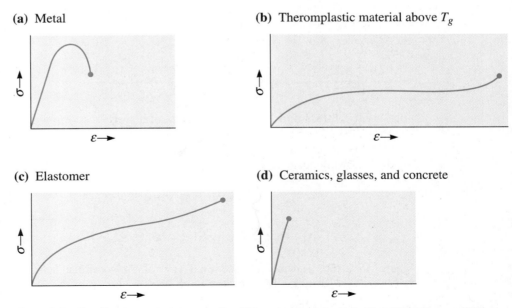

Figure 6-9   Tensile stress-strain curves for different materials. Note that these are *qualitative*.

Figure 6-9 shows *qualitatively* the stress-strain curves for a typical (a) metal, (b) a thermoplastic material, (c) an elastomer, and (d) ceramics (or glass) under relatively small strain rates. The scales in this figure are qualitative and different for each material. In practice, the actual magnitude of stresses and strains will be very different. The plastic material is assumed to be above its **glass temperature** (T_g). Metallic materials are assumed to be at room temperature. Metallic and thermoplastic materials show an initial elastic region followed by a non-linear plastic region. A separate curve for elastomers (e.g., rubber or silicones) is also included since the behavior of these materials is different from other polymeric materials. For elastomers, a large portion of the deformation is elastic and non-linear. On the other hand, ceramics and glasses show only a linear elastic region and almost no plastic deformation at room temperature.

When a tensile test is conducted, the data recorded includes load or force as a function of change in length (Δl). The change in length is typically measured using a strain gage. Table 6-1 shows the effect of the load on the changes in length of an aluminum alloy test bar. These data are then subsequently converted into stress and strain. The stress-strain curve is analyzed further to the extract properties of materials (e.g., Young's modulus, yield strength, etc.).

TABLE 6-1 ■ *The results of a tensile test of a 0.505-in. diameter aluminum alloy test bar, initial length $(l_0) = 2$ in.*

Measured Change in Length (Δl)		Calculated	
Load (lb)	(in.)	Stress (psi)	Strain (in./in.)
0	0.000	0	0
1000	0.001	5,000	0.0005
3000	0.003	15,000	0.0015
5000	0.005	25,000	0.0025
7000	0.007	35,000	0.0035
7500	0.030	37,500	0.0150
7900	0.080	39,500	0.0400
8000 (maximum load)	0.120	40,000	0.0600
7950	0.160	39,700	0.0800
7600 (fracture)	0.205	38,000	0.1025

Engineering Stress and Strain   The results of a single test apply to all sizes and cross-sections of specimens for a given material if we convert the force to stress and the distance between gage marks to strain. **Engineering stress** and **engineering strain** are defined by the following equations,

$$\text{Engineering stress} = \sigma = \frac{F}{A_0} \tag{6-4}$$

$$\text{Engineering strain} = \varepsilon = \frac{\Delta l}{l_0}, \tag{6-5}$$

where A_0 is the *original* cross-sectional area of the specimen before the test begins, l_0 is the *original* distance between the gage marks, and Δl is the change in length after

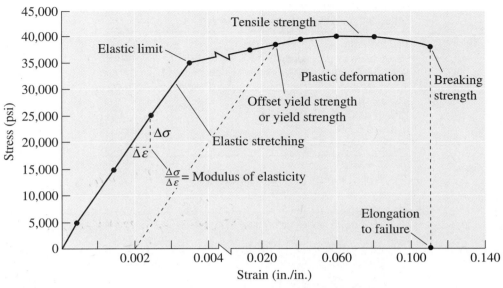

Figure 6-10   The stress-strain curve for an aluminum alloy from Table 6-1.

force F is applied. The conversions from load and sample length to stress and strain are included in Table 6-1. The stress-strain curve (Figure 6-10) is used to record the results of a tensile test.

EXAMPLE 6-1   *Tensile Testing of Aluminum Alloy*

Convert the change in length data in Table 6-1 to engineering stress and strain and plot a stress-strain curve.

SOLUTION

For the 1000-lb load:

$$\sigma = \frac{F}{A_0} = \frac{1000 \text{ lb}}{(\pi/4)(0.505 \text{ in.})^2} = \frac{1000 \text{ lb}}{0.2 \text{ in.}^2} = 5000 \text{ psi}$$

$$\varepsilon = \frac{\Delta l}{l_0} = \frac{0.001 \text{ in.}}{2.000 \text{ in.}} = 0.0005 \text{ in./in.}$$

The results of similar calculations for each of the remaining loads are given in Table 6-1 and are plotted in Figure 6-10.

Units   Many different units are used to report the results of the tensile test. The most common units for stress are pounds per square inch (psi) and MegaPascals (MPa). The units for strain include inch/inch, centimeter/centimeter, and meter/meter. The conversion factors for stress are summarized in Table 6-2. Because strain is dimensionless, no conversion factors are required to change the system of units.

TABLE 6-2 ■ *Units and conversion factors*

1 pound (lb) = 4.448 Newtons (N)
1 psi = pounds per square inch
1 MPa = MegaPascal = MegaNewtons per square meter (MN/m²)
      = Newtons per square millimeter (N/mm²) = 1,000,000 Pa
1 GPa = 1000 MPa = GigaPascal
1 ksi = 1000 psi = 6.895 MPa
1 psi = 0.006895 MPa
1 MPa = 0.145 ksi = 145 psi

EXAMPLE 6-2   *Design of a Suspension Rod*

An aluminum rod is to withstand an applied force of 45,000 pounds. To assure a sufficient safety, the maximum allowable stress on the rod is limited to 25,000 psi. The rod must be at least 150 in. long but must deform elastically no more than 0.25 in. when the force is applied. Design an appropriate rod.

SOLUTION

We can use the definition of engineering stress to calculate the required cross-sectional area of the rod:

$$A_0 = \frac{F}{\sigma} = \frac{45,000}{25,000} = 1.8 \text{ in.}^2$$

The rod could be produced in various shapes, provided that the cross-sectional area is 1.8 in.² For a round cross-section, the minimum diameter to assure that the stress is not too high is:

$$A_0 = \frac{\pi d^2}{4} = 1.8 \text{ in.}^2 \quad \text{or} \quad d = 1.51 \text{ in.}$$

The maximum allowable elastic deformation is 0.25 in. From the definition of engineering strain:

$$\varepsilon = \frac{\Delta l}{l_0} = \frac{0.25 \text{ in.}}{l_0}$$

From Figure 6-10, the strain expected for a stress of 25,000 psi is 0.0025 in./in. If we use the cross-sectional area determined previously, the maximum length of the rod is:

$$0.0025 = \frac{\Delta l}{l_0} = \frac{0.25 \text{ in.}}{l_0} \quad \text{or} \quad l_0 = 100 \text{ in.}$$

However, the minimum length of the rod is specified as 150 in. To produce a longer rod, we might make the cross-sectional area of the rod larger. The minimum strain allowed for the 150-in. rod is:

$$\varepsilon = \frac{\Delta l}{l_0} = \frac{0.25 \text{ in.}}{150 \text{ in.}} = 0.001667 \text{ in./in.}$$

The stress, from Figure 6-10, is about 16,670 psi, which is less than the maximum of 25,000 psi. The minimum cross-sectional area then is:

$$A_0 = \frac{F}{\sigma} = \frac{45,000 \text{ psi}}{16,670 \text{ lb}} = 2.70 \text{ in.}^2$$

In order to satisfy both the maximum stress and the minimum elongation requirements, cross-sectional area of the rod must be at least 2.7 in.^{2}, or a minimum diameter of 1.85 in.

6-4   Properties Obtained from the Tensile Test

Yield Strength   As we apply stress to a material, the material initially exhibits elastic deformation. The strain that develops is completely recovered when the applied stress is removed. However, as we continue to increase the applied stress the material begins to exhibit both elastic and plastic deformation. The material eventually "yields" to the applied stress. The critical stress value needed to initiate plastic deformation is defined as the **elastic limit** of the material. In metallic materials, this is usually the stress required for dislocation motion, or slip to be initiated. In polymeric materials, this stress will correspond to disentanglement of polymer molecule chains or sliding of chains past each other. The **proportional limit** is defined as the level of stress above which the relationship between stress and strain is not linear.

In most materials the elastic limit and proportional limit are quite close. However, neither the elastic limit nor the proportional limit values can be determined precisely. Measured values depend on the sensitivity of the equipment used. We, therefore, define them at an **offset strain value** (typically, but not always, 0.002 or 0.2%). We then draw a line starting with this offset value of strain and draw a line parallel to the linear portion of the engineering stress-strain curve. The stress value corresponding to the intersection of this line and the engineering stress-strain curve is defined as the **offset yield strength**, also often stated as the **yield strength**. The 0.2% offset yield strength for gray cast iron is 40,000 psi as shown in Figure 6-11(a). Engineers normally prefer to use the offset yield strength for design purposes.

For some materials the transition from elastic deformation to plastic flow is rather abrupt. This transition is known as the **yield point phenomenon**. In these materials, as the plastic deformation begins the stress value drops first from the *upper yield point* (σ_2) [Figure 6-11(b)]. The stress value then decreases and oscillates around an average value defined as the *lower yield point* (σ_1). For these materials, the yield strength is usually defined from the 0.2% strain offset as shown in Figure 6-11(a).

The stress-strain curve for certain low-carbon steels displays a double yield point [Figure 6-11(b)]. The material is expected to plastically deform at stress σ_1. However, small interstitial atoms clustered around the dislocations interfere with slip and raise the yield point to σ_2. Only after we apply the higher stress σ_2 do the dislocations slip. After slip begins at σ_2, the dislocations move away from the clusters of small atoms and continue to move very rapidly at the lower stress σ_1.

When we design parts for load-bearing applications we prefer little or no plastic deformation. As a result we must select a material such that the design stress is considerably lower than the yield strength at the temperature at which the material will be used. We can also make the component cross-section larger so that the applied force produces a stress that is well below the yield strength. On the other hand, when we want

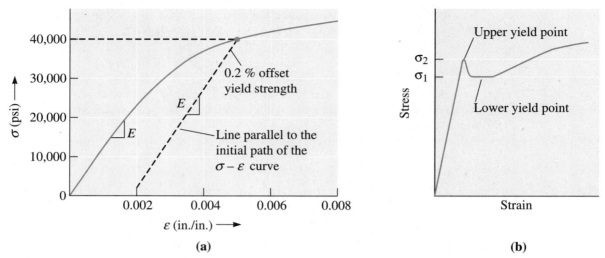

Figure 6-11   (a) Determining the 0.2% offset yield strength in gray cast iron, and (b) upper and lower yield point behavior in a low-carbon steel.

to shape materials into components (e.g., take a sheet of steel and form a car chassis), we need to apply stresses that are well above the yield strength.

Tensile Strength   The stress obtained at the highest applied force is the **tensile strength** (σ_{TS}), which is the maximum stress on the engineering stress-strain curve. In many ductile materials, deformation does not remain uniform. At some point, one region deforms more than others and a large local decrease in the cross-sectional area occurs (Figure 6-12). This locally deformed region is called a "neck." This phenomenon is known as **necking**. Because the cross-sectional area becomes smaller at this point, a lower force is required to continue its deformation, and the engineering stress, calculated from the *original* area A_0, decreases. The tensile strength is the stress at which necking begins in ductile materials. Many ductile metals and polymers show the phe-

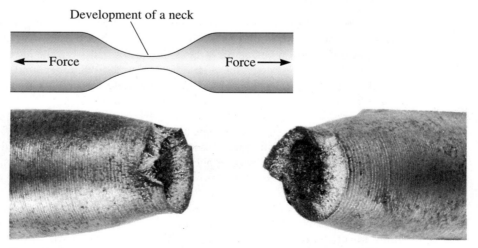

Figure 6-12   Localized deformation of a ductile material during a tensile test produces a necked region. The micrograph at the bottom shows necked region in a fractured sample.

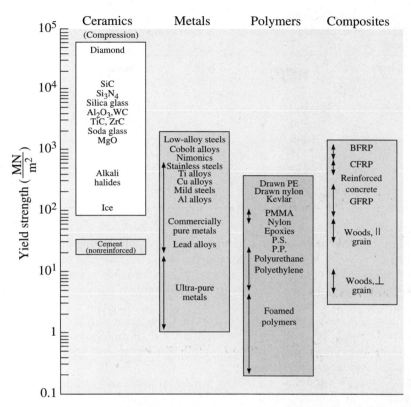

Figure 6-13 Typical yield strength values for different engineered materials. (*Source: Reprinted from* Engineering Materials I, *Second Edition, M.F. Ashby and D.R.H. Jones, 1996, Fig. 8-12, p. 85. Copyright © Butterworth-Heinemann. Reprinted with permission from Elsevier Science.*)

nomenon of necking. In compression testing, the materials will bulge, thus necking is seen only in a tensile test.

Figure 6-13 shows typical yield strength values for different engineered materials. The yield strength of pure metals is smaller. For example, ultra-pure metals have a yield strength of $\sim\left(1 - 10\frac{MN}{m^2}\right)$. On the other hand, the yield strength of alloys is higher. Strengthening in alloys is achieved using different mechanisms described before (e.g., grain size, solid solution formation, strain hardening, etc.). The yield strength of plastics and elastomers is generally lower than metals and alloys, ranging up to about $\left(10 - 100\frac{MN}{m^2}\right)$. The values for ceramics are for compressive strength (obtained using a hardness test). Tensile strength of most ceramics is much lower (~100–200 MPa). The tensile strength of glasses is about ~70 MPa and depends on surface flaws.

Elastic Properties   The modulus of elasticity, or *Young's modulus* (E), is the slope of the stress-strain curve in the elastic region. This relationship is **Hooke's Law**:[6]

$$E = \frac{\sigma}{\varepsilon} \tag{6-6}$$

The modulus is closely related to the binding energies (Figure 2-26). A steep slope in the force-distance graph at the equilibrium spacing indicates that high forces are required to separate the atoms and cause the material to stretch elastically. Thus, the material has a

TABLE 6-3 ■ *Elastic properties and melting temperature (T_m) of selected materials*

Material	T_m (°C)	E (psi)	Poisson's ratio (μ)
Pb	327	2.0×10^6	0.45
Mg	650	6.5×10^6	0.29
Al	660	10.0×10^6	0.33
Cu	1085	18.1×10^6	0.36
Fe	1538	30.0×10^6	0.27
W	3410	59.2×10^6	0.28
Al_2O_3	2020	55.0×10^6	0.26
Si_3N_4		44.0×10^6	0.24

high modulus of elasticity. Binding forces, and thus the modulus of elasticity, are typically higher for high melting point materials (Table 6-3). In metallic materials, modulus of elasticity is considered microstructure *insensitive* property since the value is dominated strongly by the strength of atomic bonds. Grain size or other microstructural features do not have a very large effect on the Young's modulus. Note that Young's modulus does depend on such factors as orientation of a single crystal material (i.e., it depends upon crystallographic direction). For ceramics, the Young's modulus depends on the level of porosity. Young's modulus of a composite depends upon the stiffness of the individual components.

Young's' modulus is a measure of the stiffness of a component. A stiff component, with a high modulus of elasticity, will show much smaller changes in dimensions if the applied stress is relatively small and, therefore, causes only elastic deformation. Figure 6-14 compares the elastic behavior of steel and aluminum. If a stress of 30,000 psi is applied to each material, the steel deforms elastically 0.001 in./in.; at the same stress, aluminum deforms 0.003 in./in. In general, most engineers view stiffness as a function of both the Young's modulus and the geometry of a component.

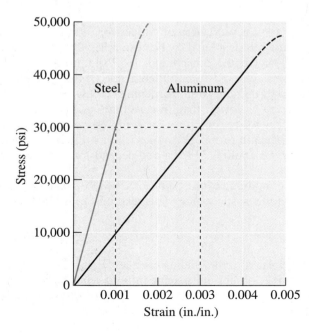

Figure 6-14
Comparison of the elastic behavior of steel and aluminum. For a given stress, aluminum deforms elastically three times as much as does steel.

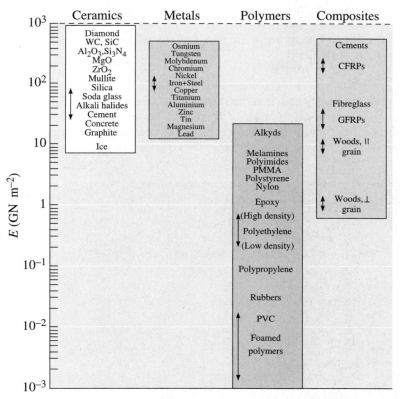

Figure 6-15   Range of elastic moduli for different engineered materials. (*Source: Reprinted from Engineering Materials I, Second Edition, M.F. Ashby and D.R.H. Jones, 1996, Fig. 3-5, p. 35, Copyright © 1996 Butterworth-Heinemann. Reprinted with permission from Elsevier Science.*)

Figure 6-15 shows the ranges of elastic moduli for different engineered materials. The modulus of elasticity of plastics is much smaller than that for metals or ceramics and glasses. For example, the modulus of elasticity of nylon is 2.7 GPa ($\sim 0.4 \times 10^6$ psi); the modulus of glass fibers is 72.4 GPa ($\sim 10.5 \times 10^6$ psi). The Young's modulus of composites such as glass fiber-reinforced composites (GFRC) or carbon fiber-reinforced composites (CFRC) lies between the values for the matrix polymer and the fiber phase (carbon or glass fibers) and depends upon their relative volume fractions. The Young's modulus of many alloys and ceramics is higher, generally ranging up to 413.7 GPa ($\sim 60 \times 10^6$ psi). Ceramics, because of the strength of ionic and covalent bonds, have the highest elastic moduli.

Poisson's ratio, μ, relates the longitudinal elastic deformation produced by a simple tensile or compressive stress to the lateral deformation that occurs simultaneously:

$$\mu = \frac{-\varepsilon_{\text{lateral}}}{\varepsilon_{\text{longitudinal}}} \tag{6-7}$$

For many metals in the elastic region the Poisson's ratio is typically about 0.3 (Table 6-3). During a tensile test the ratio increases beyond yielding to about 0.5 since during the plastic deformation volume remains constant. Some interesting structures exhibit negative Poisson's ratio.[7]

The **modulus of resilience** (E_r), the area contained under the elastic portion of a stress-strain curve, is the elastic energy that a material absorbs during loading and subsequently releases when the load is removed. For linear elastic behavior:

$$E_r = \left(\tfrac{1}{2}\right)(\text{yield strength})(\text{strain at yielding}) \qquad (6\text{-}8)$$

The ability of a spring or a golf ball to perform satisfactorily depends on a high modulus of resilience.

Tensile Toughness   The energy absorbed by a material prior to fracturing is known as **tensile toughness** and is sometimes measured as the area under the true stress-strain curve (also known as **work of fracture**). We will define true stress and true strain in Section 6-5. Since it is easier to measure engineering stress-strain, engineers often equate tensile toughness to the area under the engineering stress-strain curve.

EXAMPLE 6-3   *Young's Modulus of Aluminum Alloy*

From the data in Example 6-1, calculate the modulus of elasticity of the aluminum alloy. Use the modulus to determine the length after deformation of a bar of initial length of 50 in. Assume that a level of stress of 30,000 psi is applied.

SOLUTION

When a stress of 35,000 psi is applied, a strain of 0.0035 in./in. is produced. Thus:

$$\text{Modulus of elasticity} = E = \frac{\sigma}{\varepsilon} = \frac{35,000 \text{ psi}}{0.0035} = 10 \times 10^6 \text{ psi}$$

From Hooke's law:

$$\varepsilon = \frac{\sigma}{E} = \frac{30,000 \text{ psi}}{10 \times 10^6} = 0.0003 = \text{in./in.} = \frac{l - l_0}{l_0}$$

$$l = l_0 + \varepsilon l_0 = 50 + (0.003)(50) = 50.15 \text{ in.}$$

Ductility   **Ductility** measures the amount of deformation that a material can withstand without breaking. We can measure the distance between the gage marks on our specimen before and after the test. The **percent elongation** describes the permanent plastic deformation before failure (i.e., the elastic deformation recovered after fracture is not included). Note that the strain to failure is smaller than strain at the breaking point.

$$\% \text{ Elongation} = \frac{l_f - l_0}{l_0} \times 100 \qquad (6\text{-}9)$$

where l_f is the distance between gage marks after the specimen breaks.

A second approach is to measure the percent change in the cross-sectional area at the point of fracture before and after the test. The **percent reduction in area** describes the amount of thinning undergone by the specimen during the test:

$$\% \text{ Reduction in area} = \frac{A_0 - A_f}{A_0} \times 100 \qquad (6\text{-}10)$$

where A_f is the final cross-sectional area at the fracture surface.

Ductility is important to both designers of load-bearing components and manufacturers of components (bars, rods, wires, plates, I-beams, fibers, etc.) utilizing materials processing. The designer of a component prefers a material that displays at least some ductility, so that, if the applied stress is too high, the component can take some of the stress by deforming and not fail via a brittle fracture. Fabricators of engineering components (metallic and polymeric) want a ductile material in order to form complicated shapes without breaking the material in the process. Ductility of materials depends on temperature and strain rate.

EXAMPLE 6-4   *Ductility of an Aluminum Alloy*

The aluminum alloy in Example 6-1 has a final length after failure of 2.195 in. and a final diameter of 0.398 in. at the fractured surface. Calculate the ductility of this alloy.

SOLUTION

$$\% \text{ Elongation} = \frac{l_f - l_0}{l_0} \times 100 = \frac{2.195 - 2.000}{2.000} \times 100 = 9.75\%$$

$$\% \text{ Reduction in area} = \frac{A_0 - A_f}{A_0} \times 100$$

$$= \frac{(\pi/4)(0.505)^2 - (\pi/4)(0.398)^2}{(\pi/4)(0.505)^2} \times 100$$

$$= 37.9\%$$

The final length is less than 2.205 in. (see Table 6-1) because, after fracture, the elastic strain is recovered.

Effect of Temperature   Mechanical properties of materials depend on temperature (Figure 6-16). Yield strength, tensile strength, and modulus of elasticity decrease at higher temperatures, whereas ductility commonly increases. A materials fabricator may wish to deform a material at a high temperature (known as *hot working*) to take

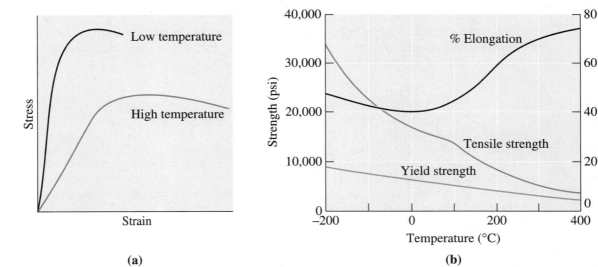

(a)   **(b)**

Figure 6-16   The effect of temperature (a) on the stress-strain curve and (b) on the tensile properties of an aluminum alloy.

advantage of the higher ductility and lower required stress. Some of you may have heard the sayings "heat and beat" or "strike while the iron is hot." The origin of these phrases is related to the ductility many metallic materials develop when heated to higher temperatures. We use the term "high temperature" here with a note of caution. Essentially, a high temperature is something that is approaching the melting temperature. Thus, 500°C is a high temperature for aluminum alloys; however, it is a relatively low temperature for the processing of steels. In metals, the yield strength decreases rapidly at higher temperatures since metallic materials can show a decreased dislocation density and an increase in grain size via grain growth (Chapter 5), or "recrystallize" into new grains that are essentially dislocation free (as described later in Chapter 7). Similarly, any strengthening that may have occurred because of formation of ultrafine precipitates (Chapter 4) may also decrease as the precipitates begin to either grow in size or dissolve into the matrix. We will discuss these effects in greater detail in later chapters. When temperatures are reduced, many, but not all, metals and alloys become brittle.

Increased temperatures also play an important role in forming polymeric materials as well as inorganic glasses. In many polymer-processing operations, such as extrusion or the stretch-blow process (Chapter 4), the increased ductility of polymers at higher temperature is advantageous. Again, a word of caution concerning the use of the term "high temperature": For polymers, the term "higher temperature" would generally mean a temperature higher than the glass temperature (T_g). In some sources, you will see T_g described as the glass transition temperature. There is no "transition" at this temperature. For our purpose, glass temperature is a temperature below which materials behave as brittle materials. Above the glass temperature plastics become ductile. Glass temperature is not a fixed temperature, but depends upon rate of cooling as well as the polymer molecular weight distribution. Many plastics are ductile at room temperature because their glass temperatures are *below* room temperature. To summarize, many polymeric materials will become harder and more brittle as they are exposed to temperatures that are below their glass temperatures. The reasons for loss of ductility at lower temperatures in polymers and metallic materials are different. However, this is a factor that played a role in failures of the *Titanic* in 1912 and the *Challenger* in 1986.

Ceramic and glassy materials are generally considered brittle at room temperatures. As the temperature increases, glasses can flow better and become more ductile. As a result, glass processing (e.g., fiber drawing or bottle manufacturing) is performed at high temperatures. Polycrystalline ceramics also can gain increased ductility at higher temperatures via mechanisms that involve grain boundary sliding and other phenomena. This is discussed in Section 6-21.

6-5   True Stress and True Strain

The decrease in engineering stress beyond the tensile strength point on an engineering stress-strain curve is related to the definition of engineering stress. We used the original area A_0 in our calculations, but this is not precise because the area continually changes. We define **true stress** and **true strain** by the following equations:

$$\text{True stress} = \sigma_t = \frac{F}{A} \tag{6-11}$$

$$\text{True strain} = \int \frac{dl}{l} = \ln\left(\frac{l}{l_0}\right) = \ln\left(\frac{A_0}{A}\right), \tag{6-12}$$

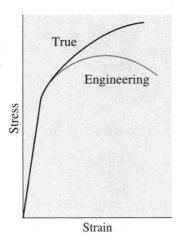

Figure 6-17
The relation between the true stress-true strain diagram and engineering stress–engineering strain diagram. The curves are identical to the yield point.

where A is the actual area at which the force F is applied. The expression $\ln(A_0/A)$ can only be used after necking begins, after which volume remains constant and $\dfrac{A_0}{l_0} = \dfrac{A}{l}$.

The true stress-strain curve is compared with the engineering stress-strain curve in Figure 6-17. True stress continues to increase after necking because, although the load required decreases, the area decreases even more.

For structural applications we often do not require true stress and true strain. When we exceed the yield strength, the material deforms. The component has failed because it no longer has the original intended shape. Furthermore, a significant difference develops between the two curves only when necking begins. But when necking begins, our component is grossly deformed and no longer satisfies its intended use. Engineers dealing with materials processing require data related to true stress and strain.

EXAMPLE 6-5   **True Stress and True Strain Calculation**

Compare engineering stress and strain with true stress and strain for the aluminum alloy in Example 6-1 at (a) the maximum load and (b) fracture. The diameter at maximum load is 0.497 in. and at fracture is 0.398 in.

SOLUTION

(a) At the tensile or maximum load:

$$\text{Engineering stress} = \frac{F}{A_0} = \frac{8000 \text{ lb}}{(\pi/4)(0.505 \text{ in.})^2} = 40{,}000 \text{ psi}$$

$$\text{True stress} = \frac{F}{A} = \frac{8000}{(\pi/4)(0.497)^2} = 41{,}237 \text{ psi}$$

$$\text{Engineering strain} = \frac{l - l_0}{l_0} = \frac{2.120 - 2.000}{2.000} = 0.060 \text{ in./in.}$$

$$\text{True strain} = \ln\left(\frac{l}{l_0}\right) = \ln\left(\frac{2.120}{2.000}\right) = 0.058 \text{ in./in.}$$

(b) At fracture:

$$\text{Engineering stress} = \frac{F}{A_0} = \frac{7600 \text{ lb}}{(\pi/4)(0.505 \text{ in.})^2} = 38{,}000 \text{ psi}$$

$$\text{True stress} = \frac{F}{A} = \frac{7600}{(\pi/4)(0.398 \text{ in.})^2} = 61{,}090 \text{ psi}$$

$$\text{Engineering strain} = \frac{\Delta l}{l_0} = \frac{0.205}{2.000} = 0.1025 \text{ in./in.}$$

$$\text{True strain} = \ln\left(\frac{A_0}{A_f}\right) = \ln\left[\frac{(\pi/4)(0.505)^2}{(\pi/4)(0.398)^2}\right]$$

$$= \ln(1.610) = 0.476 \text{ in./in.}$$

The true stress becomes much greater than the engineering stress only after necking begins.

6-6   The Bend Test for Brittle Materials

In ductile metallic materials, the engineering stress-strain curve typically goes through a maximum; this maximum stress is the tensile strength of the material. Failure occurs at a lower stress after necking has reduced the cross-sectional area supporting the load. In more brittle materials, failure occurs at the maximum load, where the tensile strength and breaking strength are the same. In brittle materials, including many ceramics, yield strength, tensile strength, and breaking strength are all the same (Figure 6-18).

In many brittle materials, the normal tensile test cannot easily be performed because of the presence of flaws at the surface. Often, just placing a brittle material in the grips of the tensile testing machine causes cracking. These materials may be tested using the **bend test** [Figure 6-19(a)]. By applying the load at three points and causing bending, a tensile force acts on the material opposite the midpoint. Fracture begins

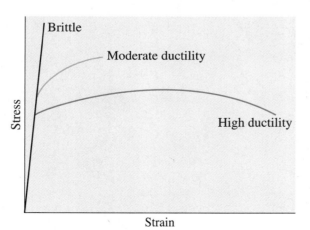

Figure 6-18
The stress-strain behavior of brittle materials compared with that of more ductile materials.

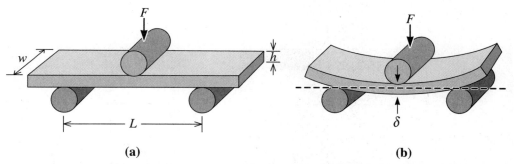

(a)   **(b)**

Figure 6-19 (a) The bend test often used for measuring the strength of brittle materials, and (b) the deflection δ obtained by bending.

at this location. The **flexural strength**, or **modulus of rupture**, describes the material's strength:

$$\text{Flexural strength for three-point bend test} = \frac{3FL}{2wh^2} = \sigma_{\text{bend}} \qquad (6\text{-}13)$$

where F is the fracture load, L is the distance between the two outer points, w is the width of the specimen, and h is the height of the specimen. The flexural strength has the units of stress and is designated by σ_{bend}. The results of the bend test are similar to the stress-strain curves; however, the stress is plotted versus deflection rather than versus strain (Figure 6-20).

The modulus of elasticity in bending, or the **flexural modulus** (E_{bend}), is calculated in the elastic region of Figure 6-20.

$$\text{Flexural modulus} = \frac{L^3 F}{4wh^3\delta} = E_{\text{bend}} \qquad (6\text{-}14)$$

where δ is the deflection of the beam when a force F is applied.

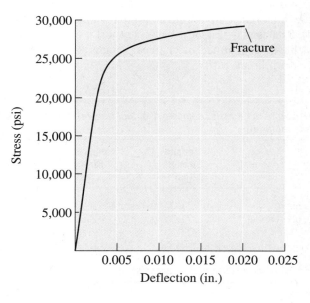

Figure 6-20
Stress-deflection curve for MgO obtained from a bend test.

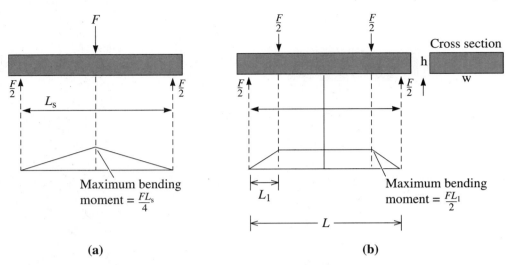

Figure 6-21   (a) Three-point and (b) four-point bend test setup.

This test can also be conducted using a setup known as the four-point bend test (Figure 6-21). The maximum stress or flexural stress for a four point bend test is given by:

$$\sigma_{bend} = \frac{3FL_1}{4wh^2} \qquad (6\text{-}15)$$

Note that while deriving Equations 6-13 through 6-15, we assume a linear stress-strain response (thus cannot be correctly applied to many polymers). The four-point bend test is better suited for testing materials containing flaws. This is because the bending moment between inner platens is constant [Figure 6-21(b)], thus samples tend to break randomly unless there is a flaw that raises the stress concentration.

Since cracks and flaws tend to remain closed in compression, brittle materials such as concrete are often designed so that only compressive stresses act on the part. Often, we find that brittle materials fail at much higher compressive stresses than tensile stresses (Table 6-4). This is why it is possible to support a fire truck on four coffee cups. However, ceramics have very limited mechanical toughness and hence when we drop a ceramic coffee cup it can break easily.

TABLE 6-4 ■ Comparison of the tensile, compressive, and flexural strengths of selected ceramic and composite materials

Material	Tensile Strength (psi)	Compressive Strength (psi)	Flexural Strength (psi)
Polyester—50% glass fibers	23,000	32,000	45,000
Polyester—50% glass fiber fabric	37,000	27,000[a]	46,000
Al_2O_3 (99% pure)	30,000	375,000	50,000
SiC (pressureless-sintered)	25,000	560,000	80,000

[a] A number of composite materials are quite poor in compression.

EXAMPLE 6-6   *Flexural Strength of Composite Materials*

The flexural strength of a composite material reinforced with glass fibers is 45,000 psi and the flexural modulus is 18×10^6 psi. A sample, which is 0.5 in. wide, 0.375 in. high, and 8 in. long, is supported between two rods 5 in. apart. Determine the force required to fracture the material and the deflection of the sample at fracture, assuming that no plastic deformation occurs.

SOLUTION

Based on the description of the sample, $w = 0.5$ in., $h = 0.375$ in., and $L = 5$ in. From Equation 6-15:

$$45,000 \text{ psi} = \frac{3FL}{2wh^2} = \frac{(3)(F \text{ lb})(5 \text{ in.})}{(2)(0.5 \text{ in.})(0.375 \text{ in.})^2} = 106.7F$$

$$F = \frac{45,000}{106.7} = 422 \text{ lb}$$

Therefore, the deflection, from Equation 6-14, is:

$$18 \times 10^6 \text{ psi} = \frac{L^3 F}{4wh^3\delta} = \frac{(5 \text{ in.})^3 (422 \text{ lb})}{(4)(0.5 \text{ in.})(0.375 \text{ in.})^3 \delta}$$

$$\delta = 0.0278 \text{ in.}$$

In this calculation, we did assume that there is no viscoelastic behavior and a linear behavior of stress versus strain.[8]

6-7   Hardness of Materials

The **hardness test** measures the resistance to penetration of the surface of a material by a hard object. Hardness as a term is not defined precisely. Hardness, depending upon the context, can represent resistance to scratching or indentation and a qualitative measure of the strength of the material. A typical Rockwell hardness tester is shown in Figure 6-22. In general, in **macrohardness** measurements the load applied is ~2 N. A variety of hardness tests have been devised, but the most commonly used are the Rockwell test and the Brinell test. Different indentors used in these tests are shown in Figure 6-23.

In the *Brinell hardness test*, a hard steel sphere (usually 10 mm in diameter) is forced into the surface of the material. The diameter of the impression, typically 2 to 6 mm, is measured and the Brinell hardness number (abbreviated as HB or BHN) is calculated from the following equation:

$$HB = \frac{2F}{\left(\pi D \left[D - \sqrt{D^2 - D_i^2} \right] \right)} \tag{6-16}$$

where F is the applied load in kilograms, D is the diameter of the indentor in millimeters, and D_i is the diameter of the impression in millimeters. The Brinell hardness has the units of stress (e.g., kg/mm^2).

Figure 6-22
Rockwell hardness tester. (*Courtesy of Newage Testing Instruments, Inc.*)

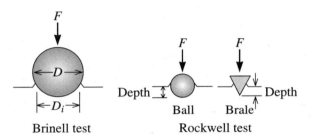

Brinell test   Rockwell test

Figure 6-23
Indentors for the Brinell and Rockwell hardness tests.

The *Rockwell hardness test* uses a small-diameter steel ball for soft materials and a diamond cone, or Brale, for harder materials. The depth of penetration of the indentor is automatically measured by the testing machine and converted to a Rockwell hardness number (HR). Since an optical measurement of the indention dimensions is not needed, the Rockwell test tends to be more popular than the Brinell test. Several variations of the Rockwell test are used, including those described in Table 6-5. A Rockwell *C* (HRC) test is used for hard steels, whereas a Rockwell *F* (HRF) test might be selected for aluminum. Rockwell tests provide a hardness number that has no units.

TABLE 6-5 ■ *Comparison of typical hardness tests*

Test	Indentor	Load	Application
Brinell	10-mm ball	3000 kg	Cast iron and steel
Brinell	10-mm ball	500 kg	Nonferrous alloys
Rockwell *A*	Brale	60 kg	Very hard materials
Rockwell *B*	1/16-in. ball	100 kg	Brass, low-strength steel
Rockwell *C*	Brale	150 kg	High-strength steel
Rockwell *D*	Brale	100 kg	High-strength steel
Rockwell *E*	1/8-in. ball	100 kg	Very soft materials
Rockwell *F*	1/16-in. ball	60 kg	Aluminum, soft materials
Vickers	Diamond pyramid	10 kg	All materials
Knoop	Diamond pyramid	500 g	All materials

Hardness numbers are used primarily as a *qualitative* basis for comparison of materials, specifications for manufacturing and heat treatment, quality control, and correlation with other properties of materials. For example, Brinell hardness is closely related to the tensile strength of steel by the relationship:

$$\text{Tensile strength (psi)} = 500 \, HB \tag{6-17}$$

where HB is in the units of kg/mm^2.

A Brinell hardness number can be obtained in just a few minutes with virtually no preparation of the specimen and without breaking the component (i.e., it is considered to be a nondestructive test), yet it provides a close approximation of the tensile strength. The Rockwell hardness number cannot be directly related to yield strength of metals and alloys, however, the test is rapid, easily performed, and popular in industry.[9]

Hardness correlates well with wear resistance. (A separate test is available for measuring the wear resistance.) A material used in crushing or grinding of ores should be very hard to assure that the material is not eroded or abraded by the hard feed materials. Similarly, gear teeth in the transmission or the drive system of a vehicle should be hard enough that the teeth do not wear out. Typically we find that polymer materials are exceptionally soft, metals and alloys have intermediate hardness, and ceramics are exceptionally hard. We use materials such as tungsten carbide-cobalt composite (WC-Co), known as "carbide," for cutting tool applications. We also use microcrystalline diamond or diamond-like carbon (DLC) materials for cutting tools and other applications.

The Knoop (HK) test is a **microhardness test**, forming such small indentations that a microscope is required to obtain the measurement. In these tests, the load applied is

Figure 6-24   Photograph of a nano-indentor—Hysitron TriboIndenter®. (*Courtesy of Hysitron Incorporated.*)

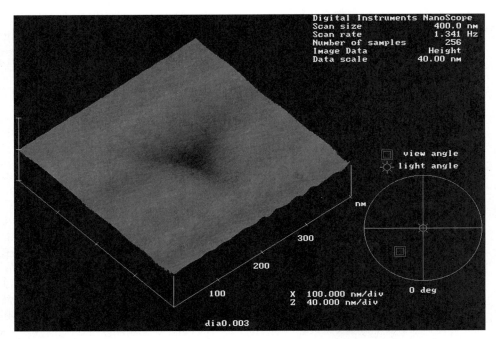

Figure 6-25   Nano-indentation on nano-scale diamond-like carbon film. (*Courtesy of Dr. Scott Mao, University of Pittsburgh.*)

less than 2 N.[10] The Vickers test, which uses a diamond pyramid indentor, can be conducted either as a macro and microhardness test. Microhardness tests are suitable for materials that may have a surface that has a higher hardness than the bulk, materials in which different areas show different levels of hardness, or on samples that are not macroscopically flat.

In Chapter 2, we described nano-structured materials and devices. For some of the nano-technology applications, measurements of hardness at a nano-scale or **nanohardness**, are important. Techniques for measuring hardness at very small length scales have become important for many applications. A nano-indentor is used for these applications (Figure 6-24). The load applied in this test is of the order of 100 μN. This technique involves creating a fine nano-scale indentation on the surface of the material being tested (Figure 6-25). Based on the applied load and the dimensions of the scratch or indentation produced, the nano-scale hardness of a material can be determined. Nano-indentation tests have been successfully used for determination of hardness of diamond-like carbon coatings deposited on magnetic hard drives. At the present time, the micro and nano-hardness testing are used primarily for research and development.

6-8   Strain Rate Effects and Impact Behavior

When a material is subjected to a sudden, intense blow, in which the strain rate ($\dot{\gamma}$ or $\dot{\varepsilon}$) is extremely rapid, it may behave in much more brittle a manner than is observed in the tensile test. This, for example, can be seen with many plastics and materials such as Silly Putty®. If you stretch a plastic such as polyethylene or Silly Putty® very slowly, the polymer molecules have time to disentangle or the chains to slide past each other and

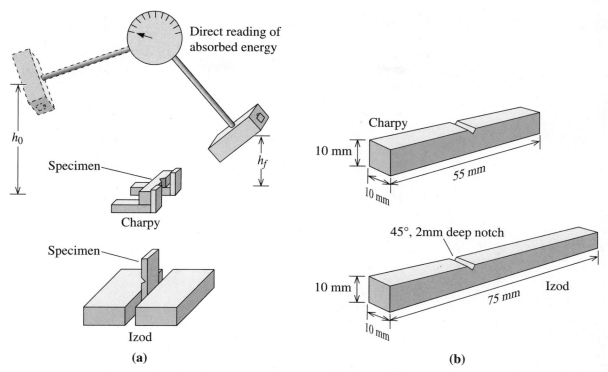

Figure 6-26 The impact test: (a) the Charpy and Izod tests, and (b) dimensions of typical specimens.

cause large plastic deformations. If, however, we apply an impact loading, there is insufficient time for these mechanisms to play a role and the materials break in a brittle manner. An **impact test** is often used to evaluate the brittleness of a material under these conditions. In contrast to the tensile test, in this test the strain rates are much higher ($\dot{\varepsilon} \sim 10^3 \, s^{-1}$).

Many test procedures have been devised, including the *Charpy* test and the *Izod* test (Figure 6-26). The Izod test is often used for plastic materials. The test specimen may be either notched or unnotched; V-notched specimens better measure the resistance of the material to crack propagation.

In the test, a heavy pendulum, starting at an elevation h_0, swings through its arc, strikes and breaks the specimen, and reaches a lower final elevation h_f. If we know the initial and final elevations of the pendulum, we can calculate the difference in potential energy. This difference is the **impact energy** absorbed by the specimen during failure. For the Charpy test, the energy is usually expressed in foot-pounds (ft · lb) or joules (J), where 1 ft · lb = 1.356 J. The results of the Izod test are expressed in units of ft · lb/in. or J/m. The ability of a material to withstand an impact blow is often referred to as the **impact toughness** of the material. As we mentioned before, in some situations, we consider the area under the true or engineering stress-strain curve as a measure of **tensile toughness**. In both cases, we are measuring the energy needed to fracture a material. The difference is that, in tensile tests, the strain rates are much smaller compared to those used in an impact test. Another difference is that in an impact test we usually deal with materials that have a notch. **Fracture toughness** of a material is defined as the ability of a material containing flaws to withstand applied load. We will discuss fracture toughness in Section 6-10.

6-9   Properties Obtained from the Impact Test

A curve showing the trends in the results of a series of impact tests performed on nylon at various temperatures is shown in Figure 6-27. In practice, the tests will be conducted at a limited number of temperatures.

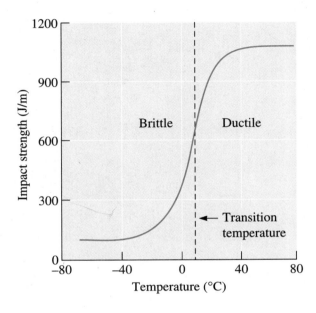

Figure 6-27
Results from a series of Izod impact tests for a super-tough nylon thermoplastic polymer.

Ductile to Brittle Transition Temperature (DBTT)   The **ductile to brittle transition temperature** is the temperature at which a material changes from ductile to brittle fracture. This temperature may be defined by the average energy between the ductile and brittle regions, at some specific absorbed energy, or by some characteristic fracture appearance. A material subjected to an impact blow during service should have a transition temperature *below* the temperature of the material's surroundings.

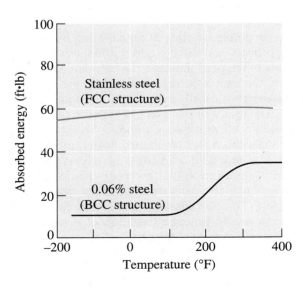

Figure 6-28
The Charpy V-notch properties for a BCC carbon steel and a FCC stainless steel. The FCC crystal structure typically leads to higher absorbed energies and no transition temperature.

Not all materials have a distinct transition temperature (Figure 6-28). BCC metals have transition temperatures, but most FCC metals do not. FCC metals have high absorbed energies, with the energy decreasing gradually and, sometimes, even increasing as the temperature decreases. As mentioned before, this transition may have contributed to the failure of the *Titanic*.

The ductile to brittle transition temperature is closely related to the glass temperature in polymers and for practical purposes is treated as the same. As mentioned before, the lower transition temperature of the polymers used in booster rocket O-rings and other factors led to the *Challenger* disaster.

Notch Sensitivity   Notches caused by poor machining, fabrication, or design concentrate stresses and reduce the toughness of materials. The **notch sensitivity** of a material can be evaluated by comparing the absorbed energies of notched versus unnotched specimens. The absorbed energies are much lower in notched specimens if the material is notch-sensitive. We will discuss in Section 6-15 how the presence of notches affect the behavior of materials subjected to cyclical stress.

Relationship to the Stress-Strain Diagram   The energy required to break a material during impact testing (i.e., the impact toughness) is not always related to the tensile toughness (i.e., the area contained within the true stress-true strain diagram (Figure 6-29)). As noted before, engineers often consider area under the engineering stress-strain curve as tensile toughness. In general, metals with both high strength and high ductility have good tensile toughness. However, this is not always the case when the strain rates are high. For example, metals that show excellent tensile toughness may show a brittle behavior under high strain rates (i.e., they may show poor impact toughness). Thus, strain rate can shift the ductile to brittle transition (DBTT). Ceramics and many composites normally have poor toughness, even though they have high strength, because they display virtually no ductility. These materials show both poor tensile toughness and poor impact toughness.

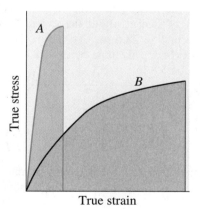

Figure 6-29
The area contained within the true stress-true strain curve is related to the tensile toughness. Although material *B* has a lower yield strength, it absorbs a greater energy than material *A*. The energies from these curves may not be the same as those obtained from impact test data.

Use of Impact Properties   Absorbed energy and DBTT are very sensitive to loading conditions. For example, a higher rate of applying energy to the specimen reduces the absorbed energy and increases the DBTT. The size of the specimen also affects the results; because it is more difficult for a thick material to deform, smaller energies are required to break thicker materials. Finally, the configuration of the notch affects the behavior; a sharp, pointed surface crack permits lower absorbed energies than does a V-notch. Because we often cannot predict or control all of these conditions, the impact test is a quick, convenient, and inexpensive way to compare different materials.

> ### EXAMPLE 6-7   *Design of a Sledgehammer*
>
> Design an 8-pound sledgehammer for driving steel fence posts into the ground.
>
> #### SOLUTION
>
> First, we must consider the design requirements to be met by the sledgehammer. A partial list would include:
>
> 1. The handle should be light in weight, yet tough enough that it will not catastrophically break.
> 2. The head must not break or chip during use, even in subzero temperatures.
> 3. The head must not deform during continued use.
> 4. The head must be large enough to assure that the user doesn't miss the fence post, and it should not include sharp notches that might cause chipping.
> 5. The sledgehammer should be inexpensive.
>
> Although the handle could be a lightweight, tough composite material (such as a polymer reinforced with Kevlar (a special polymer) fibers), a wood handle about 30 in. long would be much less expensive and would still provide sufficient toughness. As shown in a later chapter, wood can be categorized as a natural fiber-reinforced composite.
>
> To produce the head, we prefer a material that has a low transition temperature, can absorb relatively high energy during impact, and yet also has enough hardness to avoid deformation. The toughness requirement would rule out most ceramics. A face-centered cubic metal, such as FCC stainless steel or copper, might provide superior toughness even at low temperatures; however, these metals are relatively soft and expensive. An appropriate choice might be a normal BCC steel. Ordinary steels are inexpensive, have good hardness and strength, and some have sufficient toughness at low temperatures.
>
> In Appendix A, we find that the density of iron is 7.87 g/cm^3, or 0.28 lb/in.^{3} We assume that the density of steel is about the same. The volume of steel required is $V = 8$ lbs/(0.28 lb/in.^{3}) = 28.6 in.^{3} To assure that we will hit our target, the head might have a cylindrical shape, with a diameter of 2.5 in. The length of the head would then be 5.8 in.

6-10   Fracture Mechanics

Fracture mechanics is the discipline concerned with the behavior of materials containing cracks or other small flaws. The term "flaw" refers to features as small pores (holes), inclusions, or micro-cracks. The term "flaw" does *not* refer to atomic level defects such as vacancies, or dislocations. What we wish to know is the maximum stress that a material can withstand if it contains flaws of a certain size and geometry. **Fracture toughness** measures the ability of a material containing a flaw to withstand an applied load. Note that this does *not* require a high strain rate (impact).

A typical fracture toughness test may be performed by applying a tensile stress to a specimen prepared with a flaw of known size and geometry (Figure 6-30). The stress applied to the material is intensified at the flaw, which acts as a *stress raiser*. For a simple case, the *stress intensity factor*, *K*, is

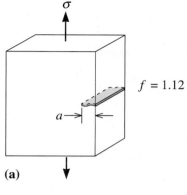

Figure 6-30
Schematic drawing of fracture toughness specimens with (a) edge and (b) internal flaws.

$$K = f\sigma\sqrt{\pi a}, \tag{6-18}$$

where f is a geometry factor for the specimen and flaw, σ is the applied stress, and a is the flaw size [as defined in Figure 6-30(b)]. Note that the analytical expression for K changes with the geometry of the flaw and specimen. If the specimen is assumed to have an "infinite" width, then $f \cong 1.0$. For a small single-edge notch [Figure 6-30(a)], $f = 1.12$.

By performing a test on a specimen with a known flaw size, we can determine the value of K that causes the flaw to grow and cause failure. This critical stress intensity factor is defined as the *fracture toughness, K_c,*

$$K_c = K \text{ required for a crack to propagate} \tag{6-19}$$

Fracture toughness depends on the thickness of the sample: As thickness increases, fracture toughness K_c decreases to a constant value (Figure 6-31). This constant is called the *plane strain fracture toughness, K_{Ic}.* It is K_{Ic} that is normally reported as the property of a material. The value of K_{Ic} does not depend upon the thickness of the sample. Table 6-6 compares the value of K_{Ic} to the yield strength of several materials. Units for fracture toughness are ksi $\sqrt{\text{in.}} = 1.0989$ MPa $\sqrt{\text{m}}$.

The ability of a material to resist the growth of a crack depends on a large number of factors:

1. Larger flaws reduce the permitted stress. Special manufacturing techniques, such as filtering impurities from liquid metals and hot pressing or hot isostatic pressing of powder particles to produce ceramic or superalloy components reduce flaw size and improve fracture toughness (Chapters 8 and 14).

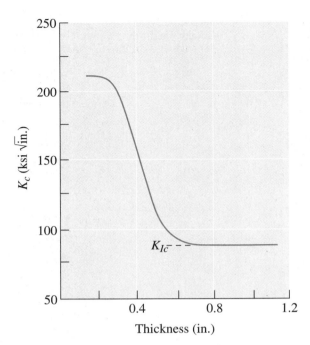

Figure 6-31
The fracture toughness K_c of a 300,000 psi yield strength steel decreases with increasing thickness, eventually leveling off at the plane strain fracture toughness K_{Ic}.

2. The ability of a material to deform is critical. In ductile metals, the material near the tip of the flaw can deform, causing the tip of any crack to become blunt, reducing the stress intensity factor, and preventing growth of the crack. Increasing the strength of a given metal usually decreases ductility and gives a lower fracture toughness. (See Table 6-6.) Brittle materials such as ceramics and many polymers have much lower fracture toughness than metals.

3. Thicker, more rigid pieces of a given material have a lower fracture toughness than thin materials.

4. Increasing the rate of application of the load, such as in an impact test, typically reduces the fracture toughness of the material.

TABLE 6-6 ■ *The plane strain fracture toughness K_{Ic} of selected materials*

Material	Fracture Toughness K_{Ic} (psi $\sqrt{in.}$)	Yield Strength or Ultimate Strength (for Brittle Solids) (psi)
Al-Cu alloy	22,000	66,000
	33,000	47,000
Ti-6% Al-4% V	50,000	130,000
	90,000	125,000
Ni-Cr steel	45,800	238,000
	80,000	206,000
Al_2O_3	1,600	30,000
Si_3N_4	4,500	80,000
Transformation toughened ZrO_2	10,000	60,000
Si_3N_4-SiC composite	51,000	120,000
Polymethyl methacrylate polymer	900	4,000
Polycarbonate polymer	3,000	8,400

5. Increasing the temperature normally increases the fracture toughness, just as in the impact test.

6. A small grain size normally improves fracture toughness, whereas more point defects and dislocations reduce fracture toughness. Thus, a fine-grained ceramic material may provide improved resistance to crack growth.

7. In certain ceramic materials we can also take advantage of stress-induced transformations that lead to compressive stresses that cause increased fracture toughness.

Fracture testing of ceramics cannot be performed easily using a sharp notch, since formation of such a notch often causes the samples to break. We can use hardness testing to gain a measure of the fracture toughness of many ceramics. When a ceramic material is indented, tensile stresses generate secondary cracks that form at the indentation and the length of secondary cracks ($2d$) provides a measure of the toughness of the ceramic material. In some cases, an indentation created using a hardness tester is used as a starter crack for the bend test. The K_{Ic} value is related to the hardness (H), Young's modulus (E), and crack dimensions by the following equation.

$$K_{Ic} = \alpha_0 \left(\frac{E}{H}\right)^{1/2} \left(\frac{P}{d^{3/2}}\right) \tag{6-20}$$

In this equation P is the indention load (in N), $2d$ is the secondary crack length (in meters), $\alpha_0 = 0.016$, a geometric parameter. In Equation 6-20, H and E have the units of N/m^2. The load, if it is in kgf, will have to be converted into N (1 kgf = 1.02 N). The resultant value of toughness is in N/m$^{3/2}$ (Pa m$^{1/2}$). Figure 6-32 shows the geometry for this equation. Figure 6-33 shows the indentation and secondary cracks in an alumina sample.

In general, this direct crack measurement method is better suited for comparison, rather than absolute measurements of fracture toughness values. The fracture toughness of many engineered materials is shown in Figure 6-34.

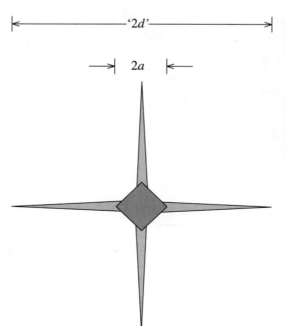

Figure 6-32
Secondary cracks developed during hardness testing can be used to assess the fracture toughness of brittle materials. Illustrated here is a schematic of such secondary cracks emanating from a diamond pyramid indenter (used in a Vicker's hardness test). The indentation length is 2a. Secondary cracks emanate from the indentation corners and have a length '2d' as measured on the sample surface. (*Source: Adapted from* Mechanical Behavior of Materials, *by T.H. Courtney, Fig. 1-21. Copyright © 2000 The McGraw-Hill Companies. Adapted with permission.*)

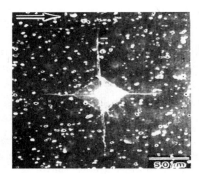

Figure 6-33
A scanning electron micrograph showing crack propagation in a *PZT* ceramic. (*Courtesy of Wang and Raj N. Singh, Ferroelectrics, 207, 555–575 (1998).*)

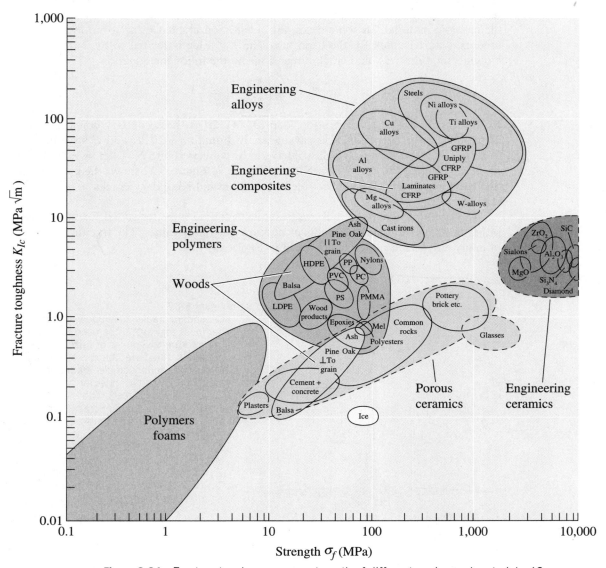

Figure 6-34 Fracture toughness versus strength of different engineered materials. (*Source: Adapted from Mechanical Behavior of Materials, by T.H. Courtney, 2000, p. 434, Fig. 9-18. Copyright © 2000 The McGraw-Hill Companies. Adapted with permission.*)

6-11   The Importance of Fracture Mechanics

The fracture mechanics approach allows us to design and select materials while taking into account the inevitable presence of flaws. There are three variables to consider: the property of the material (K_c or K_{Ic}), the stress σ that the material must withstand, and the size of the flaw a. If we know two of these variables, the third can be determined.

Selection of a Material   If we know the maximum size a of flaws in the material and the magnitude of the applied stress, we can select a material that has a fracture toughness K_c or K_{Ic} large enough to prevent the flaw from growing.

Design of a Component   If we know the maximum size of any flaw and the material (and therefore it's K_c or K_{Ic}) has already been selected, we can calculate the maximum stress that the component can withstand. Then we can design the appropriate size of the part to assure that the maximum stress is not exceeded.

Design of a Manufacturing or Testing Method   If the material has been selected, the applied stress is known, and the size of the component is fixed, we can calculate the maximum size of a flaw that can be tolerated. A nondestructive testing technique that detects any flaw greater than this critical size can help assure that the part will function safely. In addition, we find that, by selecting the correct manufacturing process, we can produce flaws that are all smaller than this critical size.

EXAMPLE 6-8   *Design of a Nondestructive Test*

A large steel plate used in a nuclear reactor has a plane strain fracture toughness of 80,000 psi $\sqrt{\text{in.}}$ and is exposed to a stress of 45,000 psi during service. Design a testing or inspection procedure capable of detecting a crack at the edge of the plate before the crack is likely to grow at a catastrophic rate.

SOLUTION

We need to determine the minimum size of a crack that will propagate in the steel under these conditions. From Equation 6-18, assuming that $f = 1.12$:

$$K_{Ic} = f\sigma\sqrt{a\pi}$$

$$80,000 = (1.12)(45,000)\sqrt{a\pi}$$

$$a = 0.8 \text{ in.}$$

A 0.8 in. deep crack on the edge should be relatively easy to detect. Often, cracks of this size can be observed visually. A variety of other tests, such as dye penetrant inspection, magnetic particle inspection, and eddy current inspection, also detect cracks much smaller than this. If the growth rate of a crack is slow and inspection is performed on a regular basis, a crack should be discovered long before reaching this critical size.

Brittle Fracture   Any crack or imperfection limits the ability of a ceramic to withstand a tensile stress. This is because a crack (sometimes called a **Griffith flaw**) concentrates and magnifies the applied stress. Figure 6-35 shows a crack of length a at the surface

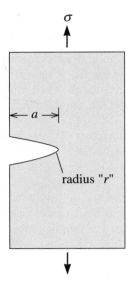

Figure 6-35
Schematic diagram of the Griffith flaw in a ceramic.

of a brittle material. The radius r of the crack is also shown. When a tensile stress σ is applied, the actual stress at the crack tip is:

$$\sigma_{\text{actual}} \cong 2\sigma\sqrt{a/r} \qquad (6\text{-}21)$$

For very thin cracks (r) or long cracks (a), the ratio $\sigma_{\text{actual}}/\sigma$ becomes large, or the stress is intensified. If the stress (σ_{actual}) exceeds the yield strength, the crack grows and eventually causes failure, even though the nominal applied stress σ is small.

In a different approach, we recognize that an applied stress causes an elastic strain, related to the modulus of elasticity, E, within the material. When a crack propagates, this strain energy is released, reducing the overall energy. At the same time, however, two new surfaces are created by the extension of the crack; this increases the energy associated with the surface. By balancing the strain energy and the surface energy, we find that the critical stress required to propagate the crack is given by the Griffith equation,

$$\sigma_{\text{critical}} \cong 2\sigma\sqrt{\frac{E\gamma}{\pi a}} \qquad (6\text{-}22)$$

where a is the length of a surface crack (or one-half the length of an internal crack) and γ is the surface energy (per unit area). Again, this equation shows that even small flaws severely limit the strength of the ceramic.

We also note that if we rearrange Equation 6-18, which described the stress intensity factor K, we obtain:

$$\sigma = \frac{K}{f\sqrt{\pi a}} \qquad (6\text{-}23)$$

This equation is similar in form to Equation 6-22. Each of these equations points out the dependence of the mechanical properties on the size of flaws present in the ceramic. Development of manufacturing processes to minimize the flaw size becomes crucial in improving the strength of ceramics.

The flaws are most important when tensile stresses act on the material. Compressive stresses try to close rather than open a crack; consequently, ceramics often have very good compressive strengths.

EXAMPLE 6-9   *Properties of SiAlON Ceramics*

Assume that an advanced ceramic, sialon (acronym for SiAlON or silicon aluminum oxynitride), has a tensile strength of 60,000 psi. Let us assume that this value is for a flaw-free ceramic. (In practice, it is almost impossible to produce flaw-free ceramics.) A thin crack 0.01 in. deep is observed before a sialon part is tested. The part unexpectedly fails at a stress of 500 psi by propagation of the crack. Estimate the radius of the crack tip.

SOLUTION

The failure occurred because the 500-psi applied stress, magnified by the stress concentration at the tip of the crack, produced an actual stress equal to the tensile strength. From Equation 6-21:

$$\sigma_{actual} = 2\sigma\sqrt{a/r}$$

$$60{,}000 \text{ psi} = (2)(500 \text{ psi})\sqrt{0.01 \text{ in.}/r}$$

$$\sqrt{0.01/r} = 60 \quad \text{or} \quad 0.01/r = 3600$$

$$r = 2.8 \times 10^{-6} \text{ in.} = 7.1 \times 10^{-6} \text{ cm} = 710 \text{ Å}$$

The likelihood of our being able to measure a radius of curvature of this size by any method of nondestructive testing is virtually zero. Therefore, although Equation 6-21 may help illustrate the factors that influence how a crack propagates in a brittle material, it does not help in predicting the strength of actual ceramic parts.

EXAMPLE 6-10   *Design of a Ceramic Support*

Design a supporting 3-in.-wide plate made of sialon, which has a fracture toughness of 9,000 psi $\sqrt{\text{in.}}$, that will withstand a tensile load of 40,000 lb. The part is to be nondestructively tested to assure that no flaws are present that might cause failure.

SOLUTION

Let's assume that we have three nondestructive testing methods available to us: X-ray radiography can detect flaws larger than 0.02 in., gamma-ray radiography can detect flaws larger than 0.008 in., and ultrasonic inspection can detect flaws larger than 0.005 in. For these flaw sizes, we must now calculate the minimum thickness of the plate that will assure that flaws of these sizes will not propagate.

From our fracture toughness equation, assuming that $f = 1$:

$$\sigma_{max} = \frac{K_{Ic}}{\sqrt{\pi a}} = \frac{F}{A}$$

$$A = \frac{F\sqrt{\pi a}}{K_{Ic}} = \frac{(40{,}000)(\sqrt{\pi})(\sqrt{a})}{9{,}000}$$

$$A = 7.88\sqrt{a} \text{ in.}^2 \quad \text{and} \quad \text{thickness} = (7.88 \text{ in.}^2/3 \text{ in.})\sqrt{a} = 2.63\sqrt{a}$$

NDT Method	Smallest Detectable Crack (in.)	Minimum Area (in.^{2})	Minimum Thickness (in.)	Maximum Stress (psi)
X-ray radiography	0.020	1.11	0.37	36,000
γ-ray radiography	0.008	0.70	0.23	57,000
Ultrasonic inspection	0.005	0.56	0.19	71,000

Our ability to detect flaws, coupled with our ability to produce a ceramic with flaws smaller than our detection limit, significantly affects the maximum stress than can be tolerated and, hence, the size of the part. In this example, the part can be smaller if ultrasonic inspection is available.

The fracture toughness is also important. Had we used Si_3N_4, with a fracture toughness of 3,000 psi $\sqrt{in.}$ instead of the sialon, we could repeat the calculations and show that, for ultrasonic testing, the minimum thickness is 0.56 in. and the maximum stress is only 24,000 psi.

6-12   Microstructural Features of Fracture in Metallic Materials

Ductile Fracture   Ductile fracture normally occurs in a **transgranular** manner (through the grains) in metals that have good ductility and toughness. Often, a considerable amount of deformation—including necking—is observed in the failed component. The deformation occurs before the final fracture. Ductile fractures are usually caused by simple overloads, or by applying too high a stress to the material.

In a simple tensile test, ductile fracture begins with the nucleation, growth, and coalescence of microvoids at the center of the test bar (Figure 6-36). **Microvoids** form when a high stress causes separation of the metal at grain boundaries or interfaces

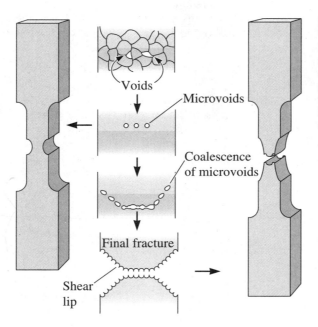

Figure 6-36
When a ductile material is pulled in a tensile test, necking begins and voids form—starting near the center of the bar—by nucleation at grain boundaries or inclusions. As deformation continues a 45° shear lip may form, producing a final cup and cone fracture.

Voids

Microvoids

Coalescence of microvoids

Final fracture

Shear lip

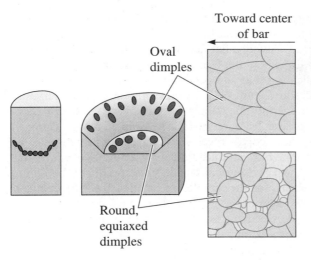

Figure 6-37
Dimples form during ductile fracture. Equiaxed dimples form in the center, where microvoids grow. Elongated dimples, pointing toward the origin of failure, form on the shear lip.

between the metal and small impurity particles (inclusions). As the local stress increases, the microvoids grow and coalesce into larger cavities. Eventually, the metal-to-metal contact area is too small to support the load and fracture occurs.

Deformation by slip also contributes to the ductile fracture of a metal. We know that slip occurs when the resolved shear stress reaches the critical resolved shear stress and that the resolved shear stresses are highest at a 45° angle to the applied tensile stress (Chapter 4, Schmid's law).

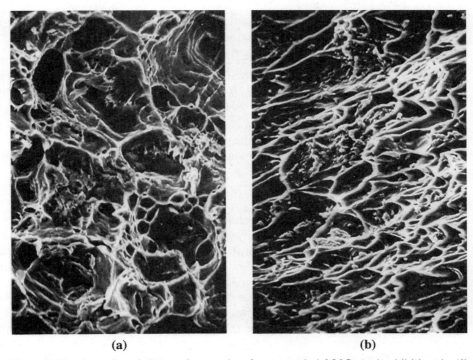

(a)   **(b)**

Figure 6-38 Scanning electron micrographs of an annealed 1018 steel exhibiting ductile fracture in a tensile test. (a) Equiaxed dimples at the flat center of the cup and cone, and (b) elongated dimples at the shear lip (×1250).

These two aspects of ductile fracture give the failed surface characteristic features. In thick metal sections, we expect to find evidence of necking, with a significant portion of the fracture surface having a flat face where microvoids first nucleated and coalesced, and a small shear lip, where the fracture surface is at a 45° angle to the applied stress. The shear lip, indicating that slip occurred, gives the fracture a cup and cone appearance (Figure 6-12 and Figure 6-37). Simple macroscopic observation of this fracture may be sufficient to identify the ductile fracture mode.

Examination of the fracture surface at a high magnification—perhaps using a scanning electron microscope—reveals a dimpled surface (Figure 6-37). The dimples are traces of the microvoids produced during fracture. Normally, these microvoids are round, or equiaxed, when a normal tensile stress produces the failure [Figure 6-38(a)]. However, on the shear lip, the dimples are oval-shaped, or elongated, with the ovals pointing toward the origin of the fracture [Figure 6-38(b)].

In a thin plate, less necking is observed and the entire fracture surface may be a shear face. Microscopic examination of the fracture surface shows elongated dimples rather than equiaxed dimples, indicating a greater proportion of 45° slip than in thicker metals.

EXAMPLE 6-11   *Hoist Chain Failure Analysis*

A chain used to hoist heavy loads fails. Examination of the failed link indicates considerable deformation and necking prior to failure. List some of the possible reasons for failure.

SOLUTION

This description suggests that the chain failed in a ductile manner by a simple tensile overload. Two factors could be responsible for this failure:

1. The load exceeded the hoisting capacity of the chain. Thus, the stress due to the load exceeded the yield strength of the chain, permitting failure. Comparison of the load to the manufacturer's specifications will indicate that the chain was not intended for such a heavy load. This is the fault of the user!

2. The chain was of the wrong composition or was improperly heat-treated. Consequently, the yield strength was lower than intended by the manufacturer and could not support the load. This may be the fault of the manufacturer!

Brittle Fracture   Brittle fracture occurs in high-strength metals and alloys or metals and alloys with poor ductility and toughness. Furthermore, even metals that are normally ductile may fail in a brittle manner at low temperatures, in thick sections, at high strain rates (such as impact), or when flaws play an important role. Brittle fractures are frequently observed when impact, rather than overload, causes failure.

In brittle fracture, little or no plastic deformation is required. Initiation of the crack normally occurs at small flaws, which cause a concentration of stress. The crack may move at a rate approaching the velocity of sound in the metal. Normally, the crack propagates most easily along specific crystallographic planes, often the {100} planes, by cleavage. In some cases, however, the crack may take an **intergranular** (along the grain boundaries) path, particularly when segregation (preferential separation of different elements) or inclusions weaken the grain boundaries.

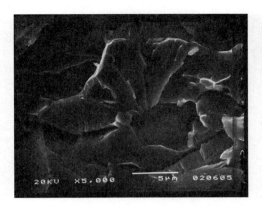

Figure 6-39
Scanning electron micrograph of a brittle fracture surface of a quenched 1010 steel (×5000). (*Courtesy of C.W. Ramsay.*)

Brittle fracture can be identified by observing the features on the failed surface. Normally, the fracture surface is flat and perpendicular to the applied stress in a tensile test. If failure occurs by cleavage, each fractured grain is flat and differently oriented, giving a crystalline or "rock candy" appearance to the fracture surface (Figure 6-39). Often, the layman claims that the metal failed because it crystallized. Of course, we know that the metal was crystalline to begin with and the surface appearance is due to the cleavage faces.

Another common fracture feature is the **Chevron pattern** (Figure 6-40), produced by separate crack fronts propagating at different levels in the material. A radiating pattern of surface markings, or ridges, fans away from the origin of the crack (Figure 6-41). The Chevron pattern is visible with the naked eye or a magnifying glass and helps us identify both the brittle nature of the failure process as well as the origin of the failure.

Figure 6-40
The Chevron pattern in a 0.5-in.-diameter quenched 4340 steel. The steel failed in a brittle manner by an impact blow.

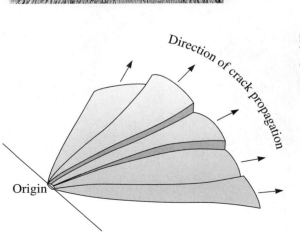

Figure 6-41
The Chevron pattern forms as the crack propagates from the origin at different levels. The pattern points back to the origin.

EXAMPLE 6-12   *Automobile Axle Failure Analysis*

An engineer investigating the cause of an automobile accident finds that the right rear wheel has broken off at the axle. The axle is bent. The fracture surface reveals a Chevron pattern pointing toward the surface of the axle. Suggest a possible cause for the fracture.

SOLUTION

The evidence suggests that the axle did not break prior to the accident. The deformed axle means that the wheel was still attached when the load was applied. The Chevron pattern indicates that the wheel was subjected to an intense impact blow, which was transmitted to the axle, causing failure. The preliminary evidence suggests that the driver lost control and crashed, and the force of the crash caused the axle to break. Further examination of the fracture surface, microstructure, composition, and properties may verify that the axle was manufactured properly.

6-13   Microstructural Features of Fracture in Ceramics, Glasses, and Composites

In ceramic materials, the ionic or covalent bonds permit little or no slip. Consequently, failure is a result of brittle fracture. Most crystalline ceramics fail by cleavage along widely spaced, closely packed planes. The fracture surface typically is smooth, and frequently no characteristic surface features point to the origin of the fracture [Figure 6-42(a)].

Glasses also fracture in a brittle manner. Frequently, a **conchoidal** fracture surface is observed. This surface contains a very smooth mirror zone near the origin of the fracture, with tear lines comprising the remainder of the surface [Figure 6-42(b)]. The tear lines point back to the mirror zone and the origin of the crack, much like the chevron pattern in metals.

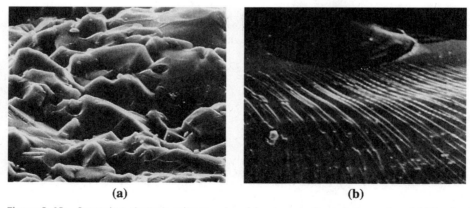

| (a) | (b) |

Figure 6-42   Scanning electron micrographs of fracture surfaces in ceramics. (a) The fracture surface of Al_2O_3, showing the cleavage faces ($\times 1250$), and (b) the fracture surface of glass, showing the mirror zone (top) and tear lines characteristic of conchoidal fracture ($\times 300$).

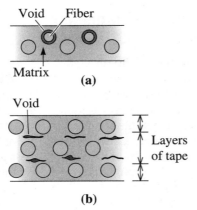

Figure 6-43
Fiber-reinforced composites can fail by several mechanisms. (a) Due to weak bonding between the matrix and fibers, fibers can pull out of the matrix, creating voids. (b) If the individual layers of the matrix are poorly bonded, the matrix may delaminate, creating voids.

Polymers can fail by either a ductile or a brittle mechanism. Below the glass temperature, thermoplastic polymers fail in a brittle manner—much like a glass. Likewise, the hard thermoset polymers, whose structure consists of inter-connected long chains of molecules, fail by a brittle mechanism. Some plastics whose structure consists of tangled but not chemically cross-linked chains, however, fail in a ductile manner above the glass temperature, giving evidence of extensive deformation and even necking prior to failure. The ductile behavior is a result of sliding of the polymer chains, which is not possible in glassy or thermosetting polymers. Thermosetting polymers have a rigid, three-dimensional cross-linked structure.

Fracture in fiber-reinforced composite materials is more complex. Typically, these composites contain strong, brittle fibers surrounded by a soft, ductile matrix, as in boron-reinforced aluminum. When a tensile stress is applied along the fibers, the soft aluminum deforms in a ductile manner, with void formation and coalescence eventually producing a dimpled fracture surface. As the aluminum deforms, the load is no longer transmitted effectively to the fibers; the fibers break in a brittle manner until there are too few of them left intact to support the final load.

Fracturing is more common if the bonding between the fibers and matrix is poor. Voids can then form between the fibers and the matrix, causing pull-out. Voids can also form between layers of the matrix if composite tapes or sheets are not properly bonded, causing **delamination** (Figure 6-43). Delamination, in this context, means the layers of different materials in a composite begin to come apart.

EXAMPLE 6-13   *Fracture in Composites*

Describe the difference in fracture mechanism between a boron-reinforced aluminum composite and a glass fiber-reinforced epoxy composite.

SOLUTION

In the boron-aluminum composite, the aluminum matrix is soft and ductile; thus we expect the matrix to fail in a ductile manner. Boron fibers, in contrast, fail in a brittle manner. Both glass fibers and epoxy are brittle; thus the composite as a whole should display little evidence of ductile fracture.

6-14 Weibull Statistics for Failure Strength Analysis

We need a statistical approach when evaluating the strength of ceramic materials. The strength of ceramics and glasses depends upon the size and distribution of sizes of flaws. In these materials, flaws originate from the processing process as the ceramics are being manufactured. These can result during machining, grinding, etc. Glasses can also develop microcracks as a result of interaction with water vapor in air. If we test alumina or other ceramic components of different sizes and geometry, we often find a large scatter in the measured values—even if their nominal composition is same. Similarly, if we are testing the strength of glass fibers of a given composition, we find that, on average, shorter fibers are stronger than longer fibers. The strength of ceramics and glasses depends upon the probability of finding a flaw that exceeds a certain critical size. For larger components or larger fibers this probability increases. As a result, the strength of larger components or fibers is likely to be lower than that of smaller components or shorter fibers. In metallic or polymeric materials, which can exhibit relatively large plastic deformations, the effect of flaws and flaw size distribution is not felt to the extent it is in ceramics and glasses. In these materials, cracks initiating from flaws get blunted by plastic deformation. Thus, for ductile materials, the distribution of strength is narrow and close to a Gaussian distribution. The strength of ceramics and glasses, however, varies considerably (i.e., if we test a large number of identical samples of silica glass or alumina ceramic, the data will show a wide scatter owing to changes in distribution of flaw sizes). The strength of brittle materials, such as ceramics and glasses, is not Gaussian; it is given by the **Weibull distribution**.[12] The Weibull distribution is an indicator of the variability of strength of materials resulting from a distribution of flaw sizes. This behavior results from critical sized flaws in materials with a distribution of flaw sizes (i.e., failure due to the weakest link of a chain).

The Weibull distribution shown in Figure 6-44 describes the fraction of samples that fail at different applied stresses. At low stresses, a small fraction of samples contain flaws large enough to cause fracture; most fail at an intermediate applied stress, and a few contain only small flaws and do not fail until large stresses are applied. To provide predictability, we prefer a very narrow distribution.

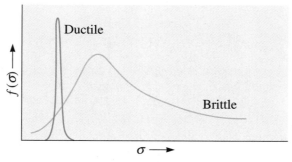

Figure 6-44 The Weibull distribution describes the fraction of the samples that fail at any given applied stress.

Consider a body of volume V with a distribution of flaws and subjected to a stress σ. If we assumed that the volume, V, was made up of n elements with volume V_0 and each element had the same flaw-size distribution, it can be shown that the survival probability, $P(V_0)$, (i.e., the probability that a brittle material will not fracture under the applied stress σ) is given by:

$$P(V_0) = \exp\left[-\left(\frac{\sigma - \sigma_u}{\sigma_0}\right)^m\right] \tag{6-24}$$

The probability of failure, $F(V_0)$, can be written as:

$$F(V_0) = 1 - P(V_0) = 1 - \exp\left[-\left(\frac{\sigma - \sigma_u}{\sigma_0}\right)^m\right] \tag{6-25}$$

In Equations 6-24 and 6-25, σ is the applied stress, σ_0 is characteristic strength (often assumed to be equal to the average strength), σ_u is the stress level below which the probability of failure is zero (i.e., the probability of survival is 1.0). In these equations, m is the Weibull modulus. In theory, Weibull modulus values can range from 0 to ∞. The Weibull modulus is a measure of the variability of the strength of the material.

The Weibull modulus m indicates the strength variability. For metals and alloys, the Weibull modulus is ~ 100. For traditional ceramics (e.g., bricks, pottery, etc.), the Weibull modulus is less than 3. Engineered ceramics, in which the processing is better controlled and hence the number of flaws is expected to be less, have a Weibull modulus of 5 to 10.

Note that for ceramics and other brittle solids, we can assume $\sigma_u = 0$. This is because there is no nonzero stress level for which we can claim a brittle material will not fail. For *brittle materials*, Equations 6-24 and 6-25 can be rewritten as follows:

$$P(V_0) = \exp\left[-\left(\frac{\sigma}{\sigma_0}\right)^m\right] \tag{6-26}$$

and

$$F(V_0) = 1 - P(V_0) = 1 - \exp\left[-\left(\frac{\sigma}{\sigma_0}\right)^m\right] \tag{6-27}$$

From Equation 6-26, for an applied stress σ of zero, the probability of survival is 1. As the applied stress σ increases, $P(V_0)$ decreases, approaching zero at very high values of applied stresses. We can also describe another meaning of the parameter σ_0. In Equation 6-26, when $\sigma = \sigma_0$, the probability of survival becomes $1/e \cong 0.37$. Therefore, σ_0 is the stress level for which the survival probability is $\cong 0.37$ or 37%. We can also state that σ_0 is the stress level for which the failure probability is $\cong 0.63$ or 63%.

Equations 6-26 and 6-27 can be modified to account for samples with different volumes. It can be shown that for an equal probability of survival samples with larger volumes will have lower strengths. This is what we mentioned before (e.g., longer glass fibers will be weaker than shorter glass fibers).

The following examples illustrate how the Weibull plots can be used for analysis of mechanical properties of materials and designing of components.

EXAMPLE 6-14   *Weibull Modulus for Steel and Alumina Ceramics*

Figure 6-45 shows the log-log plots of the probability of failure and strength of a 0.2% plain carbon steel, an alumina ceramics prepared using conventional powder processing in which alumina powders are compacted in a press and sintered into a dense mass at high temperature. Also, included is a plot for alumina ceramics prepared using special techniques that leads to much more uniform and controlled particle size. This in turn minimizes the flaws. These samples are labeled as controlled particle size (CPS). Comment on the nature of these graphs.

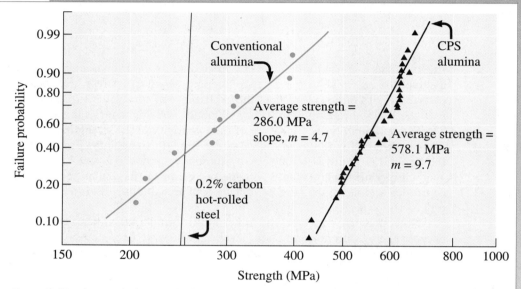

Figure 6-45   A cumulative plot (using special graph paper) of the probability that a sample will fail at any given stress yields the Weibull modulus or slope. Alumina produced by two different methods is compared with low carbon steel. Good reliability in design is obtained for a high Weibull modulus. (*Source: From* Mechanical Behavior of Materials, *by M.A. Meyers and K.K. Chawla, 1999. Copyright © 1999 Prentice-Hall. Used with permission of Pearson Education, Inc., Upper Saddle River, NJ.*)

SOLUTION

The failure probability and strength when plotted on a log-log scale result in data that can be fitted to a straight line. The slope of these lines provides us the measure of variability (i.e., the Weibull modulus).

For plain carbon steel the line is almost vertical (i.e., slope or m value is essentially approaching large values). This means that there is very little variation (5 to 10%) in the strength of different samples of the 0.2% C steel.

For alumina ceramics prepared using traditional processing, the variability is high (i.e., m is low ~4.7).

For ceramics prepared using improved and controlled processing techniques the m is higher ~9.7 indicating a more uniform distribution of flaws. The average strength is also higher (~578 MPa) suggesting lesser number of flaws that will lead to fracture.

EXAMPLE 6-15 *Strength of Ceramics and Probability of Failure*

An advanced engineered ceramic has a Weibull modulus $m = 9$. The flexural strength is 250 MPa at a probability of failure $F = 0.4$. What is the level of flexural strength if the probability of failure has to be 0.1?

SOLUTION

We assume all samples tested had the same volume, thus the size of the sample will not be a factor in this case. We can use the symbol V for sample volume instead of V_0. We are dealing with a brittle material, so we begin with Equation 6-27.

$$F(V) = 1 - P(V) = 1 - \exp\left[-\left(\frac{\sigma}{\sigma_0}\right)^m\right]$$

or

$$1 - F(V) = \exp\left[-\left(\frac{\sigma}{\sigma_0}\right)^m\right]$$

Take the logarithm of both sides to get

$$\ln[1 - F(V)] = \left[-\left(\frac{\sigma}{\sigma_0}\right)^m\right]$$

Take logarithms of both sides again,

$$\ln\{\ln[1 - F(V)]\} = -m(\ln \sigma - \ln \sigma_0) \qquad (6\text{-}28)$$

We can eliminate the minus sign on the right-hand side of Equation 6-28 by rewriting it as:

$$\ln\left[\ln\left(\frac{1}{1 - F(V)}\right)\right] = m(\ln \sigma - \ln \sigma_0) \qquad (6\text{-}29)$$

For $F = 0.4$, $\sigma = 250$ MPa, and $m = 9$, so from Equation 6-29, we have

$$\ln\left[\ln\left(\frac{1}{1 - 0.4}\right)\right] = 9(\ln 250 - \ln \sigma_0) \qquad (6\text{-}30)$$

Therefore, $\ln\{\ln 1/0.6)\} = \ln\{(\ln 1.66667)\} = \ln\{0.510826\} = -0.67173 = 9(5.52146 - \ln \sigma_0)$.

Therefore, $\ln \sigma_0 = 5.52146 + 0.07464 = 5.5961$. This gives us a value of $\sigma_0 = 269.4$ MPa. This is the characteristic strength of the ceramic, often taken as the average strength of the ceramic. For a stress level of 269.4 MPa, the probability of survival is 0.37 (or the probability of failure is 0.63). As the required probability of failure (F) goes down, the stress level to which the ceramic can be subjected (σ) also goes down.

Now, we want to determine the value of σ for $F = 0.1$. We know that $m = 9$ and $\sigma_0 = 269.4$ MPa, so we need to get the value of σ. We substitute these values into Equation 6-29:

$$\ln\left[\ln\left(\frac{1}{1 - 0.1}\right)\right] = 9(\ln \sigma - \ln 269.4)$$

$$\ln\left[\ln\left(\frac{1}{0.9}\right)\right] = 9(\ln \sigma - \ln 269.4)$$

$$\ln(\ln 1.11111) = \ln(0.105361) = -2.25037 = 9(\ln \sigma - 5.596097),$$

$$\therefore \quad -0.25004 = \ln \sigma - 5.596097, \quad \text{or}$$

$$\ln \sigma = 5.346056$$

or $\sigma = 209.8$ MPa. As expected, as we lowered the probability of failure to 0.1, we also decreased the level of stress that can be supported.

EXAMPLE 6-16   *Weibull Modulus Parameter Determination*

Seven silicon carbide specimens were tested and the following fracture strengths were obtained: 23, 49, 34, 30, 55, 43, and 40 MPa. Estimate the Weibull modulus for the data by fitting the data to Equation 6-29. Discuss the reliability of the ceramic.

SOLUTION

First, we point out that for any type of statistical analysis we need a large number of samples. Seven samples are not enough. The purpose of this example is to illustrate the calculation.

One simple, though not completely accurate, method for determining the behavior of the ceramic is to assign a numerical rank (1 to 7) to the specimens, with the specimen having the lowest fracture strength assigned the value 1.

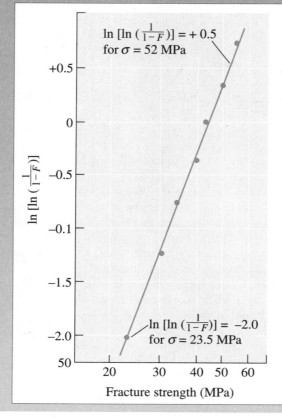

Figure 6-46
Plot of cumulative probability of failure versus fracture stress. Note the fracture strength is plotted on a log scale.

$\ln \left[\ln \left(\frac{1}{1-F} \right) \right] = +0.5$
for $\sigma = 52$ MPa

$\ln \left[\ln \left(\frac{1}{1-F} \right) \right] = -2.0$
for $\sigma = 23.5$ MPa

The total number of specimens is n (in our case, 7). The probability of failure F is then the numerical rank divided by $n + 1$ (in our case, 8). We can then plot $\ln[\ln(1/1 - F(V_0))]$ versus $\ln \sigma$. The following table and Figure 6-46 show the results of these calculations. Note that σ is plotted on a log scale.

i^{th} Specimen	σ (MPa)	$F(V_0)$	$\ln\{\ln 1/[1 - F(V_0)]\}$
1	23	1/8 = 0.125	−2.013
2	30	2/8 = 0.250	−1.246
3	34	3/8 = 0.375	−0.755
4	40	4/8 = 0.500	−0.367
5	43	5/8 = 0.625	−0.019
6	49	6/8 = 0.750	+0.327
7	55	7/8 = 0.875	+0.732

The slope of the fitted line, or the Weibull modulus m, is (using the two points indicated on the curve):

$$m = \frac{0.5 - (-2.0)}{\ln(52) - \ln(23.5)} = \frac{2.5}{3.951 - 3.157} = 3.15$$

This low Weibull modulus of 3.15 suggests that the ceramic has a highly variable fracture strength, making it difficult to use reliably in high load-bearing applications.

6-15 Fatigue

Fatigue is the lowering of strength or failure of a material due to repetitive stress which may be above or below the yield strength.[13,14] It is a common phenomenon in load-bearing components in cars and airplanes, turbine blades, springs, crankshafts and other machinery, biomedical implants, and consumer products, such as shoes, that are subjected constantly to repetitive stresses in the form of tension, compression, bending, vibration, thermal expansion and contraction, or other stresses. These stresses are often *below* the yield strength of the material! However, when the stress occurs a sufficient number of times, it causes failure by fatigue! Quite a large fraction of components found in an automobile junkyard belongs to those that failed by fatigue. The possibility of a fatigue failure is the main reason why aircraft components have a finite life. Fatigue is an interesting phenomenon in that load-bearing components can fail while the overall stress applied may not exceed the yield stress! Fatigue can occur even if the components are subjected to stress above the yield strength. A component is often subjected to the repeated application of a stress below the yield strength of the material.

There are examples of situations in which certain materials, such as thin films of ferroelectric materials, also show fatigue. This fatigue is electrical in nature and it is linked to the eventual inability of materials to show changes in electrical properties in response to the applied electric field. A detailed discussion of this is outside the scope of this book. The point is anytime we have a component that is going to be subjected to mechanical, electrical, thermal, and magnetic or other forces that are likely to be cyclical, we need to look at the effect of these external factors over a long period of time.

Fatigue failures typically occur in three stages. First, a tiny crack initiates or nucleates typically at the surface, often at a time well after loading begins. Normally, nucleation sites are at or near the surface, where the stress is at a maximum, and include surface defects such as scratches or pits, sharp corners due to poor design or manufacture, inclusions, grain boundaries, or dislocation concentrations. Next, the crack gradually propagates as the load continues to cycle. Finally, a sudden fracture of the material occurs when the remaining cross-section of the material is too small to support the applied load. Thus, components fail by fatigue because even though the overall applied stress may remain below the yield stress, at a local length scale the stress intensity exceeds the yield strength. For fatigue to occur, at least part of the stress in the material has to be tensile. We normally are concerned with fatigue of metallic and polymeric materials.

In ceramics, we normally do not consider fatigue since ceramics typically fail because of their low fracture toughness. Any fatigue cracks that may form will lower the useful life of the ceramic since it will cause the lowering of the fracture toughness. In general, we design ceramics for static (and not cyclic) loading and we factor in the Weibull modulus.

Polymeric materials also show fatigue failure. The mechanism of fatigue in polymers is different than that in metallic materials. In polymers, as the materials are subjected to repetitive stresses, considerable heating can occur near the crack tips and the inter-relationships between fatigue and another mechanism, known as **creep** (discussed in Section 6-18), affect the overall behavior.

Fatigue is also important in dealing with composites. As fibers or other reinforcing phases begin to degrade as a result of fatigue, the overall elastic modulus of the composite decreases and this weakening will be seen before the fracture due to fatigue.

Fatigue failures are often easy to identify. The fracture surface—particularly near the origin—is typically smooth. The surface becomes rougher as the original crack increases in size and may be fibrous during final crack propagation. Microscopic and macroscopic examinations reveal a fracture surface including a beach mark pattern and striations (Figure 6-47). **Beach** or **clamshell marks** (Figure 6-48) are normally formed when the load is changed during service or when the loading is intermittent, perhaps

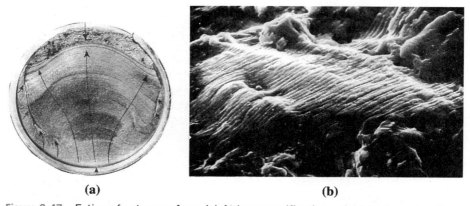

(a)   (b)

Figure 6-47 Fatigue fracture surface. (a) At low magnifications, the beach mark pattern indicates fatigue as the fracture mechanism. The arrows show the direction of growth of the crack front, whose origin is at the bottom of the photograph. (*Image* (a) is *from C.C. Cottell, "Fatigue Failures with Special Reference to Fracture Characteristics,"* Failure Analysis: The British Engine Technical Reports, *American Society for Metals, 1981, p. 318.*) (b) At very high magnifications, closely spaced striations formed during fatigue are observed (×1000).

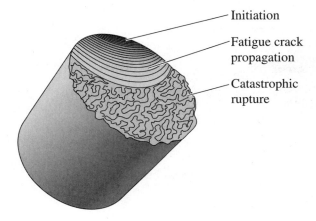

— Initiation

— Fatigue crack propagation

— Catastrophic rupture

Figure 6-48
Schematic representation of a fatigue fracture surface in a steel shaft, showing the initiation region, the propagation of fatigue crack (with beam markings), and catastrophic rupture when the crack length exceeds a critical value at the applied stress.

permitting time for oxidation inside the crack. **Striations**, which are on a much finer scale, show the position of the crack tip after each cycle. Beach marks always suggest a fatigue failure, but—unfortunately—the absence of beach marks does not rule out fatigue failure.

EXAMPLE 6-17   *Fatigue Failure Analysis of a Crankshaft*

A crankshaft in a diesel engine fails. Examination of the crankshaft reveals no plastic deformation. The fracture surface is smooth. In addition, several other cracks appear at other locations in the crankshaft. What type of failure mechanism would you expect?

SOLUTION

Since the crankshaft is a rotating part, the surface experiences cyclical loading. We should immediately suspect fatigue. The absence of plastic deformation supports our suspicion. Furthermore, the presence of other cracks is consistent with fatigue; the other cracks didn't have time to grow to the size that produced catastrophic failure. Examination of the fracture surface will probably reveal beach marks or fatigue striations.

A conventional and older method used to measure a material's resistance to fatigue is the **rotating cantilever beam test** (Figure 6-49). One end of a machined, cylindrical specimen is mounted in a motor-driven chuck. A weight is suspended from the opposite end. The specimen initially has a tensile force acting on the top surface, while the bottom surface is compressed. After the specimen turns 90°, the locations that were originally in tension and compression have no stress acting on them. After a half revolution of 180°, the material that was originally in tension is now in compression. Thus, the stress at any one point goes through a complete sinusoidal cycle from maximum tensile stress to maximum compressive stress. The maximum stress acting on this type of specimen is given by

$$\pm\sigma = \frac{32\,M}{\pi d^3} \tag{6-31a}$$

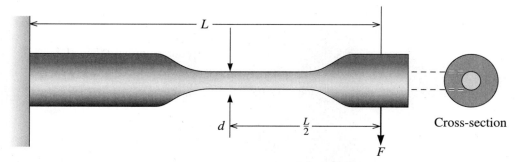

Figure 6-49   Geometry for the rotating cantilever beam specimen setup.

In this equation M is the bending moment at the cross-section, and d is the specimen diameter. The bending moment $M = F \cdot (L/2)$, and therefore,

$$\pm\sigma = \frac{16\,FL}{\pi d^3} = 5.09\frac{FL}{d^3} \tag{6-31b}$$

where L is the distance between the bending force location and the support (Figure 6-49), F is the load, and d is the diameter.

Newer machines used for fatigue testing are known as direct-loading machines. In these machines, a servo-hydraulic system, an actuator, and a control system, driven by computers, applies a desired force, deflection, displacement or strain. In some of these machines, temperature and atmosphere (e.g., humidity level) also can be controlled.

After a sufficient number of cycles in a fatigue test, the specimen may fail. Generally, a series of specimens are tested at different applied stresses. The results are presented as an **S-N curve** (also known as the **Wöhler curve**), with the stress (S) plotted versus the number of cycles (N) to failure (Figure 6-50).

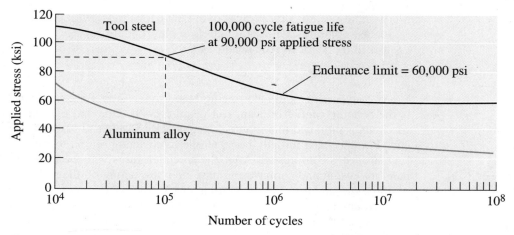

Figure 6-50   The stress-number of cycles to failure (S-N) curves for a tool steel and an aluminum alloy.

6-16 Results of the Fatigue Test

The fatigue test can tell us how long a part may survive or the maximum allowable loads that can be applied to prevent failure. The **endurance limit**, which is the stress below which there is a 50% probability that failure by fatigue will never occur, is our preferred design criterion. To prevent a tool steel part from failing (Figure 6-50), we must be sure that the applied stress is below 60,000 psi. The assumption of existence of an endurance limit is a relatively older concept. Recent research on many metals has shown that probably an endurance limit does not exist. We also need to account for the presence of corrosion, occasional overloads, and other mechanisms that may can cause the material to fail below the endurance limit. Thus, values for an endurance limit should be treated with caution.

Fatigue life tells us how long a component survives at a particular stress. For example, if the tool steel (Figure 6-50) is cyclically subjected to an applied stress of 90,000 psi, the fatigue life will be 100,000 cycles. Knowing the time associated with each cycle, we can calculate a fatigue life value in years. **Fatigue strength** is the maximum stress for which fatigue will not occur within a particular number of cycles, such as 500,000,000. The fatigue strength is necessary for designing with aluminum and polymers, which have no endurance limit.

In some materials, including steels, the endurance limit is approximately half the tensile strength. The ratio is the **endurance ratio**:

$$\text{Endurance ratio} = \frac{\text{endurance limit}}{\text{tensile strength}} \approx 0.5 \qquad (6\text{-}32)$$

The endurance ratio allows us to estimate fatigue properties from the tensile test. The endurance ratio values are ~0.3 to 0.4 for metallic materials other than low and medium strength steels. *Again, recall the cautionary note that has shown that endurance limit does not exist for many materials.*

Most materials are **notch sensitive**, with the fatigue properties particularly sensitive to flaws at the surface. Design or manufacturing defects concentrate stresses and reduce the endurance limit, fatigue strength, or fatigue life. Sometimes highly polished surfaces are prepared in order to minimize the likelihood of a fatigue failure. **Shot peening** is also a process that is used very effectively to enhance fatigue life of materials. Small metal spheres are shot at the component. This leads to a residual compressive stress at the surface similar to **tempering** of inorganic glasses (discussed later in this chapter).

EXAMPLE 6-18   *Design of a Rotating Shaft*

A solid shaft for a cement kiln produced from the tool steel in Figure 6-50 must be 96 in. long and must survive continuous operation for one year with an applied load of 12,500 lb. The shaft makes one revolution per minute during operation. Design a shaft that will satisfy these requirements.

SOLUTION

The fatigue life required for our design is the total number of cycles N that the shaft will experience in one year:

$$N = (1 \text{ cycle}/min)(60\ min/h)(24\ h/d)(365\ d/y)$$

$$N = 5.256 \times 10^5 \text{ cycles}/y$$

where y = year, d = day, and h = hour.

From Figure 6-50, the applied stress therefore must be less than about 72,000 psi. If Equation 6-31 is appropriate, then the diameter of the shaft must be:

$$\pm \sigma = \frac{16FL}{\pi d^3} = 5.09 \frac{FL}{d^3}$$

$$72{,}000 \text{ psi} = \frac{(5.09)(96 \text{ in.})(12{,}500 \text{ lb})}{d^3}$$

$$d = 4.39 \text{ in.}$$

A shaft with a diameter of 4.39 in. should operate for one year under these conditions. However, a significant margin of safety might be incorporated in the design. In addition, we might consider producing a shaft that would never fail.

Let us assume the factor of safety to be 2 (i.e., we will assume that the maximum allowed stress level will be 72,000/2 = 36,000 psi). The minimum diameter required to prevent failure would now be:

$$36{,}000 \text{ psi} = \frac{(5.09)(96 \text{ in.})(12{,}500 \text{ lb})}{d^3}$$

$$d = 5.53 \text{ in.}$$

Selection of a larger shaft reduces the stress level and makes fatigue less likely to occur or delay the failure. Other considerations might, of course, be important. High temperatures and corrosive conditions are inherent in producing cement. If the shaft is heated or attacked by the corrosive environment, fatigue is accelerated. Thus, in the applications involving fatigue of components regular inspections of the components go a long way toward avoiding a catastrophic failure.

6-17 Application of Fatigue Testing

Material components are often subjected to loading conditions that do not give equal stresses in tension and compression (Figure 6-51). For example, the maximum stress during compression may be less than the maximum tensile stress. In other cases, the loading may be between a maximum and a minimum tensile stress; here the S-N curve is presented as the stress amplitude versus number of cycles to failure. *Stress amplitude* (σ_a) is defined as half of the difference between the maximum and minimum stresses; *mean stress* (σ_m) is defined as the average between the maximum and minimum stresses:

$$\sigma_a = \frac{\sigma_{\max} - \sigma_{\min}}{2} \tag{6-33}$$

$$\sigma_m = \frac{\sigma_{\max} + \sigma_{\min}}{2} \tag{6-34}$$

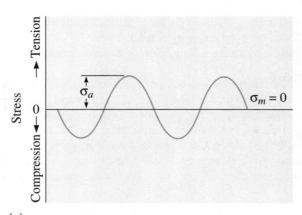

(a)

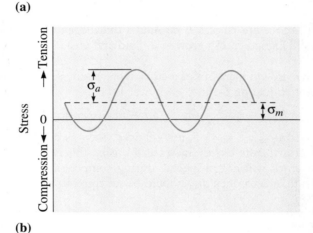

(b)

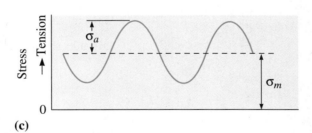

(c)

Figure 6-51
Examples of stress cycles. (a) Equal stress in tension and compression, (b) greater tensile stress than compressive stress, and (c) all of the stress is tensile.

A compressive stress is a "negative" stress. Thus, if the maximum tensile stress is 50,000 psi and the minimum stress is a 10,000 psi compressive stress, using Equations 6-33 and 6-34 the stress amplitude is 30,000 psi and the mean stress is 20,000 psi.

As the mean stress increases, the stress amplitude must decrease in order for the material to withstand the applied stresses. This condition can be summarized by the Goodman relationship:

$$\sigma_a = \sigma_{fs}\left[1 - \left(\frac{\sigma_m}{\sigma_{TS}}\right)\right] \tag{6-35}$$

where σ_{fs} is the desired fatigue strength for zero mean stress and σ_{TS} is the tensile strength of the material. Therefore, in a typical rotating cantilever beam fatigue test, where the mean stress is zero, a relatively large stress amplitude can be tolerated without fatigue. If, however, an airplane wing is loaded near its yield strength, vibrations of even a small amplitude may cause a fatigue crack to initiate and grow.

Crack Growth Rate   In many cases, a component may not be in danger of failure even when a crack is present. To estimate when failure might occur, the rate of propagation of a crack becomes important. Figure 6-52 shows the crack growth rate versus the range of the stress-intensity factor ΔK, which characterizes crack geometry and the stress amplitude. Below a threshold ΔK, a crack does not grow; for somewhat higher stress-intensities, cracks grow slowly; and at still higher stress-intensities, a crack grows at a rate given by:

$$\frac{da}{dN} = C(\Delta K)^n \tag{6-36}$$

In this equation, C and n are empirical constants that depend upon the material. Finally, when ΔK is still higher, cracks grow in a rapid and unstable manner until fracture occurs.

The rate of crack growth increases as a crack increases in size, as predicted from the stress intensity factor (Equation 6-18):

$$\Delta K = K_{max} - K_{min} = f\sigma_{max}\sqrt{\pi a} - f\sigma_{min}\sqrt{\pi a} = f\Delta\sigma\sqrt{\pi a} \tag{6-37}$$

If the cyclical stress $\Delta\sigma(\sigma_{max} - \sigma_{min})$ is not changed, then as crack length a increases, ΔK and the crack growth rate da/dN increase. In using this expression, however, one should note that a crack will not propagate during compression. Therefore, if σ_{min} is compressive, or less than zero, then σ_{min} should be set equal to zero.

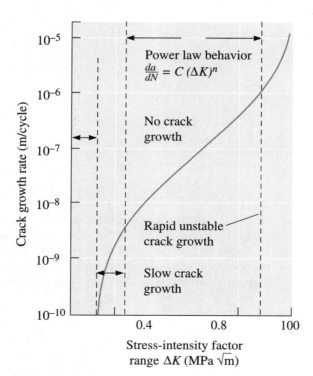

Figure 6-52
Crack growth rate versus stress-intensity factor range for a high-strength steel. For this steel, $C = 1.62 \times 10^{-12}$ and $n = 3.2$ for the units shown.

Knowledge of crack growth rate is of assistance in designing components and in nondestructive evaluation to determine if a crack poses imminent danger to the structure. One approach to this problem is to estimate the number of cycles required before failure occurs. By rearranging Equation 6-36 and substituting for ΔK:

$$dN = \frac{1}{Cf^n \Delta\sigma^n \pi^{n/2}} \frac{da}{a^{n/2}}$$

If we integrate this expression between the initial size of a crack and the crack size required for fracture to occur, we find that

$$N = \frac{2[(a_c)^{(2-n)/2} - (a_i)^{(2-n)/2}]}{(2-n)Cf^n \Delta\sigma^n \pi^{n/2}} \tag{6-38}$$

where a_i is the initial flaw size and a_c is the flaw size required for fracture. If we know the material constants n and C in Equation 6-36, we can estimate the number of cycles required for failure for a given cyclical stress (Example 6-19).

EXAMPLE 6-19   *Design of a Fatigue-Resistant Plate*

A high-strength steel plate (Figure 6-52), which has a plane strain fracture toughness of 80 MPa $\sqrt{m}$, is alternately loaded in tension to 500 MPa and in compression to 60 MPa. The plate is to survive for 10 years, with the stress being applied at a frequency of once every 5 minutes. Design a manufacturing and testing procedure that assures that the component will serve as intended.

SOLUTION

To design our manufacturing and testing capability, we must determine the maximum size of any flaws that might lead to failure within the 10-year period. The critical crack size (a_c), using the fracture toughness and the maximum stress, is:

$$K_{Ic} = f\sigma\sqrt{\pi a_c}$$
$$80 \text{ MPa}\sqrt{m} = (1)(500 \text{ MPa})\sqrt{\pi a_c}$$
$$a_c = 0.0081 \text{ m} = 8.1 \text{ mm}$$

The maximum stress is 500 MPa; however, the minimum stress is zero, not 60 MPa in compression, because cracks do not propagate in compression. Thus, $\Delta\sigma$ is:

$$\Delta\sigma = \sigma_{max} - \sigma_{min} = 500 - 0 = 500 \text{ MPa}$$

We need to determine the minimum number of cycles that the plate must withstand:

$$N = (1 \text{ cycle}/5 \text{ min})(60 \text{ min/h})(24 \text{ h/d})(365 \text{ d/y})(10 \text{ y})$$
$$N = 1,051,200 \text{ cycles}$$

If we assume that $f = 1$ for all crack lengths and note that $C = 1.62 \times 10^{-12}$ and $n = 3.2$ in Equation 6-38, then;

$$1,051,200 = \frac{2[(0.008)^{(2-3.2)/2} - (a_i)^{(2-3.2)/2}]}{(2-3.2)(1.62 \times 10^{-12})(1)^{3.2}(500)^{3.2}\pi^{3.2/2}}$$

$$1,051,200 = \frac{2[18 - a_i^{-0.6}]}{(-1.2)(1.62 \times 10^{-12})(1)(4.332 \times 10^8)(6.244)}$$

$$a_i^{-0.6} = 18 + 2764 = 2782$$

$$a_i = 1.82 \times 10^{-6} \text{ m} = 0.00182 \text{ mm for surface flaws}$$

$$2a_i = 0.00364 \text{ mm for internal flaws}$$

The manufacturing process must produce surface flaws smaller than 0.00182 mm in length. We can conduct a similar calculation for specifying a limit on edge cracks. In addition, nondestructive tests must be available to assure that cracks exceeding this length are not present.

Effect of Temperature   As the material's temperature increases, both fatigue life and endurance limit decrease. Furthermore, a cyclical temperature change encourages failure by thermal fatigue; when the material heats in a nonuniform manner, some parts of the structure expand more than others. This nonuniform expansion introduces a stress within the material, and when the structure later cools and contracts, stresses of the opposite sign are imposed. As a consequence of the thermally induced stresses and strains, fatigue may eventually occur. The frequency with which the stress is applied also influences fatigue behavior. In particular, high-frequency stresses may cause polymer materials to heat; at increased temperature, polymers fail more quickly. Chemical effects of temperature (e.g., oxidation) must also be considered.

6-18   Creep, Stress Rupture, and Stress Corrosion

If we apply stress to a material at an elevated temperature, the material may stretch and eventually fail, even though the applied stress is *less* than the yield strength at that temperature. A time dependent permanent deformation under a constant load or constant stress and at high temperatures is known as **creep**. A large number of failures occurring in components used at high temperatures can be attributed to creep or a combination of creep and fatigue. Essentially, in creep the material begins to flow slowly. Diffusion, dislocation glide or climb, or grain boundary sliding can contribute to the creep of metallic materials. Polymeric materials also show creep. In ductile metals and alloys subjected to creep, fracture is accompanied by necking, void nucleation and coalescence, or grain boundary sliding (Figure 6-53).

A material is considered failed by creep even if it has *not* actually fractured. When a material does actually creep and then ultimately break the fracture is defined as **stress rupture**. Normally, ductile stress-rupture fractures include necking and the presence of many cracks that did not have an opportunity to produce final fracture. Furthermore, grains near the fracture surface tend to be elongated. Ductile stress-rupture failures generally occur at high creep rates and relatively low exposure temperatures and have short rupture times. Brittle stress-rupture failures usually show little necking and occur more often at smaller creep rates and high temperatures. Equiaxed grains are observed near the fracture surface. Brittle failure typically occurs by formation of voids at the intersection of three grain boundaries and precipitation of additional voids along grain boundaries by diffusion processes (Figure 6-54).

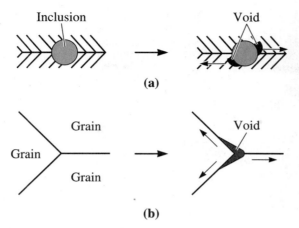

Figure 6-53
Grain boundary sliding during creep causes (a) the creation of voids at an inclusion trapped at the grain boundary, and (b) the creation of a void at a triple point where three grains are in contact.

Stress-Corrosion **Stress-corrosion** is a phenomenon in which materials react with corrosive chemicals in the environment. This leads to formation of cracks and lowering of strength. Stress-corrosion can occur at stresses well below the yield strength of the metallic, ceramic, or glassy material due to attack by a corrosive medium. In metallic materials, deep, fine corrosion cracks are produced, even though the metal as a whole shows little uniform attack. The stresses can be either externally applied or stored residual stresses. Stress-corrosion failures are often identified by microstructural examination of the nearby metal. Ordinarily, extensive branching of the cracks along grain boundaries is observed (Figure 6-55). The location at which cracks initiated may be identified by the presence of a corrosion product.

Inorganic silicate glasses are especially prone to failure by reaction with water vapor. It is well known that the strength of silica fibers or silica glass products is very high when these materials are protected from water vapor. As the fibers or silica glass components get exposed to water vapor, corrosion reactions begin leading to formation of surface flaws, which ultimately cause the cracks to grow when stress is applied. As discussed in Chapter 5, polymeric coatings are applied to optical fibers to prevent them from reacting with water vapor. For bulk glasses, special heat treatments such as tempering are used. Tempering produces an overall compressive stress on the surface of glass. Thus, even if the glass surface reacts with water vapor the cracks do not grow since the overall stress at the surface is compressive. If we create a flaw that will pene-

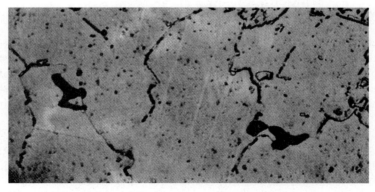

Figure 6-54 Creep cavities formed at grain boundaries in an austentic stainless steel (×500). (*From* ASM Handbook, *Vol. 7, (1972) ASM International, Materials Park, OH 44073.*)

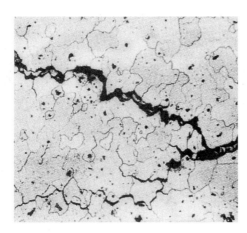

Figure 6-55
Photomicrograph of a metal near a stress-corrosion fracture, showing the many intergranular cracks formed as a result of the corrosion process (×200). (*From* ASM Handbook, *Vol. 7, (1972) ASM International, Materials Park, OH 44073.*)

trate the compressive stress region on the surface, tempered glass will shatter. Tempered glass is used widely in building and automotive applications.

EXAMPLE 6-20   *Failure Analysis of a Pipe*

A titanium pipe used to transport a corrosive material at 400°C is found to fail after several months. How would you determine the cause for the failure?

SOLUTION

Since a period of time at a high temperature was required before failure occurred, we might first suspect a creep or stress-corrosion mechanism for failure. Microscopic examination of the material near the fracture surface would be advisable. If many tiny, branched cracks leading away from the surface are noted, stress-corrosion is a strong possibility. However, if the grains near the fracture surface are elongated, with many voids between the grains, creep is a more likely culprit.

6-19   Evaluation of Creep Behavior

To determine the creep characteristics of a material, a constant stress is applied to a heated specimen in a **creep test**. As soon as the stress is applied, the specimen stretches elastically a small amount ε_0 (Figure 6-56), depending on the applied stress and the modulus of elasticity of the material at the high temperature. Creep testing can also be conducted under a constant load and is important from an engineering design viewpoint.

Dislocation Climb   High temperatures permit dislocations in a metal to **climb**. In climb, atoms move either to or from the dislocation line by diffusion, causing the dislocation to move in a direction that is perpendicular, not parallel, to the slip plane (Figure 6-57). The dislocation escapes from lattice imperfections, continues to slip, and causes additional deformation of the specimen even at low applied stresses.

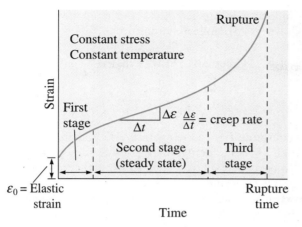

Figure 6-56
A typical creep curve showing the strain produced as a function of time for a constant stress and temperature.

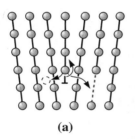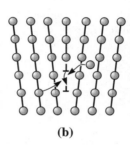

Figure 6-57
Dislocations can climb (a) when atoms leave the dislocation line to create interstitials or to fill vacancies or (b) when atoms are attached to the dislocation line by creating vacancies or eliminating interstitials.

Creep Rate and Rupture Times   During the creep test, strain or elongation is measured as a function of time and plotted to give the creep curve (Figure 6-56). In the first stage of creep of metals, many dislocations climb away from obstacles, slip, and contribute to deformation. Eventually, the rate at which dislocations climb away from obstacles equals the rate at which dislocations are blocked by other imperfections. This leads to second-stage, or steady-state, creep. The slope of the steady-state portion of the creep curve is the **creep rate**:

$$\text{Creep rate} = \frac{\Delta \text{ strain}}{\Delta \text{ time}} \tag{6-39}$$

Eventually, during third-stage creep, necking begins, the stress increases, and the specimen deforms at an accelerated rate until failure occurs. The time required for failure to occur is the **rupture time**. Either a higher stress or a higher temperature reduces the rupture time and increases the creep rate (Figure 6-58).

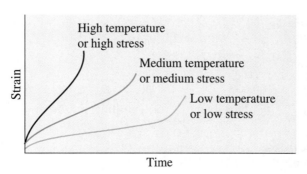

Figure 6-58
The effect of temperature or applied stress on the creep curve.

The combined influence of applied stress and temperature on the creep rate and rupture time (t_r) follows an Arrhenius relationship:

$$\text{Creep rate} = C\sigma^n \exp\left(-\frac{Q_c}{RT}\right) \tag{6-40}$$

$$t_r = K\sigma^m \exp\left(\frac{Q_r}{RT}\right) \tag{6-41}$$

where R is the gas constant, T is the temperature in Kelvin, C, K, n, and m are constants for the material. Q_c is the activation energy for creep, and Q_r is the activation energy for rupture. In particular, Q_c is related to the activation energy for self-diffusion when dislocation climb is important.

In crystalline ceramics, other factors—including grain boundary sliding and nucleation of microcracks—are particularly important. Often, a noncrystalline or glassy material is present at the grain boundaries; the activation energy required for the glass to deform is low, leading to high creep rates compared with completely crystalline ceramics. For the same reason, creep occurs at a rapid rate in ceramic glasses and amorphous polymers.

6-20   Use of Creep Data

The **stress-rupture curves**, shown in Figure 6-59(a), estimate the expected lifetime of a component for a particular combination of stress and temperature. The **Larson-Miller parameter**, illustrated in Figure 6-59(b), is used to consolidate the stress-temperature-rupture time relationship into a single curve. The Larson-Miller parameter ($L.M.$) is:

$$L.M. = \left(\frac{T}{1000}\right)(A + B \ln t) \tag{6-42}$$

where T is in Kelvin, t is the time in hours, and A and B are constants for the material.

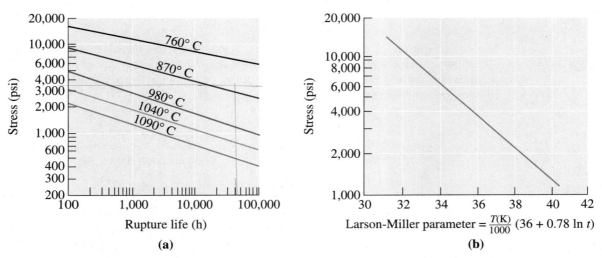

Figure 6-59 Results from a series of creep tests. (a) Stress-rupture curves for an iron-chromium-nickel alloy and (b) the Larson-Miller parameter for ductile cast iron.

EXAMPLE 6-21   *Design of Links for a Chain*

Design a ductile cast iron chain (Figure 6-60) to operate in a furnace used to fire ceramic bricks. The furnace is to operate without rupturing for five years at 600°C, with an applied load of 5000 lbs.

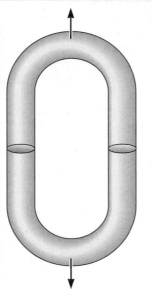

Figure 6-60
Sketch of chain link (for Example 6-21).

SOLUTION

The Larson-Miller parameter for ductile cast iron is:

$$L.M. = \frac{T(36 + 0.78 \ln t)}{1000}$$

The chain is to survive five years, or:

$$t = (24 \ h/d)(365 \ d/y)(5 \ y) = 43{,}800 \ h$$

$$L.M. = \frac{(600 + 273)[36 + 0.78 \ln(43{,}800)]}{1000} = 38.7$$

From Figure 6-59(b), the applied stress must be no more than 2000 psi.

Let us assume a **factor of safety** of 2, this will mean the applied stress should not be more than 2000/2 = 1000 psi.

The total cross-sectional area of the chain required to support the 5000 lb load is:

$$A = F/\sigma = \frac{5000 \ \text{lb}}{1000 \ \text{psi}} = 5 \ \text{in.}^2$$

The cross-sectional area of each "half" of the iron link is then 2.5 in.^{2} and, assuming a round cross-section:

$$d^2 = (4/\pi)A = (4/\pi)(2.5) = 3.18$$

$$d = 1.78 \ \text{in.}$$

6-21 Superplasticity

Many metallic and ceramic materials show unusually high plastic strains under tensile loadings at high temperatures and smaller strain rates. **Superplasticity** is large deformations in materials, resulting from high temperatures and smaller strain-rate responses. Superplasticity is the mechanical response of these materials to the applied strain rate, leading to avoidance of necking. This, in turn, results in a superplastic behavior.[15] In ceramics, fine grain size and grain boundary sliding can allow for superplastic behavior. Tetragonal polycrystalline zirconia (TZP) and many other ceramics exhibit considerable plastic deformation (strains up to 500%) at high temperatures ($\sim$1200 to 1600°C) and small strain rates ($\sim 10^{-4}$ to 10^{-5} s^{-1}). Some materials demonstrate additional plasticity as a result of changes in crystal structure. This is known as *transformation induced plasticity* and the origin of this is different from superplasticity.

SUMMARY

◈ The mechanical behavior of materials is described by their mechanical properties, which are measured with idealized, simple tests. These tests are designed to represent different types of loading conditions. The properties of a material reported in various handbooks are the results of these tests. Consequently, we should always remember that handbook values are average results obtained from idealized tests and, therefore, must be used with some care.

◈ The tensile test describes the resistance of a material to a slowly applied stress. Important properties include yield strength (the stress at which the material begins to permanently deform), tensile strength (the stress corresponding to the maximum applied load), modulus of elasticity (the slope of the elastic portion of the stress-strain curve), and % elongation and % reduction in area (both, measures of the ductility of the material).

◈ The bend test is used to determine the tensile properties of brittle materials. A modulus of elasticity and a flexural strength (similar to a tensile strength) can be obtained.

◈ The hardness test measures the resistance of a material to penetration and provides a measure of the wear and abrasion resistance of the material. A number of hardness tests, including the Rockwell and Brinell tests, are commonly used. Often the hardness can be correlated to other mechanical properties, particularly tensile strength.

◈ The impact test describes the response of a material to a rapidly applied load. The Charpy and Izod tests are typical. The energy required to fracture the specimen is measured and can be used as the basis for comparison of various materials tested under the same conditions. In addition, a transition temperature above which the material fails in a ductile, rather than a brittle, manner can be determined.

◈ Toughness refers to the ability of materials to absorb energy before they fracture. Tensile toughness is equal to the area under the true or engineering stress-true strain curve. The impact toughness is measured using the impact test. This could be very different from the tensile toughness. Fracture toughness describes how easily a crack or flaw in a material propagates. The plane strain fracture toughness K_{Ic} is a common result of these tests.

◆ Weibull statistics are used to describe and characterize the variability in the strength of brittle materials. The Weibull modulus is a measure of the variability of the strength of a material.

◆ The fatigue test permits us to understand how a material performs when a cyclical stress is applied. Knowledge of the rate of crack growth can help determine fatigue life.

◆ In engineering design of components, it is important to account for a factor of safety.

◆ Microstructural analysis of fractured surfaces can lead to better insights into the origin and cause of fracture. Different microstructural features are associated with ductile and brittle fracture as well as fatigue failure.

◆ The creep test provides information on the load-carrying ability of a material at high temperatures. Creep rate and rupture time are important properties obtained from these tests.

GLOSSARY

Anelastic (viscoelastic) material A material in which the total strain developed has elastic and viscous components. Part of the total strain recovers similar to elastic strain. Some part, though, recovers over a period of time. Examples of viscoelastic materials: polymer melts, many polymers including Silly Putty®. Typically, the term anelastic is used for metallic materials.

Apparent viscosity Viscosity obtained by dividing shear stress by the corresponding value of the shear-strain rate for that stress.

Beach or clamshell marks Patterns often seen on a component subjected to fatigue. Normally formed when the load is changed during service or when the loading is intermittent, perhaps permitting time for oxidation inside the crack.

Bend test Application of a force to the center of a bar that is supported on each end to determine the resistance of the material to a static or slowly applied load. Typically used for brittle materials.

Bingham plastic A material whose mechanical response is given by $\tau = G\gamma$, when $\tau < \tau_y$; and $\tau = \tau_y + \eta\dot{\gamma}$, when $\tau \geq \tau_y$.

Chevron pattern A common fracture feature produced by separate crack fronts propagating at different levels in the material.

Climb Movement of a dislocation perpendicular to its slip plane by the diffusion of atoms to or from the dislocation line.

Compliance Inverse of Young's modulus or modulus of elasticity.

Conchoidal fracture Fracture surface containing a very smooth mirror zone near the origin of the fracture, with tear lines comprising the remainder of the surface. This is typical of amorphous materials.

Creep A time dependent, permanent deformation at high temperatures, occurring at constant load or constant stress.

Creep rate The rate at which a material deforms when a stress is applied at a high temperature.

Creep test   Measures the resistance of a material to deformation and failure when subjected to a static load below the yield strength at an elevated temperature.

Delamination   The process by which different layers in a composite will begin to debond.

Dilatant (shear thickening)   Materials in which the apparent viscosity increases with increasing rate of shear.

Ductile to brittle transition temperature (DBTT)   The temperature below which a material behaves in a brittle manner in an impact test. The ductile to brittle switchover also depends on the strain rate.

Ductility   The ability of a material to be permanently deformed without breaking when a force is applied.

Elastic deformation   Deformation of the material that is recovered instantaneously when the applied load is removed.

Elastic limit   The magnitude of stress at which the relationship between stress and strain begins to depart from linearity.

Elastic strain   Fully and instantaneously recoverable strain in a material.

Elastomers   Natural or synthetic plastics that are comprised of molecules with spring-like coils that lead to large elastic deformations (e.g., natural rubber, silicones).

Endurance limit   An older concept that defined a stress below which a material will not fail in a fatigue test. Factors as corrosion or occasional overloading can cause materials to fail at stresses below the assumed endurance limit.

Endurance ratio   The endurance limit divided by the tensile strength of the material. The ratio is about 0.5 for many ferrous metals. See the cautionary note on endurance limit.

Engineering strain   The amount that a material deforms per unit length in a tensile test.

Engineering stress   The applied load, or force, divided by the original cross-sectional area of the material.

Extensometer   An instrument to measure change in length of a tensile specimen, thus allowing calculation of strain.

Factor of safety   The ratio of the stress level for which a component is designed to the actual stress level experienced. A factor used to design load-bearing components. For example, the maximum load a component can take is 10,000 psi. We design it (i.e., choose material, geometry, etc.) such that it can withstand 20,000 psi, in this case the factor of safety will be 2.0.

Fatigue life   The number of cycles permitted at a particular stress before a material fails by fatigue.

Fatigue strength   The stress required to cause failure by fatigue in a given number of cycles, such as 500 million cycles.

Fatigue test   Measures the resistance of a material to failure when a stress below the yield strength is repeatedly applied.

Flexural modulus   The modulus of elasticity calculated from the results of a bend test, giving the slope of the stress-deflection curve.

Flexural strength   The stress required to fracture a specimen in a bend test. Also called the modulus of rupture.

Fracture mechanics   The study of a material's ability to withstand stress in the presence of a flaw.

Fracture toughness   The resistance of a material to failure in the presence of a flaw.

Glass temperature (T_g)   A temperature below which an otherwise ductile material behaves as if it is brittle. Usually, this temperature is not fixed and is affected by processing of the material.

Griffith flaw   A crack or flaw in a material that concentrates and magnifies the applied stress.

Hardness test   Measures the resistance of a material to penetration by a sharp object. Common hardness tests include the Brinell test, Rockwell test, Knoop test, and Vickers test.

Hooke's law   The relationship between stress and strain in the elastic portion of the stress-strain curve.

Impact energy   The energy required to fracture a standard specimen when the load is applied suddenly.

Impact loading   Application of stress at a very high strain rate ($\sim{>}100 \ s^{-1}$).

Impact test   Measures the ability of a material to absorb the sudden application of a load without breaking. The Charpy and Izod tests are commonly used impact tests.

Impact toughness   Energy absorbed by a material, usually notched, during fracture, under the conditions of impact test.

Intergranular   In between grains or along the grain boundaries.

Kinematic viscosity (μ)   Ratio of viscosity and density, often expressed in centiStokes.

Larson-Miller parameter   A parameter used to relate the stress, temperature, and rupture time in creep.

Load   The force applied to a material during testing.

Materials processing   Manufacturing or fabrication methods used for shaping of materials (e.g., extrusion, forging).

Macrohardness   Overall bulk hardness of materials measured using loads ${>}2$ N.

Microhardness   Hardness of materials typically measured using loads less than 2 N using such test as Knoop (HK).

Microvoids   Development of small holes in a material. These form when a high stress causes separation of the metal at grain boundaries or interfaces between the metal and inclusions.

Modulus of elasticity (E)   Young's modulus, or the slope of the linear part of the stress-strain curve in the elastic region. It is a measure of the stiffness of a material, depends upon strength of interatomic bonds and composition, and is not strongly dependent upon microstructure.

Modulus of resilience (E_r)   The maximum elastic energy absorbed by a material when a load is applied.

Modulus of rupture   The stress required to fracture a specimen in a bend test. Also called the flexural strength.

Nano-hardness   Hardness of materials measured at 1–10 nm length scale using extremely small ($\sim 100 \ \mu$N) forces.

Necking   Local deformation causing reduction in the cross-sectional area of a tensile specimen. Many ductile materials show this behavior. The engineering stress begins to decrease at the onset of necking.

Newtonian   Materials in which the shear stress and shear strain rate are linearly related (e.g., light oil or water).

Non-Newtonian   Materials in which shear stress and shear strain rate are not linearly related, these materials are shear thinning or shear thickening (e.g., polymer melts, slurries, paints, etc.).

Notch sensitivity   Measures the effect of a notch, scratch, or other imperfection on a material's properties, such as toughness or fatigue life.

Offset strain value   A value of strain (e.g., 0.002) used to obtain the offset yield stress value.

Offset yield strength   A stress value obtained graphically that describes the stress that gives no more than a specified amount of plastic deformation. Most useful for designing components. Also, simply stated as the yield strength.

Percent elongation   The total percentage increase in the length of a specimen during a tensile test.

Percent reduction in area   The total percentage decrease in the cross-sectional area of a specimen during the tensile test.

Plastic deformation or strain   Permanent deformation of a material when a load is applied, then removed.

Poisson's ratio   The ratio between the lateral and longitudinal strains in the elastic region.

Proportional limit   A level of stress above which the relationship between stress and strain is not linear.

Pseudoplastics (shear thinning)   Materials in which the apparent viscosity decreases with increasing rate of shear.

Rheometer   An instrument used to measure rheological properties such as viscosity, yield stress, shear thinning behavior, etc. of a material.

Rheopectic behavior   Materials that show shear thickening and also an apparent viscosity that at a constant rate of shear increases with time.

Rotating cantilever beam test   An older test for fatigue testing.

Rupture time   The time required for a specimen to fail by creep at a particular temperature and stress.

S-N curve (also known as the Wöhler curve)   A graph showing stress as a function of number of cycles in fatigue.

Shear modulus (*G*)   The slope of the linear part of the shear stress-shear strain curve.

Shear-strain rate   Time derivative of shear strain. See "Strain rate."

Shear thickening (dilatant)   Materials in which the apparent viscosity increases with increasing rate of shear.

Shear thinning (pseudoplastics)   Materials in which the apparent viscosity decreases with increasing rate of shear.

Shot peening   A process in which metal spheres are shot at a component. This leads to a residual compressive stress at the surface of a component and this enhances fatigue life.

Stiffness   A qualitative measure of the elastic deformation produced in a material. A stiff material has a high modulus of elasticity. Stiffness also depends upon geometry.

Strain   Elongation change in dimension per unit length.

Strain gage   A device used for measuring change in length and hence strain.

Strain rate   The rate at which strain develops in or is applied to a material indicated as $\dot{\varepsilon}$ or $\dot{\gamma}$ for tensile and shear-strain rates, respectively. Strain rate can have an effect on whether a material would behave in a ductile or brittle fashion.

Stress   Force or load per unit area of cross-section over which the force or load is acting.

Stress-corrosion   A phenomenon in which materials react with corrosive chemicals in the environment leading to the formation of cracks and lowering of strength.

Stress relaxation   Decrease in the stress for a material held under constant strain, as a function of time, observed in viscoelastic materials. Stress relaxation is different from time dependent recovery of strain.

Stress-rupture curve   A method of reporting the results of a series of creep tests by plotting the applied stress versus the rupture time.

Striations   Patterns seen on a fractured surface of a fatigued sample. These are on a much finer scale than beach marks and show the position of the crack tip after each cycle.

Superplasticity   Large deformations in materials, resulting from high temperatures and low strain rates.

Tempering   A glass heat treatment that makes the glass safer; it does so by creating a compressive stress layer at the surface.

Tensile strength   The stress that corresponds to the maximum load in a tensile test.

Tensile test   Measures the response of a material to a slowly applied uniaxial force. The yield strength, tensile strength, modulus of elasticity, and ductility are obtained.

Tensile toughness   The area under the true stress-true strain tensile test curve, it is a measure of the energy required to cause fracture under tensile test conditions.

Thixotropic behavior   Materials that show shear thinning and also an apparent viscosity that at a constant rate of shear decreases with time.

Toughness   A qualitative measure of the energy required to cause fracture of a material. A material that resists failure by impact is said to be tough. One measure of toughness is the area under the true stress-strain curve (tensile toughness), another is the impact energy measured during an impact test (impact toughness). The ability of materials containing flaws to withstand load is known as fracture toughness.

Transgranular   Meaning across the grains (e.g., a transgranular fracture would be fracture in which cracks would go through the grains).

True strain   The strain calculated using actual and not original dimensions, given by $\varepsilon_t = \ln(l/l_0)$.

True stress   The load divided by the actual cross-sectional area of the specimen at that load.

Viscoelastic (or anelastic) material   See Anelastic material.

Viscosity (η)   Measure of resistance to flow, defined as the ratio of shear stress to shear strain rate (units Poise or Pa-s).

Viscous material   A viscous material is one in which the strain develops over a period of time and the material does not go to its original shape after the stress is removed.

Work of fracture   Area under the stress-strain curve, considered as a measure of tensile toughness.

Weibull distribution   A mathematical distribution showing the probability of failure or survival of a material as a function of the stress.

Weibull modulus (*m*)   A parameter related to the Weibull distribution. It is an indicator of the variability of the strength of materials resulting from a distribution of flaw sizes.

Wöhler curve   Graph showing fatigue stress as a function of number of cycles (also known as the S-N curve).

Yield point phenomenon   An abrupt transition, seen in some materials, from elastic deformation to plastic flow.

Yield strength   A stress value obtained graphically that describes no more than a specified amount of deformation (usually 0.002). Also known as offset yield strength.

Young's modulus   The slope of the linear part of the stress-strain curve in the elastic region, same as modulus of elasticity.

ADDITIONAL INFORMATION

1. VAUGHAN, D., *"The Challenger Launch Decision: Risky Technology, Culture and Deviance at NASA"*, 1996: University of Chicago Press.

2. HILL, S., "The Mystery of the Titanic: A Case of Brittle Fracture?" *Materials World*, 1996. 4(6): p. 334–335.

3. GANNON, R., "What Really Sank the Titanic," *Popular Science*, 1995. 246(2): p. 49–55.

4. PIEHLER, H.R., *MRS Bulletin 2000*, 2000. 25(8): p. 67–70.

5. PLAZEK, DONALD, University of Pittsburgh, *Personal communication*, 2000.

6. HOOKE, R., *In Latin* Ut Tesnio sic Vis *Meaning as the Tension Increases so Does the Strain*, 1678.

7. LAKES, R., "Foam Structures with a Negative Poisson's Ratio," *Science*, 1987. 235: p. 1038–1040.

8. NEWEY, C. and G. WEAVER, *Materials In Action Series-Materials Principles and Practice*, 1991: Butterworth-Heinemann.

9. COURTNEY, T.H., *Mechanical Behavior of Materials*, 2000: McGraw Hill.

10. MEYERS, M. and CHAWLA, K.K., *Mechanical Behavior of Materials*, 1999: Prentice Hall.

11. BENGISU, M., *Engineering Ceramics*. 2001: Springer Verlag.

12. WEIBULL, W., *App. Mech.*, 1951. 18: p. 293.

13. DOWLING, N.E., *Mechanical Behavior of Materials: Engineering Methods for Deformation, Fracture, and Fatigue*, Second Edition, 1999. Upper Saddle River, N.J.: Prentice Hall.

14. SURESH, S., *Fatigue of Materials*, 1991. Cambridge, U.K.: Cambridge University Press.

15. CASTAING, J. and DOMINGUEZ-RODRIGUEZ, A., "Ductility of Ceramics," in *The Encyclopedia of Advanced Materials*, D. BLOOR, et al., Editors. 1994. p. 659–664.

✓ PROBLEMS

Section 6-1 Technological Significance

6-1 Explain the role of mechanical properties in load-bearing applications using real-world examples.

6-2 Explain the importance of mechanical properties in functional applications (e.g., optical, magnetic, electronic, etc.) using real-world examples.

6-3 Explain the importance of understanding mechanical properties in processing of materials.

Section 6-2 Terminology for Mechanical Properties

6-4 Define "engineering stress" and "engineering strain."

6-5 Define "modulus of elasticity."

6-6 Explain the terms "elastic stress" and "elastic strain."

6-7 Define "plastic deformation" and compare it with "elastic deformation."

6-8 What is strain rate? How does it affect the mechanical behavior of polymeric and metallic materials?

6-9 Why does Silly Putty® break when you stretch it very fast?

6-10 What is a viscoelastic material? Give an example.

6-11 What is meant by the term "stress relaxation?"

6-12 Define the terms "viscosity", "apparent viscosity", and "kinematic viscosity."

6-13 In the units of Pa-s and cP, what is the viscosity of water and light oil at room temperature?

6-14 What is the kinematic viscosity of water at room temperature in cSt?

6-15 What two equations are used to describe Bingham plastic-like behavior?

6-16 What is a Newtonian material? Give an example.

6-17 What is an elastomer? Give an example.

6-18 What is meant by the terms "shear thinning" and "shear thickening" materials?

6-19 Many paints and other dispersions are not only shear thinning, but also thixotropic. What does the term "thixotropy" mean?

6-20 Draw a schematic diagram showing development of strain in an elastic and viscoelastic material. Assume that the load is applied at some time $t = 0$ and taken off at some time t.

Section 6-3 The Tensile Test: Use of the Stress-Strain Diagram

6-21 Draw qualitative engineering stress-engineering strain curves for a ductile polymer, a ductile metal, a ceramic, a glass, and natural rubber. Label carefully. Rationalize your sketch for each material.

6-22 What is necking? How does it lead to reduction in engineering stress as true stress increases?

6-23 Why do some polymers get stronger as we stretch them beyond a region where necking occurs?

6-24 A 850-lb force is applied to a 0.15-in.-diameter nickel wire having a yield strength of 45,000 psi and a tensile strength of 55,000 psi. Determine

(a) whether the wire will plastically deform; and
(b) whether the wire will experience necking.

6-25 A force of 100,000 N is applied to a 10 mm × 20 mm iron bar having a yield strength of 400 MPa and a tensile strength of 480 MPa. Determine

(a) whether the bar will plastically deform; and
(b) whether the bar will experience necking.

6-25 Calculate the maximum force that a 0.2-in.-diameter rod of Al_2O_3, having a yield strength of 35,000 psi, can withstand with no plastic deformation. Express your answer in pounds and Newtons.

6-26 A force of 20,000 N will cause a 1 cm × 1 cm bar of magnesium to stretch from 10 cm to 10.045 cm. Calculate the modulus of elasticity, both in GPa and psi.

6-27 A polymer bar's dimensions are 1 in. × 2 in. × 15 in. The polymer has a modulus of elasticity of 600,000 psi. What force is required to stretch the bar elastically to 15.25 in.?

6-28 An aluminum plate 0.5 cm thick is to withstand a force of 50,000 N with no permanent deformation. If the aluminum has a yield strength of 125 MPa, what is the minimum width of the plate?

6-29 A 3-in.-diameter rod of copper is to be reduced to a 2-in.-diameter rod by being pushed through an opening. To account for the elastic strain, what should be the diameter of the opening? The modulus of elasticity for the copper is 17×10^6 psi and the yield strength is 40,000 psi. A 0.15-cm-thick, 8-cm-wide sheet of magnesium that is originally 5 m long is to be stretched to a final length of 6.2 m. What should be the length of the sheet before the applied stress is released? The modulus of elasticity of magnesium is 45 GPa and the yield strength is 200 MPa.

6-30 A steel cable 1.25 in. in diameter and 50 ft long is to lift a 20-ton load. What is the length of the cable during lifting? The modulus of elasticity of the steel is 30×10^6 psi.

Section 6-4 Properties Obtained from the Tensile Test
and
Section 6-5 True Stress and True Strain

6-31 Define "true stress" and "true strain." Compare with engineering stress and engineering strain.

6-32 Write down the formulas for calculating the stress and strain for a sample subjected to a tensile test. Assume the sample shows necking.

6-33 The following data were collected from a standard 0.505-in.-diameter test specimen of a copper alloy (initial length (l_0) = 2.0 in.):

Load (lb)	Δl (in.)
0	00000
3,000	0.00167
6,000	0.00333
7,500	0.00417
9,000	0.0090
10,500	0.040
12,000	0.26
12,400	0.50 (maximum load)
11,400	1.02 (fracture)

After fracture, the total length was 3.014 in. and the diameter was 0.374 in. Plot the data and calculate the 0.2% offset yield strength along with

(a) the tensile strength;
(b) the modulus of elasticity;
(c) the % elongation;
(d) the % reduction in area;
(e) the engineering stress at fracture;
(f) the true stress at fracture; and
(g) the modulus of resilience.

6-34 The following data were collected from a 0.4-in.-diameter test specimen of polyvinyl chloride (l_0 = 2.0 in.):

Load (lb)	Δl (in.)
0	0.00000
300	0.00746
600	0.01496
900	0.02374
1200	0.032
1500	0.046
1660	0.070 (maximum load)
1600	0.094
1420	0.12 (fracture)

After fracture, the total length was 2.09 in. and the diameter was 0.393 in. Plot the data and calculate

(a) the 0.2% offset yield strength;
(b) the tensile strength;
(c) the modulus of elasticity;
(d) the % elongation;
(e) the % reduction in area;
(f) the engineering stress at fracture;

(g) the true stress at fracture; and
(h) the modulus of resilience.

6-35 The following data were collected from a 12-mm-diameter test specimen of magnesium (l_0 = 30.00 mm):

Load (N)	Δl (mm)
0	0.0000
5,000	0.0296
10,000	0.0592
15,000	0.0888
20,000	0.15
25,000	0.51
26,500	0.90
27,000	1.50 (maximum load)
26,500	2.10
25,000	2.79 (fracture)

After fracture, the total length was 32.61 mm and the diameter was 11.74 mm. Plot the data and calculate:

(a) the 0.2% offset yield strength;
(b) the tensile strength;
(c) the modulus of elasticity;
(d) the % elongation;
(e) the % reduction in area;
(f) the engineering stress at fracture;
(g) the true stress at fracture; and
(h) the modulus of resilience.

6-36 The following data were collected from a 20-mm-diameter test specimen of a ductile cast iron (l_0 = 40.00 mm):

Load (N)	Δl (mm)
0	0.0000
25,000	0.0185
50,000	0.0370
75,000	0.0555
90,000	0.20
105,000	0.60
120,000	1.56
131,000	4.00 (maximum load)
125,000	7.52 (fracture)

After fracture, the total length was 47.42 mm and the diameter was 18.35 mm. Plot the data and calculate:

(a) the 0.2% offset yield strength;
(b) the tensile strength;
(c) the modulus of elasticity;
(d) the % elongation;
(e) the % reduction in area;

(f) the engineering stress at fracture;

(g) the true stress at fracture; and

(h) the modulus of resilience.

Section 6-6 The Bend Test for Brittle Materials

6-37 Define the term "flexural strength" and "flexural modulus."

6-38 Why is that we often conduct a bend test on brittle materials?

6-39 A bar of Al_2O_3 that is 0.25 in. thick, 0.5 in. wide, and 9 in. long is tested in a three-point bending apparatus, with the supports located 6 in. apart. The deflection of the center of the bar is measured as a function of the applied load. The data are shown below. Determine the flexural strength and the flexural modulus.

Force (lb)	Deflection (in.)
14.5	0.0025
28.9	0.0050
43.4	0.0075
57.9	0.0100
86.0	0.0149 (fracture)

6-40 (a) A 0.4-in.-diameter, 12-in.-long titanium bar has a yield strength of 50,000 psi, a modulus of elasticity of 16×10^6 psi, and Poisson's ratio of 0.30. Determine the length and diameter of the bar when a 500-lb load is applied. (b) When a tensile load is applied to a 1.5-cm diameter copper bar, the diameter is reduced to 1.498-cm diameter. Determine the applied load, using the data in Table 6-3.

6-41 A three-point bend test is performed on a block of ZrO_2 that is 8 in. long, 0.50 in. wide, and 0.25 in. thick and is resting on two supports 4 in. apart. When a force of 400 lb is applied, the specimen deflects 0.037 in. and breaks. Calculate

(a) the flexural strength; and

(b) the flexural modulus, assuming that no plastic deformation occurs.

6-42 A three-point bend test is performed on a block of silicon carbide that is 10 cm long, 1.5 cm wide, and 0.6 cm thick and is resting on two supports 7.5 cm apart. The sample breaks when a deflection of 0.09 mm is recorded. The flexural modulus for silicon carbide is 480 GPa. Assume that no plastic deformation occurs. Calculate

(a) the force that caused the fracture; and

(b) the flexural strength.

6-43 (a) A thermosetting polymer containing glass beads is required to deflect 0.5 mm when a force of 500 N is applied. The polymer part is 2 cm wide, 0.5 cm thick, and 10 cm long. If the flexural modulus is 6.9 GPa, determine the minimum distance between the supports. Will the polymer fracture if its flexural strength is 85 MPa? Assume that no plastic deformation occurs. The flexural modulus of alumina is 45×10^6 psi and its flexural strength is 46,000 psi. (b) A bar of alumina 0.3 in. thick, 1.0 in. wide, and 10 in. long is placed on supports 7 in. apart. Determine the amount of deflection at the moment the bar breaks, assuming that no plastic deformation occurs.

6-44 Ceramics are much stronger in compression than in tension. Explain why.

6-45 Why is it that we can load a fire truck on four ceramic coffee cups? However, when we drop a ceramic cup on floor, why is it likely to break easily?

6-46 Dislocations have a major effect on the plastic deformation of metals, but do not play a major role in mechanical behavior of ceramics. Why?

6-47 What controls the strength of ceramics and glasses?

Section 6-7 Hardness of Materials

6-48 What does the term "hardness of a material" mean?

6-49 Why is hardness data difficult to correlate to mechanical properties of materials in a quantitative fashion?

6-50 What is the hardest material (natural or synthetic)? Is it diamond?

6-51 Explain the terms "macrohardness", "microhardness", and "nanohardness."

6-52 A Brinell hardness measurement, using a 10-mm-diameter indenter and a 500-kg load, produces an indentation of 4.5 mm on an aluminum plate. Determine the Brinell hardness number (HB) of the metal.

6-53 When a 3000-kg load is applied to a 10-mm-diameter ball in a Brinell test of a steel, an indentation of 3.1 mm is produced. Estimate the tensile strength of the steel.

6-54 How is nanohardness of materials measured?

Section 6-8 Strain Rate Effects and Impact Behavior
and
Section 6-9 Properties from the Impact Test

6-55 The following data were obtained from a series of Charpy impact tests performed on four steels, each having a different manganese content. Plot the data and determine

(a) the transition temperature (defined by the mean of the absorbed energies in the ductile and brittle regions); and

(b) the transition temperature (defined as the temperature that provides 50 J of absorbed energy).

Test Temperature °C	Impact Energy (J) 0.30% Mn	0.39% Mn	1.01% Mn	1.55% Mn
−100	2	5	5	15
−75	2	5	7	25
−50	2	12	20	45
−25	10	25	40	70
0	30	55	75	110
25	60	100	110	135
50	105	125	130	140
75	130	135	135	140
100	130	135	135	140

6-56 Plot the transition temperature versus manganese content and using the previous data shown discuss the effect of manganese on the toughness of steel. What would be the minimum manganese allowed in the steel if a part is to be used at 0°C?

6-57 The following data were obtained from a series of Charpy impact tests performed on four ductile cast irons, each having a different silicon content. Plot the data and determine

(a) the transition temperature (defined by the mean of the absorbed energies in the ductile and brittle regions); and

(b) the transition temperature (defined as the temperature that provides 10 J of absorbed energy).

Plot the transition temperature versus silicon content and discuss the effect of silicon on the toughness of the cast iron. What would be the maximum silicon allowed in the cast iron if a part is to be used at 25°C?

Test Temperature °C	Impact Energy (J) 2.55% Si	2.85% Si	3.25% Si	3.63% Si
−50	2.5	2.5	2	2
−5	3	2.5	2	2
0	6	5	3	2.5
25	13	10	7	4
50	17	14	12	8
75	19	16	16	13
100	19	16	16	16
125	19	16	16	16

6-58 FCC metals are often recommended for use at low temperatures, particularly when any sudden loading of the part is expected. Explain.

6-59 A steel part can be made by powder metallurgy (compacting iron powder particles and sintering to produce a solid) or by machining from a solid steel block. Which part is expected to have the higher toughness? Explain.

6-60 What is meant by the term notch sensitivity?

6-61 What is the difference between a tensile test and an impact test? Using this, explain why the toughness values measured using impact tests may not always correlate with tensile toughness measured using tensile tests.

6-62 A number of aluminum-silicon alloys have a structure that includes sharp-edged plates of brittle silicon in the softer, more ductile aluminum matrix. Would you expect these alloys to be notch-sensitive in an impact test? Would you expect these alloys to have good toughness? Explain your answers.

6-63 What is the ductile to brittle transition temperature (DBTT)?

6-64 Some polymers and some metals/alloys get brittle at low temperatures. Consult your instructor and discuss how the origin of brittleness is different in these materials.

6-65 What caused NASA's *Challenger* 1986 accident?

6-66 How is tensile toughness defined in relation to the true stress-strain diagram? How is tensile toughness related to impact toughness?

Section 6-10 Fracture Mechanics and
Section 6-11 The Importance of Fracture Mechanics

6-67 Alumina Al_2O_3 is a brittle ceramic with low toughness. Suppose that fibers of silicon carbide SiC, another brittle ceramic with low toughness, could be embedded within the alumina. Would doing this affect the toughness of the ceramic matrix composite? Explain. (These materials are discussed in later chapters.)

6-68 A ceramic matrix composite contains internal flaws as large as 0.001 cm in length. The plane strain fracture toughness of the composite is 45 MPa$\sqrt{m}$ and the tensile strength is 550 MPa. Will the stress cause the composite to fail before the tensile strength is reached? Assume that $f = 1$.

6-69 An aluminum alloy that has a plane strain fracture toughness of 25,000 psi$\sqrt{in}$. fails when a stress of 42,000 psi is applied. Observation of the fracture surface indicates that fracture began at the surface of the part. Estimate the size of the flaw that initiated fracture. Assume that $f = 1.1$.

6-70 A polymer that contains internal flaws 1 mm in length fails at a stress of 25 MPa. Determine the plane strain fracture toughness of the polymer. Assume that $f = 1$.

6-71 A ceramic part for a jet engine has a yield strength of 75,000 psi and a plane strain fracture toughness of 5,000 psi$\sqrt{\text{in.}}$ To be sure that the part does not fail, we plan to assure that the maximum applied stress is only one-third the yield strength. We use a nondestructive test that will detect any internal flaws greater than 0.05 in. long. Assuming that $f = 1.4$, does our non-destructive test have the required sensitivity? Explain.

6-72 Explain how the fracture toughness of ceramics can be obtained using hardness testing, explain why such a method provides qualitative measurements.

Section 6-12 Microstructural Features of Fracture in Metallic Materials
and
Section 6-13 Microstructural Features of Fracture in Ceramics, Glasses, and Composites

6-73 Explain the terms intergranular and intragranular fractures. Use a schematic to show grains, grain boundaries and a crack path that is typical of intergranular and intragranular fracture in materials.

6-74 What are the characteristic microstructural features associated with ductile fracture?

6-75 What are the characteristic microstructural features associated with a brittle fracture in a metallic material?

6-76 What materials typically show a conchoidal fracture?

6-77 Briefly describe how fiber-reinforced composite materials can fail.

6-78 Some aircraft components made from carbon fiber-reinforced composites have been known to fail suddenly as a result of delamination. It is difficult to see the damage to these types of materials since the damage is inside the composite. Discuss with your instructor the fracture in these materials.

6-79 Concrete has exceptional strength in compression but it fails rather easily in tension. Explain why.

6-80 What controls the strength of glasses? What can be done to enhance the strength of silicate glasses?

Section 6-14 Weibull Statistics for Failure Strength Analysis

6-81 Sketch a schematic of the strength of ceramics and that of metals and alloys as a function of probability of failure. Explain the differences you anticipate.

6-82 Why does the strength of ceramics vary considerably with the size of ceramic components?

6-83 What parameter tells us about the variability of the strength of ceramics and glasses?

6-84 Why do glass fibers of different lengths have different strengths?

6-85 Explain the significance of Weibull distribution.

Section 6-15 Fatigue
and
Section 6-16 Results of the Fatigue Test
and
Section 6-17 Application of Fatigue Testing

6-86 A cylindrical tool steel specimen that is 6 in. long and 0.25 in. in diameter rotates as a cantilever beam and is to be designed so that failure never occurs. Assuming that the maximum tensile and compressive stresses are equal, determine the maximum load that can be applied to the end of the beam. (See Figure 6-50.)

6-87 A 2-cm-diameter, 20-cm-long bar of an acetal polymer (Figure 6-61) is loaded on one end and is expected to survive one million cycles of loading, with equal maximum tensile and compressive stresses, during its lifetime. What is the maximum permissible load that can be applied?

6-88 A cyclical load of 1500 lb is to be exerted at the end of a 10-in.-long aluminum beam (Figure 6-50). The bar must survive for at least 10^6 cycles. What is the minimum diameter of the bar?

6-89 A cylindrical acetal polymer bar 20 cm long and 1.5 cm in diameter is subjected to a vibrational load at a frequency of 500 vibrations per minute, with a load of 50 N. How many hours will the part survive before breaking? (See Figure 6-61.)

6-90 Suppose that we would like a part produced from the acetal polymer shown in Figure 6-61 to survive for one million cycles under conditions that provide for equal compressive and tensile stresses. What is the fatigue strength, or maximum stress amplitude, required? What are the maximum stress, the minimum stress, and the mean stress

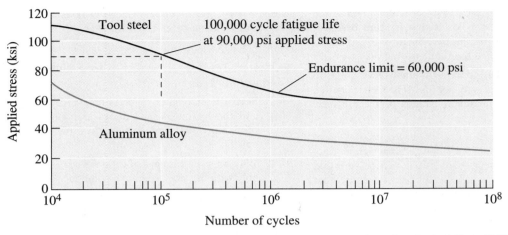

Figure 6-50 (Repeated for Problems 6-86 and 6-88) The stress-number of cycles to failure (S-N) curves for a tool steel and an aluminum alloy.

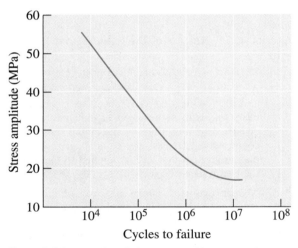

Figure 6-61 The S-N fatigue curve for an acetal polymer (for Problems 6-87, 6-89, and 6-90).

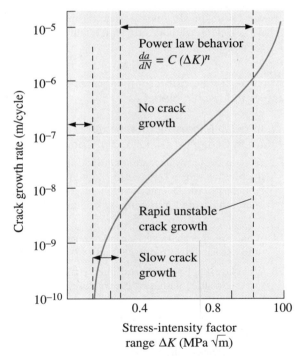

Figure 6-52 (Repeated for Problems 6-91 and 6-92) Crack growth rate versus stress-intensity factor range for a high-strength steel. For this steel, $C = 1.62 \times 10^{-12}$ and $n = 3.2$ for the units shown.

on the part during its use? What effect would the frequency of the stress application have on your answers? Explain.

6-91 The high-strength steel in Figure 6-52 is subjected to a stress alternating at 200 revolutions per minute between 600 MPa and 200 MPa (both tension). Calculate the growth rate of a surface crack when it reaches a length of 0.2 mm in both m/cycle and m/s. Assume that $f = 1.0$.

6-92 The high-strength steel in Figure 6-52, which has a critical fracture toughness of 80 MPa$\sqrt{\text{m}}$, is subjected to an alternating stress varying from -900 MPa (compression) to $+900$ MPa (tension). It is to survive for 10^5 cycles before failure

occurs. Assume that $f = 1$. Calculate:

(a) the size of a surface crack required for failure to occur; and

(b) the largest initial surface crack size that will permit this to happen.

6-93 The acrylic polymer from which Figure 6-62 was obtained has a critical fracture toughness of 2 MPa$\sqrt{m}$. It is subjected to a stress alternating between -10 and $+10$ MPa. Calculate the growth rate of a surface crack when it reaches a length of 5×10^{-6} m if $f = 1.0$.

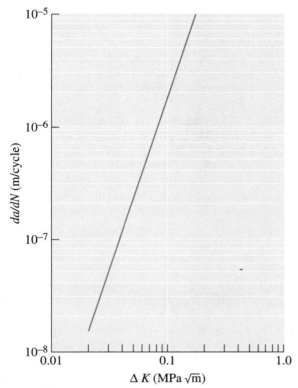

Figure 6-62 The crack growth rate for an acrylic polymer (for Problems 6-93, 6-94, 6-95).

6-94 Calculate the constants C and n in Equation 6-36 for the crack growth rate of an acrylic polymer. (See Figure 6-62.)

6-95 The acrylic polymer from which Figure 6-62 was obtained is subjected to an alternating stress between 15 MPa and 0 MPa. The largest surface cracks initially detected by nondestructive testing are 0.001 mm in length. If the critical fracture toughness of the polymer is 2 MPa$\sqrt{m}$, calculate the number of cycles required before failure occurs. Let $f = 1.0$. (*Hint*: Use the results of Problem 6-94.)

6-96 Explain how fatigue failure occurs even if the material does not see overall stress levels higher than the yield strength.

6-97 Verify what integration of $da/dN = C(\Delta K)^n$ will give Equation 6-38.

6-98 What is shot peening? What is the purpose of using this process?

Section 6-18 Creep, Stress Rupture, and Stress Corrosion

6-99 Define the term "creep" and differentiate creep from stress relaxation.

6-100 What is meant by the terms "stress rupture" and "stress corrosion?"

6-101 What is the difference between failure of a material by creep and that by stress rupture?

6-102 The activation energy for self-diffusion in copper is 49,300 cal/mol. A copper specimen creeps at 0.002 $\dfrac{\text{in./in.}}{\text{h}}$ when a stress of 15,000 psi is applied at 600°C. If the creep rate of copper is dependent on self-diffusion, determine the creep rate if the temperature is 800°C.

6-103 When a stress of 20,000 psi is applied to a material heated to 900°C, rupture occurs in 25,000 h. If the activation energy for rupture is 35,000 cal/mol, determine the rupture time if the temperature is reduced to 800°C.

6-104 The following data were obtained from a creep test for a specimen having a gage length of 2.0 in. and an initial diameter of 0.6 in. The initial stress applied to the material is 10,000 psi. The diameter of the specimen after fracture is 0.52 in.

Length Between Gage Marks (in.)	Time (h)
2.004	0
2.010	100
2.020	200
2.030	400
2.045	1000
2.075	2000
2.135	4000
2.193	6000
2.230	7000
2.300	8000 (fracture)

Determine

(a) the load applied to the specimen during the test;

(b) the approximate length of time during which linear creep occurs;

(c) the creep rate in $\dfrac{\text{in./in.}}{\text{h}}$ and in %/h; and

(d) the true stress acting on the specimen at the time of rupture.

6-105 A stainless steel is held at 705°C under different loads. The following data are obtained:

Applied Stress (MPa)	Rupture Time (h)	Creep Rate (%/h)
106.9	1200	0.022
128.2	710	0.068
147.5	300	0.201
160.0	110	0.332

Determine the exponents n and m in Equations 6-40 and 6-41 that describe the dependence of creep rate and rupture time on applied stress.

6-106 Using the data in Figure 6-59(a) for an iron-chromium-nickel alloy, determine the activation energy Q_r and the constant m for rupture in the temperature range 980 to 1090°C.

6-107 A 1-in.-diameter bar of an iron-chromium-nickel alloy is subjected to a load of 2500 lb. How many days will the bar survive without rupturing at 980°C? [See Figure 6-59(a).]

6-108 A 5 mm × 20 mm bar of an iron-chromium-nickel alloy is to operate at 1040°C for 10 years without rupturing. What is the maximum load that can be applied? [See Figure 6-59(a).]

6-109 An iron-chromium-nickel alloy is to withstand a load of 1500 lb at 760°C for 6 years. Calculate the minimum diameter of the bar. [See Figure 6-59(a).]

6-110 A 1.2-in.-diameter bar of an iron-chromium-nickel alloy is to operate for 5 years under a load of 4000 lb. What is the maximum operating temperature? [See Figure 6-59(a).]

6-111 A 1 in. × 2 in. ductile cast iron bar must operate for 9 years at 650°C. What is the maximum load that can be applied? [See Figure 6-59(b).]

6-112 A ductile cast iron bar is to operate at a stress of 6000 psi for 1 year. What is the maximum allowable temperature? [See Figure 6-59(b).]

✳ Design Problems

6-113 A hook (Figure 6-63) for hoisting containers of ore in a mine is to be designed using a non-ferrous (not based on iron) material. (A non-ferrous material is used because iron and steel could cause a spark that would ignite explosive gases in the mine.) The hook must support a

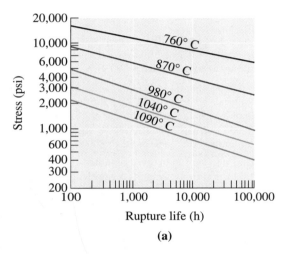

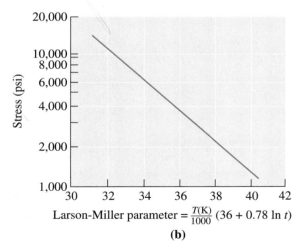

Figure 6-59 (Repeated for Problems 6-106 through 6-112) Results from a series of creep tests. (a) Stress-rupture curves for an iron-chromium-nickel alloy and (b) the Larson-Miller parameter for ductile cast iron.

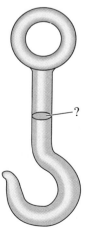

Figure 6-63 Hook (for Problem 6-113).

load of 25,000 pounds, and a factor of safety of 2 should be used. We have determined that the cross-section labeled "?" is the most critical area; the rest of the device is already well over-designed. Determine the design requirements for this device and, based on the mechanical property data given in Chapters 13 and 14 and the metal/alloy prices obtained from such sources as your local newspapers, the internet website of London Metal Exchange or *The Wall Street Journal*, design the hook and select an economical material for the hook.

6-114 A support rod for the landing gear of a private airplane is subjected to a tensile load during landing. The loads are predicted to be as high as 40,000 pounds. Because the rod is crucial and failure could lead to a loss of life, the rod is to be designed with a factor of safety of 4 (that is, designed so that the rod is capable of supporting loads four times as great as expected). Operation of the system also produces loads that may induce cracks in the rod. Our nondestructive testing equipment can detect any crack greater than 0.02 in. deep. Based on the materials given in Table 6-6, design the support rod and the material, and justify your answer.

6-115 A lightweight rotating shaft for a pump on the national aerospace plane is to be designed to support a cyclical load of 15,000 pounds during service. The maximum stress is the same in both tension and compression. The endurance limits or fatigue strengths for several candidate materials are shown below. Design the shaft, including an appropriate material, and justify your solution.

Material	Endurance Limit/ Fatigue Strength (MPa)
Al-Mn alloy	110
Al-Mg-Zn alloy	225
Cu-Be alloy	295
Mg-Mn alloy	80
Be alloy	180
Tungsten alloy	320

6-117 A ductile cast-iron bar is to support a load of 40,000 lb in a heat-treating furnace used to make malleable cast iron. The bar is located in a spot that is continuously exposed to 500°C. Design the bar so that it can operate for at least 10 years without failing.

Computer Problems

6-118 *Hardness Calculator.* Write a computer program to calculate Brinell hardness of materials. The program should ask the user to provide the diameter of the indentor (D) in millimeters, the applied load F in kilograms, and the diameter of the impression created (D_i) in millimeters. The program should then provide the Brinell hardness number value using Equation 6-16.

6-119 *Tensile Test Data and Calculations.* Write a computer program that will provide the user of this software with useful results from tensile test data. First the program should read important record keeping information such as date, name of the operator, sample number, etc. The program should also give the user an opportunity to use British or SI units. Next the program should read the sample dimensions and gage length that was used. Your software should read a series of inputs (up to a maximum of 50 sets of load and corresponding elongation values). The program should also ask and record a strain rate at which the test was conducted. The program output should provide values of engineering stress, engineering strain, true stress, true strain, and maximum load. The program should provide values of tensile strength, fracture strength, and ductility (to be measured as 100 times strain at failure). Finally, use Excel™, Origin™, or Sigma Plot™, to plot the graphs showing load versus elongation, stress versus strain, and true stress versus true strain. Using this graph calculate the 0.2% offset yield strength and Young's modulus. Test your program using any of the sets of data in the text.

6-120 *Flexural Strength or Modulus of Rupture Calculator.* Write a computer program that will ask the user to provide the value of the fracture load (F), the distance between outer points (L), width of the specimen (w), and the height (h). The program should then use Equation 6-13 to calculate the flexural strength (modulus of rupture).

Chapter

In applications, such as the chassis formation of automobiles, metals and alloys are deformed. The mechanical properties of materials change during this process due to strain hardening. The strain-hardening behavior of steels used in the fabrication of chassis influences the ability to form aerodynamic shapes. The strain-hardening behavior is also important in improving the crashworthiness of vehicles. (*Courtesy of PhotoDisc/Getty Images.*)

7

Strain Hardening and Annealing

Have You Ever Wondered?

- Why does bending a copper wire make it stronger?

- What type of steel improves the crashworthiness of cars?

- How are aluminum beverage cans made?

- Why do thermoplastics get stronger when strained?

- What is the difference between an annealed, tempered, and laminated safety glass?

- Why is it that the strength of the metallic material around a weld could be lower than that of the surrounding material?

- Can ceramic materials be stretched by several hundred percent?

In this chapter, we will learn how the strength of metals and alloys is influenced by mechanical processing and heat treatments. In Chapter 4, we learned about the different techniques that can strengthen metals and alloys (e.g., enhancing dislocation density, decreasing grain size, alloying, etc.). In this chapter, we will learn how to enhance the strength of metals and alloys using cold working, a process by which a metallic material is simultaneously deformed and strengthened. We will also see how hot working can be used to shape metals and alloys by deformation at high temperatures without strengthening. We will learn how the annealing heat treatment can be used to enhance ductility and counter the increase in hardness caused by cold working. The strengthening we obtain during cold working, which is brought about by increasing the dislocation density, is called strain hardening or work hardening. By controlling the thermo-mechanical processing (i.e., combinations of mechanical processing and heat treatment), we are able to process metallic materials into a usable shape yet still improve and control their mechanical properties.

The topics discussed in this chapter pertain particularly to metals and alloys. Strain hardening (obtained by multiplication of dislocations) requires that the material have ductility. We use strain hardening as a tool to enhance strength of a material. We have to counter the effects of strain hardening in manufacturing processes. For example, when we draw a wire or extrude a tube, strain hardening can occur and we have to ensure that the product still has acceptable ductility. Cars and trucks are made by stamping out a material known as sheet steel. This process leads to aerodynamic and aesthetically pleasing car chassis. The sheet steel used must exhibit an ability to stretch and bend easily during stamping. However, we must ultimately produce a strong steel that can withstand minor bumps and major impacts. The increase in the strength of steel as a result of strain hardening helps us in this regard. Furthermore, for better crashworthiness we must use steels that exhibit rapid strain hardening during impact loading.

What about polymers, glasses, and ceramics? Do they also exhibit strain hardening? We will show that the deformation of thermoplastic polymers often produces a strengthening effect. However, the mechanism of deformation strengthening is completely different in polymers than that in metallic materials. The strength of most brittle materials such as ceramics and glasses depends upon the flaws and flaw-size distribution (Chapter 6). Therefore, inorganic glasses and ceramics do not respond well to strain hardening. We should therefore consider different strategies to strengthen these materials. In this context, we will learn the principles of tempering and annealing of glasses. These processes make glass safer. We will also examine conditions under which ceramic materials can show large (several hundred percent) plastic deformations. Thus, all ceramic materials are not intrinsically brittle! There are conditions under which many ceramics can exhibit considerable ductility.

We begin by discussing strain hardening in metallic materials in the context of stress-strain curves.

7-1   Relationship of Cold Working to the Stress-Strain Curve

A stress-strain curve for a ductile metallic material is shown in Figure 7-1(a). If we apply a stress σ_1 that is greater than the yield strength (σ_y), it causes a permanent deformation or strain. When the stress is removed, it leaves behind a strain of ε_1. If we make a tensile test sample from the metallic material that had been previously stressed to σ_1 and retest that material, we obtain the stress-strain curve shown in Figure 7-1(b). Our new test specimen would begin to deform plastically or flow at stress level σ_1. We define **flow stress** as the stress that is needed to initiate plastic flow in previously deformed material.[1] Thus, σ_1 is now the flow stress of the material. If we continue to apply a stress until we reach σ_2, then release the stress and again retest the metallic material, the new flow stress is σ_2. Each time we apply a higher stress, the flow stress and tensile strength

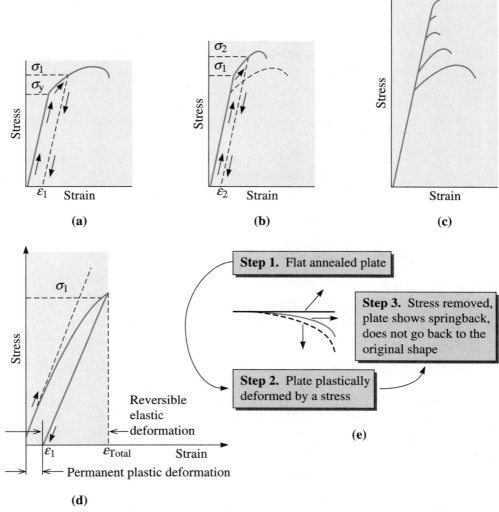

Figure 7-1 Development of strain hardening from the stress-strain diagram. (a) A specimen is stressed beyond the yield strength before the stress is removed. (b) Now the specimen has a higher yield strength and tensile strength, but lower ductility. (c) By repeating the procedure, the strength continues to increase and the ductility continues to decrease until the alloy becomes very brittle. (d) Note the total strain and the elastic strain recovery lead to remnant plastic strain and (e) illustration of springback. (*Source: Reprinted from* Engineering Materials I, *Second Edition, M.F. Ashbey, and D.R.H. Jones, 1996. Copyright © 1996 Butterworth-Heinemann. Reprinted with permission from Elsevier Science.*)

increase and the ductility decreases. We eventually strengthen the metallic material until the flow stress, tensile, and breaking strengths are equal and there is no ductility [Figure 7-1(c)]. At this point, the metallic material can be plastically deformed no further. Figures 7-1(d) and (e) are related to springback, a concept that is discussed a bit later in this section.

By applying a stress that exceeds the original yield strength of the metallic material, we have **strain hardened** or **cold worked** the metallic material, while simultaneously deforming it. This is the basis for many manufacturing techniques, such as wire drawing. Figure 7-2 illustrates several manufacturing processes that make use of both cold-

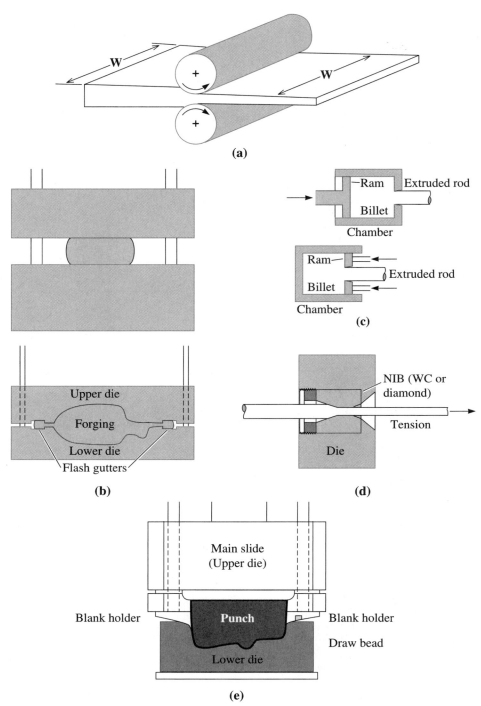

Figure 7-2 Manufacturing processes that make use of cold working as well as hot working. Common metalworking methods. (a) Rolling. (b) Forging (open and closed die). (c) Extrusion (direct and indirect). (d) Wire drawing. (e) Stamping. (Adapted from Mechanical Behavior of Materials by Meyers, M.A. & Chawla, K.K. 1999, Prentice Hall.) (*Source: Adapted from* Mechanical Behavior of Materials, *by M.A. Meyers and K.K. Chawla, p. 292, Fig. 6-1. Copyright © 1999 Prentice Hall. Adapted with permission of Pearson Education, Inc., Upper Saddle River, NJ.*)

working and hot-working processes. We will discuss the difference between **hot-working** and **cold-working** later in this chapter. Many techniques are used to simultaneously shape and strengthen a material by cold working (Figure 7-2). For example, **rolling** is used to produce metal plate, sheet, or foil. **Forging** deforms the metal into a die cavity, producing relatively complex shapes such as automotive crankshafts or connecting rods. In **drawing**, a metallic rod is pulled through a die to produce a wire or fiber. In **extrusion**, a material is pushed through a die to form products of uniform cross-sections, including rods, tubes, or aluminum trims for doors or windows. **Deep drawing** is used to form the body of aluminum beverage cans.[2] **Stretch forming** and **bending** are used to shape sheet material. Thus, cold working is an effective way of shaping metallic materials while simultaneously increasing their strength. The down side of this process is the loss of ductility. If you take a metal wire and bend it repeatedly it will harden and eventually break because of strain hardening. Strain hardening is used in many products, especially those that are not going to be exposed to very high temperatures. For example, an aluminum beverage can derives almost 70% of its strength as a result of strain hardening that occurs during its fabrication. Some of the strength of aluminum cans also come from the alloying elements (e.g., Mg) added. Note that many of the processes such as rolling can be conducted using both cold and hot working. The pros and cons of using each will be discussed later in this chapter.

Strain-Hardening Exponent (*n*) The response of a metallic material to cold working is given by the **strain-hardening exponent**, *n*, which is the slope of the plastic portion of the *true stress-true strain* curve in Figure 7-3 when a logarithmic scale is used:

$$\sigma_t = K\varepsilon_t^n \tag{7-1a}$$

or

$$\ln \sigma_t = \ln K + n \ln \varepsilon_t \tag{7-1b}$$

The constant K (strength coefficient) is equal to the stress when $\varepsilon_t = 1$. The strain-hardening exponent is relatively low for HCP metals, but is higher for BCC and, particularly, for FCC metals (Table 7-1). Metals with a low strain-hardening exponent respond poorly to cold working. If we take a copper wire and bend, the bent wire is stronger as a result of strain hardening.

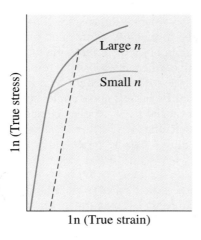

Figure 7-3
The true stress-true strain curves for metals with large and small strain-hardening exponents. Larger degrees of strengthening are obtained for a given strain for the metal with the larger *n*.

TABLE 7-1 ■ *Strain-hardening exponents and strength coefficients of typical metals and alloys*

Metal	Crystal Structure	*n*	K (psi)
Titanium	HCP	0.05	175,000
Annealed alloy steel	BCC	0.15	93,000
Quenched and tempered medium-carbon steel	BCC	0.10	228,000
Molybdenum	BCC	0.13	105,000
Copper	FCC	0.54	46,000
Cu-30% Zn	FCC	0.50	130,000
Austenitic stainless steel	FCC	0.52	220,000

Adapted from G. Dieter, Mechanical Metallurgy, *McGraw-Hill, 1961, and other sources.*

Strain-Rate Sensitivity (*m*)   The **strain-rate sensitivity** (*m*) of stress is defined as:

$$m = \left[\frac{\partial(\ln \sigma)}{\partial(\ln \dot{\varepsilon})}\right] \tag{7-2}$$

This describes how fast strain hardening occurs in response to plastic deformation. As mentioned before, the mechanical behavior of sheet steels under high strain rates ($\dot{\varepsilon}$) is important not only for shaping, but also for how well the steel will perform under high-impact loading. The crashworthiness of sheet steels is an important consideration for the automotive industry. Steels that harden rapidly under impact loading are useful in absorbing mechanical energy.

A positive value of *m* implies that material will resist necking (Chapter 6). High values of *m* and *n* mean the material can exhibit a better formability in stretching. However, these values do not affect the deep drawing characteristics. For deep drawing, the **plastic strain ratio** (*r*) is important.[3] We define plastic strain ratio as:

$$r = \frac{\varepsilon_w}{\varepsilon_t} = \frac{\ln\left(\dfrac{w}{w_0}\right)}{\ln\left(\dfrac{h}{h_0}\right)} \tag{7-3}$$

In this equation, *w* and *h* correspond to the width and thickness of the material being processed and the subscript zero indicates original dimensions. Forming limit diagrams are often used to better understand the **formability** of metallic materials. Overall, we define formability of a material as the ability of a material to maintain its integrity while being shaped. Formability of material is often described in terms of two strains—a major strain, always positive, and a minor strain that could be positive or negative. As illustrated in Figure 7-4, strain conditions on the left make circles stamped into a sample transform into ellipses; for conditions on the right, smaller circles stamped into samples become larger circles indicating stretching.[4] The forming limit diagrams illustrate the specific regions over which the material can be processed without compromising mechanical integrity.

Springback   Another point to be noted is that when a metallic material is deformed using a stress above its yield strength to a higher level (σ_1 in Figure 7-1(d)), the corresponding strain existing at stress σ_1 is obtained by dropping a perpendicular line to the

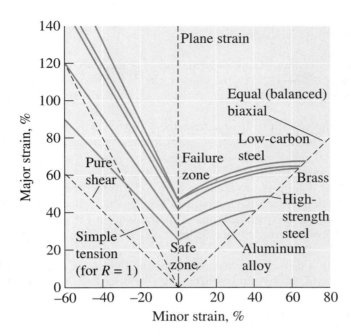

Figure 7-4
Forming limit diagram for different materials. (*Source: Reprinted from* Metals Handbook—Desk Edition, *Second Edition, ASM International, Materials Park, OH 44073, p. 146, Fig. 5 © 1998 ASM International. Reprinted by permission.*)

x-axis (point $\varepsilon_{\text{total}}$). A strain equal to $(\varepsilon_{\text{total}} - \varepsilon_1)$ is recovered since it is elastic in nature. The *elastic strain* that is recovered after a material has been *plastically* deformed is known as **springback** [Figure 7-1(e)]. The occurrence of springback is extremely important for the formation of automotive body panels from sheet steels along with many other applications. This effect is also seen in processing of polymeric materials processed, for example, by extrusion. This is because many polymers are viscoelastic, as discussed in Chapter 6.

It is possible to account for springback in designing components, however, variability in springback makes this very difficult. For example, an automotive supplier will receive coils of sheet steel from different steel manufacturers, and even though the specifications for the steel are identical, the springback variation in steels received from each manufacturer (or even for different lots from the same manufacturer) will make it harder to obtain cold worked components that have precisely the same shape and dimensions.[5]

Bauschinger Effect   Consider a material that has been subjected to tensile plastic deformation. Then, consider two separate samples (*A* and *B*) of this material that has been previously deformed. Test sample *A* in tension, and sample *B* under compression. We notice that for the deformed material the flow stress in tension ($\sigma_{\text{flow, tension}}$) for sample *A* is greater than the compressive yield strength ($\sigma_{\text{flow, compression}}$) for sample *B*. This effect, in which a material subjected to tension shows reduction in compressive strength, is known as the **Bauschinger effect**.[6] Note that we are comparing the yield strength of a material under compression and tension after the material has been subjected to plastic deformation under a tensile stress. The Bauschinger effect is also seen on stress reversal. Consider a sample deformed under compression. We can then evaluate two separate samples *C* and *D*. The sample subjected to compressive stress (*C*) shows a higher flow stress than that for the sample *D* subjected to tensile stress. The Bauschinger effect plays an important role in mechanical processing of steels and other alloys.

7-2   Strain-Hardening Mechanisms

We obtain strengthening during deformation of a metallic material by increasing the number of dislocations. Before deformation, the dislocation density is about 10^6 cm of dislocation line per cubic centimeter of metal—a relatively small concentration of dislocations.

When we apply a stress greater than the yield strength, dislocations begin to slip (Schmid's Law, Chapter 4). Eventually, a dislocation moving on its slip plane encounters obstacles that pin the ends of the dislocation line. As we continue to apply the stress, the dislocation attempts to move by bowing in the center. The dislocation may move so far that a loop is produced (Figure 7-5). When the dislocation loop finally touches itself, a new dislocation is created. The original dislocation is still pinned and can create additional dislocation loops. This mechanism for generating dislocations is called a **Frank-Read source**; Figure 7-5(e) shows an electron micrograph of a Frank-Read source.

The dislocation density may increase to about 10^{12} cm of dislocation line per cubic centimeter of metal during strain hardening. As discussed in Chapter 4, dislocation motion is the cause for the plastic flow that occurs in metallic materials; however, when we have too many dislocations, they interfere with their own motions. An analogy for this is when we have too many people in a room it is difficult for them to move around. The result is an increased strength, but reduced ductility, for metallic materials that have undergone cold working or work hardening.

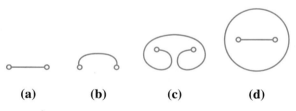

(a)   **(b)**   **(c)**   **(d)**

Figure 7-5
The Frank-Read source can generate dislocations. (a) A dislocation is pinned at its ends by lattice defects. (b) As the dislocation continues to move, the dislocation bows, eventually bending back on itself. (c) finally the dislocation loop forms, and (d) a new dislocation is created. (e) Electron micrograph of a Frank-Read source (330,000). (*Adapted from Brittain, J., "Climb Sources in Beta Prime-NiAl,"* Metallurgical Transactions, *Vol. 6A, April 1975.*)

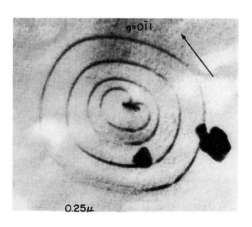

(e)

Ceramics contain dislocations and can even be strain-hardened to a small degree. However, dislocations in ceramics are normally not very mobile. Polycrystalline ceramics also contain porosity. As a result, ceramics behave as brittle materials and significant deformation and strengthening by cold working are not possible. Likewise, covalently bonded materials such as silicon (Si) are too brittle to work-harden appreciably. Glasses are amorphous and do not contain dislocations and therefore cannot be strain hardened.

Thermoplastic polymers are polymers such as polyethylene, polystyrene, and nylon. These materials consist of molecules that are long spaghetti-like chains. Thermoplastics will strengthen when they are deformed. However, this is *not* strain hardening due to dislocation multiplication but, instead, strengthening of these materials involves alignment and possibly localized crystallization of the long, chainlike molecules. When a stress greater than the yield strength is applied to thermoplastic polymers such as polyethylene, the van der Waals bonds (Chapter 3) between the molecules in different chains are broken. The chains straighten and become aligned in the direction of the applied stress (Figure 7-6). The strength of the polymer, particularly in the direction of the applied stress, increases as a result of the alignment of polymeric chains in the direction of the applied stress. As illustrated in previous chapters, polyethylene terephthalate (PET) bottles made by the blow-stretch process (Figure 3-9) involves such stress-induced crystallization. Thermoplastic polymers, get stronger as a result of local alignments of polymer chains occurring as a result of applied stress. This strength increase is seen in the stress-strain curve of typical thermoplastics (Chapter 6). Many techniques used for polymer processing are similar to those used for the fabrication of metallic materials. Extrusion, for example, is the most widely used polymer processing technique. Although many of these techniques share conceptual similarities, there are important differences between the mechanisms by which polymers become strengthened during their processing.

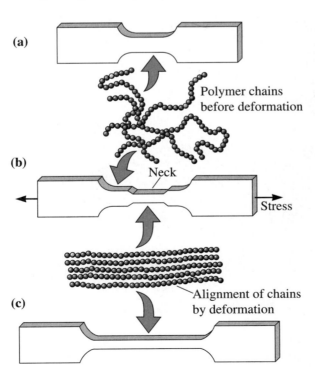

Figure 7-6
In an undeformed thermoplastic polymer tensile bar, (a) the polymer chains are randomly oriented. (b) When a stress is applied, a neck develops as chains become aligned locally. The neck continues to grow until the chains in the entire gage length have aligned. (c) The strength of the polymer is increased.

Polymer chains before deformation

Neck

Stress

Alignment of chains by deformation

7-3   Properties versus Percent Cold Work

By controlling the amount of plastic deformation, we control strain hardening. We normally measure the amount of deformation by defining the **percent cold work**:

$$\text{Percent cold work} = \left[\frac{A_0 - A_f}{A_0}\right] \times 100, \qquad (7\text{-}4)$$

where A_0 is the original cross-sectional area of the metal and A_f is the final cross-sectional area after deformation.

The effect of cold work on the mechanical properties of commercially pure copper is shown in Figure 7-7. As the cold work increases, both the yield and the tensile strength increase; however, the ductility decreases and approaches zero. The metal breaks if more cold work is attempted. Therefore, there is a maximum amount of cold work or deformation that we can perform on a metallic material before it becomes too brittle and breaks.

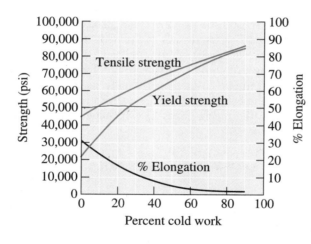

Figure 7-7
The effect of cold work on the mechanical properties of copper.

EXAMPLE 7-1   *Cold Working a Copper Plate*

A 1-cm-thick copper plate is cold-reduced to 0.50 cm, and later further reduced to 0.16 cm. Determine the total percent cold work and the tensile strength of the 0.16-cm plate. (See Figure 7-8.)

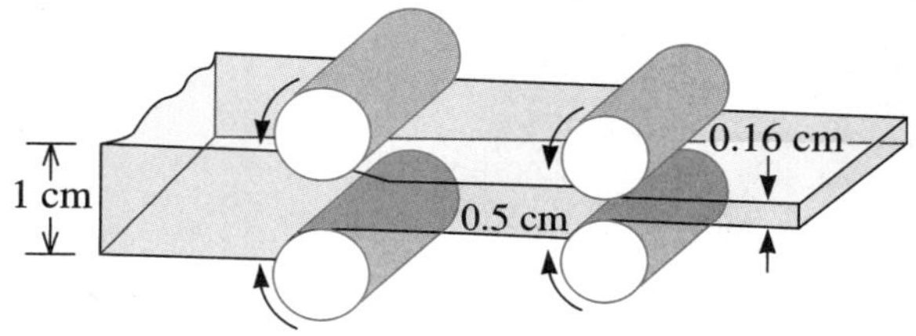

Figure 7-8
Diagram showing the rolling of a 1-cm plate to a 0.16-cm plate (for Example 7-1).

SOLUTION

Note that, because the width of the plate does not change during rolling, the cold work can be expressed as the percentage reduction in the thickness t.

We might be tempted to determine the amount of cold work accomplished in each step, that is:

$$\% \text{ CW} = \left[\frac{A_0 - A_f}{A_0}\right] \times 100 = \left[\frac{t_0 - t_f}{t_0}\right] \times 100 = \left[\frac{1 \text{ cm} - 0.50 \text{ cm}}{1 \text{ cm}}\right] \times 100$$

$$= 50\%$$

$$\% \text{ CW} = \left[\frac{A_0 - A_f}{A_0}\right] \times 100 = \left[\frac{t_0 - t_f}{t_0}\right] \times 100 = \left[\frac{0.50 \text{ cm} - 0.16 \text{ cm}}{0.50 \text{ cm}}\right] \times 100$$

$$= 68\%$$

We might then be tempted to combine the two cold work percentages ($50\% + 68\% = 118\%$) to obtain the total cold work. *This would be incorrect.*

Our definition of cold work is the percentage change between the original and final cross-sectional areas; it makes no difference how many intermediate steps are involved. Thus, the total cold work is actually

$$\% \text{ CW} = \left[\frac{t_0 - t_f}{t_0}\right] \times 100 = \left[\frac{1 \text{ cm} - 0.16 \text{ cm}}{1 \text{ cm}}\right] \times 100 = 84\%$$

and, from Figure 7-7, the tensile strength is about 82,000 psi.

We can predict the properties of a metal or an alloy if we know the amount of cold work during processing. We can then decide whether the component has adequate strength at critical locations.

When we wish to select a material for a component that requires certain minimum mechanical properties, we can design the deformation process. We first determine the necessary percent cold work and then, using the final dimensions we desire, calculate the original metal dimensions from the cold work equation.

EXAMPLE 7-2   *Design of a Cold Working Process*

Design a manufacturing process to produce a 0.1-cm-thick copper plate having at least 65,000 psi tensile strength, 60,000 psi yield strength, and 5% elongation.

SOLUTION

From Figure 7-7, we need at least 35% cold work to produce a tensile strength of 65,000 psi and 40% cold work to produce a yield strength of 60,000 psi, but we need less than 45% cold work to meet the 5% elongation requirement. Therefore, any cold work between 40% and 45% gives the required mechanical properties.

To produce the plate, a cold-rolling process would be appropriate. The original thickness of the copper plate prior to rolling can be calculated from Equation 7-4, assuming that the width of the plate does not change. Because there is a range of allowable cold work—between 40% and 45%—there is a range of initial plate thicknesses:

$$\% \, \mathrm{CW}_{\min} = 40 = \left[\frac{t_{\min} \text{ cm} - 0.1 \text{ cm}}{t_{\min} \text{ cm}} \right] \times 100, \quad \therefore t_{\min} = 0.167 \text{ cm}$$

$$\% \, \mathrm{CW}_{\max} = 45 = \left[\frac{t_{\max} \text{ cm} - 0.1 \text{ cm}}{t_{\max} \text{ cm}} \right] \times 100, \quad \therefore t_{\max} = 0.182 \text{ cm}$$

To produce the 0.1-cm copper plate, we begin with a 0.167- to 0.182-cm copper plate in the softest possible condition, then cold roll the plate 40% to 45% to achieve the 0.1 cm thickness.

7-4 Microstructure, Texture Strengthening, and Residual Stresses

During plastic deformation using cold or hot working, a microstructure consisting of grains that are elongated in the direction of the stress applied is often produced (Figure 7-9).

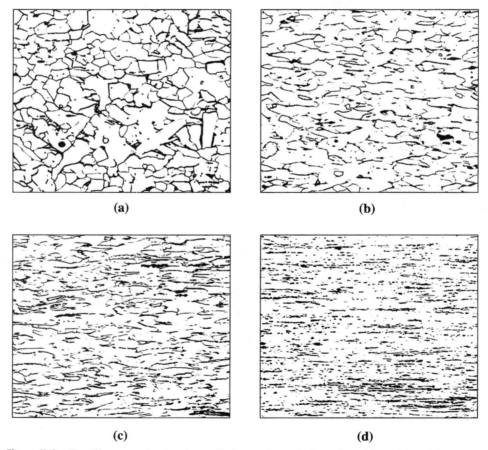

(a)   **(b)**

(c)   **(d)**

Figure 7-9 The fibrous grain structure of a low carbon steel produced by cold working: (a) 10% cold work, (b) 30% cold work, (c) 60% cold work, and (d) 90% cold work (×250). (*Source: From* ASM Handbook Vol. 9, Metallography and Microstructure, *(1985) ASM International, Materials Park, OH 44073. Used with permission.*)

Anisotropic Behavior   During deformation, the grains rotate as well as elongate, causing certain crystallographic directions and planes to become aligned with the direction in which stress is applied. Consequently, preferred orientations, or textures, develop and cause anisotropic behavior.

In processes such as wire drawing and extrusion, a **fiber texture** is produced. The term "fibers" refers to the grains in the metallic material which become elongated in a direction parallel to the axis of the wire or an extruded product. In BCC metals, $\langle 110 \rangle$ directions line up with the axis of the wire. In FCC metals, $\langle 111 \rangle$ or $\langle 100 \rangle$ directions are aligned. This gives the highest strength along the axis of the wire or the extrudate (product being extruded such as a tube), which is what we desire.

As mentioned previously, a somewhat similar effect is seen in thermoplastic materials when they are drawn into fibers or other shapes. The cause, as discussed before, is that polymer chains line up side-by-side along the length of the fiber. As in metallic materials, the strength is greatest along the axis of the polymer fiber. This type of strengthening is also seen in PET plastic bottles made using the blow-stretch process (Chapter 4). This process leads to alignment of polymer chains along the radial and length directions, leading to increased strength of PET bottles along those directions.

In processes such as rolling, grains become oriented in a preferred crystallographic direction and plane, giving a **sheet texture**. The properties of a rolled sheet or plate depend on the direction in which the property is measured. Figure 7-10 summarizes the tensile properties of a cold worked aluminum-lithium (Al-Li) alloy. For this alloy,

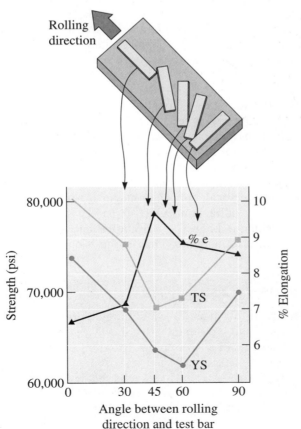

Figure 7-10
Anisotropic behavior in a rolled aluminum-lithium sheet material used in aerospace applications. The sketch relates the position of tensile bars to the mechanical properties that are obtained.

Table 7-2 ■ *Common wire drawing and extrusion, and sheet textures in materials[8]*

Crystal Structure	Wire Drawing and Extrusion (Fiber Texture) (Direction Parallel to Wire Axis)	Sheet or Rolling Texture
FCC	$\langle 111 \rangle$ and $\langle 100 \rangle$	{110} planes parallel to rolling plane $\langle 112 \rangle$ directions parallel to rolling direction
BCC	$\langle 110 \rangle$	{001} planes parallel to rolling plane $\langle 110 \rangle$ directions parallel to rolling direction
HCP	$\langle 10\bar{1}0 \rangle$	{0001} planes parallel to rolling plane $\langle 11\bar{2}0 \rangle$ directions parallel to rolling direction

strength is highest parallel to the rolling direction, whereas ductility is highest at a 45° angle to the rolling direction. The strengthening that occurs by the development of anisotropy or of a texture, is known as **texture strengthening**. As pointed out in Chapter 6, the Young's modulus of materials also depends upon crystallographic directions. For example, the Young's modulus of iron along [111] and [100] directions is ~260 and 140 GPa, respectively.[1] The dependence of yield strength on texture is even stronger. Development of texture not only has an effect on mechanical properties but also on magnetic and other properties of materials. For example, grain oriented magnetic steels made from about 3% Si and 97% Fe used in making transformer cores are textured via proper thermo-mechanical processing so as to optimize their electrical and magnetic properties. Some common fiber (wire drawing) and sheet (rolling) textures with different crystal structures are shown in Table 7-2.[8]

Texture Development in Thin Films   Orientation or crystallographic texture development also occurs in thin films and is used widely using lattice matching with substrate crystal structures or domains (Chapter 8). In the case of thin films, the texturing is often a result of the mechanisms of the growth process and not as a result of externally applied stresses. Sometimes, internally generated thermal stresses can play a role in the determination of thin-film crystallographic texture. Oriented thin films are often sought after since they can offer better electrical, optical, or magnetic properties. **Pole figure analysis**, a technique based on x-ray diffraction (XRD) (Chapter 3), or a specialized scanning electron microscopy technique known as **orientation microscopy** are used to identify textures in different engineered materials (films, sheets, single crystals, etc.).

EXAMPLE 7-3   *Design of a Stamping Process*

One method for producing fans for cooling automotive and truck engines is to stamp the blades from cold-rolled steel sheet, then attach the blades to a "spider" that holds the blades in the proper position. A number of fan blades, all produced at the same time, have failed by the initiation and propagation of a fatigue crack transverse to the axis of the blade (Figure 7-11). All other fan blades perform satisfactorily. Provide an explanation for the failure of the blades and redesign the manufacturing process to prevent these failures.

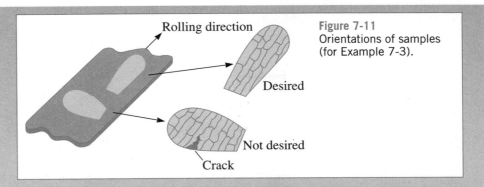

Figure 7-11
Orientations of samples
(for Example 7-3).

SOLUTION

There may be several explanations for the failure of the blades—for example, the wrong type of steel may have been selected, the dies used to stamp the blades from the sheet may be worn, or the clearance between the parts of the dies may be incorrect, producing defects that initiate fatigue failure.

The failures could also be related to the anisotropic behavior of the steel sheet caused by rolling. To achieve the best performance from the blade, the axis of the blade should be aligned with the rolling direction of the steel sheet. This procedure produces high strength along the axis of the blade and, by assuring that the grains are aligned with the blade axis, reduces the number of grain boundaries along the leading edge of the blade that might help initiate a fatigue crack. Suppose your examination of the blade using, for example, pole figure analysis, or metallographic analysis indicates that the steel sheet was aligned 90° from its usual position during stamping. Now the blade has a low strength in the critical direction and, in addition, fatigue cracks will more easily initiate and grow. This mistake in manufacturing can cause failures and injuries to mechanics performing maintenance on automobiles.

You might recommend that the manufacturing process be redesigned to assure that the blades cannot be stamped from misaligned sheet. Perhaps special guides or locking devices on the die will assure that the die is properly aligned with the sheet.

Residual Stresses A small portion of the applied stress—perhaps about 10%—is stored in the form of **residual stresses** within the structure as a tangled network of dislocations. The residual stresses increase the total energy of the structure. The presence of dislocations increases the total internal energy of the structure. The higher the extent of cold working, the higher would be the level of total internal energy of the material. Residual stresses generated by cold working may not always be desirable and can be relieved by a heat treatment known as **stress-relief anneal** (Section 7-6). As will be discussed shortly, in some instances we deliberately create residual compressive stresses at the surface of materials to enhance their mechanical properties.

The residual stresses are not uniform throughout the deformed metallic material. For example, high compressive residual stresses may be present at the surface of a rolled plate and high tensile stresses may be stored in the center. If we machine a small amount of metal from one surface of a cold-worked part, we remove metal that contains only compressive residual stresses. To restore the balance, the plate must distort. If there is a net compressive residual stress at the surface of a component, this may be beneficial from a viewpoint of mechanical properties since any crack or flaw on the

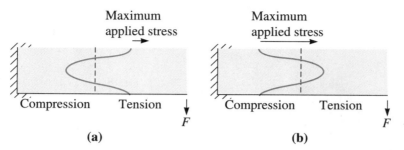

Figure 7-12 The compressive residual stresses can be harmful or beneficial. (a) A bending force applies a tensile stress on the top of the beam. Since there are already tensile residual stresses at the top, the load-carrying characteristics are poor. (b) The top contains compressive residual stresses. Now the load-carrying characteristics are very good.

surface will not likely grow. This is why any residual stresses, originating from cold work or any other source, affect the ability of the part to carry a load (Figure 7-12). If a tensile stress is applied to a material that already contains tensile residual stresses, the total stress acting on the part is the sum of the applied and residual stresses. If, however, compressive stresses are stored at the surface of a metal part, an applied tensile stress must first balance the compressive residual stresses. Now the part may be capable of withstanding a larger than normal load. In Chapter 6, we learned that fatigue is a common mechanism of failure for load-bearing components. Sometimes, components that are subject to fatigue failure can be strengthened by **shot peening**. Bombarding the surface with steel shot propelled at a high velocity introduces compressive residual stresses at the surface that increase the resistance of the metal surface to fatigue failure (Chapter 6). The following example explains the use of shot peening.

EXAMPLE 7-4 *Design of a Fatigue-Resistant Shaft*

Your company has produced several thousand shafts that have a fatigue strength of 20,000 psi. The shafts are subjected to high-bending loads during rotation. Your sales engineers report that the first few shafts placed into service failed in a short period of time by fatigue. Design a process by which the remaining shafts can be salvaged by improving their fatigue properties.

SOLUTION

Fatigue failures typically begin at the surface of a rotating part; thus, increasing the strength at the surface improves the fatigue life of the shaft. A variety of methods might be used to accomplish this.

If the shaft is made of steel, we could carburize the surface of the part (Chapter 5). In carburizing, carbon is diffused into the surface of the shaft. After an appropriate heat treatment, the higher carbon at the surface increases the strength of the surface and, perhaps more importantly, introduces *compressive* residual stresses at the surface.

We might consider cold working the shaft; cold working increases the yield strength of the metal and, if done properly, introduces compressive residual stresses. However, the cold work also reduces the diameter of the shaft and, because of the dimensional change, the shaft may not be able to perform its function.

> Another alternative would be to shot peen the shaft. Shot peening introduces local compressive residual stresses at the surface without changing the dimensions of the part. This process, which is also inexpensive, might be sufficient to salvage the remaining shafts.

Tempering and Annealing of Glasses   Residual stresses originating during the cooling of glasses also are of considerable interest. We can deal with residual stresses in glasses in two ways. First, we can reheat the glass to a high temperature known as the **annealing point** ($\sim$450°C for silicate glasses with a viscosity of $\sim$10^{13} Poise) and let it cool at a slow rate so that the outside and inside cool at about same rate. The resultant glass will have little or no stress. This process is known as **annealing** and the resultant glass that is nearly stress free is known as **annealed glass**. The purpose of annealing glasses and the process known as stress-relief annealing in metallic materials is the same (i.e., to remove or significantly lower the level of residual stress). The origin of residual stress, though, is different for these materials. Another option we have in glass processing is to conduct a heat treatment that leads to compressive stresses on the surface of a glass; this is known as **tempering**. The resultant glass is known as **tempered glass**. Tempered glass is obtained by heating glass to a temperature just below the annealing point, then, deliberately letting the surface cool more rapidly than the center. This leads to a uniform compressive stress at the surface of the glass. The center region remains under a tensile stress. It is also possible to exchange ions in the glass structure and to introduce a compressive stress. This is known as **chemical tempering**. In Chapter 6, we saw that the strength of glass depends on flaws on the surface. If we have a compressive stress at the surface of the glass and a tensile stress in the center (so as to have overall zero stress), the strength of glass is improved significantly. Any micro-cracks present will not grow readily, owing to the presence of a net compressive stress on the surface of the glass. If we, however, create a large impact, then the crack does penetrate through the region where the stresses are compressive and the glass shatters.

Tempered glass has many uses. For example, side window panes and rear windshields of cars are made using tempered glass. Applications such as fireplace screens, ovens, shelving, furniture, and refrigerators also make use of tempered glass. For automobile windshields in the front, though, we make use of **laminated safety glass**. Front windshield glass is made from two annealed glass pieces laminated inside using a plastic known as polyvinyl butyral (PVB). This plastic helps hold the glass pieces. If the windshield glass breaks, the laminated glass pieces are held together by PVB plastic. This helps minimize injuries to the driver and passengers. Also, the use of laminated safety glass reduces the chances of glass pieces cutting into the fabric of airbags that are probably being deployed simultaneously.

7-5   Characteristics of Cold Working

There are a number of advantages and limitations to strengthening a metallic material by cold working or strain hardening:

- We can simultaneously strengthen the metallic material and produce the desired final shape.

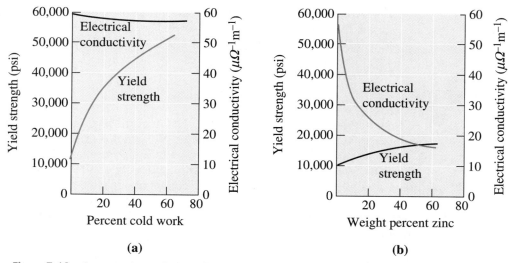

Figure 7-13   A comparison of strengthening copper by (a) cold working and (b) alloying with zinc. Note that cold working produces greater strengthening, yet has little effect on electrical conductivity.

- We can obtain excellent dimensional tolerances and surface finishes by the cold-working process.

- The cold-working process is an inexpensive method for producing large numbers of small parts, since high forces and expensive forming equipment are not needed. Also, no alloying elements are needed, which means lower-cost raw materials can be used.

- Some metals, such as HCP magnesium, have a limited number of slip systems, and are rather brittle at room temperature; thus, only a small degree of cold working can be accomplished.

- Ductility, electrical conductivity, and corrosion resistance are impaired by cold working. However, since the extent to which electrical conductivity is reduced by cold working is less than that for other strengthening processes, such as introducing alloying elements (Figure 7-13), cold working is a satisfactory way to strengthen conductor materials, such as the copper wires used for transmission of electrical power.

- Properly controlled residual stresses and anisotropic behavior may be beneficial. However, if residual stresses are not property controlled, the materials properties are greatly impaired.

- As will be seen in Section 7-6, since the effect of cold working is decreased or eliminated at higher temperatures, we cannot use cold working as a strengthening mechanism for components that will be subjected to high temperatures during application or service.

- Some deformation processing techniques can be accomplished only if cold working occurs. For example, wire drawing requires that a rod be pulled through a die to produce a smaller cross-sectional area (Figure 7-14). For a given draw force F_d, a different stress is produced in the original and final wire. The stress on the initial wire must exceed the yield strength of the metal to cause deformation. The stress on the final wire must be less than its yield strength to prevent failure. This is accomplished only if the wire strain hardens during drawing.

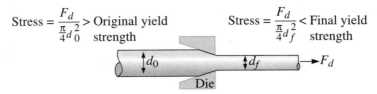

Stress $= \dfrac{F_d}{\frac{\pi}{4}d_0^2} >$ Original yield strength $\qquad$ Stress $= \dfrac{F_d}{\frac{\pi}{4}d_f^2} <$ Final yield strength

$d_0 \qquad d_f \qquad \rightarrow F_d$

Die

Figure 7-14 The wire-drawing process. The force F_d acts on both the original and final diameters. Thus, the stress produced in the final wire is greater than that in the original. If the wire did not strain harden during drawing, the final wire would break before the original wire was drawn through the die.

EXAMPLE 7-5   *Design of a Wire Drawing Process*

Design a process to produce 0.20-in. diameter copper wire. The mechanical properties of copper are to be assumed as those shown in Figure 7-7.

SOLUTION

Wire drawing is the obvious manufacturing technique for this application. To produce the copper wire as efficiently as possible, we make the largest reduction in the diameter possible. Our design must assure that the wire strain hardens sufficiently during drawing to prevent the drawn wire from breaking.

As an example calculation, let's assume that the starting diameter of the copper wire is 0.40 in. and that the wire is in the softest possible condition. The cold work is:

$$\% \text{ CW} = \left[\frac{A_0 - A_f}{A_0}\right] \times 100 = \left[\frac{(\pi/4)d_0^2 - (\pi/4)d_f^2}{(\pi/4)d_0^2}\right] \times 100$$

$$= \left[\frac{(0.40 \text{ in.})^2 - (0.20 \text{ in.})^2}{(0.40 \text{ in.})^2}\right] \times 100 = 75\%$$

From Figure 7-7, the initial yield strength with 0% cold work is 22,000 psi. The final yield strength with 75% cold work is about 77,500 psi (with very little ductility). The draw force required to deform the initial wire is:

$$F = \sigma_y A_0 = (22,000 \text{ psi})(\pi/4)(0.40 \text{ in.})^2 = 2765 \text{ lb}$$

The stress acting on the wire after passing through the die is:

$$\sigma = \frac{F_d}{A_f} = \frac{2765 \text{ lb}}{(\pi/4)(0.20 \text{ in.})^2} = 88,010 \text{ psi}$$

The applied stress of 88,010 psi is greater than the 77,500 psi yield strength of the drawn wire. Therefore, the wire breaks.

We can perform the same set of calculations for other initial diameters, with the results shown in Table 7-3 and Figure 7-15.

The graph shows that the draw stress exceeds the yield strength of the drawn wire when the original diameter is about 0.37 in. To produce the wire as efficiently as possible, the original diameter should be just under 0.37 in.

TABLE 7-3 ■ *Mechanical properties of copper wire (see Example 7-5)*

d_0 (in.)	% CW	Yield Strength of Drawn Wire (psi)	Force (lb)	Draw Stress on Drawn Wire (psi)
0.25	36	58,000	1080	34,380
0.30	56	68,000	1555	49,500
0.35	67	74,000	2117	67,390
0.40	75	77,500	2765	88,010

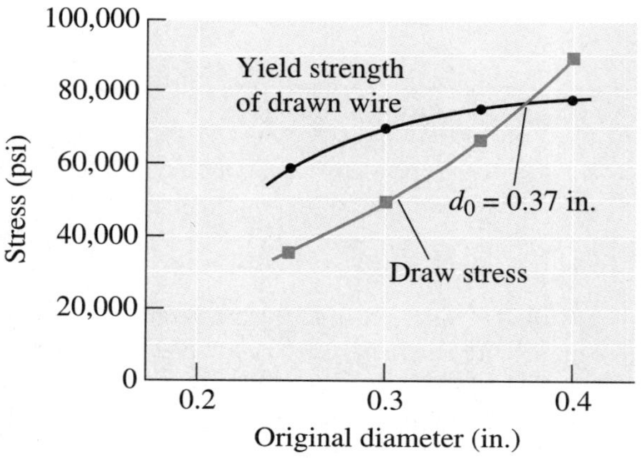

Figure 7-15
Yield strength and draw stress of wire (for Example 7-5).

7-6   The Three Stages of Annealing

Cold working is a useful strengthening mechanism. It is also a very effective tool for shaping materials using wire drawing, rolling, extrusion, etc. However, cold working leads to some effects that are sometimes undesirable. For example, the loss of ductility or development of residual stresses may not be desirable for certain applications. Since cold working or strain hardening results from increased dislocation density we can assume that any treatment to rearrange or annihilate dislocations would begin to undo the effects of cold working.

Annealing is a heat treatment used to eliminate some or all of the effects of cold working. Annealing at a low temperature may be used to eliminate the residual stresses produced during cold working without affecting the mechanical properties of the finished part. Or, annealing may be used to completely eliminate the strain hardening achieved during cold working. In this case, the final part is soft and ductile but still has a good surface finish and dimensional accuracy. After annealing, additional cold work could be done, since the ductility is restored; by combining repeated cycles of cold working and annealing, large total deformations may be achieved. There are three possible stages in the annealing process; their effects on the properties of brass are shown in Figure 7-18.

Note that the term "annealing" is also used to describe other thermal treatments. For example, ceramic glasses may be annealed, or heat treated, to eliminate residual

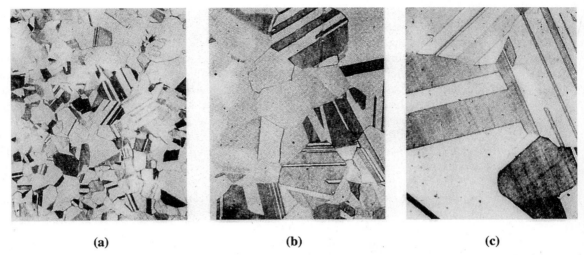

(a)   **(b)**   **(c)**

Figure 7-16   Photomicrographs showing the effect of annealing temperature on grain size in brass. Twin boundaries can also be observed in the structures. (a) Annealed at 400°C, (b) annealed at 650°C, and (c) annealed at 800°C (×75). (*Adapted from Brick, R. and Phillips, A.,* The Structure and Properties of Alloys, *1949: McGraw-Hill.*)

stresses. Cast irons and steels may be annealed to produce the maximum ductility, even though no prior cold work was done to the mateiral. These annealing heat treatments will be discussed in later chapters.

Recovery   The original cold-worked microstructure is composed of deformed grains containing a large number of tangled dislocations. When we first heat the metal, the additional thermal energy permits the dislocations to move and form the boundaries of a **polygonized subgrain structure** (Figure 7-17). The dislocation density, however, is virtually unchanged. This low-temperature treatment removes the residual stresses due to cold working without causing a change in dislocation density and is called **recovery**.

The mechanical properties of the metal are relatively unchanged because the number of dislocations is not reduced during recovery. However, since residual stresses are reduced or even eliminated when the dislocations are rearranged; recovery is often called a **stress relief anneal**. In addition, recovery restores high electrical conductivity to the metal, permitting us to manufacture copper or aluminum wire for transmission of electrical power that is strong yet still has high conductivity. Finally, recovery often improves the corrosion resistance of the material.

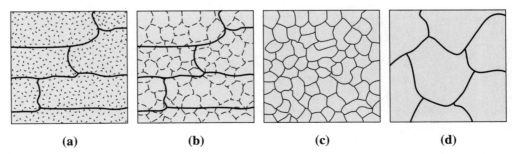

(a)   **(b)**   **(c)**   **(d)**

Figure 7-17   The effect of annealing temperature on the microstructure of cold-worked metals. (a) cold-worked, (b) after recovery, (c) after recrystallization, and (d) after grain growth.

Recrystallization   When a cold worked metallic material is heated above a certain temperature, rapid recovery eliminates residual stresses and produces the polygonized dislocation structure. New small grains then nucleate at the cell boundaries of the polygonized structure, eliminating most of the dislocations (Figure 7-17). Because the number of dislocations is greatly reduced, the recrystallized metal has low strength but high ductility. The temperature at which a microstructure of new grains that have very low dislocation density appears is known as the **recrystallization temperature**. The process of formation of new grains by heat treating a cold-worked material is known as **recrystallization**. As will be seen in Section 7-7, recrystallization temperature depends on many variables and is not a fixed temperature similar to melting temperature of elements and compounds.

Grain Growth   At still higher annealing temperatures, both recovery and recrystallization occur rapidly, producing a fine recrystallized grain structure. If the temperature is high enough, as discussed in Chapter 5, the grains begin to grow, with favored grains consuming the smaller grains (Figure 7-16). This phenomenon, called **grain growth**, is driven by the reduction in grain boundary area and was described in Chapter 5. Illustrated for a copper-zinc alloy in Figure 7-18, grain growth is almost always undesirable. Remember that grain growth will occur in most materials if they are subjected to a high enough temperature and, as such, is not related to cold working. Thus, recrystallization or recovery are not needed for grain growth to occur.

You may be aware that incandescent light bulbs contain filaments that are made from tungsten (W). High temperature causing grain growth is one of the factors that causes the filament to fail.

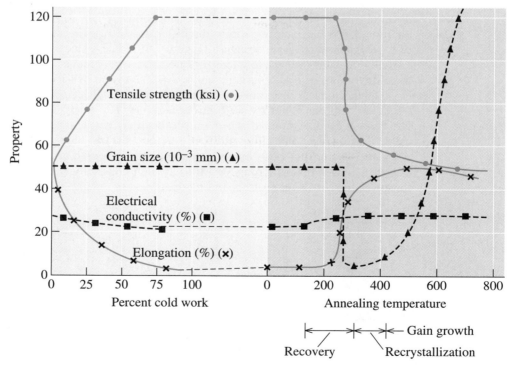

Figure 7-18   The effect of cold work on the properties of a Cu-35% Zn alloy and the effect of annealing temperature on the properties of a Cu-35% Zn alloy that is cold-worked 75%.

Ceramic materials, which normally do not show any significant strain hardening, show a considerable amount of grain growth (Chapter 5). Also, abnormal grain growth can occur in some materials as a result of a formation of liquid phase during their sintering (Chapter 14). An example of where grain growth is useful is the application of alumina ceramics for making optical materials used in lighting. In this application, we want very large grains since the scattering of light from grain boundaries has to be minimized. Some researchers have also developed methods for growing single crystals of ceramic materials using grain growth.

7-7 Control of Annealing

In many metallic material applications, we need a combination of strength and toughness. Therefore, we need to design processes that involve shaping via cold working. We then need to control the annealing process to obtain a level of ductility. To design an appropriate annealing heat treatment, we need to know the recrystallization temperature and the size of the recrystallized grains.

Recrystallization Temperature This is the temperature at which grains in the cold-worked microstructure begin to transform into new, equiaxed, and dislocation-free grains. The driving force for recrystallization is the difference between the internal energy between a cold worked material and that of a recrystallized material. It is important for us to emphasize that the recrystallization temperature is *not a* fixed temperature, like the melting temperature of a pure element, and is influenced by a variety of processing variables:

- Recrystallization temperature decreases when the amount of cold work increases. Greater amounts of cold work make the metal less stable and encourage nucleation of recrystallized grains. There is a minimum amount of cold work, about 30 to 40%, below which recrystallization will not occur.

- A smaller original cold-worked grain size reduces the recrystallization temperature by providing more sites—the former grain boundaries—at which new grains can nucleate.

- Pure metals recrystallize at lower temperatures than alloys.

- Increasing the annealing time reduces the recrystallization temperature (Figure 7-19), since more time is available for nucleation and growth of the new recrystallized grains.

- Higher melting-point range alloys have a higher recrystallization temperature. Since recrystallization is a diffusion-controlled process, the recrystallization temperature is roughly proportional to $0.4T_m$ Kelvin. Typical recrystallization temperatures for selected metals are shown in Table 7-4.

The concept of recrystallization temperature is very important since it also defines the boundary between cold working and hot working of a metallic material. If we conduct deformation (shaping) of a material above the recrystallization temperature, we refer to it as hot working. If we conduct the shaping or deformation at a temperature below the recrystallization temperature, we refer to this as cold working. Therefore, the terms "hot" and "cold" working refer to whether we are conducting the shaping process or deformation at temperatures above or below the recrystallization temperature. As can be seen from Table 7-4, for lead (Pb) or tin (Sn) deformed at 25°C, we are con-

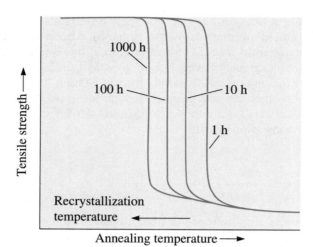

Figure 7-19
Longer annealing times reduce the recrystallization temperature. Note that the recrystallization temperature is not a fixed temperature.

ducting hot working! This is why iron (Fe) can be cold worked at room temperature but not lead (Pb). For tungsten (W) being deformed at 1000°C, we are conducting cold working! In some cases, processes conducted above 0.6 times the melting temperature (T_m) of metal (in K) are considered as hot working. Processes conducted below 0.3 times the melting temperature are considered cold working and processes conducted between 0.3 and 0.6 times T_m are considered **warm working**.[9] These descriptions of ranges that define hot, cold and warm working, however, are approximate and should be used with caution.

Recrystallized Grain Size   A number of factors influence the size of the recrystallized grains. Reducing the annealing temperature, the time required to heat to the annealing

TABLE 7-4 ■ *Typical recrystallization temperatures for selected metals*

Metal	Melting Temperature (°C)	Recrystallization Temperature (°C)
Sn	232	−4
Pb	327	−4
Zn	420	10
Al	660	150
Mg	650	200
Ag	962	200
Cu	1085	200
Fe	1538	450
Ni	1453	600
Mo	2610	900
W	3410	1200

(Source: Adapted from Structure and Properties of Engineering Materials, *by R. Brick, A. Pense, and R. Gordon, 1977. Copyright © 1977 The McGraw-Hill Companies. Adapted by permission.)*

temperature, or the annealing time reduces grain size by minimizing the opportunity for grain growth. Increasing the initial cold work also reduces final grain size by providing a greater number of nucleation sites for new grains. Finally, the presence of a second phase in the microstructure helps prevent grain growth and keeps the recrystallized grain size small.

7-8 Annealing and Materials Processing

The effects of recovery, recrystallization, and grain growth are important in the processing and eventual use of a metal or an alloy.

Deformation Processing By taking advantage of the annealing heat treatment, we can increase the total amount of deformation we can accomplish. If we are required to reduce a 5-in. thick plate to a 0.05-in. thick sheet, we can do the maximum permissible cold work, anneal to restore the metal to its soft, ductile condition, then cold work again. We can repeat the cold work-anneal cycle until we approach the proper thickness. The final cold-working step can be designed to produce the final dimensions and properties required (Example 7-6).

EXAMPLE 7-6   *Design of a Process to Produce Copper Strip*

We wish to produce a 0.1-cm-thick, 6-cm-wide copper strip having at least 60,000 psi yield strength and at least 5% elongation. We are able to purchase 6-cm-wide strip only in thicknesses of 5 cm. Design a process to produce the product we need.

SOLUTION

In Example 7-2, we found that the required properties can be obtained with a cold work of 40 to 45%. Therefore, the starting thickness must be between 0.167 cm and 0.182 cm, and this starting material must be as soft as possible—that is, in the annealed condition. Since we are able to purchase only 5-cm-thick stock, we must reduce the thickness of the 5-cm strip to between 0.167 and 0.182 cm, then anneal the strip prior to final cold working. But can we successfully cold work from 5 cm to 0.182 cm?

$$\% \, CW = \left[\frac{t_o - t_f}{t_o} \right] \times 100 = \left[\frac{5 \, cm - 0.182 \, cm}{5 \, cm} \right] \times 100 = 96.4\%$$

Based on Figure 7-7, a maximum of about 90% cold work is permitted. Therefore, we must do a series of cold work and anneal cycles. Although there are many possible combinations, one is as follows:

1. Cold work the 5-cm strip 80% to 1 cm:

$$80 = \left[\frac{t_o - t_f}{t_o} \right] \times 100 = \left[\frac{5 \, cm - t_i \, cm}{5 \, cm} \right] \times 100 \quad \text{or} \quad t_f = 1 \, cm$$

2. Anneal the 1-cm strip to restore the ductility. If we don't know the recrystallization temperature, we can use the $0.4T_m$ relationship to provide an estimate. The melting point of copper is 1085°C:

$$T_r \cong (0.4)(1085 + 273) = 543 \text{ K} = 270°C$$

3. Cold work the 1-cm-thick strip to 0.182 cm:

$$\% \text{ CW} = \left[\frac{1 \text{ cm} - 0.182 \text{ cm}}{1 \text{ cm}} \right] \times 100 = 81.8\%$$

4. Again anneal the copper at 270°C to restore ductility.

5. Finally cold work 45%, from 0.182 cm to the final dimension of 0.1 cm. This process gives the correct final dimensions and properties.

High-Temperature Service As mentioned previously, neither strain hardening nor grain size strengthening (Hall-Petch equation, Chapter 4) are appropriate for an alloy that is to be used at elevated temperatures, as in creep-resistant applications. When the cold-worked metal is placed into service at a high temperature, recrystallization immediately causes a catastrophic decrease in strength. In addition, if the temperature is high enough, the strength continues to decrease because of growth of the newly recrystallized grains.

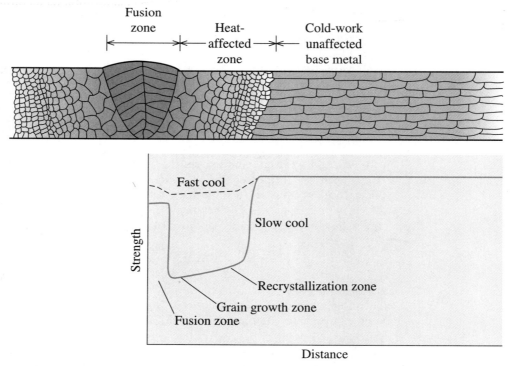

Figure 7-20 The structure and properties surrounding a fusion weld in a cold-worked metal. Note: only the right-hand side of the heat-affected zone is marked on the diagram. Note the loss in strength caused by recrystallization and grain growth in the heat-affected zone.

Joining Processes Metallic materials can be joined using processes such as welding.[10,11] When we join a cold-worked metal using a welding process, the metal adjacent to the weld heats above the recrystallization and grain growth temperatures and subsequently cools slowly. This region is called the **heat-affected zone** (HAZ). The structure and properties in the heat-affected zone of a weld are shown in Figure 7-20. The mechanical properties are reduced catastrophically by the heat of the welding process.

Welding processes, such as electron-beam welding or laser welding, which provide high rates of heat input for brief times, and, thus, subsequent fast cooling, minimize the exposure of the metallic materials to temperatures above recrystallization and minimize this type of damage. Similarly a process known as friction stir welding provides almost no HAZ and is being commercially used for welding aluminum alloys. We will discuss metal-joining processes in greater detail in Chapter 8.

7-9 Hot Working

We can deform a metal into a useful shape by hot working rather than cold working. As described previously, hot working is defined as plastically deforming the metallic material at a temperature above the recrystallization temperature. During hot working, the metallic material is continually recrystallized (Figure 7-21). As mentioned before, at room temperature lead (Pb) is well above its recrystallization temperature of $-4°C$ and therefore Pb does not strain harden and remains soft and ductile at room temperature.

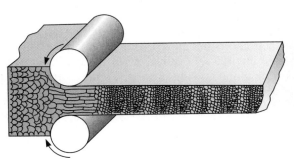

Figure 7-21
During hot working, the elongated, anisotropic grains immediately recrystallize. If the hot-working temperature is properly controlled, the final hot-worked grain size can be very fine.

Lack of Strengthening No strengthening occurs during deformation by hot working; consequently, the amount of plastic deformation is almost unlimited. A very thick plate can be reduced to a thin sheet in a continuous series of operations. The first steps in the process are carried out well above the recrystallization temperature to take advantage of the lower strength of the metal. The last step is performed just above the recrystallization temperature, using a large percent deformation in order to produce the finest possible grain size.

Hot working is well-suited for forming large parts, since the metal has a low yield strength and high ductility at elevated temperatures. In addition, HCP metals such as magnesium have more active slip systems at hot-working temperatures; the higher ductility permits larger deformations than are possible by cold working. The following example illustrates design of a hot-working process.

EXAMPLE 7-7   *Design of a Process to Produce a Copper Strip*

We want to produce a 0.1-cm-thick, 6-cm-wide copper strip having at least 60,000 psi yield strength and at least 5% elongation. We are able to purchase 6-cm-wide strip only in thicknesses of 5 cm. Design a process to produce the product we need, but in fewer steps than were required in Example 7-6.

SOLUTION

In Example 7-6, we relied on a series of cold work-anneal cycles to obtain the required thickness. We could reduce the steps by hot rolling to the required intermediate thickness:

$$\% \text{ HW} = \left[\frac{t_o - t_f}{t_o}\right] \times 100 = \left[\frac{5 \text{ cm} - 0.182 \text{ cm}}{5 \text{ cm}}\right] \times 100 = 96.4\%$$

$$\% \text{ HW} = \left[\frac{t_o - t_f}{t_o}\right] \times 100 = \left[\frac{5 \text{ cm} - 0.167 \text{ cm}}{5 \text{ cm}}\right] \times 100 = 96.7\%$$

Note that the formulas for hot and cold work are the same.

Because recrystallization occurs simultaneously with hot working, we can obtain these large deformations and a separate annealing treatment is not required. Thus our design might be:

1. Hot work the 5-cm strip 96.4% to the intermediate thickness of 0.182 cm.

2. Cold work 45% from 0.182 cm to the final dimension of 0.1 cm. This design gives the correct dimensions and properties.

Elimination of Imperfections   Some imperfections in the original metallic material may be eliminated or their effects minimized. Gas pores can be closed and welded shut during hot working—the internal lap formed when the pore is closed is eliminated by diffusion during the forming and cooling process. Composition differences in the metal can be reduced as hot working brings the surface and center of the plate closer together, thereby reducing diffusion distances.

Anisotropic Behavior   The final properties in hot-worked parts are *not* isotropic. The forming rolls or dies, which are normally at a lower temperature than the metal, cool the surface more rapidly than the center of the part. The surface then has a finer grain size than the center. In addition, a fibrous structure is produced because inclusions and second-phase particles are elongated in the working direction.

Surface Finish and Dimensional Accuracy   The surface finish is usually poorer than that obtained by cold working. Oxygen often reacts with the metal at the surface to form oxides, which are forced into the surface during forming. Hot worked steels and other metals are often subjected to a "pickling" treatment in which acids are used to dissolve the oxide scale. In some metals, such as tungsten (W) and beryllium (Be), hot-working must be done in a protective atmosphere to prevent oxidation.

The dimensional accuracy is also more difficult to control during hot working. A greater elastic strain must be considered, since the modulus of elasticity is low at hot-working temperatures. In addition, the metal contracts as it cools from the hot-working temperature. The combination of elastic strain and thermal contraction requires that the part be made oversize during deformation; forming dies must be carefully designed, and precise temperature control is necessary if accurate dimensions are to be obtained.

7-10 Superplastic Forming [SPF]

When specially heat-treated and processed, some materials can be uniformly deformed an exceptional amount—in some cases, more than 1000%.[11,12] This behavior, mentioned in Chapter 6, is called **superplasticity**. Often, superplastic forming can be coupled with diffusion bonding to produce complicated assemblies in a single step. Several conditions are required for a material to display superplastic behavior:

1. The metallic material must have a very fine grain structure, with grain diameters less than about 0.005 mm.

2. The alloy must be deformed at a high temperature, often near 0.5 to 0.65 times the absolute melting point of the alloy.

3. A very slow rate of forming, or **strain rate**, must be employed. In addition, the stress required to deform the alloy must be very sensitive to the strain rate, i.e. must have a high strain-rate sensitivity m. If necking begins to occur in a material with a high m, the necked region strains at a higher rate. The higher strain rate, in turn, strengthens the necked region, stops the necking, and permits uniform deformation to continue. Values for m for selected superplastic materials are included in Table 7-5.

4. The grain boundaries in the alloy should allow grains to slide easily over one another and rotate when stress is applied. The proper temperature and a fine grain size are necessary for **grain boundary sliding** to occur. Superplastic forming is most commonly done for metals, including alloys such as Ti-6% Al-4% V, Cu-10% Al, and Al-4.5%-Zn-4.5% Ca (Table 7-5). Complex aerospace components are often produced using the superplastic titanium alloy. Superplasticity is also found in ceramics that are normally considered brittle. A number of ceramic materials (Al_2O_3 and ZrO_2) and intermetallic compounds (e.g., Ni_3Si) show superplastic behavior.[6,13] Typically, the grain size of superplastic ceramics is ultrafine (<300 nm). One of the key factors in observing superplastic behavior is to ensure that grains remain fine and do not show grain growth when the ceramic is exposed to high temperatures. Some recently synthesized nano-scale ceramics also show superplastic behavior.

TABLE 7-5 ■ *Superplastic characteristics of several materials[13,14]*

Alloy	Superplastic Temperature	% Elongation (At Strain Rate Listed Here)	Strain Rate (s^{-1})	Strain Rate Sensitivity m	Flow Stress (MPa)
Ti-6% Al-4% V	927	1000–2000	2×10^{-4}	0.8	10
Al-4.5% Cu-0.5% Zn (Supral 100)	450	600–1000	10^{-3}	0.38	9
Al-4.5% Zn-4.5% Ca	565	500	10^{-3}	0.3	2.8
Fe-26% Cr-6.5% Ni	900	1000	5×10^{-5}	—	28
3mol.%Ytrria-97%Zirconia	1450	>160	$\sim 10^{-4}$	0.53	—
Al_2O_3 + 0.05%MgO + 0.05%Ytrria	1550	65	$\sim 10^{-4}$	—	—
80wt.% of (3mol.%Ytrria, 97%Zirconia) + 20wt.%Al_2O_3	1650	~ 500	$\sim 10^{-3}$	0.5	—

SUMMARY

◈ The properties of metallic materials can be controlled by combining plastic deformation and heat treatments.

◈ When a metallic material is deformed by cold working, strain hardening occurs as additional dislocations are introduced into the structure. Very large increases in strength may be obtained in this manner. The ductility of the strain hardened metallic material is reduced.

◈ Strain hardening, in addition to increasing strength and hardness, increases residual stresses, produces anisotropic behavior, and reduces ductility, electrical conductivity, and corrosion resistance.

◈ The amount of strain hardening is limited because of the simultaneous decrease in ductility; FCC metallic materials typically have the best response to strengthening by cold working.

◈ Wire drawing, stamping, rolling, and extrusion are some examples of manufacturing methods for shaping metallic materials. Some of the underlying principles for these processes can also be used for the manufacturing of polymeric materials.

◈ Springback and the Bauschinger effect are very important in manufacturing processes for the shaping of steels and other metallic materials. Forming limit diagrams are useful in defining shaping processes for metallic materials.

◈ The strain-hardening mechanism is not effective at elevated temperatures, where the effects of the cold work are eliminated by recrystallization.

◈ Annealing of metallic materials is a heat treatment intended to eliminate all or a portion of, the effects of strain hardening. The annealing process may involve as many as three steps.

◈ Recovery occurs at low temperatures, eliminating residual stresses and restoring electrical conductivity without reducing the strength. A "stress relief anneal" refers to recovery.

◈ Recrystallization occurs at higher temperatures and eliminates almost all of the effects of strain hardening. The dislocation density decreases dramatically during recrystallization as new grains nucleate and grow.

◈ Grain growth, which normally should be avoided, occurs at still higher temperatures. In cold-worked metallic materials, grain growth follows recovery and recrystallization. In ceramic materials, grain growth can occur due to high temperatures, longer times, or the presence of liquid phase during sintering.

◈ Hot working combines plastic deformation and annealing in a single step, permitting large amounts of plastic deformation without embrittling the material.

◈ Residual stresses in materials need to be controlled. In cold-worked metallic materials, residual stresses can be eliminated using a stress-relief anneal.

◈ Annealing of glasses leads to the removal of stresses developed during cooling. Thermal tempering of glasses is a heat treatment in which deliberate rapid cooling of the glass surface leads to a compressive stress at the surface. We use tempered or laminated glass in applications where safety is important.

◆ In metallic materials, compressive residual stresses can be introduced using shot peening. This treatment will lead to an increase in the fatigue life.

◆ Superplastic deformation provides unusually large amounts of deformation in some metallic and ceramic materials. Careful control of temperature, ultrafine grain size, and strain rate are required for superplastic forming.

GLOSSARY

Annealed glass Glass that has been treated by heating above the annealing point temperature (where the viscosity of glass becomes 10^{13} Poise) and then cooled slowly to minimize or eliminate residual stresses.

Annealing In the context of metallic material, annealing is a heat treatment used to eliminate part or all of the effects of cold working. For glasses, annealing is a heat treatment that removes thermally induced stresses.

Bauschinger effect A material previously plastically deformed under tension shows decreased flow stress under compression or vice versa.

Cold working Deformation of a metal below the recrystallization temperature. During cold working, the number of dislocations increases, causing the metal to be strengthened as its shape is changed.

Deformation processing Techniques for the manufacturing of metallic and other materials using such processes as rolling, extrusion, drawing, etc.

Drawing A deformation processing technique in which a material is pulled through an opening in a die (e.g., wire drawing).

Extrusion A deformation processing technique in which a material is pushed through an opening in a die, used for metallic and polymeric materials.

Fiber texture A preferred orientation of grains obtained during the wire drawing process. Certain crystallographic directions in each elongated grain line up with the drawing direction, causing anisotropic behavior.

Formability The ability of a material to stretch and bend without breaking. Forming diagrams describe the ability to stretch and bend materials.

Frank-Read source A pinned dislocation that, under an applied stress, produces additional dislocations. This mechanism is at least partly responsible for strain hardening.

Heat-affected zone (HAZ) The volume of material adjacent to a weld that is heated during the welding process above some critical temperature at which a change in the structure, such as grain growth or recrystallization, occurs.

Hot working Deformation of a metal above the recrystallization temperature. During hot working, only the shape of the metal changes; the strength remains relatively unchanged because no strain hardening occurs.

Laminated safety glass Two pieces of annealed glass held together by a plastic such as poly vinyl butyral (PVB). This type of glass can be used in car windshields.

Orientation microscopy A specialized technique, often based on scanning electron microscopy, used to determine the crystallographic orientation of different grains in a polycrystalline sample.

Pole figure analysis   A specialized technique based on x-ray diffraction, used for the determination of preferred orientation of thin films, sheets, or single crystals.

Polygonized structure   A subgrain structure produced in the early stages of the annealing. The subgrain boundaries are a network of dislocations rearranged during heating.

Recovery   A low-temperature annealing heat treatment designed to eliminate residual stresses introduced during deformation without reducing the strength of the cold-worked material. This is the same as a stress-relief anneal.

Recrystallization   A medium-temperature annealing heat treatment designed to eliminate all of the effects of the strain hardening produced during cold working.

Recrystallization temperature   A temperature above which essentially dislocation-free and new grains emerge from a material that was previously cold worked. This depends upon the extent of cold work, time of heat treatment, etc., and is not a fixed temperature.

Residual stresses   Stresses introduced in a material during processing. These can originate as a result of cold working or differential thermal expansion and contraction. A stress-relief anneal in metallic materials and the annealing of glasses minimize residual stresses. Compressive residual stresses deliberately introduced on the surface by the tempering of glasses or shot peening of metallic materials improve their mechanical properties.

Sheet texture   A preferred orientation of grains obtained during the rolling process. Certain crystallographic directions line up with the rolling direction, and certain preferred crystallographic planes become parallel to the sheet surface.

Shot peening   Introducing compressive residual stresses at the surface of a part by bombarding that surface with steel shot. The residual stresses may improve the overall performance of the material.

Strain hardening   Strengthening of a material by increasing the number of dislocations by deformation, or cold working. Also known as "work hardening."

Strain-hardening exponent (n)   A parameter that describes susceptibility of a material to cold working. It describes the effect that strain has on the resulting strength of the material. A material with a high strain-hardening coefficient obtains high strength with only small amounts of deformation or strain.

Strain rate   The rate at which a material is deformed.

Strain-rate sensitivity (m)   The rate at which stress develops changes as a function of strain rate. A material may behave much differently if it is slowly pressed into a shape rather than smashed rapidly into a shape by an impact blow.

Stress-relief anneal   The recovery stage of the annealing heat treatment during which residual stresses are relieved without reducing the mechanical properties of the material.

Superplasticity   The ability of a metallic or ceramic material to deform uniformly by an exceptionally large amount. Careful control over temperature, ultra-fine grain size, and strain rate are required for a material to behave in a superplastic manner.

Superplastic forming   Manufacturing processes that make use of superplasticity in ceramic or metallic materials.

Tempered glass   A glass, mainly for applications where safety is particularly important, obtained by either heat treatment and quenching or by the chemical exchange of ions. Tempering results in a net-compressive stress at the surface of the glass.

Tempering   In the context of glass making, tempering refers to a heat treatment that leads to a compressive stress on the surface of a glass. This compressive-stress layer makes tempered glass safer. In the context of processing of metallic materials, tempering refers to a heat treatment used to soften the material and to increase its toughness.

Texture strengthening   Increase in the yield strength of a material as a result of preferred crystallographic texture.

Thermomechanical processing   Processes involved in the manufacturing of metallic components using mechanical deformation and various heat treatments.

Thermoplastics   A class of polymers that consist of large, long spaghetti-like molecules that are intertwined (e.g., polyethylene, nylon, *PET*, etc.).

Warm working   A term used to indicate the processing of metallic materials in a temperature range that is between those that define cold and hot working (usually a temperature between 0.3 to 0.6 of melting temperature in K).

Work hardening   Term sometimes used instead of strain hardening or cold working to describe the effect of deformation on the strengthening of metallic materials.

ADDITIONAL INFORMATION

1. MEYERS, M.A. and K.K. CHAWLA, *Mechanical Behavior of Materials*. 1998: Prentice Hall.

2. HOSFORD, W.F. and J.L. DUNCAN, "The Aluminum Beverage Can," *Scientific American*, 1994. September: p. 48–53.

3. DAVIS, J.R., *Metals Handbook—Desk Edition*. 1998: ASM International.

4. DEARDO, A.J., I. GARCIA, and M. HUA, *Technical discussions and assistance are acknowledged*. 2002.

5. CHEN, Z., S. MAEKAWA, and T. TAKEDA, "Bauschinger Effect and Multiaxial Yield Behavior of Stress-Reversed Mild Steel," *Metallurgical and Materials Transactions*, 1999. 30A (December): p. 3069–3078.

6. CHEN, I.W., "Superplasticity in Ceramic Systems," in *Encyclopedia of Advanced Materials*, D.B. BLORR, R.J., FLEMINGS, M.C. and MAHAJAN, S., Editors. 1994. p. 2715–2717.

7. VERHOEVEN, J.D., *Fundamentals of Physical Metallurgy*. 1975: Wiley.

8. DEGARMO, E.P., J.T. BLACK, and R.A. KOSHE, *Materials and Processes in Manufacturing*, 8th ed. 1997: Prentice Hall.

9. DAVID, S.A. and T. DEBROY, "Current Issues and Problems in Welding Science," *Science*, 1992. 257: p. 497–502.

10. DEBROY, T. and S.A. DAVID, "Physical Processes in Fusion Welding," *Reviews of Modern Physics*, 1995. 67(1): p. 85–112.

11. BARNES, A.J., "Superplastic Forming: Trends and Prospects," *Aluminum World*, 2001. 1(1): p. 130–132.

12. SHERBY, O.D., T.G. NIEH, and J. WADSWORTH, "Some Thoughts on Future Directions for Research and Applications in Superplasticity," in *Material Science Forum: Superplasticity in Advanced Materials, ICSAM-97*. 1997. p. 11–20 and 243–245.

13. BENGISU, M., *Engineering Ceramics*. 2001: Springer Verlag.

14. HAMILTON, C.H., "Superplastic Forming of Metals," in *Encyclopedia of Advanced Materials*, D.B. BLORR, R.J., FLEMINGS, M.C. and MAHAJAN, S., Editors. 1994. p. 2712–2714.

✓ PROBLEMS

Section 7-1 Relationship of Cold Working to the Stress-Strain Curve

7-1   Using a stress-strain diagram, explain what the term "strain hardening" means.

7-2   What is meant by the term "springback?" What is the significance of this term from a manufacturing viewpoint?

7-3   What does the term "Bauschinger effect" mean?

7-4   What manufacturing techniques make use of the cold-working process?

7-5   A 0.505-in.-diameter metal bar with a 2-in. gage length l_0 is subjected to a tensile test. The following measurements are made in the plastic region:

Force (lb)	Change in Gage length (in.) (Δl)	Diameter (in.)
27,500	0.2103	0.4800
27,000	0.4428	0.4566
25,700	0.6997	0.4343

Determine the strain-hardening exponent for the metal. Is the metal most likely to be FCC, BCC, or HCP? Explain.

7-6   Define the following terms: strain-hardening exponent (n), strain-rate sensitivity (m), and plastic strain ratio (r). Use appropriate equations.

7-7   A 1.5-cm-diameter metal bar with a 3-cm gage length (l_0) is subjected to a tensile test. The following measurements are made:

Force (N)	Change in Gage Length (cm) (Δl)	Diameter (cm)
16,240	0.6642	1.2028
19,066	1.4754	1.0884
19,273	2.4663	0.9848

Determine the strain-hardening coefficient for the metal. Is the metal most likely to be FCC, BCC, or HCP? Explain.

7-8   What does the term "formability of a material" mean?

7-9   A true stress-true strain curve is shown in Figure 7-22. Determine the strain-hardening exponent for the metal.

7-10   A Cu-30% Zn alloy tensile bar has a strain-hardening coefficient of 0.50. The bar, which has an initial diameter of 1 cm and an initial gage length of 3 cm, fails at an engineering stress of

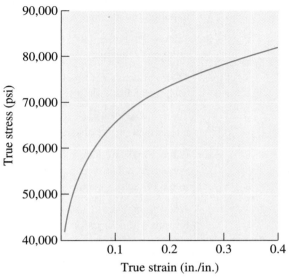

Figure 7-22   True stress-true strain curve (for Problem 7-9).

120 MPa. After fracture, the gage length is 3.5 cm and the diameter is 0.926 cm. No necking occurred. Calculate the true stress when the true strain is 0.05 cm/cm.

Section 7-2 Strain-Hardening Mechanisms

7-11   Explain why many metallic materials exhibit strain hardening.

7-12   Does a strain-hardening mechanism depend upon grain size? Does it depend upon dislocation density?

7-13   Compare and contrast strain hardening with grain size strengthening? What causes resistance to dislocation motion in each of these mechanisms?

7-14   The Frank-Read source shown in Figure 7-5(e) has created four dislocation loops from the original dislocation line. Estimate the total dislocation line present in the photograph and determine the percent increase in the length of dislocations produced by the deformation.

7-15   Strain hardening is normally not a consideration in ceramic materials. Explain why.

7-16   Thermoplastic polymers such as polyethylene show an increase in strength when subjected to stress. Explain how this strengthening occurs.

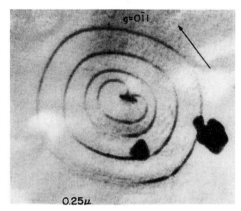

Figure 7-5   (Repeated for Problem 7-14) (e) Electron micrograph of a Frank-Read source (330,000). (*Adapted from Brittain, J., "Climb Sources in Beta Prime-NiAl,"* Metallurgical Transactions, *Vol. 6A, April 1975.*)

7-17 Bottles of carbonated beverages are made using *PET* plastic. Explain how stress-induced crystallization increases the strength of *PET* bottles made by the blow-stretch process (see Chapters 4 and 5).

Section 7-3 Properties versus Percent Cold Work

7-18 Write down the equation that defines percent cold work. Explain the meaning of each term.

7-19 A 0.25-in.-thick copper plate is to be cold worked 63%. Find the final thickness.

7-20 A 0.25-in.-diameter copper bar is to be cold worked 63%. Find the final diameter.

7-21 A 2-in.-diameter copper rod is reduced to a 1.5-in. diameter, then reduced again to a final diameter of 1 in. In a second case, the 2-in.-diameter rod is reduced in one step from a 2-in. to a 1-in. diameter. Calculate the % CW for both cases.

7-22 A 3105 aluminum plate is reduced from 1.75 in. to 1.15 in. Determine the final properties of the plate. Note 3105 designates a special composition of aluminum alloy. (See Figure 7-23.)

7-23 A Cu-30% Zn brass bar is reduced from a 1-in. diameter to a 0.45-in. diameter. Determine the final properties of the bar. (See Figure 7-24.)

7-24 A 3105 aluminum bar is reduced from a 1-in. diameter, to a 0.8-in. diameter, to a 0.6-in. diameter, to a final 0.4-in. diameter. Determine the % CW and the properties after each step of the process. Calculate the total percent cold work. Note 3105 designates a special composition of aluminum alloy. (See Figure 7-23.)

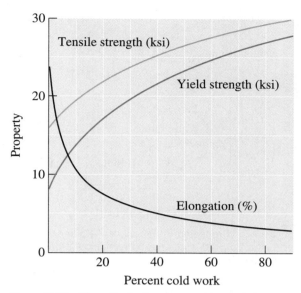

Figure 7-23   The effect of percent cold work on the properties of a 3105 aluminum alloy (for Problems 7-22 and 7-24).

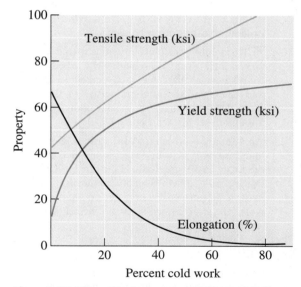

Figure 7-24   The effect of percent cold work on the properties of a Cu-30% Zn brass (for Problems 7-23 and 7-26).

7-25 We want a copper bar to have a tensile strength of at least 70,000 psi and a final diameter of 0.375 in. What is the minimum diameter of the original bar? (See Figure 7-7.)

7-26 We want a Cu-30% Zn brass plate originally 1.2-in. thick to have a yield strength greater than 50,000 psi and a % elongation of at least 10%.

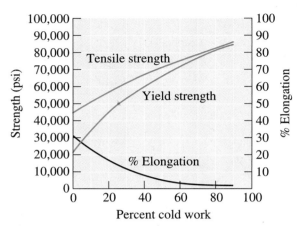

Figure 7-7 (Repeated for Problems 7-25 and 7-27) The effect of cold work on the mechanical properties of copper.

What range of final thicknesses must be obtained? (See Figure 7-24.)

7-27 We want a copper sheet to have at least 50,000 psi yield strength and at least 10% elongation, with a final thickness of 0.12 in. What range of original thickness must be used? (See Figure 7-7.)

7-28 A 3105 aluminum plate previously cold worked 20% is 2-in. thick. It is then cold worked further to 1.3 in. Calculate the total percent cold work and determine the final properties of the plate. Note 3105 designates a special composition of aluminum alloy. (See Figure 7-23.)

7-29 An aluminum-lithium (Al-Li) strap 0.25-in. thick and 2-in. wide is to be cut from a rolled sheet, as described in Figure 7-10. The strap must be able to support a 35,000-lb load without plastic deformation. Determine the range of orientations from which the strap can be cut from the rolled sheet.

Section 7-4 Microstructure, Texture Strengthening, and Residual Stresses

7-30 Does the yield strength of metallic materials depend upon the crystallographic texture materials develop during cold working? Explain.

7-31 Does the Young's modulus of a material depend upon crystallographic directions in a crystalline material? Explain.

7-32 What do the terms "fiber texture" and "sheet texture" mean?

7-33 One of the disadvantages of the cold-rolling process is the generation of residual stresses. Explain

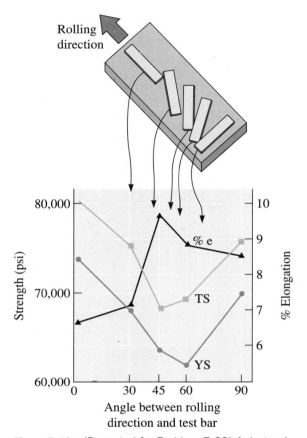

Figure 7-10 (Repeated for Problem 7-29) Anisotropic behavior in a rolled aluminum-lithium sheet material used in aerospace applications. The sketch relates the position of tensile bars to the mechanical properties that are obtained.

how we can eliminate residual stresses in cold-worked metallic materials.

7-34 What is shot peening?

7-35 What is the difference between tempering and the annealing of glasses?

7-36 How is laminated safety glass different from tempered glass? Where is laminated glass used?

7-37 What is thermal tempering? How is it different from chemical tempering? State applications of tempered glasses.

7-38 Explain factors that affect the strength of glass most. Explain how thermal tempering helps increase the strength of glass.

7-39 Residual stresses are not always undesirable. True or false? Justify your answer.

Section 7-5 Characteristics of Cold Working

7-40 Cold working cannot be used as a strengthening mechanism for materials that are going to be subjected to high temperatures during their use. Explain why.

7-41 Aluminum cans made by deep drawing derive considerable strength during their fabrication. Explain why.

7-42 Such metals as magnesium can not be effectively strengthened using cold working. Explain why.

7-43 We want to draw a 0.3-in.-diameter copper wire having yield strength of 20,000 psi into a 0.25-in.-diameter wire.

    (a) Find the draw force, assuming no friction;
    (b) Will the drawn wire break during the drawing process? Show why. (See Figure 7-7.)

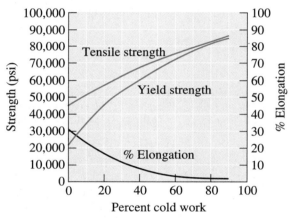

Figure 7-7 (Repeated for Problem 7-43) The effect of cold work on the mechanical properties of copper.

7-44 A 3105 aluminum wire is to be drawn to give a 1-mm-diameter wire having yield strength of 20,000 psi. Note 3105 designates a special composition of aluminum alloy.

    (a) Find the original diameter of the wire;
    (b) Calculate the draw force required; and
    (c) Determine whether the as-drawn wire will break during the process. (See Figure 7-23.)

Section 7-6 The Three Stages of Annealing

7-45 Explain the three stages of the annealing of metallic materials.

7-46 What is the driving force for recrystallization?

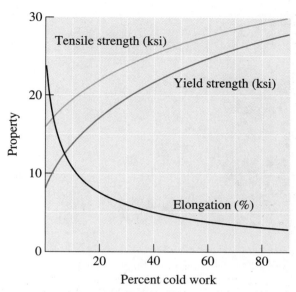

Figure 7-23 The effect of percent cold work on the properties of a 3105 aluminum alloy (for Problems 7-28 and 7-44).

7-47 In the recovery stage, the residual stresses are reduced, however, the strength of the metallic material remains unchanged. Explain why.

7-48 What is the driving force for grain growth?

7-49 Treating grain growth as the third stage of annealing, explain its effect on the strength of metallic materials.

7-50 Why is it that grain growth is usually undesirable? Cite an example where grain growth is actually useful.

7-51 What are the different ways one can encounter grain growth in ceramics?

7-52 Are annealing and recovery always prerequisites to grain growth? Explain.

7-53 A titanium alloy contains a very fine dispersion of tiny Er_2O_3 particles. What will be the effect of these particles on the grain growth temperature and the size of the grains at any particular annealing temperature? Explain.

7-54 Explain why a tungsten filament used in an incandescent light bulb ultimately fails.

7-55 The following data were obtained when a cold-worked metal was annealed.

    (a) Estimate the recovery, recrystallization, and grain growth temperatures;

  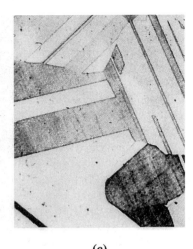

(a)   (b)   (c)

Figure 7-16   (Repeated for Problem 7-58) Photomicrographs showing the effect of annealing temperature on grain size in brass. Twin boundaries can also be observed in the structures. (a) Annealed at 400°C, (b) annealed at 650°C, and (c) annealed at 800°C (×75). (*Adapted from Brick, R. and Phillips, A.,* The Structure and Properties of Alloys, *1949: McGraw-Hill.*)

(b) Recommend a suitable temperature for a stress-relief heat treatment;

(c) Recommend a suitable temperature for a hot-working process; and

(d) Estimate the melting temperature of the alloy.

Annealing Temperature (°C)	Electrical Conductivity (ohm$^{-1} \cdot$ cm^{-1})	Yield Strength (MPa)	Grain Size (mm)
400	3.04×10^5	86	0.10
500	3.05×10^5	85	0.10
600	3.36×10^5	84	0.10
700	3.45×10^5	83	0.098
800	3.46×10^5	52	0.030
900	3.46×10^5	47	0.031
1000	3.47×10^5	44	0.070
1100	3.47×10^5	42	0.120

7-56 The following data were obtained when a cold-worked metallic material was annealed:

(a) Estimate the recovery, recrystallization, and grain growth temperatures;

(b) Recommend a suitable temperature for obtaining a high-strength, high-electrical-conductivity wire;

(c) Recommend a suitable temperature for a hot working process; and

(d) Estimate the melting temperature of the alloy.

Annealing Temperature (°C)	Residual Stresses (psi)	Tensile Strength (psi)	Grain Size (in.)
250	21,000	52,000	0.0030
275	21,000	52,000	0.0030
300	5,000	52,000	0.0030
325	0	52,000	0.0030
350	0	34,000	0.0010
375	0	30,000	0.0010
400	0	27,000	0.0035
425	0	25,000	0.0072

7-57 What is meant by the term "recrystallization?" Explain why the yield strength of a metallic material goes down during this stage of annealing.

7-58 Determine the ASTM grain size number for each of the micrographs in Figure 7-16, and plot the grain size number versus the annealing temperature.

Section 7-7 Control of Annealing

7-59 How do we distinguish between the hot working and cold working of metallic materials?

7-60 Why is it that the recrystallization temperature is not a fixed temperature for a given material?

7-61 Why does increasing the time for a heat treatment mean recrystallization will occur at a lower temperature?

7-62 Two sheets of steel have cold works of 20% and 80%, respectively. Which one would likely have a lower recrystallization temperature? Why?

7-63 What does the term "warm working" mean?

7-64 Give examples of two metallic materials for which deformation at room temperature will mean "hot working."

7-65 Give examples of two metallic materials for which mechanical deformation operation at 900°C would mean "cold working."

Section 7-8 Annealing and Materials Processing

7-66 Using the data in Table 7-4, plot the recrystallization temperature versus the melting temperature of each metal, using absolute temperatures (Kelvin). Measure the slope and compare with the expected relationship between these two temperatures. Is our approximation a good one?

TABLE 7-4 ■ *Typical recrystallization temperatures for selected metals (Repeated for Problem 7-66)*

Metal	Melting Temperature (°C)	Recrystallization Temperature (°C)
Sn	232	−4
Pb	327	−4
Zn	420	10
Al	660	150
Mg	650	200
Ag	962	200
Cu	1085	200
Fe	1538	450
Ni	1453	600
Mo	2610	900
W	3410	1200

(*Source: Adapted from* Structure and Properties of Engineering Materials, *by R. Brick, A. Pense, and R. Gordon, 1977. Copyright © 1977 The McGraw-Hill Companies. Adapted by permission.*)

7-67 We wish to produce a 0.3-in.-thick plate of 3105 aluminum having a tensile strength of at least 25,000 psi and a % elongation of at least 5%. The original thickness of the plate is 3 in. The maximum cold work in each step is 80%. Describe the cold working and annealing steps required to make this product. Compare this process with what you would recommend if you could do the initial deformation by hot working. (See Figure 7-23.)

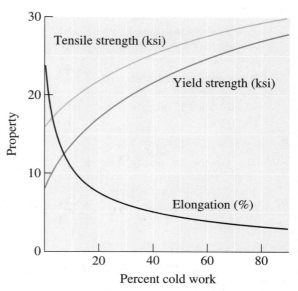

Figure 7-23 The effect of percent cold work on the properties of a 3105 aluminum alloy (for Problem 7-67).

7-68 We wish to produce a 0.2-in.-diameter wire of copper having a minimum yield strength of 60,000 psi and a minimum % elongation of 5%. The original diameter of the rod is 2 in. and the maximum cold work in each step is 80%. Describe the cold working and annealing steps required to make this product. Compare this process with what you would recommend if you could do the initial deformation by hot working. (See Figure 7-7.)

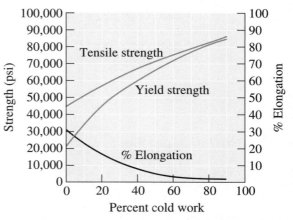

Figure 7-7 (Repeated for Problem 7-68) The effect of cold work on the mechanical properties of copper.

7-69 What is a heat-affected zone? Why do some welding processes result in a joint where the material in the heat-affected zone is weaker than the base metal?

7-70 What welding techniques can be used to avoid loss of strength in the material in the heat-affected zone? Explain why these techniques are effective.

Section 7-9 Hot Working

7-71 The amount of plastic deformation that can be performed during hot working is almost unlimited. Justify this statement.

7-72 Compare and contrast hot working and cold working.

Section 7-10 Superplastic Forming

7-73 What is meant by the term "superplasticity?"

7-74 How is it that many people regard ceramics as brittle, yet many ceramics exhibit superplasticity?

7-75 What are the microstructural requirements for metallic or ceramic materials to exhibit superplasticity?

Design Problems

7-76 Design, using one of the processes discussed in this chapter, a method to produce each of the following products. Should the process include hot working, cold working, annealing, or some combination of these? Explain your decisions.

(a) paper clips;
(b) I-beams that will be welded to produce a portion of a bridge;
(c) copper tubing that will connect a water faucet to the main copper plumbing;
(d) the steel tape in a tape measure;
(e) a head for a carpenter's hammer formed from a round rod.

7-77 We plan to join two sheets of cold-worked copper by soldering. Soldering involves heating the metal to a high enough temperature that a filler material melts and is drawn into the joint (Chapter 8). Design a soldering process that will not soften the copper. Explain. Could we use higher soldering temperatures if the sheet material were a Cu-30% Zn alloy? Explain.

7-78 We wish to produce a 1-mm-diameter copper wire having a minimum yield strength of 60,000 psi and a minimum % elongation of 5%. We start with a 20-mm-diameter rod. Design the process by which the wire can be drawn. Include all-important details and explain.

Computer Problems

7-79 *Plastic Strain Ratio.* Write a computer program that will ask the user to provide the initial and final dimensions (width and thickness) of a plate and provide the value of the plastic strain ratio.

7-80 *Design of a Wire Drawing Process.* Write a program that will effectively computerize the solution to solving Example 7-5. The program should ask the user to provide a value of the final diameter for the wire (e.g., 0.20 cm). The program should assume a reasonable value for the initial diameter (d_0) (e.g., 0.40 cm), and calculate the extent of cold work using the proper formula. Assume that the user has access to the yield strength versus % cold work curve and the user is then asked to enter the value of the yield strength for 0% cold work. Use this value to calculate the forces needed for drawing and the stress acting on the wire as it comes out of the die. The program should then ask the user to provide the value of the yield strength of the wire for the amount of cold work calculated for the assumed initial diameter and the final diameter needed. As in Example 7-5, the program should repeat these calculations until obtaining a value of d_0 that will be acceptable.

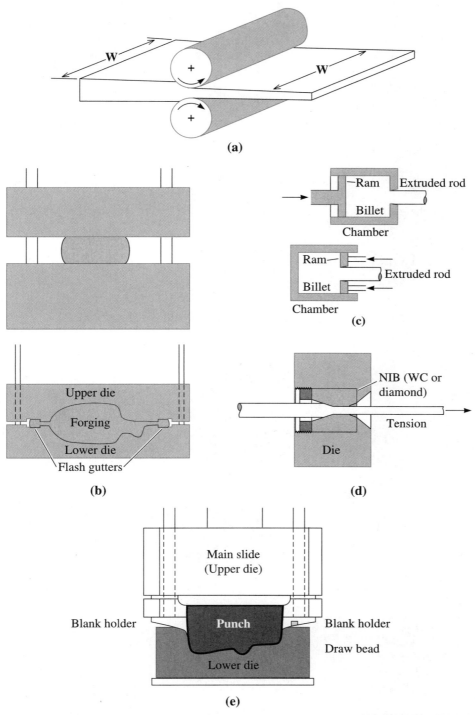

(a)

(b)

(c)

(d)

(e)

Figure 7-2 (Repeated for Problem 7-76) Manufacturing processes that make use of cold working as well as hot working. Common metalworking methods. (a) Rolling. (b) Forging (open and closed die). (c) Extrusion (direct and indirect). (d) Wire drawing. (e) Stamping. (*Source: Adapted from* Mechanical Behavior of Materials, *by M.A. Meyers and K.K. Chawla, p. 292, Fig. 6-1. Copyright © 1999 Prentice Hall. Adapted with permission of Pearson Education, Inc., Upper Saddle River, NJ.*)

Chapter

The process of nucleation and growth of one phase from another is important in the formation of snowflakes, shown here. Similarly, the nucleation and growth of small crystals from molten materials is also very important in manufacturing processes related to solidification. On the other hand, technologies related to the production of glass depend on avoiding the nucleation and growth of crystals. (*Courtesy of Getty Images.*)

8

Principles of Solidification

- *Whether water really does "freeze" at 0°C and "boil" at 100°C?*

- *How do snowmaking machines make snow?*

- *What role has the process of solidification played in the evolution of mankind?*

- *What is the process used to produce several million pounds of steels and other alloys?*

- *Is there a specific melting temperature for an alloy or a thermoplastic material?*

- *What factors determine the strength of a cast product?*

- *Why are most glasses and glass-ceramics processed by melting and casting?*

Of all the processing techniques used in the manufacturing of materials, solidification is probably the most important.[1,2] All metallic materials, as well as many ceramics, inorganic glasses, and thermoplastic polymers, are liquid or molten at some point during processing. Like water freezes to ice, molten materials solidify as they cool below their freezing temperature. In Chapter 2, we learned how materials are classified based on their atomic, ionic, or molecular order. During the solidification of materials that crystallize, the atomic arrangement of crystallized materials changes from a short-range order (SRO) to a long-range order (LRO). The solidification of crystalline materials requires two steps: In the first step, ultra-fine crystallites, known as the nuclei of a solid phase, form from the liquid. In the second step, which can overlap with the first, the ultra-fine solid crystallites begin to grow as atoms from the liquid are attached to the nuclei until no liquid remains. Some materials, such as inorganic silicate glasses, will become solid without developing a long-range order (i.e., they remain amorphous). Many polymeric materials may develop partial crystallinity during solidification or processing.

The solidification of metallic, polymeric, and ceramic materials is an important process to study because of its effect on the properties of the materials involved. In Chapter 7, we examined how strain hardening can be used to strengthen and shape metallic materials. We learned in Chapter 4 that grain size plays an important role in determining the strength of metallic materials. In this chapter, we will study the principles of solidification as they apply to pure metals. We will discuss solidification of alloys and more complex materials in subsequent chapters. We will first discuss the technological significance of solidification, and then examine the mechanisms by which solidification occurs. This will be followed by an examination of the microstructure of cast metallic materials and its effect on the material's mechanical properties. We will also examine the role of casting as a materials shaping process. We will examine how techniques such as welding, brazing, and soldering are used for joining metals. Applications of the solidification process in single crystal growth and the solidification of glasses and polymers also will be discussed.

8-1   Technological Significance

The ability to use fire to produce, melt, and cast metals such as copper, bronze, and steel indeed is regarded as an important hallmark in the evolution of mankind (Figure 8-1). The use of fire for reducing naturally occurring ores into metals and alloys led to the production of useful tools and other products. Today, thousands of years later, solidification is still considered one of the most important manufacturing processes. Several million pounds of steel, aluminum alloys, copper, and zinc are being produced through the casting process. The solidification process is also used to manufacture specific components (e.g., aluminum alloy for automotive wheels) [Figure 8-2(a)]. Industry also uses the solidification process as a **primary processing** step to produce metallic slabs or ingots (a simple, and often large casting that is processed later into useful shapes). The ingots or slabs are then hot and cold worked through **secondary processing** steps into more useful shapes (i.e., sheets, wires, rods, plates, etc.). Solidification also is applied when joining metallic materials using techniques such as welding, brazing, and soldering.

We also use solidification for processing inorganic glasses; silicate glass, for example, is processed using the float-glass process. High-quality optical fibers [Figure 8-2(b)] and other materials, such as fiberglass, also are produced from the solidification of molten glasses. During the solidification of inorganic glasses, rather than crystalline, amorphous materials are produced. In the manufacture of glass-ceramics, we first shape the materials by casting amorphous glasses, and then crystallize them using a heat treatment to enhance their strength. Many thermoplastic materials such as polyethylene, polyvinyl chloride (PVC), polypropylene, and the like are processed into useful shapes (i.e., fibers, tubes, bottles, toys, utensils, etc.) using a process that involves

Figure 8-1 An image of a bronze object. This canteen (bian hu) is from China, Warring States period, circa 3rd century BCE (bronze inlaid with silver). (*Courtesy of Freer Gallery of Art, Smithsonian Institution, Washington, D.C.*)

melting and solidification. Therefore, solidification is an extremely important technology used to control the properties of many melt-derived products as well as a tool for the manufacturing of modern engineered materials. In the sections that follow, we first discuss the nucleation and growth processes.

(a)   **(b)**

Figure 8-2 (a) Aluminum alloy wheels for automobiles, (b) optical fibers for communication. (*Courtesy of PhotoDisc/Getty Images.*)

8-2   Nucleation

In the context of solidification, the term **nucleation** refers to the formation of the first nano-sized crystallites from molten material. For example, as water begins to freeze, nano-sized ice crystals, known as nuclei, form first. In a broader sense, the term nucleation refers to the initial stage of formation of one phase from another phase. When a vapor condenses into liquid, the nanoscale sized drops of liquid that appear when the condensation begins are referred to as **nuclei**. Later, we will also see that there are many systems in which the nuclei of a solid (β) will form a material (α) (i.e., α- to β-phase transformation). For example, water droplets nucleate from water vapor. What is interesting about these transformations is that, in most engineered materials, many of them occur while the material is in solid state (i.e., there is no melting involved). Therefore, although we discuss nucleation from a solidification perspective, it is important to note that the phenomenon of nucleation is general and is associated with phase transformations.

We expect a material to solidify when the liquid cools to just below its freezing (or melting) temperature, because the energy associated with the crystalline structure of the solid is then less than the energy of the liquid. This energy difference between the liquid and the solid is the free energy per unit volume ΔG_v and is the driving force for solidification.

When the solid forms, however, a solid-liquid interface is created (Figure 8-3). A surface free energy σ_{sl} is associated with this interface; the larger the solid, the greater the increase in surface energy. Thus, the total change in energy ΔG, shown in Figure 8-4, is:

$$\Delta G = \tfrac{4}{3}\pi r^3 \Delta G_v + 4\pi r^2 \sigma_{sl} \tag{8-1}$$

where $\tfrac{4}{3}\pi r^3$ is the volume of a spherical embryo of radius r, $4\pi r^2$ is the surface area of a spherical embryo, σ_{sl} is the surface free energy of the solid liquid interface (in this case), and ΔG_v is the free energy change per unit volume, which is a negative since the phase

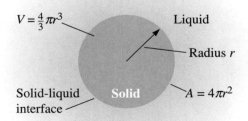

Figure 8-3
An interface is created when a solid forms from the liquid.

$V = \tfrac{4}{3}\pi r^3$ Liquid

Radius r

Solid-liquid interface   Solid $A = 4\pi r^2$

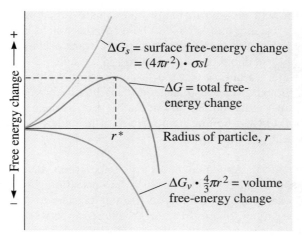

Figure 8-4
The total free energy of the solid-liquid system changes with the size of the solid. The solid is an embryo if its radius is less than the critical radius, and is a nucleus if its radius is greater than the critical radius.

transformation is assumed to be thermodynamically feasible. Note that σ_{sl} is not a strong function of r and is assumed constant. The ΔG_v also does not depend on r.

An **embryo** is a tiny particle of solid that forms from the liquid as atoms cluster together. The embryo is unstable, and may either grow into a stable nucleus or redissolve.

In Figure 8-4, the top curve shows the parabolic variation of the total surface energy $(4\pi r^2 \cdot \sigma_{sl})$. The bottom most curve shows the total volume free-energy change term $(\frac{4}{3}\pi r^3 \cdot \Delta G_v)$. The curve in the middle shows the variation of ΔG. At the temperature where the solid and liquid phases are predicted to be in thermodynamic equilibrium (i.e., at the freezing temperature), the free energy of the solid phase and that of the liquid phase are equal $(\Delta G_v = 0)$, so the total free energy change (ΔG) will be positive. When the solid is very small, with a radius less than the **critical radius** for nucleation (r^*) (Figure 8-4), further **growth** causes the total free energy to increase. The critical radius (r^*) is the minimum size of a crystal that must be formed by atoms clustering together in the liquid before the solid particle is stable and begins to grow. But instead of growing, the solid has a tendency to remelt, causing the free energy to decrease; thus, the bulk of the material remains liquid leaving just a small crystal solid. At freezing temperatures, embryos are thermodynamically unstable. So how can they grow?

The formation of embryos is a statistical process. Many embryos form and redissolve. If by chance, an embryo forms which has a radius that is larger than r^*, further growth causes the total free energy to decrease. The new solid is then stable and sustainable since nucleation has occurred, and growth of the solid particle—which is now called a nucleus—begins. At the thermodynamic melting or freezing temperatures, the probability of forming stable, sustainable nuclei is extremely small. Therefore, solidification does not begin at the thermodynamic melting or freezing temperature. If the temperature continues decreased below the equilibrium freezing temperature, the liquid phase that should have transformed into a solid becomes thermodynamically and increasingly unstable. Because the liquid is below the equilibrium freezing temperature, the liquid is considered undercooled. The **undercooling** of ΔT is the equilibrium freezing temperature minus the actual temperature of the liquid. As the extent of undercooling increases, the thermodynamic driving force for the formation of a solid phase from liquid overtakes the resistance to create a solid-liquid interface.

This phenomenon can be seen in many other phase transformations. When one solid phase (α) transforms into another solid phase (β), the system has to be cooled to a temperature that is below the thermodynamic phase transformation temperature (at which free energies of the α and β phases are equal). When a liquid is transformed into a vapor (i.e., boiling water), a bubble of vapor is created in the liquid. In order to create the transformation, though, we need to **superheat** the liquid above its boiling temperature! Therefore, we can see that liquids do not really freeze at their freezing temperature and do not really boil at their boiling point! We need to undercool the liquid for it to solidify and superheat it for it to boil!

Homogeneous Nucleation   As liquid cools to temperatures below the equilibrium freezing temperature, two factors combine to favor nucleation. First, since atoms are losing their thermal energy the probability of forming clusters to form larger embryos increases. Second, the larger volume free energy difference between the liquid and the solid reduces the critical size (r^*) of the nucleus. **Homogeneous nucleation** occurs when the undercooling becomes large enough to cause the formation of a stable nucleus.

The size of the critical radius r^* is given by

$$r^* = \frac{2\sigma_{sl}\, T_m}{\Delta H_f\, \Delta T} \tag{8-2}$$

where ΔH_f is the latent heat of fusion, T_m is the equilibrium solidification temperature in Kelvin, and $\Delta T = (T_m - T)$ is the undercooling when the liquid temperature is T. The latent heat of fusion represents the heat given up during the liquid-to-solid transformation. As the undercooling increases, the critical radius required for nucleation decreases. Table 8-1 presents values for σ_{sl}, ΔH_f, and typical undercoolings observed experimentally for homogeneous nucleation.

The following example shows how we can calculate the critical radius of the nucleus for the solidification of copper.

TABLE 8-1 ■ *Values for freezing temperature, latent heat of fusion, surface energy, and maximum undercooling for selected materials*

Metal	Freezing Temperature (T_m) (°C)	Heat of Fusion (ΔH_f) (J/cm³)	Solid-Liquid Interfacial Energy (σ_{sl}) (J/cm²)	Typical Undercooling for Homogeneous Nucleation (ΔT) (°C)
Ga	30	488	56×10^{-7}	76
Bi	271	543	54×10^{-7}	90
Pb	327	237	33×10^{-7}	80
Ag	962	965	126×10^{-7}	250
Cu	1085	1628	177×10^{-7}	236
Ni	1453	2756	255×10^{-7}	480
Fe	1538	1737	204×10^{-7}	420
NaCl	801			169
CsCl	645			152
H$_2$O	0			40

EXAMPLE 8-1 *Calculation of Critical Radius for the Solidification of Copper*

Calculate the size of the critical radius and the number of atoms in the critical nucleus when solid copper forms by homogeneous nucleation. Comment on the size of the nucleus and assumptions we made while deriving the equation for radius of nucleus.

SOLUTION

From Table 8-1:

$$\Delta T = 236°C \quad T_m = 1085 + 273 = 1358 \text{ K}$$
$$\Delta H_f = 1628 \text{ J/cm}^3$$
$$\sigma_{sl} = 177 \times 10^{-7} \text{ J/cm}^2$$
$$r^* = \frac{2\sigma_{sl}T_m}{\Delta H_f \Delta T} = \frac{(2)(177 \times 10^{-7})(1358)}{(1628)(236)} = 12.51 \times 10^{-8} \text{ cm}$$

The lattice parameter for FCC copper is $a_0 = 0.3615$ nm $= 3.615 \times 10^{-8}$ cm

$$V_{\text{unit cell}} = (a_0)^3 = (3.615 \times 10^{-8})^3 = 47.24 \times 10^{-24} \text{ cm}^3$$
$$V_{r^*} = \tfrac{4}{3}\pi r^3 = \left(\tfrac{4}{3}\pi\right)(12.51 \times 10^{-8})^3 = 8200 \times 10^{-24} \text{ cm}^3$$

The number of unit cells in the critical nucleus is

$$\frac{8200 \times 10^{-24}}{47.24 \times 10^{-24}} = 174 \text{ unit cells}$$

Since there are four atoms in each unit cell of FCC metals, the number of atoms in the critical nucleus must be:

$$(4 \text{ atoms/cell})(174 \text{ cells/nucleus}) = 696 \text{ atoms/nucleus}$$

In these types of calculations, we assume that a nucleus that is made from only a few hundred atoms still exhibits properties similar to those of bulk materials. This is not strictly correct and as such considered to be a weakness of the classical theory of nucleation.

Heterogeneous Nucleation From Table 8-1, we can see that water will not solidify into ice via homogeneous nucleation until we reach a temperature of −40°C (undercooling of 40°C)! Except in controlled laboratory experiments, homogeneous nucleation never occurs in liquids. Instead, impurities in contact with the liquid, either suspended in the liquid or on the walls of the container that holds the liquid, provide a surface on which the solid can form (Figure 8-5). Now, a radius of curvature greater than the critical radius is achieved with very little total surface between the solid and liquid. Only a few atoms must cluster together to produce a solid particle that has the required radius of curvature. Much less undercooling is required to achieve the critical size, so nucleation occurs more readily. Nucleation on preexisting surfaces is known as **heterogeneous nucleation**. This process is dependent on the contact angle (θ) for the nucleating phase and the surface on which nucleation occurs. The same type of phenomenon occurs in solid state transformations.

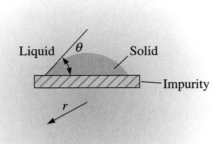

Figure 8-5
A solid forming on an impurity can assume the critical radius with a smaller increase in the surface energy. Thus, heterogeneous nucleation can occur with relatively low undercoolings.

Rate of Nucleation   The **rate of nucleation** (the number of nuclei formed per unit time) is a function of temperature. Prior to solidification, of course, there is no nucleation and, at temperatures above the freezing point, the rate of nucleation is zero. As the temperature drops, the driving force for nucleation increases. However, as the temperature becomes lower, atomic diffusion becomes slower, hence slowing the nucleation process. Thus, a typical rate of nucleation (I) reaches a maximum at some temperature below the transformation temperature (Figure 8-6). In heterogeneous nucleation, the rate of nucleation is dictated by the concentration of the nucleating agents introduced. By considering the rates of nucleation and growth, we can predict overall rate of a phase transformation.

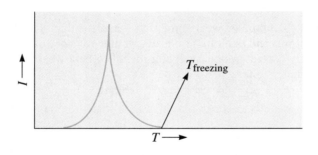

Figure 8-6
Rate of nucleation (I) as a function of temperature of the liquid (T).

8-3   Applications of Controlled Nucleation

Grain Size Strengthening   When a metal casting freezes, impurities in the melt and walls of the mold in which solidification occurs serve as heterogeneous nucleation sites. Sometimes we intentionally introduce nucleating particles into the liquid. Such practices are called **grain refinement** or **inoculation**. Chemicals added to molten metals to promote nucleation and, hence, a finer grain size, are known as grain refiners or **inoculants**. For example, a combination of 0.03% titanium (Ti) and 0.01% boron (B) is added to many liquid-aluminum alloys. Tiny particles of an aluminum titanium compound (Al_3Ti) or titanium diporide (TiB_2) form and serve as sites for heterogeneous nucleation. Grain refining or inoculation produces a large number of grains, each beginning to grow from one nucleus. The greater grain boundary area provides grain size strengthening in metallic materials. This was discussed using the Hall-Petch equation in Chapters 3 and 4.

Second-Phase Strengthening   In Chapters 4 and 5, we learned that in metallic materials, dislocation motion can be resisted by grain boundaries or the formation of ultra-fine precipitates of a second phase. Strengthening materials using ultra-fine precipitates is known as **dispersion** or **second-phase strengthening**; it is used extensively in enhancing the mechanical properties of many alloys. This process involves **solid-state phase transformations** (i.e., one solid transforming into another where the grain boundaries of the transforming phase (α), as well as atomic level defects within the grains of the parent phase, often serve as nucleation sites for heterogeneous nucleation of the new phase (β)). This nucleation phenomenon plays a critical role in strengthening mechanisms. This will be discussed in Chapters 9 and 10.

Glasses   For rapid cooling rates and/or high viscosity melts, there may be insufficient time for nuclei to form and grow. When this happens, the liquid structure is locked into place and an amorphous—or glassy—solid forms. The complex crystal structure of many ceramic and polymer materials prevents nucleation of a solid crystalline structure even at slow cooling rates. In metallic materials, however, cooling rates of 10^6 °C/s or higher may be required to suppress nucleation of the crystals. The production of metallic glasses—as well as other unique structures—by rapid cooling has been termed **rapid solidification processing**. The high cooling rates are obtained using ultra-fine metal powder particles or by forming continuous thin metallic ribbons about 0.0015 in. in thickness. Many alloy powders prepared by spraying molten metals tend to contain very small crystallites (i.e., they may be microcrystalline).

Examples of metallic glasses include complex iron-nickel-boron alloys containing chromium, phosphorus, cobalt, and other elements. Some metallic glasses may obtain strengths in excess of 500,000 psi while retaining fracture toughness of more than 10,000 psi $\sqrt{\text{in}}$. Excellent corrosion resistance, magnetic properties, and other physical properties make these materials attractive for applications concerning electrical power, aircraft engines, tools and dies, and magnetism.

Other examples of materials that make use of controlled nucleation are colored glass and **photochromic glass** (glass that can change color or tint upon exposure to sunlight). In these otherwise amorphous materials, nano-sized crystallites of different materials are deliberately nucleated. The crystals are small and, hence, do not make the glass opaque. They do have special optical properties that make the glass brightly colored or photochromic. Controlled nucleation in amorphous glasses can be used to make nano-sized crystallites of semiconductors known as **quantum dots**, which have special electrical and optical properties.[3] For example, the bandgap of quantum dot semiconductor structures is different from that of bulk semiconductor materials.

Many materials formed from a vapor phase can be cooled quickly so that they do not crystallize and, therefore, are amorphous (i.e., amorphous silicon). The point is that amorphous or non-crystalline materials do *not* always have to be formed from melts.

Glass-ceramics   The term **glass-ceramics** refers to engineered materials that begin as amorphous glasses and end up as crystalline ceramics with an ultra-fine grain size.[4] These materials are nearly free from porosity, mechanically stronger, and often much more resistant to thermal shock. Nucleation does not occur easily in silicate glasses. However, we can help by introducing nucleating agents such as titania (TiO_2) and zirconia (ZrO_2). Engineered glass-ceramics take advantage of the ease with which glasses can be melted and formed. Once a glass is formed, we can then heat it to deliberately form ultra-fine crystals, obtaining a material that has considerable mechanical toughness and thermal shock resistance. The crystallization of glass-ceramics continues until all of the material crystallizes (up to 99.9% crystallinity can be obtained). If the grain size is kept small ($\sim$50–100 nm), glass-ceramics can often be made transparent. All glasses eventually will crystallize as a result of exposure to high temperatures for long lengths of times. However, the crystallization must be controlled so every glass composition cannot be converted into a glass-ceramic. As described in Chapter 3, Donald Stookey of Corning and George Beall were two of the scientists who helped develop some of the first glass-ceramics.[4] Indeed, the process of controlled nucleation in amorphous materials forms the underpinnings for this fascinating class of materials. Specific compositions of glasses containing Li_2O-SiO_2-Al_2O_3, MgO-Al_2O_3-SiO_2, and BaO-SiO_2-Al_2O_3 are some examples of materials that can be transformed into useful glass-ceramics materials.[5]

Nucleation of Water Droplets and the Formation of Ice Crystals   It is possible to inject ultrafine crystals into clouds to promote heterogeneous nucleation of liquid water that becomes rain! Similarly, ski resorts use the process of deliberate heterogeneous nucleation for making snow with their snow making machines! Snomax™, a freeze-dried strain of the bacterium *Pseudomonas syringae* acts as a heterogeneous nucleation agent. The protein is nontoxic and nonpathogenic.[6]

8-4   Growth Mechanisms

Once the solid nuclei of a phase forms (in a liquid or another solid phase), growth begins to occur as more atoms become attached to the solid surface. In this discussion, we will concentrate on nucleation and the growth of crystals from liquid. The nature of the growth of the solid nuclei depends on how heat is removed from the molten material. Let's consider casting a molten metal in a mold, for example. We assume we have a nearly pure metal and not an alloy (as solidification of alloys is different in that in most cases it occurs over a range of temperatures). In the solidification process, two types of heat must be removed: the specific heat of the liquid and the latent heat of fusion. The **specific heat** is the heat required to change the temperature of a unit weight of the material by one degree. The specific heat must be removed first, either by radiation into the surrounding atmosphere or by conduction into the surrounding mold, until the liquid cools to its freezing temperature. This is simply a cooling of the liquid from one temperature to a temperature at which nucleation begins.

We know that to melt a solid we need to supply heat. Therefore, when solid crystals form from a liquid, heat is generated! This type of heat is called the **latent heat of fusion** (ΔH_f). The latent heat of fusion must be removed from the solid-liquid interface before solidification is completed. The manner in which we remove the latent heat of fusion determines the material's growth mechanism and final structure of a casting.

Planar Growth   When a well-inoculated liquid (i.e., a liquid containing nucleating agents) cools under equilibrium conditions, there is no need for undercooling since heterogeneous nucleation can occur. Therefore, the temperature of the liquid ahead of the **solidification front** (i.e., solid–liquid interface) is greater than the freezing temperature. The temperature of the solid is at or below the freezing temperature. During solidification, the latent heat of fusion is removed by conduction from the solid-liquid interface.

Growth direction

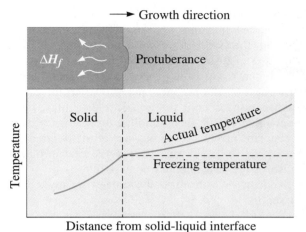

ΔH_f Protuberance

Solid | Liquid

Actual temperature

Freezing temperature

Temperature

Distance from solid-liquid interface

Figure 8-7
When the temperature of the liquid is above the freezing temperature, a protuberance on the solid-liquid interface will not grow, leading to maintenance of a planar interface. Latent heat is removed from the interface through the solid.

Any small protuberance that begins to grow on the interface is surrounded by liquid above the freezing temperature (Figure 8-7). The growth of the protuberance then stops until the remainder of the interface catches up. This growth mechanism, known as **planar growth**, occurs by the movement of a smooth solid-liquid interface into the liquid.

Dendritic Growth   When the liquid is not inoculated and the nucleation is poor, the liquid has to be undercooled before the solid forms (Figure 8-8). Under these condi-

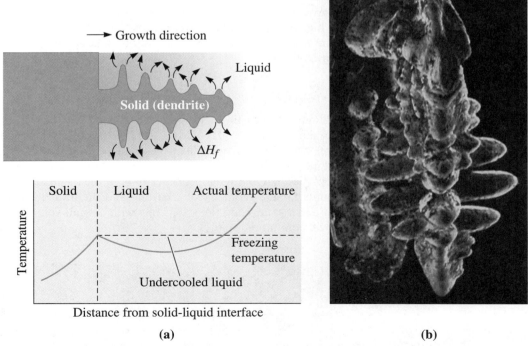

Growth direction

Liquid

Solid (dendrite)

ΔH_f

Solid | Liquid | Actual temperature

Freezing temperature

Undercooled liquid

Temperature

Distance from solid-liquid interface

(a)

(b)

Figure 8-8   (a) If the liquid is undercooled, a protuberance on the solid-liquid interface can grow rapidly as a dendrite. The latent heat of fusion is removed by raising the temperature of the liquid back to the freezing temperature. (b) Scanning electron micrograph of dendrites in steel ($\times 15$).

tions, a small solid protuberance called a **dendrite**, which forms at the interface, is encouraged to grow since the liquid ahead of the solidification front is undercooled. The word dendrite comes from the Greek word *dendron* that means tree. As the solid dendrite grows, the latent heat of fusion is conducted into the undercooled liquid, raising the temperature of the liquid toward the freezing temperature. Secondary and tertiary dendrite arms can also form on the primary stalks to speed the evolution of the latent heat. Dendritic growth continues until the undercooled liquid warms to the freezing temperature. Any remaining liquid then solidifies by planar growth. The difference between planar and dendritic growth arises because of the different sinks for the latent heat of fusion. The container or mold must absorb the heat in planar growth, but the undercooled liquid absorbs the heat in dendritic growth.

In pure metals, dendritic growth normally represents only a small fraction of the total growth and is given by:

$$\text{Dendritic fraction} = f = \frac{c\Delta T}{\Delta H_f} \tag{8-3}$$

where c is the specific heat of the liquid. The numerator represents the heat that the undercooled liquid can absorb, and the latent heat in the denominator represents the total heat that must be given up during solidification. As the undercooling ΔT increases, more dendritic growth occurs. If the liquid is well inoculated, undercooling is almost zero and growth would be mainly via the planar front solidification mechanism.

8-5   Solidification Time and Dendrite Size

The rate at which growth of the solid occurs depends on the cooling rate, or the rate of heat extraction. A higher cooling rate produces rapid solidification, or short solidification times. The time t_s required for a simple casting to solidify completely can be calculated using **Chvorinov's rule**:[7]

$$t_s = B\left(\frac{V}{A}\right)^n \tag{8-4}$$

where V is the volume of the casting and represents the amount of heat that must be removed before freezing occurs, A is the surface area of the casting in contact with the mold and represents the surface from which heat can be transferred away from the casting, n is a constant (usually about 2), and B is the **mold constant**. The mold constant depends on the properties and initial temperatures of both the metal and the mold. This rule basically accounts for the geometry of a casting and the heat transfer conditions. The rule states that under same conditions a casting with a small volume and relatively large surface area will cool more rapidly.

EXAMPLE 8-2   *Redesign of a Casting for Improved Strength*

Your company currently is producing a disk-shaped brass casting 2 in. thick and 18 in. in diameter. You believe that by making the casting solidify 25% faster, the improvement in the tensile properties of the casting will permit the casting to be made lighter in weight. Design the casting to permit this. Assume that the mold constant is 22 min/in.² for this process.

SOLUTION

One approach would be to use the same casting process, but reduce the thickness of the casting. The thinner casting would solidify more quickly and, because of the faster cooling, should have improved mechanical properties. Chvorinov's rule helps us calculate the required thickness. If d is the diameter and x is the thickness of the casting, then the volume, surface area, and solidification time of the 2-in. thick casting are:

$$V = (\pi/4)d^2x = (\pi/4)(18)^2(2) = 508.9 \text{ in.}^3$$

$$A = 2(\pi/4)d^2 + \pi dx = 2(\pi/4)(18)^2 + \pi(18)(2) = 622 \text{ in.}^2$$

$$t = B\left(\frac{V}{A}\right)^2 = (22)\left(\frac{508.9}{622}\right)^2 = 14.72 \text{ min}$$

The solidification time of the redesigned casting should be 25% shorter than the current time, or $t_r = 0.75t$, where:

$$t_r = 0.75t = (0.75)(14.72) = 11.04 \text{ min}$$

Since the casting conditions have not changed, the mold constant B is unchanged. The V/A ratio of the new casting is:

$$t_r = B\left(\frac{V}{A}\right)^2 = (22)\left(\frac{V}{A}\right)^2 = 11.04 \text{ min}$$

$$\left(\frac{V}{A}\right)^2 = 0.5018 \text{ in.}^2 \quad \text{or} \quad \frac{V}{A} = 0.708 \text{ in.}$$

If x is the required thickness for our redesigned casting, then:

$$\frac{V_r}{A_r} = \frac{(\pi/4)d^2x}{2(\pi/4)d^2 + \pi dx} = \frac{(\pi/4)(18)^2(x)}{2(\pi/4)(18)^2 + \pi(18)(x)} = 0.708 \text{ in.}$$

Therefore, $x = 1.68$ in.

This thickness provides the required solidification time, while reducing the overall weight of the casting by nearly 15%.

Solidification begins at the surface, where heat is dissipated into the surrounding mold material. The rate of solidification of a casting can be described by how rapidly the thickness d of the solidified skin grows:

$$d = k_{\text{solidification}}\sqrt{t} - c_1, \tag{8-5}$$

where t is the time after pouring, $k_{\text{solidification}}$ is a constant for a given casting material and mold, and c_1 is a constant related to the pouring temperature.

Effect on Structure and Properties   The solidification time affects the size of the dendrites. Normally, dendrite size is characterized by measuring the distance between the secondary dendrite arms (Figure 8-9). The **secondary dendrite arm spacing** (SDAS) is reduced when the casting freezes more rapidly. The finer, more extensive dendritic network serves as a more efficient conductor of the latent heat to the undercooled liquid. The SDAS is related to the solidification time by

$$\text{SDAS} = kt_s^m, \tag{8-6}$$

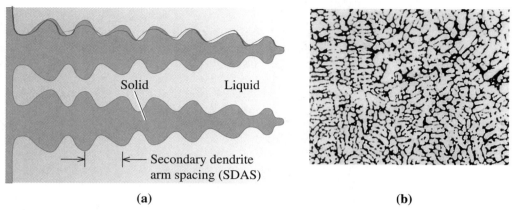

(a)   **(b)**

Figure 8-9   (a) The secondary dendrite arm spacing (SDAS). (b) Dendrites in an aluminum alloy (×50). (*From* ASM Handbook, *Vol. 9, Metallography and Microstructure (1985), ASM International, Materials Park, OH 44073-0002.*)

where m and k are constants depending on the composition of the metal. This relationship is shown in Figure 8-10 for several alloys. Small secondary dendrite arm spacings are associated with higher strengths and improved ductility (Figure 8-11).

Rapid solidification processing is used to produce exceptionally fine secondary

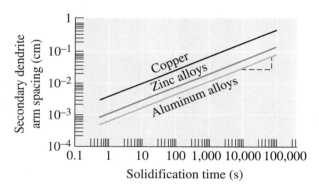

Figure 8-10
The effect of solidification time on the secondary dendrite arm spacings of copper, zinc, and aluminum.

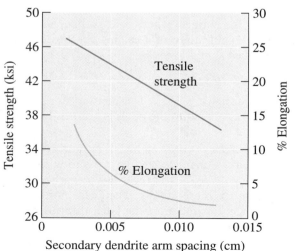

Figure 8-11
The effect of the secondary dendrite arm spacing on the properties of an aluminum casting alloy.

dendrite arm spacings; a common method is to produce very fine liquid droplets that freeze into solid particles. This process is known as spray atomization. The tiny droplets freeze at a rate of about $10^{4}°C/s$, producing powder particles that range from ~5–100 μm. This cooling rate is not rapid enough to form a metallic glass, but does produce a fine dendritic structure. By carefully consolidating the solid droplets by powder metallurgy processes, improved properties in the material can be obtained. Since the particles are derived from melt, many complex alloy compositions can be produced in the form of chemically homogenous powders.

The following three examples discuss how Chvorinov's rule, the relationship between SDAS and the time of solidification, and the SDAS and mechanical properties can be used to design a casting processes.

EXAMPLE 8-3 *Secondary Dendrite Arm Spacing for Aluminum Alloys*

Determine the constants in the equation that describe the relationship between secondary dendrite arm spacing and solidification time for aluminum alloys (Figure 8-10).

SOLUTION

We could obtain the value of SDAS at two times from the graph and calculate k and m using simultaneous equations. However, if the scales on the ordinate and abscissa are equal for powers of ten (as in Figure 8-10), we can obtain the slope m from the log-log plot by directly measuring the slope of the graph. In Figure 8-10, we can mark five equal units on the vertical scale and 12 equal units on the horizontal scale. The slope is:

$$m = \frac{5}{12} = 0.42$$

The constant k is the value of SDAS when $t_s = 1$ s, since:

$$\log SDAS = \log k + m \log t_s$$

If $t_s = 1$ s, $m \log t_s = 0$, and SDAS = k, from Figure 8-10:

$$k = 8 \times 10^{-4} \frac{cm}{s^m}$$

EXAMPLE 8-4 *Time of Solidification*

A 4-in.-diameter aluminum bar solidifies to a depth of 0.5 in. beneath the surface in 5 minutes. After 20 minutes, the bar has solidified to a depth of 1.5 in. How much time is required for the bar to solidify completely?

SOLUTION

From our measurements, we can determine the constants $k_{solidification}$ and c_1 in Equation 8-5:

$$0.5 \text{ in.} = k_{solidification} \sqrt{(5 \text{ min})} - c_1 \quad \text{or} \quad c_1 = k\sqrt{5} - 0.5$$

$$1.5 \text{ in.} = k_{solidification} \sqrt{(20 \text{ min})} - c_1 = k\sqrt{20} - (k\sqrt{5} - 0.5)$$

$$1.5 = k_{solidification}(\sqrt{20} - \sqrt{5}) + 0.5$$

$$k_{solidification} = \frac{1.5 - 0.5}{4.472 - 2.236} = 0.447 \frac{in.}{\sqrt{min}}$$

$$c_1 = (0.447)\sqrt{5} - 0.5 = 0.4995 \text{ in.}$$

Solidification is complete when $d = 2$ in. (half the diameter, since freezing is occurring from all surfaces):

$$2 = 0.447\sqrt{t} - 0.4995$$

$$\sqrt{t} = \frac{2 + 0.4995}{0.447} = 5.59$$

$$t = 31.27 \text{ min}$$

In actual practice, we would find that the total solidification time is somewhat longer than 31.27 min. As solidification continues, the mold becomes hotter and is less effective in removing heat from the casting.

EXAMPLE 8-5   *Design of an Aluminum Alloy Casting*

Design the thickness of an aluminum alloy casting whose length is 12 in. and width is 8 in., in order to produce a tensile strength of 40,000 psi. The mold constant in Chvorinov's rule for aluminum alloys cast in a sand mold is 45 min/in^2. Assume that data shown in Figures 8-10 and 8-11 can be used.

SOLUTION

In order to obtain a tensile strength of 42,000 psi, a secondary dendrite arm spacing of about 0.007 cm is required (see Figure 8-11). From Figure 8-10 we can determine that the solidification time required to obtain this spacing is about 300 s, or 5 minutes.

From Chvorinov's rule:

$$t_s = B\left(\frac{V}{A}\right)^2$$

where $B = 45$ min/in.^{2} and x is the thickness of the casting. Since the length is 12 in. and the width is 8 in.:

$$V = (8)(12)(x) = 96x$$

$$A = (2)(8)(12) + (2)(x)(8) + (2)(x)(12) = 40x + 192$$

$$5 \text{ min} = (45 \text{ min/in.}^2)\left(\frac{96x}{40x + 192}\right)^2$$

$$\frac{96x}{40x + 192} = \sqrt{(5/45)} = 0.333$$

$$96x = 13.33x + 63.9$$

$$x = 0.77 \text{ in.}$$

8-6   Cooling Curves

We can summarize our discussion at this point by examining cooling curves. A cooling curve shows how the temperature of a material (in this case, a pure metal) with time [Figure 8-12(a) and (b)]. The liquid is poured into a mold at pouring temperature, point A. The difference between the pouring temperature and the freezing temperature is the superheat. The specific heat is extracted by the mold until the liquid reaches the freezing temperature (point B). If the liquid is not well inoculated it must be under-cooled (point B to C). The slope of the cooling curve before solidification begins is the cooling rate $\dfrac{\Delta T}{\Delta t}$. As nucleation begins (point C) latent heat of fusion is given off and the temperature rises. This increase in temperature of the undercooled liquid as a result of nucleation is known as **recalescence** (point C to D). Solidification proceeds iso-thermally at the melting temperature (point D to E) as the latent heat given off from continued solidification is balanced by the heat lost by cooling. This region between

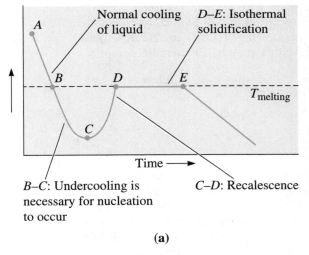

(a)

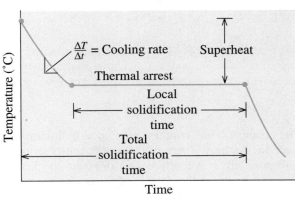

(b)

Figure 8-12
(a) Cooling curve for a pure metal that has not been well inoculated. Liquid cools as specific heat is removed (between points B and B). Undercooling is thus necessary (between points B and C). As the nucleation begins (point C), latent heat of fusion is released causing an increase in the temperature of the liquid. This process is known as recalescence (point C to point D). Metal continues to solidify at a constant temperature (T_{melting}). At point E, solidification is complete. Solid casting continues to cool from the point. (b) Cooling curve for a well inoculated, but otherwise pure, metal. No undercooling is needed. Recalescence is not observed. Solidification begins at the melting temperature.

points D and E, where the temperature is constant, is known as **thermal arrest**. A thermal arrest, or plateau, is produced because of the evolution of the latent heat of fusion balances the heat being lost because of cooling. At point E, solidification is complete and the solid casting cools from point E to room temperature.

If the liquid is well-inoculated, the extent of undercooling is usually very small. The undercooling and recalescence are very small and can be observed in cooling curves only by very careful measurements. If effective heterogeneous nuclei are present in the liquid, solidification begins at the freezing temperature [Figure 8-12(b)]. The latent heat keeps the remaining liquid at the freezing temperature until all of the liquid has solidified and no more heat can be evolved. Growth under these conditions is planar. The **total solidification time** of the casting is the time required to remove both the specific heat of the liquid and the latent heat of fusion. Measured from the time of pouring until solidification is complete, this time is given by Chvorinov's rule. The **local solidification time** is the time required to remove only the latent heat of fusion at a particular location in the casting; it is measured from when solidification begins until solidification is completed. The local solidification times (and the total solidification times) for liquids solidified via undercooled and inoculated liquids will be slightly different.

We often use the terms "melting temperature" and "freezing temperature" while discussing solidification. It would be more accurate to use the term "melting temperature" to describe when a solid turns completely into a liquid. For pure metals and compounds, this happens at a fixed temperature (assuming fixed pressure) and without superheating. "Freezing temperature" or "freezing point" can be defined as the temperature at which solidification of a material is complete.

8-7   Cast Structure

In manufacturing components by casting, molten metals are often poured into molds and permitted to solidify. The mold produces a finished shape, known as a **casting**. In other cases, the mold produces a simple shape, called an **ingot**. An ingot usually requires extensive plastic deformation before a finished product is created. A **macrostructure**, sometimes referred to as the **ingot structure**, consists of as many as three parts (Figure 8-13). (Recall that in Chapters 1 and 2 we had used the term "macrostructure" to describe the structure of a material at a macroscopic scale. Hence, the term "ingot structure" may be more appropriate.)

Chill Zone   The **chill zone** is a narrow band of randomly oriented grains at the surface of the casting. The metal at the mold wall is the first to cool to the freezing temperature. The mold wall also provides many surfaces at which heterogeneous nucleation takes place.

Columnar Zone   The **columnar zone** contains elongated grains oriented in a particular crystallographic direction. As heat is removed from the casting by the mold material, the grains in the chill zone grow in the direction opposite the heat flow, or from the coldest toward the hottest areas of the casting. This tendency usually means that the grains grow perpendicular to the mold wall.

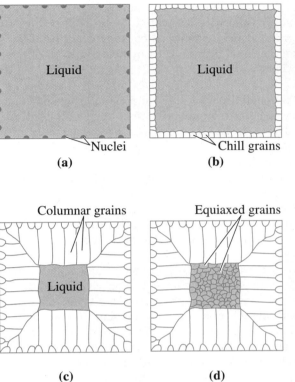

Figure 8-13
Development of the ingot structure of a casting during solidification: (a) Nucleation begins, (b) the chill zone forms, (c) preferred growth produces the columnar zone, and (d) additional nucleation creates the equiaxed zone.

Grains grow fastest in certain crystallographic directions. In metals with a cubic crystal structure, grains in the chill zone that have a $\langle 100 \rangle$ direction perpendicular to the mold wall grow faster than other less favorably oriented grains (Figure 8-14). Eventually, the grains in the columnar zone have $\langle 100 \rangle$ directions that are parallel

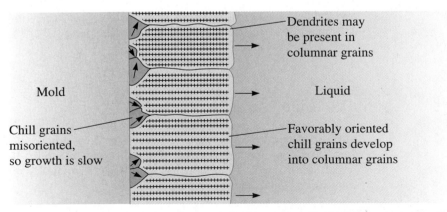

Figure 8-14 Competitive growth of the grains in the chill zone results in only those grains with favorable orientations developing into columnar grains.

to one another, giving the columnar zone anisotropic properties. This formation of the columnar zone is influenced primarily by growth—rather than nucleation—phenomena. The grains may be composed of many dendrites if the liquid is originally undercooled. Or solidification may proceed by planar growth of the columnar grains if no undercooling occurs.

Equiaxed Zone Although the solid may continue to grow in a columnar manner until all of the liquid has solidified, an equiaxed zone frequently forms in the center of the casting or ingot. The **equiaxed zone** contains new, randomly oriented grains, often caused by a low pouring temperature, alloying elements, or grain refining or inoculating agents. Small grains or dendrites in the chill zone may also be torn off by strong convection currents that are set up as the casting begins to freeze.[8] These also provide the heterogeneous nucleation sites for what ultimately become equiaxed grains. These grains grow as relatively round, or equiaxed, grains with a random orientation, and they stop the growth of the columnar grains. The formation of the equiaxed zone is a nucleation-controlled process and causes that portion of the casting to display isotropic behavior.

By understanding the factors that influence solidification in different regions, it is possible to produce castings that first form a "skin" of a chill zone and then dendrites. Metals and alloys that show this macrostructure are known as **skin-forming alloys**. We also can control the solidification such that no skin or advancing dendritic arrays of grains are seen; columnar to equiaxed switchover is almost at the mold walls. The result is casting with a macrostructure consisting predominantly of equiaxed grains. Metals and alloys that solidify in this fashion are known as **mushy-forming alloys** since the cast material seems like a mush of solid grains floating in a liquid melt. Many aluminum and magnesium alloys show this type of solidification.[8] Often, we encourage an all-equiaxed structure and thus create a casting with isotropic properties by effective grain refinement or inoculation. In a later section, we will examine one case (turbine blades) where we control solidification to encourage all columnar grains and hence anisotropic behavior.

Cast ingot structure and microstructure are important particularly for components that are directly cast into a final shape. In many situations though, as discussed in Section 8-1, metals and alloys are first cast into ingots and the ingots are subsequently subjected to thermomechanical processing (e.g., rolling, forging etc.). During these steps, the cast macrostructure is broken down and a new microstructure will emerge, depending upon the thermomechanical process used (Chapter 7).

8-8 Solidification Defects

Although there are many defects that potentially can be introduced during solidification, shrinkage and the porosity deserve special mention. If a casting contains pores (small holes), the cast component can fail catastrophically when used for load-bearing applications (e.g., turbine blades).

Shrinkage Almost all materials are more dense in the solid state than in the liquid state. During solidification, the material contracts, or shrinks, as much as 7% (Table 8-2).

TABLE 8-2 ■ *Shrinkage during solidification for selected materials*

Material	Shrinkage (%)
Al	7.0
Cu	5.1
Mg	4.0
Zn	3.7
Fe	3.4
Pb	2.7
Ga	+3.2 (expansion)
H_2O	+8.3 (expansion)
Low-carbon steel	2.5–3.0
High-carbon steel	4.0
White Cast Iron	4.0–5.5
Gray Cast Iron	+1.9 (expansion)

Note: Some data from Ref. [9]

Often, the bulk of the **shrinkage** occurs as **cavities**, if solidification begins at all surfaces of the casting, or **pipes**, if one surface solidifies more slowly than the others (Figure 8-15). The presence of such pipes can pose problems. For example, if in the production of zinc ingots a shrinkage pipe remains, water vapor can condense in it. This water can lead to an explosion if the ingot gets introduced in a furnace in which zinc is being remelted for such applications as hot-dip galvanizing.

A common technique for controlling **cavity** and **pipe shrinkage** is to place a **riser**, or an extra reservoir of metal, adjacent and connected to the casting. As the casting solidifies and shrinks, liquid metal flows from the riser into the casting to fill the shrinkage void. We need only assure that the riser solidifies after the casting; and that there is an internal liquid channel that connects the liquid in the riser to the last liquid to solidify in the casting. Chvorinov's rule can be used to help design the size of the riser. The following example illustrates how risers can be designed to compensate for shrinkage.

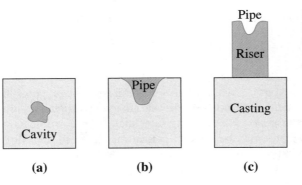

Figure 8-15
Several types of macroshrinkage can occur, including cavities and pipes. Risers can be used to help compensate for shrinkage.

EXAMPLE 8-6   *Design of a Riser for a Casting*

Design a cylindrical riser, with a height equal to twice its diameter, that will compensate for shrinkage in a 2 cm × 8 cm × 16 cm casting (Figure 8-16).

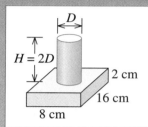

Figure 8-16
The geometry of the casting and riser (for Example 8-6).

SOLUTION

We know that the riser must freeze after the casting. To be conservative, however, we typically require that the riser take 25% longer to solidify than the casting. Therefore:

$$t_{\text{riser}} = 1.25 t_{\text{casting}} \quad \text{or} \quad \mathrm{B}\left(\frac{V}{A}\right)_r^2 = 1.25\mathrm{B}\left(\frac{V}{A}\right)_c^2$$

Subscripts r and c stand for riser and casting, respectively.

The mold constant B is the same for both casting and riser, so:

$$(V/A)_r = \sqrt{1.25(V/A)_c}$$
$$V_c = (2)(8)(16) = 256 \text{ cm}^3$$
$$A_c = (2)(2)(8) + (2)(2)(16) + (2)(8)(16) = 352 \text{ cm}^2$$

We can write equations for the volume and area of a cylindrical riser, noting that $H = 2D$:

$$V_r = (\pi/4)D^2 H = (\pi/4)D^2(2D) = (\pi/2)D^3$$
$$A_r = 2(\pi/4)D^2 + \pi DH = 2(\pi/4)D^2 + \pi D(2D) = (5\pi/2)D^2$$
$$\frac{V_r}{A_r} = \frac{(\pi/2)(D)^3}{(5\pi/2)(D)^2} = \frac{D}{5} > \sqrt{\frac{(1.25)(256)}{352}}$$
$$D = 4.77 \text{ cm} \quad H = 2D = 9.54 \text{ cm} \quad V_r = 170.5 \text{ cm}^3$$

Although the volume of the riser is less than that of the casting, the riser solidifies more slowly because of its compact shape.

Interdendritic Shrinkage This consists of small shrinkage pores between dendrites (Figure 8-17). This defect, also called **microshrinkage** or **shrinkage porosity**, is difficult to prevent by the use of risers. Fast cooling rates may reduce problems with interdendritic shrinkage; the dendrites may be shorter, permitting liquid to flow through the dendritic network to the solidifying solid interface. In addition, any shrinkage that remains may be finer and more uniformly distributed.

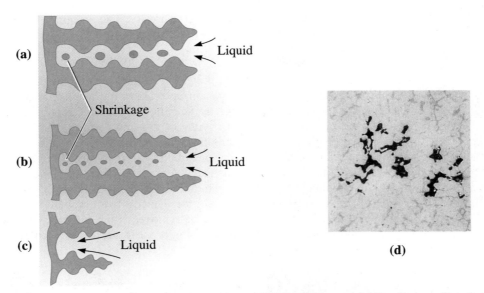

Figure 8-17 (a) Shrinkage can occur between the dendrite arms. (b) Small secondary dendrite arm spacings result in smaller, more evenly distributed shrinkage porosity. (c) Short primary arms can help avoid shrinkage. (d) Interdendritic shrinkage in an aluminum alloy is shown (×80).

Gas Porosity Many metals dissolve a large quantity of gas when they are molten. Aluminum, for example, dissolves hydrogen. When the aluminum solidifies, however, the solid metal retains in its crystal structure only a small fraction of the hydrogen since the solubility is remarkably lower (Figure 8-18). The excess hydrogen that cannot be incorporated in the solid metal or alloy crystal structure forms bubbles that may be

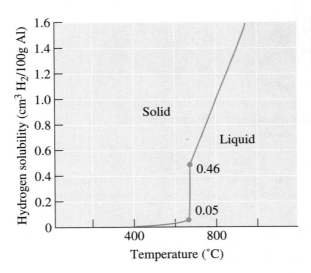

Figure 8-18
The solubility of hydrogen gas in aluminum when the partial pressure of $H_2 = 1$ atm.

trapped in the solid metal, producing **gas porosity**. The amount of gas that can be dissolved in molten metal is given by **Sievert's law**:

$$\text{Percent of gas} = K\sqrt{p_{gas}} \tag{8-7}$$

where p_{gas} is the partial pressure of the gas in contact with the metal and K is a constant which, for a particular metal-gas system, increases with increasing temperature. We can minimize gas porosity in castings by keeping the liquid temperature low, by adding materials to the liquid to combine with the gas and form a solid, or by assuring that the partial pressure of the gas remains low. The latter may be achieved by placing the molten metal in a vacuum chamber or bubbling an inert gas through the metal. Because p_{gas} is low in the vacuum, the gas leaves the metal, enters the vacuum, and is carried away. **Gas flushing** is a process in which bubbles of a gas, inert or reactive, are injected into a molten metal to remove undesirable elements from molten metals and alloys. For example, hydrogen in aluminum can be removed using nitrogen or chlorine. The following example illustrates how a degassing process can be designed.

EXAMPLE 8-7   *Design of a Degassing Process for Copper*

After melting at atmospheric pressure, molten copper contains 0.01 weight percent oxygen. To assure that your castings will not be subject to gas porosity, you want to reduce the weight percent to less than 0.00001% prior to pouring. Design a degassing process for the copper.

SOLUTION

We can solve this problem in several ways. In one approach, the liquid copper is placed in a vacuum chamber; the oxygen is then drawn from the liquid and carried away into the vacuum. The vacuum required can be estimated from Sievert's law:

$$\frac{\%O_{initial}}{\%O_{vacuum}} = \frac{K\sqrt{p_{initial}}}{K\sqrt{p_{vacuum}}} = \sqrt{\left(\frac{1\text{ atm.}}{p_{vacuum}}\right)}$$

$$\frac{0.01\%}{0.00001\%} = \sqrt{\left(\frac{1}{p_{vacuum}}\right)}$$

$$\frac{1\text{ atm.}}{p_{vacuum}} = (1000)^2 \quad \text{or} \quad p_{vacuum} = 10^{-6}\text{ atm.}$$

Another approach would be to introduce a copper-15% phosphorous alloy. The phosphorous reacts with oxygen to produce P_2O_5, which floats out of the liquid, by the reaction:

$$5O + 2P \rightarrow P_2O_5$$

Typically, about 0.01 to 0.02% P must be added to remove the oxygen.

In the manufacturing of **stainless steel**, a process known as **argon oxygen decarburization** (AOD) is used to lower the carbon content of the melt without oxidizing chromium or nickel. In this process, a mixture of argon (or nitrogen) and oxygen gases is forced into molten stainless steel. The carbon dissolved in the molten steel is oxidized by the oxygen gas via the formation of carbon monoxide (CO) gas; the CO is carried away by the inert argon (or nitrogen) gas bubbles. These processes need very careful control since some reactions (e.g., oxidation of carbon to CO) are exothermic (generate heat). For example, the goal in the AOD process is to remove carbon without oxidizing the chromium or nickel that are used as alloying elements in stainless steel.

8-9   Casting Processes for Manufacturing Components

Figure 8-19 summarizes four of the dozens of commercial casting processes. In some processes the same mold can be used; in others the mold is expendable. **Sand casting** processes include green sand molding, for which silica (SiO_2) sand grains bonded with wet clay are packed around a removable pattern. Ceramic casting processes use a fine-grained ceramic material as the mold; a slurry containing the ceramic may be poured around a reusable pattern, which is removed after the ceramic hardens. In **investment casting**, the ceramic slurry of a material such as colloidal silica (consisting of nano-sized ceramic particles) coats a wax pattern. After the ceramic hardens (i.e., the colloidal silica dispersion gels), the wax is melted and drained from the ceramic shell, leaving behind a cavity that is filled with molten metal. The investment casting process, also known as the **lost wax process**, is best suited for generating most complex shapes. Dentists and jewelers originally used the precision investment casting process. Currently, this process is used to produce such components as turbine blades, titanium heads of golf clubs, and parts for knee and hip prosthesis. In another process known as the **lost foam process**, polystyrene beads, similar to those used to make coffee cups or packaging materials, are used to produce a foam pattern. Loose sand is compacted around the pattern to produce a mold. When molten metal is poured into the mold, the polymer foam pattern melts and decomposes, with the metal taking the place of the pattern. Figure 8-20 shows an engine block produced using expandable polystyrene beads.

In the permanent mold and pressure die casting processes, a cavity is machined from metallic material. After the liquid poured into the cavity solidifies, the mold is opened, the casting is removed, and the mold is reused. The processes using metallic molds tend to give the highest strength castings because of the rapid solidification. Ceramic molds, including those used in investment casting, are good insulators and give the slowest-cooling and lowest-strength castings. Millions of truck and car pistons are made in foundries using permanent mold casting.[9] Good surface finish and dimensional accuracy are the advantages of **permanent mold casting** technique. High mold costs and limited complexity in shape are the limitations of this technique.

In **pressure die casting**, molten metallic material is forced into the mold under high pressures and is held under pressure during solidification. Many zinc, aluminum, and magnesium-based alloys are processed using pressure die casting. Extremely smooth surface finishes, very good dimensional accuracy, the ability to cast intricate shapes, and high production rates are the advantages of the pressure die casting process. Since the mold is metallic and must withstand high pressures, the dies used are expensive and the technique is limited to smaller sized components.

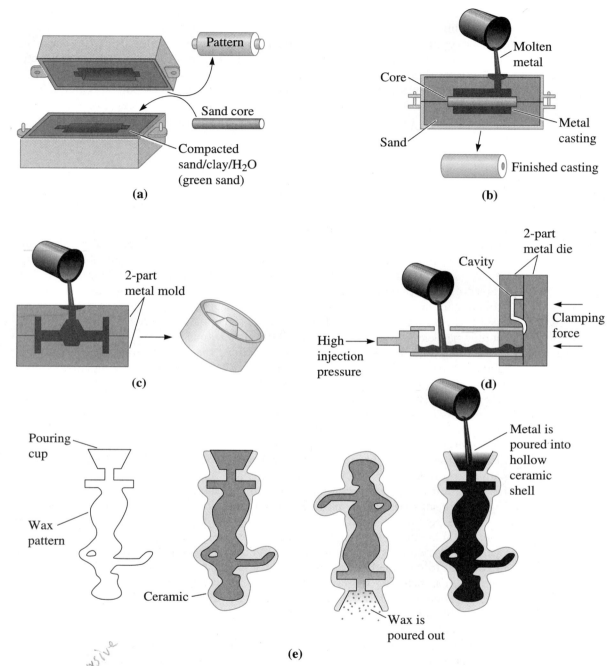

(a)

(b)

(c)

(d)

(e)

most expensive

Figure 8-19 Four typical casting processes: (a) and (b) Green sand molding, in which clay-bonded sand is packed around a pattern. Sand cores can produce internal cavities in the casting. (c) The permanent mold process, in which metal is poured into an iron or steel mold. (d) Die casting, in which metal is injected at high pressure into a steel die. (e) Investment casting, in which a wax pattern is surrounded by a ceramic; after the wax is melted and drained, metal is poured into the mold.

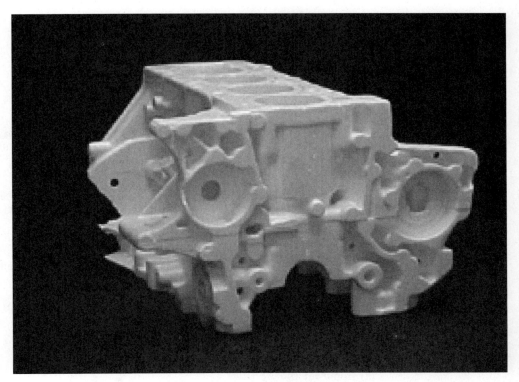

Figure 8-20   Engine block produced using the lost foam casting process. (*Courtesy of Paul Arch, Nova Chemicals.*)

8-10   Continuous Casting and Ingot Casting

As discussed in the prior section, casting is a tool used for the manufacturing of components. It is also a process for producing ingots or slabs that can be further processed into different shapes (e.g., rods, bars, wires, etc.). In the steel industry, millions of pounds of steels are produced using blast furnaces, electric arc furnaces and other processes. Figure 8-21 shows a summary of steps that go into the extraction of steels using iron ores, coke, and limestone. Although the details change, most metals and alloys (e.g., copper and zinc) are extracted from their ores using similar processes. Certain metals, such as aluminum, are produced using an electrolytic process since aluminum oxide is too stable and can not be readily reduced to aluminum metal using coke or other reducing agents.

In many cases, we begin with scrap metals and recyclable alloys. In this case, the scrap metal is melted and processed, removing the impurities and adjusting the composition. Considerable amounts of steels, aluminum, zinc, stainless steels, titanium, and many other materials are recycled every year.

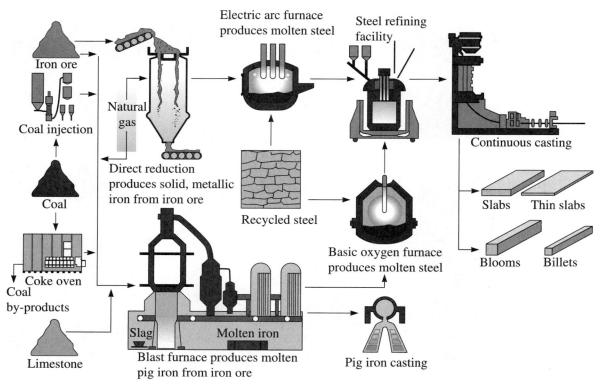

Figure 8-21   Summary of steps in the extraction of steels using iron ores, coke and limestone. (*Source: www.steel.org. Used with permission of the American Iron and Steel Institute.*)

In **ingot casting**, molten steels or alloys obtained from a furnace are cast into large molds. The resultant castings, called ingots, are then processed for conversion into useful shapes via thermomechanical processing, often at another location. In the **continuous casting** process, the idea is to go from molten metallic material to some more useful "semi-finished" shape such as a plate, slab, etc. Figure 8-22 illustrates a common method for producing steel plate and bars. The liquid metal is fed from a holding vessel (a tundish) into a water-cooled oscillating copper mold, which rapidly cools the surface of the steel. The partially solidified steel is withdrawn from the mold at the same rate that additional liquid steel is introduced. The center of the steel casting finally solidifies well after the casting exits the mold. The continuously cast material is then cut into appropriate lengths by special cutting machines.

Continuous casting is cost effective for processing many steels, stainless steels, and aluminum. Ingot casting is also cost effective and used for many steels where a continuous caster is not available or capacity is limited, and for alloys of non-ferrous metals (e.g., zinc, copper) where the volumes are relatively small and the capital expenditure needed for a continuous caster may not be justified. Also, not all alloys can be cast using the continuous casting process.

The secondary processing steps in the processing of steels and other alloys are shown in Figure 8-23.

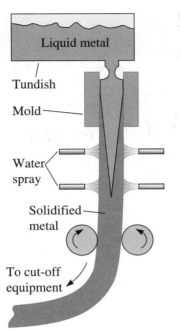

Figure 8-22
Vertical continuous casting, used in producing many steel products. Liquid metal contained in the tundish partially solidifies in a mold.

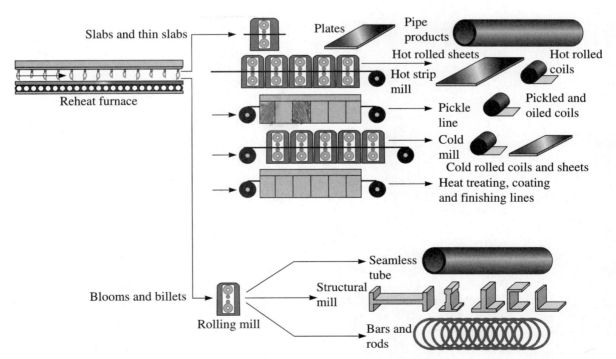

Figure 8-23 Secondary processing steps in processing of steel and alloys. (*Source: www.steel.org. Used with permission of the American Iron and Steel Institute.*)

EXAMPLE 8-8 *Design of a Continuous Casting Machine*

Figure 8-24 shows a method for continuous casting of 0.25-in.-thick, 48-in.-wide aluminum plate that is subsequently rolled into aluminum foil. The liquid aluminum is introduced between two large steel rolls that slowly turn. We want the aluminum to be completely solidified by the rolls just as the plate emerges from the machine. The rolls act as a permanent mold with a mold constant B of about 5 min/in.^{2} when the aluminum is poured at the proper superheat. Design the rolls required for this process.

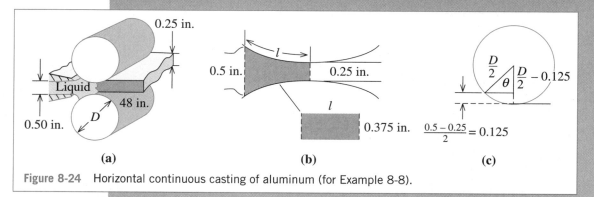

Figure 8-24 Horizontal continuous casting of aluminum (for Example 8-8).

SOLUTION

It would be helpful to simplify the geometry so that we can determine a solidification time for the casting. Let's assume that the shaded area shown in Figure 8-24(b) represents the casting and can be approximated by the average thickness times a length and width. The average thickness is (0.50 in. + 0.25 in.)/2 = 0.375 in. Then:

$$V = (\text{thickness})(\text{length})(\text{width}) = 0.375lw$$

$$A = 2(\text{length})(\text{width}) = 2lw$$

$$\frac{V}{A} = \frac{0.375lw}{2lw} = 0.1875 \text{ in.}$$

Only the area directly in contact with the rolls is used in Chvorinov's rule, since little or no heat is transferred from other surfaces. The solidification time should be:

$$t_s = B\left(\frac{V}{A}\right)^2 = (5)(0.1875)^2 = 0.175 \text{ min}$$

For the plate to remain in contact with the rolls for this period of time, the diameter of the rolls and the rate of rotation of the rolls must be designed. Figure 8-24(c) shows that the angle θ between the points where the liquid enters and exits the rolls is:

$$\cos \theta = \frac{(D/2) - 0.125}{(D/2)} = \frac{D - 0.25}{D}$$

The surface velocity of the rolls is the product of the circumference and the rate of rotation of the rolls, $v = \pi DR$, where R is in revolutions/minute. The velocity v is also the rate at which we can produce the aluminum plate. The time required for the rolls to travel the distance l must equal the required solidification time.

$$t = \frac{l}{v} = 0.175 \text{ min}$$

The length l is the fraction of the roll diameter that is in contact with the aluminum during freezing and can be given by

$$l = \frac{\pi D \theta}{360}$$

Then, by substituting for l and v in the equation for the time:

$$t = \frac{l}{v} = \frac{\pi D \theta}{360 \pi D R} = \frac{\theta}{360 R} = 0.175 \text{ min}$$

$$R = \frac{\theta}{(360)(0.175)} = 0.0159 \theta \text{ rev/min}$$

A number of combinations of D and R provide the required solidification rate. Let's calculate θ for several diameters and then find the required R.

D	θ	l	$R = 0.0159\theta$	$v = \pi D R$
24 in.	8.2771	1.7334 in.	0.1316 rev/min	9.923 in./min
36 in.	6.7560	2.1230 in.	0.1074 rev/min	12.149 in./min
48 in.	5.8502	2.4505 in.	0.0930 rev/min	14.027 in./min
60 in.	5.2322	2.7396 in.	0.0832 rev/min	15.683 in./min

As the diameter of the rolls increases, the contact area (l) between the rolls and the metal increases. This, in turn, permits a more rapid surface velocity (v) of the rolls and increases the rate of production of the plate. However, the larger diameter rolls do not need to rotate as rapidly to achieve these higher velocities.

In selecting our final design, we prefer to use the largest practical roll diameter to assure high production rates. As the rolls become more massive, however, they and their supporting equipment become more expensive.

In actual operation of such a continuous caster, faster speeds could be used, since the plate does not have to be completely solidified at the point where it emerges from the rolls.

8-11 Directional Solidification (DS), Single Crystal Growth, and Epitaxial Growth

There are some applications for which a small equiaxed grain structure in the casting is not desired. Castings used for blades and vanes in turbine engines are an example (Figure 8-25). These castings are often made of titanium, cobalt or nickel-based superalloys using precision investment casting.

In conventionally cast parts, an equiaxed grain structure is often produced. However, blades and vanes for turbine and jet engines fail along transverse grain boundaries.

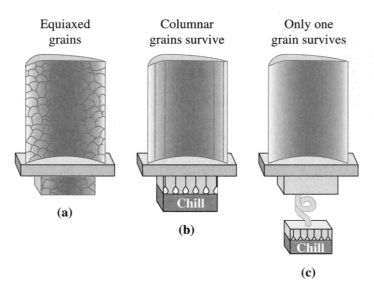

Equiaxed grains

Columnar grains survive

Only one grain survives

(a)

(b)

(c)

Chill

Chill

Figure 8-25
Controlling grain structure in turbine blades: (a) conventional equiaxed grains, (b) directionally solidified columnar grains, and (c) single crystal.

Better creep and fracture resistance are obtained using the **directional solidification** (DS) growth technique. In the DS process, the mold is heated from one end and cooled from the other, producing a columnar microstructure with all of the grain boundaries running in the longitudinal direction of the part. No grain boundaries are present in the transverse direction [Figure 8-25(b)].

Still better properties are obtained by using a **single crystal** (SC) technique. Solidification of columnar grains again begins at a cold surface; however, due to the helical connection, only one columnar grain is able to grow to the main body of the casting [Figure 8-25(c)]. The single crystal casting has no grain boundaries, so its crystallographic planes and directions can be directed in an optimum orientation.

Single Crystal Growth   One of the most important applications of solidification is the growth of single crystals. Polycrystalline materials cannot be used effectively in many electronic and optical applications. Grain boundaries and other defects interfere with the mechanisms that provide useful electrical or optical functions. For example, in order to utilize the semiconducting behavior of doped silicon, high-purity single crystals must be used. The current technology for silicon makes use of large (up to 10-in. diameter) single crystals. Typically, a large crystal of the material is grown [Figure 8-26(a)]. The large crystal is then cut into silicon wafers that are only a few millimeters thick [Figure 8-26(b)]. The **Bridgman** and **Czochralski processes** are some of the popular methods used for growing single crystals of silicon, GaAs, lithium niobate ($LiNbO_3$), and many other materials. Figure 8-26(c) shows the schematic for the Bridgman process for growing single crystals.

Crystal growth furnaces containing molten materials must be maintained at a precise and stable temperature. Often, a small crystal of a pre-determined crystallographic orientation is used as a "seed." Heat transfer is controlled so that the entire melt crystallizes into a single crystal. Typically, single crystals offer considerably improved, controllable, and predictable properties. The cost of single crystal materials though is typically higher than that of polycrystalline materials. However, with increased demand the cost of single crystals may not be a significant factor compared to the rest of the processing costs involved in making novel and useful devices.

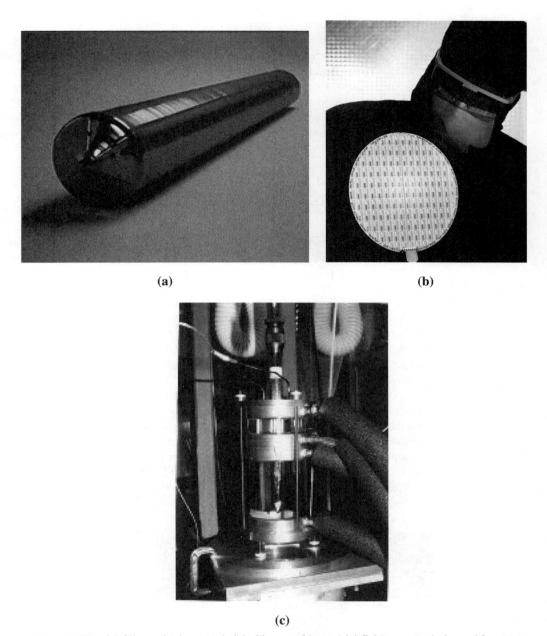

(a)

(b)

(c)

Figure 8-26   (a) Silcon single crystal, (b) silicon wafer, and (c) Bridgman technique. (*Courtesy of PhotoDisc/Getty Images.*)

Epitaxial Growth   There are probably over a hundred processes for the deposition of thin films materials. In some of these processes, there is a need to control the texture or crystallographic orientation of the polycrystalline material being deposited; others require a single crystal film oriented in a particular direction. If this is the case, we can make use of a substrate of a known orientation. If the lattice matching between the substrate and the film is good (within a few %), it is possible to grow highly oriented or single crystal thin films. This is known as **epitaxial growth**. The word *epitaxy* comes from two Greek words: *epi* (top) and *taxis* (ordered).[10] When we grow thin films of one material (i.e., Si doped with phosphorus (P)) on the substrate of the same material (silicon (Si)) the process is known as **homoepitaxy**. When highly oriented polycrystalline or single crystal thin films of one material are grown on another material that provide sites for heterogeneous nucleation, we refer to this process as **heteroepitaxy**. Many compound semiconductor devices can be grown using liquid and vapor phase epitaxial processes.[11] Solidification is *not* always required for epitaxial growth. Similarly, many materials can be grown in a highly oriented fashion (i.e., significant crystallographic texture) on amorphous substrates (e.g., ZnO on glass).

8-12 Solidification of Polymers and Inorganic Glasses

Similar to the processing of metals and alloys, the processing of thermoplastics depends critically on our ability to melt and process them via extrusion and other processes. We will discuss these processes in later chapters.

Many polymers do not crystallize, but solidify, when cooled. In these materials, the thermodynamic driving force may exist; however, the rate of nucleation of the solid may be too slow, or the complexity of the polymer chains may be so great that a crystalline solid does not form. Crystallization in polymers is almost never complete and is significantly different from that of metallic materials, requiring long polymer chains to become closely aligned over relatively large distances. By doing so, the polymer grows as **lamellar**, or plate-like, crystals (Figure 8-27). The region between each lamella contains polymer chains arranged in an amorphous manner. Amorphous regions are present between the individual lamellae, bundles of lamellae, and individual spherulites. In addition, bundles of lamellae grow from a common nucleus, but the crystallographic orientation of the lamellae within any one bundle is different from that in another. As the bundles grow, they may produce a spheroidal shape called a **spherulite**. The spherulite is composed of many individual bundles of differently oriented lamellae.

Many polymers of commercial interest develop crystallinity during their processing. Crystallinity can originate from cooling as discussed previously, or from the application

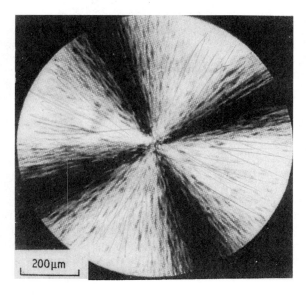

Figure 8-27
A spherulite in polystyrene (×8000). (*From R. Young and P. Lovell*, Introduction to Polymers, *2nd Ed., Chapman & Hall, 1991*).

200μm

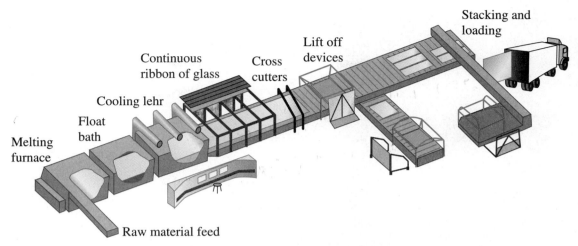

Figure 8-28 Processing scheme for float glasses. (*Source: www.glassrecruiters.com.*)

of stress. For example, in previous chapters, we have learned how PET plastic bottles are prepared using the blow-stretch process (Chapters 3 and 4) and how they can develop considerable crystallinity during formation. This crystallization is a result of the application of stress, and thus, is different from that encountered in the solidification of metals and alloys. In general, such polymers as nylon and polyethylene crystallize more easily when compared to many other thermoplastics.

Inorganic glasses, such as silicate glasses, also do not crystallize easily for kinetic reasons. While the thermodynamic driving force exists, similar to the solidification of metals and alloys; however, the melts are often too viscous and the diffusion is too slow for crystallization to proceed during solidification. The float-glass process is used to melt and cast large flat pieces of glasses (Figure 8-28). In this process, molten glass is made to float on molten tin. As discussed in Chapter 6, since the strength of inorganic glasses depends critically on surface flaws produced by the manufacturing process or the reaction with atmospheric moisture, most glasses are strengthened using tempering. When safety is not a primary concern, annealing is used to reduce stresses. Long lengths of glass fibers, such as those used with fiber optics, are produced by melting a high-purity glass rod known as a **preform**. As mentioned earlier, careful control of nucleation in glasses can lead to glass-ceramics, colored glasses, and photochromic glasses (glasses that can change their color or tint upon exposure to sunlight).

8-13   Joining of Metallic Materials

In **brazing**, an alloy, known as a filler, is used to join one metal to itself or to another metal. The brazing filler metal has a melting temperature above about 450°C. **Soldering** is a brazing process in which the filler has a melting temperature below 450°C. Lead-tin and antimony-tin alloys are the most common materials used for soldering. Currently, there is a need to develop lead-free soldering materials due to the toxicity levels of lead. Alloys being developed include those that are based on Sn-Cu-Ag.[12] In brazing and soldering, the metallic materials being joined do not melt; only the filler material melts.

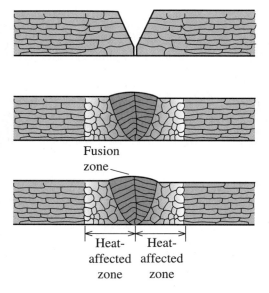

Figure 8-29
A schematic diagram of the fusion zone and solidification of the weld during fusion welding: (a) initial prepared joint, (b) weld at the maximum temperature, with joint filled with filler metal, and (c) weld after solidification.

On both brazing and soldering, the composition of the filler material is different from that of the base material being joined. Various aluminum-silicon, copper, magnesium and precious metals are used for brazing.

Solidification is also important in the joining of metals through **fusion welding**.[13,14] In the fusion-welding processes, a portion of the metals to be joined is melted and, in many instances, additional molten filler metal is added. The pool of liquid metal is called the **fusion zone** (Figures 8-29 and 8-30). When the fusion zone subsequently solidifies, the original pieces of metal are joined together. During solidification of the fusion zone, nucleation is not required. The solid simply begins to grow from existing grains, frequently in a columnar manner. Growth of the solid grains in the fusion zone from the pre-existing grains is via epitaxial growth.

The structure and properties of the fusion zone depend on many of the same variables as in a metal casting. Addition of inoculating agents to the fusion zone reduces

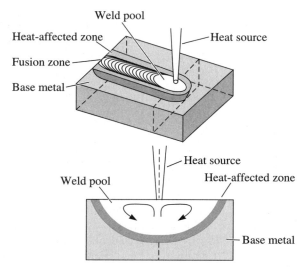

Figure 8-30
Schematic diagram showing interaction between the heat source and the base metal. Three distinct regions in the weldment are the fusion zone, the heat-affected zone, and the base metal. (*Source: Reprinted with permission from "Current Issues and Problems in Welding Science," by S.A. David and T. DebRoy, 1992*, Science, *257, pp. 497–502, Fig. 2. Copyright © 1992 American Association for the Advancement of Science.*)

the grain size. Fast cooling rates or short solidification times promote a finer microstructure and improved properties. Factors that increase the cooling rate include increased thickness of the metal, smaller fusion zones, low original metal temperatures, and certain types of welding processes. Oxyacetylene welding, for example, uses a relatively low-intensity heat source; consequently, welding times are long and the surrounding solid metal, which becomes very hot, is not an effective heat sink. Arc-welding processes provide a more intense heat source, thus reducing heating of the surrounding metal and providing faster cooling. Laser welding and electron-beam welding are exceptionally intense heat sources and produce very rapid cooling rates and potentially strong welds. The friction stir welding process has been developed for Al and Al-Li alloys for aerospace applications.

SUMMARY

◈ Transformation of a liquid to a solid is probably the most important phase transformation in applications of materials science and engineering.

◈ Solidification plays a critical role in the processing of metals, alloys, thermoplastics, and inorganic glasses. Solidification is also important in the processing of glass-ceramics and techniques used for the joining of metallic materials.

◈ Nucleation produces a critical-size solid particle from the liquid melt. Formation of nuclei is determined by the thermodynamic driving force for solidification, and is opposed by the need to create solid-liquid interface. As a result, solidification may not occur at the freezing temperature.

◈ Homogeneous nucleation requires large undercoolings of the liquid and is not observed in the normal solidification processing. By introducing foreign particles into the liquid, nuclei are provided for heterogeneous nucleation. This is done in practice by inoculation or grain refining. This process permits the grain size of the casting to be controlled.

◈ Rapid cooling of the liquid can prevent nucleation and growth, giving amorphous solids, or glasses, with unusual mechanical and physical properties. Polymeric, metallic, and inorganic materials can be made in the form of amorphous glasses.

◈ In solidification from melts, growth occurs as the nuclei grow into the liquid melt. Either planar or dendritic modes of growth may be observed. In planar growth, a smooth solid-liquid interface grows with little or no undercooling of the liquid. Special directional solidification processes take advantage of planar growth. Dendritic growth occurs when the liquid is undercooled. Rapid cooling, or a short solidification time, produces a finer dendritic structure and often leads to improved mechanical properties of a metallic casting.

◈ Chvorinov's rule, $t_s = B(V/A)^n$, can be used to estimate the solidification time of a casting. Metallic castings that have a smaller interdendritic spacing and finer grain size have higher strengths.

◈ Cooling curves can be used to determine pouring temperature, observe any undercooling, recoalescence, and time for solidification.

◈ By controlling nucleation and growth, a casting may be given a columnar grain structure, an equiaxed grain structure, or a mixture of the two. Isotropic behavior is typical of the equiaxed grains, whereas anisotropic behavior is found in columnar grains.

◇ Porosity and shrinkage are major defects that can be present in cast products. If present, it can cause cast products to fail catastrophically.

◇ In commercial solidification processing methods, defects in a casting (such as solidification shrinkage or gas porosity) can be controlled by proper design of the casting and riser system or by appropriate treatment of the liquid metal prior to casting.

◇ Sand casting, investment casting, and pressure die casting are some of the processes for casting components. Ingot casting and continuous casting are employed in the production and recycling of metals and alloys.

◇ The solidification process can be carefully controlled to produce directionally solidified materials as well as single crystals. Epitaxial processes make use of crystal structure match between substrate and the material being grown, and are useful for making electronic and other devices.

GLOSSARY

Argon oxygen decarburization (AOD) A process to refine stainless steel. The carbon dissolved in molten stainless steel is reduced by blowing argon gas mixed with oxygen.

Brazing An alloy, known as a filler, is used to join two materials to one another. The composition of the filler, which has a melting temperature above about 450°C, is quite different from the metal being joined.

Bridgman processes A process to grow semiconductor and other single crystals.

Cavities Small holes present in a casting.

Cavity shrinkage A large void within a casting caused by the volume contraction that occurs during solidification.

Chill zone A region of small, randomly oriented grains that forms at the surface of a casting as a result of heterogeneous nucleation.

Chvorinov's rule The solidification time of a casting is directly proportional to the square of the volume-to-surface area ratio of the casting.

Columnar zone A region of elongated grains having a preferred orientation that forms as a result of competitive growth during the solidification of a casting.

Continuous casting A process to convert molten metal or an alloy into a semi-finished product such as a slab.

Critical radius (r^*) The minimum size that must be formed by atoms clustering together in the liquid before the solid particle is stable and begins to grow.

Czochralski processes A process for growing single crystals. A small crystal is used as a seed.

Dendrite The treelike structure of the solid that grows when an undercooled liquid solidifies.

Directional solidification (DS) A solidification technique in which cooling in a given direction leads to preferential growth of grains in the opposite direction, leading to an anisotropic-oriented microstructure.

Dispersion strengthening Increase in strength of a metallic material by generating resistance to dislocation motion by the introduction of small clusters of a second material.

Embryo A tiny particle of solid that forms from the liquid as atoms cluster together. The embryo may grow into a stable nucleus or redissolve.

Epitaxial growth Growth of a material via epitaxy.

Epitaxy The process by which one material is made to grow in an oriented fashion using a substrate that has crystallographic match with the material being grown.

Equiaxed zone A region of randomly oriented grains in the center of a casting produced as a result of widespread nucleation.

Fusion welding Joining processes in which a portion of the materials must melt in order to achieve good bonding.

Fusion zone The portion of a weld heated to produce all liquid during the welding process. Solidification of the fusion zone provides joining.

Gas flushing A process in which a stream of gas is injected into a molten metal in order to eliminate a dissolved gas that might produce porosity.

Gas porosity Bubbles of gas trapped within a casting during solidification, caused by the lower solubility of the gas in the solid compared with that in the liquid.

Glass-ceramics Polycrystalline, ultrafine grained ceramic materials obtained by controlled crystallization of amorphous glasses.

Grain refinement The addition of heterogeneous nuclei in a controlled manner to increase the number of grains in a casting.

Growth The physical process by which a new phase increases in size. In the case of solidification, this refers to the formation of a stable solid as the liquid freezes.

Heteroepitaxy Growth of a highly oriented material onto a different substrate material. Typically, the material being grown and the substrate will have similar crystal structure and similar lattice constants.

Heterogeneous nucleation Formation of a critically sized solid from the liquid on an impurity surface.

Homoepitaxy Growth of a highly oriented material onto a crystal of the same material. The material being grown may contain useful dopants.

Homogeneous nucleation Formation of a critically sized solid from the liquid by the clustering together of a large number of atoms at a high undercooling (without an external interface).

Ingot A simple casting that is usually remelted or reprocessed by another user to produce a more useful shape.

Ingot casting The process of casting ingots. This is different from the continuous casting route.

Ingot structure The macrostructure of a casting, including the chill zone, columnar zone, and equiaxed zone.

Inoculants Materials that provide heterogeneous nucleation sites during the solidification of a material.

Inoculation   The addition of heterogeneous nuclei in a controlled manner to increase the number of grains in a casting.

Interdendritic shrinkage   Small pores between the dendrite arms formed by the shrinkage that accompanies solidification. Also known as microshrinkage or shrinkage porosity.

Investment casting   A casting process that is used for making complex shapes such as turbine blades, also known as the lost wax process.

Lamellar   A plate-like arrangement of crystals within a material.

Latent heat of fusion (ΔH_f)   The heat evolved when a liquid solidifies. The latent heat of fusion is related to the energy difference between the solid and the liquid.

Local solidification time   The time required for a particular location in a casting to solidify once nucleation has begun.

Lost foam process   A process in which a polymer foam is used as a pattern to produce a casting.

Lost wax process   A casting process in which a wax pattern is used to cast a metal.

Microshrinkage   Small, frequently isolated pores between the dendrite arms formed by the shrinkage that accompanies solidification. Also known as microshrinkage or shrinkage porosity.

Mold constant (B)   A characteristic constant in Chvorinov's rule.

Mushy-forming alloys   Alloys whose castings have a macrostructure consisting predominantly of equiaxed grains. They are known as such since the cast material seems like a mush of solid grains floating in a liquid melt.

Nucleation   The physical process by which a new phase is produced in a material. In the case of solidification, this refers to the formation of a tiny, stable solid particles in the liquid.

Nuclei   Tiny particles of solid that form from the liquid as atoms cluster together. Because these particles are large enough to be stable, nucleation has occurred and growth of the solid can begin.

Permanent mold casting   A casting process in which a mold can be used many times.

Phase diagrams   Diagrams that describe what phases will be expected to be stable, from a thermodynamic viewpoint, for any system consisting of one or more elements or compounds.

Photochromic glasses   Glassses that change color or tint upon exposure to sunlight.

Pipe shrinkage   A large conical-shaped void at the surface of a casting caused by the volume contraction that occurs during solidification.

Planar growth   The growth of a smooth solid-liquid interface during solidification, when no undercooling of the liquid is present.

Pouring temperature   The temperature of a metal or an alloy when it is poured into a mold during the casting process.

Preform   A component from which a fiber is drawn or a bottle is made.

Pressure die casting   A casting process in which molten metal/alloys is forced into a die under pressure.

Primary processing   Processes involving casting of molten metals into ingots or semi-finished useful shapes such as slabs.

Quantum dots   Nano-sized clusters of semiconductors showing properties that are different from those of bulk materials (e.g., bandgap).

Rapid solidification processing   Producing unique material structures by promoting unusually high cooling rates during solidification.

Recalescence   The increase in temperature of an undercooled liquid metal as a result of the liberation of heat during nucleation.

Riser   An extra reservoir of liquid metal connected to a casting. If the riser freezes after the casting, the riser can provide liquid metal to compensate for shrinkage.

Sand casting   A casting process using sand molds.

Secondary dendrite arm spacing (SDAS)   The distance between the centers of two adjacent secondary dendrite arms.

Secondary processing   Processes such as rolling, extrusion, etc. used to process ingots or slabs and other semi-finished shapes.

Shrinkage   Contraction of a casting during solidification.

Shrinkage porosity   Small pores between the dendrite arms formed by the shrinkage that accompanies solidification. Also known as microshrinkage or interdendritic porosity.

Sievert's law   The amount of a gas that dissolves in a metal is proportional to the partial pressure of the gas in the surroundings.

Skin forming alloys   Alloys whose microstructure shows an outer skin of small grains in the chill zone followed by dendrites.

Soldering   Soldering is a joining process in which the filler has a melting temperature below 450°C, no melting of the base materials occurs.

Solidification front   Interface between a solid and liquid.

Solidification process   Processing of materials involving solidification (e.g., single crystal growth, continuous casting, etc.).

Solid-state phase transformation   A change in phase that occurs in the solid state.

Specific heat   The heat required to change the temperature of a unit weight of the material one degree.

Spherulite   Spherical-shaped crystals produced when certain polymers solidify.

Stainless steel   A corrosion resistant alloy made from Fe-Cr-Ni-C.

Superheat   The pouring temperature minus the freezing temperature.

Thermal arrest   A plateau on the cooling curve during the solidification of a material caused by the evolution of the latent heat of fusion during solidification. This heat generation balances the heat being lost as a result of cooling.

Total solidification time   The time required for the casting to solidify completely after the casting has been poured.

Undercooling   The temperature to which the liquid metal must cool below the equilibrium freezing temperature before nucleation occurs.

1. KURZ, W., and FISHER, D.J., *Fundamentals of Solidification*, 1998: Trans-Tech.

2. BRONOWSKI, J., *The Ascent of Man*, 1973: Back Bay Books, pp. 124–131.

3. LEE, H.W.H., et al. *Quantum Confined Electron-Hole States in ZnSe Quantum Dots*, in *Mater. Res. Soc. Symp. Proc. Vol. 571—Semiconductor Quantum Dots*, 2000.

4. BEALL, G.H., "Glass-Ceramics", in *Advances in Ceramics Vol. 18—Commercial Glasses*, BOYD, D.C. and MACDOWELL, J.F. Editors. 1986: American Ceramic Society, pp. 157–173.

5. KINGERY, W.D., BOWEN, H.K., and UHLMANN, D.R., *Introduction to Ceramics*, 2nd Ed. 1976: Wiley.

6. BROWN, R., "Man-Made Snow," *Scientific American*, 1992 (January).

7. CHVORINOV, N., *Proc. Inst. British Found*, 1938–1939. 32: p. 229.

8. VERHOEVEN, J.D., *Fundamentals of Physical Metallurgy*, 1975: Wiley.

9. DEGARMO, E.P., BLACK, J.T., and KOSHE, R.A., *Materials and Processes in Manufacturing*, 8th Ed. 1997: Prentice Hall.

10. SINGH, J., *Optoelectronics: An Introduction to Materials and Devices*, 1996.

11. MAHAJAN, S. and SREE HARSHA, K.S., *Principles of Growth and Processing of Semiconductors*, 1999: McGraw Hill.

12. RAO, M., "A Primer on Lead Free Solder," *Chip Scale Review*, 2000 (March–April).

13. DEBROY, T. and DAVID, S.A., "Physical Processes in Fusion Welding," *Reviews of Modern Physics*, 1995. 67(1): pp. 85–112.

14. DAVID, S.A., and DEBROY, T., "Current Issues and Probelms in Welding Science", *Science*, 1992. 257: pp. 497–502.

PROBLEMS

Section 8-1 Technological Significance

8-1 Give examples of materials based on inorganic glasses that are made by solidification.

8-2 What do the terms "primary" and "secondary processing" mean?

8-3 Why are ceramic materials not prepared by melting and casting?

Section 8-2 Nucleation

8-4 Define the following terms: nucleation, embryo, heterogeneous nucleation, and homogeneous nucleation.

8-5 Does water freeze at 0°C and boil at 100°C? Explain.

8-6 Does ice melt at 0°C? Explain.

8-7 Assume that instead of a spherical nucleus, we had a nucleus in the form of a cube of length (x). Calculate the critical dimension x^* of the cube necessary for nucleation. Write down an equation similar to Equation 8-1 for a cubical nucleus, and derive an expression for x^* similar to Equation 8-2.

8-8 Why is it that nuclei seen experimentally are often sphere-like but faceted? Why are they sphere-like and not like cubes or other shapes?

8-9 Explain the meaning of each term in Equation 8-2.

8-10 Suppose that liquid nickel is undercooled until homogeneous nucleation occurs. Calculate

(a) the critical radius of the nucleus required; and
(b) the number of nickel atoms in the nucleus.

Assume that the lattice parameter of the solid FCC nickel is 0.356 nm.

8-11 Suppose that liquid iron is undercooled until homogeneous nucleation occurs. Calculate

(a) the critical radius of the nucleus required; and
(b) the number of iron atoms in the nucleus.

Assume that the lattice parameter of the solid BCC iron is 2.92 Å.

8-12 Suppose that solid nickel was able to nucleate homogeneously with an undercooling of only 22°C. How many atoms would have to group together spontaneously for this to occur? Assume that the lattice parameter of the solid FCC nickel is 0.356 nm.

8-13 Suppose that solid iron was able to nucleate homogeneously with an undercooling of only 15°C. How many atoms would have to group together spontaneously for this to occur? Assume that the lattice parameter of the solid BCC iron is 2.92 Å.

8-14 Calculate the fraction of solidification that occurs dendritically when iron nucleates

(a) at 10°C undercooling;

(b) at 100°C undercooling; and

(c) homogeneously.

The specific heat of iron is 5.78 J/cm$^3 \cdot$ °C.

Section 8-3 Applications of Controlled Nucleation

8-15 Explain the term inoculation.

8-16 Explain how aluminum alloys can be strengthened using small levels of titanium and boron additions.

8-17 Compare and contrast grain size strengthening and strain hardening mechanisms.

8-18 What is second-phase strengthening?

8-19 Why is it that many inorganic melts solidify into amorphous materials more easily compared to those of metallic materials?

8-20 What is a glass-ceramic? How are glass-ceramics made?

8-21 What is a photochromic glass?

8-22 What is a metallic glass?

8-23 How do machines in ski resorts make snow?

Section 8-4 Growth Mechanisms

8-24 What are the two steps encountered in the solidification of molten metals? As a function of time, can they overlap with one another?

8-25 During solidification, specific heat of the material and the latent heat of fusion need to be removed. Define each of these terms.

8-26 Describe under what conditions we expect molten metals to undergo dendritic solidification.

8-27 Describe under what conditions we expect molten metals to undergo planar front solidification.

8-28 Calculate the fraction of solidification that occurs dendritically when silver nucleates

(a) at 10°C undercooling;

(b) at 100°C undercooling; and

(c) homogeneously.

The specific heat of silver is 3.25 J/cm$^3 \cdot$ °C.

8-29 Analysis of a nickel casting suggests that 28% of the solidification process occurred in a dendritic manner. Calculate the temperature at which nucleation occurred. The specific heat of nickel is 4.1 J/cm$^3 \cdot$ °C.

Section 8-5 Solidification Time and Dendrite Size

8-30 Write down Chvorinov's rule and explain the meaning of each term.

8-31 A 2-in. cube solidifies in 4.6 min. Calculate

(a) the mold constant in Chvorinov's rule; and

(b) the solidification time for a 0.5 in. × 0.5 in. × 6 in. bar cast under the same conditions.

Assume that $n = 2$.

8-32 A 5-cm-diameter sphere solidifies in 1050 s. Calculate the solidification time for a 0.3 cm × 10 cm × 20 cm plate cast under the same conditions. Assume that $n = 2$.

8-33 Find the constants B and n in Chvorinov's rule by plotting the following data on a log-log plot:

Casting Dimensions (in.)	Solidification Time (min)
0.5 × 8 × 12	3.48
2 × 3 × 10	15.78
2.5 cube	10.17
1 × 4 × 98.13	

8-34 Find the constants B and n in Chvorinov's rule by plotting the following data on a log-log plot:

Casting Dimensions (cm)	Solidification Time (s)
1 × 1 × 6	28.58
2 × 4 × 4	98.30
4 × 4 × 4	155.89
8 × 6 × 5	306.15

8-35 A 3-in.-diameter casting was produced. The times required for the solid-liquid interface to reach different distances beneath the casting surface were measured and are shown in the following table:

Distance from Surface (in.)	Time (s)
0.1	32.6
0.3	73.5
0.5	130.6
0.75	225.0
1.0	334.9

Determine

(a) the time at which solidification begins at the surface; and

(b) the time at which the entire casting is expected to be solid.

(c) Suppose the center of the casting actually solidified in 720 s. Explain why this time might differ from the time calculated in part (b).

8-36 Figure 8-9(b) shows a photograph of an aluminum alloy. Estimate

(a) the secondary dendrite arm spacing; and

(b) the local solidification time for that area of the casting.

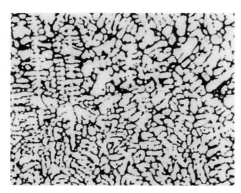

Figure 8-9 (Repeated for Problem 8-36) (b) Dendrites in an aluminum alloy (×50). (*From* ASM Handbook, *Vol. 9, Metallography and Microstructure (1985), ASM International, Materials Park, OH 44073-0002.*)

8-37 Figure 8-31 shows a photograph of FeO dendrites that have precipitated from a ceramic glass (an undercooled liquid). Estimate the secondary dendrite arm spacing.

8-38 Find the constants c and m relating the secondary dendrite arm spacing to the local solidification time by plotting the following data on a log-log plot:

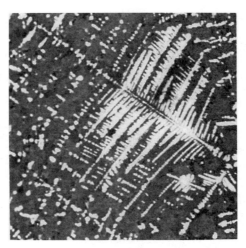

Figure 8-31 Photomicrograph of FeO dendrites in a ceramic glass (×450) (for Problem 8-37). (*Courtesy of C.W. Ramsay, University of Missouri—Rolla.*)

Solidification Time (s)	SDAS (cm)
156	0.0176
282	0.0216
606	0.0282
1356	0.0374

8-39 Figure 8-32 shows dendrites in a titanium powder particle that has been rapidly solidified. Assuming that the size of the titanium dendrites is related to solidification time by the same relationship as in aluminum, estimate the solidification time of the powder particle.

Figure 8-32 Tiny dendrites exposed within a titanium powder particle produced by rapid solidification processing (×2200) (for Problem 8-39). (*From J.D. Ayers and K. Moore, "Formation of Metal Carbide Powder by Spark Machining of Reactive Metals," in Metallurgical Transactions, Vol. 15A, June 1984, p. 1120.*)

8-40 The secondary dendrite arm spacing in an electron-beam weld of copper is 9.5×10^{-4} cm. Estimate the solidification time of the weld.

Section 8-6 Cooling Curves

8-41 Sketch a cooling curve for a pure metal and label different regions carefully.

8-42 What is meant by the term recoalescence?

8-43 What is thermal arrest?

8-44 What is the difference between local and total solidification times?

8-45 A cooling curve is shown in Figure 8-33. Determine

- (a) the pouring temperature;
- (b) the solidification temperature;
- (c) the superheat;
- (d) the cooling rate, just before solidification begins;
- (e) the total solidification time;
- (f) the local solidification time; and
- (g) the probable identity of the metal.
- (h) If the cooling curve was obtained at the center of the casting sketched in the figure, determine the mold constant, assuming that $n = 2$.

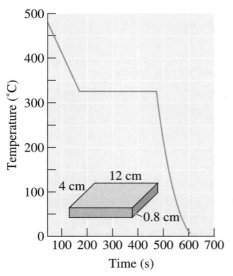

Figure 8-33 Cooling curve (for Problem 8-45).

8-46 A cooling curve is shown in Figure 8-34. Determine

- (a) the pouring temperature;
- (b) the solidification temperature;

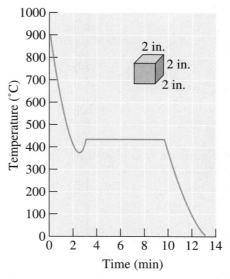

Figure 8-34 Cooling curve (for Problem 8-46).

- (c) the superheat;
- (d) the cooling rate, just before solidification begins;
- (e) the total solidification time;
- (f) the local solidification time;
- (g) the undercooling; and
- (h) the probable identity of the metal.
- (i) If the cooling curve was obtained at the center of the casting sketched in the figure, determine the mold constant, assuming that $n = 2$.

8-47 Figure 8-35 shows the cooling curves obtained from several locations within a cylindrical aluminum casting. Determine the local solidification times and the SDAS at each location, then plot the tensile strength versus distance from the casting surface. Would you recommend that the casting be designed so that a large or small amount of material must be machined from the surface during finishing? Explain.

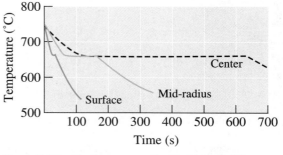

Figure 8-35 Cooling curves (for Problem 8-47).

8-48 Calculate the volume, diameter, and height of the cylindrical riser required to prevent shrinkage in a 4 in. × 10 in. × 20 in. casting if the *H/D* of the riser is 1.5.

Section 8-7 Cast Structure

8-49 What are the features expected in the macro-structure of a cast component? Explain using a sketch.

8-50 In cast materials why does solidification almost always begin at the mold walls?

8-51 Why is it that forged components do not show cast ingot structure?

Section 8-8 Solidification Defects

8-52 What type of defect in a casting can cause catastrophic failure of cast components such as turbine blades? What precautions are taken to prevent porosity in castings?

8-53 In general, compared to components prepared using forging, rolling, extrusion, etc., cast products tend to have lower fracture toughness. Explain why this may be the case.

8-54 What is a riser? Why should it freeze after the casting?

8-55 Calculate the volume, diameter, and height of the cylindrical riser required to prevent shrinkage in a 1 in. × 6 in. × 6 in. casting if the *H/D* of the riser is 1.0.

8-56 Figure 8-36 shows a cylindrical riser attached to a casting. Compare the solidification times for each casting section and the riser and determine whether the riser will be effective.

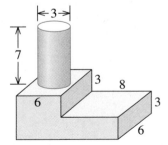

Figure 8-36   Step-block casting (for Problem 8-56).

8-57 Figure 8-37 shows a cylindrical riser attached to a casting. Compare the solidification times for each casting section and the riser and determine whether the riser will be effective.

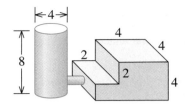

Figure 8-37   Step-block casting (for Problem 8-57).

8-58 A 4-in.-diameter sphere of liquid copper is allowed to solidify, producing a spherical shrinkage cavity in the center of the casting. Compare the volume and diameter of the shrinkage cavity in the copper casting to that obtained when a 4-in. sphere of liquid iron is allowed to solidify.

8-59 A 4-in. cube of a liquid metal is allowed to solidify. A spherical shrinkage cavity with a diameter of 1.49 in. is observed in the solid casting. Determine the percent volume change that occurs during solidification.

8-60 A 2 cm × 4 cm × 6 cm magnesium casting is produced. After cooling to room temperature, the casting is found to weigh 80 g. Determine

(a) the volume of the shrinkage cavity at the center of the casting; and

(b) the percent shrinkage that must have occurred during solidification.

8-61 A 2 in. × 8 in. × 10 in. iron casting is produced and, after cooling to room temperature, is found to weigh 43.9 lb. Determine

(a) the percent of shrinkage that must have occurred during solidification; and

(b) the number of shrinkage pores in the casting if all of the shrinkage occurs as pores with a diameter of 0.05 in.

8-62 Give examples of materials that expand upon solidification.

8-63 How can gas porosity in molten alloys be removed or minimized?

8-64 In the context of stainless steel making, what is argon oxygen decarburization?

8-65 Liquid magnesium is poured into a 2 cm × 2 cm × 24 cm mold and, as a result of directional solidification, all of the solidification shrinkage occurs along the length of the casting. Determine the length of the casting immediately after solidification is completed.

8-66 A liquid cast iron has a density of 7.65 g/cm^3. Immediately after solidification, the density of the solid cast iron is found to be 7.71 g/cm^3. Determine the percent volume change that occurs during solidification. Does the cast iron expand or contract during solidification?

8-67 From Figure 8-18, find the solubility of hydrogen in liquid aluminum just before solidification begins when the partial pressure of hydrogen is 1 atm. Determine the solubility of hydrogen (in $cm^3/100$ g Al) at the same temperature if the partial pressure were reduced to 0.01 atm.

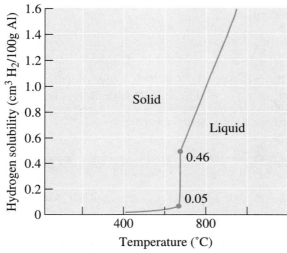

Figure 8-18 (Repeated for Problem 8-67) The solubility of hydrogen gas in aluminum when the partial pressure of $H_2 = 1$ atm.

8-68 The solubility of hydrogen in liquid aluminum at 715°C is found to be 1 $cm^3/100$ g Al. If all of this hydrogen precipitated as gas bubbles during solidification and remained trapped in the casting, calculate the volume percent gas in the solid aluminum.

Section 8-9 Casting Processes for Manufacturing Components

8-69 Explain the green sand molding process.

8-70 Why is it that castings made from pressure die casting are likely to be stronger than those made using the sand casting process?

8-71 An alloy is cast into a shape using a sand mold and a metallic mold. Which casting will be expected to be stronger and why?

8-72 What is investment casting? What are the advantages of investment casting? Explain why this process is often used to cast turbine blades.

8-73 Why is pressure a key ingredient in the pressure die casting process?

Section 8-10 Continuous Casting and Ingot Casting

8-74 What is an ore?

8-75 Explain briefly how steel is made, starting with iron ore, coke, and limestone.

8-76 Explain how scrap is used for making alloys.

8-77 What is an ingot?

8-78 Why has continuous casting of steels and other alloys assumed increased importance?

8-79 What are some of the steps that follow the continuous casting process?

Section 8-11 Directional Solidification (DS), Single Crystal Growth, and Epitaxial Growth

8-80 Define the term directional solidification.

8-81 Explain the role of nucleation and growth steps in growing single crystals.

8-82 Why do we use single crystal silicon wafers for the fabrication of microelectronic devices?

8-83 What does the term epitaxy mean?

8-84 Is epitaxial growth always from liquid? Explain.

8-85 What do the terms heteroepitaxy and homo-epitaxy mean?

Section 8-12 Solidification of Polymers and Inorganic Glasses

8-86 Why do most plastics contain amorphous and crystalline regions?

8-87 What is a spherulite?

8-88 How can processing influence crystallinity of polymers?

8-89 Explain why silicate glasses tend to form amorphous glasses, however, metallic melts typically crystallize easily.

Section 8-13 Joining of Metallic Materials

8-90 Define the terms brazing and soldering.

8-91 What is the difference between fusion welding and brazing and soldering?

8-92 What is a heat affected by zone?

8-93 Explain why, while using low intensity heat sources, the strength of the material in a weld region can be reduced.

8-94 Why do laser and electron-beam welding processes lead to stronger welds?

✳ Design Problems

8-95 Aluminum is melted under conditions that give 0.06 cm^3 H$_2$ per 100 g of aluminum. We have found that we must have no more than 0.002 cm^3 H$_2$ per 100 g of aluminum in order to prevent the formation of hydrogen gas bubbles during solidification. Design a treatment process for the liquid aluminum that will assure that hydrogen porosity does not form.

8-96 When two 0.5-in.-thick copper plates are joined using an arc-welding process, the fusion zone contains dendrites having a SDAS of 0.006 cm. However, this process produces large residual stresses in the weld. We have found that residual stresses are low when the welding conditions produce a SDAS of more than 0.02 cm. Design a process by which we can accomplish low residual stresses. Justify your design.

8-97 Design an efficient risering system for the casting shown in Figure 8-38. Be sure to include a sketch of the system, along with appropriate dimensions.

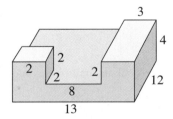

Figure 8-38   Casting to be risered (for Problem 8-97).

8-98 Design a process that will produce a steel casting having uniform properties and high strength. Be sure to include the microstructure features you wish to control and explain how you would do so.

8-99 An aluminum casting is to be injected into a steel mold under pressure (die casting). The casting is essentially a 12-in.-long, 2-in.-diameter cylinder with a uniform wall thickness, and it must have a minimum tensile strength of 40,000 psi. Based on the properties given in Figure 8-11, design the casting and process.

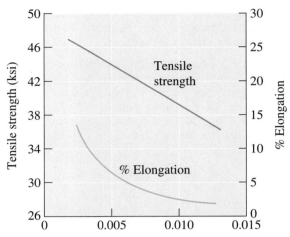

Figure 8-11   (Repeated for Problem 8-99) The effect of the secondary dendrite arm spacing on the properties of an aluminum casting alloy.

▢ Computer Problems

8-100 *Critical Radius for Homogeneous Nucleation.* Write a computer program that will allow calculation of the critical radius of nucleation (r^*). The program should ask the user to provide inputs for values for σ_{sl}, T_m, undercooling (ΔT), and enthalpy of fusion ΔH_f. Please be sure to have the correct prompts in the program to have the values entered in correct units.

8-101 *Free Energy for Formation of Nucleus of Critical Size via Heterogeneous Nucleation.* When nucleation occurs heterogeneously, the free energy for nucleus of a critical size ($\Delta G^{*\text{hetero}}$) is given by

$$\Delta G^{*\text{hetero}} = \Delta G^{*\text{homo}} f(\theta), \text{ where}$$

$$f(\theta) = \left(\frac{(2 + \cos\theta)(1 - \cos\theta)^2}{4} \right)$$

where ($\Delta G^{*\text{homo}}$) is given by $\dfrac{16\pi\sqrt{\gamma_{sl}^3}}{3\Delta G_v^2}$, which is the free energy for homogeneous nucleation of a nucleus of a critical size. If the contact angle (θ) of the phase that is nucleating on the pre-existing surface is 180°, there is no wetting, and the value of the function $f(\theta)$ is 1, and the free energy of forming a nucleus of a critical radius is same as that for homogeneous nucleation.[1] If the nucleating phase wets the solid completely (i.e., $\theta = 0$), then $f(\theta) = 0$ and there is no barrier for nucleation. Write a computer program that will ask the user to provide the values of parameters needed to calculate the free energy for formation of a nucleus via homogeneous nucleation. The program should then calculate the value of $\Delta G^{*hetero}$ as a function of the contact angle (θ) ranging from 0 to 180°. Examine the variation of the free energy values as a function of contact angle.

8-102 *Chvorinov's Rule.* Write a computer program that will calculate the time of solidification for a casting. The program should ask the user to enter the volume of the casting and surface area from which heat transfer will occur and the mold constant. The program should then use Chvorinov's rule to calculate the time of solidification.

Chapter

Josiah Willard Gibbs (1839–1903) was a brilliant American physicist and mathematician who conducted some of the most important pioneering work related to thermodynamic equilibrium. (*Courtesy of the University of Pennsylvania Library.*)

9

Solid Solutions and Phase Equilibrium

Have You Ever Wondered?

- Is it possible for the solid, liquid, and gaseous forms of a material to coexist?

- How ice skaters move so easily on ice?

- How is freeze-dried coffee made?

- What material is used to make red light-emitting diodes used in many modern product displays?

- When an alloy such as brass solidifies, which element solidifies first—copper or zinc?

We have seen that the strength of metallic materials can be enhanced using:

(a) grain-size strengthening (Hall-Petch equation);

(b) cold working or strain hardening;

(c) additions of small levels of elements; and

(d) formation of small particles of second phases.

These mechanisms cause the yield strength of metallic materials to increase because more obstacles for dislocation motion are created. In this chapter, we begin to explore the formation of how a solid solution occurs in metallic and ceramic systems. A solid solution is a solid material in which atoms or ions of elements constituting it are dispersed uniformly. The mechanical and other properties of materials can be controlled by creating point defects, such as substitutional and interstitial atoms. In metallic materials, the point defects disturb the atomic arrangement in the crystalline material and interfere with the movement of dislocations or slip. The point defects, therefore, cause the material to be solid-solution strengthened.

The introduction of alloying elements or impurities during processing changes the composition of the material and influences its solidification behavior. In this chapter, we will examine this effect by introducing the concept of an equilibrium phase diagram. For now, we consider a "phase" as a unique form in which a material exists. We will define the term "phase" more precisely later in this chapter. A phase diagram depicts the stability of different phases for a set of elements (e.g., Al and Si). From the phase diagram, we can predict how a material will solidify under equilibrium conditions. We can also predict what phases will be expected to be thermodynamically stable and in what concentrations such phases should be present.

Therefore, the major objectives of this chapter are to explore:

1. the formation of solid solutions;

2. the effects of solid-solution formation on the mechanical properties on metallic materials;

3. the conditions under which solid solutions can form;

4. the development of some basic ideas concerning phase diagrams; and

5. the solidification process in simple alloys.

We will explore how solidification of alloys occurs under equilibrium and nonequilibrium conditions. We will also see that the concept of solid-solution formation is not limited to metallic materials. We will examine how solid solutions can be formed in ceramic systems. Since the strength of ceramics depends more on flaw size distribution (and not on dislocation motion), the effect of solid-solution formation on the mechanical properties of ceramics is weak. However, ceramic solid solutions allow formation of exotic compositions whose dielectric or magnetic properties of ceramic materials can be "tuned" precisely. Similarly, we will examine how solid-solution formation is used to engineer the bandgap of semiconductors, such as those from the gallium arsenide (GaAs)-aluminum arsenide (AlAs) system. Formation of solid solutions in these types of materials leads to light emitting diodes (LEDs) of different colors. We will also point out how similar ideas are used in designing copolymers. With copolymers, we can blend properties of two or more different polymers into one material.

9-1 Phases and the Phase Diagram

Pure metallic elements have engineering applications; for example, ultra-high purity copper (Cu) or aluminum (Al) is used to make microelectronic circuitry. However, in most applications we use **alloys**. We define an "alloy" as a material that exhibits properties of a metallic material and is made from multiple elements. A **plain carbon steel** is an alloy of iron (Fe) and carbon (C). Corrosion-resistant **stainless steels** are alloys that usually contain iron (Fe), carbon (C), chromium (Cr), nickel (Ni), and some other elements. Similarly, there are alloys based on aluminum (Al), copper (Cu), cobalt (Co), nickel (Ni), titanium (Ti), zinc (Zn), and zirconium (Zr). There are two types of alloys: **single-phase alloys** and **multiple-phase alloys**. In this chapter, we will examine the behavior of single-phase alloys. As a first step, let's define a "phase" and determine how the **phase rule** helps us to determine the state—solid, liquid, or gas—in which a pure material exists.

A **phase** can be defined as any portion, including the whole, of a system which is physically homogeneous within itself and bounded by a surface so that it is mechanically separable from any other portions.[1] For example, water has three phases—liquid water, solid ice, and steam. A phase has the following characteristics:

1. the same structure or atomic arrangement throughout;
2. roughly the same composition and properties throughout; and
3. a definite interface between the phase and any surrounding or adjoining phases.

For example, if we enclose a block of ice in a vacuum chamber [(Figure 9-1(a)], the ice begins to melt and some of the water vaporizes. Under these conditions, we have three phases coexisting: solid H_2O, liquid H_2O, and gaseous H_2O. Each of these forms of H_2O is a distinct phase; each has a unique atomic arrangement, unique properties, and a definite boundary between each form. Although, in this case the phases have identical compositions.

Figure 9-1 Illustration of phases and solubility: (a) The three forms of water—gas, liquid, and solid—are each a phase. (b) Water and alcohol have unlimited solubility. (c) Salt and water have limited solubility. (d) Oil and water have virtually no solubility.

Phase Rule Josiah Willard Gibbs (1839–1903) was a brilliant American physicist and mathematician who conducted some of the most important pioneering work related to thermodynamic equilibrium.[2] His contributions originally were not well known in other parts of the world since they were published in a relatively obscure American journal. It was only after Ostwald (in 1892) and le Châelier (in 1899), translated his work in German and French, respectively, that Gibbs' work received considerable notoriety and fame. Also, some of the mathematical derivations in his original work, including that of the phase rule, were complex; most of the physical chemists at that time did not readily appreciate the significance of his work.

Gibbs developed the **phase rule** in 1875–1876. It describes the relationship between the number of components and the number of phases for a given system and the conditions that may be allowed to change (e.g., temperature, pressure, etc.). It has the general form:

$$2 + C = F + P \quad \text{(when temperature and pressure both can vary)} \tag{9-1}$$

A useful mnemonic (something that will allow you to remember) for the Gibbs phase rule is to start with a numeric and follow with the rest of the terms alphabetically (i.e., C, F, and P) using all positive signs. In the phase rule, C is the number of chemically independent components, usually elements or compounds, in the system; F is the number of degrees of freedom, or the number of variables (such as temperature, pressure, or composition), that are allowed to change independently without changing the number of phases in equilibrium; and P is the number of phases present (please do not confuse P with "pressure.") The constant "2" in Equation 9-1 implies that both the temperature and pressure are allowed to change. The term "chemically independent" refers to the number of different elements or compounds needed to specify a system. For example, water (H_2O) is considered as a one component system, since the concentrations of H and O in H_2O cannot be independently varied.

It is important to note that the Gibbs phase rule assumes thermodynamic equilibrium and, more often than not in materials processing, we encounter conditions in which equilibrium is *not* maintained. Therefore, you should not be surprised to see that the number and compositions of phases seen in practice are dramatically different from those predicted by the Gibbs phase rule.

Another point that needs to be noted is that phases do not always have to be solid, liquid, and gaseous forms of a material. An element, such as iron (Fe), can exist in FCC and BCC crystal structures. These two solid forms of iron are two different phases of iron that will be stable at different temperatures and pressure conditions. Similarly, ice, itself, can exist in several crystal structures. Therefore, when we refer to ice, strictly speaking, we must define a particular solid phase of ice. Carbon can exist in many forms (e.g., graphite or diamond). These are only two of the many possible phases of carbon.

As an example of the use of the phase rule, let's consider the case of pure magnesium (Mg). Figure 9-2 shows a **unary** ($C = 1$) phase diagram in which the lines divide the liquid, solid, and vapor phases. This unary phase diagram is also called a pressure-temperature or **P-T diagram**. In the unary phase diagram, there is only one component; in this case, magnesium (Mg). Depending on the temperature and pressure, however, there may be one, two, or even three *phases* present at any one time: solid magnesium, liquid magnesium, and magnesium vapor. Note that at atmospheric pressure (one atmosphere, given by the dashed line), the intersection of the lines in the phase diagram give the usual melting and boiling temperatures for magnesium. At very low pressures, a solid such as magnesium (Mg) can *sublime*, or go directly to a vapor form without melting when it is heated.

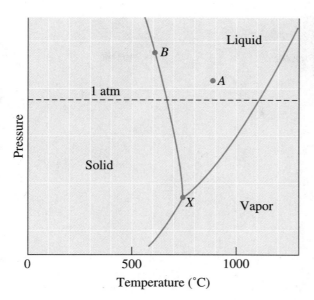

Figure 9-2
Schematic unary phase diagram for magnesium, showing the melting and boiling temperatures at one atmosphere pressure.

Suppose we have a pressure and temperature that put us at point A in the phase diagram (Figure 9-2). At this point, magnesium is all liquid. The number of phases is one (liquid). The phase rule tells us that there are two degrees of freedom. From Equation 9-1:

$$2 + C = F + P, \quad \text{therefore,} \quad 2 + 1 = F + 1 \text{ (i.e., } F = 2)$$

What does this mean? Within limits, as seen in the diagram in Figure 9-2, we can change the pressure, the temperature, or both, and still be in an all-liquid portion of the diagram. Put another way, we must fix both the temperature and the pressure to know precisely where we are in the liquid portion of the diagram.

Consider point B, the boundary between the solid and liquid portions of the diagram. The number of components, C, is still one, but at point B, the solid and liquid coexist, or the number of phases P is two. From the phase rule Equation 9-1,

$$2 + C = F + P, \quad \text{therefore,} \quad 2 + 1 = F + 2 \text{ (i.e., } F = 1)$$

or there is only one degree of freedom. For example, if we change the temperature, the pressure must also be adjusted if we are to stay on the boundary where the liquid and solid coexist. On the other hand, if we fix the pressure, the phase diagram tells us the temperature that we must have if solid and liquid are to coexist.

Finally, at point X, solid, liquid, and vapor coexist. While the number of components is still one, there are three phases. The number of degrees of freedom is

$$2 + C = F + P, \quad \text{therefore,} \quad 2 + 1 = F + 3 \text{ (i.e., } F = 0)$$

Now we have no degrees of freedom; all three phases coexist only if both the temperature and the pressure are fixed. A point on the phase diagram at which the solid, liquid, and gaseous phases coexist under equilibrium conditions is the **triple point**. In the following two examples, we see how some of these ideas underlying the Gibbs phase rule can be applied.

EXAMPLE 9-1   *Design of an Aerospace Component*

Because magnesium (Mg) is a low-density material ($\rho_{Mg} = 1.738$ g/cm^3), it has been suggested for use in an aerospace vehicle intended to enter the outer space environment. Is this a good design?

SOLUTION

In space the pressure is very low. Even at relatively low temperatures, solid magnesium can begin to change to a vapor, causing metal loss that could damage a space vehicle. In addition, solar radiation could cause the vehicle to heat, increasing the rate of magnesium loss.

A low-density material with a higher boiling point (and, therefore, lower vapor pressure at any given temperature) might be a better choice. At atmospheric pressure, aluminum boils at 2494°C and beryllium (Be) boils at 2770°C, compared with the boiling temperature of 1107°C for magnesium. Although aluminum and beryllium are somewhat denser than magnesium, either might be a better design. Given the toxic effects of Be and many of its compounds when in powder form, we may want to consider aluminum first.

Other factors to consider: In load-bearing applications, we should not only look for density but also for relative strength. Therefore, the ratio of Young's modulus to density or yield strength to density could be a better parameter to compare different materials. In this comparison, we will have to be aware that yield strength, for example, depends strongly on microstructure and Young's modulus can depend upon crystallographic directions since the strength of aluminum can be enhanced using aluminum alloys, while keeping the density about the same. This is also true for some carbon-fiber-based composites.

These materials may make a good choice. Other factors such as oxidation during reentry into Earth's atmosphere may be applicable and will also have to be considered.

EXAMPLE 9-2   *Freeze Drying Synthesis of Ceramic Superconductors*

Many ceramic materials are made into powders using different oxides and carbonates (Chapter 14). This is because ceramics melt at too high a temperature and tend to exhibit brittle behavior. For example, the synthesis process for $YBa_2Cu_3O_{7-x}$, a ceramic superconductor known as *YBCO*, involves mixing and reacting powders of yttrium oxide (Y_2O_3), copper oxide (CuO), and barium carbonate ($BaCO_3$). The barium carbonate decomposes to BaO during the high temperature reactions and reacts with yttria and copper oxide to form different phases. Often, this process, known as the "oxide mix" technique, produces ceramic powders that are relatively coarse. Some other undesired phases may also form, and deleterious impurities (from processing or raw materials) may become incorporated in the product. On the other hand, chemical techniques that make use of high-purity chemicals may offer a better product (most likely at a higher cost). One such chemical process, used in the food industry (e.g., coffee, spices, etc.) is **freeze drying**.[3] For example, freeze-dried coffee is made by extracting coffee from beans and then evaporating the water from the coffee using freeze drying.

An engineer decides to use water-based solutions to develop a freeze-drying synthesis of a *YBCO* superconductor. Using the unary phase diagram for water [Figure 9-3(a)], explain the different steps that could be followed to generate high-purity and chemically homogeneous powders of the *YBCO* superconductor.

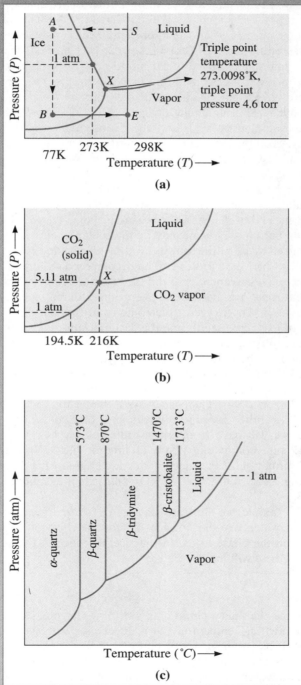

Figure 9-3
(a) Pressure-temperature diagram for H_2O. The triple point temperature is 273.0098 K and the triple point pressure is 4.6 torr. Notice the solid-liquid line sloping to the left. At normal pressure (1 atm or 760 torr), the melting temperature is 273 K. A possible scheme for freeze drying is shown as starting with point *S* and following the dashed line to the left.
(b) Pressure-temperature diagram for CO_2. Many researchers are examining the applications of super-critical CO_2 for use as a solvent for applications related to the processing of plastics and pharmaceuticals. (c) Pressure-temperature diagram for SiO_2. The dotted line shows the 1 atm pressure.

SOLUTION

Since water is a medium from which the ceramic superconductor powder is to be made, we can begin with the chemicals—known as precursors—of yttrium, copper, and barium. Typically, nitrates of many metals are water soluble and work well.

Thus, we can prepare a solution of nitrates of yttrium, copper, and barium in proper cation stoichiometry. We may have to adjust the pH of the solution so that all the nitrates go into solution. This step is important since this is where we make sure that the final ceramic powder is chemically homogeneous. Once we have a solution of mixed metal nitrates at room temperature (298 K), we can begin to lower the temperature or begin the freezing process. Follow the dashed arrows in Figure 9-3(a) starting at point S. We can lower the temperature to about 77 K using liquid nitrogen (point A). We will then be in a region where the water, nitric acid, and the dissolved nitrates are in a solid state (labeled as "ice" on the diagram). The trick now is to remove the nitric acid (HNO_3) and H_2O without causing any melting. We can do this by lowering the pressure to approximately 10^{-2} torr (point B). We then increase the temperature while maintaining the low pressure causing the ice and nitric acid to sublimate (solid $\rightarrow$ vapor). This will leave behind a dry powder of mixed metal nitrates at the end of the freeze-drying cycle (point E). In this powder, yttrium, copper, and barium will be mixed at a molecular level.

The mixed metal nitrate powder will then be heated carefully (not shown in Figure 9-3(a)) and the nitrates can be decomposed to form a ceramic powder. We will have to experiment with this process to establish what temperature and how much time are needed to form the desired superconducting phase ($YBa_2Cu_3O_{7-x}$). The oxygen stoichiometry in the material will be dependent upon the atmosphere used during the decomposition of the nitrates. If we conduct the decomposition of the nitrates in air, the oxygen stoichiometry will be that dictated by the partial pressure of oxygen in air. A temperature of about 700 to 900°C, and a total time of about 2 hours at the selected temperature should be adequate. In this process, solid-state reactions between the finely mixed precursors occur. There is always a chance that other phases may occur; oxides that are either not superconducting or other phases of oxides may form. Therefore, we will have to optimize the necessary heat treatment so that we obtain the superconducting phase of interest (i.e., $YBa_2Cu_3O_{7-x}$). We can verify the formation of the superconducting phase of the desired stoichiometry phase using x-ray diffraction (XRD) and other techniques as described in Chapter 3.

In this example, we illustrated the use of freeze drying using a unary phase diagram for water. We did not account for the effect of nitric acid and other precursors on the phase fields. In practice, a refined P-T diagram that reflects inclusion of these will have to be used.

Let's examine the unary phase diagram for water more closely. In Figure 9-3(a), notice how the melting temperature of ice decreases with increasing pressure. This is rather unusual. It has been suggested that one of the reasons why skaters can skate on ice is that the pressure from their skates actually melts the ice, thus maintaining a layer of water. This, however, has been shown *not* to be an important factor. Instead, a phenomenon known as "*surface melting*" of ice has been shown to be the main factor in providing a lubricating film of water allowing the skater to move easily across the ice.

Figure 9-3(b) shows a P-T diagram for carbon dioxide (CO_2). In recent years, there has been considerable interest in the use of supercritical carbon dioxide as a solvent for the processing of some materials, particularly plastics.

Figure 9-3(c) shows a P-T phase diagram for silica (SiO_2).[1] Different polymorphs of silica can be seen from this diagram. In Chapter 3, we saw that polymorphs are different crystal structures of the same compound. In this diagram, the stable form of silica at room temperature is α-quartz. The α-quartz transforms to β-quartz at 573°C. Upon cooling, the reverse transformation occurs. Upon heating above 870°C, the β-quartz phase is expected to transform into a β-tridymite. This change is sluggish and β-quartz may continue to exist in the form of what is termed a **metastable phase**. The metastable β-quartz can then form into metastable β-cristobalite or molten silica.

A metastable phase is a phase that is in the state of metastable equilibrium and this equilibrium can be approached usually through only one procedure.[1] Undercooled liquid metals or water represent examples of metastable phases. For example, we can obtain undercooled liquid water, a metastable phase of water, by cooling water from room temperature to a temperature below its normal freezing temperature of ∼0°C. (Chapter 8). However, we cannot heat solid ice and form metastable or undercooled water. Upon cooling from high temperatures, molten silica will form glass. Notice that the glass does not appear as a phase on the phase diagram for silica [Figure 9-3(c)] since a phase diagram deals only with equilibrium phases. When scientifically or technologically important metastable phases exist, they are sometimes shown on phase diagrams using dashed lines.

9-2   Solubility and Solid Solutions

Often, it is beneficial to know how much of each material or component we can combine without producing an additional phase. When we begin to combine different components or materials, as when we add alloying elements to a metal, solid or liquid solutions can form. For example, when we add sugar to water, we form a sugar solution. When we diffuse a small number of phosphorus (P) atoms into single crystal silicon (Si), we produce a solid solution of P in Si (Chapter 5). In other words, we are interested in the **solubility** of one material into another (e.g., sugar in water, copper in nickel, phosphorus in silicon, etc.).

Unlimited Solubility   Suppose we begin with a glass of water and a glass of alcohol. The water is one phase, and the alcohol is a second phase. If we pour the water into the alcohol and stir, only one phase is produced [Figure 9-1(b)]. The glass contains a solution of water and alcohol that has unique properties and composition. Water and alcohol are soluble into each other. Furthermore, they display **unlimited solubility**. Regardless of the ratio of water and alcohol, only one phase is produced when they are mixed together.

Similarly, if we were to mix any amounts of liquid copper and liquid nickel, only one liquid phase would be produced. This liquid alloy has the same composition and properties everywhere [Figure 9-4(a)], because nickel and copper have unlimited liquid solubility.

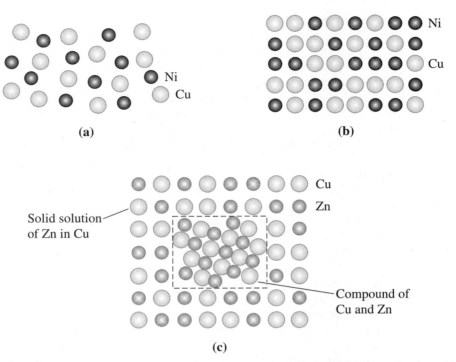

Figure 9-4 (a) Liquid copper and liquid nickel are completely soluble in each other. (b) Solid copper-nickel alloys display complete solid solubility, with copper and nickel atoms occupying random lattice sites. (c) In copper-zinc alloys containing more than 30% Zn, a second phase forms because of the limited solubility of zinc in copper.

If the liquid copper-nickel alloy solidifies and cools to room temperature while maintaining thermal equilibrium, only one solid phase is produced. After solidification, the copper and nickel atoms do not separate but, instead, are randomly located within the FCC crystal structure. Within the solid phase, the structure, properties, and composition are uniform and no interface exists between the copper and nickel atoms. Therefore, copper and nickel also have unlimited solid solubility. The solid phase is a solid solution of copper and nickel [Figure 9-4(b)].

A solid solution is *not* a mixture. A mixture contains more than one type of phase whose characteristics are retained when the mixture is formed. In contrast to this, the components of a solid solution completely dissolve in one another and do not retain their individual characteristics.

Another example of a system forming a solid solution is that of barium titanate ($BaTiO_3$) and strontium titanate ($SrTiO_3$), which are compounds found in the BaO-TiO_2-SrO ternary system. Let's consider phase relations between $BaTiO_3$ and $SrTiO_3$. If we take $BaTiO_3$ powder and begin to add $SrTiO_3$ powder to it, effectively nothing happens at room temperature since the diffusion of cations is too slow. The powders simply form a mechanical mixture. It will be hard, but not impossible, to separate out the powders again. However, if we heat this mixture of powders to a high temperature (e.g., 1000°C) and anneal it for approximately 24 hours, then a solid solution of $BaTiO_3$ and $SrTiO_3$ will form. We designate this solid solution as $(Ba_xSr_{1-x})TiO_3$, where x will be the mole fraction of $BaTiO_3$. In this new material, we will not be able to pick out a particle of $BaTiO_3$ or a particle of $SrTiO_3$. Each $(Ba_xSr_{1-x})TiO_3$ powder particle will have Ba^{+2}, Sr^{+2}, Ti^{+4} and O^{-2} ions. If the processing is handled correctly, making sure the powders are dispersed homogenously and reacted fully, each particle will have a uniform concentration of these ions (i.e., different particles selected randomly will exhibit identical chemical composition). The actual ratio of Ba^{+2} to Sr^{+2}, of course, will depend upon how much $BaTiO_3$ and $SrTiO_3$ we started with—the mole fraction of $BaTiO_3$. The solid solution of $(Ba_xSr_{1-x})TiO_3$ will have its own electrical and other properties. We could produce different solid solutions with different properties by changing the mole fraction of $BaTiO_3$ x from 0 to 1. We use solid solutions of $BaTiO_3$ with $SrTiO_3$ and other oxides to make such electronic components as capacitors. Millions of multi-layer capacitors are made each year using such materials (Chapter 18). Here the advantage of forming solid solutions is that we can "tune" the electrical properties of these materials as well as the temperature stability of the manufactured components. Ferrites, used in many similar applications, are magnetic materials that have spinel crystal structures (Chapter 3). We can also "tune" the magnetic properties of many ferrites using this solid solution approach (Chapter 19).

Many compound semiconductors that share the same crystal structures readily form solid solutions with 100% solubility. For example, we can form solid solutions of gallium arsenide (GaAs) and aluminum arsenide (AlAs). The solid solution is designated as $Ga_xAl_{1-x}As$, where x is the mole fraction of Ga, and $1 - x$ will be the mole fraction of As. As x changes from 0 to 1, the electrical properties (such as the bandgap) of semiconductors change. Thus, the formation of solid solutions is key to engineering the bandgap of different semiconductors. The bandgap is related to the color of light emitted by a diode made by using a direct bandgap semiconductor. We will examine direct and indirect bandgap semiconductors in Chapter 18. The solid-solution formation in semiconductors such as GaAs-AlAs, gallium phosphide (GaP)-indium phosphide (InP), etc. allows us to tune bandgap, thereby making it possible to make light emitting diodes of different colors, including infrared LEDs, and therefore invisible to our eye! Infrared LEDs are used in optical fiber communication systems. The most commonly used red LEDs used in displays are made using solid solutions based on the GaAs-GaP system. Solid solutions can be formed using more than two compounds or elements.

Limited Solubility   When we add a small quantity of salt (one phase) to a glass of water (a second phase) and stir, the salt dissolves completely in the water. Only one phase—salty water or brine—is found. However, if we add too much salt to the water, the excess salt sinks to the bottom of the glass [Figure 9-1(c)]. Now we have two phases—water that is saturated with salt plus excess solid salt. We find that salt has a **limited solubility** in water.

If we add a small amount of liquid zinc to liquid copper, a single liquid solution is produced. When that copper-zinc solution cools and solidifies, a single solid solution having an FCC structure results, with copper and zinc atoms randomly located at the normal lattice points. However, if the liquid solution contains more than about 30% Zn, some of the excess zinc atoms combine with some of the copper atoms to form a CuZn compound [Figure 9-4(c)]. Two solid phases now coexist: a solid solution of copper saturated with about 30% Zn plus a CuZn compound. The solubility of zinc in copper is limited. Figure 9-5 shows a portion of the Cu-Zn phase diagram illustrating the solubility of zinc in copper at low temperatures. The solubility increases with increasing temperature. This is similar to how we can dissolve more sugar or salt in water by increasing the temperature.

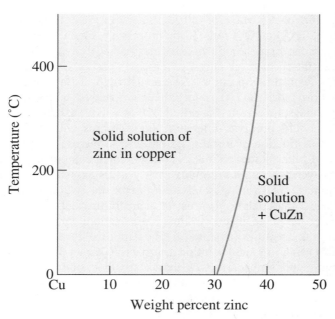

Figure 9-5
The solubility of zinc in copper. The solid line represents the solubility limit; when excess zinc is added, the solubility limit is exceeded and two phases coexist.

In Chapter 5, we examined how silicon (Si) doped with phosphorous (P), boron (B), arsenic (As), etc. can be prepared. All of these dopant elements exhibit limited solubility in Si (i.e., at small concentrations they form a solid solution with Si). Thus, solid solutions are produced even if there is limited solubility. We do not need 100% solid solubility to form solid solutions. Note that solid solutions may form either by substitutional or interstitial mechanisms. The guest atoms or ions may enter the host crystal structure at regular crystallographic positions or into the interstices.

In the extreme case, there may be almost no solubility of one material in another. This is true for oil and water [Figure 9-1(d)] or for copper-lead (Cu-Pb) alloys. Note that even though materials do not dissolve into one another they can be dispersed into one another. For example, oil-like phases and aqueous liquids can be mixed, often using surfactants (soap-like molecules), to form emulsions. Immiscibility, or lack of solubility, is seen in many molten, solid ceramic and metallic materials.

Polymeric Systems   We can process polymeric materials to enhance their usefulness using a concept similar to the formation of solid solutions in metallic and ceramic systems. We can form materials that are known as **copolymers** that consist of different monomers. For example, acrylonitrile (A), butadiene (B), and styrene (S) monomers can be made to react to form a copolymer known as ABS. This resultant copolymer is similar to a solid solution in that it has the functionalities of the three monomers from which it is derived, blending their properties. Similar to the Cu-Ni or $BaTiO_3$-$SrTiO_3$ solid solutions, we will not be able to separate out the acrylonitrile, butadiene, or styrene from an ABS plastic. Injection molding is used to convert ABS into telephones, helmets, steering wheels, and small appliance cases. Figure 9-6 illustrates the properties of different copolymers in the ABS system.[4] Note that this is *not* a phase diagram. Dylark™ is another example of a copolymer. It is formed using maleic anhydride and a styrene monomer.[5] The Dylark™ copolymer, with carbon black for UV protection, reinforced with fiberglass, and toughened with rubber is used for instrument panels in many automobiles (Chapter 15).

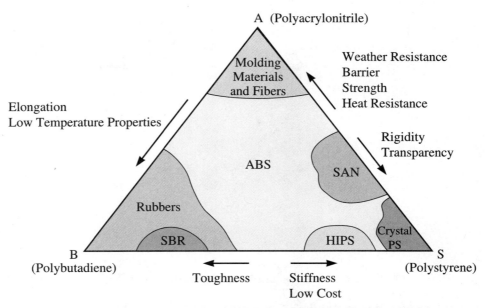

Figure 9-6   Diagram showing how the properties of copolymers formed in the ABS system vary. This is not a phase diagram. (*Source: From* Plastics, Materials and Processing, *Second Edition, by B.A. Strong, p. 223, Fig. 6-14. Copyright © 2000 Prentice Hall. Reprinted by permission of Pearson Education.*)

9-3   Conditions for Unlimited Solid Solubility

In order for an alloy system, such as copper-nickel to have unlimited solid solubility, certain conditions must be satisfied. These conditions, the **Hume-Rothery rules**, are as follows:

1. *Size factor*: The atoms or ions must be of similar size, with no more than a 15% difference in atomic radius, in order to minimize the lattice strain (i.e., to minimize, at an atomic level, the deviations caused in interatomic spacing).

2. *Crystal structure*: The materials must have the same crystal structure; otherwise, there is some point at which a transition occurs from one phase to a second phase with a different structure.

3. *Valence*: The ions must have the same valence; otherwise, the valence electron difference encourages the formation of compounds rather than solutions.

4. *Electronegativity*: The atoms must have approximately the same electronegativity. Electronegativity is the affinity for electrons (Chapter 2). If the electronegativities differ significantly, compounds form—as when sodium and chlorine ions combine to form sodium chloride.

Hume-Rothery's conditions must be met, but they are not necessarily sufficient, for two metals (e.g., Cu and Ni) or compounds (e.g., $BaTiO_3$-$SrTiO_3$) to have unlimited solid solubility.

Figure 9-7 shows schematically the two-dimensional structures of MgO and NiO. The Mg^{+2} and Ni^{+2} ions are similar in size and valence and, consequently, can replace one another in a sodium chloride (NaCl) crystal structure (Chapter 3), forming a com-

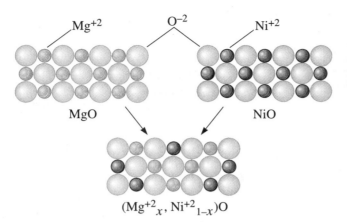

$(Mg^{+2}_x, Ni^{+2}_{1-x})O$

Figure 9-7
MgO and NiO have similar crystal structures, ionic radii, and valences; thus the two ceramic materials can form solid solutions.

plete series of solid solutions of the form $(Mg^{+2}_x Ni^{+2}_{1-x})O$, where $x = $ the mole fraction of Mg^{+2} or MgO.

The solubility of interstitial atoms is always limited. Interstitial atoms are much smaller than the atoms of the host element, thereby violating the first of Hume-Rothery's conditions.

EXAMPLE 9-3   *Ceramic Solid Solutions of MgO*

NiO can be added to MgO to produce a solid solution. What other ceramic systems are likely to exhibit 100% solid solubility with MgO?

SOLUTION

In this case, we must consider oxide additives that have metal cations with the same valence and ionic radius as the magnesium cations. The valence of the magnesium ion is +2 and its ionic radius is 0.66 Å. From Appendix B, some other possibilities in which the cation has a valence of +2 include the following:

	r (Å)	$\left[\dfrac{r_{ion} - r_{Mg^{+2}}}{r_{Mg^{+2}}} \right] \times 100\%$	Crystal Structure
Cd^{+2} in CdO	$r_{Cd^{+2}} = 0.97$	47	NaCl
Ca^{+2} in CaO	$r_{Ca^{+2}} = 0.99$	50	NaCl
Co^{+2} in CoO	$r_{Co^{+2}} = 0.72$	9	NaCl
Fe^{+2} in FeO	$r_{Fe^{+2}} = 0.74$	12	NaCl
Sr^{+2} in SrO	$r_{Sr^{+2}} = 1.12$	70	NaCl
Zn^{+2} in ZnO	$r_{Zn^{+2}} = 0.74$	12	NaCl

The percent difference in ionic radii and the crystal structures are also shown and suggest that the FeO-MgO system will probably display unlimited solid solubility. The CoO and ZnO systems also have appropriate radius ratios and crystal structures.

9-4 Solid-Solution Strengthening

In metallic materials, one of the important effects of solid-solution formation is the resultant **solid-solution strengthening**. This strengthening, via solid-solution formation, is caused by increased resistance to dislocation motion. This is one of the important reasons why brass (Cu-Zn alloy) is stronger than pure copper. We will learn later that carbon also plays another role in the strengthening of steels by forming iron carbide (Fe_3C) and other phases (Chapter 11). Jewelry could be made out of pure gold or silver. However, pure gold and pure silver are extremely soft and malleable and the jewelry pieces made will not retain their shape. This is also why jewelers add copper to gold or silver.

In the copper-nickel (Cu-Ni) system, we intentionally introduce a solid substitutional atom (nickel) into the original crystal structure (copper). The copper-nickel alloy is stronger than pure copper. Similarly, if less than 30% Zn is added to copper, the zinc behaves as a substitutional atom that strengthens the copper-zinc alloy, as compared with pure copper.

Recall from Chapter 6 that the strength of ceramics is mainly dictated by the distribution of flaws; solid-solution formation does not have a strong effect on the mechanical properties. This is similar to why strain hardening was not much of a factor in enhancing the strength of ceramics or semiconductors such as silicon (Chapter 7). As discussed before, solid-solution formation in ceramics and semiconductors (such as Si,

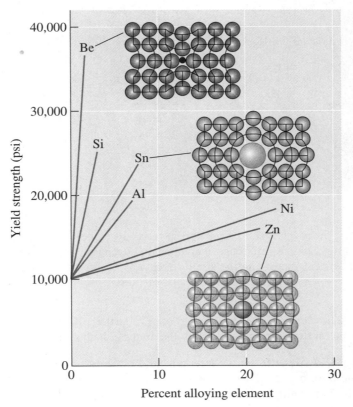

Figure 9-8
The effects of several alloying elements on the yield strength of copper. Nickel and zinc atoms are about the same size as copper atoms, but beryllium and tin atoms are much different from copper atoms. Increasing both atomic size difference and amount of alloying element increases solid-solution strengthening.

GaAs, etc.) has considerable influence on their magnetic, optical, and dielectric properties. The following discussion related to mechanical properties, therefore, applies mainly to metallic materials.

Degree of Solid-Solution Strengthening   The degree of solid-solution strengthening depends on two factors. First, a large difference in atomic size between the original (host or solvent) atom and the added (guest or solute) atom increases the strengthening effect. A larger size difference produces a greater disruption of the initial crystal structure, making slip more difficult (Figure 9-8).

Second, the greater the amount of alloying element added, the greater the strengthening effect (Figure 9-8). A Cu-20% Ni alloy is stronger than a Cu-10% Ni alloy. Of course, if too much of a large or small atom is added, the solubility limit may be exceeded and a different strengthening mechanism, **dispersion strengthening**, is produced. In dispersion strengthening, the interface between the host phase and guest phase resists dislocation motion and contributes to strengthening. This mechanism is discussed further in Chapter 10.

EXAMPLE 9-4   *Solid-Solution Strengthening*

From the atomic radii, show whether the size difference between copper atoms and alloying atoms accurately predicts the amount of strengthening found in Figure 9-8.

SOLUTION

The atomic radii and percent size difference are shown below:

Metal	Atomic Radius (Å)	$\left[\dfrac{r - r_{Cu}}{r_{Cu}}\right] \times 100\%$
Cu	1.278	0
Zn	1.332	+4.2
Al	1.432	+12.1
Sn	1.509	+18.1
Ni	1.243	−2.7
Si	1.176	−8.0
Be	1.143	−10.6

For atoms larger than copper—namely, zinc, aluminum, and tin—increasing the size difference increases the strengthening effect. Likewise for smaller atoms, increasing the size difference increases strengthening.

Effect of Solid-Solution Strengthening on Properties   The effects of solid-solution strengthening on the properties of a metallic material include the following (Figure 9-9):

1. The yield strength, tensile strength, and hardness of the alloy are greater than those of the pure metals. This is one reason why we most often use alloys rather than pure metals. For example, small concentrations of Mg are added to aluminum to provide higher strength to the aluminum alloys used in making aluminum beverage cans.[6]

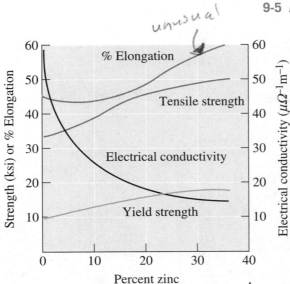

unusual

Figure 9-9
The effect of additions of zinc to copper on the properties of the solid-solution-strengthened alloy. The increase in % elongation with increasing zinc content is *not* typical of solid-solution strengthening.

2. Almost always, the ductility of the alloy is less than that of the pure metal. Only rarely, as in copper-zinc alloys, does solid-solution strengthening increase both strength and ductility.

3. Electrical conductivity of the alloy is much lower than that of the pure metal (Chapter 18). This is because electrons get more scattered off the atoms of the alloying elements. Solid-solution strengthening of copper or aluminum wires used for transmission of electrical power is not recommended because of this pronounced effect. Electrical conductivity of many alloys, although lower than pure metals, is often more stable as a function of temperature.

4. The resistance to creep, or loss of strength at elevated temperatures, is improved by solid-solution strengthening. High temperatures do not cause a catastrophic change in the properties of solid-solution-strengthened alloys. Many high-temperature alloys, such as those used for jet engines, rely partly on extensive solid-solution strengthening.

9-5   Isomorphous Phase Diagrams

A phase diagram shows the phases and their compositions at any combination of temperature and alloy composition. When only two elements or two compounds are present in a material, a **binary phase diagram** can be constructed. **Isomorphous binary phase diagrams** are found in a number of metallic and ceramic systems. In the isomorphous systems, which include the copper-nickel and NiO-MgO systems [Figure 9-10(a) and (b)], only one solid phase forms; the two components in the system display complete solid solubility. As shown in the phase diagrams for $CaO \cdot SiO_2$-$SrO \cdot SiO_2$, and thallium-lead (Tl-Pb) systems, it is possible to have phase diagrams show a minimum or maximum point, respectively [Figure 9-10(c) and (d)].[1] Notice the *x*-axis scale can represent either mole% or weight% of one of the components. We can also plot atomic% or mole fraction of one of the components. Also, notice that the CaO-SiO_2 and SrO-SiO_2 diagram could be plotted as a **ternary phase diagram**. A ternary phase diagram is a phase diagram for systems consisting of three components. Here, we represent it as a **pseudo-binary diagram** (i.e., we assume that this is a diagram

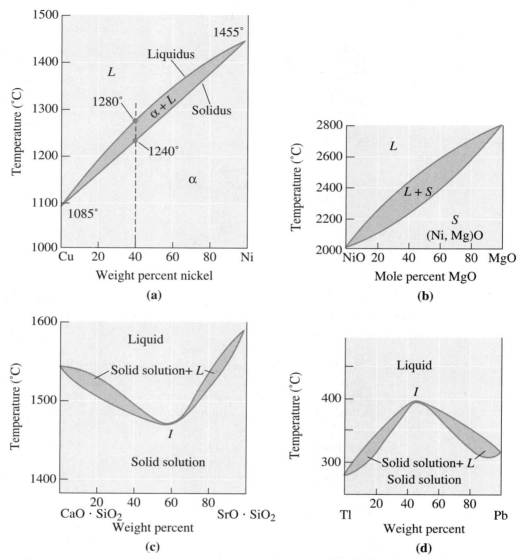

Figure 9-10 (a) The equilibrium phase diagrams for the Cu-Ni and NiO-MgO systems. (b) The liquidus and solidus temperatures are shown for a Cu-40% Ni alloy. (c) and (d) Systems with solid solution maxima and minima. (*Source: Adapted from* Introduction to Phase Equilibria, *by C.G. Bergeron, and S.H. Risbud. Copyright © 1984 American Ceramic Society. Adapted by permission.*)

that represents phase equilibria between $CaO \cdot SiO_2$ and $SrO \cdot SiO_2$). In a pseudo-binary diagram, we represent equilibria between three or more components using two compounds. Ternary phase diagrams are often encountered in ceramic and metallic systems.[1,5]

More recently, considerable developments also have occurred in the development of phase diagrams using computer databases containing thermodynamic properties of different elements and compounds. Using graphical interfaces, it is possible to generate phase diagrams for relatively complex compositions. A software package known as Thermocalc™ is one example of a commercial software package developed for this purpose. These software tools can be effective. However, considerable familiarity with the materials and training with the software is needed. Some professional societies such as the American Ceramic Society and the American Society of Metals have also published valuable and extensive collections of phase diagrams for different metal and ceramic systems.[6,7] Some of these collections are available on CD-ROM and may be available in your libraries.

There are several valuable pieces of information to be obtained from phase diagrams, as follows.

Liquidus and Solidus Temperatures   We define **liquidus temperature** as the temperature above which a material is completely liquid. The upper curve in Figure 9-10(a) represents the liquidus temperatures for copper-nickel alloys of different compositions. We must heat a copper-nickel alloy above the liquidus temperature to produce a completely liquid alloy that can then be cast into a useful shape. The liquid alloy begins to solidify when the temperature cools to the liquidus temperature. For the Cu-40% Ni alloy in Figure 9-10(a), the liquidus temperature is 1280°C.

The **solidus temperature** for the copper-nickel alloys is the temperature below which the alloy is 100% solid. The lower curve in Figure 9-10(a) represents the solidus temperatures for Cu-Ni alloys of different compositions. A copper-nickel alloy is not completely solid until the material cools below the solidus temperature. If we use a copper-nickel alloy at high temperatures, we must be sure that the service temperature is below the solidus so that no melting occurs. For the Cu-40% Ni alloy in Figure 9-10(a), the solidus temperature is 1240°C.

Copper-nickel alloys melt and freeze over a range of temperatures between the liquidus and the solidus. The temperature difference between the liquidus and the solidus is the **freezing range** of the alloy. Within the freezing range, two phases coexist: a liquid and a solid. The solid is a solution of copper and nickel atoms and is designated as the α phase. For the Cu-40% Ni alloy (α phase) in Figure 9-10(a), the freezing range is $1280 - 1240 = 40°C$. Note that pure metals solidify at a fixed temperature (i.e., the freezing range is zero degrees).

Phases Present   Often we are interested in which phases are present in an alloy at a particular temperature. If we plan to make a casting, we must be sure that the metal is initially all liquid; if we plan to heat treat an alloy component, we must be sure that no liquid forms during the process. Different solid phases have different properties. For example, BCC Fe (indicated as α phase on the iron carbon phase diagram) is magnetic. However, FCC iron (indicated as γ phase on the Fe-C diagram) is not. We will learn more about ferroelectricity in later chapters. We will learn that the tetragonal phase learn of barium titanate ($BaTiO_3$) is ferroelectric, but the cubic phase of $BaTiO_3$ is not (Chapter 18).

The phase diagram can be treated as a road map; if we know the coordinates—temperature and alloy composition—we can determine the phases present, assuming we know that thermodynamic equilibrium exists. There are many examples of technologically important situations where we do not want equilibrium phases to form. For example, in the formation of silicate glass, we want an amorphous glass and not a crystalline SiO_2 to form. When we harden steels by quenching them from a high temperature, the hardening occurs because of the formation of nonequilibrium phases. In such cases, phase diagrams will not provide all of the information we need. In these cases, we need to use special diagrams that take into account the effect of time (i.e., kinetics) on phase transformations. We will examine the use of such diagrams in later chapters.

The following two examples illustrate the applications of some of these concepts.

EXAMPLE 9-5   *NiO-MgO Isomorphous System*

From the phase diagram for the NiO-MgO binary system [Figure 9-10(b)], describe a composition that can melt at 2600°C but will not melt when placed into service at 2300°C.

SOLUTION

The material must have a liquidus temperature below 2600°C, but a solidus temperature above 2300°C. The NiO-MgO phase diagram [Figure 9-10(b)] permits us to design an appropriate composition.

To identify a composition with a liquidus temperature below 2600°C, there must be less than 65 mol% MgO in the refractory. To identify a composition with solidus temperature above 2300°C, there must be at least 50 mol% MgO present. Consequently, we can use any composition between 50 mol% MgO and 65 mol% MgO.

EXAMPLE 9-6   *Design of a Composite Material*

One method to improve the fracture toughness of a ceramic material (Chapter 6) is to reinforce the ceramic matrix with ceramic fibers. A materials designer has suggested that Al_2O_3 could be reinforced with 25% Cr_2O_3 fibers, which would interfere with the propagation of any cracks in the alumina. The resulting composite is expected to operate under load at 2000°C for several months. Criticize the appropriateness of this design.

SOLUTION

Since the composite will operate at high temperatures for a substantial period of time, the two phases—the Cr_2O_3 fibers and the Al_2O_3 matrix—must not react with one another. In addition, the composite must remain solid to at least 2000°C. The phase diagram in Figure 9-11 permits us to consider this choice for a composite.

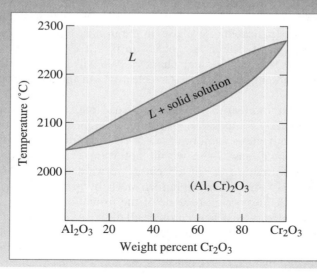

Figure 9-11
The Al_2O_3-Cr_2O_3 phase diagram (for Example 9-6).

Pure Cr_2O_3, pure Al_2O_3, and Al_2O_3-25% Cr_2O_3 have solidus temperatures above 2000°C; consequently, there is no danger of melting any of the constituents. However, Cr_2O_3 and Al_2O_3 display unlimited solid solubility. At the high service temperature, 2000°C, Al^{3+} ions will diffuse from the matrix into the fiber, replacing Cr^{3+} ions in the fibers. Simultaneously, Cr^{3+} ions will replace Al^{3+} ions in the matrix. Long before several months have elapsed, these diffusion processes cause the fibers to completely dissolve into the matrix. With no fibers remaining, the fracture toughness will again be poor.

Composition of Each Phase   For each phase we can specify a composition, expressed as the percentage of each element in the phase. Usually the composition is expressed in weight percent (wt%). When only one phase is present in the alloy or a ceramic solid solution, the composition of the phase equals the overall composition of the material. If the original composition of a single phase alloy or ceramic material changes, then the composition of the phase must also change.

However, when two phases, such as liquid and solid, coexist, their compositions differ from one another and also differ from the original overall composition. In this case, if the original composition changes slightly, the composition of the two phases is unaffected, provided that the temperature remains constant.

This difference is explained by the Gibbs phase rule. In this case, unlike the example of pure magnesium (Mg) described earlier, we keep the pressure fixed at one atmosphere, which is normal for binary phase diagrams. The phase rule given by Equation 9-1 can be rewritten as:

$$1 + C = F + P \quad \text{(for constant pressure)}, \tag{9-2}$$

where, again, C is the number of independent chemical components, P is the number of phases (*not pressure*), and F is the number of degrees of freedom. We now use number 1 instead of number 2 because we are holding the pressure constant. This reduces the number of degrees of freedom by 1. The pressure is typically, although not necessarily, one atmosphere. In a binary system, the number of components C is two; the degrees of freedom that we have include changing the temperature and changing the composition of the phases present. We can apply this form of the phase rule to the Cu-Ni system, as shown in Example 9-7.

EXAMPLE 9-7   *Gibbs Rule for Isomorphous Phase Diagram*

Determine the degrees of freedom in a Cu-40% Ni alloy at (a) 1300°C, (b) 1250°C, and (c) 1200°C.

SOLUTION

This is a binary system ($C = 2$). Two components are Cu and Ni. We will assume constant pressure. Therefore, Equation 9-2 ($1 + C = F + P$) can be used as follows:

(a) At 1300°C, $P = 1$, since only one phase (liquid) is present; $C = 2$, since both copper and nickel atoms are present. Thus:

$$1 + C = F + P \quad \therefore \ 1 + 2 = F + 1 \text{ or } F = 2$$

We must fix both the temperature and the composition of the liquid phase to completely describe the state of the copper-nickel alloy in the liquid region.

(b) At 1250°C, $P = 2$, since both liquid and solid are present; $C = 2$, since copper and nickel atoms are present. Now:

$$1 + C = F + P \quad \therefore \ 1 + 2 = F + 2 \text{ or } F = 1$$

If we fix the temperature in the two-phase region, the compositions of the two phases are also fixed. Or, if the composition of one phase is fixed, the temperature and composition of the second phase are automatically fixed.

(c) At 1200°C, $P = 1$, since only one phase, solid, is present; $C = 2$, since both copper and nickel atoms are present. Again,

$$1 + C = F + P \quad \therefore \ 1 + 2 = F + 1 \text{ or } F = 2$$

and we must fix both temperature and composition to completely describe the state of the solid.

Because there is only one degree of freedom in a two-phase region of a binary phase diagram, the compositions of the two phases are always fixed when we specify the temperature. This is true even if the overall composition of the alloy changes. Therefore, we can use a tie line to determine the composition of the two phases. A **tie line** is a horizontal line within a two-phase region drawn at the temperature of interest (Figure 9-12). In an isomorphous system, the tie line connects the liquidus and solidus points at the specified temperature. The ends of the tie line represent the compositions of the two phases in equilibrium. Tie lines are not used in single-phase regions because we do not have two phases to "tie" in.

For any alloy with overall or bulk composition lying between c_L and c_S, the composition of the liquid is c_L and the composition of the solid α is c_S.

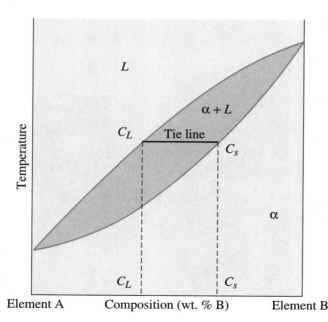

Figure 9-12
A hypothetical binary phase diagram between two elements A and B. When an alloy is present in a two-phase region, a tie line at the temperature of interest fixes the composition of the two phases. This is a consequence of the Gibbs phase rule, which provides only one degree of freedom.

The following example illustrates how the concept of tie line is used to determine the composition of different phases in equilibrium.

EXAMPLE 9-8 *Compositions of Phases in Cu-Ni Phase Diagram*

Determine the composition of each phase in a Cu-40% Ni alloy at 1300°C, 1270°C, 1250°C, and 1200°C. (See Figure 9-13.)

SOLUTION

The vertical line at 40% Ni represents the overall composition of the alloy:

- **1300°C:** Only liquid is present. The liquid must contain 40% Ni, the overall composition of the alloy.
- **1270°C:** Two phases are present. A horizontal line within the $\alpha + L$ field is drawn. The endpoint at the liquidus, which is in contact with the liquid region, is at 37% Ni. The endpoint at the solidus, which is in contact with the α region, is at 50% Ni. Therefore, the liquid contains 37% Ni and the solid contains 50% Ni.
- **1250°C:** Again two phases are present. The tie line drawn at this temperature shows that the liquid contains 32% Ni and the solid contains 45% Ni.
- **1200°C:** Only solid α is present, so the solid must contain 40% Ni.

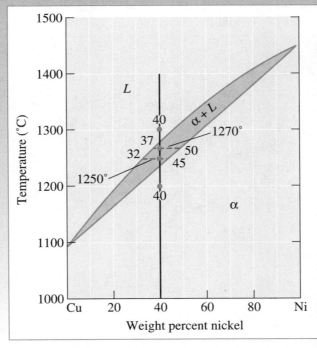

Figure 9-13
Tie lines and phase compositions for a Cu-40% Ni alloy at several temperatures (for Example 9-8).

In Example 9-8, we find that the solid α contains more nickel than the overall alloy and the liquid L contains more copper than the original alloy. Generally, the higher melting point element (in this case, nickel) is concentrated in the first solid that forms.

Amount of Each Phase (the Lever Rule)   Lastly, we are interested in the relative amounts of each phase present in the alloy. These amounts are normally expressed as weight percent (wt%). We express absolute amounts of different phases in units of mass or weight (grams, kilograms, pounds, etc.). The following example illustrates the rationale for the **lever rule**.

EXAMPLE 9-9   *Application of Lever Rule*

Calculate the amounts of α and L at 1250°C in the Cu-40% Ni alloy shown in Figure 9-14.

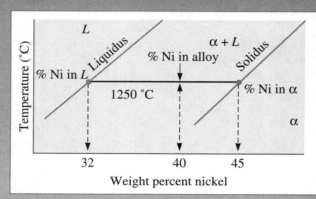

Figure 9-14
A tie line at 1250°C in the copper-nickel system that is used in Example 9-9 to find the amount of each phase.

SOLUTION

Let's say that x = mass fraction of the alloy that is solid α. Since we have only two phases, the balance of nickel must be in the liquid phase (L). Thus, the mass fraction of nickel in liquid will be $1 - x$. Consider 100 grams of the alloy. This alloy will consist of 40 grams of nickel at all temperatures. At 1250°C, let us write an equation that will represent mass balance for nickel. At 1250°C, we have x grams of the alpha phase. We have $100(1 - x)$ grams of liquid.

Total mass of nickel in 100 grams of the alloy = mass of nickel in liquid + mass of nickel in α

$$\therefore 100 \times (\% \text{ Ni in alloy}) = [(100)(1 - x)](\% \text{ Ni in L}) + (100)[x](\% \text{ Ni in } \alpha)$$

$$\therefore (\% \text{ Ni in alloy}) = (\% \text{ Ni in } L)(1 - x) + (\% \text{ Ni in } \alpha)(x)$$

By multiplying and rearranging,

$$x = \frac{(\% \text{ Ni in alloy}) - (\% \text{ Ni in } L)}{(\% \text{ Ni in } \alpha) - (\% \text{ Ni in } L)}$$

From the phase diagram at 1250°C:

$$x = \frac{40 - 32}{45 - 32} = \frac{8}{13} = 0.62$$

If we convert from mass fraction to mass percent, the alloy at 1250°C contains 62% α and 38% L. Note that the concentration of Ni in alpha phase (at 1250°C) is 45% and concentration of nickel in liquid phase (at 1250°C) is 32%.

To calculate the amounts of liquid and solid, we construct a lever on our tie line, with the fulcrum of our lever being the original composition of the alloy. The leg of the lever *opposite* to the composition of the phase, whose amount we are calculating, is divided by the total length of the lever to give the amount of that phase. In Example 9-9, note that the denominator represents the total length of the tie line and the numerator is the portion of the lever that is *opposite* the composition of the solid we are trying to calculate.

The lever rule in general can be written as:

$$\text{Phase percent} = \frac{\text{opposite arm of lever}}{\text{total length of tie line}} \times 100 \qquad (9\text{-}3)$$

We can work the lever rule in any two-phase region of a binary phase diagram. The lever rule calculation is not used in single-phase regions because the answer is trivial (there is 100% of that phase present). The lever rule is used to calculate the relative fraction or % of a phase in a two-phase mixture. The end points of the tie line we use give us the composition (i.e., chemical concentration of different components) of each phase.

The following example reinforces the application of the lever rule for calculating the amounts of phases for an alloy at different temperatures. This is one way to track the solidification behavior of alloys, something we had not seen in Chapter 8.

EXAMPLE 9-10   *Solidification of a Cu-40% Ni Alloy*

Determine the amount of each phase in the Cu-40% Ni alloy shown in Figure 9-13 at 1300°C, 1270°C, 1250°C, and 1200°C.

SOLUTION

- **1300°C:** There is only one phase, so 100% *L*.

- **1270°C:** $\%L = \dfrac{50 - 40}{50 - 37} \times 100 = 77\%$

  $\%\alpha = \dfrac{40 - 37}{50 - 37} \times 100 = 23\%$

- **1250°C:** $\%L = \dfrac{45 - 40}{45 - 32} \times 100 = 38\%$

  $\%\alpha = \dfrac{40 - 32}{45 - 32} \times 100 = 62\%$

- **1200°C:** There is only one phase, so 100% α.

Note that at each temperature, we can determine the composition of the phases in equilibrium from the ends of the tie line drawn at that temperature.

This may seem a little odd at first. How does the α phase change its composition? The liquid phase also changes its composition and the amounts of each phase change with temperature as the alloy cools from liquidus to solidus.

Sometimes we wish to express composition as atomic percent (at%) rather than weight percent (wt%). For a Cu-Ni alloy, where M_{Cu} and M_{Ni} are the molecular weights, the following equations provide examples for making these conversions:

$$at\% \ Ni = \left(\frac{\left[\dfrac{wt\% \ Ni}{M_{Ni}} \right]}{\left[\dfrac{(wt\% \ Ni)}{M_{Ni}} + \dfrac{(wt\% \ Cu)}{M_{Cu}} \right]} \right) \times 100 \qquad (9\text{-}4)$$

$$wt\% \ Ni = \left(\frac{(at\% \ Ni)/(M_{Ni})}{(at\% \ Ni)/(M_{Ni}) + (at\% \ Cu)/(M_{Cu})} \right) \times 100 \qquad (9\text{-}5)$$

9-6   Relationship Between Properties and the Phase Diagram

We have previously mentioned that a copper-nickel alloy will be stronger than either pure copper or pure nickel because of solid solution strengthening. The mechanical properties of a series of copper-nickel alloys can be related to the phase diagram as shown in Figure 9-15.

The strength of copper increases by solid-solution strengthening until about 60% Ni is added. Pure nickel is solid-solution strengthened by the addition of copper until 40%

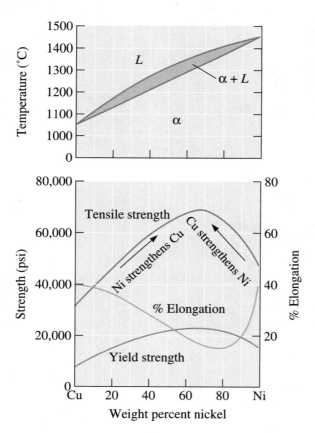

Figure 9-15
The mechanical properties of copper-nickel alloys. Copper is strengthened by up to 60% Ni and nickel is strengthened by up to 40% Cu.

Cu is added. The maximum strength is obtained for a Cu-60% Ni alloy, known as *Monel*. The maximum is closer to the pure nickel side of the phase diagram because pure nickel is stronger than pure copper.

EXAMPLE 9-11 *Design of a Melting Procedure for a Casting*

You need to produce a Cu-Ni alloy having minimum yield strength of 20,000 psi, a minimum tensile strength of 60,000 psi, and a minimum % elongation of 20%. You have in your inventory a Cu-20% Ni alloy and pure nickel. Design a method for producing castings having the required properties.

SOLUTION

From Figure 9-15, we determine the required composition of the alloy. To meet the required yield strength, the alloy must contain between 30 and 90% Ni; for the tensile strength, 33 to 90% Ni is required. The required % elongation can be obtained for alloys containing less than 60% Ni or more than 90% Ni. To satisfy all of these conditions, we could use:

$$Cu\text{-}90\% \ Ni \quad \text{or} \quad Cu\text{-}33\% \ \text{to} \ 60\% \ Ni$$

We prefer to select a low nickel content, since nickel is more expensive than copper. In addition, the lower nickel alloys have a lower liquidus, permitting castings to be made with less energy being expended. Therefore, a reasonable alloy might be Cu-35% Ni.

To produce this composition from the available melting stock, we must blend some of the pure nickel with the Cu-20% Ni ingot. Assume we wish to produce 10 kg of the alloy. Let x be the mass of Cu-20% Ni alloy we will need. The mass of pure Ni needed will be $10 - x$.

Since the final alloy consists of 35% Ni, the total mass of Ni needed will be:

$$(10 \ \text{Kg})\left(\frac{35\% \ \text{Ni}}{100\%}\right) = 3.5 \ \text{kg Ni}$$

Now let's write a mass balance for nickel. Nickel from the Cu-20% alloy + pure nickel added = total nickel in the 35% alloy being produced.

$$(x \ \text{kg})\left(\frac{20\%}{100\%}\right) + (10 - x \ \text{kg Ni})\left(\frac{100\%}{100\%}\right) = 3.5 \ \text{kg Ni}$$

$$0.2x + 10 - x = 3.5$$

$$6.5 = 0.8x$$

$$x = 8.125 \ \text{kg}$$

Therefore, we need to melt 8.125 kg of Cu-20% Ni with 1.875 kg of pure nickel to produce the required alloy. We would then heat the alloy above the liquidus temperature, which is 1250°C for the Cu-35% Ni alloy, before pouring the liquid metal into an appropriate mold.

We need to conduct such calculations for many practical situations dealing with the processing of alloys because when we make these materials, we use new and recycled materials.

9-7   Solidification of a Solid-Solution Alloy

When an alloy such as Cu-40% Ni is melted and cooled, solidification requires both nucleation and growth. Heterogeneous nucleation permits little or no undercooling, so solidification begins when the liquid reaches the liquidus temperature (Chapter 8). The phase diagram (Figure 9-16), with a tie line drawn at the liquidus temperature, tells that the *first solid to form* has a composition of Cu-52% Ni.

Two conditions are required for growth of the solid α. First, growth requires that the latent heat of fusion (ΔH_f), which evolves as the liquid solidifies, be removed from the solid–liquid interface. Second, unlike the case of pure metals, diffusion must occur so that the compositions of the solid and liquid phases follow the solidus and liquidus curves during cooling. The latent heat of fusion (ΔH_f), is removed over a range of temperatures so that the cooling curve shows a change in slope, rather than a flat plateau (Figure 9-17). Thus, as we mentioned before in Chapter 8, the solidification of alloys is different from that of pure metals.

At the start of freezing, the liquid contains Cu-40% Ni and the first solid contains Cu-52% Ni. Nickel atoms must have diffused to and concentrated at the first solid to form. But after cooling to 1250°C, solidification has advanced and the phase diagram

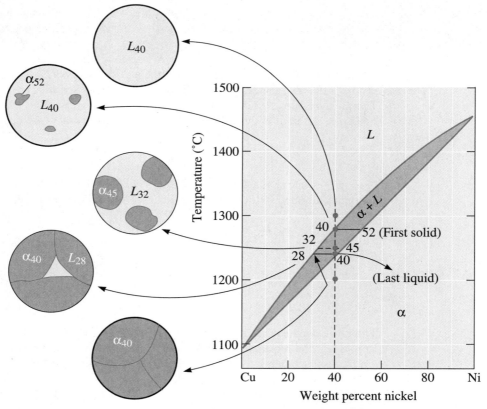

Figure 9-16   The change in structure of a Cu-40% Ni alloy during equilibrium solidification. The nickel and copper atoms must diffuse during cooling in order to satisfy the phase diagram and produce a uniform equilibrium structure.

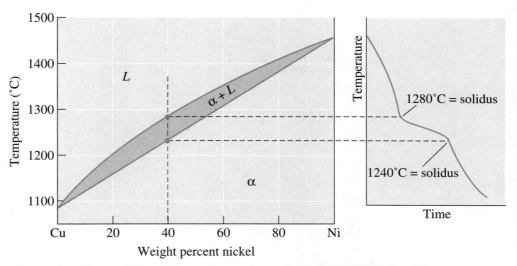

Figure 9-17 The cooling curve for an isomorphous alloy during solidification. We assume that cooling rates are small go as to allow thermal equilibrium to take place. The changes in slope of the cooling curve indicate the liquidus and solidus temperatures, in this case for a Cu-40% Ni alloy.

tells us that now all of the liquid must contain 32% Ni and all of the solid must contain 45% Ni. On cooling from the liquidus to 1250°C, some nickel atoms must diffuse from the first solid to the new solid, reducing the nickel in the first solid. Additional nickel atoms diffuse from the solidifying liquid to the new solid. Meanwhile, copper atoms have concentrated—by diffusion—into the remaining liquid. This process must continue until we reach the solidus temperature, where the last liquid to freeze, which contains Cu-28% Ni, solidifies and forms a solid containing Cu-40% Ni. Just below the solidus, all of the solid must contain a uniform concentration of 40% Ni throughout.

In order to achieve this equilibrium final structure, the cooling rate must be extremely slow. Sufficient time must be permitted for the copper and nickel atoms to diffuse and produce the compositions given by the phase diagram. In many practical casting situations, the cooling rate is too rapid to permit equilibrium. Therefore, in most castings made from alloys we expect chemical segregation. We had seen in Chapter 8 that porosity is a defect that could be present in many cast products. Another such defect often present in cast products is chemical **segregation**. This is discussed in detail in the next section.

9-8 Nonequilibrium Solidification and Segregation

In Chapter 5, we examined the thermodynamic and kinetic driving forces for diffusion. We know that diffusion occurs fastest in gases, followed by liquids, and then solids. We also saw that increasing the temperature enhances diffusion rates. When cooling is too rapid for atoms to diffuse and produce equilibrium conditions, nonequilibrium structures are produced in the casting. Let's see what happens to our Cu-40% Ni alloy on rapid cooling.

Again, the first solid, containing 52% Ni, forms on reaching the liquidus temperature (Figure 9-18). On cooling to 1260°C, the tie line tells us that the liquid contains

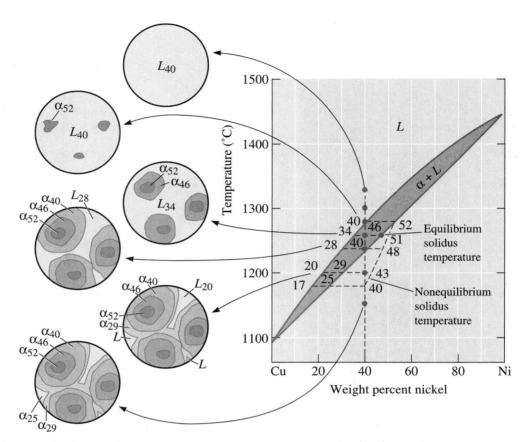

Figure 9-18   The change in structure of a Cu-40% Ni alloy during nonequilibrium solidification. Insufficient time for diffusion in the solid produces a segregated structure.

34% Ni and the solid that forms at that temperature contains 46% Ni. Since diffusion occurs rapidly in liquids, we expect the tie line to predict the liquid composition accurately. However, diffusion in solids is comparatively slow. The first solid that forms still has about 52% Ni, but the new solid contains only 46% Ni. We might find that the average composition of the solid is 51% Ni. This gives a different nonequilibrium solidus than that given by the phase diagram. As solidification continues, the nonequilibrium solidus line continues to separate from the equilibrium solidus.

When the temperature reaches 1240°C (the equilibrium solidus) a significant amount of liquid remains. The liquid will not completely solidify until we cool to 1180°C, where the nonequilibrium solidus intersects the original composition of 40% Ni. At that temperature, liquid containing 17% Ni solidifies, giving solid containing 25% Ni. The last liquid to freeze therefore contains 17% Ni, and the last solid to form contains 25% Ni. The average composition of the solid is 40% Ni, but the composition is not uniform.

The actual location of the nonequilibrium solidus line and the final nonequilibrium solidus temperature depend on the cooling rate. Faster cooling rates cause greater departures from equilibrium. The following example illustrates how we can account for the changes in composition under nonequilibrium conditions.

EXAMPLE 9-12 *Nonequilibrium Solidification of Cu-Ni Alloys*

Calculate the composition and amount of each phase in a Cu-40% Ni alloy that is present under the nonequilibrium conditions shown in Figure 9-18 at 1300°C, 1280°C, 1260°C, 1240°C, 1200°C, and 1150°C. Compare with the equilibrium compositions and amounts of each phase.

SOLUTION

We use the tie line upto the equilibrium solidus temperature to calculate composition and percentages of phases as per the lever rule. Similarly, the nonequilibrium solidus temperature curve is used to calculate percentages and concentrations of different phases formed under nonequilibrium conditions.

Temperature	Equilibrium	Nonequilibrium
1300°C	L: 40% Ni 100% L	L: 40% Ni 100% L
1280°C	L: 40% Ni 100% L	L: 40% Ni 100% L
	α: 52%, Ni ~ 0%	
1260°C	L: 34% Ni $\dfrac{46-40}{46-34} = 50\%$ L	L: 34% Ni $\dfrac{51-40}{51-34} = 65\%$ L
	α: 46% Ni $\dfrac{40-34}{46-34} = 50\%$ α	α: 51% Ni $\dfrac{40-34}{51-34} = 35\%$ α
1240°C	L: 28% Ni ~ 0% L	L: 28% Ni $\dfrac{48-40}{48-28} = 40\%$ L
	α: 40% Ni 100% α	α: 48% Ni $\dfrac{40-28}{48-28} = 60\%$ α
1200°C	α: 40% Ni 100% α	L: 20% Ni $\dfrac{43-40}{43-20} = 13\%$ L
		α: 43% Ni $\dfrac{40-20}{43-20} = 87\%$ α
1150°C	α: 40% Ni 100% α	α: 40% Ni 100% α

Microsegregation   The nonuniform composition produced by nonequilibrium solidification is known as segregation. **Microsegregation**, also known as **interdendritic segregation** and **coring**, occurs over short distances, often between small dendrite arms. The centers of the dendrites, which represent the first solid to freeze, are rich in the higher melting point element in the alloy. The regions between the dendrites are rich in the lower melting point element, since these regions represent the last liquid to freeze. The composition and properties of the phase α (in the case of Cu-Ni alloys) differ from one region to the next, and we expect the casting to have poorer properties as a result.

Microsegregation can cause **hot shortness**, or melting of the lower melting point interdendritic material at temperatures below the equilibrium solidus. When we heat the Cu-40% Ni alloy to 1225°C, below the equilibrium solidus but above the nonequilibrium solidus, the low-nickel regions between the dendrites melt.

Homogenization   We can reduce the interdendritic segregation and problems with hot shortness by means of a **homogenization heat treatment**. If we heat the casting to a

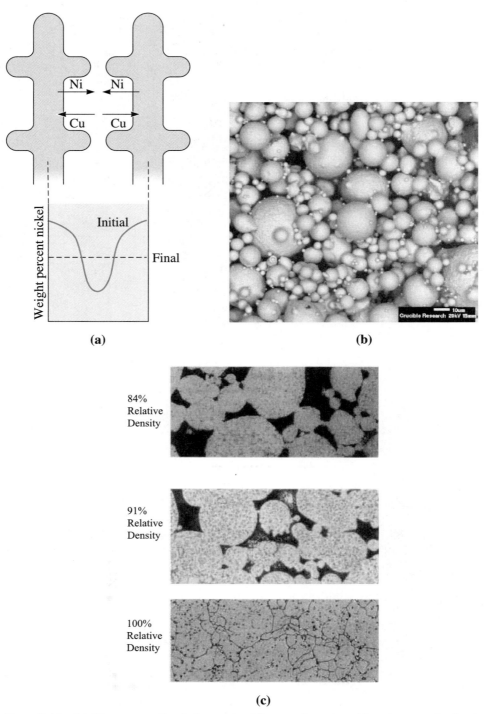

Figure 9-19 (a) Microsegregation between dendrites can be reduced by a homogenization heat treatment. Counterdiffusion of nickel and copper atoms may eventually eliminate the composition gradients and produce a homogeneous composition. (b) Spray atomized powders of superalloys. (c) Progression of densification in low carbon Astroalloy sample processed using HIP. (*Courtesy of J. Staite, Hann, B. and Rizzo, F., Crucible Compaction Metals.*)

temperature below the nonequilibrium solidus, the nickel atoms in the centers of the dendrites diffuse to the interdendritic regions; copper atoms diffuse in the opposite direction [Figure 9-19(a)]. Since the diffusion distances are relatively short, only a few hours are required to eliminate most of the composition differences. The homogenization time is related to

$$t = c\frac{(\text{SDAS})^2}{D_s} \tag{9-6}$$

where SDAS is the secondary dendrite arm spacing, D_s is the rate of diffusion of the solute in the matrix, and c is a constant. A small SDAS reduces the diffusion distance and permits short homogenization times.

Macrosegregation   There exists another type of segregation, known as **macrosegregation**, which occurs over a large distance, between the surface and the center of the casting, with the surface (which freezes first) containing slightly more than the average amount of the higher melting point metal. We cannot eliminate macrosegregation by a homogenization treatment, because the diffusion distances are too great. Macrosegregation can be reduced by hot working, which was discussed in Chapter 7. This is because in hot working we are basically breaking down the cast macrostructure.

Rapidly Solidified Powders   In applications where porosity, microsegregation, and macrosegregation must be minimized, powders of complex alloys are prepared using **spray atomization** [Figure 9-19(b)].[7] In spray atomization, homogeneous melts of complex compositions are prepared and sprayed through a ceramic nozzle. The melt stream is broken into finer droplets and quenched using argon (Ar), nitrogen (N_2) gases (gas atomization) or water (water atomization). The molten droplets solidify rapidly generating powder particles ranging from $\sim$10–100 μm in size. Since the solidification in droplets occurs very quickly, there is very little time for diffusion. Therefore, chemical segregation does not occur. Many complex nickel- and cobalt-based superalloys and stainless steel powders are examples of materials prepared using this technique. The spray atomized powders are blended and converted into desired shapes. The techniques used in processing such powders include sintering (Chapter 5), **hot pressing** (HP) and **hot isostatic pressing** (HIP). In HIP, sintering is conducted under an isostatic pressure ($\sim$25,000 psi) using, for example, argon gas. Very large ($\sim$ up to 2 foot diameter, several feet long) and smaller components can be processed using HIP. Smaller components such as disks that hold turbine blades can be machined from these. The progression of densification in a low carbon Astroalloy sample processed using spray atomized powders is shown in Figure 9-19(c).[7]

Hot pressing is sintering under a uniaxial pressure and is used in the production of smaller components of materials that are difficult to sinter otherwise (Chapter 14). The HIP and hot pressing techniques are used for both metallic and ceramic powder materials. For example, such materials as silicon carbide (SiC), silicon nitride (Si_3N_4), and lead lanthanum zirconium titanate (*PLZT*) are often prepared using hot pressing since high densities can not be obtained using conventional sintering. The pressure used provides one more driving force for densification. Note that we do *not* use spray atomization to make ceramic powders since ceramics melt at very high temperatures. Ceramic powders are most often produced using the oxide mix process (Chapter 14).

SUMMARY

◆ A phase is any portion, including the whole, of a system that is physically homogeneous within it and bounded by a surface so that it is mechanically separable from any other portions.

◆ A phase diagram typically shows phases that are expected to be present in a system under thermodynamic equilibrium conditions. Sometimes metastable phases may also be shown.

◆ Solid solutions in metallic or ceramic materials form when elements or compounds with similar crystal structure can form a phase that is chemically homogeneous.

◆ Forming solid solutions of ceramics and semiconducting materials is a useful way to "tune" the electrical, magnetic, and optical properties.

◆ In plastics, we can form copolymers by reacting different type of monomers. Forming copolymers is similar to forming solid solutions of metallic or ceramic systems.

◆ Solid-solution strengthening is accomplished in metallic materials by formation of solid solutions. The point defects created restrict dislocation motion and cause strengthening.

◆ The degree of solid-solution strengthening increases when (1) the amount of alloying element increases and (2) the atomic size difference between the host material and the alloying element increases.

◆ The amount of alloying element (or compound) that we can add to produce solid-solution strengthening is limited by the solubility of the alloying element or compound in the host material. The solubility is limited when (1) the atomic size difference is more than about 15%, (2) the alloying element (or compound) has a different crystal structure than the host element (or compound), and (3) the valence and electronegativity of the alloying element or constituent ions are different from those of host element (or compound).

◆ In addition to increasing strength and hardness, solid-solution strengthening typically decreases ductility and electrical conductivity of metallic materials. An important function of solid-solution strengthening is to provide good high-temperature properties to the alloy.

◆ The addition of alloying elements to provide solid-solution strengthening also changes the physical properties, including the melting temperature, of the alloy. The phase diagram helps explain these changes.

◆ A phase diagram in which constituents exhibit complete solid solubility is known as an isomorphous phase diagram.

◆ As a result of solid-solution formation, solidification begins at the liquidus temperature and is completed at the solidus temperature; the temperature difference over which solidification occurs is the freezing range.

◆ In two-phase regions of the phase diagram, the ends of a tie line fix the composition of each phase and the lever rule permits the amount of each phase to be calculated.

◆ Microsegregation and macrosegregation occur during solidification: Microsegregation, or coring, occurs over small distances, often between dendrites. The centers of the dendrites are rich in the higher melting point element, whereas interdendritic regions, which solidify last, are rich in the lower melting point element.

◇ Homogenization can reduce microsegregation.

◇ Macrosegregation describes differences in composition over long distances, such as between the surface and center of a casting. Hot working may reduce macrosegregation.

◇ Spray atomization is used for production of multicomponent metallic materials that need to be produced without chemical segregation.

◇ Sintering, hot pressing, and HIP are techniques that are also used for processing metallic ceramic powders. Very high densities can be obtained using hot pressing and HIP.

GLOSSARY

Alloy   A material that exhibits properties of a metallic material made from multiple elements.

Binary phase diagram   A phase diagram for a system with two components.

Copolymer   A polymer that is formed by combining two or more different types of monomers usually with the idea of blending the properties affiliated with individual polymers, example Dylark™ a copolymer of maleic anhydride and styrene.

Coring   Chemical segregation in cast products, also known as microsegregation or interdendritic segregation. The centers of the dendrites are rich in the higher melting point element, whereas interdendritic regions, which solidify last, are rich in the lower melting point element.

Dispersion strengthening   Strengthening, typically used in metallic materials, by the formation of ultra-fine dispersions of a second phase. The interface between the newly formed phase and the parent phase provides additional resistance to dislocation motion thereby causing strengthening of metallic materials (Chapter 10).

Freeze drying   A process used for processing of materials and food products in which a solution is cooled until it turns solid. Water and other chemicals are then sublimed leaving the solid behind.

Freezing range   The temperature difference between the liquidus and solidus temperatures.

Gibbs phase rule   Describes the number of degrees of freedom, or the number of variables that must be fixed to specify the temperature and composition of a phase ($2 + C = F + P$, where pressure and temperature can change, $1 + C = F + P$, where pressure or temperature is constant).

Homogenization heat treatment   The heat treatment used to reduce the microsegregation caused during nonequilibrium solidification. This heat treatment cannot eliminate macrosegregation.

Hot isostatic pressing (HIP)   Sintering of metallic or ceramic powders, conducted under an isostatic pressure.

Hot pressing (HP)   Sintering of metal or ceramic powders under a uniaxial pressure; used for production of smaller components of materials that are difficult to sinter otherwise.

Hot shortness   Melting of the lower melting point nonequilibrium material that forms due to segregation, even though the temperature is below the equilibrium solidus temperature.

Hume-Rothery rules   The conditions that an alloy or ceramic system must meet if the system is to display unlimited solid solubility. Hume-Rothery's rules are necessary but are not sufficient for materials to show unlimited solid solubility.

Interdendritic segregation   See "Coring."

Isomorphous phase diagram   A phase diagram in which components display unlimited solid solubility.

Lever rule   A technique for determining the amount of each phase in a two-phase system.

Limited solubility   When only a maximum amount of a solute material can be dissolved in a solvent material.

Liquidus   Curves on phase diagrams that describe the liquidus temperatures of all possible alloys.

Liquidus temperature   The temperature at which the first solid begins to form during solidification.

Macrosegregation   The presence of composition differences in a material over large distances caused by nonequilibrium solidification. The only way to remove this type of segregation is to break down the cast structure by hot working.

Metastable phase   A material such as undercooled liquid that exists under nonequilibrium conditions. A metastable phase usually can be formed only in one direction (e.g., supercooled water can be formed by cooling water below its freezing temperature, but not by heating ice).

Microsegregation   See "Coring."

Multiple-phase alloy   An alloy that consists of two or more phases.

Phase   Any portion including the whole of a system, which is physically homogeneous within it and bounded by a surface so that it is mechanically separable from any other portions.

Phase diagrams   Diagrams showing phases present under equilibrium conditions and the phase compositions at each combination of temperature and overall composition. Sometimes phase diagrams also indicate metastable phases.

Phase rule   See Gibbs phase rule.

P-T diagram   A diagram describing thermodynamic stability of phases under different temperature and pressure conditions (same as a unary phase diagram).

Segregation   The presence of composition differences in a material, often caused by insufficient time for diffusion during solidification.

Single-phase alloy   An alloy consisting of one phase.

Solid solution   A solid phase formed by combining multiple elements or compounds such that overall phase has uniform composition and properties that are different from those of the elements or compounds forming it.

Solid-solution strengthening   Increasing the strength of a metallic material via the formation of a solid solution.

Solidus   Curves on phase diagrams that trace the solidus temperatures for all possible alloys.

Solidus temperature   The temperature below which all liquid has completely solidified.

Solubility   The amount of one material that will completely dissolve in a second material without creating a second phase.

Spray atomization   A process in which molten alloys or metals are sprayed using a ceramic nozzle. The molten material stream is broken using a gas (e.g., Ar, N_2) or water. This leads to finer droplets that solidify rapidly forming metal or alloy powders with $\sim$10–100 μm particle size range.

Stainless steels   Corrosion-resistant alloys that usually contain iron (Fe), carbon (C), chromium (Cr), nickel (Ni), and some other elements.

Tie line   A horizontal line drawn in a two-phase region of a phase diagram to assist in determining the compositions of the two phases.

Triple point   A pressure and temperature at which three phases of a single material are in equilibrium.

Unary phase diagram   A phase diagram in which there is only one component.

Unlimited solubility   When the amount of one material that will dissolve in a second material without creating a second phase is unlimited.

ADDITIONAL INFORMATION

1. BERGERON, C.G. and RISBUD, S.H., *Introduction to Phase Equilibria*. 1984: American Ceramic Society.

2. GIBBS, W.J., *The Scientific Papers of J. Willard Gibbs: Thermodynamics* (Reprints of Papers). 1993: Ox Bow Press.

3. ASCHKENASY, H., "Freeze Drying", *Scientific American*, 1996(9).

4. STRONG, B.A., *Plastics-Materials and Processing*. Second Ed., 2000: Prentice Hall.

5. ARCH, P., *Nova Chemicals, Technical discussions and assistance is acknowledged.* 2002.

6. HOSFORD, W.F. and J.L. DUNCAN, "The Aluminum Beverage Can", *Scientific American*, 1994. September: p. 48–53.

7. STAITE, J., HANN, B., and RIZZO, F., *Technical Assistance and discussions are acknowledged.* 2002.

✓ PROBLEMS

9-1   Explain the principle of grain size strengthening. Does this mechanism work at high temperatures? Explain.

9-2   Explain the principle of strain hardening. Does this mechanism work at high temperatures? Explain.

9-3   What is the principle of solid-solution strengthening? Does this mechanism work at high temperatures? Explain.

9-4   What is the principle of dispersion strengthening?

Section 9-1 Phases and Phase Diagrams

9-5   What does the term "phase" mean?

9-6   What are the different phases of water?

9-7   Ice has been known to exist in different polymorphs. Are these different phases of water?

9-8   Write down the Gibbs phase rule, assuming temperature and pressure are allowed to change. Explain clearly the meaning of each term.

9-9   What is a phase diagram?

9-10   Explain why the P-T diagram for H_2O is considered to be a unary diagram.

9-11   The yttrium barium copper oxide (YBCO) superconductor discussed in Example 9-2 is made using oxides of yttrium, barium, and copper. Will the phase diagram for this system be unary, binary, or ternary? Explain.

9-12   What is a triple point?

9-13   What are the triple point pressures and temperatures for water and CO_2?

9-14   Explain how ultra fine powder of a yttria stabilized zirconia (YSZ) can be made using the freeze drying process.

9-15   The unary phase diagram for SiO_2 is shown in Figure 9-3(c). Locate the triple point where solid, liquid, and vapor coexist and give the temperature and the type of solid present. What do the other "triple" points indicate?

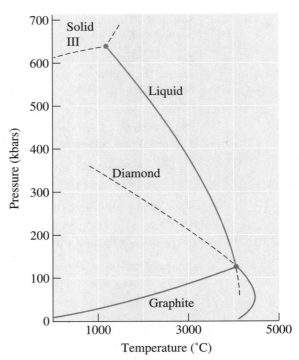

Figure 9-20   Unary phase diagram for carbon. Region for diamond formation is shown with a dotted line (for Problem 9-17). (*Source: Adapted from* Introduction to Phase Equilibria, *by C.G. Bergeron, S.H. Risbud, Fig. 2-11. Copyright © 1984 American Ceramic Society. Adapted by permission.*)

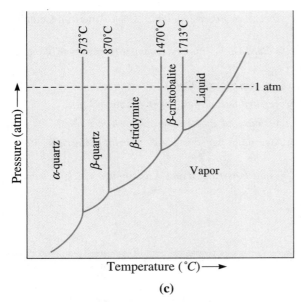

(c)

Figure 9-3   (Repeated for Problem 9-15) (c) Pressure-temperature diagram for SiO_2. The dotted line shows one atmosphere pressure.

9-16   What is a metastable phase?

9-17   Figure 9-20 shows the unary phase diagram for carbon. Based on this diagram, under what conditions can carbon in the form of graphite be converted into diamond?

9-18   Why is supercritical CO_2 being considered for applications in processing of materials and pharmaceuticals?

Section 9-2   Solubility and Solid Solutions

9-19   What is a solid solution?

9-20   How can solid solutions form in ceramic systems?

9-21   Fine powders of nickel ferrite ($NiFe_2O_4$) and zinc ferrite ($ZnFe_2O_4$) are heated at a high temperature ($\sim 1000°C$) for a long time (24 hours). Explain what may happen at the end of the process.

9-22   Explain how the dielectric and magnetic properties of materials such as $BaTiO_3$ and ferrites can be controlled using solid solution formation.

9-23   Why is it that forming solid solutions of differing compositions of GaAs-AlAs leads to LEDs of different colors?

9-24   Do we need 100% solid solubility to form a solid solution of one material in another?

9-25   Small concentrations of lead zirconate ($PbZrO_3$) are added to lead titanate ($PbTiO_3$). Draw a schematic of the resultant solid-solution crystal structure that is expected to form. This material, known as lead zirconium titanate (better known as *PZT*), has many applications ranging from spark igniters to ultrasound imaging. See Chapter 3 for information on perovskite crystal structure.

9-26 Can solid solutions be formed between three elements or three compounds?

9-27 What is a copolymer? What is the advantage to forming copolymers?

9-28 Is copolymer formation similar to solid-solution formation?

9-29 What is the ABS copolymer? State some of the applications of this material.

9-30 What is Dylark™? What are some of the applications of this material?

9-31 What other compounds are added to Dylark™? Why? Do these compounds form a solid solution or a mixture?

Section 9-3 Conditions for Unlimited Solid Solubility

9-32 Briefly state the Hume-Rothery rules and explain the rationale.

9-33 Can the Hume-Rothery rules apply to ceramic systems? Explain.

9-34 Based on Hume-Rothery's conditions, which of the following systems would be expected to display unlimited solid solubility? Explain. (a) Au-Ag; (b) Al-Cu; (c) Al-Au; (d) U-W; (e) Mo-Ta; (f) Nb-W; (g) Mg-Zn; and (h) Mg-Cd.

Section 9-4 Solid-Solution Strengthening

9-35 Suppose 1 at% of the following elements is added to copper (forming a separate alloy with each element) without exceeding the solubility limit. Which one would be expected to give the higher strength alloy? Are any of the alloying elements expected to have unlimited solid solubility in copper? (a) Au; (b) Mn; (c) Sr; (d) Si; and (e) Co.

9-36 Suppose 1 at% of the following elements is added to aluminum (forming a separate alloy with each element) without exceeding the solubility limit. Which one would be expected to give the least reduction in electrical conductivity? Are any of the alloy elements expected to have unlimited solid solubility in aluminum? (a) Li; (b) Ba; (c) Be; (d) Cd; and (e) Ga.

9-37 Which of the following oxides is expected to have the largest solid solubility in Al_2O_3? (a) Y_2O_3; (b) Cr_2O_3; and (c) Fe_2O_3.

9-38 What is the role of small concentrations of Mg in aluminum alloys used to make beverage cans?

9-39 Why do jewelers add small amounts of copper to gold and silver?

9-40 Why is it not a good idea to use solid solution strengthening as a mechanism to increase the strength of copper for electrical applications?

Section 9-5 Isomorphous Phase Diagrams

9-41 Determine the liquidus temperature, solidus temperature, and freezing range for the following NiO-MgO ceramic compositions. [See Figure 9-10(b).] (a) NiO-30 mol% MgO; (b) NiO-45 mol% MgO; (c) NiO-60 mol% MgO; and (d) NiO-85 mol% MgO.

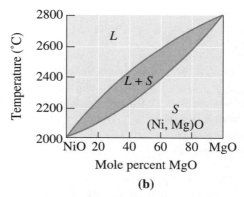

(b)

Figure 9-10 (Repeated for Problems 9-41, 9-43 and 9-46) (b) The liquidus and solidus temperatures are shown for a Cu-40% Ni alloy.

9-42 Determine the liquidus temperature, solidus temperature, and freezing range for the following MgO-FeO ceramic compositions. (See Figure 9-21.)

(a) MgO-25 wt% FeO; (b) MgO-45 wt% FeO;
(c) MgO-65 wt% FeO; (d) MgO-80 wt% FeO.

9-43 Determine the phases present, the compositions of each phase, and the amount of each phase in mol% for the following NiO-MgO ceramics at 2400°C. [See Figure 9-10(b).] (a) NiO-30 mol% MgO; (b) NiO-45 mol% MgO; (c) NiO-60 mol% MgO; and (d) NiO-85 mol% MgO.

9-44 (a) Determine the phases present, the compositions of each phase, and the amount of each phase in wt% for the following MgO-FeO ceramics at 2000°C. (See Figure 9-21.) (i) MgO-25 wt% FeO; (ii) MgO-45 wt% FeO; (iii) MgO-60 wt% FeO; and (iv) MgO-80 wt% FeO.

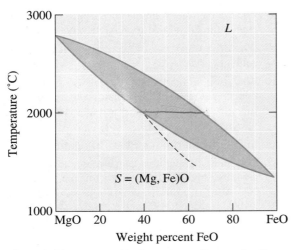

Figure 9-21 The equilibrium phase diagram for the MgO-FeO system.

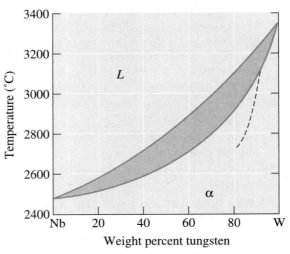

Figure 9-22 The equilibrium phase diagram for the Nb-W system.

(b) Consider an alloy of 65 wt% Cu and 35 wt% Al. Calculate the composition of the alloy in at%.

9-45 Consider a ceramic composed of 30 mol% MgO and 70 mol% FeO. Calculate the composition of the ceramic in wt%.

9-46 A NiO-20 mol% MgO ceramic is heated to 2200°C. Determine **(a)** the composition of the solid and liquid phases in both mol% and wt%; **(b)** the amount of each phase in mol% and wt%; and **(c)** assuming that the density of the solid is 6.32 g/cm³ and that of the liquid is 7.14 g/cm³, determine the amount of each phase in vol% (see Figure 9-10(b)).

9-47 A Nb-60 wt% W alloy is heated to 2800°C. Determine **(a)** the composition of the solid and liquid phases in both wt% and at%; **(b)** the amount of each phase in both wt% and at%; and **(c)** assuming that the density of the solid is 16.05 g/cm³ and that of the liquid is 13.91 g/cm³, determine the amount of each phase in vol%. (See Figure 9-22.)

9-48 How many grams of nickel must be added to 500 grams of copper to produce an alloy that has a liquidus temperature of 1350°C? What is the ratio of the number of nickel atoms to copper atoms in this alloy?

9-49 How many grams of nickel must be added to 500 grams of copper to produce an alloy that contains 50 wt% α at 1300°C?

9-50 How many grams of MgO must be added to 1 kg of NiO to produce a ceramic that has a solidus temperature of 2200°C?

9-51 How many grams of MgO must be added to 1 kg of NiO to produce a ceramic that contains 25 mol% solid at 2400°C?

9-52 We would like to produce a solid MgO-FeO ceramic that contains equal mol percentages of MgO and FeO at 1200°C. Determine the wt% FeO in the ceramic. (See Figure 9-21.)

9-53 We would like to produce a MgO-FeO ceramic that is 30 wt% solid at 2000°C. Determine the original composition of the ceramic in wt%. (See Figure 9-21.)

9-54 A Nb-W alloy held at 2800°C is partly liquid and partly solid. **(a)** If possible, determine the composition of each phase in the alloy; and **(b)** if possible, determine the amount of each phase in the alloy. (See Figure 9-22.)

9-55 A Nb-W alloy contains 55% α at 2600°C. Determine **(a)** the composition of each phase; and **(b)** the original composition of the alloy. (See Figure 9-22.)

9-56 Suppose a 1200-lb bath of a Nb-40 wt% W alloy is held at 2800°C. How many pounds of tungsten can be added to the bath before any solid forms? How many pounds of tungsten must be added to cause the entire bath to be solid? (See Figure 9-22.)

9-57 A fiber-reinforced composite material is produced, in which tungsten fibers are embedded in a Nb matrix. The composite is composed of 70 vol% tungsten. **(a)** Calculate the wt% of tungsten fibers in the composite; and **(b)** suppose the composite is heated to 2600°C and held for several years. What happens to the fibers? Explain. (See Figure 9-22.)

9-58 Suppose a crucible made of pure nickel is used to contain 500 g of liquid copper at 1150°C. Describe what happens to the system as it is held at this temperature for several hours. Explain.

Section 9-6 Relationship between Properties and the Phase Diagram

9-59 What is brass? Explain which element strengthens the matrix for this alloy.

9-60 What is the composition of Monel alloy?

Section 9-7 Solidification of a Solid-Solution Alloy

9-61 Equal moles of MgO and FeO are combined and melted. Determine **(a)** the liquidus temperature, the solidus temperature, and the freezing range of the ceramic; and **(b)** determine the phase(s) present, their composition(s), and their amount(s) at 1800°C. (See Figure 9-21.)

9-62 Suppose 75 cm³ of Nb and 45 cm³ of W are combined and melted. Determine **(a)** the liquidus temperature, the solidus temperature, and the freezing range of the alloy; and **(b)** determine the phase(s) present, their composition(s), and their amount(s) at 2800°C. (See Figure 9-22.)

9-63 A NiO-60 mol% MgO ceramic is allowed to solidify. Determine **(a)** the composition of the first solid to form; and **(b)** the composition of the last liquid to solidify under equilibrium conditions.

9-64 A Nb-35% W alloy is allowed to solidify. Determine **(a)** the composition of the first solid to form; and **(b)** the composition of the last liquid to solidify under equilibrium conditions. (See Figure 9-22.)

9-65 For equilibrium conditions and a MgO-65 wt% FeO ceramic, determine **(a)** the liquidus temperature; **(b)** the solidus temperature; **(c)** the freezing range; **(d)** the composition of the first solid to form during solidification; **(e)** the composition of the last liquid to solidify; **(f)** the phase(s) present, the composition of the phase(s), and the amount of the phase(s) at 1800°C; and **(g)** the phase(s) present, the composition of the phase(s), and the amount of the phase(s) at 1600°C. (See Figure 9-21.)

9-66 Figure 9-23 shows the cooling curve for a NiO-MgO ceramic. Determine **(a)** the liquidus temperature; **(b)** the solidus temperature; **(c)** the freezing range; **(d)** the pouring temperature; **(e)** the superheat; **(f)** the local solidification time; **(g)** the total solidification time; and **(h)** the composition of the ceramic.

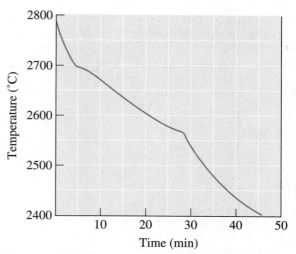

Figure 9-23 Cooling curve for a NiO-MgO ceramic (for Problem 9-66).

9-67 For equilibrium conditions and a Nb-80 wt% W alloy, determine **(a)** the liquidus temperature; **(b)** the solidus temperature; **(c)** the freezing range; **(d)** the composition of the first solid to form during solidification; **(e)** the composition of the last liquid to solidify; **(f)** the phase(s) present, the composition of the phase(s), and the amount of the phase(s) at 3000°C; and **(g)** the phases(s) present, the composition of the phase(s), and the amount of the phase(s) at 2800°C. (See Figure 9-22.)

9-68 Figure 9-24 shows the cooling curve for a Nb-W alloy. Determine **(a)** the liquidus temperature; **(b)** the solidus temperature; **(c)** the freezing range;

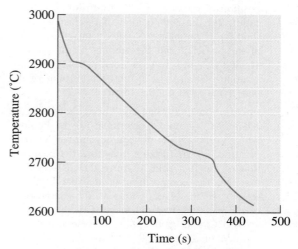

Figure 9-24 Cooling curve for a Nb-W alloy (for Problem 9-68).

(d) the pouring temperature; (e) the superheat; (f) the local solidification time; (g) the total solidification time; and (h) the composition of the alloy.

9-69 Cooling curves are shown in Figure 9-25 for several Mo-V alloys. Based on these curves, construct the Mo-V phase diagram.

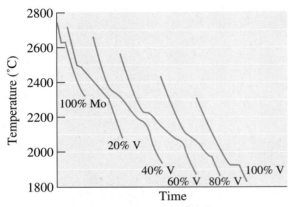

Figure 9-25   Cooling curves for a series of Mo-V alloys (for Problem 9-69).

Section 9-8 Nonequilibrium Solidification and Segregation

9-70 What are the origins of chemical segregation in cast products?

9-71 For the nonequilibrium conditions shown for the MgO-65 wt% FeO ceramic, determine (a) the liquidus temperature; (b) the nonequilibrium solidus temperature; (c) the freezing range; (d) the composition of the first solid to form during solidification; (e) the composition of the last liquid to solidify; (f) the phase(s) present, the composition of the phase(s), and the amount of the phase(s) at 1800°C; and (g) the phase(s) present, the composition of the phase(s), and the amount of the phase(s) at 1600°C. (See Figure 9-21.)

9-72 For the nonequilibrium conditions shown for the Nb-80 wt% W alloy, determine (a) the liquidus temperature; (b) the nonequilibrium solidus temperature; (c) the freezing range; (d) the composition of the first solid to form during solidification; (e) the composition of the last liquid to solidify; (f) the phase(s) present, the composition of the phase(s), and the amount of the phase(s) at 3000°C; and (g) the phase(s) present, the composition of the phase(s), and the amount of the phase(s) at 2800°C. (See Figure 9-22.)

9-73 How can microsegregation be removed?

9-74 What is macrosegregation? Is there a way to remove it without breaking up the cast structure?

9-75 What is homogenization? What type of segregation can it remove?

9-76 What is spray atomization? Can it be used for making ceramic powders?

9-77 Suppose you are asked to manufacture a critical component based on a nickel-based superalloy. The component must not contain any porosity and it must be chemically homogeneous. What manufacturing process would you use for this application? Why?

9-78 What is hot pressing? How is it different from hot isostatic pressing?

Design Problems

9-79 Homogenization of a slowly cooled Cu-Ni alloy having a secondary dendrite arm spacing of 0.025 cm requires 8 hours at 1000°C. Design a process to produce a homogeneous structure in a more rapidly cooled Cu-Ni alloy having a SDAS of 0.005 cm.

9-80 Design a process to produce a NiO-60% MgO refractory whose structure is 40% glassy phase at room temperature. Include all relevant temperatures.

9-81 Design a method by which glass beads (having a density of 2.3 g/cm³) can be uniformly mixed and distributed in a Cu-20% Ni alloy (density of 8.91 g/cm³).

9-82 Suppose that MgO contains 5 mol% NiO. Design a solidification purification method that will reduce the NiO to less than 1 mol% in the MgO.

Computer Problems

9-83 *Gibbs Phase Rule.* Write a computer program that will automate the Gibbs phase rule calculation. The program should ask the user for information on whether the pressure and temperature or only the pressure is to be held constant. The program then should use the correct equation to calculate the appropriate variable the user wants to know. The user will provide inputs for number of components. Then, if the user wishes to provide the number of phases present, the program should calculate the degrees of freedom and vice-versa.

9-84 *Conversion of wt% to at% for a Binary System.*

Write a computer program that will allow conversion of wt% into at%. The program should ask the user to provide appropriate formula weights of the elements/compounds. (See Equations 9-4 and 9-5.)

9-85 *Hume-Rothery Rules.* Write a computer program that will predict whether or not there will likely be 100% solid solubility between two elements. The program should ask the user to provide the user with information on crystal structures of the elements or compounds, and radii of different/ atoms or ions involved, valence and electronegativity values. You will have to make some assumption as to how much difference in values of electronegativity could be acceptable. The program should then use the Hume-Rothery rules and provide the user with guidance on the possibility of forming a system that shows 100% solid solubility.

Chapter

The soldering process plays a key role in processing of printed circuit boards and other devices for micro-electronics. This process often makes use of alloys based on lead and tin. Specific compositions of these alloys, known as the eutectic compositions, melt at constant temperature (like pure elements). These compositions are also mechanically strong since their microstructure comprises an intimate mixture of two distinct phases. The enhancement of mechanical properties of dispersing one phase in another phase by formation of eutectics is the central theme for this chapter. (*Courtesy of PhotoDisc/Getty Images.*)

10

Dispersion Strengthening and Eutectic Phase Diagrams

Have You Ever Wondered?

- *Why did some of the earliest glassmakers use plant ash to make glass?*

- *What alloys are most commonly used for soldering?*

- *What is a fiberglass?*

- *Is there an alloy that freezes at a constant temperature?*

- *What is Pyrex® glass?*

When the solubility of a material is exceeded by adding too much of an alloying element or compound, a second phase forms and a two-phase material is produced. The boundary between the two phases, known as the interphase interface, is a surface where the atomic arrangement is not perfect. In metallic materials, this boundary interferes with the slip or movement of dislocations, causing their strengthening. The general term for such strengthening by the introduction of a second phase is known as dispersion strengthening. In this chapter, we first discuss the fundamentals of dispersion strengthening to determine the microstructure we should aim to produce. Next, we examine the types of reactions that produce multiple-phase alloys. Finally, we examine in some detail at methods to achieve dispersion strengthening by controlling the solidification process. In this context, we will examine phase diagrams that involve the formation of multiple phases. We will concentrate on eutectic phase diagrams. In Chapter 11, we will learn about a special way of producing a dispersion of a second phase via a solid-state phase transformation sequence known as precipitation or age hardening. Precipitation hardening is a sub-set of the overall class of dispersion-strengthened materials.

10-1   Principles and Examples of Dispersion Strengthening

Most engineered materials are composed of more than one phase, and many of these materials are designed to provide improved strength. In simple **dispersion-strengthened alloys**, tiny particles of one phase, usually very strong and hard, are introduced into a second phase, which is weaker but more ductile.[1] The soft phase, usually continuous and present in larger amounts, is called the **matrix**. The hard-strengthening phase may be called the **dispersed phase** or the **precipitate**, depending on how the alloy is formed. In some cases, a phase or a mixture of phases may have a very characteristic appearance – in these cases this phase or phase mixture may be called a **microconstituent**. For dispersion strengthening to occur, the dispersed phase or precipitate must be small enough to provide effective obstacles to dislocation movement, thus providing the strengthening mechanism.

In most metal alloys, dispersion strengthening is produced by phase transformations. In this chapter, we will concentrate on a solidification transformation by which a liquid freezes to simultaneously form two solid phases. This will be called an **eutectic reaction** and is of particular importance in cast irons and many aluminum alloys. In the next chapter, we will discuss the eutectoid reaction, by which one solid phase reacts to simultaneously form two different solid phases; this reaction is key in the control of properties in steels. In Chapter 11, we will also discuss **precipitation, or age, hardening** which produces precipitates by a sophisticated heat treatment.

There are some general considerations for determining how the characteristics of the matrix and the dispersed phase affect the overall properties of an alloy (Figure 10-1).

1. The matrix should be soft and ductile, although the dispersed phase should be hard and strong. The dispersed phase particles interfere with slip, while the matrix provides at least some ductility to the overall alloy.

2. The hard dispersed phase should be discontinuous, while the soft, ductile matrix should be continuous. If the hard and brittle dispersed phase were continuous, cracks could propagate through the entire structure.

3. The dispersed phase particles should be small and numerous, increasing the likelihood that they interfere with the slip process since the area of the interphase interface is increased significantly.

4. The dispersed phase particles should be round, rather than needlelike or sharp edged, because the rounded shape is less likely to initiate a crack or to act as a notch.

5. Higher concentrations of the dispersed phase increase the strength of the alloy.

EACH OF THESE PHOTOGRAPHIC ESSAYS DEPICTS

the relationship between the structure of a material, its processing, and how the properties of the material can be utilized in a final product. The intimate structure-property-processing relationship emphasized in the text is demonstrated for several materials, including metallic alloys, ceramics, polymers, and composites.

(a)

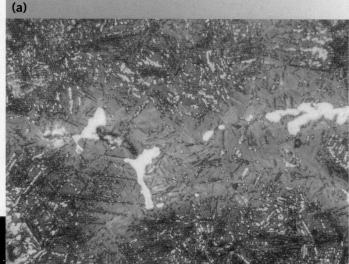

DUCTILE CAST IRON

High strength can be obtained in cast iron by heat treatment, which, in the microstructure shown, produces tiny Fe_3C precipitates in a ferritic matrix **(a)**.

Courtesy of American Foundryman's Society

Courtesy of American Foundryman's Society

(b)

The cast iron develops good ductility during its treatment with magnesium before the liquid iron is poured into molds **(b)**.

Courtesy of John Deere Company

(c)

Reliable yet economical components to be used for products that see rugged, heavy duty use, such as this farm tractor **(c)**, can be produced from metals such as ductile cast iron, discussed in Chapter 12.

FILAMENT WINDING OF COMPOSITES AND MECHANICAL MODELING

Fiber-reinforced composites often begin with a woven fabric, such as these glass fibers **(a)**, which are later impregnated with a polymer resin (see Chapter 16). Either individual fibers or these woven fabrics can then be formed into large and complex shapes, like a rocket motor casing **(b)**.

Courtesy of NASA

(a)

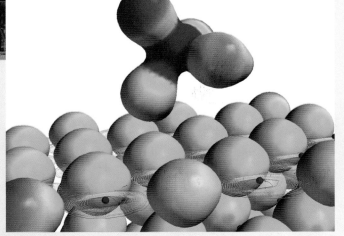

Astrid & Hanns-Frieder Michler/Science Photo Library

(b)

We use computers to design, screen and create new materials and components before their ultimate laboratory or industrial development. Image **(c)**, showing an arsenate ion absorbing onto an oxide surface, was computer generated using a form of quantum mechanical modeling. The Sandia researchers are testing new materials, called Specific Anion Nanoengineered Sorbents (SANS). To create these materials, they select mineral families with known affinities for anions, then use supercomputer modeling to rapidly simulate the arsenic-trapping aptitudes of thousands of combinations and variations of the minerals.

(c)

Photograph and information courtesy of Sandia National Laboratories

SOLID SOLUTION ALLOYS

Structure, even on an atomic scale, is crucial in developing the properties of materials. A scanning, tunneling electron micrograph reveals a missing atom, or vacancy (described in Chapter 4), in a coating on the surface of a platinum crystal **(a)**. Solid solution alloys that are intended to have good electrical conductivity and corrosion resistance, such as copper, are often engineered to have a minimum number of such defects, permitting them also to be easily joined, as in this electron beam weld used to produce a nuclear waste container **(b)**.

(a)

Courtesy of Dr. Bruce C. Schardt,
Digital Instruments

(b)

Courtesy of General Motors

(c)

PTR-Precision Technologies

On the other hand, the presence of these point defects makes it possible to control the properties of semiconducting materials (Chapters 18 and 20), such as those used to create the solar cells that power the Sunraycer **(c)**.

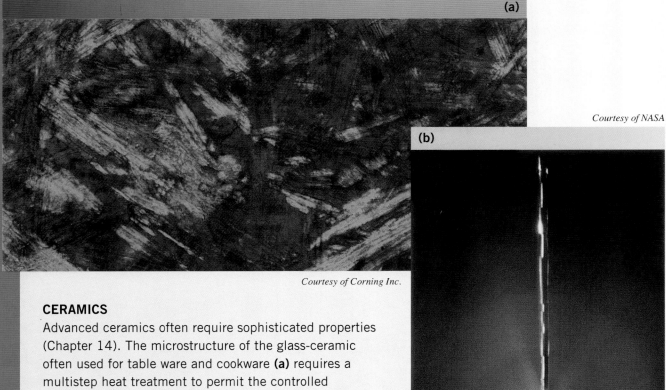

Courtesy of NASA

Courtesy of Corning Inc.

(a)

(b)

CERAMICS

Advanced ceramics often require sophisticated properties (Chapter 14). The microstructure of the glass-ceramic often used for table ware and cookware **(a)** requires a multistep heat treatment to permit the controlled precipitation of the correct number and amount of a crystalline phase from a glass matrix. Continuous single-crystal Al_2O_3 fibers to be used as reinforcements for high-temperature composites based on intermetallic compounds might be solidified from a narrow region of liquid ceramic melted by a laser **(b).**

Courtesy of Textron

(c)

Products such as silicon nitride engine blades reinforced with silicon carbide filaments **(c)** take advantage of the high-temperature properties obtained by these exotic structures and processes.

(d)

NASA's space shuttle **(d)** makes use of highly sophisticated metallic, ceramic, and polymeric materials. These materials are designed and fabricated such that they can reliably function in extreme environments encountered in space and also survive high temperatures encountered during reentry in the earth's atmosphere.

Courtesy of NASA

PHOTONICS

Although most engineered materials are polycrystalline, in many applications related to microelectronics and photonics we need single crystals.

(e)

Courtesy of NASA

Examples of materials used as single crystals include but are not limited to silicon, gallium arsenide, and lithium niobate. The growth of such crystals is often achieved using solidification of melts or from a vapor phase. This photograph **(e)** shows a mercuric iodide (HgI) single crystal grown in micro gravity using NASA's Fluids Experiment System/Vapor Crystal Growth Facility. The source material was heated, vaporized, and transported to a seed crystal where the vapor condensed. Mercury iodide crystals have practical uses as sensitive X-ray and gamma-ray detectors. In addition to their excellent optical properties, these crystals can operate at room temperature, which makes them useful for portable detector devices for nuclear power plant monitoring, natural resource prospecting, biomedical applications, and astronomical observing.

Argonne National Laboratory

Courtesy of NASA

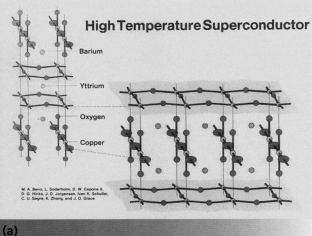

High Temperature Superconductor

Barium

Yttrium

Oxygen

Copper

M. A. Beno, L. Soderholm, D. W. Capone II,
D. G. Hinks, J. D. Jorgensen, Ivan K. Schuller,
C. U. Segre, K. Zhang, and J. D. Grace

(a)

(b)

SUPERCONDUCTORS

The complex crystal structure of the ceramic
$YBa_2Cu_3O_7$ contributes to high-temperature
superconduction in the material **(a)** discussed in
Chapter 18. This structure excludes magnetic fields
from the material (the Meissner effect), permitting
electrical currents to flow with virtually no resistance
and enabling a sample of the superconductor to be
levitated above a permanent magnet **(b)**.

Magnetic hard disks with 35 to 100
Gbit/in^2 areal densities are now
possible, thanks to the advances in
processing of magnetic materials
and development of new devices
such as the giant magnetoresistance sensors (GMR). Magnetic
hard drives **(c)** represent a
fascinating and complex system
which makes use of highly
sophisticated magnetically hard and
soft materials. The industry trend is
toward increasing the amount of
information that can be stored per
unit area. The challenge is to make
the grains of the magnetic materials
smaller without an accompanying
loss of magnetic behavior.

(c)

Courtesy of Seagate Corporation

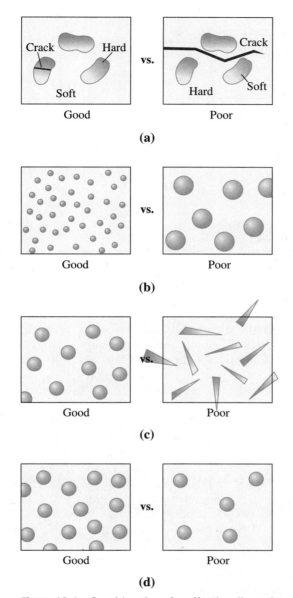

Figure 10-1   Considerations for effective dispersion strengthening: (a) The precipitate phase should be hard and discontinuous, (b) the dispersed phase particles should be small and numerous, (c) the dispersed phase particles should be round rather than needlelike, and (d) larger amounts of dispersed phase increase strengthening.

10-2   Intermetallic Compounds

An **intermetallic compound** is made up of two or more metallic elemen~~
a new phase with its own composition, crystal structure, and propertie~
compounds are almost always very hard and brittle. Intermetallics ~

TABLE 10-1 ■ *Properties of some intermetallic compounds*

Intermetallic Compound	Crystal Structure	Melting Temperature (°C)	Density $\left(\dfrac{g}{cm^3}\right)$	Young's Modulus (GPa)
FeAl	Ordered BCC	1250–1400	5.6	263
NiAl	Ordered FCC (B2)*	1640	5.9	206
Ni_3Al	Ordered FCC (L1$_2$)*	1390	7.5	337
TiAl	Ordered tetragonal (L1$_0$)*	1460	3.8	94
Ti_3Al	Ordered HCP	1600	4.2	210
$MoSi_2$	Tetragonal	2020	6.31	430

*Also known as. (*Source: Adapted from* Mechanical Behavior of Materials, *by M.A. Meyers and K.K. Chawla. Copyright © 1998 Prentice Hall. Adapted by permission of Pearson Education.*)

compounds are similar to ceramic materials in terms of their mechanical properties, however, in that they are formed by combining two or more metallic elements. Our interest in intermetallics is two-fold. First, often dispersion-strengthened alloys contain an intermetallic compound as the dispersed phase. Secondly, many intermetallic compounds, on their own (and not as second phase) are being investigated and developed for high-temperature applications.[2–4] In this section, we will discuss properties of intermetallics as stand-alone materials. In the sections that follow, we will discuss how intermetallic phases help strengthen metallic materials. Table 10-1 summarizes the properties of some intermetallic compounds.[1]

Stoichiometric intermetallic compounds have a fixed composition. Steels are often strengthened by a stoichiometric compound, iron carbide (Fe_3C) which has a fixed ratio of three iron atoms to one carbon atom. Stoichiometric intermetallic compounds are represented in the phase diagram by a vertical line [Figure 10-2(a)]. An example of a useful intermetallic compound is molybdenum disilicide ($MoSi_2$). This material is used for making heating elements of high temperature furnaces. At high temperatures (~1000–1600°C), $MoSi_2$ shows outstanding oxidation resistance. At low temperatures (~500°C and below), $MoSi_2$ is brittle and shows catastrophic oxidation known as pesting.[1]

Nonstoichiometric intermetallic compounds have a range of compositions and are sometimes called **intermediate solid solutions**. In the molybdenum-rhodium system, the γ phase is a nonstoichiometric intermetallic compound [Figure 10-2(b)]. Because the molybdenum-rhodium atom ratio is not fixed, the γ phase can contain from 45 wt% to 83 wt% Rh at 1600°C. Precipitation of the nonstoichiometric intermetallic copper aluminide $CuAl_2$ causes strengthening in a number of important aluminum alloys (Figure 11-5).[4]

Properties and Applications of Intermetallics   Intermetallic such as Ti_3Al and Ni_3Al, maintain their strength and even develop usable ductility at elevated temperatures (Figure 10-3). Lower ductility, though, has impeded further development of these materials. It has been shown that the addition of small levels of boron (B) (up to 0.2%) can enhance the ductility of polycrystalline Ni_3Al. Environmental effects also probably play a role in limiting the ductility levels in intermetallics.[3,6] Enhanced ductility levels could make it possible for intermetallics to be used in many high temperature and load-bearing applications. Ordered compounds of NiAl and Ni_3Al are also candidates for

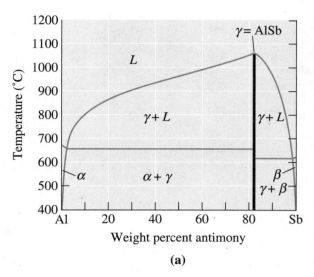

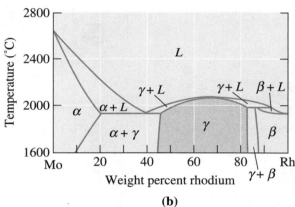

Figure 10-2
(a) The aluminum-antimony phase diagram includes a stoichiometric intermetallic compound γ.
(b) The molybdenum-rhodium phase diagram includes a nonstoichiometric intermetallic compound γ.

supersonic aircraft, jet engines, and high-speed commercial aircraft.[4] Not all applications of intermetallics are structural. Intermetallics based on silicon (e.g., platinum silicide) play a useful role in microelectronics and certain intermetallics such as Nb_3Sn are useful as superconductors (Chapter 18).

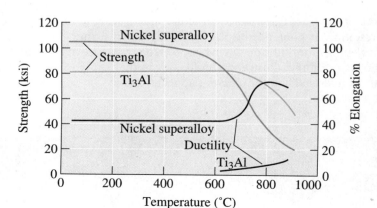

Figure 10-3
The strength and ductility of the intermetallic compound Ti_3Al compared with that of a conventional nickel superalloy. The Ti_3Al maintains its strength to higher temperatures lor̶ ̶ ̶ ̶ does the nic̶ superalloy.

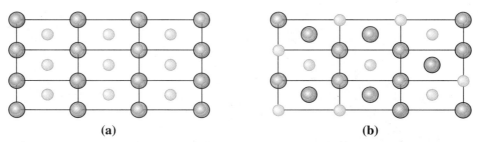

Figure 10-4   (a) In an ordered structure, the substituting atoms occupy specific lattice points, (b) while in normal structure, the constituent atoms are randomly located at different lattice points.

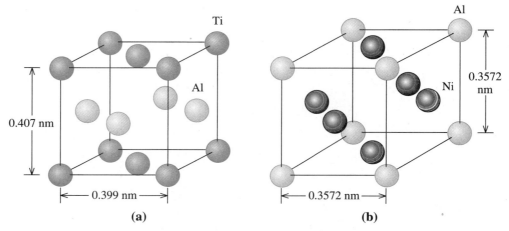

Figure 10-5   The unit cells of two intermetallic compounds: (a) TiAl has an ordered tetragonal structure, and (b) Ni_3Al has an ordered cubic structure.

The titanium aluminides, TiAl (also called the gamma (γ) alloy) and Ti_3Al (the α_2 alloy) are considered for a variety of applications, including gas turbine engines. These compounds have **ordered crystal structures**; in the ordered structure, the Ti and Al atoms occupy specific locations in the crystal structure rather than random locations as in most solid solutions (Figure 10-4). In the ordered tetragonal structure of TiAl (Figure 10-5), the titanium atoms are located at lattice points at the corners and the top and bottom faces of the unit cell, whereas aluminum atoms are located only at the other four faces of the cell. The ordered structure makes it more difficult for dislocations to move (resulting in poor ductility at low temperatures), but this also leads to a high activation energy for diffusion, giving good creep resistance at elevated temperatures.[1] The following example illustrates considerations that can go into the selection of such materials as intermetallics for different applications.

EXAMPLE 10-1   *Materials Selection for an Aerospace Vehicle*

Design a material suitable for the parts of an aerospace vehicle that reach high temperatures during re-entry from Earth orbit.

SOLUTION

The material must withstand the high temperatures generated as the vehicle enters Earth's atmosphere, where the material is exposed to oxygen. During reentry, the temperature of the material could be as high as 1600–1700°C.[4] Some ductility is needed to provide damage tolerance to the vehicle. Finally, the material should have a low density to minimize vehicle weight. The requirement for ductility suggests a metallic alloy. High-temperature metals that might be considered include tungsten (density of 19.254 g/cm^3), nickel (density of 8.902 g/cm^3), and titanium (density of 4.507 g/cm^3). Tungsten and nickel are quite dense and, as we will see in a later chapter, titanium has poor oxidation resistance at high temperatures. Given their ability to withstand oxidation, a high-temperature intermetallic compounds might be a solution. TiAl and Ni$_3$Al have good high-temperature properties and oxidation resistance and, at high temperatures, have at least some ductility. We can estimate their densities from their crystal structures.

The lattice parameters of the face-centered tetragonal TiAl unit cell are $a_0 = 3.99 \times 10^{-8}$ cm and $c_0 = 4.07 \times 10^{-8}$ cm (Figure 10-5). There are two titanium and two aluminum atoms in each unit cell. Therefore:

$$\rho_{\text{TiAl}} = \frac{(\text{Mass of Ti atoms per unit cell} + \text{Mass of Al atoms per unit cell})}{\text{Volume of the TiAl unit cell}}$$

$$\rho_{\text{TiAl}} = \frac{\dfrac{(2 \text{ Ti atoms})(\text{Atomic mass of Ti})}{N_{AV}} + \dfrac{(2 \text{ Al atoms})(\text{Atomic mass of Al})}{N_{AV}}}{(a_0^2 \times c_0)}$$

$$\rho_{\text{TiAl}} = \frac{(2)\left(47.9 \dfrac{\text{gm}}{\text{mole}}\right) + (2)\left(26.981 \dfrac{\text{gm}}{\text{mole}}\right)}{(6.023 \times 10^{23}) \cdot (3.99 \times 10^{-8} \text{ cm})^2 (4.07 \times 10^{-8} \text{ cm})}$$

$$\rho_{\text{TiAl}} = 3.84 \frac{\text{g}}{\text{cm}^3}$$

The lattice parameter for Ni$_3$Al is 3.572×10^{-8} cm [Figure 10-5(b)]; each unit cell contains three nickel atoms and one aluminum atom:

$$\rho_{\text{Ni}_3\text{Al}} = \frac{(3\text{Ni})\left(58.71 \dfrac{\text{g}}{\text{mol}}\right) + (1\text{Al})\left(26.981 \dfrac{\text{g}}{\text{mol}}\right)}{(3.572 \times 10^{-8} \text{ cm})^3 \left(6.02 \times 10^{23} \dfrac{\text{atoms}}{\text{mol}}\right)} = 7.40 \frac{\text{g}}{\text{cm}^3}$$

The TiAl is only about half the density of the Ni$_3$Al. However the specific strength (ratio of Young's modulus to density) is higher for Ni$_3$Al (see Table 10-1). Thus, Ni$_3$Al may be our choice although other properties and processing issues will have to be considered.

The emphasis for such materials used in outer space is on stiffness and not strength. Thus, carbon or aramid fiber-matrix composites (based on Kevlar^{TM}), coated properly to protect oxidation, also will be very good choices for outer space applications.[7] In addition to the simple properties considered here, manufactured materials used in space travel will have to withstand such effects as radiation, degassing of materials and resultant corona discharges, the effect of atomic oxygen (present in lower Earth orbits) on oxidation corrosion, the impact from space debris, and the stresses due to thermal expansion and contraction causing fatigue (Chapter 6).[4]

10-3   Phase Diagrams Containing Three-Phase Reactions

Many binary systems produce phase diagrams more complicated than the isomorphous phase diagrams discussed in Chapter 9. The systems we will discuss here contain reactions that involve three separate phases. Five such reactions are defined in Figure 10-6. Each of these reactions can be identified in a phase diagram by the following procedure:

1. Locate a horizontal line on the phase diagram. The horizontal line, which indicates the presence of a three-phase reaction, represents the temperature at which the reaction occurs under equilibrium conditions.

2. Locate three distinct points on the horizontal line: the two endpoints plus a third point, in between the two endpoints of the horizontal line. This third point represents the composition at which the three-phase reaction occurs. In Figure 10-6 the point in between has been shown at the center. However, on a real phase diagram this point is not necessarily at the center.

3. Look immediately above the in-between point and identify the phase or phases present; look immediately below the point in between the end points, and identify the phase or phases present. Then write in the reaction from the phase(s) above the point in between transforming to the phase(s) below the point. Compare this reaction with those in Figure 10-6 to identify the reaction.

Eutectic	$L \rightarrow \alpha + \beta$	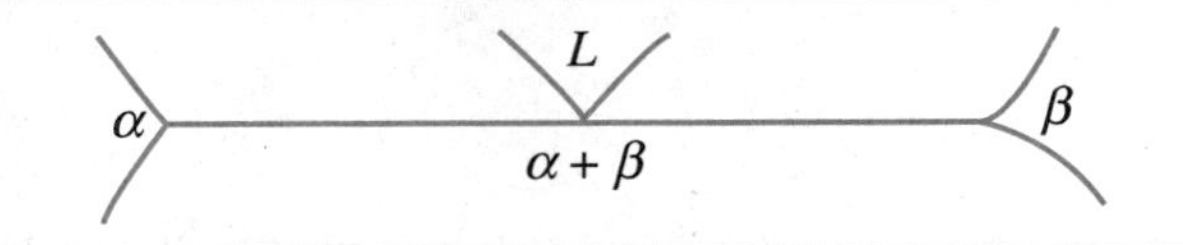
Peritectic	$\alpha + L \rightarrow \beta$	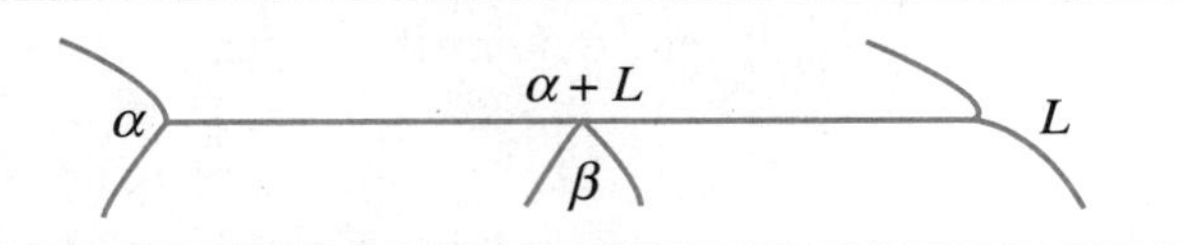
Monotectic	$L_1 \rightarrow L_2 + \alpha$	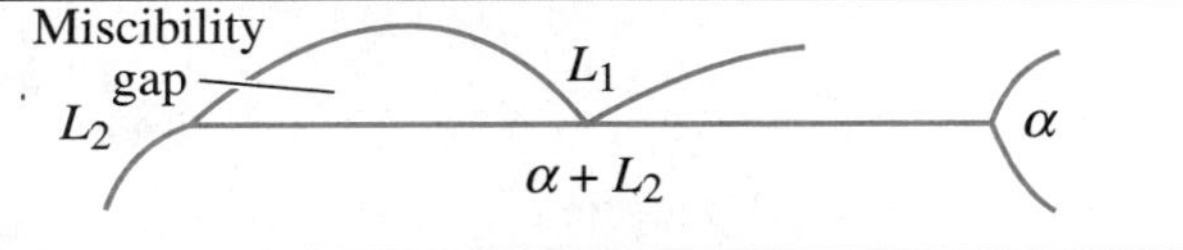
Eutectoid	$\gamma \rightarrow \alpha + \beta$	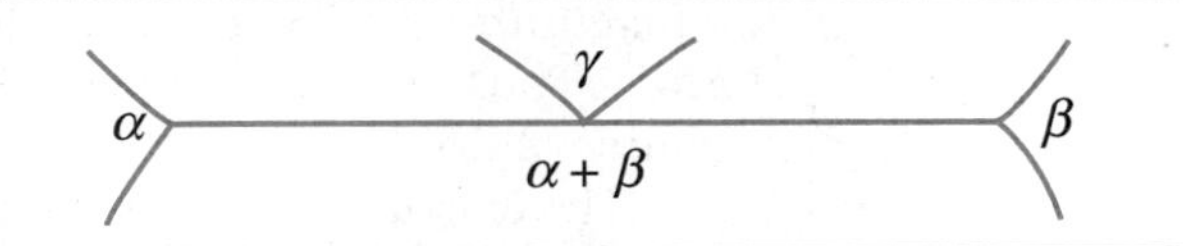
Peritectoid	$\alpha + \beta \rightarrow \gamma$	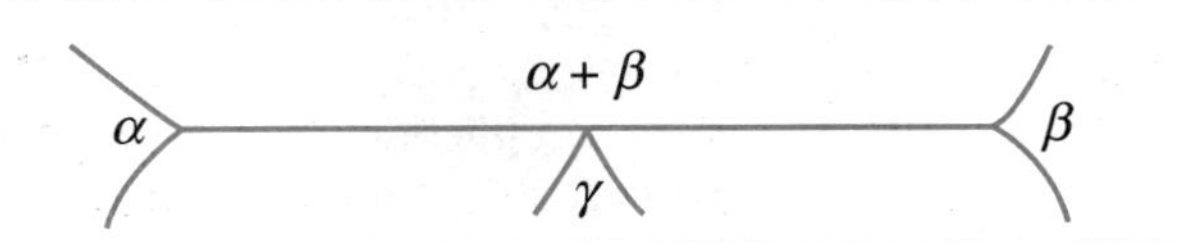

Figure 10-6   The five most important three-phase reactions in binary phase diagrams.

EXAMPLE 10-2 *Identifying Three-Phase Reactions*

Consider the binary phase diagram in Figure 10-7. Identify the three-phase reactions that occur.

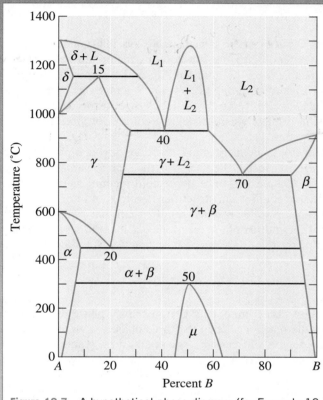

Figure 10-7 A hypothetical phase diagram (for Example 10-2).

SOLUTION

We find horizontal lines at 1150°C, 920°C, 750°C, 450°C, and 300°C:

1150°C: The in-betwen point is at 15% B. $\delta + L$ are present above the point, γ is present below. The reaction is:

$$\delta + L \rightarrow \gamma, \quad \text{a } peritectic$$

920°C: This reaction occurs at 40% B:

$$L_1 \rightarrow \gamma + L_2, \quad \text{a } monotectic$$

750°C: This reaction occurs at 70% B:

$$L \rightarrow \gamma + \beta, \quad \text{a } eutectic$$

450°C: This reaction occurs at 20% B:

$$\gamma \rightarrow \alpha + \beta, \quad \text{a } eutectoid$$

300°C: This reaction occurs at 50% B:

$$\alpha + \beta \rightarrow \mu \quad \text{or a } peritectoid$$

The eutectic, **peritectic**, and **monotectic** reactions are part of the solidification process. Alloys used for casting or soldering often take advantage of the low melting point of the eutectic reaction. The phase diagram of monotectic alloys contains a dome, or a **miscibility gap**, in which two liquid phases coexist. In the copper-lead system, the monotectic reaction produces tiny globules of dispersed lead, which improve the machinability of the copper alloy. Peritectic reactions lead to nonequilibrium solidification and segregation.

In many systems there is **metastable miscibility gap**.[8,9] In this case, the immiscibility dome extends into the sub-liquidus region. In some cases, the entire miscibility gap is metastable (i.e., the immiscibility dome is completely under the liquidus). These systems form such materials as Vycor™ and Pyrex® glasses, also known as phase-separated glasses.[10] R. Roy was the first scientist to describe the underlying science for the formation of these glasses using the concept of a metastable miscibility gap existing below the liquidus.[11]

The eutectoid and **peritectoid** reactions are completely solid-state reactions. The eutectoid reaction forms the basis for the heat treatment of several alloy systems, including steel (Chapter 11). The peritectoid reaction is extremely slow, often producing undesirable, nonequilibrium structures in alloys. As noted in Chapter 5, the rate of diffusion of atoms in solids is much smaller than that in liquids.

Each of these three-phase reactions occurs at a fixed temperature and composition. The Gibbs phase rule for a three-phase reaction is (at a constant pressure),

$$1 + C = F + P$$
$$F = 1 + C - P = 1 + 2 - 3 = 0,$$

(10-1)

since there are two components C in a binary phase diagram and three phases P are involved in the reaction. When the three phases are in equilibrium during the reaction, there are no degrees of freedom. As a result, these reactions are known as invariant. The temperature and the composition of each phase involved in the three-phase reaction are fixed. Note that of these five reactions discussed here only eutectic and eutectoid reactions can lead to dispersion strengthening.

10-4 The Eutectic Phase Diagram

The lead-tin (Pb-Sn) system contains only a simple eutectic reaction (Figure 10-8). This alloy system is the basis for the most common alloys used for soldering. As mentioned before, because of the toxicity of Pb, there is an intense effort underway to replace lead in Pb-Sn solders.[12,13] We will continue to use a Pb-Sn system, though, as a convenient way to discuss the eutectic phase diagram. Let's examine four classes of alloys in this system.

Solid Solution Alloys Alloys that contain 0 to 2% Sn behave exactly like the copper-nickel alloys; a single-phase solid solution α forms during solidification (Figure 10-9). These alloys are strengthened by solid-solution strengthening, by strain hardening, and by controlling the solidification process to refine the grain structure.

Alloys That Exceed the Solubility Limit Alloys containing between 2% and 19% Sn also solidify to produce a single solid solution α. However, as the alloy continues to cool, a solid-state reaction occurs, permitting a second solid phase (β) to precipitate from the original α phase (Figure 10-10).

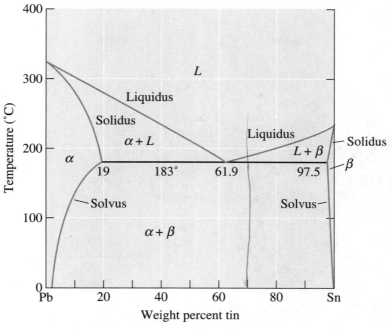

Figure 10-8 The lead-tin equilibrium phase diagram.

On this phase diagram, the α is a solid solution of tin in lead. However, the solubility of tin in the α solid solution is limited. At 0°C, only 2% Sn can dissolve in α. As the temperature increases, more tin dissolves into the lead until, at 183°C, the solubility of tin in lead has increased to 19% Sn. This is the maximum solubility of tin in lead. The

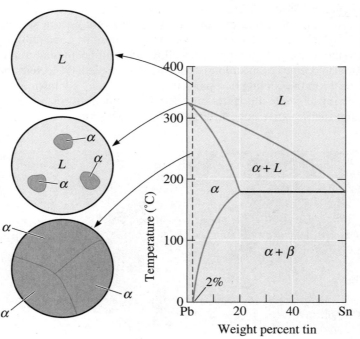

Figure 10-9
Solidification and microstructure of a Pb-2% Sn alloy. The alloy is a single-phase solid solution.

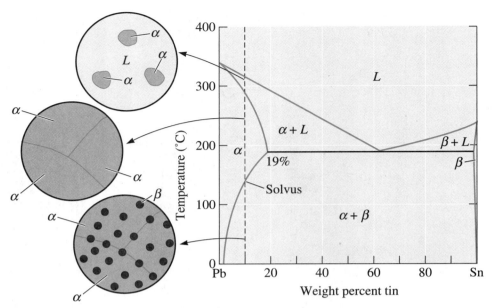

Figure 10-10 Solidification, precipitation, and microstructure of a Pb-10% Sn alloy. Some dispersion strengthening occurs as the β solid precipitates.

solubility of tin in solid lead at any temperature is given by the **solvus curve**. Any alloy containing between 2% and 19% Sn cools past the solvus, the solubility limit is exceeded, and a small amount of β forms.

We control the properties of this type of alloy by several techniques, including solid-solution strengthening of the α portion of the structure, controlling the microstructure produced during solidification, and controlling the amount and characteristics of the β phase. These types of compositions, which form a single solid phase at high temperatures and two solid phases at lower temperatures, are suitable for age or precipitate hardening. In Chapter 11, we will learn how nonequilibrium processes are needed to make precipitation hardened alloys. A component diagram (e.g., Figure 10-10) that shows a specific composition is known as an **isopleth**.[9] Determination of reactions that occur upon the cooling of a particular composition is known as an **isoplethal study**. The following example illustrates how certain calculations related to the composition of phases and their relative concentrations can be performed.

EXAMPLE 10-3   *Phases in the Lead–Tin (Pb-Sn) Phase Diagram*

Determine (a) the solubility of tin in solid lead at 100°C, (b) the maximum solubility of lead in solid tin, (c) the amount of β that forms if a Pb-10% Sn alloy is cooled to 0°C, (d) the masses of tin contained in the α and β phases, and (e) mass of lead contained in the α and β phases. Assume that the total mass of the Pb-10% Sn alloy is 100 grams.

SOLUTION

The phase diagram we need is shown in Figure 10-10. All percentages shown are weight %.

(a) The 100°C temperature intersects the solvus curve at 5% Sn. The solubility of tin (Sn) in lead (Pb) at 100°C therefore is 5%.

(b) The maximum solubility of lead (Pb) in tin (Sn), which is found from the tin-rich side of the phase diagram, occurs at the eutectic temperature of 183°C and is 97.5% Sn.

(c) At 0°C, the 10% Sn alloy is in a $\alpha + \beta$ region of the phase diagram. By drawing a tie line at 0°C and applying the lever rule, we find that:

$$\% \beta = \frac{10 - 2}{100 - 2} \times 100 = 8.2\%.$$

Note that the tie line intersects the solvus curve for solubility of Pb in Sn (on the right-hand side of the β-phase field) at a non-zero concentration of Sn. However, we can not read this accurately from the diagram; hence, we assume that the right-hand point for the tie line is 100% Sn. The % of α would be $(100 - \% \beta) = 91.8\%$. This means if we have 100 g of the 10% Sn alloy, it will consist of 8.2 g of the β phase and will consist of 91.8 g of the α phase.

(d) Note that 100 g of the alloy will consist of 10 g of Sn and 90 g of Pb. The Pb and Sn are distributed in two phases (i.e., α and β). The mass of Sn in the α phase = 2% Sn × 91.8 g of α phase = 0.02 × 91.8 g = 1.836 g. Since tin (Sn) appears in both the α and β phases, the mass of Sn in the β phase will be =(10 − 1.836) g = 8.164 g. Note that in this case the β phase at 0°C is nearly pure Sn.

(e) Let's now calculate the mass of lead in the two phases. The mass of Pb in the α phase will be equal to the mass of the α phase minus the mass of Sn in the α phase = 91.8 g − 1.836 g = 89.964 g. We could have also calculated this as:

$$\text{Mass of Pb in the } \alpha \text{ phase} = 98\% \text{ Sn} \times 91.8 \text{ g of } \alpha \text{ phase} = 0.98 \times 91.8 \text{ g}$$

$$= 89.964 \text{ g}$$

We know the total mass of the lead (90 g) and we also know the mass of lead in the α phase, therefore, the mass of Pb in the β phase = 90 − 89.964 = 0.036 g. This is consistent with what we said earlier (i.e., the β phase, in this case, is almost pure tin). The Figure 10-11 shows a summary of the calculations we have conducted here.

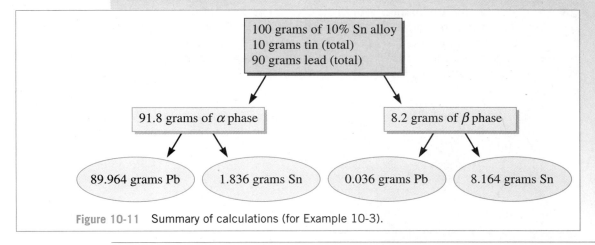

Figure 10-11 Summary of calculations (for Example 10-3).

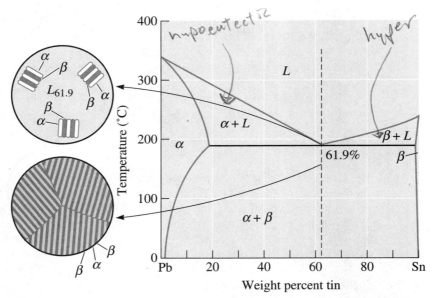

Figure 10-12   Solidification and microstructure of the eutectic alloy Pb-61.9% Sn.

Eutectic Alloys   The alloy containing 61.9% Sn has the eutectic composition (Figure 10-12). The word eutectic comes from the Greek word *eutectos* that means easily fused. Indeed, in a binary system showing one eutectic reaction, an alloy with a eutectic composition has the lowest melting temperature. This is the composition for which there is no freezing range (i.e., solidification of this alloy occurs at one temperature, 183°C in the Pb-Sn system). Above 183°C the alloy is all liquid and, therefore, must contain 61.9% Sn. After the liquid cools to 183°C, the eutectic reaction begins:

$$L_{61.9\% \text{ Sn}} \rightarrow \alpha_{19\% \text{ Sn}} + \beta_{97.5\% \text{ Sn}}$$

Two solid solutions—α and β—are formed during the eutectic reaction. The compositions of the two solid solutions are given by the ends of the eutectic line.

During solidification, growth of the eutectic requires both removal of the latent heat of fusion and redistribution of the two different atom species by diffusion. Since solidification occurs completely at 183°C, the cooling curve (Figure 10-13) is similar to that of a pure metal; that is, a thermal arrest or plateau occurs at the eutectic temperature. In Chapter 8, we had stated that alloys solidify over a range of temperatures (between the liquidus and solidus) known as the freezing range. Eutectic compositions are an exception to this rule since they transform from a liquid to a solid at a constant temperature (i.e., the eutectic temperature).

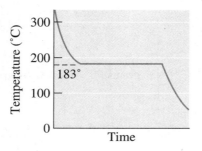

Figure 10-13
The cooling curve for an eutectic alloy is a simple thermal arrest, since eutectics freeze or melt at a single temperature.

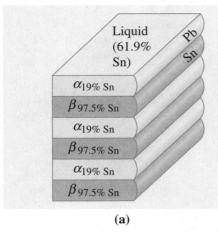

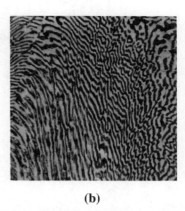

(a)   **(b)**

Figure 10-14 (a) Atom redistribution during lamellar growth of a lead-tin eutectic. Tin atoms from the liquid preferentially diffuse to the β plates, and lead atoms diffuse to the α plates. (b) Photomicrograph of the lead-tin eutectic microconstituent ($\times 400$).

As atoms are redistributed during eutectic solidification, a characteristic microstructure develops. In the lead-tin system, the solid α and β phases grow from the liquid in a **lamellar**, or plate-like, arrangement (Figure 10-14). The lamellar structure permits the lead and tin atoms to move through the liquid, in which diffusion is rapid, without having to move an appreciable distance. This lamellar structure is characteristic of numerous other eutectic systems.

The product of the eutectic reaction has a characteristic arrangement of the two solid phases called the **eutectic microconstituent**. In the Pb-61.9% Sn alloy, 100% of the eutectic microconstituent is formed, since all of the liquid goes through the reaction. The following example shows how the amount of eutectic alloy can be calculated.

EXAMPLE 10-4   *Amount of Phases in the Eutectic Alloy*

(a) Determine the amount and composition of each phase in a lead-tin alloy of eutectic composition. (b) Calculate the mass of phases present. (c) Calculate the amount of lead and tin in each phase, assuming you have 200 g of the alloy.

SOLUTION

(a) The eutectic alloy contains 61.9% Sn. We work the lever law at a temperature just below the eutectic—say, at 182°C, since that is the temperature at which the eutectic reaction is just completed. The fulcrum of our lever is 61.9% Sn. The ends of the tie line coincide approximately with the ends of the eutectic line.

$$\alpha: (Pb - 19\% \text{ Sn}) \quad \% \, \alpha = \frac{97.5 - 61.9}{97.5 - 19.0} \times 100 = 45.35\%$$

$$\beta: (Pb - 97.5\% \text{ Sn}) \quad \% \, \beta = \frac{61.9 - 19.0}{97.5 - 19.0} \times 100 = 54.65\%$$

Or we could state that the weight fraction of the α phase = 0.4535, that of the β phase is 0.5565.

A 200 g sample of the alloy would contain a total of $200 \times (0.6190) = 123.8$ g Sn and a balance of 76.2 g lead. The total mass of lead and tin cannot change as a result of conservation of mass. What changes is the mass of lead and tin in the different phases.

(b) At a temperature just below the eutectic:

The mass of the α phase in 200 g of the alloy

$$= \text{mass of the alloy} \times \text{fraction of the } \alpha \text{ phase}$$

$$= 200 \text{ g} \times 0.4535 = 90.7 \text{ g}$$

The amount of the β phase in 200 g of the alloy

$$= (\text{mass of the alloy} - \text{mass of the } \alpha \text{ phase})$$

$$= 200.0 \text{ g} - 90.7 \text{ g} = 109.3 \text{ g}$$

We could have also written this as:

Amount of β phase in 200 g of the alloy

$$= \text{mass of the alloy} \times \text{fraction of the } \beta \text{ phase}$$

$$= 200 \text{ g} \times 0.5465 = 109.3 \text{ g}$$

Thus, at a temperature just below the eutectic (i.e., at 182°C), the alloy contains 109.3 g of the β phase and 90.7 g of the α phase.

(c) Now let's calculate the masses of lead and tin in the α and β phases:

Mass of Pb in the α phase $=$ mass of the α phase in 200 g

$$\times (\text{concentration of Pb in } \alpha)$$

Mass of Pb in the α phase $= (90.7 \text{ g}) \times (1 - 0.190) = 73.467 \text{ g}$

Mass of Sn in the α phase $=$ mass of the α phase $-$ mass of Pb in the α phase

Mass of Sn in the α phase $= (90.7 - 73.467 \text{ g}) = 17.233 \text{ g}$

Mass of Pb in β phase $=$ mass of the β phase in 200 g $\times$ (wt. fraction Pb in β)

Mass of Pb in the β phase $= (109.3 \text{ g}) \times (1 - 0.975) = 2.73 \text{ g}$

Mass of Sn in the β phase $=$ total mass of Sn $-$ mass of Sn in the α phase

$$= 123.8 \text{ g} - 17.233 \text{ g} = 106.57 \text{ g}$$

Notice, that we could have obtained the same result by considering the total lead mass balance as follows:

Total mass of lead in the alloy $=$ mass of lead in the α phase

$$+ \text{mass of lead in the } \beta \text{ phase}$$

$$76.2 \text{ g} = 73.467 \text{ g} + \text{mass of lead in the } \beta \text{ phase}$$

Mass of lead in the β phase $= 76.2 - 73.467 \text{ g} = 2.73 \text{ g}$

This is same as what we calculated before. Figure 10-15 summarizes the various concentrations and masses.

This analysis confirms that most of the lead in the eutectic alloy gets concentrated in the α phase. Most of the tin gets concentrated in the β phase.

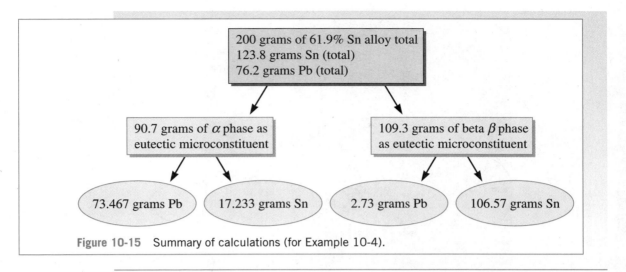

Figure 10-15 Summary of calculations (for Example 10-4).

Hypoeutectic and Hypereutectic Alloys   A **hypoeutectic alloy** is an alloy whose composition will be between that of the left-hand-side end of the tie line defining the eutectic reaction and the eutectic composition. As a hypoeutectic alloy containing between 19% and 61.9% Sn cools, the liquid begins to solidify at the liquidus temperature producing

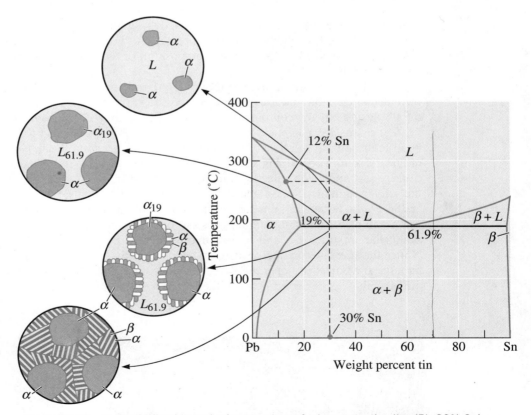

Figure 10-16 The solidification and microstructure of a hypoeutectic alloy (Pb-30% Sn).

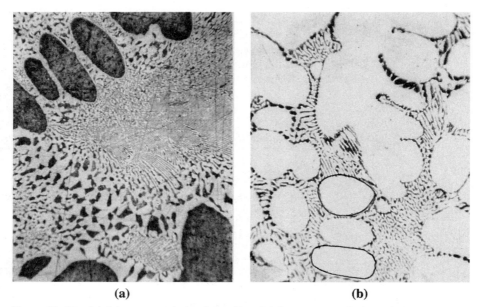

(a)   **(b)**

Figure 10-17   (a) A hypoeutectic lead-tin alloy. (b) A hypereutectic lead-tin alloy. The dark constituent is the lead-rich solid α, the light constituent is the tin-rich solid β, and the fine plate structure is the eutectic ($\times$400).

solid α. However, solidification is completed by going through the eutectic reaction (Figure 10-16). This solidification sequence occurs for compositions in which the vertical line corresponding to the original composition of the alloy crosses both the liquidus and the eutectic.

An alloy composition between that of the right-hand-side end of the tie line defining the eutectic reaction and the eutectic composition is known as a **hypereutectic alloy**. In the Pb-Sn system, any composition between 61.9% and 97.5% Sn is hypereutectic.

Let's consider a hypoeutectic alloy containing Pb-30% Sn and follow the changes in structure during solidification (Figure 10-16). On reaching the liquidus temperature of 260°C, solid α containing about 12% Sn nucleates. The solid α grows until the alloy cools to just above the eutectic temperature. At 184°C, we draw a tie line and find that the solid α contains 19% Sn and the remaining liquid contains 61.9% Sn. We note that at 184°C, the liquid contains the eutectic composition! When the alloy is cooled below 183°C, all of the remaining liquid goes through the eutectic reaction and transforms to a lamellar mixture of α and β. The microstructure shown in Figure 10-17(a) results. Notice that the eutectic microconstituent surrounds the solid α that formed between the liquidus and eutectic temperatures. The eutectic microconstituent is continuous and the primary phase is dispersed between the colonies of the eutectic microconstituent.

EXAMPLE 10-5   *Determination of Phases and Amounts in a Pb-30% Sn Hypoeutectic Alloy*

For a Pb-30% Sn alloy, determine the phases present, their amounts, and their compositions at 300°C, 200°C, 184°C, 182°C, and 0°C.

SOLUTION

Temperature (°C)	Phases	Compositions	Amounts
300	L	L: 30% Sn	$L = 100\%$
200	$\alpha + L$	L: 55% Sn	$L = \dfrac{30 - 18}{55 - 18} \times 100 = 32\%$
		α: 18% Sn	$\alpha = \dfrac{55 - 30}{55 - 18} \times 100 = 68\%$
184	$\alpha + L$	L: 61.9% Sn	$L = \dfrac{30 - 19}{61.9 - 19} \times 100 = 26\%$
		α: 19% Sn	$\alpha = \dfrac{61.9 - 30}{61.9 - 19} \times 100 = 74\%$
182	$\alpha + \beta$	α: 19% Sn	$\alpha = \dfrac{97.5 - 30}{97.5 - 19} \times 100 = 86\%$
		β: 97.5% Sn	$\beta = \dfrac{30 - 19}{97.5 - 19} \times 100 = 14\%$
0	$\alpha + \beta$	α: 2% Sn	$\alpha = \dfrac{100 - 30}{100 - 2} \times 100 = 71\%$
		β: 100% Sn	$\beta = \dfrac{30 - 2}{100 - 2} \times 100 = 29\%$

Note that in these calculations, the fractions have been rounded off to the nearest %. This can pose problems if we were to calculate masses of different phases, in that you may not be able to preserve mass balance. It is usually a good idea not to round these percentages if you are going to perform calculations concerning amounts of different phases or masses of elements in different phases.

We call the solid α phase that forms when the liquid cooled from the liquidus to the eutectic the **primary** or **proeutectic microconstituent**. This solid α did not take part in the eutectic reaction. Thus, the morphology and appearance of this α phase is very distinct from that of the α phase that appears in the eutectic microconstituent. Often we find that the amounts and compositions of the microconstituents are of more use to us than the amounts and compositions of the phases.

EXAMPLE 10-6   *Microconstituent Amount and Composition for a Hypoeutectic Alloy*

Determine the amounts and compositions of each microconstituent in a Pb-30% Sn alloy immediately after the eutectic reaction has been completed.

SOLUTION

This is a hypoeutectic composition. Therefore, the microconstituents expected are primary α and eutectic. Note that we still have only two phases (α and β). We can determine the amounts and compositions of the microconstituents if we look at how they form. The *primary* α microconstituent is all of the solid α that forms before the alloy cools to the eutectic temperature; the eutectic microconstituent is all of the liquid that goes through the eutectic reaction. At a temperature just above the eutectic—say, 184°C—the amounts and compositions of the two phases are:

$$\alpha: 19\% \text{ Sn} \quad \% \alpha = \frac{61.9 - 30}{61.9 - 19} \times 100 = 74\% = \% \text{ primary } \alpha$$

$$L: 61.9\% \text{ Sn} \quad \% L = \frac{30 - 19}{61.9 - 19} \times 100 = 26\% = \% \text{ eutectic at } 182°C$$

Thus, the primary alpha microconstituent is obtained by determining the amount of α present at the temperature just above eutectic. The amount of eutectic microconstituent at a temperature *just below* the eutectic (e.g., 182°C) is determined by calculating the amount of liquid *just above* the eutectic temperature (e.g., at 184°C), since all of this liquid of eutectic composition is transformed into the eutectic microconstituent. Note that at the eutectic temperature (183°C) the eutectic reaction is in progress (formation of the pro-eutectic α is complete); hence, the amount of the eutectic microconstituent at 183°C will change with time (starting at 0% and ending at 26% eutectic, in this case). Please be certain that you understand this example since many students tend to miss how the calculation is performed.

When the alloy cools below the eutectic to 182°C, all of the liquid at 184°C transforms to eutectic and the composition of the eutectic microconstituent is 61.9% Sn. The solid α present at 184°C remains unchanged after cooling to 182°C and is the primary microconstituent.

The cooling curve for a hypoeutectic alloy is a composite of those for solid-solution alloys and "straight" eutectic alloys (Figure 10-18). A change in slope occurs at the liquidus as primary α begins to form. Evolution of the latent heat of fusion slows the cooling rate as the solid α grows. When the alloy cools to the eutectic temperature, a thermal arrest is produced as the eutectic reaction proceeds at 183°C. The solidification sequence is similar in a hypereutectic alloy, giving the microstructure shown in Figure 10-17(b).

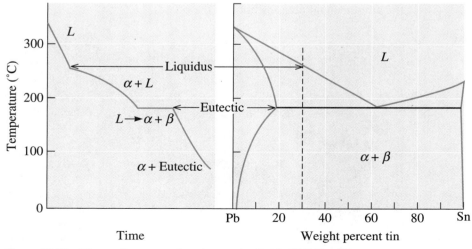

Figure 10-18   The cooling curve for a hypoeutectic Pb-30% Sn alloy.

10-5   Strength of Eutectic Alloys

Each phase in the eutectic alloy is, to some degree, solid-solution strengthened. In the lead-tin system, α, which is a solid solution of tin in lead, is stronger than pure lead (Chapter 9). Some eutectic alloys can be strengthened by cold working. We also control grain size by adding appropriate inoculants or grain refiners during solidification. Finally, we can influence the properties by controlling the amount and microstructure of the eutectic.

Eutectic Colony Size   Eutectic colonies, or grains, each nucleate and grow independently. Within each colony, the orientation of the lamellae in the eutectic microconstituent is identical. The orientation changes on crossing a colony boundary [Figure 10-19(a)]. We can refine the eutectic colonies and improve the strength of the eutectic alloy by inoculation (Chapter 8).

Interlamellar Spacing   The **interlamellar spacing** of a eutectic is the distance from the center of one α lamella to the center of the next α lamella [Figure 10-19(b)]. A small interlamellar spacing indicates that the amount of α–β interface area is large. A small interlamellar spacing therefore increases the strength of the eutectic.

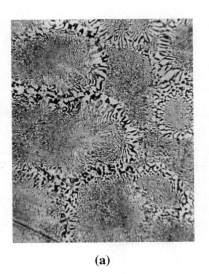

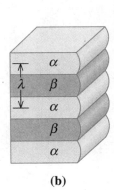

Figure 10-19
(a) Colonies in the lead-tin eutectic (×300).
(b) The interlamellar spacing in a eutectic microstructure.

(a)   **(b)**

Figure 10-20
The effect of growth rate on the interlamellar spacing in the lead-tin eutectic.

The interlamellar spacing is determined primarily by the growth rate of the eutectic,

$$\lambda = cR^{-1/2} \tag{10-2}$$

where R is the growth rate (cm/s) and c is a constant. The interlamellar spacing for the lead-tin eutectic is shown in Figure 10-20. We can increase the growth rate R, and consequently reduce the interlamellar spacing, by increasing the cooling rate or reducing the solidification time. The following example demonstrates how the solidification of a Pb-Sn alloy can be controlled.

EXAMPLE 10-7   *Design of a Directional Solidification Process*

Design a process to produce a single "grain" of Pb-Sn eutectic microconstituent in which the interlamellar spacing is 0.00034 cm.

SOLUTION

We could use a directional solidification (DS) process to produce the single grain, while controlling the growth rate to assure that the correct interlamellar spacing is achieved. To obtain $\lambda = 0.00034$ cm, we need a growth rate of 0.00025 cm/s (Figure 10-20).

Figure 10-21 shows how we might achieve this growth rate. The Pb-61.9% Sn alloy would be melted in a mold within a furnace. The mold would be withdrawn from the furnace at the rate of 0.00025 cm/s, with the mold

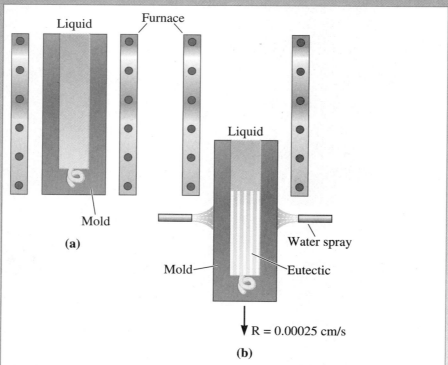

Figure 10-21 Directional solidification of a Pb-Sn eutectic alloy: (a) The metal is melted in the furnace, and (b) the mold is slowly withdrawn from the furnace and the casting is cooled (for Example 10-7).

quenched with a water spray as it emerges from the furnace. If only one eutectic colony grows through the spiral, all of the lamellae are lined up in parallel with the growth direction. If the part to be made is 10 cm long, it would take 40,000 s, or 11 h, to produce the part.

As discussed in Chapter 8, this method has been used to produce directionally solidified, high-temperature jet engines turbine blades from nickel-based superalloys.

Amount of Eutectic   We also control the properties by the relative amounts of the primary microconstituent and the eutectic. In the lead-tin system, the amount of the eutectic microconstituent changes from 0% to 100% when the tin content increases from 19% to 61.9%. With increasing amounts of the stronger eutectic microconstituent, the strength of the alloy increases (Figure 10-22). Similarly, when we increase the lead added to tin from 2.5% to 38.1% Pb, the amount of primary β in the hypereutectic alloy decreases, the amount of the strong eutectic increases, and the strength increases. When both individual phases have about the same strength, the eutectic alloy is expected to have the highest strength due to effective dispersion strengthening.

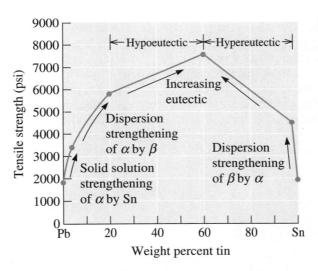

Figure 10-22
The effect of the composition and strengthening mechanism on the tensile strength of lead-tin alloys.

Microstructure of the Eutectic   Not all eutectics give a lamellar structure. The shapes of the two phases in the microconstituent are influenced by the cooling rate, the presence of impurity elements, and the nature of the alloy (Figure 10-24).

The aluminum-silicon eutectic phase diagram (Figure 10-23) forms the basis for a number of important commercial alloys. However, the silicon portion of the eutectic grows as thin, flat plates that appear needle-like in a photomicrograph [Figure 10-24(a)]. The brittle silicon platelets concentrate stresses and reduce ductility and toughness.

The eutectic microstructure in aluminum-silicon alloys is altered by modification. **Modification** causes the silicon phase to grow as thin, interconnected rods between aluminum dendrites [Figure 10-24(b)], improving both tensile strength and % elongation. In two dimensions, the modified silicon appears to be composed of tiny, round particles. Rapidly cooled alloys, such as those in die casting, are naturally modified during solidification. At slower cooling rates, however, about 0.02% Na or 0.01% Sr must be added to cause modification.

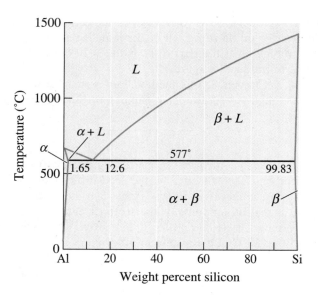

Figure 10-23
The aluminum-silicon phase diagram.

The shape of the primary phase is also important. Often the primary phase grows in a dendritic manner; decreasing the secondary dendrite arm spacing of the primary phase may improve the properties of the alloy. However, in hypereutectic aluminum-silicon alloys, coarse β is the primary phase [Figure 10-25(a)]. Because β is hard, the hypereutectic alloys are wear-resistant and are used to produce automotive engine parts. But the coarse β causes poor machinability and gravity segregation (where the primary β floats to the surface of the casting during freezing). Addition of 0.05% phosphorus (P) encourages nucleation of primary silicon, refines its size, and minimizes its deleterious qualities [Figure 10-25(b)]. The two examples following show how eutectic compositions can be designed to achieve certain levels of mechanical properties.

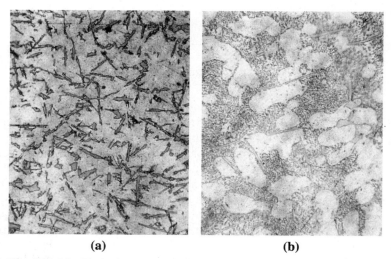

(a)   **(b)**

Figure 10-24 Typical eutectic microstructures: (a) needle-like silicon plates in the aluminum-silicon eutectic ($\times 100$), and (b) rounded silicon rods in the modified aluminum-silicon eutectic ($\times 100$).

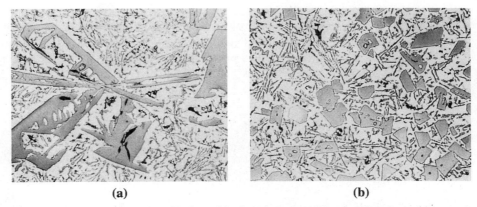

(a)   (b)

Figure 10-25 The effect of hardening with phosphorus on the microstructure of hypereutectic aluminum-silicon alloys: (a) coarse primary silicon, and (b) fine primary silicon, as refined by phosphorus addition (×75). (*From* ASM Handbook, *Vol. 7, (1972), ASM International, Materials Park, OH 44073.*)

EXAMPLE 10-8   *Design of Materials for a Wiping Solder*

One way to repair dents in a metal is to wipe a partly liquid-partly solid material into the dent, then allow this filler material to solidify. For our application, the wiping material should have the following specifications: (1) a melting temperature below 230°C, (2) a tensile strength in excess of 6000 psi, (3) be 60% to 70% liquid during application, and (4) the lowest possible cost. Design an alloy and repair procedure that will meet these specifications.

SOLUTION

Let's see if one of the Pb-Sn alloys will satisfy these conditions. First, the alloy must contain more than 40% Sn in order to have a melting temperature below 230°C (Figure 10-8). This low temperature will make it easier for the person doing the repairs to apply the filler.

Second, Figure 10-22 indicates that the tin content must lie between 23% and 80% to achieve the required 6000 psi tensile strength. In combination with the first requirement, any alloy containing between 40 and 80% Sn will be satisfactory.

Third, the cost of tin is about \$5500/ton whereas that of lead is \$550/ton. Thus, an alloy of Pb-40% Sn might be the most economical choice. There are other considerations, as well, such as: What is the geometry? Can the alloy flow well under that geometry (i.e., the viscosity of the molten metal)?

Finally, the filler material must be at the correct temperature in order to be 60% to 70% liquid. As the calculations below show, the temperature must be between 200°C and 210°C:

$$\% \, L_{200} = \frac{40 - 18}{55 - 18} \times 100 = 60\%$$

$$\% \, L_{210} = \frac{40 - 17}{50 - 17} \times 100 = 70\%$$

Our recommendation, therefore, is to use a Pb-40% Sn alloy applied at 205°C, a temperature at which there will be 65% liquid and 35% primary α. As mentioned before, we should also pay attention to the toxicity of lead and any legal liabilities the use of such materials may cause. A number of new lead free solders have been developed.[12,13]

EXAMPLE 10-9   *Design of a Wear-Resistant Part*

Design a lightweight, cylindrical component that will provide excellent wear-resistance at the inner wall, yet still have reasonable ductility and toughness overall. Such a product might be used as a cylinder liner in an automotive engine.

SOLUTION

Many wear-resistant parts are produced from steels, which have a relatively high density, but the hypereutectic Al-Si alloys containing primary β may provide the wear-resistance that we wish at one-third the weight of the steel.

Since the part to be produced is cylindrical in shape, centrifugal casting (Figure 10-26) might be a unique method for producing it. In centrifugal casting, liquid metal is poured into a rotating mold and the centrifugal force produces a hollow shape. In addition, material that has a high density is spun to the outside wall of the casting, while material that has a lower density than the liquid migrates to the inner wall.

When we centrifugally cast a hypereutectic Al-Si alloy, primary β nucleates and grows. The density of the β phase (if we assume it to be same as that of pure Si) is, according to Appendix A, 2.33 g/cm^3, compared with a density near 2.7 g/cm^3 for aluminum. As the primary β particles precipitate from the liquid, they are spun to the inner surface. The result is a casting that is composed of

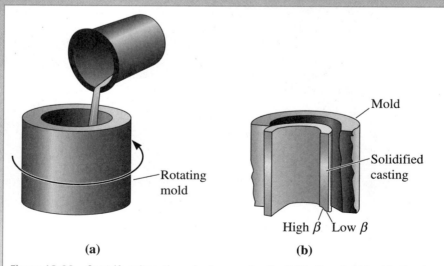

(a)   (b)

Figure 10-26   Centrifugal casting of a hypereutectic Al-Si alloy: (a) Liquid alloy is poured into a rotating mold, and (b) the solidified casting is hypereutectic at the inner diameter and eutectic at the outer diameter (for Example 10-9).

posed of eutectic microconstituent (with reasonable ductility) at the outer wall and a hypereutectic composition, containing large amounts of primary β, at the inner wall.

A typical alloy used to produce aluminum engine components is Al-17% Si. From Figure 10-23, the total amount of primary β that can form is calculated at 578°C, just above the eutectic temperature:

$$\% \text{ Primary } \beta = \frac{17 - 12.6}{99.83 - 12.6} \times 100 = 5.0\%$$

Although only 5.0% primary β is expected to form, the centrifugal action can double or triple the amount of β at the inner wall of the casting.

10-6 Eutectics and Materials Processing

Manufacturing processes take advantage of the low melting temperature associated with the eutectic reaction. The Pb-Sn alloys are the basis for a series of alloys used to produce filler materials for soldering (Chapter 8). If, for example, we wish to join copper pipe, individual segments can be joined by introducing the low-melting-point eutectic Pb-Sn alloy into the joint [Figure 10-27(a)]. The copper is heated just above the eutectic temperature. The heated copper melts the Pb-Sn alloy, which is then drawn into the thin gap by capillary action. When the Pb-Sn alloy cools and solidifies, the copper is joined. The prospects of corrosion of such pipes and the introduction of lead (Pb) into water must also be factored in.

Many casting alloys are also based on eutectic alloys. Liquid can be melted and poured into a mold at low temperatures, reducing energy costs involved in melting, minimizing casting defects such as gas porosity, and preventing liquid metal-mold reactions. Cast iron (Chapter 12) and many aluminum alloys (Chapter 13) are eutectic alloys.

Although most of this discussion has been centered around metallic materials, it is important to recognize that the eutectics are very important in many ceramic systems (Chapter 14). Formation of eutectics played a role in the successful formation of glasslike materials known as the Egyptian faience.[14] The sands of the Nile River Valley contained appreciable amounts of limestone ($CaCO_3$). Plant ash contains considerable amounts of potassium and sodium oxide is and used to cause the sand to melt at lower temperatures by the formation of the eutectics.

Silica and alumina are the most widely used ceramic materials. Figure 10-27(b) shows a phase diagram for the Al_2O_3-SiO_2.[9] Notice the eutectic at $\sim$1587°C. The dashed lines on this diagram show metastable extensions of the liquidus and metastable miscibility gaps.[8,10,15] As mentioned before, the existence of these gaps makes it possible to make technologically useful products such as the Vycor^{TM} and the Pyrex$^®$ glasses. A Vycor^{TM} glass is made by first melting (approximately at 1500°C) silica (63%), boron oxide (27%), sodium oxide (7%), and alumina (3%). The glass is then formed into desired shapes. During glass formation, the glass has phase separated (because of the metastable miscibility gap) into boron-oxide rich and silica rich regions. The boron oxide rich regions are dissolved using an acid. The porous object is sintered to form Vycor^{TM} glass that contains 95% silica, 4% boron oxide, and 1% sodium oxide.[10] It would be very difficult to achieve a high silica glass such as this without resorting to the technique described above. Pyrex$^®$ glasses contain about 80% silica,

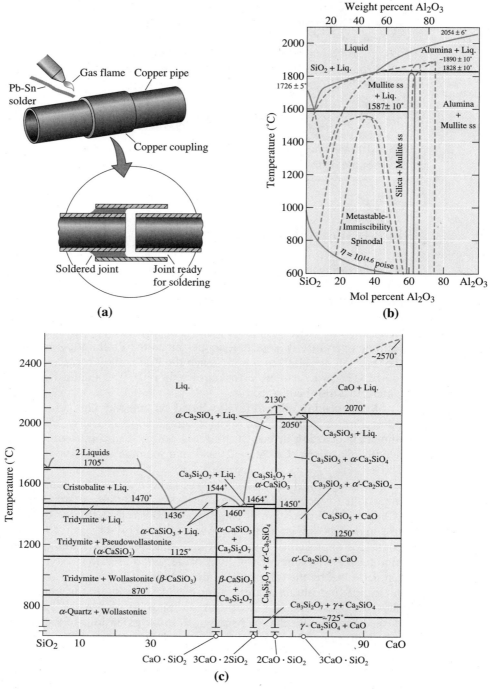

Figure 10-27 (a) A Pb-Sn eutectic alloy is often used during soldering to assemble parts. A heat source, such as a gas flame, heats both the parts and the filler material. The filler is drawn into the joint and solidifies. (b) A phase diagram for Al_2O_3-SiO_2.[9] (Adapted from *Introduction to Phase Equilibria in Ceramics*, by Bergeron, C.G. and Risbud, S.H., The American Ceramic Society, Inc., 1984, page 44.) (c) A phase diagram for the CaO-SiO_2 system. (*Source: Adapted from* Introduction to Phase Equilibria, *by C.G. Bergeron and S.H. Risbud, pp. 44 and 45, Figs. 3-36 and 3-37. Copyright © 1984 American Ceramic Society.*)

13% boron oxide, 4% sodium oxide, 2% alumina. These are used widely in making laboratory ware (i.e., beakers, etc.) and household products.

Figure 10-27(c) shows a binary phase diagram for the CaO-SiO_2 system.[9] Compositions known as the E-glass or S-glass are used to make the fibers that go into fiber-reinforced plastics. These glasses are made by melting silica sand, limestone, boric acid at about 1260°C.[7] The glass is then drawn into fibers. The E-glass (the letter "E" stands for "electrical", as the glass was originally made for electrical insulation) contains approximately 52–56 wt.% silica, 12–16% Al_2O_3, 5–10% B_2O_3, 0–5% MgO, 0–2% Na_2O, 0–2% K_2O. The "S"-glass (the letter "S" represents "strength") contains approximately 65 wt.% silica, 12–25% Al_2O_3, 10% MgO, 0–2% Na_2O, 0–2% K_2O.[16]

The determination of ceramic phase diagrams using experimental techniques involves heat treatment and the processing of components (which means crushing and grinding and mixing high purity powders to achieve uniformity) at different temperatures to ensure equilibrium. The materials equilibriated at different temperatures are then quenched rapidly to room temperature. Assuming that the structure and phase assembly that existed at the heat treatment/processing temperature are the same in the quenched material, x-ray diffraction (XRD) and electron microscopy (Chapter 3) analyses can be performed. Often, the attainment of equilibrium is very difficult in ceramic systems since the reaction kinetics tend to be slow. As a result, it is not unusual to find different variations and corrections made to phase diagrams for ceramic (and even metallic) systems. For metallic systems, the determination of cooling curves can be used to generate phase diagrams. In Chapter 9, we described the concept of cooling curves. As mentioned before in Chapter 9, many databases (e.g., those from ASM International, National Institute of Standards and Technology (NIST), and the American Ceramic Society) and computerized interfaces (e.g., ThermoCalc™) have made it easier to locate and use phase diagrams.[17]

Formation of the eutectics helps in the manufacture of inorganic glasses. Many common glasses are based on SiO_2, which melts at 1710°C. By adding Na_2O to SiO_2, a eutectic reaction is introduced, with a eutectic temperature of about 790°C. The SiO_2-Na_2O glass can be produced at a low temperature. This is why we add many oxides to form silicate glasses. Some oxides, such as sodium oxide, help melt the glass at lower temperatures (i.e., they act as a flux). A glass made from sodium oxide and silica will be water soluble and is known as "water glass." We can add lime or calcium carbonate ($CaCO_3$) to the melt. Calcium carbonate decomposes and forms CaO. The CaO also acts as a flux, producing compositions that do not dissolve in water. A soda lime glass (or A-glass) is basically a glass made using sodium oxide, calcium oxide (added as lime or $CaCO_3$), and silica in the form of sand.[16]

The eutectic reaction may be used to speed diffusion bonding or to increase the rate of sintering of compacted powders for both metal and ceramic systems. In both cases, a liquid is produced to join dissimilar materials or powder particles, even though the temperature at which the processing is done is below the melting temperature of the individual constituents involved in the process (liquid phase sintering).

Sometimes, however, the formation of the eutectic is undesirable. Because the eutectic is the last to solidify, it surrounds the primary phases. Eutectics that are brittle, therefore, embrittle the overall alloy, even if only a small percentage of the eutectic microconstituent is present in the structure. Deformation of such an alloy could cause failure through the brittle eutectic.

As another example, Al_2O_3 has a high melting point (2020°C), which makes it attractive as a refractory for containing liquid steel. The melting temperature of CaO (2570°C) is even higher. If an Al_2O_3 brick is placed into contact with a CaO brick, a series of eutectics is produced, giving a liquid with a melting temperature below the usual steel-making temperature. Thus, the refractory containing the liquid steel may fail. Similar considerations apply to silica bricks used in steel making. If a small amount of alumina is present, a eutectic liquid can form rather easily and this can lower the load-bearing capacity of the bricks considerably.

10-7   Nonequilibrium Freezing in the Eutectic System

Suppose we have an alloy, such as Pb-15% Sn, that ordinarily solidifies as a solid solution alloy. The last liquid should freeze near 230°C, well above the eutectic. However, if the alloy cools too quickly, a nonequilibrium solidus curve is produced (Figure 10-28). The primary α continues to grow until, just above 183°C, the remaining nonequilibrium liquid contains 61.9% Sn. This liquid then transforms to the eutectic microconstituent, surrounding the primary α. For the conditions shown in Figure 10-28, the amount of nonequilibrium eutectic is:

$$\% \text{ eutectic} = \frac{15 - 10}{61.9 - 10} \times 100 = 9.6\%$$

When heat treating an alloy such as Pb-15% Sn, we must keep the maximum temperature below the eutectic temperature of 183°C to prevent hot shortness or partial melting

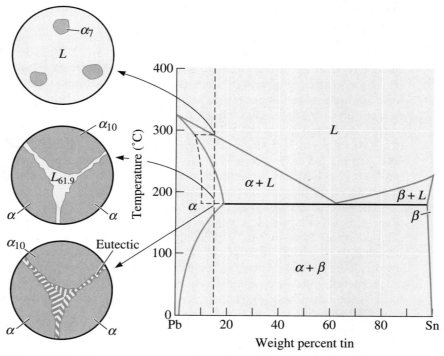

Figure 10-28   Nonequilibrium solidification and microstructure of a Pb-15% Sn alloy. A nonequilibrium eutectic microconstituent can form if the solidification is too rapid.

(Chapter 9). This concept is very important in the precipitation, or age, hardening of metallic alloys such as those in the Al-Cu system.

10-8 Ternary Phase Diagrams

Many alloy systems are based on three or even more elements. When three elements are present, we have a **ternary** alloy. For a ternary system under constant pressure a maximum of four phases can coexist and this will define an invariant point. To describe the changes in structure with temperature, we must draw a three-dimensional phase diagram. Figure 10-29 shows a hypothetical **ternary phase diagram** made up of elements *A*, *B*, and *C*. Note that two binary eutectics are present on the two visible faces of the diagram; a third binary eutectic between elements *B* and *C* is hidden on the back of the plot.

It is difficult to use the three-dimensional ternary plot; however, we can present the information from the diagram in two dimensions by any of several methods, including the liquidus plot and the isothermal plot.

Liquidus Plot We note in Figure 10-29 that the temperature at which freezing begins is shaded. We could transfer these temperatures for each composition onto a triangular diagram, as in Figure 10-30, and plot the liquidus temperatures as isothermal contours. This presentation is helpful in predicting the freezing temperature of the material. The liquidus plot also gives the identity of the primary phase that forms during solidification for any given composition.

Isothermal Plot The isothermal plot shows the phases present in the material at a particular temperature. It is useful in predicting the phases and their amounts and compositions at that temperature. Figure 10-31 shows an isothermal plot from Figure 10-29 at room temperature.

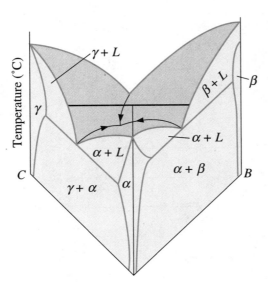

Figure 10-29
Hypothetical ternary phase diagram. Binary phase diagrams are present at the three faces.

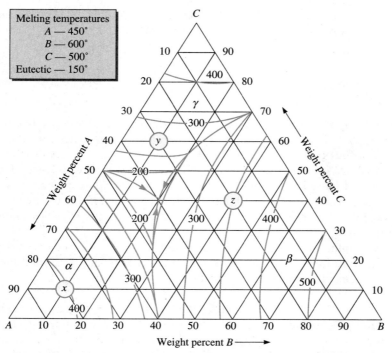

Figure 10-30 A liquidus plot for the hypothetical ternary phase diagram. The circles labeled *x*, *y*, and *z* refer to the different compositions discussed in Example 10-10.

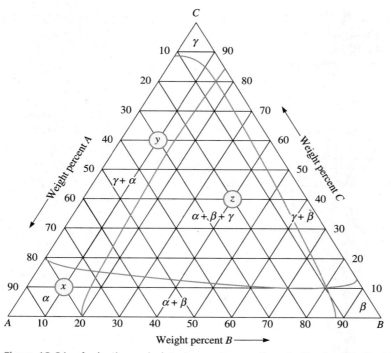

Figure 10-31 An isothermal plot at room temperature for the hypothetical ternary phase diagram. The circles labeled *x*, *y*, and *z* correspond to the different compositions discussed in Example 10-10.

EXAMPLE 10-10   *Determination of Liquidus on a Ternary Phase Diagram*

Using the ternary plots in Figures 10-30 and 10-31, determine the liquidus temperature, the primary phase that forms during solidification, and the phases at room temperature for the following materials:

Point *x*: 10% *B*, 10% *C*, balance *A*

Point *y*: 10% *B*, 60% *C*, balance *A*

Point *z*: 40% *B*, 40% *C*, balance *A*

SOLUTION

The composition 10% *B*-10% *C* balance *A* is located at point *x* in the figures; from the isotherm in this region, the liquidus temperature is 400°C. The primary phase, as indicated in the diagram, is α. The final structure (Figure 10-31) is all α.

The composition 10% *B*-60% *C* balance *A* is located at point *y*; by interpolating the isotherms in this region, the liquidus temperature is about 270°C. The primary phase that forms in this section of the diagram is γ, and the room temperature phases are α and γ.

The composition 40% *B*-40% *C* balance *A* is located at point *z*; the liquidus temperature at this point is 350°C, and the point is in the primary β region. The room temperature phases are α, β, and γ.

SUMMARY

◆ By producing a material containing two or more phases, dispersion strengthening is obtained. In metallic materials, the boundary between the phases impedes the movement of dislocations and improves strength. Introduction of multiple phases may provide other benefits, including improvement of the fracture toughness of ceramics and polymers.

◆ For optimum dispersion strengthening, particularly in metallic materials, a large number of small, hard, discontinuous dispersed phase particles should form in a soft, ductile matrix to provide the most effective obstacles to dislocations. Round dispersed phase particles minimize stress concentrations, and the final properties of the alloy can be controlled by the relative amounts of these and the matrix.

◆ Intermetallic compounds, which normally are strong but brittle, are frequently introduced as dispersed phases.

◆ Matrix structures based on intermetallic compounds have been developed to take advantage of their high-temperature properties.

◆ Phase diagrams for materials containing multiple phases normally contain one or more three-phase reactions:

   ▪ The eutectic reaction permits liquid to solidify as an intimate mixture of two phases. By controlling the solidification process, we can achieve a wide range of properties. Some of the factors that can be controlled include: the grain size or secondary dendrite arm spacings of primary microconstituents, the colony size of the eutectic microconstituent, the interlamellar spacing within the eutectic microconstituent, the microstructure, or shape, of the phases within the eutectic microconstituent, and the amount of the eutectic microconstituent that forms.

- The eutectoid reaction causes a solid to transform to a mixture of two other solids. As shown in the next chapter, heat treatments to control the eutectoid reaction provide an excellent basis for dispersion strengthening.

- In the peritectic and peritectoid reactions, two phases transform to a single phase on cooling. Dispersion strengthening does not occur, and severe segregation problems are often encountered.

- The monotectic reaction produces a mixture of a solid and liquid. Although this reaction does not provide dispersion strengthening, it does provide other benefits, such as good machinability, to some alloys.

GLOSSARY

Age hardening   A strengthening mechanism that relies on a sequence of solid-state phase transformations in generating a dispersion of ultrafine particles of a second phase. This is same as precipitation hardening. Age hardening is a form of dispersion strengthening. (Chapter 11).

Dispersion strengthening   Increasing the strength of a material by forming more than one phase. By proper control of the size, shape, amount, and individual properties of the phases, excellent combinations of properties can be obtained.

Eutectic   A three-phase invariant reaction in which one liquid phase solidifies to produce two solid phases.

Eutectic microconstituent   A characteristic mixture of two phases formed as a result of the eutectic reaction.

Eutectoid   A three-phase invariant reaction in which one solid phase transforms to two different solid phases.

Hyper-   A prefix indicating that the composition of an alloy is more than the composition at which a three-phase reaction occurs.

Hypereutectic alloys   An alloy composition between that of the right-hand-side end of the tie line defining the eutectic reaction and the eutectic composition.

Hypo-   A prefix indicating that the composition of an alloy is less than the composition at which a three-phase reaction occurs.

Hypoeutectic alloy   An alloy composition between that of the left-hand-side end of the tie line defining the eutectic reaction and the eutectic composition.

Interlamellar spacing   The distance between the center of a lamella or plate of one phase and the center of the adjoining lamella or plate of the same phase.

Intermediate solid solution   A nonstoichiometric intermetallic compound displaying a range of compositions.

Intermetallic compound   A compound formed of two or more metals that has its own unique composition, structure, and properties.

Interphase interface   The boundary between two phases in a microstructure. In metallic materials, this boundary resists dislocation motion and provides dispersion strengthening and precipitation hardening.

Isopleth   A line on a phase diagram that shows constant chemical composition.

Isoplethal study   Determination of reactions and microstructural changes that are expected while studying a particular chemical composition in a system.

Isothermal plot   A horizontal section through a ternary phase diagram showing the phases present at a particular temperature.

Lamella   A thin plate of a phase that forms during certain three-phase reactions, such as the eutectic or eutectoid.

Liquidus plot   A two-dimensional plot showing the temperature at which a three-component alloy system begins to solidify on cooling.

Matrix   The continuous solid phase in a complex microstructure. Solid dispersed phase particles may form within the matrix.

Metastable miscibility gap   A miscibility gap that extends below the liquidus or exists completely below the liquidus. Two liquids that are immiscible continue to exist as liquids and remain unmixed. These systems form the basis for Vycor™ and Pyrex® glasses.

Microconstituent   A phase or mixture of phases in an alloy that has a distinct appearance. Frequently, we describe a microstructure in terms of the microconstituents rather than the actual phases.

Miscibility gap   A region in a phase diagram in which two phases, with essentially the same structure, do not mix, or have no solubility in one another.

Modification   Addition of alloying elements, such as sodium or strontium, which change the microstructure of the eutectic microconstituent in aluminum-silicon alloys.

Monotectic   A three-phase reaction in which one liquid transforms to a solid and a second liquid on cooling.

Nonstoichiometric intermetallic compound   A phase formed by the combination of two components into a compound having a structure and properties different from either component. The nonstoichiometric compound has a variable ratio of the components present in the compound (see also Intermediate solid solution).

Ordered crystal structure   Solid solutions in which the different atoms occupy specific, rather than random, sites in the crystal structure.

Peritectic   A three-phase reaction in which a solid and a liquid combine to produce a second solid on cooling.

Peritectoid   A three-phase reaction in which two solids combine to form a third solid on cooling.

Precipitate   A solid phase that forms from the original matrix phase when the solubility limit is exceeded. We often use the term precipitate, as opposed to dispersed phase particles, for alloys formed by precipitation or age hardening. In most cases, we try to control the formation of the precipitate second phase particles to produce the optimum dispersion strengthening or age hardening.

Precipitation hardening   A strengthening mechanism that relies on a sequence of solid-state phase transformations in generating a dispersion of ultrafine precipitates of a second phase (Chapter 11). This is same as age hardening. It is a form of dispersion strengthening.

Primary microconstituent   The microconstituent that forms before the start of a three-phase reaction.

Solvus   A solubility curve that separates a single-solid phase region from a two-solid phase region in the phase diagram.

Stoichiometric intermetallic compound   A phase formed by the combination of two components into a compound having a structure and properties different from either component. The stoichiometric intermetallic compound has a fixed ratio of the components present in the compound.

Ternary alloy   An alloy formed by combining three elements or components.

Ternary phase diagram   A phase diagram between three components showing the phases present and their compositions at various temperatures. This diagram requires a three-dimensional plot or is presented as two-dimensional isothermal sections of a three-dimensional diagram.

ADDITIONAL INFORMATION

1. MEYERS, M.A. and K.K. CHAWLA, *Mechanical Behavior of Materials*. 1998: Prentice Hall.

2. WESTBROOK, J.H. and R. FLEISCHER, *Intermetallic Compounds: Principles and Practice*. 2002: Wiley.

3. POPE, D.P., et. al, *High Temperature Intermetallics*. 1997: Elsevier.

4. BANERJEE, D., "Structural Materials in Aerospace Systems," *MRS Bulletin*, 2001.

5. BAKER, I. and P.R. MUNROE, *Journal of Metals*, 1988. 2: p. 28.

6. GEORGE, E.P., C.T. LIU, and D.P. POPE, *Acta. Met*, 1996. 44: p. 1757.

7. HARPER, C.A., *Handbook of Materials Product Design*. 3rd ed. 2001: McGraw Hill.

8. RISBUD, S.H. and J.A. PASK, "On the Location of Metastable Immiscibility in the System Silicon Dioxide-aluminum Oxide," *J. Am. Ceram. Soc*, 1979. 62(3–4): p. 214–215.

9. BERGERON, C.G. and S.H. RISBUD, *Introduction to Phase Equilibria*. 1984: American Ceramic Society.

10. HALLER, W.K., "Phase Separated and Reconstructed Glasses," in *Advances in Ceramics Vol. 18-Commercial Glasses*, D.C. BOYD and J.F. MACDOWELL, Editors. 1986, American Ceramic Society. p. 51–64.

11. ROY, R., J. Am. Ceram. Soc, 1960. 430: p. 670.

12. GAYLE, F.W., et. al, "High Temperature Lead-Free Solder for Microelectronics," *Journal of Metals*, 2001. 53(6): p. 17–21.

13. RAO, M., "A Primer on Lead Free Solder," *Chip Scale Review*, 2000 (March–April).

14. HARES, G.B., "3500 Years of Glassmaking," in *Advances in Ceramics Vol. 18-Commercial Glasses*, D.C. BOYD and J.F. MACDOWELL, Editors. 1986, American Ceramic Society. p. 51–64.

15. KINGERY, W.D., H.K. BOWEN, and D.R. UHLMANN, *Introduction to Ceramics*, Second ed. 1976: Wiley.

16. AUBOURG, P.F. and W.W. WOLF, "Glass-Fibers," in *Advances in Ceramics Vol. 18-Commercial Glasses*, D.C. BOYD and J.F. MACDOWELL, Editors. 1986, American Ceramic Society. p. 51–64.

17. OKAMOTO, H., *Phase Diagrams of Dilute Binary Alloys*. 2002: ASM International.

✓ PROBLEMS

10-1 What is the principle of dispersion strengthening?

10-2 What is an interphase interface?

10-3 Can dispersion hardening be applied to ceramic materials? Explain.

Section 10-1 Principles and Examples of Dispersion Strengthening

10-4 What is a precipitate?

10-5 What is a microconstituent? Is microconstituent a phase? Explain.

10-6 What are the requirements of a matrix and precipitate for dispersion strengthening to be effective?

10-7 What is a microconstituent?

10-8 What is fiberglass? Is it a dispersion-strengthened material? Explain.

10-9 Why are needle-like precipitates not good for dispersion strengthening?

10-10 Can dispersion-strengthened materials be used at high temperatures? Can materials strengthened using cold working be used at high temperatures? Explain.

10-11 Why are small particles prefered for dispersion strengthening?

10-12 Why should the precipitate particles be hard?

Section 10-2 Intermetallic Compounds

10-13 What is an intermetallic compound? How is it different from other compounds? For example, other than the obvious difference in composition how is TiAl different from, for example, Al_2O_3?

10-14 Explain clearly the two different ways in which intermetallic compounds can be used.

10-15 What are some of the applications of $MoSi_2$? What does the term pesting mean as it applies to this material?

10-16 What are some of the major problems in the utilization of intermetallics for high-temperature applications?

10-17 Based on the material presented here, do you think the ductility of intermetallics is considered to be intrinsically limited?

10-18 What factors need to be considered for the possible use of materials for a shuttle that will visit outer space and return to earth repeatedly?

10-19 State any two non-structural applications of intermetallics.

Section 10-3 Phase Diagrams Containing Three-Phase Reactions

10-20 Define the terms eutectic, eutectoid, peritectic, peritectoid, and monotectic reactions

10-21 What is an invariant reaction? Show that for a two-component system the number of degrees of freedom for an invariant reaction is zero.

10-22 A hypothetical phase diagram is shown in Figure 10-32.

(a) Are any intermetallic compounds present? If so, identify them and determine whether they are stoichiometric or nonstoichiometric.

(b) Identify the solid solutions present in the system. Is either material A or B allotropic? Explain.

(c) Identify the three-phase reactions by writing down the temperature, the reaction in equa-

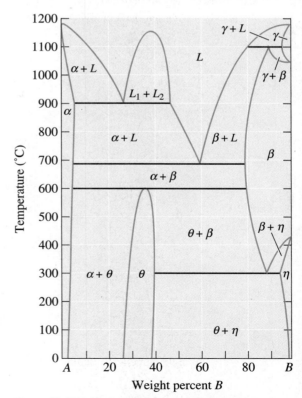

Figure 10-32 Hypothetical phase diagram (for Problem 10-22).

tion form, the composition of each phase in the reaction, and the name of the reaction.

10-23 The Cu-Zn phase diagram is shown in Figure 10-33.

(a) Are any intermetallic compounds present? If so, identify them and determine whether they are stoichiometric or nonstoichiometric.

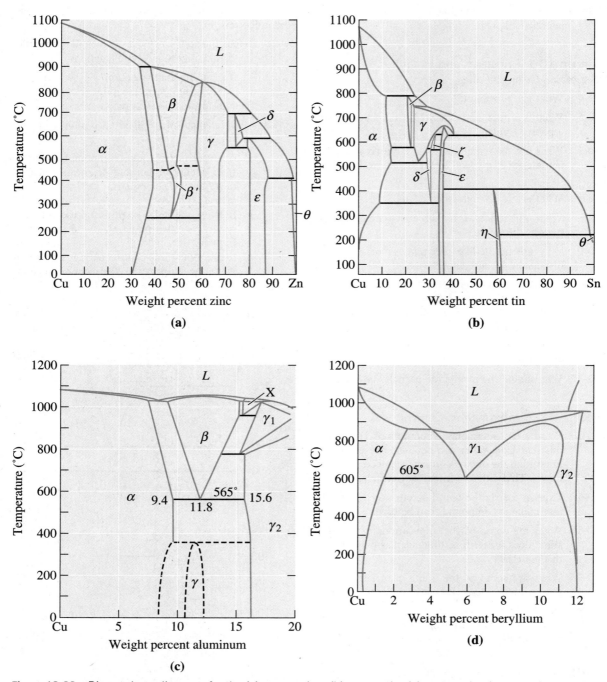

Figure 10-33 Binary phase diagrams for the (a) copper-zinc, (b) copper-tin, (c) copper-aluminum, and (d) copper-beryllium systems (for Problem 10-23).

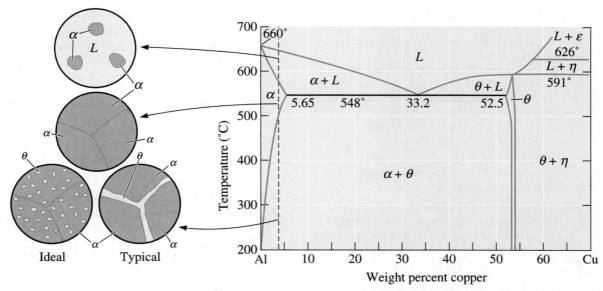

Figure 10-34 The aluminum-copper phase diagram and the microstructures that may develop during cooling of an Al-4%-Cu alloy (for Problem 10-24).

(b) Identify the solid solutions present in the system.
(c) Identify the three-phase reactions by writing down the temperature, the reaction in equation form, and the name of the reaction.

10-24 A portion of the Al-Cu phase diagram is shown in Figure 10-34.

(a) Determine the formula for the θ compound.
(b) Identify the three-phase reaction by writing down the temperature, the reaction in equation form, the composition of each phase in the reaction, and the name of the reaction.

10-25 The Al-Li phase diagram is shown in Figure 10-35.

(a) Are any intermetallic compounds present? If so, identify them and determine whether they are stoichiometric or nonstoichiometric. Determine the formula for each compound.
(b) Identify the three-phase reactions by writing down the temperature, the reaction in equation form, the composition of each phase in the reaction, and the name of the reaction.
(c) An intermetallic compound is found for 38 wt% Sn in the Cu-Sn phase diagram. Determine the formula for the compound.

10-26 An intermetallic compound is found for 10 wt% Si in the Cu-Si phase diagram. Determine the formula for the compound.

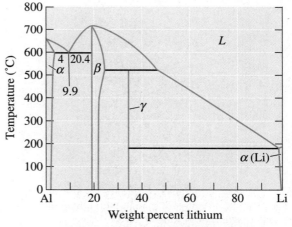

Figure 10-35 The aluminum-lithium phase diagram (for Problem 10-25).

10-27 Using the phase rule, predict and explain how many solid phases will form in a eutectic reaction in a ternary (three-component) phase diagram, assuming that the pressure is fixed.

Section 10-4 The Eutectic Phase Diagram

10-28 What is the origin of the word eutectic?

10-29 What is an isopleth? What is an isophethal study?

10-30 Consider a Pb-15% Sn alloy. During solidification, determine

(a) the composition of the first solid to form,

(b) the liquidus temperature, solidus temperature, solvus temperature, and freezing range of the alloy,

(c) the amounts and compositions of each phase at 260°C,

(d) the amounts and compositions of each phase at 183°C, and

(e) the amounts and compositions of each phase at 25°C.

10-31 Consider an Al-12% Mg alloy (Figure 10-36). During solidification, determine

(a) the composition of the first solid to form,

(b) the liquidus temperature, solidus temperature, solvus temperature, and freezing range of the alloy,

(c) the amounts and compositions of each phase at 525°C,

(d) the amounts and compositions of each phase at 450°C, and

(e) the amounts and compositions of each phase at 25°C.

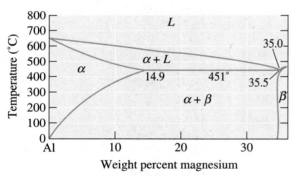

Figure 10-36   Portion of the aluminum-magnesium phase diagram (for Problem 10-31).

10-32 Consider a Pb-35% Sn alloy. Determine

(a) if the alloy is hypoeutectic or hypereutectic,

(b) the composition of the first solid to form during solidification,

(c) the amounts and compositions of each phase at 184°C,

(d) the amounts and compositions of each phase at 182°C,

(e) the amounts and compositions of each microconstituent at 182°C, and

(f) the amounts and compositions of each phase at 25°C.

10-33 Consider a Pb-70% Sn alloy. Determine

(a) if the alloy is hypoeutectic or hypereutectic,

(b) the composition of the first solid to form during solidification,

(c) the amounts and compositions of each phase at 184°C,

(d) the amounts and compositions of each phase at 182°C,

(e) the amounts and compositions of each microconstituent at 182°C, and

(f) the amounts and compositions of each phase at 25°.

10-34 Calculate the total % β and the % eutectic microconstituent at room temperature for the following lead-tin alloys: 10% Sn, 20% Sn, 50% Sn, 60% Sn, 80% Sn, and 95% Sn. Using Figure 10-22, plot the strength of the alloys versus the % β and the % eutectic and explain your graphs.

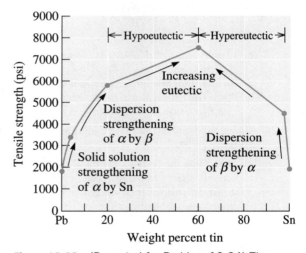

Figure 10-22   (Repeated for Problem 10-34) The effect of the composition and strengthening mechanism on the tensile strength of lead-tin alloys.

10-35 Consider an Al-4% Si alloy. (See Figure 10-23 on page 474.) Determine

(a) if the alloy is hypoeutectic or hypereutectic,

(b) the composition of the first solid to form during solidification,

(c) the amounts and compositions of each phase at 578°C,

(d) the amounts and compositions of each phase at 576°C, the amounts and compositions of each microconstituent at 576°C, and

(e) the amounts and compositions of each phase at 25°C.

10-36 Consider an Al-25% Si alloy. (See Figure 10-23 on page 474.) Determine

   (a) if the alloy is hypoeutectic or hypereutectic,
   (b) the composition of the first solid to form during solidification,
   (c) the amounts and compositions of each phase at 578°C,
   (d) the amounts and compositions of each phase at 576°C,
   (e) the amounts and compositions of each microconstituent at 576°C, and
   (f) the amounts and compositions of each phase at 25°C.

10-37 A Pb-Sn alloy contains 45% α and 55% β at 100°C. Determine the composition of the alloy. Is the alloy hypoeutectic or hypereutectic?

10-38 An Al-Si alloy contains 85% α and 15% β at 500°C. Determine the composition of the alloy. Is the alloy hypoeutectic or hypereutectic?

10-39 A Pb-Sn alloy contains 23% primary α and 77% eutectic microconstituent. Determine the composition of the alloy.

10-40 An Al-Si alloy contains 15% primary β and 85% eutectic microconstituent. Determine the composition of the alloy.

10-41 Determine the maximum solubility for the following cases:

   (a) lithium in aluminum (Figure 10-35),
   (b) aluminum in magnesium (Figure 10-37),
   (c) copper in zinc (Figure 10-33), and
   (d) carbon in γ-iron (Figure 10-38).

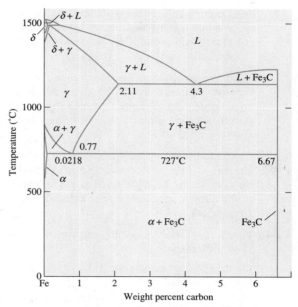

Figure 10-38 A portion of the Fe-Fe₃C phase diagram. The vertical line at 6.67% C is the stoichiometric compound Fe₃C (for Problem 10-41).

10-42 Determine the maximum solubility for the following cases:

   (a) magnesium in aluminum (Figure 10-36),
   (b) zinc in copper (Figure 10-33),
   (c) beryllium in copper (Figure 10-33), and
   (d) Al₂O₃ in MgO (Figure 10-39).

10-43 Observation of a microstructure shows that there is 28% eutectic and 72% primary β in an Al-Li alloy (Figure 10-35).

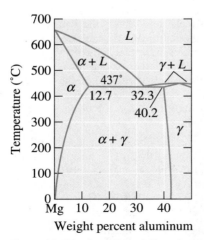

Figure 10-37 A portion of the magnesium-aluminum phase diagram (for Problem 10-41).

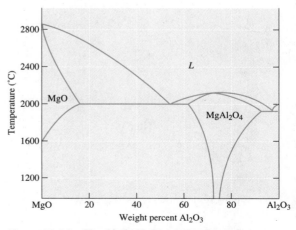

Figure 10-39 The MgO-Al₂O₃ phase diagram, showing limited solid solubility and the presence of MgAl₂O₄, or spinel (for Problem 10-42).

(a) Determine the composition of the alloy and whether it is hypoeutectic or hypereutectic.

(b) How much α and β are in the eutectic microconstituent?

10-44 Write the eutectic reaction that occurs, including the compositions of the three phases in equilibrium, and calculate the amount of α and β in the eutectic microconstituent in the Mg-Al system (Figure 10-36).

10-45 Calculate the total amount of α and β and the amount of each microconstituent in a Pb-50% Sn alloy at 182°C. What fraction of the total α in the alloy is contained in the eutectic microconstituent?

10-46 Figure 10-40 shows a cooling curve for a Pb-Sn alloy. Determine

(a) the pouring temperature,
(b) the superheat,
(c) the liquidus temperature,
(d) the eutectic temperature,
(e) the freezing range,
(f) the local solidification time,
(g) the total solidification time, and
(h) the composition of the alloy.

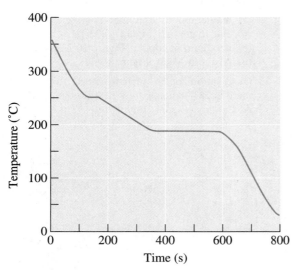

Figure 10-40 Cooling curve for a Pb-Sn alloy (for Problem 10-46).

10-47 Figure 10-41 shows a cooling curve for an Al-Si alloy. Determine

(a) the pouring temperature,
(b) the superheat,
(c) the liquidus temperature,

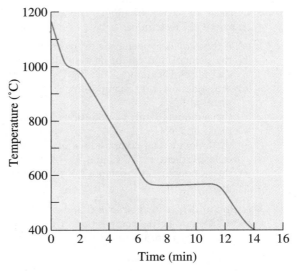

Figure 10-41 Cooling curve for an Al-Si alloy (for Problem 10-47).

(d) the eutectic temperature,
(e) the freezing range,
(f) the local solidification time,
(g) the total solidification time, and
(h) the composition of the alloy.

10-48 Draw the cooling curves, including appropriate temperatures, expected for the following Al-Si alloys:

(a) Al-4% Si,
(b) Al-12.6% Si,
(c) Al-25% Si, and
(d) Al-65% Si.

10-49 Based on the following observations, construct a phase diagram. Element *A* melts at 850°C and element *B* melts at 1200°C. Element *B* has a maximum solubility of 5% in element *A*, and element *A* has a maximum solubility of 15% in element *B*. The number of degrees of freedom from the phase rule is zero when the temperature is 725°C and there is 35% *B* present. At room temperature, 1% *B* is soluble in *A* and 7% *A* is soluble in *B*.

10-50 Cooling curves are obtained for a series of Cu-Ag alloys (Figure 10-42). Use this data to produce the Cu-Ag phase diagram. The maximum solubility of Ag in Cu is 7.9% and the maximum solubility of Cu in Ag is 8.8%. The solubilities at room temperature are near zero.

10-51 The SiO-Al$_2$O$_3$ phase diagram is included in Figure 10-27(b). A refractory is required to contain molten metal at 1900°C.

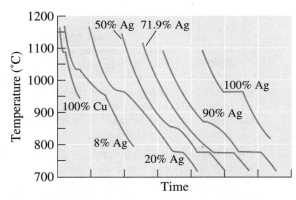

Figure 10-42 Cooling curves for a series of Cu-Ag alloys (for Problem 10-50).

(a) Will pure Al_2O_3 be a potential candidate? Explain.

(b) Will Al_2O_3 contaminated with 1% SiO_2 be a candidate? Explain.

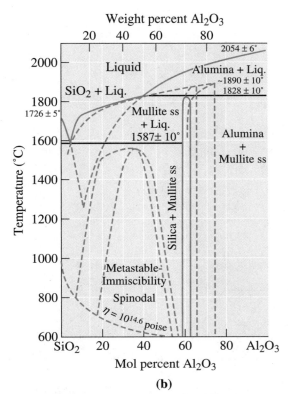

(b)

Figure 10-27 (Repeated for Problem 10-51) (b) A phase diagram for the Al_2O_3-SiO_2.[9] (*Adapted from* Introduction to Phase Equilibria *by Bergeron, C.G. and Risbud, S.H., The American Ceramic Society, Inc., 1984, page 44, Fig. 3-36.*)

Section 10-5 Strength of Eutectic Alloys

10-52 In Chapter 4, we learned that the yield strength of metallic materials can be enhanced using a finer grain size (Hall-Petch equation). On what factors does the strength of a eutectic alloy depend upon? Explain.

10-53 In regards to eutectic alloys, what does the term "modification" mean? How does it help properties of the alloy?

10-54 For the Pb-Sn system, explain why the tensile strength is maximum at the eutectic composition.

10-55 Does the shape of the proeutectic phase have an effect on the strength of eutectic alloys? Explain.

Section 10-6 Eutectics and Materials Processing

10-56 Can eutectics form in ceramic-oxide systems? Can eutectics form in polymeric or organic materials?

10-57 Explain why Pb-Sn alloys are used for soldering.

10-58 What alternative alloys are being developed to replace Pb-Sn alloys? Explain what is the motivation for this. Also, explain what factors must be considered in developing such materials.

10-59 What is water glass? What is soda lime glass? Explain the role of eutectics in the processing of glasses. What are the compositions of Vycor™ and Pyrex® glasses? Where are these materials used?

10-60 Can eutectic formation play a role in the sintering of ceramics? Explain.

10-61 Refractories used in steel making include silica brick that contain very small levels of alumina (Al_2O_3). The eutectic temperature in this system is about 1587°C. Silica melts at about 1725°C. Explain what will happen to the load bearing capacity of the bricks if a small amount of alumina gets incorporated into the silica bricks.

10-62 How can we determine phase diagram for metallic systems?

10-63 What techniques are used to establish phase diagrams for ceramic systems?

Section 10-7 Nonequilibrium Freezing in the Eutectic System

10-64 What is hot shortness? How does it affect the temperature at which eutectic alloys can be used?

Section 10-8 Ternary Phase Diagrams

10-65 Apply the Gibbs phase rule to a ternary system to show that the maximum number of phases that can coexist will be four.

10-66 Consider the ternary phase diagram shown in Figures 10-30 and 10-31. Determine the liquidus temperature, the first solid to form, and the phases present at room temperature for the following compositions:

(a) 30% *B*-20% *C*, balance *A*,
(b) 10% *B*-25% *C*, balance *A*, and
(c) 60% *B*-10% *C*, balance *A*.

10-67 Consider the ternary phase diagram shown in Figures 10-30 and 10-31. Determine the liquidus temperature, the first solid to form, and the phases present at room temperature for the following compositions:

(a) 5% *B*-80% *C*, balance *A*,
(b) 50% *B*-5% *C*, balance *A*, and
(c) 30% *B*-35% *C*, balance *A*.

10-68 Consider the liquidus plot in Figure 10-30.

(a) For a constant 20% *B*, draw a graph showing how the liquidus temperature changes from 80% 20% *B*-0% *C*, balance *A* to 0% 20% *B*-80% *C*, balance *A*.
(b) What is the composition of the ternary eutectic in this system?
(c) Estimate the temperature at which the ternary eutectic reaction occurs.

10-69 From the liquidus plot in Figure 10-30, prepare a graph of liquidus temperature versus percent *B* for a constant ratio of materials *A* and *C* (that is, from pure *B* to 50% *A*-50% *C* on the liquidus plot). Material *B* melts at 600°C.

✳ Design Problems

10-70 Design a processing method that permits a Pb-15% Sn alloy solidified under nonequilibrium conditions to be hot worked.

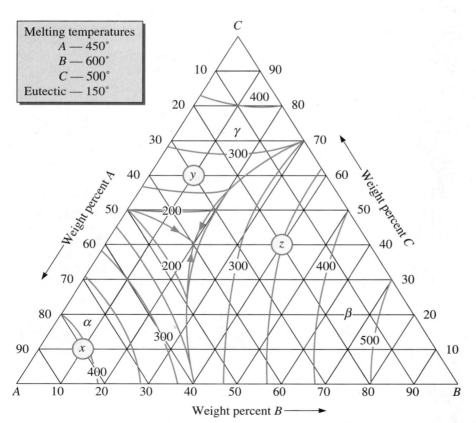

Figure 10-30 (Repeated for Problems 10-67, 10-68, and 10-69) A liquidus plot for the hypothetical ternary phase diagram.

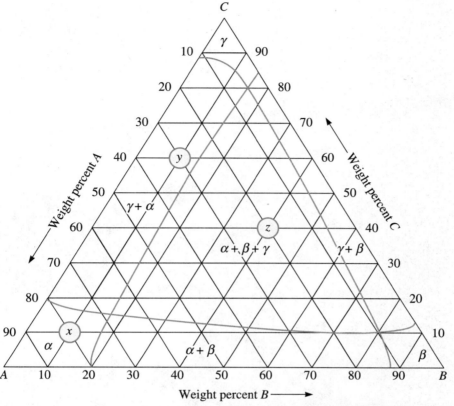

Figure 10-31 (Repeated for Problems 10-66, 10-67) An isothermal plot at room temperature for the hypothetical ternary phase diagram.

10-71 Design a eutectic diffusion bonding process to join aluminum to silicon. Describe the changes in microstructure at the interface during the bonding process.

10-72 Design a directional solidification process that will give an interlamellar spacing of 0.00005 cm in a Pb-Sn eutectic alloy. You may need to determine the constant *c* in Equation 10-2.

10-73 Design an Al-Si brazing alloy and process that will be successful in joining an Al-Mn alloy that has a liquidus of 659°C and a solidus of 656°C. Brazing, like soldering, involves introducing a liquid filler metal into a joint without melting the metals that are to be joined.

10-74 Your company would like to produce light-weight aluminum parts that have both excellent hardness and wear-resistance. The parts must have a good combination of strength, ductility, and internal integrity. Design the process flow

from the start of melting to the time that the liquid metal enters the mold cavity.

▣ Computer Problems

10-75 Write a computer program to assist you in solving problems such as those illustrated in Examples 10-3 or 10-4. The input will be, for example, the bulk composition of the alloy, the mass of the alloy, and the atomic masses of the elements (or compounds) forming the binary system. The program should then prompt the reader to provide temperature, number of phases, and the composition of the phases. The program should provide the user with the outputs for the fractions of phases present and the total masses of the phases, as well as the mass of each element in different phases. Start with a program that will provide a solution for Example 10-3 and then extend it to Example 10-4.

Chapter

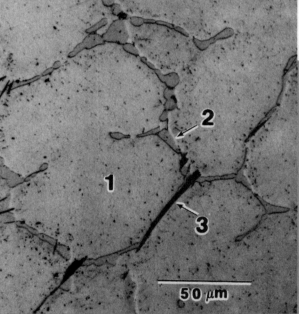

On December 17, 1903, the Wright brothers flew the first controllable airplane. This historic first flight lasted only 12 seconds and covered 120 feet, but changed the world forever. (*Courtesy of PhotoDisc/Getty Images.*)

The micrograph of the alloy used in the Wright brothers airplane is also shown. (*Courtesy Dr. Frank Gayle, NTST.*)

What was not known at the time was that the aluminum alloy engine that they used was inadvertently strengthened by precipitation hardening!

11

Dispersion Strengthening by Phase Transformations and Heat Treatment

Have You Ever Wondered?

- Who invented and flew the first controllable airplane?

- How do engineers strengthen aluminum alloys used in aircrafts?

- Can alloys remember their shape?

- Why do some steels become very hard upon quenching from high temperatures?

- What alloys are used to make orthodontic braces?

- Would it be possible to further enhance the strength and, hence, the dent resistance of sheet steels after the car chassis is made?

- What are smart materials?

In Chapter 10, we examined in detail, how second-phase particles can increase the strength of metallic materials. We also saw how dispersion-strengthened materials are prepared, and what pathways are possible for the formation of second phases during the solidification of alloys, especially eutectic alloys. In this chapter, we further discuss dispersion strengthening as we describe a variety of solid-state transformation processes including precipitation or age hardening and the eutectoid reaction. We also examine how nonequilibrium phase transformations—in particular, the martensitic reaction—can provide strengthening. Each of these dispersion-strengthening techniques requires a heat treatment.

As we discuss these strengthening mechanisms, keep in mind the characteristics that produce the most desirable dispersion strengthening as discussed in Chapter 10:

- The matrix should be relatively soft and ductile and the precipitate, or second phase, should be strong;
- the precipitate should be round and discontinuous;

- the second-phase particles should be small and numerous; and
- in general, the more precipitate we have, the stronger the alloy will be.

As in Chapter 10, we will concentrate on how these reactions influence the strength of the materials and how heat treatments can influence other properties. Since we will be dealing with solid-state phase transformations, we will begin with the nucleation and growth of second-phase particles in solid-state phase transformations.

11-1   Nucleation and Growth in Solid-State Reactions

In Chapter 8, we discussed nucleation of a solid nucleus from a melt. We also discussed the concepts of supersaturation, undercooling, and homogeneous and heterogeneous nucleation. Let's now see how these concepts apply to solid-state phase transformations such as the eutectoid reaction. In order for a precipitate of phase β to form from a solid matrix of phase α, both nucleation and growth must occur. The total change in free energy required for nucleation of a spherical solid precipitate from the matrix is:

$$\Delta G = \tfrac{4}{3}\pi r^3 \Delta G_{v(\alpha \to \beta)} + 4\pi r^2 \sigma_{\alpha\beta} + \tfrac{4}{3}\pi r^3 \varepsilon \tag{11-1}$$

The first two terms include the free energy change per unit volume (ΔG_v), and the energy change needed to create the unit area of the interface ($\sigma_{\alpha\beta}$), just as in solidification (Equation 8-1). However, the third term takes into account the **strain energy** per unit volume (ε), the energy required to permit a precipitate to fit into the surrounding matrix during the nucleation and growth of the precipitate, introduced when the precipitate forms in a solid, rigid matrix. The precipitate does not occupy the same volume that is displaced, so additional energy is required to accommodate the precipitate in the matrix.

Nucleation   As in solidification, nucleation occurs most easily on surfaces already present in the structure, thereby minimizing the surface energy term. Thus, the precipitates heterogeneously nucleate most easily at grain boundaries and other defects.

Growth   Growth of the precipitates normally occurs by long-range diffusion and redistribution of atoms. Diffusing atoms must be detached from their original locations (perhaps at lattice points in a solid solution), move through the surrounding material to the nucleus, and be incorporated into the crystal structure of the precipitate. In some cases, the diffusing atoms might be so tightly bonded within an existing phase that the detachment process limits the rate of growth. In other cases, attaching the diffusing atoms to the precipitate—perhaps because of the lattice strain—limits growth. This result sometimes leads to the formation of precipitates that have a special relationship to the matrix structure that minimizes the strain. In most cases, however, the controlling factor is the diffusion step.

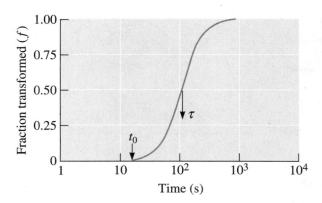

Figure 11-1
Sigmoidal curve showing the rate of transformation of FCC iron at a constant temperature. The incubation time t_0 and the time τ for 50% transformation are also shown.

Kinetics   The overall rate, or *kinetics*, of the transformation process depends on both nucleation and growth. If more nuclei are present at a particular temperature, growth occurs from a larger number of sites and the phase transformation is completed in a shorter period of time. At higher temperatures, the diffusion coefficient is higher, growth rates are more rapid, and again we expect the transformation to be completed in a shorter time, assuming an equal number of nuclei.

The rate of transformation is given by the Avrami equation (Equation 11-2), with the fraction of the transformation, f, related to time, t, by

$$f = 1 - \exp(-ct^n) \tag{11-2}$$

where c and n are constants for a particular temperature. This **Avrami relationship**, shown in Figure 11-1, produces a sigmoidal, or S-shaped, curve. This equation can describe most solid-state phase transformations.[1] An incubation time, t_0, during which no observable transformation occurs, is the time required for nucleation to occur. Initially, the transformation occurs slowly as nuclei form.

Incubation is followed by rapid growth as atoms diffuse to the growing precipitate. Near the end of the transformation, the rate again slows as the source of atoms available to diffuse to the growing precipitate is depleted. The transformation is 50% complete in time τ; the rate of transformation is often given by the reciprocal of τ:

$$\text{Rate} = 1/\tau \tag{11-3}$$

Effect of Temperature   In many phase transformations, the material undercools below the temperature at which the phase transformation occurs under equilibrium conditions. Recall from Chapter 8, the undercooling of water and other liquids and other supersaturation phenomena. Because both nucleation and growth are temperature-dependent, the rate of phase transformation depends on the undercooling (ΔT). The rate of nucleation is low for small undercoolings (since the thermodynamic driving force is low) and increases for larger undercoolings as the thermodynamic driving force increases at least up to a certain point (since diffusion becomes slower as temperature decreases). At the same time, the growth rate of the new phase decreases continuously, because of slower diffusion, as the undercooling increases. The growth rate follows an Arrhenius relationship (recall, Equation 5-1):

$$\text{Growth rate} = A \exp\left(\frac{-Q}{RT}\right) \tag{11-4}$$

where Q is the activation energy (in this case for the phase transformation), R is the gas constant, T is the temperature, and A is a constant.

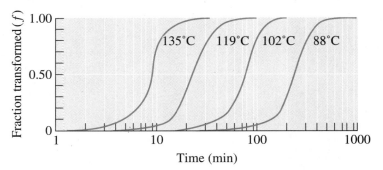

Figure 11-2   The effect of temperature on recrystallization of cold-worked copper.

Figure 11-2 shows sigmoidal curves at different temperatures for the recrystallization of copper; as the temperature increases, the rate of recrystallization of copper *increases*, because growth is the most important factor for copper.

At any particular temperature, the overall rate of transformation is the product of the nucleation and growth rates. In Figure 11-3(a), the combined effect of the nucleation and growth rates is shown. A maximum transformation rate may be observed at a critical undercooling. The time required for transformation is inversely related to the rate of transformation; Figure 11-3(b) describes the time (on a log scale) required for the transformation. This *C*-shaped curve is common for many transformations in metals, ceramics, glasses, and polymers. Notice that the time required at a temperature corresponding to the equilibrium phase transformation would be ∞ (i.e., the phase transformation will not occur). This is because there is no undercooling and, hence, the rate of homogenous nucleation is zero.

In some processes, such as the recrystallization of a cold-worked metal, we find that the transformation rate continually decreases with decreasing temperature. In this case, nucleation occurs easily, and diffusion—or growth—predominates (i.e., the growth is the rate limiting step for the transformation). The following example illustrates how the activation energy for a solid-state phase transformation such as recrystallization can be obtained from data related to the kinetics of the process.

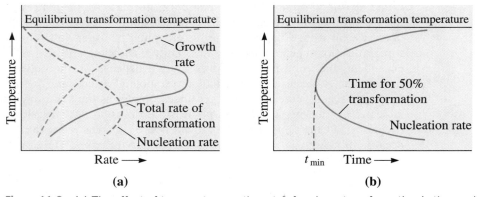

Figure 11-3   (a) The effect of temperature on the rate of a phase transformation is the product of the growth rate and nucleation rate contributions, giving a maximum transformation rate at a critical temperature. (b) Consequently, there is a minimum time (t_{min}) required for the transformation, given by the "*C*-curve".

EXAMPLE 11-1   *Activation Energy for the Recrystallization of Copper*

Determine the activation energy for the recrystallization of copper from the sigmoidal curves in Figure 11-2.

SOLUTION

The rate of transformation is the reciprocal of the time τ required for half of the transformation to occur. From Figure 11-2, the times required for 50% transformation at several different temperatures can be calculated:

$T(°C)$	T (K)	τ (min)	Rate (min^{-1})
135	408	9	0.111
119	392	22	0.045
102	375	80	0.0125
88	361	250	0.0040

The rate of transformation is an Arrhenius equation, so a plot of ln (rate) versus $1/T$ (Figure 11-4 and Equation 11-4) allows us to calculate the constants in the equation. Taking natural log of both sides of Equation 11-4:

$$\ln(\text{Growth rate}) = \ln A - \frac{Q}{RT}$$

Thus, if we plot ln(Growth rate) as a function of $1/T$, we expect a straight line that has a slope of $-Q/R$.

$$\text{slope} = \frac{-Q}{R} = \left[\frac{\Delta \ln(\text{rate})}{\Delta\left(\frac{1}{T}\right)} \right] = \frac{[(\ln(0.111) - \ln(0.004))]}{\left[\frac{1}{408} - \frac{1}{361}\right]}$$

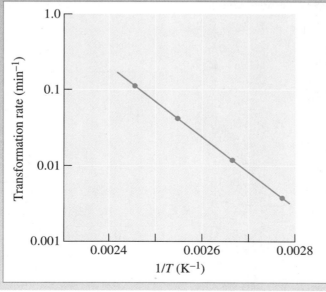

Figure 11-4
Arrhenius plot of transformation rate versus reciprocal temperature for recrystallization of copper (for Example 11-1).

$$Q/R = 10,414$$

$$Q = 20,693 \frac{\text{cal}}{\text{mol}}$$

$$\therefore 0.111 = A \exp\left(\frac{-20,693 \text{ cal/mol}}{\left(1.987 \frac{\text{cal}}{\text{deg} - \text{mol}}\right) \times (408 \text{ deg})}\right)$$

$$A = 0.111/8.21 \times 10^{-12} = 1.351 \times 10^{10} \text{ s}^{-1}$$

$$\therefore \text{rate} = 1.351 \times 10^{10} \exp\left(\frac{-20,693}{RT}\right)$$

In this particular example, the rate at which the reaction occurs *increases* as the temperature increases, indicating that the reaction may be dominated by diffusion.

11-2   Alloys Strengthened by Exceeding the Solubility Limit

In Chapter 10, we learned that lead-tin (Pb-Sn) alloys containing about 2 to 19% Sn can be dispersion-strengthened because the solubility of tin in lead is exceeded.

A similar situation occurs in aluminum-copper alloys. For example, the Al-4% Cu alloy (shown in Figure 11-5) is 100% α above 500°C. The α phase is a solid solution of aluminum containing copper up to 5.65 wt%. On cooling below the solvus temperature, a second phase, θ, precipitates. The θ phase, which is the hard, brittle intermetallic compound $CuAl_2$, provides dispersion strengthening. Applying the lever rule to the phase diagram shown Figure 11-5, we can show that at 200°C and below, in a 4% Cu

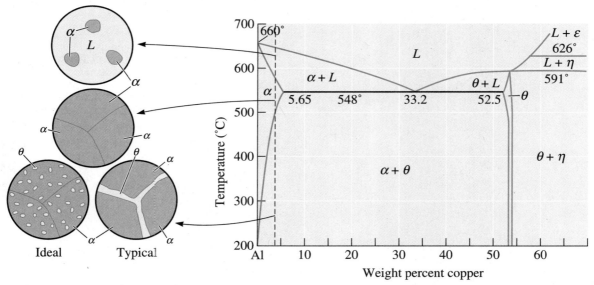

Figure 11-5 The aluminum-copper phase diagram and the microstructures that may develop during cooling of an Al-4% Cu alloy.

alloy, only about 7.5% of the final structure is θ. We must control the precipitation of the second phase to satisfy the requirements of good dispersion strengthening.

Widmanstätten Structure   The second phase may grow so that certain planes and directions in the precipitate are parallel to preferred planes and directions in the matrix, creating a basket-weave pattern known as the **Widmanstätten structure**.[2] This growth mechanism minimizes strain and surface energies and permits faster growth rates. Widmanstätten growth produces a characteristic appearance for the precipitate. When a needle-like shape is produced [Figure 11-6(a)], the Widmanstätten precipitate may encourage the nucleation of cracks, thus reducing the ductility of the material. However, some of these structures make it more difficult for cracks, once formed, to propagate, therefore providing good fracture toughness. Certain titanium alloys and ceramics obtain toughness in this way.

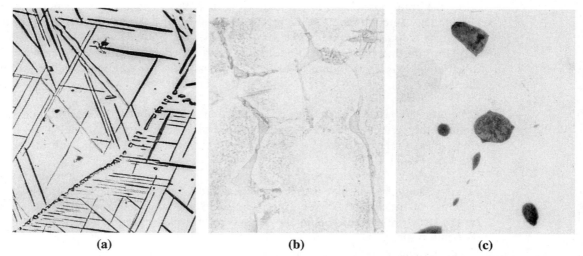

| (a) | (b) | (c) |

Figure 11-6   (a) Widmanstätten needles in a Cu-Ti alloy (×420). (*From* ASM Handbook, *Vol. 9, Metallography and Microstructure (1985), ASM International, Materials Park, OH 44073.*) (b) Continuous θ precipitate in an Al-4% Cu alloy, caused by slow cooling (×500). (c) Precipitates of lead at grain boundaries in copper (×500).

Interfacial Energy Relationships   We expect the precipitate to have a spherical shape in order to minimize surface energy. However, when the precipitate forms at an interface, the precipitate shape is also influenced by the **interfacial energy** (γ_{pm}) of the boundary between the matrix grains and the precipitate. The relationship between γ_{pm} and the interfacial energy of the matrix phase grain boundaries ($\gamma_{m,gb}$) is given by the equation that follows. This assumes the precipitate phase forms in these locations.

Assuming that the second phase is nucleating, the interfacial surface energies of the matrix-precipitate boundary (γ_{mp}) and the grain boundary energy of the matrix ($\gamma_{m,gb}$) fix a **dihedral angle** θ between the matrix-precipitate interface that, in turn, determines the shape of the precipitate (Figure 11-7). The relationship is:

$$\gamma_{m,gb} = 2\gamma_{mp} \cos \frac{\theta}{2}$$ (11-5)

Note that this equation cannot be used when the dihedral angle is 0° or 180°.

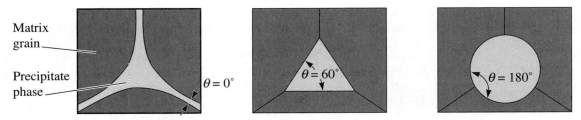

Matrix grain

Precipitate phase

Figure 11-7   The effect of surface energy and the dihedral angle on the shape of a precipitate.

If the precipitate phase completely wets the grain (similar to how water wets glass) then the dihedral angle is zero and the second phase would grow as a continuous layer along the grain boundaries of the matrix phase. If the dihedral angle is small, the precipitate may be continuous. If the precipitate is also hard and brittle, the thin film that surrounds the matrix grains causes the alloy to be very brittle [Figure 11-6(b)].

On the other hand, discontinuous and even spherical precipitates form when the dihedral angle is large [Figure 11-6(c)]. This occurs if the precipitate phase does not wet the matrix.

Coherent Precipitate   Even if we produce a uniform distribution of discontinuous precipitate, the precipitate may not significantly disrupt the surrounding matrix structure. Consequently, the precipitate blocks slip only if it lies directly in the path of the dislocation [Figure 11-8(a)].

But when a **coherent precipitate** forms, the planes of atoms in the crystal structure of the precipitate are related to—or even continuous with—the planes in the crystal structure of the matrix [Figure 11-8(b)]. Now a widespread disruption of the matrix crystal structure is created, and the movement of a dislocation is impeded even if the dislocation merely passes near the coherent precipitate. A special heat treatment, such as age hardening, may produce the coherent precipitate.

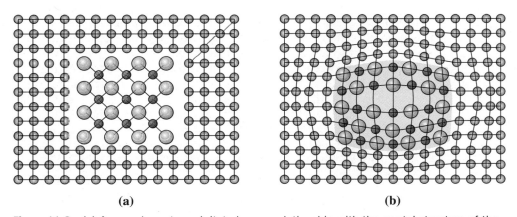

(a)   **(b)**

Figure 11-8   (a) A noncoherent precipitate has no relationship with the crystal structure of the surrounding matrix. (b) A coherent precipitate forms so that there is a definite relationship between the precipitate's and the matrix's crystal structure.

11-3   Age or Precipitation Hardening

Age hardening, or **precipitation hardening**, is produced by a sequence of phase transformations that leads to a uniform dispersion of nano-sized, coherent precipitates in a softer, more ductile matrix. The inadvertent occurrence of this process may have helped the Wright brothers, who, on December 17, 1903, made the first controllable flight that changed the world forever. Recently, Gayle and co-workers have shown that the aluminum alloy used by the Wright brothers for making the engine of the first airplane ever flown picked up copper from the casting mold.[3] The age hardening occurred inadvertently as the mold remained hot during the casting process. The formal discovery of age hardening was in 1906.[1] It also was by accident. The discovery is credited to A. Wilm who was working on solid-solution strengthening of an Al-Cu-Mn alloy. The application of age hardening started with the Wright brothers' historic flight and, even today, aluminum alloys used for aircrafts are strengthened using this technique. Age or precipitation hardening is probably one of the earliest examples of nanostructured materials that have found widespread applications.[4] A micrograph of age-hardened alloy used in the Wright brothers airplane engine is shown in the photograph that opens this chapter. Label 1 shows the matrix, labels 2 and 3 show the precipitates formed.

11-4   Applications of Age-Hardened Alloys

Before we examine the details of the mechanisms of phase transformations that are needed for age hardening to occur, let's examine some of the applications of this technique. A major advantage of precipitation hardening is that it can be used to increase the yield strength of many metallic materials via relatively simple heat treatments and without creating significant changes in density. Thus, the strength-to-density ratio of an alloy can be improved substantially using age hardening. For example, the yield strength of an aluminum alloy can be increased from about 20,000 psi to 60,000 psi as a result of age hardening.

Nickel-based super alloys (alloys based on Ni, Cr, Al, Ti, Mo, and C) are precipitation hardened by precipitation of a Ni_3Al-like γ' phase that is rich in Al and Ti.[1,2] Similarly, titanium alloys (e.g., Ti-6A-4V), stainless steels, Be-Cu and many steels are precipitation hardened and used for a variety of applications.

New sheet-steel formulations are designed such that precipitation hardening occurs in the material while the paint on the chassis is being "baked" or cured ($\sim100°C$). These bake-hardenable steels are just one example of steels that take advantage of the strengthening effect provided by age-hardening mechanisms.[5]

Figure 11-9(a) illustrates the increase in the strength of a bake-hardenable steel as result of strain hardening (Chapter 7) and precipitation hardening (in conditions that simulate baking of paint). Figure 11-9(b) shows the improvement in properties that are obtained using bake-hardenable steels. A small increase in the yield strength, after the formation process of auto body is complete, makes a big difference in the dent resistance of steels.

There are also many other steel compositions, known as high-strength, low-alloy (HSLA) or microalloyed steels, in which precipitation of nano-sized niobium carbide, titanium carbide, or nitride and other carbides or carbonitrides provides a significant increases in strength. Figure 11-9(c) shows a transmission electron micrograph of a niobium carbide precipitate formed in steel.

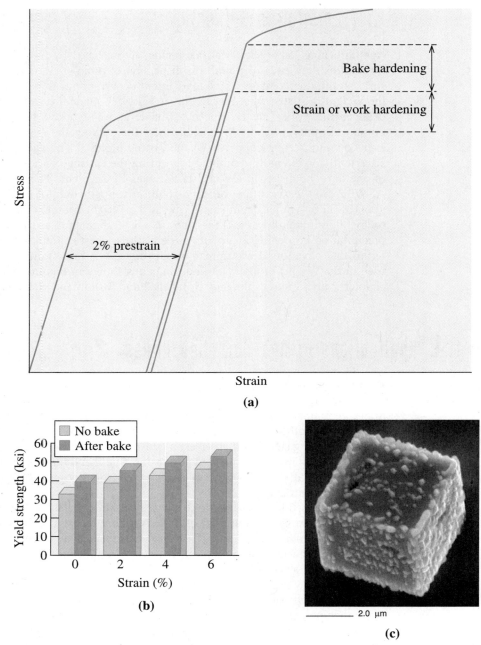

Figure 11-9 (a) A stress-strain curve showing the increase in strength of a bake-hardenable steel as a result of strain hardening and precipitation hardening. (*Source: U.S. Steel Corporation, Pittsburgh, PA.*) (b) A graph showing the increase in the yield strength of a bake hardenable steel (*Source: Bethlehem Steel, PA.*) (c) A TEM micrograph of a steel containing niobium (Nb) and manganese (Mn). The niobium react with carbon (C) and forms NbC precipitates that lead to strengthening. (*Courtesy of Dr. A.J. Deardo, Dr. I. Garcia, Dr. M. Hua, University of Pittsburgh.*)

A weakness associated with this mechanism is that age-hardened alloys can be used over a limited range of temperatures. At higher temperatures, the precipitates formed initially begin to grow and eventually dissolve if the temperatures are high enough (Section 11-8). This is where alloys in which dispersion strengthening is achieved by using a second phase that is insoluble are more effective than age-hardened alloys.

11-5   Microstructural Evolution in Age or Precipitation Hardening

How do precipitates form in precipitation hardening? How do they grow or age? Can the precipitates grow too much, or overage, so that they can not provide maximum dispersion strengthening? Answers to these questions can be found by following the microstructural evolution in the sequence of phase transformations that are necessary for age hardening.

Let's use Al-Cu as an archetypal system to illustrate these ideas. The Al-4% Cu alloy is a classical example of an age-hardenable alloy. There are three steps in the age-hardening heat treatment (Figure 11-10).

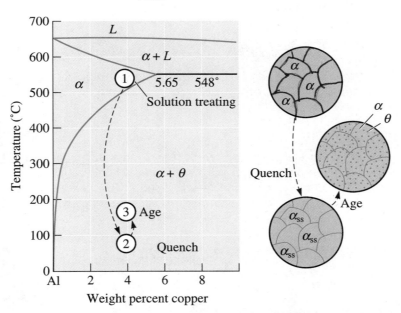

Figure 11-10   The aluminum-rich end of the aluminum-copper phase diagram showing the three steps in the age-hardening heat treatment and the microstructures that are produced.

Step 1: Solution Treatment   In the **solution treatment**, the alloy is first heated above the solvus temperature and held until a homogeneous solid solution α is produced. This step dissolves the θ phase precipitate and reduces any microchemical segregation present in the original alloy.

We could heat the alloy to just below the solidus temperature and increase the rate of homogenization. However, the presence of a nonequilibrium eutectic microconstituent may cause melting (hot shortness, Chapter 10). Thus, the Al-4% Cu alloy is solution-treated between 500°C and 548°C, that is, between the solvus and the eutectic temperatures.

Step 2: Quench   After solution treatment, the alloy, which contains only α in its structure, is rapidly cooled, or quenched. The atoms do not have time to diffuse to potential nucleation sites, so the θ does not form. After the quench, the structure is a **supersaturated solid solution** α_{ss} containing excess copper, and it is not an equilibrium structure. It is a metastable structure.[6] This situation is effectively the same as under-cooling of water and molten metals, and silicate glasses that are undercooled (Chapter 8). The only difference is we are dealing with materials in their solid state.

Step 3: Age   Finally, the supersaturated α is heated at a temperature below the solvus temperature. At this *aging* temperature, atoms diffuse only short distances. Because the supersaturated α is metastable, the extra copper atoms diffuse to numerous nucleation sites and precipitates grow. Eventually, if we hold the alloy for a sufficient time at the aging temperature, the equilibrium α + θ structure is produced. Note that even though the structure that is formed has two equilibrium phases (i.e., α + θ), the morphology of the phases is different from the structure that would have been obtained by the slow cooling of this alloy (Figure 11-5). When we go through the three steps described previously, we produce the θ phase in the form of ultrafine uniformly dispersed second-phase precipitate particles. This is what we need for effective precipitation strengthening.

The following two examples illustrate the effect of quenching on the composition of phases and a design for age-hardening treatment.

EXAMPLE 11-2   *Composition of Al-4% Cu Alloy Phases*

Compare the composition of the α solid solution in the Al-4% Cu alloy at room temperature when the alloy cools under equilibrium conditions with that when the alloy is quenched.

SOLUTION

From Figure 11-5, a tie line can be drawn at room temperature. The composition of the α determined from the tie line is about 0.02% Cu. However, the composition of the α after quenching is still 4% Cu. Since α contains more than the equilibrium copper content, the α is supersaturated with copper.

EXAMPLE 11-3   *Design of an Age-Hardening Treatment*

The magnesium-aluminum phase diagram is shown in Figure 11-11. Suppose a Mg-8% Al alloy is responsive to an age-hardening heat treatment. Design a heat treatment for the alloy.

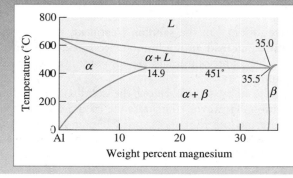

Figure 11-11
Portion of the aluminum-magnesium phase diagram.

SOLUTION

Step 1: Solution-treat at a temperature between the solvus and the eutectic to avoid hot shortness. Thus, heat between 340°C and 451°C.

Step 2: Quench to room temperature fast enough to prevent the precipitate phase β from forming.

Step 3: Age at a temperature below the solvus, that is, below 340°C, to form a fine dispersion of β phase.

Nonequilibrium Precipitates during Aging   During aging of aluminum-copper alloys, a continuous series of other precursor precipitate phases forms prior to the formation of the equilibrium θ phase. This is fairly common in precipitation-hardened alloys. The simplified diagram in Figure 11-10 does not show these intermediate phases. At the start of aging, the copper atoms concentrate on {100} planes in the α matrix and produce very thin precipitates called **Guinier-Preston** (GP) **zones**. As aging continues, more copper atoms diffuse to the precipitate and the GP-I zones thicken into thin disks, or GP-II zones. With continued diffusion, the precipitates develop a greater degree of order and are called θ'. Finally, the stable θ precipitate is produced.

The nonequilibrium precipitates—GP-I, GP-II, and θ'—are coherent precipitates. The strength of the alloy increases with aging time as these coherent phases grow in size during the initial stages of the heat treatment. When these coherent precipitates are present, the alloy is in the aged condition. Figure 11-12 shows the structure of an aged Al-Ag alloy. This important development in the microstructure evolution of precipitation-hardened alloys is the reason the time for heat treatment during aging is very important. Since the changes in the structure occur at a nano-scale, no visible change is seen in the structure of the alloy at a micro-scale. This is one reason why the mechanisms of precipitation hardening were established more firmly only in the 1950s, (when electron microscopes were starting to be used), even though the phenomenon was discovered earlier. In Chapter 3 (Figure 3-47), we examined a TEM image of an aluminum alloy that was also strengthened using precepitation hardening.

When the stable noncoherent θ phase precipitates, the strength of the alloy begins to decrease. Now the alloy is in the overaged condition. The θ still provides some dispersion strengthening, but with increasing time, the θ grows larger and even the simple dispersion-strengthening effect diminishes.

Figure 11-12
An electron micrograph of aged Al-15% Ag showing coherent γ' plates and round GP zones (×40,000). (*Courtesy of J.B. Clark.*)

11-6   Effects of Aging Temperature and Time

The properties of an age-hardenable alloy depend on both aging temperature and aging time (Figure 11-13). At 260°C, diffusion in the Al-4% Cu alloy is rapid, and precipitates quickly form. The strength reaches a maximum after less than 0.1 h exposure. Over-aging occurs if the alloy is held for longer than 0.1 h (6 minutes).

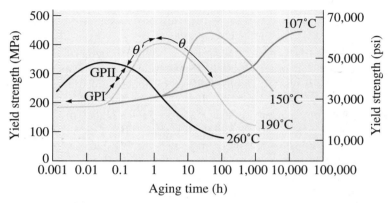

Figure 11-13   The effect of aging temperature and time on the yield strength of an Al-4% Cu alloy.

At 190°C, which is a typical aging temperature for many aluminum alloys, a longer time is required to produce the optimum strength. However, there are benefits to using the lower temperature. First, the maximum strength increases as the aging temperature decreases. Second, the alloy maintains its maximum strength over a longer period of time. Third, the properties are more uniform. If the alloy is aged for only 10 min at 260°C, the surface of the part reaches the proper temperature and strengthens, but the center remains cool and ages only slightly. The example that follows illustrates the effect of aging heat treatment time on the strength of aluminum alloys.

EXAMPLE 11-4   *Effect of Aging Heat Treatment Time on the Strength of Aluminum Alloys*

The operator of a furnace left for his hour lunch break without removing the Al-4% Cu alloy from the furnace used for the aging treatment. Compare the effect on the yield strength of the extra hour of aging for the aging temperatures of 190°C and 260°C.

SOLUTION

At 190°C, the peak strength of 400 MPa (60,000 psi) occurs at 2 h (Figure 11-13). After 3 h, the strength is essentially the same.

At 260°C, the peak strength of 340 MPa (50,000 psi) occurs at 0.06 h. However, after 1 h, the strength decreases to 250 MPa (40,000 psi).

Thus, the higher aging temperature gives lower peak strength and makes the strength more sensitive to aging time.

Aging at either 190°C or 260°C is called **artificial aging**, because the alloy is heated to produce precipitation. Some solution-treated and quenched alloys age at room temperature; this is called **natural aging**. Natural aging requires long times—often several days—to reach maximum strength. However, the peak strength is higher than that obtained in artificial aging, and no overaging occurs.

An interesting observation made by Gayle and coworkers is a striking example of the difference between natural aging and artificial aging.[3] Gayle and coworkers analyzed the aluminum alloy of the engine used in the Wright brothers' airplane. They found two interesting things. First, they found that the original alloy had undergone precipitation hardening as a result of being held in the mold for a period of time and at a temperature that was sufficient to cause precipitation hardening. Second, since when the research was done in 1903 until about 1993 (almost ninety years), the alloy had continued to age naturally! This could be seen from two different size distributions for the precipitate particles using transmission electron microscopy. In some aluminum alloys (designated as T4) used to make tapered poles or fasteners, it may be necessary to refrigerate the alloy to avoid natural aging at room temperature.[7] If not, the alloy would age at room temperature, become harder and not be workable! A SEM micrograph of the alloy used in the Wright's brothers plane appears at the beginning of this chapter.

11-7 Requirements for Age Hardening

Not all alloys are age hardenable. Four conditions must be satisfied for an alloy to have an age-hardening response during heat treatment:

catedral?

1. The alloy system must display decreasing solid solubility with decreasing temperature. In other words, the alloy must form a single phase on heating above the solvus line, then enter a two-phase region on cooling.

2. The matrix should be relatively soft and ductile, and the precipitate should be hard and brittle. In most age hardenable alloys, the precipitate is a hard, brittle intermetallic compound.

3. The alloy must be quenchable. Some alloys cannot be cooled rapidly enough to suppress the formation of the precipitate. Quenching may, however, introduce residual stresses that cause distortion of the part (Chapter 7). To minimize residual stresses, aluminum alloys are quenched in hot water, at about 80°C.

4. A coherent precipitate must form.

As mentioned before in Section 11-4, a number of important alloys, including certain stainless steels and alloys based on aluminum, magnesium, titanium, nickel, chromium, iron, and copper, meet these conditions and are age-hardenable.

11-8 Use of Age-Hardenable Alloys at High Temperatures

Based on our previous discussion, we would not select an age-hardened Al-4% Cu alloy for use at high temperatures. At service temperatures ranging from 100°C to 500°C, the alloy overages and loses its strength. Above 500°C, the second phase redissolves in the matrix and we do not even obtain dispersion strengthening. In general, the aluminum age-hardenable alloys are best suited for service near room temperature. However, some

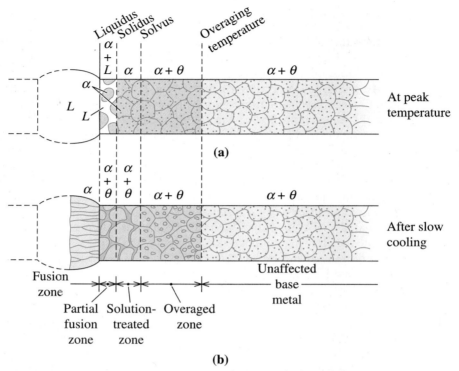

Figure 11-14   Microstructural changes that occur in age-hardened alloys during fusion welding: (a) microstructure in the weld at the peak temperature, and (b) microstructure in the weld after slowly cooling to room temperature.

magnesium alloys may maintain their strength to about 250°C and certain nickel superalloys resist overaging at 1000°C.

We may also have problems when welding age-hardenable alloys (Figure 11-14). During welding, the metal adjacent to the weld is heated. The heat-affected zone (HAZ) contains two principle zones. The lower temperature zone near the unaffected base metal is exposed to temperatures just below the solvus and may overage. The higher temperature zone is solution-treated, eliminating the effects of age hardening. If the solution-treated zone cools slowly, stable θ may form at the grain boundaries, embrittling the weld area. Very fast welding processes such as electron-beam welding, complete reheat treatment of the area after welding, or welding the alloy in the solution-treated condition improve the quality of the weld (Chapter 8). Welding of nickel-based superalloys strengthened by precipitation hardening does not pose such problems since the precipitation process is sluggish and the welding process simply acts as a solution and quenching treatment.[2] The process of friction stir welding has also been recently applied to welding of Al and Al-Li alloys for aerospace and aircraft applications.

11-9   The Eutectoid Reaction

In Chapter 10, we defined the eutectoid as a solid-state reaction in which one solid phase transforms to two other solid phases:

$$S_1 \rightarrow S_2 + S_3 \tag{11-6}$$

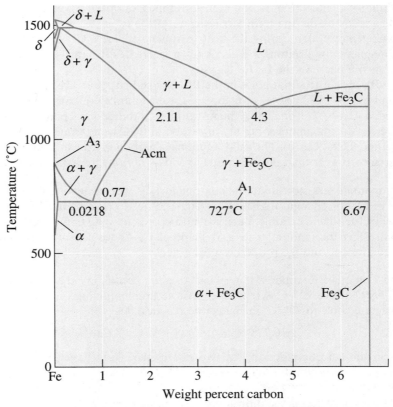

Figure 11-15 The Fe-Fe$_3$C phase diagram (a portion of the Fe-C diagram). The vertical line at 6.67% C is the stoichiometric compound Fe$_3$C.

As an example of how we can use the eutectoid reaction to control the microstructure and properties of an alloy, let's examine the technologically important portion of the iron-iron carbide (Fe-Fe$_3$C) phase diagram (Figure 11-15), which is the basis for steels and cast irons. The formation of the two solid phases (α and Fe$_3$C) permits us to obtain dispersion strengthening. The ability to control the occurrence of eutectoid reaction (this includes either making it happen, slowing it down, or avoiding it all together) is probably the most important step in the thermomechanical processing of steels. On the Fe-Fe$_3$C diagram, the eutectoid temperature is known as the A_1 temperature. The boundary between austenite (γ) and the two-phase field consisting of ferrite and ferrite and austenite is known as the A_3. The boundary between austenite (γ) and the two-phase field consisting of cementite and austenite is known as A_{cm}.

We are normally not interested in the carbon-rich end of the Fe-C phase diagram and this is why we examine the Fe-Fe$_3$C diagram as part of the Fe-C binary phase diagram.

Solid Solutions Iron goes through two allotropic transformations (Chapter 3) during heating or cooling. Immediately after solidification, iron forms a BCC structure called δ-ferrite. On further cooling, the iron transforms to a FCC structure called γ, or **austenite**. The austenite phase is named after Sir W.C. Robert-Austen who first reported it.[7] Finally, iron transforms back to the BCC structure at lower temperatures; this structure is called α, or **ferrite**. Both of the ferrites (α and δ) and the austenite are solid solutions of interstitial carbon atoms in iron (Example 4-3). Normally, when no specific reference is made, the term ferrite refers to the α ferrite, since this is the phase we encounter more often during the heat treatment of steels. Certain ceramic materials used in magnetic applications are also known as ferrites (Chapter 14) but are not related to the ferrite phase in the Fe-Fe$_3$C system.

Because interstitial holes in the FCC crystal structure are somewhat larger than the holes in the BCC crystal structure, a greater number of carbon atoms can be accommodated in FCC iron. Thus, the maximum solubility of carbon in austenite is 2.11% C, whereas the maximum solubility of carbon in BCC iron is much lower (i.e., ~0.0218% C in α and 0.09% C in δ). The solid solutions of carbon in iron are relatively soft and ductile, but are stronger than pure iron due to solid-solution strengthening by the carbon.

Compounds   A stoichiometric compound Fe$_3$C, or **cementite**, forms when the solubility of carbon in solid iron is exceeded. The Fe$_3$C contains 6.67% C, is extremely hard and brittle (like a ceramic material), and is present in all commercial steels. By properly controlling the amount, size, and shape of Fe$_3$C, we control the degree of dispersion strengthening and the properties of the steel.

The Eutectoid Reaction   If we heat an alloy containing the eutectoid composition of 0.77% C above 727°C, we produce a structure containing only austenite grains. When austenite cools to 727°C, the eutectoid reaction begins:

$$\gamma(0.77\% \text{ C}) \rightarrow \alpha(0.0218\% \text{ C}) + \text{Fe}_3\text{C}(6.67\% \text{ C}) \tag{11-7}$$

As in the eutectic reaction, the two phases that form have different compositions, so atoms must diffuse during the reaction (Figure 11-16). Most of the carbon in the austenite diffuses to the Fe$_3$C, and most of the iron atoms diffuse to α. This redistribution of atoms is easiest if the diffusion distances are short, which is the case when the α and Fe$_3$C grow as thin lamellae, or plates.

Pearlite   The lamellar structure of α and Fe$_3$C that develops in the iron-carbon system is called **pearlite**, which is a microconstituent in steel. This was so named because a polished and etched pearlite shows the colorfulness of mother-of-pearl.[1] The lamellae

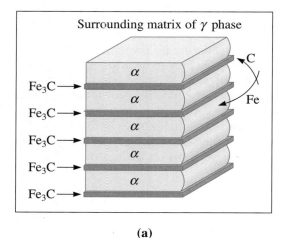
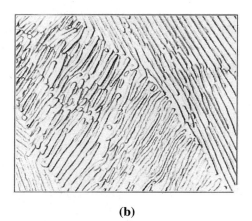

(a)   (b)

Figure 11-16   Growth and structure of pearlite: (a) redistribution of carbon and iron, and (b) photomicrograph of the pearlite lamellae (×2000). (*From* ASM Handbook, *Vol. 7, (1972), ASM International, Materials Park, OH 44073.*)

in pearlite are much finer than the lamellae in the lead-tin eutectic because the iron and carbon atoms must diffuse through solid austenite rather than through liquid. One way to think about pearlite is to consider it as a metal-ceramic nano-composite. The following example shows the calculation of the amounts of the phases in the pearlite microconstituent.

EXAMPLE 11-5   *Phases and Composition of Pearlite*

Calculate the amounts of ferrite and cementite present in pearlite.

SOLUTION

Since pearlite must contain 0.77% C, using the lever rule:

$$\% \; \alpha = \frac{6.67 - 0.77}{6.67 - 0.0218} \times 100 = 88.7\%$$

$$\% \; Fe_3C = \frac{0.77 - 0.0218}{6.67 - 0.0218} \times 100 = 11.3\%$$

In Example 11-5, we saw that most of the pearlite is composed of ferrite. In fact, if we examine the pearlite closely, we find that the Fe_3C lamellae are surrounded by α. The pearlite structure therefore, provides dispersion strengthening—the continuous ferrite phase is relatively soft and ductile and the hard, brittle cementite is dispersed. The next example illustrates the similarity and differences between composites and pearlites.

EXAMPLE 11-6   *Tungsten Carbide (WC)-Cobalt (Co) Composite and Pearlite*

Tungsten carbide-cobalt composites, known as cemented carbides or carbides, are used as bits for cutting tools and drills (Chapter 1). What features are similar between these "cemented carbides" and pearlite, a microconstituent in steels? What are some of the major differences?

SOLUTION

Pearlite is very similar to the tungsten carbide-cobalt (WC-Co) composites known as carbides. You may recall from earlier chapters that WC-Co are ceramic-metal composites (known as cermets) and used for bits for cutting tools, drill bits, etc. In both materials, we take advantage of the toughness of one phase (ferrite or cobalt metal, in the case of pearlite in steel and WC-Co, respectively) and the hard ceramic like phase (WC and Fe_3C, in WC-Co and steel, respectively). The metallic phase helps with ductility and the hard phase helps with strength. The difference is, WC and Co are two separate compounds that are sintered together using the powder metallurgy route. Pearlite is a microconstituent made up of two phases derived from same two elements (Fe-C). Another difference is in pearlite, the phases are formed via a eutectoid reaction. No such reaction occurs in the formation of WC-Co composites. Typically, WC-Co microstructure consists mainly of WC grains that are "glued" by cobalt grains. In pearlite, the metal-like ferrite phase dominates.

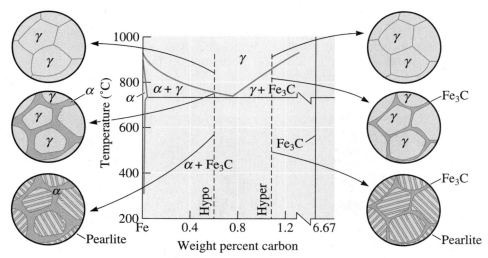

Figure 11-17   The evolution of the microstructure of hypoeutectoid and hypereutectoid steels during cooling, in relationship to the Fe-Fe$_3$C phase diagram.

Primary Microconstituents   Hypoeutectoid steels contain less than 0.77% C, and hypereutectoid steels contain more than 0.77% C. Ferrite is the primary or proeutectoid microconstituent in hypoeutectoid alloys, and cementite is the primary or proeutectoid microconstituent in hypereutectoid alloys. If we heat a hypoeutectoid alloy containing 0.60% C above 750°C, only austenite remains in the microstructure. Figure 11-17 shows what happens when the austenite cools. Just below 750°C, ferrite nucleates and grows, usually at the austenite grain boundaries. Primary ferrite continues to grow until the temperature falls to 727°C. The remaining austenite at that temperature is now surrounded by ferrite and has changed in composition from 0.60% C to 0.77% C. Subsequent cooling to below 727°C causes all of the remaining austenite to transform to pearlite by the eutectoid reaction. The structure contains two phases—ferrite and cementite—arranged as two microconstituents—primary ferrite and pearlite. The final microstructure contains islands of pearlite surrounded by the primary ferrite [Figure 11-18(a)]. This structure permits the alloy to be strong, due to the dispersion-strengthened pearlite, yet ductile, due to the continuous primary ferrite.

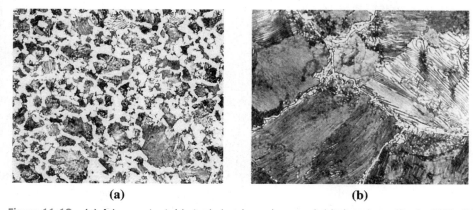

(a)   (b)

Figure 11-18   (a) A hypoeutectoid steel showing primary α (white) and pearlite (×400). (b) A hypereutectoid steel showing primary Fe$_3$C surrounding pearlite (×800). (*From* ASM Handbook, *Vol. 7, (1972), ASM International, Materials Park, OH 44073.*)

In hypereutectoid alloys, the primary phase is Fe_3C, which forms at the austenite grain boundaries. After the austenite cools through the eutectoid reaction, the steel contains hard, brittle cementite surrounding islands of pearlite [Figure 11-18(b)]. Now, because the hard, brittle microconstituent is continuous, the steel is also brittle. Fortunately, we can improve the microstructure and properties of the hypereutectoid steels by heat treatment. The following example shows the calculation for the composition of phases and microconstituent in a plain carbon steel.

EXAMPLE 11-7   *Phases in Hypoeutectoid Plain Carbon Steel*

Calculate the amounts and compositions of phases and microconstituents in a Fe-0.60% C alloy at 726°C.

SOLUTION

The phases are ferrite and cementite. Using a tie line and working the lever law at 726°C, we find:

$$\alpha \ (0.0218\% \ C) \quad \% \ \alpha = \left[\frac{6.67 - 0.60}{6.67 - 0.0218} \right] \times 100 = 91.3\%$$

$$Fe_3C \ (6.67\% \ C) \quad \% \ Fe_3C = \left[\frac{0.60 - 0.0218}{6.67 - 0.0218} \right] \times 100 = 8.7\%$$

The microconstituents are primary ferrite and pearlite. If we construct a tie line just above 727°C, we can calculate the amounts and compositions of ferrite and austenite just before the eutectoid reaction starts. All of the austenite at that temperature will have eutectoid composition (i.e., it will contain 0.77% C) and will transform to pearlite; all of the proeutectoid ferrite will remain as primary ferrite.

$$\text{Primary } \alpha : 0.0218\% \ C \quad \% \ \text{Primary } \alpha = \left[\frac{0.77 - 0.60}{6.67 - 0.0218} \right] \times 100 = 22.7\%$$

$$\text{Austentite just above 727°C} = \text{Pearlite} : 0.77\% \ C$$

$$\% \ \text{Pearlite} = \left[\frac{0.60 - 0.0218}{6.67 - 0.0218} \right] \times 100 = 77.3\%$$

11-10   Controlling the Eutectoid Reaction

We control dispersion strengthening in the eutectoid alloys in much the same way that we did in eutectic alloys (Chapter 10).

Controlling the Amount of the Eutectoid   By changing the composition of the alloy, we change the amount of the hard second phase. As the carbon content of steel increases towards the eutectoid composition of 0.77% C, the amounts of Fe_3C and pearlite increase, thus increasing the strength. However, this strengthening effect eventually peaks and the properties level out or even decrease when the carbon content is too high (Table 11-1).

TABLE 11-1 ■ *The effect of carbon on the strength of steels*

	Slow Cooling (Coarse Pearlite)			Fast Cooling (Fine Pearlite)		
Carbon %	Yield Strength (psi)	Tensile Strength (psi)	% Elongation	Yield Strength (psi)	Tensile Strength (psi)	% Elongation
0.20	42,750	57,200	36.5	50,250	64,000	36.0
0.40	51,250	75,250	30.0	54,250	85,500	28.0
0.60	54,000	90,750	23.0	61,000	112,500	18.0
0.80	54,500	89,250	25.0	76,000	146,500	11.0
0.95	55,000	95,250	13.0	72,500	147,000	9.5

After Metals Progress Materials and Processing Databook, *1981.*

Controlling the Austenite Grain Size   Pearlite grows as grains or *colonies*. Within each colony, the orientation of the lamellae is identical. The colonies nucleate most easily at the grain boundaries of the original austenite grains. We can increase the number of pearlite colonies by reducing the prior austenite grain size, usually by using low temperatures to produce the austenite. Typically, we can increase the strength of the alloy by reducing the initial austenite grain size, thus increasing the number of colonies.

Controlling the Cooling Rate   By increasing the cooling rate during the eutectoid reaction, we reduce the distance that the atoms are able to diffuse. Consequently, the lamellae produced during the reaction are finer or more closely spaced. By producing fine pearlite, we increase the strength of the alloy (Table 11-1 and Figure 11-19).

Controlling the Transformation Temperature   The solid-state eutectoid reaction is rather slow, and the steel may cool below the equilibrium eutectoid temperature before

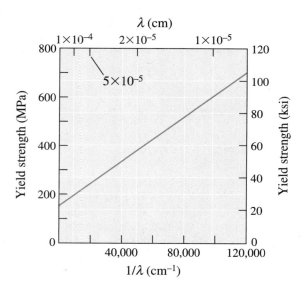

Figure 11-19
The effect of interlamellar spacing (λ) of on the yield strength of pearlite.

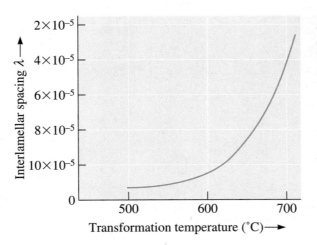

Figure 11-20
The effect of the austenite transformation temperature on the
interlamellar spacing in pearlite.

the transformation begins (i.e., the austenite phase can be undercooled). Lower transformation temperatures give a finer, stronger structure (Figure 11-20), influence the time
required for transformation, and even alter the arrangement of the two phases. This information is contained in the **time-temperature-transformation** (TTT) **diagram** (Figure
11-21). This diagram, also called the **isothermal transformation** (IT) diagram or the C-
curve, permits us to predict the structure, properties, and heat treatment required in
steels.

The shape of the TTT diagram is a consequence of the kinetics of the eutectoid reaction and is similar to the diagram shown by the Avrami relationship (Figure 11-3). At
any particular temperature, a sigmoidal curve describes the rate at which the austenite

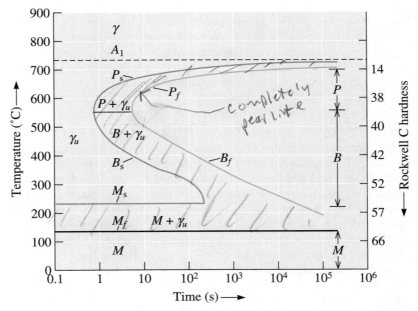

Figure 11-21   The time-temperature-transformation (TTT) diagram for an eutectoid steel.

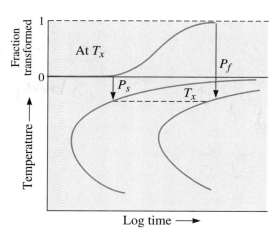

Figure 11-22
The sigmoidal curve is related to the start and finish times on the TTT diagram for steel. In this case, austenite is transforming to pearlite.

transforms to a mixture of ferrite and cementite (Figure 11-22). An incubation time is required for nucleation. The P_s (pearlite start) curve represents the time at which austenite starts to transform to ferrite and cementite via the eutectoid transformation. The sigmoidal curve also gives the time at which the transformation is completed; this time is given by the P_f (pearlite finish) curve. When the temperature decreases from 727°C, the rate of nucleation increases, while the rate of growth of the eutectoid decreases. As in Figure 11-3, a maximum transformation rate, or minimum transformation time, is found; the maximum rate of transformation occurs near 550°C for an eutectoid steel (Figure 11-21).

Two types of microconstituents are produced as a result of the transformation. Pearlite (P) forms above 550°C, and bainite (B) forms at lower temperatures:

1. *Nucleation and growth of phases in pearlite*: If we quench to just below the eutectoid temperature, the austenite is only slightly undercooled. Long times are required before stable nuclei for ferrite and cementite form. After the phases that form pearlite nucleate, atoms diffuse rapidly and *coarse* pearlite is produced; the transformation is complete at the pearlite finish (P_f) time. Austenite quenched to a lower temperature is more highly undercooled. Consequently, nucleation occurs more rapidly and the P_s is shorter. However, diffusion is also slower, so atoms diffuse only short distances and *fine* pearlite is produced. Even though growth rates are slower, the overall time required for the transformation is reduced because of the shorter incubation time. Finer pearlite forms in shorter times as we reduce the isothermal transformation temperature to about 550°C, which is the *nose*, or *knee*, of the TTT curve (Figure 11-21).

2. *Nucleation and growth of phases in bainite*: At a temperature just below the nose of the TTT diagram, diffusion is very slow and total transformation times increase. In addition, we find a different microstructure! At low transformation temperatures, the lamellae in pearlite would have to be extremely thin and, consequently, the boundary area between the ferrite and Fe_3C lamellae would be very large. Because of the energy associated with the ferrite-cementite interface, the total energy of the steel would have to be very high. The steel can reduce its internal energy by permitting the cementite to precipitate as discrete, rounded particles in a ferrite matrix. This new microconstituent, or arrangement of ferrite and cementite, is called bainite. It is named so in honor of E.C. Bain who first reported it in 1930.[8] Transformation begins at a bainite start (B_s) time and ends at a bainite finish (B_f) time.

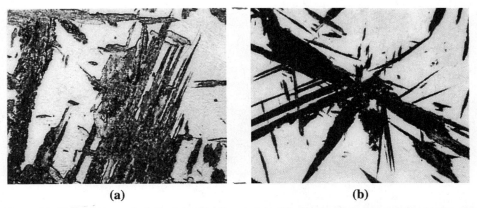

Figure 11-23   (a) Upper bainite (gray, feathery plates) (×600). (b) Lower bainite (dark needles) (×400). (*From* ASM Handbook, *Vol. 8, (1973), ASM International, Materials Park, OH 44073.*)

The times required for austenite to begin and finish its transformation to bainite increase and the bainite becomes finer as the transformation temperature continues to decrease. The bainite that forms just below the nose of the curve is called coarse bainite, upper bainite, or feathery bainite. The bainite that forms at lower temperatures is called fine bainite, lower bainite, or acicular bainite. Figure 11-23 shows typical microstructures of bainite. Note that the morphology of bainite depends on the heat treatment used.

Figure 11-24 shows the effect of transformation temperature on the properties of eutectoid (0.77% C) steel. As the temperature decreases, there is a general trend toward higher strength and lower ductility due to the finer microstructure that is produced. The following two examples illustrate how we can design heat treatments of steels to produce desired microstructures and properties.

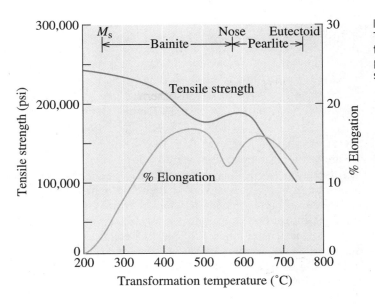

Figure 11-24
The effect of transformation temperature on the properties of an eutectoid steel.

EXAMPLE 11-8   *Design of a Heat Treatment to Generate Pearlite Microstructure*

Design a heat treatment to produce the pearlite structure shown in Figure 11-16(b).

SOLUTION

First, we need to determine the interlamellar spacing of the pearlite. If we count the number of lamellar spacings in the upper right of Figure 11-16(b), remembering that the interlamellar spacing is measured from one α plate to the next α plate, we find 14 spacings over a 2-cm distance. Due to the $\times 2000$ magnification, this 2-cm distance is actually 0.001 cm. Thus:

$$\lambda = \left[\frac{0.001 \text{ cm}}{14 \text{ spacings}} \right] = 7.14 \times 10^{-5} \text{ cm}$$

If we assume that the pearlite is formed by an isothermal transformation, we find from Figure 11-20 that the transformation temperature must have been approximately 700°C. From the TTT diagram (Figure 11-21), our heat treatment must have been:

1. Heat the steel to about 750°C and hold—perhaps for 1 h—to produce all austenite. A higher temperature may cause excessive growth of austenite grains.

2. Quench to 700°C and hold for at least 10^5 s (the P_f time). We assume here that the steel cools instantly to 700°C. In practice, this does not happen and thus the transformation does not occur at one temperature. We may need to use the continuous cooling transformation diagrams to be more precise (Chapter 12).

3. Cool to room temperature.

The steel should have a hardness of HRC 14 (Figure 11-21) and a yield strength of about 200 MPa (30,000 psi), as shown in Figure 11-19.

EXAMPLE 11-9   *Heat Treatment to Generate Bainite Microstructure*

Excellent combinations of hardness, strength, and toughness are obtained from bainite. One heat treatment facility austenitized an eutectoid steel at 750°C, quenched and held the steel at 250°C for 15 min, and finally permitted the steel to cool to room temperature. Was the required bainitic structure produced?

SOLUTION

Let's examine the heat treatment using Figure 11-21. After heating at 750°C, the microstructure is 100% γ. After quenching to 250°C, unstable austenite remains for slightly more than 100 s, when fine bainite begins to grow. After 15 min, or 900 s, about 50% fine bainite has formed and the remainder of the steel still contains unstable austenite. As we will see later, the unstable austenite transforms to martensite when the steel is cooled to room temperature and the final structure is a mixture of bainite and hard, brittle martensite. The heat treatment was not successful! The heat treatment facility should have held the steel at 250°C for at least 10^4 s, or about 3 h.

11-11 The Martensitic Reaction and Tempering

Martensite is a phase that forms as the result of a diffusionless solid-state transformation. In this transformation there is no diffusion and, hence, it does not follow the Avrami transformation kinetics. The growth rate in martensitic transformations (also known as **displacive** or **athermal transformations**) is so high that nucleation becomes the controlling step. The phase that forms upon the quenching of steels was named "martensite" by Floris Osmond in 1895 in honor of German metallurgist Adolf Martens.[1] Similar martensitic phase transformations occur in other systems as well.

Cobalt, for example, transforms from a FCC to a HCP crystal structure by a slight shift in the atom locations that alters the stacking sequence of close-packed planes. Because the reaction does not depend on diffusion, the martensite reaction is an athermal transformation—that is, the reaction depends only on the temperature, not on the time. The martensite reaction often proceeds rapidly, at speeds approaching the velocity of sound in the material.

Many other alloys such as Cu-Zn-Al and Cu-Al-Ni and Ni-Ti show martensitic phase transformations. These transformations can also be driven by the application of mechanical stress. Other than martensite that forms in certain type of steels, the Ni-Ti alloy, known as nitinol (which stands for Nickel Titanium Naval Ordinance Laboratory, developed by the U.S. Naval Ordinance Laboratory in the 1940s) is perhaps the best-known example of alloys that make use of martensitic phase transformations. These materials can remember their shape and are known as shape-memory alloys (SMAs).[9] We will discuss applications of these materials in Section 11-12.

Martensite in Steels   In steels with less than about 0.2% C, the FCC austenite transforms to a supersaturated BCC martensite structure. In higher carbon steels, the martensite reaction occurs as FCC austenite transforms to BCT (body centered tetragonal)

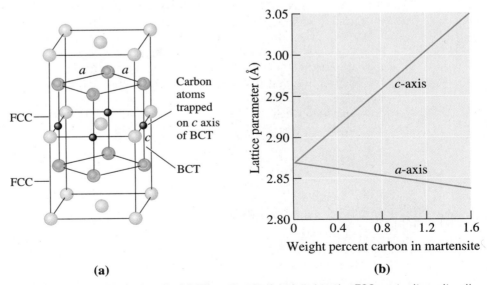

(a)   **(b)**

Figure 11-25   (a) The unit cell of BCT martensite is related to the FCC austenite unit cell. (b) As the percentage of carbon increases, more interstitial sites are filled by the carbon atoms and the tetragonal structure of the martensite becomes more pronounced.

martensite. The relationship between the FCC austenite and the BCT martensite [Figure 11-25(a)] shows that carbon atoms in the 1/2, 0, 0 type of interstitial sites in the FCC cell can be trapped during the transformation to the body-centered structure, causing the tetragonal structure to be produced. As the carbon content of the steel increases, a greater number of carbon atoms are trapped in these sites, thereby increasing the difference between the a- and c-axes of the martensite [Figure 11-25(b)].

The steel must be quenched, or rapidly cooled, from the stable austenite region to prevent the formation of pearlite, bainite, or primary microconstituents. The martensite reaction begins in an eutectoid steel when austenite cools below 220°C, the martensite start (M_s) temperature (Figure 11-21). The amount of martensite increases as the temperature decreases. When the temperature passes below the martensite finish temperature (M_f), the steel should contain 100% martensite. At any intermediate temperature, the amount of martensite does not change as the time at that temperature increases.

Owing to the conservation of mass, the composition of martensite must be the same as that of the austenite from which it forms. There is no long-range diffusion during the transformation that can change the composition. Thus, in iron-carbon alloys, the initial austenite composition and the final martensite composition are the same. The following example illustrates how heat treatment is used to produce a dual phase steel.

EXAMPLE 11-10   *Design of a Heat Treatment for a Dual Phase Steel*

Unusual combinations of properties can be obtained by producing a steel whose microstructure contains 50% ferrite and 50% martensite; the martensite provides strength and the ferrite provides ductility and toughness. Design a heat treatment to produce a dual phase steel in which the composition of the martensite is 0.60% C.

SOLUTION

To obtain a mixture of ferrite and martensite, we need to heat-treat a hypoeutectoid steel into the $\alpha + \gamma$ region of the phase diagram. The steel is then quenched, permitting the γ portion of the structure to transform to martensite.

The heat treatment temperature is fixed by the requirement that the martensite contain 0.60% C. From the solubility line between the γ and the $\alpha + \gamma$ regions, we find that 0.60% C is obtained in austenite when the temperature is about 750°C. To produce 50% martensite, we need to select a steel that gives 50% austenite when the steel is held at 750°C. If the carbon content of the steel is x, then:

$$\% \, \gamma = \left[\frac{(x - 0.02)}{(0.60 - 0.02)} \right] \times 100 = 50 \quad \text{or} \quad x = 0.31\% \text{ C}$$

Our final design is:

1. Select a hypoeutectoid steel containing 0.31% C.

2. Heat the steel to 750°C and hold (perhaps for 1 h, depending on the thickness of the part) to produce a structure containing 50% ferrite and 50% austenite, with 0.60% C in the austenite.

3. Quench the steel to room temperature. The austenite transforms to martensite, also containing 0.60% C.

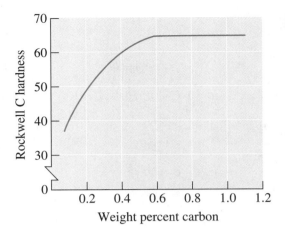

Figure 11-26
The effect of carbon content on the hardness of martensite in steels.

Properties of Steel Martensite   Martensite in steels is very hard and brittle, just like ceramics. The BCT crystal structure has no close-packed slip planes in which dislocations can easily move. The martensite is highly supersaturated with carbon, since iron normally contains less than 0.0218% C at room temperature, and martensite contains the amount of carbon present in the steel. Finally, martensite has a fine grain size and an even finer substructure within the grains.

The structure and properties of steel martensites depend on the carbon content of the alloy (Figure 11-26). When the carbon content is low, the martensite grows in a "lath" shape, composed of bundles of flat, narrow plates that grow side by side [Figure 11-27(a)]. This martensite is not very hard. At a higher carbon content, plate martensite grows, in which flat, narrow plates grow individually rather than as bundles [Figure 11-27(b)]. The hardness is much greater in the higher carbon, plate martensite structure, partly due to the greater distortion, or large c/a ratio, of the crystal structure.

Tempering of Steel Martensite   Martensite is not an equilibrium phase. This is why it does not appear on the $Fe-Fe_3C$ phase diagram (Figure 11-15). When martensite in a steel is heated below the eutectoid temperature, the thermodynamically stable α and Fe_3C phases precipitate. This process is called **tempering**. The decomposition of

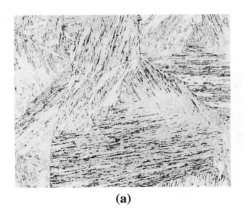

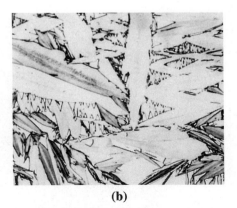

(a)   (b)

Figure 11-27   (a) Lath martensite in low-carbon steel (×80). (b) Plate martensite in high-carbon steel (×400). (*From* ASM Handbook, *Vol. 8, (1973), ASM International, Materials Park, OH 44073.*)

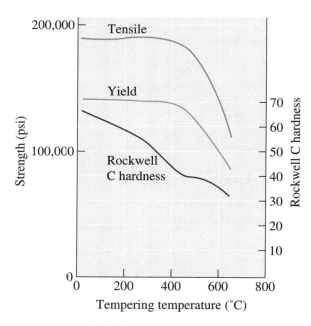

Figure 11-28
Effect of tempering temperature on the properties of an eutectoid steel.

martensite in steels causes the strength and hardness of the steel to decrease while the ductility and impact properties are improved (Figure 11-28). Note that the term tempering here is different from the term we used for tempering of silicate glasses. In both tempering of glasses and tempering of steels, however, the key result is an increase in the toughness of the material.

At low tempering temperatures, the martensite may form two transition phases—a lower carbon martensite and a very fine nonequilibrium ε-carbide, or $Fe_{2.4}C$. The steel is still strong, brittle, and perhaps even harder than before tempering. At higher temperatures, the stable α and Fe_3C form and the steel becomes softer and more ductile. If the steel is tempered just below the eutectoid temperature, the Fe_3C becomes very coarse and the dispersion-strengthening effect is greatly reduced. By selecting the appropriate tempering temperature, a wide range of properties can be obtained. The product of the tempering process is a microconstituent called tempered martensite (Figure 11-29).

Martensite in Other Systems   The characteristics of the martensite reaction are different in other alloy systems. For example, martensite can form in iron-based alloys that

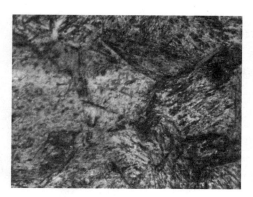

Figure 11-29
Tempered martensite in steel (×500). (*From ASM Handbook, Vol. 9, Metallography and Microstructure (1985), ASM International Materials Park, OH 44073.*)

contain little or no carbon by a transformation of the FCC crystal structure to a BCC crystal structure. In certain high-manganese steels and stainless steels, the FCC structure changes to a HCP crystal structure during the martensite transformation. In addition, the martensitic reaction occurs during the transformation of many polymorphic ceramic materials, including ZrO_2, and even in some crystalline polymers. Thus the terms martensitic reaction and martensite are rather generic. In the context of steel properties, microstructure, and heat treatment, the term "martensite" refers to the hard and brittle bct phase obtained upon the quenching of steels.

The properties of martensite in other alloys are also different from the properties of steel martensite. In titanium alloys, BCC titanium transforms to a HCP martensite structure during quenching. However, the titanium martensite is softer and weaker than the original structure. The martensite that forms in other alloys can also be tempered. The martensite produced in titanium alloys can be reheated to permit the precipitation of a second phase. Unlike the case of steel, however, the tempering process *increases*, rather than decreases, the strength of the titanium alloy.

11-12 The Shape-Memory Alloys [SMAs]

Shape-memory effect is a unique property possessed by some alloys that undergo the martensitic reaction.[10] This effect was discovered originally in Au-Cd alloys in 1951 and received more prominence after its discovery in Ni-Ti alloys in 1963. A Ni-50 at% Ti alloy and several copper-based alloys can be processed using a sophisticated thermomechanical treatment to produce a martensitic structure. At the end of the treatment process, the metal is deformed to a predetermined shape. The metal can then be deformed into a second shape, but when the temperature is increased, the metal changes back to its original shape! Orthodontal braces, blood clot filters, engines, antennas for cellular phones, frames for eyeglasses and actuators for smart systems have been developed using these materials. Flaps that change direction of airflow depending upon temperature have been developed and used for air conditioners.

More recently, a special class of materials known as ferromagnetic shape-memory alloys also has been developed. Examples of ferromagnetic shape-memory alloys include Ni_2MnGa, Fe-Pd, and Fe_3Pt.[10] Unlike Ni-Ti, these materials show a shape-memory effect in response to a magnetic field. Most commercial shape-memory alloys including Ni-Ti are nonmagnetic. We discussed in Chapter 6 that many polymers are viscoelastic and the viscous component is recovered over time. Thus, many polymers do have a memory of their shape! Recently, researchers have developed new shape-memory plastics.[11]

Shape-memory alloys exhibit a memory that can be triggered by stress or temperature change.[12] **Smart materials** are materials that can sense an external stimulus (such as stress, temperature change, magnetic field, etc.) and undergo some type of change. Actively smart materials can even initiate a response (i.e., they function as a sensor and an actuator). Shape-memory alloys are a family of passively smart materials in that they merely sense a change in stress or temperature.

Shape-memory alloys also show a superelastic behavior.[13] Recoverable strains up to 10% are possible. This is why shape-memory alloys have been used so successfully in such applications as orthodontic wires, eyeglass frames, underwires for brassieres, and antennas for cellular phones. In these applications we make use of the superelastic (and not the shape memory) effect. The following examples illustrate the use of shape-memory alloys in different fields of application.

EXAMPLE 11-11   *Design of a Coupling for Tubing*

At times, you need to join titanium tubing in the field. Design a method for doing this quickly.[14]

SOLUTION

Titanium is quite reactive and, unless special welding processes are used, may be contaminated. In the field, we may not have access to these processes. Therefore, we wish to make the joint without resorting to high-temperature processes.

We can take advantage of the shape-memory effect for this application (Figure 11-30). Ahead of time, we can set a Ni-Ti coupling into a small diameter, then deform it into a larger diameter in the martensitic state. In the field, the coupling, which is in the martensitic state, is slipped over the tubing and heated (at a low enough temperature so that the titanium tubing is not contaminated). The coupling contracts back to its predetermined shape as a result of the shape-memory effect, producing a strong mechanical bond to join the tubes.

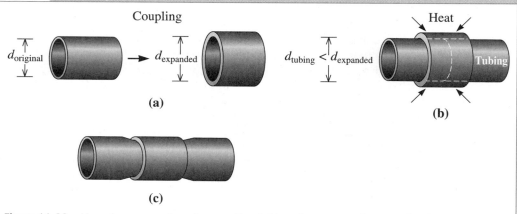

Figure 11-30   Use of memory alloys for coupling tubing: A memory alloy coupling is expanded (a) so it fits over the tubing (b). When the coupling is reheated, it shrinks back to its original diameter (c), squeezing the tubing for a tight fit (for Example 11-11).

EXAMPLE 11-12   *Selection of Material for a Self-Expandable Cardiovascular Stent*

Close to a half million people in the world have coronary stents. These are mostly made from 316 stainless steel, but some are made from platinum. How would you go about designing a material for making a cardiovascular stent?[15] A conventional stent is essentially a slotted tube that is inserted into an artery. This procedure is typically done by doctors after a procedure known as an angioplasty is conducted.

SOLUTION

We can consider using Ni-Ti shape-memory alloys. A narrow stent made from a Ni-Ti shape-memory alloy can be inserted into the artery that needs to be opened. This will be a self-expanding stent that opens up when it reaches a certain temperature (e.g., the internal temperature of the body). The shape-memory alloy composition will have to be chosen such that the stent will open at body temperature (37°C). These types of stents have been developed and tested for many other applications involving gastroentrology, urology, and what is known as a vena cava filter. Other considerations would be: Is Ni-Ti biocompatible? Corrosion resistance will also have to be taken into account. Would it be durable?

Currently, the cardiovascular stents are made mostly from 316 stainless steel. This material is nonferromagnetic. Suppose the patient needs a magnetic resonance imaging (MRI) scan in which the body and the implants would be subjected to very intense (~1.5 Tesla) magnetic flux densities. If the stent material is a ferromagnetic shape-memory alloy, the stent could become dislodged. Therefore, these materials must be ruled out.

Thus, a number of issues such as the use of a nonmagnetic material, biocompatibility, corrosion resistance, mechanical properties, and durability must be considered carefully, especially for biomedical applications. Most likely, the cost of the material or stent would also play a small part in the total cost of the procedure.

SUMMARY

◆ Solid-state phase transformations, which have a profound effect on the structure and properties of a material, can be controlled by proper heat treatments. These heat treatments are designed to provide an optimum distribution of two or more phases in the microstructure. Dispersion strengthening permits a wide variety of structures and properties to be obtained.

◆ These transformations typically require both nucleation and growth of new phases from the original structure. The kinetics of the phase transformation help us understand the mechanisms that control the reaction and the rate at which the reaction occurs, enabling us to design the heat treatment to produce the desired microstructure. Reference to appropriate phase diagrams also helps us select the necessary compositions and temperatures.

◆ Age hardening, or precipitation hardening, is one powerful method for controlling the optimum dispersion strengthening in many metallic alloys. In age hardening, a very fine widely dispersed coherent precipitate is allowed to precipitate by a heat treatment that includes (a) solution treating to produce a single-phase solid solution, (b) quenching to retain that single phase, and (c) aging to permit a precipitate to form. In order for age hardening to occur, the phase diagram must show decreasing solubility of the solute in the solvent as the temperature decreases.

◆ Aluminum alloys, stainless steels, and bake-hardenable steels are just a few examples of materials that are strengthened using precipitation hardening.

◆ The eutectoid reaction can be controlled to permit one type of solid to transform to two different types of solid. The kinetics of the reaction depends on the nucleation of the new solid phases and the diffusion of the different atoms in the material to permit the growth of the new phases.

◇ The most widely used eutectoid reaction occurs in producing steels from iron-carbon alloys: Either pearlite or bainite can be produced as a result of the eutectoid reaction in steel. In addition, primary ferrite or primary cementite may be present, depending on the carbon content of the alloy. The trick is to formulate a microstructure that consists of a right mix of metal-like phases that are tough and ceramic-like phases that are hard and brittle.

◇ Factors that influence the mechanical properties of the microconstituent produced by the eutectoid reaction include (a) the composition of the alloy (amount of eutectoid microconstituent), (b) the grain size of the original solid, the eutectoid microconstituent, and any primary microconstituents, (c) the fineness of the structure within the eutectoid microconstituent (interlamellar spacing), (d) the cooling rate during the phase transformation, and (e) the temperature at which the transformation occurs (the amount of undercooling).

◇ A martensitic reaction occurs with no long-range diffusion. Again, the best known example occurs in steels:

    ○ The amount of martensite that forms depends on the temperature of the transformation (an athermal reaction).

    ○ Martensite is very hard and brittle, with the hardness determined primarily by the carbon content.

    ○ The amount and composition of the martensite are the same as the austenite from which it forms.

◇ Martensite can be tempered. During tempering, a dispersion-strengthened structure is produced. In steels, tempering reduces the strength and hardness but improves the ductility and toughness.

◇ Since optimum properties are obtained through heat treatment, we must remember that the structure and properties may change when the material is used at or exposed to elevated temperatures. Overaging or overtempering occur as a natural extension of the phenomena governing these transformations when the material is placed into service.

◇ Shape-memory alloys (e.g., Ni-Ti) are a class of smart materials that can remember their shape. They also exhibit superelastic behavior. These materials depend upon the martensitic phase transformations that occur as a result of stress or temperature change. Shape-memory alloys are useful for biomedical and other applications.

GLOSSARY

Age hardening   A special dispersion-strengthening heat treatment. By solution treatment, quenching, and aging, a coherent precipitate forms that provides a substantial strengthening effect. Also known as precipitation hardening, it is a form of dispersion strengthening.

Artificial aging   Reheating a solution-treated and quenched alloy to a temperature below the solvus in order to provide the thermal energy required for a precipitate to form.

Athermal transformation   When the amount of the transformation depends only on the temperature, not on the time (same as martensitic transformation or displacive transformation).

Austenite   The name given to the FCC crystal structure of iron.

Avrami relationship   Describes the fraction of a transformation that occurs as a function of time. This describes most solid-state transformations that involve diffusion, thus martensitic transformations are not described.

Bainite   A two-phase microconstituent, containing ferrite and cementite, that forms in steels that are isothermally transformed at relatively low temperatures.

Bake-hardenable steels   These are steels that can show an increase in their yield stress as a result of precipitation hardening that can occur at fairly low temperatures ($\sim 100°C$), conditions that simulate baking of paints on cars. This additional increase leads to better dent resistance.

Cementite   The hard, brittle ceramic-like compound Fe_3C that, when properly dispersed, provides the strengthening in steels.

Coherent precipitate   A precipitate whose crystal structure and atomic arrangement have a continuous relationship with the matrix from which the precipitate is formed. The coherent precipitate provides excellent disruption of the atomic arrangement in the matrix and provides excellent strengthening.

Dihedral angle   The angle that defines the shape of a precipitate particle in the matrix. The dihedral angle is determined by the relative surface energies of the grain boundary energy of the matrix and the matrix-precipitate interfacial energy.

Displacive transformation   A phase transformation that occurs via small displacements of atoms or ions and without diffusion. Same as athermal or martensitic transformation.

Ferrite   The name given to the BCC crystal structure of iron that can occur as α or δ. This is not to be confused with ceramic ferrites which are magnetic materials with spinel or inverse spinel structure.

Guinier-Preston (GP) zones   Tiny clusters of atoms that precipitate from the matrix in the early stages of the age-hardening process. Although the GP zones are coherent with the matrix, they are too small to provide optimum strengthening.

Interfacial energy   The energy associated with the boundary between two phases.

Isothermal transformation   When the amount of a transformation at a particular temperature depends on the time permitted for the transformation.

Martensite   A metastable phase formed in steel and other materials by a diffusionless, athermal transformation.

Martensitic transformation   A phase transformation that occurs without diffusion. Same as athermal or displacive transformation. These occur in steels, Ni-Ti and many ceramic materials.

Natural aging   When a coherent precipitate forms from a solution-treated and quenched age-hardenable alloy at room temperature, providing optimum strengthening.

Pearlite   A two-phase lamellar microconstituent, containing ferrite and cementite, that forms in steels cooled in a normal fashion or isothermally transformed at relatively high temperatures.

Shape-memory effect   The ability of certain materials to develop microstructures that, after being deformed, can return the material to its initial shape when heated (e.g. Ni-Ti alloys).

Smart materials   Materials that can sense an external stimulus (e.g., stress, pressure, temperature change, magnetic field, etc.) and initiate a response. Passively smart materials can sense external stimulus, actively smart materials have sensing and actuation capabilities.

Solution treatment   The first step in the age-hardening heat treatment. The alloy is heated above the solvus temperature to dissolve any second phase and to produce a homogeneous single-phase structure.

Strain energy   The energy required to permit a precipitate to fit into the surrounding matrix during nucleation and growth of the precipitate.

Superelastic behavior   Shape-memory alloys deformed above a critical temperature show a large reversible elastic deformation as a result of a stress-induced martensitic transformation.

Supersaturated solid solution   The solid solution formed when a material is rapidly cooled from a high-temperature single-phase region to a low-temperature two-phase region without the second phase precipitating. Because the quenched phase contains more alloying element than the solubility limit, it is supersaturated in that element.

Tempering   A low-temperature heat treatment used to reduce the hardness of martensite by permitting the martensite to begin to decompose to the equilibrium phases. This leads to increased toughness.

TTT diagram   The time-temperature-transformation diagram describes the time required at any temperature for a phase transformation to begin and end. The TTT diagram assumes that the temperature is constant during the transformation.

Widmanstätten structure   The precipitation of a second phase from the matrix when there is a fixed crystallographic relationship between the precipitate and matrix crystal structures. Often needle-like or plate-like structures form in the Widmanstätten structure.

ADDITIONAL INFORMATION

1. VERHOEVEN, J.D., *Fundamentals of Physical Metallurgy,* 1975: Wiley.

2. WEIDMANN, G., P. LEWIS, and N. REID, *Structural Materials (Materials in Action Series)*, ed. and N. REID. 1990: The Open University, Butterworth Heinemann.

3. GAYLE, F.W. and M. GOODWAY, "Precipitation Hardening in the First Aerospace Aluminum Alloy: The Wright Flyer Crankcase," *Science*, 1994. 266(5187): p. 1015–1017.

4. HORNBOGEN, E., "Precipitation Hardening—The Oldest Nanotechnology," *Metall* (Isernhagen, Germany), 2001. 55(9): p. 522–526.

5. BELANGER, P.J., et al. "Evolution of IF-Based Steel for Exposed Applications at Daimler-Chrysler in North America," in *IF Steels 2000 Proceedings*, Pittsburgh: Iron and Steel Society, Warrendale, Pa. 2000.

6. BERGERON, C.G. and S.H. RISBUD, *Introduction to Phase Equilibria*. 1984: American Ceramic Society.

7. HARPER, C.A., *Handbook of Materials Product Design*. 3rd ed. 2001: McGraw Hill.

8. BAIN, E.C. and E.S. DAVENPORT, *Trans. AIME*, 1930. 90: p. 117.

9. FRIEND, C.M. and N.B. MORGAN, "Shape-Memory Alloys in Medicine—from New Prosthetic Devices to "Cuddly" Instruments," in *Medical Applications for Shape-Memory Alloys (SMA)*. 1999.

10. KAKESHITA, T. and K. ULLAKKO, "Giant Magnetostriction in Ferromagnetic Shape-Memory Alloys," *MRS Bulletin*, 2002. 27(2): p. 91.

11. LENDELIN, A., A. SCHMIDT, and R. LANGER, *Proc. Natl. Acade. Sci. USA*, 2001. 98: p. 842.

12. OTSUKA, K. and WAYMAN, C.M., *Shape-Memory Materials*. First ed. 1998, Cambridge: Cambridge University Press.

13. OTSUKA, K. and T. KAKESHITA, "Science and Technology of Shape-Memory Alloys: New Developments," *MRS Bulletin*, 2002. 27(2).

14. HARISSON, J.D. and D.E. HODGSON, eds., *Shape-Memory Effects in Alloys.* ed. J. PERKINS. 1975, Plenum Publisher, New York. 517.

15. JOST, C. and V. KUMAR, "Are Current Cardiovascular Stents MRI Safe?" *J. Invas. Cardiology*, 1998. 10(8): p. 477–479.

✓ PROBLEMS

Section 11-1 Nucleation and Growth in Solid-State Reactions

11-1 How is the equation for nucleation of a phase in the solid state different from that for a liquid to solid transformation?

11-2 Determine the constants c and n in Equation 11-2 that describe the rate of crystallization of polypropylene at 140°C. (See Figure 11-31.)

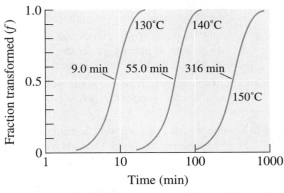

Figure 11-31 The effect of temperature on the crystallization of polypropylene (for Problems 11-2, 11-4 and 11-106).

11-3 Determine the constants c and n in Equation 11-2 that describe the rate of recrystallization of copper at 135°C. (See Figure 11-2.)

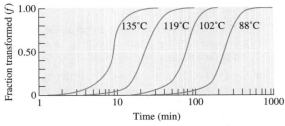

Figure 11-2 (Repeated for Problem 11-3) The effect of temperature on recrystallization of cold-worked copper.

11-4 Determine the activation energy for crystallization of polypropylene, using the curves in Figure 11-36 (on page 539).

11-5 Most solid-state phase transformations follow the Avrami equation. True or False. Discuss briefly.

11-6 What step controls the rate of recrystallization of a cold-worked metal?

Section 11-2 Alloys Strengthened By Exceeding the Solubility Limit

11-7 What are the different ways by which a second phase can be made to precipitate in a two-phase microstructure?

11-8 Explain, when cooled slowly, why is it that the second phase in Al-4% Cu alloys nucleates and grows along the grain boundaries. Is this usually desirable?

11-9 What do the terms "coherent" and "incoherent" precipitates mean?

11-10 What properties of the precipitate phase are needed for precipitation hardening? Why?

Section 11-3 Age or Precipitation Hardening

11-11 What is the principle of precipitation hardening?

11-12 What is the difference between precipitation hardening and dispersion strengthening?

11-13 What is a supersaturated solution? How do we obtain supersaturated solutions during precipitation hardening? Why is the formation of a supersaturated solution necessary?

11-14 Why do the precipitates formed during precipitation hardening form throughout the microstructure and not just at grain boundaries?

11-15 On aging for longer times, why do the second-phase precipitates grow? What is the driving force? Compare this with driving forces for grain growth and solid-state sintering.

11-16 (a) Recommend an artificial age-hardening heat treatment for a Cu-1.2% Be alloy. (See Figure 11-34 on page 537.) Include appropriate temperatures.

(b) Compare the amount of the γ_2 precipitate that forms by artificial aging at 400°C with the amount of the precipitate that forms by natural aging.

11-17 Suppose that age hardening is possible in the Al-Mg system. (See Figure 11-11.)

(a) Recommend an artificial age-hardening heat treatment for each of the following alloys, and

(b) compare the amount of the β precipitate that forms from your treatment of each alloy.
(i) Al-4% Mg   (ii) Al-6% Mg
(iii) Al-12% Mg

(c) Testing of the alloys after the heat treatment reveals that little strengthening occurs as a result of the heat treatment. Which of the requirements for age hardening is likely not satisfied?

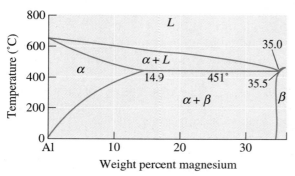

Figure 11-11   (Repeated for Problem 11-17) Portion of the aluminum-magnesium phase diagram.

11-18 An Al-2.5% Cu alloy is solution-treated, quenched, and overaged at 230°C to produce a stable microstructure. If the spheroidal θ precipitates so that the form has a diameter of 9000 Å and a density of 4.26 g/cm^3, determine the number of precipitate particles per cm^3.

Section 11-4 Applications of Age-Hardened Alloys

11-19 Why is precipitation hardening an attractive mechanism of strengthening for aircraft materials?

11-20 Why are most precipitation-hardened alloys suitable only for low-temperature applications?

11-21 What are bake-hardenable steels?

11-22 When an automobile chassis is formed using a sheet of bake-hardenable steel, what mechanism of strengthening contributes to the final properties?

11-23 What are micro-alloyed or high-strength, low-alloy steels?

Section 11-5 Microstructural Evolution in Age or Precipitation Hardening

11-24 Explain the three basic steps encountered during precipitation hardening.

11-25 Explain how hot shortness can occur in precipitation-hardened alloys.

11-26 In precipitation hardening, does the phase that provides strengthening form directly from the supersaturated matrix phase? Explain.

Section 11-6 Effects of Aging Temperature and Time

11-27 What is aging? Why is this step needed in precipitation hardening?

11-28 What is overaging?

11-29 What do the terms "natural aging" and "artificial aging" mean?

11-30 In the plane flown by the Wright brothers, how was the alloy precipitation strengthened?

11-31 Why did the work of Dr. Gayle and coworkers reveal two sets of precipitates in the alloy that was used to make the Wright brothers' plane?

11-32 What analytical techniques would you use to characterize the nano-sized precipitates in an alloy?

11-33 Why do we have to keep some aluminum alloys at low temperatures until they are ready for forming steps?

11-34 What phases cause the increase in the yield strength of bake-hardenable steels?

11-35 Why not simply use steel that has ~10% higher strength to begin with, instead of using bake-hardenable steels?

Section 11-7 Requirements for Age Hardening

11-36 Can all alloy compositions be strengthened using precipitation hardening? Can we use this mechanism for the strengthening of ceramics, glasses, or polymers?

11-37 A conductive copper wire is to be made. Would you choose precipitation hardening as a way of strengthening this wire? Explain.

11-38 Figure 11-32 shows a hypothetical phase diagram. Determine whether each of the following alloys might be good candidates for age hardening, and explain your answer. For those alloys that might be good candidates, describe the heat treatment required, including recommended temperatures.

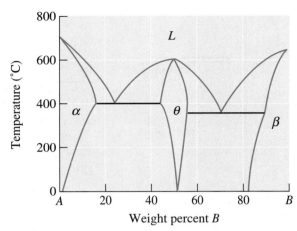

Figure 11-32 Hypothetical phase diagram (for Problem 11-38).

(a) *A*-10% *B* (b) *A*-20% *B* (c) *A*-55% *B*
(d) *A*-87% *B* (e) *A*-95% *B*

Section 11-8 Use of Age-Hardenable Alloys at High Temperatures

11-39 What is the major limitation of the use for precipitation-hardened alloys? Can dispersion-strengthened materials such as TD-nickel (Chapter 16) be used at high temperatures?

11-40 Why is it that certain aluminum (not nickel-based) alloys strengthened using age hardening can lose their strength on welding?

11-41 Would you choose a precipitation-hardened alloy to make an aluminum alloy baseball bat?

11-42 What type of dispersion strengthened alloys can retain their strength up to ~1000°C?

Section 11-9 The Eutectoid Reaction

11-43 Write down the eutectoid reaction in Fe-Fe$_3$C system.

11-44 Sketch the microstructure of pearlite formed by the slow cooling of a steel with the eutectoid composition.

11-45 Compare and contrast eutectic and eutectoid reactions.

11-46 What are the solubilities of carbon in α, δ, and γ forms of iron?

11-47 Define the following terms: ferrite, austenite, pearlite, and cementite.

11-48 The pearlite microstructure is similar to a ceramic-metal nanocomposite. True or False. Comment.

11-49 What do the terms "hypoeutectoid" and "hypereutectoid" steels mean?

11-50 What is the difference between a microconstituent and a phase?

11-51 Figure 11-1 shows the sigmoidal curve for the transformation of austenite. Determine the constants c and n in Equation 11-2 for this reaction. By comparing this figure with the TTT diagram (Figure 11-21), estimate the temperature at which this transformation occurred.

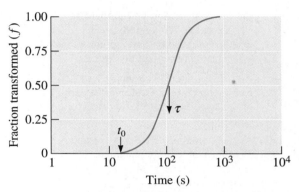

Figure 11-1 (Repeated for Problem 11-51) Sigmoidal curve showing the rate of transformation of FCC iron at a constant temperature. The incubation time t_0 and the time τ for 50% transformation are also shown.

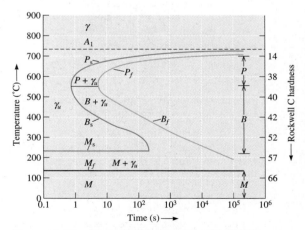

Figure 11-21 (Repeated for Problem 11-51) The time-temperature-transformation (TTT) diagram for an eutectoid steel.

11-52 For an Fe-0.35% C alloy, determine

(a) the temperature at which austenite first begins to transform on cooling,

(b) the primary microconstituent that forms,

(c) the composition and amount of each phase present at 728°C,

(d) the composition and amount of each phase present at 726°C, and

(e) the composition and amount of each micro-constituent present at 726°C.

11-53 For an Fe-1.15% C alloy, determine

(a) the temperature at which austenite first begins to transform on cooling,

(b) the primary microconstituent that forms,

(c) the composition and amount of each phase present at 728°C,

(d) the composition and amount of each phase present at 726°C, and

(e) the composition and amount of each micro-constituent present at 726°C.

11-54 A steel contains 8% cementite and 92% ferrite at room temperature. Estimate the carbon content of the steel. Is the steel hypoeutectoid or hyper-eutectoid?

11-55 A steel contains 18% cementite and 82% ferrite at room temperature. Estimate the carbon con-tent of the steel. Is the steel hypoeutectoid or hypereutectoid?

11-56 A steel contains 18% pearlite and 82% primary ferrite at room temperature. Estimate the car-bon content of the steel. Is the steel hypo-eutectoid or hypereutectoid?

11-57 A steel contains 94% pearlite and 6% primary cementite at room temperature. Estimate the carbon content of the steel. Is the steel hypo-eutectoid or hypereutectoid?

11-58 A steel contains 55% α and 45% γ at 750°C. Es-timate the carbon content of the steel.

11-59 A steel contains 96% γ and 4% Fe_3C at 800°C. Estimate the carbon content of the steel.

11-60 A steel is heated until 40% austenite, with a carbon content of 0.5%, forms. Estimate the temperature and the overall carbon content of the steel.

11-61 A steel is heated until 85% austenite, with a carbon content of 1.05%, forms. Estimate the temperature and the overall carbon content of the steel.

11-62 Determine the eutectoid temperature, the com-position of each phase in the eutectoid reac-tion, and the amount of each phase present in the eutectoid microconstituent for the following systems. For the metallic systems, comment on whether you expect the eutectoid microconstitu-ent to be ductile or brittle.

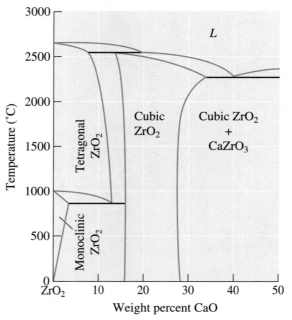

Figure 11-33 The ZrO_2-CaO phase diagram. A polymorphic phase transformation occurs for pure ZrO_2. Adding 16 to 26% CaO produces a single cubic zirconia phase at all temperatures (for Problem 11-62).

(a) ZrO_2-CaO (See Figure 11-33.)

(b) Cu-Al at 11.8% Al [See Figure 11-34(c).]

(c) Cu-Zn at 47% Zn [See Figure 11-34(a).]

(d) Cu-Be [See Figure 11-34(d).]

Section 11-10 Controlling the Eutectoid Reaction

11-63 Why are the distances between lamellae formed in an eutectoid reaction typically separated by distances smaller than those formed in eutectic reactions?

11-64 Compare the interlamellar spacing and the yield strength when an eutectoid steel is isothermally transformed to pearlite at

(a) 700°C, and

(b) 600°C.

11-65 Why is it that a eutectoid steel exhibits different yield strengths and % elongations, depending upon if it was cooled slowly or relatively fast?

11-66 What is a TTT diagram?

11-67 Sketch and label clearly different parts of a TTT diagram for a plain-carbon steel with 0.77% carbon.

11-68 On the TTT diagram what is the difference be-tween the γ and γ_u phases?

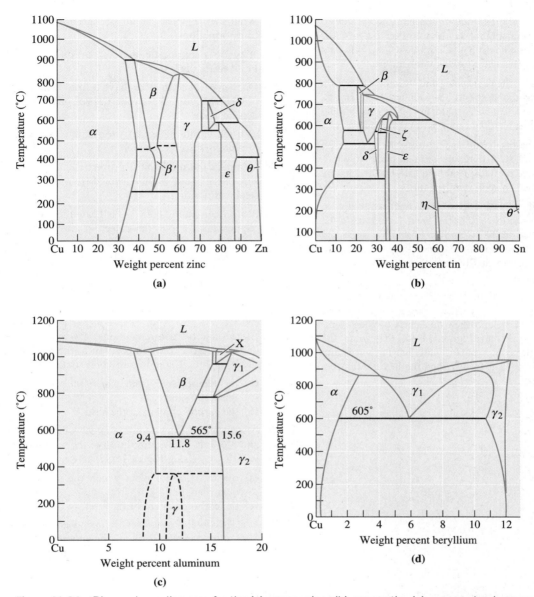

Figure 11-34 Binary phase diagrams for the (a) copper-zinc, (b) copper-tin, (c) copper-aluminum, and (d) copper-berylium systems (for Problems 11-16 and 11-62).

11-69 How is it that bainite and pearlite do not appear in the Fe-Fe$_3$C diagram? Are these phases or microconstituents?

11-70 Why is it that we cannot make use of TTT diagrams for describing heat treatment profiles in which samples are getting cooled over a period of time (i.e., why are TTT diagrams suitable for only following isothermal transformations)?

11-71 What is bainite? Why do steels containing bainite exhibit higher levels of toughness?

11-72 An isothermally transformed eutectoid steel is found to have a yield strength of 410 MPa. Estimate

(a) the transformation temperature, and
(b) the interlamellar spacing in the pearlite.

11-73 Determine the required transformation temperature and microconstituent if an eutectoid steel is to have the following hardness values:

(a) HRC 38   (b) HRC 42
(c) HRC 48   (d) HRC 52

11-74 Describe the hardness and microstructure in an eutectoid steel that has been heated to 800°C for 1 h, quenched to 350°C and held for 750 s, and finally quenched to room temperature.

11-75 Describe the hardness and microstructure in an eutectoid steel that has been heated to 800°C, quenched to 650°C, held for 500 s, and finally quenched to room temperature.

11-76 Describe the hardness and microstructure in an eutectoid steel that has been heated to 800°C, quenched to 300°C and held for 10 s, and finally quenched to room temperature.

11-77 Describe the hardness and microstructure in an eutectoid steel that has been heated to 800°C, quenched to 300°C and held for 10 s, quenched to room temperature, and then reheated to 400°C before finally cooling to room temperature again.

11-78 A steel containing 0.3% C is heated to various temperatures above the eutectoid temperature, held for 1 h, and then quenched to room temperature. Using Figure 11-35, determine the amount, composition, and hardness of any martensite that forms when the heating temperature is:

(a) 728°C   (b) 750°C
(c) 790°C   (d) 850°C

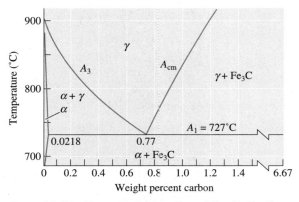

Figure 11-35   The eutectoid portion of the Fe-Fe$_3$C phase diagram (for Problems 11-78, 11-86, 11-87 and 11-88).

Section 11-11 The Martensitic Reaction and Tempering

11-79 What is the difference between solid-state phase transformations such as the eutectoid reaction and martensitic phase transformation?

11-80 What is the difference between isothermal and athermal transformation?

11-81 What step controls the rate of martensitic phase transformations?

11-82 Why does martensite not appear on the Fe-Fe$_3$C phase diagram?

11-83 Does martensite in steels have a fixed composition? What do the properties of martensite depend upon?

11-84 Can martensitic phase transformations occur in other alloys and ceramics?

11-85 Compare the mechanical properties of martensite, pearlite, and bainite formed from eutectoid steel composition.

11-86 A steel containing 0.95% C is heated to various temperatures above the eutectoid temperature, held for 1 h, and then quenched to room temperature. Using Figure 11-35 determine the amount and composition of any martensite that forms when the heating temperature is:

(a) 728°C   (b) 750°C
(c) 780°C   (d) 850°C

11-87 A steel microstructure contains 75% martensite and 25% ferrite; the composition of the martensite is 0.6% C. Using Figure 11-35, determine

(a) the temperature from which the steel was quenched, and
(b) the carbon content of the steel.

11-88 A steel microstructure contains 92% martensite and 8% Fe$_3$C; the composition of the martensite is 1.10% C. Using Figure 11-35, determine

(a) the temperature from which the steel was quenched, and
(b) the carbon content of the steel.

11-89 A steel containing 0.8% C is quenched to produce all martensite. Estimate the volume change that occurs, assuming that the lattice parameter of the austenite is 3.6 Å. Does the steel expand or contract during quenching?

11-90 Describe the complete heat treatment required to produce a quenched and tempered eutectoid steel having a tensile strength of at least 125,000 psi. Include appropriate temperatures.

11-91 Describe the complete heat treatment required to produce a quenched and tempered eutectoid steel having a HRC hardness of less than 50. Include appropriate temperatures.

11-92 In eutectic alloys, the eutectic microconstituent is generally the continuous one, but in the eutectoid structures, the primary microconstituent is normally continuous. By describing the changes that occur with decreasing temperature in each reaction, explain why this difference is expected.

11-93 What is the tempering of steels? Why is tempering necessary?

11-94 What phases are formed by the decomposition of martensite?

11-95 What is tempered martensite?

11-96 If tempering results in the decomposition of martensite, why should we form martensite in the first place?

11-97 Describe the changes in properties that occur upon the tempering of an eutectoid steel.

Section 11-12 Shape-Memory Alloys (SMAs)

11-98 What is the principle by which shape-memory alloys display a memory effect?

11-99 Give examples of materials that display a shape-memory effect.

11-100 What are ferromagnetic shape-memory alloys? How are they different from conventional shape-memory alloys? Give examples.

11-101 Do the phase transformations from parent phase to martensitic phase and vice-versa occur at the same temperature? Explain.

11-102 Explain the origin of superelastic behavior of shape-memory alloys. How is this different from the superplastic behavior we have seen before in certain metallic and ceramic materials?

11-103 What are some of the applications of shape-memory alloys?

11-104 What factors and properties must be considered in the use of shape-memory alloys for biomedical applications such as cardiovascular stents?

❊ Design Problems

11-105 You wish to attach aluminum sheets to the frame of the twenty-fourth floor of a skyscraper. You plan to use rivets made of an age-hardenable aluminum, but the rivets must be soft and ductile in order to close. After the sheets are attached, the rivets must be very strong. Design a method for producing, using, and strengthening the rivets.

11-106 Design a process to produce a polypropylene polymer with a structure that is 75% crystalline. Figure 11-36 will provide appropriate data.

11-107 An age-hardened, Al-Cu bracket is used to hold a heavy electrical-sensing device on the outside of a steel-making furnace. Temperatures may exceed 200°C. Is this a good de-

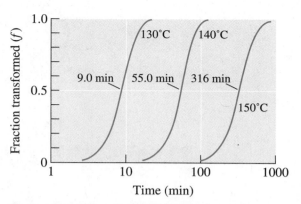

Figure 11-36 The effect of temperature on the crystallization of polypropylene (for Problems 11-4 and 11-106).

sign? Explain. If it is not, design an appropriate bracket and explain why your choice is acceptable.

11-108 You use an arc-welding process to join an eutectoid steel. Cooling rates may be very high following the joining process. Describe what happens in the heat-affected area of the weld and discuss the problems that might occur. Design a joining process that may minimize these problems.

▣ Computer Problems

11-109 *Calculation of Phases and Microconstituents in Plain-Carbon Steels* Write a computer program that will calculate the amount of phases in plain-carbon steels in compositions that range from 0 to 1.5% carbon. Assume room temperature for your calculations. The program should ask the user to provide a value for the carbon concentration. The program should make use of this input and provide the amount of phases (ferrite, Fe_3C) in the steel using the lever rule (similar to Example 11-7). The program should also tell the user (as part of the output) whether the steel is eutectoid, hypoeutectoid, or hypereutectoid. Depending upon the composition, the program should also make available the amounts of microconstituents present (e.g., how much pearlite). The program should also print an appropriate descriptive message that describes the expected microstructure (assume slow cooling). For example, if the composition chosen is that of a hypoeutectoid steel, the program should output a message stating that the microstructure will primarily consist of α grains and pearlite.

Appropriate design of heat treatments enables the engineer to control the microstructure and mechanical properties of metal alloys. In this example, the polymorphic behavior of a nonferrous titanium alloy permits α plates of titanium to form in a matrix of β titanium. The plate-like structure interferes with the growth of cracks, therefore improving the fracture toughness of the alloy. (ASM Handbook, *Vol. 2, Properties and Selection: Nonferrous Alloys and Special-Purpose Alloys (1990), ASM International, Materials Park, OH 44073.*)

PART 3

Engineering Materials

The mechanical properties of materials can be predicted and controlled by understanding the atomic bonding, atomic arrangement, and strengthening mechanisms discussed in the previous sections. This fact is particularly evident in Chapters 12 and 13, in which the ideas of solid solution strengthening, strain hardening, and dispersion strengthening are applied to the ferrous and nonferrous alloys.

The discussion of ceramics and polymers in Chapters 14 and 15 emphasizes the importance of atomic bonding and atomic arrangement. The mechanical properties of ceramics and polymers are explained in these chapters by mechanisms that do not involve dislocation movement.

Composite materials are even more difficult to categorize because of the many types and intended uses of the materials, as pointed out in Chapter 16. Many composites, are designed to provide special characteristics that go beyond conventional methods for controlling the structure-property relationship. Construction materials such as wood and concrete, described in Chapter 17, are special types of "composite" materials.

Frequently we find that complex parts and structures are composed of materials from several or even all of these groups. Each group has its own unique set of properties that best suit the individual application.

Chapter

Steels constitute the most widely used family of materials for structural, load-bearing applications. Most buildings, bridges, tools, automotives, and numerous other applications make use of ferrous alloys. With a range of heat treatments that can provide a wide assortment of microstructures and properties, steels are probably the most versatile family of engineered materials.

Gold is for the mistress—silver for the maid
Copper for the craftsman cunning at his trade.
"Good!" said the Baron. sitting in his hall,
"But Iron—Cold Iron—is master of them all!"

—Rudyard Kipling
(Courtesy of PhotoDisc/Getty Images.)

12

Ferrous Alloys

Have You Ever Wondered?

- *What is the most widely used engineered material?*

- *What makes stainless steels "stainless?"*

- *What is the difference between cast iron and steel?*

- *Are stainless steels magnetic?*

- *Is a tin can made out of tin?*

- *Is high-purity iron powder used as a supplement in breakfast cereals?*

Ferrous alloys, which are based on iron-carbon alloys, include plain-carbon steels, alloy and tool steels, stainless steels, and cast irons. These are the most widely used materials in the world. In the history of civilization, these materials made their mark by defining the *Iron Age.*[1] Steels typically are produced in two ways: by re-fining iron ore or by recycling scrap steel (Figure 12-1).

In producing primary steel, iron ore (processed to contain 50 to 70% iron oxide, Fe_2O_3 or Fe_3O_4) is heated in a *blast furnace* in the presence of coke (a form of carbon) and oxygen (Figure 12-1). Coke has a dual role: it is a fuel for

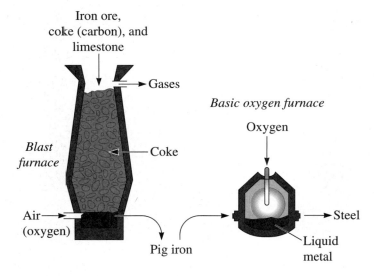

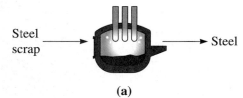

(a)

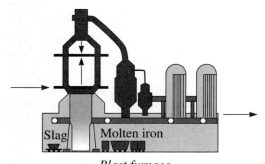

Blast furnace

Produces molten pig
iron from iron ore

(b)

Figure 12-1   (a) In a blast furnace, iron ore is reduced using coke (carbon) and air to produce
liquid pig iron. The high-carbon content in the pig iron is reduce by introducing oxygen into the
basic oxygen furnace to produce liquid steel. An electric arc furnace can be used to produce
liquid steel by melting scrap. (b) Schematic of a blast furnace operation. (*Source:
www.steel.org. Used with permission of the American Iron and Steel Institute.*)

the blast furnace and it is also a reducing agent. The coke is burned using a blast of air (sometimes enriched with oxygen). The coke reduces the iron oxide into a crude molten iron known as **pig iron**. At about $\sim 1600°C$, this material contains about 95% iron; 4% carbon; 0.3 to 0.9% silicon; 0.5% Mn; and 0.025 to 0.05% of sulfur, phosphorus, and titanium. Carbon monoxide and carbon dioxide are produced as gaseous by-products. Limestone ($CaCO_3$) is added as a fluxing agent to help remove impurities. The limestone decomposes and forms CaO. The calcium oxide forms eutectics with silica and other oxides contained as impurities in the ore concentrate, which help produce a molten **slag**. Slag is a by-product of the blast furnace process. It contains silica, CaO, and other impurities in the form of a silicate melt.

Because the liquid pig iron contains a large amount of carbon, oxygen is blown into it in the *basic oxygen furnace* (BOF) to eliminate the excess carbon and produce liquid steel. Steel has a carbon content up to a maximum of $\sim 2\%$. Steel processing occurs at a very large scale. About 300 tons of pig iron can be refined into molten steel in about 30 minutes! Steel is also produced by recycling steel scrap. The scrap is often melted in an **electric arc furnace** in which the heat of the arc melts the scrap. Many alloy and specialty steels, such as stainless steels, are produced using electric melting. Molten steels (including stainless steels) often undergo further refining using such processes as ladle refining, argon oxygen decarburization (AOD), and the like (Chapter 8). The goal here is to reduce the levels of impurities such as phosphorus, sulfur, etc. and to bring the carbon to a desired level. This basic process is used to oxidize impurities such as P, S, Si, Mn, etc. and eventually transfer them to the slag. The trick is not to oxidize the iron and other desirable alloying elements. For example, in stainless steel making, the goal is not to oxidize chromium and nickel, sending these expensive elements to slag. Thus, the oxidation of impurities has to be performed under controlled conditions. The steels refined this way are often described as *clean steels*.

Molten steel is poured into molds to produce finished steel castings; or cast continuously into shapes that are later processed through metal-forming techniques such as rolling or forging (Chapter 7). In the latter case, the steel is either poured into large ingot molds or is continuously cast into regular shapes.

All of the strengthening mechanisms discussed in the previous chapter apply to at least some of the ferrous alloys. In this chapter, we will discuss how to use the eutectoid reaction to control the structure and properties of steels through heat treatment and alloying. We will also examine two special classes of ferrous alloys: stainless steels and cast irons.

12-1   Designations and Classification of Steels

The dividing point between "steels" and "cast irons" is 2.11% C, where the eutectic reaction becomes possible. For steels, we concentrate on the eutectoid portion of the diagram (Figure 12-2) in which the solubility lines and the eutectoid isotherm are specially identified. The A_3 shows the temperature at which ferrite starts to form on cooling; the A_{cm} shows the temperature at which cementite starts to form; and the A_1 is the eutectoid temperature.

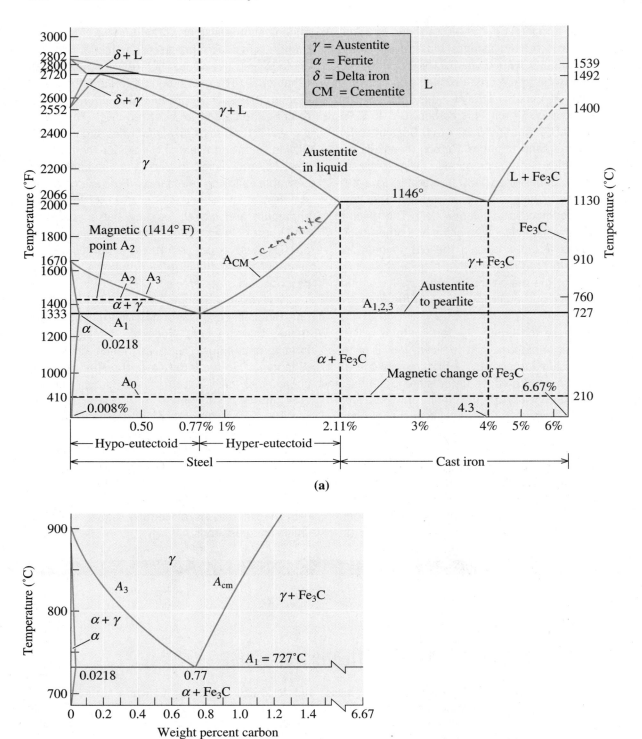

Figure 12-2 (a) The eutectoid portion of the Fe-Fe$_3$C phase diagram. (b) An expanded version of the Fe-C diagram, adapted from several sources.

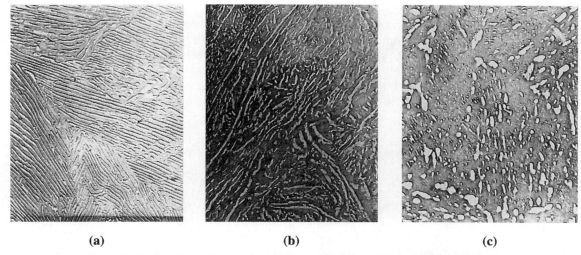

(a)   **(b)**   **(c)**

Figure 12-3   Electron micrographs of (a) pearlite, (b) bainite, and (c) tempered martensite, illustrating the differences in cementite size and shape among these three microconstituents (×7500). (*From* The Making, Shaping, and Treating of Steel, 10th Ed. *Courtesy of the Association of Iron and Steel Engineers.*)

Almost all of the heat treatments of steel are directed toward producing the mixture of ferrite and cementite that gives the proper combination of properties. Figure 12-3 shows the three important microconstituents, or arrangements of ferrite and cementite, that are usually sought. Pearlite is a microconstituent consisting of a lamellar mixture of ferrite and cementite. In bainite, which is obtained by transformation of austenite at a large undercooling, the cementite is more rounded than in pearlite. Tempered martensite, a mixture of very fine and nearly round cementite in ferrite, forms when martensite is reheated following its formation.

Designations   The AISI (American Iron and Steel Institute) and SAE (Society of Automotive Engineers) [3] provide designation systems (Table 12-1) that use a four- or

TABLE 12-1 ■ *Compositions of selected AISI-SAE steels*

AISI-SAE Number	% C	% Mn	% Si	% Ni	% Cr	Others
1020	0.18–0.23	0.30–0.60				
1040	0.37–0.44	0.60–0.90				
1060	0.55–0.65	0.60–0.90				
1080	0.75–0.88	0.60–0.90				
1095	0.90–1.03	0.30–0.50				
1140	0.37–0.44	0.70–1.00				0.08–0.13% S
4140	0.38–0.43	0.75–1.00	0.15–0.30		0.80–1.10	0.15–0.25% Mo
4340	0.38–0.43	0.60–0.80	0.15–0.30	1.65–2.00	0.70–0.90	0.20–0.300% Mo
4620	0.17–0.22	0.45–0.65	0.15–0.30	1.65–2.00		0.20–0.30% Mo
52100	0.98–1.10	0.25–0.45	0.15–0.30		1.30–1.60	
8620	0.18–0.23	0.70–0.90	0.15–0.30	0.40–0.70	0.40–0.60	0.15–0.25% Y
9260	0.56–0.64	0.75–1.00	1.80–2.20			

five-digit number. The first two numbers refer to the major alloying elements present, and the last two or three numbers refer to the percentage of carbon. An AISI 1040 steel is a plain-carbon steel with 0.40% C. An SAE 10120 steel is a plain-carbon steel containing 1.20% C. An AISI 4340 steel is an alloy steel containing 0.40% C. Note that the American Society for Testing of Materials (ASTM) has a different way of classifying steels. The ASTM has a list of specifications that describe steels suitable for different applications. The example below illustrates the use of AISI numbers.

EXAMPLE 12-1 *Design of a Method to Determine AISI Number*

An unalloyed steel tool used for machining aluminum automobile wheels has been found to work well, but the purchase records have been lost and you do not know the steel's composition. The microstructure of the steel is tempered martensite, and assume that you cannot estimate the composition of the steel from the structure. Design a treatment that may help determine the steel's carbon content.

SOLUTION

Assume that there is no access to equipment that would permit you to analyze the chemical composition directly. Since the entire structure of the steel is a very fine tempered martensite, we can do a simple heat treatment to produce a structure that can be analyzed more easily. This can be done in two different ways.

The first way is to heat the steel to a temperature just below the A_1 temperature and hold for a long time. The steel overtempers and large Fe_3C spheres form in a ferrite matrix. We then estimate the amount of ferrite and cementite and calculate the carbon content using the lever law. If we measure 16% Fe_3C using this method, the carbon content is:

$$\% \ Fe_3C = \left[\frac{(x - 0.0218)}{(6.67 - 0.0218)}\right] \times 100 = 16 \quad \text{or} \quad x = 1.086\% \ C$$

A better approach, however, is to heat the steel above the A_{cm} to produce all austenite. If the steel then cools slowly, it transforms to pearlite and a primary microconstituent. If, when we do this, we estimate that the structure contains 95% pearlite and 5% primary Fe_3C, then:

$$\% \ Pearlite = \left[\frac{(6.67 - x)}{(6.67 - 0.77)}\right] \times 100 = 95 \quad \text{or} \quad x = 1.065\% \ C$$

The carbon content is on the order of 1.065 to 1.086%, consistent with a 10110 steel. In this procedure, we assumed that the weight and volume percentages of the microconstituents are the same, which is nearly the case in steels.

Classifications Steels can be classified based on their composition or the way they have been processed. Carbon steels contain up to ~2% carbon. These steels may also contain other elements, such as Si (maximum 0.6%), copper (up to 0.6%), and Mn (up to 1.65%). Decarburized steels contain less than 0.005% C. Ultra-low carbon steels contain a maximum of 0.03% carbon. They also contain very low levels of other elements such as Si and Mn. Low-carbon steels contain 0.04 to 0.15% carbon. These low-carbon steels are used for making car bodies and hundreds of other applications. Mild steel contains 0.15 to 0.3% carbon. This steel is used in buildings, bridges, piping, etc. Medium-carbon steels contain 0.3 to 0.6% carbon. These are used in making machinery, tractors, mining equipment, etc. High-carbon steels contain above 0.6% carbon. These are used in making springs, railroad car wheels, and the like. Note that cast irons are Fe-C alloys containing 2 to 4% carbon.

Alloy steels are compositions that contain more significant levels of alloying elements. We will discuss the effect of alloying elements later in this chapter. They improve the hardenability (Sections 12-4 and 12-5) of steels. The AISI defines alloy steels as steels that exceed in one or more of these elements: ≥1.65% Mn, 0.6% Si, 0.6% Cu. The total carbon content is up to 1% and the total alloying elements content is below 5%. A material is also an alloy steel if a definite concentration of alloying elements, such as Ni, Cr, Mo, Ti, etc., is specified. These steels are used for making tools (hammers, chisels, etc.) and also in making parts such as axles, shafts, and gears.

Certain specialty steels may consist of higher levels of sulfur (>0.1%) or lead (~0.15–0.35%) to provide machinability. These, however, can not be welded easily. Recently, researchers have developed "green steel" in which lead, an environmental toxin, was replaced with tin (Sn) and/or antimony (Sb). Steels can also be classified based on their processing. For example, the term "concast steels" refers to continuously cast steels. Galvanized steels have a zinc coating for corrosion resistance (Chapter 5). Similarly, tin-plated steel is used to make corrosion-resistant tin cans and other products. Tin is deposited using electroplating—a process known as "continuous web electrodeposition."[3] "E-steels" are steels that are melted using an electric furnace, while "B-steels" contain a small (0.0005 to 0.003%), yet significant, concentration of boron. Recently, a "germ-resistant" coated stainless steel has been developed.[4] There is no need to remember all of these classifications. It is important to understand the composition of steel and its microstructure after processing.

12-2 Simple Heat Treatments

Four simple heat treatments—process annealing, annealing, normalizing, and spheroidizing—are commonly used for steels (Figure 12-4). These heat treatments are

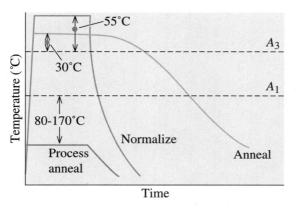
(a) Hypoeutectoid

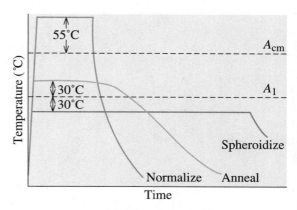
(b) Hypereutectoid

Figure 12-4   Schematic summary of the simple heat treatments for (a) hypoeutectoid steels and (b) hypereutectoid steels.

used to accomplish one of three purposes: (1) eliminating the effects of cold work, (2) controlling dispersion strengthening, or (3) improving machinability.

Process Annealing—Eliminating Cold Work   The recrystallization heat treatment used to eliminate the effect of cold working in steels with less than about 0.25% C is called a **process anneal**. The process anneal is done 80°C to 170°C below the A_1 temperature. The intent of the process anneal treatment for steels is similar to the annealing of inorganic glasses in that the main idea is to significantly reduce or eliminate residual stresses.

Annealing and Normalizing—Dispersion Strengthening   Steels can be dispersion-strengthened by controlling the fineness of pearlite. The steel is initially heated to produce homogeneous austenite (FCC γ phase), a step called **austenitizing**. **Annealing**, or a full anneal, allows the steel to cool slowly in a furnace, producing coarse pearlite. **Normalizing** allows the steel to cool more rapidly, in air, producing fine pearlite. Figure 12-5 shows the typical properties obtained by annealing and normalizing plain-carbon steels.

For annealing, austenitizing of hypoeutectoid steels is conducted about 30°C above the A_3, producing 100% γ. However, austenitizing of a hypereutectoid steel is done at about 30°C above the A_1, producing austenite and Fe$_3$C. This process prevents the formation of a brittle, continuous film of Fe$_3$C at the grain boundaries that occurs on slow cooling from the 100% γ region. In both cases, the slow furnace cool and coarse pearlite provide relatively low strength and good ductility.

For normalizing, austenitizing is done at about 55°C above the A_3 or A_{cm}; the steel is then removed from the furnace and cooled in air. The faster cooling gives fine pearlite and provides higher strength.

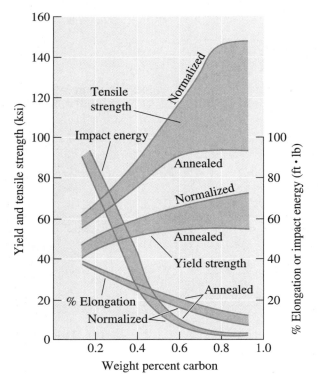

Figure 12-5
The effect of carbon and heat treatment on the properties of plain-carbon steels.

Spheroidizing—Improving Machinability Steels, which contain a large concentration of Fe_3C, have poor machining characteristics. It is possible to transform the morphology of Fe_3C using *spheroidizing*. During the spheroidizing treatment, which requires several hours at about 30°C below the A_1, the Fe_3C phase morphology changes into large, spherical particles in order to reduce boundary area. The microstructure, known as **spheroidite**, has a continuous matrix of soft, machinable ferrite (Figure 12-6). After machining, the steel is given a more sophisticated heat treatment to produce the required properties. A similar microstructure occurs when martensite is tempered just below the A_1 for long periods of time. As noted before, alloying elements such as Pb and S are also added to improve machinability of steels and, more recently, lead-free "green steels" that have very good machinability have been developed.[5,6]

The example below shows how different heat treatment conditions can be developed for a given composition of steel.

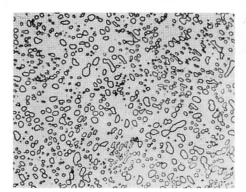

Figure 12-6
The microstructure of spheroidite, with Fe_3C particles dispersed in a ferrite matrix (×850). (*From* ASM Handbook, *Vol. 7, (1972), ASM International, Materials Park, OH 44073.*)

EXAMPLE 12-2 *Determination of Heat Treating Temperatures*

Recommend temperatures for the process annealing, annealing, normalizing, and spheroidizing of 1020, 1077, and 10120 steels.

SOLUTION

From Figure 12-2, we find the critical A_1, A_3, or A_{cm}, temperatures for each steel. We can then specify the heat treatment based on these temperatures.

Steel Type	1020	1077	10120
Critical temperatures	$A_1 = 727°C$ $A_3 = 830°C$	$A_1 = 727°C$	$A_1 = 727°C$ $A_{cm} = 895°C$
Process annealing	727 − (80 to 170) = 557°C to 647°C	Not done	Not done
Annealing	830 + 30 = 860°C	727 + 30 = 757°C	727 + 30 = 757°C
Normalizing	830 + 55 = 885°C	727 + 55 = 782°C	895 + 55 = 950°C
Spheroidizing	Not done	727 − 30 = 697°C	727 − 30 = 697°C

Isothermal Heat Treatments

The effect of transformation temperature on the properties of a 1080 (eutectoid) steel was discussed in Chapter 11. As the isothermal transformation temperature decreases, pearlite becomes progressively finer before bainite begins to form. At very low temperatures, martensite is obtained.

Austempering and Isothermal Annealing The isothermal transformation heat treatment used to produce bainite, called **austempering**, simply involves austenitizing the steel, quenching to some temperature below the nose of the TTT curve, and holding at that temperature until all of the austenite transforms to bainite (Figure 12-7).

Annealing and normalizing are usually used to control the fineness of pearlite. However, pearlite formed by an **isothermal anneal** (Figure 12-7) may give more uniform properties, since the cooling rates and microstructure obtained during annealing and normalizing vary across the cross section of the steel. *Note that the TTT diagrams only describe isothermal heat treatments (i.e., we assume that the sample begins and completes heat treatment at a given temperature).* Thus, we cannot exactly describe heat treatments by superimposing cooling curves on a TTT diagram such as those shown in Figure 12-7 (Section 12-4).

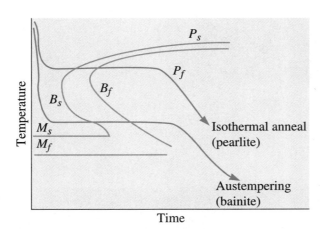

Figure 12-7
The austempering and isothermal anneal heat treatments in a 1080 steel.

Effect of Changes in Carbon Concentration on the TTT Diagram In either a hypoeutectoid or a hypereutectoid steel, the TTT diagram must reflect the possible formation of a primary phase. The isothermal transformation diagrams for a 1050 and a 10110 steel are shown in Figure 12-8. The most remarkable change is the presence of a "wing" which begins at the nose of the curve and becomes asymptotic to the A_3 or A_{cm} temperature. The wing represents the ferrite start (F_s) time in hypoeutectoid steels or the cementite start (C_s) time in hypereutectoid steels.

When a 1050 steel is austenitized, quenched, and held between the A_1 and the A_3, primary ferrite nucleates and grows. Eventually, an equilibrium amount of ferrite and austenite result. Similarly, primary cementite nucleates and grows to its equilibrium amount in a 10110 steel held between the A_{cm} and A_1 temperatures.

If an austenitized 1050 steel is quenched to a temperature between the nose and the A_1 temperatures, primary ferrite again nucleates and grows until reaching the equilibrium amount. The remainder of the austenite then transforms to pearlite. A similar situation, producing primary cementite and pearlite, is found for the hypereutectoid steel.

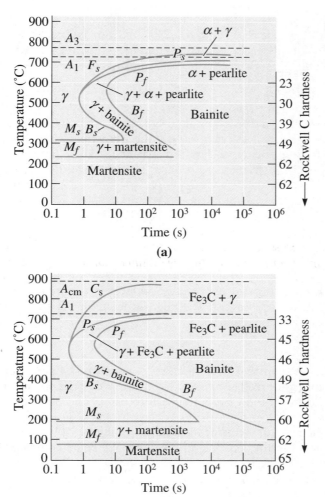

If we quench the steel below the nose of the curve, only bainite forms, regardless of the carbon content of the steel. If the steels are quenched to temperatures below the M_s, martensite will form. The following example shows how the phase diagram and TTT diagram can guide development of the heat treatment of steels.

EXAMPLE 12-3   *Design of a Heat Treatment for an Axle*

A heat treatment is needed to produce a uniform microstructure and hardness of HRC 23 in a 1050 steel axle.

SOLUTION

We might attempt this task in several ways. We could austenitize the steel, then cool at an appropriate rate by annealing or normalizing to obtain the correct hardness. By doing this, however, we find that the structure and hardness vary from the surface to the center of the axle.

A better approach is to use an isothermal heat treatment. From Figure 12-8, we find that a hardness of HRC 23 is obtained by transforming austenite to a mixture of ferrite and pearlite at 600°C. From Figure 12-2, we find that the A_3 temperature is 770°C. Therefore, our heat treatment is:

1. Austenitize the steel at $770 + (30 \text{ to } 55) = 805°C$ to $825°C$, holding for 1 h and obtaining 100% γ.

2. Quench the steel to 600°C and hold for a minimum of 10 s. Primary ferrite begins to precipitate from the unstable austenite after about 1.0 s. After 1.5 s, pearlite begins to grow, and the austenite is completely transformed to ferrite and pearlite after about 10 s. After this treatment, the microconstituents present are:

$$\text{Primary } \alpha = \left[\frac{(0.77 - 0.5)}{(0.77 - 0.0218)} \right] \times 100 = 36\%$$

$$\text{Pearlite} = \left[\frac{(0.5 - 0.0218)}{(0.77 - 0.0218)} \right] \times 100 = 64\%$$

3. Cool in air-to-room temperature, preserving the equilibrium amounts of primary ferrite and pearlite. The microstructure and hardness are uniform because of the isothermal anneal.

Interrupting the Isothermal Transformation   Complicated microstructures are produced by interrupting the isothermal heat treatment. For example, we could austenitize the 1050 steel (Figure 12-9) at 800°C, quench to 650°C and hold for 10 s (permitting some ferrite and pearlite to form), then quench to 350°C and hold for 1 h (3600 s). Whatever unstable austenite remained before quenching to 350°C transforms to bainite. The final structure is ferrite, pearlite, and bainite. We could complicate the treatment further by interrupting the treatment at 350°C after 1 min (60 s) and quenching. Any austenite remaining after 1 min at 350°C forms martensite. The final structure now contains ferrite, pearlite, bainite, and martensite. Note that each time we change the temperature, we start at zero time! In practice, temperatures can not be changed

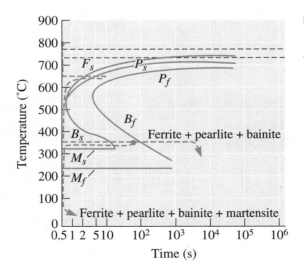

Figure 12-9
Producing complicated structures by interrupting the isothermal heat treatment of a 1050 steel.

Figure 12-10
Dark feathers of bainite surrounded by light martensite, obtained by interrupting the isothermal transformation process (×1500). (ASM Handbook, *Vol. 9 Metallography and Microstructure (1985), ASM International, Materials Park, OH 44073.*)

instantaneously (i.e., we cannot go instantly from 800 to 650 or 650 to 350°C). This is why it is better to use the continuous cooling transformation (CCT) diagrams.

Figure 12-10 shows the structure obtained by interrupting the transformation to bainite of a 0.5% C steel by quenching the remaining austenite to martensite. Because such complicated mixtures of microconstituents produce unpredictable properties, these structures are seldom produced intentionally.

12-4   Quench and Temper Heat Treatments

Quenching hardens most steels and tempering increases the toughness. This has been known for perhaps thousands of years. For example, a series of such heat treatments has been used for making Damascus steel and Japanese Samurai swords.[1,7] We can obtain an exceptionally fine dispersion of Fe_3C (known as tempered martensite) if we first quench the austenite to produce martensite, then temper. During tempering, an

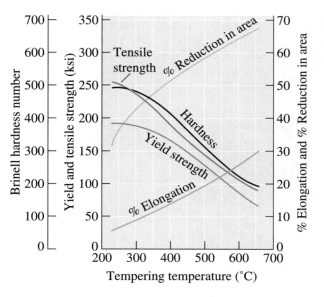

Figure 12-11
The effect of tempering temperature on the mechanical properties of a 1050 steel.

intimate mixture of ferrite and cementite forms from the martensite, as discussed in Chapter 11. The tempering treatment controls the final properties of the steel (Figure 12-11 on page 555). Note that this is different from a spheroidizing heat treatment (Figure 12-6). The following example shows how a combination of heat treatments is used to obtain steels with desired properties.

EXAMPLE 12-4   *Design of a Quench and Temper Treatment*

A rotating shaft that delivers power from an electric motor is made from a 1050 steel. Its yield strength should be at least 145,000 psi, yet it should also have at least 15% elongation in order to provide toughness. Design a heat treatment to produce this part.

SOLUTION

We are not able to obtain this combination of properties by annealing or normalizing (Figure 12-5). However a quench and temper heat treatment produces a microstructure that can provide both strength and toughness. Figure 12-11 shows that the yield strength exceeds 145,000 psi if the steel is tempered below 460°C, whereas the elongation exceeds 15% if tempering is done above 425°C. The A_3 temperature for the steel is 770°C. A possible heat treatment is:

1. Austenitize above the A_3 temperature of 770°C for 1 h. An appropriate temperature may be $770 + 55 = 825°C$.

2. Quench rapidly to room temperature. Since the M_f is about 250°C, martensite will form.

3. Temper by heating the steel to 440°C. Normally, 1 h will be sufficient if the steel is not too thick.

4. Cool to room temperature.

Retained Austenite   There is a large volume expansion when martensite forms from austenite. As the martensite plates form during quenching, they surround and isolate small pools of austenite (Figure 12-12), which deform to accommodate the lower-density martensite. However, for the remaining pools of austenite to transform, the surrounding martensite must deform. Because the strong martensite resists the transformation, either the existing martensite cracks or the austenite remains trapped in the structure as **retained austenite**. Retained austenite can be a serious problem. Martensite

Figure 12-12
Retained austenite (white) trapped between martensite needles (black) (×1000). (*From* ASM Handbook, *Vol. 8, (1973), ASM International, Materials Park, OH 44073.*)

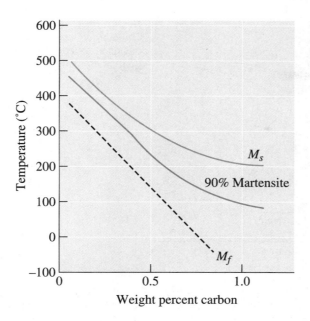

Figure 12-13
Increasing carbon reduces the M_s and M_f temperatures in plain-carbon steels.

softens and becomes more ductile during tempering. After tempering, the retained austenite cools below the M_s and M_f temperatures and transforms to martensite, since the surrounding **tempered martensite** can deform. But now the steel contains more of the hard, brittle martensite! A second tempering step may be needed to eliminate the martensite formed from the retained austenite. Retained austenite is also more of a problem for high-carbon steels. The martensite start and finish temperatures are reduced when the carbon content increases (Figure 12-13). High-carbon steels must be refrigerated to produce all martensite.

Residual Stresses and Cracking   Residual stresses are also produced because of the volume change or because of cold working. A stress-relief anneal can be used to remove or minimize residual stresses due to cold working. Stresses are also induced because of thermal expansion and contraction. In steels, there is one more mechanism that causes

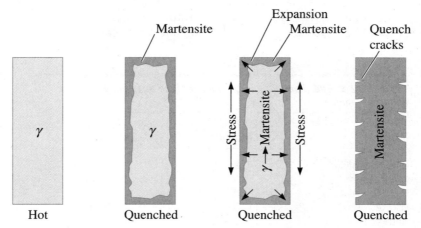

Figure 12-14   Formation of quench cracks caused by residual stresses produced during quenching. The figure illustrates the development of stresses as the austenite transforms to martensite during cooling.

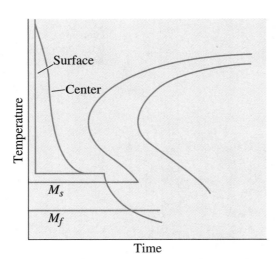

Figure 12-15
The marquenching heat treatment, designed to reduce residual stresses and quench cracking.

stress. When steels are quenched, the surface of the quenched steel cools rapidly and transforms to martensite. When the austenite in the center later transforms, the hard surface is placed in tension, while the center is compressed. If the residual stresses exceed the yield strength, **quench cracks** form at the surface (Figure 12-14). However, if we first cool to just above the M_s and hold until the temperature equalizes in the steel, subsequent quenching permits all of the steel to transform to martensite at about the same time. This heat treatment is called **marquenching** or **martempering** (Figure 12-15). Note that as discussed below, strictly speaking, the CCT diagrams should be used to examine non-isothermal heat treatments.

Quench Rate   In using the TTT diagram, we assumed that we could cool from the austenitizing temperature to the transformation temperature instantly. Because this does not occur in practice, undesired microconstituents may form during the quenching process. For example, pearlite may form as the steel cools past the nose of the curve, particularly because the time of the nose is less than one second in plain-carbon steels.

The rate at which the steel cools during quenching depends on several factors. First, the surface always cools faster than the center of the part. In addition, as the size of the part increases, the cooling rate at any location is slower. Finally, the cooling rate depends on the temperature and heat transfer characteristics of the quenching medium (Table 12-2). Quenching in oil, for example, produces a lower H coefficient, or slower

TABLE 12-2 ■ *The H coefficient, or severity of the quench, for several quenching media*

Medium	H Coefficient	Cooling Rate at the Center of a 1-in. Bar (°C/s)
Oil (no agitation)	0.25	18
Oil (agitation)	1.0	45
H_2O (no agitation)	1.0	45
H_2O (agitation)	4.0	190
Brine (no agitation)	2.0	90
Brine (agitation)	5.0	230

cooling rate, than quenching in water or brine. The H coefficient is equivalent to the heat transfer coefficient. Agitation helps break the vapor blanket (e.g., when water is the quenching medium) and improves overall heat transfer rate by bringing cooler liquid into contact with the parts being quenched.

Continuous Cooling Transformation Diagrams We can develop a continuous cooling transformation (CCT) diagram by determining the microstructures produced in the steel at various rates of cooling. The CCT curve for a 1080 steel is shown in Figure 12-16. The CCT diagram differs from the TTT diagram in that longer times are required for transformations to begin and no bainite region is observed.

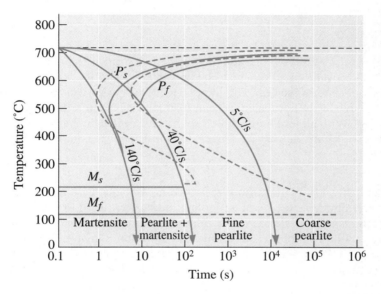

Figure 12-16
The CCT diagram (solid lines) for a 1080 steel compared with the TTT diagram (dashed lines).

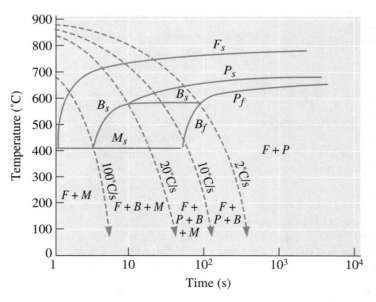

Figure 12-17
The CCT diagram for a low-alloy, 0.2% C steel.

If we cool a 1080 steel at 5°C/s, the CCT diagram tells us that we obtain coarse pearlite; we have annealed the steel. Cooling at 35°C/s gives fine pearlite and is a normalizing heat treatment. Cooling at 100°C/s permits pearlite to start forming, but the reaction is incomplete and the remaining austenite changes to martensite. We obtain 100% martensite and thus are able to perform a quench and temper heat treatment, only if we cool faster than 140°C/s. Other steels, such as the low-carbon steel in Figure 12-17 on page 559, have more complicated CCT diagrams. In various handbooks, you can find a compilation of TTT and CCT diagrams for different grades of steels.

12-5   Effect of Alloying Elements

Alloying elements are added to steels to (a) provide solid-solution strengthening of ferrite, (b) cause the precipitation of alloy carbides rather than that of Fe_3C, (c) improve corrosion resistance and other special characteristics of the steel, and (d) improve **hardenability**. The term hardenability describes the ease with which steels can form martensite. This relates to how easily we can form martensite in a thick section of steel that is quenched. With a more hardenable steel we can "get away" with relatively slow cooling rate and still form martensite. Improving hardenability is most important in alloy and tool steels.

Hardenability   In plain-carbon steels, the nose of the TTT and CCT curves occurs at very short times; hence, very fast cooling rates are required to produce all martensite. In thin sections of steel, the rapid quench produces distortion and cracking. In thick steels, we are unable to produce martensite. All common alloying elements in steel shift the TTT and CCT diagrams to longer times, permitting us to obtain all martensite even in thick sections at slow cooling rates. Figure 12-18 shows the TTT and CCT curves for a 4340 steel.

Plain-carbon steels have low hardenability—only very high cooling rates produce all martensite. Alloy steels have high hardenability—even cooling in air may produce martensite. Hardenability does not refer to the hardness of the steel. A low-carbon, high-alloy steel may easily form martensite but, because of the low-carbon content, the martensite is not hard.

Effect on the Phase Stability   When alloying elements are added to steel, the binary Fe-Fe_3C stability is affected and the phase diagram is altered (Figure 12-19). Alloying elements reduce the carbon content at which the eutectoid reaction occurs and change the A_1, A_3, and A_{cm} temperatures. A steel containing only 0.6% C is hypoeutectoid and would operate at 700°C without forming austenite; the otherwise same steel containing 6% Mn is hypereutectoid and austenite forms at 700°C.

Shape of the TTT Diagram   Alloying elements may introduce a "bay" region into the TTT diagram, as in the case of the 4340 steel (Figure 12-18). The bay region is used as the basis for a thermomechanical heat treatment known as **ausforming**. A steel can be austenitized, quenched to the bay region, plastically deformed, and finally quenched to produce martensite (Figure 12-20). Steels subjected to this treatment are known as *ausformed steels*.

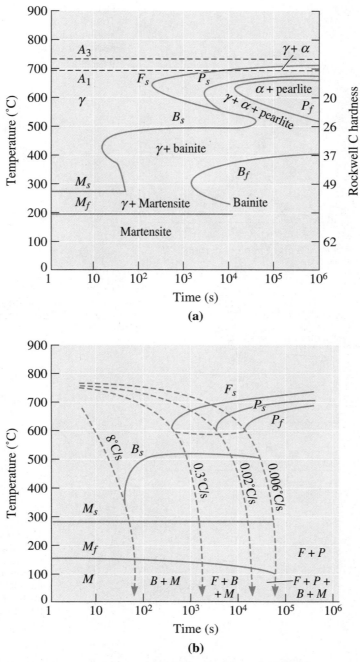

Figure 12-18 (a) TTT and (b) CCT curves for a 4340 steel.

Tempering Alloying elements reduce the rate of tempering compared with that of a plain-carbon steel (Figure 12-21). This effect may permit the alloy steels to operate more successfully at higher temperatures than plain-carbon steels since overaging will not occur during service.

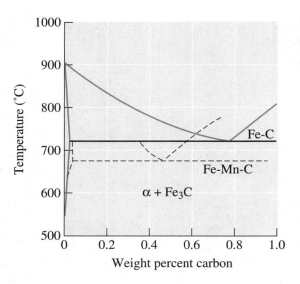

Figure 12-19
The effect of 6% manganese on the stability ranges of the phases in the eutectoid portion of the Fe-Fe$_3$C phase diagram.

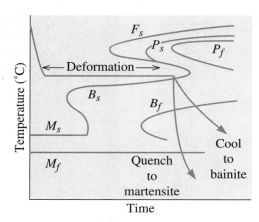

Figure 12-20
When alloying elements introduce a bay region into the TTT diagram, the steel can be ausformed.

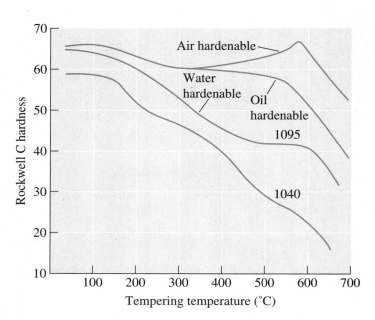

Figure 12-21
The effect of alloying elements on the phases formed during the tempering of steels. The air-hardenable steel shows a secondary hardening peak.

12-6   Application of Hardenability

A **Jominy test** (Figure 12-22) is used to compare hardenabilities of steels. A steel bar 4-in. long and 1 in. in diameter is austenitized, placed into a fixture, and sprayed at one end with water. This procedure produces a range of cooling rates—very fast at the quenched end, almost air cooling at the opposite end. After the test, hardness measurements are made along the test specimen and plotted to produce a **hardenability curve** (Figure 12-23). The distance from the quenched end is the **Jominy distance** and is related to the cooling rate (Table 12-3).

Virtually any steel transforms to martensite at the quenched end. Thus, the hardness at zero Jominy distance is determined solely by the carbon content of the steel. At larger Jominy distances, there is a greater likelihood that bainite or pearlite will form instead of martensite. An alloy steel with a high hardenability (such as 4340) maintains a rather flat hardenability curve; a plain-carbon steel (such as 1050) has a curve that drops off quickly. The hardenability is determined primarily by the alloy content of the steel.

We can use hardenability curves in selecting or replacing steels in practical applications. The fact that two different steels cool at the same rate if quenched under iden-

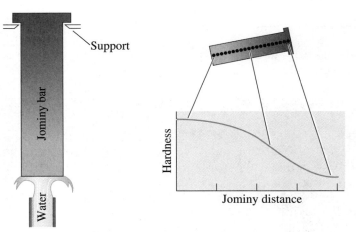

Figure 12-22
The set-up for the Jominy test used for determining the hardenability of a steel.

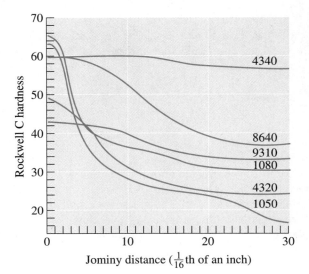

Figure 12-23
The hardenability curves for several steels.

TABLE 12-3 ■ *The relationship between cooling rate and Jominy distance*

Jominy Distance (in.)	Cooling Rate (°C/s)
$\frac{1}{16}$	315
$\frac{2}{16}$	110
$\frac{3}{16}$	50
$\frac{4}{16}$	36
$\frac{5}{16}$	28
$\frac{6}{16}$	22
$\frac{7}{16}$	17
$\frac{8}{16}$	15
$\frac{10}{16}$	10
$\frac{12}{16}$	8
$\frac{16}{16}$	5
$\frac{20}{16}$	3
$\frac{24}{16}$	2.8
$\frac{28}{16}$	2.5
$\frac{36}{16}$	2.2

tical conditions helps in this selection process. The Jominy test data are used as shown in the following example.

EXAMPLE 12-5   *Design of a Wear-Resistant Gear*

A gear made from 9310 steel, which has an as-quenched hardness at a critical location of HRC 40, wears at an excessive rate. Tests have shown that an as-quenched hardness of at least HRC 50 is required at that critical location. Design a steel that would be appropriate.

SOLUTION

We know that if different steels of the same size are quenched under identical conditions, their cooling rates or Jominy distances are the same. From Figure 12-23, a hardness of HRC 40 in a 9310 steel corresponds to a Jominy distance of 10/16 in. (10°C/s). If we assume the same Jominy distance, the other steels shown in Figure 12-23 have the following hardnesses at the critical location:

1050   HRC 28

1080   HRC 36

4320   HRC 31

8640   HRC 52

4340   HRC 60

Both the 8640 and 4340 steels are appropriate. The 4320 steel has too low a carbon content ever to reach HRC 50; the 1050 and 1080 have enough carbon, but the hardenability is too low. In Table 12-1, we find that the 86xx steels contain less alloying elements than the 43xx steels; thus the 8640 steel is probably less expensive than the 4340 steel and might be our best choice. We must also consider other factors such as durability.

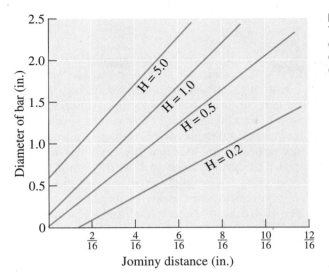

Figure 12-24
The Grossman chart used to determine the hardenability at the center of a steel bar for different quenchants.

In another simple technique, we utilize the severity of the quench and the Grossman chart (Figure 12-24) to determine the hardness at the *center* of a round bar. The bar diameter and H coefficient, or severity of the quench in Table 12-2, give the Jominy distance at the center of the bar. We can then determine the hardness from the hardenability curve of the steel. (See Example 12-6.)

EXAMPLE 12-6   *Design of a Quenching Process*

Design a quenching process to produce a minimum hardness of HRC 40 at the center of a 1.5-in. diameter 4320 steel bar.

SOLUTION

Several quenching media are listed in Table 12-2. We can find an approximate H coefficient for each of the quenching media, then use Figure 12-24 to estimate the Jominy distance in a 1.5-in. diameter bar for each media. Finally, we can use the hardenability curve (Figure 12-23) to find the hardness in the 4320 steel. The results are listed below.

	H Coefficient	Jominy Distance	HRC
Oil (no agitation)	0.25	11/16	30
Oil (agitation)	1.00	6/16	39
H_2O (no agitation)	1.00	6/16	39
H_2O (agitation)	4.00	4/16	44
Brine (no agitation)	2.00	5/16	42
Brine (agitation)	5.00	3/16	46

The last three methods, based on brine or agitated water, are satisfactory. Using an unagitated brine quenchant might be least expensive, since no extra equipment is needed to agitate the quenching bath. However, H_2O is less corrosive than the brine quenchant.

12-7   Specialty Steels

There are many special categories of steels, including tool steels, interstitial-free steels, high-strength-low-alloy (*HSLA*) steels, dual-phase steels, and maraging steels.

Tool steels are usually high-carbon steels that obtain high hardnesses by a quench and temper heat treatment. Their applications include cutting tools in machining operations, dies for die casting, forming dies, and other uses in which a combination of high strength, hardness, toughness, and temperature resistance is needed.

Alloying elements improve the hardenability and high-temperature stability of the tool steels. The water-hardenable steels such as 1095 must be quenched rapidly to produce martensite and also soften rapidly even at relatively low temperatures. Oil-hardenable steels form martensite more easily, temper more slowly, but still soften at high temperatures. The air-hardenable and special tool steels may harden to martensite while cooling in air. In addition, these steels may not soften until near the A_1 temperature. In fact, the highly alloyed tool steels may pass through a **secondary hardening peak** near 500°C as the normal cementite dissolves and hard alloy carbides precipitate (Figure 12-21). The alloy carbides are particularly stable, resist growth or spheroidization, and are important in establishing the high-temperature resistance of these steels.

High-strength-low-alloy (*HSLA*) steels are low-carbon steels containing small amounts of alloying elements. The *HSLA* steels are specified on the basis of yield strength, with grades up to 80,000 psi; the steels contain the least amount of alloying element that still provides the proper yield strength without heat treatment. In these steels, careful processing permits precipitation of carbides and nitrides of Nb, V, Ti, or Zr, which provide dispersion strengthening and a fine grain size (Figure 11-9c).

Dual-phase steels contain a uniform distribution of ferrite and martensite, with the dispersed martensite providing yield strengths of 60,000 to 145,000 psi. These low-carbon steels do not contain enough alloying elements to have good hardenability using the normal quenching processes. But when the steel is heated into the ferrite-plus-austenite portion of the phase diagram, the austenite phase becomes enriched in carbon, which provides the needed hardenability. During quenching, only the austenite portion transforms to martensite (Figure 12-25).

Maraging steels are low-carbon, highly alloyed steels. The steels are austenitized and quenched to produce a soft martensite that contains less than 0.3% C. When the martensite is aged at about 500°C, intermetallic compounds such as Ni_3Ti, Fe_2Mo, and Ni_3Mo precipitate.

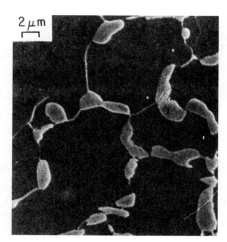

2 μm

Figure 12-25
Microstructure of a dual-phase steel, showing islands of light martensite in a ferrite matrix (×2500). (*From G. Speich, "Physical Metallurgy of Dual-Phase Steels,"* Fundamentals of Dual-Phase Steels, *The Metallurgical Society of AIME, 1981.*)

Interstitial-free steels are steels containing Nb and Ti. They react with C and S to form precipitates of carbides and sulfides. Thus, virtually no carbon remains in the ferrite. These steels are very formable and attractive for the automobile industry.

Grain-oriented steels containing silicon are used as soft magnetic materials and are used in transformer cores. Nearly pure iron powder (known as carbonyl iron), obtained by the decomposition of iron pentacarbonyl $(Fe(CO)_5)$ and sometimes a reducing heat treatment, is used to make magnetic materials. It is also used as an additive for food supplements in breakfast cereals and other iron-fortified food products under the name reduced iron.

As mentioned before, many steels are also coated, usually to provide good corrosion protection. *Galvanized* steel is coated with a thin layer of zinc (Chapter 5), *terne* steel is coated with lead, and other steels are coated with aluminum or tin.

12-8   Surface Treatments

We can, by proper heat treatment, produce a structure that is hard and strong at the surface, so that excellent wear and fatigue resistance are obtained, but at the same time gives a soft, ductile, tough core that provides good resistance to impact failure. We have seen principles of carburizing in Chapter 5, when we discussed diffusion. In this section, we see this and similar other processes.

Selectively Heating the Surface   We could begin by rapidly heating the surface of a medium-carbon steel above the A_3 temperature (the center remains below the A_1). After the steel is quenched, the center is still a soft mixture of ferrite and pearlite, while the surface is martensite (Figure 12-26). The depth of the martensite layer is the **case depth**. Tempering produces the desired hardness at the surface. We can provide local heating of the surface by using a gas flame, an induction coil, a laser beam, or an electron beam. We can, if we wish, harden only selected areas of the surface that are most subject to failure by fatigue or wear.

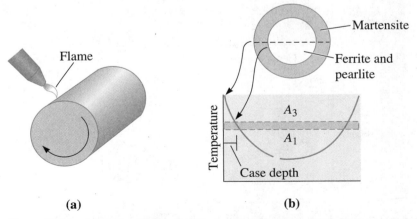

Figure 12-26   (a) Surface hardening by localized heating. (b) Only the surface heats above the A_1 temperature and is quenched to martensite.

Carburizing and Nitriding   These techniques involve controlled diffusion of carbon and nitrogen, respectively (Chapter 5). For best toughness, we start with a low-carbon steel.

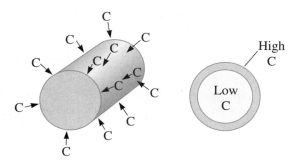

Figure 12-27
Carburizing of a low-carbon steel to produce a high-carbon, wear-resistant surface.

In **carburizing**, carbon is diffused into the surface of the steel at a temperature above the A_3 (Figure 12-27). A high-carbon content is produced at the surface due to rapid diffusion and the high solubility of carbon in austenite. When the steel is then quenched and tempered, the surface becomes a high-carbon tempered martensite, while the ferritic center remains soft and ductile. The thickness of the hardened surface, again called the case depth, is much smaller in carburized steels than in flame- or induction-hardened steels.

Nitrogen provides a hardening effect similar to that of carbon. In **cyaniding**, the steel is immersed in a liquid cyanide bath which permits both carbon and nitrogen to diffuse into the steel. In **carbonitriding**, a gas containing carbon monoxide and ammonia is generated and both carbon and nitrogen diffuse into the steel. Finally, only nitrogen diffuses into the surface from a gas in nitriding. Nitriding is carried out below the A_1 temperature.

In each of these processes, compressive residual stresses are introduced at the surface, providing excellent fatigue resistance (Chapter 7) in addition to the good combination of hardness, strength, and toughness. The following example explains considerations that go into considering heat treatments such as quenching and tempering and surface hardening.

EXAMPLE 12-7   *Design of Surface-Hardening Treatments for a Drive Train*

Design the materials and heat treatments for an automobile axle and drive gear (Figure 12-28).

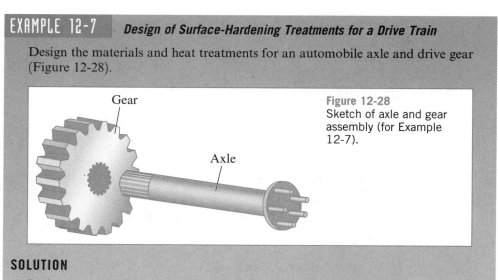

Figure 12-28
Sketch of axle and gear assembly (for Example 12-7).

SOLUTION

Both parts require good fatigue resistance. The gear also should have a good hardness to avoid wear, and the axle should have good overall strength to withstand bending and torsional loads. Both parts should have good toughness. Finally, since millions of these parts will be made, they should be inexpensive.

Quenched and tempered alloy steels might provide the required combination of strength and toughness; however, the alloy steels are expensive. An alternative approach for each part is described below.

The axle might be made from a forged 1050 steel containing a matrix of ferrite and pearlite. The axle could be surface-hardened, perhaps by moving the axle through an induction coil to selectively heat the surface of the steel above the A_3 temperature (about 770°C). After the coil passes any particular location of the axle, the cold interior quenches the surface to martensite. Tempering then softens the martensite to improve ductility. This combination of carbon content and heat treatment meets our requirements. The plain-carbon steel is inexpensive; the core of ferrite and pearlite produces good toughness and strength; and the hardened surface provides good fatigue and wear resistance.

The gear is subject to more severe loading conditions, for which the 1050 steel does not provide sufficient toughness, hardness, and wear resistance. Instead, we might carburize a 1010 steel for the gear. The original steel contains mostly ferrite, providing good ductility and toughness. By performing a gas carburizing process above the A_3 temperature (about 860°C), we introduce about 1.0% C in a very thin case at the surface of the gear teeth. This high-carbon case, which transforms to martensite during quenching, is tempered to control the hardness. Now we obtain toughness due to the low-carbon ferrite core, wear resistance due to the high-carbon surface, and fatigue resistance due to the high-strength surface containing compressive residual stresses introduced during carburizing. In addition, the plain-carbon 1010 steel is an inexpensive starting material that is easily forged into a near-net shape prior to heat treatment.

12-9   Weldability of Steel

In Chapter 8, we discussed welding and other joining processes.[8,9] We noted that steels are the most widely used structural materials. In bridges, buildings, and many other applications, steels must be welded. The structural integrity of steel structures not only depends upon the strength of the steel but also the strength of the welded joints. This is why the weldability of steel is always an important consideration.

Many low-carbon steels weld easily. Welding of medium- and high-carbon steels is comparatively more difficult since martensite can form in the heat-affected zone rather easily, thereby causing a weldment with poor toughness. Several strategies such as preheating the material or minimizing incorporation of hydrogen have been developed to counter these problems. The incorporation of hydrogen causes the steel to become brittle. In low-carbon steels, the strength of the welded regions in these materials is higher than the base material. This is due to the finer pearlite microstructure that forms during cooling of the heat-affected zone. Retained austenite along ferrite grain boundaries also limits recrystallization and thus helps retain a fine grain size, which contributes to the strength of the welded region.[3] During welding, the metal nearest the weld heats above the A_1 temperature and austenite forms (Figure 12-29). During cooling, the austenite in this heat-affected zone transforms to a new structure, depending on the cooling rate and the CCT diagram for the steel. Plain low-carbon steels have such a low hardenability

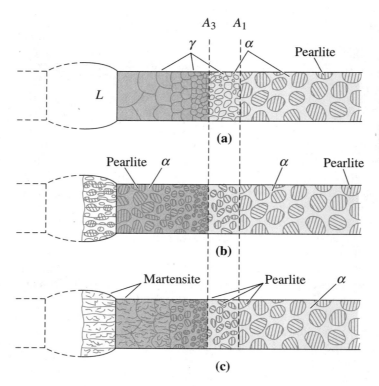

Figure 12-29
The development of the heat-affected zone in a weld: (a) the structure at the maximum temperature, (b) the structure after cooling in a steel of low hardenability, and (c) the structure after cooling in a steel of high hardenability.

that normal cooling rates seldom produce martensite. However, an alloy steel may have to be preheated to slow down the cooling rate or post-heated to temper any martensite that forms.

A steel that is originally quenched and tempered has two problems during welding. First, the portion of the heat-affected zone that heats above the A_1 may form martensite after cooling. Second, a portion of the heat-affected zone below the A_1 may overtemper. Normally, we should not weld a steel in the quenched and tempered condition. The example below shows how the heat-affected zone microstructure can be accounted for using CCT diagrams.

EXAMPLE 12-8   *Structures of Heat-Affected Zones*

Compare the structures in the heat-affected zones of welds in 1080 and 4340 steels if the cooling rate in the heat-affected zone is 5°C/s.

SOLUTION

From the CCT diagrams, Figures 12-16 and 12-18, the cooling rate in the weld produces the following structures:

1080: 100% pearlite

4340: Bainite and martensite

The high hardenability of the alloy steel reduces the weldability, permitting martensite to form and embrittle the weld.

12-10   Stainless Steels

Stainless steels are selected for their excellent resistance to corrosion.[10] Harry Brearly was working on an alloy containing 0.35% carbon and 14% chromium (today known as 420 stainless steel).[3] He was looking for a way to "etch" the sample using acids so that he could see the microstructure. He had difficulty finding a suitable chemical agent to achieve this. While others had experienced this "problem", Mr. Brearly saw the opportunity in having a material that was corrosion resistant!

All true stainless steels contain a minimum of about 11% Cr, which permits a thin, protective surface layer of chromium oxide to form when the steel is exposed to oxygen.

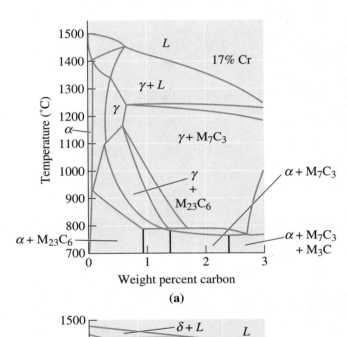

Figure 12-30
(a) The effect of 17% chromium on the iron-carbon phase diagram. At low-carbon contents, ferrite is stable at all temperatures. (b) A section of the iron-chromium-nickel-carbon phase diagram at a constant 18% Cr-8% Ni. At low-carbon contents, austenite is stable at room temperature.

TABLE 12-4 ■ *Typical compositions and properties of stainless steels*

Steel	% C	% Cr	% Ni	Others	Tensile Strength (psi)	Yield Strength (psi)	% Elongation	Condition
Austenitic:								
201	0.15	17	5	6.5% Mn	95,000	45,000	40	Annealed
304	0.08	19	10		75,000	30,000	30	Annealed
					185,000	140,000	9	Cold-worked
304L	0.03	19	10		75,000	30,000	30	Annealed
316	0.08	17	12	2.5% Mo	75,000	30,000	30	Annealed
321	0.08	18	10	0.4% Ti	85,000	35,000	55	Annealed
347	0.08	18	11	0.8% Nb	90,000	35,000	50	Annealed
Ferritic:								
430	0.12	17			65,000	30,000	22	Annealed
442	0.12	20			75,000	40,000	20	Annealed
Martensitic:								
416	0.15	13		0.6% Mo	180,000	140,000	18	Quenched and tempered
431	0.20	16	2		200,000	150,000	16	Quenched and tempered
440C	1.10	17		0.7% Mo	285,000	275,000	2	Quenched and tempered
Precipitation hardening:								
17-4	0.07	17	4	0.4% Nb	190,000	170,000	10	Age-hardened
17-7	0.09	17	7	1.0% Al	240,000	230,000	6	Age-hardened

The chromium is what makes stainless steels stainless. Chromium is also a *ferrite stabilizing element*. Figure 12-30(a) illustrates the effect of chromium on the iron-carbon phase diagram. Chromium causes the austenite region to shrink, while the ferrite region increases in size. For high-chromium, low-carbon compositions, ferrite is present as a single phase up to the solidus temperature.

There are several categories of stainless steels based on crystal structure and strengthening mechanism. Typical properties are included in Table 12-4.

Ferritic Stainless Steels Ferritic stainless steels contain up to 30% Cr and less than 0.12% C. Because of the BCC structure, the ferritic stainless steels have good strengths and moderate ductilities derived from solid-solution strengthening and strain hardening. Ferritic stainless steels are magnetic. They are not heat treatable. They have excellent corrosion resistance, moderate formability and are relatively inexpensive.

Martensitic Stainless Steels From Figure 12-30(a), we find that a 17% Cr-0.5% C alloy heated to 1200°C forms 100% austenite, which transforms to martensite on quenching in oil. The martensite is then tempered to produce high strengths and hardnesses [Figure 12-31(a)]. The chromium content is usually less than 17% Cr; otherwise, the austenite field becomes so small that very stringent control over both austenitizing temperature and carbon content is required. Lower chromium contents also permit the carbon content to vary from about 0.1% to 1.0%, allowing martensites of different hardnesses to be produced. The combination of hardness, strength, and corrosion resistance makes the alloys attractive for applications such as high-quality knives, ball bearings, and valves.

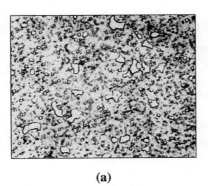
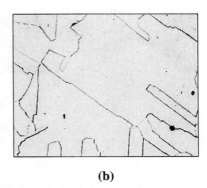

(a)   (b)

Figure 12-31 (a) Martensitic stainless steel containing large primary carbides and small carbides formed during tempering (×350). (b) Austenitic stainless steel (×500). (*From* ASM Handbook, *Vols. 7 and 8, (1972, 1973), ASM International, Materials Park, OH 44073.*)

Austenitic Stainless Steels Nickel, which is an *austenite stabilizing element*, increases the size of the austenite field, while nearly eliminating ferrite from the iron-chromium-carbon alloys [Figure 12-30(b)]. If the carbon content is below about 0.03%, the carbides do not form and the steel is virtually all austenite at room temperature [Figure 12-31(b)].

The FCC austenitic stainless steels have excellent ductility, formability, and corrosion resistance. Strength is obtained by extensive solid-solution strengthening, and the austenitic stainless steels may be cold worked to higher strengths than the ferritic stainless steels. These are nonmagnetic, which is an advantage for many applications. For example, as seen in Chapter 11, cardiovascular stents are often made from 316 stainless steels. The steels have excellent low-temperature impact properties, since they have no transition temperature. Furthermore, the austenitic stainless steels are not ferromagnetic. Unfortunately, the high-nickel and chromium contents make the alloys expensive. The 304 alloy containing 18% Cr and 8% nickel (also known as 18-8 stainless) is the most widely used grade of stainless steel. Although stainless, this alloy can undergo **sensitization**. When heated to a temperature of ~480–860°C, chromium carbides precipitate along grain boundaries rather than within grains. This causes chromium depletion in the interior of the grains and this will cause the stainless steel to corrode very easily. This is known as sensitization.

Precipitation-Hardening (PH) Stainless Steels The precipitation-hardening (or PH) stainless steels contain Al, Nb, or Ta and derive their properties from solid-solution strengthening, strain hardening, age hardening, and the martensitic reaction. The steel is first heated and quenched to permit the austenite to transform to martensite. Reheating permits precipitates such as Ni_3Al to form from the martensite. High-mechanical properties are obtained even with low-carbon contents.

Duplex Stainless Steels In some cases, mixtures of phases are deliberately introduced into the stainless steel structure. By appropriate control of the composition and heat treatment, a **duplex stainless steel** containing approximately 50% ferrite and 50% austenite can be produced. This combination provides a set of mechanical properties, corrosion resistance, formability, and weldability not obtained in any one of the usual stainless steels.

Most stainless steels are recyclable and the following example shows how differences in properties can be used to separate different types of stainless steels.

EXAMPLE 12-9	*Design of a Test to Separate Stainless Steels*

In order to efficiently recycle stainless steel scrap, we wish to separate the high-nickel stainless steel from the low-nickel stainless steel. Design a method for doing this.

SOLUTION

Performing a chemical analysis on each piece of scrap is tedious and expensive. Sorting based on hardness might be less expensive; however, because of the different types of treatments—such as annealing, cold working, or quench and tempering—the hardness may not be related to the steel composition.

The high-nickel stainless steels are ordinarily austenitic, whereas the low-nickel alloys are ferritic or martensitic. An ordinary magnet will be attracted to the low-nickel ferritic and martensitic steels, but will not be attracted to the high-nickel austenitic steel. We might specify this simple and inexpensive magnetic test for our separation process.

12-11 Cast Irons

Cast irons are iron-carbon-silicon alloys, typically containing 2–4% C and 0.5–3% Si, that pass through the eutectic reaction during solidification. The microstructures of the five important types of cast irons are shown schematically in Figure 12-32.

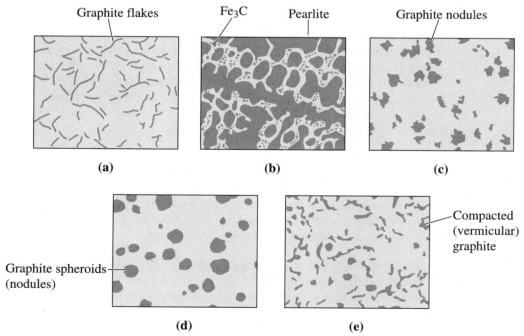

Figure 12-32 Schematic drawings of the five types of cast iron: (a) gray iron, (b) white iron, (c) malleable iron, (d) ductile iron, and (e) compacted graphite iron.

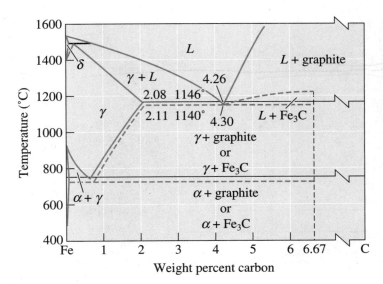

Figure 12-33
The iron-carbon phase diagram showing the relationship between the stable iron-graphite equilibria (solid lines) and the metastable iron-cementite reactions (dashed lines).

Eutectic Reaction in Cast Irons Based on the Fe-Fe$_3$C phase diagram (dashed lines in Figure 12-33), the eutectic reaction that occurs in Fe-C alloys at 1140°C is:

$$L \rightarrow \gamma + Fe_3C \tag{12-1}$$

This reaction produces *white cast iron,* with a microstructure composed of Fe$_3$C and pearlite. The Fe-Fe$_3$C system, however, is really a metastable phase diagram.[11] Under truly equilibrium conditions, the eutectic reaction is:

$$L \rightarrow \gamma + graphite \tag{12-2}$$

The Fe-C phase diagram is shown as solid lines in Figure 12-33. When the stable $L \rightarrow \gamma$ + graphite eutectic reaction occurs at 1146°C, gray, ductile, or compacted-graphite cast iron forms.

In Fe-C alloys, the liquid easily undercools 6°C (the temperature difference between the stable and metastable eutectic temperatures) and white iron forms. Adding about 2% silicon to the iron increases the temperature difference between the eutectics, permitting larger undercoolings to be tolerated and more time for the stable graphite eutectic to nucleate and grow. Silicon, therefore, is a *graphite stabilizing* element. Elements such as chromium and bismuth have the opposite effect and encourage white cast iron. We can also introduce inoculants, such as silicon (as Fe-Si ferrosilicon), to encourage the nucleation of graphite, or we can reduce the cooling rate of the casting to provide more time for the growth of graphite.

Silicon also reduces the amount of carbon contained in the eutectic. We can take this effect into account by defining the **carbon equivalent** (CE):

$$CE = \% \, C + \tfrac{1}{3}\% \, Si \tag{12-3}$$

The eutectic composition is always near 4.3% CE. A high-carbon equivalent encourages the growth of the graphite eutectic.

Eutectoid Reaction in Cast Irons The matrix structure and properties of each type of cast iron are determined by how the austenite transforms during the eutectoid reaction. In the Fe-Fe$_3$C phase diagram used for steels, the austenite transformed to ferrite and cementite, often in the form of pearlite. However, silicon also encourages the *stable*

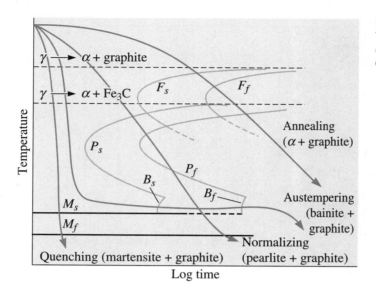

Figure 12-34
The transformation diagram for austenite in a cast iron.

eutectoid reaction:

$$\gamma \rightarrow \alpha + \text{graphite} \qquad (12\text{-}4)$$

Under equilibrium conditions, carbon atoms diffuse from the austenite to existing graphite particles, leaving behind the low-carbon ferrite. The transformation diagram (Figure 12-34) describes how the austenite might transform during heat treatment. **Annealing** (or furnace cooling) of cast iron gives a soft ferritic matrix (not coarse pearlite as in steels!). Normalizing, or air cooling, gives a pearlitic matrix. The cast irons can also be austempered to produce bainite, or can be quenched to martensite and tempered. Austempered ductile iron, with strengths of up to 200,000 psi, is used for high-performance gears.

Gray cast iron contains small, interconnected graphite flakes that cause low strength and ductility. This is the most widely used cast iron and is named for the dull gray color of the fractured surface. Gray cast iron contains many clusters, or **eutectic cells**, of interconnected graphite flakes (Figure 12-35). The point at which the flakes are connected

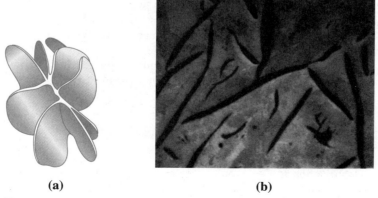

(a)   **(b)**

Figure 12-35   (a) Sketch and (b) photomicrograph of the flake graphite in gray cast iron (×100).

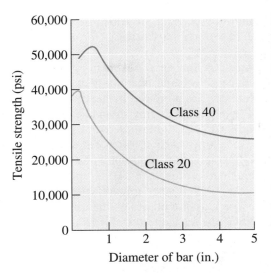

Figure 12-36
The effect of the cooling rate or casting size on the tensile properties of two gray cast irons.

is the original graphite nucleus. Inoculation helps produce smaller eutectic cells, thus improving strength. The gray irons are specified by a class number of 20 to 80. A class 20 gray iron has a nominal tensile strength of 20,000 psi. In thick castings, however, coarse graphite flakes and a ferrite matrix produce tensile strengths as low as 12,000 psi (Figure 12-36), whereas in thin castings, fine graphite and pearlite form and give tensile strengths near 40,000 psi. Higher strengths are obtained by reducing the carbon equivalent, by alloying, or by heat treatment. The graphite flakes concentrate stresses and cause low strength and ductility, but gray iron has a number of attractive properties, including high-compressive strength, good machinability, good resistance to sliding wear, good resistance to thermal fatigue, good thermal conductivity, and good vibration damping.

White cast iron is a hard, brittle alloy containing massive amounts of Fe_3C. A fractured surface of this material appears white, hence the name. A group of highly alloyed white irons are used for their hardness and resistance to abrasive wear. Elements

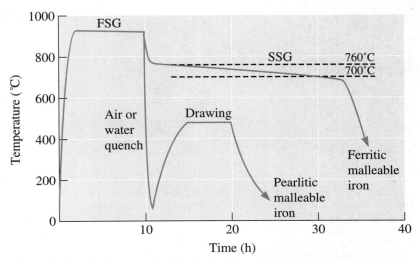

Figure 12-37   The heat treatments for ferritic and pearlitic malleable irons.

such as chromium, nickel, and molybdenum are added so that, in addition to the alloy carbides formed during solidification, martensite is formed during subsequent heat treatment.

Malleable cast iron formed by the heat treatment of white cast iron, produces rounded clumps of graphite. It exhibits better ductility than gray or white cast irons. It is also very machinable. Malleable iron is produced by heat treating unalloyed 3% carbon equivalent (2.5% C, 1.5% Si) white iron. During the malleabilizing heat treatment, the cementite formed during solidification is decomposed and graphite clumps, or nodules, are produced. The nodules, or temper carbon, often resemble popcorn. The rounded graphite shape permits a good combination of strength and ductility. The production of malleable iron requires several steps (Figure 12-37). Graphite nodules nucleate as the white iron is slowly heated. During **first stage graphitization** (FSG), cementite decomposes to the stable austenite and graphite phases as the carbon in Fe_3C diffuses to the graphite nuclei. Following FSG, the austenite transforms during cooling. Figure 12-38

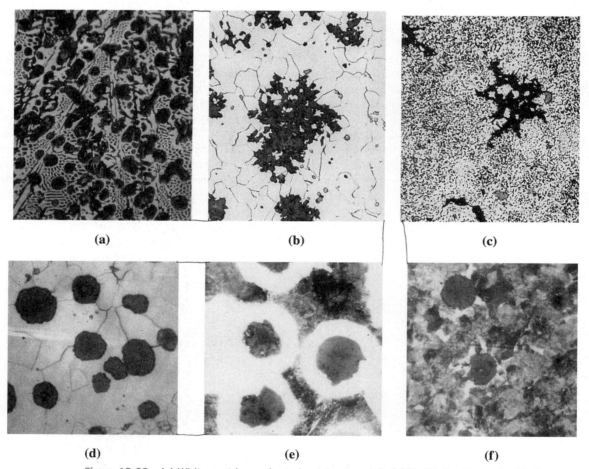

(a)　　　　　　　**(b)**　　　　　　　**(c)**

(d)　　　　　　　**(e)**　　　　　　　**(f)**

Figure 12-38　(a) White cast iron prior to heat treatment ($\times100$). (b) Ferritic malleable iron with graphite nodules and small MnS inclusions in a ferrite matrix ($\times200$). (c) Pearlitic malleable iron drawn to produce a tempered martensite matrix ($\times500$). (*Images (b) and (c) are from* Metals Handbook, *Vols. 7 and 8, (1972, 1973), ASM International, Materials Park, OH 44073.*) (d) Annealed ductile iron with a ferrite matrix ($\times250$). (e) As-cast ductile iron with a matrix of ferrite (white) and pearlite ($\times250$). (f) Normalized ductile iron with a pearlite matrix ($\times250$).

shows the microstructures of the original white iron (a) and the two types of malleable iron that can be produced (b and c). To make *ferritic malleable iron*, the casting is cooled slowly through the eutectoid temperature range to cause **second stage graphitization** (SSG). The ferritic malleable iron has good toughness compared with that of other irons because its low-carbon equivalent reduces the transition temperature below room temperature. *Pearlitic malleable iron* is obtained when austenite is cooled in air or oil to form pearlite or martensite. In either case, the matrix is hard and brittle. The iron is then drawn at a temperature below the eutectoid. **Drawing** tempers the martensite or spheroidizes the pearlite. A higher drawing temperature decreases strength and increases ductility and toughness.

Ductile or **nodular, cast iron** contains spheroidal graphite particles. Ductile iron is produced by treating liquid iron with a carbon equivalent of near 4.3% with magnesium, which causes spheroidal graphite (called nodules) to grow during solidification, rather than during a lengthy heat treatment. Several steps are required to produce this iron. These include desulfurization, **nodulizing**, and inoculation. In desulfurization, any sulfur and oxygen in the liquid metal is removed by adding desulfurizing agents such as calcium oxide (CaO). In nodulizing, Mg is added, usually in a dilute form such as an MgFeSi alloy. If pure Mg is added, the nodulizing reaction is very violent, since the boiling point of Mg is much lower than the temperature of the liquid iron, and most of the Mg will be lost. A residual of about 0.03% Mg must be in the liquid iron after treatment in order for spheroidal graphite to grow. Finally, inoculation with FeSi compounds to cause heterogeneous nucleation of the graphite is essential; if inoculation is not effective, white iron will form instead of ductile iron. The nodulized and inoculated iron must then be poured into molds within a few minutes to avoid fading. **Fading** oc-

TABLE 12-5 ■ *Typical properties of cast irons*

	Tensile Strength (psi)	Yield Strength (psi)	% E	Notes
Gray irons:				
Class 20	12,000–40,000	—	—	
Class 40	28,000–54,000	—	—	
Class 60	44,00–66,000	—	—	
Malleable irons:				
32510	50,000	32,500	10	Ferritic
35018	53,000	35,000	18	Ferritic
50005	70,000	50,000	5	Pearlitic
70003	85,000	70,000	3	Pearlitic
90001	105,000	90,000	1	Pearlitic
Ductile irons:				
60–40–18	60,000	40,000	18	Annealed
65–45–12	65,000	45,000	12	As-cast ferritic
80–55–06	80,000	55,000	6	As-cast pearlitic
100–70–03	10,000	70,000	3	Normalized
120–90–02	120,000	90,000	2	Quenched and tempered
Compacted graphite irons:				
Low strength	40,000	28,000	5	90% Ferritic
High strength	65,000	55,000	1	80% Pearlitic

curs by the gradual, nonviolent loss of Mg due to vaporization and/or reaction with oxygen, resulting in flake or compacted graphite instead of spheroidal graphite. In addition, the inoculant effect will also fade, resulting in white iron.

Compared with gray iron, ductile cast iron has excellent strength and ductility. Due to the higher silicon content (typically around 2.4%) in ductile irons compared with 1.5% Si in malleable irons, the ductile irons are stronger but not as tough as malleable irons.

Compacted graphite cast iron contains rounded but interconnected graphite also produced during solidification. The graphite shape in compacted graphite cast iron is intermediate between flakes and spheres, with numerous rounded rods of graphite that are interconnected to the nucleus of the eutectic cell. This compacted graphite, sometimes called **vermicular graphite**, also forms when ductile iron fades. The compacted graphite permits strengths and ductilities that exceed those of gray cast iron, but allows the iron to retain good thermal conductivity and vibration damping properties. The treatment for the compacted graphite iron is similar to that for ductile iron; however, only about 0.015% Mg is introduced during nodulizing. A small amount of titanium (Ti) is added to assure the formation of the compacted graphite.

Typical properties of cast irons are given in Table 12-5.

SUMMARY

◆ The properties of steels, determined by dispersion strengthening, depend on the amount, size, shape, and distribution of cementite. These factors are controlled by alloying and heat treatment.

◆ A process anneal recrystallizes cold-worked steels.

◆ Spheroidizing produces large, spheroidal Fe_3C and good machinability in high-carbon steels.

◆ Annealing, involving a slow furnace cool after austenitizing, giving a coarse pearlitic structure containing lamellar Fe_3C.

◆ Normalizing, involving an air cool after austenitizing, gives a fine pearlitic structure and higher strength compared with annealing.

◆ In isothermal annealing, pearlite with a uniform interlamellar spacing is obtained by transforming the austenite at a constant temperature.

◆ Austempering is used to produce bainite, containing rounded Fe_3C, by an isothermal transformation.

◆ Quench and temper heat treatments require the formation and decomposition of martensite, providing exceptionally fine dispersions of round Fe_3C.

◆ We can better understand the mechanics of the heat treatments by use of TTT diagrams, CCT diagrams, and hardenability curves.

◆ The TTT diagrams describe how austenite transforms to pearlite and bainite at a constant temperature.

◆ The CCT diagrams describe how austenite transforms during continuous cooling. These diagrams give the cooling rates needed to obtain martensite in quench and temper treatments.

◇ The hardenability curves compare the ease with which different steels transform to martensite.

◇ Alloying elements increase the times required for transformations in the TTT diagrams, reduce the cooling rates necessary to produce martensite in the CCT diagrams, and improve the hardenability of the steel.

◇ Specialty steels and heat treatments provide unique properties or combinations of properties. Of particular importance are surface-hardening treatments, such as carburizing, that produce an excellent combination of fatigue and impact resistance. Stainless steels, which contain a minimum of 12% Cr, have excellent corrosion resistance.

◇ Cast irons, by definition, undergo the eutectic reaction during solidification. Depending on the composition and treatment, either γ and Fe_3C or γ and graphite form during freezing.

◇ White cast iron, with good wear-resistance, is obtained when Fe_3C forms during the eutectic reaction. Malleable cast iron, with good strength, ductility, and toughness, is produced by heat treating white cast iron to form rounded graphite. By producing graphite directly during solidification, gray cast iron, ductile cast iron, and compacted graphite cast iron are produced. Because graphite flakes form in gray iron, its strength and ductility are limited. The graphite spheres that form in ductile iron as a result of the addition of magnesium permit good strength and ductility. Compacted graphite iron has intermediate properties.

GLOSSARY

Annealing (cast iron)   A heat treatment used to produce a ferrite matrix in a cast iron by austenitizing, then furnace cooling.

Annealing (steel)   A heat treatment used to produce a soft, coarse pearlite in steel by austenitizing, then furnace cooling.

Ausforming   A thermomechanical heat treatment in which austenite is plastically deformed below the A_1 temperature, then permitted to transform to bainite or martensite.

Austempering   The isothermal heat treatment by which austenite transforms to bainite.

Austenitizing   Heating a steel or cast iron to a temperature where homogeneous austenite can form. Austenitizing is the first step in most of the heat treatments for steel and cast irons.

Carbon equivalent   Carbon plus one-third of the silicon in a cast iron.

Carbonitriding   Hardening the surface of steel with carbon and nitrogen obtained from a special gas atmosphere.

Carburizing   A group of surface-hardening techniques by which carbon diffuses into steel.

Case depth   The depth below the surface of a steel at which hardening occurs by surface hardening and carburizing processes.

Cast iron   Ferrous alloys containing sufficient carbon so that the eutectic reaction occurs during solidification.

Compacted-graphite cast iron   A cast iron treated with small amounts of magnesium and titanium to cause graphite to grow during solidification as an interconnected, coral-shaped precipitate, giving properties midway between gray and ductile iron.

Cyaniding   Hardening the surface of steel with carbon and nitrogen obtained from a bath of liquid cyanide solution.

Drawing   Reheating a malleable iron in order to reduce the amount of carbon combined as cementite by spheroidizing pearlite, tempering martensite, or graphitizing both.

Dual-phase steels   Special steels treated to produce martensite dispersed in a ferrite matrix.

Ductile cast iron   Cast iron treated with magnesium to cause graphite to precipitate during solidification as spheres, permitting excellent strength and ductility. Also known as nodular cast iron.

Duplex stainless steel   A special class of stainless steels containing a microstructure of ferrite and austenite.

Electric arc furnace   A furnace used to melt steel scrap using electricity. Often, specialty steels are made using electric arc furnaces.

Eutectic cell   A cluster of graphite flakes produced during solidification of gray iron that are all interconnected to a common nucleus.

Fading   The loss of the nodulizing or inoculating effect in cast irons as a function of time, permitting undesirable changes in microstructure and properties.

First stage graphitization   The first step in the heat treatment of a malleable iron, during which the massive carbides formed during solidification are decomposed to graphite and austenite.

Gray cast iron   Cast iron which, during solidification, contains graphite flakes, causing low strength and poor ductility. This is the most widely used type of cast iron.

Hardenability   The ease with which a steel can be quenched to form martensite. Steels with high hardenability form martensite even on slow cooling.

Hardenability curves   Graphs showing the effect of the cooling rate on the hardness of as-quenched steel.

Hot metal   The molten iron produced in a blast furnace, also known as pig iron. It contains about 95% iron, 4% carbon, 0.3–0.9% silicon, 0.5% Mn, and 0.025–0.05% each sulfur, phosphorus, and titanium.

Inoculation   The addition of an agent to the molten cast iron that provides nucleation sites at which graphite precipitates during solidification.

Interstitial free steels   These are steels containing Nb and Ti. They react with C and S to form precipitates of carbides and sulfides, leaving the ferrite nearly free of interstial elements.

Isothermal annealing   Heat treatment of a steel by austenitizing, cooling rapidly to a temperature between the A_1 and the nose of the TTT curve, and holding until the austenite transforms to pearlite.

Jominy distance   The distance from the quenched end of a Jominy bar. The Jominy distance is related to the cooling rate.

Jominy test   The test used to evaluate hardenability. An austenitized steel bar is quenched at one end only, thus producing a range of cooling rates along the bar.

Malleable cast iron Cast iron obtained by a lengthy heat treatment, during which cementite decomposes to produce rounded clumps of graphite. Good strength, ductility, and toughness are obtained as a result of this structure.

Maraging steels A special class of alloy steels that obtain high strengths by a combination of the martensitic and age-hardening reactions.

Marquenching Quenching austenite to a temperature just above the M_S and holding until the temperature is equalized throughout the steel before further cooling to produce martensite. This process reduces residual stresses and quench cracking. Also known as *martempering*.

Nitriding Hardening the surface of steel with nitrogen obtained from a special gas atmosphere.

Nodulizing The addition of magnesium to molten cast iron to cause the graphite to precipitate as spheres rather than as flakes during solidification.

Normalizing A simple heat treatment obtained by austenitizing and air cooling to produce a fine pearlitic structure. This can be done for steels and cast irons.

Pig iron The molten iron produced in a blast furnace also known as hot metal. It contains about 95% iron, 4% carbon, 0.3–0.9% silicon, 0.5% Mn, and 0.025–0.05% each sulfur, phosphorus, and titanium.

Process anneal A low-temperature heat treatment used to eliminate all or part of the effect of cold working in steels.

Quench cracks Cracks that form at the surface of a steel during quenching due to tensile residual stresses that are produced because of the volume change that accompanies the austenite-to-martensite transformation.

Retained austenite Austenite that is unable to transform into martensite during quenching because of the volume expansion associated with the reaction.

Second stage graphitization The second step in the heat treatment of malleable irons that are to have a ferritic matrix. The iron is cooled slowly from the first stage graphitization temperature so that austenite transforms to ferrite and graphite rather than to pearlite.

Secondary hardening peak Unusually high hardness in a steel tempered at a high temperature caused by the precipitation of alloy carbides.

Sensitization When heated to a temperature of ~480–860°C, chromium carbides precipitate along grain boundaries rather than within grains, causing chromium depletion in the interior. This causes the stainless steel to corrode very easily.

Slag A byproduct of a melting or refining process, usually containing many impurities and other oxides.

Spheroidite A microconstituent containing coarse spheroidal cementite particles in a matrix of ferrite, permitting excellent machining characteristics in high-carbon steels.

Stainless steels A group of ferrous alloys that contain at least 11% Cr, providing extraordinary corrosion resistance.

Tempered martensite The microconstituent of ferrite and cementite formed when martensite is tempered.

Tool steels A group of high-carbon steels that provide combinations of high hardness, toughness, or resistance to elevated temperatures.

Vermicular graphite   The rounded, interconnected graphite that forms during the solidification of cast iron. This is the intended shape in compacted-graphite iron, but it is a defective shape in ductile iron.

White cast iron   Cast iron that produces cementite rather than graphite during solidification. The white irons are hard and brittle.

ADDITIONAL INFORMATION

1. BRONOWSKI, J., *The Ascent of Man*. 1973: Back Bay Books. 124–131.

2. FRUEHAN, R.J., *The Making, Shaping and Treating of Steel (Steel Making and Refining)*. 11th ed, ed. R.J. FRUEHAN. 1998: The AISE Steel Foundation.

3. HARPER, C.A., *Handbook of Materials Product Design*. 3rd ed. 2001: McGraw Hill.

4. BERENDOSOHN, R., "Germfree House," *Popular Mechanics*, 2002 (May): p. 104–106.

5. DEARDO, A.J. and I. GARCIA, *US Patent 6,200,395: Free-Machining Steels Containing Tin Antimony and/or Arsenic*. 2001: University of Pittsburgh.

6. DEARDO, A.J., I. GARCIA, and M. HUA, *Technical discussions and assistance are acknowleded*. 2002.

7. EYLON, D. and H.H. SUZUKI, "On the Use of Lamination in the Making of Iron and Steel Swords," in *The Fifth International Conference on the Beginning of Metals and Alloys*. 2002. Kyongju, Korea.

8. DAVID, S.A. and T. DEBROY, "Current Issues and Problems in Welding Science," *Science*, 1992. 257: p. 497–502.

9. DEBROY, T. and S.A. DAVID, "Physical Processes in Fusion Welding," *Reviews of Modern Physics*, 1995. 67(1): p. 85–112.

10. PATIL, B., "Stainless Steel Making," in *The Making, Shaping and Treating of Steel (Steel Making and Refining)*, R.J. FRUEHAN, Editor. 1998: The AISE Steel Foundation.

11. BERGERON, C.G. and S.H. RISBUD, *Introduction to Phase Equilibria*. 1984: American Ceramic Society.

✓ PROBLEMS

12-1 Explain how steel is made from iron ore and scrap steel.

Section 12-1 Designations and Classification of Steels

12-2 What is the difference between cast iron and steels?

12-3 What do A_1, A_3, and A_{cm} temperatures refer to? Are these temperatures constant?

12-4 Calculate the amounts of ferrite, cementite, primary microconstituent, and pearlite in the following steels:

(a) 1015   (b) 1035   (c) 1095   (d) 10130

12-5 Estimate the AISI-SAE number for steels having the following microstructures:

(a) 38% pearlite-62% primary ferrite
(b) 93% pearlite-7% primary cementite
(c) 97% ferrite-3% cementite
(d) 86% ferrite-14% cementite

12-6 Complete the following table:

	1035 Steel	10115 Steel
A_1 temperature		
A_3 or A_{cm} temperature		
Full annealing temperature		
Normalizing temperature		
Process annealing temperature		
Spheroidizing temperature		

12-7 What do the terms low-, medium-, and high-carbon steels mean?

Section 12-2 Simple Heat Treatments

Section 12-3 Isothermal Heat Treatments

12-8 Explain the following heat treatments: (a) process anneal, (b) austenitizing, (c) annealing, (d) normalizing, and (e) quenching.

12-9 Explain why, strictly speaking, TTT diagrams can be used for isothermal treatments only.

12-10 In a pearlitic 1080 steel, the cementite platelets are 4×10^{-5} cm thick, and the ferrite platelets are 14×10^{-5} cm thick. In a spheroidized 1080 steel, the cementite spheres are 4×10^{-3} cm in diameter. Estimate the total interface area between the ferrite and cementite in a cubic centimeter of each steel. Determine the percentage reduction in surface area when the pearlitic steel is spheroidized. The density of ferrite is 7.87 g/cm^3 and that of cementite is 7.66 g/cm^3.

12-11 Describe the microstructure present in a 1050 steel after each step in the following heat treatments:

   (a) heat at 820°C, quench to 650°C and hold for 90 s, and quench to 25°C

   (b) heat at 820°C, quench to 450°C and hold for 90 s, and quench to 25°C

   (c) heat at 820°C, and quench to 25°C

   (d) heat at 820°C, quench to 720°C and hold for 100 s, and quench to 25°C

   (e) heat at 820°C, quench to 720°C and hold for 100 s, quench to 400°C and hold for 500 s, and quench to 25°C

   (f) heat at 820°C, quench to 720°C and hold for 100 s, quench to 400°C and hold for 10 s, and quench to 25°C

   (g) heat at 820°C, quench to 25°C, heat to 500°C and hold for 10^3 s, and air cool to 25°C

12-12 Describe the microstructure present in a 10110 steel after each step in the following heat treatments

   (a) heat to 900°C, quench to 400°C and hold for 10^3 s, and quench to 25°C

   (b) heat to 900°C, quench to 600°C and hold for 50 s, and quench to 25°C

   (c) heat to 900°C, and quench to 25°C

   (d) heat to 900°C, quench to 300°C and hold for 200 s, and quench to 25°C

   (e) heat to 900°C, quench to 675°C and hold for 1 s, and quench to 25°C

   (f) heat to 900°C, quench to 675°C and hold for 1 s, quench to 400°C and hold for 900 s, and slowly cool to 25°C

   (g) heat to 900°C, quench to 675°C and hold for 1 s, quench to 300°C and hold for 10^3 s, and air cool to 25°C

   (h) heat to 900°C, quench to 300°C and hold for 100 s, quench to 25°C, heat to 450°C for 3600 s, and cool to 25°C

12-13 Recommend appropriate isothermal heat treatments to obtain the following, including appropriate temperatures and times:

   (a) an isothermally annealed 1050 steel with HRC 23,

   (b) an isothermally annealed 10110 steel with HRC 40,

   (c) an isothermally annealed 1080 steel with HRC 38,

   (d) an austempered 1050 steel with HRC 40,

   (e) an austempered 10110 steel with HRC 55, and

   (f) an austempered 1080 steel with HRC 50.

12-14 Compare the minimum times required to isothermally anneal the following steels at 600°C. Discuss the effect of the carbon content of the steel on the kinetics of nucleation and growth during the heat treatment.

   (a) 1050   (b) 1080   (c) 10110

Section 12-4 Quench and Temper Heat Treatments

12-15 Explain the following terms: (a) quenching, (b) tempering, (c) retained austenite, and (d) marquenching/martempering.

12-16 We wish to produce a 1050 steel that has a Brinell hardness of at least 330 and an elongation of at least 15%.

   (a) Recommend a heat treatment, including appropriate temperatures, that permits this to be achieved. Determine the yield strength and tensile strength that are obtained by this heat treatment.

   (b) What yield and tensile strength would be obtained in a 1080 steel by the same heat treatment?

   (c) What yield strength, tensile strength and % elongation would be obtained in the 1050 steel if it were normalized?

12-17 We wish to produce a 1050 steel that has a tensile strength of at least 175,000 psi and a reduction in area of at least 50%.

   (a) Recommend a heat treatment, including appropriate temperatures, that permits this to be achieved. Determine the Brinell hardness number, % elongation, and yield strength that are obtained by this heat treatment.

(b) What yield strength and tensile strength would be obtained in a 1080 steel by the same heat treatment?

(c) What yield strength, tensile strength, and % elongation would be obtained in the 1050 steel if it were annealed?

12-18 A 1030 steel is given an improper quench and temper heat treatment, producing a final structure composed of 60% martensite and 40% ferrite. Estimate the carbon content of the martensite and the austenitizing temperature that was used. What austenitizing temperature would you recommend?

12-19 A 1050 steel should be austenitized at 820°C, quenched in oil to 25°C, and tempered at 400°C for an appropriate time.

(a) What yield strength, hardness, and % elongation would you expect to obtain from this heat treatment?

(b) Suppose the actual yield strength of the steel is found to be 125,000 psi. What might have gone wrong in the heat treatment to cause this low strength?

(c) Suppose the hardness is found to be HB 525. What might have gone wrong in the heat treatment to cause this high hardness?

12-20 A part produced from a low-alloy, 0.2% C steel (Figure 12-17) has a microstructure containing ferrite, pearlite, bainite, and martensite after quenching. What microstructure would be obtained if we used a 1080 steel? What microstructure would be obtained if we used a 4340 steel?

12-21 Fine pearlite and a small amount of martensite are found in a quenched 1080 steel. What microstructure would be expected if we used a low-alloy, 0.2% C steel? What microstructure would be expected if we used a 4340 steel?

12-22 Explain the origin of quench cracking.

12-23 What different media are used for quenching steels? Explain why agitated media provide better heat-transfer rates.

Section 12-5 Effect of Alloying Elements

Section 12-6 Application of Hardenability

12-24 Explain briefly why alloying elements are used in steels.

12-25 Explain the difference between hardenability and hardness. Explain using a sketch how hardenability of steels is measured.

12-26 We have found that a 1070 steel, when austenitized at 750°C, forms a structure containing pearlite and a small amount of grain-boundary ferrite that gives acceptable strength and ductility. What changes in the microstructure, if any, would be expected if the 1070 steel contained an alloying element such as Mo or Cr? Explain.

12-27 Using the TTT diagrams, compare the hardenabilities of 4340 and 1050 steels by determining the times required for the isothermal transformation of ferrite and pearlite (F_s, P_s, and P_f) to occur at 650°C.

12-28 We would like to obtain a hardness of HRC 38 to 40 in a quenched steel. What range of cooling rates would we have to obtain for the following steels? Are some steels inappropriate?

(a) 4340 (b) 8640 (c) 9310
(d) 4320 (e) 1050 (f) 1080

12-29 A steel part must have an as-quenched hardness of HRC 35 in order to avoid excessive-wear rates during use. When the part is made from 4320 steel, the hardness is only HRC 32. Determine the hardness if the part were made under identical conditions, but with the following steels. Which, if any, of these steels would be better choices than 4320?

(a) 4340 (b) 8640 (c) 9310
(d) 1050 (e) 1080

12-30 A part produced from a 4320 steel has a hardness of HRC 35 at a critical location after quenching. Determine

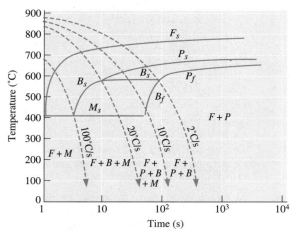

Figure 12-17   (Repeated for Problem 12-20) The CCT diagram for a low-alloy, 0.2% C steel.

(a) the cooling rate at that location, and

(b) the microstructure and hardness that would be obtained if the part were made of a 1080 steel.

12-31 A 1080 steel is cooled at the fastest possible rate that still permits all pearlite to form. What is this cooling rate? What Jominy distance, and hardness are expected for this cooling rate?

12-32 Determine the hardness and microstructure at the center of a 1.5-in.-diameter 1080 steel bar produced by quenching in

(a) unagitated oil,

(b) unagitated water, and

(c) agitated brine.

12-33 A 2-in.-diameter bar of 4320 steel is to have a hardness of at least HRC 35. What is the minimum severity of the quench (H coefficient)? What type of quenching medium would you recommend to produce the desired hardness with the least chance of quench cracking?

12-34 A steel bar is to be quenched in agitated water. Determine the maximum diameter of the bar that will produce a minimum hardness of HRC 40 if the bar is:

(a) 1050 (b) 1080 (c) 4320

(d) 8640 (e) 4340

12-35 The center of a 1-in.-diameter bar of 4320 steel has a hardness of HRC 40. Determine the hardness and microstructure at the center of a 2-in. bar of 1050 steel quenched in the same medium.

Section 12-7 Specialty Steels

12-36 Explain what is meant by the following terms: (a) tool steels, (b) dual-phase steels, (c) maraging steels, and (d) high-strength-low-alloy steels.

12-37 What material is used to fortify breakfast cereals with iron?

Section 12-8 Surface Treatments

12-38 What is the principle of the surface hardening of steels using carburizing and nitriding?

12-39 A 1010 steel is to be carburized using a gas atmosphere that produces 1.0% C at the surface of the steel. The case depth is defined as the distance below the surface that contains at least 0.5% C. If carburizing is done at 1000°C, determine the time required to produce a case depth of 0.01 in. (See Chapter 5 for a review.)

12-40 A 1015 steel is to be carburized at 1050°C for 2 h using a gas atmosphere that produces 1.2% C at the surface of the steel. Plot the percent carbon versus the distance from the surface of the steel. If the steel is slowly cooled after carburizing, determine the amount of each phase and microconstituent at 0.002-in. intervals from the surface. (See Chapter 5.)

12-41 Why is it that the strength of the heat-affected zone is higher for low-carbon steels? What is the role of retained austenite, in this case?

12-42 Why is it easier to weld low-carbon steels, however, it is difficult to weld high-carbon steels?

12-43 A 1050 steel is welded. After cooling, hardnesses in the heat-affected zone are obtained at various locations from the edge of the fusion zone. Determine the hardnesses expected at each point if a 1080 steel were welded under the same conditions. Predict the microstructure at each location in the as-welded 1080 steel.

Distance from Edge of Fusion Zone	Hardness in 1050 Weld
0.05 mm	HRC 50
0.10 mm	HRC 40
0.15 mm	HRC 32
0.20 mm	HRC 28

Section 12-10 Stainless Steels

12-44 What is a stainless steel? Why are stainless steels stainless?

12-45 We wish to produce a martensitic stainless steel containing 17% Cr. Recommend a carbon content and austenitizing temperature that would permit us to obtain 100% martensite during the quench. What microstructure would be produced if the martensite were then tempered until the equilibrium phases formed?

12-46 What is sensitization?

12-47 What are the different classes of stainless steels? Which one is most widely used?

12-48 Occasionally, when an austenitic stainless steel is welded, the weld deposit may be slightly magnetic. Based on the Fe-Cr-Ni-C phase diagram [Figure 12-30(b)], what phase would you expect is causing the magnetic behavior? Why might this phase have formed? What could you do to restore the nonmagnetic behavior?

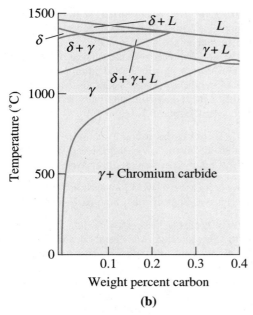

(b)

Figure 12-30 (Repeated for Problem 12-48) (b) A section of the iron-chromium-nickel-carbon phase diagram at a constant 18% Cr-8% Ni. At low-carbon contents, austenite is stable at room temperature.

Section 12-11 Cast Irons

12-49 Define cast iron using the Fe-Fe$_3$C phase diagram.

12-50 What are the different types of cast irons? Explain using a sketch.

12-51 A tensile bar of a class 40 gray iron casting is found to have a tensile strength of 50,000 psi. Why is the tensile strength greater than that given by the class number? What do you think is the diameter of the test bar?

12-52 You would like to produce a gray iron casting that freezes with no primary austenite or graphite. If the carbon content in the iron is 3.5%, what percentage of silicon must you add?

12-53 Compare the expected hardenabilities of a plain-carbon steel, a malleable cast iron, and a ductile cast iron. Explain why you expect different hardenabilities.

✳ Design Problems

12-54 We would like to produce a 2-in.-thick steel wear plate for a rock-crushing unit. To avoid frequent replacement of the wear plate, the hardness should exceed HRC 38 within 0.25 in. of the steel surface. However, the center of the plate should have a hardness of no more than HRC 32 to assure some toughness. We have only a water quench available to us. Design the plate, assuming that we only have the steels given in Figure 12-23 available to us.

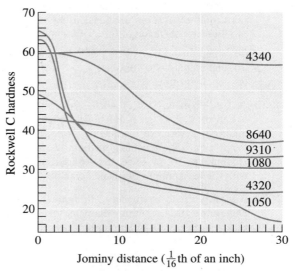

Figure 12-23 (Repeated for Problem 12-54) The hardenability curves for several steels.

12-55 A quenched and tempered 10110 steel is found to have surface cracks that cause the heat-treated part to be rejected by the customer. Why did the cracks form? Design a heat treatment, including appropriate temperatures and times that will minimize these problems.

12-56 Design a corrosion-resistant steel to use for a pump that transports liquid helium at 4 K in a superconducting magnet.

12-57 Design a heat treatment for a hook made of 1-in.-diameter steel rod having a microstructure containing a mixture of ferrite, bainite, and martensite after quenching. Estimate the mechanical properties of your hook.

12-58 Design an annealing treatment for a 1050 steel. Be sure to include details of temperatures, cooling rates, microstructures, and properties.

12-59 Design a process to produce a 0.5-cm-diameter steel shaft having excellent toughness, yet excellent wear and fatigue resistance. The surface hardness should be at least HRC 60, and the hardness 0.01 cm beneath the surface should be approximately HRC 50. Describe the process, including details of the heat-treating atmosphere, the composition of the steel, temperatures, and times.

Computer Problems

12-60 *Empirical Relationships for Transformations in Steels*. Heat treatment of steels depends upon the ability of knowing the A_1, A_3, A_{cm}, M_s, and M_f temperatures. Steel producers often provide

their customers empirical formulas that define these temperatures as a function of alloying element concentrations. For example, one such formula described in the literature is:

$$M_s(\text{in } °C) = 561 - 474(\% \text{ C}) - 33(\% \text{ Mn})$$
$$- 17(\% \text{ Ni}) - 17(\% \text{ Cr})$$
$$- 21(\% \text{ Mo})$$

Write a computer program that will ask the user to provide information for the concentrations of different alloying elements and provide the M_s and M_f temperatures. Assume that the M_f temperature is 215°C below M_s.

Chapter

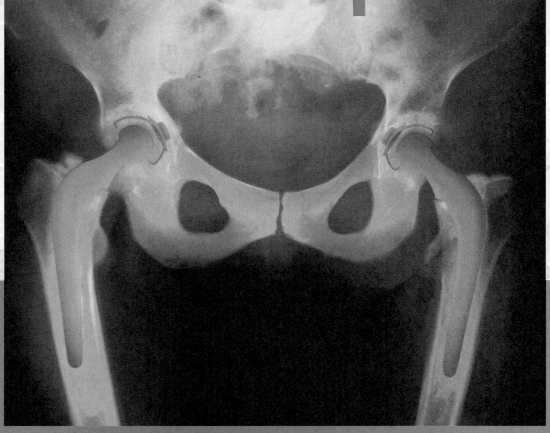

Every year, approximately 500,000 people receive a hip implant such as that shown in the photo. The implant must be made of a biocompatible material that has strong fracture toughness, an excellent fatigue life, be corrosion resistant, and posses a stiffness that is similar to that of bone. Nonferrous alloys based on titanium are used for this application. (*Courtesy of PhotoDisc/Getty Images.*)

13

Nonferrous Alloys

Have You Ever Wondered?

- In the history of mankind, which came first: copper or steel?

- Why does the Statue of Liberty have a green patina?

- What materials are used to manufacture biomedical implants for hip prostheses?

- Would it be possible to make aluminum cars?

- What fuels the solid rocket boosters that propel the space shuttle at a speed of 3000 miles an hour?

- What materials are used as "catalysts" in the automobile catalytic converter? What do they "convert?"

Nonferrous alloys (i.e., alloys of elements other than iron) include, but are not limited to, alloys based on aluminum, copper, nickel, cobalt, zinc, precious metals (such as Pt, Au, Ag, Pd), and other metals (e.g., Nb, Ta, W). In this chapter, we will briefly explore the properties and applications of Cu, Al, and Ti alloys in load-bearing applications. We will not discuss the electronic, magnetic, and other applications of nonferrous alloys.

In many applications, weight is a critical factor. To relate the strength of the material to its weight, a **specific strength**, or strength-to-weight ratio, is defined:

$$\text{Specific strength} = \frac{\text{strength}}{\text{density}} \quad (13\text{-}1)$$

Table 13-1 compares the specific strength of steel, some high-strength nonferrous alloys, and polymer-matrix composites. Another factor to consider in designing with nonferrous metals is their cost, which also varies considerably. Table 13-1 gives the approximate price of differnt materials. One should note, however, that the price of the material is only a small portion of the cost of a part. Fabrication and finishing, not to mention marketing and distribution, often contribute much more to the overall cost of a part. Composites based on carbon and other fibers also have significant advantages with respect to their specific strength. However, their properties could be anisotropic and the temperature at which they can be used is limited. In practice, to overcome the anisotropy, composites are often made in many layers. The directions of fibers are changed in different layers so as to minimize the anisotropy in properties.

TABLE 13-1 ■ *Specific strength and cost of nonferrous alloys, steels, and polymer composites*

Metal	Density g/cm³	Density (lb/in.³)	Tensile Strength (psi)	Specific Strength (in.)	Cost per lb ($)[c]
Aluminum	2.70	(0.097)	83,000	8.6×10^5	0.60
Beryllium	1.85	(0.067)	55,000	8.2×10^5	350.00
Copper	8.93	(0.322)	150,000	4.7×10^5	0.71
Lead	11.36	(0.410)	10,000	0.2×10^5	0.45
Magnesium	1.74	(0.063)	55,000	8.7×10^5	1.50
Nickel	8.90	(0.321)	180,000	5.6×10^5	4.10
Titanium	4.51	(0.163)	160,000	9.8×10^5	4.00
Tungsten	19.25	(0.695)	150,000	2.2×10^5	4.00
Zinc	7.13	(0.257)	75,000	2.9×10^5	0.40
Steels	~7.87	(0.284)	200,000	7.0×10^5	0.10
Aramid/epoxy (Kevlar, vol. fraction of fibers 0.6, longitudinal tension)	1.4	(0.05)	200,000	4.0×10^6	—
Aramid/epoxy (Kevlar, vol. fraction of fibers 0.6, transverse tension)[a]	1.4	(0.05)	4,300	0.86×10^4	—
Glass/epoxy (Vol. fraction of E-glass fibers 0.6, longitudinal tension)[b]	2.1	(0.075)	150,000	2.0×10^6	—
Glass/epoxy (Vol. fraction of E-glass fibers 0.6, transverse tension)	2.1	(0.075)	7,000	9.3×10^4	—

[a] *Data for composites from Harper, C.A., Handbook of Materials Product Design, 3rd ed. 2001: McGraw-Hill. Commodity composites are relatively inexpensive; high-performance composites are expensive.*
[b] *Properties of composites are highly anisotropic. This is taken care of during fabrication though.*
[c] *Costs based on average prices for the years 1998 to 2002.*

13-1   Aluminum Alloys

Aluminum is the third most plentiful element on earth (next to oxygen and silicon), but, until the late 1800s, was expensive and difficult to produce. The six-pound cap installed

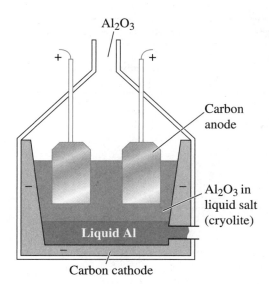

Figure 13-1
Production of aluminum in an electrolytic cell.

on the top of the Washington Monument in 1884 was one of the largest aluminum parts made at that time.[2] Development of electrical power and the **Hall-Heroult process** [3] for electrolytically reducing alumina (Al_2O_3) to liquid Al metal (Figure 13-1) allowed aluminum to become one of the most widely used and inexpensive engineering materials. In this process, alumina is converted into a material known as molten cryolite (Na_3AlF_6). Then, in an electrolytic cell, the following reactions occur:

$$Al^{+3} \text{ (in melt)} + 3 \text{ e}^- \rightarrow Al \text{ (molten) (occurs at the negatively charged cathode)}$$

$$2O^{-2} \text{ (in melt)} + C \text{ (solid, anode)} \rightarrow CO_2 \text{ (g)} + 4 \text{ e}^- \text{ (at the positively charged anode)}$$

Molten aluminum is formed at the cathode. The anode, made out of graphite, is consumed during the reaction.

General Properties and Uses of Aluminum   Aluminum has a density of 2.70 g/cm^3, or one-third the density of steel, and a modulus of elasticity of 10×10^6 psi. Although aluminum alloys have lower tensile properties compared with those of steel, their specific strength (or strength-to-weight ratio) is excellent. The Wright brothers used an Al-Cu alloy for their engine for this very reason (Chapter 11).[4] Aluminum can be formed easily, it has high thermal and electrical conductivity, and does not show a ductile-to-brittle transition at low temperatures.[5] It is nontoxic and can be recycled with only about 5% of the energy that was needed to make it from alumina. This is why the recycling of aluminum is so successful. Aluminum's beneficial physical properties include nonmagnetic behavior and its resistance to oxidation and corrosion. However, an aluminum does not display a true endurance limit (Chapter 6), so failure by fatigue may eventually occur, even at low stresses. Because of its low-melting temperature, aluminum does not perform well at elevated temperatures. Finally, aluminum alloys have low hardness, leading to poor wear resistance. Aluminum responds readily to strengthening mechanisms. Table 13-2 compares the strength of pure annealed aluminum with that of alloys strengthened by various techniques. The alloys may be 30 times stronger than pure aluminum.

About 25% of the aluminum produced today is used in the transportation industry, another 25% is used for the manufacture of beverage cans and other packaging, about 15% is used in construction, 15% in electrical applications, and 20% in other applica-

TABLE 13-2 ■ *The effect of strengthening mechanisms in aluminum and aluminum alloys*

Material	Tensile Strength (psi)	Yield Strength (psi)	% Elongation	Ratio of Alloy-to-Metal Yield Strengths
Pure Al	6,500	2,500	60	1
Commercially pure Al (at least 99% pure)	13,000	5,000	45	2.0
Solid-solution-strengthened Al alloy	16,000	6,000	35	2.4
Cold-worked Al	24,000	22,000	15	8.8
Dispersion-strengthened Al alloy	42,000	22,000	35	8.8
Age-hardened Al alloy	83,000	73,000	11	29.2

tions.[6] About 200 pounds of aluminum was used in an average car made in the United States in 1996.[1] Aluminum reacts with oxygen, even at room temperature, to produce an extremely thin aluminum-oxide layer that protects the underlying metal from many corrosive environments. We should be careful, though, not to generalize this behavior. For example, aluminum powder (because it has a high surface area), when present in the form of an oxidizer, such as ammonium perchlorate and iron oxide as catalysts, serves as the fuel for solid rocket boosters (*SRBs*). These boosters use ~200,000 lbs of atomized aluminum powder every time the space shuttle takes off and can generate enough force for the shuttle to reach a speed of ~3000 miles per hour. New develop-

TABLE 13-3 ■ *Designation system for aluminum alloys*

Wrought alloys:

1xxx[a]	Commercially pure Al (>99% Al)	Not age-hardenable
2xxx	Al-Cu and Al-Cu-Li	Age-hardenable
3xxx	Al-Mn	Not age-hardenable
4xxx	Al-Si and Al-Mg-Si	Age-hardenable if magnesium is present
5xxx	Al-Mg	Not age-hardenable
6xxx	Al-Mg-Si	Age-hardenable
7xxx	Al-Mg-Zn	Age-hardenable
8xxx	Al-Li, Sn, Zr, or B	Age-hardenable
9xxx	Not currently used	

Casting alloys:

1xx.x.[b]	Commercially pure Al	Not age-hardenable
2xx.x.	Al-Cu	Age-hardenable
3xx.x.	Al-Si-Cu or Al-Mg-Si	Some are age-hardenable
4xx.x.	Al-Si	Not age-hardenable
5xx.x.	Al-Mg	Not age-hardenable
7xx.x.	Al-Mg-Zn	Age-hardenable
8xx.x.	Al-Sn	Age-hardenable
9xx.x.	Not currently used	

[a] *The first digit shows the main alloying element, the second digit shows modification, and the last two digits shows the decimal % of the Al concentration (e.g., 1060: will be 99.6% Al alloy).*
[b] *Last digit indicates product form, 1 or 2 is ingot (depends upon purity) and 0 is for casting.*

ments related to aluminum include the development of aluminum alloys containing higher Mg concentrations for use in making automobiles. There is also interest in developing processes that will transform molten Al into sheet or other solid products.

Designation Aluminum alloys can be divided into two major groups: wrought and casting alloys, depending on their method of fabrication. **Wrought alloys**, which are shaped by plastic deformation, have compositions and microstructures significantly different from casting alloys, reflecting the different requirements of the manufacturing process. Within each major group we can divide the alloys into two subgroups: heat-treatable and nonheat-treatable alloys.

Aluminum alloys are designated by the numbering system shown in Table 13-3. The first number specifies the principle alloying elements, and the remaining numbers refer to the specific composition of the alloy.

The degree of strengthening is given by the **temper designation** T or H, depending on whether the alloy is heat-treated or strain-hardened (Table 13-4). Other designations indicate whether the alloy is annealed (O), solution-treated (W), or used in the as-fabricated condition (F). The numbers following the T or H indicate the amount of strain hardening, the exact type of heat treatment, or other special aspects of the processing of the alloy. Typical alloys and their properties are included in Table 13-5.

Wrought Alloys The 1xxx, 3xxx, 5xxx, and most of the 4xxx wrought alloys are not age-hardenable. The 1xxx and 3xxx alloys are single-phase alloys except for the presence of small amounts of inclusions or intermetallic compounds (Figure 13-2). Their

---

TABLE 13-4 ■ *Temper designations for aluminum alloys*

F   As-fabricated (hot-worked, forged, cast, etc.)
O   Annealed (in the softest possible condition)
H   Cold-worked
    H1x—cold-worked only. (x refers to the amount of cold work and strengthening.)
      H12—cold work that gives a tensile strength midway between the O and H14 tempers.
      H14—cold work that gives a tensile strength midway between the O and H18 tempers.
      H16—cold work that gives a tensile strength midway between the H14 and H18 tempers.
      H18—cold work that gives about 75% reduction.
      H19—cold work that gives a tensile strength greater than 2000 psi of that obtained by the H18 temper.
    H2x—cold-worked and partly annealed.
    H3x—cold-worked and stabilized at a low temperature to prevent age hardening of the structure.
W   Solution-treated
T   Age-hardened
    T1—cooled from the fabrication temperature and naturally aged.
    T2—cooled from the fabrication temperature, cold-worked, and naturally aged.
    T3—solution-treated, cold-worked, and naturally aged.
    T4—solution-treated and naturally aged.
    T5—cooled from the fabrication temperature and artificially aged.
    T6—solution-treated and artificially aged.
    T7—solution-treated and stabilized by overaging.
    T8—solution-treated, cold-worked, and artificially aged.
    T9—solution-treated, artificially aged, and cold-worked.
    T10—cooled from the fabrication temperature, cold-worked, and artificially aged.

TABLE 13-5 ■ *Properties of typical aluminum alloys*

Alloy		Tensile Strength (psi)	Yield Strength (psi)	% Elongation	Applications
Non heat-treatable wrought alloys:					
1100-O	>99% Al	13,000	5,000	40	Electrical components, foil,
1100-H18		24,000	22,000	10	food processing,
3004-O	1.2% Mn-1.0% Mg	26,000	10,000	25	beverage can bodies,
3004-H18		41,000	36,000	9	architectural uses,
4043-0	5.2% Si	21,000	10,000	22	filler metal for welding,
4043-H18		41,000	39,000	1	beverage can tops, and
5182-O	4.5% Mg	42,000	19,000	25	marine components
5182-H19		61,000	57,000	4	
Heat-treatable wrought alloys:					
2024-T4	4.4% Cu	68,000	47,000	20	Truck wheels, aircraft skins,
2090-T6	2.4% Li-2.7% Cu	80,000	75,000	6	pistons, canoes, railroad
4032-T6	12% Si-1% Mg	55,000	46,000	9	cars, and aircraft frames
6061-T6	1% Mg-0.6% Si	45,000	40,000	15	
7075-T6	5.6% Zn-2.5% Mg	83,000	73,000	11	
Casting alloys:					
201-T6	4.5% Cu	70,000	63,000	7	Transmission housings,
319-F	6% Si-3.5% Cu	27,000	18,000	2	general purpose castings,
356-T6	7% Si-0.3% Mg	33,000	24,000	3	aircraft fittings, motor
380-F	8.5% Si-3.5% Cu	46,000	23,000	3	housings, automotive
390-F	17% Si-4.5% Cu	41,000	35,000	1	engines, food-handling
443-F	5.2% Si (sand cast)	19,000	8,000	8	equipment, and marine
	(permanent mold)	23,000	9,000	10	fittings
	(die cast)	33,000	16,000	9	

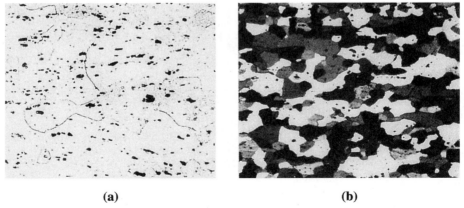

       (a)   **(b)**

Figure 13-2   (a) FeAl$_3$ inclusions in annealed 1100 aluminum (×350). (b) Mg$_2$Si precipitates in annealed 5457 aluminum alloy (×75). (*From* ASM Handbook, *Vol. 7, (1972), ASM International, Materials Park, OH 44073.*)

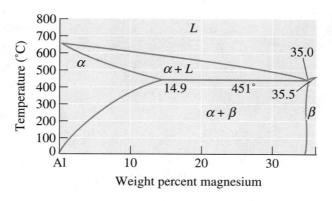

Figure 13-3
Portion of the aluminum-magnesium phase diagram.

properties are controlled by strain hardening, solid-solution strengthening, and grain-size control. However, because the solubilities of the alloying elements in aluminum are small at room temperature, the degree of solid-solution strengthening is limited. The 5xxx alloys contain two phases at room temperature—α, a solid solution of magnesium in aluminum, and Mg_2Al_3, a hard, brittle intermetallic compound (Figure 13-3). The aluminum-magnesium alloys are strengthened by a fine dispersion of Mg_2Al_3, as well as by strain hardening, solid-solution strengthening, and grain-size control.[5] However, because Mg_2Al_3 is not coherent, age-hardening treatments are not possible. The 4xxx series alloys also contain two phases, α and nearly pure silicon, β (Chapter 10). Alloys that contain both silicon and magnesium can be age hardened by permitting Mg_2Si to precipitate. The 2xxx, 6xxx, and 7xxx alloys are age-hardenable alloys. Although excellent specific strengths are obtained for these alloys, the amount of precipitate that can form is limited. In addition, they cannot be used at temperatures above approximately 175°C in the aged condition. Alloy 2024 is the most widely used aircraft alloy.[1] There is also an interest in the development of precipitation-hardened Al-Li alloys due to their high Young's modulus and low density. However, high-processing costs, anisotropic properties, and lower fracture toughness have proved to be limiting factors. Al-Li alloys are used to make space shuttle fuel tanks.

Casting Alloys   Many of the common aluminum casting alloys shown in Table 13-5 contain enough silicon to cause the eutectic reaction, giving the alloys low melting points, good fluidity, and good castability. **Fluidity** is the ability of the liquid metal to flow through a mold without prematurely solidifying, and **castability** refers to the ease with which a good casting can be made from the alloy.

The properties of the aluminum-silicon alloys are controlled by solid-solution strengthening of the α aluminum matrix, dispersion strengthening by the β phase, and solidification, which controls the primary grain size and shape as well as the nature of the eutectic microconstituent. Fast cooling obtained in die casting or permanent-mold casting (Chapter 8) increases strength by refining grain size and the eutectic microconstituent (Figure 13-4). Grain refinement using boron and titanium additions, modification using sodium or strontium to change the eutectic structure, and hardening with phosphorus to refine the primary silicon (Chapter 8) are all done in certain alloys to improve the microstructure and, thus, the degree of dispersion strengthening. Many alloys also contain copper, magnesium, or zinc, thus permitting age hardening. The following examples illustrate applications of aluminum alloys.

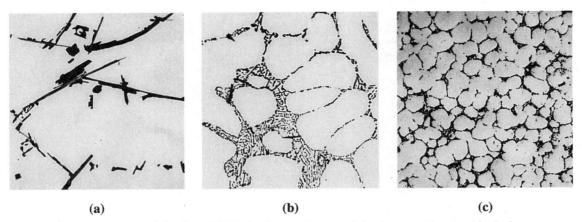

(a)   **(b)**   **(c)**

Figure 13-4 (a) Sand-cast 443 aluminum alloy containing coarse silicon and inclusions. (b) Permanent-mold 443 alloy containing fine dendrite cells and fine silicon due to faster cooling. (c) Die-cast 443 alloy with a still finer microstructure (×350). (*From* ASM Handbook, *Vol. 7, (1972), ASM International, Materials Park, OH 44073.*)

EXAMPLE 13-1   *Strength-to-Weight Ratio in Design*

A steel cable 0.5 in. in diameter has a yield strength of 70,000 psi. The density of steel is about 7.87 g/cm^3. Based on the data in Table 13-5, determine (a) the maximum load that the steel cable can support, (b) the diameter of a cold-worked aluminum-manganese alloy (3004-H 18) required to support the same load as the steel, and (c) the weight per foot of the steel cable versus the aluminum alloy cable.

SOLUTION

a. Load $= F = (\sigma_y \times A) = 70,000 \left(\dfrac{\pi}{4}\right)(0.5 \text{ in.})^2 = 13,744 \text{ lb}$

b. The yield strength of the aluminum alloy is 36,000 psi. Thus:

$$A = \frac{\pi}{4}d^2 = \frac{F}{\sigma_y} = \frac{13,744}{36,000} = 0.38 \text{ in.}^2$$

$$\therefore d = 0.697 \text{ in.}$$

Density of steel $= \rho = 7.87 \text{ g/cm}^3 = 0.284 \text{ lb/in.}^3$

Density of aluminum $= \rho = 2.70 \text{ g/cm}^3 = 0.097 \text{ lb/in.}^3$

c. Weight of steel $= Al\rho = \dfrac{\pi}{4}(0.5 \text{ in.})^2(12)(0.284) = 0.669 \text{ lb/ft}$

Weight of aluminum $= Al\rho = \dfrac{\pi}{4}(0.697)^2(12)(0.097) = 0.444 \text{ lb/ft}$

Although the yield strength of the aluminum is lower than that of the steel and the cable must be larger in diameter, the aluminum cable weighs only about half as much as the steel cable. When comparing materials, a proper factor-of-safety should also be included during design.

EXAMPLE 13-2 *Design of an Aluminum Recycling Process*

Design a method for recycling aluminum alloys used for beverage cans.

SOLUTION

Recycling aluminum is advantageous because only a fraction (about 5%) of the energy required to produce aluminum from Al_2O_3 is required.[1] However, recycling beverage cans does present several difficulties.

The beverage cans are made of two aluminum alloys (3004 for the main body, and 5182 for the lids) having different compositions (Table 13-5).[6] The 3004 alloy has the exceptional formability needed to perform the deep-drawing process; the 5182 alloy is harder and permits the pull-tops to function properly. When the cans are remelted, the resulting alloy contains both Mg and Mn and is not suitable for either application.

One approach to recycling the cans is to separate the two alloys from the cans. The cans are shredded, then heated to remove the lacquer that helps protect the cans during use. We could then further shred the material at a temperature where the 5182 alloy begins to melt. The 5182 alloy has a wider freezing range than the 3004 alloy and breaks into very small pieces; the more ductile 3004 alloy remains in larger pieces. The small pieces of 5182 can therefore be separated by passing the material through a screen. The two separated alloys can then be melted, cast, and rolled into new can stock.

An alternative method would be to simply remelt the cans. Once the cans have been remelted, we could bubble chlorine gas through the liquid alloy. The chlorine reacts selectively with the magnesium, removing it as a chloride. The remaining liquid can then be adjusted to the proper composition and be recycled as 3004 alloy.

EXAMPLE 13-3 *Design/Materials Selection for a Cryogenic Tank*

Design the material to be used to contain liquid hydrogen fuel for the space shuttle.

SOLUTION

Liquid hydrogen is stored below −253°C; therefore, our tank must have good cryogenic properties. The tank is subjected to high stresses, particularly when the plane is inserted into orbit, and it should have good fracture toughness to minimize the chances of catastrophic failure. Finally, it should be light in weight to permit higher payloads or less fuel consumption.

Lightweight aluminum would appear to be a good choice. Aluminum does not show a ductile to brittle transition. Because of its good ductility, we expect aluminum to also have good fracture toughness, particularly when the alloy is in the annealed condition.

One of the most common cryogenic aluminum alloys is 5083-O. Aluminum-lithium alloys are also being considered for low-temperature applications to take advantage of their even lower density.

EXAMPLE 13-4   *Design of a Casting Process for Wheels*

Design a casting process to produce automotive wheels having reduced weight and consistent and uniform properties.

SOLUTION

Many automotive wheels are produced by permanent mold casting of 356 aluminum alloy. In permanent mold casting, liquid aluminum is introduced into a heated cast iron mold and solidifies. Risers may be needed to assure that shrinkage voids are not created. This need may require that the wheel be designed to assure good casting characteristics, rather than to minimize weight. The casting may also cool at very different rates, producing differences in microstructure, such as secondary dendrite arm spacing, and properties throughout the wheel.

An alternative process might be to use the **thixocasting process** in which the material is stirred during solidification, producing a partly liquid, partly solid structure that behaves as a solid when no external force is applied, yet flows as a liquid under pressure. We would select an alloy with a wide-freezing range so that a significant portion of the solidification process occurs by the growth of dendrites. A hypoeutectic aluminum-silicon alloy might be appropriate. In the thixocasting process, the dendrites are broken up by stirring during solidification. The billet is later reheated to cause melting of just the eutectic portion of the alloy, and it is then forced into the mold in its semi-solid condition at a temperature below the liquidus temperature. When the alloy again solidifies, the primary aluminum phase will be uniform, round grains (rather than dendrites) surrounded by a continuous matrix of eutectic. Because approximately half of the alloy is already solid at the time of injection, the total amount of shrinkage is small, reducing the possibility of internal defects. This result also reduces the requirement for risers, which in turn provides more freedom in designing the wheel for its eventual function rather than for its ease of manufacture.

13-2   Magnesium and Beryllium Alloys

Magnesium, which is often extracted electrolytically from concentrated magnesium chloride in seawater, is lighter than aluminum with a density of 1.74 g/cm^3, and it melts at a slightly lower temperature than aluminum.[3] In many environments, the corrosion resistance of magnesium approaches that of aluminum; however, exposure to salts, such as that near a marine environment, causes rapid deterioration. Although magnesium alloys are not as strong as aluminum alloys, their specific strengths are comparable. Consequently, magnesium alloys are used in aerospace applications, high-speed machinery, and transportation and materials handling equipment.

Magnesium, however, has a low modulus of elasticity (6.5×10^6 psi) and poor resistance to fatigue, creep, and wear. Magnesium also poses a hazard during casting and machining, since it combines easily with oxygen and burns. Finally, the response of magnesium to strengthening mechanisms is relatively poor.

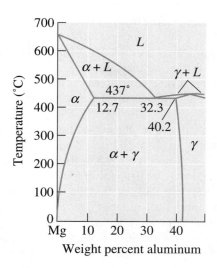

Figure 13-5
The magnesium-aluminum phase diagram.

Structure and Properties   Magnesium, which has the HCP structure, is less ductile than aluminum. However, the alloys do have some ductility because alloying increases the number of active slip planes. Some deformation and strain hardening can be accomplished at room temperature, and the alloys can be readily deformed at elevated temperatures. Strain hardening produces a relatively small effect in pure magnesium because of the low strain-hardening coefficient (Chapters 6 and 7).

As in aluminum alloys, the solubility of alloying elements in magnesium at room temperature is limited, causing only a small degree of solid-solution strengthening. However the solubility of many alloying elements increases with temperature, as shown in the Mg-Al phase diagram (Figure 13-5). Therefore, alloys may be strengthened by either dispersion strengthening or age hardening. Some age-hardened magnesium alloys, such as those containing Zr, Th, Ag, or Ce, have good resistance to overaging at temperatures as high as 300°C. Alloys containing up to 9% Li have exceptionally light weight. Properties of typical magnesium alloys are listed in Table 13-6.

TABLE 13-6 ■ *Properties of typical magnesium alloys*

Alloy	Composition	Tensile Strength (psi)	Yield Strength (psi)	% Elongation
Pure Mg:				
Annealed		23,000	13,000	3–15
Cold-worked		26,000	17,000	2–10
Casting alloys:				
AM 100-T6	10% Al-0.1% Mn	40,000	22,000	1
AZ81A-T4	7.6% Al-0.7% Zn	40,000	12,000	15
ZK61A-T6	6% Zn-0.7% Zr	45,000	28,000	10
Wrought alloys:				
AZ80A-T5	8.5% Al-0.5% Zn	55,000	40,000	7
ZK40A-T5	4% Zn-0.45% Zr	40,000	37,000	4
HK31A-H24	3% Th-0.6% Zr	38,000	30,000	8

Advanced magnesium alloys include those with very low levels of impurities and those containing large amounts (>5%) of cerium and other rare earths. These alloys form a protective MgO film that improves corrosion resistance. Rapid solidification processing permits larger amounts of alloying elements to be dissolved in the magnesium, further improving corrosion resistance. Improvements in strength, particularly at high temperatures, can be obtained by introducing ceramic particles or fibers such as silicon carbide into the metal.

Beryllium is lighter than aluminum, with a density of 1.848 g/cm^3, yet it is stiffer than steel, with a modulus of elasticity of 42×10^6 psi. Beryllium alloys, which have yield strengths of 30,000 to 50,000 psi, have high specific strengths and maintain both strength and stiffness to high temperatures. Instrument grade beryllium is used in inertial guidance systems where the elastic deformation must be minimal; structural grades are used in aerospace applications; and nuclear applications take advantage of the transparency of beryllium to electromagnetic radiation. Unfortunately, beryllium is expensive (Table 13-1), brittle, reactive, and toxic. Its production is quite complicated and hence the applications of Be alloys are very limited. Beryllium oxide (BeO) which is also toxic *in a powder form*, is used to make high-thermal conductivity ceramics.

13-3   Copper Alloys

Copper occurs in nature as elemental copper and was extracted successfully from rock long before iron, since the relatively lower temperatures required for the extraction could be achieved more easily.[7] Copper is typically produced by a pyrometallurgical (high-temperature) process. The copper ore containing high-sulfur contents is concentrated, then converted into a molten immiscible liquid containing copper sulfide-iron sulfide and is known as a copper matte. This is done in a flash smelter.[8] In a separate reactor, known as a copper converter, oxygen introduced to the matte converts the iron sulfide to iron oxide and the copper sulfide to an impure copper called **blister copper**, which is then purified electrolytically. Other methods for copper extraction include leaching copper from low-sulfur ores with a weak acid, then electrolytically extracting the copper from the solution.[3]

Copper-based alloys have higher densities than that for steels. Although the yield strength of some alloys is high, their specific strength is typically less than that of aluminum or magnesium alloys. These alloys have better resistance to fatigue, creep, and wear than the lightweight-aluminum and magnesium alloys. Many of these alloys have excellent ductility, corrosion resistance, electrical and thermal conductivity, and most can easily be joined or fabricated into useful shapes. Applications for copper-based alloys include electrical components (such as wire), pumps, valves, and plumbing parts, where these properties are used to advantage.

Copper alloys are also unusual in that they may be selected to produce an appropriate decorative color. Pure copper is red; zinc additions produce a yellow color, and nickel produces a silver color. Copper can corrode easily; forming a basic copper sulfate ($CuSO_4 \cdot 3Cu(OH)_2$). This is a green compound that is insoluble in water (but soluble in acids). This green patina provides an attractive finish for many applications. The Statue of Liberty is green because of the green patina of the oxidized copper skin that covers the steel structure.

The wide variety of copper-based alloys take advantage of all of the strengthening mechanisms that we have discussed. The effects of these strengthening mechanisms on the mechanical properties are summarized in Table 13-7.

TABLE 13-7 ■ *Properties of typical copper alloys obtained by different strengthening mechanisms*

Material	Tensile Strength (psi)	Yield Strength (psi)	% Elongation	Strengthening Mechanism
Pure Cu, annealed	30,300	4,800	60	None
Commercially pure Cu, annealed to coarse grain size	32,000	10,000	55	Solid solution
Commercially pure Cu, annealed to fine grain size	34,000	11,000	55	Grain size
Commercially pure Cu, cold-worked 70%	57,000	53,000	4	Strain hardening
Annealed Cu-35% Zn	47,000	15,000	62	Solid solution
Annealed Cu-10% Sn	66,000	28,000	68	Solid solution
Cold-worked Cu-35% Zn	98,000	63,000	3	Solid solution + strain hardening
Age-hardened Cu-2% Be	190,000	175,000	4	Age hardening
Quenched and tempered Cu-Al	110,000	60,000	5	Martensitic reaction
Cast manganese bronze	71,000	28,000	30	Eutectoid reaction

Copper containing less than 1% impurities is used for electrical and micro-electronics applications. Small amounts of cadmium, silver, and Al_2O_3 improve their hardness without significantly impairing conductivity. The single-phase copper alloys are strengthened by cold working. Examples of this effect are shown in Table 13-7. The FCC copper has excellent ductility and a high strain-hardening coefficient.

Solid-Solution-Strengthened Alloys A number of copper-based alloys contain large quantities of alloying elements, yet remain single phase. Important binary phase diagrams are shown in Figure 13-6. The copper-zinc, or **brass**, alloys with less than 40% Zn form single-phase solid solutions of zinc in copper. The mechanical properties—even elongation—increase as the zinc content increases. These alloys can be cold formed into rather complicated yet corrosion-resistant components. **Bronzes** are generally considered alloys of copper containing tin and can certainly contain other elements. Manganese bronze is a particularly high-strength alloy containing manganese as well as zinc for solid-solution strengthening.

Tin bronzes, often called phosphor bronzes, may contain up to 10% Sn and remain single phase. The phase diagram predicts that the alloy will contain the Cu_3Sn (ε) compound. However, the kinetics of the reaction are so slow that the precipitate may not form.

Alloys containing less than about 9% Al or less than 3% Si are also single phase. These aluminum bronzes and silicon bronzes have good forming characteristics and are often selected for their good strength and excellent toughness.

Age-Hardenable Alloys A number of copper-base alloys display an age-hardening response, including zirconium-copper, chromium-copper, and beryllium-copper. The copper-beryllium alloys are used for their high strength, their high stiffness (making them useful as springs and fine wires), and their nonsparking qualities (making them useful for tools to be used near flammable gases and liquids).

Phase Transformations Aluminum bronzes that contain over 9% Al can form β phase on heating above 565°C, the eutectoid temperature [Figure 13-6(c)]. On subsequent

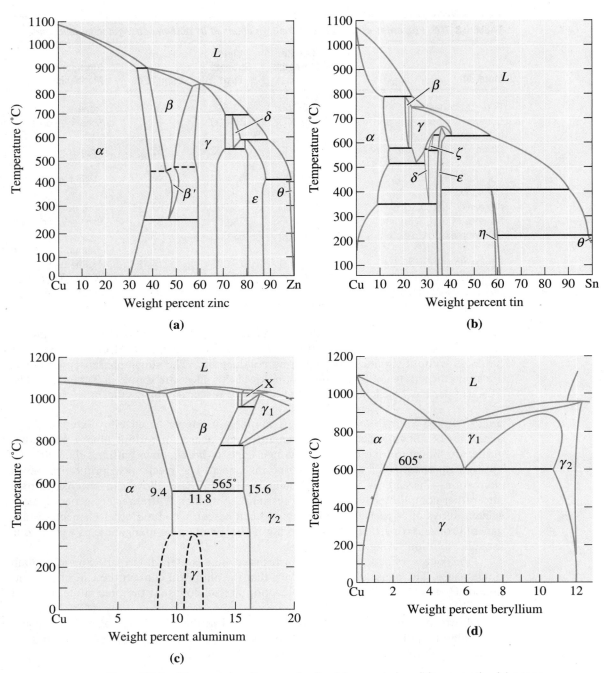

Figure 13-6   Binary phase diagrams for the (a) copper-zinc, (b) copper-tin, (c) copper-aluminum, and (d) copper-beryllium systems.

cooling, the eutectoid reaction produces a lamellar structure (like pearlite) that contains a brittle γ_2 compound. The low-temperature peritectoid reaction, $\alpha + \gamma_2 \rightarrow \gamma$, normally does not occur. The eutectoid product is relatively weak and brittle, but we can rapidly quench the β to produce martensite, or β', which has high strength and low ductility. When β' is subsequently tempered, a combination of high strength, good ductility, and excellent toughness is obtained as fine platelets of α precipitate from the β'.

Leaded-Copper Alloys   Virtually any of the wrought alloys may contain up to 4.5% Pb. The lead forms a monotectic reaction with copper and produces tiny lead spheres as the last liquid to solidify. The lead improves machining characteristics. Use of leaded-copper alloys, however, has a major environmental impact and, consequently, new alloys that are lead free have been developed. The following two examples illustrate the use of copper-based alloys.

EXAMPLE 13-5   *Design/Materials Selection for an Electrical Switch*

Design the contacts for a switch or relay that opens and closes a high-current electrical circuit.

SOLUTION

When the switch or relay opens and closes, contact between the conductive surfaces can cause wear and result in poor contact and arcing. A high hardness would minimize wear, but the contact materials must allow the high current to pass through the connection without overheating or arcing.

Therefore, our design must provide for both good electrical conductivity and good wear resistance. A relatively pure copper alloy dispersion strengthened with a hard phase that does not disturb the copper lattice would, perhaps, be ideal. In a $Cu-Al_2O_3$ alloy, the hard ceramic-oxide particles provide wear resistance but do not interfere with the electrical conductivity of the copper matrix.

EXAMPLE 13-6   *Design of a Heat Treatment for a Cu-Al Alloy Gear*

Design the heat treatment required to produce a high-strength aluminum-bronze gear containing 10% Al.

SOLUTION

The aluminum bronze can be strengthened by a quench and temper heat treatment. We must heat above 900°C to obtain 100% β for a Cu-10% Al alloy [Figure 13-6(c)]. The eutectoid temperature for the alloy is 565°C. Therefore, our recommended heat treatment is:

1. Heat the alloy to 950°C and hold to produce 100% β.
2. Quench the alloy to room temperature to cause β to transform to martensite, β', which is supersaturated in copper.
3. Temper below 565°C; a temperature of 400°C might be suitable. During tempering, the martensite transforms to α and γ_2. The amount of the γ_2 that forms at 400°C is:

$$\% \, \gamma_2 = \frac{10 - 9.4}{15.6 - 9.4} \times 100 = 9.7\%$$

4. Cool rapidly to room temperature so that the equilibrium γ does not form.

Note that if tempering were carried out below about 370°C, γ would form rather than γ_2.

13-4   Nickel and Cobalt Alloys

Nickel and cobalt alloys are used for corrosion protection and for high-temperature resistance, taking advantage of their high melting points and high strengths. Nickel is FCC and has good formability; cobalt is an allotropic metal, with an FCC structure above 417°C and an HCP structure at lower temperatures. Special cobalt alloys are used for exceptional wear resistance and, because of resistance to human body fluids, for prosthetic devices. Typical alloys and their applications are listed in Table 13-8.

TABLE 13-8 ■ *Compositions, properties, and applications for selected nickel and cobalt alloys*

Material	Tensile Strength (psi)	Yield Strength (psi)	% Elongation	Strengthening Mechanism	Applications
Pure Ni (99.9% Ni)	50,000	16,000	45	Annealed	Corrosion resistance
	95,000	90,000	4	Cold-worked	Corrosion resistance
Ni-Cu alloys:					
Monel 400 (Ni-31.5% Cu)	78,000	39,000	37	Annealed	Valves, pumps, heat exchangers
Monel K-500 (Ni-29.5% Cu-2.7% Al-0.6% Ti)	150,000	110,000	30	Aged	Shafts, springs, impellers
Ni superalloys:					
Inconel 600 (Ni-15.5% Cr-8% Fe)	90,000	29,000	49	Carbides	Heat-treatment equipment
Hastelloy B-2 (Ni-28% Mo)	130,000	60,000	61	Carbides	Corrosion resistance
DS-Ni (Ni-2% ThO$_2$)	71,000	48,000	14	Dispersion	Gas turbines
Fe-Ni superalloys:					
Incoloy 800 (Ni-46% Fe-21% Cr)	89,000	41,000	37	Carbides	Heat exchangers
Co superalloys:					
Stellite 6B (60% Co-30% Cr-4.5% W)	177,000	103,000	4	Carbides	Abrasive wear resistance

In Chapter 8, we saw how rapidly solidified powders of nickel- and cobalt-based superalloys can be formed using spray atomization followed by hot isostatic pressing. These materials are used to make the rings that retain turbine blades, as well as for turbine blades for aircraft engines.[9] In Chapter 11, we discussed applications of shape-memory alloys based on Ni-Ti.[10] Iron, nickel and cobalt are magnetic. Certain Fe-Ni- and Fe-Co-based alloys form very good magnetic materials (Chapter 19).[11,12] A Ni-36% Fe alloy (Invar) displays practically no expansion during heating; this effect is exploited in producing bimetallic composite materials. Cobalt is used in WC-Co cutting tools.

Nickel and Monel   Nickel and its alloys have excellent corrosion resistance and forming characteristics. When copper is added to nickel, the maximum strength is obtained near 60% Ni. A number of alloys, called Monels, with approximately this composition are used for their strength and corrosion resistance in salt water and at elevated temperatures. Some of the Monels contain small amounts of aluminum and titanium. These alloys show an age-hardening response by the precipitation of γ', a coherent Ni_3Al or Ni_3Ti precipitate which nearly doubles the tensile properties. The precipitates resist overaging at temperatures up to 425°C (Figure 13-7).

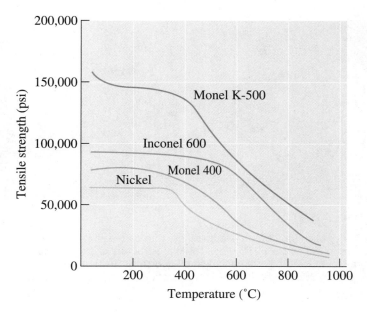

Figure 13-7
The effect of temperature
on the tensile strength of
several nickel-based
alloys.

Superalloys **Superalloys** are nickel, iron-nickel, and cobalt alloys that contain large amounts of alloying elements intended to produce a combination of high strength at elevated temperatures, resistance to creep at temperatures up to 1000°C, and resistance to corrosion. These excellent high-temperature properties are obtained even though the melting temperatures of the alloys are about the same as that for steels. Typical applications include vanes and blades for turbine and jet engines, heat exchangers, chemical reaction vessel components, and heat-treating equipment.

To obtain high strength and creep resistance, the alloying elements must produce a strong, stable microstructure at high temperatures. Solid-solution strengthening, dispersion strengthening, and precipitation hardening are generally employed.

Solid-Solution Strengthening Large additions of chromium, molybdenum, and tungsten and smaller additions of tantalum, zirconium, niobium, and boron provide solid-solution strengthening. The effects of solid-solution strengthening are stable and, consequently, make the alloy resistant to creep, particularly when large atoms such as molybdenum and tungsten (which diffuse slowly) are used.

Carbide-Dispersion Strengthening All alloys contain a small amount of carbon which, by combining with other alloying elements, produces a network of fine, stable carbide particles. The carbide network interferes with the dislocation movement and prevents grain boundary sliding. The carbides include TiC, BC, ZrC, TaC, Cr_7C_3, $Cr_{23}C_6$, Mo_6C, and W_6C, although often they are more complex and contain several alloying elements. Stellite 6B, a cobalt-based superalloy, has unusually good wear resistance at high temperatures due to these carbides.

Precipitation Hardening Many of the nickel and nickel-iron superalloys that contain aluminum and titanium form the coherent precipitate $\gamma'(Ni_3Al$ or $Ni_3Ti)$ during aging. The γ' particles (Figure 13-8) have a crystal structure and lattice parameter similar to that of the nickel matrix; this similarity leads to a low-surface energy and minimizes

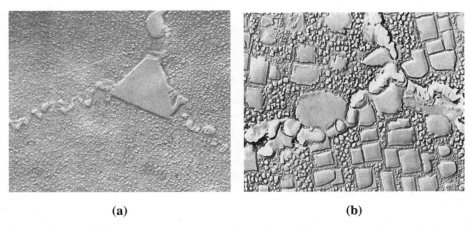

(a)   **(b)**

Figure 13-8   (a) Microstructure of a superalloy, with carbides at the grain boundaries and γ' precipitates in the matrix ($\times$15,000). (b) Microstructure of a superalloy aged at two temperatures, producing both large and small cubical γ' precipitates ($\times$10,000). (*ASM Handbook, Vol. 9, Metallography and Microstructure (1985), ASM International, Materials Park, OH 44073.*)

overaging of the alloys, providing good strength and creep resistance even at high temperatures.

By varying the aging temperature, precipitates of various sizes can be produced. Small precipitates, formed at low-aging temperatures, can grow between the larger precipitates produced at higher temperatures, therefore increasing the volume percentage of the γ' and further increasing the strength [Figure 13-8(b)].

The high-temperature use of the superalloys can be improved when a ceramic or coating is used. One method for doing this is to first coat the superalloy with a metallic bond coat composed of a complex NiCoCrAlY alloy, then apply an outer thermal barrier coating of a stabilized ZrO_2-ceramic (Figure 5-5). The coating helps reduce oxidation of the superalloy and permits jet engines to operate at higher temperatures and with greater efficiency. The next example shows the application of a nickel-based superalloy.

EXAMPLE 13-7   *Design/Materials Selection for a High-Performance Jet Engine Turbine Blade*

Design a nickel-based superalloy for producing turbine blades for a gas turbine aircraft engine that will have a particularly long creep-rupture time at temperatures approaching 1100°C.

SOLUTION

First, we need a very stable microstructure. Addition of aluminum or titanium permits the precipitation of up to 60 vol% of the γ' phase during heat treatment and may permit the alloy to operate at temperatures approaching 0.85 times the absolute melting temperature. Addition of carbon and alloying elements such as Ta and Hf permits the precipitation of alloy carbides that prevent grain boundaries from sliding at high temperatures. Other alloying elements, including molybdenum and tungsten, provide solid-solution strengthening.

Second, we might produce a directionally solidified or even single-crystal turbine blade (Chapter 8). In directional solidification, only columnar grains form during freezing, eliminating transverse grain boundaries that might nucleate cracks. In a single crystal, no grain boundaries are present. We might use the investment casting process, being sure to pass the liquid superalloy through a filter to trap any tiny inclusions before the metal enters the ceramic investment mold.

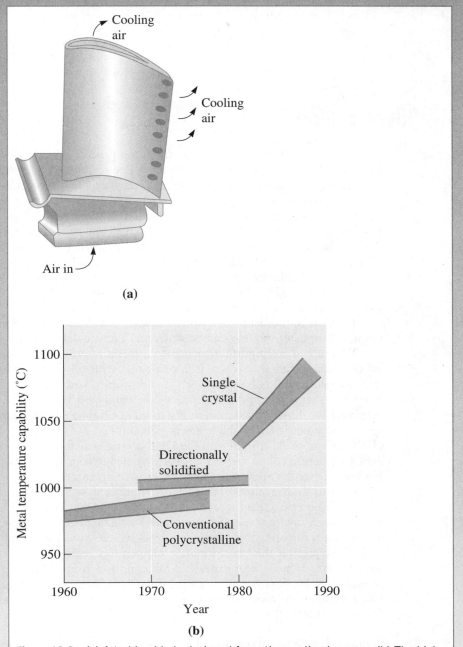

Figure 13-9 (a) A turbine blade designed for active cooling by a gas. (b) The high-temperature capability of superalloys has increased with improvements in manufacturing methods (for Example 13-7).

We would then heat treat the casting to assure that the carbides and γ' precipitate with the correct size and distribution. Multiple aging temperatures might be used to assure that the largest possible volume percent γ' is formed.

Finally, the blade might contain small cooling channels along its length. Air for combustion in the engine can pass through these channels, providing active cooling to the blade, before reacting with fuel in the combustion chamber. Figure 13-9 shows the improvements in performance that can be obtained using these design methods.

13-5 Titanium Alloys

Titanium is produced from TiO_2 by the Kroll process. The TiO_2 is converted to $TiCl_4$ (titanium tetra chloride, also informally known as *tickle*!), which is subsequently reduced to titanium metal by sodium or magnesium. The resultant titanium sponge is then consolidated, alloyed as necessary, and processed using vacuum arc melting. Recently, a new process for the production of titanium sponge directly from TiO_2 has been reported.[13] Titanium provides excellent corrosion resistance, high specific strength, and good high-temperature properties. Strengths up to 200,000 psi, coupled with a density of 4.505 g/cm^3, provide excellent mechanical properties. An adherent, protective TiO_2 film provides excellent resistance to corrosion and contamination below 535°C. Above 535°C, the oxide film breaks down and small atoms such as carbon, oxygen, nitrogen, and hydrogen embrittle the titanium.

Titanium's excellent corrosion resistance provides applications in chemical processing equipment, marine components, and biomedical implants such as hip prostheses. Titanium is an important aerospace material, finding applications as airframe and jet engine components. When it is combined with niobium, a superconductive intermetallic compound is formed; when it is combined with nickel, the resulting alloy displays the shape-memory effect; when it is combined with aluminum, a new class of intermetallic alloys is produced, as discussed in Chapter 10. Titanium alloys are used for sports equipment such as the heads of golf clubs.

Titanium is allotropic, with the HCP crystal structure (α) at low temperatures and a BCC structure (β) above 882°C. Alloying elements provide solid-solution strengthening and change the allotropic transformation temperature. The alloying elements can be divided into four groups (Figure 13-10). Additions such as tin and zirconium provide solid-solution strengthening without affecting the transformation temperature. Aluminum, oxygen, hydrogen, and other alpha-stabilizing elements increase the temperature at which α transforms to β. Beta stabilizers such as vanadium, tantalum, molybdenum, and niobium lower the transformation temperature, even causing β to be stable at room temperature. Finally, manganese, chromium, and iron produce a eutectoid reaction, reducing the temperature at which the α–β transformation occurs and producing a two-phase structure at room temperature. Several categories of titanium and its alloys are listed in Table 13-9.

Commercially Pure Titanium   Unalloyed titanium is used for its superior corrosion resistance. Impurities, such as oxygen, increase the strength of the titanium (Figure 13-11) but reduce corrosion resistance. Applications include heat exchangers, piping, reactors, pumps, and valves for the chemical and petrochemical industries.

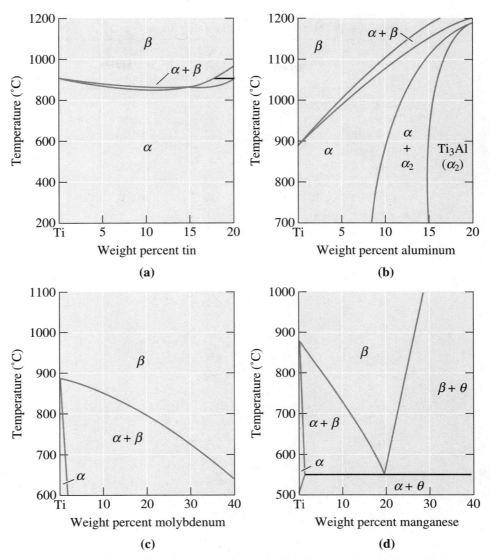

Figure 13-10 Portions of the phase diagrams for (a) titanium-tin, (b) titanium-aluminum, (c) titanium-molybdenum, and (d) titanium-manganese.

Alpha Titanium Alloys   The most common of the all-alpha alloys contains 5% Al and 2.5% Sn, which provide solid-solution strengthening to the HCP alpha. The alpha alloys are annealed at high temperatures in the β region. Rapid cooling gives an acicular, or Widmanstätten, α-grain structure (Figure 13-12) that provides good resistance to fatigue.[14] Furnace cooling gives a more platelike α structure that provides better creep resistance.

Beta Titanium Alloys   Although large additions of vanadium or molybdenum produce an entirely β structure at room temperature, none of the beta alloys are actually alloyed to that extent. Instead, they are rich in β stabilizers, so that rapid cooling produces a metastable structure composed of all β. Strengthening is obtained both from the large amount of solid-solution-strengthening alloying elements and by aging the metastable β

TABLE 13-9 ■ *Properties of selected titanium alloys*

Material	Tensile Strength (psi)	Yield Strength (psi)	% Elongation
Commercially pure Ti:			
99.5% Ti	35,000	25,000	24
99.0% Ti	80,000	70,000	15
Alpha Ti alloys:			
5% Al-2.5% Sn	125,000	113,000	15
Beta Ti alloys:			
13% V-11% Cr-3% Al	187,000	176,000	5
Alpha-beta Ti alloys:			
6% Al-4% V	150,000	140,000	8

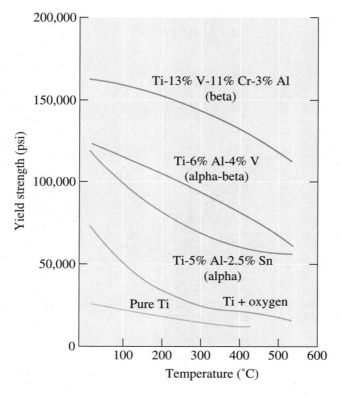

Figure 13-11
The effect of temperature on the yield strength of selected titanium alloys.

structure to permit α to precipitate. Applications include high-strength fasteners, beams, and other fittings for aerospace applications.

Alpha-Beta Titanium Alloys With proper balancing of the α and β stabilizers, a mixture of α and β is produced at room temperature. Ti-6% Al-4% V, an example of this approach, is by far the most common of all the titanium alloys. Because the alloys contain two phases, heat treatments can be used to control the microstructure and properties.

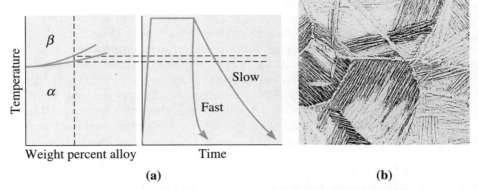

(a)   **(b)**

Figure 13-12   (a) Annealing and (b) microstructure of rapidly cooled alpha titanium (×100). Both the grain boundary precipitate and the Widmanstätten plates are alpha. (*From* ASM Handbook, *Vol. 7, (1972), ASM International, Materials Park, OH 44073.*)

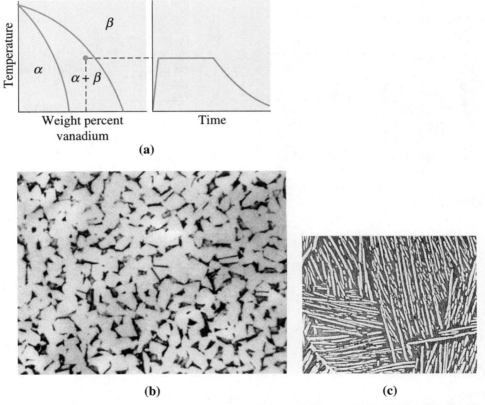

(a)

(b)   **(c)**

Figure 13-13   Annealing of an alpha-beta titanium alloy. (a) Annealing is done just below the α–β transformation temperature, (b) slow cooling gives equiaxed α grains (×250), and (c) rapid cooling yields acicular α grains (×2500). (*From* Metals Handbook, *Vol. 7, (1972), ASM International, Materials Park, OH 44073.*)

Annealing provides a combination of high ductility, uniform properties, and good strength. The alloy is heated just below the β-transus temperature, permitting a small amount of α to remain and prevent grain growth (Figure 13-13). Slow cooling causes equiaxed α grains to form; the equiaxed structure provides good ductility and formability while making it difficult for fatigue cracks to nucleate. Faster cooling, particularly from above the α-β transus temperature, produces an acicular—or "basketweave"—alpha phase (Figure 13-13). Although fatigue cracks may nucleate more easily in this structure, cracks must follow a tortuous path along the boundaries between α and β. This condition results in a low-fatigue crack growth rate, good fracture toughness, and good resistance to creep.

Two possible microstructures can be produced when the β phase is quenched from a high temperature. The phase diagram in Figure 13-14 includes a dashed martensite start line, which provides the basis for a quench and temper treatment. The β transforms to titanium martensite (α') in an alloy that crosses the M_s line on cooling. The titanium martensite is a relatively soft supersaturated phase. When α' is reheated, tempering occurs by the precipitation of β from the supersaturated α':

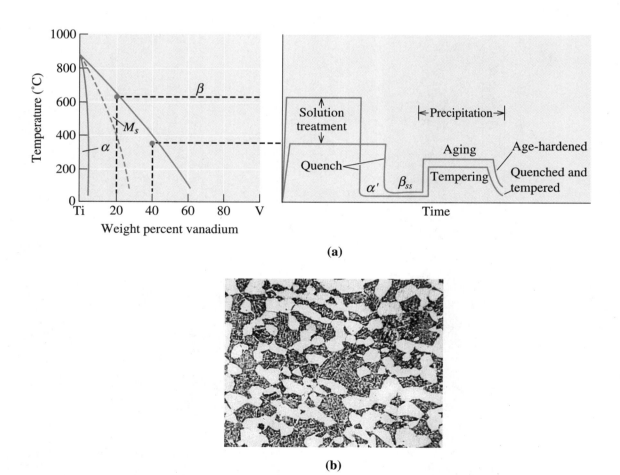

(a)

(b)

Figure 13-14 (a) Heat treatment and (b) microstructure of the alpha-beta titanium alloys. The structure contains primary α (large white grains) and a dark β matrix with needles of α formed during aging (×250). (*From* ASM Handbook, *Vol. 7, (1972), ASM International, Materials Park, OH 44073.*)

$$\alpha' \rightarrow \alpha + \beta \text{ precipitates}$$

Fine β precipitates initially increase the strength compared with the α', opposite to what is found when a steel martensite is tempered. However, softening occurs when tempering is done at too high a temperature.

More highly alloyed α-β compositions are age-hardened. When the β in these alloys is quenched, β_{ss}, which is supersaturated in titanium, remains. When β_{ss} is aged, α precipitates in a Widmanstätten structure, (Figure 13-14):

$$\beta_{ss} \rightarrow \beta + \alpha \text{ precipitates}$$

The formation of this structure leads to improved strength and fracture toughness. Components for airframes, rockets, jet engines, and landing gear are typical applications for the heat-treated alpha-beta alloys. Some alloys, including the Ti-6% Al-4% V alloy, are superplastic and can be deformed as much as 1000%. This alloy is also used for making implants for human bodies. (See the photograph at the beginning of this Chapter.) Titanium alloys are considered **biocompatible** (i.e., they are not rejected by the body). By developing porous coatings of bone-like ceramic compositions known as hydroxyapatite, it may be possible to make titanium implants **bioactive** (i.e., the natural bone can grow into the hydroxyapatite coating). The following three examples illustrate applications of titanium alloys.

EXAMPLE 13-8 *Design of a Heat Exchanger*

Design a 5-ft-diameter, 30-ft-long heat exchanger for the petrochemical industry (Figure 13-15).

Ti tubes

Figure 13-15
Sketch of a heat exchanger using titanium tubes (for Example 13-8).

SOLUTION

The heat exchanger must meet several design criteria. It must have good corrosion resistance to handle aggressive products of the chemical refinery; it must operate at relatively high temperatures; it must be easily formed into the sheet and tubes from which the heat exchanger will be fabricated; and it must have good weldability for joining the tubes to the body of the heat exchanger.

Provided that the maximum operating temperature is below 535°C so that the oxide film is stable, titanium might be a good choice to provide corrosion resistance at elevated temperatures. A commercially pure titanium provides the best corrosion resistance.

Pure titanium also provides superior forming and welding characteristics and would, therefore, be our most logical selection. If pure titanium does not provide sufficient strength, an alternative is an alpha titanium alloy, still providing good corrosion resistance, forming characteristics, and weldability but also somewhat improved strength.

EXAMPLE 13-9   *Design of a Connecting Rod*

Design a high-performance connecting rod for the engine of a racing automobile (Figure 13-16).

Figure 13-16
Sketch of connecting rod (for Example 13-9).

SOLUTION

A high-performance racing engine requires materials that can operate at high temperatures and stresses while minimizing the weight of the engine. In normal automobiles, the connecting rods are often a forged steel or a malleable cast iron. We might be able to save considerable weight by replacing these parts with titanium.

To achieve high strengths, we might consider an alpha-beta titanium alloy. Because of its availability, the Ti-6% Al-4% V alloy is a good choice. The alloy is heated to about 1065°C, which is in the all-β portion of the phase diagram. On quenching, a titanium martensite forms; subsequent tempering produces a microstructure containing β precipitates in an α matrix.

When the heat treatment is performed in the all-β region, the tempered martensite has an acicular structure, which reduces the rate of growth of any fatigue cracks that might develop.

EXAMPLE 13-10   *Materials for Hip Prosthesis*

What type of a material would you choose for an implant to be used for a total hip replacement implant?

SOLUTION

A hip prosthesis is intended to replace part of the worn out or damaged femur bone. The implant has a metal head and fits down the cavity of the femur. We need to consider the following factors: biocompatibility, corrosion resistance, high-fracture toughness, excellent fatigue life (so that implants last for many years since it is difficult to do the surgery as patients get older), and wear resistance. We also need to consider the stiffness. If the alloy chosen is too stiff compared to the bone, most of the stress will be carried by the implant. This leads to weakening of the remaining bone and, in turn, can make the implant loose. Thus, we need a material that has a high tensile strength, corrosion resistance, biocompatibility, and fracture toughness. These requirements suggest 316 stainless steel or Ti-6% Al-4% V. Neither of these materials are ferromagnetic and both are opaque to x-rays. This is good for magnetic resonance and x-ray imaging. Titanium alloys are not very hard and can wear out. Stainless steels are harder, but they are much stiffer than bone. Titanium is biocompatible and would be a better choice. Perhaps a composite material in which the stem is made from a Ti-6% Al-4% V alloy and a head that is made from a wear-resistant, corrosion resistant, and fractured tough ceramic, such as alumina, may be an answer. The inside of the socket could be made from an ultra-high-density (ultra-high molecular weight) polyethylene that has a very low-friction coefficient. The surface of the implant could be made porous so as to encourage the bone to grow. Another option is to coat the implant with a material like porous hydroxyapatite to encourage bone growth. A titanium alloy implant is shown as the chapter opening photo for this chapter.

13-6   Refractory and Precious Metals

The **refractory metals**, which include tungsten, molybdenum, tantalum, and niobium (or columbium), have exceptionally high-melting temperatures (above 1925°C) and, consequently, have the potential for high-temperature service. Applications include filaments for light bulbs, rocket nozzles, nuclear power generators, tantalum- and niobium-based electronic capacitors, and chemical processing equipment. The metals, however, have a high density, limiting their specific strengths (Table 13-10).

TABLE 13-10 ■ *Properties of some refractory metals*

| Metal | Melting Temperature (°C) | Density (g/cm³) | T = 1000°C | | Transition Temperature (°C) |
			Tensile Strength (psi)	Yield Strength (psi)	
Nb	2468	8.57	17,000	8,000	−140
Mo	2610	10.22	50,000	30,000	30
Ta	2996	16.6	27,000	24,000	−270
W	3410	19.25	66,000	15,000	300

Oxidation   The refractory metals begin to oxidize between 200°C and 425°C and are rapidly contaminated or embrittled. Consequently, special precautions are required during casting, hot working, welding, or powder metallurgy. The metals must also be protected during service at elevated temperatures. For example, the tungsten filament in a light bulb is protected by a vacuum.

For some applications, the metals may be coated with a silicide or aluminide coating. The coating must (a) have a high melting temperature, (b) be compatible with the refractory metal, (c) provide a diffusion barrier to prevent contaminants from reaching the underlying metal, and (d) have a coefficient of thermal expansion similar to that of the refractory metal. Coatings are available that protect the metal to about 1650°C. In some applications, such as capacitors for cellular phones, the formation of oxides is useful since we want to make use of the oxide as a nonconducting material (Chapter 18).

Forming Characteristics   The refractory metals, which have the BCC crystal structure, display a ductile-to-brittle transition temperature. Because the transition temperatures for niobium and tantalum are below room temperature, these two metals can readily be formed. However, annealed molybdenum and tungsten normally have a transition temperature above room temperature, causing them to be brittle. Fortunately, if these metals are hot worked to produce a fibrous microstructure, the transition temperature is lowered and the forming characteristics are improved.

Alloys   Large increases in both room-temperature and high-temperature mechanical properties are obtained by alloying. Tungsten alloyed with hafnium, rhenium, and carbon can operate up to 2100°C. These alloys typically are solid-solution strengthened; in fact, tungsten and molybdenum form a complete series of solid solutions, much like copper and nickel. Some alloys, such as W-2% ThO_2, are dispersion strengthened by oxide particles during their manufacture by powder metallurgy processes. Composite materials, such as niobium reinforced with tungsten fibers, may also improve high-temperature properties.

Precious Metals   These include gold, silver, palladium, platinum, and rhodium. As their name suggests, these are precious and expensive. From an engineering viewpoint, these materials resist corrosion and make very good conductors of electricity. As a result, alloys of these materials are often used as electrodes for devices. These electrodes are formed using a thin-film deposition (e.g., sputtering or electroplating) or screen printing of metal powder dispersions/pastes. Nano-sized particles of Pt/Rh/Pd (loaded onto a ceramic support) are also used as catalysts in automobiles. These metals facilitate the oxidation of CO to CO_2 and NO_x to N_2 and O_2. They are also used as catalysts in petroleum refining.

SUMMARY

◆ The "light metals" include low-density alloys based on aluminum, magnesium, and beryllium. Aluminum alloys have a high specific strength due to their low density and, as a result, find many aerospace applications. Excellent corrosion resistance and electrical conductivity of aluminum also provide for a vast number of applications. The most important and powerful strengthening mechanism in these alloys is age hardening. Aluminum and magnesium are limited to use at low temperatures because of the loss of their mechanical properties as a result of overaging or recrystallization. Copper alloys (brasses and bronzes) are also used in many structural and other applications. Beryllium, on the other hand, has an exceptional strength-to-weight ratio, maintains its strength at high temperatures, and is unusually stiff. It is toxic and expensive. Titanium alloys have intermediate densities and temperature resistance, along with excellent corrosion resistance, leading to applications in aerospace, chemical processing, and biomedical devices. These nonferrous alloys show a powerful response to strengthening by age hardening and, in some cases, quench and temper heat treatments.

◆ Nickel and cobalt alloys, including superalloys, provide good properties at even higher temperatures. Combined with their good corrosion resistance, these alloys find many applications in aircraft engines and chemical processing equipment. Strengthening even at high temperatures is usually obtained by age hardening, solid-solution strengthening, and dispersion strengthening due to alloy carbides.

◆ Refractory metals are able to operate at the highest temperatures, although they may have to be protected from oxidation by appropriate atmospheres or coatings. Precious metals are used as conductors in electronic devices, and as catalysts for automobiles and in petroleum refining.

GLOSSARY

Bioactive   A material that is not rejected by the human body, and eventually becomes part of the body (e.g., hydroxyapatite).

Biocompatible   A material that is not rejected by the human body.

Blister copper   An impure form of copper obtained during the copper refining process.

Brass   A group of copper-based alloys, normally containing zinc as the major alloying element.

Bronze   Generally, copper alloys containing tin, can contain other elements.

Castability   The ease with which a metal can be poured into a mold to make a casting without producing defects or requiring unusual or expensive techniques to prevent casting problems.

Fluidity   The ability of liquid metal to fill a mold cavity without prematurely freezing.

Hall-Heroult process   An electrolytic process by which aluminum is extracted from its ore.

Monel   The copper-nickel alloy, containing approximately 60% Ni, that gives the maximum strength in the binary alloy system.

Nonferrous alloy   An alloy based on some metal other than iron.

Refractory metals   Metals having a melting temperature above 1925°C.

Specific strength   The ratio of strength to density. Also called the strength-to-weight ratio.

Superalloys   A group of nickel, iron-nickel, and cobalt-based alloys that have exceptional heat resistance, creep resistance, and corrosion resistance.

Temper designation   A shorthand notation using letters and numbers to describe the processing of an alloy. H tempers refer to cold-worked alloys; T tempers refer to age-hardening treatments.

Thixocasting   A process by which a material is stirred during solidification, producing a partly liquid, partly solid structure that behaves as a solid when no external force is applied, yet flows as a liquid under pressure.

Wrought alloys   Alloys that are shaped by a deformation process.

ADDITIONAL INFORMATION

1. HARPER, C.A., *Handbook of Materials Product Design*. 3rd ed. 2001: McGraw Hill.

2. BINCZEWSKI, "The Point of a Monument: A History of the Aluminum Cap of the Washington Monument," *Journal of Metals*, 1995. 47(11).

3. EVANS, J.W. and L.C. DE JONGHE, *The Production of Inorganic Materials*. 1991: Prentice Hall.

4. GAYLE, F.W. and M. GOODWAY, "Precipitation Hardening in the First Aerospace Aluminum Alloy: The Wright Flyer Crankcase," *Science*, 1994. 266(5187): p. 1015–1017.

5. SAMPLE, V., Alcoa, *Technical discussions are acknowledged*. 2002.

6. HOSFORD, W.F. and J.L. DUNCAN, "The Aluminum Beverage Can," *Scientific American*, 1994. September: p. 48–53.

7. BRONOWSKI, J., *The Ascent of Man*. 1973: Back Bay Books. p. 124–131.

8. DAVENPORT, W.G., et al., *Flash Smelting: Analysis, Control and Optimization*. 2001: Warrendale, PA., The Minerals, Metals & Materials Society.

9. STAITE, J., B. HANN, and F. RIZZO, Crucible Compaction Metals, Pittsburgh, PA. *Technical assistance and discussions are acknowledged*. 2002.

10. OTSUKA, K. and T. KAKESHITA, "Science and Technology of Shape-Memory Alloys: New Developments," *MRS Bulletin*, 2002. 27(2).

11. Bozorth, R.M., *Ferromagnetism.* 1978: New York, IEEE Press.

12. Chen, C.W., *Magnetisn and Metallurgy of Soft Magnetic Materials.* 2nd ed. 1986: Dover Publication.

13. Chen, G.Z., D.J. Fray, and T.W. Farthing, *Nature*, 2000. 407: p. 361.

14. Weidmann, G., P. Lewis, and N. Reid, *Structural Materials (Materials in Action Series)*, ed. N. Reid. 1990: The Open University, Butterworth Heinemann.

✓ PROBLEMS

13-1 In some cases, we may be more interested in cost per unit volume than in cost per unit weight. Rework Table 13-1 to show the cost in terms of $/cm^3. Does this change/alter the relationship between the different materials?

Section 13-1 Aluminum Alloys

13-2 Assuming that the density remains unchanged, compare the specific strength of the 2090-T6 aluminum alloy to that of a die-cast 443-F aluminum alloy. If you considered the actual density, do you think the difference between the specific strengths would increase or become smaller? Explain.

13-3 Explain why aluminum alloys containing more than about 15% Mg are not used.

13-4 Explain how aluminum metal is made from alumina using the Hall-Heroult process.

13-5 What are some of the major applications of aluminum and its alloys?

13-6 Explain why aluminum cans are recycled effectively.

13-7 Would you expect a 2024-T9 aluminum alloy to be stronger or weaker than a 2024-T6 alloy? Explain.

13-8 Estimate the tensile strength expected for the following aluminum alloys:

(a) 1100-H14   (b) 5182-H12   (c) 3004-H16

13-9 Suppose, by rapid solidification from the liquid state, that a supersaturated Al-7% Li alloy can be produced and subsequently aged. Compare the amount of β that will form in this alloy with that formed in a 2090 alloy.

13-10 Determine the amount of Mg_2Al_3 (β) expected to form in a 5182-O aluminum alloy (See Figure 13-5).

13-11 Based on the phase diagrams, which of the following alloys would be most suited for thixocasting? Explain your answer. (See Figure 13-3 and phase diagrams from Chapters 10 and 11.)

(a) Al-12% Si   (b) Al-1% Cu   (c) Al-10% Mg

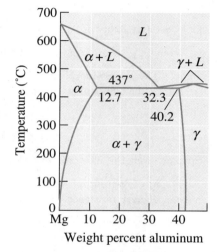

Figure 13-5 (Repeated for Problem 13-10) The magnesium-aluminum phase diagram.

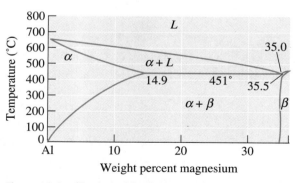

Figure 13-3 (Repeated for Problem 13-11) Portion of the aluminum-magnesium phase diagram.

Section 13-2 Magnesium and Beryllium Alloys

13-12 From the data in Table 13-6, estimate the ratio by which the yield strength of magnesium can be increased by alloying and heat treatment and compare with that of aluminum alloys.

13-13 Suppose a 24-in.-long round bar is to support a load of 400 lb without any permanent deformation. Calculate the minimum diameter of the bar if it is made of

(a) AZ80A-T5 magnesium alloy, and
(b) 6061-T6 aluminum alloy.

Calculate the weight of the bar and the approximate cost (based on pure Al and Mg) in each case.

13-14 A 10-m rod 0.5 cm in diameter must elongate no more than 2 mm under load. Determine the maximum force that can be applied if the rod is made of:

(a) aluminum   (b) magnesium   (c) beryllium

Section 13-3 Copper Alloys

13-15 (a) Explain how pure copper is made. (b) What are some of the important properties of copper? (c) What is brass? (d) What is bronze? (e) Why is Statue of Liberty green?

13-16 We say that copper can contain up to 40% Zn or 9% Al and still be single phase. How do we explain this statement in view of the phase diagrams in Figure 13-6?

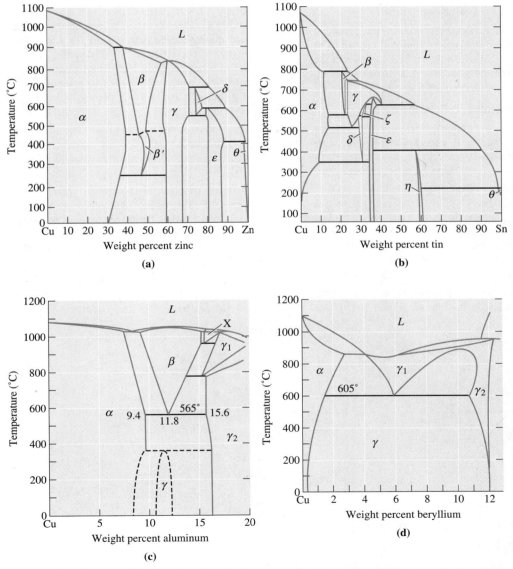

Figure 13-6 (Repeated for Problem 13-16) Binary phase diagrams for the (a) copper-zinc, (b) copper-tin, (c) copper-aluminum, and (d) copper-beryllium systems.

13-17 Compare the percentage increase in the yield strength of commercially pure annealed aluminum, magnesium, and copper by strain hardening. Explain the differences observed.

13-18 We would like to produce a quenched and tempered aluminum bronze containing 13% Al. Recommend a heat treatment, including appropriate temperatures. Calculate the amount of each phase after each step of the treatment.

13-19 A number of casting alloys have very high lead contents; however, the Pb content in wrought alloys is comparatively low. Why isn't more lead added to the wrought alloys? What precautions must be taken when a leaded wrought alloy is hot worked or heat treated?

13-20 Would you expect the fracture toughness of quenched and tempered aluminum bronze to be high or low? Would there be a difference in the resistance of the alloy to crack nucleation compared with crack growth? Explain.

Section 13-4 Nickel and Cobalt Alloys

13-21 Based on the photomicrograph in Figure 13-8(a), would you expect the γ' precipitate or the carbides to provide a greater strengthening effect in superalloys at low temperatures? Explain.

13-22 The density of Ni_3Al is 7.5 g/cm^3. Suppose a Ni-5 wt% Al alloy is heat treated so that all of the aluminum reacts with nickel to produce Ni_3Al. Determine the volume percentage of the Ni_3Al precipitate in the nickel matrix.

13-23 Figure 13-8(b) shows a nickel superalloy containing two sizes of γ' precipitates. Which precipitate likely formed first? Which precipitate

formed at the higher temperature? What does our ability to perform this treatment suggest concerning the effect of temperature on the solubility of Al and Ti in nickel? Explain.

Section 13-5 Titanium Alloys

13-24 When steel is joined using arc welding, only the liquid-fusion zone must be protected by a gas or flux. However, when titanium is welded, both the front and back sides of the welded metal must be protected. Why must these extra precautions be taken when joining titanium?

13-25 Both a Ti-15% V alloy and a Ti-35% V alloy are heated to a temperature at which all β just forms. They are then quenched and reheated to 300°C. Describe the changes in microstructure during the heat treatment for each alloy, including the amount of each phase. What is the matrix and what is the precipitate in each case? Which is an age-hardening process? Which is a quench and temper process? [See Figure 13-14(a)].

13-26 The θ phase in the Ti-Mn phase diagram has the formula MnTi. Calculate the amount of α and θ in the eutectoid microconstituent. [See Figure 13-10(d).]

13-27 How is titanium metal made? Explain the terms "biocompatible" and "bioactive" materials. What would be the advantages in using a Ti-6% Al-4% V alloy for hip prosthesis applications? What would be some of the difficulties?

13-28 Determine the specific strength of the strongest Al, Mg, Cu, Ti, and Ni alloys. Use the densities of the pure metals, in $lb/in.^3$ in your calculations. Try to explain their order.

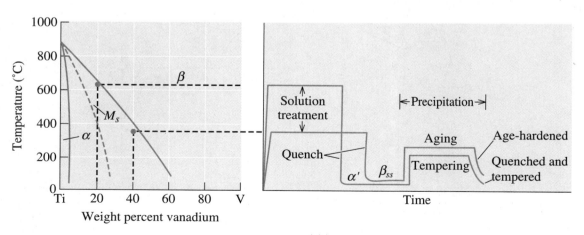

(a)

Figure 13-14 (Repeated for Problem 13-25) (a) Heat treatment.

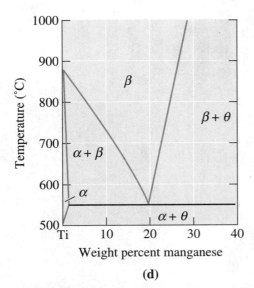

(d)

Figure 13-10 (Repeated for Problem 13-26)
(d) titanium-manganese.

13-29 Based on the phase diagrams, estimate the solubilities of Ni, Zn, Al, Sn, and Be in copper at room temperature. Are these solubilities expected in view of Hume-Rothery's conditions for solid solubility? Explain.

Section 13-6 Refractory and Precious Metals

13-30 What is a refractory metal or an alloy? What is a precious metal?

13-31 The temperature of a coated tungsten part is increased. What happens when the protective coating on a tungsten part expands more than the tungsten? What happens when the protective coating on a tungsten part expands less than the tungsten?

13-32 What are some of the applications of W?

13-33 For what applications are Pt, Rh, Pd, Ag used?

13-34 What metals are used as catalysts for catalytic converters in automobiles? What reactions do these metals catalyze?

✳ Design Problems

13-35 A part for an engine mount for a private aircraft must occupy a volume of 60 cm^3 with a minimum thickness of 0.5 cm and a minimum width of 4 cm. The load on the part during service may be as much as 75,000 N. The part is expected to remain below 100°C during service. Design a material and its treatment that will perform satisfactorily in this application.

13-36 You wish to design the rung on a ladder. The ladder should be light in weight so that it can be easily transported and used. The rungs on the ladder should be 0.25 in. × 1 in. and are 12-in. long. Design a material and its processing for the rungs.

13-37 We have determined that we need an alloy having a density of 2.3 ± 0.05 g/cm^3 that must be strong, yet still have some ductility. Design a material and its processing that might meet these requirements.

13-38 We wish to design a mounting device that will position and aim a laser for precision cutting of a composite material. What design requirements might be important? Design a material and its processing that might meet these requirements.

13-39 Design a nickel-titanium alloy that will produce 60 volume percent Ni$_3$Ti precipitate in a pure-nickel matrix.

13-40 An actuating lever in an electrical device must open and close almost instantly and carry a high current when closed. What design requirements would be important for this application? Design a material and its processing to meet these requirements.

13-41 A fan blade in a chemical plant must operate at temperatures as high as 400°C under rather corrosive conditions. Occasionally, solid material is ingested and impacts the fan. What design requirements would be important? Design a material and its processing for this application.

▭ Computer Problems

13-42 *Database Identification System for Alloys.* Write a computer program that will ask the user to input a three- or four-digit code for aluminum alloys (Table 13-3). You will have to ask the user to provide one digit at a time or figure out a way of comparing different digits in a string. This will be followed by a letter and a number (e.g., T and 4). The program should then provide the user with some more detailed information about the type of alloy. For example, if the user enters 2024 and then T4, the program should provide an output that will specify that **(a)** the alloy is wrought type, **(b)** Cu is the major alloying element (since the first digit is 2), and **(c)** it is naturally aged. Do not make the program too complex. The main idea here is for you to see how databases for alloys are designed.

Ceramics play a critical role in a wide array of technologies related to electronic, magnetic, optical, and energy. Many advanced ceramics play a very important role in providing thermal insulation and high-temperature properties. Applications of advanced ceramics range from credit cards, houses for silicon chips, tiles for the space shuttle, medical imaging, optical fibers that enable communication, and safe and energy efficient glasses. Traditional ceramics play a very important role as refractories for metals processing and consumer applications. (*Courtesy of NASA.*)

14

Ceramic Materials

Have You Ever Wondered?

- What is the magnetic strip on your credit card is made from?

- What material is used to protect the space shuttle from high temperatures during re-entry?

- What ceramic material is commonly added to paints?

- What ceramic material is found in bone and teeth?

- What are spark plugs made from?

The goal of this chapter is to examine more closely the synthesis, processing, and applications of ceramic materials. Ceramics have been used for many thousands of years.[1] Most ceramics exhibit good strength under compression; however, they exhibit virtually no ductility under tension. The family of ceramic materials includes polycrystalline and single-crystal inorganic materials, amorphous inorganic glasses, and glass-ceramics.[2] We have discussed many of these materials in previous chapters.

In Chapters 2 and 3, we learned about the bonding in ceramic materials, the crystal structures of technologically useful ceramics, the existence of ionic defects (Kroger-Vink notation for defect chemistry), and the arrangements of ions in glasses. In Chapter 4 and other chapters, we examined the process for making float glass. In Chapter 8, we examined how the principles of nucleation and growth are applied to prepare glass-ceramics. In Chapter 10, we examined phase diagrams for such ceramic systems as Al_2O_3-SiO_2.[3] We will now use these previously developed ideas to further explore the world of ceramics.

This chapter focuses on the synthesis, processing, and applications of ceramics, especially those of advanced or high-tech ceramics. We will also recapitulate processing and applications of inorganic glasses and glass-ceramics. We begin with a discussion that summarizes the classification and applications of ceramics.

14-1 | Applications of Ceramics

There are many different ways to classify ceramics. One way is to define ceramics based on their class of chemical compounds (e.g., oxides, carbides, nitrides, sulfides, fluorides, etc.). Another way, which we will use here, is to classify ceramics by their major function.

Ceramics are used in a wide range of technologies such as refractories, spark plugs, dielectrics in capacitors, sensors, abrasives, magnetic recording media, etc. The space shuttle makes use of ~25,000 reusable, lightweight, highly porous ceramic tiles that protect the aluminum frame from the heat generated during re-entry into the Earth's atmosphere. These tiles are made from high-purity silica fibers and colloidal silica coated with a borosilicate glass. Ceramics can also appear in nature as oxides and in natural materials; the human body has the amazing ability of making hydroxyapatite, a ceramic found in bones and teeth. Ceramics are also used as coatings. **Glazes** are ceramic coatings applied to glass objects; **enamels** are ceramic coatings applied to metallic objects. Let's follow the classification shown in Table 14-1 and take note of different applications. Alumina and silica are the most widely used ceramic materials and, as you will notice, there are numerous applications listed in Table 14-1 that depend upon the use of these two ceramics.

The following is a brief summary of applications of some of the more widely used ceramic materials:

- *Alumina* (Al_2O_3) is used to contain molten metal or in applications where a material must operate at high temperatures, but where high strength is also required. Alumina is also used as a low dielectric constant substrate for electronic packaging that houses silicon chips. One classical application is for insulators in spark plugs. Some unique applications are also being found in dental and medical use. Chromium-doped alumina is used for making lasers. Fine particles of alumina are also used as catalyst support.

- *Aluminum nitride* (AlN) provides good electrical insulation, but has a high thermal conductivity. Because its coefficient of thermal expansion is similar to that of silicon, AlN is a suitable replacement for Al_2O_3 as a substrate material for integrated circuits. Cracking is minimized and electrical insulation is obtained; yet heat generated by the electronic circuit can be quickly removed. It is also better suited than many competing materials for use in electrical circuits operating at a high frequency.

TABLE 14-1 ■ *Functional classification of ceramics*

Function	Application	Examples of Ceramics
Electrical	Capacitor dielectrics	$BaTiO_3$, $SrTiO_3$, Ta_2O_5
	Microwave dielectrics	$Ba(Mg_{1/3}Ta_{2/3})O_3$, $Ba(Zn_{1/3}Ta_{2/3})O_3$
		$BaTi_4O_9$, $Ba_2Ti_9O_{20}$, $Zr_xSn_{1-x}TiO_4$, Al_2O_3
	Conductive oxides	In-doped SnO_2 (*ITO*)
	Superconductors	$YBa_2Cu_3O_{7-x}$ (*YBCO*)
	Electronic packaging	Al_2O_3
	Insulators	Porcelain
	Solid-oxide fuel cells	ZrO_2, $LaCrO_3$
	Piezoelectric	$Pb(Zr_xTi_{1-x})O_3$(*PZT*), $Pb(Mg_{1/3}Nb_{2/3})O_3$
	Electro-optical	*PLZT*, $LiNbO_3$
Magnetic	Recording media	γ-Fe_2O_3, CrO_2 ("chrome" cassettes)
	Ferrofluids, credit cards	Fe_3O_4
	Circulators, isolators,	Nickel zinc ferrite
	Inductors, magnets	Manganese zinc ferrite
Optical	Fiber optics	Doped SiO_2
	Glasses	SiO_2 based
	Lasers	Al_2O_3, yttrium aluminum garnate (*YAG*)
	Lighting	Al_2O_3, glasses.
Automotive	Oxygen sensors, fuel cells	ZrO_2
	Catalyst support	Cordierite
	Spark plugs	Al_2O_3
	Tires	SiO_2
	Windshields/windows	SiO_2 based glasses
Mechanical/Structural	Cutting tools	WC-Co cermets
		SiAlON
		Al_2O_3
	Composites	SiC, Al_2O_3, silica glass fibers
	Abrasives	SiC, Al_2O_3, diamond, BN, $ZrSiO_4$
Biomedical	Implants	Hydroxyapatite
	Dentistry	Porcelain, Al_2O_3
	Ultrasound imaging	*PZT*
Construction	Buildings	Concrete
		Glass
		Sanitaryware
Others	Defense applications	*PZT*, B_4C
	Armor materials	
	Sensors	SnO_2
	Nuclear	UO_2
		Glasses for waste disposal
	Metals processing	Alumina and silica-based refractories, oxygen sensors, casting molds, etc.
Chemical	Catalysis	Various oxides (Al_2O_3, ZrO_2, ZnO, TiO_2)
	Air, liquid filtration	
	Sensors	
	Paints, rubber	
Domestic	Tiles, sanitaryware, Whiteware, kitchenware, Pottery, art, jewelry	Clay, alumina, and silica-based ceramics, glass-ceramics, diamond, ruby, cubic zirconia and other crystals

Acronyms are indicated in italics.

- *Barium titanate* ($BaTiO_3$) is the most widely used electronic ceramic. Several million capacitors are made using this material. It has a high dielectric constant that makes it possible to create smaller capacitors that can hold considerable amounts of charge (Chapter 18).

- *Boron carbide* (B_4C) is very hard, yet unusually lightweight. Boron carbide is the third hardest material known—after diamond and cubic boron nitride (*CBN*). In addition to its use as nuclear shielding, it finds uses in applications requiring excellent abrasion resistance and as a portion of bulletproof armor plate, although it has rather poor properties at high temperatures.

- *Cordierite* ($2MgO$-$2Al_2O_3$-$5SiO_2$) is useful as an electronic ceramic. It is also used to make a honeycomb structure used in the catalysts in a catalytic converter. This support contains a dispersion of nano-sized metal particles such as Pt, Rh, etc. that serve as catalysts.

- *Diamond* (C) is the hardest naturally occurring material. Industrial diamonds are used as abrasives for grinding and polishing. Diamond and diamond-like coatings prepared using chemical vapor deposition processes are used to make abrasion-resistant coatings for many different applications (e.g., cutting tools). It is, of course, also used in jewelry.

- *Lead zirconium titanate* (*PZT*) This is the most widely used piezoelectric material (Chapter 18). Applying pressure or stress leads to the development of a voltage in this material. In addition, the application of a voltage leads to the development of a strain in this material. As a result, *PZT* is used in many applications such as gas igniters, submarines for detection of underwater objects, and ultrasound imaging.

- *Silica* (SiO_2) is probably the most widely used ceramic material. Silica is an essential ingredient in glasses and many glass ceramics. Silica-based materials are used in thermal insulation, refractories, abrasives, as fiber-reinforced composites, laboratory glassware, etc. In the form of long continuous fibers, silica is used to make optical fibers for communications. Powders made using fine particles of silica are used in tires, paints, and many other applications.

- *Silicon carbide* (SiC) provides outstanding oxidation resistance at temperatures even above the melting point of steel. SiC often is used as a coating for metals, carbon-carbon composites, and other ceramics to provide protection at these extreme temperatures. SiC is also used as an abrasive in grinding wheels and as particulate and fibrous reinforcement in both metal matrix and ceramic matrix composites. It is also used to make heating elements for furnaces. SiC is a semiconductor and is a very good candidate for high-temperature electronics.

- *Silicon nitride* (Si_3N_4) has properties similar to those of SiC, although its oxidation resistance and high-temperature strength are somewhat lower. Both silicon nitride and silicon carbide are likely candidates for components for automotive and gas turbine engines, permitting higher operating temperatures and better fuel efficiencies with less weight than traditional metals and alloys.

- *Sialon* is an acronym that stands for *si*licon *al*uminum *oxyn*itride. It is formed when aluminum and oxygen are partially substituted for silicon and nitrogen in silicon nitride. The general form of the material is $Si_{6-z}Al_zO_zN_{8-z}$; when $z = 3$, the formula is $Si_3Al_3O_3N_5$. Sialon crystals are typically embedded in a glassy phase based on Y_2O_3. The glassy phase is then allowed to devitrify (crystallize) by a heat treatment to improve the creep resistance. The result is a ceramic that is relatively lightweight, with a low coefficient of thermal expansion, good fracture toughness, and a higher strength than many of the other common advanced ceramics. Sialon may find applications in cutting tools, engine components, and other applications involving both high temperatures and demanding wear conditions.

▪ *Titanium Dioxide* (TiO_2) is used to make electronic ceramics such as $BaTiO_3$. The largest use, though, is as a white pigment to make paints. Titania is used in certain glass ceramics as a nucleating agent. Fine particles of TiO_2 are used to make suntan lotions that provide protection against ultraviolet rays.

▪ *Titanium boride* (TiB_2) is a good conductor of both electricity and heat. In addition, it provides excellent toughness. TiB_2, along with boron carbide, silicon carbide, and alumina, finds application in producing armor.

▪ *Uranium dioxide* (UO_2) is widely used as a nuclear reactor fuel. This material has exceptional dimensional stability because its crystal structure can accommodate the products of the fission process.

▪ *Yttrium Aluminum Garnet* (*YAG*, $Y_3Al_5O_{12}$) crystals are used as a host for making the Nd-YAG lasers.

▪ *Zinc Oxide* (ZnO) is used as an accelerator in the vulcanization of rubbers used in tires, for example (Chapter 15). It is also used in paints, surge protection devices, medicated body powders, and skin ointments.

▪ *Zirconia* (ZrO_2) is used to make many other ceramics such as zircon. Zirconia is also used to make oxygen gas sensors that are used in automotives and to measure dissolved oxygen in molten steels. Zirconia is used as an additive in many electronic ceramics as well as a refractory material. The cubic form of zirconia single crystals is used to make jewelry items. Fuel cells based on zirconia will likely appear in cars by year 2015.

14-2   Properties of Ceramics

The properties of some ceramics are summarized in Table 14-2.[4,5] Mechanical properties of some structural ceramics are summarized in Table 14-3.

Take note of the high-melting temperatures and high-compressive strengths of ceramics. As mentioned in Chapter 6, the weight of an entire firetruck can be supported on four ceramic coffee cups. We should also remember that the tensile and flexural strength values show considerable strength since the strength of ceramics is dependent

TABLE 14-2 ▪ *Properties of commonly encountered polycrystalline ceramics*

Material	Melting Point (°C)	Thermal Expansion Coefficient ($\times 10^{-6}$ cm/cm)/°C	Knoop Hardness (HK) (100 g)
Al_2O_3	2000	~6.8	2100
BN	2732	0.57[a], −0.46[b]	5000
SiC	2700	~3.7	2500
Diamond		1.02	7000
Mullite	1810	4.5	—
TiO_2	1840	8.8	—
Cubic ZrO_2	2700	10.5	—

[a] *Perpendicular to pressing direction.*
[b] *Parallel to pressing direction.*

TABLE 14-3 ■ *Mechanical properties of selected advanced ceramics*

Material	Density (g/cm³)	Tensile Strength (psi)	Flexural Strength (psi)	Compressive Strength (psi)	Young's Modulus (psi)	Fracture Toughness (psi √in.)
Al_2O_3	3.98	30,000	80,000	400,000	56×10^6	5,000
SiC (sintered)	3.1	25,000	80,000	560,000	60×10^6	4,000
Si_3N_4 (reaction bonded)	2.5	20,000	35,000	150,000	30×10^6	3,000
Si_3N_4 (hot pressed)	3.2	80,000	130,000	500,000	45×10^6	5,000
Sialon	3.24	60,000	140,000	500,000	45×10^6	9,000
ZrO_2 (partially stabilized)	5.8	65,000	100,000	270,000	30×10^6	10,000
ZrO_2 (transformation toughened)	5.8	50,000	115,000	250,000	29×10^6	11,000

on the distribution of flaw sizes and is not affected by dislocation motion. We discussed the Weibull distribution and the strength of ceramics and glasses in Chapter 6.[6,7] Also note that, contrary to common belief, ceramics are not always brittle. Under smaller strain rates and at high temperatures, many ceramics with a very fine grain size indeed show superplastic behavior.[8] This was discussed in Chapter 7.

14-3   Synthesis of Ceramic Powders

Ceramic materials melt at high temperatures and they exhibit a brittle behavior under tension. As a result, the casting and thermomechanical processing, used widely for metals, alloys, and thermoplastics, cannot be applied when processing ceramics. Inorganic glasses, though, make use of lower melting temperatures due to the formation of eutectics in the float-glass process.[9] Since melting, casting, and thermomechanical processing is not a viable option for polycrystalline ceramics, we typically process ceramics into useful shapes starting with ceramic powders. A "powder" is a collection of fine particles. The step of making a ceramic powder is defined here as the **synthesis** of ceramics. We begin with a ceramic powder and get it ready for shaping by crushing, grinding, separating impurities, blending different powders, drying, and **spray drying** to form soft agglomerates. Different techniques such as compaction, **tape casting**, extrusion, and **slip casting** are then used to convert properly processed powders into a desired shape to form what is known as a **green ceramic**. A green ceramic is a ceramic that has not yet been sintered. The steps of converting a ceramic powder (or mixture of powders) into a useful shape are known as **powder processing**. The green ceramic is then consolidated further using a high-temperature treatment known as sintering or firing. In this process, the green ceramic is heated to a high temperature, using a controlled heat-treatment and atmosphere, so that a dense material is obtained. The ceramic may be then subjected to additional operations such as grinding, polishing, or machining as needed for the final application. In some cases, leads will be attached, electrodes will be deposited, or coatings may have to be deposited. These general steps encountered in the synthesis and processing of ceramics are summarized in Figure 14-1.

Next, we will describe the traditional synthesis techniques for some commonly encountered ceramics. These techniques used for making ceramic powders are also referred to as the "oxide mix" process.

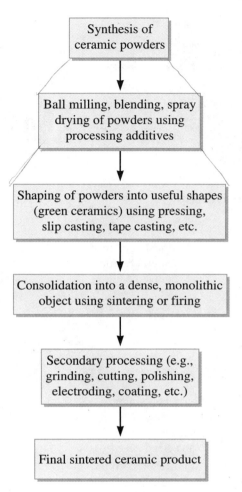

Figure 14-1
Typical steps encountered in the processing of
ceramics.

Crushing and grinding are conducted to reduce the particle size of the minerals mined and to physically "liberate" the minerals of interest from the rest of the "gangue" material. Gangue material is a collection of minerals that has no economical value.

Ball milling is a way of further reducing the size of minerals and blending the different powders used to make certain ceramics. This consists of a cylindrical vessel containing grinding media (such as steel, alumina, or zirconia spheres). Collisions between the grinding media and the mineral (or ceramic) lead to particle size reduction. Figure 14-2 includes a schematic of a ball mill.[10,11] In the processing of electronic magnetic ceramics, special attention has to be paid to ensure that impurities do not appear from the wear of the grinding media. For example, if alumina is used in the grinding of barium polytitanate powders, impurities that adversely affect the properties are formed as a result of alumina pick-up during ball milling.[12] However, inexpensive steel balls can be used for the grinding of nickel zinc ferrites, as long as we keep track of how much iron gets incorporated by the wear of the grinding media, since iron oxide is a component of the final ceramic.

Leaching is a process in which acids or alkalis are used to dissolve a mineral. This is conducted typically to get the metal of interest in solution.

Calcination refers to the heating of a mineral or an intermediate product in order to decompose it (e.g., heating a carbonate to remove CO_2 and form an oxide) or remove

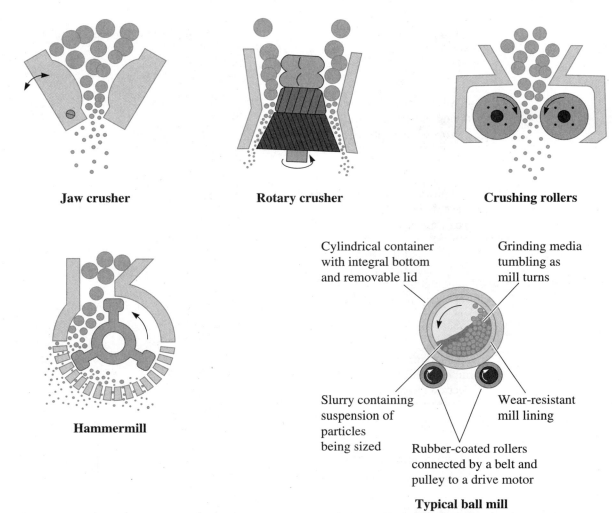

Jaw crusher

Rotary crusher

Crushing rollers

Hammermill

Cylindrical container with integral bottom and removable lid

Grinding media tumbling as mill turns

Slurry containing suspension of particles being sized

Wear-resistant mill lining

Rubber-coated rollers connected by a belt and pulley to a drive motor

Typical ball mill

Figure 14-2   Schematic of the jaw, rotary, crushing rollers, and hammermill crushing equipment and ball mill (grinding) equipment. (*Jaw, rotary, crushing, and hammermill: Source: From* Principles of Ceramics Processing, *Second Edition, by J.S. Reed, p. 314, Figs. 17-1 and 17-2. Copyright © 1995 John Wiley & Sons, Inc. Reprinted by permission. Ball mill grinding: Source: From* Modern Ceramic Engineering, *by D.W. Richerson, p. 387, Fig. 9-3. Copyright © 1992 Marcel Dekker, Inc.*)

the water after hydration. Calcination can also refer to heating a mixture of chemicals (e.g., oxides or carbonates) where, in some decomposition, reactions may occur and chemical reactions between solids can lead to the formation of ceramic phases of interest. For example, $BaCO_3$ and TiO_2 are calcined at ~1000°C to form $BaTiO_3$. Carbon dioxide is formed as a byproduct.

Other synthesis techniques can produce the following materials:

- *Aluminum oxide (Al_2O_3):* Chemically, corundum, ruby, and sapphire are all aluminum oxide or Al_2O_3. Alumina is extracted from bauxite ore using the Bayer process. In this process, bauxite ore is leached or digested using NaOH. Impurities are separated out by filtration. Very fine crystals of a mineral called gibbsite are added that causes heterogeneous nucleation (Chapter 8) to precipitate the rest of the aluminum in the solution as gibbsite. The gibbsite precipitated is washed to remove sodium and then calcined at ~1100–1200°C to form alpha alumina (α-Al$_2$O$_3$). Higher calcination temperatures (1600°C) form tabular alumina. Lower sodium grades are preferred for electronic applications since sodium ions can provide partial ionic conductivity.

■ *Silicon carbide (SiC):* In the Acheson process, coke and silica are heated using electrodes. The temperature is about 2200°C and leads to the formation of SiC.

■ *Titanium dioxide (TiO$_2$):* Titania is made using either the chloride or sulfate process. In the sulfuric acid process, ilmenite (FeTiO$_3$), is leached to form soluble titanium oxysulfate (TiOSO$_4$). This is then hydrolyzed to form titanium oxyhydroxide (TiO(OH)$_2$), which is calcined at 1000°C to form TiO$_2$. In the chloride process, the titanium ore is converted into titanium tetrachloride (TiCl$_4$), which in turn is converted into TiO$_2$. The level and type of impurities in TiO$_2$ are different. We can select a proper type of TiO$_2$ for a given application.

■ *Zinc Oxide (ZnO):* High-purity ZnO powder used in the rubber industry and in the manufacture of surge protection and other devices is made using the oxidation of molten zinc metal.[13]

■ *Zirconium Oxide (ZrO$_2$):* The mineral zircon (ZrSiO$_4$) is dissolved using NaOH. The resultant sodium zirconate is converted into zirconia. It is also possible to convert zircon into zirconium oxychloride (ZrOCl$_2$) or zirconium sulfate, which is then converted to zirconia.[10,14,15]

Conventional vs. Chemical Techniques for Powder Synthesis   Ceramic compositions can be formed using alumina, zirconia, or other powders. Typically, the conventional processes of ball milling and calcination at high temperatures will produce the desired powders. For example, to make BaTiO$_3$ powder, we start with barium carbonate (BaCO$_3$) and TiO$_2$ batched in correct amounts. These are calcined at ~1000°C by solid-state reactions to form BaTiO$_3$. Many other complex ferrites and multicomponent ceramics are formed in similar ways. For example, a nickel zinc ferrite can be formed through the reactions of iron oxide, zinc oxide, and nickel oxide. Powders made using these conventional processes require close attention as to the level of impurities. Typically, the powders made using conventional techniques tend to be coarse (up to approximately 10 μm) and they can often contain other phases in smaller concentration. The major advantage of using conventional synthesis processes to make ceramic powders is the low cost.

Chemical processes, on the other hand, can provide ultrafine, high-purity, and chemically homogenous powders efficiently. These techniques include **sol-gel process**, hydrothermal synthesis, chemical precipitation, and freeze drying (Chapter 9). Although the purity of powders made using these chemical techniques is high, they often are not cost effective. The use of chemical techniques is often preferred for applications where the difference in performance is justified by the increased cost. High-purity raw materials are readily available at lower costs, maybe as byproducts of some other processes. The sol-gel process, for example, is used to make large quantities of high quality ceramic abrasives and other materials in the form of powders and fibers.[16,17] A brief description of the sol-gel process is as follows. A sol is a dispersion of colloidal matter. This is converted into a gel and ultimately into a useful product such as a thin film, powder, or porous and monolithic ceramic. Polymeric metallorganic precursors or colloidal oxide (or hydroxide) particles are used to generate single or multi-component ceramic materials at low temperatures.

No matter how the ceramic powder is made, the average particle size, particle size distribution, particle shape, type, and concentration of impurities, crystalline phases present, and the powder surface area are some of the properties of powders that are routinely measured or determined. Properties of final sintered ceramics depend critically on these characteristics of ceramic powders, as well as powder processing. The following examples illustrates the formulation of barium titanate.

EXAMPLE 14-1   *Formation of a Barium Titanate Formulation*

We want to make 1000 kilograms of $BaTiO_3$ ceramic from $BaCO_3$ and TiO_2. How much barium carbonate and titanium dioxide should be ball milled and calcined?

SOLUTION

Even though the actual formation of $BaTiO_3$ involves many different intermediate phases, we can write the final reaction as:

$$BaCO_3(s) + TiO_2(s) \rightarrow BaTiO_3(s) + CO_2(g)$$

Thus, one mole of $BaCO_3$ reacts with one mole of TiO_2 to form one mole of $BaTiO_3$ and one mole of CO_2.

Molecular weights are: $BaCO_3$: $137(Ba) + 12(C) + (3 \times 16(O)) = 197$ g/mol

TiO_2: $48(Ti) + 2 \times 16 = 80$ g/mol

$BaTiO_3$: $137(Ba) + 48(Ti) + (3 \times 16(O)) = 233$ g/mol

CO_2: $12(C) + 2 \times 16(O) = 44$ g/mol

Thus, 197 g of $BaCO_3$ reacts with 80 g of TiO_2 to form 233 g of $BaTiO_3$ and 44 g of CO_2.

Since 233 g of $BaTiO_3$ is produced using 197 g of $BaCO_3$, for 1000 kilograms of $BaTiO_3$, we will need 845.5 kilograms of $BaCO_3$.

Similarly, since 233 g of $BaTiO_3$ is produced using 80 grams of TiO_2, for 1000 kilograms of $BaTiO_3$, we will need 343.3 kilograms of TiO_2.

In these calculations, we assumed that the raw materials contained no other impurities. In practice, we will have to determine the % TiO_2 in the titanium dioxide powder. Similarly, the actual concentration of $BaCO_3$ in the barium carbonate powder will have to be determined and accounted for. The concentrations of other impurities will also have to be monitored.

14-4   Powder Processing

Ceramic powders prepared using conventional or chemical techniques are shaped using the techniques shown in Figure 14-3. We emphasize that very similar processes are used for processing metal and alloy powders, a route known as **powder metallurgy**. Powders consist of particles that are loosely bonded, and powder processing involves the consolidation of these powders into a desired shape. Often, the ceramic powders prepared need to be converted into soft agglomerates by spraying a slurry of the powder through a nozzle into a chamber (spray dryer) in the presence of hot air. This process leads to the formation of soft agglomerates that flow into the dies used for powder compaction; this is known as *spray drying*.

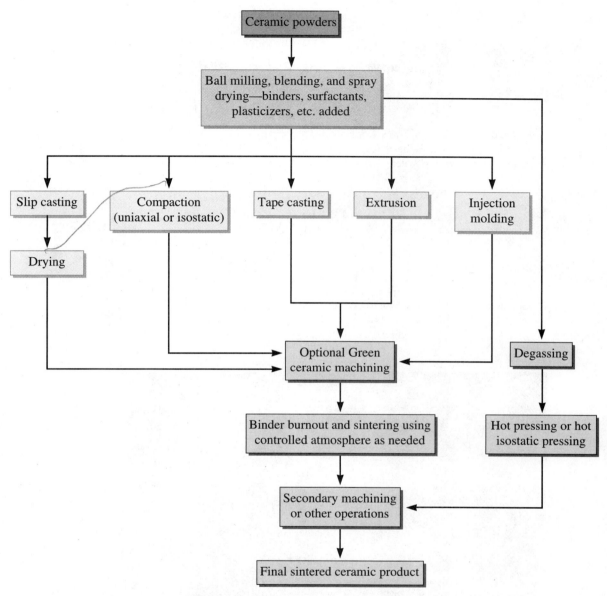

Figure 14-3 Different techniques for processing of advanced ceramics.

Compaction and Sintering One of the most cost-effective ways to produce thousands of relatively simple shapes of relatively small pieces (<6 inches) is compaction and sintering. Many electronic and magnetic ceramics, WC-Co cutting tool bits, and other materials are processed using this technique. The biggest advantage to sintering is the reduction in the surface area of a powder (Chapter 5). Fine powders can be spray dried, forming soft agglomerates that flow and compact well. The different steps of uniaxial compaction, in which the compacting force is applied in one direction, are shown in Figure 14-4(a). As an example, the microstructure of a barium magnesium tantalate ceramic prepared using compaction and sintering is shown in Figure 14-4(b). Sintering involves different mass transport mechanisms [Figure 14-4(c)].[18] With sintering, the

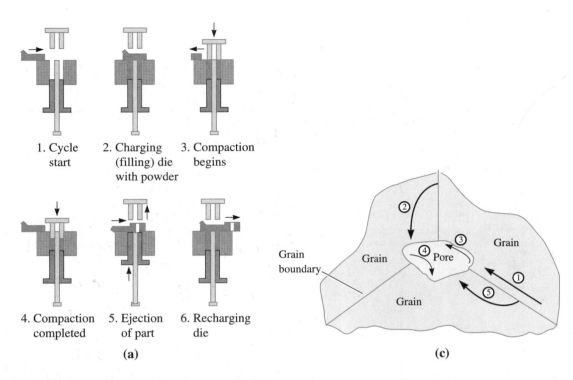

1. Cycle start
2. Charging (filling) die with powder
3. Compaction begins
4. Compaction completed
5. Ejection of part
6. Recharging die

(a)

(b)

Figure 14-4   (a) Uniaxial powder compaction showing the die-punch assembly during different stages. Typically, for small parts these stages are completed in less than a minute. (*Source: From* Materials and Processes in Manufacturing, *Eighth Edition, by E.P. DeGarmo, J.T. Black, and R.A. Koshe, Fig. 16-4. Copyright © 1997 Prentice Hall. Reprinted by permission Pearson Educaiton, Inc.*) (b) Microstructure of a barium magnesium tantalate (BMT) ceramic prepared using compaction and sintering. (*Courtesy Heather Shivey.*) (c) Different diffusion mechanisms involved in sintering. The grain boundary and bulk diffusion (1, 2 and 5) to the neck contribute to densification. Evaporation-condensation (4) and surface diffusion (3) do not contribute to densification. (*Source: From* Physical Ceramics: Principles for Ceramic Science and Engineering, *by Y.M. Chiang, D. Birnie, and W.D. Kingery, Fig. 5-40. Copyright © 1997 John Wiley & Sons, Inc. This material is used by permission of John Wiley & Sons, Inc.*)

grain boundary and bulk (volume) diffusion contribute to densification (increase in density). Surface diffusion and evaporation condensation can cause grain growth, but they do not cause densification.[19]

The compaction process can be completed within one minute for smaller parts; thus, uniaxial compaction is well suited for making a large number of smaller and simple shapes. Compaction is used to create what we call "green ceramics"; these have respectable strengths and can be handled and machined. In some cases, very large pieces (up to a few feet in diameter and six to eight feet long) can be produced using a process called **cold isostatic pressing** (CIP) where pressure is applied using oil. Such large pieces are then sintered with or without pressure. Cold isostatic pressing is used for achieving a higher green ceramic density or where the compaction of more complex shapes is required.

Binders (polymers that can help form a network of particles) are often added to the powders to enhance green ceramic density. The green parts are loaded into trays where they contact a ceramic material with which they do not react and are then sent to batch or continuous furnaces for sintering under controlled conditions. The binders used typically are burnt at lower temperatures using smaller heating rates ($\sim$300 to 500°C at a rate of 5°C/h). Attention must be paid to the left over residues (e.g., Na_2O) and gaseous products formed during binder burnout. Sintering sometimes is conducted under a controlled atmosphere at a given temperature and time profile (the atmosphere during sintering may change the material's composition). The temperature at which sintering occurs is dictated by the kinetics of diffusion, defect chemistry, average particle size of powder particles, and the presence of any sintering additives added or those that are inadvertently present (as impurities). Typically with sintering, a linear shrinkage of about 10 to 20% and density levels of approximately 95% (of the theoretical density) are quite common. Higher levels of density (>99%) are more difficult to achieve. The fact that the final sintered dimensions are smaller than those at the start of the process and that some machining may be needed must be factored into the design of ceramic materials. The cooling rate for sintered ceramics has to be monitored to avoid microcracking of parts due to thermal expansion and contraction stresses, especially in larger parts. Solid-state reactions (such as the incorporation of atmospheric oxygen onto vacant oxygen sites, ordering of cations, or the crystallization of glassy phases) can also occur during cooling after sintering and attention needs to be paid to these processes as well. Any glassy phases that remain often are detrimental to the mechanical properties of ceramics.

In some cases, parts may be produced under conditions in which sintering is conducted using applied pressure. This technique, known as **hot pressing**, is used for refractory and covalently bonded ceramics that do not show good pressureless sintering behavior. Similarly, large pieces of metals and alloys compacted using CIP can be sintered under pressure in a process known as **hot isostatic pressing** (HIP) (Chapter 5). In hot pressing or HIP, the applied pressure acts against the internal pore pressure and enhances densification without causing grain growth.[20] Hot pressing or hot isostatic pressing also are used for making ceramics or metallic where very little or almost no porosity is required. Some recent innovative processes that make use of microwaves (similar to the way food gets heated in microwave oven) have also been developed for the drying and sintering of ceramic materials.

Some ceramics, such as silicon nitride (Si_3N_4), are produced by **reaction bonding**. Silicon is formed into a desired shape and then reacted with nitrogen to form the nitride. Reaction bonding, which can be done at lower temperatures, provides better dimensional control compared with hot pressing; however, lower densities and mechanical properties are obtained. As a comparison, the effect of processing on silicon nitride ceramics is shown in Table 14-4.

TABLE 14-4 ■ *Properties of Si$_3$N$_4$ processed using different techniques*

Process	Compressive Strength (psi)	Flexural Strength (psi)
Slip casting	20,000	10,000
Reaction bonding	112,000	30,000
Hot pressing	50,000	125,000

Tape Casting   A technique known as *tape casting* is used for the production of thin ceramic tapes (~3 to 100 μm) (Figure 14-5). This technique is also known as the *doctor blade* process. In tape casting, a slurry consisting of ceramic particles is formed in water or an organic liquid such as methyl ethyl ketone (MEK).[21] Since the removal of organic solvents and binders poses problems related to volatile organics and other polycyclic aromatic compounds (which are cancer causing), the tape casting industry has been moving to water-based slurries. The slurry also contains surfactants that enable proper dispersion of ceramic particles. A polymer binder is also added. The function of a binder is to "bind" or hold ceramic particles in a polymer tape. To lower the glass temperature of the binder and, therefore, to make the tape more flexible, a plasticizer is added (Chapter 15). The slurry containing ceramic particles, solvent, plasticizers, and binders is then made to flow under a blade and onto a plastic substrate (e.g., Mylar^{TM}). The shear thinning slurry spreads under the blade. The tape is then dried using clean hot air.

A tape casting machine can make tapes that are 3 to 4 feet wide and from about 3 to 100 μm in thickness. The tape is made continuously and wound onto a roll at the other end. Tape casting machines can be up to 60 to 70 feet long. The tapes made are then machined (i.e., they are cut and diced into desired shapes) and holes may be drilled to form conductive pathways necessary for certain electronic devices. Electrodes on the surfaces of the tapes can be applied using metal-powder inks, depending upon the application.

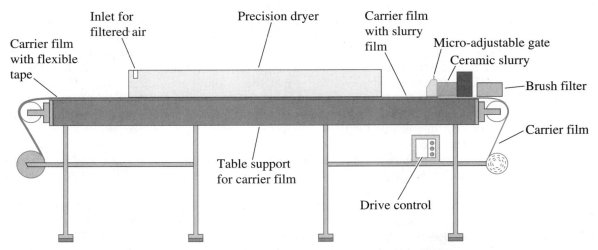

Figure 14-5   Schematic of a tape casting machine. (*Source: From* Principles of Ceramics Processing, *Second Edition, by J.S. Reed, p. 532, Fig. 26-6. Copyright © 1995 John Wiley & Sons, Inc. Reprinted by permission.*)

After these operations, the tape is subjected to binder burnout and sintering operations. Many commercially important electronic packages based on alumina substrates and millions of barium titanate capacitors are made using this type of tape casting process. A modern multi-layer capacitor based on $BaTiO_3$ consists of about 100 layers of thin sheets of ceramic dielectric (each layer about 5 to 10 μm thick) separated by electrode layers.

Slip Casting   This technique typically uses an aqueous slurry of ceramic powder. The slurry, known as the *slip*, is poured into a plaster of Paris ($CaSO_4 : 2H_2O$) mold (Figure 14-6).[11] As the water from the slurry begins to move out by capillary action, a thick mass builds along the mold wall. When sufficient product thickness is built, the rest of the slurry is poured out (this is called *drain casting*). It is also possible to continue to pour more slurry in to form a solid piece (this is called *solid casting*) (Figure 14-6).[10]

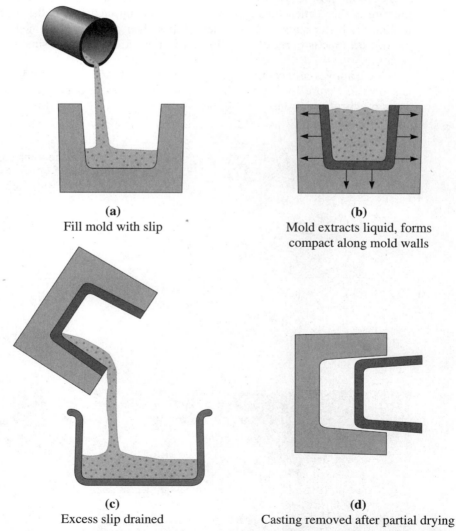

(a)
Fill mold with slip

(b)
Mold extracts liquid, forms
compact along mold walls

(c)
Excess slip drained

(d)
Casting removed after partial drying

Figure 14-6   Steps in slip casting of ceramics. (*Source: From* Modern Ceramic Engineering, *by D.W. Richerson, p. 462, Fig. 10-34. Copyright © 1992 Marcel Dekker. Reprinted by permission.*)

Pressure may also be used to inject the slurry into polymer molds. The green ceramic is then dried and "fired" or sintered at a high temperature. Slip casting is widely used to make ceramic art (figurines and statues), sinks and other ceramic sanitaryware.

Extrusion and Injection Molding These are popular techniques used for making furnace tubes, bricks, tiles, and insulators. The idea behind the extrusion process is to use a viscous, dough-like mixture of ceramic particles containing a binder and other additives. This mixture has a clay-like consistency, which is then fed to an extruder where it is mixed well in a pug mill, sheared, deaerated, and then injected into a die where a continuous shape of green ceramic is produced by the extruder. This material is cut at appropriate lengths and then dried and sintered. Cordierite ceramics used for making catalyst honeycomb structures are also made using the extrusion process.

Injection molding of ceramics is similar to injection molding of polymers (Chapter 15). Ceramic powder is mixed with a thermoplastic plasticizer and other additives. The mixture is then taken through an extruder and injected into a die. Ceramic injection molding is better suited for complex shapes. The polymer contained in the injection-molded ceramic is burnt off and the rest of the ceramic body is sintered at a high temperature.

An examples of a ceramic microstructure of a PLZT ceramic made by hot pressing is shown in Figure 14-7. The following examples illustrate the use of these concepts in some practical situations.

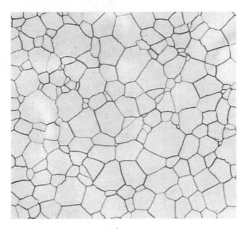

Figure 14-7
Magnetic ferrites made using compaction and sintering. (*Courtesy of Dr. Gene Haertling.*)

EXAMPLE 14-2 *Sintering of Lead Zirconium Titanate (PZT) Ceramics for Barbeque Grill Gas Igniters*

A manufacturer of PZT ceramics wants to make 100,000 cylindrical pieces (5 mm in diameter and 10 mm in height) per day. These pieces are to be used for a gas igniter. Since PZT is piezoelectric, it can be squeezed to generate a charge and the charge can be used to make a spark (Chapter 18). The manufacturer has access to several different types of powders (e.g., traditional, hydrothermal, and sol-gel). The manufacturer also can invest in various processing techniques (e.g., tape casting, slip casting, uniaxial pressing, or sintering). What type of synthesis and processing techniques would you suggest for the manufacturing of these elements?

SOLUTION

Lead zirconium titanate (PZT) contains oxides of three different elements. Chemical processing techniques, such as sol-gel, can give us powders that have better purity and chemical homogeneity and finer size. Conventional powders prepared using mixed oxides will have more impurities and these powders will be coarser. The application of PZT here is for making gas igniter pieces, so to keep the cost low, we will choose conventionally prepared powders. We can guarantee that the powders provide the necessary piezoelectric properties using pilot tests.

Powders can be ball milled as necessary to break down agglomerates. An organic medium can be used since PZT powder particles may cause undesirable surface reactions with water. Zirconia is used as a grinding media to minimize contamination due to other grinding media. The uptake of ZrO_2 will have to be monitored. Spray drying can be used to form soft agglomerates that provide higher green densities. The need for these steps must be determined during pilot testing.

For powder processing, we choose uniaxial pressing and sintering since the pieces are cylindrical and can be produced easily using this technique. We can also consider slip casting as a possible process, however, the reactivity of PZT powders, the possible incorporation of lead (Pb) in discharge waters, the cost of treatments, and the need to achieve higher levels of density make this an unfavorable processing choice. In this application, since we ultimately want an electric voltage to be developed across the PZT, it is important to make sure the density is high. We have to account for sintering shrinkage and the possibility that machining may be needed (e.g., grinding of surfaces). Thus, the die size must be larger than the final part dimensions required.

In sintering PZT, we must carefully consider using binders to achieve higher green density. When PZT is heated in open air, lower volatility components can begin to evaporate (e.g., PbO). This can cause the formation of undesirable phases and slow down the sintering process significantly. This, in turn, can have a deleterious effect on the properties of PZT. We can minimize evaporation of PbO by conducting sintering in a self-contained atmosphere that is rich in PbO vapor. This can be achieved by conducting sintering in closed crucibles in the presence of PbO powder.

Following sintering, the PZT ceramics may have to undergo some machining. Electrodes must be deposited and other steps (e.g., a process known as poling) will be required to make sure the desired piezoelectric properties can be obtained. Thus, discussions with device engineers and product engineers must be conducted first to make sure these and any other issues are addressed. For example, because of the toxicity of lead, it would be very important to make sure that all powder handling (synthesis, processing, machining, etc.) steps are very carefully monitored. Employees must also be asked to use proper equipment to eliminate exposure to powder particles to protect their health. Lead exposure must be monitored carefully during the manufacturing of such products.

EXAMPLE 14-3   *Sintering of Microwave Dielectrics*

A manufacturer of barium zinc tantalate ($Ba(Zn_{1/3}Ta_{2/3})O_3$ or BZT ceramics) produces cylindrical, puck-shaped devices known as dielectric resonators. The particular pucks made in this case are 2 inches in diameter and $\frac{1}{2}$ inch tall and they have a $\frac{1}{2}$ inch hole at the center. The sintering of these pieces is conducted at 1500°C, using additives. The manufacturing process used to make these pieces is well established. Suddenly, one day, a batch of the dielectric resonators comes out of the furnace with cracks and must be rejected. The plant manager determines that the material has reached 98% of the theoretical target level of densification. What could be the reason for the cracks?

SOLUTION

Problems such as these have to be traced back through the manufacturing process starting with the raw materials, blending, compaction, and the use of binders. In a quality manufacturing operation (such as the one having a ISO 9000 certification), all raw materials, manufacturing steps, procedures, and suppliers must be traceable. Since the resonators have reached the intended level of densification, most likely the composition of the batch was fine.

In this case, the raw materials were checked and did not have any unusual levels of known impurities. In addition, the blending of powders was done correctly. The parts were also compacted properly. Discussions with the furnace operator revealed that he had increased the speed of the furnace conveyer belt so that he could get off work an hour early, causing the parts to cool too quickly. The resultant dielectric resonators developed microcracks because of rapid cooling from the sintering temperature. The problem, in this case, turned out to be due to a human error. Often, in the ceramic processing of materials, problems can be traced to changes in raw materials, changes in suppliers, and other unforeseen events.

14-5   Characteristics of Sintered Ceramics

For sintered ceramics, the average grain size, grain size distribution, and the level and type of porosity are important. Similarly, depending upon the application, second phases in the microstructure could occur as separate grains of components dissolved in solid solutions of the matrix, so second phases at grain boundaries also become important. In the case of extruded ceramics, orientation effects also can be important.

Grains and Grain Boundaries   The average grain size often closely related to the primary particle size (Figures 14-4 and 14-7). An exception to this is if there is grain growth due to long sintering times or exaggerated or abnormal grain growth (Chapter 5). Typically, ceramics with a small grain size are stronger than coarse-grained ceramics. Finer grain sizes help reduce stresses that develop at grain boundaries due to anisotropic expansion and contraction. Normally, starting with finer ceramic raw materials produces a fine grain size. Magnetic, dielectric, and optical properties of ceramic materials depend upon the average grain size and, in these applications, grain size must be controlled properly. Although we have not discussed this here in detail, in certain applications; it is important to use single crystals of ceramic materials so as to avoid the deleterious grain boundaries that are always present in polycrystalline ceramics. In some other applications, such as for electrical surge-protection devices based on ZnO, it is important to have grain boundaries with a different chemical composition than that of the grain. This composition and its structure are such that selective electrical breakdown of the material along the grain boundaries at a certain voltage is allowed. This helps protect equipment connected to the surge protector since the electrical current passes through the ZnO device instead of the equipment. In these cases, special additives such as Sb_2O_5 or Bi_2O_3 are added to ZnO formulations.

Porosity Pores represent the most important defect in polycrystalline ceramics. The presence of pores is usually detrimental to the mechanical properties of bulk ceramics, since pores provide a pre-existing location from which a crack can grow. The presence of pores is one of the reasons why ceramics show such brittle behavior under tensile loading. Since there is a distribution of pore sizes and the overall level of porosity changes, the mechanical properties of ceramics vary. This variability is measured using the Weibull statistics (Chapter 6).[6] The presence of pores, on the other hand, may be useful for increasing the resistance thermal shock. In certain applications, such as filters for hot metals and alloys or for liquids or gases, the presence of interconnected pores is desirable.

Pores in a ceramic may be either interconnected or closed. The **apparent porosity** measures the interconnected pores and determines the permeability, or the ease with which gases and fluids seep through the ceramic component. The apparent porosity is determined by weighing the dry ceramic (W_d), then reweighing the ceramic both when it is suspended in water (W_s) and after it is removed from the water (W_w). Using units of grams and cm^3:

$$\text{Apparent porosity} = \frac{W_w - W_d}{W_w - W_s} \times 100 \tag{14-1}$$

The **true porosity** includes both interconnected and closed pores. The true porosity, which better correlates with the properties of the ceramic, is:

$$\text{True porosity} = \frac{\rho - B}{\rho} \times 100, \tag{14-2}$$

where

$$B = \frac{W_d}{W_w - W_s} \tag{14-3}$$

B is the **bulk density** and ρ is the true density or specific gravity of the ceramic. The bulk density is the weight of the ceramic divided by its volume. The following example illustrates how porosity levels in ceramics are determined.

EXAMPLE 14-4   *Silicon Carbide Ceramics*

Silicon carbide particles are compacted and fired at a high temperature to produce a strong ceramic shape. The specific gravity of SiC is 3.2 g/cm^3. The ceramic shape subsequently is weighed when dry (360 g), after soaking in water (385 g), and while suspended in water (224 g). Calculate the apparent porosity, the true porosity, and the fraction of the pore volume that is closed.

SOLUTION

$$\text{Apparent porosity} = \frac{W_w - W_d}{W_w - W_s} \times 100 = \frac{385 - 360}{385 - 224} \times 100 = 15.5\%$$

$$\text{Bulk density} = B = \frac{W_d}{W_w - W_s} = \frac{360}{385 - 224} = 2.24$$

$$\text{True porosity} = \frac{\rho - B}{\rho} \times 100 = \frac{3.2 - 2.24}{3.2} \times 100 = 30\%$$

The closed-pore percentage is the true porosity minus the apparent porosity, or $30 - 15.5 = 14.5\%$. Thus:

$$\text{Fraction closed pores} = \frac{14.5}{30} = 0.483$$

14-6   Inorganic Glasses

In Chapter 3, we discussed amorphous materials such as glasses, gels, and thin amorphous films. We also discussed the concepts of short- versus long-range order in terms of atomic or ionic arrangements in noncrystalline materials. The most important of the noncrystalline materials are glasses, especially those based on silica. Of course, there are glasses based on other compounds (e.g., sulfides, fluorides, and different alloys). A glass is a metastable material that has hardened and become rigid without crystallizing. A glass in some ways resembles an undercooled liquid. Below the **glass temperature** (Figure 14-8), the rate of volume contraction on cooling is reduced and the material can be considered a "glass" rather than an "undercooled liquid." Joining silica tetrahedra or other ionic groups produces a solid, but noncrystalline framework structure produces the glassy structures (Chapter 3).

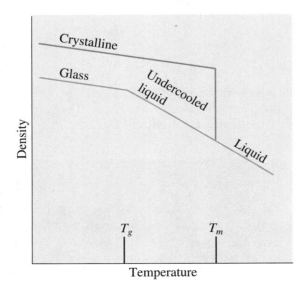

Figure 14-8
When silica crystallizes on cooling, an abrupt change in the density is observed. For glassy silica, however, the change in slope at the glass temperature indicates the formation of a glass from the undercooled liquid. Glass does not have a fixed T_m or T_g. Crystalline materials have a fixed T_m and they do not have a T_g.

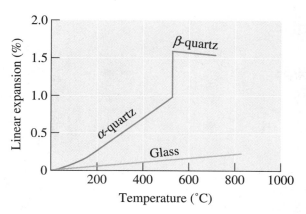

Figure 14-9
The expansion of quartz. In addition to the regular—almost linear— expansion, a large, abrupt expansion accompanies the α- to β-quartz transformation. However, glasses expand uniformly.

Silicate Glasses   The silicate glasses are the most widely used. *Fused silica*, formed from pure SiO_2, has a high-melting point, and the dimensional changes during heating and cooling are small (Figure 14-9). Generally, however, the silicate glasses contain additional oxides (Table 14-5). While oxides such as silica behave as **glass formers**, an **intermediate** oxide (such as lead oxide or aluminum oxide) does not form a glass by itself but is incorporated into the network structure of the glass formers. A third group of oxides, the **modifiers**, break up the network structure and eventually cause the glass to devitrify, or crystallize.

TABLE 14-5 ■ *Division of the oxides into glass formers, intermediates, and modifiers*

Glass Formers	Intermediates	Modifiers
B_2O_3	TiO_2	Y_2O_3
SiO_2	ZnO	MgO
GeO_2	PbO_2	CaO
P_2O_5	Al_2O_3	PbO
V_2O_3	BeO	Na_2O

Modified Silicate Glasses   Modifiers break up the silica network if the oxygen-to-silicon ratio (O:Si) increases significantly. When Na_2O is added, for example, the sodium ions enter holes within the network rather than becoming part of the network. However, the oxygen ion that enters with the Na_2O does become part of the network (Figure 14-10). When this happens, there are not enough silicon ions to combine with the extra oxygen ions and keep the network intact. Eventually, a high O:Si ratio causes the remaining silica tetrahedra to form chains, rings, or compounds, and the silica no longer transforms to a glass. When the O:Si ratio is above about 2.5, silica glasses are difficult to form; above a ratio of three, a glass forms only when special precautions are taken, such as the use of rapid cooling rates.

Modification also lowers the melting point and viscosity of silica, making it possible to produce glass at lower temperatures. The effect of Na_2O additions to silica is shown in Figure 14-11. As you can see, the addition of Na_2O produces eutectics with very low melting temperatures. Adding CaO, which reduces the solubility of the glass in water, further modifies these glasses. The example that follows shows how to design a glass.

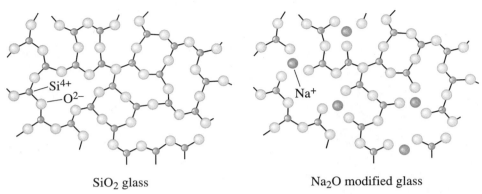

SiO$_2$ glass   Na$_2$O modified glass

Figure 14-10 The effect of Na$_2$O on the silica glass network. Sodium oxide is a modifier, disrupting the glassy network and reducing the ability to form a glass.

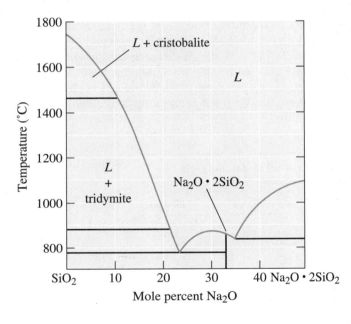

Figure 14-11
The SiO$_2$-Na$_2$O phase diagram. Additions of soda (Na$_2$O) to silica dramatically reduce the melting temperature of silica, by forming eutectics.

EXAMPLE 14-5 *Design of a Glass*

We produce good chemical resistance in a glass when we introduce B$_2$O$_3$ into silica. To assure that we have good glass-forming tendencies, we wish the O:Si ratio to be no more than 2.5, but we also want the glassware to have a low-melting temperature to make the glass-forming process easier and more economical. Design such a glass.

SOLUTION

Because B$_2$O$_3$ reduces the melting temperature of silica, we would like to add as much as possible. We also, however, want to assure that the O:Si ratio is no more than 2.5, so the amount of B$_2$O$_3$ is limited. As an example, let us determine the amount of B$_2$O$_3$ we must add to obtain exactly an O:Si ratio of 2.5. Let f_B be the mole fraction of B$_2$O$_3$ added to the glass, and $(1 - f_B)$ be the mole fraction of SiO$_2$:

$$\frac{O}{Si} = \frac{\left(3\dfrac{O \text{ ions}}{B_2O_3}\right)(f_B) + \left(2\dfrac{O \text{ ions}}{SiO_2}\right)(1-f_B)}{\left(1\dfrac{Si \text{ ion}}{SiO_2}\right)(1-f_B)} = 2.5$$

$$3f_B + 2 - 2f_B = 2.5 - 2.5f_B \quad \text{or} \quad f_B = 0.143$$

Therefore, we must produce a glass containing no more than 14.3 mol% B_2O_3. In weight percent:

$$\text{wt\% } B_2O_3 = \frac{(f_B)(69.62 \text{ g/mol})}{(f_B)(69.62 \text{ g/mol}) + (1-f_B)(60.08 \text{ g/mol})} \times 100$$

$$\text{wt\% } B_2O_3 = \frac{(0.143)(69.62)}{(0.143)(69.62) + (0.857)(60.08)} \times 100 = 16.2$$

14-7 Processing and Applications of Glasses

Glasses are manufactured into useful articles at a high temperature with viscosity controlled so that the glass can be shaped without breaking. Figure 14-12 helps us understand the processing in terms of the viscosity ranges.

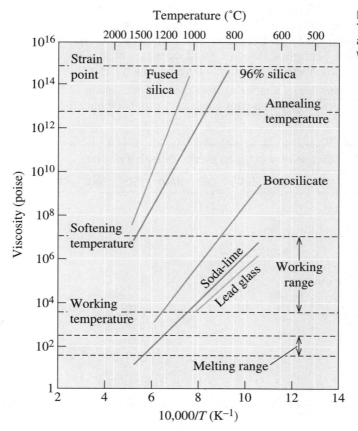

Figure 14-12
The effect of temperature and composition on the viscosity of glass.

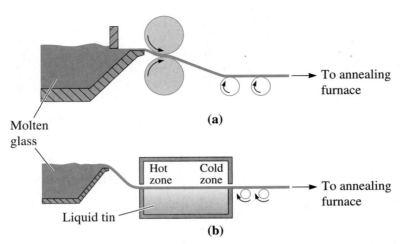

Figure 14-13 Techniques for manufacturing sheet and plate glass: (a) rolling and (b) floating the glass on molten tin.

1. *Liquid range.* Sheet and plate glass are produced when the glass is in the molten state. Techniques include rolling the molten glass through water-cooled rolls or floating the molten glass over a pool of liquid tin (Figure 14-13). The liquid-tin process produces an exceptionally smooth surface on the glass. The development of the float-glass process was a genuine breakthrough in the area of glass processing. Basic float-glass composition has been essentially unchanged for many years (Table 14-6).[9]

Some glass shapes, including large optical mirrors, are produced by casting the molten glass into a mold, then assuring that cooling is as slow as possible to minimize residual stresses and avoid cracking of the glass part. Glass fibers may be produced by drawing the liquid glass through small openings in a platinum die [Figure 14-14(c)]. Typically, many fibers are produced simultaneously for a single die.

2. *Working range.* Shapes such as those of containers or light bulbs can be formed by pressing, drawing, or blowing glass into molds (Figure 14-14). A hot *gob* of liquid

TABLE 14-6 ■ *Compositions of typical glasses (in weight percent)*

Glass	SiO_2	Al_2O_3	CaO	Na_2O	B_2O_3	MgO	PbO	Others
Fused silica	99							
Vycor™	96				4			
Pyrex™	81	2		4	12			
Glass jars	74	1	5	15		4		
Window glass	72	1	10	14		2		
Plate glass/Float glass	73	1	13	13				
Light bulbs	74	1	5	16		4		
Fibers	54	14	16		10	4		
Thermometer	73	6		10	10			
Lead glass	67			6			17	10% K_2O
Optical flint	50			1			19	13% BaO, 8% K_2O, ZnO
Optical crown	70			8		10		2% BaO, 8% K_2O
E-glass fibers	55	15	20		10			
S-glass fibers	65	25				10		

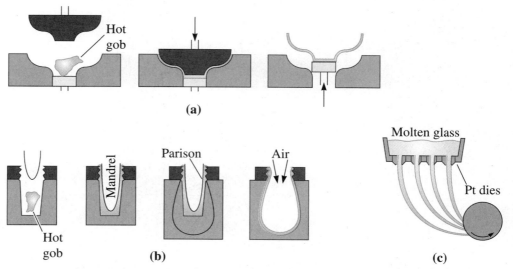

Figure 14-14 Techniques for forming glass products: (a) pressing, (b) press and blow process, and (c) drawing of fibers.

glass may be preformed into a crude shape (**parison**), then pressed or blown into a heated die to produce the final shape. The glass is heated to the working range so that the glass is formable, but not "runny."

3. *Annealing range.* Some ceramic parts may be annealed to reduce residual stresses introduced during forming. Large glass castings, for example, are often annealed and slowly cooled to prevent cracking. Some glasses may be heat treated to cause **devitrification**, or the precipitation of a crystalline phase from the glass.

We discussed the tempering of glasses in and some of the earlier chapters. **Tempered glass** is produced by quenching the surface of plate glass with air, causing the surface layers to cool and contract. When the center cools, its contraction is restrained by the already rigid surface, which is placed in compression (Figure 14-15). Tempered glass is capable of withstanding much higher tensile stresses and impact blows than untempered glass. Tempered glass is used in car and home windows, shelving for refrigerators, ovens, furniture, and many other applications where safety is important. Laminated glass, consisting of two annealed glass pieces with a polymer (such as polyvinyl butyral or PVB) in between, is used to make car windshields.

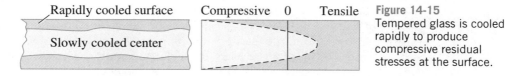

Figure 14-15
Tempered glass is cooled rapidly to produce compressive residual stresses at the surface.

Glass Compositions and Phase Diagrams   Pure SiO_2 must be heated to very high temperatures to obtain viscosities that permit economical forming. Most commercial glasses are based on silica; modifiers such as soda (Na_2O) are added to break down the network structure and form eutectics with low-melting temperatures, whereas lime (CaO) is added to reduce the solubility of the glass in water. The most common commercial glass contains approximately 75% SiO_2, 15% Na_2O, and 10% CaO and is

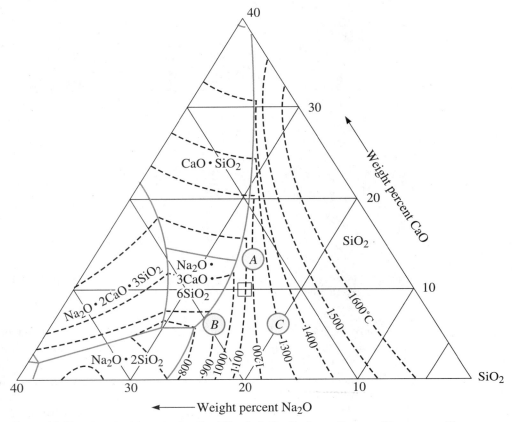

Figure 14-16   The liquidus plot for the SiO_2-CaO-Na_2O phase diagram. The compositions corresponding to points *A*, *B*, and *C* refer to those discussed in Example 14-6.

known as the soda line glass. Figure 14-16 is a liquidus plot of a portion of the ternary phase diagram showing the effect of both soda and lime on silica; the square shows that the SiO_2-15% Na_2O-10% CaO composition has a liquidus temperature of about 1100°C. Although inexpensive to produce, the soda-lime glasses have poor resistance to chemical attack and thermal stresses. Table 14-6 compares the compositions of several typical glasses.

Borosilicate glasses, which contain about 15% B_2O_3, have excellent chemical and dimensional stability. Their uses include laboratory glassware (Pyrex^{TM}) and containers for the disposal of high-level radioactive nuclear waste. Calcium aluminoborosilicate glass—or E-glass—is used as a general-purpose fiber for composite materials, such as fiberglass. Aluminosilicate glass, with 20% Al_2O_3 and 12% MgO, and high-silica glasses, with 3% B_2O_3, are excellent for high-temperature resistance and for protection against heat or thermal shock. The S-glass, a magnesium aluminosilicate, is used to produce high-strength fibers for composite materials. Fused silica, or virtually pure SiO_2, has the best resistance to high temperature, thermal shock, and chemical attack, although it is also expensive. Particularly high-quality fused silica is used for fiber-optic systems.

Special optical qualities can also be obtained, including sensitivity to light. Photochromic glass, which is darkened by the ultraviolet portion of sunlight, is used for sunglasses. Photosensitive glass darkens permanently when exposed to ultraviolet light;

if only selected portions of the glass are exposed and then immersed in hydrofluoric acid, etchings can be produced. Polychromatic glasses are sensitive to all light, not just ultraviolet radiation. Similarly, nanosized crystals of semiconductors such as cadmium sulfide (CdS) are nucleated in silicate glasses in a process known as *striking*. These glasses exhibit lively colors and also have useful optical properties. Some of the semiconductor crystals such as these are known as quantum dots and are candidate materials for electro-optical devices.[22]

EXAMPLE 14-6   *Design of a Soda-Lime Glass*

Design a soda-lime glass capable of being cast at a temperature of 1000°C.

SOLUTION

For casting, the glass should be heated above its liquidus temperature. Its viscosity should be low enough so that it flows easily into the mold. Therefore, if we want to pour the liquid at 1000°C, we might select a glass composition that has a lower liquidus-say 900°C. Suppose we used the following compositions (indicated on the silica-soda-lime phase diagram, Figure 14-16):

Glass A: 74% SiO_2-13% CaO-13% Na_2O

Glass B: 74% SiO_2-6% CaO-20% Na_2O

Glass C: 80% SiO_2-7% CaO-13% Na_2O

From the liquidus plot, we find that glass A has a liquidus of 1200°C, glass B has a liquidus of 900°C, and glass C has a liquidus of 1300°C. Of these three glasses, glass B is our obvious choice.

Of course, other compositions might also have a liquidus of 900°C. Increasing the CaO slightly still gives a glass with the required liquidus but processing will be more difficult; decreasing the CaO also gives a glass with the required liquidus, but the solubility of the glass in water is higher.

14-8   Glass-Ceramics

Glass-ceramics are crystalline materials that are derived from amorphous glasses. Usually, glass-ceramics have a substantial level of crystallinity ($\sim$>70–99%).[2] As mentioned in Chapters 3 and 8, the formation of glass-ceramics was discovered serendipitously by Don Stookey.[23] With glass-ceramics, we can take advantage of the formability and density of glass. A product that contains very low porosity can be obtained by producing a shape with conventional glass-forming techniques, such as pressing or blowing. Glass, however, has poor creep resistance. We then crystallize the glass using heterogeneous nucleation by such oxides as TiO_2 and/or ZrO_2. These oxides react with the glass and with each other and provide the nuclei that ultimately lead to glass crystallization. Phase separation of glasses (Chapter 10) plays an important role in formation of the nuclei. In some commercial glass-ceramics (e.g., Visionware™) nanosized crystallites are formed and the resultant material remains optically transparent.

The first step in producing a glass-ceramic is to assure that crystallization does not occur during cooling from the forming temperature. A continuous and isothermal cooling transformation diagram, much like the CCT and TTT diagrams for steels, can be seen for silicate-based glasses. Figure 14-17(a) shows a TTT diagram for a glass. If glass cools too slowly, a transformation line is crossed; nucleation and growth of the crystals begin, but in an uncontrolled manner. Addition of modifying oxides to glass, much like addition of alloying elements to steel, shifts the transformation curve to longer times and prevents devitrification even at slow cooling rates. As noted in previous chapters, strictly speaking we should make use of CCT (and not TTT) diagrams for this discussion.

Nucleation of the crystalline phase is controlled in two ways. First, the glass contains agents, such as TiO_2, that react with other oxides and form phases that provide the nucleation sites. Second, a heat treatment is designed to provide the appropriate number of nuclei; the temperature should be relatively low in order to maximize the rate of nucleation [Figure 14-17(b)]. However, the overall rate of crystallization depends on

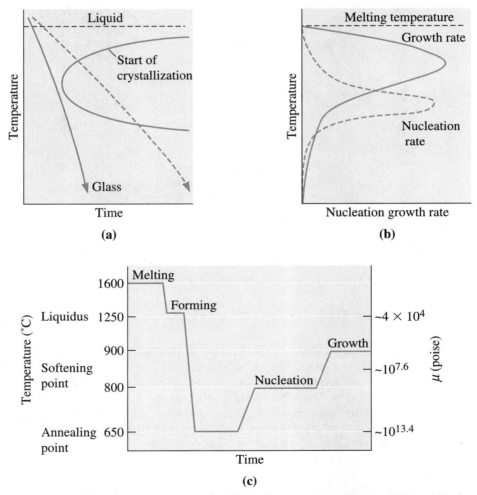

Figure 14-17 Producing a glass-ceramic: (a) Cooling must be rapid to avoid the start of crystallization. Isothermal and continuous cooling curves for lunar glass. (b) The rate of nucleation of precipitates is high at low temperatures, whereas the rate of growth of the precipitates is high at higher temperatures. (c) A typical heat-treatment profile for glass-ceramics fabrication, illustrated here for Li_2O-Al_2O_3-SiO_2 glasses.

the growth rate of the crystals once nucleation occurs; higher temperatures are required to maximize the growth rate. Consequently, a heat-treatment schedule similar to that shown in Figure 14-17(c) for the Li_2O-Al_2O_3-SiO_2 glass-ceramics (Pyroceram™) can be used.[18] The low-temperature step provides nucleation sites, and the high-temperature step speeds the rate of growth of the crystals; as much as 99% of the part may crystallize.

This special structure of glass-ceramics can provide good mechanical strength and toughness, often with a low coefficient of thermal expansion and high-temperature corrosion resistance. Perhaps the most important glass-ceramic is based on the Li_2O-Al_2O_3-SiO_2 system.[3] These materials are used for cooking utensils (Corning Ware™) and ceramic tops for stoves. Other glass-ceramics are used in communication, computer, and optical applications.

14-9 Processing and Applications of Clay Products

Crystalline ceramics are often manufactured into useful articles by preparing a shape, or compact, composed of the raw materials in a fine powder form. The powders are then bonded by chemical reaction, partial or complete **vitrification** (melting), or sintering.

Clay products form a group of traditional ceramics used for producing pipe, brick, cooking ware, and other common products. Clay, such as kaolinite, and water serve as the initial binder for the ceramic powders, which are typically silica. Other materials, such as feldspar [$(K, Na)_2O \cdot Al_2O_3 \cdot 6SiO_2$] serve as fluxing (glass-forming) agents during later heat treatment.

Forming Techniques for Clay Products   The powders, clay, flux, and water are mixed and formed into a shape (Figure 14-18). Dry or semi-dry mixtures are mechanically pressed into "green" (unbaked) shapes of sufficient strength to be handled. For more uniform compaction of complex shapes, isostatic pressing may be done; the powders are placed into a rubber mold and subjected to high pressures through a gas or liquid medium. Higher moisture contents permit the powders to be more plastic or formable. **Hydroplastic forming** processes, including extrusion, jiggering, and hand working, can be applied to these plastic mixes. Ceramic slurries containing large amounts of organic plasticizers, rather than water, can be injected into molds.

Still higher moisture contents permit the formation of a slip, or pourable slurry, containing fine ceramic powder. The slip is poured into a porous mold. The water in the slip nearest to the mold wall is drawn into the mold, leaving behind a soft solid which has a low-moisture content. When enough water has been drawn from the slip to produce a desired thickness of solid, the remaining liquid slip is poured from the mold, leaving behind a hollow shell. Slip casting is used in manufacturing washbasins and other commercial products. After forming, the ceramic bodies—or greenware—are still weak, contain water or other lubricants, and are porous, and subsequent drying and firing are required.

Drying and Firing of Clay Products   During drying, excess moisture is removed and large dimensional changes occur (Figure 14-19). Initially, the water between the clay platelets—the interparticle water—evaporates and provides most of the shrinkage. Relatively little dimensional change occurs as the remaining water between the pores evaporates. The temperature and humidity are carefully controlled to provide uniform drying throughout the part, thus minimizing stresses, distortion, and cracking.

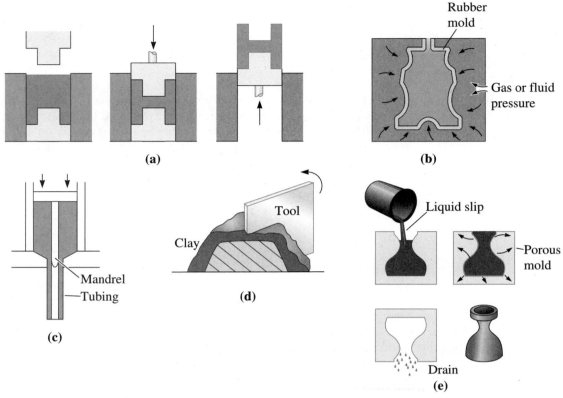

Figure 14-18 Processes for shaping crystalline ceramics: (a) pressing, (b) isostatic pressing, (c) extrusion, (d) jiggering, and (e) slip casting.

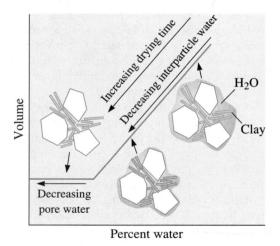

Figure 14-19
The change in the volume of a ceramic body as moisture is removed during drying. Dimensional changes cease after the interparticle water is gone.

The rigidity and strength of a ceramic part are obtained during **firing**. During heating, the clay dehydrates, eliminating the hydrated water that is part of the kaolinite crystal structure, and vitrification, or melting, begins (Figure 14-20). Impurities and the fluxing agent react with the ceramic particles (SiO_2) and clay, producing a low-melting-point liquid phase at the grain surfaces. The liquid helps eliminate porosity and, after

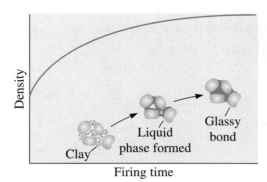

Figure 14-20
During firing, clay and other fluxing materials react with coarser particles to produce a glassy bond and reduce porosity.

cooling, changes to a rigid glass that binds the ceramic particles. This glassy phase provides a **ceramic bond**, but it also causes additional shrinkage of the entire ceramic body.

The grain size of the final part is determined primarily by the size of the original powder particles. Furthermore, as the amount of flux increases, the melting temperature decreases; more glass forms, and the pores become rounder and smaller. A smaller initial grain size accelerates this process by providing more surface area at which vitrification can occur.

Applications of Clay Products   Many structural clay products and whitewares are produced using these processes. Brick and tile used for construction are pressed or extruded into shape, dried, and fired to produce the ceramic bond. Higher firing temperatures or finer original particle sizes produce more vitrification, less porosity, and higher density. The higher density improves mechanical properties but reduces the insulating qualities of the brick or tile.

Earthenware are porous clay bodies fired at relatively low temperatures. Little vitrification occurs; the porosity is very high and interconnected, and earthenware ceramics may leak. Consequently, these products must be covered with an impermeable glaze.

Higher firing temperatures, which provide more vitrification and less porosity, produce stoneware. The stoneware, which is used for drainage and sewer pipe, contains only 2% to 4% porosity. Ceramics known as china and porcelain require even higher firing temperatures to cause complete vitrification and virtually no porosity.

14-10   Refractories

Refractory materials are important components of the equipment used in the production, refining, and handling of metals and glasses, for constructing heat-treating furnaces, and for other high-temperature processing equipment. The **refractories** must survive at high temperatures without being corroded or weakened by the surrounding environment. Typical refractories are composed of coarse-oxide particles bonded by a finer refractory material. The finer material melts during firing, providing bonding. In some cases, refractory bricks contain about 20% to 25% apparent porosity to provide improved thermal insulation.

Refractories are often divided into three groups—acid, basic, and neutral—based on their chemical behavior (Table 14-7).

TABLE 14-7 ■ Compositions of typical refractories (weight percents)

Refractory	SiO$_2$	Al$_2$O$_3$	MgO	Fe$_2$O$_3$	Cr$_2$O$_3$
Acidic					
Silica	95–97				
Superduty firebrick	51–53	43–44			
High-alumina firebrick	10–45	50–80			
Basic					
Magnesite			83–93	2–7	
Olivine	43		57		
Neutral					
Chromite	3–13	12–30	10–20	12–25	30–50
Chromite-magnesite	2–8	20–24	30–39	9–12	30–50

From Ceramic Data Book, *Cahners Publishing Co., 1982.*

Acid Refractories   Common acidic refractories include silica, alumina, and fireclay (an impure kaolinite). Pure silica is sometimes used to contain molten metal. In some applications, the silica may be bonded with small amounts of boron oxide, which melts and produces the ceramic bond. When a small amount of alumina is added to silica, the refractory contains a very low-melting-point eutectic microconstituent (Figure 14-21) and is not suited for refractory applications at temperatures above about 1600°C, a temperature often required for steel making. However, when larger amounts of alumina are added, the microstructure contains increasing amounts of mullite, 3Al$_2$O$_3$ · 2SiO$_2$, which has a high melting temperature. These fireclay refractories are generally relatively weak, but they are inexpensive. Alumina concentrations above about 50% constitute the high-alumina refractories.

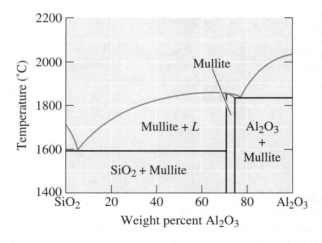

Figure 14-21
A simplified SiO$_2$-Al$_2$O$_3$ phase diagram, the basis for alumina silicate refractories.

Basic Refractories   A number of refractories are based on MgO (magnesia, or periclase). Pure MgO has a high melting point, good refractoriness, and good resistance to attack by the basic environments often found in steel-making processes. Olivine refractories contain forsterite, or Mg$_2$SiO$_4$, and also have high melting points (Figure 14-22). Other magnesia refractories may include CaO or carbon. Typically, the basic refractories are more expensive than the acid refractories.

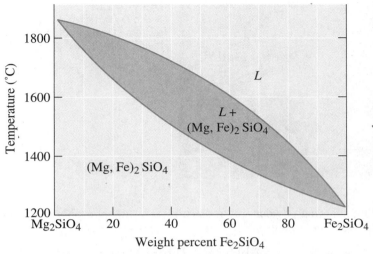

Figure 14-22 The Mg_2SiO_4-Fe_2SiO_4 phase diagram, showing complete solid solubility.

Neutral Refractories These refractories, which include chromite and chromite-magnesite, might be used to separate acid and basic refractories, preventing them from attacking one another.

Special Refractories Carbon, or graphite, is used in many refractory applications, particularly when oxygen is not present. Other refractory materials include zirconia (ZrO_2), zircon ($ZrO_2 \cdot SiO_2$), and a variety of nitrides, carbides, and borides. Most of the carbides, such as TiC and ZrC, do not resist oxidation well, and their high-temperature applications are best suited to reducing conditions. However, silicon carbide is an exception; when SiC is oxidized at high temperatures, a thin layer of SiO_2 forms at the surface, protecting the SiC from further oxidation up to about 1500°C. Nitrides and borides also have high-melting temperatures and are less susceptible to oxidation. Some of the oxides and nitrides are candidates for use in jet engines.

14-11 Other Ceramic Materials

In addition to their use in producing construction materials, appliances, structural materials, and refractories, ceramics find a host of other applications, including the following.

Cements Ceramic raw materials are joined using a binder that does not require firing or sintering in a process called **cementation**. A chemical reaction converts a liquid resin to a solid that joins the particles. In the case of sodium silicate, the introduction of CO_2 gas acts as a catalyst to dehydrate the sodium silicate solution into a glassy material:

$$x\text{Na}_2\text{O} \cdot y\text{SiO}_2 \cdot \text{H}_2\text{O} + \text{CO}_2 \rightarrow \text{glass (not balanced)}$$

Figure 14-23 shows silica sand grains used to produce molds for metal casting. The liquid sodium silicate coats the sand grains and provides bridges between the sand grains. Introduction of the CO_2 converts the bridges to a solid, joining the sand grains.

Figure 14-23
A photograph of silica sand grains bonded with sodium silicate through the cementation mechanism (×60).

Fine alumina-powder solutions catalyzed with phosphoric acid produce an aluminum phosphate cement:

$$Al_2O_3 + 2H_3PO_4 \rightarrow 2AlPO_4 + 3H_2O$$

When alumina particles are bonded with the aluminum phosphate cement, refractories capable of operating at temperatures as high as 1650°C are produced.

Plaster of paris, or gypsum, is another material that is hardened by a cementation reaction:

$$CaSO_4 \cdot \tfrac{1}{2}H_2O + \tfrac{3}{2}H_2O \rightarrow CaSO_4 \cdot 2H_2O$$

When the liquid slurry reacts, interlocking solid crystals of gypsum ($CaSO_4 \cdot 2H_2O$) grow, with very small pores between the crystals. Larger amounts of water in the original slurry provide more porosity, but they also decrease the strength of the final plaster. One of the important uses of this material is for construction of walls in buildings.

The most common and important of the cementation reactions occurs in Portland cement, which is used to produce concrete.

Coatings   Ceramics are often used to provide protective coatings to other materials. Common commercial coatings include glazes and enamels. Glazes are applied to the surface of a ceramic material to seal a permeable clay body, to provide protection and decoration, or for special purposes. Enamels are applied to metal surfaces. The enamels and glazes are clay products that vitrify easily during firing. A common composition is $CaO \cdot Al_2O_3 \cdot 2SiO_2$.

Special colors can be produced in glazes and enamels by the addition of other minerals. Zirconium silicate gives a white glaze, cobalt oxide makes the glaze blue, chromium oxide produces green, lead oxide gives a yellow color, and a red glaze may be produced by adding a mixture of selenium and cadmium sulfides.

One of the problems encountered with a glaze or enamel is surface cracking, or crazing, which occurs when the glaze has a coefficient of thermal expansion different than that of the underlying material. This is frequently the most important factor in determining the composition of the coating.

Special coatings are used for advanced ceramics and high-service temperature metals. SiC coatings are applied to carbon-carbon composite materials to improve their oxidation resistance. Zirconia coatings are applied to nickel-based superalloys to provide thermal barriers that protect the metal from melting or adverse reactions.

Thin Films and Single Crystals   Thin films of many complex and multi-component ceramics are produced using different techniques such as sputtering, sol-gel, and chemical-vapor deposition (CVD). Usually, the thickness of such films is less than 0.05 to 10 μm, and more likely greater than 2 μm. Many functional electronic ceramic thin films are prepared and integrated onto silicon wafers, glasses, and other substrates. For example, the magnetic strips on credit cards use gamma iron oxide (γ-Fe_2O_3 or Fe_3O_4) thin films for storing data. Indium tin oxide (ITO), a conductive and transparent material, is coated on glass and used in applications such as touch-screen displays (Table 14-1). Many other coatings are used on glass to make the glass energy efficient. Recently, a self-cleaning glass using a TiO_2 coating has been developed.[9,24] Similarly, thin films of ceramics, such as PZT, PLZT, and $BaTiO_3$-$SrTiO_3$ solid solutions, can be prepared and used. Often, the films develop an orientation, or texture, which may be advantageous for a given application. Single crystals of many ceramics (e.g., SiO_2 or quartz, lithium niobate ($LiNbO_3$), sapphire, or yttrium aluminum garnate) are used in many electrical and electro-optical applications. These crystals are grown from melts using techniques similar to those described in Chapter 8.

Fibers   Fibers are produced from ceramic materials for several uses: as a reinforcement in composite materials, for weaving into fabrics, or for use in fiber-optic systems. Borosilicate glass fibers, the most commonly produced fibers, provide strength and stiffness in fiberglass. Fibers can be produced from a variety of other ceramics, including alumina, silicon carbide, silica, and boron carbide. The sol-gel process is also used to produce commercial fibers for many applications.[17]

A special type of fibrous material is the silica tile used to provide the thermal protection system for NASA's space shuttle. Silica fibers are bonded with colloidal silica to produce an exceptionally lightweight tile with densities as low as 0.144 g/cm^3; the tile is coated with special high-emissivity glazes to permit protection up to 1300°C.

Joining and Assembly of Ceramic Components   Ceramics are often made as monolithic components rather than assemblies of numerous components.[25] When two ceramic parts are placed in contact under a load, stress concentrations at the brittle surface are created, leading to an increased probability of failure. In addition, methods for joining ceramic parts into a larger assembly are limited. The brittle ceramics cannot be joined by fusion welding or deformation bonding processes. At low temperatures, adhesive bonding using polymer materials may be accomplished; ceramic cements may be used at higher temperatures. Diffusion bonding and brazing can be used to join ceramics and to join ceramics to metals.

SUMMARY

◆ Ceramics are inorganic materials that have high hardness and high melting points. These include single crystal and polycrystalline ceramics, glasses, and glass-ceramics. Typical ceramics are electrical and thermal insulators with good chemical stability and good strength in compression.

◆ Polycrystalline ceramics exhibit a brittle behavior, partly because of porosity. Because most polycrystalline ceramics cannot plastically deform (unless special conditions with respect to temperature and strain rates are created), the porosity limits the ability of the material to withstand a tensile load.

◆ Ceramics play a critical role in a wide array of technologies related to electronic, magnetic, optical, and energy. Many advanced ceramics play a very important role in providing thermal insulation and high-temperature properties. Applications of advanced ceramics range from credit cards, houses for silicon chips, tiles for the space shuttle, medical imaging, optical fibers that enable communication, and safe and energy efficient glasses. Traditional ceramics play a very important role as refractories for metals processing and consumer applications.

◇ Traditional ceramic-powder synthesis involves the treatment of ores using processes such as crushing and grinding, followed by ball milling, and calcination. These processes are inexpensive and widely used for making many ceramic powders.

◇ Specialized chemical precursor-based processes (e.g., precipitation, sol-gel, etc.) are used to make higher-purity and chemically homogeneous powders.

◇ Ceramic processing is commonly conducted using compaction and sintering. For specialized applications, isostatic compaction, hot pressing, and hot isostatic pressing (HIP) are used, especially to achieve higher-densification levels.

◇ Tape casting, slip casting, extrusion, and injection molding are some of the other techniques used to form green ceramics into different shapes. These processes are then followed by a burnout step in which binders and plasticizers are burnt off, and the resultant ceramic is sintered.

◇ The driving force to sintering is a reduction in the particle surface area. Bulk diffusion and grain-boundary diffusion are used for densification. Evaporation, condensation, and surface diffusion contribute to grain growth but not to densification. Sintering of ceramics often involves the formation of a liquid phase due to the dopants added or impurities present.

◇ Many silicates and other ceramics form glasses rather easily, since the kinetics of crystallization are sluggish. Glasses can be formed in the form of sheets using float-glass or fibers and other shapes. Silicate glasses are used in a significant number of applications that include window glass, windshields, fiber optics, and fiberglass. The strength of the glass is controlled by its surface flaws. Tempering is used to introduce a compressive surface layer on the glass. This glass is used where safety is important. Annealing is used to remove stresses from glass. Windshields are made using laminated glass. Many different types of coatings can be deposited on glasses to make them more useful.

◇ Glass-ceramics are formed using controlled crystallization of inorganic glasses. These materials are used widely for kitchenware and many other applications.

◇ Ceramics, in the form of fibers, thin films, coatings, and single crystals, have many different applications.

GLOSSARY

Apparent porosity   The percentage of a ceramic body that is composed of interconnected porosity.

Bulk density   The mass of a ceramic body per unit volume, including closed and interconnected porosity.

Calcination   Heating of chemicals to decompose and or react with different chemicals; used in traditional synthesis of ceramics.

Cementation   Bonding ceramic raw materials into a useful product, using binders that form a glass or gel without firing at high temperatures.

Ceramic   An inorganic material with high melting temperature. Usually hard and brittle.

Ceramic bond   Bonding ceramic materials by permitting a glassy product to form at high-firing temperatures.

Cermet   A ceramic-metal composite (e.g., WC-Co) providing a good combination of hardness with other properties such as toughness.

Cold isostatic pressing (CIP)   A powder-shaping technique in which hydrostatic pressure is applied during compaction. This is used for achieving a higher green ceramic density or compaction of more complex shapes.

Devitrification   The crystallization of glass.

Enamel   A ceramic coating on metal.

Firing   Heating a ceramic body at a high temperature to cause a ceramic bond to form.

Flux   Additions to ceramic raw materials that reduce the melting temperature.

Glass   An amorphous inorganic material derived by cooling of a melt.

Glass-ceramics   Ceramic shapes formed in the glassy state and later allowed to crystallize during heat treatment to achieve improved strength and toughness.

Glass formers   Oxides with a high-bond strength that easily produce a glass during processing.

Glass temperature   The temperature below which an undercooled liquid becomes a glass.

Glaze   A ceramic coating applied to glass. The glaze contains glassy and crystalline ceramic phases.

Green ceramic   A ceramic that has been shaped into a desired form but has not yet been sintered.

Hot isostatic pressing (HIP)   A powder-processing technique in which large pieces of metals, alloys and ceramics can be produced using sintering under a hydrostatic pressure generated by a gas.

Hot pressing   A ceramic processing technique in which sintering is conducted under uniaxial pressure.

Hydroplastic forming   A number of processes by which a moist ceramic clay body is formed into a useful shape.

Injection molding   A processing technique in which a thermoplastic mass (loaded with ceramic powder) is mixed in an extruder-like setup and then injected into a die to form complex parts. In the case of ceramics, the thermoplastic is burnt off. Also, widely used for thermoplastics (Chapter 15).

Intermediates   Oxides that, when added to a glass, help to extend the glassy network; although the oxides normally do not form a glass themselves.

Laminated glass   Annealed glass with a polymer (e.g., polyvinyl butyral, *PVB*) sandwiched in between, used for car windshields.

Leaching   A process in which acids or alkalis are used to dissolve a mineral, conducted typically to get the metal or mineral of interest in solution.

Parison   A crude glassy shape that serves as an intermediate step in the production of glassware. The parison is later formed into a finished product.

Powder   A collection of fine particles.

Powder metallurgy Powder processing routes used for converting metal and alloy powders into useful shapes.

Powder processing Unit operations conducted to convert powders into useful shapes (e.g., pressing, tape casting, etc.).

Reaction bonding A ceramic processing technique by which a shape is made using one material that is later converted into a ceramic material by reaction with a gas.

Refractories A group of ceramic materials capable of withstanding high temperatures for prolonged periods of time.

Slip A liquid slurry that is poured into a mold. When the slurry begins to harden at the mold surface, the remaining liquid slurry is decanted, leaving behind a hollow ceramic casting.

Slip casting Forming a hollow ceramic part by introducing a pourable slurry into a mold. The water in the slurry is extracted into the porous mold, leaving behind a drier surface. Excess slurry can then be decanted.

Sol-gel process A method that uses a sol (dispersion of colloidal particles or molecules) that is converted into a gel and ultimately into a useful product such as a thin film, powder, or porous and monolithic ceramic. Polymeric metallorganic or colloidal oxide (hydroxide) particles are used to generate single or multi-component ceramic materials at low temperatures.

Spray drying A slurry of a ceramic powder is sprayed into a large chamber in the presence of hot air. This leads to the formation of soft agglomerates that can flow well into the dies used during powder compaction.

Synthesis Steps conducted to make a ceramic powder.

Tape casting A process for making thin sheets of ceramics using a ceramic slurry consisting of binders, plasticizers, etc. The slurry is cast with the help of a blade onto a plastic substrate. The resultant green tape is then dried, cut, and machined and used to make electronic ceramic and other devices.

Tempered glass A high-strength glass that has a surface layer where the stress is compressive, induced thermally during cooling or by the chemical diffusion of ions.

True porosity The percentage of a ceramic body that is composed of both closed and interconnected porosity.

Viscous flow Deformation of a glassy material at high temperatures.

Vitrification Melting, or formation, of a glass.

ADDITIONAL INFORMATION

1. BRONOWSKI, J., *The Ascent of Man*. 1973: Back Bay Books. 124–131.

2. HÖLAND, W. and G.H. BEALL, *Glass-Ceramic Technology*. 2002: American Ceramic Society.

3. BERGERON, C.G. and S.H. RISBUD, *Introduction to Phase Equilibria*. 1984: American Ceramic Society.

4. BENGISU, M., *Engineering Ceramics*. 2001: Springer Verlag.

5. HARPER, C.A., *Handbook of Materials Product Design*. 3rd ed. 2001: McGraw Hill.

6. MEYERS, M.A. and K.K. CHAWLA, *Mechanical Behavior of Materials*. 1998: Prentice Hall.

7. DOWLING, N.E., *Mechanical Behavior of Materials: Engineering Methods for Deformation, Fracture, and Fatigue*. 2nd ed. 1999: Upper Saddle River, N.J., Prentice Hall.

8. CHEN, I.W., "Superplasticity in Ceramic Systems," *Encyclopedia of Advanced Materials*, D.B. BLORR, R.J., FLEMINGS, M.C. and MAHAJAN, S., Editor. 1994. p. 2715–2717.

9. GREENBERG, CHARLES, PPG Industries. *Technical discussions are acknowledged.* 2002.

10. REED, J.S., *Principles of Ceramics Processing.* 2nd ed. 1995: John Wiley and Sons.

11. RICHERSON, D.W., *Modern Ceramic Engineering.* 1992: Dekker.

12. MHAISALKAR, S. and D.W. READY, *Personal communication.* 1989.

13. YOUNG, C. and B. DUGAN, *Zinc Coporation of America, technical discussions and assistance are acknowledged.* 2002.

14. STEVENS, R., *Zirconia and Zirconia Ceramics.* 2nd ed. 1986: Magnesium Elektron Ltd.

15. NETTLESHIP, I. and R. STEVENS, "Tetragonal Zirconia Polycrystal (TZP)—A Review," *Int. J. High Technol. Ceram*, 1987. 3(1): p. 1–32.

16. PHULÉ, P.P. and S.H. RISBUD, "Synthesis of Ceramics in the BaO-TiO$_2$ System." *Journal of Materials Science*, 1990.

17. PHULÉ, P.P. and T. WOOD, "Sol-Gel Processing of Ceramics," *Encyclopedia of Advanced Materials*, M. FLEMINGS, Editor. 2001: Permgaon.

18. CHIANG, Y.M., D. BIRNIE, and W.D. KINGERY, *Physical Ceramics: Principles for Ceramic Science and Engineering.* 1997: Wiley.

19. BARSOUM, M., *Fundamentals of Ceramics.* 1997: McGraw Hill.

20. WILKISNON, D.S. and M.F. ASHBY, *Sintering and Catalysis*, G.C. KUCZYNSKI, Editor. 1976: New York, Plenum. p. 476.

21. MISTLER, R.E. and E.R. TWINAME, *Tape Casting: Theory and Practice.* 2002: American Ceramic Society.

22. LEE, H., RISBUD, S.H., et al. "Quantum Confined Electron-Hole States in ZnSe Quantum Dots." *Mater. Res. Soc. Symp. Proc. Vol. 571 Semiconductor Quantum Dots.* 2000.

23. STOOKEY, DONALD, "Profiles in Ceramics," *Ceramic Bulletin.* 2001.

24. GREENBERG, C., *Discussion on self-cleaning glasses and use of thin-film coatings is acknowledged.* 2002.

25. REIMANIS, I., C.J. HENAGER, and A. TOMSIA. "Ceramic Joining Ceramic Transactions, Volume 77." *Proceedings of the Ceramic Joining Symposium at the 98th Annual Meeting of the American Ceramic Society*, 1996: Indianapolis, Indiana, American Ceramic Society.

26. YINNON, H., D.R. UHLMANN, and N.J. KRIDEL, Editors, *Glass-Science and Technology.* Vol. 1. 1983: Academic Press.

✓ PROBLEMS

Section 14-1 Applications of Ceramics

14-1 What are the primary types of atomic bonds in ceramics?

14-2 Explain the meaning of following terms: ceramics, inorganic glasses, and glass-ceramics.

14-3 Explain why ceramics typically are processed as powders. How is this similar to or different from the processing of metals?

14-4 What do the terms "glaze" and "enamel" mean?

14-5 What material is used to make the tiles that provide thermal protection in NASA's space shuttle?

14-6 Which ceramic materials are most widely used?

14-7 Explain how ceramic materials can be classified in different ways.

14-8 State any one application of the following ceramics: (a) alumina, (b) silica, (c) barium titanate, (d) zirconia, (e) boron carbide, and (f) diamond.

14-9 What is a piezoelectric material? State an example of a piezoelectric ceramic and give some applications.

14-10 State applications of silica in the forms of a fiber and a fine particle.

14-11 What does the term sialon mean? What is WC-Co used for? What are the roles of WC and Co?

14-12 What are some applications of magnetic ceramics?

14-13 How does zirconia serve as a sensor for measuring oxygen in car exhaust or in molten steel?

Section 14-2 Properties of Ceramics

14-14 What are some of the typical characteristics of ceramic materials?

14-15 Why is the tensile strength of ceramics much lower than that under a compressive stress?

14-16 Plastic deformation due to dislocation motion is important in metals; however, this is not a very important consideration for the properties of ceramics and glasses. Explain.

14-17 Can ceramic materials show superplastic behavior or are they always brittle? Explain.

14-18 Explain why the strength of ceramics tends to show a wide scatter in their mechanical properties.

Section 14-3 Synthesis of Ceramic Powders

14-19 Explain the role of crushing and grinding in the synthesis of ceramic powders.

14-20 What is ball milling? Explain using a sketch.

14-21 For ball milling of nickel zinc ferrite, steel spheres are used as a grinding medium. However, zirconia spheres are used to grind powders of $BaTiO_3$. Explain why this may be the case.

14-22 What does the term "leaching" mean?

14-23 What is calcination?

14-24 How are the following ceramic powders made? (a) alumina, (b) titanium dioxide, (c) zirconium dioxide, and (d) zinc oxide.

14-25 You want to use alumina to make spark plugs. Assuming alumina is made using the Bayer process, what impurity would you be concerned about most? Explain.

14-26 Compare chemical and conventional (oxide-mix) processes for manufacturing ceramic powders.

14-27 A manufacturer of *PZT* ceramics wants to make 500 kilograms of $Pb(Zr_{0.4}Ti_{0.6})O_3$ ceramic powder. How much PbO, ZrO_2, and TiO_2

would be needed? Assume that these precursors have no impurities.

14-28 Barium titanate can be made using a hydrothermal process in which TiO_2 is reacted with barium ions in an alkaline solution of pH ~10. Assuming there is a practically unlimited supply of Ba^{+2} in the aqueous solution, how much TiO_2 would be needed to make 10 kilograms of hydrothermal $BaTiO_3$. Assume the following reaction leads to the formation of hydrothermal $BaTiO_3$.

$$Ba(OH)_2 + TiO_{2-} \rightarrow BaTiO_3 + H_2O$$

14-29 Explain how one can make yttria-stabilized zirconia powder using the freeze-drying process (see Chapter 9 for help).

Section 14-4 Powder Processing

14-30 What the biggest advantage to sintering?

14-31 What the biggest advantage to grain growth?

14-32 What mechanisms of diffusion play the most important role in the solid-state sintering of ceramics?

14-33 What is liquid-phase sintering?

14-34 Explain the use of the following processes (use a sketch as needed): (a) uniaxial compaction and sintering, (b) hot pressing, (c) HIP, and (d) tape casting.

14-35 What are the applications of slip casting and tape casting?

14-36 Why is it that the atmosphere used in the sintering of ceramics often needs to be controlled?

Section 14-5 Characteristics of Sintered Ceramics

14-37 What are some of the important characteristics of sintered ceramics?

14-38 For observing the grain size of metallic materials, it is possible to etch the grain boundaries using acids and alkalis. This, however, is difficult for ceramic materials. Explain how polishing, grinding, and subsequent exposure to a high temperature can be used to affect the surface microstructure of ceramics. (Note: This technique is known as thermal grooving or thermal etching.)

14-39 What typical density levels are obtained in sintered ceramics?

14-40 Zinc oxide is used to make electrical surge protection devices known as varistors (variable resistance devices). A zinc oxide manufacturer is assessing different ZnO powders. The company finds that the starting ZnO powder has a size of about ~1 μm. In making the electrical surge protection devices, bismuth oxide, antimony oxide, and other oxides are added in small quantities. The average grain size of the sintered ceramic is ~30 μm. What does this say about the possible mechanism of sintering?

14-41 What do the terms "apparent porosity" and "true porosity" of ceramics mean?

14-42 The specific gravity of Al_2O_3 is 3.96 g/cm^3. A ceramic part is produced by sintering alumina powder. It weighs 80 g when dry, 92 g after it has soaked in water, and 58 g when suspended in water. Calculate the apparent porosity, the true porosity, and the closed pores.

14-43 Silicon carbide (SiC) has a specific gravity of 3.1 g/cm^3. A sintered SiC part is produced, occupying a volume of 500 cm^3 and weighing 1200 g. After soaking in water, the part weighs 1250 g. Calculate the bulk density, the true porosity, and the volume fraction of the total porosity that consists of closed pores.

14-44 Explain how polymeric foams (similar to Styrofoam^{TM}) can be used to make ceramic foams that could be used for the filtration of molten metals.

Section 14-6 Inorganic Glasses

Section 14-7 Processing and Applications of Glasses

14-45 What is the main reason why glass formation is easy in silicate systems?

14-46 Can glasses be formed using metallic materials?

14-47 Define the terms "glass formers", "intermediates", and "modifiers".

14-48 What does the term "glass temperature" mean? Is this a fixed temperature for a given composition of glass?

14-49 Explain the float-glass process for making glass.

14-50 What is the role of lime and soda in a soda-lime glass?

14-51 Explain how phase separation and metastable immiscibility are used to make Vycor^{TM} glass (see earlier discussions in Chapters 9 to 12).

14-52 Chemically speaking, what is Pyrex$^®$ glass? What is it used for?

14-53 What does the letter "E" in E-glass stand for? What does the letter "S" in S-glass stand for? What are the E- and S-glasses used for?

14-54 Calculate the O:Si ratio when 20 wt% Na_2O is added to SiO_2. Explain whether this material will provide good glass-forming tendencies. Above what temperature must the ceramic be heated to be all-liquid?

14-55 How many grams of BaO can be added to 1 kg of SiO_2 before the O:Si ratio exceeds 2.5 and glass-forming tendencies are poor? Compare this with the case when Li_2O is added to SiO_2.

14-56 Calculate the O:Si ratio when 30 wt% Y_2O_3 is added to SiO_2. Will this material provide good glass-forming tendencies?

14-57 Lead can be introduced into a glass either as PbO (where the Pb has a valence of +2) or as PbO_2 (where the Pb has a valence of +4). Such leaded glasses are used to make what is marketed as "crystal glass" for dinnerware. These glasses are also used to make glasses for computer and TV CRT screens. Draw a sketch (similar to Figure 14-10) showing the effect of each of these oxides on the silicate network.

SiO$_2$ glass   Na$_2$O modified glass

Figure 14-10 (Repeated for Problem 14-57) The effect of Na$_2$O on the silica glass network. Sodium oxide is a modifier, disrupting the glassy network and reducing the ability to form a glass.

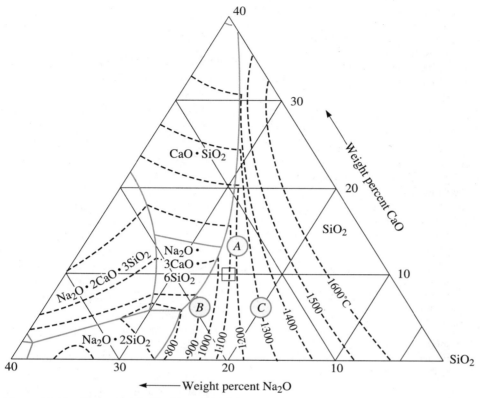

Figure 14-16 (Repeated for Problem 14-58.) The liquidus plot for the SiO_2-CaO-Na_2O phase diagram.

Which oxide is a modifier and which is an intermediate?

14-58 A glass composed of 65 mol% SiO_2, 20 mol% CaO, and 15 mol% Na_2O is prepared. Calculate the O:Si ratio and determine whether the material has good glass-forming tendencies. Estimate the liquidus temperature of the material using Figure 14-16.

Section 14-8 Glass-Ceramics

14-59 How is a glass-ceramic different from a glass and a ceramic?

14-60 What are the advantages of using glass-ceramics as compared to either glasses or ceramics?

14-61 What is the role of heterogeneous nucleation and phase separation in the formation of glass-ceramics?

14-62 Draw a typical heat-treatment profile encountered in processing of glass-ceramics.

14-63 Compare the kinetics of crystallization of glass with the TTT and CCT curves encountered in the heat treatment of steels.

14-64 What are some of the important applications of glass-ceramics?

Section 14-11 Other Ceramic Materials

14-65 In tape casting and injection molding, we use polymeric molecules as binders. What is the binder in forming clays?

14-66 What are some of the applications of clay products?

14-67 What different types of refractories are used in the manufacturing of metals?

14-68 What is cement?

14-69 What is the role of a yttria stabilized zirconia (*YSZ*) coating used on certain turbine blades? Why does the zirconia have to be "stabilized?"

14-70 What is the difference between a thin film and a coating?

14-71 What materials are used for making a conductive and transparent glass coating?

14-72 Give examples of some ceramic materials that are used to make lasers.

14-73 What are some applications of ceramic fibers?

✳ Design Problems

14-74 Design a silica soda-lime glass that can be cast at 1300°C and that will have a O:Si ratio of less than 2.3 to assure good glass-forming tendencies. To assure adequate viscosity, the casting temperature should be at least 100°C above the liquidus temperature.

◻ Computer Problems

14-75 *Sintering Profile.* Write a computer program that will ask the user to provide temperatures during different stages of sintering. The program should ask the user to provide starting temperatures, temperatures for binder burn out, sintering-hold temperatures, and final temperatures. A time for sintering must also be provided by the user. The program should ask the user to provide the heating rates for the binder burn out and the stage between binder burn out and sintering, as well as the cooling rate. The program should then provide temperatures for any given time and declare the sintering cycle stage in the furnace.

14-76 *Apparent and Bulk Density.* Write a program that will ask the user to provide the different information needed to calculate apparent and bulk density of ceramic parts. The program should also calculate the interconnected and closed porosity.

Chapter

Polymers are used in an amazing array of technologies, including automobiles, microelectronics, and composites. This photograph is of Charles Goodyear who invented the vulcanization process. Similar to many other very important events in the history of science and technology, serendipity played a key role in this invention. (*Courtesy of Goodyear Inc.*)

15

Polymers

Have You Ever Wondered?

- What are compact disks (CDs) made from?

- Who was Charles Goodyear?

- What is Silly Putty® made from?

- What polymer is used in chewing gum?

- Which was the first synthetic fiber ever made?

- Why are some plastics "dishwasher safe" and some not?

- What are bulletproof vests made from?

- Why are nylon strings in a tennis rackets initially stretched to provide a tension higher than that desired ultimately?

- What polymer is used for non-stick cookware?

The suffix *mer* means a "unit." In this context, the term mer refers to a unit group of atoms or molecules that defines a characteristic arrangement for a polymer. A polymer can be thought of as a material made by combining several mers or units. Polymers are materials consisting of giant or macromolecules, chain-like molecules having average molecular weights from 10,000 to more than 1,000,000 g/mol built by joining many mers or units through chemical bonding. Molecular weight is defined as the sum of atomic masses in each molecule. Most polymers, solids or liquids, are carbon-based organic; however, they can be inorganic (e.g., silicones based on a Si-O network).

Plastics are materials that are composed principally of naturally occurring and modified or artificially made polymers often containing additives such as fibers, fillers, pigments, and the like that further enhance their properties.[1] Plastics include thermoplastics (commodity and engineering), thermoset materials, and elastomers (natural or synthetic). In this book, we use the terms plastic and polymers interchangably. *Polymerization* is the process by which small molecules consisting of one unit (known as a **monomer**) or a few units (known as **oligomers**) are chemically joined to create these giant molecules. Polymerization normally begins with the production of long chains in which the atoms are strongly joined by covalent bonding. Plastics are used in an amazing number of applications including clothing, toys, home appliances, structural and decorative items, coatings, paints, adhesives, automobile tires, biomedical materials, car bumpers and interiors, foams, and packaging. Polymers are often used in composites, both as fibers and as a matrix. Liquid crystal displays (LCDs) are based on polymers.[2] We also use polymers in photochromic lenses. Plastics are often used to make electronic components because of their insulating ability and low dielectric constant. More recently, significant developments have occurred in the area of flexible electronic devices based on the useful piezoelectricity, semiconductivity, optical and electro-optical properties seen in some polymers.[3] Polymers such as polyvinyl acetate (PVA) are water-soluble. Many such polymers can be dissolved in water or organic solvents to be used as binders, surfactants, or plasticizers in processing ceramics, semiconductors, and as additives to many consumer products. Polyvinyl butyral (PVB), a polymer, makes up part of the laminated glass used for car windshields. Polymers are probably used in more technologies than any other class of materials.

Commercial—or standard commodity—polymers are lightweight, corrosion-resistant materials with low strength and stiffness, and they are not suitable for use at high temperatures. These polymers are, however, relatively inexpensive and are readily formed into a variety of shapes, ranging from plastic bags to mechanical gears to bathtubs. *Engineering polymers* are designed to give improved strength or better performance at elevated temperatures. These materials are produced in relatively small quantities and often are expensive. Some of the engineering polymers can perform at temperatures as high as 350°C; others—usually in a fiber form—have strengths that are greater than that of steel.

Polymers also have many useful physical properties. Some polymers such as acrylics like Plexiglas™ and Lucite™ are transparent and can substitute for glasses. Although most polymers are electrical insulators, special polymers (such as the acetals) and polymer-based composites possess useful electrical conductivity. Teflon™ has a low coefficient of friction and is the coating for nonstick cookware. Polymers also resist corrosion and chemical attack.

15-1   Classification of Polymers

Polymers are classified in several ways: by how the molecules are synthesized, by their molecular structure, or by their chemical family. One way to classify polymers is to state if the polymer is a **linear polymer** or a **branched polymer** (Figure 15-1). A linear polymer consists of spaghetti-like molecular chains. In a branched polymer, there are primary polymer chains and secondary (offshoots) of smaller chains that stem from

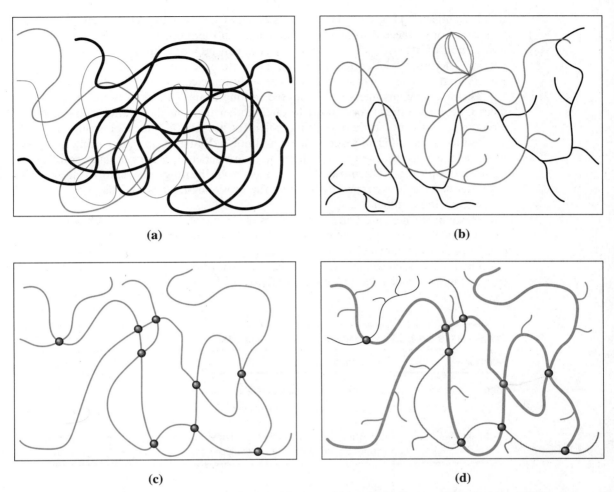

Figure 15-1 Schematic showing linear and branched polymers. Note that branching can occur in any type of polymer (e.g., thermoplastics, thermosets, and elastomers). (a) Linear unbranched polymer: notice chains are not straight lines and not connected. Different polymer chains are shown using different shades and design to show clearly that each chain is not connected to another. (b) Linear branched polymer: chains are not connected, however they have branches. (c) Thermoset polymer without branching: chains are connected to one another by covalent bonds but they do not have branches. Joining points are highlighted with solid circles. (d) Thermoset polymer that has branches and chains that are interconnected via covalent bonds. Different chains and branches are shown in different shades for better contrast. Places where chains are actually chemically bonded are shown with filled circles.

these main chains. Note that even though we say "linear", the chains are actually not in the form of straight lines. A better method to describe polymers is in terms of their mechanical and thermal behavior. Table 15-1 compares the three major polymer categories.

Thermoplastics are composed of long chains produced by joining together monomers; they typically behave in a plastic, ductile manner. The chains may or may not have branches. Individual chains are intertwined. There are relatively weak van der Waals bonds between atoms of different chains. This is somewhat similar to a few trees that are tangled up together. The trees may or may not have branches, each tree is on its own and not connected to another. The chains in thermoplastics can be untangled by application of a tensile stress. Thermoplastics can be amorphous or crystalline. Upon heating, thermoplastics soften and melt. They are processed into shapes by heating to elevated temperatures. Thermoplastics are easily recycled.

Thermosetting polymers are composed of long chains (linear or branched) of molecules that are strongly cross-linked to one another to form three-dimensional network structures. Network or thermosetting polymers are like a bunch of strings that are knotted to one another in several places and not just tangled up. Each string may have other side strings attached to it. Thermosets are generally stronger, but more brittle, than thermoplastics. Thermosets do not melt upon heating but begin to decompose. They cannot easily be reprocessed after the cross-linking reaction has occurred and hence recycling is difficult.

Elastomers These are known as rubbers. They have an elastic deformation >200%. These may be thermoplastics or lightly cross-linked thermosets. The polymer chains consist of coil-like molecules that can reversibly stretch by applying a force.

Thermoplastic elastomers are a special group of polymers. They have the processing ease of thermoplastics and the elastic behavior of elastomers.

TABLE 15-1 ■ *Comparison of the three polymer categories*

Behavior	General Structure	Example
Thermoplastic	Flexible linear chains (straight or branched)	Polyethylene
Thermosetting	Rigid three-dimensional network (chains may be linear or branched)	Polyurethanes
Elastomers	Thermoplastics or lightly cross-linked thermosets, consist of spring-like molecules	Natural rubber

Representative Structures   Figure 15-2 shows three ways we could represent a segment of polyethylene, the simplest of the thermoplastics. The polymer chain consists of a backbone of carbon atoms; two hydrogen atoms are bonded to each carbon atom in the chain. The chain twists and turns throughout space. In the figure, polyethylene has no branches and hence is a linear thermoplastic. The simple two-dimensional model in Figure 15-2(c) includes the essential elements of the polymer structure and will be used to describe the various polymers. The single lines (—) between the carbon atoms and between the carbon and hydrogen atoms represent a single covalent bond. Two parallel lines (=) represent a double covalent bond between atoms. A number of polymers include ring structures, such as the benzene ring found in polystyrene and other polymers (Figure 15-3).

In the structure shown in Figure 15-2 (c) if we replace one of the hydrogen atoms in CH_2 with (CH_3), a benzene ring, or chlorine, we get the structure of polypropylene, polystyrene, and polyvinyl chloride (PVC). If we replaced all H in CH_2 groups with fluorine (F), we get the structure of polytetrafluoroethylene or Teflon^{TM}. Like many other discoveries Teflon^{TM}, was also discovered by accident.[4,5] Many polymer structures can thus be derived from the structure of polyethylene. The following example shows how different types of polymers are used.

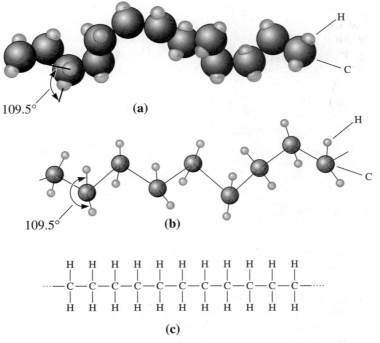

Figure 15-2   Three ways to represent the structure of polyethylene: (a) a solid three-dimensional model, (b) a three-dimensional "space" model, and (c) a simple two-dimensional model.

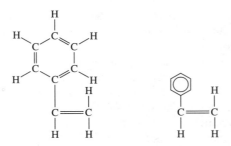

Figure 15-3
Two ways to represent the benzene ring. In this case, the benzene ring is shown attached to a pair of carbon atoms, producing styrene.

EXAMPLE 15-1   *Design/Materials Selection for Polymer Components*

Design the type of polymer material you might select for the following applications: a surgeon's glove, a beverage container and a pulley.

SOLUTION

The glove must be capable of stretching a great deal in order to slip onto the surgeon's hand, yet it must conform tightly to the hand to permit the maximum sensation of touch during surgery. A material that undergoes a large amount of elastic strain—particularly with relatively little applied stress—might be appropriate; this requirement describes an elastomer.

The beverage container should be easily and economically produced. It should have some ductility and toughness so that it does not accidentally shatter and leak the contents. If the beverage is carbonated, diffusion of CO_2 is a major concern (Chapter 5). A thermoplastic such as polyethylene terephthalate (PET) will have the necessary formability and ductility needed for this application.

The pulley will be subjected to some stress and wear as a belt passes over it. A relatively strong, rigid, hard material is required to prevent wear, so a thermosetting polymer might be most appropriate.

15-2   Addition Polymerization

Addition polymerization and **condensation polymerization** are the two main ways to conduct "polymerization" (creating a polymer). The polymers derived from these processes are known as addition and condensation polymers, respectively. The formation of the most common polymer, polyethylene (PE), from ethylene molecules is an example of addition or chain-growth polymerization. Ethylene, a gas, is the monomer (single unit) and has the formula C_2H_4. The two carbon atoms are joined by a double covalent bond. Each carbon atom shares two of its electrons with the second carbon atom, and two hydrogen atoms are bonded to each of the carbon atoms (Figure 15-4).

In presence of an appropriate combination of heat, pressure, and catalysts, the double bond between the carbon atoms is broken and replaced with a single covalent bond. The ends of the monomer are now *free radicals*; each carbon atom has an unpaired electron that it may share with other free radicals. Addition polymerization occurs because the original monomer contains a double covalent bond between the carbon atoms. The double bond is an **unsaturated bond**. After changing to a single bond, the carbon atoms are still joined, but they become active; other *repeat units* or *mers* can be added to produce the polymer chain. In ethylene there are two locations (each carbon atom) at which molecules can be attached. Thus, ethylene is **bifunctional**, and only chains form. The **functionality** is the number of sites at which new molecules can be attached to the repeat unit of the polymer. If there are three or more sites at which molecules can be attached, a three-dimensional network forms.

To begin the addition polymerization process, an initiator is added to the monomer (Figure 15-5). The initiator, which is the "on" switch, forms free radicals with a reactive

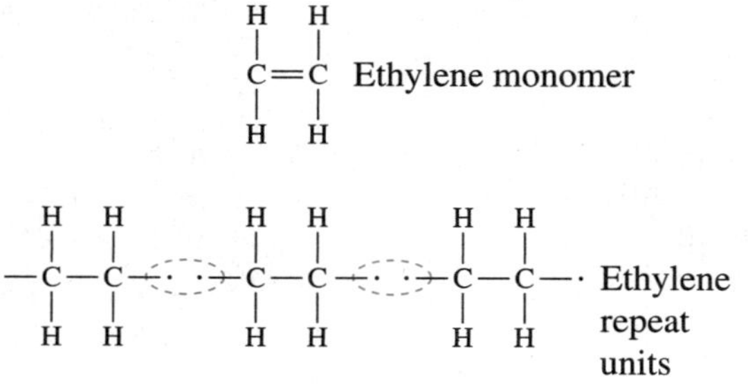

Figure 15-4
The addition reaction for producing polyethylene from ethylene molecules. The unsaturated double bond in the monomer is broken to produce active sites, which then attract additional repeat units to either end to produce a chain.

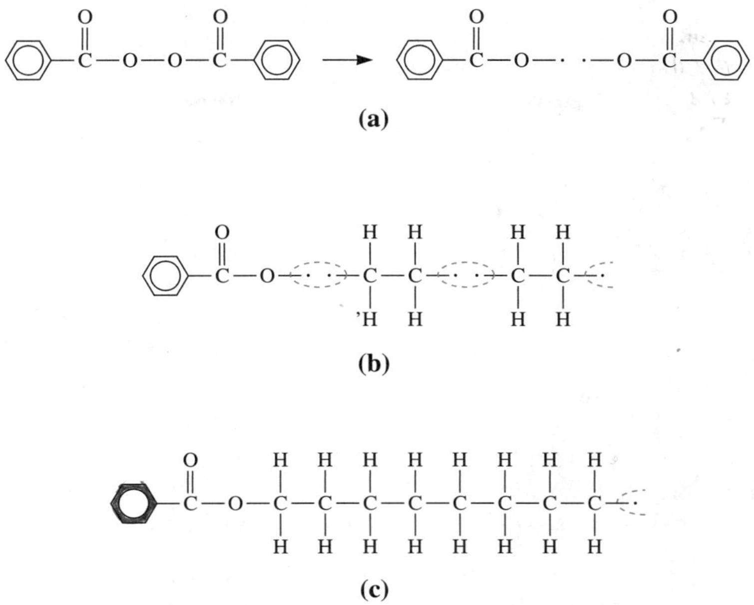

(a)

(b)

(c)

Figure 15-5   Initiation of a polyethylene chain by chain-growth may involve (a) producing free radicals from initiators such as benzoyl peroxide, (b) attachment of a polyethylene repeat unit to one of the initiator radicals, and (c) attachment of additional repeat units to propagate the chain.

site that attracts one of the carbon atoms of an ethylene monomer. When this reaction occurs, the reactive site is transferred to the other carbon atom in the monomer and a chain begins to form. A second repeat unit of ethylene can be attached at this new site, and the chain begins to lengthen. This process continues until a long polyethylene chain—an addition polymer—is formed. Because the initiators—which are often per-oxides—react with one another as well as with the monomer, their lifetimes are rela-tively short. A common initiator is benzoyl peroxide (Figure 15-5). Once the chain is initiated, repeat units are added onto each chain at a high rate, with perhaps several thousand additions each second. When polymerization is nearly complete, the few remaining monomers must diffuse a long distance before reaching an active site at the end of a chain; consequently, the growth rate decreases.

Termination of Addition Polymerization   We need polymers that have a controlled average molecular weight and a molecular weight distribution. Thus, the polymeriza-tion reactions must have an "off" switch as well! The chains may be terminated by two mechanisms (Figure 15-6). First, the ends of two growing chains may be joined. This process, called *combination*, creates a single large chain from two smaller chains. Sec-ond, the active end of one chain may remove a hydrogen atom from a second chain by a process known as *disproportionation*. This reaction terminates two chains, rather than combining two chains into one larger chain. Sometimes, compounds known as termi-nators are added to end the polymerization reactions. In general, for thermoplastics, the higher the average molecular weight the higher will be the melting temperature and the higher will be the Young's modulus of the polymer (Section 15-4). The following ex-ample illustrates an addition polymerization reaction.

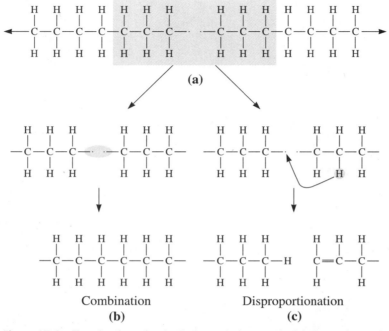

Figure 15-6 Termination of polyethylene chain growth: (a) the active ends of two chains come into close proximity, (b) the two chains undergo combination and become one large chain, and (c) rearrangement of a hydrogen atom and creation of a double covalent bond by disproportionation cause termination of two chains.

EXAMPLE 15-2   *Calculation of Initiator Required*

Calculate the amount of benzoyl peroxide initiator required to produce 1 kg of polyethylene with an average molecular weight of 200,000 g/mol. What is the degree of polymerization? Assume that 20% of the initiator is actually effective and that all termination occurs by the combination mechanism.

SOLUTION

For 100% efficiency, we need one molecule of benzoyl peroxide for each polyethylene chain. One of the free radicals would initiate one chain, the other free radical a second chain; then, the two chains combine into one larger one. The molecular weight of ethylene = (2 C)(12) + (4 H)(1) = 28 g/mol. Therefore, the degree of polymerization is:

$$\frac{200{,}000 \text{ g/mol}}{28 \text{ g/mol}} = 7143 \text{ ethylene molecules per average chain}$$

$$\frac{(1000 \text{ g polyethylene})(6.02 \times 10^{23} \text{ monomers/mol})}{28 \text{ g/mol}} = 215 \times 10^{23} \text{ monomers}$$

The combination mechanism requires the number of benzoyl peroxide molecules to be:

$$\frac{215 \times 10^{23} \text{ ethylene molecules}}{7143 \text{ ethylenes/chain}} = 0.03 \times 10^{23} \text{ chains}$$

The molecular weight of benzoyl peroxide is $(14\ C)(12) + (10\ H)(1) + (4\ O)(16) = 242$ g/mol. Therefore, the amount of initiator needed to form the ends of the chains is:

$$\frac{(0.03 \times 10^{23})(242 \text{ g/mol})}{6.02 \times 10^{23}} = 1.206 \text{ g}$$

However, only 20% of the initiator actually is effective; the rest recombines or combines with other molecules and does not cause initiation of a chain. Thus, we need five times this amount, or 6.03 g of benzoyl peroxide per kilogram of polyethylene.

15-3   Condensation Polymerization

Polymer chains can also form by condensation reactions, or *step-growth* polymerization, producing structures and properties that resemble those of addition polymers. In condensation polymerization, a relatively small molecule (such as water, ethanol, methanol etc.) is formed as a result of the polymerization reaction. This mechanism may often involve different monomers as starting or precursor molecules. The polymerization of dimethyl terephthalate and ethylene glycol (also used as radiator coolant) to produce *polyester* is an important example (Figure 15-7). During polymerization, a hydrogen atom on the end of the ethylene glycol monomer combines with an OCH_3 (methoxy) group from the dimethyl terephthalate. A byproduct, methyl alcohol (CH_3OH), is "condensed" off and the two monomers combine to produce a larger molecule. Each of

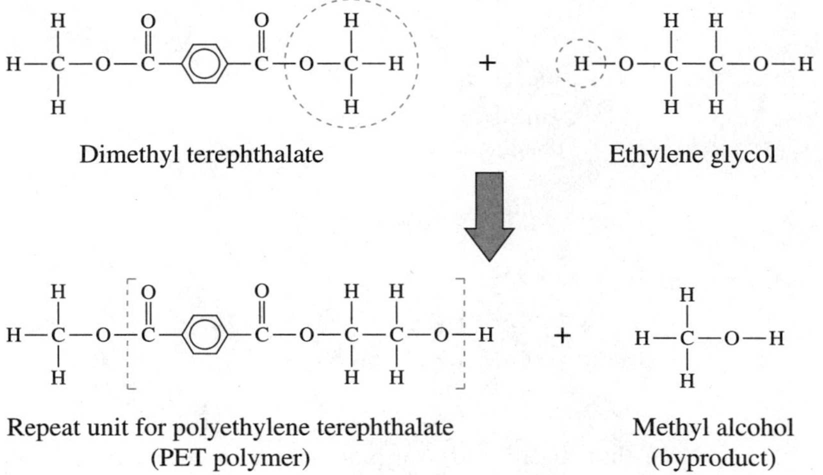

Dimethyl terephthalate   Ethylene glycol

Repeat unit for polyethylene terephthalate   Methyl alcohol
(PET polymer)   (byproduct)

Figure 15-7   The condensation reaction for polyethylene terephthalate (PET), a common polyester. The OCH3 group and a hydrogen atom are removed from the monomers, permitting the two monomers to join and producing methyl alcohol as a byproduct.

the monomers in this example are bifunctional, and the condensation polymerization can continue by the same reaction. Eventually, a long polymer chain—a polyester—is produced. The repeat unit for this polyester consists of two original monomers: one ethylene glycol and one dimethyl terephthalate. The length of the polymer chain depends on the ease with which the monomers can diffuse to the ends and undergo the condensation reaction. Chain growth ceases when no more monomers reach the end of the chain to continue the reaction. Condensation polymerization reactions also occur in sol-gel processing of ceramic materials.[6]

The following example describes discovery of nylon and calculations related to condensation polymerization.

EXAMPLE 15-3   *Condensation Polymerization of 6,6-Nylon*

Nylon was first reported by Wallace Hume Carothers, of du Pont in about 1934. In 1939, du Pont's Charles Stine reported the discovery of this first synthetic fiber to a group of 3000 women gathered for the New York World's Fair. The first application was nylon stockings that were strong. Today nylon is used in hundreds of applications. Prior to nylon, Carothers had discovered neoprene (an elastomer).

The linear polymer 6,6-nylon is to be produced by combining 1000 g of hexamethylene diamine with adipic acid. A condensation reaction then produces the polymer. Show how this reaction occurs and determine the byproduct that forms. How many grams of adipic acid are needed, and how much 6,6-nylon is produced, assuming 100% efficiency?

SOLUTION

The molecular structures of the monomers are shown below. The linear nylon chain is produced when a hydrogen atom from the hexamethylene diamine combines with an OH group from adipic acid to form a water molecule.

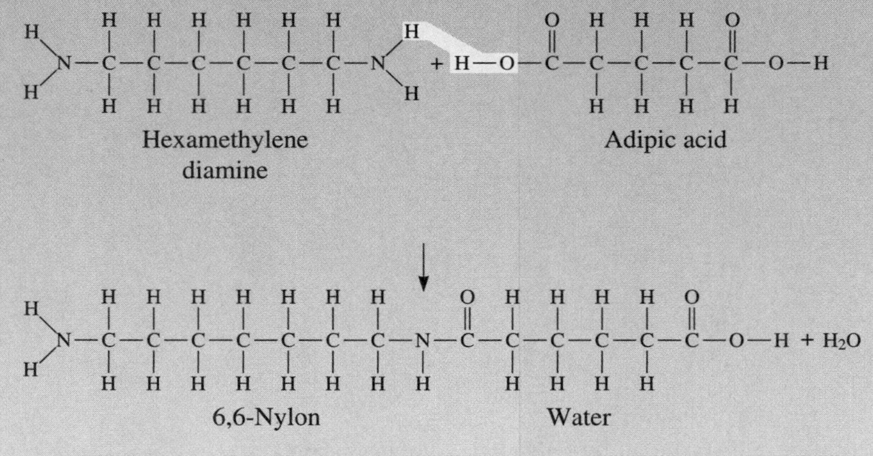

Note that the reaction can continue at both ends of the new molecule; consequently, long chains may form. This polymer is called 6,6-nylon because both monomers contain six carbon atoms.

We can determine that the molecular weight of hexamethylene diamine is 116 g/mol, of adipic acid is 146 g/mol, and of water is 18 g/mol. The number of moles of hexamethylene diamine added (calculated below) is equal to the number of moles of adipic acid:

$$\frac{1000 \text{ g}}{116 \text{ g/mol}} = 8.621 \text{ moles} = \frac{x \text{ g}}{146 \text{ g/mol}}$$

$$x = 1259 \text{ g of adipic acid required}$$

The number of moles of water lost is also 8.621:

$$y = (8.621 \text{ moles})(18 \text{ g/mol}) = 155.2 \text{ g H}_2\text{O}$$

But each time one more monomer is attached, another H_2O is released. Therefore, the total amount of nylon produced is

$$1000 \text{ g} + 1259 \text{ g} - 2(155.2 \text{ g}) = 1948.6 \text{ g}.$$

15-4   Degree of Polymerization

Polymers, unlike organic or inorganic compounds, do not have a fixed molecular weight. For example, polyethylene may have a molecular weight that ranges from ~25,000 to 6 million! The average length of a linear polymer is represented by the **degree of polymerization**, or the number of repeat units in the chain. The degree of polymerization can also be defined as:

$$\text{Degree of polymerization} = \frac{\text{average molecular weight of polymer}}{\text{molecular weight of repeat unit}} \quad (15\text{-}1)$$

If the polymer contains only one type of monomer, the molecular weight of the repeat unit is that of the monomer. If the polymer contains more than one type of monomer, the molecular weight of the repeat unit is the sum of the molecular weights of the monomers, less the molecular weight of the byproduct.

The length of the chains in a linear polymer varies considerably. Some may be quite short due to early termination; other may be exceptionally long. We can define an average molecular weight in two ways.

The *weight average molecular weight* is obtained by dividing the chains into size ranges and determining the fraction of chains having molecular weights within that range. The weight average molecular weight $\overline{M}_w$ is

$$\overline{M}_w = \sum f_i M_i, \quad (15\text{-}2)$$

where M_i is the mean molecular weight of each range and f_i is the weight fraction of the polymer having chains within that range.

The *number average molecular weight* $\overline{M}_n$ is based on the number fraction, rather than the weight fraction, of the chains within each size range. It is always smaller than the weight average molecular weight and is given by

$$\overline{M}_n = \sum x_i M_i, \quad (15\text{-}3)$$

where M_i is again the mean molecular weight of each size range, but x_i is the fraction of the total number of chains within each range. Either $\overline{M}_x$ or $\overline{M}_n$ can be used to calculate the degree of polymerization.

The two following examples illustrate these concepts.

EXAMPLE 15-4   *Degree of Polymerization for 6,6-Nylon*

Calculate the degree of polymerization if 6,6-nylon has a molecular weight of 120,000 g/mol.

SOLUTION

The reaction by which 6,6-nylon is produced was described in Example 15-3. Hexamethylene diamine and adipic acid combine and release a molecule of water. When a long chain forms, there is, on an average, one water molecule released for each reacting molecule. The molecular weights are 116 g/mol for hexamethylene diamine, 146 g/mol for adipic acid, and 18 g/mol for water. The repeat unit for 6,6-nylon is:

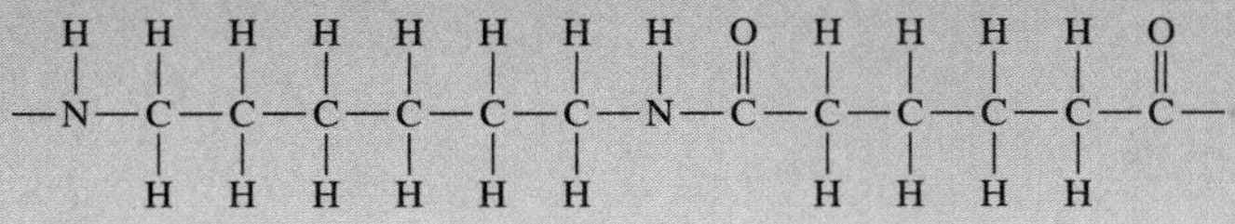

The molecular weight of the repeat unit is the sum of the molecular weights of the two monomers, minus that of the two water molecules that are evolved:

$$M_{\text{repeat unit}} = 116 + 146 - 2(18) = 226 \text{ g/mol}$$

$$\text{Degree of polymerization} = \frac{120,000}{226} = 531$$

The degree of polymerization refers to the total number of repeat units in the chain. The chain contains 531 hexamethylene diamine and 531 adipic acid molecules.

EXAMPLE 15-5   *Number and Weight Average Molecular Weights*

We have a polyethylene sample containing 4000 chains with molecular weights between 0 and 5000 g/mol, 8000 chains with molecular weights between 5000 and 10,000 g/mol, 7000 chains with molecular weights between 10,000 and 15,000 g/mol, and 2000 chains with molecular weights between 15,000 and 20,000 g/mol. Determine both the number and weight average molecular weights.

SOLUTION

First we need to determine the number fraction x_i and weight fraction f_i for each of the four ranges. For x_i, we simply divide the number in each range by 21,000, the total number of chains. To find f_i, we first multiply the number of chains by the mean molecular weight of the chains in each range, giving the "weight" of each group, then find f_i by dividing by the total weight of 192.5×10^6. We can then use Equations 15-2 and 15-3 to find the molecular weights.

Number of Chains	Mean M per Chain	x_i	x_iM_i	Weight	f_i	f_iM_i
4000	2500	0.191	477.5	10×10^6	0.0519	129.75
8000	7500	0.381	2857.5	60×10^6	0.3118	2338.50
7000	12,500	0.333	4162.5	87.5×10^6	0.4545	5681.25
2000	17,500	0.095	1662.5	35×10^6	0.1818	3181.50
$\sum = 21,000$		$\sum = 1.00$	$\sum = 9160$	$\sum = 192.5 \times 10^6$	$\sum = 1$	$\sum = 11,331$

$$\overline{M}_n = \sum x_iM_i = 9160 \text{ g/mol}$$

$$\overline{M}_w = \sum f_iM_i = 11,331 \text{ g/mol}$$

The weight average molecular weight is larger than the number average molecular weight.

15-5 Typical Thermoplastics

Some of the mechanical properties of typical thermoplastics are shown in Table 15-2.

TABLE 15-2 ■ *Properties of selected thermoplastics*

	Tensile Strength (psi)	% Elongation	Elastic Modulus (psi)	Density (g/cm³)	Izod Impact (ft lb/in.)
Polyethylene (PE):					
Low-density	3,000	800	40,000	0.92	9.0
High-density	5,500	130	180,000	0.96	4.0
Ultrahigh molecular weight	7,000	350	100,000	0.934	30.0
Polyvinyl chloride (PVC)	9,000	100	600,000	1.40	
Polypropylene (PP)	6,000	700	220,000	0.90	1.0
Polystyrene (PS)	8,000	60	450,000	1.06	0.4
Polyacrylonitrile (PAN)	9,000	4	580,000	1.15	4.8
Polymethyl methacrylate (PMMA) (acrylic, Plexiglas)	12,000	5	450,000	1.22	0.5
Polychlorotrifluoroethylene	6,000	250	300,000	2.15	2.6
Polytetrafluoroethylene (PTFE, Teflon)	7,000	400	80,000	2.17	3.0
Polyoxymethylene (POM) (acetal)	12,000	75	520,000	1.42	2.3
Polyamide (PA) (nylon)	12,000	300	500,000	1.14	2.1
Polyester (PET)	10,500	300	600,000	1.36	0.6
Polycarbonate (PC)	11,000	130	400,000	1.20	16.0
Polyimide (PI)	17,000	10	300,000	1.39	1.5
Polyetheretherketone (PEEK)	10,200	150	550,000	1.31	1.6
Polyphenylene sulfide (PPS)	9,500	2	480,000	1.30	0.5
Polyether sulfone (PES)	12,200	80	350,000	1.37	1.6
Polyamide-imide (PAI)	27,000	15	730,000	1.39	4.0

Because bonding within the chains is stronger, rotation and sliding of the chains is more difficult, leading to higher strengths, higher stiffnesses, and higher melting points than the simpler addition polymers (Table 15-3). In some cases, good impact properties can be gained from these complex chains, with polycarbonates being particularly remarkable. Polycarbonates (Lexan™, Merlon™, and Sparlux™) are used to make bulletproof windows, compact disks for data storage, and in many other applications.

TABLE 15-3 ■ Repeat units and applications for selected addition thermoplastics

Polymer	Repeat Unit	Application	Polymer	Repeat Unit	Application
Polyethylene (PE)		Packing films, wire insulation, squeeze bottles, tubing, household items	Polyacrylonitrile (PAN)		Textile fibers, precursor for carbon fibers, food container
Polyvinyl chloride (PVC)		Pipe, valves, fittings, floor tile, wire insulation, vinyl automobile roofs	Polymethyl methacrylate (PMMA) (acrylic-Plexiglas)		Windows, windshields, coatings, hard contact lenses, lighted signs
Polypropylene (PP)		Tanks, carpet fibers, rope, packaging			
Polystyrene (PS)		Packaging and insulation foams, lighting panels, appliance components, egg cartons	Polychlorotri-fluoroethylene		Valve components, gaskets, tubing, electrical insulation
			Polytetrafluoro-ethylene (Teflon) (PTFE)		Seals, valves, nonstick coatings

Thermoplastics with Complex Structures   A large number of polymers, which typically are used for special applications and in relatively small quantities, are formed from complex monomers, often by the condensation mechanism. Oxygen, nitrogen, sulfur, and benzene rings (or aromatic groups) may be incorporated into the chain. Table 15-4 shows the repeat units and typical applications for a number of these complex polymers. Polyoxymethylene, or acetal, is a simple example in which the backbone of the polymer chain contains alternating carbon and oxygen atoms. A number of these polymers, including polyimides and polyetheretherketone (PEEK), are important aerospace materials.

TABLE 15-4 ■ *Repeat units and applications for complex thermoplastics*

Polymer	Repeat Unit	Applications
Polyoxymethylene (acetal)(POM)		Plumbing fixtures, pens, bearings, gears, fan blades
Polyamide (nylon) (PA)		Bearings, gears, fibers, rope, automotive components, electrical components
Polyester (PET)		Fibers, photographic film, recording tape, boil-in-bag containers, beverage containers
Polycarbonate (PC)		Electrical and appliance housings, automotive components, football helmets, returnable bottles
Polyimide (PI)		Adhesives, circuit boards, fibers for space shuttle
Polyetheretherketone (PEEK)		High-temperature electrical insulation and coatings
Polyphenylene sulfide (PPS)		Coatings, fluid-handling components, electronic components, hair dryer components
Polyether sulfone (PES)		Electrical components, coffeemakers, hair dryers, microwave oven components
Polyamide-imide (PAI)		Electronic components, aerospace and automotive applications

15-6 Structure—Property Relationships in Thermoplastics

Degree of Polymerization In general, for a given type of thermoplastic (e.g., poly-ethylene) the tensile strength, creep resistance, impact toughness, wear resistance, and melting temperature all increase with increasing average molecular weight or degree of polymerization. The increases in these properties are not linear.[1] As the average molecular weight increases, the melting temperature increases and this makes the processing more difficult. In fact, we can make use of a bimodal molecular weight distribution in polymer processing. One component has lower molecular weight and helps melting. In certain applications (e.g., polystyrene for coffee cups), it is important to ensure that the residual monomer concentration is extremely small (<ppm) because of the toxicity concerns.

Effect of Side Groups   In polyethylene, the linear chains easily rotate and slide when stress is applied, and no strong polar bonds are formed between the chains; thus, polyethylene has a low strength.

Vinyl compounds have one of the hydrogen atoms replaced with a different atom or atom group. When R in the side group is chlorine, we produce polyvinyl chloride (PVC); when the side group is CH_3, we produce polypropylene (PP); addition of a benzene ring as a side group gives polystyrene (PS); and a CN group produces polyacrylonitrile (PAN). Generally, a head-to-tail arrangement of the repeat units in the polymers is obtained (Figure 15-8). When two of the hydrogen atoms are replaced, the monomer is a *vinylidene compound*, important examples of which include polyvinylidene chloride (the basis for Saran Wrap™) and polymethyl methacrylate (acrylics such as Lucite™ and Plexiglas™).

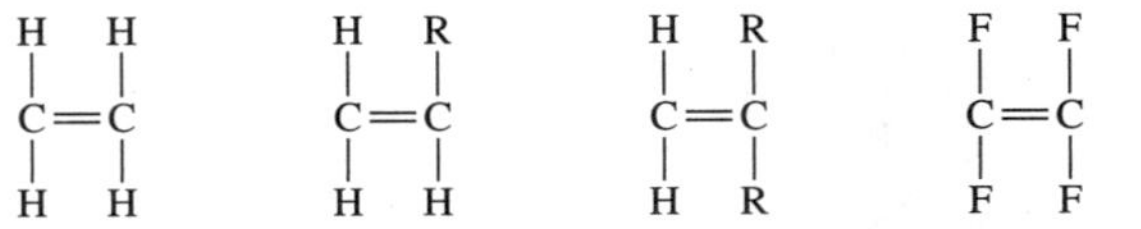

| Ethylene | Vinyl compound | Vinylidene compound | Tetrafluoro- ethylene |

The effects of adding other atoms or atom groups to the carbon backbone in place of hydrogen atoms are illustrated by the typical properties given in Table 15-3. Larger atoms such as chlorine or groups of atoms such as methyl (CH_3) and benzene make it more difficult for the chains to rotate, uncoil, disentangle, and deform by viscous flow when a stress is applied or when the temperature is increased. This condition leads to higher strength, stiffness, and melting temperature than those for polyethylene. The chlorine atom in PVC and the carbon-nitrogen group in PAN are strongly attracted by hydrogen bonding to hydrogen atoms on adjacent chains. This, for example, is the reason why PVC is more rigid than many other polymers. The way to get around the rigidity of PVC is to add low molecular weight compounds such as phthalate esters, known as plasticizers. When PVC contains these compounds, the glass temperature is lowered. This makes PVC more ductile and workable and is known as vinyl (not to be confused with the vinyl group mentioned here and in other places). PVC is used to make three-ring binders, pipes, tiles, and clear Tygon™ tubing.

In polytetrafluoroethylene (PTFE or Teflon™), all four hydrogen atoms in the polyethylene structure are replaced by fluorine. The monomer again is symmetrical, and

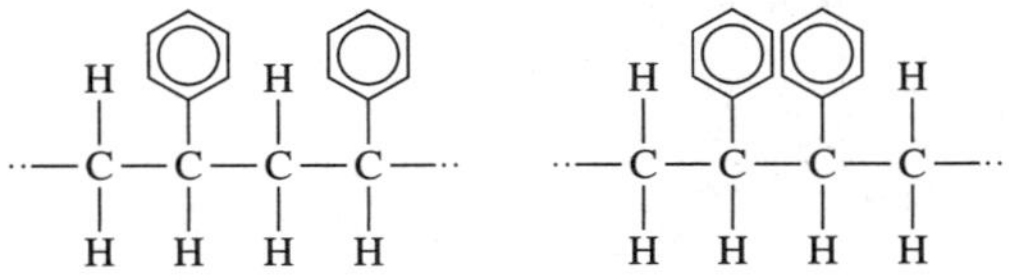

Head-to-tail   Head-to-head

Figure 15-8
Head-to-tail versus head-to-head arrangement of repeat units. The head-to-tail arrangement is most typical.

the strength of the polymer is not much greater than that of polyethylene. However, the C-F bond permits PTFE to have a high melting point with the added benefit of low friction, nonstick characteristics that make the polymer useful for bearings and cookware. Teflon™ was invented by accident by Roy Plunkett, who was working with tetrafluoroethylene gas. He found a tetrafluoroethylene gas cylinder that had no pressure (and, thus, seemed empty) but was heavier than usual. The gas inside had polymerized into solid Teflon™!

Branching prevents dense packing of the chains, thereby reducing the density, stiffness, and strength of the polymer. Low-density (LD) polyethylene, which has many branches, is weaker than high-density (HD) polyethylene, which has virtually no branching (Table 15-2).

Crystallization and Deformation   **Crystallinity** is important in polymers since it affects mechanical and optical properties. Crystallinity evolves in the processing of polymers as a result of temperature changes and applied stress (e.g., formation of PET bottles discussed in previous chapters). If crystalline regions become too large, they begin to scatter light and make the plastic translucent. Of course, in certain special polymers, localized regions crystallize in response to an applied electric field and this is the principle by which the liquid crystal displays work. We will discuss this in detail in the following section. As we have discussed previously, encouraging crystallization of the polymer also helps to increase density, resistance to chemical attack, and mechanical properties—even at higher temperatures—because of the stronger bonding between the chains. In addition, deformation straightens and aligns the chains, producing a preferred orientation. Deformation of a polymer is often used in producing fibers having mechanical properties in the direction of the fiber that exceed those of many metals and ceramics. This texture strengthening (Chapter 7), in fact, played a key role in the discovery of nylon fibers. In the previous chapters we have seen how PET bottles develop a biaxial texture and strength along the radial and length direction.

Tacticity   When a polymer is formed from nonsymmetrical repeat units, the structure and properties are determined by the location of the nonsymmetrical atoms or atom

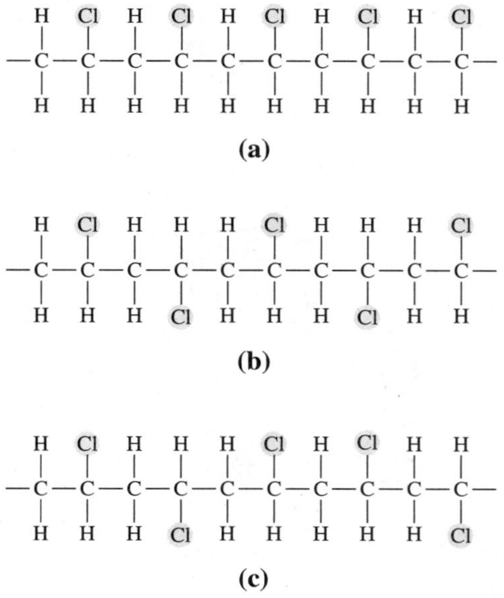

Figure 15-9
Three possible arrangements of nonsymmetrical monomers: (a) isotactic, (b) syndiotactic, and (c) atactic.

groups. This condition is called **tacticity**, or stereoisomerism. In the syndiotactic arrangement, the atoms or atom groups alternatively occupy positions on opposite sides of the linear chain. The atoms are all on the same side of the chain in *isotactic* polymers, whereas the arrangement of the atoms is random in *atactic* polymers (Figure 15-9 on page 685).

The atactic structure, which is the least regular and least predictable, tends to give poor packing, low density, low strength and stiffness, and poor resistance to heat or chemical attack. Atactic polymers are more likely to have an amorphous structure with a relatively high glass temperature. An important example of the importance of tacticity occurs in polypropylene. Atactic polypropylene is an amorphous wax-like polymer with poor mechanical properties, whereas isotactic polypropylene may crystallize and is one of the most widely used commercial polymers.

Copolymers Similar to the concept of solid solutions or the idea of composites, linear addition chains composed of two or more types of molecules can be arranged to form copolymers. This is a very powerful way to blend properties of different polymers. The arrangement of the monomers in a copolymer may take several forms (Figure 15-10). These include alternating, random, block, and grafted copolymers. ABS, composed of acrylonitrile, butadiene (a synthetic elastomer), and styrene, is one of the most common polymer materials (Figure 15-11). See the phase stability diagram of ABS in Chapter 9. Styrene and acrylonitrile form a linear copolymer (SAN) that serves as a matrix. Styrene and butadiene also form a linear copolymer, BS rubber, which acts as the filler

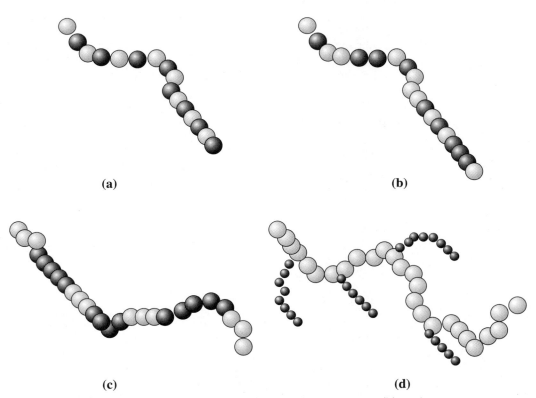

(a)   **(b)**

(c)   **(d)**

Figure 15-10 Four types of copolymers: (a) alternating monomers, (b) random monomers, (c) block copolymers, and (d) grafted copolymers. Circles of different colors or sizes represent different monomers.

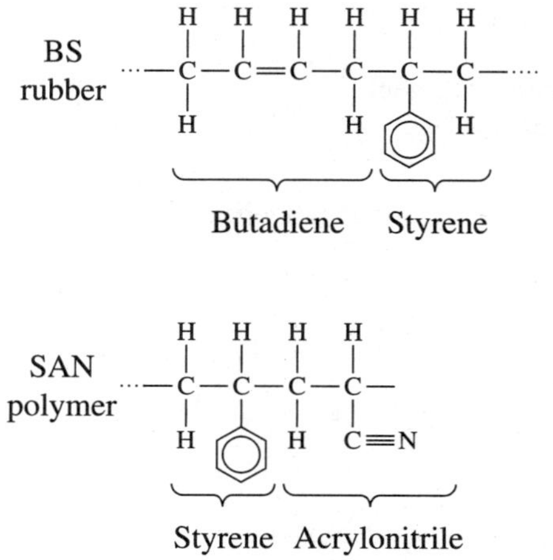

BS
rubber

Butadiene   Styrene

SAN
polymer

Styrene   Acrylonitrile

Figure 15-11
Copolymerization produces the polymer ABS, which is really made up of two copolymers, SAN and BS, grafted together.

material. The combination of the two copolymers gives ABS an excellent combination of strength, rigidity, and toughness. Another common copolymer contains repeat units of ethylene and propylene. Whereas polyethylene and polypropylene are both easily crystallized, the copolymer remains amorphous. When this copolymer is cross-linked, it behaves as an elastomer. Dylark™ is a copolymer of maleic anhydride and styrene. Styrene provides toughness, while maleic anhydride provides high-temperature properties. Carbon black (for protection from ultraviolet rays and enhancing stiffness), rubber (for toughness), and glass fibers (for stiffness) are added to the Dylark™ copolymer. It is used to make instrument panels for car dashboards. The Dylark™ plastic is then coated with vinyl that provides a smooth and soft finish.

Blending and Alloying   We can improve the mechanical properties of many of the thermoplastics by blending or alloying. By mixing an immiscible elastomer with the thermoplastic, we produce a two-phase polymer, as we found in ABS. The elastomer does not enter the structure as a copolymer but, instead, helps to absorb energy and improve toughness. Polycarbonates used to produce transparent aircraft canopies are also toughened by elastomers in this manner.

Liquid Crystalline Polymers   Some of the complex thermoplastic chains become so stiff that they act as rigid rods, even when heated above the melting point. These materials are **liquid crystalline polymers** (LCPs).[2] Some aromatic polyesters and aromatic polyamides (or **aramids**) are examples of liquid crystalline polymers and are used as high-strength fibers (as discussed in Chapter 16). Kevlar™ an aromatic polyamide, is the most familiar of the LCPs, and is used as a reinforcing fiber for aerospace applications and for bulletproof vests. Liquid crystal polymers are, of course, used to make electronic displays.

15-7   Effect of Temperature on Thermoplastics

Properties of thermoplastics change depending upon temperature. We need to know how these changes occur because this can help us (a) better design components, and (b) guide the type of processing techniques that need to be used. Several critical temperatures and structures, summarized in Figures 15-12 and 15-13, may be observed.

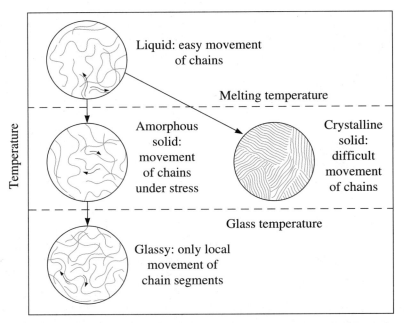

Figure 15-12   The effect of temperature on the structure and behavior of thermoplastics.

Thermoplastics can be amorphous or crystalline once they cool below the melting temperature, (Figure 15-12). Most often, engineered thermoplastics consist of regions that are amorphous and crystalline. The crystallinity in thermoplastics can be introduced by temperature (slow cooling) or by the application of stress that can untangle chains (stress-induced crystallization, Chapter 6). Similar to dispersion strengthening of metallic materials, the formation of crystalline regions in an otherwise amorphous matrix helps increase the strength of thermoplastics. In typical thermoplastics, bonding within the chains is covalent, but the long coiled chains are held to one another by weak van der Waals bonds and by entanglement. When a tensile stress is applied to the thermoplastic, the weak bonding between the chains can be overcome and the chains can rotate and slide relative to one another. The ease with which the chains slide depends on both temperature and the polymer structure.

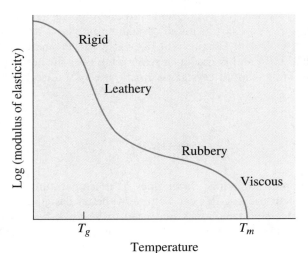

Figure 15-13
The effect of temperature on the modulus of elasticity for an amorphous thermoplastic. Note that T_g and T_m are not fixed.

Degradation Temperature At very high temperatures, the covalent bonds between the atoms in the linear chain may be destroyed, and the polymer may burn or char. In thermoplastics decomposition occurs in the liquid state, in thermosets the decomposition occurs in the solid state. This temperature T_d (not shown in Figure 15-13), is the **degradation** (or decomposition) **temperature**. When plastics burn, they create smoke, and that is dangerous. Some materials (such as limestone, talc, alumina etc.) added to thermoplastics are thermal or heat stabilizers. They absorb heat and protect the polymer matrix. Fire retardant additives, such as hydrated alumina, antimony compounds, or halogen compounds (e.g., $MgBr$, PCl_5) are added to retard the flammability of polymers. Some additives retard fire by excluding oxygen but generate dangerous gases and are not appropriate for certain applications.

Exposure to other forms of chemicals or energy (e.g., oxygen, ultraviolet radiation, and attack by bacteria) also cause a polymer to degrade or **age** slowly, even at low temperatures. Carbon black (up to ~3%) is one of the commonly used additives that helps improve the resistance of plastics to ultraviolet degradation.

Liquid Polymers Thermoplastics usually do not melt at a precise temperature. Instead there is usually a range of temperatures over which melting occurs. The approximate melting ranges of typical polymers are included in Table 15-5. At or above the melting temperature T_m, bonding between the twisted and intertwined chains is weak. If a force is applied, the chains slide past one another and the polymer flows with virtually no elastic strain. The strength and modulus of elasticity are nearly zero and the polymer is

TABLE 15-5 ■ *Melting and glass temperature ranges (°C) for selected thermoplastics and elastomers*

Polymer	Melting Temperature	Glass Temperature (T_g)	Processing Temperature
Addition polymers			
Low-density (LD) polyethylene	98–115	−90 to −25	149–232
High-density (HD) polyethylene	130–137	−110	177–260
Polyvinyl chloride	175–212	87	
Polypropylene	160–180	−25 to −20	190–288
Polystyrene	240	85–125	
Polyacrylonitrile	320	107	
Polytetrafluoroethylene (Teflon)	327		
Polychlorotrifluoroethylene	220		
Polymethyl methacrylate (acrylic)		90–105	
Acrylonitrile butadiene styrene (ABS)	110–125	100	177–260
Condensation polymers			
Acetal	181	−85	
6,6-nylon	243–260	49	260–327
Cellulose acetate	230		
Polycarbonate	230	149	271–300
Polyester	255	75	
Polyethylene terephthalate (PET)	212–265	66–80	227–349
Elastomers			
Silicone		−123	
Polybutadiene	120	−90	
Polychloroprene	80	−50	
Polyisoprene	30	−73	

suitable for casting and many forming processes. Most thermoplastic melts are shear thinning (i.e., their apparent viscosity decreases within an increase in the steady state shear rate (Chapter 6)).

Rubbery and Leathery States Below the melting temperature, the polymer chains are still twisted and intertwined. These polymers have an amorphous structure. Just below the melting temperature, the polymer behaves in a *rubbery* manner. When stress is applied, both elastic and plastic deformation of the polymer occurs. When the stress is removed, the elastic deformation is quickly recovered, but the polymer is permanently deformed due to the movement of the chains. Large permanent elongations can be achieved, permitting the polymer to be formed into useful shapes by molding and extrusion.

At lower temperatures, bonding between the chains is stronger, the polymer becomes stiffer and stronger, and a **leathery** behavior is observed. Many of the commercial polymers, including polyethylene, have a useable strength in this condition.

Glassy State Below the **glass temperature** T_g, the linear amorphous polymer becomes hard, brittle, and glass-like (Chapter 6).[7] This is again not a fixed temperature but a range of temperatures. When the polymer cools below the glass temperature, certain properties—such as density or modulus of elasticity—change at a different rate (Figure 15-14).

Although glassy polymers have poor ductility and formability, they do have good strength, stiffness, and creep resistance. A number of important polymers, including polystyrene and polyvinyl chloride, have glass temperatures above room temperature (Table 15-5).

The glass temperature is typically about 0.5 to 0.75 times the absolute melting temperature T_m. Polymers such as polyethylene, which have no complicated side groups attached to the carbon backbone, have low glass temperatures (even below room temperature) compared with polymers such as polystyrene, which have more complicated side groups.

As pointed out in Chapter 6, many thermoplastics become brittle at lower temperatures. The brittleness of the polymer used for some of the O-rings ultimately caused the 1986 *Challenger* disaster.[8] The lower temperatures that existed during the launch time caused the embrittlement of the rubber O-rings used for the booster rockets.

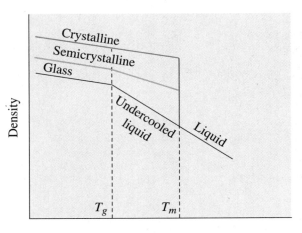

Figure 15-14
The relationship between the density and the temperature of the polymer shows the melting and glass temperatures. Note that T_g and T_m are not fixed; rather, they are ranges of temperatures.

Observing and Measuring Crystallinty in Polymers Many thermoplastics partially crystallize when cooled below the melting temperature, with the chains becoming closely aligned over appreciable distances. A sharp increase in the density occurs as the coiled and intertwined chains in the liquid are rearranged into a more orderly, close-packed structure (Figure 15-14).

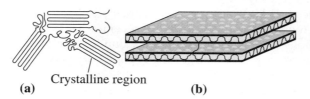

(a) Crystalline region   **(b)**

Figure 15-15
The folded chain model for crystallinity in polymers, shown in (a) two dimensions and (b) three dimensions.

One model describing the arrangement of the chains in a crystalline polymer is shown in Figure 15-15. In this *folded chain* model, the chains loop back on themselves, with each loop being approximately 100 carbon atoms long. The folded chain extends in three dimensions, producing thin plates or lamellae. The crystals can take various forms, with the spherulitic shape shown in Figure 15-16 being particularly common. The crystals have a unit cell that describes the regular packing of the chains. The crystal structure for polyethylene, shown in Figure 3-41, describes one such unit cell. Crystal structures for several polymers are described in Table 15-6. Some polymers are polymorphic, having more than one crystal structure.

Even in crystalline polymers, there are always thin regions between the lamellae, as well as between spherulites, that are amorphous transition zones. The weight percentage of the structure that is crystalline can be calculated from the density of the polymer

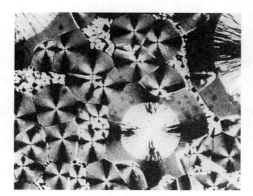

Figure 15-16
Photograph of spherulitic crystals in an amorphous matrix of nylon (×200). (*From R. Brick, A. Pense and R. Gordon*, Structure and Properties of Engineering Materials, *4th Ed., McGraw-Hill, 1977.*)

TABLE 15-6 ■ *Crystal structures of several polymers*

Polymer	Crystal Structure	Lattice Parameters (nm)
Polyethylene	Orthorhombic	$a_0 = 0.742$ $b_0 = 0.495$ $c_0 = 0.255$
Polypropylene	Orthorhombic	$a_0 = 1.450$ $b_0 = 0.569$ $c_0 = 0.740$
Polyvinyl chloride	Orthorhombic	$a_0 = 1.040$ $b_0 = 0.530$ $c_0 = 0.510$
Polyisoprene (cis)	Orthorhombic	$a_0 = 1.246$ $b_0 = 0.886$ $c_0 = 0.810$

$$\% \text{ Crystalline} = \frac{\rho_c}{\rho} \frac{(\rho - \rho_a)}{(\rho_c - \rho_a)} \times 100, \tag{15-4}$$

where ρ is the measured density of the polymer, ρ_a is the density of amorphous polymer, and ρ_c is the density of completely crystalline polymer. Similarly, x-ray diffraction (XRD) can be used to measure the level of crystallinity and determination of lattice constants for single crystal polymers.[9] (See Chapter 3.)

As the side groups get more complex, it becomes harder to crystallize thermoplastics. For example, polyethylene (H as side group) can be crystallized more easily than polystyrene (benzene ring as side group). High-density polyethylene (HDPE) has a higher level of crystallinty and, therefore, a higher density (0.97 g/cc). Low-density polyethylene (LDPE) has a density of 0.92 g/cc. The crystallinity and, hence, the density in LDPE is lower, since the polymer is branched. Thus, branched polymers show lower levels of crystallinity. A completely crystalline polymer would not display a glass temperature; however, the amorphous regions in semicrystalline polymers do change to a glassy material below the glass temperature (Figure 15-14). Such polymers as acetal, nylon, HDPE, and polypropylene are referred to as crystalline even though the level of crystallinity may be moderate.[1] The following examples show how properties of plastics can be accounted for in different applications.

EXAMPLE 15-6   *Design of a Polymer Insulation Material*

A storage tank for liquid hydrogen will be made of metal, but we wish to coat the metal with a 3-mm thickness of a polymer as an intermediate layer between the metal and additional insulation layers. The temperature of the intermediate layer may drop to −80°C. Design a material for this layer.

SOLUTION

We want the material to have reasonable ductility. As the temperature of the tank changes, stresses develop in the coating due to differences in thermal expansion, and we do not want the polymer to fail due to these stresses. A material that has good ductility and/or can undergo large elastic strains is needed. We therefore would prefer either a thermoplastic that has a glass temperature below −80°C or an elastomer, also with a glass temperature below −80°C. Of the polymers listed in Table 15-2, thermoplastics such as polyethylene and acetal are satisfactory. Suitable elastomers include silicone and polybutadiene.

We might prefer one of the elastomers, for they can accommodate thermal stress by elastic, rather than plastic, deformation.

EXAMPLE 15-7   *Impact-Resistant Polyethylene*

A new grade of flexible, impact-resistant polyethylene for use as a thin film requires a density of 0.88 to 0.915 g/cm^3. Design the polyethylene required to produce these properties. The density of amorphous polyethylene is about 0.87 g/cm^3.

SOLUTION

To produce the required properties and density, we must control the percent crystallinity of the polyethylene. We can use Equation 15-4 to determine the crystallinity that corresponds to the required density range. To do so, however, we must know the density of completely crystalline polyethylene. We can use the data in Table 15-3 to calculate this density if we recognize that there are two polyethylene repeat units in each unit cell (see Example 3-16):

$$\rho_c = \frac{(4\ C)(12) + (8\ H)(1)}{(7.42)(4.95)(2.55)(10^{-24})(6.02 \times 10^{23})} = 0.9932 \text{ g/cm}^3$$

We know that $\rho_a = 0.87$ g/cm^3 and that ρ varies from 0.88 to 0.915 g/cm^3. The required crystallinity then varies from:

$$\% \text{ crystalline} = \frac{(0.9932)(0.88 - 0.87)}{(0.88)(0.9932 - 0.87)} \times 100 = 9.2$$

$$\% \text{ crystalline} = \frac{(0.9932)(0.915 - 0.87)}{(0.915)(0.9932 - 0.87)} \times 100 = 39.6$$

Therefore, we must be able to process the polyethylene to produce a range of crystallinity between 9.2 and 39.6%.

15-8 Mechanical Properties of Thermoplastics

Most thermoplastics (molten and solid) exhibit a non-Newtonian and **viscoelastic** behavior (Chapter 6). The behavior is non-Newtonian (i.e., the stress and strain are not linearly related for most parts of the stress-strain curve). The viscoelastic behavior means when an external force is applied to a thermoplastic polymer, both elastic and plastic (or viscous) deformation occurs. The mechanical behavior is closely tied to the manner in which the polymer chains move relative to one another under load. Deformation is more complicated in thermoplastics. The deformation process depends on both time and the rate at which the load is applied. Figure 15-17 shows a stress-strain curve for 6,6-nylon.

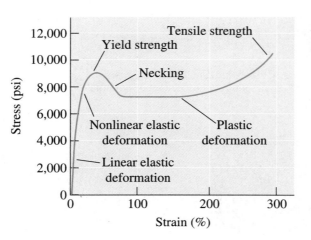

Figure 15-17
The stress-strain curve for 6,6-nylon, a typical thermoplastic polymer.

Elastic Behavior Elastic deformation in thermoplastics is the result of two mechanisms. An applied stress causes the covalent bonds within the chain to stretch and distort, allowing the chains to elongate elastically. When the stress is removed, recovery from this distortion is almost instantaneous. This behavior is similar to that in metals and ceramics, which also deform elastically by the stretching of metallic, ionic, or covalent bonds. But in addition, entire segments of the polymer chains may be distorted; when the stress is removed, the segments move back to their original positions only over a period of time—often hours or even months. This time-dependent, or viscoelastic, behavior may contribute to some nonlinear elastic behavior.

Plastic Behavior of Amorphous Thermoplastics These polymers deform plastically when the stress exceeds the yield strength. Unlike deformation in the case of metals, however, plastic deformation is not a consequence of dislocation movement. Instead, chains stretch, rotate, slide, and disentangle under load to cause permanent deformation. The drop in the stress beyond the yield point can be explained by this phenomenon. Initially, the chains may be highly tangled and intertwined. When the stress is sufficiently high, the chains begin to untangle and straighten. Necking also occurs, permitting continued sliding of the chains at a lesser stress. Eventually, however, the chains become almost parallel and close together; stronger van der Waals bonding between the more closely aligned chains requires higher stresses to complete the deformation and fracture process (Figure 15-18). This type of crystallization due to orientation played an important role in the discovery of nylon as a material to make strong fibers.

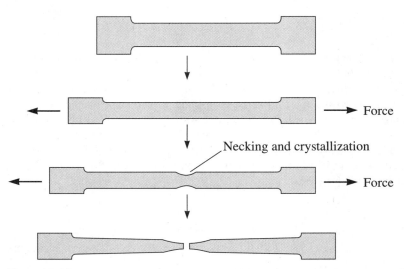

Figure 15-18 Necks are not stable in amorphous polymers, because local alignment strengthens the necked region and reduces its rate of deformation.

EXAMPLE 15-8 *Comparing Mechanical Properties of Thermoplastics*

Compare the mechanical properties of LD polyethylene, HD polyethylene, polyvinyl chloride, polypropylene, and polystyrene, and explain their differences in terms of their structures.

SOLUTION

Let us look at the maximum tensile strength and modulus of elasticity for each polymer.

Polymer	Tensile Strength (psi)	Modulus of Elasticity (ksi)	Structure
LD polyethylene	3000	40	Highly branched, amorphous structure with symmetrical monomers
HD polyethylene	5500	180	Amorphous structure with symmetrical monomers but little branching
Polypropylene	6000	220	Amorphous structure with small methyl side groups
Polystyrene	8000	450	Amorphous structure with benzene side groups
Polyvinyl chloride	9000	600	Amorphous structure with large chlorine atoms as side groups

We can conclude that:

1. Branching, which reduces the density and close packing of chains, reduces the mechanical properties of polyethylene.

2. Adding atoms or atom groups other than hydrogen to the chain increases strength and stiffness. The methyl group in polypropylene provides some improvement, the benzene ring of styrene provides higher properties, and the chlorine atom in polyvinyl chloride provides a large increase in properties.

Creep and Stress Relaxation   Thermoplastics also exhibit creep, a time-dependent permanent deformation with constant stress or load (Figures 15-19 and 15-20).

They also show stress relaxation (i.e., under a constant strain the stress level decreases with time) (Chapter 6). Stress relaxation, like creep, is a consequence of the viscoelastic behavior of the polymer. Perhaps the most familiar example of this behavior is a rubber band (an elastomer) stretched around a pile of books. Initially, the tension in the rubber band is high, when the rubber band is taut. After several weeks, the strain in the rubber band is unchanged (it still completely encircles the books), but the stress will have decreased—that is, the band is no longer taut. Similarly, the nylon strings in tennis rackets are pulled at a higher tension initially since this tension (i.e., stress) decreases with time.

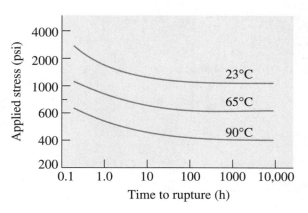

Figure 15-19
The effect of temperature on the stress-rupture behavior of high-density polyethylene.

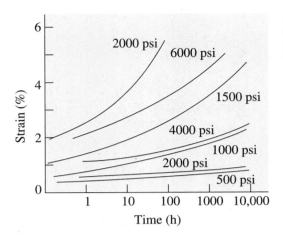

Figure 15-20
Creep curves for acrylic (PMMA) (colored lines) and polypropylene (black lines) at 20°C and several applied stresses.

In a simple model, the rate at which stress relaxation occurs is related to the **relaxation time** λ, which is considered a property of the polymer (more complex models consider a distribution of relaxation times). The stress after time t is given by

$$\sigma = \sigma_0 \exp(-t/\lambda) \qquad (15\text{-}5)$$

where σ_0 is the original stress. The *relaxation time*, in turn, depends on the viscosity and, thus, the temperature:

$$\lambda = \lambda_0 \exp(Q/RT) \qquad (15\text{-}6)$$

where σ_0 is a constant and Q is the activation energy related to the ease with which polymer chains slide past each other. Relaxation of the stress occurs more *rapidly* at *higher temperatures* and for polymers with a low viscosity.

The following example shows how stress relaxation can be accounted for while designing with polymers.

EXAMPLE 15-9   *Design of Initial Stress in a Polymer*

A band of polyisoprene is to hold together a bundle of steel rods for up to one year. If the stress on the band is less than 1500 psi, the band will not hold the rods tightly. Design the initial stress that must be applied to a polyisoprene band when it is slipped over the steel. A series of tests showed that an initial stress of 1000 psi decreased to 980 psi after six weeks.

SOLUTION

Although the strain of the elastomer band may be constant, the stress will decrease over time due to stress relaxation. We can use Equation 15-5 and our initial tests to determine the relaxation time for the polymer:

$$\sigma = \sigma_0 \exp\left(-\frac{t}{\lambda}\right)$$

$$980 = 1000 \exp\left(-\frac{6}{\lambda}\right)$$

$$-\frac{6}{\lambda} = \ln\left(\frac{980}{1000}\right) = \ln(0.98) = -0.0202$$

$$\lambda = \frac{6}{0.0202} = 297 \text{ weeks}$$

Now that we know the relaxation time, we can determine the stress that must be initially placed onto the band in order that it still be stressed to 1500 psi after 1 year (52 weeks).

$$1500 = \sigma_0 \exp(-52/297) = \sigma_0 \exp(-0.175) = 0.839\sigma_0$$

$$\sigma_0 = \frac{1500}{0.839} = 1788 \text{ psi}$$

The polyisoprene band must be made significantly undersized so it can slip over the materials it is holding together with a tension of 1788 psi. After one year, the stress will still be 1500 psi.

One more practical measure for high temperature and creep properties of a polymer is the **heat deflection temperature** or **heat distortion temperature** under load, which is the temperature at which a given deformation of a beam occurs for a standard load. A high-deflection temperature indicates good resistance to creep and permits us to compare various polymers. The deflection temperatures for several polymers are shown in Table 15-7, which gives the temperature required to cause a 0.01 in. deflection for a 264 psi load at the center of a bar resting on supports 4 in. apart. A polymer is "dishwasher safe" if it has a heat distortion temperature greater than $\sim 50°C$.

TABLE 15-7 ■ *Deflection temperatures for selected polymers for a 264-psi load*

Polymer	Deflection Temperature (°C)
Polyester	40
Polyethylene (ultra-high density)	40
Polypropylene	60
Phenolic	80
Polyamide (6,6-nylon)	90
Polystyrene	100
Polyoxymethylene (acetal)	130
Polyamide-imide	280
Epoxy	290

Impact Behavior Viscoelastic behavior also helps us understand the impact properties of polymers. At very high rates of strain, as in an impact test, there is insufficient time for the chains to slide and cause plastic deformation. For these conditions, the thermoplastics behave in a brittle manner and have poor impact values. As discussed here and in Chapter 6, polymers may have a transition temperature. At low temperatures, brittle behavior is observed in an impact test, whereas more ductile behavior is observed at high temperatures, where the chains move more easily. These effects of temperature and strain rate are similar to those seen in metals that exhibit a ductile-to-brittle transition temperature; however, the mechanisms are different.

Deformation of Crystalline Polymers   A number of polymers are used in the crystalline state. As we discussed earlier, however, the polymers are never completely crystalline. Instead, small regions—between crystalline lamellae and between crystalline spherulites—are amorphous transition regions. Polymer chains in the crystalline region extend into these amorphous regions as tie chains. When a tensile load is applied to the polymer, the crystalline lamellae within the spherulites slide past one another and begin to separate as the tie chains are stretched. The folds in the lamellae tilt and become aligned with the direction of the tensile load. The crystalline lamellae break into smaller units and slide past one another, until eventually the polymer is composed of small aligned crystals joined by tie chains and oriented parallel to the tensile load. The spherulites also change shape and become elongated in the direction of the applied stress. With continued stress, the tie chains disentangle or break, causing the polymer to fail.

Crazing   Crazing occurs in thermoplastics when localized regions of plastic deformation occur in a direction perpendicular to that of the applied stress. In transparent thermoplastics, such as some of the glassy polymers, the craze produces a translucent or opaque region that looks like a crack. The craze can grow until it extends across the entire cross-section of the polymer part. But the craze is not a crack, and, in fact, it can continue to support an applied stress. The process is similar to that for the plastic deformation of the polymer, but the process can proceed even at a low stress over an extended length of time. Crazing can lead to brittle fracture of the polymer and is often assisted by the presence of a solvent (known as solvent crazing).[1]

Blushing   Blushing or whitening refers to failure of a plastic because of localized crystallization (due to repeated bending, for example) that ultimately causes voids to form.

15-9   Elastomers (Rubbers)

A number of natural and synthetic polymers called *elastomers* display a large amount (>200%) of elastic deformation when a force is applied. Rubber bands, automobile tires, O-rings, hoses, and insulation for electrical wires are common uses for these materials. Crude natural rubber could erase pencil marks; hence, elastomers got the name rubber.

Geometric Isomers   Some monomers that have different structures, even though they have the same composition, are called **geometric isomers**. Isoprene, or natural rubber, is an important example (Figure 15-21). The monomer includes two double bonds between carbon atoms; this type of monomer is called a **diene**. Polymerization occurs by breaking both double bonds, creating a new double bond at the center of the molecule and active sites at both ends.

In the *trans* form of isoprene, the hydrogen atom and the methyl group at the center of the repeat unit are located on opposite sides of the newly formed double bond. This arrangement leads to relatively straight chains; the polymer crystallizes and forms a hard rigid polymer called *gutta percha*. This is used to make golf balls and shoe soles.

In the *cis* form, however, the hydrogen atom and the methyl group are located on the same side of the double bond. This different geometry causes the polymer chains to develop a highly coiled structure, preventing close packing and leading to an amorphous, rubbery polymer. If a stress is applied to the cis-isoprene, the polymer behaves in

Cis

Trans

Monomer   Repeat unit   Polymer

Figure 15-21   The cis and trans structures of isoprene. The cis form is useful for producing the isoprene elastomer.

a viscoelastic manner. The chains uncoil and bonds stretch, producing elastic deformation, but the chains also slide past one another, producing nonrecoverable plastic deformation. The polymer behaves as a thermoplastic rather than an elastomer (Figure 15-22).

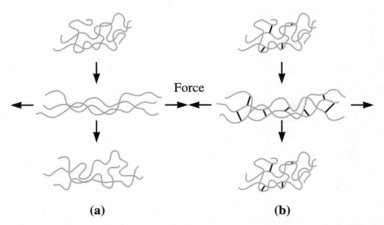

Force

(a)   (b)

Figure 15-22   (a) When the elastomer contains no cross-links, the application of a force causes both elastic and plastic deformation; after the load is removed, the elastomer is permanently deformed. (b) When cross-linking occurs, the elastomer still may undergo large elastic deformation; however, when the load is removed, the elastomer returns to its original shape.

Cross-Linking   We prevent viscous plastic deformation while retaining large elastic deformation by **cross-linking** the chains. **Vulcanization**, which uses sulfur atoms, is a common method for cross-linking. (Vulcan was the Roman god of fire.) This process was invented, accidentally, by Charles Goodyear, who, after many years of hard work

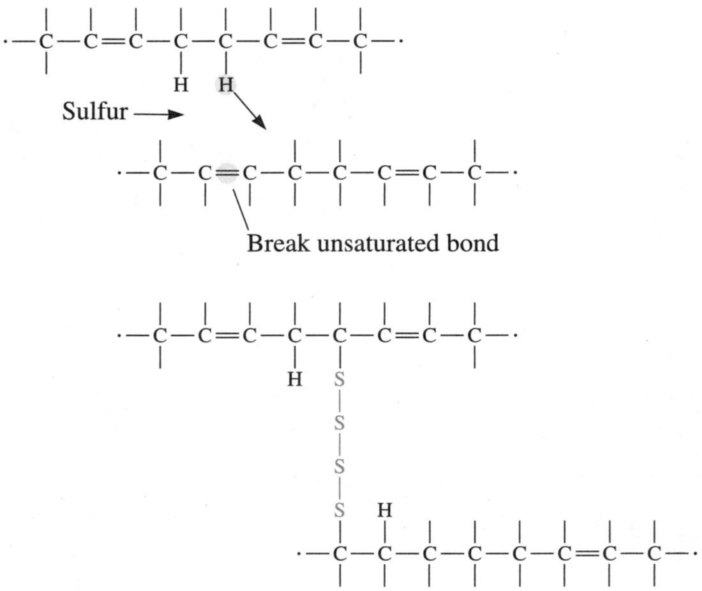

$$\cdot-\overset{|}{\underset{|}{C}}-\overset{|}{\underset{|}{C}}=\overset{|}{\underset{|}{C}}-\overset{|}{\underset{|}{C}}-\overset{|}{\underset{|}{C}}-\overset{|}{\underset{|}{C}}=\overset{|}{\underset{|}{C}}-\overset{|}{\underset{|}{C}}-\cdot$$

Sulfur ⟶

Break unsaturated bond

Figure 15-23
Cross-linking of polyisoprene chains may occur by introducing strands of sulfur atoms. Sites for attachment of the sulfur strands occur by rearrangement or loss of a hydrogen atom and the breaking of an unsaturated bond.

trying to make more natural rubber less sticky and more durable, accidentally dropped a mixture of India rubber and sulfur on a hot stove. However, it was Thomas Hancock (the person who patented the process in England after evaluating samples Goodyear had sent), who named the process vulcanization. Figure 15-23 describes how strands of sulfur atoms can link the polymer chains as the polymer is processed and shaped at temperatures of about 120 to 180°C. The cross-linking steps may include rearranging a hydrogen atom and replacing one or more of the double bonds with single bonds. The cross-linking process is not reversible; consequently, the elastomer cannot easily be recycled.

The stress-strain curve for an elastomer is shown in Figure 15-24. Virtually all of the curve represents elastic deformation; thus, elastomers display a nonlinear elastic behavior. Initially, the modulus of elasticity decreases because of the uncoiling of the chains. However, after the chains have been extended, further elastic deformation occurs by the stretching of the bonds, leading to a higher modulus of elasticity.

The number of cross-links determines the elasticity of the rubber, or the amount of sulfur added to the material. Low sulfur additions leave the rubber soft and flexible, as in elastic bands or rubber gloves. Increasing the sulfur content restricts the uncoiling of

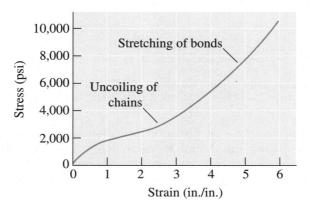

Figure 15-24
The stress-strain curve for an elastomer. Virtually all of the deformation is elastic; therefore, the modulus of elasticity varies as the strain changes.

the chains and the rubber becomes harder, more rigid, and brittle, as in rubber used for motor mounts. Typically, 0.5 to 5% sulfur is added to provide cross-linking in elastomers. Many more *efficient vulcanizing* (EV) systems, which are sulfur free, also have been developed and used in recent years.[10]

Typical Elastomers Elastomers, which are amorphous polymers, do not easily crystallize during processing. They have a low glass temperature, and chains can easily be deformed elastically when a force is applied. The typical elastomers (Tables 15-8 and

TABLE 15-8 ■ *Repeat units and applications for selected elastomers*

Polymer	Repeat Unit	Applications
Polyisoprene		Tires, golf balls, shoe soles
Polybutadiene (or butadiene rubber or Buna-S)		Industrial tires, toughening other elastomers, inner tubes of tires, weatherstripping, steam hoses
Polyisobutylene (or butyl rubber)		
Polychloroprene (Neoprene)		Hoses, cable sheathing
Butadiene-styrene (BS or SBR rubber)		Tires
Butadiene-acrylonitrile (Buna-N)		Gaskets, fuel hoses
Silicones		Gaskets, seals

TABLE 15-9 ■ *Properties of selected elastomers*

	Tensile Strength (psi)	% Elongation	Density (g/cm^3)
Polyisoprene	3000	800	0.93
Polybutadiene	3500		0.94
Polyisobutylene	4000	350	0.92
Polychloroprene (Neoprene)	3500	800	1.24
Butadiene-styrene (BS or SBR rubber)	3000	2000	1.0
Butadiene-acrylonitrile	700	400	1.0
Silicones	1000	700	1.5
Thermoplastic elastomers	5000	1300	1.06

15-9) meet these requirements. Polyisoprene is a natural rubber. Polychloroprene, or Neoprene, is a common material for hoses and electrical insulation. Many of the important synthetic elastomers are copolymers. Polybutadiene (butadiene rubber or Buna-S) is similar to polyisoprene, but the repeat unit has four carbon atoms consisting of one double bond. This is a relatively low-cost rubber, though the resistance to solvents is poor. As a result, it is used as a toughening material to make other elastomers. Butadiene-styrene rubber (BSR or BS), which is also one of the components of ABS (Figure 15-11), is used for automobile tires. Butyl rubber is different from polybutadiene (butadiene rubber). Butyl rubber, or polyisobutadiene, is used to make inner tubes for tires, vibration mounts, and as weather-stripping material. Silicones are another important elastomer based on chains composed of silicon and oxygen atoms. Silly Putty® was invented by James Wright of General Electric. It is made using hydroxyl terminated polydimethyl siloxane, boric oxide, and some other additives. At slower strain rates you can stretch it significantly, while you pull it fast it snaps (Chapter 7). The silicone rubbers (also known as polysiloxanes) provide high-temperature resistance, permitting use of the elastomer at temperatures as high as 315°C. Low molecular weight silicones form liquids and are known as silicon oils. Silicones can also be purchased as a two-part system that can be molded and cured. Chewing gum contains a base that is made from natural rubber, styrene butadiene, or polyvinyl acetate (PVA).

Thermoplastic Elastomers (TPEs) This is a special group of polymers that do not rely on cross-linking to produce a large amount of elastic deformation. Figure 15-25 shows the structure of a styrene-butadiene block copolymer engineered so that the styrene repeat units are located only at the ends of the chains. Approximately 25% of the chain is composed of styrene. The styrene ends of several chains form spherical-shaped domains. The styrene has a high glass temperature; consequently, the domains are strong and rigid and tightly hold the chains together. Rubbery areas containing butadiene repeat units are located between the styrene domains; these portions of the polymer have a glass temperature below room temperature and therefore behave in a soft, rubbery manner. Elastic deformation occurs by recoverable movement of the chains; however, sliding of the chains at normal temperatures is prevented by the styrene domains.

The styrene-butadiene block copolymers differ from the BS rubber discussed earlier in that cross-linking of the butadiene monomers is not necessary and, in fact, is undesirable. When the thermoplastic elastomer is heated, the styrene heats above the glass temperature, the domains are destroyed, and the polymer deforms in a viscous

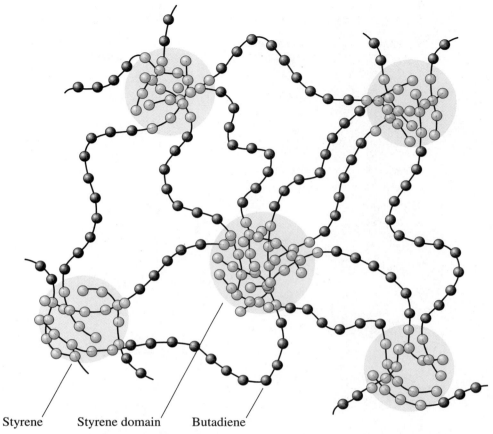

Styrene   Styrene domain   Butadiene

Figure 15-25 The structure of the styrene-butadiene (SB) copolymer in a thermoplastic elastomer. The glassy nature of the styrene domains provides elastic behavior without cross-linking of the butadiene.

manner—that is, it behaves as any other thermoplastic, making fabrication very easy. When the polymer cools, the domains reform and the polymer reverts to its elastomer characteristics. The thermoplastic elastomers consequently behave as ordinary thermoplastics at elevated temperatures and as elastomers at low temperatures. This behavior also permits the thermoplastic elastomers to be more easily recycled than conventional elastomers. A useful fluoroelastomer for high temperature and corrosive environments is Viton™. It is used for seals, O-rings, etc.

15-10   Thermosetting Polymers

Thermosets are highly cross-linked polymer chains that form a three-dimensional network structure. Because the chains cannot rotate or slide, these polymers possess good strength, stiffness, and hardness. However, thermosets also have poor ductility and impact properties and a high glass temperature. In a tensile test, thermosetting polymers display the same behavior as a brittle metal or ceramic.

Thermosetting polymers often begin as linear chains. Depending on the type of repeat units and the degree of polymerization, the initial polymer may be either a solid

TABLE 15-10 ■ *Functional units and applications for selected thermosets*

Polymer	Functional Units	Typical Applications
Phenolics		Adhesives, coatings, laminates
Amines		Adhesives, cookware, electrical moldings
Polyesters		Electrical moldings, decorative laminates, polymer matrix in fiberglass
Epoxies		Adhesives, electrical moldings, matrix for composites
Urethanes		Fibers, coatings, foams, insulation
Silicone		Adhesives, gaskets, sealants

or a liquid resin; in some cases, a two- or three-part liquid resin is used (as in the case of the two tubes of epoxy glue that we often use). Heat, pressure, mixing of the various resins, or other methods initiate the cross-linking process. Cross-linking is not reversible; once formed, the thermosets cannot be reused or conveniently recycled.

The functional groups for a number of common thermosetting polymers are summarized in Table 15-10, and representative properties are given in Table 15-11.

Phenolics   Phenolics, the most commonly used thermosets, are often used as adhesives, coatings, laminates, and molded components for electrical or motor applications.

Bakelite^{TM} is one of the common phenolic thermosets. A condensation reaction joining phenol and formaldehyde molecules produces the initial linear phenolic resin (Figure 15-26 on page 705). The oxygen atom in the formaldehyde molecule reacts with a hydrogen atom on each of two phenol molecules, and water is evolved as the by-product. The two phenol molecules are then joined by the carbon atom remaining in the formaldehyde.

This process continues until a linear phenol-formaldehyde chain is formed. However, the phenol is trifunctional. After the chain has formed, there is a third location on each phenol ring that provides a site for cross-linking with the adjacent chains.

TABLE 15-11 ■ *Properties of typical thermosetting polymers*

	Tensile Strength (psi)	% Elongation	Elastic Modulus (psi)	Density (g/cm³)
Phenolics	9,000	2	1300	1.27
Amines	10,000	1	1600	1.50
Polyesters	13,000	3	650	1.28
Epoxies	15,000	6	500	1.25
Urethanes	10,000	6		1.30
Silicone	4,000	0	1200	1.55

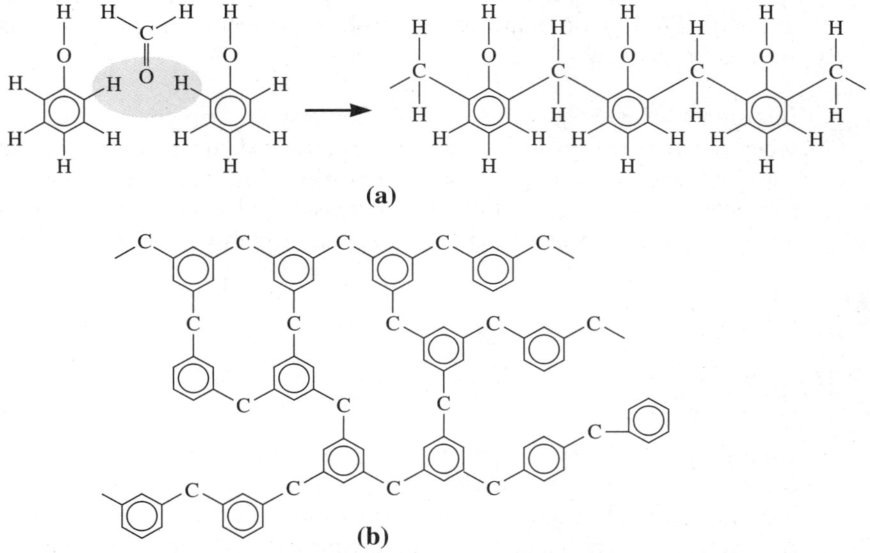

Figure 15-26 Structure of a phenolic. In (a), two phenol rings are joined by a condensation reaction through a formaldehyde molecule. Eventually, a linear chain forms. In (b), excess formaldehyde serves as the cross-linking agent, producing a network, thermosetting polymer.

Amines Amino resins, produced by combining urea or melamine monomers with formaldehyde, are similar to the phenolics. The monomers are joined by a formaldehyde link to produce linear chains. Excess formaldehyde provides the cross-linking needed to give strong, rigid polymers suitable for adhesives, laminates, molding materials for cookware, and electrical hardware such as circuit breakers, switches, outlets, and wall plates.

Urethanes Depending on the degree of cross-linking, the urethanes behave as thermosetting polymers, thermoplastics, or elastomers. These polymers find application as fibers, coatings, and foams for furniture, mattresses, and insulation.

Polyesters Polyesters form chains from acid and alcohol molecules by a condensation reaction, giving water as a byproduct. When these chains contain unsaturated bonds, a styrene molecule may provide cross-linking. Polyesters are used as molding or casting materials for a variety of electrical applications, decorative laminates, boats and other marine equipment, and as a matrix for composites such as fiberglass.

Epoxies   Epoxies are thermosetting polymers formed from molecules containing a tight C—O—C ring. During polymerization, the C—O—C rings are opened and the bonds are rearranged to join the molecules. The most common of the commercial epoxies is based on bisphenol A, to which have been added two epoxide units. These molecules are polymerized to produce chains and then co-reacted with curing agents that provide cross-linking. Epoxies are used as adhesives, rigid molded parts for electrical applications, automotive components, circuit boards, sporting goods and a matrix for high-performance fiber-reinforced composite materials for aerospace.

Polyimides   Polyimides display a ring structure that contains a nitrogen atom. One special group, the bismaleimides (BMI), is important in the aircraft and aerospace industry. They can operate continuously at temperatures of 175°C and do not decompose until reaching 460°C.

Interpenetrating Polymer Networks   Some special polymer materials can be produced when linear thermoplastic chains are intertwined through a thermosetting framework, forming **interpenetrating polymer networks**. For example, nylon, acetal, and polypropylene chains can penetrate into a cross-linked silicone thermoset. In more advanced systems, two interpenetrating thermosetting framework structures can be produced.

15-11   Adhesives

Adhesives are polymers used to join other polymers, metals, ceramics, composites, or combinations of these materials. The adhesives are used for a variety of applications. The most critical of these are the "structural adhesives," which find use in the automotive, aerospace, appliance, electronics, construction, and sporting equipment areas.

Chemically Reactive Adhesives   These adhesives include polyurethane, epoxy, silicone, phenolics, anaerobics, and polyimides. One-component systems consist of a single polymer resin cured by exposure to moisture, heat, or—in the case of anaerobics—the absence of oxygen. Two-component systems (such as epoxies) cure when two resins are combined.

Evaporation or Diffusion Adhesives   The adhesive is dissolved in either an organic solvent or water and is applied to the surfaces to be joined. When the carrier evaporates, the remaining polymer provides the bond. Water-base adhesives are preferred from the standpoint of environmental and safety considerations. The polymer may be completely dissolved in water or may consist of latex, or a stable dispersion of polymer in water. A number of elastomers, vinyls, and acrylics are used.

Hot-Melt Adhesives   These thermoplastics and thermoplastic elastomers melt when heated. On cooling, the polymer solidifies and joins the materials. Typical melting temperatures of commercial hot-melts are about 80°C to 110°C, which limits the elevated-temperature use of these adhesives. High-performance hot-melts, such as polyamides and polyesters, can be used up to 200°C.

Pressure-Sensitive Adhesives   These adhesives are primarily elastomers or elastomer copolymers produced as films or coatings. Pressure is required to cause the polymer to stick to the substrate. They are used to produce electrical and packaging tapes, labels, floor tiles, wall coverings, and wood-grained textured films.

Conductive Adhesives   A polymer adhesive may contain a filler material such as silver, copper, or aluminum flakes or powders to provide electrical and thermal conductivity. In some cases, thermal conductivity is desired but electrical conductivity is not wanted; alumina, boron nitride, and silica may be used as fillers to provide this combination of properties.

15-12 Additives for Plastics

Most plastics contain additives that impart special characteristics to the polymeric-based material.

Fillers   Fillers are added for a variety of purposes. One of the best known examples is the addition of carbon black to rubber to improve the strength and wear resistance of tires. It is also used to enhance protection against ultraviolet rays (e.g. Dylark^{TM}).[11] Some fillers, such as short fibers or flakes of inorganic materials (e.g. $CaCO_3$ and SiO_2), improve the mechanical properties of the polymers (Chapter 14). Each automobile tire contains about $\frac{1}{2}$ lb of nano-sized silica particles for enhancing the stiffness of the elastomer. Other fillers, called **extenders**, permit a large volume of a polymer material to be produced with relatively little actual resin, thus reducing cost. Calcium carbonate, silica, talc, and clay are frequently used extenders.

Pigments   Used to produce colors in polymers and paints, pigments are finely ground particles, such as TiO_2, that are uniformly dispersed in the polymer.

Stabilizers   Stabilizers prevent deterioration of the polymer due to environmental effects. Heat stabilizers are required in processing polyvinyl chloride; otherwise, hydrogen and chlorine atoms may be removed as hydrochloric acid, causing the polymer to be embrittled. Stabilizers also prevent the deterioration of polymers due to ultraviolet radiation.

Antistatic Agents   Most polymers, because they are poor conductors, build up a charge of static electricity. Antistatic agents attract moisture from the air to the polymer surface, improving the surface conductivity of the polymer and reducing the likelihood of a spark or discharge.

Flame Retardants   Because they are organic materials, most polymers are flammable. Additives that contain chlorine, bromine, phosphorus, or metallic salts reduce the likelihood that combustion will occur or spread.

Plasticizers   Low molecular weight molecules or chains called **plasticizers** reduce the glass temperature and provide internal lubrication, thereby improving the forming characteristics of the polymer. Plasticizers are particularly important for polyvinyl chloride, which has a glass temperature well above room temperature.

Reinforcements   The strength and rigidity of polymers are improved by introducing glass, polymer, or carbon filaments as **reinforcements**. For example, fiberglass consists of short filaments of glass in a polymer matrix.

Catalysts   In the processing of elastomers, zinc oxide (ZnO) is used to expedite the vulcanization process for making tires (Chapter 14). The ZnO expedites the cross-linking process, but does not participate in the formation of cross-links.

15-13   Polymer Processing and Recycling

There are a number of methods for producing polymer shapes, including molding, extrusion, and manufacture of films and fibers. The techniques used to form the polymers depend to a large extent on the nature of the polymer—in particular, whether it is thermoplastic or thermosetting. The greatest variety of techniques are used to form the thermoplastics. The polymer is heated to near or above the melting temperature so that it becomes rubbery or liquid. The polymer is then formed in a mold or die to produce the required shape. Thermoplastic elastomers can be formed in the same manner. In these processes, scrap can be easily recycled and waste is minimized. Fewer forming techniques are used for the thermosetting polymers because, once cross-linking has occurred, the thermosetting polymers are no longer capable of being formed. Elastomers are processed in high-shear equipment such as a Banbury mixer. Carbon black and other additives are added. The heating from viscoelastic deformation can begin to cross-link the material prematurely. After the mixing step, a curing agent (e.g., zinc oxide) is added. The material discharged from the mixer is pliable and is processed using a short extruder, molded using a two-roll mill, or applied on parts by dip coating. This processing of elastomers is known as **compounding** of rubber.

The following are some of the techniques mainly used for processing of polymers; most of these, you will note, apply only to thermoplastics.

Extrusion   This is the most widely used technique for processing thermoplastics. Extrusion can serve two purposes. First, it provides a way to form certain simple shapes continuously. Second, extrusion provides an excellent mixer for additives (e.g., carbon black, fillers, etc.) when processing polymers that ultimately may be processed using some other process. A screw mechanism consisting of one or a pair of screws (twin screw) forces heated thermoplastic (either solid or liquid) and additives through a die opening to produce solid shapes, films, sheets, tubes, pipes, and even plastic bags (Figure 15-27). An industrial extruder can be up to 60 to 70 feet long, 2 feet in diameter, and consist of different heating or cooling zones. Since thermoplastics show shear thinning behavior and are viscoelastic, the control of both temperature and viscosity is critical in polymer extrusion. One special extrusion process for producing films is illustrated in Figure 15-28. Extrusion also can be used to coat wires and cables with either thermoplastics or elastomers.

Blow Molding   A hollow preform of a thermoplastic called a **parison** is introduced into a die by gas pressure and expanded against the walls of the die (Figure 15-29). This process is used to produce plastic bottles, containers, automotive fuel tanks, and other hollow shapes.

Injection Molding   Thermoplastics heated above the melting temperature using an extruder are forced into a closed die to produce a molding. This process is similar to

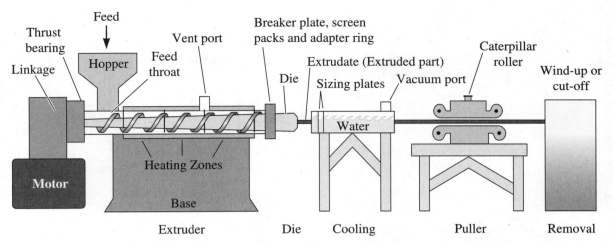

Figure 15-27 Schematic of an extruder used for polymer processing. (*Source: Adapted from* Plastics: Materials and Processing, Second Edition, *by A. Brent Strong, p. 382, Fig. 11-1. Copyright © 2000 Prentice Hall. Adapted with permission of Pearson Education, Inc., Upper Saddle River, NJ.*)

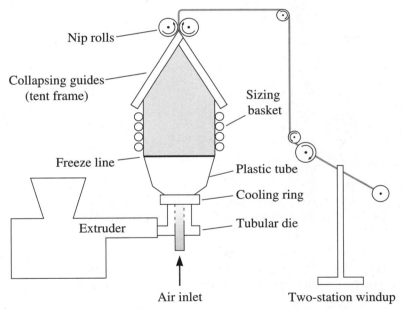

Figure 15-28 One technique by which polymer films (used in the manufacture of garbage bags, for example) can be produced. The film is extruded in the form of a bag, which is separated by air pressure until the polymer cools. (*Source: Adapted from* Plastics: Materials and Processing, Second Edition, *by A. Brent Strong, p. 397, Fig. 11-8. Copyright © 2000 Prentice Hall. Adapted with permission of Pearson Education, Inc., Upper Saddle River, NJ.*)

die casting of molten metals. A plunger or a special screw mechanism applies pressure to force the hot polymer into the die. A wide variety of products, ranging from cups, combs, and gears to garbage cans, can be produced in this manner.

Thermoforming Thermoplastic polymer sheets heated to the plastic region can be formed over a die to produce such diverse products as egg cartons and decorative panels. The forming can be done using matching dies, a vacuum, or air pressure.

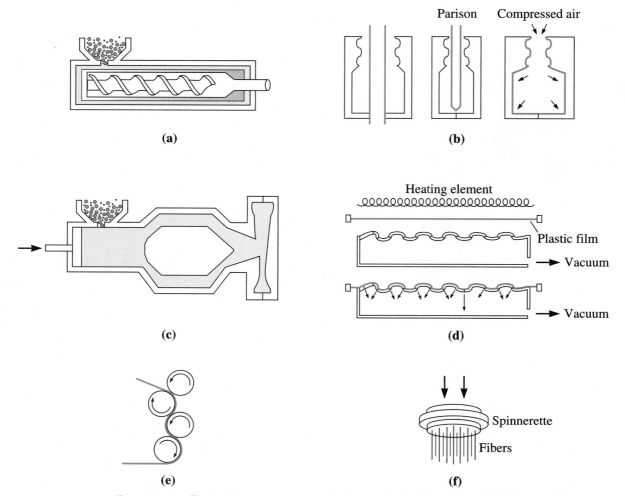

Figure 15-29 Typical forming processes for thermoplastic: (a) extrusion, (b) blow molding, (c) injection molding, (d) thermoforming, (e) calendaring, and (f) spinning.

Calendaring   In a calendar, molten plastic is poured into a set of rolls with a small opening. The rolls, which may be embossed with a pattern, squeeze out a thin sheet of the polymer—often, polyvinyl chloride. Typical products include vinyl floor tile and shower curtains.

Spinning   Filaments, fibers, and yarns may be produced by spinning. The molten thermoplastic polymer is forced through a die containing many tiny holes. The die, called a **spinnerette**, can rotate and produce a yarn. For some materials, including nylon, the fiber may subsequently be stretched to align the chains parallel to the axis of the fiber; this process increases the strength of the fibers.

Casting   Many polymers can be cast into molds and permitted to solidify. The molds may be plate glass for producing individual thick plastic sheets or moving stainless steel belts for continuous casting of thinner sheets. *Rotational molding* is a special casting process in which molten polymer is poured into a mold rotating about two axes. Centrifugal action forces the polymer against the walls of the mold, producing a thin shape such as a camper top.

Compression Molding   Thermoset moldings are most often formed by placing the solid material before cross-linking into a heated die. Application of high pressure and temperature causes the polymer to melt, fill the die, and immediately begin to harden. Small electrical housings as well as fenders, hoods, and side panels for automobiles can be produced by this process (Figure 15-30).

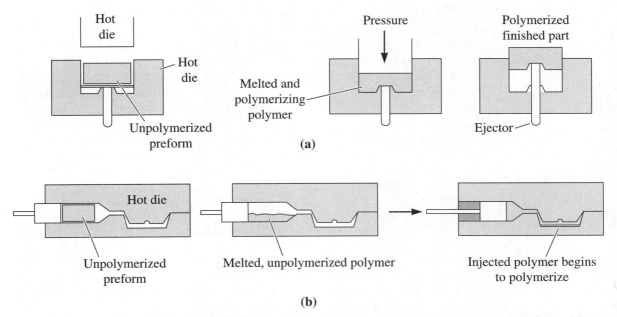

Figure 15-30   Typical forming processes for thermosetting polymers: (a) compression molding and (b) transfer molding.

Transfer Molding   A double chamber is used in the transfer molding of thermosetting polymers. The polymer is heated under pressure in one chamber. After melting, the polymer is injected into the adjoining die cavity. This process permits some of the advantages of injection molding to be used for thermosetting polymers (Figure 15-30).

Reaction Injection Molding (RIM)   Thermosetting polymers in the form of liquid resins are first injected into a mixer and then directly into a heated mold to produce a shape. Forming and curing occur simultaneously in the mold. In reinforced-reaction injection molding (RRIM), a reinforcing material consisting of particles or short fibers is introduced into the mold cavity and is impregnated by the liquid resins to produce a composite material. Automotive bumpers, fenders, and furniture parts are made using this process.

Foams   Foamed products can be produced in polystyrene, urethanes, polymethyl methacrylate, and a number of other polymers. The polymer is produced in the form of tiny beads, often containing a blowing agent such as pentane. During the pre-expansion process, the bead increases in diameter by as many as 50 times. The pre-expanded beads are then injected into a die, with the individual beads fusing together, using steam, to form exceptionally lightweight products with densities of perhaps only 0.02 g/cm^3. Expandable polystyrene (EPS) cups, packaging, and insulation are some of the applications for foams. Engine blocks for Saturn cars are made using a pattern made from expanded polystyrene beads.[11]

EXAMPLE 15-10   *Insulation Boards for Houses*

You want to design a material that can be used for making insulation boards that are approximately 4 ft wide and 8 ft tall. The material must provide good thermal insulation. What material would you choose?

SOLUTION

Glasses tend to be good insulators of heat. However, they will be heavy, more expensive, and prone to fracture. Polymers are lightweight, can be produced inexpensively, and they can be good thermal insulators. We can use foamed polystyrene since the air contained in the pores adds significantly to their effectiveness as thermal insulators. For better mechanical properties, we may want to produce foams that have relatively high density (compared to foams that are used to make coffee cups). Finally, from a safety viewpoint, we want to be sure that some fire and flame retardants are added to the foams. Such panels are made using expanded polystyrene beads containing pentane. A molding process is used to make the foams. The sheets can be cut into required sizes using a heated metal wire.

Recycling of Plastics   Recycling is a very important issue and a full discussion of the entire process is outside the scope of this book. But recycling plays an important role in our everyday lives. Material is recycled in many ways. For example, part of the polymer that is scrap from a manufacturing process (known as regrind) is used by recycling plants. The recycling of thermoplastics is relatively easy and practiced widely. Note that many of the everyday plastic products you encounter (bags, soda bottles, yogurt containers, etc.) have numbers stamped on them. For PET products (recycling symbol "PETE", because of trademark issues), the number is 1. For HDPE vinyl (recycling symbol V), LDPE, PP, and PS the numbers are 2, 3, 4, 5 and 6, respectively. Other plastics are marked number 7.

Thermosets and elastomers are more difficult to recycle, although they can still be used. For example, tires can be shredded and used to make safer playground surfaces or roads.

Despite enormous recycling efforts, a large portion of the materials in a landfill today are plastics (the largest is that of paper). Given the limited amount of petroleum, the threat of global warming, and a need for a cleaner and safer environment, careful use and recycling makes sense for all materials.

SUMMARY

◇ Polymers are made from large macromolecules produced by the joining of smaller molecules, called monomers, using addition or condensation polymerization reactions. *Plastics* are materials that are based on polymeric compounds, and they contain many other additives that improve their properties. Compared with most metals and ceramics, plastics have low strength, stiffness, and melting temperatures; however, they also have a low density and good chemical resistance. Plastics are used in a most diverse number of technologies.

◇ Thermoplastics have chains that are not chemically bonded to each other, permitting the material to be easily formed into useful shapes, to have good ductility, and to be economically recycled. Thermoplastics can have an amorphous structure, which provides low strength and good ductility when the ambient temperature is above the glass temperature. The polymers are more rigid and brittle when the temperature falls below the glass temperature. Many thermoplastics can also partially crystallize during cooling or by application of a stress. This increases their strength.

◆ The thermoplastic chains can be made more rigid and stronger by using nonsymmetrical monomers that increase the bonding strength between the chains and make it more difficult for the chains to disentangle when stress is applied. In addition, many monomers produce more rigid chains containing atoms or groups of atoms other than carbon; this structure also produces high-strength thermoplastics.

◆ Elastomers are thermoplastics or lightly cross-linked thermosets that exhibit greater than 200% elastic deformation. Chains are eventually cross-linked using vulcanization. The cross-linking makes it possible to obtain very large elastic deformations without permanent plastic deformation. Increasing the number of cross-links increases the stiffness and reduces the amount of elastic deformation of the elastomers.

◆ Thermoplastic elastomers combine features of both thermoplastics and elastomers. At high temperatures, these polymers behave as thermoplastics and are plastically formed into shapes; at low temperatures, they behave as elastomers.

◆ Thermosetting polymers are highly cross-linked into a three-dimensional network structure. Typically, high glass temperatures, good strength, and brittle behavior are found. Once cross-linking occurs, these polymers cannot be easily recycled.

◆ Manufacturing processes depend on the behavior of the polymers. Processes such as extrusion, injection molding, thermoforming, casting, drawing, and spinning are made possible by the viscoelastic behavior of the thermoplastics. The non-reversible behavior of bonding in thermosetting polymers limits their processing to fewer techniques, such as compression molding, transfer molding and reaction-injection molding.

GLOSSARY

Addition polymerization Process by which polymer chains are built up by adding monomers together without creating a byproduct.

Aging Slow degradation of polymers as a result of exposure to low levels of heat, oxygen, bacteria, or ultraviolet rays.

Aramids Polyamide polymers containing aromatic groups of atoms in the linear chain.

Blushing A thermoplastic bent repeatedly leads to small volumes of the material crystallizing; this leads to voids that ultimately cause the material to fail.

Branched polymer Any polymer consisting of chains that consist of a main chain and secondary chains that branch off from the main chain.

Compounding Processing of elastomers in device known as a Banbury mixer followed by forming using extrusion, molding, or dip coating.

Condensation polymerization A polymerization mechanism in which a small molecule (e.g., water, methanol, etc.) is condensed out as a byproduct.

Copolymer An addition polymer produced by joining more than one type of monomer.

Crazing Localized plastic deformation in a polymer. A craze may lead to the formation of cracks in the material.

Cross-linking   Attaching chains of polymers together to produce a three-dimensional network polymer.

Deflection temperature   The temperature at which a polymer will deform a given amount under a standard load.

Degradation temperature   The temperature above which a polymer burns, chars, or decomposes.

Degree of polymerization   The average molecular weight of the polymer divided by the molecular weight of the monomer.

Diene   A group of monomers that contain two double-covalent bonds. These monomers are often used in producing elastomers.

Elastomers   These are polymers (thermoplastics or lightly cross-linked thermosets) that have an elastic deformation > 200%.

Extenders   Additives or fillers to polymers that provide bulk at a low cost.

Functionality   The number of sites on a monomer at which polymerization can occur.

Geometric isomer   A molecule that has the same composition as, but a structure different from, a second molecule.

Glass temperature   The temperature range below which the amorphous polymer assumes a rigid glassy structure.

Interpenetrating polymer networks   Polymer structures produced by intertwining two separate polymer structures or networks.

Linear polymer   Any polymer in which molecules are in the form of spaghetti-like chains.

Liquid-crystalline polymers   Exceptionally stiff polymer chains that act as rigid rods, even above their melting point.

Mer   A unit group of atoms and molecules that defines a characteristic arrangement for a polymer. A polymer can be thought off as a material made by combining several mers or units.

Monomer   The molecule from which a polymer is produced.

Oligomer   Low molecular weight molecules, these may contain two (dimers) or three (trimers) mers.

Parison   A hot glob of soft or molten polymer that is blown or formed into a useful shape.

Plastic   A predominantly polymeric material comprising of other additives.

Plasticizer   An additive that, by decreasing the glass temperature, improves the formability of a polymer.

Polymer   Polymers are materials made from giant (or macromolecular), chain-like molecules having average molecular weights from 10,000 to more than 1,000,000 g/mol built by the joining of many mers or units by chemical bonds. Polymers are usually, but not always, carbon based.

Reinforcement   Additives to polymers designed to provide significant improvement in strength. Fibers are typical reinforcements.

Relaxation time   A property of a polymer that is related to the rate at which stress relaxation occurs.

Repeat unit   The repeating structural unit from which a polymer is built. Also called a *mer*.

Spinnerette An extrusion die containing many small openings through which hot or molten polymer is forced to produce filaments. Rotation of the spinnerette twists the filaments into a yarn.

Stress relaxation A reduction of the stress acting on a material over a period of time at a constant strain due to viscoelastic deformation.

Tacticity Describes the location in the polymer chain of atoms or atom groups in nonsymmetrical monomers.

Thermoplastic elastomers Polymers that behave as thermoplastics at high temperatures, but as elastomers at lower temperatures.

Thermoplastics Linear or branched polymers in which chains of molecules are not interconnected to one another.

Thermosetting polymers Polymers that are heavily cross-linked to produce a strong three dimensional network structure.

Unsaturated bond The double- or even triple-covalent bond joining two atoms together in an organic molecule. When a single covalent bond replaces the unsaturated bond, polymerization can occur.

Viscoelasticity The deformation of a material by elastic deformation and viscous flow of the material when stress is applied.

Vulcanization Cross-linking elastomer chains by introducing sulfur or other chemicals.

ADDITIONAL INFORMATION

1. STRONG, B.A., *Plastics-Materials and Processing*. Second ed. 2000: Prentice Hall.
2. COLLINGS, P.J. and M. HIRD, *Introduction to Liquid Crystals: Chemistry and Physics*. 1997: Taylor and Francis.
3. KULKARNI, V.G., *Polyanilines: Progress in Processing and Properties*, ACS, (1999).
4. ROBERTS, R.M., *Serendipity: Accidental Discoveries in Science*. 1989: John Wiley and Sons.
5. DRONBY, G.J., *Technology of Fluoropolymers*, CRS Press (2000).
6. PHULÉ, P.P. and T. WOOD, *Sol-Gel Processing of Ceramics*, in *Encyclopedia of Advanced Materials*, D. BLOOR, S. MAHAJAN, and M. FLEMINGS, Editors. 2001: Permgaon.
7. PLAZEK, D.J. 2002, *Personal Communication*.
8. VAUGHAN, D., *The Challenger Launch Decision: Risky Technology, Culture, and Deviance at NASA*. 1996: Chicago Press.
9. MAGILL, J., *Technical discussions are acknowledged*. 2002.
10. HARPER, C.A., *Handbook of Materials Product Design*. 3rd ed. 2001: McGraw Hill.
11. ARCH, P., *Nova Chemicals, Technical discussions and assistance is acknowledged*. 2002.

✓ PROBLEMS

Section 15-1 Classification of Polymers

15-1 What are linear and branched polymers? Can thermoplastics be branched?

15-2 Define (a) a thermoplastic, (b) thermosetting plastics, (c) elastomers, and (d) thermoplastic elastomers.

15-3 What electrical and optical applications are polymers used for? Explain using examples.

15-4 What are the major advantages of plastics compared to ceramics, glasses, and metallic materials?

Sections 15-2 Addition Polymerization

Sections 15-3 Condensation Polymerization

15-5 What do the terms condensation polymerization, addition polymerization, initiator, and terminator mean?

15-6 Suppose that 20 g of benzoyl peroxide are introduced to 5 kg of propylene monomer (see Table 15-4). If 30% of the initiator groups are effective, calculate the expected degree of polymerization and the molecular weight of the polypropylene polymer if:

(a) All of the termination of the chains occurs by combination and all of the termination occurs by disproportionation.

(b) Suppose hydrogen peroxide (H_2O_2) is used as the initiator for 10 kg of vinyl chloride monomer (see Table 15-3). Show schematically how the hydrogen peroxide will initiate the polymer chains. Calculate the amount of hydrogen peroxide (assuming that it is 10% effective) required to produce a degree of polymerization of 4000 if termination of the chains occurs by combination and termination occurs by disproportionation.

15-7 The formula for formaldehyde is HCHO. Draw the structure of the formaldehyde molecule and repeat unit. Does formaldehyde polymerize to produce an acetal polymer (see Table 15-4) by the addition mechanism or the condensation mechanism? Try to draw a sketch of the reaction and the acetal polymer by both mechanisms.

15-8 You would like to combine 5 kg of dimethyl terephthalate with ethylene glycol to produce polyester (PET). Calculate the amount of ethylene glycol required, the amount of byproduct evolved, and the amount of polyester produced.

15-9 Would you expect polyethylene to polymerize at a faster or slower rate than polymethyl methacrylate? Explain. Would you expect polyethylene to polymerize at a faster or slower rate than a polyester? Explain.

15-10 You would like to combine 10 kg of ethylene glycol with terephthalic acid to produce a polyester. The monomer for terephthalic acid is shown below:

$$H—O—\overset{\overset{\displaystyle O}{\|}}{C}—\bigcirc\!\!\!\!-\!\!\!\!\bigcirc—\overset{\overset{\displaystyle O}{\|}}{C}—O—H$$

Determine the byproduct of the condensation reaction and calculate the amount of tereph-

thalic acid required, the amount of byproduct evolved, and the amount of polyester produced.

Section 15-4 Degree of Polymerization

15-11 Explain why low-density polyethylene is good to make grocery bags, however, super high molecular weight polyethylene must be used where strength and very high wear resistance is needed.

15-12 Explain why, in polymer processing, a bimodal molecular weight distribution is often necessary.

15-13 The molecular weight of polymethyl methacrylate (see Table 15-3) is 250,000 g/mol. If all of the polymer chains are the same length,

(a) calculate the degree of polymerization, and
(b) the number of chains in 1 g of the polymer.

15-14 Calculate (a) the degree of polymerization of polytetrafluoroethylene (see Table 15-3) is 7500. (b) If all of the polymer chains are the same length, calculate

(i) the molecular weight of the chains, and
(ii) the total number of chains in 1000 g of the polymer.

15-15 A polyethylene rope weighs 0.25 lb per foot. If each chain contains 7000 repeat units,

(a) calculate the number of polyethylene chains in a 10-ft length of rope, and
(b) the total length of chains in the rope, assuming that carbon atoms in each chain are approximately 0.15 nm apart.

15-16 A common copolymer is produced by including both ethylene and propylene monomers in the same chain. Calculate the molecular weight of the polymer produced using 1 kg of ethylene and 3 kg of propylene, giving a degree of polymerization of 5000.

15-17 Analysis of a sample of polyacrylonitrile (see Table 15-3) shows that there are six lengths of chains, with the following number of chains of each length. Determine

(a) the weight average molecular weight and degree of polymerization, and
(b) the number average molecular weight and degree of polymerization.

Number of Chains	Mean Molecular Weight of Chains (g/mol)
10,000	3,000
18,000	6,000
17,000	9,000
15,000	12,000
9,000	15,000
4,000	18,000

15-18 Explain why you would prefer that the number average molecular weight of a polymer be as close as possible to the weight average molecular weight.

Section 15-5 Typical Thermoplastics

Section 15-6 Structure–Property Relationships in Thermoplastics

Section 15-7 Effect of Temperature on Thermoplastics

Section 15-8 Mechanical Properties of Thermoplastics

15-19 Explain what the following terms mean: decomposition temperature, heat distortion temperature, glass temperature, and melting temperature. Why is it that thermoplastics do not have a fixed melting or glass temperature?

15-20 Using Table 15-5, plot the relationship between the glass temperatures and the melting temperatures of the addition thermoplastics. What is the approximate relationship between these two critical temperatures? Do the condensation thermoplastics and the elastomers also follow the same relationship?

15-21 List the addition polymers in Table 15-5 that might be good candidates for making the bracket that holds the rearview mirror onto the outside of an automobile, assuming that temperatures frequently fall below zero degrees Celsius. Explain your choices.

15-22 Based on Table 15-5, which of the elastomers might be suited for use as a gasket in a pump for liquid CO_2 at $-78°C$? Explain.

15-23 How do the glass temperatures of polyethylene, polypropylene, and polymethyl methacrylate compare? Explain their differences, based on the structure of the monomer.

15-24 Which of the addition polymers in Table 15-5 are used in their leathery condition at room temperature? How is this condition expected to affect their mechanical properties compared with those of glassy polymers?

15-25 The density of polypropylene is approximately 0.89 g/cm^3. Determine the number of propylene repeat units in each unit cell of crystalline polypropylene.

15-26 The density of polyvinyl chloride is approximately 1.4 g/cm^3. Determine the number of vinyl chloride repeat units, hydrogen atoms, chlorine atoms, and carbon atoms in each unit cell of crystalline PVC.

15-27 A polyethylene sample is reported to have a density of 0.97 g/cm^3. Calculate the percent crystallinity in the sample. Would you expect that the structure of this sample has a large or small amount of branching? Explain.

15-28 Amorphous polyvinyl chloride is expected to have a density of 1.38 g/cm^3. Calculate the % crystallization in PVC that has a density of 1.45 g/cm^3. (*Hint*: Find the density of completely crystallized PVC from its lattice parameters, assuming four repeat units per unit cell.)

15-29 What factors influence the crystallinity of polymers? Explain the development and role of crystallinity in PET and nylon.

15-30 Describe the relative tendencies of the following polymers to crystallize. Explain your answer.

(a) branched polyethylene versus linear polyethylene,

(b) polyethylene versus polyethylene-polypropylene copolymer,

(c) isotactic polypropylene versus atactic polypropylene, and

(d) polymethyl methacrylate versus acetal (polyoxymethylene).

15-31 Explain the meaning of these terms-creep, stress relaxation, crazing, blushing, environmental stress cracking, and aging of polymers.

15-32 A stress of 2500 psi is applied to a polymer serving as a fastener in a complex assembly. At a constant strain, the stress drops to 2400 psi after 100 h. If the stress on the part must remain above 2100 psi in order for the part to function properly, determine the life of the assembly.

15-33 A stress of 1000 psi is applied to a polymer that operates at a constant strain; after six months, the stress drops to 850 psi. For a particular application, a part made of the same polymer must maintain a stress of 900 psi after 12 months. What should be the original stress applied to the polymer for this application?

15-34 Data for the rupture time of polyethylene are shown in Figure 15-19. At an applied stress of 700 psi, the figure indicates that the polymer ruptures in 0.2 h at 90°C but survives 10,000 h at 65°C. Assuming that the rupture time is related to the viscosity, calculate the activation energy for the viscosity of the polyethylene and estimate the rupture time at 23°C.

15-35 For each of the following pairs, recommend the one that will most likely have the better impact properties at 25°C. Explain each of your choices.

(a) polyethylene versus polystyrene,

(b) low-density polyethylene versus high-density polyethylene, and

(c) polymethyl methacrylate versus polytetra-fluoroethylene.

Section 15-9 Elastomers (Rubbers)

15-36 Who invented the vulcanization process? What role does sulfur play?

15-37 What is natural rubber?

15-38 The polymer ABS can be produced with varying amounts of styrene, butadiene, and acrylonitrile monomers, which are present in the form of two copolymers: BS rubber and SAN.

(a) How would you adjust the composition of ABS if you wanted to obtain good impact properties?

(b) How would you adjust the composition if you wanted to obtain good ductility at room temperature?

(c) How would you adjust the composition if you wanted to obtain good strength at room temperature?

15-39 Figure 15-24 shows the stress-strain curve for an elastomer. From the curve, calculate and plot the modulus of elasticity versus strain and explain the results.

15-40 The maximum number of cross-linking sites in polyisoprene is the number of unsaturated bonds in the polymer chain. If three sulfur atoms are in each cross-linking sulfur strand, calculate the amount of sulfur required to provide cross-links at every available site in 5 kg of polymer and the wt% S that would be present in the elastomer. Is this typical?

15-41 Suppose we vulcanize polychloroprene, obtaining the desired properties by adding 1.5% sulfur by weight to the polymer. If each cross-linking strand contains an average of four sulfur atoms, calculate the fraction of the unsaturated bonds that must be broken.

15-42 The monomers for adipic acid, ethylene glycol, and maleic acid are shown below. These monomers can be joined into chains by condensation reactions, then cross-linked by breaking unsaturated bonds and inserting a styrene molecule as the cross-linking agent. Show how a linear chain composed of these three monomers can be produced.

Adipic acid

Maleic acid

Ethylene glycol

Section 15-10 Thermosetting Polymers

15-43 Explain the term thermosetting polymer. Why can't a thermosetting polymer be produced using just adipic acid and ethylene glycol?

15-44 Show how styrene provides cross-linking between the linear chains.

15-45 If 50 g of adipic acid, 100 g of maleic acid, and 50 g of ethylene glycol are combined, calculate the amount of styrene required to completely cross-link the polymer.

15-46 How much formaldehyde is required to completely cross-link 10 kg of phenol to produce a thermosetting phenolic polymer? How much byproduct is evolved?

15-47 Explain why the degree of polymerization is not usually used to characterize thermosetting polymers.

15-48 Defend or contradict the choice to use the following materials as hot-melt adhesives for an application in which the assembled part is subjected to impact-type loading:

(a) polyethylene,

(b) polystyrene,

(c) styrene-butadiene thermoplastic elastomer,

(d) polyacrylonitrile, and

(e) polybutadiene.

15-49 Compare and contrast properties of thermoplastics, thermosetting materials, and elastomers.

Section 15-11 Adhesives

15-50 Many paints are based on polymeric materials. Explain why plasticizers are added to paints. What must happen to the plasticizers after the paint is applied?

15-51 You want to extrude a complex component from an elastomer. Should you vulcanize the rubber before or after the extrusion operation? Explain.

15-52 Suppose a thermoplastic polymer can be produced in sheet form either by rolling (deformation) or by continuous casting (with a rapid cooling rate). In which case would you expect to obtain the higher strength? Explain.

Section 15-12 Additives for Plastics

15-53 What are some of the common fillers added to plastics? What is the role of carbon black, silica, and calcium carbonate?

15-54 What are fire-resistant plastics? What are self-extinguishing plastics?

15-55 What material is added to stiffen rubber used for tires?

15-56 Why are plasticizers needed for PVC? How is the role of plasticizers used in polymer processing similar to that of plasticizers used in ceramic tape casting?

Section 15-13 Polymer Processing and Recycling

15-57 Draw a schematic of an extruder. Explain how extrusion is used in polymer processing.

15-58 How are thermoplastics processed?

15-59 What is rubber compounding?

15-60 What is expanded polystyrene (EPS)? How can this material be used to make coffee cups and a complex casting, insulation boards, or packaging for shipping TVs and VCRs?

15-61 Which plastics are easiest to recycle? Which plastics are the most difficult?

✳ Design Problems

15-62 Figure 15-31 shows the behavior of polypropylene, polyethylene, and acetal at two temperatures. We would like to produce a 12-in.-long rod of a polymer that will operate at 40°C for 6 months under a constant load of 500 lb. Design the material and size of the rod such that no more than 5% Elongation will occur by creep.

15-63 Design a polymer material that might be used to produce a 3-in.-diameter gear to be used to transfer energy from a low-power electric motor. What are the design requirements? What class of polymers (thermoplastics, thermosets, elastomers) might be most appropriate? What particular polymer might you first consider? What additional information concerning the application and polymer properties do you need to know to complete your design?

15-64 Design a polymer material and a forming process to produce the case for a personal computer. What are the design and forming requirements? What class of polymers might be most

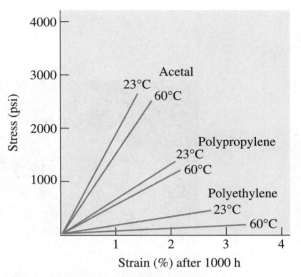

Figure 15-31 The effect of applied stress on the percent creep strain for three polymers (for Problem 15-62).

appropriate? What particular polymer might you first consider? What additional information do you need to know?

15-65 What kind of polymer can be used to line the inside of the head of hip prosthesis implant? Discuss what requirements would be needed for this type of polymer.

▣ Computer Problems

15-66 *Polymer Molecular Weight Distribution.* The following data were obtained for polyethylene. Determine the average molecular weight and degree of polymerization.

Molecular Weight Range (g/mol)	f_i	x_i
0–3,000	0.01	0.03
3,000–6,000	0.08	0.10
6,000–9,000	0.19	0.22
9,000–12,000	0.27	0.36
12,000–15,000	0.23	0.19
15,000–18,000	0.11	0.07
18,000–21,000	0.06	0.02
21,000–24,000	0.05	0.01

Write a computer program or use a spreadsheet program to solve this problem.

Composites are formed by incorporating multiple phase components in a material in such a way that the properties of the resultant material are unique and not attainable otherwise. Fiberglass is an archetypal composite material. Composites tell us the success stories of materials that work in teams and materials that work in synergy.

The two photographs above illustrate how nature makes composites. The abalone shell is made from calcium carbonate. The SEM fractured surface of the nacre (mother-of-pearl) section of red abalone (*Haliotis rufescens*) displays a tortuous pattern due to nanolayered nature of the hybrid composite of aragonitic (CaCO3-orthorhombic) platelets surrounded by a thin-film (10 nm) proteinaceous organic matrix. (*Courtesy of Professor Mehmet Sarikaya, University of Washington, Seattle.*) Although it is difficult to copy exactly the processes that occur in nature when making engineered composites, these materials certainly provide the inspiration for developing and understanding composites. Today, composites are used in so many applications and technologies that it would be almost impossible to think about a world without them. (*Abalone shell image courtesy of PhotoDisc/Getty Images.*)

16

Composites: Teamwork and Synergy in Materials

Have You Ever Wondered?

- What are some of the naturally occurring composites?

- Why is abalone shell, made primarily of calcium carbonate, so much stronger than chalk, which is also made of calcium carbonate?

- What sporting gear applications make use of composites?

- Why are composites finding increased uses in aircrafts and automobiles?

Composites are produced when two or more materials or phases are used together to give a combination of properties that cannot be attained otherwise. Composite materials may be selected to give unusual combinations of stiffness, strength, weight, high-temperature performance, corrosion resistance, hardness, or conductivity.[1,2] Composites highlight how different materials can work in synergy. Abalone shell, wood, bone, and teeth are examples of naturally occurring composites. Microstructures of selected composites are shown in Figure 16-1. An example of a material that is a composite at a macro-scale would be steel-reinforced concrete. Composites that are at

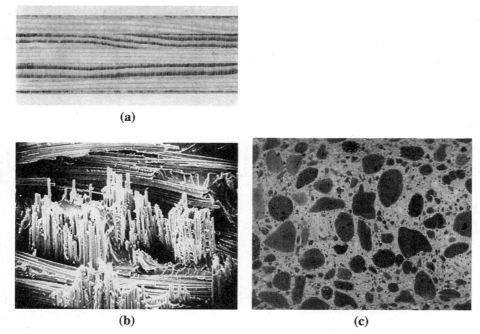

Figure 16-1 Some examples of composite materials: (a) plywood is a laminar composite of layers of wood veneer, (b) fiberglass is a fiber-reinforced composite containing stiff, strong glass fibers in a softer polymer matrix ($\times$175), and (c) concrete is a particulate composite containing coarse sand or gravel in a cement matrix (reduced 50%).

a micro-scale include such materials as carbon or glass-fiber reinforced plastics (CFRP or GFRP). These composites offer significant gains in specific strengths and are finding increasing uses in airplanes, electronic components, automotives and sporting equipment.[3]

As mentioned in Chapters 11 and 12, dispersion-strengthened (like steels) and precipitation-hardened alloys are examples of traditional materials that are **nanocomposites**. In a nanocomposite, the **dispersed phase** consists of nanoscale particles and is distributed in a **matrix phase**. Essentially, the same concept has been applied in developing what are described as **hybrid organic-inorganic nanocomposites**. These are materials in which the molecular, or microstructure, of the composites consists of an inorganic part or block and an organic block.[4,5] The idea is not too different from the formation of block copolymers (Chapter 15). These are often made using the sol-gel process as mentioned in Chapter 14.[6,7] These and other functional composites can provide unusual combinations of

electronic, magnetic, or optical properties. For example, a porous dielectric material prepared using phase-separated inorganic glasses exhibits a dielectric constant that is lower than that for the same material with no porosity. A transparent polymer, that is responsive to magnetic field, can also be prepared using composites. Space shuttle tiles are made from silica fibers are lightweight because they consists of air and silica fibers and exhibit a very small thermal conductivity. The two phases in these examples would be ceramic and air. Many glass-ceramics are nano-scale composites of different ceramic phases. Many plastics can be considered composites as well. For example, Dylark™ is a composite of maleic anhydride-styrene copolymer. It contains carbon black for stiffness and protection against ultraviolet rays. It also contains glass fibers for increased Young's modulus and rubber for toughness.[8] Silver-filled epoxies provide thermal conductivities higher than those for expoxies. Some dielectric materials are made using multiple phases such that the overall dielectric properties of interest (e.g., the dielectric constant) do not change appreciably with temperature (within a certain range). Similarly, many composites are prepared using magnetic and optical materials. Some composite structures may consist of different materials arranged in different layers. This leads to what are known as functionally graded materials and structures. For example, yttria stabilized zirconia (YSZ) coating on a turbine blade will have other layers in between that provide bonding with the turbine blade material. The YSZ coating itself is made using a plasma spray or other technique and contains certain levels of porosity which are essential for providing protection against high temperatures.[9] Similarly, different coatings on glass are examples of composite structures.[10] Thus, the *concept* of using composites is a generic one and can be applied at the macro, micro, and nano length-scales.

In composites, the properties and volume fractions of individual phases are important. The connectivity of phases is also very important. Usually the **matrix phase** is the continuous phase and the other phase is said to be the dispersed phase. Thus, terms such as "metal-matrix" indicate a metallic material used to form the continuous phase.

Connectivity describes how the two or more phases are connected in the composite.

Newnham has described a connectivity model for describing connectivities for functional composites.[11] Composites are often classified based on the shape or nature of the dispersed phase (e.g., particle-reinforced, whisker-reinforced, or fiber-reinforced composites). **Whiskers** are like fibers; but their length is much smaller. The bonding between the particles, whiskers, or fibers and the matrix is also very important. In structural composites, polymeric molecules known as "coupling agents" are used. These molecules form bonds with the dispersed phase and become integrated into the continuous matrix phase as well.

In this chapter, we will primarily focus on composites used in structural or mechanical applications. Composites can be placed into three categories—particulate, fiber, and laminar—based on the shapes of the materials (Figure 16-1). Concrete, a mixture of cement and gravel, is a particulate composite; fiberglass, containing glass fibers embedded in a polymer, is a fiber-reinforced composite; and plywood, having alternating layers of wood veneer, is a laminar composite. If the reinforcing particles are uniformly distributed, particulate composites have isotropic properties; fiber composites may be either isotropic or anisotropic; laminar composites always display anisotropic behavior.

16-1 Dispersion-Strengthened Composites

A special group of dispersion-strengthened nanocomposite materials containing particles 10 to 250 nm in diameter is classified as particulate composites. These **dispersoids**, usually a metallic oxide, are introduced into the matrix by means other than traditional phase transformations (Chapters 11 and 12). Even though the small particles are not coherent with the matrix, they block the movement of dislocations and produce a pronounced strengthening effect.

At room temperature, the dispersion-strengthened composites may be weaker than traditional age-hardened alloys, which contain a coherent precipitate. However, because the composites do not catastrophically soften by overaging, overtempering, grain growth, or coarsening of the dispersed phase, the strength of the composite decreases

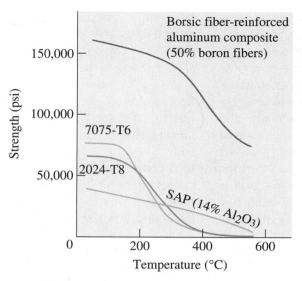

Figure 16-2
Comparison of the yield strength of dispersion-strengthened sintered aluminum powder (SAP) composite with that of two conventional two-phase high-strength aluminum alloys. The composite has benefits above about 300°C. A fiber-reinforced aluminum composite is shown for comparison.

only gradually with increasing temperature (Figure 16-2). Furthermore, their creep resistance is superior to that of metals and alloys.

The dispersoid must have a low solubility in the matrix and must not chemically react with the matrix, but a small amount of solubility may help improve the bonding between the dispersant and the matrix. Copper oxide (Cu_2O) dissolves in copper at high temperatures; thus, the Cu_2O-Cu system would not be effective. However, Al_2O_3 does not dissolve in aluminum; the Al_2O_3-Al system does give an effective dispersion-strengthened material.

Examples of Dispersion-Strengthened Composites Table 16-1 lists some materials of interest. Perhaps the classic example is the sintered aluminum powder (SAP) composite. SAP has an aluminum matrix strengthened by up to 14% Al_2O_3. The composite is formed by powder metallurgy. In one method, aluminum and alumina powders are blended, compacted at high pressures, and sintered. In a second technique, the aluminum powder is treated to add a continuous-oxide film on each particle. When the powder is compacted, the oxide film fractures into tiny flakes that are surrounded by the aluminum metal during sintering.

TABLE 16-1 ■ *Examples and applications of selected dispersion-strengthened composites*

System	Applications
Ag-CdO	Electrical contact materials
Al-Al_2O_3	Possible use in nuclear reactors
Be-BeO	Aerospace and nuclear reactors
Co-ThO_2, Y_2O_3	Possible creep-resistant magnetic materials
Ni-20% Cr-ThO_2	Turbine engine components
Pb-PbO	Battery grids
Pt-ThO_2	Filaments, electrical components
W-ThO_2, ZrO_2	Filaments, heaters

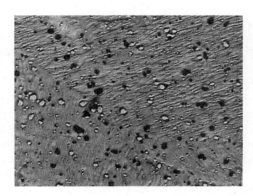

Figure 16-3
Electron micrograph of TD-nickel. The dispersed ThO_2 particles have a diameter of 300 nm or less (×2000). (*From* Oxide Dispersion Strengthening, *p. 714, Gordon and Breach, 1968.* © *AIME.*)

Another important group of dispersion-strengthened composites includes thoria-dispersed metals such as TD-nickel (Figure 16-3). TD-nickel can be produced by internal oxidation. Thorium is present in nickel as an alloying element. After a powder compact is made, oxygen is allowed to diffuse into the metal, react with the thorium, and produce thoria (ThO_2). The following example illustrates calculations related to a dispersion-strengthened composite.

EXAMPLE 16-1   *TD-Nickel Composite*

Suppose 2 wt% ThO_2 is added to nickel. Each ThO_2 particle has a diameter of 1000 Å. How many particles are present in each cubic centimeter?

SOLUTION

The densities of ThO_2 and nickel are 9.69 and 8.9 g/cm^3, respectively. The volume fraction is:

$$f_{ThO_2} = \frac{\dfrac{2}{9.69}}{\dfrac{2}{9.69} + \dfrac{98}{8.9}} = 0.0184$$

Therefore, there is 0.0184 cm^3 of ThO_2 per cm^3 of composite. The volume of each ThO_2 sphere is:

$$V_{ThO_2} = \tfrac{4}{3}\pi r^3 = \tfrac{4}{3}\pi(0.5 \times 10^{-5}\text{ cm})^3 = 0.52 \times 10^{-15}\text{ cm}^3$$

$$\text{Concentration of } ThO_2 \text{ particles} = \frac{0.0184}{0.52 \times 10^{-15}} = 35.4 \times 10^{12}\text{ particles/cm}^3$$

16-2   Particulate Composites

The particulate composites contain large amounts of coarse particles that do not effectively block slip. The particulate composites are designed to produce unusual combinations of properties rather than to improve strength.

Rule of Mixtures   Certain properties of a particulate composite depend only on the relative amounts and properties of the individual constituents. The **rule of mixtures** can accurately predict these properties. The density of a particulate composite, for example, is

$$\rho_c = \sum (f_i \cdot \rho_i) = f_1\rho_1 + f_2\rho_2 + \cdots + f_n\rho_n, \tag{16-1}$$

where ρ_c is the density of the composite, $\rho_1, \rho_2, \ldots, \rho_n$ are the densities of each constituent in the composite, and $f_1, f_2, \ldots, f_n$ are the volume fractions of each constituent. Note that the connectivity of different phases (i.e., how the dispersed phase is arranged with respect to the continuous phase) is also very important.[11]

Cemented Carbides   **Cemented carbides**, or cermets, contain hard ceramic particles dispersed in a metallic matrix (Chapter 1). Tungsten carbide inserts used for cutting tools in machining operations are typical of this group. Tungsten carbide (WC) is a hard, stiff, high-melting-temperature ceramic. To improve toughness, tungsten carbide particles are combined with cobalt powder and pressed into powder compacts. The compacts are heated above the melting temperature of the cobalt. The liquid cobalt surrounds each of the solid tungsten carbide particles (Figure 16-4). After solidification, the cobalt serves as the binder for tungsten carbide and provides good impact resistance. Other carbides, such as TaC and TiC, may also be included in the cermet. The following example illustrates the calculation of density for cemented carbide.

Figure 16-4
Microstructure of tungsten carbide—20% cobalt-cemented carbide (×1300). (*From* Metals Handbook, *Vol. 7, 8th Ed., American Society for Metals, 1972.*)

EXAMPLE 16-2   *Cemented Carbides*

A cemented carbide cutting tool used for machining contains 75 wt% WC, 15 wt% TiC, 5 wt% TaC, and 5 wt% Co. Estimate the density of the composite.

SOLUTION

First, we must convert the weight percentages to volume fractions. The densities of the components of the composite are:

$$\rho_{WC} = 15.77 \text{ g/cm}^3 \quad \rho_{TiC} = 4.94 \text{ g/cm}^3$$

$$\rho_{TaC} = 14.5 \text{ g/cm}^3 \quad \rho_{Co} = 8.90 \text{ g/cm}^3$$

$$f_{WC} = \frac{75/15.77}{75/15.77 + 15/4.94 + 5/14.5 + 5/8.9} = \frac{4.76}{8.70} = 0.547$$

$$f_{TiC} = \frac{15/4.94}{8.70} = 0.349$$

$$f_{TaC} = \frac{5/14.5}{8.70} = 0.040$$

$$f_{Co} = \frac{5/8.90}{8.70} = 0.064$$

From the rule of mixtures, the density of the composite is

$$\rho_c = \sum (f_i \rho_i) = (0.547)(15.77) + (0.349)(4.94) + (0.040)(14.5)$$
$$+ (0.064)(8.9)$$
$$= 11.50 \text{ g/cm}^3$$

Abrasives   Grinding and cutting wheels are formed from alumina (Al_2O_3), silicon carbide (SiC), and cubic boron nitride (CBN). To provide toughness, the abrasive particles are bonded by a glass or polymer matrix. Diamond abrasives are typically bonded with a metal matrix. As the hard particles wear, they fracture or pull out of the matrix, exposing new cutting surfaces.

Electrical Contacts   Materials used for electrical contacts in switches and relays must have a good combination of wear resistance and electrical conductivity. Otherwise, the contacts erode, causing poor contact and arcing. Tungsten-reinforced silver provides this combination of characteristics. A tungsten powder compact is made using conventional powder metallurgical processes (Figure 16-5) to produce high interconnected

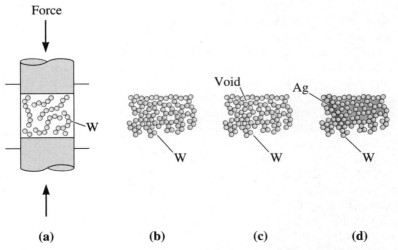

Figure 16-5   The steps in producing a silver-tungsten electrical composite: (a) Tungsten powders are pressed, (b) a low-density compact is produced, (c) sintering joins the tungsten powders, and (d) liquid silver is infiltrated into the pores between the particles.

porosity. Liquid silver is then vacuum infiltrated to fill the interconnected voids. Both the silver and the tungsten are continuous. Thus, the pure silver efficiently conducts current while the hard tungsten provides wear resistance.

EXAMPLE 16-3 *Silver-Tungsten Composite*

A silver-tungsten composite for an electrical contact is produced by first making a porous tungsten powder metallurgy compact, then infiltrating pure silver into the pores. The density of the tungsten compact before infiltration is 14.5 g/cm^3. Calculate the volume fraction of porosity and the final weight percent of silver in the compact after infiltration.

SOLUTION

The densities of pure tungsten and pure silver are 19.3 g/cm^3 and 10.49 g/cm^3. We can assume that the density of a pore is zero, so from the rule of mixtures:

$$\rho_c = f_W \rho_W + f_{pore} \rho_{pore}$$

$$14.5 = f_W (19.3) + f_{pore}(0)$$

$$f_W = 0.75$$

$$f_{pore} = 1 - 0.75 = 0.25$$

After infiltration, the volume fraction of silver equals the volume fraction of pores:

$$f_{Ag} = f_{pore} = 0.25$$

$$\text{wt\% Ag} = \frac{(0.25)(10.49)}{(0.25)(10.49) + (0.75)(19.3)} \times 100 = 15.3\%$$

This solution assumes that all of the pores are open, or interconnected.

Polymers Many engineering polymers that contain fillers and extenders are particulate composites. A classic example is carbon black in vulcanized rubber. Carbon black consists of tiny carbon spheroids only 5 to 500 nm in diameter. The carbon black improves the strength, stiffness, hardness, wear resistance, resistance to degradation due to ultraviolet rays, and heat resistance of the rubber. As mentioned in Chapter 15, nanoparticles of silica are added to rubber tire to enhance their stiffness.[12,13]

Extenders, such as calcium carbonate ($CaCO_3$), solid glass spheres, and various clays, are added so that a smaller amount of the more expensive polymer is required. The extenders may stiffen the polymer, increase the hardness and wear resistance, increase thermal conductivity, or improve resistance to creep; however, strength and ductility normally decrease (Figure 16-6). Introducing hollow glass spheres may impart the same changes in properties while significantly reducing the weight of the composite. Other special properties can be obtained. Elastomer particles are introduced into polymers to improve toughness (e.g., Dylark™). Polyethylene may contain metallic powders, such as lead, to improve the absorption of fission products in nuclear applications. The design of a polymer composite is illustrated in the example that follows.

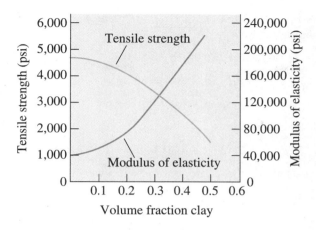

Figure 16-6
The effect of clay on the properties of polyethylene.

EXAMPLE 16-4 *Design of a Particulate Polymer Composite*

Design a clay-filled polyethylene composite suitable for injection molding of inexpensive components. The final part must have a tensile strength of at least 3000 psi and a modulus of elasticity of at least 80,000 psi. Polyethylene costs approximately 50 cents per pound and clay costs approximately 5 cents per pound. The density of polyethylene is 0.95 g/cm^3 and that of clay is 2.4 g/cm^3.

SOLUTION

From Figure 16-6, a volume fraction of clay below 0.35 is required to maintain a tensile strength greater than 3000 psi, whereas a volume fraction of at least 0.2 is needed for the minimum modulus of elasticity. For lowest cost, we use the maximum allowable clay, or a volume fraction of 0.35.

In 1000 cm^3 of composite parts, there are 350 cm^3 of clay and 650 cm^3 of polyethylene in the composite, or:

$$\frac{(350 \text{ cm}^3)(2.4 \text{ g/cm}^3)}{454 \text{ g/lb}} = 1.85 \text{ lb clay}$$

$$\frac{(650 \text{ cm}^3)(0.95 \text{ g/cm}^3)}{454 \text{ g/lb}} = 1.36 \text{ lb polyethylene}$$

The cost of materials is:

$$(1.85 \text{ lb clay})(\$0.05/\text{lb}) = \$0.0925$$

$$(1.36 \text{ lb PE})/(\$0.50/\text{lb}) = \$0.68$$

$$\text{total} = \$0.7725 \text{ per 1000 cm}^3$$

Suppose that weight is critical. The composite's density is:

$$\rho_c = (0.35)(2.4) + (0.65)(0.95) = 1.4575 \text{ g/cm}^3$$

We may wish to sacrifice some of the economic savings in order to obtain lighter weight. If we use only 0.2 volume fraction clay, then (using the same method as above) we find that we need 1.06 lb clay and 1.67 lb polyethylene. The cost of materials is now:

$$(1.06 \text{ lb clay})(\$0.05/\text{lb}) = \$0.053$$

$$(1.67 \text{ lb PE})(\$0.50/\text{lb}) = \$0.835$$

$$\text{total} = \$0.89 \text{ per } 1000 \text{ cm}^3$$

The density of the composite is:

$$\rho_c = (0.2)(2.4) + (0.8)(0.95) = 1.24 \text{ g/cm}^3$$

The material costs about 10% more, but there is a weight savings of more than 10%.

Cast Metal Particulate Composites Aluminum castings containing dispersed SiC particles for automotive applications, including pistons and connecting rods, represent an important commercial application for particulate composites (Figure 16-7). With special processing, the SiC particles can be wet by the liquid, helping to keep the ceramic particles from sinking during freezing.

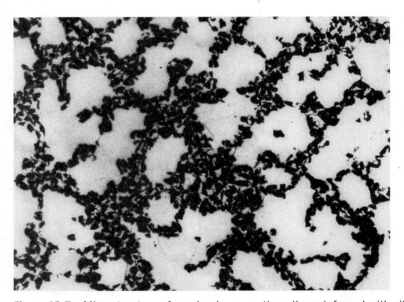

Figure 16-7 Microstructure of an aluminum casting alloy reinforced with silicon carbide particles. In this case, the reinforcing particles have segregated to interdendritic regions of the casting (×125). (*Courtesy of David Kennedy, Lester B. Knight Cost Metals Inc.*)

16-3 Fiber-Reinforced Composites

Most fiber-reinforced composites provide improved strength, fatigue resistance, Young's modulus, and strength-to-weight ratio by incorporating strong, stiff, but brittle fibers into a softer, more ductile matrix. The matrix material transmits the force to the fibers, which carry most of the applied force. The matrix also provides protection for the fiber surface and minimizes diffusion of species such as oxygen or moisture that can degrade the mechanical properties of fibers. The strength of the composite may be high at both room temperature and elevated temperatures (Figure 16-2).

Many types of reinforcing materials are employed. Straw has been used to strengthen mud bricks for centuries. Steel-reinforcing bars are introduced into concrete structures. Glass fibers in a polymer matrix produce fiberglass for transportation and aerospace applications. Fibers made of boron, carbon, polymers (e.g., aramids, Chapter 15), and ceramics provide exceptional reinforcement in advanced composites based on matrices of polymers, metals, ceramics, and even intermetallic compounds.

The Rule of Mixtures in Fiber-Reinforced Composites As for particulate composites, the rule of mixtures always predicts the density of fiber-reinforced composites:

$$\rho_c = f_m \rho_m + f_f \rho_f, \tag{16-2}$$

where the subscripts m and f refer to the matrix and the fiber. Note that $f_m = 1 - f_f$.

In addition, the rule of mixtures accurately predicts the electrical and thermal conductivity of fiber-reinforced composites along the fiber direction if the fibers are *continuous* and *unidirectional*:

$$K_c = f_m K_m + f_f K_f \tag{16-3}$$

$$\sigma_c = f_m \cdot \sigma_m + f_f \cdot \sigma_f \tag{16-4}$$

where K is the thermal conductivity and σ is the electrical conductivity. Thermal or electrical energy can be transferred through the composite at a rate that is proportional to the volume fraction of the conductive material. In a composite with a metal matrix and ceramic fibers, the bulk of the energy would be transferred through the matrix; in a composite consisting of a polymer matrix containing metallic fibers, energy would be transferred through the fibers.

When the fibers are not continuous or unidirectional, the simple rule of mixtures may not apply. For example, in a metal fiber-polymer matrix composite, electrical conductivity would be low and would depend on the length of the fibers, the volume fraction of the fibers, and how often the fibers touch one another. This is expressed using the concept of connectivity of phases.[11]

Modulus of Elasticity The rule of mixtures is used to predict the modulus of elasticity when the fibers are continuous and unidirectional. Parallel to the fibers, the modulus of elasticity may be as high as:

$$E_{c,\|} = f_m \cdot E_m + f_f \cdot E_f \tag{16-5}$$

However, when the applied stress is very large, the matrix begins to deform and the stress-strain curve is no longer linear (Figure 16-8). Since the matrix now contributes little to the stiffness of the composite, the modulus can be approximated by:

$$E_{c,\|} = f_f \cdot E_f, \tag{16-6}$$

When the load is applied perpendicular to the fibers, each component of the composite acts independently of the other. The modulus of the composite is now:

$$\frac{1}{E_{c,\perp}} = \frac{f_m}{E_m} + \frac{f_f}{E_f} \tag{16-7}$$

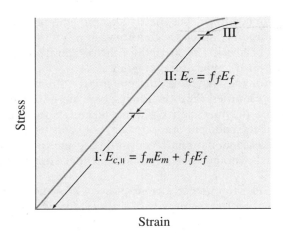

Figure 16-8
The stress-strain curve for a fiber-reinforced composite. At low stresses (region I), the modulus of elasticity is given by the rule of mixtures. At higher stresses (region II), the matrix deforms and the rule of mixtures is no longer obeyed.

Again, if the fibers are not continuous and unidirectional, the rule of mixtures does not apply.

The following examples further illustrate these concepts.

EXAMPLE 16-5   *Rule of Mixtures for Composites: Stress Parallel to Fibers*

Derive the rule of mixtures (Equation 16-5) for the modulus of elasticity of a fiber-reinforced composite when a stress (σ) is applied along the axis of the fibers.

SOLUTION

The total force acting on the composite is the sum of the forces carried by each constituent:

$$F_c = F_m + F_f$$

Since $F = \sigma A$:

$$\sigma_c A_c = \sigma_m A_m + \sigma_m A_f$$

$$\sigma_c = \sigma_m \left(\frac{A_m}{A_c}\right) + \sigma_f \left(\frac{A_f}{A_c}\right)$$

If the fibers have a uniform cross-section, the area fraction equals the volume fraction f:

$$\sigma_c = \sigma_m \cdot f_m + \sigma_f \cdot f_f$$

From Hooke's law, $\sigma = \varepsilon E$. Therefore:

$$E_{c,\parallel}\varepsilon_c = E_m \varepsilon_m f_m + E_f \varepsilon_f f_f$$

If the fibers are rigidly bonded to the matrix, both the fibers and the matrix must stretch equal amounts (iso-strain conditions):

$$\varepsilon_c = \varepsilon_m = \varepsilon_f$$

$$E_{c,\parallel} = f_m E_m + f_f E_f$$

EXAMPLE 16-6   *Modulus of Elasticity for Composites: Stress Perpendicular to Fibers*

Derive the equation for the modulus of elasticity of a fiber-reinforced composite when a stress is applied perpendicular to the axis of the fiber (Equation 16-7).

SOLUTION

In this example, the strains are no longer equal; instead, the weighted sum of the strains in each component equals the total strain in the composite, whereas the stresses in each component are equal (iso-stress conditions):

$$\varepsilon_c = f_m \varepsilon_m + f_f \varepsilon_f$$

$$\frac{\sigma_c}{E_c} = f_m \left(\frac{\sigma_m}{E_m}\right) + f_f \left(\frac{\sigma_f}{E_f}\right)$$

Since $\sigma_c = \sigma_m = \sigma_f$:

$$\frac{1}{E_{c,\perp}} = \frac{f_m}{E_m} + \frac{f_f}{E_f}$$

Strength of Composites   The tensile strength of a fiber-reinforced composite (TS_c) depends on the bonding between the fibers and the matrix. However, the rule of mixtures is sometimes used to approximate the tensile strength of a composite containing continuous, parallel fibers:

$$TS_c = f_f TS_f + f_m \sigma_m, \tag{16-8}$$

where TS_f is the tensile strength of the fiber and σ_m is the stress acting on the matrix when the composite is strained to the point where the fiber fractures. Thus, σ_m is *not* the actual tensile strength of the matrix. Other properties, such as ductility, impact properties, fatigue properties, and creep properties, are difficult to predict even for unidirectionally aligned fibers.

EXAMPLE 16-7   *Boron Aluminum Composites*

Boron coated with SiC (or Borsic) reinforced aluminum containing 40 vol% fibers is an important high-temperature, lightweight composite material. Estimate the density, modulus of elasticity, and tensile strength parallel to the fiber axis. Also estimate the modulus of elasticity perpendicular to the fibers.

SOLUTION

The properties of the individual components are shown below.

Material	Density (g/cm^3)	Modulus of Elasticity (psi)	Tensile Strength (psi)
Fibers	2.36	55,000,000	400,000
Aluminum	2.70	10,000,000	5,000

From the rule of mixtures:

$$\rho_c = (0.6)(2.7) + (0.4)(2.36) = 2.56 \text{ g/cm}^3$$

$$E_{c,\parallel} = (0.6)(10 \times 10^6) + (0.4)(55 \times 10^6) = 28 \times 106 \text{ psi}$$

$$TS_c = (0.6)(5,000) + (0.4)(400,000) = 163,000 \text{ psi}$$

Perpendicular to the fibers:

$$\frac{1}{E_{c,\perp}} = \frac{0.6}{10 \times 10^6} + \frac{0.4}{55 \times 10^6} = 0.06727 \times 10^{-6}$$

$$E_{c,\perp} = 14.9 \times 10^6 \text{ psi}$$

The actual modulus and strength parallel to the fibers are shown in Figure 16-9. The calculated modulus of elasticity (28×10^6 psi) is exactly the same as the measured modulus. However, the estimated strength (163,000 psi) is substantially higher than the actual strength (about 130,000 psi). We also note that the modulus of elasticity is very anisotropic, with the modulus perpendicular to the fiber being only half the modulus parallel to the fibers.

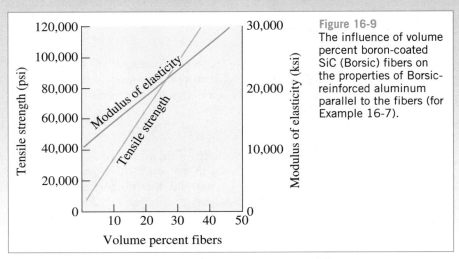

Figure 16-9
The influence of volume percent boron-coated SiC (Borsic) fibers on the properties of Borsic-reinforced aluminum parallel to the fibers (for Example 16-7).

EXAMPLE 16-8 | **Nylon-Glass Fiber Composites**

Glass fibers in nylon provide reinforcement. If the nylon contains 30 vol% E-glass, what fraction of the applied force is carried by the glass fibers?

SOLUTION

The modulus of elasticity for each component of the composite is:

$$E_{\text{glass}} = 10.5 \times 10^6 \text{ psi} \quad E_{\text{nylon}} = 0.4 \times 10^6 \text{ psi}$$

Both the nylon and the glass fibers have equal strain if bonding is good, so:

$$\varepsilon_c = \varepsilon_m = \varepsilon_f$$

$$\varepsilon_m = \frac{\sigma_m}{E_m} = \varepsilon_f = \frac{\sigma_f}{E_f}$$

$$\frac{\sigma_f}{\sigma_m} = \frac{E_f}{E_m} = \frac{10.5 \times 10^6}{0.4 \times 10^6} = 26.25$$

$$\text{Fraction} = \frac{F_f}{F_f + F_m} = \frac{\sigma_f A_f}{\sigma_f A_f + \sigma_m A_m} = \frac{\sigma_f(0.3)}{\sigma_m(0.3) + \sigma_m(0.7)}$$

$$= \frac{0.3}{0.3 + 0.7(\sigma_m/\sigma_f)} = \frac{0.3}{0.3 + 0.7(1/26.25)} = 0.92$$

Almost all of the load is carried by the glass fibers.

16-4 Characteristics of Fiber-Reinforced Composites

Many factors must be considered when designing a fiber-reinforced composite, including the length, diameter, orientation, amount, and properties of the fibers; the properties of the matrix; and the bonding between the fibers and the matrix.

Fiber Length and Diameter Fibers can be short, long, or even continuous. Their dimensions are often characterized by the **aspect ratio** l/d, where l is the fiber length and d is the diameter. Typical fibers have diameters varying from 10 μm (10×10^{-4} cm) to 150 μm (150×10^{-4} cm).

The strength of a composite improves when the aspect ratio is large. Fibers often fracture because of surface imperfections. Making the diameter as small as possible gives the fiber less surface area and, consequently, fewer flaws that might propagate during processing or under a load. We also prefer long fibers. The ends of a fiber carry less of the load than the remainder of the fiber; consequently, the fewer the ends, the higher the load-carrying ability of the fibers (Figure 16-10).

In many fiber-reinforced systems, discontinuous fibers with an aspect ratio greater than some critical value are used to provide an acceptable compromise between pro-

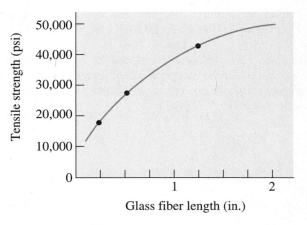

Figure 16-10
Increasing the length of chopped E-glass fibers in an epoxy matrix increases the strength of the composite. In this example, the volume fraction of glass fibers is about 0.5.

cessing ease and properties. A critical fiber length l_c, for any given fiber diameter d can be determined:

$$l_c = \frac{TS_f d}{2\tau_i} \tag{16-9}$$

where TS_f is the strength of the fiber and τ_i is related to the strength of the bond between the fiber and the matrix, or the stress at which the matrix begins to deform. If the fiber length l is smaller than l_c, little reinforcing effect is observed; if l is greater than about $15l_c$, the fiber behaves almost as if it were continuous. The strength of the composite can be estimated from

$$\sigma_c = f_f TS_f \left(1 - \frac{l_c}{2l}\right) + f_m \sigma_m \tag{16-10}$$

where σ_m is the stress on the matrix when the fibers break.

Amount of Fiber A greater volume fraction of fibers increases the strength and stiffness of the composite, as we would expect from the rule of mixtures. However, the maximum volume fraction is about 80%, beyond which fibers can no longer be completely surrounded by the matrix.

Orientation of Fibers The reinforcing fibers may be introduced into the matrix in a number of orientations. Short, randomly oriented fibers having a small aspect ratio—typical of fiberglass—are easily introduced into the matrix and give relatively isotropic behavior in the composite.

Long, or even continuous, unidirectional arrangements of fibers produce anisotropic properties, with particularly good strength and stiffness parallel to the fibers. These fibers are often designated as 0° plies, indicating that all of the fibers are aligned with the direction of the applied stress. However, unidirectional orientations provide poor properties if the load is perpendicular to the fibers (Figure 16-11).

One of the unique characteristics of fiber-reinforced composites is that their properties can be tailored to meet different types of loading conditions. Long, continuous fibers can be introduced in several directions within the matrix (Figure 16-12); in orthogonal arrangements (0°/90° plies), good strength is obtained in two perpendicular directions. More complicated arrangements (such as 0°/±45°/90° plies) provide reinforcement in multiple directions.

Fibers can also be arranged in three-dimensional patterns. In even the simplest of fabric weaves, the fibers in each individual layer of fabric have some small degree of orientation in a third direction. Better three-dimensional reinforcement occurs when the fabric layers are knitted or stitched together. More complicated three-dimensional weaves (Figure 16-13 on page 738) can also be used.

Fiber Properties In most fiber-reinforced composites, the fibers are strong, stiff, and lightweight. If the composite is to be used at elevated temperatures, the fiber should also have a high melting temperature. Thus the specific strength and specific modulus of the fiber are important characteristics:

$$\text{Specific strength} = \frac{TS}{\rho} \tag{16-11}$$

$$\text{Specific modulus} = \frac{E}{\rho} \tag{16-12}$$

where TS is the tensile strength, ρ is the density, and E is the modulus of elasticity.

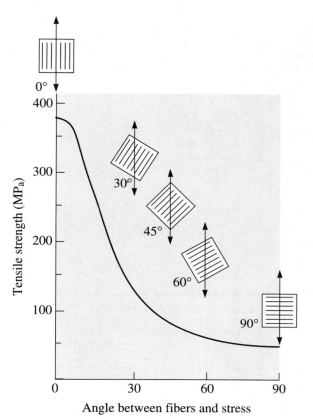

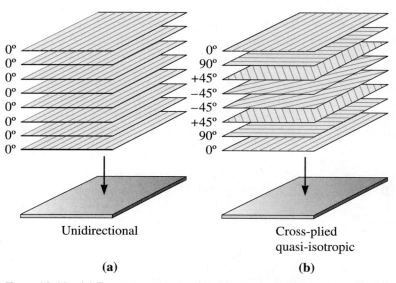

Figure 16-12 (a) Tapes containing aligned fibers can be joined to produce a multi-layered unidirectional composite structure. (b) Tapes containing aligned fibers can be joined with different orientations to produce a quasi-isotropic composite. In this case, a 0°/±45°/90° composite is formed.

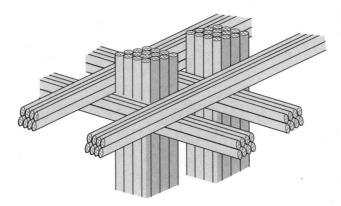

Figure 16-13
A three-dimensional weave for fiber-reinforced composites.

TABLE 16-2 ■ *Properties of selected reinforcing materials**

Material	Density (g/cm³)	Tensile Strength (ksi)	Modulus of Elasticity ($\times 10^6$ psi)	Melting Temperature (°C)	Specific Modulus ($\times 10^7$ in.)	Specific Strength ($\times 10^6$ in.)
Polymers:						
Kevlar™	1.44	650	18.0	500	34.7	12.5
Nylon	1.14	120	0.4	249	1.0	2.9
Polyethylene	0.97	480	25.0	147	7.1	13.7
Metals:						
Be	1.83	185	44.0	1277	77.5	2.8
Boron	2.36	500	55.0	2030	64.7	4.7
W	19.40	580	59.0	3410	8.5	0.8
Glass:						
E-glass	2.55	500	10.5	<1725	11.4	5.6
S-glass	2.50	650	12.6	<1725	14.0	7.2
Carbon:						
HS (high strength)	1.75	820	40.0	3700	63.5	13.0
HM (high modulus)	1.90	270	77.0	3700	112.0	3.9
Ceramics:						
Al_2O_3	3.95	300	55.0	2015	38.8	2.1
B_4C	2.36	330	70.0	2450	82.4	3.9
SiC	3.00	570	70.0	2700	47.3	5.3
ZrO_2	4.84	300	50.0	2677	28.6	1.7
Whiskers:						
Al_2O_3	3.96	3000	62.0	1982	43.4	21.0
Cr	7.20	1290	35.0	1890	13.4	4.9
Graphite	1.66	3000	102.0	3700	170.0	50.2
SiC	3.18	3000	70.0	2700	60.8	26.2
Si_3N_4	3.18	2000	55.0		47.8	17.5

$^*1\frac{gm}{cm^3} = 0.0361\frac{lb}{in.^3}$

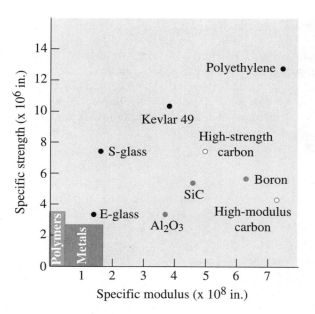

Figure 16-14
Comparison of the specific strength and specific modulus of fibers versus metals and polymers.

Properties of typical fibers are shown in Table 16-2 and Figure 16-14. Note in Table 16-2, the density is in g/cm^3. Also, note that $1\dfrac{gm}{cm^3} = 0.0361\dfrac{lb}{in.^3}$. The highest specific modulus is usually found in materials having a low atomic number and covalent bonding, such as carbon and boron. These two elements also have a high strength and melting temperature.

Aramid fibers, of which Kevlar™ is the best known example, are aromatic polyamide polymers strengthened by a backbone containing benzene rings (Figure 16-15) and are examples of liquid-crystalline polymers in that the polymer chains are rod-like and very stiff. Specially prepared polyethylene fibers are also available. Both the aramid and polyethylene fibers have excellent strength and stiffness but are limited to low-temperature use. Because of their lower density, polyethylene fibers have superior specific strength and specific modulus.

Ceramic fibers and whiskers, including alumina, glass, and silicon carbide, are strong and stiff. Glass fibers, which are the most commonly used, include pure silica, S-glass (25% Al_2O_3, 100% MgO, balance SiO_2), and E-glass (18% CaO, 15% Al_2O_3, balance SiO_2). Although they are considerably denser than the polymer fibers, the ceramics can be used at much higher temperatures. Beryllium and tungsten, although metallically bonded, have a high modulus that makes them attractive fiber materials for certain applications. The following example discusses issues related to designing with composites.

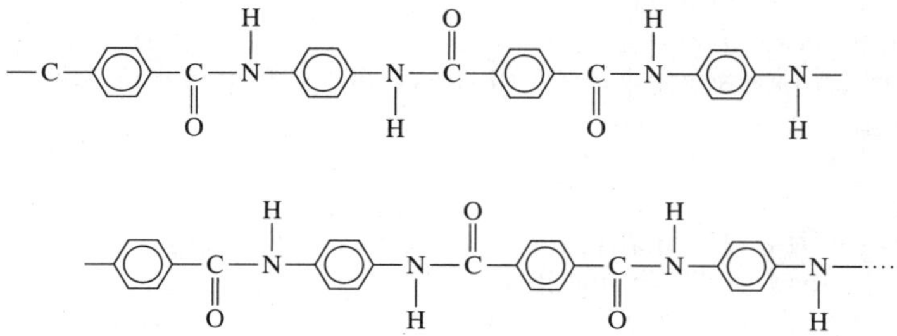

Figure 16-15
The structure of Kevlar™. The fibers are joined by secondary bonds between oxygen and hydrogen atoms on adjoining chains.

EXAMPLE 16-9   *Design of an Aerospace Composite*

We are now using a 7075-T6 aluminum alloy (modulus of elasticity of 10×10^6 psi) to make a 500-pound panel on a commercial aircraft. Experience has shown that each pound reduction in weight on the aircraft reduces the fuel consumption by 500 gallons each year. Design a material for the panel that will reduce weight, yet maintain the same specific modulus, and will be economical over a 10-year lifetime of the aircraft.

SOLUTION

There are many possible materials that might be used to provide a weight savings. As an example, let's consider using a boron fiber-reinforced Al-Li alloy in the T6 condition. Both the boron fiber and the lithium alloying addition increase the modulus of elasticity; the boron and the Al-Li alloy also have densities less than that of typical aluminum alloys.

The specific modulus of the current 7075-T6 alloy is:

$$\text{Specific modulus} = \frac{(10 \times 10^6 \text{ psi})}{\left[\dfrac{\left(2.7 \dfrac{\text{g}}{\text{cm}^3}\right)\left(2.54 \dfrac{\text{cm}}{\text{in.}}\right)^3}{454 \left(\dfrac{\text{g}}{\text{lb}}\right)}\right]}$$

$$= 1.03 \times 10^8 \text{ in.}$$

The density of the boron fibers is approximately 2.36 g/cm^3 (0.085 lb/in.^{3}) and that of a typical Al-Li alloy is approximately 2.5 g/cm^3 (0.09 lb/in.^{3}). If we use 0.6 volume fraction boron fibers in the composite, then the density, modulus of elasticity, and specific modulus of the composite are:

$$\rho_c = (0.6)(0.085) + (0.4)(0.09) = 0.087 \text{ lb/in.}^3$$

$$E_c = (0.6)(55 \times 10^6) + (0.4)(11 \times 10^6) = 37 \times 10^6 \text{ psi}$$

$$\text{Specific modulus} = \frac{37 \times 10^6}{0.087} = 4.25 \times 10^8 \text{ in.}$$

If the specific modulus is the only factor influencing the design of the component, the thickness of the part might be reduced by 75%, giving a component weight of 125 pounds rather than 500 pounds. The weight savings would then be 375 pounds, or (500 gal/lb)(375 lb) = 187,500 gal per year. At about $2.00 per gallon, about $375,000 in fuel savings could be realized each year, or $3.75 million over the 10-year aircraft lifetime.

This is certainly an optimistic comparison, since strength or fabrication factors may not permit the part to be made as thin as suggested. In addition, the high cost of boron fibers (over $300/lb) and higher manufacturing costs of the composite compared with those of 7075 aluminum would reduce cost savings.

Matrix Properties   The matrix supports the fibers and keeps them in the proper position, transfers the load to the strong fibers, protects the fibers from damage during manufacture and use of the composite, and prevents cracks in the fiber from propagating throughout the entire composite. The matrix usually provides the major control over electrical properties, chemical behavior, and elevated-temperature use of the composite.

Polymer matrices are particularly common. Most polymer materials—both thermoplastics and thermosets—are available in short glass fiber-reinforced grades. These composites are formed into useful shapes by the processes described in Chapter 15. Sheet-molding compounds (SMCs) and bulk-molding compounds (BMCs) are typical of this type of composite. Thermosetting aromatic polyimides are used for somewhat higher temperature applications.

Metal-matrix composites include aluminum, magnesium, copper, nickel, and intermetallic compound alloys reinforced with ceramic and metal fibers. A variety of aerospace and automotive applications are satisfied by the MMCs. The metal matrix permits the composite to operate at high temperatures, but producing the composite is often more difficult and expensive than producing the polymer-matrix materials.

The ceramic-matrix composites (CMCs) have good properties at elevated temperatures and are lighter in weight than the high-temperature metal-matrix composites. In a later section, we discuss how to develop toughness in CMCs.

Bonding and Failure Particularly in polymer and metal-matrix composites, good bonding must be obtained between the various constituents. The fibers must be firmly

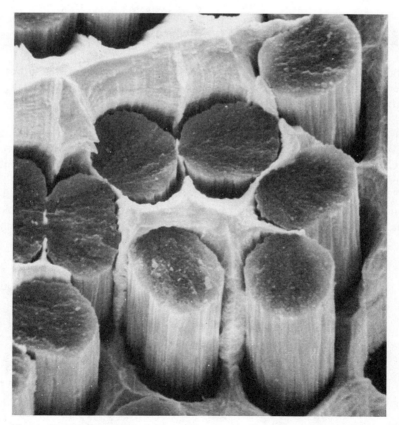

Figure 16-16 Scanning electron micrograph of the fracture surface of a silver-copper alloy reinforced with carbon fibers. Poor bonding causes much of the fracture surface to follow the interface between the metal matrix and the carbon tows (×3000). (*From* Metals Handbook, *American Society for Metals, Vol. 9, 9th Ed., 1985.*)

bonded to the matrix material if the load is to be properly transmitted from the matrix to the fibers. In addition, the fibers may pull out of the matrix during loading, reducing the strength and fracture resistance of the composite if bonding is poor. Figure 16-16 illustrates poor bonding of carbon fibers in a copper matrix. In some cases, special coatings may be used to improve bonding. Glass fibers are coated with a silane coupling or "keying" agent (called **sizing**) to improve bonding and moisture resistance in fiberglass composites. Carbon fibers are similarly coated with an organic material to improve bonding. Boron fibers have been coated with silicon carbide or boron nitride to improve bonding with an aluminum matrix; in fact, these fibers have been called Borsic fibers to reflect the presence of the silicon carbide (SiC) coating.

Another property that must be considered when combining fibers into a matrix is the similarity between the coefficients of thermal expansion for the two materials. If the fiber expands and contracts at a rate much different from that of the matrix, fibers may break or bonding can be disrupted, causing premature failure.

In many composites, individual plies or layers of fabric are joined. Bonding between these layers must also be good or another problem—**delamination**—may occur. Delamination has been suspected to be a cause in some accidents involving airplanes using composite-based structures. The layers may tear apart under load and cause failure. Using composites with a three-dimensional weave will help prevent delamination.

16-5   Manufacturing Fibers and Composites

Producing a fiber-reinforced composite involves several steps, including producing the fibers, arranging the fibers into bundles or fabrics, and introducing the fibers into the matrix.

Making the Fiber   Metallic fibers, glass fibers, and many polymer fibers (including nylon, aramid, and polyacrylonitrile) can be formed by drawing processes, as described in Chapter 7 (wire drawing of metal) and Chapter 15 (using the spinnerette for polymer fibers).

Boron, carbon, and ceramics are too brittle and reactive to be worked by conventional drawing processes. Boron fiber is produced by **chemical vapor deposition** (CVD) [Figure 16-17(a)]. A very fine, heated tungsten filament is used as a substrate, passing

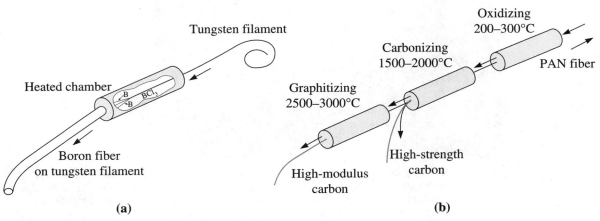

(a)   **(b)**

Figure 16-17   Methods for producing (a) boron and (b) carbon fibers.

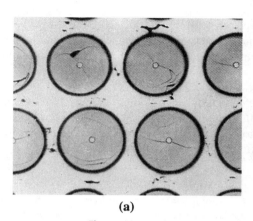

(a)

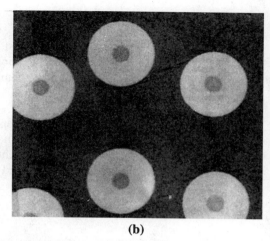

(b)

Figure 16-18 Photomicrographs of two fiber-reinforced composites: (a) In Borsic fiber-reinforced aluminum, the fibers are composed of a thick layer of boron deposited on a small-diameter tungsten filament (×1000). (*From* Metals Handbook, *American Society for Metals, Vol. 9, 9th Ed., 1985.*) (b) In this microstructure of a ceramic-fiber–ceramic-matrix composite, silicon carbide fibers are used to reinforce a silicon nitride matrix. The SiC fiber is vapor-deposited on a small carbon precursor filament (×125). (*Courtesy of Dr. R.T. Bhatt, NASA Lewis Research Center.*)

through a seal into a heated chamber. Vaporized boron compounds such as BCl_3 are introduced into the chamber, decompose, and permit boron to precipitate onto the tungsten wire (Figure 16-18). SiC fibers are made in a similar manner, with carbon fibers as the substrate for the vapor deposition of silicon carbide.

Carbon fibers are made by **carbonizing**, or pyrolizing, an organic filament, which is more easily drawn or spun into thin, continuous lengths [Figure 16-17(b)]. The organic filament, known as a **precursor**, is often rayon (a cellulosic polymer), polyacrylonitrile (PAN), or pitch (various aromatic organic compounds). High temperatures decompose the organic polymer, driving off all of the elements but carbon. As the carbonizing temperature increases from 1000°C to 3000°C, the tensile strength decreases while the modulus of elasticity increases (Figure 16-19). Drawing the carbon filaments at critical

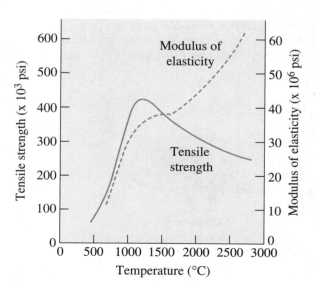

Figure 16-19
The effect of heat-treatment temperature on the strength and modulus of elasticity of carbon fibers.

times during carbonizing may produce desirable preferred orientations in the final carbon filament.

Whiskers are single crystals with aspect ratios of 20 to 1000. Because the whiskers contain no mobile dislocations, slip cannot occur and they have exceptionally high strengths. Because of the complex processing required to produce fibers, their cost may be quite high.

Arranging the Fibers   Exceptionally fine filaments are bundled together as rovings, yarns, or tows. In **yarns**, as many as 10,000 filaments are twisted together to produce the fiber. A **tow** contains a few hundred to more than 100,000 untwisted filaments (Figure 16-20). **Rovings** are untwisted bundles of filaments, yarns, or tows.

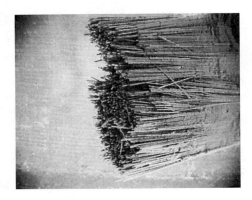

Figure 16-20
A scanning electron micrograph of a carbon tow containing many individual carbon filaments (×200).

Often, fibers are chopped into short lengths of 1 cm or less. These fibers, also called **staples**, are easily incorporated into the matrix and are typical of the sheet-molding and bulk-molding compounds for polymer-matrix composites. The fibers often are present in the composite in a random orientation.

Long or continuous fibers for polymer-matrix composites can be processed into mats or fabrics. **Mats** contain non-woven, randomly oriented fibers loosely held together by a polymer resin. The fibers can also be woven, braided, or knitted into two-dimensional or three-dimensional fabrics. The fabrics are then impregnated with a polymer resin. The resins at this point in the processing have not yet been completely polymerized; these mats or fabrics are called **prepregs**.

When unidirectionally aligned fibers are to be introduced into a polymer matrix, **tapes** may be produced. Individual fibers can be unwound from spools onto a mandrel,

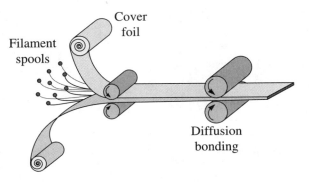

Figure 16-21
Production of fiber tapes by encasing fibers between metal cover sheets by diffusion bonding.

which determines the spacing of the individual fibers, and prepregged with a polymer resin. These tapes, only one fiber diameter thick, may be up to 48 in. wide. Figure 16-21 illustrates that tapes can also be produced by covering the fibers with upper and lower layers of metal foil that are then joined by diffusion bonding.

Producing the Composite   A variety of methods for producing composite parts are used, depending on the application and materials. Short fiber-reinforced composites are normally formed by mixing the fibers with a liquid or plastic matrix, then using relatively conventional techniques such as injection molding for polymer-base composites or casting for metal-matrix composites. Polymer matrix composites can also be produced by a spray-up method, in which short fibers mixed with a resin are sprayed against a form and cured.

Special techniques, however, have been devised for producing composites using continuous fibers, either in unidirectionally aligned, mat, or fabric form (Figure 16-22). In hand lay-up techniques, the tapes, mats, or fabrics are placed against a form, saturated with a polymer resin, rolled to assure good contact and freedom from porosity, and finally cured. Fiberglass car and truck bodies might be made in this manner, which is generally slow and labor intensive.

Tapes and fabrics can also be placed in a die and formed by bag molding. High-pressure gases or a vacuum are introduced to force the individual plies together so that good bonding is achieved during curing. Large polymer matrix components for the skins of military aircraft have been produced by these techniques. In matched die molding, short fibers or mats are placed into a two-part die; when the die is closed, the composite shape is formed.

Filament winding is used to produce products such as pressure tanks and rocket motor castings (Figure 16-23). Fibers are wrapped around a form or mandrel to gradually build up a hollow shape that may be even several feet in thickness. The filament

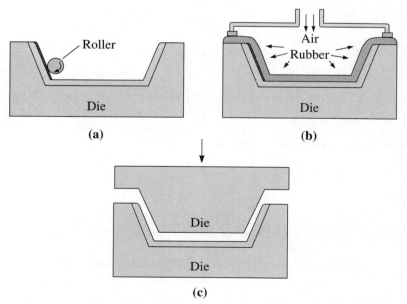

Figure 16-22   Producing composite shapes in dies by (a) hand lay-up, (b) pressure bag molding, and (c) matched die molding.

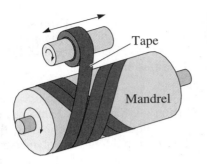

Figure 16-23
Producing composite shapes by filament winding.

can be dipped in the polymer-matrix resin prior to winding, or the resin can be impregnated around the fiber during or after winding. Curing completes the production of the composite part.

Pultrusion is used to form a simple-shaped product with a constant cross section, such as round, rectangular, pipe, plate, or sheet shapes (Figure 16-24). Fibers or mats are drawn from spools, passed through a polymer resin bath for impregnation, and gathered together to produce a particular shape before entering a heated die for curing. Curing of the resin is accomplished almost immediately, so a continuous product is produced. The pultruded stock can subsequently be formed into somewhat more complicated shapes, such as fishing poles, golf club shafts, and ski poles.

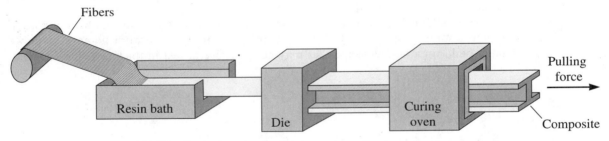

Figure 16-24   Producing composite shapes by pultrusion.

Metal-matrix composites with continuous fibers are more difficult to produce than are the polymer-matrix composites. Casting processes that force liquid around the fibers using capillary rise, pressure casting, vacuum infiltration, or continuous casting are used. Various solid-state compaction processes can also be used.

16-6   Fiber-Reinforced Systems and Applications

Before completing our discussion of fiber-reinforced composites, let's look at the behavior and applications of some of the most common of these materials. Figure 16-25 compares the specific modulus and specific strength of several composites with those of metals and polymers. Note that the values in this figure are lower than those in Figure 16-14, since we are now looking at the composite, not just the fiber.

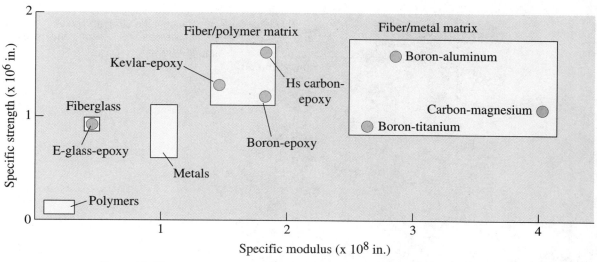

Figure 16-25 A comparison of the specific modulus and specific strength of several composite materials with those of metals and polymers.

Advanced Composites The term advanced composites is often used when the composite is intended to provide service in very critical applications, as in the aerospace industry (Table 16-3). The advanced composites normally are polymer–matrix composites reinforced with high-strength polymer, metal, or ceramic fibers. Carbon fibers are used extensively where particularly good stiffness is required; aramid—and, to an even greater extent, polyethylene—fibers are better suited to high-strength applications in which toughness and damage resistance are more important. Unfortunately, the polymer fibers lose their strength at relatively low temperatures, as do all of the polymer matrices (Figure 16-26).

TABLE 16-3 ■ *Examples of fiber-reinforced materials and applications*

Material	Applications
Borsic aluminum	Fan blades in engines, other aircraft and aerospace applications
Kevlar™-epoxy and Kevlar™-polyester	Aircraft, aerospace applications (including space shuttle), boat hulls, sporting goods (including tennis rackets, golf club shafts, fishing rods), flak jackets
Graphite-polymer	Aerospace and automotive applications, sporting goods
Glass-polymer	Lightweight automotive applications, water and marine applications, corrosion-resistant applications, sporting goods equipment, aircraft and aerospace components

The advanced composites are also frequently used for sporting goods. Tennis rackets, golf clubs, skis, ski poles, and fishing poles often contain carbon or aramid fibers because the higher stiffness provides better performance. In the case of golf clubs, carbon fibers allow less weight in the shaft and therefore more weight in the head. Fabric reinforced with polyethylene fibers is used for lightweight sails for racing yachts.

A unique application for aramid fiber composites is armor. Tough Kevlar™ composites provide better ballistic protection than do other materials, making them suitable for lightweight, flexible bulletproof clothing.

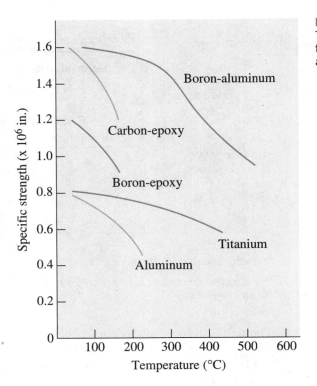

Figure 16-26
The specific strength versus temperature for several composites and metals.

Hybrid composites are composed of two or more types of fibers. For instance, Kevlar™ fibers may be mixed with carbon fibers to improve the toughness of a stiff composite, or Kevlar™ may be mixed with glass fibers to improve stiffness. Particularly good tailoring of the composite to meet specific applications can be achieved by controlling the amounts and orientations of each fiber.

Tough composites can also be produced if careful attention is paid to the choice of materials and processing techniques. Better fracture toughness in the rather brittle composites can be obtained by using long fibers, amorphous (such as PEEK and PPS) rather than crystalline or cross-linked matrices, thermoplastic-elastomer matrices, or interpenetrating network polymers.

Metal-Matrix Composites   These materials, strengthened by metal or ceramic fibers, provide high-temperature resistance. Aluminum reinforced with borsic fibers has been used extensively in aerospace applications, including struts for the space shuttle. Copper-based alloys have been reinforced with SiC fibers, producing high-strength propellers for ships.

Aluminum is commonly used in metal-matrix composites. Al_2O_3 fibers reinforce the pistons for some diesel engines; SiC fibers and whiskers are used in aerospace applications, including stiffeners and missile fins; and carbon fibers provide reinforcement for the aluminum antenna mast of the Hubble telescope. Polymer fibers, because of their low melting or degradation temperatures, are not normally used in a metallic matrix. *Polymets*, however, are produced by hot extruding aluminum powder and high-melting-temperature liquid-crystalline polymers. A reduction of 1000 to 1 during the extrusion process elongates the polymer into aligned filaments and bonds the aluminum powder particles into a solid matrix.

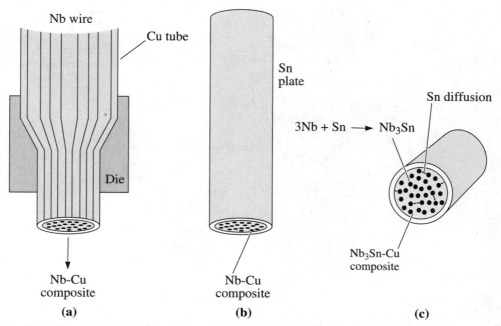

Figure 16-27   The manufacture of composite superconductor wires: (a) Niobium wire is surrounded with copper during forming. (b) Tin is plated onto Nb-Cu composite wire. (c) Tin diffuses to niobium to produce the Nb_3Sn-Cu composite.

Metal-matrix composites may find important applications in components for rocket or aircraft engines. Superalloys reinforced with metal fibers (such as tungsten) or ceramic fibers (such as SiC or B_4N) maintain their strength at higher temperatures, permitting jet engines to operate more efficiently. Similarly, titanium and titanium aluminides reinforced with SiC fibers are considered for turbine blades and disks.

A unique application for metal-matrix composites is in the superconducting wire required for fusion reactors. The intermetallic compound Nb_3Sn has good superconducting properties but is very brittle (Chapter 18). To produce Nb_3Sn wire, pure niobium wire is surrounded by copper as the two metals are formed into a wire composite (Figure 16-27). The niobium-copper composite wire is then coated with tin. The tin diffuses through the copper and reacts with the niobium to produce the intermetallic compound. Niobium-titanium systems are also used.

Ceramic-Matrix Composites   Composites containing ceramic fibers in a ceramic matrix are also finding applications. Two important uses will be discussed to illustrate the unique properties that can be obtained with these materials.

Carbon-carbon (C-C) composites are used for extraordinary temperature resistance in aerospace applications. Carbon-carbon composites can operate at temperatures of up to 3000°C and, in fact, are stronger at high temperatures than at low temperatures (Figure 16-28). Carbon-carbon composites are made by forming a polyacrylonitrile or carbon fiber fabric into a mold, then impregnating the fabric with an organic resin, such as a phenolic. The part is pyrolyzed to convert the phenolic resin to carbon. The composite, which is still soft and porous, is impregnated and pyrolyzed several more times, continually increasing the density, strength, and stiffness. Finally, the part is coated with silicon carbide to protect the carbon-carbon composite from oxidation. Strengths of 300,000 psi and stiffnesses of 50×10^6 psi can be obtained. Carbon-carbon composites

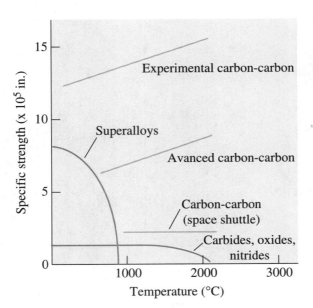

Figure 16-28
A comparison of the specific strength of various carbon-carbon composites with that of other high-temperature materials relative to temperature.

have been used as nose cones and leading edges of high-performance aerospace vehicles such as the space shuttle, and as brake discs on racing cars and commercial jet aircraft.

Ceramic-fiber–ceramic-matrix composites provide improved strength and fracture toughness compared with conventional ceramics (Table 16-4). Fiber reinforcements improve the toughness of the ceramic matrix in several ways. First, a crack moving through the matrix encounters a fiber; if the bonding between the matrix and the fiber is poor, the crack is forced to propagate around the fiber in order to continue the fracture process. In addition, poor bonding allows the fiber to begin to pull out of the matrix [Figure 16-29(a)]. Both processes consume energy, thereby increasing fracture toughness. Finally, as a crack in the matrix begins, unbroken fibers may bridge the crack, providing a compressive stress that helps keep the crack from opening [Figure 16-29(b)].

TABLE 16-4 ■ *Effect of SiC-reinforcement fibers on the properties of selected ceramic materials*

Material	Flexural Strength (psi)	Fracture Toughness (psi $\sqrt{in.}$)
Al_2O_3	80,000	5,000
Al_2O_3/SiC	115,000	8,000
SiC	72,000	4,000
SiC/SiC	110,000	23,000
ZrO_2	30,000	5,000
ZrO_2/SiC	65,000	20,200
Si_3N_4	68,000	4,000
Si_3N_4/SiC	115,000	51,000
Glass	9,000	1,000
Glass/SiC	120,000	17,000
Glass ceramic	30,000	2,000
Glass ceramic/SiC	120,000	16,000

(b)

(a)

Figure 16-29 Two failure modes in ceramic–ceramic composites: (a) Extensive pull-out of SiC fibers in a glass matrix provides good composite toughness (×20). (*From Metals Handbook, American Society for Metals, Vol. 9, 9th Ed., 1985.*) (b) Bridging of some fibers across a crack enhances the toughness of a ceramic-matrix composite (unknown magnification). (*From Journal of Metals, May 1991.*)

Unlike polymer and metal matrix composites, poor bonding—rather than good bonding—is required! Consequently, control of the interface structure is crucial. In a glass-ceramic (based on $Al_2O_3 \cdot SiO_2 \cdot Li_2O$) reinforced with SiC fibers, an interface layer containing carbon and NbC is produced that makes debonding of the fiber from the matrix easy. If, however, the composite is heated to a high temperature, the interface is oxidized; the oxide occupies a large volume, exerts a clamping force on the fiber, and prevents easy pull-out. Fracture toughness is then decreased. The following example illustrates some of the cost and property issues that come up while working with composites.

EXAMPLE 16-10   *Design of a Composite Strut*

Design a unidirectional fiber-reinforced epoxy-matrix strut having a round cross-section. The strut is 10 ft long and, when a force of 500 pounds is applied, it should stretch no more than 0.10 in. We want to assure that the stress acting on the strut is less than the yield strength of the epoxy matrix, 12,000 psi. If the fibers should happen to break, the strut will stretch an extra amount but may not catastrophically fracture. Epoxy costs about \$0.80/lb and has a modulus of elasticity of 500,000 psi.

SOLUTION

Suppose that the strut were made entirely of epoxy (that is, no fibers):

$$\varepsilon_{max} = \frac{0.10 \text{ in.}}{120 \text{ in.}} = 0.00083 \text{ in./in.}$$

$$\sigma_{max} = E\varepsilon = (500{,}000)(0.00083) = 415 \text{ psi}$$

$$A_{strut} = \frac{F}{\sigma} = \frac{500}{415} = 1.2 \text{ in.}^2 \quad \text{or} \quad d = 1.24 \text{ in.}$$

Since $\rho_{epoxy} = 1.25 \text{ g/cm}^3 = 0.0451 \text{ lb/in.}^3$:

$$\text{Weight}_{strut} = (0.0451)(\pi)(1.24/2)^2(120) = 6.54 \text{ lb}$$

$$\text{Cost}_{strut} = (6.54 \text{ lb})(\$0.80/\text{lb}) = \$5.23$$

With no reinforcement, the strut is large and heavy; the materials cost is high due to the large amount of epoxy needed.

In a composite, the maximum strain is still 0.00083 in./in. If we make the strut as small as possible—that is, it operates at 12,000 psi—then the minimum modulus of elasticity E_c of the composite is:

$$E_c > \frac{\sigma}{\varepsilon_{max}} = \frac{12{,}000}{0.00083} = 14.5 \times 10^6 \text{ psi}$$

Let's look at several possible composite systems. The modulus of glass fibers is less than 14.5×10^6 psi; therefore, glass reinforcement is not a possible choice.

For high modulus carbon fibers, $E = 77 \times 10^6$ psi; the density is $1.9 \text{ g/cm}^3 = 0.0686 \text{ lb/in.}^3$, and the cost is about \$30/lb. The minimum volume fraction of carbon fibers needed to give a composite modulus of 14.5×10^6 psi is:

$$E_c = f_C(77 \times 10^6) + (1 - f_C)(0.5 \times 10^6) > 14.5 \times 10^6$$

$$f_C = 0.183$$

The volume fraction of epoxy remaining is 0.817. An area of 0.817 times the total cross-sectional area of the strut must support a 500-lb load with no more than 12,000 psi if all of the fibers should fail:

$$A_{epoxy} = 0.817 A_{total} = \frac{F}{\sigma} = \frac{500 \text{ lb}}{12,000 \text{ psi}} = 0.0416 \text{ in.}^2$$

$$A_{total} = \frac{0.0416}{0.817} = 0.051 \text{ in.}^2 \quad \text{or} \quad d = 0.255 \text{ in.}$$

$$\text{Volume}_{strut} = A_{total} = (0.051 \text{ in.}^2)(120 \text{ in.}) = 6.12 \text{ in.}^3$$

$$\text{Weight}_{strut} = \rho V = [(0.0686)(0.183) + (0.0451)(0.817)](6.12)$$

$$= 0.302 \text{ lb}$$

$$\text{Weight fraction carbon} = \frac{(0.183)(1.9 \text{ g/cm}^3)}{(0.183)(1.9) + (0.817)(1.25)} = 0.254$$

Weight carbon $= (0.254)(0.302 \text{ lb}) = 0.077$

Weight epoxy $= (0.746)(0.302 \text{ lb}) = 0.225$

Cost of each strut $= (0.077)(\$30) + (0.225)(\$0.80) = \$2.49$

The carbon-fiber reinforced strut is less than one-quarter the diameter of an all-epoxy structure, with only 5% of the weight and half of the cost.

We might also repeat these calculations using Kevlar™ fibers, with a modulus of 18×10^6 psi, a density of 1.44 g/cm^3 = 0.052 lb/in.^{3}, and a cost of about \$20/lb. By doing so, we would find that a volume fraction of 0.8 fibers is required. Note that 0.8 volume fraction is at the maximum of fiber volume that can be incorporated into a matrix. We would also find that the required diameter of the strut is 0.515 in. and that the strut weighs 1.263 lb and costs \$20.94. The modulus of the Kevlar™ is not high enough to offset its high cost.

Although the carbon fibers are the most expensive, they permit the lightest weight and the lowest material cost strut. (This calculation does not, however, take into consideration the costs of manufacturing the strut.) Our design, therefore, is to use a 0.255-in.-diameter strut containing 0.183 volume fraction high modulus carbon fiber.

16-7 Laminar Composite Materials

Laminar composites include very thin coatings, thicker protective surfaces, claddings, bimetallics, laminates, and a host of other applications. In addition, the fiber-reinforced composites produced from tapes or fabrics can be considered partly laminar. Many laminar composites are designed to improve corrosion resistance while retaining low cost, high strength, or light weight. Other important characteristics include superior wear or abrasion resistance, improved appearance, and unusual thermal expansion characteristics.

Rule of Mixtures Some properties of the laminar composite materials parallel to the lamellae are estimated from the rule of mixtures. The density, electrical and thermal conductivity, and modulus of elasticity parallel to the lamellae can be calculated with little error using following formulas:

$$\text{Density} = \rho_{c,\parallel} = \sum (f_i \rho_i) \tag{16-13}$$

$$\text{Electrical conductivity} = \sigma_{c,\parallel} = \sum (f_i \sigma_i)$$

$$\text{Thermal conductivity} = K_{c,\parallel} = \sum (f_i K_i) \tag{16-14}$$

$$\text{Modulus of elasticity} = E_{c,\parallel} = \sum (f_i E_i)$$

The laminar composites are very anisotropic. The properties perpendicular to the lamellae are:

$$\text{Electrical conductivity} = \frac{1}{\sigma_{c,\perp}} = \sum \left(\frac{f_i}{\sigma_i} \right)$$

$$\text{Thermal conductivity} = \frac{1}{K_{c,\perp}} = \sum \left(\frac{f_i}{K_i} \right) \tag{16-15}$$

$$\text{Modulus of elasticity} = \frac{1}{E_{c,\perp}} = \sum \left(\frac{f_i}{E_i} \right)$$

However, many of the really important properties, such as corrosion and wear resistance, depend primarily on only one of the components of the composite, so the rule of mixtures is not applicable.

Producing Laminar Composites   Several methods are used to produce laminar composites, including a variety of deformation and joining techniques used primarily for metals. (Figure 16-30).

Individual plies are often joined by *adhesive bonding*, as is the case in producing plywood. Polymer-matrix composites built up from several layers of fabric or tape prepregs are also joined by adhesive bonding; a film of unpolymerized material is placed between each layer of prepreg. When the layers are pressed at an elevated temperature, polymerization is completed and the prepregged fibers are joined to produce composites that may be dozens of layers thick.

Most of the metallic laminar composites, such as claddings and bimetallics, are produced by *deformation bonding*, such as hot- or cold-roll bonding. The pressure exerted by the rolls breaks up the oxide film at the surface, brings the surfaces into

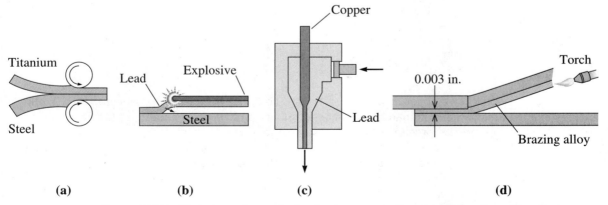

| (a) | (b) | (c) | (d) |

Figure 16-30   Techniques for producing laminar composites: (a) roll bonding, (b) explosive bonding, (c) coextrusion, and (d) brazing.

atom-to-atom contact, and permits the two surfaces to be joined. Explosive bonding can also be used. An explosive charge provides the pressure required to join metals. This process is particularly well suited for joining very large plates that will not fit into a rolling mill. Very simple laminar composites, such as coaxial cable, are produced by coextruding two materials through a die in such a way that the soft material surrounds the harder material. Metal conductor wire can be coated with an insulating thermoplastic polymer in this manner.

Brazing can join composite plates (Chapter 8). The metallic sheets are separated by a very small clearance—preferably, about 0.003 in.—and heated above the melting temperature of the brazing alloy. The molten brazing alloy is drawn into the thin joint by capillary action.

16-8 Examples and Applications of Laminar Composites

The number of laminar composites is so varied and their applications and intentions are so numerous that we cannot make generalizations concerning their behavior. Instead we will examine the characteristics of a few commonly used examples.

Laminates Laminates are layers of materials joined by an organic adhesive. In laminated safety glass, a plastic adhesive, such as polyvinyl butyral (PVB), joins two pieces of glass; the adhesive prevents fragments of glass from flying about when the glass is broken (Chapter 14). Laminates are used for insulation in motors, for gears, for printed circuit boards, and for decorative items such as Formica® countertops and furniture.

Microlaminates include composites composed of alternating layers of aluminum sheet and fiber-reinforced polymer. *Arall* (aramid-aluminum laminate) and *Glare* (glass-aluminum laminate) have been developed as possible skin materials for aircraft. In Arall, an aramid fiber such as Kevlar™ is prepared as a fabric or unidirectional tape, impregnated with an adhesive, and laminated between layers of aluminum alloy (Figure 16-31). The composite laminate has an unusual combination of strength, stiffness, corrosion resistance, and light weight. Fatigue resistance is improved, since the interface between the layers may block cracks. Compared with polymer-matrix composites, the microlaminates have good resistance to lightning-strike damage (which is important in aerospace applications), are formable and machinable, and are easily repaired.

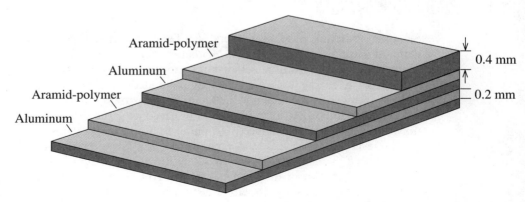

Figure 16-31 Schematic diagram of an aramid-aluminum laminate, Arall, which has potential for aerospace applications.

Clad Metals   Clad materials are metal-metal composites. A common example of **cladding** is United States silver coinage. A Cu-80% Ni alloy is bonded to both sides of a Cu-20% Ni alloy. The ratio of thicknesses is about 1/6:2/3:1/6. The high-nickel alloy is a silver color, while the predominantly copper core provides low cost.

Clad materials provide a combination of good corrosion resistance with high strength. *Alclad* is a clad composite in which commercially pure aluminum is bonded to higher-strength aluminum alloys. The pure aluminum protects the higher-strength alloy from corrosion. The thickness of the pure aluminum layer is about 1% to 15% of the total thickness. Alclad is used in aircraft construction, heat exchangers, building construction, and storage tanks, where combinations of corrosion resistance, strength, and light weight are desired.

Bimetallics   Temperature indicators and controllers take advantage of the different coefficients of thermal expansion of the two metals in laminar composite. If two pieces of metal are heated, the metal with the higher coefficient of thermal expansion becomes longer. If the two pieces of metal are rigidly bonded together, the difference in their coefficients causes the strip to bend and produce a curved surface. The amount of movement depends on the temperature. By measuring the curvature or deflection of the strip, we can determine the temperature. Likewise, if the free end of the strip activates a relay, the strip can turn on or off a furnace or air conditioner to regulate temperature. Metals selected for **bimetallics** must have (a) very different coefficients of thermal expansion, (b) expansion characteristics that are reversible and repeatable, and (c) a high modulus of elasticity, so that the bimetallic device can do work. Often the low-expansion strip is made from Invar, an iron-nickel alloy, whereas the high-expansion strip may be brass, Monel, or pure nickel.

Bimetallics can act as circuit breakers as well as thermostats; if a current passing through the strip becomes too high, heating causes the bimetallic to deflect and break the circuit.

Multilayer capacitors   Similar geometry is used to make billions of multi-layer capacitors. Their structure is comprisesd of thin sheets of $BaTiO_3$-based ceramics separated by Ag/Pd or Ni electrodes (Chapter 18).

16-9   Sandwich Structures

Sandwich materials have thin layers of a facing material joined to a lightweight filler material, such as a polymer foam. Neither the filler nor the facing material is strong or rigid, but the composite possesses both properties. A familiar example is corrugated cardboard. A corrugated core of paper is bonded on either side to flat, thick paper. Neither the corrugated core nor the facing paper is rigid, but the combination is.

Another important example is the honeycomb structure used in aircraft applications. A **honeycomb** is produced by gluing thin aluminum strips at selected locations. The honeycomb material is then expanded to produce a very low-density cellular panel that, by itself, is unstable (Figure 16-32). When an aluminum facing sheet is adhesively bonded to either side of the honeycomb, however, a very stiff, rigid, strong, and exceptionally lightweight sandwich with a density as low as 0.04 g/cm^3 is obtained.

The honeycomb cells can have a variety of shapes, including hexagonal, square, rectangular, and sinusoidal, and they can be made from aluminum, fiberglass, paper,

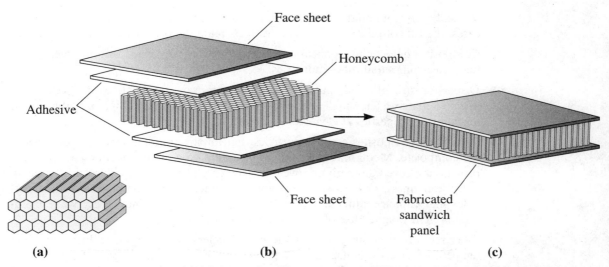

Figure 16-32 (a) A hexagonal cell honeycomb core, (b) can be joined to two face sheets by means of adhesive sheets, (c) producing an exceptionally lightweight yet stiff, strong honeycomb sandwich structure.

aramid polymers, and other materials. The honeycomb cells can be filled with foam or fiberglass to provide excellent sound and vibration absorption. Figure 16-33 describes one method by which the honeycomb can be fabricated.

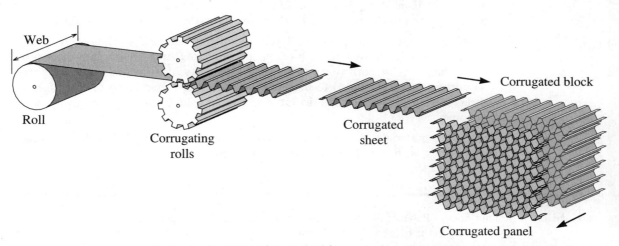

Figure 16-33 In the corrugation method for producing a honeycomb core, the material (such as aluminum) is corrugated between two rolls. The corrugated sheets are joined together with adhesive and then cut to the desired thickness.

SUMMARY

◆ Composites are composed of two or more materials or phases joined or connected in such a way so as to give a combination of properties that cannot be attained otherwise. The volume fraction and the connectivity of the phases or materials in a composite and the nature of the interface between the dispersed phase and matrix are very important in determining the properties.

◆ Virtually any combination of metals, polymers, and ceramics is possible. In many cases, the rule of mixtures can be used to estimate the properties of the composite.

◆ Composites have many applications in construction, aerospace, automotive, sports, microelectronics and other industries.

◆ Dispersion-strengthened materials contain exceptionally small oxide particles in a metal matrix. The small stable dispersoids interfere with slip, providing good mechanical properties at elevated temperatures.

◆ Particulate composites contain particles that impart combinations of properties to the composite. Metal-matrix composites contain ceramic or metallic particles that provide improved strength and wear resistance and assure good electrical conductivity, toughness, or corrosion resistance. Polymer-matrix composites contain particles that enhance stiffness, heat resistance, or electrical conductivity while maintaining light weight, ease of fabrication, or low cost.

◆ Fiber-reinforced composites provide improvements in strength, stiffness, or high-temperature performance in metals and polymers and impart toughness to ceramics. Fibers typically have low densities, giving high specific strength and specific modulus, but they often are very brittle. Fibers can be continuous or discontinuous. Discontinuous fibers with a high aspect ratio l/d produce better reinforcement.

◆ Fibers are introduced into a matrix in a variety of orientations. Random orientations and isotropic behavior are obtained using discontinuous fibers; unidirectionally aligned fibers produce composites with anisotropic behavior, with large improvements in strength and stiffness parallel to the fiber direction. Properties can be tailored to meet the imposed loads by orienting the fibers in multiple directions.

◆ Laminar composites are built of layers of different materials. These layers may be sheets of different metals, with one metal providing strength, and the other providing hardness or corrosion resistance. Layers may also include sheets of fiber-reinforced material bonded to metal or polymer sheets or even to fiber-reinforced sheets having different fiber orientations. The laminar composites are always anisotropic.

◆ Sandwich materials, including honeycombs, are exceptionally lightweight laminar composites, with solid facings joined to an almost hollow core.

GLOSSARY

Aramid fibers   Polymer fibers, such as Kevlar™, formed from polyamides, which contain the benzene ring in the backbone of the polymer.

Aspect ratio   The length of a fiber divided by its diameter.

Bimetallic   A laminar composite material produced by joining two strips of metal with different thermal expansion coefficients, making the material sensitive to temperature changes.

Brazing   A process in which a liquid filler metal is introduced by capillary action between two solid-based materials that are to be joined. Solidification of the brazing alloy provides the bond.

Carbonizing   Driving off the non-carbon atoms from a polymer fiber, leaving behind a carbon fiber of high strength. Also known as pyrolizing.

Cemented carbides   Particulate composites containing hard ceramic particles bonded with a soft metallic matrix. The composite combines high hardness and cutting ability, yet still has good shock resistance.

Chemical vapor deposition   Method for manufacturing materials by condensing the material from a vapor onto a solid substrate.

Cladding   A laminar composite produced when a corrosion-resistant or high-hardness layer of a laminar composite formed onto a less expensive or higher-strength backing.

Delamination   Separation of individual plies of a fiber-reinforced composite.

Dispersed phase   The phase or phases that are dispersed in a continuous matrix of the composite.

Dispersoids   Tiny oxide particles formed in a metal matrix that interfere with dislocation movement and provide strengthening, even at elevated temperatures.

Filament winding   Process for producing fiber-reinforced composites in which continuous fibers are wrapped around a form or mandrel. The fibers may be prepregged or the filament-wound structure may be impregnated to complete the production of the composite.

Honeycomb   A lightweight but stiff assembly of aluminum strip joined and expanded to form the core of a sandwich structure.

Hybrid organic-inorganic composites   Nanocomposites consisting of structures that are partly organic and partly inorganic, often made using sol-gel processing.

Matrix phase   The continuous phase in a composite. The composites are named after the continuous phase (e.g., polymer-matrix composites).

Nanocomposite   A material in which the dispersed phase is nano-sized and is distributed in the continuous matrix.

Precursor   A starting chemical (e.g., a polymer fiber) that is carbonized to produce carbon fibers.

Prepregs   Layers of fibers in unpolymerized resins. After the prepregs are stacked to form a desired structure, polymerization joins the layers together.

Pultrusion   A method for producing composites containing mats or continuous fibers.

Rovings   Bundles of less than 10,000 filaments.

Rule of mixtures   The statement that the properties of a composite material are a function of the volume fraction of each material in the composite.

Sandwich   A composite material constructed of a lightweight, low-density material surrounded by dense, solid layers. The sandwich combines overall light weight with excellent stiffness.

Sizing   Coating glass fibers with an organic material to improve bonding and moisture resistance in fiberglass.

Specific modulus   The modulus of elasticity divided by the density.

Specific strength   The tensile or yield strength of a material divided by the density.

Staples   Fibers chopped into short lengths.

Tapes   Single-filament-thick strips of prepregs, with the filaments either unidirectional or woven fibers. Several layers of tapes can be joined to produce composite structures.

Tow   A bundle of more than 10,000 filaments.

Whiskers   Very fine fibers grown in a manner that produces single crystals with no mobile dislocations, thus giving nearly theoretical strengths.

Yarns   Continuous fibers produced from a group of twisted filaments.

ADDITIONAL INFORMATION

1. MEYERS, M.A. and K.K. CHAWLA, *Mechanical Behavior of Materials*. 1998: Prentice Hall.

2. HARPER, C.A., *Handbook of Materials Product Design*. 3rd ed. 2001: McGraw Hill.

3. FROES, F.H., "Materials for Sports," *MRS Bulletin*, 1998. 23(3).

4. LOY, D.A., "Hybrid Organic-Inorganic Materials," *MRS Bulletin*, 2001. 26(5): p. 364.

5. VAIA, R.A. and E.P. GIANNELIS, "Polymer Nanocomposites: Status and Opportunities," *MRS Bulletin*, 2001. 26(5): p. 394.

6. ARKLES, B., "Commercial Applications of Sol-Gel Derived Hybrid Materials," *MRS Bulletin*, 2001. 26(5): p. 402.

7. PHULÉ, P.P. and T. WOOD, "Sol-Gel Processing of Ceramics," *Encylopedia of Advanced Materials*, M. FLEMINGS, Editor. 2001: Pergamon.

8. ARCH, P., *Nova Chemicals, Technical discussions and assistance are acknowledged*. 2002.

9. SAMPATH, S. AND McCUNE, R., "Thermal Spray Processing of Materials," *MRS Bulletin*, 2000. Vol. 25, No. 7, p. 12.

10. Greenberg, *Personal communication*. 2002.

11. NEWNHAM, R.E., "Nonmechanical Properties of Composites", in *Concise Encyclopedia of Composite Materials*, edited by A. Kelly, pp. 205–211, Pergamon Press, Oxford (1989).

12. ROSATO, D., D. DIMATTIA, and D. ROSATO, *Designing with Plastics and Composites*. 1991: New York, NY, Van Nostrand Reinhold.

13. STRONG, B.A., *Plastics-Materials and Processing*. Second ed. 2000: Prentice Hall.

✓ PROBLEMS

Section 16-1 Dispersion-Strengthened Composites

16-1 What is a composite?

16-2 What do the properties of composite materials depend upon?

16-3 Give examples of applications where composites are used for load bearing applications.

16-4 Give two examples of applications where composites are used for non-structural applications.

16-5 What is a dispersion-strengthened composite? How is it different from a particle-reinforced composite?

16-6 What is a nanocomposite? How can steels containing ferrite and martensite be described as composites? Explain.

16-7 Nickel containing 2 wt% thorium is produced in powder form, consolidated into a part, and sintered in the presence of oxygen, causing all of the thorium to produce ThO_2 spheres 80 nm in diameter. Calculate the number of spheres per cm^3. The density of ThO_2 is 9.86 g/cm^3.

16-8 Spherical aluminum powder (SAP) 0.002 mm in diameter is treated to create a thin oxide layer and is then used to produce a SAP dispersion-strengthened material containing 10 vol% Al_2O_3. Calculate the average thickness of the oxide film prior to compaction and sintering of the powders into the part.

16-9 Yttria (Y_2O_3) particles 750 Å in diameter are introduced into tungsten by internal oxidation. Measurements using an electron microscope show that there are 5×10^{14} oxide particles per cm^3. Calculate the wt% Y originally in the alloy. The density of Y_2O_3 is 5.01 g/cm^3.

16-10 With no special treatment, aluminum is typically found to have an Al_2O_3 layer that is 3 nm thick. If spherical aluminum powder prepared with a total diameter of 0.01 mm is used to produce the SAP dispersion-strengthened material, calculate the volume percent Al_2O_3 in the material and the number of oxide particles per cm^3. Assume that the oxide breaks into disk-shaped flakes 3 nm thick and 3×10^{-4} mm in diameter. Compare the number of oxide particles per cm^3 with the number of solid solution atoms per cm^3 when 3 at% of an alloying element is added to aluminum.

Section 16-2 Particulate Composites

16-11 What is a particulate composite?

16-12 What is a cermet? What is the role of WC and Co in a cermet?

16-13 Calculate the density of a cemented carbide, or cermet, based on a titanium matrix if the composite contains 50 wt% WC, 22 wt% TaC, and 14 wt% TiC. (See Example 16-2 for densities of the carbides.)

16-14 A typical grinding wheel is 9 in. in diameter, 1 in. thick, and weighs 6 lb. The wheel contains SiC (density of 3.2 g/cm^3) bonded by silica glass (density of 2.5 g/cm^3); 5 vol% of the wheel is porous. The SiC is in the form of 0.04 cm cubes. Calculate

   (a) the volume fraction of SiC particles in the wheel and

   (b) the number of SiC particles lost from the wheel after it is worn to a diameter of 8 in.

16-15 An electrical contact material is produced by infiltrating copper into a porous tungsten carbide (WC) compact. The density of the final composite is 12.3 g/cm^3. Assuming that all of the pores are filled with copper, calculate

   (a) the volume fraction of copper in the composite,

   (b) the volume fraction of pores in the WC compact prior to infiltration, and

   (c) the original density of the WC compact before infiltration.

16-16 An electrical contact material is produced by first making a porous tungsten compact that weighs 125 g. Liquid silver is introduced into the compact; careful measurement indicates that 105 g of silver is infiltrated. The final density of the composite is 13.8 g/cm^3. Calculate the volume fraction of the original compact that is interconnected porosity and the volume fraction that is closed porosity (no silver infiltration).

16-17 How much clay must be added to 10 kg of polyethylene to produce a low-cost composite having a modulus of elasticity greater than 120,000 psi and a tensile strength greater than 2000 psi? The density of the clay is 2.4 g/cm^3 and that of polyethylene is 0.92 g/cm^3.

16-18 We would like to produce a lightweight epoxy part to provide thermal insulation. We have available hollow glass beads for which the outside diameter is 1/16 in. and the wall thickness is 0.001 in. Determine the weight and number of beads that must be added to the epoxy to produce a one-pound composite with a density of 0.65 g/cm^3. The density of the glass is 2.5 g/cm^3 and that of the epoxy is 1.25 g/cm^3.

Section 16-4 Characteristics of Fiber-Reinforced Composites

16-19 What is a fiber-reinforced composite?

16-20 What fiber-reinforcing materials are commonly used?

16-21 In a fiber-reinforced composite, what is the role of the matrix?

16-22 What do the terms CFRP and GFRP mean?

16-23 Explain briefly how the volume of fiber, fiber orientation, and fiber strength and modulus affect the properties of fiber-reinforced composites.

16-24 Five kg of continuous boron fibers are introduced in a unidirectional orientation into 8 kg of an aluminum matrix. Calculate

   (a) the density of the composite,

   (b) the modulus of elasticity parallel to the fibers, and

   (c) the modulus of elasticity perpendicular to the fibers.

16-25 We want to produce 10 lbs. of a continuous unidirectional fiber-reinforced composite of HS carbon in a polyimide matrix that has a modulus of elasticity of at least 25×10^6 psi parallel to the fibers. How many pounds of fibers are required? See Chapter 15 for properties of polyimide.

16-26 We produce a continuous unidirectionally reinforced composite containing 60 vol% HM carbon fibers in an epoxy matrix. The epoxy has a tensile strength of 15,000 psi. What fraction of the applied force is carried by the fibers?

16-27 A polyester matrix with a tensile strength of 13,000 psi is reinforced with Al$_2$O$_3$ fibers. What vol% fibers must be added to insure that the fibers carry 75% of the applied load?

16-28 An epoxy matrix is reinforced with 40 vol% E-glass fibers to produce a 2-cm-diameter composite that is to withstand a load of 25,000 N. Calculate the stress acting on each fiber.

16-29 A titanium alloy with a modulus of elasticity of 16×10^6 psi is used to make a 1000-lb part for a manned space vehicle. Determine the weight of a part having the same modulus of elasticity parallel to the fibers, if the part is made of

(a) aluminum reinforced with boron fibers and

(b) polyester (with a modulus of 650,000 psi) reinforced with high-modulus carbon fibers.

(c) Compare the specific modulus for all three materials.

16-30 Short, but aligned, Al_2O_3 fibers with a diameter of 20 μm are introduced into a 6,6-nylon matrix. The strength of the bond between the fibers and the matrix is estimated to be 1000 psi. Calculate the critical fiber length and compare with the case when 1-μm alumina whiskers are used instead of the coarser fibers. What is the minimum aspect ratio in each case?

16-31 We prepare several epoxy-matrix composites using different lengths of 3-μm-diameter ZrO_2 fibers and find that the strength of the composite increases with increasing fiber length up to 5 mm. For longer fibers, the strength is virtually unchanged. Estimate the strength of the bond between the fibers and the matrix.

Section 16-5 Manufacturing Fibers and Composites

16-32 Explain briefly how boron and carbon fibers are made.

16-33 Explain briefly how continuous-glass fibers are made.

16-34 What is a coupling agent? What is "sizing" as it relates to the production of glass fibers?

16-35 What is the difference between a fiber and a whisker?

16-36 In one polymer-matrix composite, as produced, discontinuous glass fibers are introduced directly into the matrix; in a second case, the fibers are first "sized." Discuss the effect this difference might have on the critical fiber length and strength of the composite.

16-37 A Borsic fiber-reinforced aluminum composite is shown in Figure 16-18. Estimate the volume fractions of tungsten, boron, and the matrix for this composite. Calculate the modulus of elasticity parallel to the fibers for this composite. What would the modulus be if the same size boron fiber could be produced without the tungsten precursor?

16-38 A silicon nitride matrix reinforced with silicon carbide fibers containing a HS-carbon precursor is shown in Figure 16-18. Estimate the volume fractions of the SiC, Si_3N_4, and carbon in this composite. Calculate the modulus of elasticity parallel to the fibers for this composite. What would the modulus be if the same size SiC

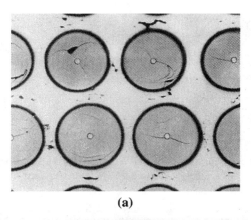

(a)

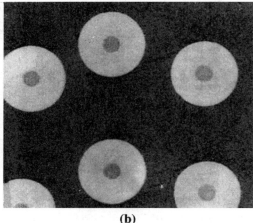

(b)

Figure 16-18 (Repeated for Problems 16-37 and 16-38.) Photomicrographs of two fiber-reinforced composites: (a) In Borsic fiber-reinforced aluminum, the fibers are composed of a thick layer of boron deposited on a small-diameter tungsten filament (×1000). (*From* Metals Handbook, *American Society for Metals, Vol. 9, 9th Ed., 1985.*) (b) In this microstructure of a ceramic-fiber–ceramic-matrix composite, silicon carbide fibers are used to reinforce a silicon nitride matrix. The SiC fiber is vapor-deposited on a small carbon precursor filament (×125). (*Courtesy of Dr. R.T. Bhatt, NASA Lewis Research Center.*)

fiber could be produced without the carbon precursor?

16-39 Explain why bonding between carbon fibers and an epoxy matrix should be excellent, whereas bonding between silicon nitride fibers and a silicon carbide matrix should be poor.

Section 16-6 Fiber-Reinforced Systems and Applications

16-40 Explain briefly in what sporting equipment composite materials are used. What is the main reason why composites are used in these applications?

16-41 A polyimide matrix is to be reinforced with 70 vol% carbon fibers to give a minimum modulus of elasticity of 40×10^6 psi. Recommend a process for producing the carbon fibers required. Estimate the tensile strength of the fibers that are produced.

16-42 What are the advantages of using ceramic-matrix composites?

Section 16-7 Laminar Composite Materials

Section 16-8 Examples and Applications of Laminar Composites

Section 16-9 Sandwich Structures

16-43 What is a laminar composite?

16-44 A microlaminate, Arall, is produced using 5 sheets of 0.4-mm-thick aluminum and 4 sheets of 0.2-mm-thick epoxy reinforced with unidirectionally aligned Kevlar™ fibers. The volume fraction of Kevlar™ fibers in these intermediate sheets is 55%. Calculate the modulus of elasticity of the microlaminate parallel and perpendicular to the unidirectionally aligned Kevlar™ fibers. What are the principle advantages of the Arall material compared with those of unreinforced aluminum?

16-45 A laminate composed of 0.1-mm-thick aluminum sandwiched around a 2-cm-thick layer of polystyrene foam is produced as an insulation material. Calculate the thermal conductivity of the laminate parallel and perpendicular to the layers. The thermal conductivity of aluminum is $0.57 \dfrac{\text{cal}}{\text{cm} \cdot \text{s} \cdot \text{K}}$ and that of the foam is $0.000077 \dfrac{\text{cal}}{\text{cm} \cdot \text{s} \cdot \text{K}}$.

16-46 A 0.01-cm-thick sheet of a polymer with a modulus of elasticity of 0.7×10^6 psi is sandwiched between two 4-mm-thick sheets of glass with a modulus of elasticity of 12×10^6 psi. Calculate the modulus of elasticity of the composite parallel and perpendicular to the sheets.

16-47 A U.S. quarter is $\frac{15}{16}$ in. in diameter and is about $\frac{1}{16}$ in. thick. Copper costs about $1.10 per pound and nickel costs about $4.10 per pound. Compare the material cost in a composite quarter versus a quarter made entirely of nickel.

16-48 Calculate the density of a honeycomb structure composed of the following elements: The two 2-mm-thick cover sheets are produced using an epoxy matrix prepreg containing 55 vol% E-glass fibers. The aluminum honeycomb is 2 cm thick; the cells are in the shape of 0.5 cm squares and the walls of the cells are 0.1 mm thick. Estimate the density of the structure. Compare the weight of a 1 m × 2 m panel of the honeycomb compared with a solid aluminum panel of the same dimensions.

Design Problems

16-49 Design the materials and processing required to produce a discontinuous, but aligned, fiber-reinforced fiberglass composite that will form the hood of a sports car. The composite should provide a density of less than 1.6 g/cm^3 and a strength of 20,000 psi. Be sure to list all of the assumptions you make in creating your design.

16-50 A 3-ft inside-diameter spherical tank is to be designed to store liquid chlorine. The tank must have a modulus of elasticity in the tangential direction of at least 15×10^6 psi, it should have a thermal conductivity in the radial direction of no more than 0.006 cal/cm $\cdot$ s $\cdot$ K, and it should weigh no more than 170 lbs. Using only the materials listed in Table 21-3, design a material and tank thickness that will be suitable. Estimate the cost of materials in your tank to assure that it is not prohibitively expensive.

16-51 Design an electrical-contact material and a method for producing the material that will result in a density of no more than 6 g/cm^3, yet at least 50 vol% of the material will be conductive.

16-52 What factors will have to be considered in designing a bicycle frame using an aluminum frame and a frame made using C-C composite?

Computer Problems

16-53 *Properties of Laminar Composite.* Write a computer program (or use a spreadsheet software) to calculate the properties of a laminar composite using the rule of mixture. For example, if the user provides the value of the thermal conductivity of each phase and the corresponding volume fraction, the program should provide the value of the effective thermal conductivity. Properties parallel and perpendicular to the lamellae should be calculated.

Chapter

Construction materials, such as steels, concrete, wood, and glasses, along with composites, such as fiberglass, play a major role in the development and maintenance of a nation's infrastructure. Although many of our current construction materials are well developed and the technologies for producing them are mature, several new directions can be seen in construction materials. The use of smart sensors and actuators that can help control structures or monitor the health of structures is one such area. The increased use of composites and polymeric adhesives is one more dimension that has developed considerably. Processes such as welding and galvanizing as well as phenomena, such as corrosion, play important roles in the reliability and durability of structures. (*Courtesy of PhotoDisc/Getty Images.*)

17

Construction Materials

Have You Ever Wondered?

- *What are the most widely used manufactured construction materials?*

- *What is the difference between cement and concrete?*

- *What construction material is a composite made by nature?*

- *What is reinforced concrete?*

A number of important materials are used in the construction of buildings, highways, bridges, and much of our country's infrastructure. In this chapter, we look at three of the most important of these materials: wood, concrete, and asphalt. The field of construction materials is indeed very important for engineers, especially to civil engineers and highway engineers. The properties and processing of steels used for making reinforced concrete, ceramics (e.g., sand, lime, concrete), plastics (e.g., epoxies and polystyrene foams), glasses, and composites (e.g., fiberglass) play a very important role in the development and use of construction materials. Another area in civil engineering that is becoming increasingly important is related to the use of sensors and actuators in buildings and bridges. Advanced materials developed for microelectronic and optical applications play an important role in this area.

For example, smart bridges and buildings that make use of optical-fiber sensors currently are being developed.[1] These sensors can monitor the health of the structures on a continuous basis and thus can provide early warnings of any potential problems. Similarly, researchers in areas of smart structures are also working on many other ideas using sensors that can detect such things as the formation of ice. If ice is detected, the system can start to spray salt water to prevent or delay freezing. There are also applications of advanced materials that detect snow or ice. These are used on steep driveways in commercial parking garages where activation of the snow/ice sensor initiates the heating of that part of the driveway.[2] Similarly, we now have

many smart coatings on glasses that can deflect heat and make buildings energy efficient. There are also new coatings that have resulted in self-cleaning glasses.[3,4] New technologies are also being implemented to develop "green buildings."

There are many other areas of materials that relate to structures. For example, the corrosion of bridges and the limitations it poses on the bridge's life expectancy is a major cost for any nation. Strategies using galvanized steels and the proper paints to protect against corrosion now play a major role in development. Similarly, many material-joining techniques, such as welding, play a very important role in the construction of bridges and buildings. Long-term environmental impacts of the materials used must be considered (e.g., what are the best materials to use for water pipes, insulation, fire retardancy, etc.?). The goal of this chapter is to present a summary of the material properties of wood, concrete, and asphalt.

17-1   The Structure of Wood

Wood, a naturally occurring composite, is one of our most familiar materials. Although it is not a "high-tech" material, we are literally surrounded by it in our homes and value it for its beauty. In addition, wood is a strong, lightweight material that still dominates much of the construction industry.

We can consider wood to be a complex fiber-reinforced composite composed of long, unidirectionally aligned, tubular polymer cells in a polymer matrix. Furthermore, the polymer tubes are composed of bundles of partially crystalline, cellulose fibers aligned at various angles to the axes of the tubes. This arrangement provides excellent tensile properties in the longitudinal direction.

Wood consists of four main constituents. **Cellulose** fibers make up about 40% to 50% of wood. Cellulose is a naturally occurring thermoplastic polymer with a degree of polymerization of about 10,000. The structure of cellulose is shown in Figure 17-1. About 25% to 35% of a tree is **hemicellulose**, a polymer having a degree of polymerization of about 200. Another 20% to 30% of a tree is **lignin**, a low molecular weight, organic cement that bonds the various constituents of the wood. Finally, **extractives** are organic impurities such as oils, which provide color to the wood or act as preservatives against the environment and insects, and inorganic minerals such as silica, which help dull saw blades during the cutting of wood. As much as 10% of the wood may be extractives.

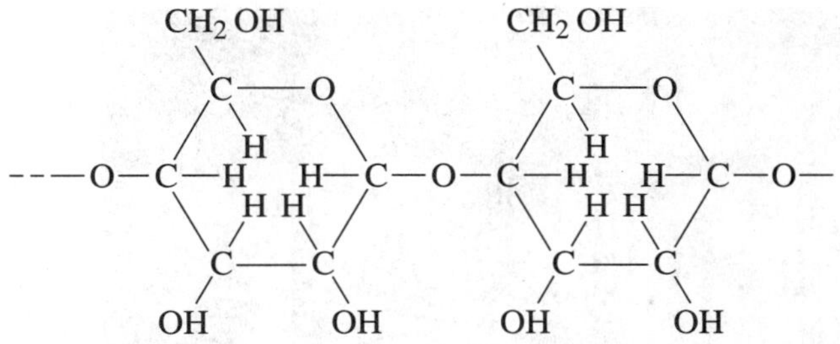

CH₂OH, CH₂OH

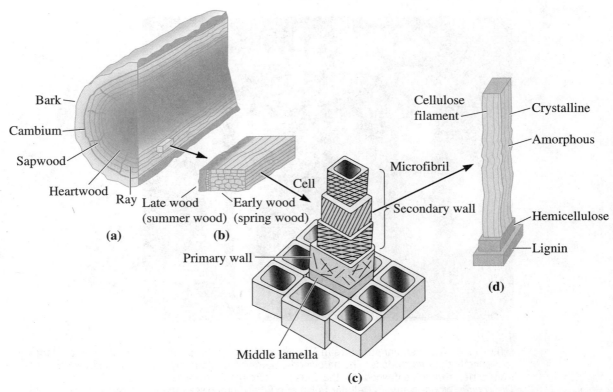

Figure 17-1
The structure of the cellulose
filaments in wood.

There are three important levels in the structure of wood: the fiber structure, the cell structure, and the macrostructure (Figure 17-2).

Fiber Structure The basic component of wood is cellulose, $C_6H_{10}O_5$, arranged in polymer chains that form long fibers. Much of the fiber length is crystalline, with the crystalline regions separated by small lengths of amorphous cellulose. A bundle of cellulose chains are encased in a layer of randomly oriented, amorphous hemicellulose chains. Finally, the hemicellulose is covered with lignin. The entire bundle, consisting of cellulose chains, hemicellulose chains, and lignin, is called a **microfibril**; it can have a virtually infinite length.

Figure 17-2 The structure of wood: (a) the macrostructure, including a layer structure outlined by the annual growth rings, (b) detail of the cell structure within one annual growth ring, (c) the structure of a cell, including several layers composed of microfibrils of cellulose fibers, hemicellulose fibers, and lignin, and (d) the microfibril's aligned, partly crystalline cellulose chains.

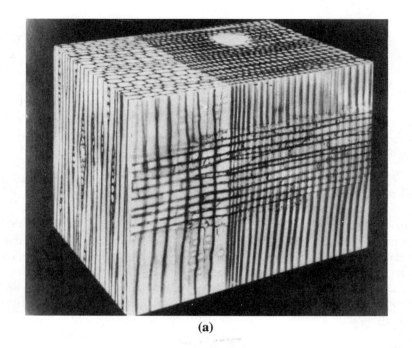

(a)

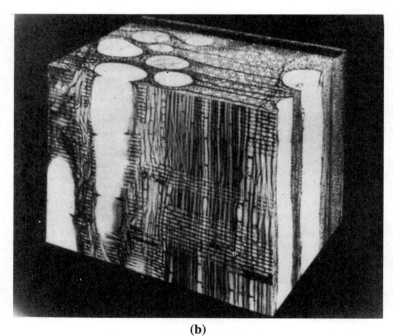

(b)

Figure 17-3   The cellular structure in (a) softwood and (b) hardwood. Softwoods contain larger, longer cells than hardwoods. The hardwoods, however, contain large-diameter vessels. Water is transported through softwoods by the cells and through hardwoods by the vessels. (*From J.M. Dinwoodie*, Wood: Nature's Cellular Polymeric Fiber-Composite, *The Institute of Metals, 1989.*)

Cell Structure   The tree is composed of elongated cells, often having an aspect ratio of 100 or more, that constitute about 95% of the solid material in wood. The hollow cells are composed of several layers built up from the microfibrils. The first, or primary, wall of the cell contains randomly oriented microfibrils. As the cell walls thicken, three more distinct layers are formed. The outer and inner walls contain microfibrils oriented in two directions that are not parallel to the cell. The middle wall, which is the thickest, contains microfibrils that are unidirectionally aligned, usually at an angle not parallel to the axis of the cell.

Macrostructure   A tree is composed of several layers. The outer layer, or *bark*, protects the tree. The **cambium**, just beneath the bark, contains new growing cells. The **sapwood** contains a few hollow living cells that store nutrients and serve as the conduit for water. Finally, the **heartwood**, which contains only dead cells, provides most of the mechanical support for the tree.

The tree grows when new elongated cells develop in the cambium. Early in the growing season, the cells are large; later they have a smaller diameter, thicker walls, and a higher density. This difference between the early (or *spring*) wood and the late (or *summer*) wood permits us to observe annual growth rings. In addition, some cells grow in a radial direction; these cells, called *rays*, provide the storage and transport of food.

Hardwood versus Softwood   The hardwoods are deciduous trees such as oak, ash, hickory, elm, beech, birch, walnut, and maple.[6] In these trees, the elongated cells are relatively short, with a diameter of less than 0.1 mm and a length of less than 1 mm. Contained within the wood are longitudinal pores, or vessels, which carry water through the tree (Figure 17-3).

The softwoods are the conifers, evergreens such as pine, spruce, hemlock, fir, spruce, and cedar, and have similar structures. In softwoods, the cells tend to be somewhat longer than in the hardwoods. The hollow center of the cells is responsible for transporting water. In general, the density of softwoods tends to be lower than that of hardwoods because of a greater percentage of void space.

17-2   Moisture Content and Density of Wood

The material making up the individual cells in virtually all woods has essentially the same density—about 1.45 g/cm^3. However, wood contains void space that causes the actual density to be much lower. The density of wood depends primarily on the species of the tree (or the amount of void space peculiar to that species) and the percentage of water in the wood (which depends on the amount of drying and on the relative humidity to which the wood is exposed during use). Completely dry wood varies in density from about 0.3 to 0.8 g/cm^3, with hardwoods having higher densities than softwoods. However, the measured density is normally higher due to the water contained in the wood. The percentage water is given by:

$$\% \text{ Water} = \frac{\text{weight of water}}{\text{weight of dry wood}} \times 100 \tag{17-1}$$

On the basis of this definition, it is possible to describe a wood as containing more than 100% water. The water is contained both in the hollow cells or vessels, where it is not tightly held, and in the cellulose structure in the cell walls, where it is more tightly bonded to the cellulose fibers. While a large amount of water is stored in a live tree, the

TABLE 17-1 ■ *Properties of typical woods*

Wood	Density (for 12% Water) (g/cm³)	Modulus of Elasticity (psi)
Cedar	0.32	1,100,000
Pine	0.35	1,200,000
Fir	0.48	2,000,000
Maple	0.48	1,500,000
Birch	0.62	2,000,000
Oak	0.68	1,800,000

amount of water in the wood after the tree is harvested depends, eventually, on the humidity to which the wood is exposed; higher humidity increases the amount of water held in the cell walls. The density of a wood is usually given at a moisture content of 12%, which corresponds to 65% humidity. The density and modulus of elasticity parallel to the grain of several common woods are included in Table 17-1 for this typical water content.

The following example illustrates the calculation for the density of the wood.

EXAMPLE 17-1 *Density of Dry and Wet Wood*

A green wood has a density of 0.86 g/cm³ and contains 175% water. Calculate the density of the wood after it has completely dried.

SOLUTION

A 100-cm³ sample of the wood would weigh 86 g. From Equation 17-1, we can calculate the weight of the dry wood to be:

$$\% \text{ Water} = \frac{\text{weight of water}}{\text{weight of dry wood}} \times 100 = 175$$

$$= \frac{\text{green weight} - \text{dry weight}}{\text{dry weight}} \times 100 = 175$$

$$\text{Dry weight of wood} = \frac{(100)(\text{green weight})}{275}$$

$$= \frac{(100)(86)}{275} = 31.3 \text{ g}$$

$$\text{Density of dry wood} = \frac{31.3 \text{ g}}{100 \text{ cm}^3} = 0.313 \text{ g/cm}^3$$

17-3 Mechanical Properties of Wood

The strength of a wood depends on its density, which in turn depends on both the water content and the type of wood. As a wood dries, water is eliminated first from the vessels

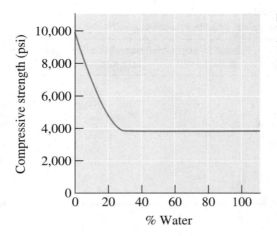

Figure 17-4
The effect of the percentage of water in a typical wood on the compressive strength parallel to the grain.

and later from the cell walls. As water is removed from the vessels, practically no change in the strength or stiffness of the wood is observed (Figure 17-4). However, on continued drying to less than about 30% water, there is water loss from the actual cellulose fibers. This loss permits the individual fibers to come closer together, increasing the bonding between the fibers and the density of the wood and, thereby, increasing the strength and stiffness of the wood.

The type of wood also affects the density. Because they contain less of the higher-density late wood, softwoods typically are less dense and therefore have lower strengths than hardwoods. In addition, the cells in softwoods are larger, longer, and more open than those in hardwoods, leading to lower density.

The mechanical properties of wood are highly anisotropic. In the longitudinal direction (Figure 17-5), an applied tensile load acts parallel to the microfibrils and cellulose chains in the middle section of the secondary wall. These chains are strong—because they are mostly crystalline—and are able to carry a relatively high load. However, in the radial and tangential directions, the weaker bonds between the microfibrils and cellulose fibers may break, resulting in very low tensile properties. Similar behavior is observed in compression and bending loads. Because of the anisotropic behavior, most lumber is cut in a tangential-longitudinal or radial-longitudinal manner. These cuts maximize the longitudinal behavior of the wood.

Wood has poor properties in compression and bending (which produces a combination of compressive and tensile forces). In compression, the fibers in the cells tend

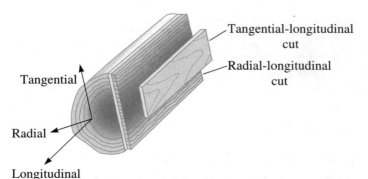

Tangential-longitudinal cut

Radial-longitudinal cut

Tangential

Radial

Longitudinal

Figure 17-5
The different directions in a log. Because of differences in cell orientation and the grain, wood displays anisotropic behavior.

TABLE 17-2 ■ *Anisotropic behavior of several woods (at 12% moisture)*

	Tensile Strength Longitudinal (psi)	Tensile Strength Radial (psi)	Compressive Strength Longitudinal (psi)	Compressive Strength Radial (psi)
Beech	12,500	1,010	7,300	1,010
Elm	17,500	660	5,520	690
Maple	15,700	1,100	7,830	1,470
Oak	11,300	940	6,200	810
Cedar	6,600	320	6,020	920
Fir	11,300	390	5,460	610
Pine	10,600	310	4,800	440
Spruce	8,600	370	5,610	580

to buckle, causing the wood to deform and break at low stresses. Unfortunately, most applications for wood place the component in compression or bending and therefore do not take full advantage of the engineering properties of the material. Similarly, the modulus of elasticity is highly anisotropic; the modulus perpendicular to the grain is about 1/20th that given in Table 17-1 parallel to the grain. Table 17-2 compares the tensile and compressive strengths parallel and perpendicular to the cells for several woods.

Clear wood has a specific strength and specific modulus that compare well with those of other common construction materials (Table 17-3). Wood also has good toughness, largely due to the slight misorientation of the cellulose fibers in the middle layer of the secondary wall. Under load, the fibers straighten, permitting some ductility and energy absorption. The mechanical properties of wood also depend on imperfections in the wood. Clear wood, free of imperfections such as knots, may have a longitudinal tensile strength of 10,000 to 20,000 psi. Less-expensive construction lumber, which usually contains many imperfections, may have a tensile strength below 5000 psi. The knots also disrupt the grain of the wood in the vicinity of the knot, causing the cells to be aligned perpendicular to the tensile load.

TABLE 17-3 ■ *Comparison of the specific strength and specific modulus of wood with those of other common construction materials*

Material	Specific Strength ($\times 10^5$ in.)	Specific Modulus ($\times 10^7$ in.)
Clear wood	7.0	9.5
Aluminum	5.0	10.5
1020 steel	2.0	10.5
Copper	1.5	5.5
Concrete	0.6	3.5

After F.F. Wangaard, "Wood: Its Structure and Properties," J. Educ. Models for Mat. Sci. and Engr., Vol. 3, No. 3, 1979.

17-4   Expansion and Contraction of Wood

Like other materials, wood changes dimensions when heated or cooled. Dimensional changes in the longitudinal direction are very small in comparison with those in metals, polymers, and ceramics. However, the dimensional changes in the radial and tangential directions are greater than those for most other materials.

In addition to dimensional changes caused by temperature fluctuations, the moisture content of the wood causes significant changes in dimension. Again, the greatest changes occur in the radial and tangential directions, where the moisture content affects the spacing between the cellulose chains in the microfibrils. The change in dimensions Δx in wood in the radial and tangential directions is approximated by

$$\Delta x = x_0[c(M_f - M_i)], \tag{17-2}$$

where x_0 is the initial dimension, M_i is the initial water content, M_f is the final water content, and c is a coefficient that describes the dimensional change and can be measured in either the radial or the tangential direction. Table 17-4 includes the dimensional coefficients for several woods. In the longitudinal direction, no more than 0.1 to 0.2% change is observed.

TABLE 17-4 ■ *Dimensional coefficient c (in./in. · % H$_2$O) for several woods*

Wood	Radial	Tangential
Beech	0.00190	0.00431
Elm	0.00144	0.00338
Maple	0.00165	0.00353
Oak	0.00183	0.00462
Cedar	0.00111	0.00234
Fir	0.00155	0.00278
Pine	0.00141	0.00259
Spruce	0.00148	0.00263

During the initial drying of wood, the large dimensional changes perpendicular to the cells may cause warping and even cracking. In addition, when the wood is used, its water content may change, depending on the relative humidity in the environment. As the wood gains or loses water during use, shrinkage or swelling continues to occur. If a wood construction does not allow movement caused by moisture fluctuations, warping and cracking can occur—a particularly severe condition in large expanses of wood, such as the floor of a large room. Excessive expansion may cause large bulges in the floor; excessive shrinkage may cause large gaps between individual planks of the flooring.

17-5   Plywood

The anisotropic behavior of wood can be reduced, and wood products can be made in larger sizes, by producing plywood. Thin layers of wood, called **plies**, are cut from logs—normally, softwoods. The plies are stacked together with the grains between adjacent plies oriented at 90° angles; usually an odd number of plies are used. Assuring that these angles are as precise as possible is important to assure that the plywood does not warp or twist when the moisture content in the material changes. The individual plies are generally bonded to one another using a thermosetting phenolic resin. The resin is introduced between the plies, which are then pressed together while hot to cause the resin to polymerize.

Similar wood products are also produced as "laminar" composite materials. The facing (visible) plies may be of a more expensive hardwood, with the center plies of less expensive softwood. Wood particles can be compacted into sheets and laminated between two wood plies, producing a particle board. Wood plies can be used as the facings for honeycomb materials.

17-6 Concrete Materials

An **aggregate** is a combination of gravel, sand, crushed stones, or slag. A **mortar** is made by mixing cement, water, air, and fine aggregate. Concrete contains all of the ingredients of the mortar and coarse aggregates. **Cements** are inorganic materials that set and harden after being mixed into a paste using water. **Concrete** is a particulate composite in which both the particulate and the matrix are ceramic materials. In concrete, sand and a coarse aggregate are bonded in a matrix of Portland cement. A cementation reaction between water and the minerals in the cement provides a strong matrix that holds the aggregate in place and provides good compressive strength to the concrete.

Cements   They are classified as hydraulic and nonhydraulic. **Hydraulic cements** set and harden under water. **Nonhydraulic cements** (e.g., lime, CaO) cannot harden under water and require air for hardening.[6] Portland cement is the most widely used and manufactured construction material. It was patented by Joseph Aspdin in 1824 and is named as such after the limestone cliffs on the Isle of Portland in England.[7]

Hydraulic cement is made out of calcium silicates, with an approximate composition of CaO ($\sim$60 to 65%), SiO_2 ($\sim$20 to 25%), and iron oxide and alumina ($\sim$7 to 12%). The cement binder, which is very fine in size, is composed of various ratios of $3CaO \cdot Al_2O_3$, $2CaO \cdot SiO_2$, $3CaO \cdot SiO_2$, $4CaO \cdot Al_2O_3 \cdot Fe_2O_3$, and other minerals. In the cement terminology, CaO, SiO_2, Al_2O_3, and Fe_2O_3 are often indicated as C, S, A, and F, respectively. Thus, C_3S means $3CaO$-SiO_2. When water is added to the cement, a hydration reaction occurs, producing a solid gel that bonds the aggregate particles. Possible reactions include

$$3CaO \cdot Al_2O_3 + 6H_2O \rightarrow Ca_3Al_2(OH)_{12} + \text{heat}$$

$$2CaO \cdot SiO_2 + xH_2O \rightarrow Ca_2SiO_4 \cdot xH_2O + \text{heat}$$

$$3CaO + SiO_2 + (x+1)H_2O \rightarrow Ca_2SiO_4 \cdot xH_2O + Ca(OH)_2 + \text{heat}$$

After hydration, the cement provides the bond for the aggregate particles. Consequently, enough cement must be added to coat all of the aggregate particles. The cement typically constitutes on the order of 15 vol% of the solids in the concrete.

The composition on the cement helps determine the rate of curing and the final properties of the concrete. For example, $3CaO \cdot Al_2O_3$ and $3CaO \cdot SiO_2$ produce rapid setting but low strengths. The $2CaO \cdot Al_2O_3$ reacts more slowly during hydration,

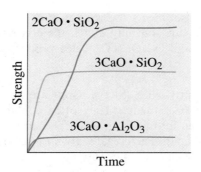

Figure 17-6
The rate of hydration of the minerals in Portland cement.

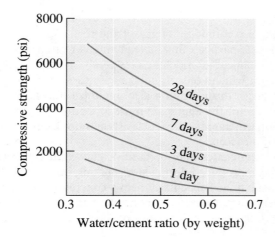

Figure 17-7
The compressive strength of concrete increases with time. After 28 days, the concrete approaches its maximum strength.

but eventually produces higher strengths (Figure 17-6). Nearly complete curing of the concrete is normally expected within 28 days (Figure 17-7), although some additional curing may continue for years.

There are about 10 general types of cements used. Some are shown in Table 17-5. In large structures such as dams, curing should be slow in order to avoid excessive heating caused by the hydration reaction. These cements typically contain low percentages of $3CaO \cdot SiO_2$, such as in Types II and IV. Some construction jobs, however, require that concrete forms be removed and reused as quickly as possible; cements for these purposes may contain large amounts of $3CaO \cdot SiO_2$, as in Type III.

type I A — type 5–10

TABLE 17-5 ■ *Types of Portland cements*

| | **Approximate Composition** | | | | |
	$3C \cdot S$	$2C \cdot S$	$3C \cdot A$	$4C \cdot A \cdot F$	**Characteristics**
Nbrmd = Type I	55	20	12	9	General purpose
Type II	45	30	7	12	Low rate of heat generation, moderate resistance to sulfates
high early strength = Type III	65	10	12	8	Rapid setting
Type IV	25	50	5	13	Very low rate of heat generation
Type V	40	35	3	14	Good sulfate resistance

The composition of the cement also affects the resistance of the concrete to the environment. For example, sulfates in the soil may attack the concrete; using higher proportions of $4CaO \cdot Al_2O_3 \cdot Fe_2O_3$ and $2CaO \cdot SiO_2$ helps produce concretes more resistant to sulfates, as in Type V.

Sand   Chemically sand is predominantly silica (SiO_2). Sands are fine minerals, typically of the order of 0.1 to 1.0 mm in diameter. They often contain at least some adsorbed water, which should be taken into account when preparing a concrete mix. The sand helps fill voids between the coarser aggregate, giving a high-packing factor, reducing the amount of open (or interconnected) porosity in the finished concrete, and reducing problems with disintegration of the concrete due to repeated freezing and thawing during service.

Aggregate   Coarse aggregate is composed of gravel and rock. Aggregate must be clean, strong, and durable. Aggregate particles that have an angular rather than a round shape provide strength due to mechanical interlocking between particles, but angular particles also provide more surface on which voids or cracks may form. It is normally preferred that the aggregate size be large; this condition also minimizes the surface area at which cracks or voids form. The size of the aggregate must, of course, be matched to the size of the structure being produced; aggregate particles should not be any larger than about 20% of the thickness of the structure.

In some cases, special aggregates may be used. Lightweight concretes can be produced by using mineral slags, produced during steel-making operations; these concretes have improved thermal insulation. Particularly heavy concretes can be produced using dense minerals or even metal shot; these heavy concretes can be used in building nuclear reactors to better absorb radiation. The densities of several aggregates are included in Table 17-6.

TABLE 17-6 ■ *Characteristics of concrete materials*

Material	True Density	
Cement	190 lb/ft^3	where 1 sack = 94 lb
Sand	160 lb/ft^3	
Aggregate	170 lb/ft^3	Normal
	80 lb/ft^3	Lightweight slag
	30 lb/ft^3	Lightweight vermiculite
	280 lb/ft^3	Heavy Fe_3O_4
	390 lb/ft^3	Heavy ferrophosphorous
Water	62.4 lb/ft^3	with 7.48 gal/ft^3

17-7   Properties of Concrete

Many factors influence the properties of concrete. Some of the most important are the water-cement ratio, the amount of air entrainment, and the type of aggregate.

Water-Cement Ratio   The ratio of water to cement affects the behavior of concrete in several ways:

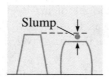

Figure 17-8
The slump test, in which deformation of a concrete shape under its own weight is measured, is used to describe the workability of concrete mix.

1. A minimum amount of water must be added to the cement to assure that all of it undergoes the hydration reaction. Too little water therefore causes low strength. Normally, however, other factors such as workability place the lower limit on the water-cement ratio.

2. A high water-cement ratio improves the **workability** of concrete—that is, how easily the concrete slurry can fill all of the space in the form. Air pockets or interconnected porosity caused by poor workability reduce the strength and durability of the concrete structure. Workability can be measured by the *slump test*. For example, a wet concrete shape 12-in. tall is produced (Figure 17-8) and is permitted to stand under its own weight. After some period of time, the shape deforms. The reduction in height of the form is the **slump**. A minimum water-cement ratio of about 0.4 (by weight) is usually required for workability. A larger slump, caused by a higher water-cement ratio, indicates greater workability. Slumps of 1 to 6 in. are typical; high slumps are needed for pouring narrow or complex forms, while low slumps may be satisfactory for large structures such as dams.

3. Increasing the water-cement ratio beyond the minimum required for workability decreases the compressive strength of the concrete. This strength is usually measured by determining the stress required to crush a concrete cylinder 6 in. in diameter and 12-in. tall. Figure 17-9 shows the effect of the water-cement ratio on concrete's strength.

4. High water-cement ratios increase the shrinkage of concrete during curing, creating a danger of cracking.

Because of the different effects of the water-cement ratio, a compromise between strength, workability, and shrinkage may be necessary. A weight ratio of 0.45 to 0.55 is typical. To maintain good workability, organic plasticizers may be added to the mix with little effect on strength.

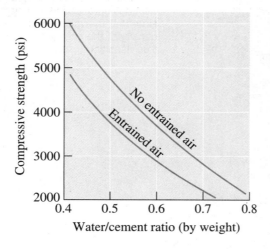

Figure 17-9
The effect of the water-cement ratio and entrained air on the 28-day compressive strength of concrete.

Air-Entrained Concrete Almost always, a small amount of air is entrained into the concrete during pouring. For coarse aggregate, such as 1.5-in. rock, 1% by volume of the concrete may be air. For finer aggregate, such as 0.5-in. gravel, 2.5% air may be trapped.

We sometimes intentionally entrain air into concrete—sometimes as much as 8% for fine gravel. The entrained air improves workability of the concrete and helps minimize problems with shrinkage and freeze–thaw conditions. However, air-entrained concrete has a lower strength (Figure 17-9).

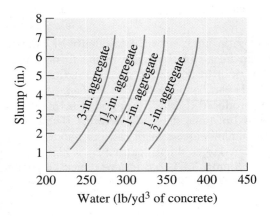

Figure 17-10
The amount of water per cubic yard of concrete required to give the desired workability (or slump) depends on the size of the coarse aggregate.

Type and Amount of Aggregate The size of the aggregate affects the concrete mix. Figure 17-10 shows the amount of water per cubic yard of concrete required to produce the desired slump, or workability; more water is required for smaller aggregates. Figure 17-11 shows the amount of aggregate that should be present in the concrete mix. The volume ratio of aggregate in the concrete is based on the bulk density of the aggregate, which is about 60% of the true density shown in Table 17-6.

The examples that follow show how to calculate the mixture contents for a concrete mixture.

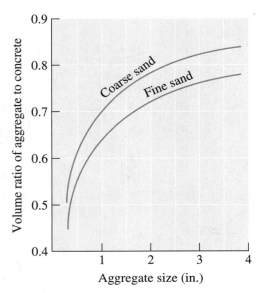

Figure 17-11
The volume ratio of aggregate to concrete depends on the sand and aggregate sizes. Note that the volume ratio uses the bulk density of the aggregate—about 60% of the true density.

17.6

EXAMPLE 17-2   *Composition of Concrete*

Determine the amounts of water, cement, sand, and aggregate in 5 cubic yards of concrete, assuming that we want to obtain a water/cement ratio of 0.4 (by weight) and that the cement/sand/aggregate ratio is 1:2.5:4 (by weight). A "normal" aggregate will be used, containing 1% water, and the sand contains 4% water. Assume that no air is entrained into the concrete.

SOLUTION

One method by which we can calculate the concrete mix is to first determine the volume of each constituent based on one sack (94 lb) of cement. We should remember that after the concrete is poured, there are no void spaces between the various constituents; therefore, we need to consider the true density—not the bulk density—of the constituents in our calculations.

For each sack of cement we use, the volume of materials required is:

$$\text{Cement} = \frac{94 \text{ lb/sack}}{190 \text{ lb/ft}^3} = 0.495 \text{ ft}^3$$

$$\text{Sand} = \frac{2.5 \times 94 \text{ lb cement}}{160 \text{ lb/ft}^3} = 1.469 \text{ ft}^3$$

$$\text{Gravel} = \frac{4 \times 94 \text{ lb cement}}{170 \text{ lb/ft}^3} = 2.212 \text{ ft}^3$$

$$\text{Water} = \frac{0.4 \times 94 \text{ lb cement}}{62.4 \text{ lb/ft}^3} = 0.603 \text{ ft}^3$$

Total volume of concrete = 4.779 ft^3/sack of cement

Therefore, in 5 cubic yards (or 135 ft^3), we need:

$$\text{Cement} = \frac{135 \text{ ft}^3}{4.779 \text{ ft}^3/\text{sack}} = 28 \text{ sacks}$$

Sand = (28 sacks)(94 lb/sack)(2.5 sand/cement) = 6580 lb

Gravel = (28 sacks)(94 lb/sack)(4 gravel/cement) = 10,528 lb

Water = (28 sacks)(94 lb/sack)(0.4 water/cement) = 1054 lb

But the sand contains 4% water and the gravel contains 1% water. To obtain the weight of the *wet* sand and gravel, we must adjust for the water content of each:

Sand = (6580 lb dry)(1.04) = 6843 lb and water = 263 lb

Gravel = (10,528 lb dry)(1.01) = 10,633 lb and water = 105 lb

Therefore, we actually only need to add:

Water = 1054 lb − 263 lb − 105 lb = 686 lb

$$= \frac{(686 \text{ lb})(7.48 \text{ gal/ft}^3)}{62.4 \text{ lb/ft}^3} = 82 \text{ gal}$$

Accordingly, we recommend that 28 sacks of cement, 6843 lb of sand, and 10,633 lb of gravel be combined with 82 gal of water.

EXAMPLE 17-3   *Design of a Concrete Mix for a Retaining Wall*

Design a concrete mix that will provide a 28-day compressive strength of 4000 psi in a concrete intended for producing a 5-in.-thick retaining wall 6-ft high. We expect to have about 2% air entrained in the concrete, although we will not intentionally entrain air. Our aggregate contains 1% moisture, and we have only coarse sand containing 5% moisture available.

SOLUTION

Some workability of the concrete is needed to assure that the form will fill properly with the concrete. A slump of 3 in. might be appropriate for such an application. The wall thickness is 5 in. To help minimize cost, we would use a large aggregate: A 1-in. diameter aggregate size would be appropriate (about 1/5 of the wall thickness).

To obtain the desired workability of the concrete using 1-in. aggregate, we should use about 320 lb of water per cubic yard (Figure 17-10).

To obtain the 4000 psi compressive strength after 28 days (assuming no intentional entrained air), we need a water/cement weight ratio of 0.57 (Figure 17-9).

Consequently, the weight of cement required per cubic yard of concrete is (320 lb water/0.57 water-cement) = 561 lb cement.

Because our aggregate size is 1 in. and we have only coarse sand available, the volume ratio of the aggregate to the concrete is 0.7 (Figure 17-11). Thus, the amount of aggregate required per yard of concrete is 0.7 yd^3; however, this amount is in terms of the bulk density of the aggregate. Because the bulk density is about 60% of the true density, the actual volume occupied by the aggregate in the concrete is 0.7 yd^3 × 0.6 = 0.42 yd^3.

Let's determine the volume of each constituent per cubic yard of concrete in order to calculate the amount of sand required:

$$\text{Water} = 320 \text{ lb}/62.4 \text{ lb/ft}^3 = 5.13 \text{ ft}^3$$

$$\text{Cement} = 561 \text{ lb}/190 \text{ lb/ft}^3 = 2.95 \text{ ft}^3$$

$$\text{Aggregate} = 0.42 \text{ yd}^3 × 27 \text{ ft}^3/\text{yd}^3 = 11.34 \text{ ft}^3$$

$$\text{Air} = 0.02 × 27 \text{ ft}^3 = 0.54 \text{ ft}^3$$

$$\text{Sand} = 27 - 5.13 - 2.95 - 11.34 - 0.54 = 7.04 \text{ ft}^3$$

Or converting to other units, assuming that the aggregate and sand are dry:

$$\text{Water} = 5.13 \text{ ft}^3 × 7.48 \text{ gal/ft}^3 = 38.4 \text{ gal}$$

$$\text{Cement} = 561 \text{ lb}/94 \text{ lb/sack} = 6 \text{ sacks}$$

$$\text{Aggregate} = 11.34 \text{ ft}^3 × 170 \text{ lb/ft}^3 = 1928 \text{ lb}$$

$$\text{Sand} = 7.04 \text{ ft}^3 × 160 \text{ lb/ft}^3 = 1126 \text{ lb}$$

However, the aggregate and the sand are wet. Thus, the actual amounts of aggregate and sand needed are:

$$\text{Aggregate} = 1928 × 1.01 = 1948 \text{ lb} \ (20 \text{ lb water})$$

$$\text{Sand} = 1126 × 1.05 = 1182 \text{ lb} \ (56 \text{ lb water})$$

The actual amount of water needed is:

$$\text{Water} = 38.4 \text{ gal} - \frac{(20 + 56 \text{ lb})(7.48 \text{ gal/ft}^3)}{62.4 \text{ lb/ft}^3} = 29.3 \text{ gal}$$

Thus, for each cubic yard of concrete, we will combine 6 sacks of cement, 1948 lb of aggregate, 1182 lb of sand, and 29.3 gal of water. This should give us a slump of 3 in. (the desired workability) and a compressive strength of 4000 psi after 28 days.

17-8   Reinforced and Prestressed Concrete

Concrete, like other ceramic-based materials, develops good compressive strength. Due to the porosity and interfaces present in the brittle structure, however, it has very poor tensile properties. Several methods are used to improve the load-bearing ability of concrete in tension.

Reinforced Concrete   Steel rods (known as rebars), wires, or mesh are frequently introduced into concrete to provide improvement in resisting tensile and bending forces. The tensile stresses are transferred from the concrete to the steel, which has good tensile properties. The steel takes the tensile part of the load. Polymer fibers, which are less likely to corrode, can also be used as reinforcement. In flexural stresses, the steel supports the part that is in tension; the part that is under compression is supported by the concrete.

Prestressed Concrete   Instead of simply being laid as reinforcing rods in a form, the steel can initially be pulled in tension between an anchor and a jack, thus remaining under tension during the pouring and curing of the concrete. After the concrete sets, the tension on the steel is released. The steel then tries to relax from its stretched condition, but the restraint caused by the surrounding concrete places the concrete in compression. Now higher tensile and bending stresses can be applied to the concrete because of the compressive residual stresses introduced by the pretensioned steel. In order to permit the external tension to be removed in a timely manner, the early-setting Type III cements are often used for these applications.

Poststressed Concrete   An alternate method of placing concrete under compression is to place hollow tubes in the concrete before pouring. After the concrete cures, steel rods or cables running through the tubes can then be pulled in tension, acting against the concrete. As the rods are placed in tension, the concrete is placed in compression. The rods or cables are then secured permanently in their stretched condition.

17-9   Asphalt

Asphalt is a composite of aggregate and **bitumen**, which is a thermoplastic polymer most frequently obtained from petroleum. Asphalt is an important material for paving roads. The properties of the asphalt are determined by the characteristics of the aggregate and binder, their relative amounts, and additives.

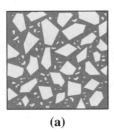   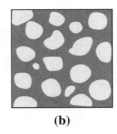

(a)   **(b)**

Figure 17-12
The ideal structure of asphalt (a) compared with the undesirable structure (b) in which round grains, a narrow distribution of grains, and excess binder all reduce the strength of the final material.

The aggregate, as in concrete, should be clean and angular and should have a distribution of grain sizes to provide a high-packing factor and good mechanical interlocking between the aggregate grains (Figure 17-12). The binder, composed of thermoplastic chains, bonds the aggregate particles. The binder has a relatively narrow useful temperature range, being brittle at sub-zero temperatures and beginning to melt at relatively low temperatures. Additives such as gasoline or kerosene can be used to modify the binder, permitting it to liquefy more easily during mixing and causing the asphalt to cure more rapidly after application.

The ratio of binder to aggregate is important. Just enough binder should be added so that the aggregate particles touch, but voids are minimized. Excess binder permits viscous deformation of the asphalt under load. Approximately 5% to 10% bitumen is present in a typical asphalt. Some void space is also required—usually, about 2 to 5%. When the asphalt is compressed, the binder can squeeze into voids, rather than be squeezed from the surface of the asphalt and lost. Too much void space, however, permits water to enter the structure; this increases the rate of deterioration of the asphalt and may embrittle the binder. Emulsions of asphalt are used for sealing driveways. The aggregate for asphalt is typically sand and fine gravel. However, there is some interest in using recycled glass products as the aggregate. **Glasphalt** provides a useful application for crushed glass. Similarly, there are also applications for materials developed using asphalt and shredded rubber tires.

SUMMARY

◆ Construction materials are composite materials occurring in nature or produced from natural-occurring materials. These materials play a vital role in the infrastructure of any nation. Concrete, wood, asphalt, glasses, composites, and steels are some of the most commonly encountered materials.

◆ Advanced materials used to make sensors and actuators also play an increasingly important role in monitoring the structural health of buildings and bridges. There are also new coatings and thin films on glasses that have contributed to more energy-efficient buildings.

◆ Wood is a natural fiber-reinforced polymer composite material. Cellulose fibers constitute aligned cells that provide excellent reinforcement in longitudinal directions in wood, but give poor strength and stiffness in directions perpendicular to the cells and fibers. The properties of wood therefore are highly anisotropic and depend on the species of the tree and the amount of moisture present in the wood. Wood has good tensile strength but poor compressive strength.

◈ Concrete is a particulate composite. In concrete, ceramic particles such as sand and gravel are used as filler in a ceramic-cement matrix. The water-cement ratio is a particularly important factor governing the behavior of the concrete. This behavior can be modified by entraining air and by varying the composition of the cement and aggregate materials. Concrete has good compressive strength but poor tensile strength.

◈ Asphalt also is a particulate composite, using the same type of aggregates as in concrete, but an organic, polymer binder.

GLOSSARY

Aggregate An aggregate is a combination of gravel, sand, crushed stones, or slag.

Bitumen The organic binder, composed of low melting point polymers and oils, for asphalt.

Cambium The layer of growing cells in wood.

Cellulose With a high degree of polymerization, a naturally occurring polymer fiber that is the major constituent of wood.

Cements Cements are inorganic materials that set and harden after being mixed into a paste using water.

Extractives Impurities in wood.

Glasphalt Asphalt in which the aggregate includes recycled glass.

Heartwood The center of a tree, comprised of dead cells, which provides mechanical support to a tree.

Hemicellulose With a low degree of polymerization, a naturally occurring polymer fiber that is an important constituent of wood.

Hydraulic cement A cement that sets and hardens under water.

Lignin The polymer cement in wood that bonds the cellulose fibers in the wood cells.

Microfibril Bundles of cellulose and other polymer chains that serve as the fiber reinforcement in wood.

Mortar A mortar is made by mixing cement, water, air, and fine aggregates. Concrete contains all of the ingredients of the mortar and coarse aggregates.

Nonhydraulic cements Cements that cannot harden under water and require air for hardening.

Plies The individual sheet of wood veneer from which plywood is constructed.

Portland cement An hydraulic cement produced by pulverizing clinker made out of calcium silicates; approximate composition is CaO ($\sim 60-65\%$), SiO_2 ($\sim 20-25\%$), and iron oxide and alumina ($\sim 7-12\%$).

Sapwood Hollow, living cells in wood that store nutrients and conduct water.

Slump The decrease in height of a standard concrete form when the concrete settles under its own weight.

Workability The ease with which a concrete slurry fills all of the space in a form.

ADDITIONAL INFORMATION

1. CHONG, K., *Technical discussions are acknowledged*. 2001.

2. DHEKNEY, V., *Discussions are acknowledged*. 2002.

3. GREENBERG, C.B., "Thin Films on Float Glass: The Extraordinary Possibilities," J. of Indu. Chemi, 2001. 40(1): p. 26–32.

4. GREENBERG, C., *Discussion on self-cleaning glasses and use of thin-film coatings is acknowledged*. 2002.

5. BRANTLEY, R.L. and R.Y. BRANTLEY, *Building Materials Technology*. 1996: McGraw Hill.

6. DERUCHER, K.N., G.P. KORFIATIS, and A.S. EZELDIN, *Materials for Civil and Highway Engineers*. 1998: Upper Saddle River, NJ, Prentice Hall.

7. MAMLOUK, M.S. and J.P. ZANIEWSKI, *Materials for Civil and Construction Engineers*. 1999: Addison-Wesley.

✓ PROBLEMS

Section 17-1 The Structure of Wood

Section 17-2 Moisture Content and Density of Wood

Section 17-3 Mechanical Properties of Wood

Section 17-4 Expansion and Contraction of Wood

Section 17-5 Plywood

17-1 A sample of wood with the dimensions 3 in. × 4 in. × 12 in. has a dry density of 0.35 g/cm³.

    (a) Calculate the number of gallons of water that must be absorbed by the sample to contain 120% water.

    (b) Calculate the density after the wood absorbs this amount of water.

17-2 The density of a sample of oak is 0.90 g/cm³. Calculate

    (a) the density of completely dry oak, and
    (b) the percent water in the original sample.

17-3 Boards of maple 1 in. thick, 6 in. wide, and 16 ft long are used as the flooring for a 60 ft × 60 ft hall. The boards were cut from logs with a tangential-longitudinal cut. The floor is laid when the boards have a moisture content of 12%. After some particularly humid days, the moisture content in the boards increases to 45%. Determine the dimensional change in the flooring parallel to the boards and perpendicular to the boards. What will happen to the floor? How can this problem be corrected?

17-4 A wall 30 feet long is built using radial-longitudinal cuts of 5-inch wide pine, with the boards arranged in a vertical fashion. The wood contains a moisture content of 55% when the wall is built; however, the humidity level in the room is maintained to give 45% moisture in the wood. Determine the dimensional changes in the wood boards, and estimate the size of the gaps that will be produced as a consequence of these changes.

Section 17-6 Concrete Materials

Section 17-7 Properties of Concrete

Section 17-8 Reinforced and Prestressed Concrete

Section 17-9 Asphalt

17-5 We have been asked to prepare 100 yd³ of normal concrete using a volume ratio of cement-sand-coarse aggregate of 1:2:4. The water-cement ratio (by weight) is to be 0.5. The sand contains 6 wt% water and the coarse aggregate contains 3 wt% water. No entrained air is expected.

    (a) Determine the number of sacks of cement that must be ordered, the tons of sand and aggregate required, and the amount of water needed.

    (b) Calculate the total weight of the concrete per cubic yard.

    (c) What is the weight ratio of cement-sand-coarse aggregate?

17-6 We plan to prepare 10 yd³ of concrete using a 1:2.5:4.5 weight ratio of cement-sand-coarse aggregate. The water-cement ratio (by weight) is 0.45. The sand contains 3 wt% water, the coarse aggregate contains 2 wt% water, and 5% entrained air is expected. Determine the number of sacks of cement, tons of sand and coarse aggregate, and gallons of water required.

✳ Design Problems

17-7 A wooden structure is functioning in an environment controlled at 65% humidity. Design a wood support column that is to hold a compressive load of 20,000 lb. The distance from the top to the bottom of the column should be 96 ± 0.25 in. when the load is applied.

17-8 Design a wood floor that will be 50 ft by 50 ft and will be in an environment in which humidity changes will cause a fluctuation of plus or minus 5% water in the wood. We want to minimize any buckling or gap formation in the floor.

17-9 We would like to produce a concrete that is suitable for use in building a large structure in a sulfate environment. For these situations, the maximum water-cement ratio should be 0.45 (by weight). The compressive strength of the concrete after 28 days should be at least 4000 psi. We have an available coarse aggregate containing 2% moisture in a variety of sizes, and both fine and coarse sand containing 4% mois-

ture. Design a concrete that will be suitable for this application.

17-10 We would like to produce a concrete sculpture. The sculpture will be as thin as 3 in. in some areas and should be light in weight, but it must have a 28-day compressive strength of at least 2000 psi. Our available aggregate contains 1% moisture and our sands contain 5% moisture. Design a concrete that will be suitable for this application.

17-11 The binder used in producing asphalt has a density of about 1.3 g/cm^3. Design an asphalt, including the weight and volumes of each constituent, that might be suitable for use as pavement. Assume that the sands and aggregates are the same as those for a normal concrete.

▪ Computer Problems

17-12 Write a computer program that will calculate the amounts of water, cement, sand, and aggregate for any number of cubic yards of concrete specified by the customer. Assume that the water-to-cement weight ratio is 0.4. The ratio of cement/sand/aggregate is to be assumed 1:2.5:4.0. The program should provide the number of concrete sacks needed (assume one sack is 94 lb.). The amount of water provided to the user should be in gallons. The amount of sand should be in pounds.

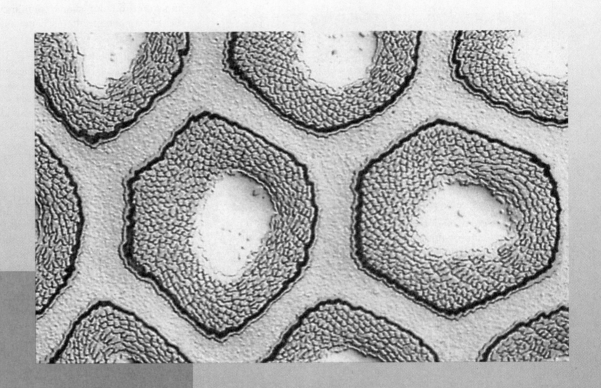

The structure of a Nb_3Sn superconductor wire for high-field magnetic applications includes many complex fibers within the wire. In this photomicrograph taken at a magnification of 1000×, fibers are embedded in a continuous copper matrix. The white core of each fiber is tin, and the "pebbly" layer around the tin core contains over 100 tiny niobium filaments. The dark ring around each fiber is vanadium; the light ring around the vanadium is niobium. These two rings keep the copper from diffusing into the fiber. During later heat treatment, the niobium filaments react with the tin, converting the niobium filaments to Nb_3Sn filaments (*From* ASM Handbook, *Vol. 2, Properties and Selection: Nonferrous Alloys and Special-Purpose Alloys (1990), ASM International, Materials Park, OH 44073.*)

PART 4

Physical Properties of Engineering Materials

The physical behavior of materials is described by a variety of electrical, magnetic, optical, and thermal properties. Most of these properties are determined by the atomic structure, atomic arrangement, and crystal structure of the material. In Chapter 18, we find that the atomic structure—in particular, the energy gap between the electrons in the valence and conduction bands —helps us to divide materials into conductors, semiconductors, and insulators. Atomic structure is responsible for the ferromagnetic behavior discussed in Chapter 19 and explains many optical (Chapter 20) and thermal (Chapter 21) properties. Modern aspects of the materials science and technology of materials for magnetics, microelectronics, and photonics are discussed in this part.

Chapter

Electronic materials include insulators, semiconductors, conductors, and superconductors. This family of materials has truly revolutionized the world. From spark plugs made from alumina, and copper wires for electrical transmission to components for wireless communications, high-powered magnets used in magnetic resonance imaging, capacitors, inductors, solar cells, active matrix displays, silicon, and gallium arsenide-based computer chips, electronic materials are found in countless numbers of applications. New advances in the materials sciences have led to several breakthroughs in the development of new electronic materials. Thanks to synthesis-processing-property relationships developed using a materials science approach, we now have ceramics that are not just excellent insulators, but also semiconductors and superconductors. Similarly, we now have polymers that are semiconductive and, more recently, a superconductive polymer has also been claimed to have been discovered. (*Courtesy of PhotoDisc/Getty Images.*)

18

Electronic Materials

Have You Ever Wondered?

- Can we make lightweight and flexible electronic circuits using plastics?

- Why are car manufacturers considering switching over to a 42-volt battery system?

- What materials are used to create some of the strongest electromagnets used in magnetic resonance imaging?

- What is the difference between a conductor and superconductor?

- How does a microwave oven heat food?

- How does a spark igniter in a gas grill work?

- Can ceramics be semiconductors or superconductors?

- What material is used to generate the high fidelity sound in a speaker tweeter?

From novel polymers exhibiting semiconductive behavior to the ceramic superconductors or amorphous silicon solar cells to the copper and aluminum wires used in power transmission and the porcelain used in electrical insulation, electronic materials have played a key role in the development of all computer-based and infor- mation-related technologies. Figure 18-1 gives a broad classification of some of the different electronic materials. Semiconductor circuitry, integrated onto silicon (Si) or gallium arsenide (GaAs) chips, plays a very important role in many of our modern devices, such as personal digital assistants, computers, cell phones, TVs,

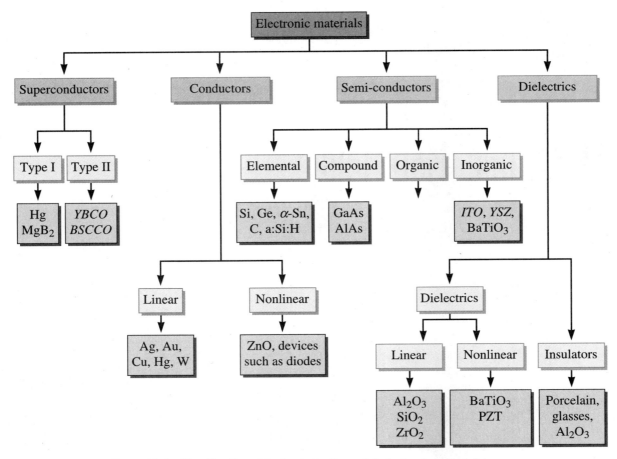

Figure 18-1 Classification of technologically useful electronic materials.

computers and VCRs. Some other materials which are used to make light-emitting diodes (LEDs), lasers, etc., exhibit electro-optic effects, while amorphous silicon (a:Si:H) materials are used in photovoltaic applications. We use porcelains, alumina, and other ceramics, as well as plastics, in applications where electrical insulating properties are important. The term "electronic materials" is used in a generic sense here. It includes materials used in microelectronics as well as those used in power transmission, electrical insulation, and other larger-scale applications.

In Chapter 2, we pointed out that, in applications of electronic materials, other properties (such as mechanical) are important and must be considered when designing components. In Chapter 3, we examined the crystal structures

or atomic arrangements in such materials as Si, GaAs, amorphous silicon, some ceramic superconductors, alumina, and barium titanate. In Chapters 4 and 5, we discussed important ideas relating to intrinsic and extrinsic semiconductors in relation to point defects and the basic concept of a *p-n* junction. We discussed how point defects created using dopants can lead to conductivity in oxides such as indium tin oxide *(ITO)* and yttria stabilized zirconia *(YSZ)*. In Chapter 8, we discussed the crystal growth of semiconductors and other materials. A review of these concepts covered in these earlier chapters would help you to better understand our discussions in this chapter.

To effectively select and use materials for electrical power and microelectronic applications, we first must understand how properties, such as electrical conductivity, are controlled. We must also realize that electrical behavior is influenced by the microstructure and processing of a material, and the environment to which a material is exposed.

18-1   Ohm's Law and Electrical Conductivity

Most of us are familiar with the common form of Ohm's law,

$$V = IR \qquad (18-1)$$

where V is the voltage (volts, V), I is the current (amps, A), and R is the resistance (ohms, Ω) to the current flow. This law is applicable to most but not all materials. The resistance (R) of a resistor is a characteristic of the size, shape, and properties of the materials used,

$$R = \rho \frac{l}{A} = \frac{l}{\sigma A} \qquad (18-2)$$

where l is the length (cm) of the resistor, A is the cross-sectional area (cm^2) of the resistor, ρ is the electrical resistivity (ohm $\cdot$ cm or $\Omega \cdot$ cm), and σ, which is the reciprocal of ρ, is the electrical conductivity (ohm$^{-1} \cdot$ cm^{-1}). The magnitude of the resistance depends upon the dimensions of the resistor, as well as the microstructure and composition of the material. The resistivity or conductivity does not depend on the dimensions of the material. Thus, resistivity or conductivity allows us to compare different materials. For example silver is a better conductor than copper. Resistivity is a **microstructure-sensitive property**, similar to yield strength. The resistivity of pure copper is much smaller than that of commercially pure copper, because impurities in commercially pure copper scatter electrons and contribute to increased resistivity. Similarly, the resistivity of annealed, pure copper is slightly lower than that of cold-worked, pure copper because of the scattering effect associated with dislocations. There are examples where we find that resistivity does appear to depend upon dimensions. For example, if we examine the resistivity of barium titanate (BaTiO$_3$) in bulk (say a disk that is $\frac{1}{2}$ cm thick and few cm in diameter) and in a thin film (say 100 nm thick), often we will find that the resistivity of bulk BaTiO$_3$ is higher than that of thin-film BaTiO$_3$. Thus, we do not have one value of resistivity for BaTiO$_3$! These effects are usually related to microstructure and chemical homogeneity, the presence of pores, the magnitude of applied electric field, and so on. In general, we would not be surprised to see BaTiO$_3$ has a higher resistivity than copper, but it is important to note that the resistivity of BaTiO$_3$ can vary significantly depending upon the presence of impurities, dopants, pores, and other microstructural features.

In components designed to conduct electrical energy, minimizing power losses is important, not only to conserve energy, but also to minimize heating. The electrical power (P, in watts) lost when a current flows through a resistance is given by:

$$P = VI = I^2R \qquad (18-3)$$

A high resistance R results in larger power losses. These electrical losses are also known as Joule heating losses.

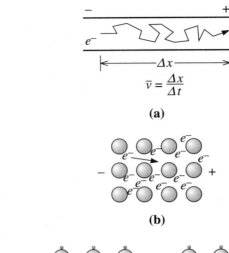

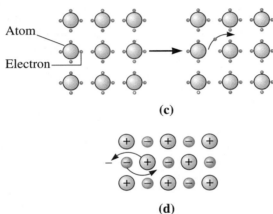

Figure 18-2
(a) Charge carriers, such as electrons, are deflected by atoms or defects and take an irregular path through a conductor. The average rate at which the carriers move is the drift velocity $\bar{v}$. (b) Valence electrons in the metallic bond move easily. (c) Covalent bonds must be broken in semiconductors and insulators for an electron to be able to move. (d) Entire ions must diffuse to carry charge in many ionically bonded materials.

A second form of Ohm's law is obtained if we combine Equations 18-1 and 18-2 to give:

$$\frac{I}{A} = \sigma \frac{V}{l}$$

If we define I/A as the **current density** J (A/cm^2) and V/l as the **electric field** E (V/cm), then:

$$J = \sigma E \qquad (18\text{-}4)$$

We can also determine that the current density J is

$$J = n \cdot q \cdot \bar{v},$$

where n is the number of charge carriers (carriers/cm^3), q is the charge on each carrier (1.6×10^{-19} C), and $\bar{v}$ is the average **drift velocity** (cm/s) at which the charge carriers move [Figure 18-2(a)]. Thus:

$$\sigma E = nq\bar{v} \quad \text{or} \quad \sigma = nq\frac{\bar{v}}{E}$$

Recall, that in Chapter 5, we discussed the difference between drift and diffusion of carriers. Diffusion occurs as a result of temperature and concentration gradients, and drift occurs as a result of an applied electric or magnetic field. Please review these concepts to better appreciate the following discussion.

The term $\bar{v}/E$ is called the **mobility** μ $\left(\dfrac{cm^2}{V \cdot s}\right)$ of the carriers (in the case of metals this is the mobility of electrons):

$$\mu = \frac{\bar{v}}{E}$$

Finally:

$$\sigma = n \cdot q \cdot \mu \tag{18-5a}$$

The charge q is a constant; from inspection of Equation 18-5a, we find that we can control the electrical conductivity of materials by (1) controlling the number of charge carriers in the material or (2) controlling the mobility—or ease of movement—of the charge carriers. The mobility is particularly important in metals, whereas the number of carriers is more important in semiconductors and insulators.

Electrons are the charge carriers in metals. In semiconductors, electrons and holes are both carriers of electricity. Thus, Equation 18-5a can be modified as follows for expressing conductivity of semiconductors:

$$\sigma = n \cdot q \cdot \mu_n + p \cdot q \cdot \mu_p \tag{18-5b}$$

In this equation, μ_n and μ_p are the mobilities of electrons and holes, respectively. The terms n and p represent the concentrations of free electrons and holes in the semiconductor.

In ceramics, when conduction does occur, it can be the result of electrons that "hop" from one defect to another or the movement of ions [Figure 18-2(b)–(d)]. The mobility depends on atomic bonding, imperfections, microstructure, and, in ionic compounds, diffusion rates.

Because of these effects, the electrical conductivity of materials varies tremendously, as illustrated in Table 18-1. These values are approximate and are for high-purity materials at 300 K (unless noted otherwise). Note that the values of conductivity for metallic and semiconducting materials depend very strongly on temperature. Thus, the data in Table 18-1 are to be used for comparing different classes of materials. Table 18-2 includes some useful units and relationships.

Electronic materials can be classified as (a) superconductors, (b) normal conductors, (c) semiconductors, and (d) dielectrics or insulators, depending upon the magnitude of their electrical conductivity. Materials with conductivity less than 10^{-12} $\Omega^{-1} \cdot cm^{-1}$ or resistivity greater than 10^{12} $\Omega - cm$, are considered insulating or dielectric. Materials with conductivity greater than 10^3 $\Omega^{-1} \cdot cm^{-1}$ or resistivity less than 10^{-3} $\Omega^{-1} \cdot cm^{-1}$ are considered conductors. Materials with conductivity less than 10^3 $\Omega^{-1} \cdot cm^{-1}$ but greater than 10^{-12} $\Omega^{-1} \cdot cm^{-1}$ are considered semiconductors. (These are approximate ranges of values.)

We use the term "dielectric" for materials that are used in applications where the dielectric constant is important. The **dielectric constant** (k) of a material is a microstructure-sensitive property relating to the material's ability to store an electrical charge. We use the term "insulator" to describe the ability of a material to stop the flow of DC or AC current, as opposed to its ability to store a charge. A measure of how good of an insulator a material is the maximum electric field it can support without an electrical break down.

The following examples illustrate how some of the concepts related to resistance, resistivity, mobility, etc., can be applied.

TABLE 18-1 ■ *Electrical conductivity of selected materials at* T = 300 *K**

Material	Conductivity (ohm^{-1} · cm^{-1})
Superconductors	
Hg, Nb$_3$Sn, YBa$_2$Cu$_3$O$_{7-x}$ MgB$_2$	Infinite (under certain conditions such as low temperatures)
Metals	
Alkali metals:	
Na	2.13×10^5
K	1.64×10^5
Alkali earth metals:	
Mg	2.25×10^5
Ca	3.16×10^5
Group 3B metals:	
Al	3.77×10^5
Ga	0.66×10^5
Transition metals:	
Fe	1.00×10^5
Ni	1.46×10^5
Group 1B metals:	
Cu	5.98×10^5
Ag	6.80×10^5
Au	4.26×10^5
Semiconductors	
Group 4B elements:	
Si	5×10^{-6}
Ge	0.02
α-Sn	0.9×10^5
Compound semiconductors	
GaAs	2.5×10^{-9}
AlAs	0.1
SiC	10^{-10}
Ionic Conductors	
Indium tin oxide (*ITO*)	
Yttria-stabilized zirconia (*YSZ*)	
Insulators, Linear and Nonlinear Dielectrics	
Polymers:	
Polyethylene	10^{-15}
Polytetrafluorethylene	10^{-18}
Polystyrene	10^{-17} to 10^{-19}
Epoxy	10^{-12} to 10^{-17}
Ceramics:	
Alumina (Al$_2$O$_3$)	10^{-14}
Silicate glasses	10^{-17}
Boron nitride (BN)	10^{-13}
Barium titanate (BaTiO$_3$)	10^{-14}
C (diamond)	$<10^{-18}$

**Unless specified otherwise, assumes high purity material.*

TABLE 18-2 ■ *Some useful relationships, constants, and units*

Electron volt = 1 eV = 1.6×10^{-19} Joule = 1.6×10^{-12} erg
1 amp = 1 coulomb/second
1 volt = 1 amp · ohm
$k_B T$ at room temperature (300 K) = 0.0259 eV
c = speed of light 2.998×10^{-8} m/s
ε_0 = perimitivity of free space = 8.85×10^{-12} F/m
q = charge on electron = 1.6×10^{-19} C
Avogadro's number N_A = 6.023×10^{23}
k_B = Boltzmann's constant = 8.63×10^{-5} eV/K = 1.38×10^{-23} J/K
h = Planck's constant 6.63×10^{-34} J-s = 4.14×10^{-15} eV-s

EXAMPLE 18-1   *Design of a Transmission Line*

Design an electrical transmission line 1500 m long that will carry a current of 50 A with no more than 5×10^5 W loss in power. The electrical conductivity of several materials is included in Table 18-1.

SOLUTION

Electrical power is given by the product of the voltage and current or:

$$P = VI = I^2 R = (50)^2 R = 5 \times 10^5 \text{ watts}$$

$$R = 200 \text{ ohms}$$

From Equation 18-2:

$$A = \frac{l}{R \cdot \sigma} = \frac{(1500 \text{ m})(100 \text{ cm/m})}{(200 \text{ ohms})\sigma} = \frac{750}{\sigma}$$

Let's consider three metals—aluminum, copper, and silver—that have excellent electrical conductivity. The table below includes appropriate data and some characteristics of the transmission line for each metal.

	σ (ohm^{-1} · cm^{-1})	A (cm^2)	Diameter (cm)
Aluminum	3.77×10^5	0.00199	0.050
Copper	5.98×10^5	0.00125	0.040
Silver	6.80×10^5	0.00110	0.037

Any of the three metals will work, but cost is a factor as well. Aluminum will likely be the most economical choice (Chapter 13), even though the wire has the largest diameter. However, other factors, such as whether the wire can support itself between transmission poles, also contribute to the final choice.

EXAMPLE 18-2   *A 42-volt Battery System for Cars*

Some manufacturers of automotives are considering using a 42-volt battery system instead of the standard 12/14-volt battery system. Why do you think this change is being considered?

SOLUTION

In some applications, such as automobiles or aircrafts, it is important to reduce the conductor's overall weight. For cars, this means an increase in the use of formable, lightweight, high-strength steels and composites. But there is also a need for increased overall power as newer cars come equipped with many electronic features such as power windows, power doors, power liftgates, and advanced lighting, braking, and steering systems. Increasing the voltage means less current is needed ($V = IR$), so the total weight of the copper wires can be reduced. Less weight in both the wiring and the motors contributes to fuel efficiency. Increased peak power requirements, ride control, and advanced electronic systems are just some of the other reasons to consider the introduction of a 42-volt battery system.

EXAMPLE 18-3   *Drift Velocity of Electrons in Copper*

Assuming that all of the valence electrons contribute to current flow, (a) calculate the mobility of an electron in copper and (b) calculate the average drift velocity for electrons in a 100-cm copper wire when 10 V are applied.

SOLUTION

1. The valence of copper is one: therefore, the number of valence electrons equals the number of copper atoms in the material. The lattice parameter of copper is 3.6151×10^{-8} cm and, since copper is FCC, there are 4 atoms/unit cell. From Table 18-1, the resistivity $= 1/\sigma = 1/5.98 \times 10^5 = 1.67 \times 10^{-6} \ \Omega \cdot$ cm:

$$n = \frac{(4 \ \text{atoms/cell})(1 \ \text{electron/atom})}{(3.6151 \times 10^{-8} \ \text{cm})^3} = 8.466 \times 10^{22} \ \text{electrons/cm}^3$$

$$q = 1.6 \times 10^{-19} \ \text{C}$$

$$\mu = \frac{\sigma}{nq} = \frac{1}{\rho n q} = \frac{1}{(1.67 \times 10^{-6})(8.466 \times 10^{22})(1.6 \times 10^{-19})}$$

$$= 44.2 \ \frac{\text{cm}^2}{\Omega \cdot \text{C}} = 44.2 \ \frac{\text{cm}^2}{\text{V} \cdot \text{s}}$$

2. The electric field is:

$$E = \frac{V}{l} = \frac{10}{100} = 0.1 \ \text{V/cm}$$

The mobility is 44.2 cm^2/V $\cdot$ s; therefore:

$$\bar{v} = \mu E = (44.2)(0.1) = 4.42 \ \text{cm/s}$$

18-2   Band Structures of Solids

In Chapter 2, we noted that the electrons in a single atom occupy discrete energy levels. The Pauli exclusion principle permits each energy level to contain only two electrons. For example, the 2s level of a single atom contains one energy level and two electrons. The 2p level contains three energy levels and a total of six electrons.

When N atoms come together to produce a solid, the Pauli principle still requires that only two electrons in the entire solid have the same energy. When a solid is formed, the different split energy levels of electrons come together to form continuous bands of energies (Figure 18-3). Consequently, the 2s band in a solid contains N discrete energy levels and 2N electrons, two in each energy level. Each of the 2p levels contains N energy levels and 2N electrons. Since the three 2p bands actually overlap, we could alternately describe a single, broad 2p band containing 3N energy levels and 6N electrons.

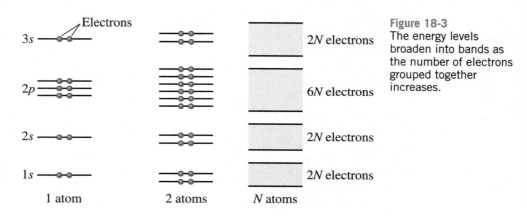

Figure 18-3
The energy levels broaden into bands as the number of electrons grouped together increases.

Band Structure of Sodium   Figure 18-4 shows an idealized picture of the band structure in sodium, which has an electronic structure of $1s^2 2s^2 2p^6 3s^1$. The energies within the bands depend on the spacing between the atoms; the vertical line represents the

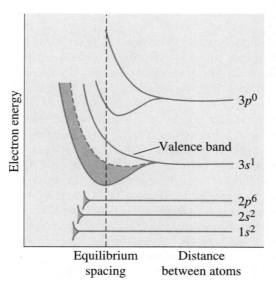

Figure 18-4
The simplified band structure for sodium. The energy levels broaden into bands. The 3s band, which is only half filled with electrons, is responsible for conduction in sodium.

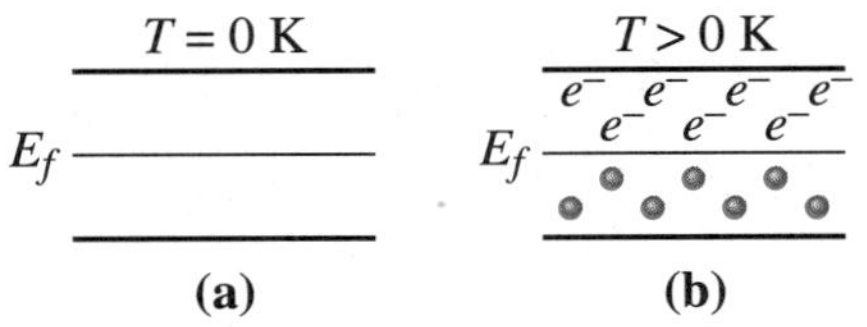

$T = 0 \text{ K}$ _____

E_f ————————

————————

(a)

$T > 0 \text{ K}$ ————————

$\overline{e^- \quad e^- \quad e^- \quad e^-}$
$e^- \quad e^- \quad e^-$

E_f ————————
○ ● ○ ● ○ ●
○ ● ○ ● ○ ●

(b)

Figure 18-5
(a) At absolute zero, all of the electrons in the outer energy level have the lowest possible energy. (b) When the temperature is increased, some electrons are excited into unfilled levels. Note that the Fermi energy is unchanged.

equilibrium interatomic spacing of the atoms in solid sodium. The 3s energy levels are the **valence band**. The empty 3p energy levels, which are separated from the 3s band by an energy gap, form the **conduction band**.

Sodium and other alkali metals in column 1A of the periodic table have only one electron in the outermost s level. The 3s valence band in sodium is half filled and, at absolute zero, only the lowest energy levels are occupied. The *Fermi energy* (E_f) is the energy at which half of the possible energy levels in the band are occupied by electrons.

But when the temperature of the metal increases, some electrons gain energy and are excited into the empty energy levels in the valence band (Figure 18-5). This condition creates an equal number of empty energy levels, or **holes**, vacated by the excited electrons [Figure 18-5(b)]. Only a small increase in energy is required to cause excitation of electrons. Both the excited electrons and the newly created holes can then carry an electrical charge.

Band Structure of Magnesium and Other Metals   Magnesium and other metals in of the periodic table have two electrons in their outermost s band. These metals have a high conductivity because the p band overlaps the s band at the equilibrium interatomic spacing. This overlap permits electrons to be excited into the large number of unoccupied energy levels in the combined 3s and 3p band. Overlapping 3s and 3p bands in aluminum and other metals in column 3B provide a similar effect.

In the transition metals, including scandium through nickel, an unfilled 3d band overlaps the 4s band. This overlap provides energy levels into which electrons can be excited; however, complex interactions between the bands prevent the conductivity from being as high as in some of the better conductors. However, in copper, the inner 3d band is full, and the atom core tightly holds these electrons. Consequently, there is little interaction between the electrons in the 4s and 3d bands, and copper has a high conductivity. A similar situation is found for silver and gold.

Band Structure of Semiconductors and Insulators   The elements in Group 4—carbon (diamond), silicon, germanium, and tin—contain two electrons in their outer p shell and have a valence of four. Based on our discussion in the previous section, we might expect these elements to have a high conductivity due to the unfilled p band, but this behavior is not observed!

These elements are covalently bonded; consequently, the electrons in the outer s and p bands are rigidly bound to the atoms. The covalent bonding produces a complex change in the band structure, or **hybridization**. The 2s and 2p levels of the carbon atoms in diamond can contain up to eight electrons, but there are only four valence electrons available. When carbon atoms are brought together to form solid diamond, the 2s and 2p levels interact and produce two bands (Figure 18-6). Each hybrid band can contain 4N electrons. Since there are only 4N electrons available, the lower (or valence) band is completely filled, whereas the upper (or conduction) band is empty.

A large **energy gap** or **bandgap** (E_g) separates the electrons from the conduction band in diamond $(E_g \sim 5.5 \text{ eV})$. Few electrons possess sufficient energy to jump the forbidden zone to the conduction band. Consequently, diamond has an electrical conduc-

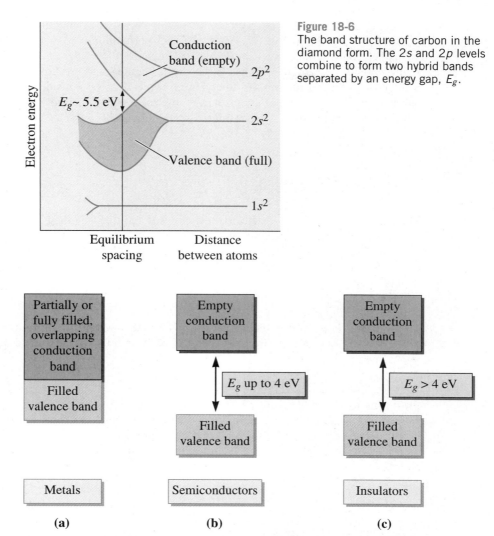

Figure 18-6
The band structure of carbon in the diamond form. The 2s and 2p levels combine to form two hybrid bands separated by an energy gap, E_g.

Figure 18-7 Schematic of band structures for (a) metals, (b) semiconductors, and (c) dielectrics or insulators. (Temperature is 0 K.)

tivity of less than 10^{-18} ohm$^{-1} \cdot$ cm^{-1}. Other covalently and ionically bonded materials have a similar band structure and, like diamond, behave as nonconductors of electricity. Increasing the temperature or an applied voltage supplies the energy required for electrons to overcome the energy gap. For example, the electrical conductivity of boron nitride increases from about 10^{-13} at room temperature to 10^{-4} ohm$^{-1} \cdot$ cm^{-1} at 800°C.

Thus, an important distinction between metals and semiconductors is that the conductivity of semiconductors *increases* with temperature, as more and more electrons can make it to the conduction band from the valence band. In other words, an increasing number of electrons from the covalent bonds in a semiconductor is freed and becomes available for conduction. The conductivity of most metals, on the other hand, *decreases* with increasing temperature. This is because the number of electrons that are already available begin to scatter more (i.e., increasing temperature reduces mobility).

Although germanium (Ge), silicon (Si), and α-Sn have the same crystal structure and band structure as diamond, the energy gap is smaller. In fact, the energy gap (E_g)

in α-Sn is so small ($E_g = 0.1$ eV) that α-Sn behaves as a metal. The energy gap is somewhat larger in silicon ($E_g = 1.1$ eV) and germanium ($E_g = 1.43$ eV)—these elements behave as semiconductors. Table 18-1 includes the electrical conductivity of these four elements. Figure 18-7 shows a schematic of the band structures of metals, semiconductors, and insulators or dielectrics. Typically, we consider materials with a bandgap greater than 4.0 eV as insulators, dielectrics, or nonconductors; materials with a bandgap less than 4 eV are considered semiconductors.

18-3   Conductivity of Metals and Alloys

The conductivity of a pure, defect-free metal is determined by the electronic structure of the atoms. But we can change the conductivity by influencing the mobility, μ, of the carriers. The mobility is proportional to the average drift velocity, $\bar{v}$, which is low if the electrons collide with imperfections in the lattice. The **mean free path** of electrons (λ_e) is defined as:

$$\lambda_e = \tau \bar{v} \tag{18-5c}$$

The average time between collisions is τ. The mean free path defines the average distance between collisions; a longer mean free path permits high mobilities and high conductivities.

Temperature Effect   When the temperature of a metal increases, thermal energy causes the atoms to vibrate (Figure 18-8). At any instant, the atom may not be in its equilibrium position, and it therefore interacts with and scatters electrons. The mean free path decreases, the mobility of electrons is reduced, and the resistivity increases. The change in resistivity of a pure metal as a function of temperature can be estimated from the equation:

$$\rho = \rho_{RT}(1 + \alpha_R \Delta T), \tag{18-6}$$

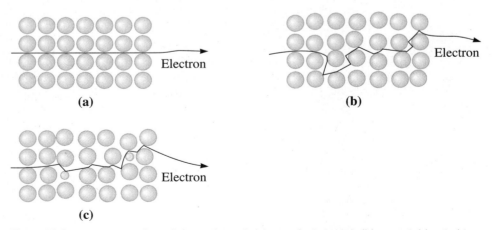

(a)

(b)

(c)

Figure 18-8   Movement of an electron through (a) a perfect crystal, (b) a crystal heated to a high temperature, and (c) a crystal containing atomic level defects. Scattering of the electrons reduces the mobility and conductivity.

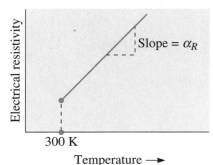

Figure 18-9
The effect of temperature on the electrical resistivity of a metal with a perfect crystal structure. The slope of the curve is the temperature resistivity coefficient.

where ρ is the resistivity at any temperature T, ρ_{RT} is the resistivity at room temperature (i.e., 25°C), $\Delta T = (T - T_{RT})$ is the difference between the temperature of interest and room temperature, and α_R is the *temperature resistivity coefficient*. The relationship between resistivity and temperature is linear over a wide temperature range (Figure 18-9). Examples of the temperature resistivity coefficient are given in Table 18-3.[1]

TABLE 18-3 ■ *The temperature resistivity coefficient α_R for selected metals[1]*

Metal	Room Temperature Resistivity (ohm · cm)	Temperature Resistivity Coefficient (α_R) (ohm/ohm · °C)
Be	4.0×10^{-6}	0.0250
Mg	4.45×10^{-6}	0.0037
Ca	3.91×10^{-6}	0.0042
Al	2.65×10^{-6}	0.0043
Cr	12.90×10^{-6} (0°C)	0.0030
Fe	9.71×10^{-6}	0.0065
Co	6.24×10^{-6}	0.0053
Ni	6.84×10^{-6}	0.0069
Cu	1.67×10^{-6}	0.0043
Ag	1.59×10^{-6}	0.0041
Au	2.35×10^{-6}	0.0035
Pd	10.8×10^{-6}	0.0037
W	5.3×10^{-6} (27°C)	0.0045
Pt	9.85×10^{-6}	0.0039

(Source: Reprinted by permission from Handbook of Electronic Materials, *by P.S. Neelakanta, pp. 215–216, Table 9-1. Copyright © 1995 CRC Press, Boca Raton, Florida.)*

The following example illustrates how resistivity of pure copper can be calculated.

EXAMPLE 18-4 *Resistivity of Pure Copper*

Calculate the electrical conductivity of pure copper at (a) 400°C and (b) −100°C.

SOLUTION

Since the conductivity of pure copper is $5.98 \times 10^5 \ \Omega^{-1} \cdot cm^{-1}$ the resistivity of copper at room temperature is 1.67×10^{-6} ohm · cm. The temperature resistivity coefficient is 0.0068 ohm · cm/°C. (See Table 18-1).

1. At 400°C:

$$\rho = \rho_{RT}(1 + \alpha_R \Delta T) = (1.67 \times 10^{-6})[1 + 0.0068(400 - 25)]$$

$$\rho = 5.929 \times 10^{-6} \text{ ohm} \cdot \text{cm}$$

$$\sigma = 1/\rho = 1.69 \times 10^{5} \text{ ohm}^{-1} \cdot \text{cm}^{-1}$$

2. At −100°C:

$$\rho = (1.67 \times 10^{-6})[1 + 0.0068(-100 - 25)] = 0.251 \times 10^{-6} \text{ ohm} \cdot \text{cm}$$

$$\sigma = 39.8 \times 10^{5} \text{ ohm}^{-1} \cdot \text{cm}^{-1}$$

Effect of Atomic Level Defects Imperfections in crystal structures scatter electrons, reducing the mobility and conductivity of the metal [Figure 18-8(c)]. For example, the increase in the resistivity due to solid solution atoms is

$$\rho_d = b(1 - x)x \qquad (18\text{-}7)$$

where ρ_d is the increase in resistivity due to the defects, x is the atomic fraction of the impurity or solid solution atoms present, and b is the defect resistivity coefficient. In a similar manner, vacancies, dislocations, and grain boundaries reduce the conductivity of the metal. Each defect contributes to an increase in the resistivity of the metal. Thus, the overall resistivity is

$$\rho = \rho_T + \rho_d \qquad (18\text{-}8)$$

where ρ_d equals the contributions from all of the defects. Equation 18-8 is known as the **Matthiessen's rule**. The effect of the defects is *independent* of temperature (Figure 18-10).

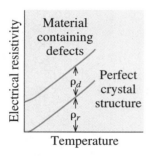

Figure 18-10
The electrical resistivity of a metal is composed of a constant defect contribution ρ_d and a variable temperature contribution ρ_T.

Effect of Processing and Strengthening Strengthening mechanisms and metal processing techniques affect the electrical properties of a metal in different ways (Table 18-4). Solid-solution strengthening is *not* a good way to obtain high strength in metals intended to have high conductivities. The mean free paths are very short due to the random distribution of the interstitial or substitutional atoms. Figure 18-11 shows the effect of zinc and other alloying elements on the conductivity of copper; as the amount of alloying element increases, the conductivity decreases substantially.

Age hardening and dispersion strengthening reduce the conductivity to an extent that is less than solid-solution strengthening, since there is a longer mean free path between precipitates, as compared with the path between point defects. Strain hardening and grain-size control have even less effect on conductivity (Figure 18-11 and Table

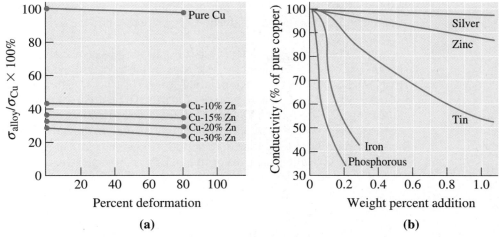

Figure 18-11 (a) The effect of solid-solution strengthening and cold working on the electrical conductivity of copper, and (b) the effect of addition of selected elements on the electrical conductivity of copper.

18-4). Since dislocations and grain boundaries are further apart than solid-solution atoms, there are large volumes of metal that have a long mean free path. Consequently, cold working is an effective way to increase the strength of a metallic conductor without seriously impairing the electrical properties of that material. In addition, the effects of

TABLE 18-4 ■ *The effect of alloying, strengthening, and processing on the electrical conductivity of copper and its alloys*

Alloy	$\dfrac{\sigma_{alloy}}{\sigma_{Cu}} \times 100$	Remarks
Pure annealed copper	100	Few defects to scatter electrons; the mean free path is long.
Pure copper deformed 80%	98	Many dislocations, but because of the tangled nature of the dislocation networks, the mean free path is still long.
Dispersion-strengthened Cu-0.7% Al_2O_3	85	The dispersed phase is not as closely spaced as solid-solution atoms, nor is it coherent, as in age hardening. Thus, the effect on conductivity is small.
Solution-treated Cu-2% Be	18	The alloy is single phase; however, the small amount of solid-solution strengthening from the supersaturated beryllium greatly decreases conductivity.
Aged Cu-2% Be	23	During aging, the beryllium leaves the copper lattice to produce a coherent precipitate. The precipitate does not interfere with conductivity as much as the solid-solution atoms.
Cu-35% Zn	28	This alloy is solid-solution strengthened by zinc, which has an atomic radius near that of copper. The conductivity is low, but not as low as when beryllium is present.

cold working on conductivity can be eliminated by the low temperature recovery heat treatment, in which good conductivity is restored while the strength is retained.

Conductivity of Alloys Alloys typically have higher resistivities than pure metals because of the scattering of the electrons in the added alloys. For example, the resistivity of pure Cu at room temperature is $\sim 1.7 \times 10^{-6}$ $\Omega \cdot$ cm and that of pure gold is $\sim 1.56 \times 10^{-6}$ $\Omega \cdot$ cm. The resistivity of a 50% Au-50% Cu alloy at room temperature is much higher, $\sim 15 \times 10^{-6}$ $\Omega \cdot$ cm. Ordering of atoms in alloys by heat treatment can decrease their resistivity. Compared to pure metals, the resistivity of alloys tends to be stable in regards to temperature variation. Alloys such as nichrome ($\sim$80% Ni-20% Cr) can be used as heating elements or to make thermocouple wires. Certain alloys of Bi-Sn-Pb-Cd are used to make electrical fuses due to their low melting ranges.

18-4 Superconductivity

A **superconductor** is a material that exhibits zero electrical resistance under certain conditions and expels a magnetic field completely (i.e., a superconductor is also a perfect diamagnet (see Chapter 19)). This is known as the Meissner or Meissnner-Ochsenfeld effect.[2]

We have discussed many examples of materials and phenomena discoveries that were the result of serendipity. The discovery of the superconductivity is one more such example. In 1911, Heike Kammerlingh Onnes found that the resistance of mercury vanished when it was cooled to liquid helium (He) temperatures ($\sim$4 K).[3] Onnes was actually working on the liquefaction of helium and was not investigating superconductivity. Following this discovery, many pure metals (e.g., Nb) have been found to be superconductive. Another breakthrough in superconductivity came, again as a result of serendipity, when Bendorz and Müller discovered the first high T_c ceramic superconductor based on a La-Sr-Cu oxide. This was followed by the discovery of the $YBa_2Cu_3O_{7-x}$ *(YBCO)* superconductor by Chu and his coworkers.[4]

The origin of **superconductivity** is related to electron-phonon coupling and the resultant pairing of conduction electrons. This is explained using Bardeen, Cooper, and Schrifer or BCS theory. According to this theory, superconductiviy occurs as a result of the formation of pairs of electrons known as Cooper pairs. This theory explains the properties of superconductivity in metallic superconductors; however, there has been no theory put forward to date to fully explain the existence of superconductivity in the ceramic materials.

Superconductivity in materials disappears above a certain temperature known as the critical temperature (T_c). The change from normal conduction to superconductivity occurs abruptly at this critical temperature T_c (Figure 18-12) the superconductivity also disappears when a certain level of magnetic field is present. There also exists a critical current density T_c above which superconducting state can not be maintained.

The applied magnetic field (H) induces a magnetic-flux density (B). The unit for B is Tesla. We sometimes represent the critical field in the form of a critical-flux density (B).

In the Type I superconductors, superconductivity is stable up to a magnetic field, H_c [Figure 18-13(a)]. Thus, for superconductivity to exist, the temperature must be below T_c and the magnetic field must be below a certain value (H_c). In order for a material to be a superconductor, the magnetic field must be excluded from the conduc-

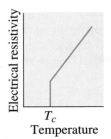

Figure 18-12
The electrical resistivity of a superconductor becomes zero below some critical temperature T_c.

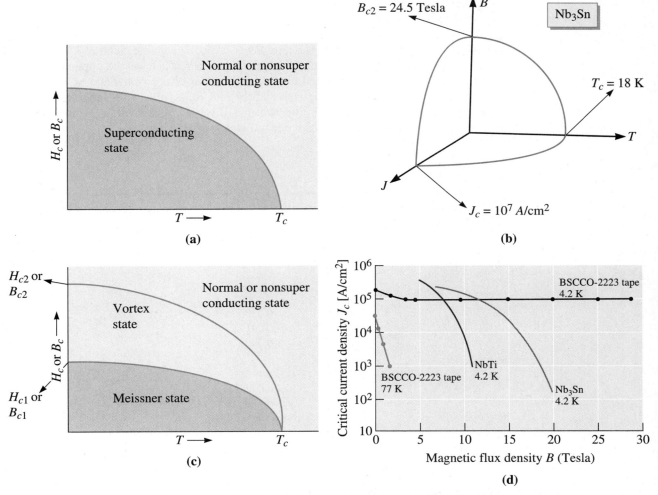

Figure 18-13 (a) The effect of a magnetic field on the temperature below which superconductivity occurs for a Type I superconductor. (b) The critical surface of Nb_3Sn, a Type II superconductor. The superconducting envelope or surface showing the combined effects of temperature, magnetic field, and current density on a Nb_3Sn superconductor. Conditions within the envelope produce superconductivity. (c) The effect of a magnetic field on the temperature below which superconductivity occurs for a Type II superconductor. (d) Critical-current density as a function of magnetic-flux density (in Tesla) created by the applied magnetic fields (for different superconductors). (*Source: From* "High Tc Super-conductors for Magnet and Energy Technology," *by B.R. Lehndorff. In* Springer Tracts in Modern Physics, *p. 6, Fig. 2-1. Copyright © 2001 Springer.*)

tor. This explains the ability of superconductors to levitate above a magnet. For Type I superconductors, which include tungsten and tin, increasing the magnetic field decreases the temperature required for superconductivity. For a magnetic field greater than H_c, lines of magnetic field enter the conductor and completely suppress superconductivity. The effect of the temperature on the critical field is given by

$$H_c = H_0 \left[1 - \left(\frac{T}{T_c} \right)^2 \right] \tag{18-9}$$

where H_0 is the critical magnetic field at 0 K [Figure 18-13(a)]. Combinations of temperatures and magnetic fields within the envelope created by H_c and T_c provide superconductivity in Type I materials. Most Type I superconductors lose their superconducting behavior, even in very weak magnetic fields.

In Type II superconductors, when a certain lower-critical magnetic field H_{c1} is applied, the magnetic flux can penetrate into the superconductor. Then, as the applied magnetic field increases to a higher value, known as the upper-critical magnetic field H_{c2}, the superconductivity disappears [Figure 18-13(b) and (c)]. Table 18-5 includes a list of some superconductors and their T_c and B_c (or B_{c2}) values. As shown in Figure 18-13(c) B_c and B_{c_2} are flux densities at which normal state appears.

TABLE 18-5 ■ *The critical temperature, magnetic field, and current density for superconductivity of selected materials*

Material	T_c (K)	B_c or B_{c2} (Tesla)	J_c (A/cm^2) at 0 K
Type I superconductors:			
W	0.015	—	
Al	1.180	0.0099	
Sn	3.720	0.0300	
Hg	4.150	0.0410	
Pb	7.190	0.0800	
Type II superconductors:			
Nb	9.250	0.1990	$\sim 10^5$
Nb$_3$Sn	18.050	24.5 at 0 K	10^7
Nb$_3$Ge	23.2	38	
MgB$_2$ (Ref. [5])	39	~ 10 at 4.2 K	
Ceramic superconductors:			
(La, Sr)$_2$CuO$_4$	40.000		
YBa$_2$Cu$_3$O$_{7-x}$			
(YBCO)	93.000	300	10^4–10^7
TlBa$_2$Ca$_3$Cu$_4$O$_{11}$	122.000		
Bi$_2$Sr$_2$Ca$_2$Cu$_3$O$_{10}$			
(BSCCO2223)	110		
Bi$_2$Sr$_2$CaCu$_2$O$_8$			
(BSCCO2212)	92		
HgBa$_2$Ca$_2$Cu$_3$O$_8$	133–135		
Hg$_{0.8}$Tl$_{0.2}$Ba$_2$Ca$_2$Cu$_3$O$_{8.33}$	138 (Highest recorded at ambient pressure)		

What Limits the Current Density in Superconductors? We defined superconductors as materials that show zero electrical resistance under certain conditions. This would suggest that an unlimited amount of current should flow through a superconductive wire, since no resistive heating of the wire occurs. With superconductors, the higher the value of the magnetic field present, the lower the critical current density (J_c) or maximum amount of current per unit area [Figure 18-13(b) and (d)].[7] Even at zero magnetic fields, the superconductor cannot carry an infinite amount of current because, in polycrystalline superconductors, the current cannot flow without resistance from one grain to another. Thus, grain boundaries and grain orientation or texture (Chapter 7) are very important in both the bulk and thin-film forms of superconductors.[8] The motion of the magnetic-flux lines limits the current transport within the grains.

Certain atomic level defects in superconductors can pin the magnetic vortices created and these, in turn, help *increase* the maximum current carrying capacity of the superconductor. In addition to the transport of grains across grain boundaries (known as the "weak link" effect), the transport of current within the grains, or the transport of current, may also be limited by percolation. This problem arises if there are pores or grains of second phases that can block the supercurrent. Thus, the current density in superconductors is indeed a microstructure-sensitive property.

Conventional ceramic and metallurgical processing techniques or chemical routes, (such as precipitation (Chapter 14) or freeze drying (Chapter 9)) as well as many thin-film processes, such as ion-beam-assisted deposition (IBAD) and laser ablation, have been used for processing ceramic superconductors into a thin-film form, sintered products, and long length wires.

Magnesium diboride (MgB_2), a previously known compound, was found to be a Type II superconductor in 2001.[5] MgB_2 is similar to the traditional metallic superconductors, however its T_c and B_{c_2} are relatively high and it does not seem to suffer from the "weak link" effects seen in high T_c ceramic superconductors. Recently, scientists from Bell Laboratories have claimed that poylthiophene an organic compound, shows superconductivity ($T_c = 2.6$ K).

Applications of Superconductors Under certain conditions superconductors offer zero electrical resistance, which means we can consider superconductors for electrical applications (motors, cables for transmission of power, etc.). Electronic circuits have also been built using superconductors and powerful superconducting electromagnets are used in magnetic resonance imaging (MRI). Also, very low electrical-loss components, known as filters, based on ceramic superconductors have been developed for wireless communications.

Until about 1986, liquid helium ($T = 4$ K) was required to cool the materials below the critical temperature and, consequently, the applications for superconductors were limited. However, a group of ceramic materials (such as those shown in Table 18-5) were discovered whose critical temperatures exceed 77 K, permitting cooling by relatively inexpensive liquid nitrogen. These materials include $YBa_2Cu_3O_{7-x}$, where x indicates that some oxygen ions are missing from the complicated perovskite crystal structure. Similar behavior has been found for a variety of other ceramics, including some (such as $TlBa_2Ca_3Cu_4O_{11}$) that have critical temperatures above 100 K. New discoveries, such as the superconductivity in MgB_2 and carbon nanotubes (Figure 18-14), may also bring new applications previously not considered possible. The following example illustrates designing for superconductors.

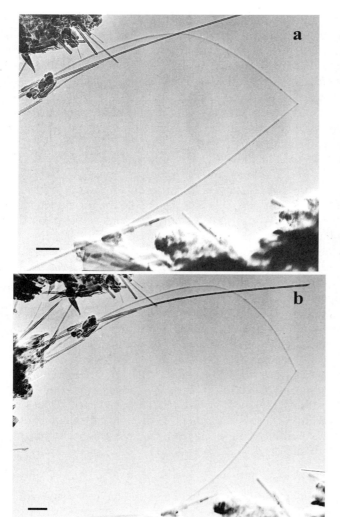

Figure 18-14
Transmission electron micrographs of carbon nanotubes *(CNT)*. The micrographs shows two carbon nanotubes that are bent considerably (shows they are highly elastic). The ends of these nanotubes were "welded" using a nanomanipulator in the transmission electron microscope. These materials are promising for microelectronics and other applications. (*Micrograph Courtesy of Professor Vinayak Dravid, Northwestern University*).

EXAMPLE 18-5   *Design of a Superconductor System*

Design the limiting magnetic field that will permit niobium to serve as a superconductor at liquid helium temperatures. Assume $H_0 = 1970$ oersteds.

SOLUTION

From Table 18-5, $T_c = 9.25$ K. We are given the value of and $H_0 = 1970$ oersted. Since the operating temperature will be 4 K, we need to determine the maximum permissible magnetic field, using Equation 18-9:

$$H_c = H_0[1 - (T/T_c)^2]$$

$$H_c = 1970[1 - (4/9.25)^2] = (1970)(0.813) = 1602 \text{ oersted}$$

Therefore, the magnetic field (H) must remain below 1602 oersted in order for the niobium to remain superconductive.

18-5   Conductivity in Other Materials

Electrical conductivity in most ceramics and polymers is normally low. However, special materials provide limited or even good conduction. In Chapter 4, we saw how the Kröger-Vink notation can be used to explain defect chemistry in ceramic materials. Using dopants, it is possible to take many ceramics (e.g., $BaTiO_3$, TiO_2, ZrO_2) that are normally insulating and convert them into conductive oxides.[9] The conduction in these materials can occur as a result of movement of ions or electrons and holes.

Conduction in Ionic Materials   Conduction in ionic materials often occurs by movement of entire ions, since the energy gap is too large for electrons to enter the conduction band. Therefore, most ionic materials behave as insulators.

In ionic materials, the mobility of the charge carriers, or ions, is

$$\mu = \frac{ZqD}{k_B T}, \tag{18-10}$$

where D is the diffusion coefficient, k_B is Boltzmann's constant (Table 18-2), T is the absolute temperature, q is the electronic charge (Table 18-2), and Z is the valence of the ion. The mobility is many orders of magnitude lower than the mobility of electrons; hence, the conductivity is very small:

$$\sigma = n \cdot Z \cdot q \cdot \mu \tag{18-11}$$

Impurities and vacancies increase conductivity; vacancies are necessary for diffusion in substitutional types of crystal structures, and impurities can diffuse and help carry the current. High temperatures increase conductivity because the rate of diffusion increases. The following example illustrates the estimation of mobility and conductivity in MgO.

EXAMPLE 18-6   *Ionic Conduction in MgO*

Suppose that the electrical conductivity of MgO is determined primarily by the diffusion of the Mg^{2+} ions. Estimate the mobility of the Mg^{2+} ions and calculate the electrical conductivity of MgO at 1800°C.

SOLUTION

From Figure 5-9, the diffusion coefficient of Mg^{2+} ions in MgO at 1800°C is 10^{-10} cm²/s. For MgO, $Z = 2$/ion, $q = 1.6 \times 10^{-19}$ C, $k_B = 1.38 \times 10^{-23}$ J/K (from Table 18-2), and $T = 2073$ K:

$$\mu = \frac{ZqD}{k_B T} = \frac{(2)(1.6 \times 10^{-19})(10^{-10})}{(1.38 \times 10^{-23})(2073)} = 1.12 \times 10^{-9} \ C \cdot cm^2/J \cdot s$$

Since the charge in coulomb is equivalent to Ampere · second, and Joules is equivalent to Amp · sec · volts:

$$\mu = 1.12 \times 10^{-9} \ cm^2/V \cdot s$$

MgO has the NaCl structure, with four magnesium ions per unit cell. The lattice parameter is 3.96×10^{-8} cm, so the number of Mg^{2+} ions per cubic centimeter is:

$$n = \frac{4 \text{ Mg}^{2+} \text{ ions/cell}}{(3.96 \times 10^{-8} \text{ cm})^3} = 6.4 \times 10^{22} \text{ ions/cm}^3$$

$$\sigma = nZq\mu = (6.4 \times 10^{22})(2)(1.6 \times 10^{-19})(1.12 \times 10^{-9})$$

$$= 22.94 \times 10^{-6} \text{ C} \cdot \text{cm}^2/\text{cm}^3 \cdot \text{V} \cdot \text{s}$$

Since charge in coulombs (C) is equivalent to ampere · second (A · S) and voltage is equivalent to ampere · ohm (A · Ω),

$$\sigma = 2.294 \times 10^{-5} \text{ ohm}^{-1} \cdot \text{cm}^{-1}$$

Applications of Ionically Conductive Oxides The most widely used conductive and transparent oxide is indium tin oxide *(ITO)*, used as a transparent conductive coating on plate glass.[10] Other applications of *ITO* include touch screen displays for computers and devices such as automated teller machines. Other conductive oxides include ytrria-stabilized zirconia *(YSZ)*, which is used as a solid electrolyte in solid oxide fuel cells.[11] Lithium cobalt oxide is used as a solid electrolyte in lithium ion batteries.[12] It is important to remember that, although most ceramic materials behave as electrical insulators, by properly engineering the point defects in ceramics it is possible to convert many of them into semiconductors.

Conduction in Polymers Because their valence electrons are involved in covalent bonding, polymers have a band structure with a large energy gap, leading to low-electrical conductivity. Polymers are frequently used in applications that require electrical insulation to prevent short circuits, arcing, and safety hazards. Table 18-1 includes the conductivity of four common polymers. In some cases, however, the low conductivity is a hindrance. For example, static electricity can accumulate on housings for electronic equipment, making the polymer transparent to electromagnetic radiation that damages the internal solid-state devices. If lightning strikes the polymer-matrix composite wing of an airplane, severe damage can occur. We can solve these problems by two approaches: (1) introducing an additive to the polymer to improve the conductivity, and (2) creating polymers that inherently have good conductivity. In some applications, such as liquid crystal displays (LCDs), we need polymers that respond to application of an electronic field (Chapter 15).

Adding ionic compounds to the polymer can reduce resistivity because the ions migrate to the polymer surface and attract moisture, which, in turn, dissipates static charges. Introducing conductive filler materials, such as carbon black, can also dissipate static charge. Polymer-matrix composites containing carbon or nickel-plated carbon fibers combine high stiffness with improved conductivity; hybrid composites containing metal fibers, along with normal carbon, glass, or aramid fibers, also produce lightning-safe aircraft skins. Figure 18-15 shows that when enough carbon fibers are introduced to nylon in order to assure fiber-to-fiber contact, the resistivity is reduced by nearly 13 orders of magnitude. Conductive fillers and fibers are also used to produce polymers that shield against electromagnetic radiation.

Some polymers inherently have good conductivity as a result of doping or processing techniques. When acetal polymers are doped with agents such as arsenic pentafluoride, electrons or holes are able to jump freely from one atom to another along the backbone of the chain, increasing the conductivity to near that of metals. Some polymers, such as polyphthalocyanine, can be cross-linked by special curing processes to raise the conductivity to as high as 10^2 ohm^{-1} · cm^{-1}, a process that permits the

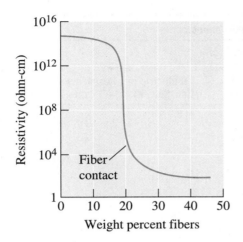

Figure 18-15
Effect of carbon fibers on the electrical resistivity of nylon.

polymer to behave as a semiconductor. Because of the cross-linking, electrons can move more easily from one chain to another.

Recently, advances have been made when polyanilines were found to exhibit semiconductivity, and many semiconducting polymers were transformed into light emitting diodes (LEDs).[13] Alan J. Heeger, Alan G. MacDiarmid, and Hideki Shirakawa won the 2000 Nobel prize in Chemistry for their discovery of conducting polymers. As mentioned in the previous section, recently a superconducting polymer (poylthiophene) has also been claimed.

Other exciting developments in polymeric materials used in electronics include the development of carbon nanotubes by Ijima.[14] A carbon nanotube is a small single or multiwalled structure made from carbon (Figure 18-14). These materials are being developed for electronics (they can be made either conducting or insulating), hydrogen storage, and other applications. Some have suggested that carbon nanotubes are 50 to 100 times stronger than steels. Such comparisons, however, do not account for the distribution of flaw sizes and are thus not appropriate.

18-6 Semiconductors

In previous chapters (e.g., Chapter 5), we discussed the concepts related to intrinsic and extrinsic semiconductors. An **intrinsic semiconductor** is one whose properties are not controlled by impurities or dopants. An **extrinsic semiconductor** (*n* or *p* type) is preferred for devices since its properties are temperature stable and can be controlled using ion implantation or diffusion of dopants (Chapter 5). Elemental semiconductors are elements such as germanium or silicon. Compound semiconductors are formed from elements in groups 2B and 6B of the periodic table (e.g., CdS, CdSe, CdTe, HgCdTe, etc.) and are known as II–VI (two–six) semiconductors. They can also be formed by combining elements from groups 3B and 5B of the periodic table (e.g., GaN, GaAs, AlAs, AlP, InP, etc.).[15] These are known as III–V (three–five) semiconductors. Semiconductor materials, including silicon and germanium, provide the building blocks for many electronic devices. These materials have an easily controlled electrical conductivity and, when properly combined, can act as switches, amplifiers, or information

Table 18-6 ■ *Properties of commonly encountered semiconductors*

Semiconductor	Bandgap eV	Mobility of Electrons (μ_n) $\dfrac{cm^2}{V\text{-}s}$	Mobility of Holes (μ_p) $\dfrac{cm^2}{V\text{-}s}$	Dielectric Constant (k)	Resistivity $\Omega \cdot cm$	Density $\dfrac{gm}{cm^3}$	Melting Temperature °C
Silicon (Si)	1.11	1350	480	11.8	2.5×10^5	2.33	1415
Amorphous Silicon (a:Si:H)	1.70	1	10^{-2}	~11.8	10^{10}	~2.30	—
Germanium (Ge)	0.67	3900	1900	16.0	43	5.32	936
SiC (α)	2.86	500		10.2	10^{10}	3.21	2830
Gallium Arsenide (GaAs)	1.43	8500	400	13.2	4×10^8	5.31	1238
Diamond	~5.50	1800	1500	5.7	$>10^{18}$	3.52	~4200
α-Sn	0.10	2000	1000	—	10^{-4}	5.80	232

storage devices. The properties of some of the commonly encountered semiconductors are included in Table 18-6.

The energy gap E_g between the valence and conduction bands in semiconductors is relatively small (Figure 18-7). As a result, some electrons possess enough thermal energy to exceed the gap and enter the conduction band. The excited electrons leave behind unoccupied energy levels, or holes, in the valence band. When an electron moves to fill a hole, another hole is created from the original electron source; consequently, the holes appear to act as positively charged electrons and carry an electrical charge. When a voltage is applied to the material, the electrons in the conduction band accelerate toward the positive terminal, while holes in the valence band move toward the negative terminal (Figure 18-16). Current is, therefore, conducted by the movement of both electrons and holes.

The conductivity is determined by the number of electron-hole pairs,

$$\sigma = n \cdot q \cdot \mu_n + p \cdot q \cdot \mu_p \tag{18-12}$$

where n is the concentration of electrons in the conduction band, p is the concentration of holes in the valence band, and μ_n and μ_p are the mobilities of electrons and holes, respectively (Table 18-6). This equation is the same as 18-5b.

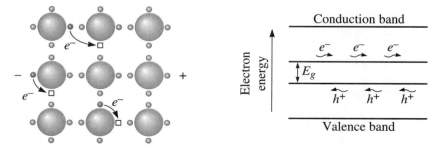

Figure 18-16   When a voltage is applied to a semiconductor, the electrons move through the conduction band, while the electron holes move through the valence band in the opposite direction.

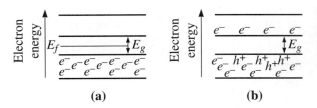

Figure 18-17
The distribution of electrons and holes in the valence and conduction bands (a) at absolute zero and (b) at an elevated temperature.

In intrinsic semiconductors, for every conduction electron promoted to the conduction band, there is a hole left in the valence band, therefore:

$$n_i = p_i$$

Therefore, the conductivity of an intrinsic semiconductor is:

$$\sigma = qn_i(\mu_e + \mu_h) \tag{18-13}$$

In intrinsic semiconductors, we control the number of charge carriers and, hence, the electrical conductivity by controlling the temperature. At absolute zero temperature, all of the electrons are in the valence band, whereas all of the levels in the conduction band are unoccupied [Figure 18-17(a)]. As the temperature increases, there is a greater probability that an energy level in the conduction band is occupied (and an equal probability that a level in the valence band is unoccupied, or that a hole is present) [Figure 18-17(b)]. The number of electrons in the conduction band, which is equal to the number of holes in the valence band, is given by

$$n = n_i = p_i = n_0 \exp\left(\frac{-E_g}{2k_B T}\right) \tag{18-14a}$$

where n_0 is related to the density of states and the effective masses of electrons and holes.

$$n_0 = 2\left(\frac{2\pi k_B T}{h^2}\right)^{3/2} (m_n^* m_p^*)^{3/4} \tag{18-14b}$$

In this equation, k and h are Boltzmann and Planck's constants and m_n^* and m_p^* are the effective masses of electrons and holes in the semiconductor respectively. For Ge, Si, and GaAs, the room temperature values of n_i are: 2.5×10^{13}, 1.5×10^{10}, and 2×10^6 electrons/cm^3, respectively. The $n_i \times p_i$ product remains constant at any given temperature and a given semiconductor. This allows us to calculate n_i or p_i values at different temperatures.

Higher temperatures permit more electrons to cross the forbidden zone and, hence, the conductivity increases:

$$\sigma = n_0 q(\mu_e + \mu_h) \exp\left(\frac{-E_g}{2k_B T}\right) \tag{18-15}$$

Note that both n and σ are related to temperature by an Arrhenius equation, rate $= A \exp\left(\frac{-Q}{RT}\right)$. As the temperature increases, the conductivity of a semiconductor also increases because more charge carriers are present, whereas the conductivity of a metal decreases due to the lower mobility of the charge carriers. The example that follows shows the calculation for carrier concentration in an intrinsic semiconductors.

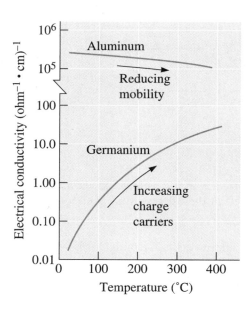

Figure 18-18
The electrical conductivity versus temperature for intrinsic semiconductors compared with metals. Note the break in y-axis scale.

EXAMPLE 18-7   *Carrier Concentrations in Intrinsic Ge*

For germanium at 25°C, estimate (a) the number of charge carriers, (b) the fraction of the total electrons in the valence band that are excited into the conduction band, and (c) the constant n_0.

SOLUTION

From Table 18-6, $\rho = 43\ \Omega\text{-cm}$, $\therefore\ \sigma = 0.023\ \Omega^{-1} \cdot \text{cm}^{-1}$

Also from Table 18-6, $E_g = 0.67$ eV

$$\mu_n = 3900\,\frac{\text{cm}^2}{\text{V} \cdot \text{s}} \quad \mu_p = 1900\,\frac{\text{cm}^2}{\text{V} \cdot \text{s}}$$

$2k_B T = (2)(8.63 \times 10^{-5}\ \text{eV/K})(273 + 25) = 0.0514\ \text{eV}$ at $T = 25°C$

1. From Equation 18-12:

$$n = \frac{\sigma}{q(\mu_n + \mu_p)} = \frac{0.023}{(1.6 \times 10^{-19})(3900 + 1900)} = 2.5 \times 10^{13}\,\frac{\text{electrons}}{\text{cm}^3}$$

There are 2.5×10^{13} electrons/cm^3 and 2.5×10^{13} holes/cm^3 helping to conduct a charge in germanium at room temperature.

2. The lattice parameter of diamond cubic germanium is 5.6575×10^{-8} cm. The total number of electrons in the valence band of germanium is:

$$\text{Total electrons} = \frac{(8\ \text{atoms/cell})(4\ \text{electrons/atom})}{(5.6575 \times 10^{-8}\ \text{cm})^3}$$

$$= 1.77 \times 10^{23}$$

$$\text{Fraction excited} = \frac{2.5 \times 10^{13}}{1.77 \times 10^{23}} = 1.41 \times 10^{-10}$$

3. From Equation 18-14(a):

$$n_0 = \frac{n}{\exp(-E_g/2k_B T)} = \frac{2.5 \times 10^{13}}{\exp(-0.67/0.0514)}$$

$$= 1.14 \times 10^{19} \text{ carriers/cm}^3$$

Extrinsic Semiconductors The temperature dependence of conductivity in intrinsic semiconductors is nearly exponential, but this is not useful for practical applications. We cannot accurately control the behavior of an intrinsic semiconductor because slight variations in temperature can significantly change the conductivity. However, by intentionally adding a small number of impurity atoms to the material (called **doping**), we can produce an extrinsic semiconductor. The conductivity of the extrinsic semiconductor depends primarily on the number of impurity, or dopant, atoms and, in a certain temperature range, is independent of temperature. This ability to have a tunable yet temperature independent conductivity is the reason why we almost always use extrinsic semiconductors to make devices.

n-type Semiconductors Suppose we add an impurity atom such as antimony (which has a valence of five) to silicon or germanium. Four of the electrons from the antimony atom participate in the covalent-bonding process, while the extra electron enters an energy level in a donor state just below the conduction band (Figure 18-19). Since the extra electron is not tightly bound to the atoms, only a small increase in energy, E_d, is required for the electron to enter the conduction band. (E_d is often defined as the energy difference between the top of the valence band and the donor band. In this case, the energy increase required would be defined as $E_g - E_d$.) The energy gap controlling conductivity is now E_d rather than E_g (Table 18-7). No corresponding holes are created when the donor electrons enter the conduction band.

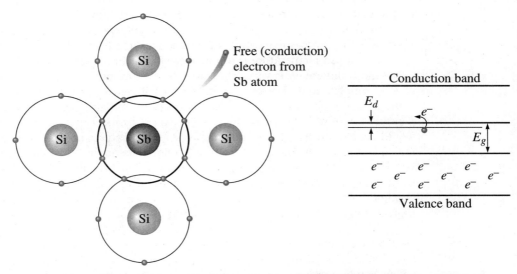

Figure 18-19 When a dopant atom with a valence greater than four is added to silicon, an extra electron is introduced and a donor energy state is created. Now electrons are more easily excited into the conduction band.

TABLE 18-7 ■ *The donor and acceptor energy gaps (in electron volts) when silicon and germanium semiconductors are doped*

Dopant	Silicon		Germanium	
	E_d	E_a	E_d	E_a
P	0.045		0.0120	
As	0.049		0.0127	
Sb	0.039		0.0096	
B		0.045		0.0104
Al		0.057		0.0102
Ga		0.065		0.0108
In		0.160		0.0112

Some intrinsic semiconduction still occurs, with a few electrons gaining enough energy to jump the large E_g gap. However, this becomes important only at higher temperatures. In an extrinsic semiconductor, there has to be overall electrical neutrality. Thus, the number of donor atoms (N_d) and holes per unit volume (p_0) (both are positively charged) is equal to the number of acceptor atoms (N_a) and electrons per unit volume (n_0) (both are negatively charged).

$$p_0 + N_d = n_0 + N_a$$

In this equation, n_0 and p_0 are the concentrations of electrons and holes in an extrinsic semiconductor.

If the extrinsic semiconductor is heavily *n*-type doped (i.e., $N_d \gg n_i$) and therefore $n_0 \sim N_d$. Similarly, if there is a heavily acceptor-doped (*p*-type) semiconductor then $N_a \gg p_i$ and hence $p_0 \sim N_a$. This is important since this says that by adding a considerable amount of dopant we can, dominate the conductivity of a semiconductor by controlling the dopant concentration. This is why extrinsic semiconductors are most useful for making controllable devices such as transistors.

The changes in carrier concentration with temperature are shown in Figure 18-20. From this the approximate conductivity changes in an extrinsic semiconductor are easy to follow. When temperature is too low, the donor or acceptor ions are not ionized and hence the conductivity is very small. As temperature begins to increase, ionization process is complete and electrons (or holes) contributed by the donors (or acceptors) become available for conduction. The conductivity remains nearly independent of temperature (region labeled as extrinsic). The value of conductivity at which plateau occurs would depend on the level of doping. When temperatures become too high, the behavior now approaches that of an intrinsic semiconductor since the effect of dopants is essentially lost. In this analysis, we have not accounted for the effects of dopant concentrations on the mobility of electrons and holes and the temperature dependence of bandgap. At very high temperatures (not shown in Figure 18-20) the conductivity *decreases* again as scattering of carriers dominates.

p-Type Semiconductors   When we add an impurity such as gallium or boron, which has a valence of three, to Si or Ge, there are not enough electrons to complete the covalent bonding process. An electron hole is created in the valence band that can be filled by electrons from other locations in the band (Figure 18-21). The holes act as acceptors of electrons. These hole sites have a somewhat higher than normal energy and create an

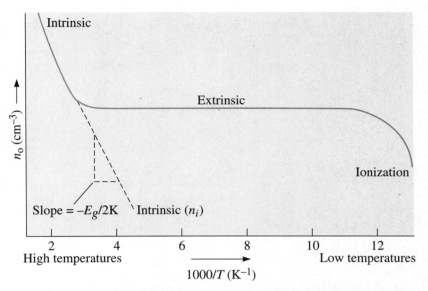

Figure 18-20 The effect of temperature on the carrier concentration of an *n*-type semiconductor. At low temperatures, the donor or acceptor atoms are not ionized. As temperature increases, the ionization process is complete and the carrier concentration increases to a level that is dictated by the level of doping. The conductivity then essentially remains unchanged, until the temperature becomes too high and the thermally generated carriers begin to dominate. The effect of dopants is lost at very high temperatures and the semiconductor essentially shows "intrinsic" behavior.

acceptor level of possible electron energies just above the valence band (Table 18-7). An electron must gain an energy of only E_a in order to create a hole in the valence band. The hole then moves and carries the charge. Now we have a *p*-type semiconductor as illustrated in the following example.

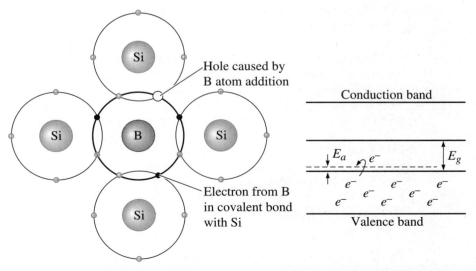

Figure 18-21 When a dopant atom with a valence of less than four is substituted into the silicon structure, a hole is created in the structure and an acceptor energy level is created just above the valence band. Little energy is required to excite the holes into motion.

EXAMPLE 18-8   *Design of a Semiconductor*

Design a *p*-type semiconductor based on silicon, which provides a constant conductivity of 100 $ohm^{-1} \cdot cm^{-1}$ over a range of temperatures. Comment on the level of purity needed.

SOLUTION

In order to obtain the desired conductivity, we must dope the silicon with atoms having a valence of +3, adding enough dopant to provide the required number of charge carriers. If we assume that the number of intrinsic carriers is small, then

$$\sigma = N_d q \mu_p,$$

where $\sigma = 100$ $ohm^{-1} \cdot cm^{-1}$ and $\mu_p = 480$ $cm^2/V \cdot s$. If we remember that coulomb can be expressed as ampere-seconds and voltage can be expressed as ampere-ohm, the number of charge carriers required is:

$$N_d = \frac{\sigma}{q\mu_p} = \frac{100}{(1.6 \times 10^{-19})(480)} = 1.30 \times 10^{18} \text{ donor atoms/cm}^3$$

Let's compare the concentration of donor atoms in Si with the concentration of Si atoms. Assume that the lattice constant of Si remains unchanged as a result, of doping:

$$N_d = \frac{(1 \text{ electron/dopant atom})(x \text{ dopant atom/Si atom})(8 \text{ Si atoms/unit cell})}{(5.4307 \times 10^{-8} \text{ cm})^3}$$

$$x = (1.3 \times 10^{18})(5.4307 \times 10^{-8})^3/8 = 26 \times 10^{-6} \text{ dopant atom/Si atom}$$

or 26 dopant atoms/10^6 Si atoms

Possible dopants include boron, aluminum, gallium, and indium (Chapter 5). High-purity chemicals and clean room conditions are essential for processing since we need 26 dopant atoms (apples) in a million silicon atoms (oranges).

Many other materials that are normally insulating (because the bandgap is too large) can be turned into semiconductors by doping. Examples of this include $BaTiO_3$, ZnO, TiO_2, and many other oxides. Thus, the concept of *n*- and *p*-type dopants is not limited to Si, Ge, GaAs, etc. We can dope $BaTiO_3$, for example, and make *n*- or *p*-type $BaTiO_3$. Such materials are useful for many sensor applications such as **thermistors**.[17]

Direct and Indirect Bandgap Semiconductors   In a direct bandgap semiconductor, an electron can be promoted from the conduction band to the valence band without changing the momentum of the electron. An example of a direct bandgap semiconductor is GaAs. When the excited falls back into the valence band, electrons and holes combine to produce light. Thus,

$$\text{electron} + \text{hole} \dashrightarrow h\nu \tag{18-16a}$$

This is known as **radiative recombination**. Thus, direct bandgap materials such as GaAs and solid solutions of these (e.g., GaAs-AlAs, etc.) are used to make light-emitting diodes (LEDs) of different colors. The bandgap of semiconductors can be tuned using solid solutions (Chapter 9). The change in bandgap produces a change in the wavelength (i.e., the frequency of the color (v) is related to the bandgap E_g as $E_g = h v$, where h is the Planck's constant). Since an optical effect is obtained using an electronic material, often the direct bandgap materials are known as optoelectronic materials. Many lasers and LEDs have been developed using these materials. LEDs that emit light in the infrared range are used in optical-fiber communication systems to convert light waves into electrical pulses. Different colored lasers, such as the newest blue laser using GaN, have been developed using direct bandgap materials.[18,19]

In an indirect bandgap semiconductor (e.g., Si, Ge, and GaP) the electron-hole recombination is very efficient and the electrons cannot be promoted to the valence band without a change in momentum. As a result, in materials that have an indirect bandgap (e.g., silicon), we cannot get light emission. Instead, electrons and holes combine to produce heat that is dissipated within the material. This is known as **nonradiative recombination**.

$$\text{electron} + \text{hole} \dashrightarrow \text{heat} \qquad (18\text{-}16b)$$

Note that both direct and indirect bandgap materials can be doped to form n- or p-type semiconductors.

The example that follows illustrates the use of radiative combination in light-emitting diodes.

EXAMPLE 18-9   *Creating the Color of a Light-Emitting Diode Display*

A light-emitting diode display made using a GaAs-GaP solid solution of composition 0.4 GaP-0.6 GaAs has a direct bandgap of 1.9 eV. What will be the color this LED display?

SOLUTION

In direct bandgap materials, a forward bias (an external voltage) is used to inject carriers that recombine via radiative recombination and give off light. The wavelength of the light emitted is related to the bandgap by:

$$E_g = h v = (h c / \lambda)$$

where h = Plancks constant (6.63×10^{-34} J-s, 4.41×10^{-15} eV-s), c = speed of light (2.998×10^8 m/s), λ = wavelength of light (m). We can put in the values of E_g, c and h and calculate the wavelength λ.

You can show that if the bandgap is in eV and λ is to be in μm, then:

$$E_g \text{ (in eV)} = 1.24 / \lambda \text{ (in } \mu\text{m)}$$

$$1.9 \text{ eV} = 1.24 / \lambda \text{ (in } \mu\text{m)}$$

Therefore, $\lambda = 0.652$ μm or 652 nm. This is the wavelength of red light, and, therefore, the LED display would be red in color.

18-7   Applications of Semiconductors

Semiconductors are at the heart of all computer chips. We make diodes, transistors, lasers, and LEDs using semiconductors. Silicon is the workhorse of very large scale integrated (VLSI) circuits. These consist of millions of transistors integrated into an area as small as 1 cm². The *p-n* junction is used to make diodes and many other devices such as transistors. Electrically, the *p-n* junction is conducting when the *p*-side is connected to a positive voltage (known as **forward bias**). When a negative bias is applied to the *p*-side of a *p-n* junction (**reverse bias**) the *p-n* junction does not carry much current. When no bias is applied, there is no current flowing through the *p-n* junction. The forward current can be up to a few milliamperes, while the reverse-bias current is few nano-amperes. In a *p-n* junction under no bias, the diffusion and drift currents due to electron and hole motions cancel out (Chapter 5). Under forward bias, the depletion layer width (W) becomes much smaller, the magnitude of the internal potential barrier becomes smaller (from V_0 to $V_0 - V_f$) and considerable current begins to flow. Under reverse bias, the width of the depletion layer increases potential barrier for current flow

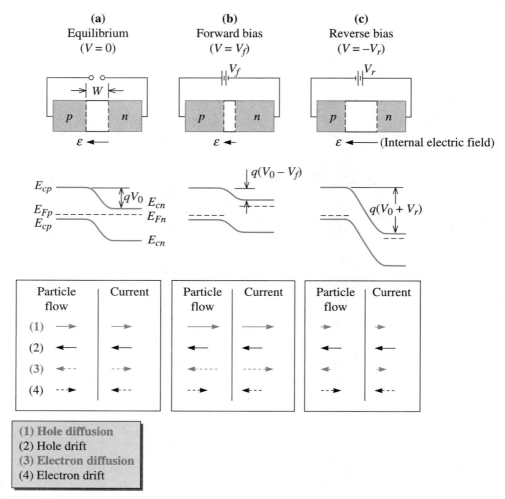

(1) **Hole diffusion**
(2) **Hole drift**
(3) **Electron diffusion**
(4) **Electron drift**

Figure 18-22   Behavior of a *p-n* junction device: (a) When no bias is applied electron and hole currents due to drift and diffusion cancel out and there is no net current and a built-in potential V_0 exists. (b) Under a forward bias causes the potential barrier to be reduced to $V_f - V_0$ and the depletion region becomes smaller causing a current to flow. (c) Under reverse bias, the potential barrier increases to $V_r + V_0$ and very little current flows. The internal electric field that develops is shown as E. (*Source: From* Solid State Electronic Devices, Third Edition, *by. B.G. Streetman, Fig. 5-10. Copyright © 2000 Prentice Hall. Reprinted by permission Pearson Education, Inc., Upper Saddle River, NJ.*)

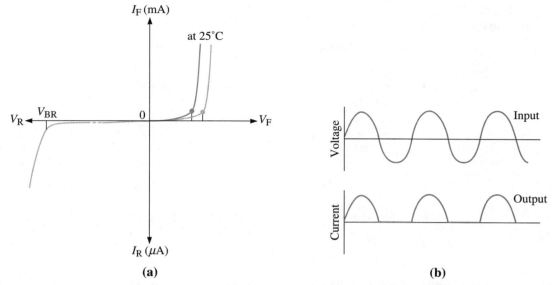

Figure 18-23 (a) The current-voltage characteristic for a *p-n* junction. Note the different scales in the first and third quadrants. (b) If an alternating signal is applied, rectification occurs and only half of the input signal passes the rectifier. (*Source: From* Electronic Devices, Sixth Edition, *by T.L. Floyd, Fig. 1-30. Copyright © 2002 Prentice Hall. Reprinted by permission Pearson Education, Inc., Upper Saddle River, NJ.*)

increases (from V_0 to $V_0 + V_r$) and very little current flows under reverse bias (Figure 18-22). In this diagram, ε is the internal electric field, E_F is the Fermi energy level. Subscripts v and c refer to valence and conduction bands, respectively. The drift currents remain unaffected, however diffusion currents become affected as a result of bias (increasing under forward bias and decreasing under reverse bias).

The *V-I* characteristics of a *p-n* junction are shown in Figure 18-23(a). Because the *p-n* junction permits current to flow in only one direction, it passes only half of an alternating current, therefore converting the alternating current to direct current [Figure 18-23(b)]. These junctions are called *rectifier diodes*. Typically, a small leakage current is produced for reverse bias due to the movement of thermally activated electrons and holes. When the reverse bias becomes too large, however, any carriers that do leak through the insulating barrier of the junction are highly accelerated, excite other charge carriers, and cause a high current in the reverse direction. The breakdown under reverse bias can be due to the tunneling of electrons. This is known as Zener breakdown, which occurs in heavily doped *p-n* junctions with smaller depletion-layer widths. Zener diodes are used to protect electrical circuits, since they can be designed to intentionally breakdown at a given voltage. Zener diodes are always used in reverse bias. For lightly doped *p-n* junctions, breakdown can occur through the **avalanche breakdown** mechanism. Essentially, when the voltage across *p-n* junctions becomes too high, electron-hole pairs are created by impact ionization. This breakdown will damage the *p-n* junction and is normally not useful.

There are two types of transistors (transfer resistors) fabricated using *p-n* junctions.

Bipolar Junction Transistors A **transistor** can be used as a switch or an amplifier. One example is the *bipolar junction transistor* (BJT), which is often used in the central processing units of computers because of their rapid switching response. A bipolar junction transistor is a sandwich of either *n-p-n* or *p-n-p* semiconductor materials. There

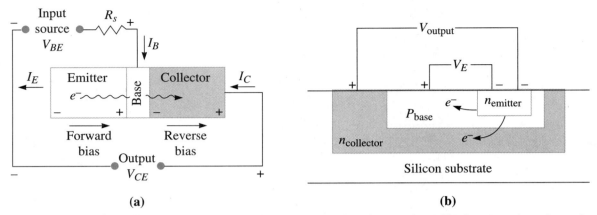

Figure 18-24   (a) A circuit for an *n-p-n* bipolar junction transistor. The input creates a forward and reverse bias that causes electrons to move from the emitter, through the base, and into the collector, creating an amplified output. (b) Sketch of the cross-section of the transistor.

are three zones in the transistor: the emitter, the base, and the collector. As in the *p-n* junction, electrons are initially concentrated in the *n*-type material and holes are concentrated in the *p*-type material.

Figure 18-24 shows an *n-p-n* transistor and its electrical circuit, both schematically and as it might appear when implanted in a silicon chip. The electrical signal to be amplified is connected between the base and the emitter, with a small voltage between these two zones. The output from the transistor, or the amplified signal, is connected between the emitter and the collector and operates at a higher voltage. The circuit is connected so that a forward bias is produced between the emitter and the base (the positive voltage is at the *p*-type base), while a reverse bias is produced between the base and the collector (with the positive voltage at the *n*-type collector). The forward bias causes electrons to leave the emitter and enter the base.

Electrons and holes attempt to recombine in the base; however, if the base is exceptionally thin and lightly doped, or if the recombination time τ is long, almost all of the electrons pass through the base and enter the collector. The reverse bias between the base and collector accelerates the electrons through the collector, the circuit is completed, and an output signal is produced. The current through the collector is given by

$$I_c = I_0 \exp\left(\frac{V_E}{B}\right), \tag{18-17}$$

where I_0 and B are constants and V_E is the voltage between the emitter and the base. If the input voltage V_E is increased, a very large current I_c is produced.

Field Effect Transistors   A second type of transistor, which is more often used for storing data in computer memories, is the *field effect transistor* (FET), which behaves in a somewhat different manner than the bipolar junction transistors. Figure 18-25 shows an example of a metal oxide semiconductor (MOS) field effect transistor, in which two *n*-type regions are formed within a *p*-type substrate. One of the *n*-type regions is called the source; the second is called the drain. A third component of the transistor is a conductor called a gate, that is, separated from the semiconductor by a thin insulating layer of SiO_2. A potential is applied between the gate and the source, with the gate region

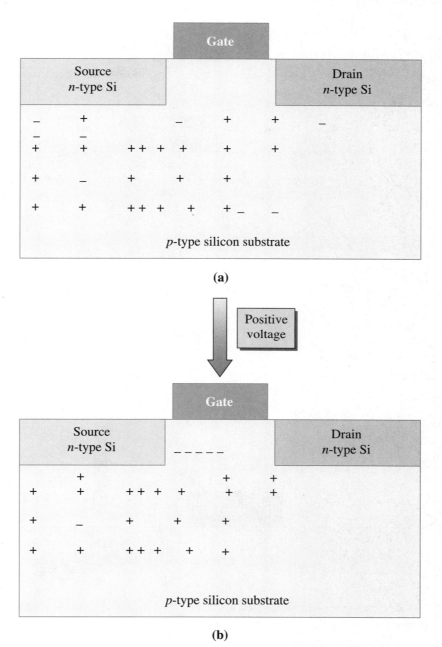

(a)

(b)

Figure 18-25 Operation of a *n-p-n* transistor. (a) Schematic of a *n-p-n* transistor. The dashes indicate electrons, the + sign indicates holes. Electrons in source and gate regions are not shown. (b) The gate has a positive voltage. Electrons in the *p*-type substrate move to the region between the source and drain. (c) When positive voltage is applied to a drain, electrons from the source move to the drain via the channel region between the source and drain (created by a positive gate voltage). The transistor is now "on." (d) When a gate voltage is removed. The electron channel between the source is broken and the transistor is turned "off." (*Source: www.intel.com.*)

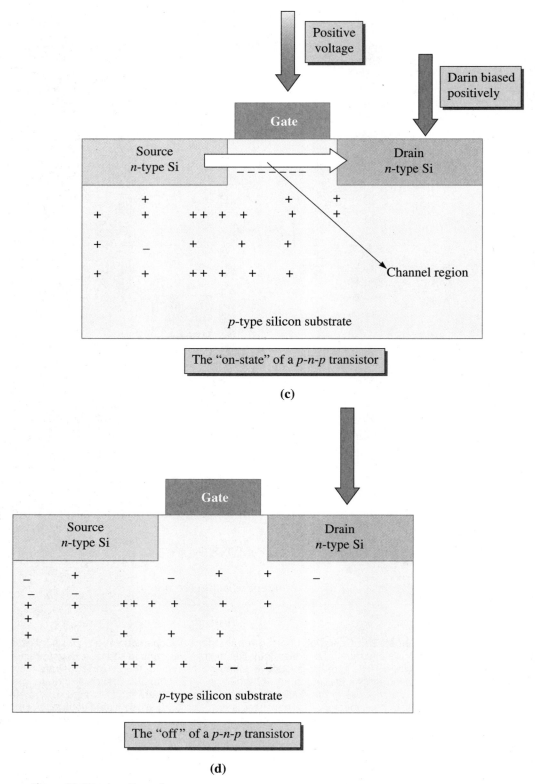

Figure 18-25 (continued)

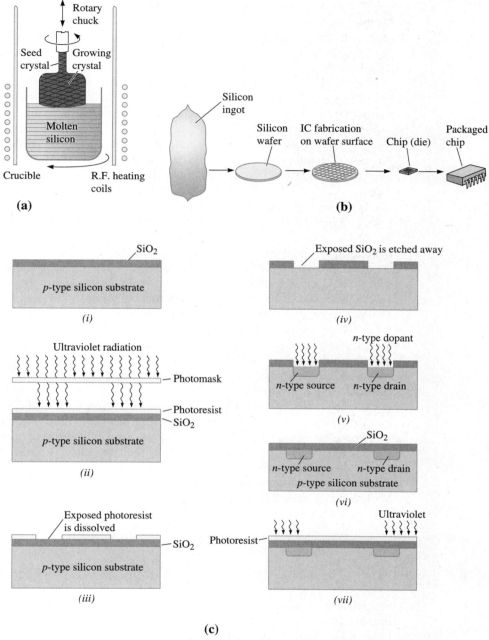

Figure 18-26 (a) Czochralski growth technique for growing single crystals of silicon. (*Source: From* Microchip Fabrication, Third Edition, *by P. VanZant, Fig. 3-7. Copyright © 1997 The McGraw-Hill Companies. Reprinted with permission The McGraw-Hill Companies.*) (b) Overall steps encountered in the processing of semiconductors. (*Source: From* Fundamentals of Modern Manufacturing, *by M.P. Groover, p. 849, Fig. 34-3. Copyright © 1996 Prentice Hall.*) (c) Production of an FET semiconductor device: (i) A *p*-type silicon substrate is oxidized. (ii) Photolithography, ultraviolet radiation passing through a photomask, exposes a portion of the photoresist layer. (iii) The exposed photoresist is dissolved. (iv) The exposed silica is removed by etching. (v) An *n*-type dopant is introduced to produce the source and drain. (vi) The silicon is again oxidized. (vii) Photolithography is repeated to introduce other components, including electrical leads, for the device.

being positive. The potential draws electrons to the vicinity of the gate, but the electrons cannot enter the gate because of the silica. The concentration of electrons beneath the gate makes this region more conductive, so that a large potential between the source and drain permits electrons to flow from the source to the drain, producing an amplified signal (on state). By changing the input voltage between the gate and the source, the number of electrons in the conductive path changes, thus also changing the output signal. When no voltage is applied to the gate, no electrons are attracted to the region between the source and the drain and there is no current flow from the source to the drain (off-state).

The field effect transistors are generally less expensive to produce than the bipolar junction transistors. Because FETs occupy less space, they are preferred in microelectronic integrated circuits, where a million transistors are present in a single silicon chip.

In addition to these, many other devices such as solar cells, photovoltaic panels, lasers, photodiodes, etc., are prepared using different semiconductors.

Manufacture and Fabrication of Semiconductor Devices   In order to produce electronic components that require little power, operate very rapidly, yet are inexpensive, microelectronic integrated circuits formed on silicon chips may contain as many as 1,000,000 transistors or other devices, each having dimensions less than 10^{-6} cm (1 μm). In order to produce these circuits, special technologies are required. The starting point for the most common devices is pure, single-crystal silicon (Si). These crystals are often grown using the Czochralski growth technique [Figure 18-26(a)].[20] A seed crystal is used to grow large (up to 10 inch diameter) silicon single crystals. Float zone and liquid encapsulated Czochralski techniques are also used.[16,20] We prefer single crystals because electrical properties of uniformly doped and essentially dislocation-free single crystals are much better defined than those of polycrystalline silicon. The overall steps encountered in semiconductor processing are shown in Figure 18-26(b).

The steps described in Figure 18-26(c) summarize the production of an FET transistor. Diffusion and ion implantation are used to create p- and n-type regions in selected areas. Almost all processing of semiconductors is conducted in clean rooms to eliminate contamination due to air- and liquid-borne particles. For the same reason, the highest purity chemicals and solutions are used in semiconductor processing.

18-8   Insulators and Dielectric Properties

Materials used to insulate an electric field from its surroundings are required in a large number of electrical and electronic applications. Electrical insulators obviously must have a very low conductivity, or high resistivity, to prevent the flow of current. Insulators must also be able to withstand intense electric fields. Insulators are produced from ceramic and polymeric materials in which there is a large energy gap between the valence and conduction bands. However, the high-electrical resistivity of these materials is not always sufficient. At high voltages, a catastrophic breakdown of the insulator may occur, similar to what happens in the p-n diodes at a large reverse bias, and current may flow. In order to select an insulating material properly, we must understand how the material stores, as well as conducts, electrical charge. Porcelain, alumina, cordierite, mica, and some glasses and plastics are used as insulators. The resistivity of most of these is $>10^{14}$ Ω-cm and the breakdown electric fields are $\sim$5 to 15 kV/mm.[21]

18-9   Polarization in Dielectrics

When we apply stress to a material, some level of strain develops. Similarly, when we subject materials to an electric field, the atoms, molecules, or ions respond to the applied electric field (E). Thus, the material is said to be polarized. A dipole is a pair of opposite charges separated by a certain distance. If one charge of $+q$ is separate from another charge of $-q$ and if d is the distance between these charges, the dipole moment is $q \times d$. The magnitude of polarization is given by $p = z \cdot q \cdot d$, where z is the number of charge centers that are displaced per cubic meter. In this equation, q is the electronic charge and d is the displacement between positive and negative charges forming the dipole.

Any mechanism that causes a separation of charges (e.g., nucleus and electronic cloud) or any mechanism that causes a change in the separation of charges that are already present (e.g., movement or vibration of ions in an ionic material) causes polarization. The more the availability of polarization pathways or mechanisms, the greater is the polarizability or dielectric susceptibility of the material. A mechanical analog for this is: if we apply a stress to a material, the strain (equivalent of polarization) can occur by elastic deformation, movement of twins (in certain materials only), or plastic deformation (movement of dislocations).

There are four primary mechanisms causing polarization: (1) electronic polarization, (2) ionic polarization, (3) molecular polarization, and (4) space charge (Figure 18-27). Their occurrence depends upon the electrical frequency of the applied field, just like the mechanical behavior of materials depends on the strain rate (Chapters 6 and 7). If we apply a very rapid rate of strain, such mechanisms as plastic deformation do not get a chance to let the material deform. Similarly, if we apply a rapidly alternating electric field, some polarization mechanisms may not get a chance to induce polarization in the material.

Polarization mechanisms play two important roles. First, if we make a **capacitor** from a material, the polarization mechanisms allow the charge to be stored, since the dipoles created in the material (as a result of polarization) can bind a certain portion of the charge on the electrodes of the capacitor. Thus, the higher the dielectric polarization, the higher the dielectric constant (k) of the material. The dielectric constant is a measure of the ability of a material to store electrical charge. It is defined as the ratio of capacitance of a capacitor filled with dielectric and that has vacuum. This charge storage, in some ways, is similar to the elastic strain developing in a material subjected to stress. The second important role is that when polarization sets in, charges move (ions or electronic clouds are displaced). If the electrical field oscillates, the charges move back and forth. These displacements are extremely small (typically <1 Å); however if they occur enough times they cause **dielectric losses**, which are the part of the electrical energy lost in causing these displacements of electronic clouds or ions. This lost energy appears as heat. The dielectric loss is similar to the viscous deformation of a material. If we want to store charges, as in a capacitor, dielectric loss is not good. However, if we want to use microwaves to heat up our lunches, dielectric losses that occur in water contained in the food are indeed good! The dielectric losses are often measured by a parameter known as **tan δ**. When we are interested in extremely low loss materials, such as those used in microwave communications (Chapter 14), we refer to a parameter known as the **quality factor** ($Q_d \sim 1/\tan \delta$). The dielectric constant and dielectric losses depend strongly on electrical frequency and temperature. This will be discussed a little later.

Electronic polarization is omnipresent since all materials contain atoms. The electronic cloud gets displaced from the nucleus in response to the field seen by the atoms.

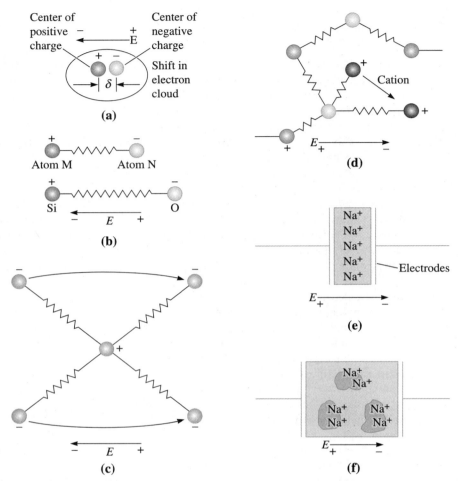

Figure 18-27 Polarization mechanisms in materials: (a) electronic, (b) atomic or ionic, (c) high-frequency dipolar or orientation (present in ferroelectrics), (d) low-frequency dipolar (present in linear dielectrics and glasses), (e) interfacial-space charge at electrodes, and (f) interfacial-space charge at heterogeneities such as grain boundaries. (*Source: From Principles of Electronic Ceramics, L.L. Hench and J.K. West, p. 188, Fig. 5-2. Copyright © 1990 Wiley Interscience. Reprinted by permission. This material is used by permission of John Wiley & Sons, Inc.*)

The separation of charges, the nucleus and the electronic cloud, creates a dipole moment [Figure 18-27(a)]. This mechanism can survive at the highest electrical frequencies ($\sim 10^{15}$ Hz) since an electronic cloud can be displaced rapidly, back and forth, as the electrical field switches. Larger atoms and ions have higher electronic polarizability (tendency to undergo polarization) since the electronic cloud is farther away from the nucleus and held less tightly. This polarization mechanism is also linked closely to the refractive index of materials since light is an electromagnetic wave where the electric field oscillates at a very high frequency ($\sim 10^{14} - 10^{16}$ Hz). The higher the electronic polarizability, the higher the refractive index. We use this mechanism in making "lead crystal", which is really an amorphous glass that contains up to 30% PbO. The large lead ions (Pb^{+2}) are highly polarizable due to the electronic polarization mechanisms and provide a high-refractive index when high enough concentrations of lead oxide are present in the glass.

EXAMPLE 18-10	*Electronic Polarization in Copper*

Suppose that the average displacement of the electrons relative to the nucleus in a copper atom is 1×10^{-8} Å when an electric field is imposed on a copper plate. Calculate the electronic polarization.

SOLUTION

The atomic number of copper is 29, so there are 29 electrons in each copper atom. The lattice parameter of copper is 3.6151 Å. Thus

$$Z = \frac{(4 \text{ atoms/cell})(29 \text{ electrons/atom})}{(3.6151 \times 10^{-10} \text{ m})^3} 2.46 \times 10^{30} \text{ electrons/m}^3$$

$$P = Zqd = \left(2.46 \times 10^{30} \frac{\text{electrons}}{\text{m}^3} \right)$$

$$\times \left(1.6 \times 10^{-19} \frac{\text{C}}{\text{electron}} \right) (10^{-8} \text{ Å})(10^{-10} \text{ m/Å})$$

$$= 3.94 \times 10^{-7} \text{ C/m}^2$$

Ionic Polarization   When an ionically bonded material is placed in an electric field, the bonds between the ions are elastically deformed. Depending on the direction of the field, cations and anions move either closer together or further apart. These temporarily induced dipoles provide polarization and may change the overall dimensions of the material. This mechanism can be in effect up to 10^{12}–10^{13} Hz. Based on this mechanism, we can interpret the infrared absorption or transmission in different materials. Ionic polarization will be absent in covalently bonded materials such as Si, Ge, diamond, etc.

Dipolar or Orientation Polarization   In Chapter 2, we learned that certain materials such as water are polar in nature. Each water molecule has a permanent dipole moment built into it. When a field is applied, the dipoles rotate to line up with the imposed field. Thus, when water is subjected to an electric field, in addition to the electronic polarization, the water will have a special dipolar or orientation polarization mechanism with a relatively high dielectric constant (k ∼ 78).

Some other materials, known as ferroelectrics, by virtue of their crystal structure, exhibit reversible and spontaneous dielectric polarization. The tetragonal polymorph of barium titanate ($BaTiO_3$) and PZT are good examples of ferroelectric materials. Certain polymers, such polyvinylidene fluoride (PVDF), also exhibit ferroelectricity.[22] Barium titanate has an asymmetrical tetragonal structure at room temperature (Figure 18-28). The titanium ion is displaced slightly from the center of the unit cell and the oxygen ions are displaced slightly in the opposite directions from their face-centered positions, causing the crystal to be tetragonal and permanently polarized. This special polarization mechanism leads to a very high dielectric constant in $BaTiO_3$, $PbTiO_3$, and other ferroelectric materials.

Space Charges   A charge may develop at interfaces between phases within a material, normally as a result of the presence of impurities or grain boundaries. Some materials can also contain relatively mobile ions such as lithium ions (Li^+). At high temperatures (that promote diffusion) and very low-electrical frequencies (∼up to 10^2 Hz, so the charges can drift a considerable distance), the interfacial polarization mechanism can

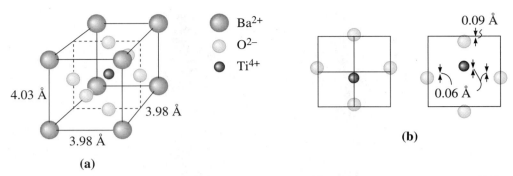

(a)

Figure 18-28 (a) The oxygen ions are at face centers, Ba^{+2} ions are at cube corners and Ti^{+4} is at cube center in cubic $BaTiO_3$. (b) In tetragonal $BaTiO_3$, the Ti^{+4} is off-center and the unit cell has a net polarization.

appear in some materials (e.g., $LiTaO_3$, $LiNbO_3$, and certain glasses containing Li_2O or Na_2O). The charge piles up at the grain boundaries or near the electrodes when the material is placed in an electric field. This type of polarization is not an important factor in most common dielectrics used in capacitors and other applications.

Frequency and Temperature Dependence of Dielectric Constant and Dielectric Losses
You may know from elementary physics that capacitance (C) is given by:

$$C = Q/A \qquad (18\text{-}18)$$

where Q is the charge on the plates and A is the area of the plates (Figure 18-29). As the material undergoes polarization, it can bind a certain amount of charge on the surface of the material. The higher the polarization, the higher the bound charge on the surface and the higher the dielectric constant (k) of the material. If we fill the space between the parallel plates (surface area A and separated by a distance t) with a material, then the dielectric constant (k) or relative permittivity ε_r is given by:

$$C = \frac{k\varepsilon_0 A}{t} \qquad (18\text{-}19)$$

The constant ε_0 is the **permittivity** of a vacuum and is 8.85×10^{-12} F/m. As noted before, k is a measure of the ability of a material to store electrical charge.

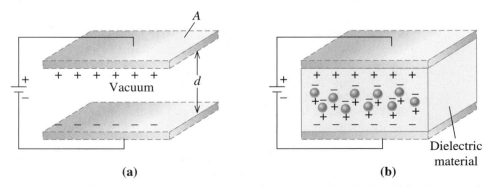

Figure 18-29 A charge can be stored at the conductor plates in a vacuum (a). However, when a dielectric is placed between the plates (b), the dielectric polarizes and additional charge is stored.

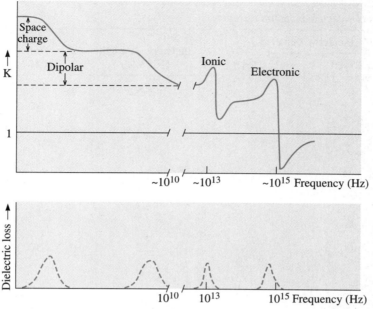

Figure 18-30 Frequency dependence of polarization mechanisms. On top is the change in the dielectric constant with increasing frequency, and the bottom curve represents the dielectric loss. (*Source: From* Electroceramics: Material, Properties, Applications, *by A.J. Moulson and J.M. Herbert, p. 68, Fig. 2-38. Copyright © 1990 Chapman and Hall. Reprinted with kind permission of Kluwer Academic Publishers and the author.*)

The dielectric constant (k) depends upon how susceptible the material is to the applied electric field. The dielectric constant of materials depends upon the composition, microstructure, electrical frequency, and temperature (Figure 18-30). The capacitance depends upon the dielectric constant, area of the electrodes, and the separation between the electrodes. Capacitors in parallel provide added capacitance (just like resistances add in a series). This is the reason why multi-layer capacitors consist of 100 or more layers interconnected in parallel. These are typically based on $BaTiO_3$-based formulations and are prepared using a tape-casting process (Chapter 14). Silver-palladium or nickel is used as electrode layers. We choose $BaTiO_3$ as the base material since it has an unusually high dielectric constant (k ~ 2000–5000 at ~1 kHz and room temperature). Most other ceramics and polymers have a lower dielectric constant since they do not have the special ferroelectric polarization mechanism that $BaTiO_3$ and other ferroelectrics have. Thus, the dielectric constant of most plastics is low (~2–10). The loss tangents of most plastics are small (tan δ ~ 0.0003–0.001). Many nonferroelectric ceramics, also known as **linear dielectrics**, (e.g., Al_2O_3, SiO_2, etc.) have a dielectric constant of ~5–10. The dielectric losses are also relatively smaller (tan δ ~ 0.0001–0.001).

For a given material, as the electrical frequency increases, more and more polarization mechanisms begin to drop out. This is because, if the electrical field oscillates too rapidly, the species contributing cannot really move back and forth. Thus, as frequency increases the overall dielectric constant begins to decrease. This is known as *dielectric relaxation*. But dielectric losses show a different behavior. When the frequency is too slow for a given polarization mechanism, the dielectric loss is small since the movement of charged species does not occur often enough. When the frequency is too high, the charged species cannot keep up with the electric field. This is why the dielectric losses show a peak or resonance for each of the polarization mechanisms. In real mate-

TABLE 18-8 ■ *Properties of selected dielectric materials*

Material	Dielectric Constant (at 60 Hz)	Dielectric Constant (at 10^6 Hz)	Dielectric Strength (10^6 V/m)	tan δ (at 10^6 Hz)	Resistivity (ohm · cm)
Polyethylene	2.3	2.3	20	0.00010	$>10^{16}$
Teflon	2.1	2.1	20	0.00007	10^{18}
Polystyrene	2.5	2.5	20	0.00020	10^{18}
PVC	3.5	3.2	40	0.05000	10^{12}
Nylon	4.0	3.6	20	0.04000	10^{15}
Rubber	4.0	3.2	24		
Phenolic	7.0	4.9	12	0.05000	10^{12}
Epoxy	4.0	3.6	18		10^{15}
Paraffin wax		2.3	10		10^{13}–10^{19}
Fused silica	3.8	3.8	10	0.00004	10^{11}–10^{12}
Soda-lime glass	7.0	7.0	10	0.00900	10^{15}
Al_2O_3	9.0	6.5	6	0.00100	10^{11}–10^{13}
TiO_2		14–110	8	0.00020	10^{13}–10^{18}
Mica		7.0	40		10^{13}
$BaTiO_3$		2000–5000	12	~0.0001	10^{8}–10^{15}
Water		78.3			10^{14}

rials, there is an assortment of bonds (e.g., Si-O, Al-O, Na-O, etc. in a silicate glass). Therefore, even for the same polarization mechanism these peaks are smeared since many such peaks occur at different frequencies. The dielectric properties of some materials are shown in Table 18-8.

For electrical insulation (e.g., spark plugs, maximum cables, etc.), the **dielectric strength** (i.e., the electric field value that can be supported prior to electrical breakdown) is important.

Linear and Nonlinear Dielectrics   The dielectric constant, as expected, is related to the polarization that can be achieved in the material. We can show that the dielectric polarization induced in a material depends upon the applied electric field (E) and the dielectric constant (k).

$$P = (k - 1)\varepsilon_0 E \text{ (for linear dielectrics)} \tag{18-20}$$

where E is the strength of the electric field (V/m). For materials that polarize easily, both the dielectric constant and the capacitance are large and, in turn, a large quantity of charge can be stored. In addition, Equation 18-20 suggests that polarization increases, at least until all of the dipoles are aligned, as the voltage (expressed by the strength of the electric field) increases. The quantity (k − 1) is known as dielectric susceptibility (χ_e). The dielectric constant of vacuum is one, or the dielectric susceptibility is zero. This makes sense since a vacuum does not contain any atoms or molecules.

In **linear dielectrics** P is linearly related to E and k is constant. This is similar to how stress and strain are linearly related to elastic solids (Chapter 6). In linear dielectrics, the k or χ_e remain constant with changing E. In polar materials, such as $BaTiO_3$, the dielectric constant changes with E and hence Equation 18-20 cannot be used. These materials in which P and E are not related by a straight line are known as **nonlinear dielectrics** or ferroelectrics. These materials are similar to elastomers for which stress and strain are not linearly related and hence cannot be defined as a unique value of the Young's modulus (Chapter 6).

18-10 Electrostriction, Piezoelectricity, Pyroelectricity, and Ferroelectricity

When any material undergoes polarization, its ions and electronic clouds are displaced, causing the development of a mechanical strain in the material. This effect is seen in all materials subjected to an electric field and is known as the *electrostriction*.

The strain (x) generated using an electric field (E) is given by:

$$x = \xi E^2 \tag{18-21}$$

The term ξ represents the electrostriction coefficient and E is the magnitude of the electric field. Of the total 32 classes made from triclinic, monoclinic, orthorhombic, and tetragonal crystal systems, 20 have a center of symmetry. This means that if we apply a mechanical stress, there is no dipole moment generated since ionic movements are symmetric. Of the 21 that remain, 20 point groups, which lack a center of symmetry, exhibit the development of dielectric polarization when subjected to stress. These materials are known as *piezoelectric*. (The word *piezo* means pressure.) When these materials are stressed, they develop a voltage. This development of a voltage upon the application of stress is known as the **direct** or **motor piezoelectric effect** (Figure 18-31). This effect helps us make devices such as spark igniters, which are often made using lead zirconium titanate (PZT). This effect is also used, for example, in detecting submarines and other objects under water.

Conversely, when an electrical voltage is applied, a piezoelectric material shows the development of strain. This is known as the **converse** or **generator piezoelectric effect**. This effect is used in making actuators. For example, this movement can be used to generate ultrasound waves that are used in medical imaging, as well as such applications as ultrasonic cleaners or toothbrushes. Sonic energy can also be created using piezoelectrics to make the high-fidelity "tweeter" found in most speakers. In addition to $Pb_xZr_{-x}TiO_3$ *(PZT)*, other piezoelectrics include SiO_2 (for making quartz crystal oscillators), ZnO, and polyinylidene fluoride *(PVDF)*. Many naturally occurring materials such as bone and silk are also piezoelectric.[22]

The "*d*" constant for a piezoelectric is defined as the ratio of strain to electric field.

$$\varepsilon = d \cdot E \tag{18-22a}$$

The "*g*" constant for a piezoelectric is defined as the ratio of voltage generated to stress applied:

$$\varepsilon = g \cdot X \tag{18-22b}$$

In these equations, E is the electric field (V/m), X is the applied stress (Pa), ε is the strain, and d is the piezoelectric coefficient.

The d and g piezoelectric coefficients are related by the dielectric constant as follows:

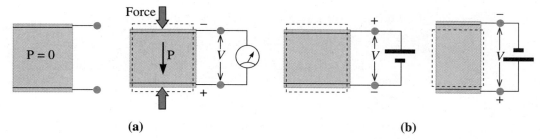

Figure 18-31 The (a) direct and (b) converse piezoelectric effect. In the direct piezoelectric effect, applied stress causes a voltage to appear. In the converse effect (b), an applied voltage leads to development of strain.

TABLE 18-9 ■ *The piezoelectric coefficients* d *and* g *for selected materials*

Material	*d* Coefficient (pC/N)	*g* Coefficient (mV/N)
Quartz (SiO_2)	2.3	50×10^{-3}
$BaTiO_3$*	190	12×10^{-3}
PZT*	268 to 480	12×10^{-3} to 35×10^{-3}
$PbNb_2O_6$*	80	
$PbTiO_3$	47	
$LiNbO_3$	6	
$LiTaO_3$	5.7	

Assumes strain and poling axis along same direction.

$$g = \frac{d}{k\varepsilon_0} \tag{18-23}$$

where k is the dielectric constant and ε_0 is the permittivity of free space (8.85×10^{-12} F/m). Typical values for *d* and *g* coefficients are given in Table 18-9.

The following example illustrates the design for a spark igniter.

EXAMPLE 18-11   *Design of a Spark Igniter*

A PZT spark igniter is made using a disk that has a 5-mm diameter and 20-mm height. Calculate the voltage generated if the *g* coefficient for PZT used is $35 \times 10^{-3} \dfrac{V - m}{N}$. Assume that a compressive force of 10,000 N is applied on the circular face.

SOLUTION

The *g* coefficient can be defined as:

$$g = \frac{E \left(\dfrac{V}{m}\right)}{X \left(\dfrac{N}{m^2}\right)}$$

In this equation, X is the applied stress, E is the electric field generated.

$$\text{The stress in this case} = \frac{10,000 \text{ N}}{\left(\dfrac{\pi}{4}\right)(5 \times 10^{-2} \text{ m})^2} = 0.0509 \times 10^8 \text{ N/m}^2$$

Therefore, the electric field E generated $= g \cdot X = (0.0509 \times 10^8 \text{ N/m}^2) \cdot \left(35 \times 10^{-3} \dfrac{V - m}{N}\right) = 1.782 \times 10^5$ V/m.

This field appears across a height (H) of 20 mm. The voltage (V) generated will be:

$$V = E \times H = (1.782 \times 10^5 \text{ V/m})(20 \times 10^{-3} \text{ m}) = 35.65 \times 10^2 = 3565 \text{ Volts.}$$

This voltage ultimately appears through the leads across a small air gap (~0.5 mm) and creates a small spark that can ignite a barbeque grill!

Of the 20 point groups of materials that exhibit the piezoelectric effect, 10 have a unique axis. A point group describes symmetry of the lattice. These develop spontaneous polarization and are known as *polar crystals*. Since the magnitude of spontaneous polarization depends upon temperature, a change in temperature leads to the development of charge. This is known as the **pyroelectric effect**. Each of the 10 classes mentioned consist of materials that exhibit pyroelectricity. Pyroelectric materials (e.g., lithium tantalate, and $LiTaO_3$) have been used to develop uncooled infrared detection and imaging for such items as night vision cameras used by the law enforcement.

We define **ferroelectrics** as materials that show the development of a spontaneous and reversible dielectric polarization (P_s). Ferroelectrics are pyroelectrics in that they develop a spontaneous polarization, however, that spontaneous polarization is reversible using an electric field. Examples include tetragonal polymorphs barium titanate ($BaTiO_3$), ($PbTiO_3$), and the like. These materials are important because the ferroelectric polarization substantially increases their dielectric constant. As a result, these materials are used widely for making multi-layer capacitors. Similarly, thin films of BST (barium strontium titanate) have been used to develop capacitors for microelectronics. The geometries of different capacitors are shown in Figure 18-32. Thin films of PZT and $BaTiO_3$-$SrTiO_3$ have applications as memory elements or actuators.[28]

We can also make a volumetrically efficient capacitor using $BaTiO_3$ based materials. Figure 18-33(a) shows the different polymorphs of $BaTiO_3$ and the corresponding changes in the lattice constants and changes in the dielectric constant along different axes.[23] The hysteresis loop for single crystalline and polycrystalline $BaTiO_3$ are shown in Figure 18-33(b) and (c). Single or polycrystalline ferroelectrics consist of regions in which the polarization is in one direction. The effect of grain size and temperature on k of $BaTiO_3$ is shown in Figure 18-34(a). These regions are known as *ferroelectric domains* [Figure 18-34(b)]. In a virgin ferroelectric material, different domains are randomly aligned. As an electric field is applied, some of the domains begin to become aligned with the electric field leading to an increase in the polarization. If a sufficiently high electric field is applied, almost all domains become aligned with the field and the polarization reaches a maximum. This is known as *saturation polarization*. If the field is increased beyond this, the material can break down electrically. When the applied field direction is decreased to zero, a certain fraction of the domains remains in the original direction. This leads to a remnant polarization (P_r). This effect has been used to develop ferroelectric thin-film-based memories. An electric field in the opposite direction has to be applied to cause the overall polarization to go to zero.

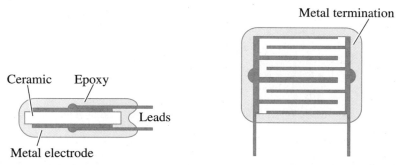

Figure 18-32 Examples of ceramic capacitors. (a) Single-layer ceramic capacitor (disk capacitors). (b) Multilayer ceramic capacitor (stacked ceramic layers). (*Source: From* Principles of Electrical Engineering Materials and Devices, *by S.O. Kasap, p. 559, Fig. 7-29. Copyright © 1997 Irwin. Reprinted by permission of The McGraw-Hill Companies.*)

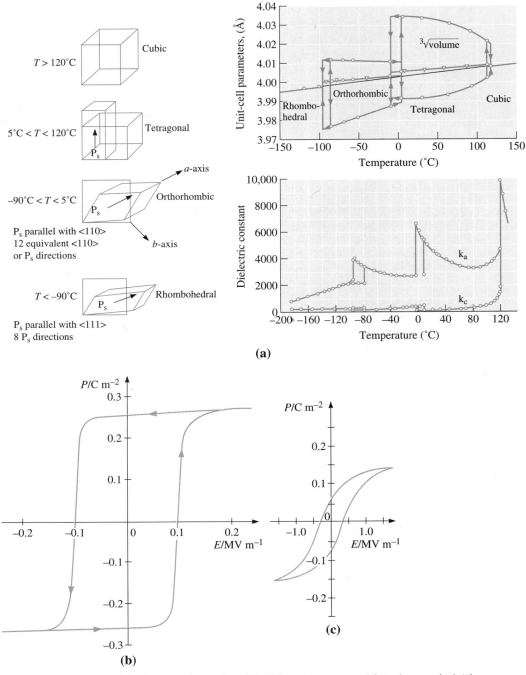

Figure 18-33   (a) Different polymorphs of $BaTiO_3$ and accompanying changes in lattice constants and dielectric constants. (*Source: From* Principles of Electronic Ceramics, *L.L. Hench and J.K. West, p. 247, Fig. 6-5. Copyright © 1990 Wiley Interscience. This material is used by permission of John Wiley & Sons, Inc.*). The ferroelectric hysteresis loops for (b) single crystal. (*Source: From* Electroceramics: Material, Properties, Applications, *by A.J. Moulson and J.M. Herbert, p. 76, Fig. 2-46. Copyright © 1990 Chapman and Hall. Reprinted with kind permission of Kluwer Academic Publishers and the author.*) (c) Polycrystalline $BaTiO_3$ showing the influence of the electric field on polarization. (*Source: From* Principles of Electrical Engineering Materials and Devices, *by S.O. Kasap, p. 565, Fig. 7-35. Copyright © 1997 Irwin. Reprinted by permission of The McGraw-Hill Companies.*)

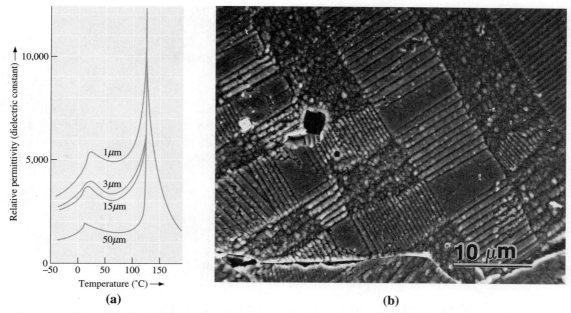

Figure 18-34   (a) The effect of temperature and grain size on the dielectric constant of barium titanate. Above the Curie temperature, the spontaneous polarization is lost due to a change in crystal structure and barium titanate is in the paraelectric state. The grain size dependence shows that similar to yield-strength dielectric constant is a microstructure sensitive property. (*Source: From* Electroceramics: Material, Properties, Applications, *by A.J. Moulson and J.M. Herbert, p. 78, Fig. 2-48. Copyright © 1990 Chapman and Hall. Reprinted with kind permission of Kluwer Academic Publishers and the author.*) (b) Ferroelectric domains can be seen in the microstructure of polycrystalline $BaTiO_3$. (*Courtesy of Dr. Rodney Roseman, University of Cincinnati.*)

Certain ferroelectrics, such as PZT, exhibit a strong piezoelectric effect, but in order to maximize the piezoelectric effect (e.g., the development of strain or voltage), piezoelectric materials are deliberately "poled" using an electric field to align all domains in one direction.

The dielectric constant of ferroelectrics goes through a maximum near a temperature known as the *Curie temperature* [(Figure 18-34(a)]. However, above this Curie temperature (T_c) of ~130°C, the titanium ion in $BaTiO_3$ rattles so fast around the cube center that, for all practical purposes, it is at the center of the cube (i.e., the structure becomes symmetric). Thus, above T_c, $BaTiO_3$ and other ferroelectrics become **paraelectric**. Note that even in the paraelectric state, the dielectric constant of $BaTiO_3$ remains quite high. However, since the cubic crystal structure has a center of symmetry, it is nonpiezoelectric. This is why we make use of the paraelectric state of $BaTiO_3$ when making single and multi-layer capacitors. This way any vibration or shock can not generate spurious voltages because of the piezoelectric effect. Since the Curie transition occurs at a high temperature, we use additives to $BaTiO_3$ that help shift the Curie transition temperature to lower in order to make the ferroelectric to paraelectric transition flatter. Many materials such as $Pb(Mg_{1/3}Nb_{2/3})$ or *PMN* make a much more gradual ferroelectric to paraelectric transition and are known as *relaxor ferroelectrics*. These materials show very high dielectric constants (up to 20,000) and good piezoelectric behavior, so they are used to make capacitors and piezoelectric devices. Note that all piezoelectric materials are not necessarily ferroelectric (e.g., SiO_2) although the converse is true; all ferroelectric materials are piezoelectric. The following examples illustrate the principles and applications of $BaTiO_3$.

EXAMPLE 18-12   *Polarization in Barium Titanate*

Calculate the maximum polarization per cubic centimeter and the maximum charge that can be stored per square centimeter for barium titanate.

SOLUTION

The product of the charge and the distance between the charges gives the strength of the dipoles. In $BaTiO_3$, the separations are the distances that the Ti^{4+} and O^{2-} ions are displaced from the normal lattice points (Figure 18-28). The charge on each ion is the product of q and the number of excess or missing electrons. Thus, the dipole moments are:

$$Ti^{4+}: (1.6 \times 10^{-19})(4 \text{ electrons/ion})(0.06 \times 10^{-8} \text{ cm})$$

$$= 0.384 \times 10^{-27} \ C \cdot cm/ion$$

$$O^{2-}_{(top)}: (1.6 \times 10^{-19})(2 \text{ electrons/ion})(0.09 \times 10^{-8} \text{ cm})$$

$$= 0.288 \times 10^{-27} \ C \cdot cm/ion$$

$$O^{2-}_{(side)}: (1.6 \times 10^{-19})(2 \text{ electrons/ion})(0.06 \times 10^{-8} \text{ cm})$$

$$= 0.192 \times 10^{-27} \ C \cdot cm/ion$$

Each oxygen ion is shared with another unit cell, so the total dipole moment in the unit cell is:

$$\begin{aligned} \text{Dipole moment} = &(1 \ Ti^{4+}/\text{cell})(0.384 \times 10^{-27}) \\ &+ (1 \ O^{2-} \text{ from top and bottom of cell})(0.288 \times 10^{-27}) \\ &+ (2 \ O^{2-} \text{ from four sides of cell})(0.192 \times 10^{-27}) \\ = &1.056 \times 10^{-27} \ C \cdot cm/\text{cell} \end{aligned}$$

The polarization per cubic centimeter is:

$$P = \frac{1.056 \times 10^{-27} \ C \cdot cm/\text{cell}}{(3.98 \times 10^{-8} \text{ cm})^2(4.03 \times 10^{-8} \text{ cm})}$$

$$= 1.65 \times 10^{-5} \ C/cm^2$$

The total charge on a $BaTiO_3$ crystal 1 cm $\times$ 1 cm is:

$$Q = PA = (1.65 \times 10^{-5} \ C/cm^2)(1 \text{ cm})^2$$

$$= 1.65 \times 10^{-5} \ C$$

EXAMPLE 18-13   *Design of a Multi-layer Capacitor*

A multi-layer capacitor is to be designed using a $BaTiO_3$-based formulation containing $SrTiO_3$. The dielectric constant of the material is 3000. (a) Calculate the capacitance of a multi-layer capacitor consisting of 100 layers connected in parallel using Ni electrodes. The sides of the capacitor are 10 $\times$ 5 mm and the thickness of each layer is 10 μm. (b) What is the role of $SrTiO_3$? (c) What processing technique will be used to make these?

SOLUTION

(a) The capacitance of a parallel plate capacitor is given by:

$$C = \frac{\varepsilon_0 k A}{t}$$

The permittivity of free space is ε_0 8.85×10^{-12} F/m. The relative permittivity of $BaTiO_3$ formulation ε_r or dielectric constant (k) is 3000.
Capacitance per layer will be:

$$C_{layer} = \frac{(8.85 \times 10^{-12} \text{ F/m})(3000)(10 \times 10^{-2} \text{ m})(5 \times 10^{-2} \text{ m})}{10 \times 10^{-6} \text{ m}}$$

Note the conversion to SI units.

$$C_{layer} = 13.27 \times 10^{-6} \text{ F} = 13.27 \text{ } \mu F.$$

We have 100 layers connected in parallel. Capacitances add up in this arrangement. All layers have same geometric dimensions in this case.

$$C_{total} = (\text{number of layers}) \cdot (\text{capacitance per layer})$$

$$C_{total} = (13.27 \text{ } \mu F)(100) = 1327 \text{ } \mu F$$

(b) The role of $SrTiO_3$ addition to the formulation is to shift the Curie temperature below room temperature. In this way, the capacitor dielectric is in the high dielectric constant, but paraelectric state. Thus, if it is subjected to stress due to vibrations, it will not generate any spurious voltages due to the piezoelectric effect present in the tetragonal ferroelectric state.

(c) Tape casting will be the best way to make such thin-interconnected layers (Chapter 14).

SUMMARY

◆ Electronic materials include insulators, dielectrics, conductors, semiconductors, and superconductors. These materials can be classified using their band structures. Electronic materials have enabled many technologies ranging from spark plugs to high-voltage line insulators to solar cells, computer chips, and many sensors and actuators.

◆ Important properties of conductors include the level of conductivity and the temperature dependence of conductivity. In pure metals, the resistivity increases with temperature. Resistivity is very sensitive to impurities and microstructural defects such as grain boundaries. Resistivity of alloys is typically higher and dominated by alloying elements and is relatively stable with respect to temperature.

◆ Superconductors are materials that offer zero resistance to the flow of electrical current and exclude a magnetic field. The superconductive effect is seen at low temperatures. In most Type I superconductors, the superconductive effect is lost even under very small magnetic-flux densities. In Type II superconductors, the flux can penetrate while the material remains superconducting and, above a certain critical field (H_{c2}), the superconductivity is lost. In Type I superconductors, the formation of electron pairs is an accepted mechanism. The mechanism is unclear in Type II superconductors. Nb_3Sn and Hg, are examples of Type I superconductors, while *YBCO* and *BSCCO* and MgB_2 are examples of Type II superconductors. In Type II superconductors, the current transport is limited by several microstructural factors. Superconductors are useful for microelectronic devices, components for wireless communications, and to make such items as the power magnets used in MRI imaging.

◆ Semiconductors have conductivities between insulators and conductors. Resistivity, a dielectric constant, the mobilities of electrons, and holes characterize semiconductors. Semiconductors can be classified as elemental (Si, Ge) or compound (SiC, GaAs). Both of these can be intrinsic or extrinsic (*n*- or *p*-type). Some semiconductors are direct (e.g., GaAs) or indirect (e.g., Si) bandgaps. The level of conductivity is controlled by the type of semiconductor and dopants. Extrinsic semiconductors are widely used because of their ability to tune the level of conductivity and temperature stability. Crystals of semiconductors are converted into computer chips using clean room conditions and ultra-clean chemicals. These direct bandgap materials are useful for making lasers, LEDs, and the like. Silicon is used to make transistors, diodes, and integrated circuits.

◆ Similar to semiconductors, conductive *n*- or *p*-type oxides can be formed. These materials include ITO, ZrO_2, and lithium cobalt oxide. They are used in displays, fuel cells, and batteries, respectively. Polymers can also exhibit semiconductivity. Such polymers could be the basis for lightweight and flexible electronic components.

◆ Dielectrics have large bandgaps and do not conduct electricity. With insulators, the focus is on breakdown voltage or field. With other dielectrics, the emphasis is on the dielectric constant, $\tan \delta$, frequency, and temperature dependence. Polarization mechanisms in materials dictate this dependence. In piezoelectrics, the application of stress results in the development of a voltage; the application of a voltage causes strain. In pyroelectrics, temperature change causes polarization. Ferroelectrics are materials that show a reversible and spontaneous polarization. $BaTiO_3$, *PZT*, PVDF are examples of ferroelectrics that are also piezoelectrics. Ferroelectrics exhibit a large dielectric constant, even in the paraelectric state, and are often used to make capacitors. Piezoelectrics are used in many sensor and actuator applications such as ultrasonic cleaners, ultrasound imaging, spark igniters, micro-actuators, and detection systems for submarines.

GLOSSARY

Avalanche breakdown   The reverse-bias voltage that causes a large current flow in a lightly doped *p-n* junction.

Bandgap   The energy between the top of the valence band and the bottom of the conduction band that a charge carrier must obtain before it can transfer a charge.

Capacitor   A microelectronic device, constructed from alternating layers of a dielectric and a conductor, that is capable of storing a charge. These can be single layer or multi-layer devices.

Conduction band   The unfilled energy levels into which electrons can be excited to provide conductivity.

Curie temperature   The temperature at which a ferroelectric to paraelectric transition occurs.

Current density   The current flowing through per unit cross-sectional area.

Defect semiconductors   Compounds, such as ZnO and $BaTiO_3$, that contain defects that provide semiconductivity.

Dielectric constant   The ratio of the permittivity of a material to the permittivity of a vacuum, thus describing the relative ability of a material to polarize and store a charge; the same as relative permittivity.

Dielectric loss   A measure of how much electrical energy is lost due to motion of charge entities that respond to the electric field via different polarization mechanisms. This energy appears as heat.

Dielectric strength   The maximum electric field that can be maintained between two conductor plates without causing a breakdown.

Doping   Deliberate addition of controlled amounts of other elements to increase the number of charge carriers in a semiconductor.

Drift velocity   The average rate at which electrons or other charge carriers move through a material under the influence of an electric or magnetic field.

Electric field   The voltage gradient or volts per unit length.

Electrostriction   The dimensional change that occurs in any material when an electric field is acting on it.

Energy gap (Bandgap)   The energy between the top of the valence band and the bottom of the conduction band that a charge carrier must obtain before it can transfer a charge.

Extrinsic semiconductor   A semiconductor prepared by adding dopants, which determine the number and type of charge carriers. Extrinsic behavior can also be seen due to impurities.

Ferroelectric   A material that shows spontaneous and reversible dielectric polarization.

Forward bias   Connecting a *p-n* junction device so that the *p*-side is connected to positive. Enhanced diffusion occurs as the energy barrier is lowered, permitting a considerable amount of current can flow under forward bias.

Holes   Unfilled energy levels in the valence band. Because electrons move to fill these holes, the holes move and produce a current.

Hybridization   When valence and conduction bands are separated by an energy gap, leading to the semiconductive behavior of silicon and germanium.

Hysteresis loop   The loop traced out by the nonlinear polarization in a ferroelectric material as the electric field is cycled. A similar loop occurs in certain magnetic materials.

Intrinsic semiconductor   A semiconductor in which properties are controlled by the element or compound that makes the semiconductor and not by dopants or impurities.

Linear dielectrics   Materials in which the dielectric polarization is linearly related to the electric field; the dielectric constant is not dependent on the electric field.

Matthiessen's rule   The resistivity of a metallic material is given by the addition of a base resistivity that accounts for the effect of temperature (ρ_T), and a temperature independent term that reflects the effect of atomic level defects, including impurities forming solid solutions (ρ_d).

Mean free path   The average distance that electrons can move without being scattered by other atoms.

Microstructure-sensitive property   Properties that depend on the microstructure of a material (e.g., conductivity, dielectric constant, or yield strength).

Mobility The ease with which a charge carrier moves through a material.

Nonradiative recombination Combining of electrons and holes leading to generation of heat; this occurs mainly in indirect bandgap materials such as Si.

Nonlinear dielectrics Materials in which dielectric polarization is not linearly related to the electric field (e.g., ferroelectric). These have a field-dependent dielectric constant.

p-n Junction A device made by creating an *n*-type region in a *p*-type material (or vice versa). A *p-n* junction behaves as a diode and multiple *p-n* junctions function as transistors. It is also the basis of LEDs and solar cells.

Permittivity The ability of a material to polarize and store a charge within it.

Piezoelectrics Materials that develop voltage upon the application of a stress and develop strain when an electric field is applied.

Polarization Movement of charged entities (i.e., electron cloud, ions, dipoles, and molecules) in response to an electric field.

Pyroelectric The ability of a material to spontaneously polarize and produce a voltage due to changes in temperature.

Radiative recombination Recombination of holes and electrons that leads to emission of light; this occurs in direct bandgap materials.

Rectifier A *p-n* junction device that permits current to flow in only one direction in a circuit.

Reverse bias Connecting a junction device so that the *p*-side is connected to a negative terminal; very little current flows through a *p-n* junction under reverse bias.

Superconductivity Flow of current through a material that has no resistance to that flow.

Superconductor A material that exhibits zero electrical resistance under certain conditions.

Thermistor A semiconductor device that is particularly sensitive to changes in temperature, permitting it to serve as an accurate measure of temperature.

Transducer A device that receives one type of input (such as strain or light) and provides an output that may be of a different type (such as an electrical signal).

Transistor A semiconductor device that can be used to amplify electrical signals.

Valence band The energy levels filled by electrons in their lowest energy states.

Zener diode A heavily doped *p-n* junction device which breaks down via a tunneling mechanism, used as a device to limit current.

ADDITIONAL INFORMATION

1. NEELKANTA, P.S., *Handbook of Electromagnetic Materials*. 1995: Boca Raton, CRC Press.

2. CHU, P.C.W., "High Temperature Superconductors," *Scientific American*, 1995. 163 (September): p. 163.

3. ONNES, K.H., "Further Experiments with Liquid Helium," *Comm. Phys. Lab. Univ. Leiden*, 1911. 120b.

4. WU, M.K., et. al, "Superconductivity at 93 K in New Mixed Phase Y-Ba-Cu-O Compound at Ambient Pressure," *Physics Rev. Letters*, 1987. 65: p. 908.

5. NAGAMATSU, J., *Nature*, 2001. 410 (March 1): p. 63.

6. COOLEY, L.D., et. al, "Potential Applications of Magnesium Diboride for Acclerator Magnet Applications," in *Proc. of the 2001 Particle Acclerator Conf.*, 2001. Chicago: IEEE.

7. LEHNDORFF, B.R., "High T_c Superconductors for Magnet and Energy Technology," *Springer Tracts in Modern Physics*. 2001: Springer.

8. MANNHART, J. and P. CHAUDHARI, "High-T_c Bicrystal Grain Boundaries," *Physics Today*, 2001: (November) p. 48–53.

9. OISHI, Y. and K. ANDO, *Transport in Nonstoichiomertic Compounds*, G. SIMKOVICH and U.S. STUBICAN, Editors, Plenum. p. 188–202.

10. GINLEY, D.S. and C. BRIGHT, "Transparent Conducting Oxides," *MRS Bulletin*, 2000. 25(8): p. 15.

11. SINGHAL, S.C., "Science and Technology of Solid-Oxide Fuel Cells," *MRS Bulletin*, 2000. 25(3): p. 16.

12. *Discussions with Professor P.N. Kumta, Carnegie Mellon University are acknowledged* (2002).

13. KULKARNI, V.G., "Polyanilines: Progress in Processing and Applications," in *ACS Symp. Series*. 1999: American Chemical Society.

14. IIJIMA, S., "Helical Microtubules of Graphitic Carbon," *Nature*, 1991. 354: p. 56.

15. MAHAJAN, S. and K.S. SREE HARSHA, *Principles of Growth and Processing of Semiconductors*. 1999: McGraw-Hill.

16. STREETMAN, B.G., *Solid-State Electronic Devices*. 3rd ed., 2000: Prentice Hall.

17. MOULSON, A.J. and J.M. HERBERT, *Electroceramics: Materials, Properties, Applications*. 1990: Chapman and Hall.

18. SINGH, J., *Optoelectronics: An Introduction to Materials and Devices*. 1996.

19. BHATTACHARYA, P., *Semiconductor Optoelectronic Devices*. 1997, Upper Saddle River, New Jersey: Prentice Hall.

20. VAN ZANT, P., *Microchip Fabrication*. 3rd ed., 1997: McGraw-Hill.

21. BUCHANAN, R.C., *Ceramic Materials for Electronics*. 2nd ed., 1991: Marcel Dekker.

22. XU, Y., *Ferroelectric Materials*. 1991: North-Holland.

23. HENCH, L.L. and J.K. WEST, *Principles of Electronic Ceramics*. 1990: Wiley Interscience.

24. KASAP, S.O., *Principles of Electrical Engineering Materials and Devices*. 1997: Irwin.

25. FLOYD, T.L., *Electronic Devices*. 2002: Upper Saddle River, New Jersey, Prentice Hall.

26. GROOVER, M.P., *Fundamentals of Modern Manufacturing*. 1996: Prentice Hall.

27. ALLURI, P., P. MAJHI, D. TANG, and S.K. DEY, *Integrated Ferroelectrics* vol. 21. (1998): p. 305.

✔ PROBLEMS

Section 18-1 Ohm's Law and Electrical Conductivity

Section 18-2 Band Structures of Solids

Section 18-3 Conductivity of Metals and Alloys

18-1 A current of 10 A is passed through a 1-mm-diameter wire 1000 m long. Calculate the power loss if the wire is made of

(a) aluminum,
(b) silicon, and
(c) silicon carbide. (See Table 18-1.)

18-2 A 0.5-mm-diameter fiber 1 cm in length made of boron nitride is placed into a 120-V circuit. Using Table 18-1, calculate

(a) the current flowing in the circuit, and
(b) the number of electrons passing through the boron-nitride fiber per second.
(c) What would the current and number of electrons be if the fiber were made of magnesium instead of boron nitride?

18-3 Do all materials and devices obey Ohm's law? (*Hint*: Consider such devices as materials used for lightning arrestors based on ZnO, *p-n* junctions, etc.)

18-4   The power lost in a 2-mm-diameter copper wire is to be less than 250 W when a 5-A current is flowing in the circuit. What is the maximum length of the wire?

18-5   A current density of 100,000 A/cm^2 is applied to a gold wire 50 m in length. The resistance of the wire is found to be 2 ohm. Calculate the diameter of the wire and the voltage applied to the wire.

18-6   We would like to produce a 5000-ohm resistor from boron-carbide fibers having a diameter of 0.1 mm. What is the required length of the fibers?

18-7   Suppose we estimate that the mobility of electrons in silver is 75 $cm^2/V \cdot s$. Estimate the fraction of the valence electrons that are carrying an electrical charge.

18-8   A current density of 5000 A/cm^2 is applied to a magnesium wire. If half of the valence electrons serve as charge carriers, calculate the average drift velocity of the electrons.

18-9   We apply a voltage of 10 V to an aluminum wire 2 mm in diameter and 20 m long. If 10% of the valence electrons carry the electrical charge, calculate the average drift velocity of the electrons in km/h and miles/h.

18-10   In a welding process, a current of 400 A flows through the arc when the voltage is 35 V. The length of the arc is about 0.1 in. and the average diameter of the arc is about 0.18 in. Calculate the current density in the arc, the electric field across the arc, and the electrical conductivity of the hot gases in the arc during welding.

18-11   Draw a schematic of the band structures of a dielectric, a semiconductor, and a metallic material. Use this to explain why the conductivity of pure metals decreases with increasing temperature, while the opposite is true for semiconductors and dielectrics.

18-12   Calculate the electrical conductivity of nickel at $-50°C$ and at $+500°C$.

18-13   The electrical resistivity of pure chromium is found to be 18×10^{-6} ohm $\cdot$ cm. Estimate the temperature at which the resistivity measurement was made.

18-14   After finding the electrical conductivity of cobalt at 0°C, we decide we would like to double that conductivity. To what temperature must we cool the metal?

18-15   From Figure 18-11(b), estimate the defect-resistivity coefficient for tin in copper.

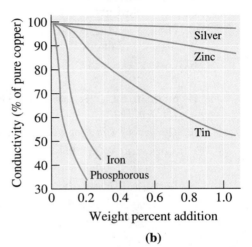

(b)

Figure 18-11   (Repeated for Problems 18-15 and 18-17) (b) the effect of selected elements on the electrical conductivity of copper.

18-16   The electrical resistivity of a beryllium alloy containing 5 at% of an alloying element is found to be 50×10^{-6} ohm $\cdot$ cm at 400°C. Determine the contributions to resistivity due to temperature and due to impurities by finding the expected resistivity of pure beryllium at 400°C, the resistivity due to impurities, and the defect-resistivity coefficient. What would be the electrical resistivity if the beryllium contained 10 at% of the alloying element at 200°C?

18-17   Is Equation 18-7 valid for the copper-zinc system? If so, calculate the defect resistivity coefficient for zinc in copper. (See Figure 18-11).

Section 18-4   Superconductivity

18-18   Define the following terms: a superconductor, Meissner effect, Type I superconductor, and Type II superconductor.

18-19 GaV_3 is to operate as a superconductor in liquid helium (at 4 K). The T_c is 16.8 K and H_0 is 350,000 oersted. What is the maximum magnetic field that can be applied to the material?

18-20 Nb_3Sn and GaV_3 are candidates for a superconductive application when the magnetic field is 150,000 oersted. Which would require the lower temperature in order to be superconductive?

18-21 A filament of Nb_3Sn 0.05 mm in diameter operates in a magnetic field of 1000 oersted at 4 K. What is the maximum current that can be applied to the filament in order for the material to behave as a superconductor?

18-22 Assume that most of the electrical charge transferred in MgO is caused by the diffusion of Mg^{2+} ions. Determine the mobility and electrical conductivity of MgO at 25°C and at 1500°C. (See Table 5-1.)

18-23 Assume that most of the electrical charge transferred in Al_2O_3 is caused by the diffusion of Al^{3+} ions. Determine the mobility and electrical conductivity of Al_2O_3 at 500°C and at 1500°C. (See Table 5-1 and Example 14-1.)

18-24 What is a "weak link" problem in Type II ceramic superconductors? What limits the current density in these materials? What is different about MgB_2?

18-25 Explain how long wires of YBCO superconductors can be made.

18-26 State any three applications of superconducting materials.

Section 18-5 Conductivity in Other Materials

18-27 Calculate the electrical conductivity of a fiber-reinforced polyethylene part that is reinforced with 20 vol% of continuous, aligned nickel fibers.

18-28 What are ionic conductors? What are their applications?

18-29 How do the touch screen displays on some computers work?

18-30 Can polymers be semiconducting? What would be the advantages in using these instead of silicon?

Section 18-6 Semiconductors

Section 18-7 Applications of Semiconductors

18-31 Explain the following terms: semiconductor, intrinsic semiconductor, extrinsic semiconductor, elemental semiconductor, compound semiconductor, direct and indirect bandgap semiconductor.

18-32 What is radiative and nonradiative recombination? What types of materials are used to make LEDs?

18-33 For germanium, silicon, and tin, compare, at 25°C, the number of charge carriers per cubic centimeter, the fraction of the total electrons in the valence band that are excited into the conduction band, and constant n_0.

18-34 For germanium, silicon, and tin, compare the temperature required to double the electrical conductivity from the room temperature value.

18-35 Determine the electrical conductivity of silicon when 0.0001 at% antimony is added as a dopant and compare it to the electrical conductivity when 0.0001 at% indium is added.

18-36 We would like to produce an extrinsic germanium semiconductor having an electrical conductivity of 2000 $ohm^{-1} \cdot cm^{-1}$. Determine the amount of phosphorous and the amount of gallium required.

18-37 Estimate the electrical conductivity of silicon doped with 0.0002 at% arsenic at 600°C, which is above the plateau in the conductivity-temperature curve.

18-38 Determine the amount of arsenic that must be combined with 1 kg of gallium to produce a *p*-type semiconductor with an electrical conductivity of 500 $ohm^{-1} \cdot cm^{-1}$ at 25°C. The lattice parameter of GaAs is about 5.65 Å, and GaAs has the zinc blende structure.

18-39 A ZnO crystal is produced in which one interstitial Zn atom is introduced for every 500 Zn lattice sites. Estimate the number of charge carriers per cubic centimeter and the electrical conductivity at 25°C.

18-40 Each Fe^{3+} ion in FeO serves as an acceptor site for an electron. If there is one vacancy per 750 unit cells of the FeO crystal (with the sodium chloride structure), determine the number of possible charge carriers per cubic centimeter. The lattice parameter of FeO is 0.429 nm.

18-41 When a voltage of 5 mV is applied to the emitter of a transistor, a current of 2 mA is produced. When the voltage is increased to 8 mV, the current through the collector rises to 6 mA. By what percentage will the collector current increase when the emitter voltage is doubled from 9 mV to 18 mV?

18-42 Design a light emitting diode that will emit at 1.12 micrometers. Is this wavelength in the visible range? What will be the application of this type of LED?

18-43 How can we make LEDs that emit white light (i.e., light that looks like sun light)?

18-44 Outline the process for manufacturing semiconductor chips.

18-45 How are amorphous silicon solar cells made?

Section 18-9 Polarization in Dielectrics

18-46 In mechanical properties we have seen that stress (a cause) produces strain (an effect). What is the electrical analog of this?

18-47 In mechanical properties, elastic modulus represents the elastic energy stored and viscous dissipation represents mechanical energy lost in deformation. What is the electrical analog for this?

18-48 Calculate the displacement of the electrons or ions for the following conditions:

(a) electronic polarization in nickel of 2×10^{-7} C/m^2,

(b) electronic polarization in aluminum of 2×10^{-8} C/m^2,

(c) ionic polarization in NaCl of 4.3×10^{-8} C/m^2, and

(d) ionic polarization in ZnS of 5×10^{-8} C/m^2.

18-49 A 2-mm-thick alumina dielectric is used in a 60-Hz circuit. Calculate the voltage required to produce a polarization of 5×10^{-7} C/m^2.

18-50 Suppose we are able to produce a polarization of 5×10^{-5} C/cm^2 in a cube (5 mm side) of barium titanate. What voltage is produced?

18-51 Calculate the thickness of polyethylene required to store the maximum charge in a 24,000-V circuit without breakdown.

18-52 What polarization mechanism will be present in (a) alumina, (b) copper, (c) silicon, and (d) barium titanate?

18-53 What is lead crystal? Why does it have a high refractive index? Is this a crystalline material?

Section 18-10 Electrostriction, Piezoelectricity, Pyroelectricity, and Ferroelectricity

18-54 Define the following terms: electrostriction, piezoelectricity (define both its direct and converse effects), pyroelectricity, and ferroelectricity.

18-55 What are ferroelectric domains? Sketch and explain the variation of dielectric constant of barium titanate as a function of temperature.

18-56 Why is pure BaTiO$_3$ not suitable for making capacitors?

18-57 Calculate the capacitance of a parallel-plate capacitor containing 5 layers of mica, where each mica sheet is 1 cm $\times$ 2 cm $\times$ 0.005 cm.

18-58 A multi-layer capacitor is to be designed using a relaxor ferroelectric formulation based on lead magnesium niobate *(PMN)*. The apparent dielectric constant of the material is 20,000. Calculate the capacitance of a multi-layer capacitor consisting of 10 layers connected in parallel using Ni electrodes. The sides of the capacitor are 10 $\times$ 10 mm and the thickness of each layer is 20 μm.

18-59 Why does the dielectric constant of most materials go down with increasing electrical frequency?

18-60 Determine the number of Al$_2$O$_3$ sheets, each 1.5 cm $\times$ 1.5 cm $\times$ 0.001 cm, required to obtain a capacitance of 0.0142 μF in a 10^6 Hz parallel-plate capacitor.

18-61 We would like to construct a barium titanate device with a 0.1-in. diameter that will produce a voltage of 250 V when a 5-pound force is applied. How thick should the device be?

18-62 A force of 20 lb is applied to the face of a 0.5 cm $\times$ 0.5 cm $\times$ 0.1 cm thick quartz crystal. Determine the voltage produced by the force. The modulus of elasticity of quartz is 10.4×10^6 psi.

18-63 Determine the strain produced when a 300 V signal is applied to a barium titanate wafer 0.2 cm $\times$ 0.2 cm $\times$ 0.01 cm thick.

18-64 Figure 18-35 shows the hysteresis loops for two ferroelectric materials: Determine the voltage required to eliminate polarization in a 0.1-cm-thick dielectric made from material A.

18-65 From Figure 18-35, determine the thickness of a dielectric made from material B if 10 V is required to eliminate polarization.

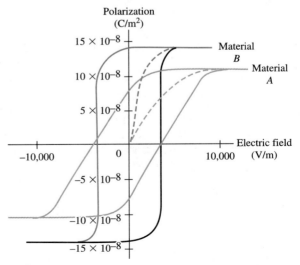

Figure 18-35 Ferroelectric hysteresis loops (for Problems 18-64–18-67).

18-66 Using Figure 18-35, what electric field is required to produce a polarization of 8×10^{-8} C/m^2 in material A, and what is the dielectric constant at this polarization?

18-67 An electric field of 2500 V/m is applied to material B in Figure 18-35. Determine the polarization and the dielectric constant at this electric field.

✳ Design Problems

18-68 We would like to produce a 100-ohm resistor using a thin wire of a material. Design such a device.

18-69 Design a capacitor that is capable of storing 1 μF when 100 V is applied.

18-70 Design an epoxy-matrix composite that has a modulus of elasticity of at least 35×10^6 psi and an electrical conductivity of at least 1×10^5 ohm$^{-1} \cdot$ cm^{-1}.

18-71 Design a semiconductor thermistor that will activate a cooling system when the ambient temperature reaches 500°C.

18-72 Design a dielectric device that will detect whether sand is at a particular level in a sand storage tank.

18-73 Design a *PZT* piezoelectric part that will produce 25,000 V when subjected to a stress of 5 psi, assuming piezoelectric properties of *PZT* listed in Table 18-9.

▢ Computer Problems

18-74 *Design of Multi-layer Capacitors.* Write a computer program that can be used to calculate the capacitance of a multi-layer capacitor. The program, for example, should ask the user to provide values of the dielectric constant, thickness of the layers, and dimensions of the layer. The program should also be flexible in that if the user provides an intended value of capacitance and other dimensions, the program should offer a value of the dielectric constant for the formulation to be used.

Chapter

A magnetic hard drive is the heart of personal and laptop computers. These disks use magnetic materials that are unique in that information can be easily written to them, but information cannot be easily erased. The hard drive system is complex in that it makes use of nanostructured and nanoscale thin-film magnetic materials for information storage. (*Courtesy of Seagate.*)

19

Magnetic Materials

Have You Ever Wondered?

- *What materials are used to make audio and video cassettes?*

- *What affects the "lifting strength" of a magnet?*

- *What role do magnetic materials play in DNA sequencing?*

- *What are "soft" and "hard" magnetic materials?*

- *Are there "nonmagnetic" materials?*

- *What is "spintronics?"*

- *Could there be materials that develop mechanical strain upon the application of a magnetic field?*

- *Who was the only husband and wife team to win the Nobel prize?*

Every material in the world responds to the presence of a magnetic field. Magnetic materials are used to operate such things as electrical motors, generators, and transformers. Much of the data storage technology (computer hard disks, computer disks, video and audio cassettes, and the like) is based on magnetic particles. Magnetic materials are also used in loudspeakers, telephones, CD players, telephones, televisions, and video recorders. Superconductors can also be viewed as magnetic materials; their applications were discussed in Chapter 18. Magnetic materials, such as iron oxide (Fe_3O_4) particles, are used to make exotic compositions of "liquid magnets" or ferrofluids. The same iron oxide particles are also used to bind DNA molecules, cells, and proteins. This is the basis for magnetic separation techniques used in bioengineering and DNA sequencing. The most widely used magnetic materials are based on ferromagnetic metals and alloys such as iron, nickel, and cobalt or ferrimagnetic ceramics, including various ferrites and garnets. Alloys of Fe-Nd-B and Sm-Co make very strong "permanent" magnetic materials and are used in a wide variety of applications. Magnetic materials are used in many sensors and actuators, and some materials, known as magnetostrictive materials, develop a strain on the application of a magnetic field.

Magnetic behavior is determined primarily by the electronic structure of a material, which provides magnetic dipoles. Interactions between these dipoles determine the type of magnetic behavior that is observed. Magnetic behavior can be controlled by composition, microstructure, and the processing of basic materials. In this chapter, we look at the fundamental basis for responses of certain materials to the presence of magnetic fields. We will also examine the properties and applications of different types of magnetic materials. Much of the terminology we will be using in describing ferromagnetic and ferrimagnetic materials is similar to that used for ferroelectric materials discussed in Chapter 18.

19-1   Classification of Magnetic Materials

Strictly speaking, there is no such thing as a "nonmagnetic" material. Every material in this world consists of atoms; atoms consist of electrons spinning around them similar to a current-carrying loop that generates a magnetic field. Thus, every material responds to a magnetic field. The manner in which this response of electrons and atoms in a material is scaled determines whether a material will be strongly or weakly magnetic. Examples of **ferromagnetic** materials are materials such as Fe, Ni, Co, and some of their alloys. Examples of **ferrimagnetic** materials include many ceramic materials such as nickel zinc ferrite and manganese zinc ferrite. If someone uses the term "nonmagnetic", they mean that the material is neither ferromagnetic nor ferrimagnetic. These "nonmagnetic" materials are further classified as **diamagnetic** (e.g., superconductors) or **paramagnetic**. In some cases, we also encounter materials that are **antiferromagnetic** or **superparamagnetic**. We will discuss these different classes of materials and their applications a bit later in the chapter. Ferromagnetic and ferrimagnetic materials are usually further classified as either soft or hard magnetic materials. High-purity iron or plain carbon steels are examples of a magnetically soft material as they can become magnetized, but when the magnetizing source is removed, these materials lose their magnet-like behavior.

Permanent magnets or **hard magnetic materials** retain their magnetization. These are pure "magnets." Many ceramic ferrites are used to make inexpensive refrigerator magnets. A hard magnetic material does not lose its magnetic behavior easily.

19-2   Magnetic Dipoles and Magnetic Moments

The magnetic behavior of materials can be traced to the structure of atoms. Electrons in atoms have a planetary motion in that they go around the nucleus. The orbital motion

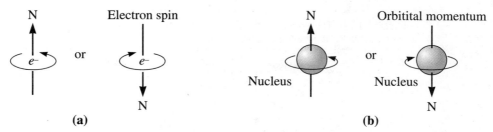

Figure 19-1 Origin of magnetic dipoles: (a) The spin of the electron produces a magnetic field with a direction dependent on the quantum number m_s. (b) Electrons orbiting around the nucleus create a magnetic field around the atom.

of the electron around the nucleus and the spin of the electron about its own axis (Figure 19-1) cause separate magnetic moments. These two motions (i.e., spin and orbital) contribute to the magnetic behavior of materials. When the electron spins, there is a magnetic moment associated with that motion. The magnetic moment of an electron due to its spin is known as the **Bohr magneton** (M_B). This is a fundamental constant and is defined as:

$$M_B = \text{Bohr magneton} = \frac{q\text{h}}{4\pi m_e} = 9.274 \times 10^{-24} \text{ A} \cdot \text{m}^2 \tag{19-1}$$

where q is the charge on the electron, h is Planck's constant, and m_e is the mass of the electron. This moment is directed along the axis of electron spin.

The nucleus of the atom consists of protons and neutrons. These also have a spin. However, the overall magnetic moment due to their spin is much smaller than that for electrons. We normally do not encounter the effects of a magnetic moment of a nucleus with the exception of such applications as nuclear magnetic resonance (NMR).

We can view electrons in materials as small elementary magnets. If the magnetic moments due to electrons in materials could line up in the same direction, the world would be a magnetic place! However this, as you know, is not the case. Thus, there must be some mechanism by which the magnetic moments associated with electron spin and their orbital motion get canceled in most materials, leaving behind only a few materials that are "magnetic." There are the two effects that, fortunately, make most materials in the world not be "magnetic."

First, we must consider the magnetic moment of atoms: According to the Pauli exclusion principle, two electrons within the same energy level (orbital state) must have opposite spins. This means their electron-spin-derived magnetic moments are opposite (one can be considered "up ↑" and the other one "down ↓") and cancel out. The second effect is that, the orbital moments of electrons also cancel each other out. Thus, in a completely filled shell, all electron-spin and orbital moments will cancel out. This is why atoms of most elements do not have a net magnetic moment. Some elements, such as transition elements (3d, 4d, 5d partially filled), the lanthanides (4f partially filled), and actinides (5f partially filled), have a net magnetic moment since some of their levels have an unpaired electron.

Certain elements, such as the transition metals, have an inner energy level that is not completely filled. The elements scandium (Sc) through copper (Cu), whose electronic structures are shown in Table 19-1, are typical. Except for chromium and copper, the valence electrons in the 4s level are paired; the unpaired electrons in chromium and copper are canceled by interactions with other atoms. Copper also has a completely filled 3d shell and thus does not display a net magnetic moment.

TABLE 19-1 ■ *The electron spins in the 3d energy level in transition metals, with arrows indicating the direction of spin*

Metal	3d					4s
Sc	↑					↑↓
Ti	↑	↑				↑↓
V	↑	↑	↑			↑↓
Cr	↑	↑	↑	↑	↑	↑
Mn	↑	↑	↑	↑	↑	↑↓
Fe	↑↓	↑	↑	↑	↑	↑↓
Co	↑↓	↑↓	↑	↑	↑	↑↓
Ni	↑↓	↑↓	↑↓	↑	↑	↑↓
Cu	↑↓	↑↓	↑↓	↑↓	↑↓	↑

The electrons in the 3*d* level of the remaining transition elements do not enter the shells in pairs. Instead, as in manganese, the first five electrons have the same spin. Only after half of the 3*d* level is filled do pairs with opposing spins form. Therefore, each atom in a transition metal has a permanent magnetic moment, which is related to the number of unpaired electrons. Each atom behaves as a magnetic dipole.

In many elements, these magnetic moments exist for free individual atoms, however, when the atoms form crystalline materials, these moments are "quenched" or canceled out. Thus, a number of materials made from elements whose atoms have a net magnetic moment do not exhibit magnetic behavior. For example, the Fe^{+2} ion has a net magnetic moment of $4\mu_B$ (4 times the magnetic moment of an electron), however $FeCl_2$ crystals are not magnetic.

The response of the atom to an applied magnetic field depends on how the magnetic dipoles represented by each atom react to the field. Most of the transition elements (e.g., Cu, Ti) react in such a way that the sum of the individual atoms' magnetic moments is zero. However, the atoms in nickel (Ni), iron (Fe), and cobalt (Co) undergo an exchange interaction, whereby the orientation of the dipole in one atom influences the surrounding atoms to have the same dipole orientation, producing a desirable amplification of the effect of the magnetic field. In the case of Fe, Ni, and Co, the magnetic moments of the atoms line up in the same directions, and these materials are known as ferromagnetic.

In certain materials, such as BCC chromium (Cr), the magnetic moments of atoms at the center of the unit cell are opposite in direction to those of the atoms at the corners of the unit cell, thus, the net moment is zero. Materials in which there is a complete cancellation of the magnetic moments of atoms or ions are known as anti-ferromagnetic.

Materials in which magnetic moments of different atoms or ions do not completely cancel out are known as ferrimagnetic materials. We will discuss these materials in a later section.

19-3   Magnetization, Permeability, and the Magnetic Field

Let's examine the relationship between the magnetic field and magnetization. Figure 19-2 depicts a coil having *n* turns. When an electric current is passed through the coil, a magnetic field *H* is produced, with the strength of the field given by

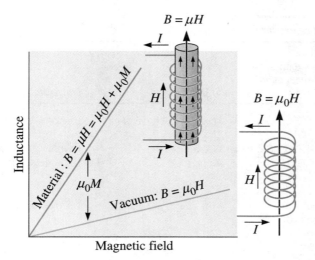

Figure 19-2
A current passing through a coil sets up a magnetic field H with a flux density B. The flux density is higher when a magnetic core is placed within the coil.

$$H = \frac{nI}{l} \tag{19-2}$$

where n is the number of turns, l is the length of the coil (m), and I is the current (A). The units of H are therefore ampere $\cdot$ turn/m, or simply A/m. An alternate unit for magnetic field is the oersted, obtained by multiplying A/m by $4\pi \times 10^{-3}$ (see Table 19-2).

When a magnetic field is applied in a vacuum, lines of magnetic flux are induced. The number of lines of flux, called the flux density, or *inductance B*, is related to the applied field by

$$B = \mu_0 H \tag{19-3}$$

where B is the inductance, H is the magnetic field, and μ_0 is a constant called the **magnetic permeability of vacuum**. If H is expressed in units of oersted, then B is in gauss and μ_0 is 1 gauss/oersted. In an alternate set of units, H is in A/m, B is in tesla (also called weber/m^2), and μ_0 is $4\pi \times 10^{-7}$ weber/A $\cdot$ m (also called henry/m).

When we place a material within the magnetic field, the magnetic-flux density is determined by the manner in which induced and permanent magnetic dipoles interact with the field. The flux density now is

$$B = \mu H, \tag{19-4}$$

where μ is the permeability of the material in the field. If the magnetic moments reinforce the applied field, then $\mu > \mu_0$, a greater number of lines of flux that can accomplish work are created, and the magnetic field is magnified. If the magnetic moments oppose the field, however, $\mu < \mu_0$.

We can describe the influence of the magnetic material by the relative permeability μ_r, where:

$$\mu_r = \frac{\mu}{\mu_0} \tag{19-5}$$

A large relative permeability means that the material amplifies the effect of the magnetic field. Thus, the relative permeability has the same importance that conductivity has in dielectrics. A material with higher magnetic permeability (e.g., iron) will carry magnetic flux more readily. We will learn later that the permeability of ferromagnetic or

TABLE 19-2 ■ *Units, conversions, and values for magnetic materials*

	Gaussian and cgs emu (*Electromagnetic Units*)	SI Units	Conversion
Inductance or magnetic flux density (B)	Gauss (G)	Tesla (or weber (Wb)/m^2)	1 tesla = 10^4 G, Wb/m^2
Magnetic flux (ϕ)	Maxwell (Mx), G-cm^2	Wb, volt-second	1 Wb = 10^8 G-cm^2
Magnetic potential difference or magnetic electromotive force (U, F)	Gilbert (Gb)	Ampere (A)	1 A = $4\pi \times 10^{-1}$ Gb
Magnetic field strength, magnetizing force (H)	Oersted (Oe), Gilbert (Gb)/cm	A/m	1 A/m = $4\pi \times 10^{-3}$ Oe
(Volume) magnetization (M)	emu/cm^3	A/m	1 A/m = 10^{-3} emu/cm^3
(Volume) magnetization ($4\pi M$)	G	A/m	1 A/m = $4\pi \times 10^{-3}$ G
Magnetic polarization or intensity of magnetization (J or I)	emu/cm^3	T, Wb/m^2	1 tesla = $(1/4\pi) \times 10^4$ emu/cm^3
(Mass) magnetization (σ, M)	emu/g	A-m^2/kg Wb-m/kg	1 1 Wb-m/kg = $(1/4\pi) \times 10^7$ emu/g
Magnetic moment (m)	emu, erg/G	A-m^2, Joules (J) per tesla (J/T)	1 J/T = 10^3 emu
Magnetic dipole moment (j)	emu, erg/G	Wb-m	1 Wb-m = $(1/4\pi) \times 10^{10}$ emu
Magnetic permeability (μ)	Dimensionless	Wb/A · m, (henry (H)/m)	1 Wb/A · m = $(1/4\pi) \times 10^7$
Magnetic permeability of free space (μ_0)	1 gauss/oersted	$\mu_0 = (4\pi) \times 10^{-7}$ H/m (value)	
Relative permeability (μ_r)	Not defined	Dimensionless	
(Volume) energy density, energy product (W)	erg/cm^3	J/m^3	1 J/m^3 = 10 erg/cm^3

ferrimagnetic materials is not constant and depends on the value of the applied magnetic field (H).

The **magnetization** M represents the increase in the inductance due to the core material, so we can rewrite the equation for inductance as:

$$B = \mu_0 H + \mu_0 M \tag{19-6}$$

The first part of this equation is simply the effect of the applied magnetic field. The second part is the effect of the magnetic material that is present. This is similar to our discussion on dielectric polarization and the mechanical behavior of materials. In materials, stress causes strain, electric field induces (E) dielectric polarization (P), and a magnetic field (H) causes magnetization ($\mu_0 M$) that contributes to the total flux density B.

The **magnetic susceptibility** χ_m, which is the ratio between magnetization and the applied field, gives the amplification produced by the material:

$$\chi_m = \frac{M}{H} \tag{19-7}$$

Both μ_r and χ_m refer to the degree to which the material enhances the magnetic field and are therefore related by:

$$\mu_r = 1 + \chi_m \tag{19-8}$$

As noted before, the μ_r and, therefore, the χ_m values for ferromagnetic and ferrimagnetic materials depend on the applied field (H). For ferromagnetic and ferrimagnetic materials, the term $\mu_0 M \gg \mu_0 H$. Thus, for these materials:

$$B \cong \mu_0 M \qquad (19\text{-}9)$$

We sometimes interchangeably refer to either inductance or magnetization. Normally, we are interested in producing a high-inductance B or magnetization M. This is accomplished by selecting materials that have a high relative permeability or magnetic susceptibility.

The following example shows how these concepts can be applied for comparing actual and theoretical magnetizations in pure iron.

EXAMPLE 19-1   *Theoretical and Actual Saturation Magnetization in Fe*

Calculate the maximum, or saturation, magnetization that we expect in iron. The lattice parameter of BCC iron is 2.866 Å. Compare this value with 2.1 tesla (a value of saturation flux density experimentally observed for pure Fe.)

SOLUTION

Based on the unpaired electronic spins, we expect each iron atom to have four electrons that act as magnetic dipoles. The number of atoms per m^3 in BCC iron is:

$$\text{Number of Fe atoms/m}^3 = \frac{2 \text{ atoms/cell}}{(2.866 \times 10^{-10} \text{ m})^3} = 8.48 \times 10^{28}$$

The maximum volume magnetization (M_{sat}) is the total magnetic moment per unit volume:

$$M_{sat} = \left[8.48 \times 10^{28} \frac{\text{atoms}}{\text{m}^3}\right] [9.27 \times 10^{-24} \text{ Am}^2] \left[4 \frac{\text{Bohr magnetons}}{\text{atom}}\right]$$

$$M_{sat} = 3.15 \times 10^6 \frac{\text{A}}{\text{m}}$$

To convert the value of saturation magnetization M into saturation flux density B in tesla, we need the value of $\mu_0 M$. In ferromagnetic materials $\mu_0 M \gg \mu_0 H$ and therefore, $B \cong \mu_0 M$.

Saturation induction in tesla $= B_{sat} = \mu_0 M_{sat}$.

$$B_{sat} = \left(4\pi \times 10^{-7} \frac{\text{Wb}}{\text{m} \cdot \text{A}}\right) \left(3.15 \times 10^6 \frac{\text{A}}{\text{m}}\right)$$

$$B_{sat} = 3.95 \frac{\text{Wb}}{\text{m}^2} = 3.95 \text{ tesla}$$

This is almost two times the experimentally observed value of 2.1 tesla. Reversing our calculations, we can show that the each iron atom contributes only about 2.2 Bohr magneton and not 4. This is the difference between behavior of individual atoms and their behavior in a crystalline solid. It can be shown that in the case of iron, the difference is due to the $3d$ electron orbital moment being quenched in the crystal.[3]

19-4   Diamagnetic, Paramagnetic, Ferromagnetic, Ferrimagnetic, and Superparamagnetic Materials

As mentioned before, there is no such thing as a "nonmagnetic" material. All materials respond to magnetic fields. When a magnetic field is applied to a material, several types of behavior are observed (Figure 19-3).

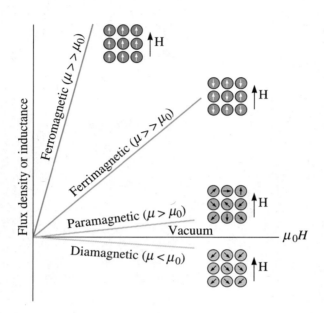

Figure 19-3
The effect of the core material on the flux density. The magnetic moment opposes the field in diamagnetic materials. Progressively stronger moments are present in paramagnetic, ferrimagnetic, and ferromagnetic materials for the same applied field.

Diamagnetic Behavior   A magnetic field acting on any atom induces a magnetic dipole for the entire atom by influencing the magnetic moment caused by the orbiting electrons. These dipoles oppose the magnetic field, causing the magnetization to be less than zero. This behavior, called **diamagnetism**, gives a relative permeability of about 0.99995 (or a negative susceptibility approximately -10^{-6}, note the negative sign). Materials such as copper, silver, silicon, gold, and alumina are diamagnetic at room temperature. Superconductors are perfect diamagnets ($\chi_m = -1$); they lose their superconductivity at higher temperatures or in the presence of a magnetic field (Chapter 18). In a diamagnetic material the magnetization (M) direction is opposite to the direction of applied field (H).

Paramagnetism   When materials have unpaired electrons, a net magnetic moment due to electron spin is associated with each atom. When a magnetic field is applied, the dipoles line up with the field, causing a positive magnetization. However, because the dipoles do not interact, extremely large magnetic fields are required to align all of the dipoles. In addition, the effect is lost as soon as the magnetic field is removed. This effect, called **paramagnetism**, is found in metals such as aluminum, titanium, and alloys of copper. The magnetic susceptibility (χ_m) of paramagnetic materials is positive and lies between 10^{-4} and 10^{-5}. Ferromagnetic and ferrimagnetic materials above the Curie temperature also exhibit paramagnetic behavior.

Ferromagnetism   **Ferromagnetic** behavior is caused by the unfilled energy levels in the $3d$ level of iron, nickel, and cobalt. Similar behavior is found in a few other materials, including gadolinium (Gd). In ferromagnetic materials, the permanent unpaired dipoles easily line up with the imposed magnetic field due to the exchange interaction, or mutual reinforcement of the dipoles. Large magnetizations are obtained even for small magnetic fields, giving large susceptibilities approaching 10^6. Similar to ferroelectrics (Chapter 18), the susceptibility of ferromagnetic materials depends upon the intensity of the applied magnetic field. This is similar to the mechanical behavior of elastomers whose modulus of elasticity depends upon the level of strain. Above the Curie temperature, ferromagnetic materials behave as paramagnetic materials and their susceptibility is given by the following equation known as the Curie-Weiss law:

$$\chi_m = \frac{C}{(T - T_c)} \qquad (19\text{-}10)$$

In this equation C is a constant that depends upon the material, T_c is the Curie temperature and T is the temperature above T_c. Essentially the same equation also describes the change in dielectric permittivity above the Curie temperature of ferroelectrics (Chapter 18). Similar to ferroelectrics, ferromagnetic materials show the formation of hystereis loop domains and magnetic domains. These materials will be discussed in the next section.

Antiferromagnetism In materials such as manganese, chromium, MnO, and NiO, the magnetic moments produced in neighboring dipoles line up in opposition to one another in the magnetic field, even though the strength of each dipole is very high. This effect is illustrated for MnO in Figure 19-4. These materials are **antiferromagnetic** and have zero magnetization. The magnetic susceptibility is positive and small. In addition, CoO and $MnCl_2$ are examples of antiferromagnetic materials.

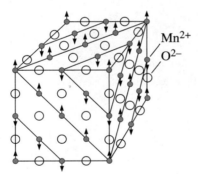

Mn^{2+}
O^{2-}

Figure 19-4
The crystal structure of MnO consists of alternating layers of {111} type planes of oxygen and manganese ions. The magnetic moments of the manganese ions in every other (111) plane are oppositely aligned. Consequently, MnO is antiferromagnetic.

Ferrimagnetism In ceramic materials, different ions have different magnetic moments. In a magnetic field, the dipoles of cation *A* may line up with the field, while dipoles of cation *B* oppose the field. But because the strength or number of dipoles are not equal, a net magnetization results. The **ferrimagnetic** materials can provide good amplification of the imposed field. We will look at a group of ceramics called ferrites that display this behavior in a later section. These materials show a large and magnetic field dependent magnetic susceptibility similar to ferromagnetic materials. They also show Curie-Weiss behavior (similar to ferromagnetic materials) at temperatures above the Curie transition temperature. Most ferrimagnetic materials are ceramics and are good insulators of electricity. Thus, in these materials, electrical losses (known as eddy current losses) are much smaller compared to those in metallic ferromagnetic materials. Therefore, ferrites are used in many high-frequency applications.

Superparamagnetism   When the size or grain size of ferromagnetic and ferrimagnetic materials gets below a certain critical size, these materials behave as if they are paramagnetic. The magnetic dipole energy of each particle becomes comparable to the thermal energy. This small magnetic moment changes its direction randomly (as a result of the thermal energy). Thus, the material behaves as if it has no net magnetic moment. This is known as superparamagnetism. Thus, if we produce iron oxide (Fe_3O_4) particles in a 3–5 nm size, they behave as superparamagnetic materials. Such iron oxide superparamagnetic particles are used to form dispersions in aqueous or organic carrier phases or to form "liquid magnets" or ferrofluids.[4,5] The particles in the fluid move in response to a gradient in the magnetic field. However, since the particles form a stable sol, the entire dispersion moves and, hence, the material behaves as a liquid magnet. Such materials are used as seals in computer hard drives and in loudspeakers as heat transfer (cooling) mediums. The permanent magnet used in the loudspeaker holds the liquid magnets in place. Superparamagnetic particles of iron oxide (Fe_3O_4) can also be coated with different chemicals and used to separate DNA molecules, proteins, and cells from other molecules.

The following example illustrates how to select a material for a given application.

EXAMPLE 19-2   *Design/Materials Selection for a Solenoid*

We want to produce a solenoid coil that produces an inductance of at least 2000 gauss when a 10-mA current flows through the conductor. Due to space limitations, the coil should be composed of 10 turns over a 1 cm length. Select a core material for the coil.

SOLUTION

First, we can determine the magnetic field H produced by the coil. From Equation 19-2:

$$H = \frac{nI}{l} = \frac{(10)(0.01\ \text{A})}{0.01\ \text{m}} = 10\ \text{A/m}$$

$$H = (10\ \text{A/m})(4\pi \times 10^{-3}\ \text{oersted/A/m}) = 0.126\ \text{oersted}$$

If the inductance B must be at least 2000 gauss, then the permeability of the core material must be:

$$\mu = \frac{B}{H} = \frac{2000}{0.126} = 15{,}873\ \text{gauss/oersted}$$

The relative permeability of the core material must be at least:

$$\mu_r = \frac{\mu}{\mu_0} = \frac{15{,}873}{1} = 15{,}873$$

If we examine the magnetic materials listed in Table 19-4, we find that 4-79 permalloy has a maximum relative permeability of 80,000 and might be a good selection for the core material.

19-5   Domain Structure and the Hysteresis Loop

From a phenomenological viewpoint, ferromagnetic materials are similar to ferro-electrics. A single crystal of iron or a polycrystalline piece of low-carbon steel is ferro-magnetic; however, they ordinarily do not show a net magnetization. Within the single crystal or polycrystalline structure of a ferromagnetic or ferrimagnetic material, a sub-structure composed of magnetic domains is produced, even in the absence of an external field. This happens because the presence of many domains in the material, arranged so that the net magnetization is zero, minimizes the magneto-static energy. **Domains** are regions in the material in which all of the dipoles are aligned in a certain direction. In a material that has never been exposed to a magnetic field, the individual domains have a random orientation. The net magnetization in the virgin ferromagnetic or ferrimagnetic material as a whole is zero [Figure 19-5(a)]. Similar to ferroelectrics, application of a magnetic field (poling) will coerce many of the magnetic domains to line up along with the magnetic field direction.

Boundaries, called **Bloch walls**, separate the individual magnetic domains. The Bloch walls are narrow zones in which the direction of the magnetic moment gradually and continuously changes from that of one domain to that of the next [Figure 19-5(b)]. The domains are typically very small, about 0.005 cm or less, while the Bloch walls are about 100 nm thick.

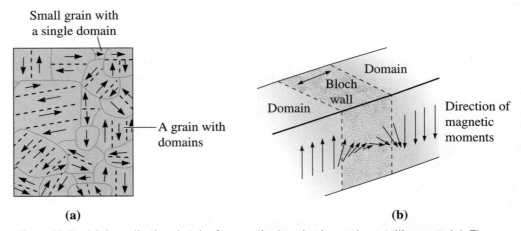

(a)   **(b)**

Figure 19-5   (a) A qualitative sketch of magnetic domains in a polycrystalline material. The dashed lines show demarcation between different magnetic domains; the dark curves show the grain boundaries. (b) The magnetic moments in adjoining atoms change direction continuously across the boundary between domains.

Movement of Domains in a Magnetic Field   When a magnetic field is imposed on the material, domains that are nearly lined up with the field grow at the expense of unaligned domains. In order for the domains to grow, the Bloch walls must move; the field provides the force required for this movement. Initially the domains grow with difficulty, and relatively large increases in the field are required to produce even a little magnetization. This condition is indicated in Figure 19-6 by a shallow slope, which is the initial permeability of the material. Thus, similar to ferroelectrics and elastomers, ferromagnetic and ferrimagnetic materials are nonlinear. As the field increases in strength, favorably oriented domains grow more easily, with permeability increasing as well. A maximum permeability can be defined as shown in the figure. Eventually, the

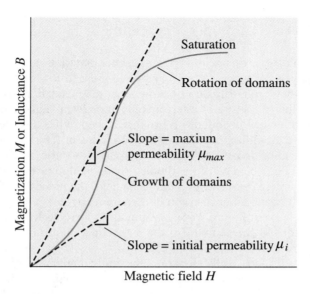

Figure 19-6
When a magnetic field is first applied to a magnetic material, magnetization initially increases slowly, then more rapidly as the domains begin to grow. Later, magnetization slows, as domains must eventually rotate to reach saturation. Notice the permeability values depend upon the magnitude of *H*.

unfavorably oriented domains disappear and rotation completes the alignment of the domains with the field. The **saturation magnetization**, produced when all of the domains are oriented along with the magnetic field, is the greatest amount of magnetization that the material can obtain. Under these conditions the permeability of these materials becomes quite small.

Effect of Removing the Field   When the field is removed, the resistance offered by the domain walls prevents regrowth of the domains into random orientations. As a

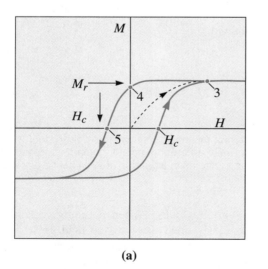

(a)

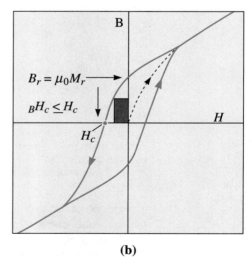

(b)

Figure 19-7   (a) The ferromagnetic hysteresis *M-H* loop showing the effect of the magnetic field on inductance or magnetization. The dipole alignment leads to saturation magnetization (point 3), a remanance (point 4), and a coercive field (point 5). (b) The corresponding *B-H* loop. Notice the end of the *B-H* loop, the *B* value does not saturate since $B = \mu_0 H + \mu_0 M$. (*Source: Adapted form Permanent Magnetism, by R. Skomski and J.M.D. Coey, p. 3, Fig. 1-1. Edited by J.M.D. Coey and D.R. Tilley. Copyright © 1999 Institute of Physics Publishing. Adapted by permission.*)

result, many of the domains remain oriented near the direction of the original field and a residual magnetization, known as the **remanence**, (M_r), is present in the material. The value of B_r (usually in Tesla) is known as the retentivity of the magnetic material. The material acts as a permanent magnet. Figure 19-7(a) shows this effect in the magnetization-field curve. Notice that the *M-H* loop shows saturation, but the *B-H* loop does not. The magnetic field needed to bring the induced magnetization to zero is the coercivity of the material. This is a microstructure-sensitive property.

For magnetic recording materials, Fe, γ-Fe_2O_3, Fe_3O_4, and needle-shaped CrO_2 particles are used. The elongated shape of magnetic particles leads to higher coercivity (H_c). The dependence of coercivity (H_c) on shape of a particle or grain is known as **magnetic shape anisotropy**. The coercivity of recording materials needs to be smaller than that for permanent magnets since data written onto a magnetic data storage medium should be erasable. On the other hand, the coercivity values should be higher than soft magnetic materials since we want to retain the information stored. Such materials are described as magnetically semi-hard.

Effect of Reversing the Field If we now apply a field in the reverse direction, the domains grow with an alignment in the opposite direction. A coercive field H_c (or coercivity) is required to force the domains to be randomly oriented and cancel one another's effect. Further increases in the strength of the field eventually align the domains to saturation in the opposite direction.

As the field continually alternates, the magnetization versus field relationship traces out a **hysteresis loop**. The hysteris loop is shown an both the *B-H* and *M-H* plots. The area contained within the hysteresis loop is related to the energy consumed during one cycle of the alternating field. The shaded area shown in Figure 19-7(b) is the largest *B-H* product and is known as the power of magnetic material.

19-6 The Curie Temperature

When the temperature of a ferromagnetic or ferrimagnetic material is increased, the added thermal energy increases the mobility of the domains, making it easier for them to become aligned, but also preventing them from remaining aligned when the field is

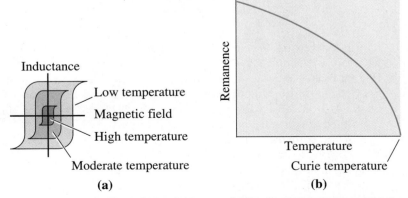

Figure 19-8 The effect of temperature on (a) the hysteresis loop and (b) the remanance. Ferromagnetic behavior disappears above the Curie temperature.

TABLE 19-3 ■ *Curie temperatures for selected materials*

Material	Curie Temperature (°C)
Gadolinium	16
$Nd_2Fe_{12}B$	312
Nickel	358
$BaO \cdot 6Fe_2O_3$	469
Co_5Sm	747
Iron	771
Alnico 1	780
Cunico	855
Alnico 5	900
Cobalt	1117

removed. Consequently, saturation magnetization, remanance, and the coercive field are all reduced at high temperatures (Figure 19-8). If the temperature exceeds the **Curie temperature** (T_c), ferromagnetic or ferrimagnetic behavior is no longer observed. Instead, the material behaves as a paramagnetic material. The Curie temperature (Table 19-3), which depends on the material, can be changed by alloying elements. Marie and Pierre Curie, French scientists, (the only husband and wife to win a Nobel prize; Marie Curie actually won two Nobel prizes) did research on magnets and the Curie temperature refers to their name. The dipoles can still be aligned in a magnetic field above the Curie temperature, but they become randomly aligned when the field is removed.

The following example shows how the T_c is an important consideration for permanent magnet design.

EXAMPLE 19-3 | *Design/Materials Selection for a High-Temperature Magnet*

Select a permanent magnet for an application in an aerospace vehicle that must re-enter Earth's atmosphere. During re-entry, the magnet may be exposed to magnetic fields as high as 600 oersted and may briefly reach temperatures as high as 500°C. We want the material to have the highest power possible and to maintain its magnetization after re-entry.

SOLUTION

It is first necessary to select potential materials having sufficient coercive field H_c and Curie temperature that re-entry will not demagnetize them. From Table 19-3, we can eliminate materials such as gadolinium, nickel, $Nd_2Fe_{12}B$, and the ceramic ferrites, since their Curie temperatures are below 500°C. Other materials from Table 19-4, such as steel and Alnico 1, can be eliminated because their coercive fields are below 600 oersted.

Of the permanent magnetic materials remaining in Table 19-3, Alnico 12 has the lowest power and can be eliminated. Thus, our choice is between Alnico 5 and Co_5Sm. The Co_5Sm has four times the power of the Alnico 5 and, based on performance, might be our best choice. Some of the benefits of Co_5Sm, however, are offset by its higher cost.

19-7 Applications of Magnetic Materials

Ferromagnetic and ferrimagnetic materials are classified as magnetically soft or magnetically hard depending upon the shape of the hysteresis loop [Figure 19-9(a)]. Generally, if the coercivity value is $\sim>10^4$ A · m^{-1}, we consider the material as magnetically hard. If the coercivity values are less than 10^3 A · m^{-1} we consider the materials as magnetically soft. Figure 19-9(b) shows classification of different commercially important magnetic materials.[6] Note that the coercivity is a very strongly *microstructure-sensitive* property, however, for a material of a given composition, the saturation magnetization is constant (i.e., it is not microstructure dependent). This is similar to the way yield strength of metallic materials is dependent strongly on the microstructure, how-

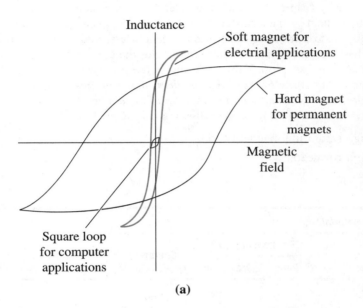

(a)

Figure 19-9
(a) Comparison of the hysteresis loops for three applications of ferromagnetic and ferrimagnetic materials. (b) Saturation magnetization and coercivity values for different magnetic materials. (*Source: Adapted from "Magnetic Materials: An Overview, Basic Concepts, Magnetic Measurements, Magnetostrictive Materials," by G.Y. Chin et al. In D. Bloor, M. Flemings, and S. Mahajan (Eds.),* Encyclopedia of Advanced Materials, *Vol. 1, 1994, p. 1423, Fig. 1. Copyright © 1994 Pergamon Press. Reprinted by permission of the editor.*)

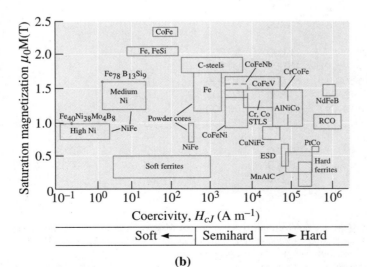

(b)

ever, the Young's modulus does not depend on it very strongly. Many factors such as the structure of grain boundaries, and the presence of pores or surface layers on particles, affect the coercivity values. The coercivity of single crystals depends strongly on crystallographic directions. There are certain directions along which it is easy to align the magnetic domains. There are other directions along which the coercivity is much higher. Coercivity of magnetic particles also depends upon shape of the particles. This is why in magnetic recording media we use acicular and not spherical particles. This effect is also used in Fe-Si steels, which are textured or grain oriented so as to minimize energy losses during the operation of an electrical transformer.

Let's look at some applications for magnetic materials.

Soft Magnetic Materials Ferromagnetic materials are often used to enhance the magnetic flux density (B) produced when an electric current is passed through the material. The magnetic field is then expected to do work. Applications include cores for electromagnets, electric motors, transformers, generators, and other electrical equipment. Because these devices utilize an alternating field, the core material is continually cycled through the hysteresis loop. Table 19-4 shows the properties of selected soft, or electrical, magnetic materials.[6] *Note that in these materials the value of relative magnetic permeability depends strongly on the strength of the applied field.*

These materials often have the following characteristics.

1. High-saturation magnetization.
2. High permeability.

TABLE 19-4 ■ Soft magnetic materials

Name	Composition	Permeability (μ_r) Initial	Maximum	Coercivity (H_c) (A · m^{-1})	Retentivity (B_r) (T)	B_{max} (T)	Resistivity ($\mu\Omega$ · m)
Ingot Iron	99.8% Fe	150	5000	80	0.77	2.14	0.10
Low-carbon steel	99.5% Fe	200	4000	100		2.14	1.12
Silicon iron, unoriented	Fe-3% Si	270	8000	60		2.01	0.47
Silicon iron, grain-oriented	Fe-3% Si	1400	50,000	7	1.20	2.01	0.50
4750 alloy	Fe-48% Ni	11,000	80,000	2		1.55	0.48
4-79 permalloy	Fe-4% Mo-79% Ni	40,000	200,000	1		0.80	0.58
Superalloy	Fe-5% Mo-80% Ni	80,000	450,000	0.4		0.78	0.65
2V-Permendur	Fe-2% V-49% Co	800	450,000	0.4		0.78	0.65
Supermendur	Fe-2% V-49% Co		100,000	16	2.00	2.30	0.40
Metglas[a] 2650SC	Fe$_{81}$B$_{13.5}$Si$_{3.5}$C$_2$		300,000	3	1.46	1.61	1.35
Metglas[a] 2650S-2	Be$_{78}$B$_{13}$S$_9$		600,000	2	1.35	1.56	1.37
MnZn Ferrite	H5C2[b]	10,000		7	0.09	0.40	1.5×10^5
MnZn Ferrite	H5E[b]	18,000		3	0.12	0.44	5×10^4
NiZn Ferrite	K5[b]	290		80	0.25	0.33	2×10^{12}

[a] Allied Corporation trademark
[b] TDK ferrite code
(Source: Adapted from "Magnetic Materials: An Overview, Basic Concepts, Magnetic Measurements, Magnetostrictive Materials," by G.Y. Chin et al. In R. Bloor, M. Flemings, and S. Mahajan (Eds.), Encyclopedia of Advanced Materials, Vol. 1, 1994, p. 1424, Table 1. Copyright © 1994 Pergamon Press. Reprinted with permission of the editor.)

3. Small coercive field.

4. Small remanance.

5. Small hysteresis loop.

6. Rapid response to high-frequency magnetic fields.

7. High-electrical resistivity.

High saturation magnetization permits a material to do work, while high permeability permits saturation magnetization to be obtained with small imposed magnetic fields. A small coercive field also indicates that domains can be reoriented with small magnetic fields. A small remanance is desired so that almost no magnetization remains when the external field is removed. These characteristics also lead to a small hysteresis loop, therefore minimizing energy losses during operation.

If the frequency of the applied field is so high that the domains cannot be realigned in each cycle, the device may heat due to dipole friction. In addition, higher frequencies naturally produce more heating because the material cycles through the hysteresis loop more often, losing energy during each cycle. For high-frequency applications, materials must permit the dipoles to be aligned at exceptionally rapid rates.

Energy can also be lost by heating if eddy currents are produced. During operation, electrical currents can be induced into the magnetic material. These currents produce power losses and joule, or I^2R heating (Chapter 18). Eddy current losses are particularly severe when the material operates at high frequencies. If the electrical resistivity is high, eddy current losses can be held to a minimum. Soft magnets produced from ferrimagnetic ceramic materials have a high resistivity and therefore are less likely to heat than metallic ferromagnetics materials. Recently, a class of smart materials, known as magnetorheological or MR fluids based on soft magnetic carbonyl iron (Fe) particles, has been introduced in various applications related to vibration control such as Delphi's Magne-Ride™ system.[7,8] These materials are like magnetic paints and can be made to absorb energy from shocks and vibrations by turning on a magnetic field. The stiffening of MR fluids is controllable and reversible. Some of the new models of Cadillac and Corvette offer a suspension based on these smart materials.

Data Storage Materials Magnetic materials are used for data storage. Memory is stored by magnetizing the material in a certain direction. For example, if the "north" pole is up, the bit of information stored is 1. If the "north" pole is down, then a 0 is stored.

For this application, materials with a square hysteresis loop, a low remanance, a low saturation magnetization, and a low coercive field are preferable. Hard ferrites based on Ba, CrO_2, acicular iron particles, and γ-Fe_2O_3 satisfy these requirements. The stripe on credit cards and bank machine cards, as well as many audio-cassettes, are made using γ-Fe_2O_3 or Fe_3O_4 particles. The square loop assures that a bit of information placed in the material by a field remains stored; a steep and abrupt change in magnetization is required to remove the information from storage in the ferromagnet. Furthermore, the magnetization produced by small external fields keeps the coercive field (H_c), saturation magnetization, and remanance (B_r) low.

The B_r and H_c values of some typical magnetic recording materials are shown in Table 19-5.[9]

Many new alloys based on Co-Pt-Ta-Cr have been developed for the manufacture of hard disks. Computer hard disks are made using sputtered thin films of these materials. As discussed in earlier chapters, many different alloys, such as those based on nanostructured Fe-Pt and Fe-Pd, are being developed for data storage applica-

TABLE 19-5 ■ *Typical magnetic recording materials*[16]

	Particle Length μm	Aspect Ratio	Magnetization (B_r)		Coercivity (H_c)		Surface Area m^2/g	Curie temp. (T_c) °C
			Wb/m^2	emu/cc	kA/m	Oe		
γ-Fe$_2$O$_3$	0.20	5:1	0.44	350	22–34	420	15–30	600
Co-γ-Fe$_2$O$_3$	0.20	6:1	0.48	380	30–75	940	20–35	700
CrO$_2$	0.20	10:1	0.50	400	30–75	950	18–55	125
Fe	0.15	10:1	1.40[a]	1100[a]	56–176	2200	20–60	770
Barium Ferrite	0.05	0.02 μm thick	0.40	320	56–240	3000	20–25	350

[a] *For overcoated, stable particles use only 50 to 80% of these values due to reduced magnetic particle volume*
(Source: From The Complete Handbook of Magnetic Recording, Fourth Edition, by F. Jorgensen, p. 324, Table 11-1. Copyright © 1996 Reprinted by permission of The McGraw-Hill Companies.)

tions. More recently, a technology known as *spintronics (spin-based electronics)* has evolved.[10] In spintronics, the main idea is to make use of the spin of electrons as a way of affecting the flow of electrical current (known as spin-polarized current) to make devices such as field effect transistors (FET).[11] The spin of the electrons (up or down) is also being considered as a way of storing information. A very successful example of a real-world spintronic-based device is as a giant magnetoresistance (GMR) sensor that is used for reading information from computer hard disks.[12]

Permanent Magnets   Finally, magnetic materials are used to make strong permanent magnets (Table 19-6). Strong permanent magnets, often called **hard magnets**, require the following:

1. High remanance (stable domains).
2. High permeability.
3. High coercive field.
4. Large hysteresis loop.
5. High power (or BH product).

Some of these are excellent examples of applications of intermetallics (Chapters 10–12). The *record* for any energy product is obtained for Nd$_2$Fe$_{14}$B magnets with an energy product of $\sim$445 kJm^{-3} ($\sim$56 Mega-Gauss-Oersteds (MGOe)).[13] These mag-

TABLE 19-6 ■ *Selected properties of hard, permanent, or magnetic materials*

Material	Common Name	$\mu_o M_r$ (T)	$\mu_o H_c$ (T)	$(BH)_{max}$ (kJ·m^{-3})	T_c (°C)
Fe-Co	Co-steel	1.07	0.02	6	887
Fe-Co-Al-Ni	Alnico-5	1.05	0.06	44	880
BaFe$_{12}$O$_{19}$	Ferrite	0.42	0.31	34	469
SmCo$_5$	Sm-Co	0.87	0.80	144	723
Nd$_2$Fe$_{14}$B	Nd-Fe-B	1.23	1.21	290–445	312

(Source: Adapted from Permanent Magnetism, by R. Skomski and J.M.D. Coey, p. 23, Table 1-2. Edited by J.M.D. Coey and D.R. Tilley. Copyright © 1999 Institute of Physics Publishing. Adapted by permission.)

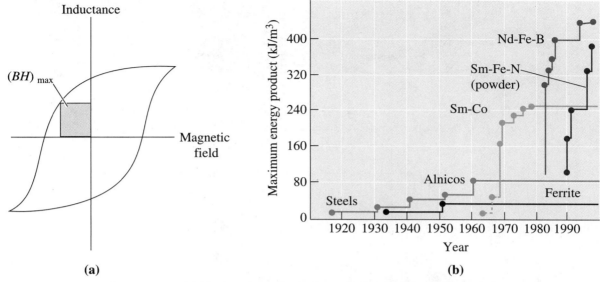

(a)   **(b)**

Figure 19-10 (a) The largest rectangle drawn in the second or fourth quadrant of the *B-H* curve gives the maximum *BH* product. $(BH)_{max}$ is related to the power, or energy, required to demagnetize the permanent magnet. (b) Development of permanent magnet materials, maximum energy product is shown on the *y*-axis. (*Source: Adapted from* Permanent Magnetism, *by R. Skomski and J.M.D. Coey, p. 23, Table 1-2. Edited by J.M.D. Coey and D.R. Tilley. Copyright © 1999 Institute of Physics Publishing. Adapted by permission.*)

nets are made in the form of a powder by the rapid solidification of a molten alloy. Powders are either bonded in a polymer matrix or by hot pressing, producing bulk materials. The energy product increases when the sintered magnet is "oriented" or poled. Corrosion resistance, brittleness, and a relatively low Curie temperature of $\sim 312°C$ are some of the limiting factors of these extraordinary materials.

The **power** of the magnet is related to the size of the hysteresis loop, or the maximum product of *B* and *H*. The area of the largest rectangle that can be drawn in the second or fourth quadrants of the *B-H* curve is related to the energy required to demagnetize the magnet [Figure 19-10(a) and Figure 19-7(b)]. For the product to be large, both the remanance and the coercive field should be large.

In many applications, we need to calculate the lifting power of a permanent magnet. The magnetic force obtainable using a permanent magnet is given by

$$F = \frac{\mu_0 M^2 A}{2} \tag{19-11}$$

In this equation A is the area of cross section, M is the magnetization, and μ_0 is the magnetic permeability of free space.

One of the most successful examples of the contributions by materials scientists and engineers in this area is the development of strong rare earth magnets. The progress made in the development of strong permanent magnets is illustrated in Figure 19-10(b).[2] Permanent magnets are used in many applications including loudspeakers, motors, generators, holding magnets, mineral separation, and bearings. Typically, they offer a nonuniform magnetic field. However, it is possible to use Halbach geometric arrangements that lead to relatively uniform magnetic fields.[14] The following examples illustrate applications of some of these concepts related to permanent magnetic materials.

EXAMPLE 19-4   *Energy Product for Permanent Magnets*

Determine the power, or *BH* product, for the magnetic material whose properties are shown in Figure 19-11.

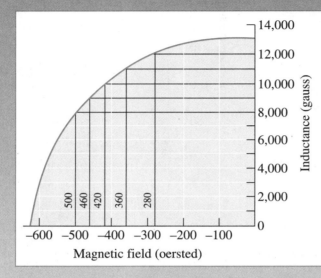

Figure 19-11
The fourth quadrant of the *B-H* curve for a permanent magnetic material (for Example 19-4).

SOLUTION

Several rectangles have been drawn in the fourth quadrant of the *B-H* curve. The *BH* product in each is:

$$BH_1 = (12,000)(280) = 3.4 \times 10^6 \text{ gauss} \cdot \text{oersted}$$

$$BH_2 = (11,000)(360) = 4.0 \times 10^6 \text{ gauss} \cdot \text{oersted}$$

$$BH_3 = (10,000)(420) = 4.2 \times 10^6 \text{ gauss} \cdot \text{oersted} = \text{maximum}$$

$$BH_4 = (9,000)(460) = 4.1 \times 10^6 \text{ gauss} \cdot \text{oersted}$$

$$BH_5 = (8,000)(500) = 4.0 \times 10^6 \text{ gauss} \cdot \text{oersted}$$

Thus, the power is about 4.2×10^6 gauss $\cdot$ oersted.

EXAMPLE 19-5   *Design/Selection of Magnetic Materials*

Select an appropriate magnetic material for the following applications: a high-electrical-efficiency motor, a magnetic device to keep cupboard doors closed, a magnet used in an ammeter or voltmeter, and magnetic resonance imaging.

SOLUTION

High-electrical-efficiency motor: To minimize hysteresis losses, we might use an oriented silicon iron, taking advantage of its anisotropic behavior and its small hysteresis loop. Since the iron-silicon alloy is electrically conductive, we would produce a laminated structure, with thin sheets of the silicon iron sandwiched between a nonconducting dielectric material. Sheets thinner than about 0.5 mm might be recommended.

Magnet for cupboard doors: The magnetic latches used to fasten cupboard doors must be permanent magnets; however, low cost is a more important design feature than high power. An inexpensive ferritic steel or a low-cost ferrite would be recommended.

Magnets for an ammeter or voltmeter: For these applications, Alnico alloys are particularly effective. We find that these alloys are among the least sensitive to changes in temperature, assuring accurate current or voltage readings over a range of temperatures.

Magnetic resonance imaging: One of the applications for MRI is in medical diagnostics. In this case, we want a very powerful magnet. A $Nd_2Fe_{12}B$ magnetic material, which has an exceptionally high *BH* product, might be recommended for this application. We can also make use of very strong electromagnets made using superconductors (Chapter 18).

The example that follows shows how the lifting power of a permanent magnet can be calculated.

EXAMPLE 19-6   *Lifting Power of a Magnet*

Calculate the force in kN for one square meter area of a permanent magnet whose saturation magnetization is 1.61 tesla.

SOLUTION

As noted before, the attractive force from a permanent magnet is given by:

$$F = \frac{\mu_0 M^2 A}{2}$$

We have been given the value of $\mu_0 M = 1.61$ tesla. We can rewrite the equation that provides the force due to a permanent magnet as follows.

$$F = \frac{\mu_0 M^2 A}{2} = \frac{(\mu_0 M)^2 A}{2\mu_0}$$

$$\therefore \frac{F}{A} = \frac{(1.61 \ T)^2}{2\left(4\pi \times 10^{-7} \frac{H}{m}\right)} = 1031.4 \frac{kN}{m^2}$$

Note that the force in this case will be 1031 kN since the area (*A*) has been specified as 1 m^2. This translates into $\sim$10.31 kilograms/cm^2.

19-8   Metallic and Ceramic Magnetic Materials

Let's look at typical alloys and ceramic materials used in magnetic applications and discuss how their properties and behavior can be enhanced. Some polymeric materials have shown magnetic activity however the Curie transition temperatures of these materials are too low compared to those for metallic and ceramic magnetic materials.[15]

Magnetic Metals Pure iron, nickel, and cobalt are not usually used for electrical applications because they have high electrical conductivities and relatively large hysteresis loops, leading to excessive power loss. They are, however, relatively poor permanent magnets; the domains are easily reoriented and both the remanance and the *BH* product are small compared with those of more complex alloys. Some change in the magnetic properties is obtained by introducing defects into the structure. Dislocations, grain boundaries, boundaries between multiple phases, and point defects help pin the domain boundaries, therefore keeping the domains aligned when the original magnetizing field is removed.

Iron-Nickel Alloys Some iron-nickel alloys, such as Permalloy, have high permeabilities, making them useful as soft magnets. One example of an application for these magnets is the "head" that stores or reads information on a computer disk (Figure 19-12). As the disk rotates beneath the head, a current produces a magnetic field in the head. The magnetic field in the head, in turn, magnetizes a portion of the disk. The direction of the field produced in the head determines the orientation of the magnetic particles embedded in the disk and, consequently, stores information. The information can be retrieved by again spinning the disk beneath the head. The magnetized region in the disk induces a current in the head; the direction of the current depends on the direction of the magnetic field in the disk.

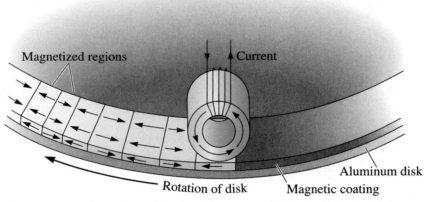

Figure 19-12 Information can be stored or retrieved from a magnetic disk by use of an electromagnetic head. A current in the head magnetizes domains in the disk during storage; the domains in the disk induce a current in the head during retrieval.

Silicon Iron These are processed into grain-oriented steels. Introduction of 3 to 5% Si into iron produces an alloy that, after proper processing, is useful in electrical applications such as motors and generators. We take advantage of the anisotropic magnetic behavior of silicon iron to obtain the best performance. As a result of rolling and subsequent annealing, a sheet texture is formed in which the $\langle 100 \rangle$ directions in each grain are aligned. Because the silicon iron is most easily magnetized in $\langle 100 \rangle$ directions, the field required to give saturation magnetization is very small, and both a small hysteresis loop and a small remanance are observed (Figure 19-13).[3] This type of anisotropy is known as **magnetocrystalline anisotropy**.

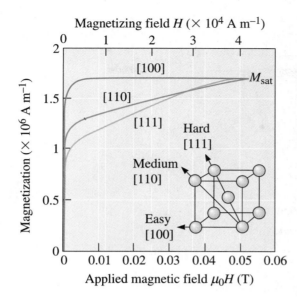

Magnetizing field H ($\times 10^4$ A m^{-1})

Figure 19-13
The initial magnetization curve for iron is highly anisotropic; magnetization is easiest when the $\langle 100 \rangle$ directions are aligned with the field and hardest along [111]. (*Source: From* Principles of Electrical Engineering Materials and Devices, *by S.O. Kasap, p. 623, Fig. 8-24. Copyright © 1997 Irwin. Reprinted by permission of The McGraw-Hill Companies.*)

Composite Magnets   Composite materials are used to reduce eddy current losses. Thin sheets of silicon iron are laminated with sheets of a dielectric material. The laminated layers are then built up to the desired overall thickness. The laminant increases the resistivity of the composite magnets and makes them successful at low and intermediate frequencies.

At very high frequencies, losses are more significant because the domains do not have time to realign. In this case, a composite material containing domain-sized magnetic particles in a polymer matrix may be used. The particles, or domains, rotate easily in the soft polymer, while eddy current losses are minimized because of the high resistivity of the polymer.

Metallic Glasses   Amorphous metallic glasses, often complex iron-boron alloys, are produced by employing extraordinarily high cooling rates during solidification (rapid-solidification processing, Chapter 8). The metallic glasses are produced in the form of thin tapes, which are stacked together to produce larger materials. These materials behave as soft magnets with a high-magnetic permeability; the absence of grain boundaries avoids magnetocrystalline anisotropy and permits easy movement of the domains, while a high electrical resistivity minimizes eddy current losses.

Magnetic Tape   Magnetic materials for information storage must have a square loop and a low coercive field, permitting very rapid transmission of information. Magnetic tape for audio or video applications is produced by evaporating, sputtering, or plating particles of a magnetic material such as γ-Fe_2O_3, or CrO_2 onto a polyester tape.[16]

Both floppy disks and hard disks for computer data storage are produced in a similar manner. In a hard disk, magnetic particles are embedded in a polymer film on a flat aluminum substrate. Because of the polymer matrix and the small particles, the domains can rotate quickly in response to a magnetic field. These materials are summarized in Table 19-5.

Complex Metallic Alloys for Permanent Magnets   Improved permanent magnets are produced by making the grain size so small that only one domain is present in each

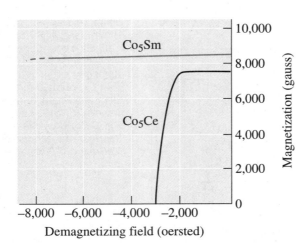

grain. Now the boundaries between domains are grain boundaries rather than Bloch walls. The domains can change their orientation only by rotating, which requires greater energy than domain growth. Two techniques are used to produce these magnetic materials: phase transformations and powder metallurgy. Alnico, one of the most common of the complex metallic alloys, has a single-phase BCC structure at high temperatures. But when Alnico slowly cools below 800°C, a second BCC phase rich in iron and cobalt precipitates. This second phase is so fine that each precipitate particle is a single domain, producing a very high remanance, coercive field, and power. Often the alloys are permitted to cool and transform while in a magnetic field to align the domains as they form.

A second technique—powder metallurgy—is used for a group of rare earth metal alloys including samarium-cobalt. A composition giving Co_5Sm, an intermetallic compound, has a high BH product (Figure 19-14) due to unpaired magnetic spins in the $4f$ electrons of samarium. The brittle intermetallic is crushed and ground to produce a fine powder in which each particle is a domain. The powder is then compacted while in an imposed magnetic field to align the powder domains. Careful sintering to avoid growth of the particles produces a solid-powder metallurgy magnet. Another rare earth magnet based on neodymium, iron, and boron has a BH product 45 mega-gauss-oersted (MGO_e). In these materials, a fine-grained intermetallic compound, $Nd_2Fe_{14}B$, provides the domains, and a fine HfB_2 precipitate prevents movement of the domain walls.

Ferrimagnetic Ceramic Materials   Common magnetic ceramics are the ferrites, which have a spinel crystal structure [Figures 19-15 and 14-1(c)]. These ferrites have nothing to do with the *ferrite phase* we encountered in studying the Fe-C phase diagram (Chapters 10–12). Ferrites are used in wireless communications and in microelectronics in such applications as inductors.[17,18] Ferrite powders are made using ceramic processing techniques (such as powder synthesis via solid state reactions, ball milling, compaction, and sintering as described in Chapter 14).

We can understand the behavior of these ceramic magnets by looking at magnetite, Fe_3O_4. Magnetite contains two different iron ions, Fe^{2+} and Fe^{3+}, so we could rewrite the formula for magnetite as $Fe^{2+}Fe_2^{3+}O_4^{2-}$. The magnetite, or spinel, crystal structure is based on an FCC arrangement of oxygen ions, with iron ions occupying selected interstitial sites. Although the spinel unit cell actually contains eight of the FCC arrangements, we need examine only one of the FCC subcells:

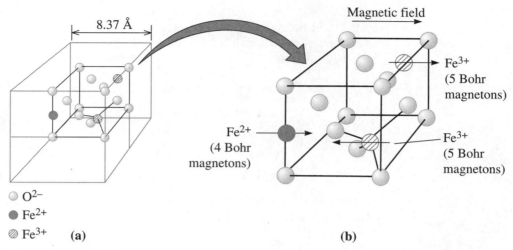

Figure 19-15 (a) The structure of magnetite, Fe_3O_4. (b) The subcell of magnetite. The magnetic moments of ions in the octahedral sites line up with the magnetic field, but the magnetic moments of ions in tetrahedral sites oppose the field. A net magnetic moment is produced by this ionic arrangement.

1. Four oxygen ions are in the FCC positions of the subcell.

2. Octahedral sites, which are surrounded by six oxygen ions, are present at each edge and the center of the subcell. One Fe^{2+} and one Fe^{3+} ion occupy octahedral sites.

3. Tetrahedral sites have indices in the subcell such as 1/4, 1/4, 1/4. One Fe^{3+} ion occupies one of the tetrahedral sites.

4. When Fe^{2+} ions form, the two $4s$ electrons of iron are removed, but all of the $3d$ electrons remain. Because there are four unpaired electrons in the $3d$ level of iron, the magnetic strength of the Fe^{2+} dipole is four Bohr magnetons. However, when Fe^{3+} forms, both $4s$ electrons and one of the $3d$ electrons are removed. The Fe^{3+} ion has five unpaired electrons in the $3d$ level and, thus, has a strength of five Bohr magnetons.

5. The ions in the tetrahedral sites of the magnetite line up so that their magnetic moments oppose the applied magnetic field, but the ions in the octahedral sites reinforce the field [Figure 19-15(b)]. Consequently, the Fe^{3+} ion in the tetrahedral site neutralizes the Fe^{3+} ion in the octahedral site (the Fe^{3+} ions have antiferromagnetic behavior). However, the Fe^{2+} ion in the octahedral site is not opposed by any other ion, and it therefore reinforces the magnetic field. The following example shows how we can calculate the magnetization in Fe_3O_4, which is one of the ferrites.

EXAMPLE 19-7   *Magnetization in Magnetite (Fe₃O₄)*

Calculate the total magnetic moment per cubic centimeter in magnetite. Calculate the value of the saturation flux density (B_{sat}) for this material.

SOLUTION

In the subcell [Figure 19-15(b)], the total magnetic moment is four Bohr magnetons, obtained from the Fe^{2+} ion, since the magnetic moments from the two Fe^{3+} ions located at tetrahedral and octahedral sites are canceled by each other.

In the unit cell overall, there are eight subcells, so the total magnetic moment is 32 Bohr magnetons per cell.

The size of the unit cell, with a lattice parameter of 8.37×10^{-8} cm, is:

$$V_{\text{cell}} = (8.37 \times 10^{-8})^3 = 5.86 \times 10^{-22} \text{ cm}^3$$

The magnetic moment per cubic centimeter is:

$$\text{Total moment} = \frac{32 \text{ Bohr magnetons/cell}}{5.86 \times 10^{-22} \text{ cm}^3/\text{cell}} = 5.46 \times 10^{22} \text{ magnetons/cm}^3$$

$$= (5.46 \times 10^{22})(9.27 \times 10^{-24} \text{ A} \cdot \text{m}^2/\text{magneton})$$

$$= 0.51 \text{ A} \cdot \text{m}^2/\text{cm}^3 = 5.1 \times 10^5 \text{ A/m}^2/\text{m}^3 = 5.1 \times 10^5 \text{ A/m}$$

This expression represents the magnetization M at saturation (M_{sat}). The value of $B_{\text{sat}} \simeq \mu_0 M_{\text{sat}}$ will be $= (4\pi \times 10^{-7})(5.1 \times 10^5) = 0.64$ Tesla.

When ions are substituted for Fe^{2+} ions in the spinel structure, the magnetic behavior may be changed. Ions that may not produce ferromagnetism in a pure metal may contribute to ferrimagnetism in the spinels, as shown by the magnetic moments in Table 19-7. Soft electrical magnets are obtained when the Fe^{2+} ion is replaced by various mixtures of manganese, zinc, nickel, and copper. The nickel and manganese ions have magnetic moments that partly cancel the effect of the two iron ions, but a net ferrimagnetic behavior, with a small hysteresis loop, is obtained. The high electrical resistivity of these ceramic compounds helps minimize eddy currents and permits the materials to operate at high frequencies. Ferrites used in computer applications may contain additions of manganese, magnesium, or cobalt to produce a square hysteresis loop behavior.

Another group of soft ceramic magnets is based on garnets, which include yttria iron garnet, $Y_3Fe_5O_{12}$ (YIG). These complex oxides, which may be modified by substituting aluminum or chromium for iron or by replacing yttrium with lanthanum or praesydium, behave much like the ferrites. Another garnet, based on gadolinium and gallium, can be produced in the form of a thin film. Tiny magnetic domains can be produced in the garnet film; these domains, or *magnetic bubbles*, can then serve as storage units for computers. Once magnetized, the domains do not lose their memory in case of a sudden power loss.

TABLE 19-7 ■ *Magnetic moments for ions in the spinel structure*

Ion	Bohr Magnetons
Fe^{3+}	5
Mn^{2+}	5
Fe^{2+}	4
Co^{2+}	3
Ni^{2+}	2
Cu^{2+}	1
Zn^{2+}	0

Hard ceramic magnets used as permanent magnets include another complex oxide family, the hexagonal ferrites. The hexagonal ferrites include $SrFe_{12}O_{19}$, $BaFe_{12}O_{19}$, and $PbFe_{12}O_{19}$.

The example that follows shows the selection of ceramic magnet.

EXAMPLE 19-8   *Design/Materials Selection for a Ceramic Magnet*

Design a cubic ferrite magnet that has a total magnetic moment per cubic meter of 5.5×10^5 A/m.

SOLUTION

We found in Example 19-7 that the magnetic moment per cubic meter for Fe_3O_4 is 5.1×10^5 A/m. To obtain a higher saturation magnetization, we must replace Fe^{2+} ions with ions having more Bohr magnetons per atom. One such possibility (Table 19-7) is Mn^{2+}, which has five Bohr magnetons.

Assuming that the addition of Mn ions does not appreciably affect the size of the unit cell, we find from Example 19-7 that:

$$V_{cell} = 5.86 \times 10^{-22} \text{ cm}^3 = 5.86 \times 10^{-28} \text{ m}^3$$

Let x be the fraction of Mn^{2+} ions that have replaced the Fe^{2+} ions, which have now been reduced to $1 - x$. Then, the total magnetic moment is:

Total moment

$$= \frac{(8 \text{ subcells})[(x)(5 \text{ magnetons}) + (1 - x)(4 \text{ magnetons})](9.27 \times 10^{-24} \text{ A} \cdot \text{m}^2)}{5.86 \times 10^{-28} \text{ m}^3}$$

$$= \frac{(8)(5x + 4 - 4x)(9.27 \times 10^{-24})}{5.86 \times 10^{-28}} = 5.5 \times 10^5$$

$$x = -4 + 4.346 = 0.346$$

Therefore we need to replace 34.6 at% of the Fe^{2+} ions with Mn^{2+} ions to obtain the desired magnetization.

Magnetostriction   Certain materials can develop strain when their magnetic state is changed. This effect is used in actuators. The magnetostrictive effect can be seen either by changing the magnetic field or by changing the temperature.[6] Iron, nickel, Fe_3O_4, $TbFe_2$, DyFe, and $SmFe_2$ are examples of some materials that show this effect. Terfenol-D, is named after its constituents, terbium (Tb), iron (Fe) and dysprosium (Dy) and its developer, the Naval Ordnance Laboratory (NOL), is one of the best known magnetostrictive materials whose composition is $\sim Tb_x Dy_{1-x} Fe_y$ ($0.27 < x < 0.30$, $1.9 < y < 2$). The magnetostriction phenomenon is analogous to electrostriction. Recently, some ferromagnetic alloys that also show magnetostriction have been developed.

 SUMMARY   ◆ All materials interact with magnetic fields. The magnetic properties of materials are related to the interaction of magnetic dipoles with a magnetic field. The magnetic dipoles originate with the electronic structure of the atom, causing several types of behavior.

◇ Magnetic materials have enabled numerous technologies that range from high-intensity superconducting magnets for MRI, semi-hard materials used in magnetic data storage, permanent magnets used in loud speakers, motors, generators, to superparamagnetic materials used to make ferrofluids and magnetic separation of DNA molecules and cells.

◇ In diamagnetic materials, the magnetic dipoles oppose the applied magnetic field.

◇ In paramagnetic materials, the magnetic dipoles weakly reinforce the applied magnetic field, increasing the net magnetization or inductance.

◇ Ferromagnetic and ferrimagnetic materials are magnetically nonlinear. Their permeability depends strongly on the applied magnetic field. In ferromagnetic materials (such as iron, nickel, and cobalt), the magnetic dipoles strongly reinforce the applied magnetic field, producing large net magnetization or inductance. In ferrimagnetic materials, some magnetic dipoles reinforce the field, whereas others oppose the field. However, a net increase in magnetization or inductance occurs. Magnetization may remain even after the magnetic field is removed. Increasing the temperature above the Curie temperature destroys the ferromagnetic or ferrimagnetic behavior.

◇ The structure of ferromagnetic and ferrimagnetic materials includes domains, within which all of the magnetic dipoles are aligned. When a magnetic field is applied, the dipoles become aligned with the field, increasing the magnetization to its maximum, or saturation, value. When the field is removed, some alignment of the domains may remain, giving a remanant magnetization.

◇ For soft magnetic materials, little remanance exists, only a small coercive field is required to remove any alignment of the domains, and little energy is consumed in reorienting the domains when an alternating magnetic field is applied.

◇ For hard, or permanent, magnetic materials, the domains remain almost completely aligned when the field is removed, large coercive fields are required to randomize the domains, and a large hysteresis loop is observed. This condition provides the magnet with a high power.

◇ Magnetostriction is the development of strain in response to the applied magnetic field or a temperature change that induces a magnetic transformation. Terfenol type magnetostrictive materials have been developed for actuator applications.

GLOSSARY

Antiferromagnetism Arrangement of magnetic moments such that the magnetic moments of atoms or ions cancel out causing zero net magnetization.

Bloch walls The boundaries between magnetic domains.

Bohr magneton The strength of a magnetic moment of an electron (μ_B) due to electron spin.

Coercivity The magnetic field needed to coerce or force the domains in a direction opposite to the magnetization direction. This is a microstructure sensitive property.

Curie temperature The temperature above (T_c) which ferromagnetic or ferrimagnetic materials become paramagnetic.

Diamagnetism   The effect caused by the magnetic moment due to the orbiting electrons, which produces a slight opposition to the imposed magnetic field.

Domains   Small regions within a single or polycrystalline material in which all of the magnetization directions are aligned.

Ferrimagnetism   Magnetic behavior obtained when ions in a material have their magnetic moments aligned in an antiparallel arrangement such that the moments do not completely cancel out and a net magnetization remains.

Ferromagnetism   Alignment of the magnetic moments of atoms in the same direction so that a net magnetization remains after the magnetic field is removed.

Hard magnet   Ferromagnetic or ferrimagnetic material that has a coercivity $> 10^4$ A $\cdot$ m^{-1}.

Hysteresis loop   The loop traced out by magnetization in a ferromagnetic or ferrimagnetic material as the magnetic field is cycled.

Magnetic moment   The strength of the magnetic field associated with a magnetic dipole.

Magnetic permeability   The ratio between inductance or magnetization and magnetic field. It is a measure of the ease with which magnetic flux lines can "flow" through a material.

Magnetic susceptibility   The ratio between magnetization and the applied field.

Magnetization   The total magnetic moment per unit volume.

Magnetocrystalline anisotropy   In single crystals, the coercivity depends upon crystallographic direction creating easy and hard axes of magnetization.

Paramagnetism   The net magnetic moment caused by the alignment of the electron spins when a magnetic field is applied.

Permanent magnet   A hard magnetic material.

Power   The strength of a permanent magnet as expressed by the maximum product of the inductance and magnetic field.

Remanance   The polarization or magnetization that remains in a material after it has been removed from the field. The remanance is due to the permanent alignment of the dipoles.

Saturation magnetization   When all of the dipoles have been aligned by the field, producing the maximum magnetization.

Shape anisotropy   The dependence of coercivity on the shape of magnetic particles.

Soft magnet   Ferromagnetic or ferrimagnetic material that has a coercivity $\leq 10^3$ Am^{-1}.

Superparamagnetism   In the nano-scale regime, materials that are ferromagnetic or ferrimagnetic but behave in a paramagnetic manner (because of their nano-sized grains or particles).

ADDITIONAL INFORMATION

1. Jacoby, M., "Data Storage: New Materials Push the Limits," *Chemical and Engineering News*, 2000 (June 12, 2000): p. 41.

2. Skomski, R. and J.M.D. Coey, *Permanent Magnetism*, J.M.M. Coey and D.R. Tilley, Editors. 1999: Institute of Publishing.

3. Kasap, S.O., *Principles of Electrical Engineering Materials and Devices*. 1997: Irwin.

4. RAJ, K., R.E. ROSENSWEIG, and L.M. AZIZ, *Stable Polysiloxane Ferrofluid Compositions and Methods of Making Same*. 1998: U.S. Patent 5851416, Ferrofluidics Corporation.

5. ROSENSWEIG, R.E. and K. RAJ, *Personal communication* 2000.

6. CHIN, G.Y., et. al, "Magnetic Materials: An Overview, Basic Concepts, Magnetic Measurements, Magnetostrictive Materials," in *The Encyclopedia of Advanced Materials*, D. BLOOR, M. FLEMINGS, and S. MAHAJAN, Editors. 1994: Pergmaon. p. 649–654.

7. PHULÉ, P.P. and Ginder, J.M., "Magnetorheological (MR) Fluids: Principles and Applications," *Smart Materials Bulletin*, 2001 (February).

8. FOISTER, R., R. IYENGAR, and D. CARLSON, *Discussions with Dr. Foister, Dr. Iyengar of Delphi and Dr. Carlson of the Lord Corporation are acknowledged.* 2002.

9. JORGENSEN, F., *The Complete Handbook of Magnetic Recording*. 4th ed. 1996: McGraw Hill.

10. VARSHNEY, U., *National Science Foundation, Technical discussions are acknowledged.* 2001.

11. AWSCHALOM, D.D., M.E. FLATTE, and N. SAMARTH, "Spintronics," *Scientific American*, 2002.

12. *Discussions with Dr. John Barnard and Dr. Tim Klemmer are acknowledged.* 2002.

13. *Discussions with Dr. S.G. Sankar of Advanced Materials Corporation are acknowledged.* 2002.

14. HALBACH, K. "High-Performance Permanent Magnetic Materials," in *MRS Symposium Proceedings*. 1987: Anaheim, CA, Materials Research Society.

15. MILLER, J.S., E., A.J., "Molecule-Based Magnets—An Overview," *MRS Bulletin*, 2000 (November 2000): p. 21–30.

16. ONODERA, S., H. KONDO, and T. KAWANA, *Materials for Magnetic-Tape Media*. MRS Bulletin, 1996. September: p. 35–41.

17. BUCHANAN, R.C., *Ceramic Materials for Electronics*. 2nd ed. 1991: Marcel Dekker.

18. HUTH, J., *Spang Magnetics, Technical discussions are acknowledged.* 2001.

✓ PROBLEMS

Section 19-1 Classification of Magnetic Materials

Section 19-2 Magnetic Dipoles and Magnetic Moments

19-1 State any four real-world applications of different magnetic materials.

19-2 Explain the following statement "Strictly speaking, there is no such thing as a nonmagnetic material."

19-3 Normally we disregard the magnetic moment of the nucleus. In what application does the nuclear magnetic moment become important?

19-4 What two motions of electrons are important in determining the magnetic properties of materials?

19-5 Explain why only a handful of solids exhibit ferromagnetic or ferrimagnetic behavior.

19-6 Calculate and compare the maximum magnetization we would expect in iron, nickel, cobalt, and gadolinium. There are seven electrons in the $4f$ level of gadolinium. Compare the calculated values with the experimentally observed values.

Section 19-3 Magnetization, Permeability, and the Magnetic Field

Section 19-4 Diamagnetic, Paramagnetic, Ferromagnetic, Ferrimagnetic, and Superparamagnetic Materials

Section 19-5 Domain Structure and Hysteresis Loop

19-7 Define the following terms: magnetic induction, magnetic field, magnetic susceptibility and magnetic permeability.

19-8 Define the following terms: ferromagnetic, ferrimagnetic, diamagnetic, paramagnetic, superparamagnetic, and antiferromagnetic materials.

19-9 What is a ferromagnetic material? What is a ferrimagnetic material? Explain and provide examples of each type of material.

19-10 How does the permeability of ferromagnetic and ferrimagnetic materials change with temperature when the temperature is greater than Curie temperature?

19-11 An alloy of nickel and cobalt is to be produced to give a magnetization of 2×10^6 A/m. The crystal structure of the alloy is FCC with a lattice parameter of 0.3544 nm. Determine the atomic percent cobalt required, assuming no interaction between the nickel and cobalt.

19-12 Estimate the magnetization that might be produced in an alloy containing nickel and 70 at% copper, assuming that no interaction occurs.

19-13 An Fe-80% Ni alloy has a maximum permeability of 300,000 when an inductance of 3500 gauss is obtained. The alloy is placed in a 20-turn coil that is 2 cm in length. What current must flow through the conductor coil to obtain this field?

19-14 An Fe-49% Ni alloy has a maximum permeability of 64,000 when a magnetic field of 0.125 oersted is applied. What inductance is obtained and what current is needed to obtain this inductance in a 200-turn, 3-cm-long coil?

19-15 Draw a schematic of the *B-H* and *M-H* loops for a typical ferromagnetic material. What is the difference between these two loops?

19-16 Is the magnetic permeability of ferromagnetic or ferrimagnetic material constant? Explain.

19-17 From a phenomenological viewpoint, what are the similarities between elastomers, ferromagnetic and ferrimagnetic materials, and ferroelectrics?

19-18 What is the major differences between ferromagnetic and ferrimagnetic materials?

19-19 Compare the electrical resistivities of ferromagnetic metals and ferrimagnetic ceramics.

19-20 Why are eddy current losses important design factors in ferromagnetic materials but less important in ferrimagnetic materials?

19-21 Which element has the highest saturation magnetization? What alloys have the highest saturation magnetization of all materials?

19-22 What material has the highest energy product of all magnetic materials?

19-23 Is coercivity of a material a microstructure sensitive property? Is remanance a microstructure sensitive property? Explain.

19-24 Is saturation magnetization of a material a microstructure-sensitive property? Explain.

19-25 Can the same material have different hysteresis loops? Explain.

19-26 The following data describe the effect of the magnetic field on the inductance in a silicon steel. Calculate the initial permeability and the maximum permeability for the material.

H (A/m)	*B* (tesla)
0.00	0
20	0.08
40	0.30
60	0.65
80	0.85
100	0.95
150	1.10
250	1.25

19-27 A magnetic material has a coercive field of 167 A/m, a saturation magnetization of 0.616 tesla, and a residual inductance of 0.3 tesla. Sketch the hysteresis loop for the material.

19-28 A magnetic material has a coercive field of 10.74 A/m, a saturation magnetization of 2.158 tesla, and a remanance induction of 1.183 tesla. Sketch the hysteresis loop for the material.

19-29 Using Figure 19-16, determine the following properties of the magnetic material: remanance, saturation magnetization, coercive field, initial permeability, maximum permeability and power (maximum *BH* product).

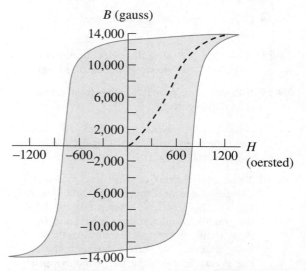

Figure 19-16 Hysteresis curve for a hard magnetic material (for Problem 19-29).

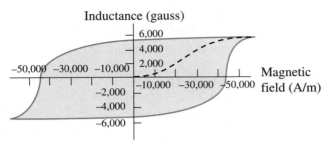

Inductance (gauss)

Figure 19-17   Hysteresis curve for a hard magnetic material (for Problem 19-30).

19-30 Using Figure 19-17, determine the following properties of the magnetic material: remanance saturation magnetization, coercive field, initial permeability, maximum permeability, and power (maximum *BH* product).

Section 19-6 The Curie Temperature

Section 19-7 Applications of Magnetic Materials

Section 19-8 Metallic and Ceramic Magnetic Materials

19-31 Define the terms soft and hard magnetic materials. Draw a typical *M-H* loop for each material.

19-32 What important characteristics are associated with soft magnetic materials.

19-33 Are materials used for magnetic data storage (e.g., audio cassettes) magnetically hard or soft? Explain.

19-34 Give examples of materials used in magnetic recording.

19-35 What are the advantages of using Fe-Nd-B magnets? What are some of their disadvantages?

19-36 Estimate the power of the Co_5Ce material shown in Figure 19-14.

19-37 What advantages does the Fe-3% Si material have compared with permalloy for use in electric motors?

19-38 The coercive field for pure iron is related to the grain size of the iron by the relationship $H_c = 1.83 + 4.14/\sqrt{A}$, where A is the area of the grain in two dimensions (mm²) and H_c is in A/m. If only the grain size influences the 99.95% iron (coercivity 0.9 oersted), estimate the size of the grains in the material. What happens to the coercivity value when the iron is annealed to increase the grain size?

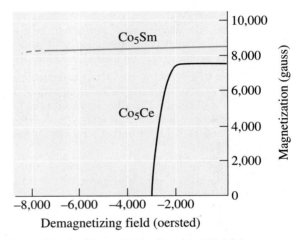

Figure 19-14   (Repeated for Problem 19-36.) Demagnetizing curves for Co_5Sm and Co_5Ce, representing a portion of the hysteresis loop.

19-39 Calculate the attractive force per meter square from a permanent magnet whose saturation magnetization is 1.0 tesla.

19-40 Suppose we replace 10% of the Fe^{2+} ions in magnetite with Cu^{2+} ions. Determine the total magnetic moment per cubic centimeter.

19-41 Suppose that the total magnetic moment per cubic meter in a spinel structure in which Ni^{2+} ions have replaced a portion of the Fe^{2+} ions is 4.6×10^5 A/m. Calculate the fraction of the Fe^{2+} ions that have been replaced and the wt% Ni present in the spinel.

19-42 What is magnetostriction? How is this similar to electrostriction? How is it different from the piezoelectric effect?

19-43 State examples of materials that show the magnetostriction effect.

19-44 What is spintronics? Give an example of a spintronics-based device used in personal and laptop computers.

Design Problems

19-45 Design a solenoid no longer than 1 cm that will produce an inductance of 3000 gauss.

19-46 Design a permanent magnet that will have a remanance of at least 5000 gauss that will not be demagnetized if exposed to a temperature of 400°C or to a magnetic field of 1000 oersted, and that has good magnetic power.

19-47 Design a spinel-structure ferrite that will produce a total magnetic moment per cubic meter of 5.6×10^5 A/m.

19-48 Design a spinel-structure ferrite that will produce a total magnetic moment per cubic meter of 4.1×10^5 A/m.

Computer Problems

19-49 *Converting Magnetic Units.* Write a computer program that will convert magnetic units from the cgs or Gaussian system to the SI system. For example, if the user provides a value of flux density in Gauss the program should provide a value in Wb/m^2 or tesla.

Chapter

Optical materials play a critical role in the infrastructure of our communications and information technology systems. There are currently more than ten million kilometers of optical fiber cable installed worldwide. Optical fibers for communications and medical applications, lasers for medical and manufacturing applications, micro-machined mirror arrays, light-emitting diodes, and solar cells have enabled a wide range of new technologies. This photograph shows a micromachined array of silicon pillars formed by anisotropic etching. These types of structures are known as *photonic crystals*. Development of such novel devices may lead to new approaches as to how we can manipulate photons in an active way and better integrate semiconductor materials and optical materials technologies. (*Courtesy of PhotoDisc/Getty Images.*)

20

Photonic Materials

Have You Ever Wondered?

- *Why does the sky appear blue?*

- *How does an optical fiber work?*

- *What factors control the transmission and absorption of light in different materials?*

- *How many miles of optical fibers have been installed worldwide to date?*

- *What does the acronym LASER stand for?*

- *What is a ruby laser made from?*

- *Does the operation of a fluorescent tube light involve phosphorescence?*

Photonic or optical materials have had a significant impact on the development of the communications infrastructure and the information technology. Photonic materials have also played a key role in many other technologies related to medicine, manufacturing, and astronomy, just to name a few. Today, more than 10 million kilo-meters (~6.25 million miles) of optical fiber has been installed worldwide.[1] The term "optoelectronics" refers to the science and technology that combines electronic and optical materials. Examples of these would include light emitting diodes (LEDs), solar cells, and semiconductor lasers. We discussed some of these in Chapter 18.

Starting with simple mirrors, prisms, and lenses to the latest photonic band gap materials, the field of optical materials and devices has advanced at a very rapid pace.[2–4] The goal of this chapter is to present a summary of fundamental principles that have guided applications of optical materials.

Optical properties of materials are related to the interaction of a material with electromagnetic radiation in the form of waves or particles of energy called **photons**. This radiation may have characteristics that fall in the visible light spectrum, or may be invisible to the human eye. In this chapter, we explore two avenues by which we can use the optical properties of materials: emission of photons from materials and interaction of photons with materials. Several sources cause the emission of photons having a certain frequency, wavelength, and energy. For example, gamma (γ) rays are produced by changes in the structure of the nucleus of the atom; x-rays, ultraviolet radiation, and the visible spectrum are produced by changes in the electronic structure of the atom. Microwaves and radio waves are low-energy, long-wavelength radiation caused by the vibration of atoms or the crystal structure.

When photons interact with a material, a variety of optical effects can be produced, including absorption, transmission, scattering, diffraction, reflection, refraction, and the generation of voltage. Examining these phenomena enables us not only to better understand the behavior of materials but also to use them to produce such things as aircraft that cannot be detected by radar; lasers for medical use, communications, or manufacturing; fiber-optic devices; light-emitting diodes; solar cells; and analytical instruments for determining crystal structure or material composition.

20-1   The Electromagnetic Spectrum

Light is energy, or radiation, in the form of waves or particles called photons that can be emitted from a material. The important characteristics of the photons—their energy E, wavelength λ, and frequency v—are related by the equation

$$E = hv = \frac{hc}{\lambda}, \tag{20-1}$$

where c is the speed of light (in vacuum the speed c_0 is 3×10^{10} cm/s) and h is Planck's constant (6.62×10^{-34} J · s). Since there are 1.6×10^{-19} J per electron volt (eV), h also is given by 4.14×10^{-15} eV · s. This equation permits us to consider the photon either as a particle of energy E or as a wave with a characteristic wavelength and frequency.

The spectrum of electromagnetic radiation is shown in Figure 20-1.[5] Gamma and x-rays have a very short wavelength, or a high frequency, and possess very high energies; microwaves and radio waves possess very low energies; and visible light represents only a very narrow portion of the electromagnetic spectrum. Figure 20-1 also shows the response of the human eye to different colors. Bandgaps (E_g) of semiconductors in eV and corresponding wavelengths of light are also shown. As discussed in Chapter 18, these relationships are used to make LEDs of different colors.

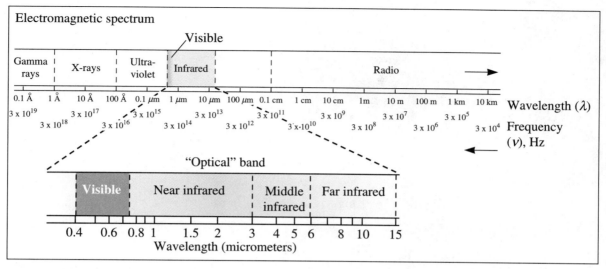

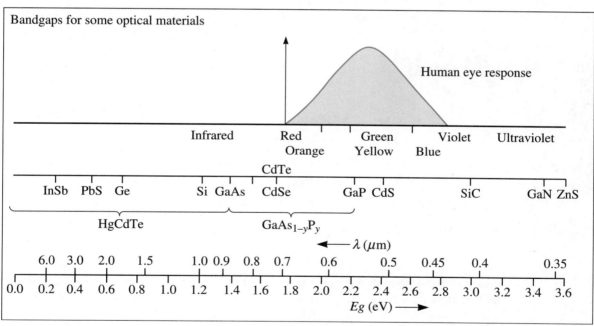

Figure 20-1   The electromagnetic spectrum of radiation; the bandgaps and cutoff frequencies for some optical materials are also shown. (*Source: From* Optoelectronics: An Introduction to Materials and Devices, *by J. Singh. Copyright © 1996 The McGraw-Hill Companies. Reprinted by permission of The McGraw-Hill Companies.*)

20-2 Refraction, Reflection, Absorption, and Transmission

All materials interact in some way with light. Photons cause a number of optical phenomena when they interact with the electronic or crystal structure of a material (Figure 20-2). If incoming photons interact with valence electrons, several things may happen. The photons may give up their energy to the material, in which case *absorption*

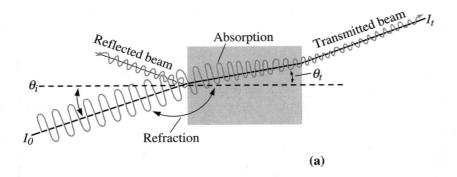

(a)

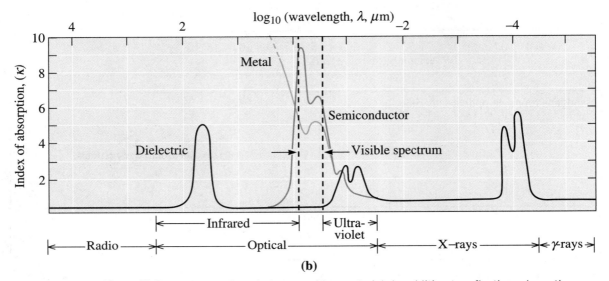

(b)

Figure 20-2   (a) Interaction of photons with a material. In addition to reflection, absorption, and transmission, the beam changes direction, or is refracted. The change in direction is given by the index of refraction n. (b) The absorption index (κ) as a function of wavelength.

occurs. Or the photons may give up their energy, but photons of identical energy are immediately emitted by the material; in this case, *reflection* occurs. Finally, the photons may not interact with the electronic structure of the material; in this case, *transmission* occurs. Even in transmission, however, photons are changed in velocity and *refraction* occurs. A very small intensity of the incident light may be scattered with a slightly different frequency (raman scattering).

As Figure 20-2 illustrates, an incident beam of intensity I_0 may be partly reflected, partly absorbed, and partly transmitted. The intensity of the incident beam therefore can be expressed as

$$I_0 = I_r + I_a + I_t \tag{20-2}$$

where I_r is the portion of the beam that is reflected, I_a as the portion that is absorbed, and I_t is the portion finally transmitted through the material. Reflection may occur at both the front and back surfaces of the material. Figure 20-2(a), shows reflection only at the front surface. Also, reflection occurs at a certain angle with respect to the normal of the surface (specular reflection) and also in many other directions (diffuse reflection, not shown in Figure 20-2(a)). Several factors are important in determining the behavior of the photon, with the energy required to excite an electron to a higher energy state being of particular importance.

Let's examine each of these four phenomena. We begin with refraction, since it is related to reflection and transmission.

Refraction   Even when a photon is transmitted, the photon causes polarization of the electrons in the material and, by interacting with the polarized material, loses some of its energy. The speed of light (c) can be related to the ease with which a material polarizes both electrically (permittivity ε) and magnetically (permeability μ), or

$$c = \frac{1}{\sqrt{\mu \cdot \varepsilon}}$$
(20-3)

Generally, optical materials are not magnetic and the permeability can be neglected. Because the speed of the photons decreases, the beam of photons changes direction when it enters the material [Figure 20-2(a)]. Suppose photons traveling in a vacuum impinge on a material. If θ_i and θ_t, respectively, are the angles that the incident and refracted beams make with the normal of the surface of the material, then:

$$n = \frac{c_0}{c} = \frac{\lambda_{\text{vacuum}}}{\lambda} = \frac{\sin \theta_i}{\sin \theta_t}$$
(20-4)

The ratio n is the **index of refraction**, c_0 is the speed of light in a vacuum (3×10^8 m/s), and c is the speed of light in the material. Typical values of the index of refraction for several materials are listed in Table 20-1.

TABLE 20-1 ■ *Index of refraction of selected materials for photons of wavelength 5890 Å*

Material	Index of Refraction (n)
Air	1.00
Ice	1.309
Water	1.333
Teflon™	1.35
SiO_2 (glass)	1.46
Polymethyl methacrylate	1.49
Typical silicate glasses	~1.50
Polyethylene	1.52
Sodium chloride (NaCl)	1.54
SiO_2 (quartz)	1.55
Epoxy	1.58
Polystyrene	1.60
TiO_2	1.74
Sapphire (Al_2O_3)	1.8
Leaded glasses (crystal)	2.50
Rutile (TiO_2)	2.6
Diamond	2.417
Silicon	3.49
Gallium arsenide	3.35
Indium phosphide	3.21
Germanium	4.0

We can also define a complex refractive index (n^*). This includes κ, a parameter known as the absorption index:

$$n^* = n(1 - i\kappa) \tag{20-5}$$

In Equation 20-5, $i = \sqrt{-1}$ is the imaginary number. The absorption index is defined as:

$$\kappa = \frac{\alpha\lambda}{4\pi n} \tag{20-6}$$

In Equation 20-6, α is the linear absorption coefficient, λ is the wavelength of light, n is the refractive index. Figure 20-2(b) shows the variation in index of absorption with the frequency of electromagnetic waves.[6] Drawing an analogy, the refractive index is similar to the dielectric constant of materials and the absorption index is similar to the dielectric loss factor. Similarly, drawing an analogy with the mechanical properties of materials, the refractive index is similar to Young's modulus and the absorption index is similar to the viscous deformation.

If the photons are traveling in Material 1, instead of in a vacuum, and then pass into Material 2, the velocities of the incident and refracted beams depend on the ratio between their indices of refraction, again causing the beam to change direction:

$$\frac{c_1}{c_2} = \frac{n_2}{n_1} = \frac{\sin \theta_i}{\sin \theta_t} \tag{20-7a}$$

Equation 20-7a is also known as Snell's law.

When a ray of light enters from a material with refractive index (n_1) into a material of refractive index (n_2), and if $n_1 > n_2$, the ray is bent away from the normal and toward the boundary surface [Figure 20-3(a)].[1] A beam traveling through Material 1 is reflected rather than transmitted if the angle θ_t becomes 90°.

More interaction of the photons with the electronic structure of the material occurs when the material is easily polarized. We saw different dielectric polarization mechanisms in Chapter 18. Among these, the electronic polarization (i.e., displacement of the electronic cloud around the atoms and ions) is the one that controls the refractive index of materials. Consequently, we expect to find a relationship between the index of re-

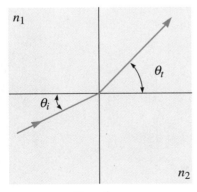

Boundary

(a)

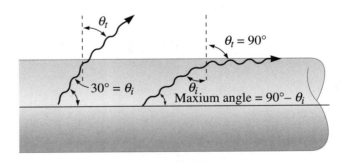

(b)

Figure 20-3   (a) When a ray of light enters from material 1 into material 2, if the refractive index of material 1 (n_1) is greater than that of material 2 (n_2), then the ray bends away from the normal and toward the boundary surface.[1, 9] (b) Diagram a light beam in glass fiber for Example 20-1.

fraction (n) and the high frequency dielectric constant (k_∞) of the material. From Equation 20-4, and for nonferromagnetic or nonferrimagnetic materials:

$$n = \frac{c}{v} = \sqrt{\frac{\mu\varepsilon}{\mu_0\varepsilon_0}} \cong \sqrt{\frac{\varepsilon}{\varepsilon_0}} = \sqrt{k_\infty} \qquad (20\text{-}7b)$$

In Equation 20-7b, c equals the speed of light, v is the frequency of light, μ and μ_0 are the magnetic permeabilities and vacuum of the material through which light is traveling, respectively, and ε and ε_0 are dielectric permittivity of material and vacuum, respectively. As we discussed in Chapter 18, the material known as "crystal" silicate glass, which is actually lead oxide (~up to 30%), has a high index of refraction ($n \sim 2.5$) since Pb^{+2} ions have very high electronic polarizability. We use a similar strategy to dope silica fibers to enhance the refractive index of the core of optical fibers as compared to their cladding region. This helps keep the light (and hence information) in the core of the optical fiber. The difference in the high-frequency dielectric constant (which is related to the refractive index) and the low-frequency dielectric constant, in fact, is a measure of the other polarization mechanisms that are contributing to the dielectric constant.

The refractive index, n, is not a constant for a particular material. The frequency, or wavelength, of the photons affects the index of refraction. **Dispersion** of a material is defined as the variation of the refractive index with wavelength:

$$(\text{Dispersion})_\lambda = \frac{dn}{d\lambda} \qquad (20\text{-}7c)$$

This dependence of the refractive index on wavelength is nonlinear. The dispersion within a material means light pulses of different wavelengths, starting at the same time at the end of an optical fiber, will arrive at different times at the other end. Thus, material dispersion plays an important role in fiber optics. This is one of the reasons why we prefer to use a single wavelength source of light for fiber-optic communications. Dispersion also causes chromatic aberration in optical lenses.

Since dielectric polarization (P) is equal to the dipole moment per unit volume and the high frequency dielectric constant is related to the refractive index, we expect that, for the same material, a denser form or polymorph will have a higher refractive index (compare the refractive indices of ice and water, or glass and quartz).

The following example illustrates how an optical fiber is designed to minimize the optical losses during insertion, followed by an example calculating the index of refraction.

EXAMPLE 20-1 *Design of a Fiber Optic System*

Optical fibers are commonly made from high-purity silicate glasses. They consist of a core that has refractive index (~1.48) that is higher than a region called cladding (refractive index ~ 1.46). This is why even a simple glass fiber in air (refractive index 1.0) can serve as an optical fiber. In designing a fiber optic transmission system, we plan to introduce a beam of photons from a laser into a glass fiber whose index of refraction of is 1.5. Design a system to introduce the beam with a minimum of leakage of the beam from the fiber.

SOLUTION

To prevent leakage of the beam, we need the total internal reflection and thus the angle θ_t must be at least 90°. Suppose that the photons enter at a 60° angle to the axis of the fiber. From Figure 20-3(b), we find that $\theta_i = 90 - 60 = 30°$. If we let the glass be Material 1 and if the glass fiber is in air ($n = 1.0$), then from Equation 20-7a:

$$\frac{n_2}{n_1} = \frac{\sin \theta_i}{\sin \theta_t} \quad \text{or} \quad \frac{1}{1.5} = \frac{\sin 30}{\sin \theta_t}$$

$$\sin \theta_t = 1.5 \sin 30 = 1.5(0.50) = 0.75 \quad \text{or} \quad \theta_t = 48.6°$$

Because θ_t is less than 90°, photons escape from the fiber. To prevent transmission, we must introduce the photons at a shallower angle, giving $\theta_t = 90°$.

$$\frac{1}{1.5} = \frac{\sin \theta_i}{\sin \theta_t} = \frac{\sin \theta_i}{\sin 90°} = \sin \theta_i$$

$$\sin \theta_i = 0.6667 \quad \text{or} \quad \theta_i = 41.8°$$

If the angle between the beam and the axis of the fiber is $90 - 41.8 = 48.2°$ or less, the beam is reflected.

   If the fiber were immersed in water ($n = 1.333$), then:

$$\frac{1.333}{1.5} = \frac{\sin \theta_i}{\sin \theta_t} = \frac{\sin \theta_i}{\sin 90°} = \sin \theta_i$$

$$\sin \theta_i = 0.8887 \quad \text{or} \quad \theta_i = 62.7°$$

In water, the photons would have to be introduced at an angle of less than $90 - 62.7 = 27.3°$ in order to prevent transmission.

EXAMPLE 20-2   *Light Transmission in Polyethylene*

Suppose a beam of photons in a vacuum strikes a sheet of polyethylene at an angle of 10° to the normal of the surface of the polymer. Calculate the index of refraction of polyethylene and find the angle between the incident beam and the beam as it passes through the polymer.

SOLUTION

The index of refraction is related to the high-frequency dielectric constant. For this material the high-frequency dielectric constant $k_\infty = 2.3$:

$$n = \sqrt{k_\infty} = \sqrt{2.3} = 1.52$$

The angle θ_t is:

$$n = \frac{\sin \theta_i}{\sin \theta_t}$$

$$\sin \theta_t = \frac{\sin \theta_i}{n} = \frac{\sin 10°}{1.52} = \frac{0.174}{1.52} = 0.114$$

$$\theta_t = 6.56°$$

Reflection   When a beam of photons strikes a material, the photons interact with the valence electrons and give up their energy. In metals, the valence bands are unfilled and the radiation of almost any wavelength excites the electrons into higher energy levels. One might expect that, if the photons are totally absorbed, no light would be reflected and the metal would appear to be black. In aluminum or silver, however, photons of almost identical wavelength are immediately reemitted as the excited electrons return to their lower energy levels—that is, reflection occurs. Since virtually the entire visible spectrum is reflected, these metals have a white, or silvery, color.

The **reflectivity** R gives the fraction of the incident beam that is reflected and is related to the index of refraction. If the material is in a vacuum or in air:

$$R = \left(\frac{n-1}{n+1}\right)^2 \tag{20-8}$$

If the material is in some other medium with an index of refraction of n_i, then:

$$R = \left(\frac{n-n_i}{n+n_i}\right)^2 \tag{20-9}$$

These equations apply to the reflection from a single surface and assume normal (perpendicular to the surface) incidence. The value of R depends upon the angle of incidence. Materials with a high index of refraction have a higher reflectivity than materials with a low index. Because the index of refraction varies with the wavelength of the photons, so does the reflectivity.

In metals, the reflectivity is typically on the order of 0.9 to 0.95, whereas the reflectivity of typical glasses is nearer to 0.05. The high reflectivity of metals is one reason that they are *opaque*.

There are many applications where we want materials to have very good reflectivity. Examples include mirrors and certain types of coatings on glasses. In fact, these coatings must also be designed such that much of a certain part of the electro-magnetic spectrum (e.g., infrared, the part that produces heat) must be reflected. Many such coatings have been developed for glasses.[7] There are also many applications where the reflectivity must be extremely limited. Such coatings are known as antireflective (AR) coatings. These coatings are used for glasses, in automobile rear view mirrors, on windows, or for the glass of picture frames so that you see through the glass without seeing your own reflection.

Absorption   That portion of the incident beam that is not reflected by the material is either absorbed or transmitted through the material. The fraction of the beam that is absorbed is related to the thickness of the material and the manner in which the photons interact with the material's structure. The intensity of the beam after passing through the material is given by

$$I = I_0 \exp(-\alpha x), \tag{20-10a}$$

where x is the path through which the photons move (usually the thickness of the material), α is the **linear absorption coefficient** of the material for the photons, I_0 is the intensity of the beam after reflection at the front surface, and I is the intensity of the beam when it reaches the back surface. Equation 20-10a is also known as Bouguer's law.

Absorption in materials occurs by several mechanisms. In *Rayleigh scattering*, the photon interacts with the electrons orbiting an atom and is deflected without any change in photon energy; this outcome is more significant for high atomic number atoms and low photon energies. The blue color in the sunlight gets scattered more than other colors in the visible spectrum and this makes the sky look blue. There is also another type of scattering from particles much larger than the wavelength of light; this type of scattering, known as the Tyndall effect, is the reason why clouds (which consist of relatively larger water droplets) look white. *Compton scattering* is caused by an interaction between the photon and orbiting electrons, causing the electron to be ejected from the atom and, consequently, consuming some of the energy of the photon. Again,

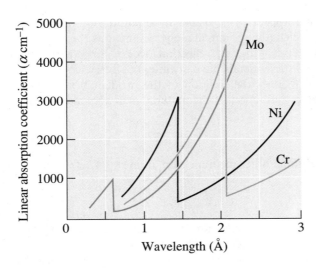

Figure 20-4
The linear absorption coefficient relative to wavelengths for several metals. Note the sudden decrease in the absorption coefficient for wavelengths greater than the absorption edge.

higher atomic number atoms and low photon energies cause more scattering. The *photoelectric effect* occurs when the energy of the photon is consumed by breaking the bond between an electron and its nucleus. As the energy of the photon increases (or the wavelength decreases, see Figures 20-1 and 20-4), less absorption occurs until the photon has energy equal to that of the binding energy. At this energy, the absorption coefficient increases significantly. The energy, or wavelength, at which this occurs is called the **absorption edge**. In Figure 20-4, the abrupt change in the absorption coefficient corresponds to the energy required to remove an electron from the K shell of the atom; this absorption edge is important in certain x-ray analytical techniques.

In some cases, the effect of scattering can be written as:

$$I = I_0 \exp(-\alpha_i + \alpha_s)x, \qquad (20\text{-}10b)$$

In Equation 20-10b, α_i is what we previously termed as α, the intrinsic absorption coefficient, and α_s is the scattering coefficient.

Examples of a portion of the characteristic spectra for several elements are included in Table 20-2. The K_α, K_β, and L_α lines correspond to the wavelengths of radiation emitted from transitions of electrons between shells, as discussed a little later in this section.

TABLE 20-2 ■ *Characteristic emission lines and absorption edges for selected elements*

Metal	K_a (Å)	K_β (Å)	L_a (Å)	Absorption Edge (Å)
Al	8.337	7.981	—	7.951
Si	7.125	6.768	—	6.745
S	5.372	5.032	—	5.018
Cr	2.291	2.084	—	2.070
Mn	2.104	1.910	—	1.896
Fe	1.937	1.757	—	1.743
Co	1.790	1.621	—	1.608
Ni	1.660	1.500	—	1.488
Cu	1.542	1.392	13.357	1.380
Mo	0.711	0.632	5.724	0.620
W	0.211	0.184	1.476	0.178

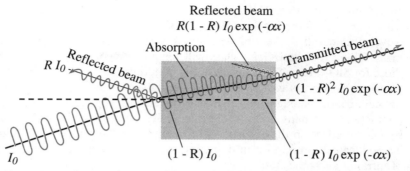

Figure 20-5 Fractions of the original beam that are reflected, absorbed, and transmitted.

Transmission The fraction of the beam that is not reflected or absorbed is transmitted through the material. Using the following steps, we can determine the fraction of the beam that is transmitted (see Figure 20-5):

1. If the incident intensity is I_0, then the loss due to reflection at the front face of the material is RI_0. The fraction of the incident beam that actually enters the material is $I_0 - RI_0 = (1 - R)I_0$:

$$I_{\text{reflected at front surface}} = RI_0$$

$$I_{\text{after reflection}} = (1 - R)I_0$$

2. A portion of the beam that enters the material is lost by absorption. The intensity of the beam after passing through a material having a thickness x is:

$$I_{\text{after absorption}} = (1 - R)I_0 \exp(-\alpha x)$$

3. Before the partially absorbed beam exits the material, reflection occurs at the back surface. The fraction of the beam that reaches the back surface and is reflected is:

$$I_{\text{reflected at back surface}} = R(1 - R)I_0 \exp(-\alpha x)$$

4. Consequently, the fraction of the beam that is completely transmitted through the material is:

$$I_{\text{transmitted}} = I_{\text{after absorption}} - I_{\text{reflected at back}}$$
$$= (1 - R)I_0 \exp(-\alpha x) - R(1 - R)I_0 \exp(-\alpha x)$$
$$= (1 - R)(1 - R)I_0 \exp(-\alpha x) \tag{20-11}$$
$$I_t = I_0(1 - R)^2 \exp(-\alpha x)$$

Again, however, the intensity of the transmitted beam depends on the wavelength of the photons in the beam. Suppose that a beam of white light (containing photons of all wavelengths in the visible spectrum) impinges on a material. If the same fraction of the photons having different wavelengths is absorbed, reflected, and therefore transmitted, the transmitted beam would also be white light, or colorless. This is what we find in materials such as diamond. If, however, longer wavelength photons (red, orange, etc.) are absorbed more than the shorter wavelength photons, we expect the transmitted light to appear blue or green.

There are two cutoff frequencies in dielectric materials. The short, wavelength-side absorption edge is determined by the bandgap (E_g) of the dielectric. Thus, many of the dielectrics (e.g., sapphire, TiO_2, NaCl, etc.) are transparent to the white light. The long, wavelength-side absorption band edge is determined by the interaction of relatively low energy photons with the lattice vibrations or phonons. Thus, the transmission in dielectrics is limited on the higher wavelength side by the lattice vibrations in the crystal. The wavelength cutoff in the long wavelength side increases with increasing ionic or atomic mass. Thus, KCl and GeO_2 glass will remain transparent to higher wavelengths than those for SiO_2 glass. Thus, for dielectric materials, there exists a "window" in which the transmission occurs easily. This will be between wavelengths that correspond to the optical absorption edge and the higher wavelength edge dictated by the phonon-photon interactions. In many cases, though, porosity, second phases, or impurities can have a significant effect on the absorption of radiation in this window of optical transparency. These ideas are important in designing materials that are transparent to certain ranges of infrared and other radiations.

The intensity of the transmitted beam also depends on microstructural features. Porosity in ceramics scatters photons; even a small amount of porosity (less than 1 volume percent) may make the ceramic opaque. For example, alumina that has relatively low density is opaque, however, a high-density alumina is optically transparent. High-density alumina is often used in the manufacture of lamp envelopes. Crystalline precipitates, particularly those that have a much different index of refraction than the matrix material, also cause scattering. These crystalline *opacifiers* cause a glass that normally may have excellent transparency to become translucent or even opaque. Typically, smaller pores or precipitates cause a greater reduction in the transmission of light.

In metals, the absorption coefficient tends to be large [Figure 20-6(a)], particularly in the visible light spectrum. Because there is no energy gap in metals, virtually any photon has sufficient energy to excite an electron into a higher energy level, thus absorbing the energy of the excited photon. Dielectrics, on the contrary, possess a large energy gap between the valence and conduction bands. If the energy of the incident

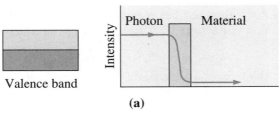

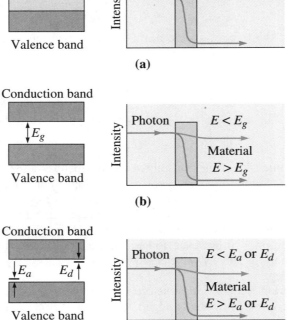

Figure 20-6
Relationships between absorption and the energy gap: (a) metals, (b) Dielectrics and intrinsic semiconductors, and (c) extrinsic semiconductors.

photons is less than the energy gap, no electrons gain enough energy to escape the valence band and, therefore, absorption does not occur [Figure 20-6(b)]. When the photons do not interact with imperfections in the material, the material is said to be *transparent*. This is the case for glass, many high-purity crystalline ceramics, and amorphous polymers such as acrylics, polycarbonates, and polysulfones.

In intrinsic semiconductors, absorption occurs when the photons have energies exceeding the energy gap E_g, whereas transmission occurs for less energetic photons [Figure 20-6(b)]. In intrinsic semiconductors, the energy gap is typically smaller than that in insulators. In extrinsic semiconductors, we have donor or acceptor energy levels that provide additional energy levels for absorption. In extrinsic semiconductors, absorption occurs when photons have energies greater than E_a or E_d [Figure 20-6(c)]. Therefore, semiconductors are opaque to short-wavelength radiation but transparent to long-wavelength photons. For example, silicon and germanium appear opaque to visible light, but they are transparent to the longer wavelength infrared radiation. Many of the narrow bandgap semiconductors (e.g., HgCdTe) are used for detection of infrared radiation. The bandgap of these semiconductors is comparable to the energy hν of the infrared radiation. These detector materials have to be cooled to lower temperatures (e.g., liquid nitrogen) since the thermal energy of electrons at room temperatures is otherwise enough to saturate the conduction band.

Photoconduction occurs in semiconducting materials if the semiconductor is part of an electrical circuit. If the energy of an incoming photon is sufficient, an electron is excited into the conduction band, or a hole is created in the valence band and the electron or hole then carries a charge through the circuit [Figure 20-7(a)]. The maximum wavelength of the incoming photon required to produce photoconduction is related to the energy gap in the semiconducting material.

$$\lambda_{\max} = \frac{hc}{E_g} \tag{20-12}$$

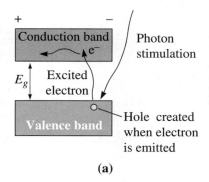

(a)

Figure 20-7
(a) Photoconduction in semiconductors involves the absorption of a stimulus by exciting electrons from the valence band to the conduction band. Rather than dropping back to the valence band to cause emission, the excited electrons carry a charge through an electrical circuit. (b) A solar cell takes advantage of this effect.

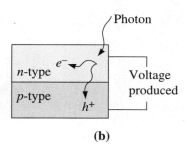

(b)

We can use this principle for photodetectors or "electric eyes" that open or close doors or switches when a beam of light focused on a semiconducting material is interrupted. The current is also intensified.

Solar cells also use the absorption of light to generate voltage [Figure 20-7(b)]. Essentially the charge carriers generated by optical absorption are separated, and this leads to the development of a voltage. This voltage causes a current flow in an external circuit. Solar cells are *p-n* junctions designed so that photons excite electrons into the conduction band. The electrons move to the *n*-side of the junction, while holes move to the *p*-side of the junction. This movement produces a contact voltage due to the charge imbalance. If the junction device is connected to an electric circuit, the junction acts as a battery to power the circuit. Solar cells make use of antireflective coatings so that maximum key elements of the solar spectrum are captured.

LEDs As discussed in Chapter 18, the light that is absorbed by a direct band gap semiconductor causes electrons to be promoted to the conduction band. When these electrons fall back into the valence bandgap they combine with the holes causing emission of light. Many semiconductor solid solutions can be used so as to have different bandgaps, producing LEDs of different colors. This phenomenon is also used in light-emitting diodes and semiconductor lasers (Figure 20-8). The design of LEDs is discussed a little later in this chapter.

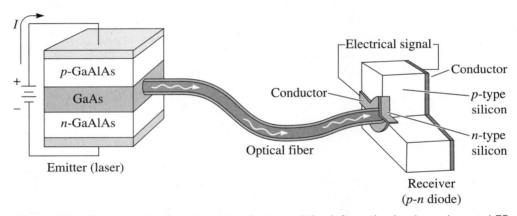

Figure 20-8   Elements of a photonic system for transmitting information involves a laser or LED to generate photons from an electrical signal, optical fibers to transmit the beam of photons efficiently, and an LED receiver to convert the photons back into an electrical signal.

The following examples illustrate applications of many of these concepts related to the absorption and transmission of light.

EXAMPLE 20-3   *Determining Critical Energy Gaps*

Determine the critical energy gaps that provide complete transmission and complete absorption of photons in the visible spectrum.

SOLUTION

The visible light spectrum varies from 4×10^{-5} cm to 7×10^{-5} cm. The minimum E_g required to assure that no photons in the visible spectrum are absorbed is:

$$E_g = \frac{hc}{\lambda} = \frac{(6.62 \times 10^{-34} \text{ J} \cdot \text{s})(3 \times 10^{10} \text{ cm/s})}{(4 \times 10^{-5} \text{ cm})(1.6 \times 10^{-19} \text{ J/eV})}$$

$$= 3.1 \text{ eV}$$

The maximum E_g below which all of the photons in the visible spectrum are absorbed is:

$$E_g = \frac{hc}{\lambda} = \frac{(6.62 \times 10^{-34} \text{ J} \cdot \text{s})(3 \times 10^{10} \text{ cm/s})}{(7 \times 10^{-5} \text{ cm})(1.6 \times 10^{-19} \text{ J/eV})}$$

$$= 1.8 \text{ eV}$$

For materials with an intermediate E_g, a portion of the photons in the visible spectrum will be absorbed.

EXAMPLE 20-4   *Design of a Radiation Shield*

A material has a reflectivity of 0.15 and an absorption coefficient (α) of 100 cm^{-1}. Design a shield that will permit only 1% of the incident radiation from being transmitted through the material.

SOLUTION

From Equation 20-11, the fraction of the incident intensity that will be transmitted is:

$$\frac{I_t}{I_0} = (1 - R)^2 \exp(-\alpha x)$$

$$0.01 = (1 - 0.15)^2 \exp(-100x)$$

$$\frac{0.01}{(0.85)^2} = 0.01384 = \exp(-100x)$$

$$\ln(0.01384) = -4.28 = -100x$$

$$x = 0.0428 \text{ cm}$$

The material should have a thickness of 0.0428 cm in order to transmit 1% of the incident radiation.

If we wished, we could determine the amount of radiation lost in each step:

Reflection at the front face: $I_r = RI_0 = 0.15I_0$.

Energy after reflection: $I = I_0 - 0.15I_0 = 0.85I_0$

Energy after absorption: $I_a = (1 - R)I_0 \exp[(-100)(0.0428)] = 0.0118I_0$

Energy due to absorption: $0.85I_0 - 0.0118I_0 = 0.838I_0$

Reflection at the back face: $I_r = R(1 - R)I_0 \exp(-\alpha x)$

$$= (0.15)(1 - 0.15)I_0 \exp[-(100)(0.0428)] = 0.0018I_0$$

20-3   Selective Absorption, Transmission, or Reflection

Unusual optical behavior is observed when photons are selectively absorbed, transmitted, or reflected. We have already seen that semiconductors transmit long-wavelength photons but absorb short-wavelength radiation. There are a variety of other cases in which similar selectivity produces unusual optical properties.

In certain materials, replacement of normal ions by transition or rare earth elements produces a *crystal field*, which creates new energy levels within the structure. This phenomenon occurs when Cr^{3+} ions replace Al^{3+} ions in Al_2O_3. The new energy levels absorb visible light in the violet and green-yellow portions of the spectrum. Red wavelengths are transmitted, giving the reddish color in ruby. In addition, the chromium ion replacement creates an energy level that permits luminescence (discussed later) to occur when the electrons are excited by a stimulus. Lasers produced from chromium-doped ruby produce a characteristic red beam because of this.

Glasses can also be doped with ions that produce selective absorption and transmission (Table 20-3). For example, photochromic glass, used for sunglasses, contains silver compounds. The glass darkens in sunlight but becomes transparent in darkness. In bright light, the silver ions in the glass gain an electron through excitation by the photons and are reduced from Ag ions to nano-sized clusters of metallic silver atoms. Thus, absorption of photons occurs. When the incoming light diminishes in intensity, the silver reverses to silver ions and no absorption occurs. More recently photochromic plastics have also been developed.

TABLE 20-3 ■ *Effect of ions on colors produced in glasses*

Ion	Color
Cr^{2+}	Blue
Cr^{3+}	Green
Cu^{2+}	Blue-green
Mn^{2+}	Orange
Fe^{2+}	Blue-green
U^{6+}	Yellow

Electron or hole traps, called *F-centers*, can also be present in crystals. When fluorite (CaF_2) is produced so that there is excess calcium, a fluoride ion vacancy is produced. To maintain electrical neutrality, an electron is trapped in the vacancy, producing energy levels that absorb all visible photons—with the exception of purple.

Polymers—particularly those containing an aromatic ring in the backbone can have complex covalent bonds that produce an energy level structure that causes selective absorption. For this reason, chlorophyll in plants appears green and hemoglobin in blood appears red. The following example illustrates the design of a "stealthy" aircraft.

| EXAMPLE 20-5 | *Design of a "Stealthy" Aircraft* |

Design an aircraft that cannot be detected by radar.

SOLUTION

Radar is electromagnetic radiation, typically in the microwave portion of the spectrum. To detect an aircraft by radar, some portion of the microwave must be reflected from the aircraft and returned to a radar receiver. Several approaches can be considered to reduce the radar signature from an aircraft:

1. We might make the aircraft from materials that are transparent to radar. Many polymers, polymer-matrix composites, and ceramics satisfy this requirement.

2. We might design the aircraft so that the radar signal is reflected at severe angles from the source. This could be done with flat surfaces at more than about a 30° angle from the horizontal, or by assuring that all surfaces are curved to help reflect the radar at a variety of angles.

3. The internal structure of the aircraft also can be made to absorb the radar. For example, use of a honeycomb material in the wings may cause the radar waves to be repeatedly reflected within the material. The honeycomb can also be filled with an absorbing material to hasten the dissipation of the radar energy.

4. We might make the aircraft less visible by selecting materials that have electronic transitions of the same energy as the radar. Carbon fibers in a polymer matrix and carbon-carbon composites at high-temperature locations in the aircraft accomplish this end. Coatings containing ferrimagnetic ceramic ferrites on an aircraft skin absorb radar waves and convert the energy to heat, but these materials are very heavy. Dielectrics, which typically are lighter in weight, can be engineered to absorb radiation of the appropriate frequency.

20-4 Examples and Use of Emission Phenomena

Let's look at some particular examples of emission phenomena which, by themselves, provide some familiar and important functions.

Gamma Rays—Nuclear Interactions   Gamma rays, which are very high-energy photons, are emitted during the radioactive decay of unstable nuclei of certain atoms. The energy of the gamma rays therefore depends on the structure of the atom nucleus and varies for different materials. The gamma rays produced from a material have fixed wavelengths. When cobalt 60 decays, for example, gamma rays having energies of 1.17×10^6 and 1.33×10^6 eV (or wavelengths of 1.06×10^{-10} cm and 0.93×10^{-10} cm) are emitted. The gamma rays can be used as the radiation source to detect defects lying within a material (a nondestructive test).

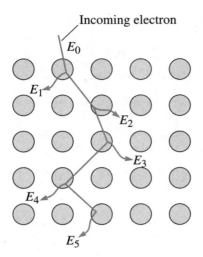

Incoming electron

E_0

E_1

E_2

E_3

E_4

E_5

$$E_1 + E_2 + E_3 + E_4 + E_5 = E_0$$

Figure 20-9
When an accelerated electron strikes and interacts with a material, its energy may be reduced in a series of steps. In the process, several photons of different energies E_1 to E_5 are emitted, each with a unique wavelength.

X-rays—Inner Electron Shell Interactions **X-rays**, which have somewhat lower energy than gamma rays, are produced when electrons in the inner shells of an atom are stimulated. The stimulus could be high-energy electrons or other x-rays. When stimulation occurs, x-rays of a wide range of energies are emitted. Both a continuous and a characteristic spectrum of x-rays are produced.

Suppose that a high-energy electron strikes a material. As the electron decelerates, energy is given up and emitted as photons. Each time the electron strikes an atom, more of its energy is given up. Each interaction, however, may be more or less severe, so the electron gives up a different fraction of its energy each time and produces photons of different wavelengths (Figure 20-9). A **continuous spectrum** is produced (the smooth portion of the curves in Figure 20-10). If the electron were to lose all of its energy in one impact, the minimum wavelength of the emitted photons would be equivalent to the original energy of the stimulus. The minimum wavelength of x-rays produced is called

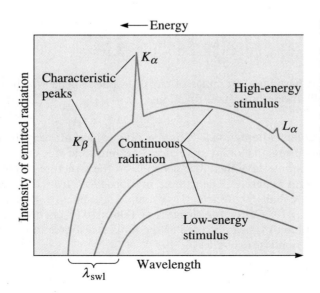

Figure 20-10
The continuous and characteristic spectra of radiation emitted from a material. Low-energy stimuli produce a continuous spectrum of low-energy, long-wavelength photons. A more intense, higher energy spectrum is emitted when the stimulus is more powerful until, eventually, characteristic radiation is observed.

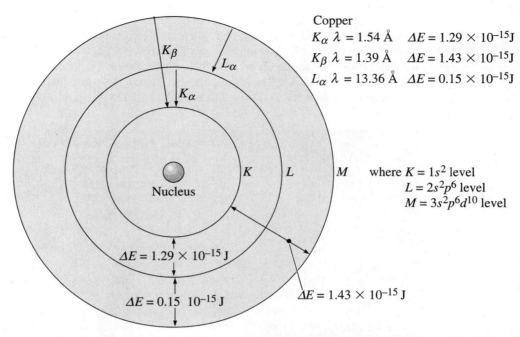

Copper
$K_\alpha \; \lambda = 1.54 \; \text{Å}$ $\Delta E = 1.29 \times 10^{-15} \text{J}$
$K_\beta \; \lambda = 1.39 \; \text{Å}$ $\Delta E = 1.43 \times 10^{-15} \text{J}$
$L_\alpha \; \lambda = 13.36 \; \text{Å}$ $\Delta E = 0.15 \times 10^{-15} \text{J}$

where $K = 1s^2$ level
$L = 2s^2 p^6$ level
$M = 3s^2 p^6 d^{10}$ level

Figure 20-11 Characteristic x-rays are produced when electrons change from one energy level to a lower energy level, as illustrated here for copper. The energy and wavelength of the x-rays are fixed by the energy differences between the energy levels.

the **short wavelength limit** λ_{swl}. The short wavelength limit decreases, and the number and energy of the emitted photons increase, when the energy of the stimulus increases.

The incoming stimulus may also have sufficient energy to excite an electron from an inner-energy level into an outer-energy level. The excited electron is not stable and, to restore equilibrium, electrons from a higher level fill the empty inner level. This process leads to the emission of a **characteristic spectrum** of x-rays that is different for each type of atom.

The characteristic spectrum is produced because there are discrete energy differences between any two energy levels. When an electron drops from one level to a second level, a photon having that particular energy and wavelength is emitted. This effect is illustrated in Figure 20-11. We typically refer to the energy levels by the K, L, M, . . . designation, as described in Chapter 2. If an electron is excited from the K shell, electrons may fill that vacancy from any outer shell. Normally, electrons in the closest shells fill the vacancies. Thus, photons with energy $\Delta E = E_K - E_L$ (K_α x-rays) or $\Delta E = E_K - E_M$ (K_β x-rays) are emitted. When an electron from the M shell fills the L shell, a photon with energy $\Delta E = E_L - E_M$ (L_α x-rays) is emitted; it has a long wavelength, or low energy. Note that we need a more energetic stimulus to produce K_α x-rays than that required for L_α x-rays.

As a consequence of the emission of photons having a characteristic wavelength, a series of peaks is superimposed on the continuous spectrum (Figure 20-10). The wavelengths at which these peaks occur are peculiar to the type of atom (Table 20-2). Thus, each element produces a different characteristic spectrum, which serves as a "fingerprint" for that type of atom. If we match the emitted characteristic wavelengths with those expected for various elements, the identity of the material can be determined. We can also measure the intensity of the characteristic peaks. By comparing measured intensities with standard intensities, we can estimate the percentage of each type of atom in the material and, hence, we can estimate the composition of the material. The energy or wavelength of the x-rays emitted when an electron beam impacts a sample (such as that in a scanning or transmission electron microscope) can be analyzed to get chemical information of the make up of a sample. This technique is known as energy dispersive x-ray analysis (EDXA). The examples that follow illustrate the application of x-ray emissions as used in XRD (x-ray diffraction) and EDXA analytical techniques.

EXAMPLE 20-6   *Design/Materials Selection for an X-ray Filter*

Design a filter that preferentially absorbs K_β x-rays from the nickel spectrum but permits K_α x-rays to pass with little absorption. This type of filter is used in x-ray diffraction (XRD) analysis of materials.

SOLUTION

When determining a crystal structure or identifying unknown materials using various x-ray diffraction techniques, we prefer to use x-rays of a single wavelength. If both K_α and K_β characteristic peaks are present and interact with the material, analysis becomes much more difficult.

However, we can use the selective absorption, or the existence of the absorption edge, to isolate the K_α peak. Table 20-2 includes the information that we need. If a filter material is selected; such that the absorption edge

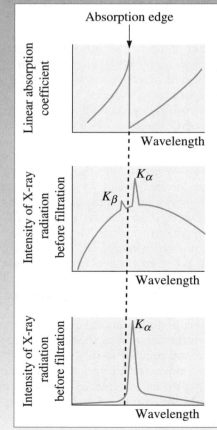

Figure 20-12
Elements have a selective lack of absorption of certain wavelengths. If a filter is selected with an absorption edge between the K_α and K_β peaks of an x-ray spectrum, all x-rays except K_α are absorbed (for Example 20-6).

lies between the K_α and K_β wavelengths, then the K_β is almost completely absorbed, whereas the K_α is almost completely transmitted. In nickel, $K_\alpha = 1.660$ Å and $K_\beta = 1.500$ Å. A filter with an absorption edge between these characteristic peaks will work. Cobalt, with an absorption edge of 1.608 Å, would be our choice. Figure 20-12 shows how this filtering process occurs.

EXAMPLE 20-7   *Design of an X-ray Filter*

Design a filter to transmit at least 95% of the energy of a beam composed of zinc K_α x-rays, using aluminum as the shielding material. (The aluminum has a linear absorption coefficient of 108 cm^{-1}.) Assume no loss to reflection.

SOLUTION

Assuming that no losses are caused by the reflection of x-rays from the aluminum, we need simply to design the thickness of the aluminum required to transmit 95% of the incident intensity. The final intensity will therefore be $0.95I_0$. Thus, from Equation 20-10(a),

$$\ln\left(\frac{0.95I_0}{I_0}\right) = -(108)(x)$$

$$\ln(0.95) = -0.051 = -108x$$

$$x = \frac{-0.051}{-108} = 0.00047 \text{ cm}$$

We would like to roll the aluminum to a thickness of 0.00047 cm or less. The filter could be thicker if a material were selected that has a lower linear-absorption coefficient for zinc K_α x-rays.

EXAMPLE 20-8   *Generation of X-rays for XRD*

Suppose an electron accelerated at 5000 V strikes a copper target. Will K_α, K_β, or L_α x-rays be emitted from the copper target?

SOLUTION

The electron must possess enough energy to excite an electron to a higher level, or its wavelength must be less than that corresponding to the energy difference between the shells:

$$E = (5000 \text{ eV})(1.6 \times 10^{-19} \text{ J/eV}) = 8 \times 10^{-16} \text{ J}$$

$$\lambda = \frac{hc}{E} = \frac{(6.62 \times 10^{-34})(3 \times 10^{10})}{8 \times 10^{-16}}$$

$$= 2.48 \times 10^{-8} \text{ cm} = 2.48 \text{ Å}$$

For copper, K_α is 1.542 Å, K_β is 1.392 Å, and L_α is 13.357 Å (Table 20-2). Therefore, the L_α peak may be produced, but K_α and K_β will not.

EXAMPLE 20-9 *Energy Dispersive X-ray Analysis (EDXA)*

The micrograph in Figure 20-13 was obtained using a scanning electron microscope at a magnification of 1000. The beam of electrons in the SEM was directed at the three different phases, creating x-rays and producing the characteristic peaks. From the energy spectra, determine the probable identity of each phase. Assume each region represents a different phase.

SOLUTION

All three phases have an energy peak of about 1.5 keV = 1500 eV, which corresponds to a wavelength of:

$$\lambda = \frac{hc}{E} = \frac{(6.62 \times 10^{-34} \text{ J} \cdot \text{s})(3 \times 10^{10} \text{ cm/s})}{(1500 \text{ eV})(1.6 \times 10^{-19} \text{ J/eV})(10^{-8} \text{ cm/A})} = 8.275 \text{ Å}$$

In a similar manner, energies and wavelengths can be found for the other peaks. These wavelengths are compared with those in Table 20-2, and the

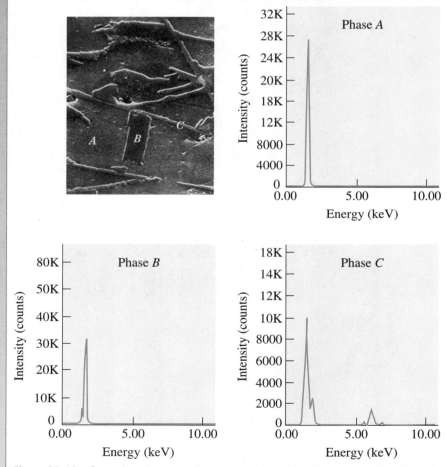

Figure 20-13 Scanning electron micrograph of a multiple-phase material. The energy distribution of emitted radiation from the three phases marked *A*, *B*, and *C* is shown. The identity of each phase is determined in Example 20-9.

identity of the elements in each phase can be found, as summarized in the table.

Phase	Peak Energy	λ	λ (Table 20-2)	Line
A	1.5 keV	8.275 Å	8.337 Å	K_αAl
B	1.5 keV	8.275 Å	8.337 Å	K_αAl
	1.7 keV	7.30 Å	7.125 Å	K_αSi
C	1.5 keV	8.275 Å	8.337 Å	K_αAl
	1.7 keV	7.30 Å	7.125 Å	K_αSi
	5.8 keV	2.14 Å	2.104 Å	K_αMn
	6.4 keV	1.94 Å	1.937 Å	K_αFe
	7.1 keV	1.75 Å	1.757 Å	K_βFe

Thus, phase *A* appears to be an aluminum matrix, phase *B* appears to be a silicon needle (perhaps containing some aluminum), and phase *C* appears to be an Al-Si-Mn-Fe compound. Actually, this is an aluminum-silicon alloy. The stable phases are aluminum and silicon, with inclusions forming when manganese and iron are present as impurities.

Luminescence—Outer Electron Shell Interactions   Whereas x-rays are produced by electron transitions in the inner-energy levels of an atom, **luminescence** is the conversion of radiation or other forms of energy to visible light. Luminescence occurs when the incident radiation excites electrons from the valence band, through the energy gap, and into the conduction band. The excited electrons remain in the higher energy levels only briefly. When the electrons drop back to the valence band, photons are emitted. If the wavelength of these photons is in the visible light range, luminescence occurs.

Luminescence does not occur in metals. Electrons are merely excited into higher energy levels within the unfilled valence band and, when the excited electron returns to the lower energy level, the photon that is produced has a very small energy and a wavelength longer than that of visible light [Figure 20-14(a)].

In certain ceramics and semiconductors, however, the energy gap between the valence and conduction bands is such that an electron dropping through this gap produces a photon in the visible range. Two different effects are observed in these luminescent materials: fluorescence and phosphorescence. In **fluorescence**, all of the excited electrons drop back to the valence band and the corresponding photons are emitted within a very short time ($\sim 10^{-8}$ seconds) after the stimulus is removed [Figure 20-14(b)]. One wavelength, corresponding to the energy gap E_g, predominates. Fluorescent dyes and microscopy are used in many advanced techniques in biochemistry and biomedical engineering.[8] X-ray fluorescence (XRF) is widely used for the chemical analysis of materials.

Phosphorescent materials have impurities that introduce a donor level within the energy gap [Figure 20-14(c)]. The stimulated electrons first drop into the donor level and are trapped. The electrons must then escape the trap before returning to the valence band. There is a delay before the photons are emitted. When the source is removed, electrons in the traps gradually escape and emit light over some additional period of time. The intensity of the luminescence is given by

$$\ln\left(\frac{I}{I_0}\right) = -\frac{t}{\tau},$$

(20-13)

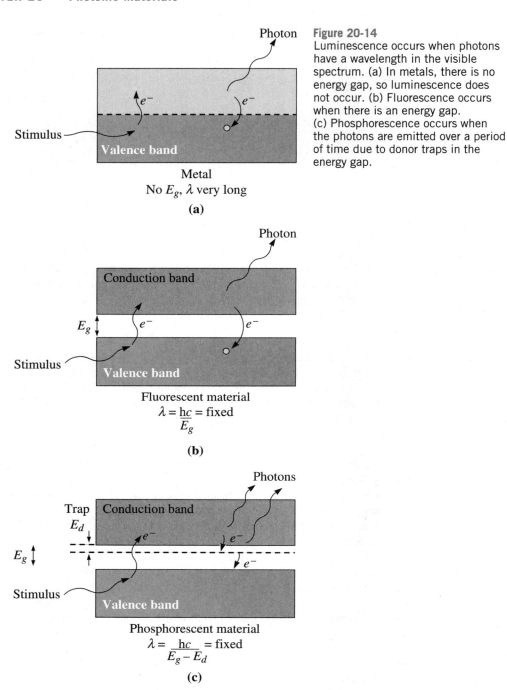

Figure 20-14
Luminescence occurs when photons have a wavelength in the visible spectrum. (a) In metals, there is no energy gap, so luminescence does not occur. (b) Fluorescence occurs when there is an energy gap. (c) Phosphorescence occurs when the photons are emitted over a period of time due to donor traps in the energy gap.

where τ is the **relaxation time**, a constant for the material. After time t following removal of the source, the intensity of the luminescence is reduced from I_0 to I. Phosphorescent materials are very important in the operation of television screens. In this case, the relaxation time must not be too long or the images begin to overlap. In color television, three types of phosphorescent materials are used; the energy gaps are engineered so that red, green, and blue colors are produced. Oscilloscope and radar screens rely on the same principle. Fluorescent lamps consist of mercury vapor. The mercury vapor, in the presence of an electrical arc, produces ultraviolet light. This ultraviolet light leads to fluorescence. The inside of the glass of these lamps is coated with a phosphorescent material. The role of this material is to convert the small wavelength ultraviolet radiation into visible light. The fluorescent decay is temperature independent. The range of times is from 5×10^{-9} seconds to about 2 seconds. The following example illustrates the selection of a phosphor for a television screen.

EXAMPLE 20-10   *Design/Materials Selection for a Television Screen*

Select a phosphor material that will produce a blue image on a television screen.

SOLUTION

Photons having energies that correspond to the color blue have wavelengths of about 4.5×10^{-5} cm (Figure 20-1). The energy of the emitted photons therefore is:

$$E = \frac{hc}{\lambda} = \frac{(4.14 \times 10^{-15} \text{ eV} \cdot \text{s})(3 \times 10^{10} \text{ cm/s})}{4.5 \times 10^{-5} \text{ cm}}$$

$$= 2.76 \text{ eV}$$

Table 18-6 includes energy gaps for a variety of materials. None of the materials listed has an E_g of 2.76 eV, but ZnS has an E_g of 3.54 eV. If a suitable dopant were introduced to provide a trap $3.54 - 2.76 = 0.78$ eV below the conduction band, phosphorescence would occur.

We would also need information concerning the relaxation time to assure that phosphorescence would not persist long enough to distort the image. Typical phosphorescent materials for television screens might include $CaWO_4$, which produces photons with a wavelength of 4.3×10^{-5} cm (blue). This material has a relaxation time of 4×10^{-6} s. ZnO doped with excess zinc produces photons with a wavelength of 5.1×10^{-5} cm (green), whereas $Zn_3(PO_4)_2$ doped with manganese gives photons with a wavelength of 6.45×10^{-5} cm (red).

Light-Emitting Diodes—Electroluminescence   Luminescence can be used to advantage in creating **light-emitting diodes** (LEDs). LEDs are used to provide the display for watches, clocks, calculators, and other electronic devices. The stimulus for these devices is an externally applied voltage, which causes electron transitions and **electroluminescence**. LEDs are *p-n* junction devices engineered so that the E_g is in the visible spectrum (often red). A voltage applied to the diode in the forward-bias direction causes holes and electrons to recombine at the junction and emit photons (Figure 20-15). GaAs, GaP, GaAlAs, and GaAsP are typical materials for LEDs.

Lasers—Amplification of Luminescence   The **laser** (*l*ight *a*mplification by *s*timulated *e*mission of *r*adiation) is another example of a special application of luminescence. In certain materials, electrons excited by a stimulus (such as the flash tube shown in Figure 20-16) produce photons which, in turn, excite additional photons of identical wave-

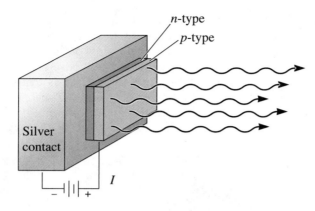

Figure 20-15
Diagram of a light-emitting diode (LED). A forward-bias voltage across the *p-n* junction produces photons.

length.[9] Consequently, a large amplification of the photons emitted in the material occurs. By proper choice of stimulant and material, the wavelength of the photons can be in the visible range. The output of the laser is a beam of photons that are parallel, of the same wavelength, and coherent. In a *coherent* beam, the wavelike nature of the photons is in phase, so that destructive interference does not occur. Lasers are useful in heat treating and melting of metals, welding, surgery, mapping, transmission and processing of information, and a variety of other applications, including reading of the compact disks used to produce noise-free stereo recordings.

A variety of materials are used to produce lasers. Ruby, which is single crystal Al_2O_3 doped with a small amount of Cr_2O_3 (emits at 6943 Å), and yttrium aluminum garnet ($Y_3Al_5O_{12}$ YAG) doped with neodymium (Nd) (emits at 1.06 μm) are two common solid-state lasers. Other lasers are based on CO_2 gas.

Semiconductor lasers, such as those based on GaAs solid solutions which have an energy gap corresponding to a wavelength in the visible range, are also used (Figure 20-1). Figure 20-17 illustrates how a semiconductor laser might operate. When a voltage

The laser process

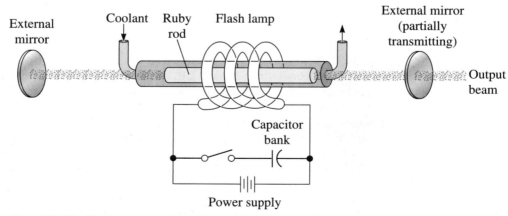

Figure 20-16   The laser converts a stimulus into a beam of coherent photons. The mirror on one side is 100% reflecting, the mirror on the right transmits partially. (*Source: From* Optical Materials: An Introduction to Selection and Application, *by S. Musikant, p. 201, Fig. 10-1. Copyrigt © 1985 Marcel Dekker, Inc.*)

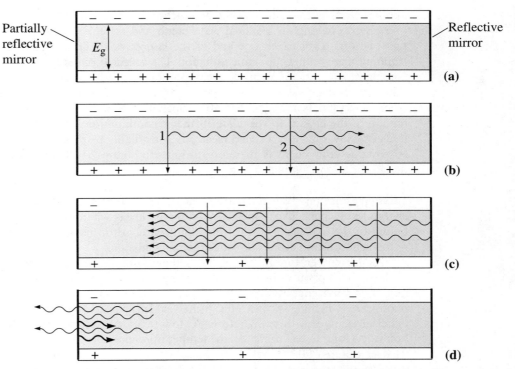

Partially reflective mirror

E_g

Reflective mirror

(a)

1

2

(b)

(c)

(d)

Figure 20-17 Creation of a laser beam from a semiconductor: (a) Electrons are excited into the conduction band by an applied voltage. (b) Electron 1 recombines with a hole to produce a photon. The photon stimulates the emission of photon 2 by a second recombination. (c) Photons reflected from the mirrored end stimulate even more photons. (d) A fraction of the photons are emitted as a laser beam, while the rest are reflected to stimulate more recombinations.

applied to the device excites the semiconductor, electrons jump from the valence band to the conduction band, leaving behind holes in the valence band. When an electron collapses back to the valence band and recombines with a hole, a photon having an energy and wavelength equivalent to the energy gap is produced. This photon stimulates

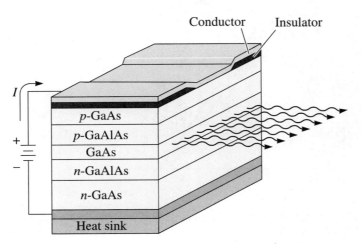

Conductor   Insulator

I

p-GaAs

p-GaAlAs

GaAs

n-GaAlAs

n-GaAs

Heat sink

Figure 20-18
Schematic cross-section of a GaAs laser. Because the surrounding p- and n-type GaAlAs layers have a higher energy gap and a lower index of refraction than GaAs, the photons are trapped in the active GaAs layer.

another electron to drop from the conduction band to the valence band, creating a second photon having an identical wavelength and a frequency that is in phase with the first photon. A mirror at one end of the laser crystal completely reflects the photons, trapping them within the semiconductor. The reflected photons stimulate even more recombinations, until an intense wave of photons is produced. The photons then reach the other end of the crystal, which is only partly mirrored. A fraction of the photons emerge from the crystal as a monochromatic, coherent laser beam, while the rest of the photons remain in the crystal to stimulate further recombinations. The applied voltage assures that a steady source of excited electrons is available to produce additional photons; a continuous laser beam is therefore produced. Figure 20-18 schematically depicts one design for a semiconductor laser.

Thermal Emission   When a material is heated, electrons are thermally excited to higher energy levels, particularly in the outer energy levels where the electrons are less tightly bound to the nucleus. The electrons immediately drop back to their normal levels and release photons, an event known as **thermal emission**.

As the temperature increases, thermal agitation increases, and the maximum energy of the emitted photons increases. A continuous spectrum of radiation is emitted, with a minimum wavelength and an intensity distribution dependent on the temperature. The photons may include wavelengths in the visible spectrum; consequently, the color of the material changes with temperature. At low temperatures, the wavelength of the radiation is too long to be visible. As the temperature increases, emitted photons have shorter wavelengths. At 700°C we begin to see a reddish tint; at 1500°C the orange and red wavelengths are emitted (Figure 20-19). Higher temperatures produce all wavelengths in the visible range, and the emitted spectrum is white light. By measuring the intensity of a narrow band of the emitted wavelengths with a pyrometer, we can estimate the temperature of the material.

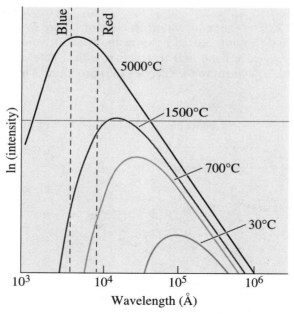

Figure 20-19   Intensity in relation to wavelengths of photons emitted thermally from a material. As the temperature increases, more photons are emitted from the visible spectrum.

20-5 Fiber Optic Communication System

Alexander Graham Bell and William Wheeler were men ahead of their times. In 1880, Bell invented a light-based communication system known as the photophone and Wheeler was granted a patent (U.S. Patent 247,229, granted September 20, 1881) for a system that used pipes to light distant rooms.[1,10,11] These two inventions were the predecessors of the fiber-optic communications systems that exist today. The other key invention that helped commercialize fiber optics technologies was the invention of the laser in 1960. The laser provided a monochromatic source of light so fiber optics could be used effectively. Another advancement came when high-purity silica glass fibers became available. These fibers provided very small optical losses and were essential for carrying information over longer distances without the need for equipment to boost the signal. Optical fibers are also free from electromagnetic interference (EMI) since they carry signal as light, not radio waves.

A fiber optic system transmits a light signal generated from some other source, such as an electrical signal. The fiber optic system transmits the light to a receiver using an optical fiber, processes the data received, and converts the data to a usable form (Figure 20-20). Photonic materials are required for this process. Most of the principles and materials presently used in photonic systems have already been introduced in the previous sections. Let's review these materials in the context of an actual system, pointing out some of the special requirements that are needed (Figure 20-20).

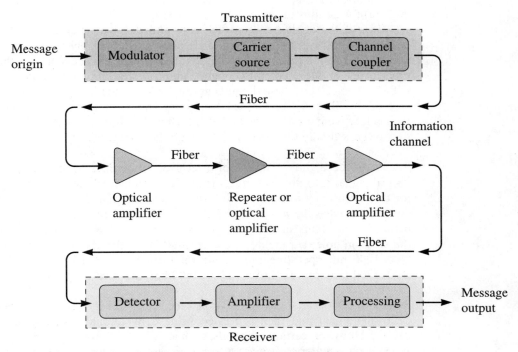

Figure 20-20 Schematic of a fiber-optic based communication system.

Generating the Signal In order to best transmit and process information, the light should be coherent and monochromatic (to minimize dispersion). Thus, a laser is an ideal method for generating the photons. The Group III–V semiconductors include

GaAs, GaAlAs (750–900 nm), and InGaAsP (1300 nm). Most long-distance systems use the InGaAsP sources that emit at 1300 nm, which is in the infrared range. A voltage can energize lasers built from these materials, thereby generating a laser beam. Light-emitting diodes are a second device that can be used to produce the photons.

By varying the voltages applied to these devices, the intensity of the photon beam can be varied. The intensity of the beam can then be exploited to convey information.

Transmitting the Beam   Optical fibers transmit the information. Waveguides, composed of glass fibers, transmit the light from the source to the receiver. In order for the optical fibers to transmit light long distances efficiently, the glass must have exceptional transparency and must not leak any light. Most optical fibers are made from ultra high-purity silica (SiO_2). The diameter of the fibers is 0.25–0.5 mm (including the plastic coating that prevents diffusion of water vapor and mechanical damage to the fiber surface) (Chapter 5). Water molecules can introduce hydroxyl groups (OH^-) on the glass surface and increase optical losses. Also, as discussed in earlier chapters, water vapor causes corrosion of the glass fiber surface and this introduces microcracks that lower the strength of the glass fibers. The diameter of the fiber itself is about ~125 μm. The core consists of a material that has a higher refractive index ($n_{core} = 1.48$) than the cladding ($n_{cladding} = 1.46$). The cladding is also made from high-purity silica and may be anywhere from about 9 to 85 μm in diameter. The core is typically doped with GeO_2 to enhance the refractive index. Some fibers based on plastics and other types of glasses (e.g., fluorides) have been developed recently. These advanced materials provide a better index of refraction which helps keep the beam within the central core. See Figure 20-21(a) for an illustration of the beam's path.

Even if the angle between the beam and the fiber is small, the distance that the beam must travel through a composite fiber is considerably longer than the fiber itself due to the sharp reflections that are involved at the distinct boundary between the two types of glass. More complex, fibers contain a core glass that is doped at the surface with B_2O_3 or GeO_2; these dopants gradually lower the index of refraction near the surface of the fiber. These Graded Refractive Index, or GRIN, fibers gradually change their refractive index [Figure 20-21(b)] thereby helping rays that enter at different angles to travel essentially the same path through the fiber [Figure 20-21(c)].[11]

Receiving the Signal   The job of a receiver in the fiber optic system is to convert the optical signal into an electronic signal. Semiconductor photodiodes typically are used for this job as they respond much faster to respond than solar cell-type detectors. When a photon reaches the *p-n* diode, an electron is excited into the conduction band, leaving behind a hole. Thus, electron-hole pairs are created. If a voltage is applied to the diode, the electron-hole pair creates a current that is amplified and further processed. Silicon (400–1000 nm operating range), Ge (600–1000 nm), GaAs (800–1000 nm), InGaAs (100–1700 nm), and InGaAsP (1100–1600 nm) are used for making such detectors.

Processing the Signal   Normally, the received signal is converted immediately into an electronic signal and then processed using conventional silicon-based semiconductor devices. However, certain materials, such as $LiNbO_3$, have a nonlinear optical response. When such a material receives a beam of photons, it acts as a transistor and amplifies the signal or acts as a switch (or a logic gate in a computer) to control the path of the beam. Development of such photonic transistors could one day lead to an optically based computer.

A fiber optic system also has of repeaters, regenerators, and optical amplifiers. One limitation of a fiber optics-based communication system is that even though the information travels between two points at the speed of light, the optical signal ultimately

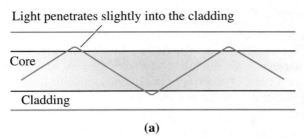

Light penetrates slightly into the cladding

Core

Cladding

(a)

Figure 20-21
Different types of optical fibers. (a) A step index glass fiber, in which the index of refraction is slightly different in each glass. (b) The profile of a refractive index in a graded refractive index (GRIN) fiber. (c) The path of rays entering at different angles.[1,5]

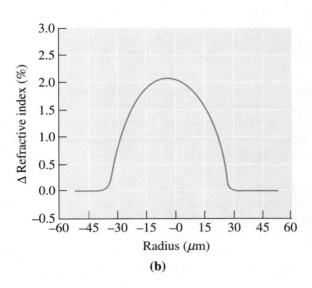

(b)

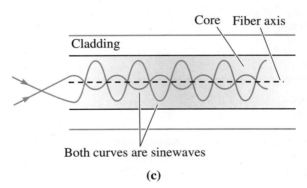

Cladding

Core   Fiber axis

Both curves are sinewaves

(c)

needs to be converted into an electronic form, thus slowing down the overall rate of information transfer. Thus, optical fibers are considered "passive" components. This is why there is a considerable amount of interest in developing active optical devices.

Photonic Bandgap Materials   This is one area of active research.[2] In **photonic bandgap materials** (PBG), there is a range of frequencies that cannot be transmitted. This is the equivalent of an optical insulator. These materials are produced by micromachining silicon or other materials; colloidal particles are used since the spacing required for these crystals is of the length scale $\cong$ or <1 μm. It is possible that these types of materials could lead to active optical devices equivalent to a transistor.

SUMMARY

◇ The optical properties of materials include the refractive index, absorption coefficient, and dispersion. These are determined by the interaction of electromagnetic radiation, or photons with materials. Refractive index of materials depends primarily upon the extent of electronic polarization and is therefore related to the high-frequency dielectric constant of materials.

◇ As a result of the interaction between light and materials, refraction, reflection, transmission, scattering, and diffraction can occur. These phenomena are used in a wide variety of applications of photonic materials. These applications include fiber optics for communication and lasers for surgery and welding. Devices that use optoelectronic effects include LEDs, solar cells, photodiodes, etc. Other applications include phosphors for fluorescent lights, televisions, and many analytical techniques.

◇ Emission of photons occurs by transitions in the atom structure of the material. Fluorescence, phosphorescence, electroluminescence (used in light-emitting diodes), and lasers are examples of luminescence. Photons are emitted by thermal excitation, with photons in the visible portion of the spectrum produced when the temperature is sufficiently high. X-ray emission from materials is used in EDXA and XRF analysis.

◇ A fiber optic system uses an optical fiber as one of its key elements. Information, in the form of a lightwave, is kept in the optical fiber using total internal reflection. The core has a higher refractive index than the cladding. Either a step or graded refractive index profile can be used. Optical fibers are coated with polymers to avoid the degradation of fibers from water vapor and mechanical stress.

◇ A fiber optic system uses semiconductor lasers to generate light, couplers, fibers, photodiodes. The system also makes use of repeaters, regenerators, etc. Over 10 million kilometers of optical fibers are currently used, including cables that cross the Atlantic Ocean.

GLOSSARY

Absorption edge   The wavelength at which the absorption characteristics of a material abruptly change.

Characteristic spectrum   The spectrum of radiation emitted from a material. It shows peaks at fixed wavelengths corresponding to particular energy level differences corresponding to the atomic structure of the material.

Continuous spectrum   Radiation emitted from a material having all wavelengths longer than a critical short wavelength limit.

Dispersion   Frequency dependence of the refractive index.

Electroluminescence   Use of an applied electrical signal to stimulate photons from a material.

Fluorescence   Emission of light obtained typically within $\sim 10^{-8}$ seconds.

Index of refraction   Relates the change in velocity and direction of radiation as it passes through a transparent medium (also known as refractive index).

Laser   The acronym stands for light amplification by stimulated emission of radiation. A beam of monochromatic coherent radiation produced by the controlled emission of photons.

Light-emitting diodes (LEDs)   Electronic *p-n* junction devices that convert an electrical signal into visible light.

Linear absorption coefficient   Describes the ability of a material to absorb radiation.

Luminescence   Conversion of radiation to visible light.

Phosphorescence   Emission of radiation from a material after the stimulus is removed.

Photoconduction   Production of a voltage due to the stimulation of electrons into the conduction band by light radiation.

Photonic bandgap materials   These are structures produced using micromachined silicon or colloidal particles, such that there is a range of frequencies that cannot be transmitted through the structure.

Photons   Energy or radiation produced from atomic, electronic, or nuclear sources that can be treated as particles or waves.

Reflectivity   The percentage of incident radiation that is reflected.

Refractive index   See Index of refraction.

Relaxation time   The time required for $1/e$ of the electrons to drop from the conduction band to the valence band in luminescence.

Short wavelength limit   The shortest wavelength or highest energy radiation emitted from a material under particular conditions.

Solar cell   A *p-n* junction device that creates a voltage due to excitation by photons.

Thermal emission   Emission of photons from a material due to excitation of the material by heat.

X-rays   Electromagnetic radiation in the wavelength range ~1 to 100 Å.

ADDITIONAL INFORMATION

1. PALAIS, J.C., *Fiber Optic Communications*. 4th ed. 1998.

2. POLMAN, A. and P. WILTZIUS, "Materials Science Aspects of Photonic Crystals," *MRS Bulletin*, 2001. 26(8): p. 608–610.

3. COLLINGS, P.J. and M. HIRD, *Introduction to Liquid Crystals: Chemistry and Physics*. 1997: Taylor and Francis.

4. BUCHANAN, R.C., *Ceramic Materials for Electronics*. 2nd ed. 1991: Marcel Dekker.

5. SINGH, J., *Optoelectronics: An Introduction to Materials and Devices*. 1996: McGraw Hill.

6. KINGERY, W.D., H.K. BOWEN, and D.R. UHLMANN, *Introduction to Ceramics*. 2nd ed. 1976: Wiley.

7. GREENBERG, C.B., "Thin Films on Float Glass: The Extraordinary Possibilities," *J. of Indu. Chemi*, 2001. 40(1): p. 26–32.

8. MUJUMDAR, R. and S. MUJUMDAR, *Technical discussions are acknowledged*. 2002.

9. MUSIKANT, S., *Optical Materials: An Introduction to Selection and Application*. 1985: New York, Marcel Dekker.

10. MIMMS, F.M., "Alexander Graham Bell and the Photophone: The Centennial of the Invention of Light Wave Communications 1880–1980," *Opt. News*, 1980. 6(1): p. 8–16.

11. HECHT, J., *Understanding Fiber Optics*. 1993: Sams Publishing.

PROBLEMS

Section 20-1 The Electromagnetic Spectrum

Section 20-2 Refraction, Reflection, Absorption, and Transmission

20-1 State the definitions of refractive index and absorption coefficient. Compare these with the definitions of dielectric constant, loss factor, Young's modulus, and viscous deformation.

20-2 What is Snell's law? Illustrate using a diagram.

20-3 What does the index of refraction of a material depend upon?

20-4 What is "lead crystal?" What makes the refractive index of this material so much higher than that of ordinary silicate glass?

20-5 What polarization mechanism affects the refractive index?

20-6 Why is the refractive index of ice smaller than that of water?

20-7 What is dispersion? What is the importance of dispersion in fiber optic systems?

20-8 What factors limit the transmission of light through dielectric materials?

20-9 What is the principle by which LEDs and solar cells work?

20-10 A beam of photons strikes a material at an angle of 25°C to the normal of the surface. Which, if any, of the materials listed in Table 20-1 could cause the beam of photons to continue at an angle of 18 to 20° from the normal of the material's surface?

20-11 A laser beam passing through air strikes a 5-cm-thick polystyrene block at a 20° angle to the normal of the block. By what distance is the beam displaced from its original path when the beam reaches the opposite side of the block?

20-12 A beam of photons passes through air and strikes a soda-lime glass that is part of an aquarium containing water. What fraction of the beam is reflected by the front face of the glass? What fraction of the remaining beam is reflected by the back face of the glass?

20-13 We find that 20% of the original intensity of a beam of photons is transmitted from air through a 1-cm-thick material having a dielectric constant of 2.3 and back into air. Determine the fraction of the beam that is

(a) reflected at the front surface,

(b) absorbed in the material, and

(c) reflected at the back surface.

Determine the linear-absorption coefficient of the photons in the material.

20-14 A beam of photons in air strikes a composite material consisting of a 1-cm-thick sheet of polyethylene and a 2-cm-thick sheet of soda-lime glass. The incident beam is 10° from the normal of the composite. Determine the angle of the beam with respect to the normal of the beam

(a) passes through the polyethylene,

(b) passes through the glass, and

(c) passes through air on the opposite side of the composite.

By what distance is the beam displaced from its original path when it emerges from the composite?

20-15 A glass fiber ($n = 1.5$) is coated with Teflon. Calculate the maximum angle that a beam of light can deviate from the axis of the fiber without escaping from the inner portion of the fiber.

20-16 A material has a linear-absorption coefficient of 591 cm^{-1} for photons of a particular wavelength. Determine the thickness of the material required to absorb 99.9% of the photons.

Section 20-3 Selective Absorption, Transmission, or Reflection

20-17 What is a photochromic glass?

20-18 How are colored glasses produced?

20-19 What is ruby crystal made from?

Section 20-4 Examples and Uses of Emission Phenomena

20-20 What is the principle of energy dispersive x-ray analysis (EDXA)?

20-21 What is fluorescence? What is phosphorescence?

20-22 What is XRF?

20-23 How does a fluorescent lamp work?

20-24 What is electroluminescence?

20-25 Calcium tungstate (CaWO$_4$) has a relaxation time of 4×10^{-6} s. Determine the time required for the intensity of this phosphorescent material to decrease to 1% of the original intensity after the stimulus is removed.

20-26 The intensity of a radiation from phosphorescent material is reduced to 90% of its original intensity after 1.95×10^{-7} s. Determine the time required for the intensity to decrease to 1% of its original intensity.

20-27 What is a laser?

20-28 In a solid-state laser, why is one mirror partially transmitting?

20-29 What color is emitted from a ruby laser and a Nd:YAG laser?

20-30 By appropriately doping yttrium aluminum garnet with neodymium, electrons are excited within the $4f$ energy shell of the Nd atoms. Determine the approximate energy transition if the Nd:YAG serves as a laser, producing a wavelength of 532 nm. What color would the laser beam possess?

20-31 Determine whether an incident beam of photons with a wavelength of 7500 Å will cause luminescence in the following materials (see Chapter 18):

(a) ZnO   (b) GaP   (c) GaAs
(d) GaSb   (e) PbS

20-32 Determine the wavelength of photons produced when electrons excited into the conduction band of indium-doped silicon:

(a) drop from the conduction band to the acceptor band, and

(b) then drop from the acceptor band to the valence band (see Chapter 18).

20-33 Which, if any, of the semiconducting compounds listed in Chapter 18 are capable of producing an infrared laser beam?

20-34 What type of electromagnetic radiation (ultraviolet, infrared, visible) is produced from

(a) pure germanium, and

(b) germanium doped with phosphorus? (See Chapter 18.)

20-35 Which, if any, of the dielectric materials listed in Chapter 18 would reduce the speed of light in air from 3×10^{10} cm/s to less than 0.5×10^{10} cm/s?

20-36 What filter material would you use to isolate the K_α peak of the following x-rays: iron, manganese, nickel? Explain your answer.

20-37 What voltage must be applied to a tungsten filament to produce a continuous spectrum of x-rays having a minimum wavelength of 0.09 nm?

20-38 A tungsten filament is heated with a 12,400 V power supply. What is

(a) the wavelength, and

(b) the frequency of the highest-energy x-rays that are produced?

20-39 What is the minimum voltage required to produce K_α x-rays in nickel?

20-40 Based on the characteristic x-rays that are emitted, determine the difference in energy between electrons in tungsten for

(a) the K and L shells,

(b) the K and M shells, and

(c) the L and M shells.

20-41 Figure 20-22 shows the results of an x-ray fluorescent analysis, in which the energy of x-rays emitted from a material are plotted relative to the wavelength of the x-rays. Determine

(a) the accelerating voltage used to produce the, exciting x-rays, and

(b) the identity of the elements in the sample.

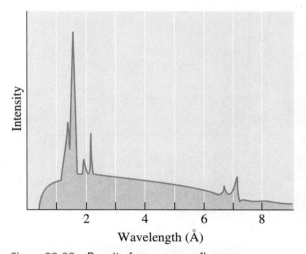

Figure 20-22 Results from an x-ray fluorescence analysis of an unknown metal sample (for Problem 20-41).

20-42 Figure 20-23 shows the energies of x-rays produced from an energy-dispersive analysis of radiation emitted from a specimen in a scanning electron microscope. Determine the identity of the elements in the sample.

20-43 Figure 20-24 shows the intensity of the radiation obtained from a copper x-ray generating tube as a function of wavelength. The accompanying table shows the linear absorption coefficient for a nickel filter for several wavelengths. If the Ni filter is 0.005 cm thick, calculate and plot the intensity of the transmitted x-ray beam versus wavelength.

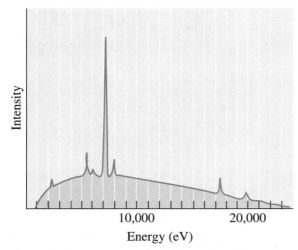

Figure 20-23 X-ray emission spectrum (for Problem 20-42).

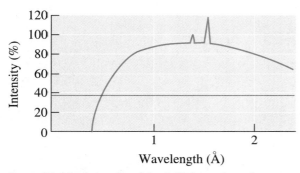

Figure 20-24 Intensity of the initial spectrum from a copper x-ray source before filtering (for Problem 20-43).

Wavelength (Å)	Linear Absorption Coefficient (cm^{-1})
0.711	422
1.436	2900
1.542	440
1.659	543
1.790	670
1.937	830
2.103	1030
2.291	1300

Section 20-5 Fiber Optic Communication System

20-44 What is the principle by which information is transmitted via an optical fiber?

20-45 Draw a sketch of a step-index optical fiber. Explain how total internal reflection keeps the light inside the fiber.

20-46 What is a GRIN fiber? Explain how the profile of refractive index in a GRIN fiber allows different incident angles.

20-47 What material is used to make most optical fibers? What is the cladding made out of? What material is used to enhance the refractive index of the core?

20-48 Why are optical fibers coated with a polymer?

20-49 What role does water vapor play in determining the optical losses and mechanical strength of optical fibers?

20-50 In a fiber optics system, how is an electrical signal converted or modulated into an optical wave?

20-51 In a fiber optics system, how is the optical signal converted into an electrical signal?

20-52 Is an optical fiber an active or passive device? Explain.

20-53 What is a photonic bandgap material?

Design Problems

20-54 Nickel x-rays are to be generated inside a container, with the x-rays being emitted from the container through only a small slot. Design a container that will assure that no more than 0.01% of the K_α nickel x-rays escape through the rest of the container walls, yet 95% of the K_α nickel x-rays pass through a thin window covering the slot. The following data give the mass-absorption coefficients of several metals for nickel K_α x-rays. The mass absorption coefficient α_m is α/ρ, where α is the linear mass-absorption coefficient and ρ is the density of the filter material.

Material	α_m (cm^2/g)
Be	1.8
Al	58.4
Ti	247.0
Fe	354.0
Co	54.4
Cu	65.0
Sn	322.0
Ta	200.0
Pb	294.0

20-55 Design a method by which a photoconductive material might be used to measure the temperature of a material from the material's thermal emission.

20-56 Design a method, based on a material's refractive characteristics, that will cause a beam of photons originally at a 2° angle to the normal of the material to be displaced from its original path by 2 cm at a distance of 50 cm from the material.

20-57 Amorphous selenium is used in photocopiers. Conduct a literature search and find out how amorphous selenium works in this application.

20-58 Design a 3-ft-diameter satellite housing an infrared detector that can be placed into a low Earth orbit and that will not be detected by radar.

Computer Problems

20-59 *Calculating Power in Decibels.* In an optical communication system or electrical power transmission system, the power or signal often is transferred between several components. The decibel (dB) is a convenient unit to measure the relative power levels. If the input power to a device is P_1 and the output power is P_2, then P_2/P_1 is the ratio of power transmitted, thus representing efficiency. This ratio in decibels is written as:

$$dB = 10 \log \frac{P_2}{P_1}.$$

Power must be expressed in similar units. Write a computer program that will calculate the dB value for the transmission of power between two components of a fiber optic system (e.g., light source to fiber). Then extend this calculation to three components (e.g., light transmitted from source to fiber, and then fiber to a detector).

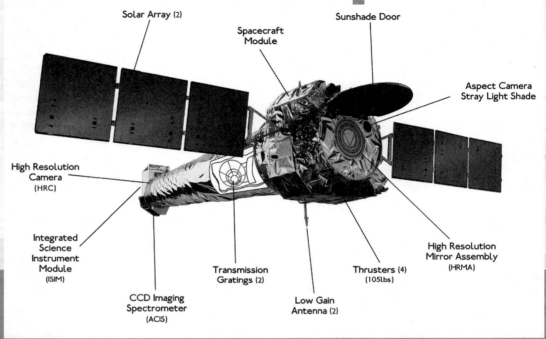

Solar Array (2)

Spacecraft Module

Sunshade Door

Aspect Camera Stray Light Shade

High Resolution Camera (HRC)

Integrated Science Instrument Module (ISIM)

CCD Imaging Spectrometer (ACIS)

Transmission Gratings (2)

Low Gain Antenna (2)

Thrusters (4) (105lbs)

High Resolution Mirror Assembly (HRMA)

Thermal properties of materials play a very important role in many technologies. From refractories used in the production of metallic materials to thermal barrier coatings (TBCs) for turbine blades, many applications require materials that minimize heat transfer. There are other applications where thermal management will be of paramount importance. For example, it is expected that a computer chip manufactured in year 2010 will have a billion transistors and generate more power (on a square inch basis) than a nuclear reactor!

In these applications high thermal conductivity materials are needed. The thermal expansion of materials also plays a key role in many situations; such expansion can lead to the development of stresses that lead to material failure. Materials scientists and engineers have created many novel materials that have a negative or near zero thermal expansion coefficient. This is an important characteristic in optical materials such as those used in telescope mirrors whose focusing will change depending upon thermal expansion and contraction. This photograph shows an artist's rendering of the Chandra x-ray telescope. The mirror substrate of NASA's Chandra and Hubble telescopes has been made from a glass-ceramic material known as Zerodur™, manufactured by Schott Glass Technologies. Notice the solar panels as well. Zerodur™ is just one of the many examples of how the science and engineering of materials has had an impact on other scientific disciplines, creating knowledge and benefits to society. (*Courtesy of NASA.*)

21

Thermal Properties of Materials

Have You Ever Wondered?

- What material has the highest thermal conductivity?

- Why, although we normally use ceramics for insulation, there are many ceramics that have high-thermal conductivity?

- Are there materials that have zero or negative thermal-expansion coefficients?

- What material is used to make the Chandra x-ray telescope mirror substrate?

In previous chapters, we described how a material's properties can change with temperature. In many cases, we found that a material's mechanical and physical properties depend on the temperature at which the material serves or may be subjected to during processing. An appreciation of the thermal properties of materials is helpful in understanding the mechanical failure of materials such as ceramics, thermal barriers, and fibers when the temperature changes; in designing processes in which materials must be heated; or in selecting materials to transfer heat rapidly, as in semiconductor-based electronic devices.

Thermal management has become a very important issue in electronic packaging materials. It has been estimated that, by the year 2010, a single computer chip will hold a billion transistors generating 1000 watts of power. This is more heat per square inch than that of a nuclear reactor.[1]

In metallic materials, electrons transfer heat. In ceramic materials, the conduction of heat involves phonons. In certain other applications, such as thermal barrier coatings or space shuttle tiles, we want to minimize the heat transfer through the material. Heat transfer is also important in many applications ranging from, for example, polystyrene foam cups used for hot beverages to sophisticated coatings on glasses to make energy efficient buildings.[2,3] In this chapter, we will discuss heat capacity, thermal expansion properties, and the thermal conductivity of materials.

21-1   Heat Capacity and Specific Heat

In Chapter 20, we noted that optical behavior depends on how photons are produced and interact with a material. The photon is treated as a particle with a particular energy or as electromagnetic radiation having a particular wavelength or frequency. Some of the thermal properties of materials can also be characterized in the same dual manner, as are optical properties. However, these properties are determined by the behavior of **phonons**, rather than photons.

At absolute zero, the atoms in a material have a minimum energy. However, when heat is supplied, the atoms gain thermal energy and vibrate at a particular amplitude and frequency. The vibration of each atom is transferred to the surrounding atoms and produces an elastic wave called a phonon. The energy of the phonon can be expressed in terms of the wavelength or frequency, just as in Equation 20-1:

$$E = \frac{hc}{\lambda} = h\nu \tag{21-1}$$

The energy required to change the temperature of the material one degree is the **heat capacity** or **specific heat**.

The heat capacity is the energy required to raise the temperature of one mole of a material by one degree. The specific heat is defined as the energy needed to increase the temperature of one grain of a material by 1°C. The heat capacity can be expressed either at constant pressure, C_p, or at a constant volume, C_v. At high temperatures, the heat capacity for a given volume of material approaches

$$C_p = 3R \simeq 6 \frac{cal}{mol \cdot K}, \tag{21-2}$$

where R is the gas constant (1.987 cal/mol). However, as shown in Figure 21-1, heat capacity is not a constant. The heat capacity of metals approaches $6 \frac{cal}{mol \cdot K}$ near room temperature, but this value is not reached in ceramics until near 1000°C.

The relationship between specific heat and heat capacity is:

$$\text{Specific heat} = C_p = \frac{\text{heat capacity}}{\text{atomic weight}} \tag{21-3}$$

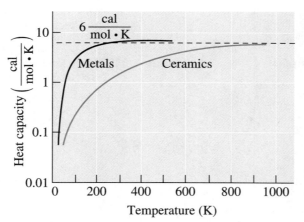

Figure 21-1
Heat capacity as a function of temperature for metals and ceramics.

In most engineering calculations, specific heat is used more conveniently than heat capacity. The specific heat of typical materials is given in Table 21-1. Neither the heat capacity nor the specific heat depends significantly on the structure of the material; thus, changes in dislocation density, grain size, or vacancies have little effect.

The most important factor affecting specific heat is the lattice vibrations or phonons. However, other factors affect the heat capacity; one striking example occurs in ferromagnetic materials such as iron (Figure 21-2). An abnormally high heat capacity is observed in iron at the Curie temperature, where the normally aligned magnetic moments of the iron atoms are randomized and the iron becomes paramagnetic. Heat capacity also depends on the crystal structure, as shown in Figure 21-2 for iron.

The following examples illustrate the use of specific heat.

TABLE 21-1 ■ *The specific heat of selected materials at 27°C*

Material	Specific Heat $\left(\dfrac{cal}{g \cdot K}\right)$	Material	Specific Heat $\left(\dfrac{cal}{g \cdot K}\right)$
Metals:		Ceramics:	
Al	0.215	Al_2O_3	0.200
Cu	0.092	Diamond	0.124
B	0.245	SiC	0.250
Fe	0.106	Si_3N_4	0.170
Pb	0.038	SiO_2 (silica)	0.265
Mg	0.243	Polymers:	
Ni	0.106	High-density polyethylene	0.440
Si	0.168	Low-density polyethylene	0.550
Ti	0.125	6,6-nylon	0.400
W	0.032	Polystyrene	0.280
Zn	0.093	Other:	
		Water	1.000
		Nitrogen	0.249

Note: 1 $\dfrac{cal}{g \cdot K}$ = 4184 $\dfrac{J}{kg \cdot K}$

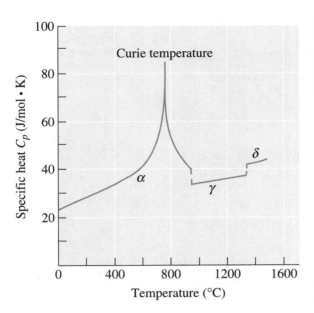

Figure 21-2
The effect of temperature on the specific heat of iron. Both the change in crystal structure and the change from ferromagnetic to paramagnetic behavior are indicated.

EXAMPLE 21-1 *Specific Heat of Tungsten*

How much heat must be supplied to 250 g of tungsten to raise its temperature from 25°C to 650°C?

SOLUTION

The specific heat of tungsten is $0.032 \dfrac{cal}{g \cdot K}$. Thus:

$$\text{Heat required} = (\text{specific heat})(\text{mass})(\Delta T)$$
$$= (0.032 \text{ cal/g} \cdot K)(250 \text{ g})(650 - 25)$$
$$= 5000 \text{ cal}$$

If no losses occur, 5000 cal (or 20,920 J) must be supplied to the tungsten. A variety of processes might be used to heat the metal. We could use a gas torch, we could place the tungsten in an induction coil to induce eddy currents, we could pass an electrical current through the metal, or we could place the metal into an oven heated by SiC resistors.

EXAMPLE 21-2 *Specific Heat of Niobium*

Suppose the temperature of 50 g of niobium increases 75°C when heated for a period. Estimate the specific heat and determine the heat in calories required.

SOLUTION

The atomic weight of niobium is 92.91 g/mol. We can use Equation 21-3 to estimate the heat required to raise the temperature of one gram by one °C:

$$C_p \approx \frac{6}{92.91} = 0.0646 \text{ cal/g} \cdot {}^{\circ}\text{C}$$

Thus the total heat required is:

$$\text{Heat} = \left(0.0646 \frac{\text{cal}}{\text{g} \cdot {}^{\circ}\text{C}} \right) (50 \text{ g})(75^{\circ}\text{C}) = 242 \text{ cal}$$

21-2 Thermal Expansion

An atom that gains thermal energy and begins to vibrate behaves as though it has a larger atomic radius. The average distance between the atoms and therefore the overall dimensions of the material increase. The change in the dimensions of the material Δl per unit length is given by the **linear coefficient of thermal expansion** α:

$$\alpha = \frac{l_f - l_0}{l_0(T_f - T_0)} = \frac{\Delta l}{l_0 \Delta T} = \frac{\varepsilon}{\Delta T} \tag{21-4}$$

where T_0 and T_f are the initial and final temperatures and l_0 and l_f are the initial and final dimensions of the material and ε is the strain. A *volume* coefficient of thermal expansion (α_v) can also be defined to describe the change in volume when the temperature of the material is changed. If the material is isotropic, $\alpha_v = 3\alpha$. An instrument known as a dilatometer is used to measure the thermal-expansion coefficient. It is also possible to trace thermal expansion using XRD. Coefficients of thermal expansion for several materials are included in Table 21-2.

The coefficient of thermal expansion of a material is related to the strength of the atomic bonds. In order for the atoms to move from their equilibrium positions, energy must be introduced into the material. If a very deep energy trough caused by strong atomic bonding is characteristic of the material, the atoms separate to a lesser degree and the material has a low linear coefficient of expansion (Chapter 2). This relationship also suggests that materials having a high melting temperature—also due to strong atomic attractions—have low coefficients of thermal expansion (Figure 21-3). Consequently, lead (Pb) has a much larger coefficient than that for high melting point metals such as tungsten (W). Most ceramics, which have strong ionic or covalent bonds, have low coefficients compared with metals. Certain glasses, such as fused silica, also have a poor packing factor, which helps accommodate thermal energy with little dimensional change. Although bonding within the chains of polymers is covalent, the secondary bonds holding the chains together are weak, leading to high coefficients. Polymers that contain strong cross-linking typically have lower coefficients than linear polymers such as polyethylene.

Several precautions must be taken when calculating dimensional changes in materials:

1. The expansion characteristics of some materials, particularly single crystals or materials having a preferred orientation, are anisotropic.

2. Allotropic materials have abrupt changes in their dimensions when the phase transformation occurs (Figure 21-4). These abrupt changes contribute to the cracking of refractories on heating or cooling and quench cracks in steels.

3. The linear coefficient of expansion continually changes with temperature. Normally, α either is listed in handbooks as a complicated temperature-dependent function or is given as a constant for only a particular temperature range.

TABLE 21-2 ■ The linear coefficient of thermal expansion at room temperature for selected materials

Material	Linear Coefficient of Thermal Expansion ($\times 10^{-6}$ 1/0°C)
Al	25.0
Cu	16.6
Fe	12.0
Ni	13.0
Pb	29.0
Si	3.0
W	4.5
1020 steel	12.0
3003 aluminum alloy	23.2
Gray iron	12.0
Invar (Fe-36% Ni)	1.54
Stainless steel	17.3
Yellow brass	18.9
Epoxy	55.0
6,6-nylon	80.0
6,6-nylon—33% glass fiber	20.0
Polyethylene	100.0
Polyethylene—30% glass fiber	48.0
Polystyrene	70.0
Al_2O_3	6.7
Fused silica	0.55
Partially stabilized ZrO_2	10.6
SiC	4.3
Si_3N_4	3.3
Soda-lime glass	9.0

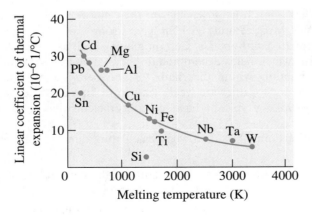

Figure 21-3
The relationship between the linear coefficient of thermal expansion and the melting temperature in metals at 25°C. Higher melting point metals tend to expand to a lesser degree.

4. Interaction of the material with electric or magnetic fields produced by magnetic domains may prevent normal expansion until temperatures above the Curie temperature are reached. This is the case for Invar, a Fe-36% Ni alloy, which undergoes practically no dimensional changes at temperatures below the Curie temperature (about 200°C). This makes Invar attractive as a material for bimetallics (Figure 21-4).

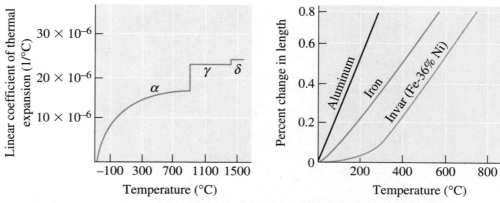

Figure 21-4 (a) The linear coefficient of thermal expansion of iron changes abruptly at temperatures where an allotropic transformation occurs. (b) The expansion of Invar is very low due to the magnetic properties of the material at low temperatures.

The thermal expansion of engineered materials can be tailored using multi-phase materials. Upon heating, one phase can show thermal expansion while the other phase can show thermal contraction. Thus, the overall material can show a zero or negative thermal-expansion coefficient. Zerodur™ is a glass-ceramic material that can be controlled to have zero or slightly negative thermal expansion developed by Schott Glass Technologies. It consists of a ~70–80 wt% crystalline phase that has a high-quartz structure. The remainder is a glassy phase. The negative thermal expansion coefficient of the glassy phase and the positive thermal expansion of the crystalline phase cancel for each other out, leading to a zero-thermal-expansion material. Zerodur™ has been used as the mirror substrate on the Hubble telescope and Chandra x-ray telescope. A dense, optically transparent, and zero-thermal-expansion material is necessary in these applications since any changes in dimensions as a result of the changes in the temperatures in space will make it difficult to focus the telescopes properly. Zerodur™ is one of the examples of how engineered materials have helped astronomers and society learn about far away galaxies. Many ceramic materials based on sodium zirconium phosphate (NZP) that have a near-zero-thermal-expansion coefficient have also been developed by materials scientists.[4–6]

The following examples show the use of the linear coefficients of thermal expansion.

EXAMPLE 21-3   *Bonding and Thermal Expansion*

Explain why, in Figure 21-3, the linear coefficients of thermal expansion for silicon and tin do not fall on the curve. How would you expect germanium to fit into this figure?

SOLUTION

Both silicon and tin are covalently bonded. The strong covalent bonds are more difficult to stretch than the metallic bonds (a deeper trough in the energy-separation curve), so these elements have a lower coefficient. Since germanium also is covalently bonded, its thermal expansion should be less than that predicted by Figure 21-3.

EXAMPLE 21-4   *Design of a Pattern for a Casting Process*

Design the dimensions for a pattern that will be used to produce a rectangular-shaped aluminum casting having dimensions at 25°C of 25 cm × 25 cm × 3 cm.

SOLUTION

To produce a casting having particular final dimensions, the mold cavity into which the liquid aluminum is to be poured must be made oversize. After the liquid solidifies, which occurs at 660°C for pure aluminum, the solid casting contracts as it cools to room temperature. If we calculate the amount of contraction expected, we can make the original pattern used to produce the mold cavity that much larger.

The linear coefficient of thermal expansion for aluminum is 25×10^{-6} 1/°C. The temperature change from the freezing temperature to 25°C is $660 - 25 = 635$°C. The change in any dimension is given by:

$$\Delta l = l_0 - l_f = \alpha l_0 \Delta T$$

For the 25-cm dimensions, $l_f = 25$ cm. We wish to find l_0:

$$l_0 - 25 = (25 \times 10^{-6})(l_0)(635)$$

$$l_0 - 25 = 0.15875 l_0$$

$$0.984 l_0 = 25$$

$$l_0 = 25.40 \text{ cm}$$

For the 3-cm dimensions, $l_f = 3$ cm.

$$l_0 - 3 = (25 \times 10^{-6})(l_0)(635)$$

$$l_0 - 3 = 0.015875 l_0$$

$$0.984 l_0 = 3$$

$$l_0 = 3.05 \text{ cm}$$

If we design the pattern to the dimensions 25.40 cm × 25.40 cm × 3.05 cm, the casting should contract to the required dimensions.

When an isotropic material is slowly and uniformly heated, the material expands uniformly without creating any residual stress. If, however, the material is restrained from moving, the dimensional changes may not be possible and, instead, stresses develop. These **thermal stresses** are related to the coefficient of thermal expansion, the modulus of elasticity E of the material, and the temperature change ΔT:

$$\sigma_{\text{thermal}} = \alpha E \Delta T \tag{21-5}$$

Thermal stresses can arise from a variety of sources. In large rigid structures such as bridges, restraint may develop as a result of the design. Some bridges are designed in sections and are comprised of steel plates in between, so that the sections move relative to one another during seasonal temperature changes.

When materials are joined—for example, coating cast iron bathtubs with a ceramic enamel or coating superalloy turbine blades with a yttria stabilized zirconia (YSZ) thermal barrier—changes in temperature cause different amounts of contraction or expansion in the different materials. This disparity leads to thermal stresses that may cause the protective coating to spall off. Careful matching of the thermal properties of the coating to those of the substrate material is necessary to prevent coating cracking (if the coefficient of the coating is less than that of the underlying substrate) or spalling (flaking of the coating due to a high-expansion coefficient). Thermal barrier coatings have porosity, which increases their thermal protection ability, and the porosity helps with reducing the level of stress and improving thermal-shock resistance.

A similar situation may occur in composite materials. Brittle fibers that have a lower coefficient than the matrix may be stretched to the breaking point when the temperature of the composite increases.

Thermal stresses may even develop in a nonrigid, isotropic material if the temperature is not uniform. In producing tempered glass (Chapter 14), the surface is cooled more rapidly than the center, permitting the surface to initially contract. When the center cools later, its contraction is restrained by the rigid surface, placing compressive residual stresses on the surface.

EXAMPLE 21-5   *Design of a Protective Coating*

A ceramic enamel is to be applied to a 1020 steel plate. The ceramic has a fracture strength of 4000 psi, a modulus of elasticity of 15×10^6 psi, and a coefficient of thermal expansion of $10 \times 10^{-6} \frac{1}{°C}$. Design the maximum temperature change that can be allowed without cracking the ceramic.

SOLUTION

Because the enamel is bonded to the 1020 steel, it is essentially restrained. If only the enamel heated (and the steel remained at a constant temperature), the maximum temperature change would be:

$$\sigma_{\text{thermal}} = \alpha E \Delta T = \sigma_{\text{fracture}}$$

$$\left(10 \times 10^{-6} \frac{1}{°C} \right) (15 \times 10^6 \text{ psi}) \Delta T = 4000 \text{ psi}$$

$$\Delta T = 26.7°C$$

However, the steel also expands. Its coefficient of thermal expansion (Table 21-2) is $12 \times 10^{-6} \frac{1}{°C}$ and its modulus of elasticity is 30×10^6 psi. Since the steel expands more than the enamel, a stress is still introduced into the enamel. The net coefficient of expansion is

$$\Delta \alpha = 12 \times 10^{-6} - 10 \times 10^{-6} = 2 \times 10^{-6} \text{ 1/°C:}$$

$$\sigma = (2 \times 10^{-6})(15 \times 10^6) \Delta T = 4000$$

$$\Delta T = 133°C$$

In order to permit greater temperature variations, we might select an enamel that has a higher coefficient of thermal expansion, an enamel that has a lower modulus of elasticity (so that greater strains can be permitted before the stress reaches the fracture stress), or an enamel that has a higher strength.

21-3   Thermal Conductivity

The **thermal conductivity** K is a measure of the rate at which heat is transferred through a material. The treatment of thermal conductivity is similar to that of diffusion (Chapter 5). Thermal conductivity, similar to the diffusion coefficient, is a microstructure sensitive property. The conductivity relates the heat Q transferred across a given plane of area A per second when a temperature gradient $\Delta T/\Delta x$ exists (Figure 21-5):

$$\frac{Q}{A} = K\frac{\Delta T}{\Delta x} \tag{21-6}$$

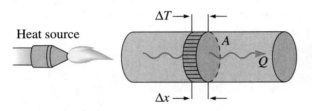

Figure 21-5
When one end of a bar is heated, a heat flux Q/A flows toward the cold end at a rate determined by the temperature gradient produced in the bar.

TABLE 21-3 ■ *Typical values of room temperature thermal conductivity of selected materials*

Material	Thermal Conductivity ($W \cdot m^{-1} \cdot K^{-1}$)	Material	Thermal Conductivity ($W \cdot m^{-1} \cdot K^{-1}$)
Pure Metals:		Ceramics:	
Ag	430		
Al	238	Al_2O_3	16–40
Cu	400	Carbon (diamond)	2000
Fe	79	Carbon (graphite)	335
Mg	100	Fireclay	0.26
Ni	90		
Pb	35	Silicon carbide	up to 270
Si	150	AlN	up to 270
		Si_3N_4	up to 150
Ti	22	Soda-lime glass	0.96–1.7
W	171	Vitreous silica	1.4
Zn	117	Vycor™ glass	12.5
Zr	23	ZrO_2	4.2
Alloys:		Polymers:	
1020 steel	100	6,6-nylon	0.25
3003 aluminum alloy	280	Polyethylene	0.33
304 stainless steel	30	Polyimide	0.21
Cementite	50	Polystyrene	0.13
		Polystyrene foam	0.029
Cu-30% Ni	50	Teflon	0.25
Ferrite	75		
Gray iron	79.5		
Yellow brass	221		

Note: 1 cal/cm $\cdot$ s $\cdot$ K = 418.4 W $\cdot$ m^{-1} $\cdot$ K^{-1}

Note that the thermal conductivity K plays the same role in heat transfer that the diffusion coefficient D does in mass transfer. Among all metals, silver (Ag) has the highest thermal conductivity at room temperature ($430 \text{ W} \cdot \text{m}^{-1} \cdot \text{K}^{-1}$). Copper is next with a thermal conductivity of $400 \text{ W} \cdot \text{m}^{-1} \cdot \text{K}^{-1}$. In general, metals have higher thermal conductivity than ceramics. However, diamond, a ceramic material, has a very high thermal conductivity of $2000 \text{ W} \cdot \text{m}^{-1} \cdot \text{K}^{-1}$. Values for the thermal conductivity of some materials are included in Table 21-3.

Thermal energy is transferred by two important mechanisms: transfer of free electrons and lattice vibrations (or phonons). Valence electrons gain energy, move toward the colder areas of the material, and transfer their energy to other atoms. The amount of energy transferred depends on the number of excited electrons and their mobility; these, in turn, depend on the type of material, lattice imperfections, and temperature. In addition, thermally induced vibrations of the atoms transfer energy through the material.

Metals Because the valence band is not completely filled in metals, electrons require little thermal excitation in order to move and contribute to the transfer of heat. Since the thermal conductivity of metals is due primarily to the electronic contribution, we expect a relationship between thermal and electrical conductivities:

$$\frac{K}{\sigma T} = L = 5.5 \times 10^{-9} \frac{\text{cal} \cdot \text{ohm}}{\text{s} \cdot \text{K}^2}, \tag{21-7}$$

where L is the **Lorentz constant**. This relationship is followed to a limited extent in many metals.

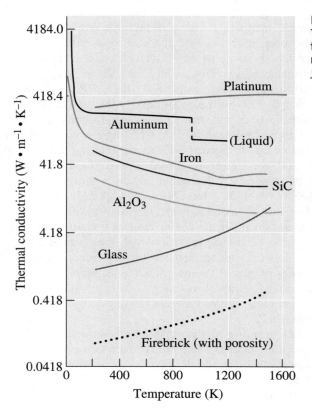

Figure 21-6
The effect of temperature on the thermal conductivity of selected materials. Note the log scale on the *y*-axis.

When the temperature of the material increases, two off-setting factors affect thermal conductivity. Higher temperatures are expected to increase the energy of the electrons, creating more "carriers" and increasing the contribution from lattice vibration; these effects increase the thermal conductivity. However, the greater lattice vibration scatters the electrons, reducing their mobility, and therefore tends to decrease the thermal conductivity. The combined effect of these factors leads to very different behavior for different metals. For iron, the thermal conductivity initially decreases with increasing temperature (due to the lower mobility of the electrons), then increases slightly (due to more lattice vibration). The conductivity *decreases* continuously when aluminum is heated but *increases* continuously when platinum is heated (Figure 21-6).

Thermal conductivity in metals also depends on crystal structure defects, microstructure, and processing. Thus, cold-worked metals, solid-solution-strengthened metals, and two-phase alloys might display lower conductivities compared with their defect-free counterparts.

Ceramics   Contrary to popular conception, ceramics do not have intrinsically low thermal conductivities. It is true that the experimentally observed thermal conductivity of many ceramics is low as compared to metals. The energy gap in ceramics is too large for many electrons to be excited into the conduction band except at very high temperatures. Consequently, the transfer of heat in ceramics occurs primarily by lattice vibrations (or phonons). Since the electronic contribution is absent, the thermal conductivity of most ceramics is much lower than that of metals. However, the main reason why the experimentally observed conductivity of ceramics is low is the level of porosity. Other factors also influence the thermal conductivity of ceramics. Crystal structure defects and porosity increase scattering and reduce conductivity. The best insulating brick, for example, contains a large fraction of porosity.

There are ceramics, such as AlN have respectable levels of thermal conductivities. As noted before, diamond has a very high thermal conductivity. Materials with a close-packed structure and high modulus of elasticity produce high-energy phonons that encourage high thermal conductivities. It has been shown that many ceramics with a diamond-like crystal structure (e.g., AlN, SiC, BeO, BP, GaN, Si, AlP) have high thermal conductivities.[7] For example, SiC and AlN with thermal conductivities of $\sim$270 W m^{-1} K^{-1} have been reported. The thermal conductivity of polycrystalline ceramic materials is typically lower than that of single crystals. The thermal conductivity of ceramics is low due to porosity. Effective sintering aids help decrease density (reducing porosity) and, formation of desired second phases at grain boundaries, via control of texture or orientation. Recently, silicon nitride ceramics, with a thermal conductivity of $\sim$150 W $\cdot$ m^{-1} $\cdot$ K^{-1} have been developed.[7]

Glasses have low thermal conductivity. The amorphous loosely packed structure minimizes the points at which silicate chains contact one another, making it more difficult for the phonons to be transferred. However, the thermal conductivity increases as the temperature increases; higher temperatures produce more energetic phonons and more rapid transfer of heat. In some applications, such as window glasses, we use double panes of glass and separate them using a gas (e.g., Ar) to provide better thermal insulation. We can also make use of different coatings on glass to make buildings and cars more energy efficient.[2,3]

The more ordered structure of the crystalline ceramics, as well as glass-ceramics that contain large amounts of crystalline precipitates, cause less scattering of phonons. Compared with glasses, these materials have a higher thermal conductivity. As the temperature increases, however, scattering becomes more pronounced and the thermal conductivity decreases (as shown for alumina and silicon carbide in Figure 21-6). At still higher temperatures, heat transfer by radiation becomes significant, and the conductivity may increase.

Some ceramics have thermal conductivities approaching that of metals. Although advanced ceramics such as AlN and SiC are good thermal conductors, they are also electrical insulators; therefore, these materials are good candidates for use in electronic packaging substrates where heat dissipation is needed.

Semiconductors   Heat is conducted in semiconductors by both phonons and electrons. At low temperatures, phonons are the principal carriers of energy, but at higher temperatures, electrons are excited through the small energy gap into the conduction band and thermal conductivity increases significantly.

Polymers   The thermal conductivity of polymers is very low—even in comparison with silicate glasses. Vibration and movement of the molecular polymer chains transfer energy. Increasing the degree of polymerization, increasing the crystallinity, minimizing branching, and providing extensive cross-linking all produce a more rigid structure and provide for higher thermal conductivity.

The thermal conductivity of many engineered materials depends upon the volume fractions of different phases and their connectivity. Silver-filled epoxies are used in many heat-transfer applications related to microelectronics. Unusually good thermal insulation is obtained using polymer foams, often produced from polystyrene or polyurethane. Styrofoam™ coffee cups are a typical product. The next example illustrates the design of a window glass for thermal conductivity.

EXAMPLE 21-6   *Design of a Window Glass*

Design a glass window 4 ft × 4 ft square that separates a room at 25°C from the outside at 40°C and allows no more than 5×10^6 cal of heat to enter the room each day. Assume thermal conductivity of glass is 0.96 W · m^{-1} · K^{-1} or $0.023 \dfrac{\text{cal}}{\text{cm} \cdot \text{s} \cdot \text{K}}$.

SOLUTION

From Equation 21-6:

$$\frac{Q}{A} = K\frac{\Delta T}{\Delta x},$$

where Q/A is the heat transferred per second through the window.

$$1 \text{ day} = (24 \text{ h/day})(3600 \text{ s/h}) = 8.64 \times 10^4 \text{ s}$$

$$A = [(4 \text{ ft})(12 \text{ in./ft})(2.54 \text{ cm/in.})]^2 = 1.486 \times 10^4 \text{ cm}^2$$

$$Q = \frac{(5 \times 10^6 \text{ cal/day})}{8.64 \times 10^4 \text{ s/day}} = 57.87 \text{ cal/s}$$

$$\frac{Q}{A} = \frac{57.87 \text{ cal/s}}{1.486 \times 10^4 \text{ cm}^2} = 0.00389 \frac{\text{cal}}{\text{cm}^2 \cdot \text{s}}$$

$$\frac{Q}{A} = 0.00389 \frac{\text{cal}}{\text{cm}^2 \cdot \text{s}} = \left(0.0023 \frac{\text{cal}}{\text{cm}^2 \cdot \text{s} \cdot \text{K}}\right)(40 - 25°\text{C})/\Delta x$$

$$\Delta x = 8.87 \text{ cm} = \text{thickness}$$

> The glass would have to be exceptionally thick to prevent the desired maximum heat flux. We might do several things to reduce the heat flux. Although all of the silicate glasses have similar thermal conductivities, we might use instead a transparent polymer material (such as polymethyl methacrylate). The polymers have thermal conductivities approximately one order of magnitude smaller than the ceramic glasses. We could also use a double-paned glass, with the glass panels separated either by a gas (air or Ar have a very low thermal conductivity) or a sheet of transparent polymer.

21-4 Thermal Shock

Stresses leading to the fracture of brittle materials can be introduced thermally as well as mechanically. When a piece of material is cooled quickly, a temperature gradient is produced. This gradient can lead to different amounts of contraction in different areas. If residual tensile stresses become high enough, flaws may propagate and cause failure. Similar behavior can occur if a material is heated rapidly. This failure of a material caused by stresses induced by sudden changes in temperature is known as **thermal shock**. Thermal shock behavior is affected by several factors:

1. *Coefficient of thermal expansion:* A low coefficient minimizes dimensional changes and reduces the ability to withstand thermal shock.

2. *Thermal conductivity:* The magnitude of the temperature gradient is determined partly by the thermal conductivity of the material. A high thermal conductivity helps to transfer heat and reduce temperature differences quickly in the material.

3. *Modulus of elasticity:* A low modulus of elasticity permits large amounts of strain before the stress reaches the critical level required to cause fracture.

4. *Fracture stress:* A high stress required for fracture permits larger strains. The fracture stress for a particular material is high if flaws are small and few in number.

5. *Phase transformations:* Additional dimensional changes can be caused by phase transformations. Transformation of silica from quartz to cristobalite, for example, introduces residual stresses and increases problems with thermal shock. Similarly, we cannot use pure $PbTiO_3$ ceramics since the stresses induced during the cubic to tetragonal transformation will cause the ceramic to fracture.

One method for measuring the resistance to thermal shock is to determine the maximum temperature difference that can be tolerated during a quench without affecting the mechanical properties of the material. Pure (fused) silica glass has a thermal-shock resistance of about 3000°C. Figure 21-7 shows the effect of quenching temperature difference on the modulus of rupture in sialon ($Si_3Al_3O_3N_5$) after quenching; no cracks and therefore no change in the properties of the ceramic are evident until the quenching-temperature difference approaches 950°C. Other ceramics have poorer resistance. Shock resistance for partially stabilized zirconia (PSZ) and Si_3N_4 is about 500°C; for SiC 350°C; and for Al_2O_3 and ordinary glass, about 200°C.

Another way to evaluate the resistance of a material to thermal shock is by the thermal shock parameter (R' or R)

$$R' = \text{Thermal shock parameter} = \frac{\sigma_f K(1 - \nu)}{E \cdot \alpha} \qquad (21\text{-}8a)$$

or

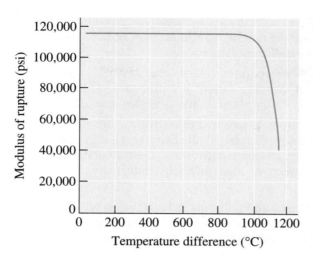

Figure 21-7
The effect of quenching temperature difference on the modulus of rupture of sialon. The thermal shock resistance of the ceramic is good up to about 950°C.

$$R = \frac{\sigma_f \cdot (1 - v)}{E \cdot \alpha} \qquad (21\text{-}8b)$$

where σ_f is the fracture stress of the material, v is the Poisson's ratio, K is the thermal conductivity, E is the modulus of elasticity, and α is the linear coefficient of thermal expansion. Equation 21-8b is used in situations where the heat transfer rate is essentially infinite. A higher value of thermal shock parameter means better resistance to thermal shock. Thermal shock parameter represents the maximum temperature change that can occur without fracturing the material.

Thermal shock is usually not a problem in most metals because metals normally have sufficient ductility to permit deformation rather than fracture. As mentioned before, many ceramic compositions developed as zero-thermal-expansion ceramics (based on sodium zirconium phosphates) and many glass-ceramics have excellent thermal-shock resistance.[8,9]

SUMMARY

◈ The thermal properties of materials can be at least partly explained by the movement of electrons and phonons.

◈ Heat capacity and specific heat represent the quantity of energy required to raise the temperature of a given amount of material by one degree, and they are influenced by temperature, crystal structure, and bonding.

◈ The coefficient of thermal expansion describes the dimensional changes that occur in a material when its temperature changes. Strong bonding leads to a low coefficient of expansion. High-melting-point metals and ceramics have low coefficients, whereas low-melting-point metals and polymers have high coefficients.

◈ The thermal expansion of many engineered materials can be made to be near zero by properly controlling the combination of different phases that show positive and negative thermal expansion coefficients. Zerodur™ glass-ceramics and NZP ceramics are examples of such materials.

◈ Because of thermal expansion, stresses develop in a material when the temperature changes. Care in design, processing, or materials selection is required to prevent failure due to thermal stresses.

◇ Heat is transferred in materials by both phonons and electrons. Thermal conductivity depends on the relative contributions of each of these mechanisms, as well as on the structure and temperature.

◇ Phonons play an important role in the thermal properties of ceramics, semiconductors, and polymers. Disordered structures, such as ceramic glasses or amorphous polymers, scatter phonons and cause low conductivity.

◇ Thermal conductivity is sensitive to microstructure. Most commonly encountered ceramics have a low thermal conductivity because of porosity and defects. Proper processing using sintering additives and the formation of desirable second phases and control of texture, high thermal conductivity AlN, SiC, Si_3N_4 have been developed.

◇ Crystalline ceramics and polymers have higher conductivities than their glassy or amorphous counterparts. Second phases, including porosity, have a significant effect on thermal conductivity in nonmetallic materials.

◇ Electronic contributions are most important in metals; consequently, lattice imperfections that scatter electrons reduce conductivity. The effect of temperature is less easy to characterize, since increasing temperature increases phonon energy but also increases the scattering of both phonons and electrons.

GLOSSARY

Heat capacity   The energy required to raise the temperature of one mole of a material by one degree.

Linear coefficient of thermal expansion   Describes the amount by which each unit length of a material changes when the temperature of the material changes by one degree.

Lorentz constant   The constant that relates electrical and thermal conductivity.

Phonon   A packet of elastic waves. It is characterized by its energy, wavelength, or frequency, which transfers energy through a material.

Specific heat   The energy required to raise the temperature of one gram of a material by one degree.

Thermal conductivity   A microstructure-sensitive property that measures the rate at which heat is transferred through a material.

Thermal shock   Failure of a material caused by stresses introduced by sudden changes in temperature.

Thermal stresses   Stresses introduced into a material due to differences in the amount of expansion or contraction that occur because of a temperature change.

ADDITIONAL INFORMATION

1. JOHNSON, G., "At Los Alamos, Two Visions of Supercomputing," *The New York Times*. 2002. p. 1,4.

2. GREENBERG, C.B., "Thin Films on Float Glass: The Extraordinary Possibilities," *J. of Indu. Chemi*, 2001. 40(1): p. 26–32.

3. GREENBERG, C., *Discussion on self-cleaning glasses and use of thin-film coatings is acknowledged.* 2002.

4. LIMAYE, S.Y., D.K. AGRAWAL, and H.A. McKINSTRY, "Synthesis and Thermal Expansion of MZr4(PO4)6, (M = Mg, Ca, Sr, Ba)," *Journal of the American Ceramic Society*, 1987. 70: p. C232–36.

5. ROY, R., et al., "A New Structural Family of Near Zero Expansion Ceramics," *Materials Research Bulletin*, 1984. 19: p. 471–477.

6. ROY, R., The American Ceramic Society. "Low Thermal Expansion Ceramics: A Retrospective," in *Low-Expansion Materials*. 1995: American Ceramic Society.

7. WATARI, K. and S.L. SHINDE, "High Thermal Conductivity Materials," *MRS Bulletin*, 2001 (June): p. 440–442.

8. LIMAYE, S., *Personal Communication.*

9. BEALL, G.H., "Glass-Ceramics," in *Advances in Ceramics Vol. 18—Commercial Glasses*, D.C. BOYD and J.F. MacDOWELL, Editors. 1986: American Ceramic Society. p. 157–173.

✓ PROBLEMS

Section 21-1 Heat Capacity and Specific Heat

21-1 State any two applications where a high thermal conductivity is desirable.

21-2 State any two applications where a very low thermal conductivity of a material is highly desirable.

21-3 Calculate the heat (in calories and joules) required to raise the temperature of 1 kg of the following materials by 50°C

    (a) lead (b) nickel (c) Si_3N_4 (d) 6,6-nylon

21-4 Calculate the temperature of a 100-g sample of the following materials (originally at 25°C) when 3000 calories are introduced:

    (a) tungsten (b) titanium
    (c) Al_2O_3 (d) low-density polyethylene

21-5 An alumina insulator for an electrical device is also to serve as a heat sink. A 10°C temperature rise in an alumina insulator 1 cm × 1 cm × 0.02 cm is observed during use. Determine the thickness of a high-density polyethylene insulator that would be needed to provide the same performance as a heat sink. The density of alumina is 3.96 g/cm^3.

21-6 A 200-g sample of aluminum is heated to 400°C and is then quenched into 2000 cm^3 of water at 20°C. Calculate the temperature of the water after the aluminum and water reach equilibrium. Assume no temperature loss from the system.

Section 21-2 Thermal Expansion

21-7 A 2-m-long soda-lime glass sheet is produced at 1400°C. Determine its length after it cools to 25°C.

21-8 A copper casting is to be produced having the final dimensions of 1 in. × 12 in. × 24 in. Determine the size of the pattern that must be used to make the mold into which the liquid copper is poured during the manufacturing process.

21-9 An aluminum casting is made using the permanent mold process. In this process, the liquid aluminum is poured into a gray cast iron mold that is heated to 350°C. We wish to produce an aluminum casting that is 15 in. long at 25°C. Calculate the length of the cavity that must be machined into the gray cast iron mold.

21-10 We coat a 100-cm-long, 2-mm-diameter copper wire with a 0.5-mm-thick epoxy insulation coating. Determine the length of the copper and the coating when their temperature increases from 25°C to 250°C. What is likely to happen to the epoxy coating as a result of this heating?

21-11 We produce a 10-in.-long bimetallic composite material composed of a strip of yellow brass bonded to a strip of Invar. Determine the length to which each material would expand when the temperature increases from 20°C to 150°C. Draw a sketch showing what will happen to the shape of the bimetallic strip.

21-12 Give examples of materials that have negative or near-zero thermal expansion coefficients.

21-13 What is Zerodur™? What are some of the properties and applications of this material?

Section 21-3 Thermal Conductivity

Section 21-4 Thermal Shock

21-14 Define the terms thermal conductivity and thermal shock of materials.

21-15 Thermal conductivity of most ceramics is low. True or False? Explain.

21-16 Is the thermal conductivity of materials a microstructure sensitive property? Explain.

21-17 A nickel engine part is coated with SiC to provide corrosion resistance at high temperatures. If no residual stresses are present in the part at 20°C, determine the thermal stresses that develop when the part is heated to 1000°C during use. (See Table 14-3.)

21-18 Alumina fibers 2 cm long are incorporated into an aluminum matrix. Assuming good bonding between the ceramic fibers and the aluminum, estimate the thermal stresses acting on the fiber when the temperature of the composite increases 250°C. Are the stresses on the fiber tensile or compressive? (See Table 14-3.)

21-19 A 24-in.-long copper bar with a yield strength of 30,000 psi is heated to 120°C and immediately fastened securely to a rigid framework. Will the copper deform plastically during cooling to 25°C? How much will the bar deform if it is released from the framework after cooling?

21-20 Repeat Problem 21-19, but using a silicon carbide rod rather than a copper rod. (See Table 14-3.)

21-21 A 3-cm-plate of silicon carbide separates liquid aluminum (held at 700°C) from a water-cooled steel shell maintained at 20°C. Calculate the heat Q transferred to the steel per cm^2 of silicon carbide each second.

21-22 A sheet of 0.01-in. polyethylene is sandwiched between two 3 ft × 3 ft × 0.125 in. sheets of soda-lime glass to produce a window. Calculate (a) the heat lost through the window each day when room temperature is 25°C and the outside air is 0°C and (b) the heat entering through the window each day when room temperature is 25°C and the outside air is 40°C.

21-23 We would like to build a heat-deflection plate that permits heat to be transferred rapidly parallel to the sheet but very slowly perpendicular to the sheet. Consequently, we incorporate 1 kg of copper wires, each 0.1 cm in diameter, into 5 kg of a polyimide polymer matrix. Estimate the thermal conductivity parallel and perpendicular to the sheet.

21-24 Suppose we just dip a 1-cm-diameter, 10-cm-long rod of aluminum into one liter of water at 20°C. The other end of the rod is in contact with a heat source operating at 400°C. Determine the length of time required to heat the water to 25°C if 75% of the heat is lost by radiation from the bar.

21-25 Write down the equations that define the thermal shock resistance (TSR) parameter. Based on this, what can you say about materials that show a near-zero thermal-expansion coefficient.

21-26 Determine the thermal shock parameter for silicon nitride, hot pressed silicon carbide, and alumina and compare it with the thermal-shock resistance as defined by the maximum quenching temperature difference. (See Table 14-3.)

21-27 Gray cast iron has a higher thermal conductivity than ductile or malleable cast iron. Review Chapter 12 and explain why this difference in conductivity might be expected.

✳ Design Problems

21-28 A chemical-reaction vessel contains liquids at a temperature of 680°C. The wall of the vessel must be constructed so that the outside wall operates at a temperature of 35°C or less. Design the vessel wall and appropriate materials if $Q_{maximum} = 6000$ cal/s.

21-29 Liquid copper is held in a silicon nitride vessel. The inside diameter of the vessel is 3 in. The outside of the vessel is in contact with copper that contains cooling channels through which water at 20°C flows at the rate of 50 liters per minute. The copper is to remain at a temperature below 25°C. Design a system that will accomplish this end.

21-30 Design a metal panel coated with glass enamel capable of thermal cycling between 20°C and 150°C. The glasses generally available are expected to have a tensile strength of 5,000 psi and a compressive strength of 50,000 psi.

21-31 What design constraints exist in selecting materials for a turbine blade for a jet engine that is capable of operating at high temperatures?

21-32 Consider the requirements of a low dielectric constant, low dielectric loss, good mechanical strength, and high thermal conductivity. What materials would you consider for electronic packaging substrates?

Computer Problems

21-33 *Thermal-Shock-Resistance Parameters.* Write a computer program or use a spreadsheet software that will provide the value of thermal-shock-resistance parameters (TSR) when the values of fracture stress, thermal conductivity, Young's modulus, Poisson's ratio, and the thermal-expansion coefficient are provided.

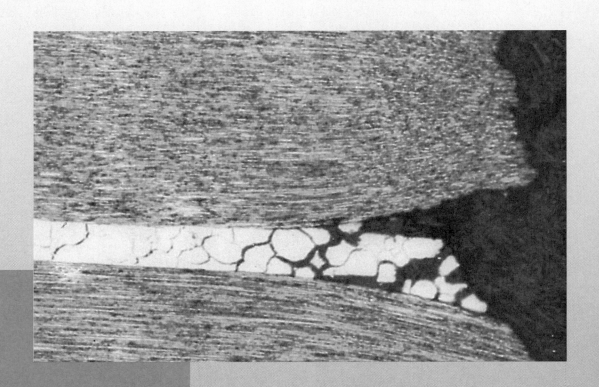

Fracture mechanisms differ in the two components of a nickel-base superalloy reinforced with tungsten fibers. The tungsten fibers (dark) failed in a ductile manner, with necking of the fibers at the fracture surface. The superalloy matrix (white) failed in a brittle manner at the boundaries between the original powder particles used to produce the composite. (Magnification ×100) (*Courtesy of* ASM Handbook, *Vol. 12, Fractography (1987), ASM International, Materials Park, OH 44073.*)

PART 5

Protection Against Deterioration and Failure of Materials

The failure of materials by corrosion and wear is discussed in this part. The failure due to fatigue and creep was discussed in Chapter 6. We again find that deterioration or failure is related to the structure, properties, and processing of materials. In Chapter 22, corrosion and wear are examined; electrochemical corrosion is found to be particularly important. In addition to examining the mechanism for corrosion, we look at techniques for controlling and preventing damage to a material by these processes.

Most developed nations spend about 6% of their total gross domestic product in addressing corrosion-related issues. In the United States, this amounts to about $550 billion. The process of corrosion affects many important areas of technology, ranging from the construction and manufacturing of bridges, airplanes, ships, and microelectronics; the food industry; space exploration; fiber optics networks; and nuclear and other power utilities, just to name a few. Corrosion can be prevented or contained using different techniques that include the use of coatings, and anodic or cathodic protection.

The photograph here shows corrosion of an aluminum guardrail, possibly because of incorrect heat treatment. Guardrails are now often made using steels that are protected with a zinc coating obtained using hot-die galvanizing. (*Courtesy of Dr. David Burleigh, New Mexico Tech.*)

22

Corrosion and Wear

Have You Ever Wondered?

- *Why does iron rust?*

- *What does the acronym "WD-40™" stand for?*

- *Is the process of corrosion ever useful?*

- *Why do household water-heater tanks contain Mg rods?*

- *Do metals like aluminum and titanium undergo corrosion?*

- *How does wear affect the useful life of different components such as crankshafts?*

- *What process makes use of mechanical erosion and chemical corrosion in the manufacture of semiconductor chips?*

The composition and physical integrity of a solid material is altered in a corrosive environment. In chemical corrosion, a corrosive liquid dissolves the material. In electrochemical corrosion, metal atoms are removed from the solid material as the result of an electric circuit that is produced. Metals and certain ceramics react with a gaseous environment, usually at elevated temperatures, and the material may be destroyed by the formation of oxides or other compounds. Polymers degrade when exposed to oxygen at elevated temperatures. Materials may be altered when exposed to radiation or even bacteria. Finally, a variety of wear and wear-corrosion mechanisms alter the shape of materials. According to a study concluded in 2001, in the United States, combating corrosion costs about 6% of the gross domestic product (GDP). This amount, which includes direct and indirect costs, was approximately $550 billion dollars in 1998.[1]

The corrosion process occurs in order to lower the free energy of a system. The corrosion process occurs over a period of time and can occur either at high or low temperatures. The trick to preventing corrosion is to either come up with a coating that can avoid contact with the reactant or to apply a counter driving force to slow down the kinetics. Chemical corrosion is an important consideration in many sectors including transportation (bridges, pipelines, cars, airplanes, trains, and ships), utilities (electrical, water, telecommunications, and nuclear power plants), and production and manufacturing (food industry, microelectronics, and petroleum refining).

Corrosion must not only be monitored but also be factored into the design of different components. A recent incident occurred at a nuclear power plant where significant corrosion led to a hole (approximately 6 inches deep, 5 inches long, and 7 inches wide) in a holding tank where corrosion had occurred at an approximate rate of 2 inches per year. The corrosion was due to leaking boric acid that was used as a coolant bath for uranium rods. If it were not detected and if it were not for the design of the stainless-steel plate that held up under corrosive conditions and high pressures, a major nuclear accident could have occurred.[2]

The process of corrosion, which is the deterioration of a metallic material by a reaction with environmental chemicals, also can be useful. For example, in the state-of-the art polishing of semiconductor wafers, a process known as chemical mechanical polishing (CMP) makes use of corrosion and erosion. Similarly, the purification of metals, such as copper, makes use of an electrochemical dissolution process, which is similar to aqueous corrosion. Thus, the underlying processes that cause corrosion and erosion are *not* always detrimental. In nature, corrosion and erosion play a major role in the formation of canyons and caverns.

The goal of this chapter is to introduce the principles and mechanisms by which corrosion and wear occur under different conditions. This includes the aqueous corrosion of metals, the oxidation of metals, the corrosion of ceramics, and the degradation of polymers. We will offer a summary of different technologies that are used to prevent or minimize corrosion and associated problems.[3–5]

22-1   Chemical Corrosion

In **chemical corrosion**, or direct dissolution, a material dissolves in a corrosive liquid medium. The material continues to dissolve until either it is consumed or the liquid is saturated. An example is copper-based alloys develop a green patina, due to the formation of copper carbonate and copper hydroxides. This is why, for example, the Statue of Liberty looks greenish. The chemical corrosion of copper, tantalum, silicon, silicon dioxide, and other materials can be achieved under extremely well-controlled conditions. In the processing of silicon wafers, for example, a process known as chemical mechanical polishing uses a silica slurry to provide mechanical erosion. This process creates extremely flat surfaces that are suitable for the processing of silicon wafers. Chemical corrosion also occurs in nature. For example, the chemical corrosion of rocks by carbonic acid (H_2CO_3) and the mechanical erosion of wind and water play an important role in the formation of canyons and caverns.

Liquid Metal Attack   Liquid metals first attack a solid at high-energy locations, such as grain boundaries. If these regions continue to be attacked preferentially, cracks eventually grow (Figure 22-1). Often this form of corrosion is complicated by the presence of fluxes that accelerate the attack or by electrochemical corrosion. Aggressive metals such as liquid lithium can also attack ceramics.

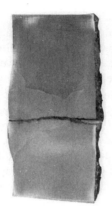

Figure 22-1
Molten lead is held in thick steel pots during refining. In this case, the molten lead has attacked a weld in a steel plate and cracks have developed. Eventually, the cracks propagate through the steel, and molten lead leaks from the pot.

Selective Leaching   One particular element in an alloy may be selectively dissolved, or leached, from the solid. **Dezincification** occurs in brass containing more than 15% Zn. Both copper and zinc are dissolved by aqueous solutions at elevated temperatures; the zinc ions remain in solution while the copper ions are replated onto the brass (Figure 22-2). Eventually, the brass becomes porous and weak.

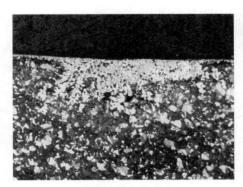

Figure 22-2
Photomicrograph of a copper deposit in brass, showing the effect of dezincification ($\times 50$).

Graphitic corrosion of gray cast iron occurs when iron is selectively dissolved in water or soil, leaving behind interconnected graphite flakes and a corrosion product. Localized graphitic corrosion often causes leakage or failure of buried gray iron gas lines, leading to explosions.

Dissolution and Oxidation of Ceramics   Ceramic refractories used to contain molten metal during melting or refining may be dissolved by the slags that are produced on the metal surface. For example, an acid (high SiO_2) refractory is rapidly attacked by a basic (high CaO or MgO) slag. A glass produced from SiO_2 and Na_2O is rapidly attacked by water; CaO must be added to the glass to minimize this attack. Nitric acid may selectively leach iron or silica from some ceramics, reducing their strength and density. As noted in Chapters 6 and 14, the strength of silicate glasses depends on flaws that are often created by corrosive interactions with water.

Chemical Attack on Polymers   Compared to metals and oxide ceramics, plastics are considered corrosion resistant. Teflon™ and Viton™ are some of the most corrosion-resistant materials and are used in many applications, including the chemical processing industry. These and other polymeric materials can withstand the presence of many acids, bases, and organic liquids. However, aggressive solvents often diffuse into low-molecular-weight thermoplastic polymers. As the solvent is incorporated into the polymer, the smaller solvent molecules force apart the chains, causing swelling. The strength of the bonds between the chains decreases. This leads to softer, lower-strength polymers with a low transition temperature. In extreme cases, the swelling leads to stress cracking.

Thermoplastics may also be dissolved into a solvent. Prolonged exposure causes a loss of material and weakening of the polymer part. This process occurs most easily when the temperature is high and when the polymer has a low molecular weight, is highly branched and amorphous, and is not cross-linked. The structure of the monomer is also important; the CH_3 groups on the polymer chain in polypropylene are more easily removed from the chain than are chloride or fluoride ions in polyvinyl chloride (PVC) or polytetrafluoroethylene (Teflon™). Teflon has exceptional resistance to chemical attack by almost all solvents.

22-2   Electrochemical Corrosion

Electrochemical corrosion, the most common form of attack of metals, occurs when metal atoms lose electrons and become ions. As the metal is gradually consumed by this process, a byproduct of the corrosion process is typically formed. Electrochemical corrosion occurs most frequently in an aqueous medium, in which ions are present in water, soil, or moist air. In this process, an electric circuit is created and the system is called an **electrochemical cell**. Corrosion of a steel pipe or a steel automobile panel, creating holes in the steel and rust as the byproduct, are examples of this reaction.

Although responsible for corrosion, electrochemical cells may also be useful. By deliberately creating an electric circuit, we can *electroplate* protective or decorative coatings onto materials. In some cases, electrochemical corrosion is even desired. For example, in etching a polished metal surface with an appropriate acid, various features in the microstructure are selectively attacked, permitting them to be observed. In fact, most of the photographs of metal and alloy microstructures in this text were obtained in this way, thus enabling, for example, the observation of pearlite in steel or grain boundaries in copper.

Components of an Electrochemical Cell   There are four components of an electrochemical cell (Figure 22-3):

1. The **anode** gives up electrons to the circuit and corrodes.
2. The **cathode** receives electrons from the circuit by means of a chemical, or cathode, reaction. Ions that combine with the electrons produce a byproduct at the cathode.

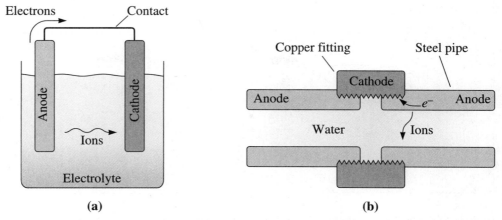

Figure 22-3 The components in an electrochemical cell: (a) a simple electrochemical cell and (b) a corrosion cell between a steel water pipe and a copper fitting.

3. The anode and cathode must be electrically connected, usually by physical contact, to permit the electrons to flow from the anode to the cathode and continue the reaction.

4. A liquid **electrolyte** must be in contact with both the anode and the cathode. The electrolyte is conductive, thus completing the circuit. It provides the means by which metallic ions leave the anode surface and assures that ions move to the cathode to accept the electrons.

This description of an electrochemical cell defines electrochemical corrosion; if the direction of the current is reversed, metal ions are deposited onto the cathode and **electroplating** occurs.

Anode Reaction   The anode, which is a metal, undergoes an **oxidation reaction** by which metal atoms are ionized. The metal ions enter the electrolytic solution, while the electrons leave the anode through the electrical connection:

$$M \rightarrow M^{n+} + ne^- \tag{22-1}$$

Because metal ions leave the anode, the anode corrodes, or oxidizes.

Cathode Reaction in Electroplating   In electroplating, a cathodic **reduction reaction**, which is the reverse of the anode reaction, occurs at the cathode:

$$M^{n+} + ne^- \rightarrow M \tag{22-2}$$

The metal ions, either intentionally added to the electrolyte or formed by the anode reaction, combine with electrons at the cathode. The metal then plates out and covers the cathode surface.

Cathode Reactions in Corrosion   Except in unusual conditions, plating of a metal does not occur during electrochemical corrosion. Instead, the reduction reaction forms a gas, solid, or liquid byproduct at the cathode (Figure 22-4).

1. *The hydrogen electrode:* In oxygen-free liquids, such as hydrochloric acid (HCl) or stagnant water, hydrogen gas may be evolved at the cathode:

$$2H^+ + 2e^- \rightarrow H_2\uparrow \text{ (gas)} \tag{22-3}$$

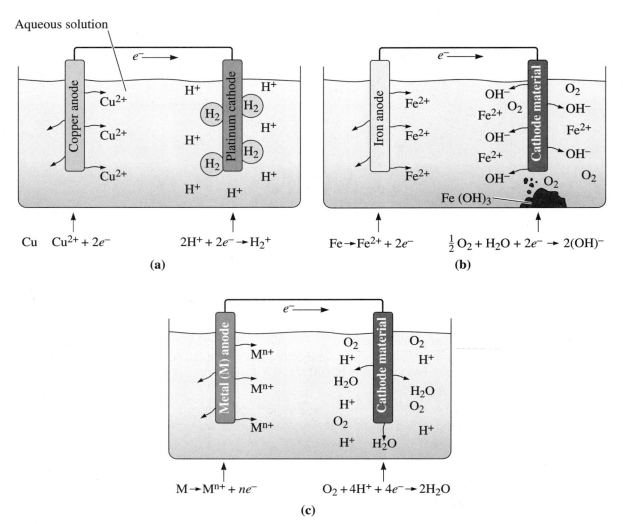

Figure 22-4 The anode and cathode reactions in typical electrolytic corrosion cells: (a) the hydrogen electrode, (b) the oxygen electrode, and (c) the water electrode.

If zinc were placed in such an environment, we would find that the overall reaction is:

$$Zn \rightarrow Zn^{2+} + 2e^- \text{ (anode reaction)}$$

$$2H^+ + 2e^- \rightarrow H_2\uparrow \text{ (cathode reaction)} \tag{22-4}$$

$$Zn + 2H^+ \rightarrow Zn^{2+} + H_2\uparrow \text{ (overall reaction)}$$

The zinc anode gradually dissolves, and hydrogen bubbles continue to evolve at the cathode.

2. *The oxygen electrode:* In aerated water, oxygen is available to the cathode, and hydroxyl, or $(OH)^-$, ions form:

$$\frac{1}{2}O_2 + H_2O + 2e^- \rightarrow 2(OH)^- \tag{22-5}$$

The oxygen electrode enriches the electrolyte in $(OH)^-$ ions. These ions react with positively charged metallic ions and produce a solid product. In the case of rusting of iron:

$$Fe \rightarrow Fe^{2+} + 2e^- \text{ (anode reaction)}$$

$$\left. \begin{array}{l} \frac{1}{2}O_2 + H_2O + 2e^- \rightarrow 2(OH)^- \\[1em] Fe^{2+} + 2(OH)^- \rightarrow Fe(OH)_2 \end{array} \right\} \text{ (cathode reactions)} \qquad (22\text{-}6)$$

$$Fe + \frac{1}{2}O_2 + H_2O \rightarrow Fe(OH)_2 \text{ (overall reaction)}$$

The reaction continues as the $Fe(OH)_2$ reacts with more oxygen and water:

$$2Fe(OH)_2 + \frac{1}{2}O_2 + H_2O \rightarrow 2Fe(OH)_3 \qquad (22\text{-}7)$$

$Fe(OH)_3$ is commonly known as *rust*.

3. *The water electrode:* In oxidizing acids, the cathode reaction produces water as a byproduct:

$$O_2 + 4H^+ + 4e^- \rightarrow 2H_2O \qquad (22\text{-}8)$$

If a continuous supply of both oxygen and hydrogen is available, the water electrode produces neither a buildup of solid rust nor a high concentration or dilution of ions at the cathode.

22-3   The Electrode Potential in Electrochemical Cells

In electroplating, an imposed voltage is required to cause a current to flow in the cell. But in corrosion, a potential naturally develops when a material is placed in a solution. Let's see how the potential required to drive the corrosion reaction develops.

Electrode Potential   When a pure metal is placed in an electrolyte, an **electrode potential** is developed that is related to the tendency of the material to give up its electrons. However, the driving force for the oxidation reaction is offset by an equal but opposite driving force for the reduction reaction. No net corrosion occurs. Consequently, we cannot measure the electrode potential for a single electrode material.

Electromotive Force Series   To determine the tendency of a metal to give up its electrons, we measure the potential difference between the metal and a standard electrode using a half-cell (Figure 22-5). The metal electrode to be tested is placed in a 1-Molar (M) solution of its ions. A standard reference electrode is also placed in a 1-M solution of its ions. The word "standard" is used in a thermodynamic sense. The two electrolytes are in electrical contact but are not permitted to mix with one another. Each electrode establishes its own electrode potential. By measuring the voltage between the two electrodes when the circuit is open, we obtain the potential difference. The potential of the hydrogen electrode, which is taken as our standard reference electrode, is arbitrarily set equal to zero volts. If the metal has a greater tendency to give up electrons than hydrogen, then the potential of the metal is negative—the metal is anodic with respect to the hydrogen electrode.

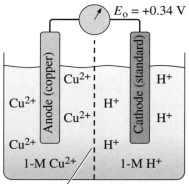

$E_o = +0.34$ V

Figure 22-5
The half-cell used to measure the electrode potential of copper under standard conditions. The electrode potential of copper is the potential difference between it and the standard hydrogen electrode in an open circuit. Since E_0 is greater than zero, copper is cathodic compared with the hydrogen electrode.

Screen that permits transfer of charge but not mixing of electrolytes

The **electromotive force** (or **emf**) **series** shown in Table 22-1 compares the standard electrode potential E_0 for each metal with that of the hydrogen electrode under standard conditions of 25°C and 1-M solution of ions in the electrolyte. Note that the measurement of the potential is made when the electric circuit is open. The voltage difference begins to change as soon as the circuit is closed.

The more negative the value of potential for the oxidation of metal, the more electropositive is the metal; this means the metal will have a higher tendency to undergo an oxidation reaction. For example, alkali and alkaline earth metals (e.g., Li, K, Ba, Sr, and Mg) are so reactive that they have to be kept under conditions that prevent any contact with oxygen. On the other hand, metals that are toward the bottom of the chart

TABLE 22-1 ■ *The electromotive force (emf) series for selected elements and reactions*

		Metal	Electrode Potential E_0 (Volts)
Anodic	↑	$Li \rightarrow Li^+ + e^-$	−3.05
		$Mg \rightarrow Mg^{2+} + 2e^-$	−2.37
		$Al \rightarrow A^{3+} + 3e^-$	−1.66
		$Ti \rightarrow Ti^{2+} + 2e^-$	−1.63
		$Mn \rightarrow Mn^{2+} + 2e^-$	−1.63
		$Zn \rightarrow Zn^{2+} + 2e^-$	−0.76
		$Cr \rightarrow Cr^{3+} + 3e^-$	−0.74
		$Fe \rightarrow Fe^{2+} + 2e^-$	−0.44
		$Ni \rightarrow Ni^{2+} + 2e^-$	−0.25
		$Sn \rightarrow Sn^{2+} + 2e^-$	−0.14
		$Pb \rightarrow Pb^{2+} + 2e^-$	−0.13
		$H_2 \rightarrow 2H^+ + 2e^-$	0.00
		$Cu \rightarrow Cu^{2+} + 2e^-$	+0.34
		$4(OH)^- \rightarrow O_2 + 2H_2O + 4e^-$	+0.40
		$Ag \rightarrow Ag^+ + e^-$	+0.80
		$Pt \rightarrow Pt^{4+} + 4e^-$	+1.20
		$2H_2O \rightarrow O_2 + 4H^+ + 4e^-$	+1.23
Cathodic	↓	$Au \rightarrow Au^{3+} + 3e^-$	+1.50

(e.g., Ag, Au, and Pt) will not tend to react with oxygen. This is why we call them "noble metals." Metals such as Fe, Cu, and Ni have intermediate reactivities. However, this is not the only consideration. Aluminum, for example, has a strongly negative standard electrode potential but does react easily with oxygen to form aluminum oxide; it also reacts easily with fluoride to form aluminum fluoride. Both of these compounds form a tenacious and impervious layer that helps stop further corrosion. Titanium also reacts readily with oxygen. The quickly formed titanium oxide creates a barrier that prevents the further diffusion of species and, thus, avoids further oxidation. This is why both aluminum and titanium are highly reactive, but can resist corrosion exceptionally well. Note that the emf series tells us about the thermodynamic feasibility and driving force, it does *not* tell us about the kinetics of the reaction.

Effect of Concentration on the Electrode Potential The electrode potential depends on the concentration of the electrolyte. At 25°C, the **Nernst equation** gives the electrode potential in nonstandard solutions:

$$E = E_0 + \frac{0.0592}{n} \log(C_{\text{ion}}), \tag{22-9}$$

where E is the electrode potential in a solution containing a concentration C_{ion} of the metal in molar units, n is the valence of the metallic ion, and E_0 is the standard electrode potential in a 1-M solution. Note that when $C_{\text{ion}} = 1$, $E = E_0$. The example that follows illustrates the calculation of the electrode potential.

EXAMPLE 22-1 *Half-Cell Potential for Copper*

Suppose 1 g of copper as Cu^{2+} is dissolved in 1000 g of water to produce an electrolyte. Calculate the electrode potential of the copper half-cell in this electrolyte.

SOLUTION

From chemistry, we know that a standard 1-M solution of Cu^{2+} is obtained when we add 1 mol of Cu^{2+} (an amount equal to the atomic mass of copper) to 1000 g of water. The atomic mass of copper is 63.54 g/mol. The concentration of the solution when only 1 g of copper is added must be:

$$C_{\text{ion}} = \frac{1}{63.54} = 0.0157 \text{ M}$$

From the Nernst equation, with $n = 2$ and $E_0 = +0.34$ V:

$$E = E_0 + \frac{0.0592}{n} \log(C_{\text{ion}}) = 0.34 + \frac{0.0592}{2} \log(0.0157)$$

$$= 0.34 + (0.0296)(-1.8) = 0.29 \text{ V}$$

Rate of Corrosion or Plating The amount of metal plated on the cathode in electroplating, or removed from the metal by corrosion, can be determined from **Faraday's equation**,

$$w = \frac{ItM}{n\text{F}}, \tag{22-10}$$

where w is the weight plated or corroded (g), I is the current (A), M is the atomic mass of the metal, n is the valence of the metal ion, t is the time (s), and F is Faraday's constant (96,500 C). This law basically states that one gram equivalent of a metal will be deposited by 96,500 C of charge. Often the current is expressed in terms of current density, $i = I/A$, so Equation 22-10 becomes

$$w = \frac{iAtM}{n\text{F}}, \tag{22-11}$$

where the area A (cm^2) is the surface area of the anode or cathode.

The following examples illustrate the use of Faraday's equation to calculate the current density.

EXAMPLE 22-2   *Design of a Copper Plating Process*

Design a process to electroplate a 0.1-cm-thick layer of copper onto a 1 cm × 1 cm cathode surface.

SOLUTION

In order for us to produce a 0.1-cm-thick layer on a 1 cm^2 surface area, the weight of copper must be:

$$\rho_{Cu} = 8.96 \text{ g/cm}^3 \quad A = 1 \text{ cm}^2$$

$$\text{Volume of copper} = (1 \text{ cm}^2)(0.1 \text{ cm}) = 0.1 \text{ cm}^3$$

$$\text{Weight of copper} = (8.96 \text{ g/cm}^3)(0.1 \text{ cm}^3) = 0.896 \text{ g}$$

From Faraday's equation, where $M_{Cu} = 63.54$ g/mol and $n = 2$:

$$It = \frac{wn\text{F}}{M} = \frac{(0.896)(2)(96,500)}{63.54} = 2722 \text{ A} \cdot \text{s}$$

Therefore, we might use several different combinations of current and time to produce the copper plate:

Current	Time
0.1 A	27,220 s = 7.6 h
1.0 A	2,722 s = 45.4 min
10.0 A	272.2 s = 4.5 min
100.0 A	27.22 s = 0.45 min

Our choice of the exact combination of current and time might be made on the basis of the rate of production and quality of the copper plate. Low currents require very long plating times, perhaps making the process economically unsound. High currents, however, may reduce plating efficiencies. The plating effectiveness may depend on the composition of the electrolyte containing the copper ions, as well as on any impurities or additives that are present. Currents such as 10 A or 100 A are too high—they can initiate other side reactions that are not desired. The deposit may also not be uniform and smooth. Additional background or experimentation may be needed to obtain the most economical and efficient plating process. A current of ~1 A and a time of ~45 minutes are not uncommon in electroplating operations.

EXAMPLE 22-3   *Corrosion of Iron*

An iron container 10 cm × 10 cm at its base is filled to a height of 20 cm with a corrosive liquid. A current is produced as a result of an electrolytic cell, and after 4 weeks, the container has decreased in weight by 70 g. Calculate (1) the current and (2) the current density involved in the corrosion of the iron.

SOLUTION

1. The total exposure time is:

$$t = (4 \text{ wk})(7 \text{ d/wk})(24 \text{ h/d})(3600 \text{ s/h}) = 2.42 \times 10^6 \text{ s}$$

From Faraday's equation, using $n = 2$ and $M = 55.847$ g/mol:

$$I = \frac{wn\text{F}}{tM} = \frac{(70)(2)(96,500)}{(2.42 \times 10^6)(55.847)}$$

$$= 0.1 \text{ A}$$

2. The total surface area of iron in contact with the corrosive liquid and the current density are:

$$A = (4 \text{ sides})(10 \times 20) + (1 \text{ bottom})(10 \times 10) = 900 \text{ cm}^2$$

$$i = \frac{I}{A} = \frac{0.1}{900} = 1.11 \times 10^{-4} \text{ A/cm}^2$$

EXAMPLE 22-4   *Copper-Zinc Corrosion Cell*

Suppose that in a corrosion cell composed of copper and zinc, the current density at the copper cathode is 0.05 A/cm^2. The area of both the copper and zinc electrodes is 100 cm^2. Calculate (1) the corrosion current, (2) the current density at the zinc anode, and (3) the zinc loss per hour.

SOLUTION

1. The corrosion current is:

$$I = i_{\text{Cu}} A_{\text{Cu}} = (0.05 \text{ A/cm}^2)(100 \text{ cm}^2) = 5 \text{ A}$$

2. The current in the cell is the same everywhere. Thus:

$$i_{\text{Zn}} = \frac{I}{A_{\text{Zn}}} = \frac{5}{100} = 0.05 \text{ A/cm}^2$$

3. The atomic mass of zinc is 65.38 g/mol. From Faraday's equation:

$$w_{\text{zinc loss}} = \frac{ItM}{n\text{F}} = \frac{\left(5 \dfrac{\text{A}}{\text{cm}^2}\right)(3600 \text{ s/h})(65.38 \text{ g/mol})}{(2)(96,500 \text{ C})}$$

$$= 6.1 \text{ g/h}$$

22-4   The Corrosion Current and Polarization

To protect metals from corrosion, we wish to make the current as small as possible. Unfortunately, the corrosion current is very difficult to measure, control, or predict. Part of this difficulty can be attributed to various changes that occur during operation of the corrosion cell. A change in the potential of an anode or cathode, which in turn affects the current in the cell, is called **polarization**. There are three important kinds of polarization: (1) activation, (2) concentration, and (3) resistance polarization.

Activation Polarization   This kind of polarization is related to the energy required to cause the anode or cathode reactions to occur. If we can increase the degree of polarization, these reactions occur with greater difficulty and the rate of corrosion is reduced. Small differences in composition and structure in the anode and cathode materials dramatically change the activation polarization. Segregation effects in the electrodes cause the activation polarization to vary from one location to another. These factors make it difficult to predict the corrosion current.

Concentration Polarization   After corrosion begins, the concentration of ions at the anode or cathode surface may change. For example, a higher concentration of metal ions may be produced at the anode if the ions are unable to diffuse rapidly into the electrolyte. Hydrogen ions may be depleted at the cathode in a hydrogen electrode, or a high $(OH)^-$ concentration may develop at the cathode in an oxygen electrode. When this situation occurs, either the anode or cathode reaction is stifled, because fewer electrons are released at the anode or accepted at the cathode.

In any of these examples, the current density, and thus the rate of corrosion, decreases because of concentration polarization. Normally, the polarization is less pronounced when the electrolyte is highly concentrated, the temperature is increased, or the electrolyte is vigorously agitated. Each of these factors increases the current density and encourages electrochemical corrosion.

Resistance Polarization   This type of polarization is caused by the electrical resistivity of the electrolyte. If a greater resistance to the flow of the current is offered, the rate of corrosion is reduced. Again, the degree of resistance polarization may change as the composition of the electrolyte changes during the corrosion process.

22-5   Types of Electrochemical Corrosion

In this section, we will look at some of the more common forms taken by electrochemical corrosion. First, there is *uniform attack*. When a metal is placed in an electrolyte, some regions are anodic to other regions. However, the location of these regions moves and even reverses from time to time. Since the anode and cathode regions continually shift, the metal corrodes uniformly.

Galvanic attack occurs when certain areas always act as anodes, whereas other areas always act as cathodes. These electrochemical cells are called galvanic cells and can be separated into three types: **composition cells**, **stress cells**, and **concentration cells**.

Composition Cells   Composition cells, or *dissimilar metal corrosion*, develop when two metals or alloys, such as copper and iron, form an electrolytic cell. Because of the effect

TABLE 22-2 ■ *The galvanic series in seawater*

Anodic		Anodic	
	Magnesium and Mg alloys		Lead
	Zinc		Tin
	Galvanized steel		Cu-40% Zn brass
	5052 aluminum		Nickel-based alloys (active)
	3003 aluminum		Copper
	1100 aluminum		Cu-30% Ni alloy
	Alclad		Nickel-based alloys (passive)
	Cadmium		Stainless steels (passive)
	2024 aluminum		Silver
	Low-carbon steel		Titanium
	Cast iron		Graphite
	50% Pb-50% Sn solder		Gold
	316 stainless steel (active)	Cathodic	Platinum

After ASM Metals Handbook, Vol. 10, 8th Ed., 1975.

of alloying elements and electrolyte concentrations on polarization, the emf series may not tell us which regions corrode and which are protected. Instead, we use a **galvanic series**, in which the different alloys are ranked according to their anodic or cathodic tendencies in a particular environment (Table 22-2). We may find a different galvanic series for seawater, freshwater, and industrial atmospheres.

The following example is an illustration of the galvanic series.

EXAMPLE 22-5   *Corrosion of a Soldered Brass Fitting*

A brass fitting used in a marine application is joined by soldering with lead-tin solder. Will the brass or the solder corrode?

SOLUTION

From the galvanic series, we find that all of the copper-based alloys are more cathodic than a 50% Pb-50% Sn solder. Thus, the solder is the anode and corrodes. In a similar manner, the corrosion of solder can contaminate water in freshwater plumbing systems with lead.

Composition cells also develop in two-phase alloys, where one phase is more anodic than the other. Since ferrite is anodic to cementite in steel, small microcells cause steel to galvanically corrode (Figure 22-6). Almost always, a two-phase alloy has less resistance to corrosion than a single-phase alloy of a similar composition.

Intergranular corrosion occurs when the precipitation of a second phase or segregation at grain boundaries produces a galvanic cell. In zinc alloys, for example, impurities such as cadmium, tin, and lead segregate at the grain boundaries during solidification. The grain boundaries are anodic compared with the remainder of the grains, and corrosion of the grain boundary metal occurs (Figure 22-7). In austenitic stainless steels, chromium carbides can precipitate at grain boundaries [Figure 22-6(b)]. The formation of the carbides removes chromium from the austenite adjacent to the boundaries. The low-chromium ($<12\%$ Cr) austenite at the grain boundaries is anodic to the remainder

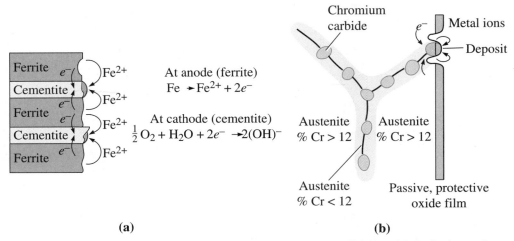

(a)   **(b)**

Figure 22-6 Example of microgalvanic cells in two-phase alloys: (a) In steel, ferrite is anodic to cementite. (b) In austenitic stainless steel, precipitation of chromium carbide makes the low Cr austenite in the grain boundaries anodic.

Figure 22-7
Photomicrograph of intergranular corrosion in a zinc die casting. Segregation of impurities to the grain boundaries produces microgalvanic corrosion cells ($\times 50$).

of the grain and corrosion occurs at the grain boundaries. In certain cold-worked aluminum alloys, grain boundaries corrode rapidly due to the presence of detrimental precipitates. This causes the grains of aluminum to peel back like the pages of a book or leaves. This is known as exfoliation.[8] See the cover photograph of this chapter for an illustration of this phenomenon.

Stress Cells Stress cells develop when a metal contains regions with different local stresses. The most highly stressed, or high-energy, regions act as anodes to the less-stressed cathodic areas (Figure 22-8). Regions with a finer grain size, or a higher density of grain boundaries, are anodic to coarse-grained regions of the same material. Highly cold-worked areas are anodic to less cold-worked areas.

Stress corrosion occurs by galvanic action, but other mechanisms, such as the adsorption of impurities at the tip of an existing crack, may also occur. Failure occurs as a result of corrosion and an applied stress. Higher applied stresses reduce the time required for failure.

Fatigue failures are also initiated or accelerated when corrosion occurs. *Corrosion fatigue* can reduce fatigue properties by initiating cracks, perhaps by producing pits or crevices, and by increasing the rate at which the cracks propagate.

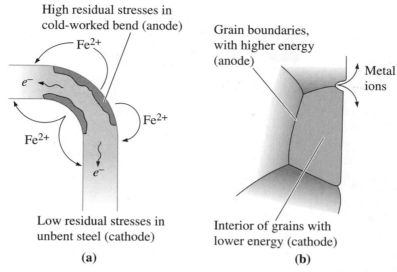

Figure 22-8 Examples of stress cells. (a) Cold work required to bend a steel bar introduces high residual stresses at the bend, which then is anodic and corrodes. (b) Because grain boundaries have a high energy, they are anodic and corrode.

EXAMPLE 22-6   *Corrosion of Cold-Drawn Steel*

A cold-drawn steel wire is formed into a nail by additional deformation, producing the point at one end and the head at the other. Where will the most severe corrosion of the nail occur?

SOLUTION

Since the head and point have been cold-worked an additional amount compared with the shank of the nail, the head and point serve as anodes and corrode most rapidly.

Concentration Cells   Concentration cells develop due to differences in the concentration of the electrolyte (Figure 22-9). According to the Nernst equation a difference in metal ion concentration causes a difference in electrode potential. The metal in contact with the most concentrated solution is the cathode; the metal in contact with the dilute solution is the anode.

The oxygen concentration cell (often referred to as **oxygen starvation**) occurs when the cathode reaction is the oxygen electrode, $H_2O + \frac{1}{2}O_2 + 4e^- \rightarrow 4(OH)^-$. Electrons flow from the low-oxygen region, which serves as the anode, to the high-oxygen region, which serves as the cathode.

Deposits, such as rust or water droplets, shield the underlying metal from oxygen. Consequently, the metal under the deposit is the anode and corrodes. This causes one form of pitting corrosion. Waterline corrosion is similar. Metal above the waterline is exposed to oxygen, while metal beneath the waterline is deprived of oxygen; hence, the metal underwater corrodes. Normally, the metal far below the surface corrodes more slowly than metal just below the waterline due to differences in the distance that electrons must travel. Because cracks and crevices have a lower oxygen concentration than

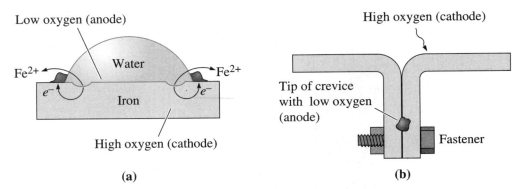

Figure 22-9 Concentration cells: (a) Corrosion occurs beneath a water droplet on a steel plate due to low oxygen concentration in the water. (b) Corrosion occurs at the tip of a crevice because of limited access to oxygen.

the surrounding base metal, the tip of a crack or crevice is the anode, causing **crevice corrosion**.

Pipe buried in soil may corrode because of differences in the composition of the soil. Velocity differences may cause concentration differences. Stagnant water contains low oxygen concentrations, whereas fast-moving, aerated water contains higher oxygen concentrations. Metal near stagnant water is anodic and corrodes. The following is an example of corrosion due to water.

EXAMPLE 22-7   *Corrosion of Crimped Steel*

Two pieces of steel are joined mechanically by crimping the edges. Why would this be a bad idea if the steel is then exposed to water? If the water contains salt, would corrosion be affected?

SOLUTION

By crimping the steel edges, we produce a crevice. The region in the crevice is exposed to less air and moisture, so it behaves as the anode in a concentration cell. The steel in the crevice corrodes.

Salt in the water increases the conductivity of the water, permitting electrical charge to be transferred at a more rapid rate. This causes a higher current density and, thus, faster corrosion due to less resistance polarization.

Microbial Corrosion   Various microbes, such as fungi and bacteria, create conditions that encourage electrochemical corrosion. Particularly in aqueous environments, these organisms grow on metallic surfaces. The organisms typically form colonies that are not continuous. The presence of the colonies and the byproducts of the growth of the organisms produce changes in the environment and, hence the rate at which it occurs.

Some bacteria reduce sulfates in the environment, producing sulfuric acid, which, in turn, attacks metal. The bacteria may be either aerobic (which thrive when oxygen is available) or anaerobic (which do not need oxygen to grow). Such bacteria cause attacks of a variety of metals, including steels, stainless steels, aluminum, and copper, and some ceramics and concrete. A common example occurs in aluminum fuel tanks for aircraft. When the fuel, typically kerosene, is contaminated with moisture, bacteria

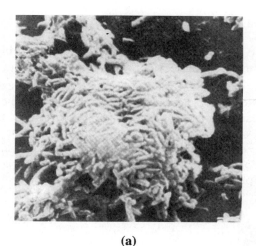

(a)

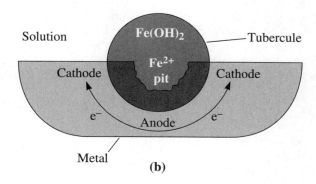

(b)

grow and excrete acids. The acids attack the aluminum, eventually causing the fuel tank to leak.

The growth of colonies of organisms on a metal surface leads to the development of oxygen concentration cells (Figure 22-10). Areas beneath the colonies are anodic, whereas unaffected areas are cathodic. In addition, the colonies of organisms reduce the rate of diffusion of oxygen to the metal, and—even if the oxygen does diffuse into the colony—the organisms tend to consume the oxygen. The concentration cell produces pitting beneath the regions covered with the organisms. Growth of the organisms, which may include products of the corrosion of the metal, produces accumulations (called **tubercules**) that may plug pipes, clog water-cooling systems in nuclear reactors, submarines, or chemical reactors.

22-6 Protection Against Electrochemical Corrosion

A number of techniques are used to combat corrosion, including design, coatings, inhibitors, cathodic protection, passivation, and materials selection.

Design Proper design of metal structures can slow or even avoid corrosion. Some of the steps that should be taken to combat corrosion are as follows:

1. Prevent the formation of galvanic cells. This can be achieved by using similar metals or alloys. For example, steel pipe is frequently connected to brass plumbing fixtures, producing a galvanic cell that causes the steel to corrode. By using intermediate plastic fittings to electrically insulate the steel and brass, this problem can be minimized.

2. Make the anode area much larger than the cathode area. For example, copper rivets can be used to fasten steel sheet. Because of the small area of the copper rivets, a limited cathode reaction occurs. The copper accepts few electrons, and the steel anode reaction proceeds slowly. If, on the other hand, steel rivets are used for joining copper sheet, the small steel anode area gives up many electrons, which are accepted by the large copper cathode area; corrosion of the steel rivets is then very rapid. This is illustrated in the following example.

EXAMPLE 22-8   *Effect of Areas on Corrosion Rate for Copper-Zinc Couple*

Consider a copper-zinc corrosion couple. If the current density at the copper cathode is 0.05 A/cm^2, calculate the weight loss of zinc per hour if (1) the copper cathode area is 100 cm^2 and the zinc anode area is 1 cm^2 and (2) the copper cathode area is 1 cm^2 and the zinc anode area is 100 cm^2.

SOLUTION

1. For the small zinc anode area:

$$I = i_{Cu}A_{Cu} = (0.05 \text{ A/cm}^2)(100 \text{ cm}^2) = 5 \text{ A}$$

$$w_{Zn} = \frac{ItM}{nF} = \frac{(5)(3600)(65.38)}{(2)(96,500)} = 6.1 \text{ g/h}$$

2. For the large zinc anode area:

$$I = i_{Cu}A_{Cu} = (0.05 \text{ A/cm}^2)(1 \text{ cm}^2) = 0.05 \text{ A}$$

$$w_{Zn} = \frac{ItM}{nF} = \frac{(0.05 \text{ A/cm}^2)(3600 \text{ s})\left(65.38 \frac{\text{gm}}{\text{mol}}\right)}{(2)(96,500 \text{ C})} = 0.061 \text{ g/h}$$

The rate of corrosion of the zinc is reduced significantly when the zinc anode is much larger than the cathode.

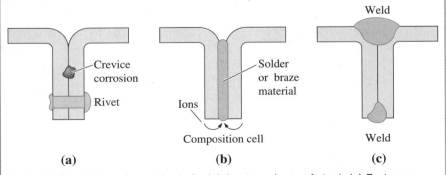

Figure 22-11 Alternative methods for joining two pieces of steel: (a) Fasteners may produce a concentration cell, (b) brazing or soldering may produce a composition cell, and (c) welding with a filler metal that matches the base metal may avoid the formation of galvanic cells.

3. Design components so that fluid systems are closed, rather than open, and so that stagnant pools of liquid do not collect. Partly filled tanks undergo water-line corrosion. Open systems continuously dissolve gas, providing ions that participate in the cathode reaction and encourage concentration cells.

4. Avoid crevices between assembled or joined materials (Figure 22-11). Welding may be a better joining technique than brazing, soldering, or mechanical fastening. Galvanic cells develop in brazing or soldering, since the filler metals have a different composition from the metal being joined. Mechanical fasteners produce crevices that lead to concentration cells. However, if the filler metal is closely matched to the base metal, welding may prevent these cells from developing.

5. In some cases, the rate of corrosion cannot be reduced to a level that will not interfere with the expected lifetime of the component. In such cases, the assembly should be designed in such a manner that the corroded part can easily and economically be replaced.

Coatings Coatings are used to isolate the anode and cathode regions. Coatings also prevent diffusion of oxygen or water vapor that initiate corrosion or oxidation. Temporary coatings, such as grease or oil, provide some protection but are easily disrupted. Organic coatings, such as paint, or ceramic coatings, such as enamel or glass, provide better protection. However, if the coating is disrupted, a small anodic site is exposed that undergoes rapid, localized corrosion.

Metallic coatings include tin-plated and hot-dip galvanized (zinc-plated) steel (Figure 22-12).[7] This was discussed in earlier chapters on ferrous materials. A continuous coating of either metal isolates the steel from the electrolyte. However, when the coating is scratched, exposing the underlying steel, the two coatings behave differently. The zinc continues to be effective, because zinc is anodic to steel. Since the area of the exposed steel cathode is small, the zinc coating corrodes at a very slow rate and the steel remains protected. However, steel is anodic to tin, so a tiny steel anode is created when tinplate is scratched and rapid corrosion of the steel subsequently occurs.

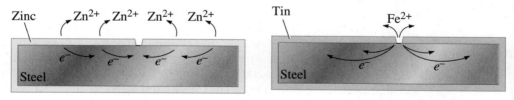

Figure 22-12 Zinc-plated steel and tin-plated steel are protected differently. Zinc protects steel even when the coating is scratched, since zinc is anodic to steel. Tin does not protect steel when the coating is disrupted, since steel is anodic with respect to tin.

Chemical conversion coatings are produced by a chemical reaction with the surface. Liquids such as zinc acid orthophosphate solutions form an adherent phosphate layer on the metal surface. The phosphate layer is, however, rather porous and is more often used to improve paint adherence. Stable, adherent, nonporous, non-conducting oxide layers form on the surface of aluminum, chromium, and stainless steel. These oxides exclude the electrolyte and prevent the formation of galvanic cells. Components such as reaction vessels can also be lined with corrosion-resistant Teflon™ or other plastics.

Inhibitors   When added to the electrolyte solution, some chemicals migrate preferentially to the anode or cathode surface and produce concentration or resistance polarization—that is, they are **inhibitors**. Chromate salts perform this function in automobile radiators. A variety of chromates, phosphates, molybdates, and nitrites produce protective films on anodes or cathodes in power plants and heat exchangers, thus stifling the electrochemical cell. Although the exact contents and the mechanism by which it works is not clear, the popular lubricant WD-40^{TM} works from the actions of inhibitors. The name "WD-40^{TM}" was designated water displacement (WD) that worked on the fortieth experimental attempt!

Cathodic Protection   We can protect against corrosion by supplying the metal with electrons and forcing the metal to be a cathode (Figure 22-13). Cathodic protection can use a sacrificial anode or an impressed voltage.

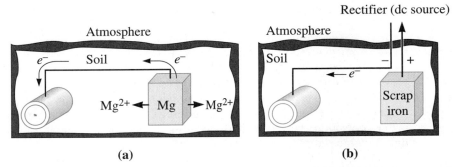

Figure 22-13   Cathodic protection of a buried steel pipeline: (a) A sacrificial magnesium anode assures that the galvanic cell makes the pipeline the cathode. (b) An impressed voltage between a scrap iron auxiliary anode and the pipeline assures that the pipeline is the cathode.

A **sacrificial anode** is attached to the material to be protected, forming an electrochemical circuit. The sacrificial anode corrodes, supplies electrons to the metal, and thereby prevents an anode reaction at the metal. The sacrificial anode, typically zinc or magnesium, is consumed and must eventually be replaced. Applications include preventing the corrosion of buried pipelines, ships, off-shore drilling platforms, and water heaters. A magnesium rod is used in many water heaters. The Mg serves as an anode and undergoes dissolution, thus protecting the steel from corroding.

An **impressed voltage** is obtained from a direct current source connected between an auxiliary anode and the metal to be protected. Essentially, we have connected a battery so that electrons flow to the metal, causing the metal to be the cathode. The auxiliary anode, such as scrap iron, corrodes.

Passivation or Anodic Protection   Metals near the anodic end of the galvanic series are active and serve as anodes in most electrolytic cells. However, if these metals are made passive or more cathodic, they corrode at slower rates than normal. **Passivation** is accomplished by producing strong anodic polarization, preventing the normal anode reaction; thus the term anodic protection.

We cause passivation by exposing the metal to highly concentrated oxidizing solutions. If iron is dipped in very concentrated nitric acid, the iron rapidly and uniformly corrodes to form a thin, protective iron hydroxide coating. The coating protects the iron from subsequent corrosion in nitric acid.

We can also cause passivation by increasing the potential on the anode above a critical level. A passive film forms on the metal surface, causing strong anodic polarization, and the current decreases to a very low level. Passivation of aluminum is called **anodizing**, and a thick oxide coating is produced. This oxide layer can be dyed to produce attractive colors. The Ta_2O_5 oxide layer formed on tantalum wires is used to make capacitors.

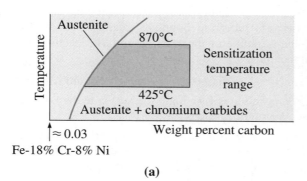

(a)

Figure 22-14
(a) Intergranular corrosion takes place in austenitic stainless steel. (b) Slow cooling permits chromium carbides to precipitate at grain boundaries. (c) A quench anneal to dissolve the carbides may prevent intergranular corrosion.

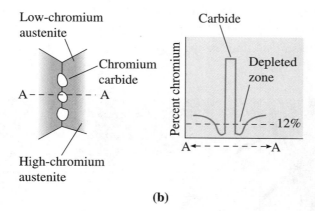

(b)

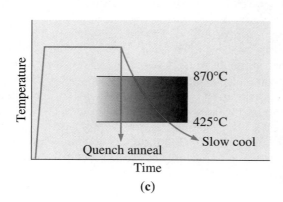

(c)

Materials Selection and Treatment Corrosion can be prevented or minimized by selecting appropriate materials and heat treatments. In castings, for example, segregation causes tiny, localized galvanic cells that accelerate corrosion. We can improve corrosion resistance with a homogenization heat treatment. When metals are formed into finished shapes by bending, differences in the amount of cold work and residual stresses cause local stress cells. These may be minimized by a stress-relief anneal or a full recrystallization anneal.

The heat treatment is particularly important in austenitic stainless steels (Figure 22-14). When the steel cools slowly from 870 to 425°C, chromium carbides precipitate at the grain boundaries. The austenite at the grain boundaries may contain less than 12% chromium, which is the minimum required to produce a passive oxide layer. The steel is **sensitized**. Because the grain boundary regions are small and highly anodic, rapid corrosion of the austenite at the grain boundaries occurs. We can minimize the problem by several techniques.

1. If the steel contains less than 0.03% C, the chromium carbides do not form.

2. If the percent chromium is very high, the austenite may not be depleted to below 12% Cr, even if the chromium carbides form.

3. Addition of titanium or niobium ties up the carbon as TiC or NbC, preventing the formation of chromium carbide. The steel is said to be **stabilized**.

4. The sensitization temperature range—425°C to 870°C—should be avoided during manufacture and service.

5. In a **quench anneal** heat treatment, the stainless steel is heated above 870°C, causing the chromium carbides to dissolve. The structure, now containing 100% austenite, is rapidly quenched to prevent formation of carbides.

The following examples illustrate the design of a corrosion protection system:

EXAMPLE 22-9   *Design of a Corrosion Protection System*

Steel troughs are located in a field to provide drinking water for a herd of cattle. The troughs frequently rust through and must be replaced. Design a system to prevent or delay this problem.

SOLUTION

The troughs are likely a low-carbon, unalloyed steel containing ferrite and cementite, producing a composition cell. The waterline in the trough, which is partially filled with water, provides a concentration cell. The trough is also exposed to the environment and the water is contaminated with impurities. Consequently, corrosion of the unprotected steel tank is to be expected.

Several approaches might be used to prevent or delay corrosion. We might, for example, fabricate the trough using stainless steel or aluminum. Either would provide better corrosion resistance than the plain carbon steel, but both are considerably more expensive than the current material.

We might suggest using cathodic protection; a small magnesium anode could be attached to the inside of the trough. The anode corrodes sacrificially and prevents corrosion of the steel. This would require that the farm operator regularly check the tank to be sure that the anode is not completely consumed. We also want to be sure that magnesium ions introduced into the water are not a health hazard.

Another approach would be to protect the steel trough using a suitable coating. Painting the steel (that is, introducing a protective polymer coating) and, using a tin-plated steel, provides protection as long as the coating is not disrupted.

The most likely approach is to use a galvanized steel, taking advantage of the protective coating and the sacrificial behavior of the zinc. Corrosion is very slow due to the large anode area, even if the coating is disrupted. Furthermore, the galvanized steel is relatively inexpensive, readily available, and does not require frequent inspection.

EXAMPLE 22-10 *Design of a Stainless-Steel Weldment*

A piping system used to transport a corrosive liquid is fabricated from 304 stainless steel. Welding of the pipes is required to assemble the system. Unfortunately, corrosion occurs and the corrosive liquid leaks from the pipes near the weld. Identify the problem and design a system to prevent corrosion in the future.

SOLUTION

Table 12-4 shows that the 304 stainless steel contains 0.08% C, causing the steel to be sensitized if it is improperly heated or cooled during welding. Figure 22-15 shows the maximum temperatures reached in the fusion and heat-affected zones during welding. A portion of the pipe in the HAZ heats into the sensitization temperature range, permitting chromium carbides to precipitate. If the cooling rate of the weld is very slow, the fusion zone and other areas of the

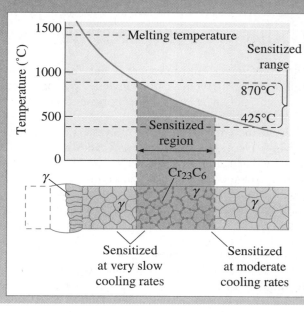

Figure 22-15
The peak temperature surrounding a stainless-steel weld and the sensitized structure produced when the weld slowly cools (for Example 22-10).

heat-affected zone may also be affected. Sensitization of the weld area, therefore, is the likely reason for corrosion of the pipe in the region of the weld.

Several solutions to the problem may be considered. We might use a welding process that provides very rapid rates of heat input, causing the weld to heat and cool very quickly. If the steel is exposed to the sensitization temperature range for only a brief time, chromium carbides may not precipitate. Joining processes such as laser welding or electron-beam welding are high-rate-of-heat-input processes, but they are expensive. In addition, electron beam welding requires the use of a vacuum, and it may not be feasible to assemble the piping system in a vacuum chamber.

We might heat treat the assembly after the weld is made. By performing a quench anneal, any precipitated carbides are re-dissolved during the anneal and do not re-form during quenching. However, it may be impossible to perform this treatment on a large assembly.

We might check the original welding procedure to determine if the pipe was preheated before joining in order to minimize the development of stresses due to the welding process. If the pipe were preheated, sensitization would be more likely to occur. We would recommend that any preheat procedure be suspended.

Perhaps our best design is to use a stainless steel that is not subject to sensitization. For example, carbides do not precipitate in a 304L stainless steel, which contains less than 0.03% C. The low-carbon stainless steels are more expensive than the normal 304 steel; however, the extra cost does prevent the corrosion process and still permits us to use conventional joining techniques.

22-7   Microbial Degradation and Biodegradable Polymers

Attack by a variety of insects and microbes is one form of "corrosion" of polymers. Relatively simple polymers (such as polyethylene, polypropylene, and polystyrene), high-molecular-weight polymers, crystalline polymers, and thermosets are relatively immune to attack.

However, certain polymers—including polyesters, polyurethanes, cellulosics, and plasticized polyvinyl chloride (which contains additives that reduce the degree of polymerization)—are particularly vulnerable to microbial degradation. These polymers can be broken into low-molecular-weight molecules by radiation or chemical attack until they are small enough to be ingested by the microbes.

We take advantage of microbial attack by producing *biodegradable* polymers, thus helping to remove the material from the waste stream. Biodegradation requires the complete conversion of the polymer to carbon dioxide, water, inorganic salts, and other small byproducts produced by the ingestion of the material by bacteria. Polymers such as cellulosics can easily be broken into molecules with low molecular weights and are therefore biodegradable. In addition, special polymers are produced to degrade rapidly; a copolymer of polyethylene and starch is one example. Bacteria attack the starch portion of the polymer and reduce the molecular weight of the remaining polyethylene.

22-8   Oxidation and Other Gas Reactions

Materials of all types may react with oxygen and other gases. These reactions can, like corrosion, alter the composition, properties, or integrity of a material. As mentioned before, metals such as Al and Ti react with oxygen very readily.

Oxidation of Metals   Metals may react with oxygen to produce an oxide at the surface. We are interested in three aspects of this reaction: the ease with which the metal oxidizes, the nature of the oxide film that forms, and the rate at which **oxidation** occurs.

The ease with which oxidation occurs is given by the standard free energy of formation for the oxide (Figure 22-16). This is known as an Ellingham diagram. There is a large driving force for the oxidation of magnesium and aluminum, but there is little tendency for the oxidation of nickel or copper. This is illustrated in the following example.

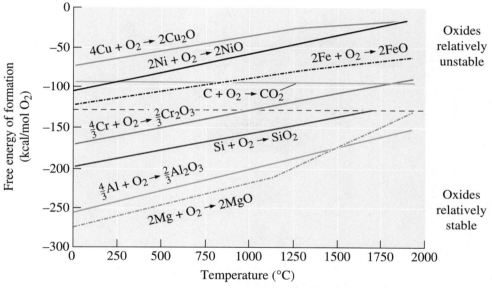

Figure 22-16   The standard free energy of formation of selected oxides as a function of temperature. A large negative free energy indicates a more stable oxide.

EXAMPLE 22-11   *Chromium-Based Steel Alloys*

Explain why we should not add alloying elements such as chromium to pig iron before the pig iron is converted to steel in a basic oxygen furnace at 1700°C.

SOLUTION

In a basic oxygen furnace, we lower the carbon content of the metal from about 4% to much less than 1% by blowing pure oxygen through the molten metal. If chromium were already present before the steel making began, chromium would oxidize before the carbon (Figure 22-16), since chromium oxide has a lower free energy of formation (or is more stable) than carbon dioxide (CO_2). Thus, any expensive chromium added would be lost before the carbon was removed from the pig iron.

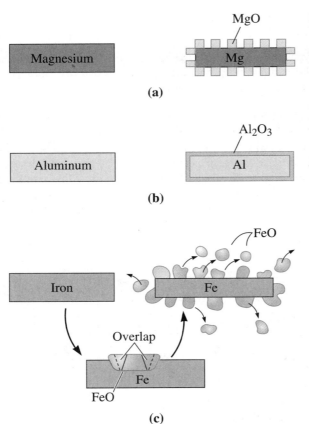

MgO

Magnesium

Mg

(a)

Aluminum

Al₂O₃

Al

(b)

FeO

Iron

Fe

Overlap

Fe

FeO

(c)

Figure 22-17
Three types of oxides may form, depending on the volume ratio between the metal and the oxide: (a) magnesium produces a porous oxide film, (b) aluminum forms a protective, adherent, nonporous oxide film, and (c) iron forms an oxide film that spalls off the surface and provides poor protection.

The type of oxide film influences the rate at which oxidation occurs (Figure 22-17). For the **oxidation reaction**

$$nM + mO_2 \rightarrow M_nO_{2m}, \tag{22-12}$$

the **Pilling-Bedworth (P-B) ratio** is:

$$\text{P-B ratio} = \frac{\text{oxide volume per metal atom}}{\text{metal volume per metal atom}} = \frac{(M_{\text{oxide}})(\rho_{\text{metal}})}{n(M_{\text{metal}})(\rho_{\text{oxide}})}, \tag{22-13}$$

where M is the atomic or molecular mass, ρ is the density, and n is the number of metal atoms in the oxide, as defined in Equation 22-12.

If the Pilling-Bedworth ratio is less than one, the oxide occupies a smaller volume than the metal from which it formed; the coating is therefore porous and oxidation continues rapidly—typical of metals such as magnesium. If the ratio is one to two, the volumes of the oxide and metal are similar and an adherent, nonporous, protective film forms—typical of aluminum and titanium. If the ratio exceeds two, the oxide occupies a large volume and may flake from the surface, exposing fresh metal that continues to oxidize—typical of iron. Although the Pilling-Bedworth equation historically has been used to characterize oxide behavior, many exceptions to this behavior are observed. The use of the P-B equation is illustrated in the following example.

EXAMPLE 22-12 *Pilling-Bedworth Ratio*

The density of aluminum is 2.7 g/cm^3 and that of Al_2O_3 is about 4 g/cm^3. Describe the characteristics of the aluminum-oxide film. Compare with the oxide film that forms on tungsten. The density of tungsten is 19.254 g/cm^3 and that of WO_3 is 7.3 g/cm^3.

SOLUTION

For $2Al + 3/2O_2 \rightarrow Al_2O_3$, the molecular weight of Al_2O_3 is 101.96 and that of aluminum is 26.981.

$$\text{P-B} = \frac{M_{Al_2O_3}\rho_{Al}}{nM_{Al}\rho_{Al_2O_3}} = \frac{\left(101.96\frac{gm}{mol}\right)(2.7 \text{ g/cm}^3)}{(2)\left(26.981\frac{gm}{mol}\right)(4 \text{ g/cm}^3)} = 1.28$$

For tungsten, $W + 3/2O_2 \rightarrow WO_3$. The molecular weight of WO_3 is 231.85 and that of tungsten is 183.85:

$$\text{P-B} = \frac{M_{WO_3}\rho_W}{nM_W\rho_{WO_3}} = \frac{\left(231.85\frac{gm}{mol}\right)(19.254 \text{ g/cm}^3)}{(1)\left(183.85\frac{gm}{mol}\right)(7.3 \text{ g/cm}^3)} = 3.33$$

Since P-B $\simeq$ 1 for aluminum, the Al_2O_3 film is nonporous and adherent, providing protection to the underlying aluminum. However, P-B > 2 for tungsten, so the WO_3 should be nonadherent and nonprotective.

The rate at which oxidation occurs depends on the access of oxygen to the metal atoms. A linear rate of oxidation occurs when the oxide is porous (as in magnesium) and oxygen has continued access to the metal surface:

$$y = kt, \tag{22-14}$$

where y is the thickness of the oxide, t is the time, and k is a constant that depends on the metal and temperature.

A parabolic relationship is observed when diffusion of ions or electrons through a nonporous oxide layer is the controlling factor (Chapter 5). This relationship is observed in iron, copper, and nickel:

$$y = \sqrt{kt} \tag{22-15}$$

Finally, a logarithmic relationship is observed for the growth of thin-oxide films that are particularly protective, as for aluminum and possibly chromium:

$$y = k \ln(ct + 1), \tag{22-16}$$

where k and c are constants for a particular temperature, environment, and composition. The example that follows shows the calculation for the time required for a nickel sheet to oxidize completely.

EXAMPLE 22-13 *Parabolic Oxidation Curve for Nickel*

At 1000°C, pure nickel follows a parabolic oxidation curve given by the constant $k = 3.9 \times 10^{-12}$ cm^2/s in an oxygen atmosphere. If this relationship is not affected by the thickness of the oxide film, calculate the time required for a 0.1-cm nickel sheet to oxidize completely.

SOLUTION

Assuming that the sheet oxidizes from both sides:

$$y = \sqrt{kt} = \sqrt{\left(3.9 \times 10^{-12} \frac{cm^2}{s}\right)(t\ s)} = \frac{0.1\ cm}{2\ sides} = 0.05\ cm$$

$$t = \frac{(0.05\ cm)^2}{3.9 \times 10^{-12}\ cm^2/s} = 6.4 \times 10^8\ s = 20.3\ years$$

Temperature also affects the rate of oxidation. In many metals, the rate of oxidation is controlled by the rate of diffusion of oxygen or metal ions through the oxide. If oxygen diffusion is more rapid, oxidation occurs between the oxide and the metal; if the metal ion diffusion is more rapid, oxidation occurs at the oxide-atmosphere interface. Consequently, we would expect oxidation rates to follow an Arrhenius relationship, increasing exponentially as the temperature increases.

Oxidation and Thermal Degradation of Polymers   Polymers degrade when heated and/or exposed to oxygen. A polymer chain may be ruptured, producing two macroradicals. In rigid thermosets, the macroradicals may instantly recombine (a process called the *cage* effect), resulting in no net change in the polymer. However, in the more flexible thermoplastics—particularly for amorphous rather than crystalline polymers—recombination does not occur and the result is a decrease in the molecular weight, the viscosity, and the mechanical properties of the polymer. Depolymerization continues as the polymer is exposed to the high temperature. Polymer chains can also *unzip*. In this case, individual monomers are removed one after another from the ends of the chain, gradually reducing the molecular weight of the remaining chains. As the degree of polymerization decreases, the remaining chains become more heavily branched or cyclization may occur. In *cyclization*, the two ends of the same chain may be bonded together to form a ring.

Polymers also degrade by the loss of side groups on the chain. Chloride ions (in polyvinyl chloride) and benzene rings (in polystyrene) are lost from the chain, forming byproducts. For example, as polyvinyl chloride is degraded, hydrochloric acid (HCl) is produced. Hydrogen atoms are bonded more strongly to the chains; thus, polyethylene does not degrade as easily as PVC or PS. Fluoride ions (in Teflon™) are more difficult to remove than hydrogen atoms, providing Teflon™ with its high-temperature resistance.

22-9   Wear and Erosion

Wear and erosion remove material from a component by mechanical attack of solids or liquids. Corrosion and mechanical failure also contribute to this type of attack.

Adhesive Wear   **Adhesive wear**—also known as scoring, galling, or seizing—occurs when two solid surfaces slide over one another under pressure. Surface projections, or asperities, are plastically deformed and eventually welded together by the high local pressures (Figure 22-18). As sliding continues, these bonds are broken, producing cavities on one surface, projections on the second surface, and frequently tiny, abrasive particles—all of which contribute to further wear of the surfaces.

Bond

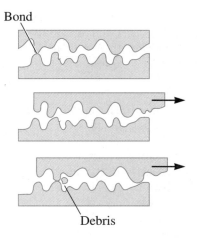

Debris

Many factors may be considered in trying to improve the wear resistance of materials. Designing components so that loads are small and surfaces are smooth, and continual lubrication is possible helps prevent adhesions that cause the loss of material.

The properties and microstructure of the material are also important. Normally, if both surfaces have high hardnesses, the wear rate is low. High strength, to help resist the applied loads, and good toughness and ductility, which prevent the tearing of material from the surface, may be beneficial. Ceramic materials, with their exceptional hardness, are expected to provide good adhesive wear resistance.

Wear resistance of polymers can be improved if the coefficient of friction is reduced by the addition of polytetrafluoroethylene (Teflon^{TM}) or if the polymer is strengthened by the introduction of reinforcing fibers such as glass, carbon, or aramid.

Abrasive Wear   When material is removed from a surface by contact with hard particles, **abrasive wear** occurs. The particles either may be present at the surface of a second material or may exist as loose particles between two surfaces (Figure 22-19). This type of wear is common in machinery such as plows, scraper blades, crushers, and grinders used to handle abrasive materials and may also occur when hard particles are unintentionally introduced into moving parts of machinery. Abrasive wear is also used for grinding operations to remove material intentionally. In many automotive applications (e.g., dampers, gears, pistons, and cylinders) abrasive wear behavior is a major concern.[9]

Materials with a high hardness, good toughness, and high hot strength are most resistant to abrasive wear. Typical materials used for abrasive-wear applications include quenched and tempered steels; carburized or surface-hardened steels; cobalt alloys such

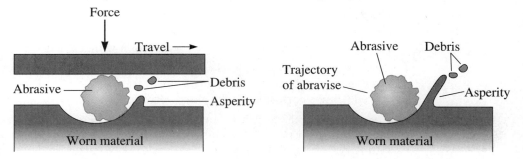

Figure 22-19   Abrasive wear, caused by either trapped or free-flying abrasives, produces troughs in the material, piling up asperities that may fracture into debris.

as Stellite; composite materials, including tungsten carbide cermets; white cast irons; and hard surfaces produced by welding. Most ceramic materials also resist wear effectively because of their high hardness; however, their brittleness may sometimes limit their usefulness in abrasive wear conditions.

Liquid Erosion   The integrity of a material may be destroyed by erosion caused by high pressures associated with a moving liquid. The liquid causes strain hardening of the metal surface, leading to localized deformation, cracking, and loss of material. Two types of liquid erosion deserve mention.

Cavitation occurs when a liquid containing a dissolved gas enters a low-pressure region. Gas bubbles, which precipitate and grow in the liquid in the low pressure environment, collapse when the pressure subsequently increases. The high-pressure, local shock wave that is produced by the collapse may exert a pressure of thousands of atmospheres against the surrounding material. Cavitation is frequently encountered in propellors, dams and spillways, and hydraulic pumps.

Liquid impingement occurs when liquid droplets carried in a rapidly moving gas strike a metal surface. High localized pressures develop because of the initial impact and the rapid lateral movement of the droplets from the impact point along the metal surface. Water droplets carried by steam may erode turbine blades in steam generators and nuclear power plants.

Liquid erosion can be minimized by proper materials selection and design. Minimizing the liquid velocity, assuring that the liquid is deaerated, selection of hard, tough materials to absorb the impact of the droplets, and coating the material with an energy-absorbing elastomer all may help minimize erosion.

SUMMARY

◆ About 6% of the Gross Domestic Product in the United States is used in combating corrosion. Corrosion causes deterioration of all types of materials. Designers and engineers must know how corrosion occurs in order to consider suitable designs, materials selection, or protective measures.

◆ In chemical corrosion, a material is dissolved in a solvent, resulting in the loss of material. All materials—metals, ceramics, polymers, and composites—are subject to this form of attack. Choice of appropriate materials having a low solubility in a given solvent or use of inert coatings on materials helps avoid or reduce chemical corrosion.

◆ Electrochemical corrosion requires that a complete electric circuit develop. The anode corrodes and a byproduct such as rust forms at the cathode. Because an electric circuit is required, this form of corrosion is most serious in metals and alloys.

◆ Composition cells are formed by the presence of two different metals, two different phases within a single alloy, or even segregation within a single phase.

◆ Stress cells form when the level of residual or applied stresses vary within the metal; the regions subjected to the highest stress are the anode and, consequently, they corrode. Stress corrosion cracking and corrosion fatigue are examples of stress cells.

◆ Concentration cells form when a metal is exposed to a nonuniform electrolyte; for example, the portion of a metal exposed to the lowest oxygen content corrodes. Pitting corrosion, waterline corrosion, and crevice corrosion are examples of these cells. Microbial corrosion, in which colonies of organisms such as bacteria grow on the metal surface, is another example of a concentration cell.

◈ Electrochemical corrosion can be minimized or prevented by using electrical insulators to break the electric circuit, designing and manufacturing assemblies without crevices, designing assemblies so that the anode area is unusually large compared with that of the cathode, using protective and even sacrificial coatings, inhibiting the action of the electrolyte, supplying electrons to the metal by means of an impressed voltage, using heat treatments that reduce residual stresses or segregation, and a host of other actions.

◈ Oxidation degrades most materials. While an oxide coating provides protection for some metals such as aluminum, most materials are attacked by oxygen. Diffusion of oxygen and metallic atoms is often important; therefore oxidation occurs most rapidly at elevated temperatures.

◈ Other factors, such as attack by microbes, damage caused by radiation, and wear or erosion of a material, may also cause a material to deteriorate.

GLOSSARY

Abrasive wear   Removal of material from surfaces by the cutting action of particles.

Adhesive wear   Removal of material from surfaces of moving equipment by momentary local bonding, then bond fracture, at the surfaces.

Anode   The location at which corrosion occurs as electrons and ions are given up in an electrochemical cell.

Anodizing   An anodic protection technique in which a thick oxide layer is deliberately produced on a metal surface.

Cathode   The location at which electrons are accepted and a byproduct is produced during corrosion.

Cavitation   Erosion of a material surface by the pressures produced when a gas bubble collapses within a moving liquid.

Chemical corrosion   Removal of atoms from a material by virtue of the solubility or chemical reaction between the material and the surrounding liquid.

Composition cells   Electrochemical corrosion cells produced between two materials having a different composition. Also known as galvanic cells.

Concentration cells   Electrochemical corrosion cells produced between two locations on a material at which the composition of the electrolyte is different.

Crevice corrosion   A special concentration cell in which corrosion occurs in crevices because of the low concentration of oxygen.

Dezincification   A special chemical corrosion process by which both zinc and copper atoms are removed from brass, but the copper is replated back onto the metal.

Electrochemical cell   A cell in which electrons and ions can flow by separate paths between two materials, producing a current which, in turn, leads to corrosion or plating.

Electrochemical corrosion   Corrosion produced by the development of a current in an electrochemical cell that removes ions from the material.

Electrode potential   Related to the tendency of a material to corrode. The potential is the voltage produced between the material and a standard electrode.

Electrolyte   The conductive medium through which ions move to carry current in an electrochemical cell.

emf series   The arrangement of elements according to their electrode potential, or their tendency to corrode.

Faraday's equation   The relationship that describes the rate at which corrosion or plating occurs in an electrochemical cell.

Galvanic series   The arrangement of alloys according to their tendency to corrode in a particular environment.

Graphitic corrosion   A special chemical corrosion process by which iron is leached from cast iron, leaving behind a weak, spongy mass of graphite.

Impressed voltage   A cathodic protection technique by which a direct current is introduced into the material to be protected, thus preventing the anode reaction.

Inhibitors   Additions to the electrolyte that preferentially migrate to the anode or cathode, cause polarization, and reduce the rate of corrosion.

Intergranular corrosion   Corrosion at grain boundaries because grain boundary segregation or precipitation produces local galvanic cells.

Liquid impingement   Erosion of a material caused by the impact of liquid droplets carried by a gas stream.

Nernst equation   The relationship that describes the effect of electrolyte concentration on the electrode potential in an electrochemical cell.

Oxidation   Reaction of a metal with oxygen to produce a metallic oxide. This normally occurs most rapidly at high temperatures.

Oxidation reaction   The anode reaction by which electrons are given up to the electrochemical cell.

Oxygen starvation   In the concentration cell, low-oxygen regions of the electrolyte cause the underlying material to behave as the anode and to corrode.

Passivation   Producing strong anodic polarization by causing a protective coating to form on the anode surface and to thereby interrupt the electric circuit.

Pilling-Bedworth ratio   Describes the type of oxide film that forms on a metal surface during oxidation.

Polarization   Changing the voltage between the anode and cathode to reduce the rate of corrosion. *Activation* polarization is related to the energy required to cause the anode or cathode reaction; *concentration* polarization is related to changes in the composition of the electrolyte; and *resistance* polarization is related to the electrical resistivity of the electrolyte.

Quench anneal   The heat treatment used to dissolve carbides and prevent intergranular corrosion in stainless steels.

Reduction reaction   The cathode reaction by which electrons are accepted from the electrochemical cell.

Sacrificial anode   Cathodic protection by which a more anodic material is connected electrically to the material to be protected. The anode corrodes to protect the desired material.

Sensitization   Precipitation of chromium carbides at the grain boundaries in stainless steels, making the steel sensitive to intergranular corrosion.

Stabilization   Addition of titanium or niobium to a stainless steel to prevent intergranular corrosion.

Stress cells   Electrochemical corrosion cells produced by differences in imposed or residual stresses at different locations in the material.

Stress corrosion Deterioration of a material in which an applied stress accelerates the rate of corrosion.

Tubercule Accumulations of microbial organisms and corrosion byproducts on the surface of a material.

ADDITIONAL INFORMATION

1. KOCH, G.H., et al., "Corrosion Cost and Preventive Strategies in the United States," *CC Technologies and NACE International*, 2001.
2. USBORNE, D., "US Nuclear Plant was Close to Disaster," *The Independent*, 2002.
3. JONES, D.A., *Principles and Prevention of Corrosion*. 1995: Prentice Hall.
4. REVIE, R. and H.H. UHLIG, *Uhlig's Corrosion Handbook*. 2nd ed. 2000: Wiley Interscience.
5. ROBERGE, P.R., *Handbook of Corrosion Engineering*. 1999: McGraw Hill.
6. BABU, S.V., K.C. CADIEN, and H. YANO, "Chemical-Mechanical Polishing 2001, Advances and Future Challenges," *MRS Proceedings*, 2001.
7. YOUNG, C. and B. DUGAN, "Zinc Corporation of America," *Technical discussions and assistance are acknowledged*, 2002.
8. BURLEIGH, D., New Mexico Tech. University, *Technical discussions are acknowledged*, 2002.
9. GANGOPADHYAY, A. and MCWATT, D., Ford Motor Company, *Technical discussions are acknowledged*, 2002.

PROBLEMS

Section 22-1 Chemical Corrosion

Section 22-2 Electrochemical Corrosion

Section 22-3 The Electrode Potential in Electrochemical Cells

Section 22-4 The Corrosion Current and Polarization

22-1 A gray cast iron pipe is used in the natural gas distribution system for a city. The pipe fails and leaks, even though no corrosion noticeable to the naked eye has occurred. Offer an explanation for why the pipe failed.

22-2 A brass plumbing fitting produced from a Cu-30% Zn alloy operates in the hot water system of a large office building. After some period of use, cracking and leaking occur. On visual examination, no metal appears to have been corroded. Offer an explanation for why the fitting failed.

22-3 Suppose 10 g of Sn^{2+} are dissolved in 1000 ml of water to produce an electrolyte. Calculate the electrode potential of the tin half-cell.

22-4 A half-cell produced by dissolving copper in water produces an electrode potential of $+0.32$ V. Calculate the amount of copper that must have been added to 1000 ml of water to produce this potential.

22-5 An electrode potential in a platinum half-cell is 1.10 V. Determine the concentration of Pt^{4+} ions in the electrolyte.

22-6 A current density of 0.05 A/cm^2 is applied to a 150-cm^2 cathode. What period of time is required to plate out a 1-mm-thick coating of silver onto the cathode?

22-7 We wish to produce 100 g of platinum per hour on a 1000 cm^2 cathode by electroplating. What plating current density is required? Determine the current required.

22-8 A 1-m-square steel plate is coated on both sides with a 0.005-cm-thick layer of zinc. A current density of 0.02 A/cm^2 is applied to the plate in an aqueous solution. Assuming that the zinc corrodes uniformly, determine the length of time required before the steel is exposed.

22-9 A 2-in.-inside-diameter, 12-ft-long copper distribution pipe in a plumbing system is accidentally connected to the power system of a manufacturing plant, causing a current of 0.05 A to flow through the pipe. If the wall thickness of the pipe is 0.125 in., estimate the time required before the pipe begins to leak, assuming a uniform rate of corrosion.

22-10 A steel surface 10 cm × 100 cm is coated with a 0.002-cm-thick layer of chromium. After one year of exposure to an electrolytic cell, the chromium layer is completely removed. Calculate the current density required to accomplish this removal.

22-11 A corrosion cell is composed of a 300 cm^2 copper sheet and a 20 cm^2 iron sheet, with a current density of 0.6 A/cm^2 applied to the copper. Which material is the anode? What is the rate of loss of metal from the anode per hour?

22-12 A corrosion cell is composed of a 20 cm^2 copper sheet and a 400 cm^2 iron sheet, with a current density of 0.7 A/cm^2 applied to the copper. Which material is the anode? What is the rate of loss of metal from the anode per hour?

Section 22-6 Protection Against Electrochemical Corrosion

Section 22-7 Microbial Degradation and Biodegradable Polymers

22-13 Alclad is a laminar composite composed of two sheets of commercially pure aluminum (alloy 1100) sandwiched around a core of 2024 aluminum alloy. Discuss the corrosion resistance of the composite. Suppose that a portion of one of the 1100 layers was machined off, exposing a small patch of the 2024 alloy. How would this affect the corrosion resistance? Explain. Would there be a difference in behavior if the core material were 3003 aluminum? Explain.

22-14 The leaf springs for an automobile are formed from a high-carbon steel. For best corrosion resistance, should the springs be formed by hot working or cold working? Explain. Would corrosion still occur even if you use the most desirable forming process? Explain.

22-15 Several types of metallic coatings are used to protect steel, including zinc, lead, tin, cadmium, aluminum, and nickel. In which of these cases will the coating provide protection even when the coating is locally disrupted? Explain.

22-16 An austenitic stainless steel corrodes in all of the heat-affected zone (HAZ) surrounding the fusion zone of a weld. Explain why corrosion occurs and discuss the type of welding process

or procedure that might have been used. What might you do to prevent corrosion in this region?

22-17 A steel is securely tightened onto a bolt in an industrial environment. After several months, the nut is found to contain numerous cracks, even though no externally applied load acts on the nut. Explain why cracking might have occurred.

22-18 The shaft for a propeller on a ship is carefully designed so that the applied stresses are well below the endurance limit for the material. Yet after several months, the shaft cracks and fails. Offer an explanation for why failure might have occurred under these conditions.

22-19 An aircraft wing composed of carbon fiber-reinforced epoxy is connected to a titanium forging on the fuselage. Will the anode for a corrosion cell be the carbon fiber, the titanium, or the epoxy? Which will most likely be the cathode? Explain.

22-20 The inside surface of a cast iron pipe is covered with tar, which provides a protective coating. Acetone in a chemical laboratory is drained through the pipe on a regular basis. Explain why, after several weeks, the pipe begins to corrode.

22-21 A cold-worked copper tube is soldered, using a lead-tin alloy, into a steel connector. What types of electrochemical cells might develop due to this connection? Which of the materials would you expect to serve as the anode and suffer the most extensive damage due to corrosion? Explain.

22-22 Pure tin is used to provide a solder connection for copper in many electrical uses. Which metal will most likely act as the anode?

22-23 Sheets of annealed nickel, cold-worked nickel, and recrystallized nickel are placed into an electrolyte. Which would be most likely to corrode? Which would be least likely to corrode? Explain.

22-24 A pipeline carrying liquid fertilizer crosses a small creek. A large tree washes down the creek and is wedged against the steel pipe. After some time, a hole is produced in the pipe at the point where the tree touches the pipe, with the diameter of the hole larger on the outside of the pipe than on the inside of the pipe. The pipe then leaks fertilizer into the creek. Offer an explanation for why the pipe corroded.

22-25 Two sheets of a 1040 steel are joined together with an aluminum rivet (Figure 22-20). Discuss

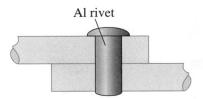

Figure 22-20 Two steel sheets joined by an aluminum rivet (for Problem 22-25).

	Oxide Density (g/cm^3)
Mg-MgO	3.60
Na-Na$_2$O	2.27
Ti-TiO$_2$	5.10
Fe-Fe$_2$O$_3$	5.30
Ce-Ce$_2$O$_3$	6.86
Nb-Nb$_2$O$_5$	4.47
W-WO$_3$	7.30

the possible corrosion cells that might be created as a result, of this joining process. Recommend a joining process that might minimize some for these cells.

22-26 Figure 22-21 shows a cross-section through an epoxy-encapsulated integrated circuit, including a small microgap between the copper lead frame and the epoxy polymer. Suppose chloride ions from the manufacturing process penetrate the package. What types of corrosion cells might develop? What portions of the integrated circuit are most likely to corrode?

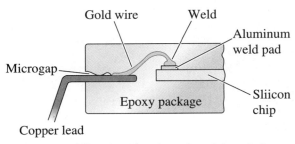

Figure 22-21 Cross-section through an integrated circuit showing the external lead connection to the chip (for Problem 22-26).

22-27 A current density of 0.1 A/cm^2 is applied to the iron in an iron-zinc corrosion cell. Calculate the weight loss of zinc per hour

(a) if the zinc has a surface area of 10 cm^2 and the iron has a surface area of 100 cm^2, and

(b) if the zinc has a surface area of 100 cm^2 and the iron has a surface area of 10 cm^2.

Section 22-8 Oxidation and Other Gas Reactions

Section 22-9 Wear and Erosion

22-28 Determine the Pilling-Bedworth ratio for the following metals and predict the behavior of the oxide that forms on the surface. Is the oxide protective, does it flake off the metal, or is it permeable? (See Appendix A for the metal density.)

22-29 Oxidation of most ceramics is not considered to be a problem. Explain.

22-30 A sheet of copper is exposed to oxygen at 1000°C. After 100 h, 0.246 g of copper are lost per cm^2 of surface area; after 250 h, 0.388 g/cm^2 are lost; and after 500 h, 0.550 g/cm^2 are lost. Determine whether oxidation is parabolic, linear, or logarithmic, then determine the time required for a 0.75-cm sheet of copper to be completely oxidized. The sheet of copper is oxidized from both sides.

22-31 At 800°C, iron oxidizes at a rate of 0.014 g/cm^2 per hour; at 1000°C, iron oxidizes at a rate of 0.0656 g/cm^2 per hour. Assuming a parabolic oxidation rate, determine the maximum temperature at which iron can be held if the oxidation rate is to be less than 0.005 g/cm^2 per hour.

✳ Design Problems

22-32 A cylindrical steel tank 3 ft in diameter and 8 ft long is filled with water. We find that a current density of 0.015 A/cm^2 acting on the steel is required to prevent corrosion. Design a sacrificial anode system that will protect the tank.

22-33 The drilling platforms for offshore oil rigs are supported on large steel columns resting on the bottom of the ocean. Design an approach to assure that corrosion of the supporting steel columns does not occur.

22-34 A storage building is to be produced using steel sheet for the siding and roof. Design a corrosion-protection system for the steel.

22-35 Design the materials for the scraper blade for a piece of earthmoving equipment.

▪ Computer Problems

22-36 Write a computer program that will calculate the thickness of a coating deposited using different inputs provided by the user (e.g., density and valence of the metal, area or dimensions of the part being plated, desired time of plating, and current).

Appendix A: Selected Physical Properties of Metals

Metal		Atomic Number	Crystal Structure	Lattice Parameters (Å)	Atomic Mass g/mol	Density (g/cm³)	Melting Temperature (°C)
Aluminum	Al	13	FCC	4.04958	26.981	2.699	660.4
Antimony	Sb	51	hex	$a = 4.307$ $c = 11.273$	121.75	6.697	630.7
Arsenic	As	33	hex	$a = 3.760$ $c = 10.548$	74.9216	5.778	816
Barium	Ba	56	BCC	5.025	137.3	3.5	729
Beryllium	Be	4	hex	$a = 2.2858$ $c = 3.5842$	9.01	1.848	1290
Bismuth	Bi	83	hex	$a = 4.546$ $c = 11.86$	208.98	9.808	271.4
Boron	B	5	rhomb	$a = 10.12$ $\alpha = 65.5°$	10.81	2.3	2300
Cadmium	Cd	48	HCP	$a = 2.9793$ $c = 5.6181$	112.4	8.642	321.1
Calcium	Ca	20	FCC	5.588	40.08	1.55	839
Cerium	Ce	58	HCP	$a = 3.681$ $c = 11.857$	140.12	6.6893	798
Cesium	Cs	55	BCC	6.13	132.91	1.892	28.6
Chromium	Cr	24	BCC	2.8844	51.996	7.19	1875
Cobalt	Co	27	HCP	$a = 2.5071$ $c = 4.0686$	58.93	8.832	1495
Copper	Cu	29	FCC	3.6151	63.54	8.93	1084.9
Gadolinium	Gd	64	HCP	$a = 3.6336$ $c = 5.7810$	157.25	7.901	1313
Gallium	Ga	31	ortho	$a = 4.5258$ $b = 4.5186$ $c = 7.6570$	69.72	5.904	29.8
Germanium	Ge	32	FCC	5.6575	72.59	5.324	937.4
Gold	Au	79	FCC	4.0786	196.97	19.302	1064.4
Hafnium	Hf	72	HCP	$a = 3.1883$ $c = 5.0422$	178.49	13.31	2227
Indium	In	49	tetra	$a = 3.2517$ $c = 4.9459$	114.82	7.286	156.6
Iridium	Ir	77	FCC	3.84	192.9	22.65	2447
Iron	Fe	26	BCC FCC BCC	2.866 3.589	55.847 (>912°C) (>1394°C)	7.87	1538
Lanthanum	La	57	HCP	$a = 3.774$ $c = 12.17$	138.91	6.146	918
Lead	Pb	82	FCC	4.9489	207.19	11.36	327.4
Lithium	Li	3	BCC	3.5089	6.94	0.534	180.7

Metal		Atomic Number	Crystal Structure	Lattice Parameters (Å)	Atomic Mass g/mol	Density (g/cm³)	Melting Temperature (°C)
Magnesium	Mg	12	HCP	$a = 3.2087$ $c = 5.209$	24.312	1.738	650
Manganese	Mn	25	cubic	8.931	54.938	7.47	1244
Mercury	Hg	80	rhomb		200.59	13.546	−38.9
Molybdenum	Mo	42	BCC	3.1468	95.94	10.22	2610
Nickel	Ni	28	FCC	3.5167	58.71	8.902	1453
Niobium	Nb	41	BCC	3.294	92.91	8.57	2468
Osmium	Os	76	HCP	$a = 2.7341$ $c = 4.3197$	190.2	22.57	2700
Palladium	Pd	46	FCC	3.8902	106.4	12.02	1552
Platinum	Pt	78	FCC	3.9231	195.09	21.45	1769
Potassium	K	19	BCC	5.344	39.09	0.855	63.2
Rhenium	Re	75	HCP	$a = 2.760$ $c = 4.458$	186.21	21.04	3180
Rhodium	Rh	45	FCC	3.796	102.99	12.41	1963
Rubidium	Rb	37	BCC	5.7	85.467	1.532	38.9
Ruthenium	Ru	44	HCP	$a = 2.6987$ $c = 4.2728$	101.07	12.37	2310
Selenium	Se	34	hex	$a = 4.3640$ $c = 4.9594$	78.96	4.809	217
Silicon	Si	14	FCC	5.4307	28.08	2.33	1410
Silver	Ag	47	FCC	4.0862	107.868	10.49	961.9
Sodium	Na	11	BCC	4.2906	22.99	0.967	97.8
Strontium	Sr	38	FCC BCC	6.0849 4.84 (>557°C)	87.62	2.6	768
Tantalum	Ta	73	BCC	3.3026	180.95	16.6	2996
Technetium	Tc	43	HCP	$a = 2.735$ $c = 4.388$	98.9062	11.5	2200
Tellurium	Te	52	hex	$a = 4.4565$ $c = 5.9268$	127.6	6.24	449.5
Thorium	Th	90	FCC	5.086	232	11.72	1775
Tin	Sn	50	FCC	6.4912	118.69	5.765	231.9
Titanium	Ti	22	HCP BCC	$a = 2.9503$ $c = 4.6831$ 3.32 (>882°C)	47.9	4.507	1668
Tungsten	W	74	BCC	3.1652	183.85	19.254	3410
Uranium	U	92	ortho	$a = 2.854$ $b = 5.869$ $c = 4.955$	238.03	19.05	1133
Vanadium	V	23	BCC	3.0278	50.941	6.1	1900
Yttrium	Y	39	HCP	$a = 3.648$ $c = 5.732$	88.91	4.469	1522
Zinc	Zn	30	HCP	$a = 2.6648$ $c = 4.9470$	65.38	7.133	420
Zirconium	Zr	40	HCP BCC	$a = 3.2312$ $c = 5.1477$ 3.6090 (>862°C)	91.22	6.505	1852

Appendix B: The Atomic and Ionic Radii of Selected Elements

Element	Atomic Radius (Å)	Valence	Ionic Radius (Å)
Aluminum	1.432	+3	0.51
Antimony	1.45	+5	0.62
Arsenic	1.15	+5	2.22
Barium	2.176	+2	1.34
Beryllium	1.143	+2	0.35
Bismuth	1.60	+5	0.74
Boron	0.46	+3	0.23
Bromine	1.19	−1	1.96
Cadmium	1.49	+2	0.97
Calcium	1.976	+2	0.99
Carbon	0.77	+4	0.16
Cerium	1.84	+3	1.034
Cesium	2.65	+1	1.67
Chlorine	0.905	−1	1.81
Chromium	1.249	+3	0.63
Cobalt	1.253	+2	0.72
Copper	1.278	+1	0.96
Fluorine	0.6	−1	1.33
Gallium	1.218	+3	0.62
Germanium	1.225	+4	0.53
Gold	1.442	+1	1.37
Hafnium	1.55	+4	0.78
Hydrogen	0.46	+1	1.54
Indium	1.570	+3	0.81
Iodine	1.35	−1	2.20
Iron	1.241 (BCC)	+2	0.74
	1.269 (FCC)	+3	0.64
Lanthanum	1.887	+3	1.016
Lead	1.75	+4	0.84
Lithium	1.519	+1	0.68
Magnesium	1.604	+2	0.66
Manganese	1.12	+2	0.80
		+3	0.66
Mercury	1.55	+2	1.10
Molybdenum	1.363	+4	0.70
Nickel	1.243	+2	0.69
Niobium	1.426	+4	0.74
Nitrogen	0.71	+5	0.15

Element	Atomic Radius (Å)	Valence	Ionic Radius (Å)
Oxygen	0.60	−2	1.32
Palladium	1.375	+4	0.65
Phosphorus	1.10	+5	0.35
Platinum	1.387	+2	0.80
Potassium	2.314	+1	1.33
Rubidium	2.468	+1	0.70
Selenium	1.15	−2	1.91
Silicon	1.176	+4	0.42
Silver	1.445	+1	1.26
Sodium	1.858	+1	0.97
Strontium	2.151	+2	1.12
Sulfur	1.06	−2	1.84
Tantalum	1.43	+5	0.68
Tellurium	1.40	−2	2.11
Thorium	1.798	+4	1.02
Tin	1.405	+4	0.71
Titanium	1.475	+4	0.68
Tungsten	1.371	+4	0.70
Uranium	1.38	+4	0.97
Vanadium	1.311	+3	0.74
Yttrium	1.824	+3	0.89
Zinc	1.332	+2	0.74
Zirconium	1.616	+4	0.79

Note that 1 Å = 10^{-8} cm = 0.1 nanometer (nm)

Appendix C: The Electronic Configuration for Each of the Elements

Atomic Number	Element	K 1s	L 2s	L 2p	M 3s	M 3p	M 3d	N 4s	N 4p	N 4d	N 4f	O 5s	O 5p	O 5d	P 6s	P 6p
1	Hydrogen	1														
2	Helium	2														
3	Lithium	2	1													
4	Beryllium	2	2													
5	Boron	2	2	1												
6	Carbon	2	2	2												
7	Nitrogen	2	2	3												
8	Oxygen	2	2	4												
9	Fluorine	2	2	5												
10	Neon	2	2	6												
11	Sodium	2	2	6	1											
12	Magnesium	2	2	6	2											
13	Aluminum	2	2	6	2	1										
14	Silicon	2	2	6	2	2										
15	Phosphorus	2	2	6	2	3										
16	Sulfur	2	2	6	2	4										
17	Chlorine	2	2	6	2	5										
18	Argon	2	2	6	2	6										
19	Potassium	2	2	6	2	6		1								
20	Calcium	2	2	6	2	6		2								
21	Scandium	2	2	6	2	6	1	2								
22	Titanium	2	2	6	2	6	2	2								
23	Vanadium	2	2	6	2	6	3	2								
24	Chromium	2	2	6	2	6	5	1								
25	Manganese	2	2	6	2	6	5	2								
26	Iron	2	2	6	2	6	6	2								
27	Cobalt	2	2	6	2	6	7	2								
28	Nickel	2	2	6	2	6	8	2								
29	Copper	2	2	6	2	6	10	1								
30	Zinc	2	2	6	2	6	10	2								
31	Gallium	2	2	6	2	6	10	2	1							
32	Germanium	2	2	6	2	6	10	2	2							
33	Arsenic	2	2	6	2	6	10	2	3							
34	Selenium	2	2	6	2	6	10	2	4							
35	Bromine	2	2	6	2	6	10	2	5							

Atomic Number	Element	K 1s	L		M			N				O			P	
			2s	2p	3s	3p	3d	4s	4p	4d	4f	5s	5p	5d	6s	6p
36	Krypton	2	2	6	2	6	10	2	6							
37	Rubidium	2	2	6	2	6	10	2	6		1					
38	Strontium	2	2	6	2	6	10	2	6		2					
39	Yttrium	2	2	6	2	6	10	2	6	1	2					
40	Zirconium	2	2	6	2	6	10	2	6	2	2					
41	Niobium	2	2	6	2	6	10	2	6	4	1					
42	Molybdenum	2	2	6	2	6	10	2	6	5	1					
43	Technetium	2	2	6	2	6	10	2	6	6	1					
44	Ruthenium	2	2	6	2	6	10	2	6	7	1					
45	Rhodium	2	2	6	2	6	10	2	6	8	1					
46	Palladium	2	2	6	2	6	10	2	6	10						
47	Silver	2	2	6	2	6	10	2	6	10		1				
48	Cadmium	2	2	6	2	6	10	2	6	10		2				
49	Indium	2	2	6	2	6	10	2	6	10		2	1			
50	Tin	2	2	6	2	6	10	2	6	10		2	2			
51	Antimony	2	2	6	2	6	10	2	6	10		2	3			
52	Tellurium	2	2	6	2	6	10	2	6	10		2	4			
53	Iodine	2	2	6	2	6	10	2	6	10		2	5			
54	Xenon	2	2	6	2	6	10	2	6	10		2	6			
55	Cesium	2	2	6	2	6	10	2	6	10		2	6		1	
56	Barium	2	2	6	2	6	10	2	6	10		2	6		2	
57	Lanthanum	2	2	6	2	6	10	2	6	10	1	2	6		2	
	⋮	⋮	⋮	⋮	⋮	⋮	⋮	⋮	⋮	⋮	⋮	⋮	⋮		⋮	
71	Lutetium	2	2	6	2	6	10	2	6	10	14	2	6	1	2	
72	Hafnium	2	2	6	2	6	10	2	6	10	14	2	6	2	2	
73	Tantalum	2	2	6	2	6	10	2	6	10	14	2	6	3	2	
74	Tungsten	2	2	6	2	6	10	2	6	10	14	2	6	4	2	
75	Rhenium	2	2	6	2	6	10	2	6	10	14	2	6	5		
76	Osmium	2	2	6	2	6	10	2	6	10	14	2	6	6		
77	Iridium	2	2	6	2	6	10	2	6	10	14	2	6	9		
78	Platinum	2	2	6	2	6	10	2	6	10	14	2	6	9	1	
79	Gold	2	2	6	2	6	10	2	6	10	14	2	6	10	1	
80	Mercury	2	2	6	2	6	10	2	6	10	14	2	6	10	2	
81	Thallium	2	2	6	2	6	10	2	6	10	14	2	6	10	2	1
82	Lead	2	2	6	2	6	10	2	6	10	14	2	6	10	2	2
83	Bismuth	2	2	6	2	6	10	2	6	10	14	2	6	10	2	3
84	Polonium	2	2	6	2	6	10	2	6	10	14	2	6	10	2	4
85	Astatine	2	2	6	2	6	10	2	6	10	14	2	6	10	2	5
86	Radon	2	2	6	2	6	10	2	6	10	14	2	6	10	2	6

Answers to Selected Problems

CHAPTER 2

2-6 6.69×10^{21} atoms.

2-7 9.79×10^{27} atoms/ton.

2-8   (a) 5.98×10^{23} atoms. (b) 0.994 mol.

2-10   3.

2-12 142×10^{23} carriers.

2-19   0.086.

2-32 Al_2O_3, with strong ionic bonding, has lower coefficient.

2-34   Weak secondary bonds between chains.

CHAPTER 3

3-25   (a) 1.426×10^{-8} cm. (b) 1.4447×10^{-8} cm.

3-27   (a) 5.3355 Å. (b) 2.3103 Å.

3-29   FCC.

3-31   BCT.

3-33   (a) 8 atoms/cell. (b) 0.387.

3-39   0.6% contraction.

3-51 A: $[00\bar{1}]$. B: $[1\bar{2}0]$. C: $[\bar{1}11]$. D: $[2\bar{1}\bar{1}]$.

3-53 A: $(1\bar{1}1)$. B: (030). C: $(10\bar{2})$.

3-55 A: $[1\bar{1}0]$ or $[1\bar{1}00]$. B: $[11\bar{1}]$ or $[11\bar{2}3]$. C: $[011]$ or $[\bar{1}2\bar{1}3]$.

3-57 A: $(1\bar{1}01)$. B: (0003). C: $(1\bar{1}00)$.

3-63 $[\bar{1}10]$, $[1\bar{1}0]$, $[101]$, $[\bar{1}0\bar{1}]$, $[011]$, $[0\bar{1}\bar{1}]$.

3-65   Tetragonal—4; orthorhombic—2; cubic—12.

3-67   (a) (111). (c) $(0\bar{1}2)$.

3-69   [100]: 0.35089 nm, 2.85 nm^{-1}, 0.866. [110]: 0.496 nm, 2.015 nm^{-1}, 0.612. [111]: 0.3039 nm, 3.291 nm^{-1}, 1.

3-71   (100): 1.617×10^{-17}/cm^2, 0.7854. (110): 1.144×10^{-17}/cm^2, 0.555. (111): 1.867×10^{-17}/cm^2, 0.907.

3-73   4,563,000.

3-79   (a) 0.2797 Å. (b) 0.629 Å.

3-87   (a) 6. (c) 8. (e) 4. (g) 6.

3-89   Fluorite. (a) 5.2885 Å. (b) 12.13 g/cm^3. (c) 0.624.

3-91   Cesium chloride. (a) 4.1916 Å. (b) 4.8 g/cm^3. (c) 0.693.

3-93   (111): 0.202 (Mg). (222): 0.806 (O).

3-101   8 SiO_2, 8 Si ions, 16 O ions.

3-105   0.40497 nm.

3-107   (a) BCC. (c) 0.23 nm.

CHAPTER 4

4-1 5.1×10^{19} vacancies/cm^3.

4-3   (a) 0.002375. (b) 1.61×10^{20} vacancies/cm^3.

4-5   (a) 1.157×10^{20} vacancies/cm^3. (b) 0.532 g/cm^3.

4-7   0.345.

4-9   8.265 g/cm^3.

4-11   (a) 0.0081. (b) one H atom per 123 unit cells.

4-13   (a) 0.0534 defects/unit cell. (b) 2.52×10^{20} defects/cm^3.

4-22   (a) $[0\bar{1}1]$, $[01\bar{1}]$, $[\bar{1}10]$, $[1\bar{1}0]$, $[\bar{1}01]$, $[10\bar{1}]$.

4-24 $(0\bar{1}1)$, $(01\bar{1})$, $(1\bar{1}0)$, $(\bar{1}10)$, $(10\bar{1})$, $(\bar{1}01)$.

4-26   Expected: $\mathbf{b} = 2.863$ Å, $d = 2.338$ Å. (110)[111]: $\mathbf{b} = 7.014$ Å, $d = 2.863$ Å. Ratio $= 0.44$.

4-49   (a) $K = 19.4$ psi/$\sqrt{\text{m}}$, $\sigma_0 = 60{,}300$ psi. (b) 103,680 psi.

4-51   (a) 128 grains/in^2. (b) 1,280,000 grains/in^2.

4-53   3.6.

4-60   284 Å.

CHAPTER 5

5-12 1.08×10^9 jumps/s.

5-24   (a) 59,230 cal/mol. (b) 0.055 cm^2/s.

5-42   (a) -0.02495 at% Sb/cm. (b) -1.246×10^{19} Sb/cm$^3 \cdot$ cm.

5-44   (a) -1.969×10^{11} H atoms/cm$^3 \cdot$ cm. (b) 3.3×10^7 H atoms/cm$^2 \cdot$ s.

5-47   0.001245 g/h.

5-48 $-198°$C.

5-60 $D_0 = 3.47 \times 10^{-16}$ cm^2/s versus $D_{Al} = 2.48 \times 10^{-13}$ cm^2/s.

5-62 $D_H = 1.07 \times 10^{-4}$ cm^2/s versus $D_N = 3.9 \times 10^{-9}$ cm^2/s.

5-66   0.01 cm: 0.87% C. 0.05 cm: 0.43% C. 0.10 cm: 0.18% C.

5-68   907°C.

5-70   0.53% C.

5-72   2.9 min.

5-74   12.8 min.

5-86   667°C.

5-88   51,000 cal/mol; yes.

CHAPTER 6

6-24   (a) Deforms. (b) Does not neck.

6-26   1100 lb or 4891 N.

6-28   20,000 lb.

6-30   1.995 in.

6-32   50.0543 ft.

6-36   (a) 11,600 psi. (b) 12,729 psi. (c) 603,000 psi. (d) 4.5%. (e) 3.5%. (f) 11,297 psi. (g) 11,706 psi. (h) 76.4 psi.

6-38 (a) 274 MPa. (b) 417 MPa. (c) 172 GPa. (d) 18.55%. (e) 15.8%. (f) 397.9 MPa. (g) 473 MPa. **(h)** 0.17 MPa.

6-42 $l = 12.00298$ in. $d = 0.39997$ in.

6-44 (a) 76,800 psi. (b) 22.14×10^6 psi.

6-46 41 mm; will not fracture.

6-56 29.8.

6-61 No transition temperature.

6-65 Not notch-sensitive; poor toughness.

6-71 No; stress required to propagate the crack is 20 times greater than the tensile strength.

6-73 0.99 MPa $\sqrt{m}$.

6-89 15.35 lb.

6-91 $d = 1.634$ in.

6-93 22 MPa; max $= +22$ MPa, min $= -22$ MPa, mean $= 0$ MPa; reduce fatigue strength due to heating.

6-94 (a) 2.5 mm. (b) 0.0039 mm.

6-97 $C = 2.047 \times 10^{-3} \cdot n = 3.01$.

6-106 101,660 h.

6-108 $n = 6.86$. $m = -6.9$.

6-110 29 days.

6-112 0.52 in.

6-114 2000 lb.

CHAPTER 7

7-5 $n = 0.12$; BCC.

7-9 $n = 0.15$.

7-14 Total $= 0.0109$ mm, % increase $= 32,700\%$.

7-19 0.152 in.

7-22 26,000 psi tensile, 22,000 psi yield, 5% elongation.

7-24 First step: 36% CW giving 26,000 psi tensile, 23,000 psi yield, 6% elongation. Second step: 64% CW giving 30,000 psi tensile, 27,000 psi yield, 3% elongation. Third step: 84% CW giving 32,000 psi tensile, 29,000 psi yield, 2% elongation.

7-26 0.78 to 0.96 in.

7-28 48% CW: 28,000 psi tensile, 25,000 psi yield, 4% elongation.

7-43 (a) 1414 lb. (b) Will not break.

7-53 Increase grain growth temperature and keep grain size small.

7-55 (a) 550°C, 750°C, 950°C. (b) 700°C. (c) 900°C. (d) 2285°C.

7-66 Slope $= 0.4$.

7-68 CW 75% from 2 to 1 in., anneal. CW 75% from 1 to 0.5 in., anneal. CW 72.3% from 0.5 to 0.263 in., anneal. CW 42% from 0.263 to 0.2 in. or hot work 98.3% from 2 to 0.263 in., then CW 42% from 0.263 to 0.2 in.

CHAPTER 8

8-10 (a) 6.65 Å. (b) 109 atoms.

8-12 1.136×10^6 atoms.

8-14 (a) 0.0333. (b) 0.333. (c) All.

8-29 1265°C.

8-32 31.15 s.

8-34 $B = 305$ s/cm^2, $n = 1.58$.

8-36 (a) 4.16×10^{-3} cm. (b) 90 s.

8-38 $c = 0.0032$ s, $m = 0.34$.

8-40 0.023 s.

8-46 (a) 900°C. (b) 420°C. (c) 480°C. (d) 312°C/min. (e) 9.7 min. (f) 8.1 min. (g) 60°C. (h) Zinc. (i) 87.3 min/in^2.

8-48 $D = 6.67$ in., $H = 10$ in., $V = 349$ in^3.

8-56 V/A (riser) $= 0.68$, V/A (middle) $= 1.13$, V/A (end) $= 0.89$; not effective.

8-58 $D_{Cu} = 1.48$ in. $D_{Fe} = 1.30$ in.

8-60 (a) 46 cm^3. (b) 4.1%.

8-65 23.04 cm.

8-67 0.046 cm^3/100 g Al.

CHAPTER 9

9-34 (a) Yes. (c) No. (e) No. (g) No.

9-36 *Cd* should give smallest decrease in conductivity; none should give unlimited solid solubility.

9-41 (a) 2330°C, 2150°C, 180°C, (c) 2570°C, 2380°C, 190°C.

9-43 (a) 100% L containing 30% MgO. (b) 70.8% L containing 38% MgO, 29.2% S containing 62% MgO. (c) 8.3% L containing 38% MgO, 91.7% S containing 62% MgO. (d) 100% S containing 85% MgO.

9-45 44.1 at% Cu − 55.9 at% Al.

9-47 (a) *L*: 15 mol% MgO or 8.69 wt% MgO. S: 38 mol% MgO or 24.85 wt% MgO. (b) 78.26 mol% L or 80.1 wt% L; 21.74 mol% S or 19.9 wt% MgO. (c) 78.1 vol% L, 21.9 vol% S.

9-49 750 g Ni, Ni/Cu $= 1.62$.

9-51 332 g MgO.

9-53 64.1 wt% FeO.

9-55 (a) 49 wt% W in L, 70 wt% W in α. (b) Not possible.

9-57 212 lb W; 1200 lb W.

9-59 Ni dissolves. When liquid reaches 10% Ni, the bath begins to freeze.

9-63 (a) 2900°C, 2690°C, 210°C. (b) 60% L containing 49% W, 40% α containing 70% W.

9-65 (a) 55% W. (b) 18% W.

9-69 (a) 2900°C. (b) 2710°C. (c) 190°C. (d) 2990°C. (e) 90°C. (f) 300 s. (g) 340 s. (h) 60% W.

9-71 (a) 2000°C. (b) 1450°C. (c) 550°C. (d) 40% FeO. (e) 92% FeO. (f) 65.5% L containing 75% FeO, 34.5% S containing 46% FeO. **(g)** 30.3% L containing 88% FeO, 69.7% S containing 55% FeO.

9-72 **(a)** 3100°C. **(b)** 2720°C. **(c)** 380°C. **(d)** 90% W. **(e)** 40% W. **(f)** 44.4% L containing 70% W, 55.6% α containing 88% W. **(g)** 9.1% L containing 50% W, 90.9% α containing 83% W.

CHAPTER 10

10-22 **(a)** θ. **(b)** α, β, γ, η. **(c)** 1100°C: peritectic. 900°C: monotectic. 680°C: eutectic. 600°C: peritectoid. 300°C: eutectoid.

10-24 **(a)** $CuAl_2$. **(b)** 548°C, eutectic, L $\to \alpha + \theta$, 33.2% Cu in L, 5.65% Cu in α, 52.5% Cu in θ.

10-26 $SnCu_3$.

10-28 3 solid phases.

10-32 **(a)** 2.5% Mg. **(b)** 600°C, 470°C, 400°C, 130°C. **(c)** 74% α containing 7% Mg, 26% L containing 26% Mg. **(d)** 100% α containing 12% Mg. **(e)** 67% α containing 1% Mg, 33% β containing 34% Mg.

10-34 **(a)** Hypereutectic. **(b)** 98% Sn. **(c)** 22.8% β containing 97.5% Sn, 77.2% L containing 61.9% Sn. **(d)** 35% α containing 19% Sn, 65% β containing 97.5% Sn. **(e)** 22.8% primary β containing 97.5% Sn, 77.2% eutectic containing 61.9% Sn. **(f)** 30% α containing 2% Sn, 70% β containing 100% Sn.

10-36 **(a)** Hypoeutectic. **(b)** 1% Si. **(c)** 78.5% α containing 1.65% Si, 21.5% L containing 12.6% Si. **(d)** 97.6% α containing 1.65% Si, 2.4% β containing 99.83% Si. **(e)** 78.5 primary α containing 1.65% Si, 21.5% eutectic containing 12.6% Si. **(f)** 96% α containing 0% Si, 4% β containing 100% Si.

10-38 Hypoeutectic.

10-40 52% Sn.

10-42 **(a)** 4% Li. **(c)** 3% Cu.

10-44 Hypereutectic. **(b)** 64% α, 36% β.

10-46 0.54.

10-48 **(a)** 1150°C. **(b)** 150°C. **(c)** 1000°C. **(d)** 577°C. **(e)** 423°C. **(f)** 10.5 min. **(g)** 11.5 min. **(h)** 45% Si.

10-52 **(a)** Yes, $T_m = 2040$°C > 1900°C. **(b)** No, forms 5% L.

10-68 **(a)** 390°C, γ, $\gamma + \alpha$. **(b)** 330°C, β, $\alpha + \beta$. **(c)** 290°C, β, $\alpha + \beta + \gamma$.

CHAPTER 11

11-2 $c = 6.47 \times 10^{-6}$, $n = 2.89$.

11-4 59,525 cal/mol.

11-17 For Al $-$ 4% Mg: solution treat between 210 and 451°C, quench, age below 210°C. For Al $-$ 12% Mg: solution treat between 390 and 451°C, quench, age below 390°C.

11-38 **(a)** Solution treat between 290 and 400°C, quench, age below 290°C. **(c)** Not good candidate. **(e)** Not good candidate.

11-52 **(a)** 795°C. **(b)** Primary ferrite. **(c)** 56.1% ferrite containing 0.0218% C and 43.9% austenite containing 0.77% C. **(d)** 95.1% ferrite containing 0.0218% C and 4.9% cementite containing 6.67% C. **(e)** 56.1% primary ferrite containing 0.0218% C and 43.9% pearlite containing 0.77% C.

11-54 0.53% C, hypoeutectoid.

11-56 0.156% C, hypoeutectoid.

11-58 0.281% C.

11-60 760°C, 0.212% C.

11-62 **(a)** 900°C: 12% CaO in tetragonal, 3% CaO in monoclinic, 14% CaO in cubic; 18% monoclinic and 82% cubic. **(c)** 250°C; 47% Zn in β', 36% Zn in α, 59% Zn in γ; 52.2% α, 47.8% γ.

11-72 **(a)** 615°C. **(b)** 1.67×10^{-5} cm.

11-74 Bainite with HRC 47.

11-76 Martensite with HRC 66.

11-78 37.2% martensite with 0.77% C and HRC 65. **(c)** 84.8% martensite with 0.35% C and HRC 58.

11-87 **(a)** 750°C. **(b)** 0.455% C.

11-89 3.06% expansion.

11-91 Austenitize at 750°C, quench, temper above 330°C.

CHAPTER 12

12-4 **(a)** 97.2% ferrite, 2.2% cementite, 82.9% primary ferrite, 17.1% pearlite. **(c)** 85.8% ferrite, 14.2% cementite, 3.1% primary cementite, 96.9% pearlite.

12-6 For 1035: $A_1 = 727$°C; $A_3 = 790$°C; anneal = 820°C; normalize = 845°C; process anneal = 557–647°C; not usually spheroidized.

12-11 **(a)** Ferrite and pearlite. **(c)** Martensite. **(e)** Ferrite and bainite. **(g)** Tempered martensite.

12-13 **(a)** Austenitize at 820°C, hold at 600°C for 10 s, cool. **(c)** Austenitize at 780°C, hold at 600°C for 10 s, cool. **(e)** Austenitize at 900°C, hold at 320°C for 5000 s, cool.

12-16 **(a)** Austenitize at 820°C, quench, temper between 420 and 480°C; 150,000 to 180,000 psi tensile, 140,000 to 160,000 psi yield. **(b)** 175,000 to 180,000 psi tensile, 130,000 to 135,000 psi yield. **(c)** 100,000 psi tensile, 65,000 yield, 20% elongation.

12-18 0.48% C in martensite; austenitized at 770°C; should austenitize at 860°C.

12-20 1080: fine pearlite. 4340: martensite.

12-26 May become hypereutectoid, with grain boundary cementite.

12-28 Not applicable. **(c)** 8 to 10°C/s. **(e)** 32 to 36°C/s.

12-30 **(a)** 16°C/s. **(b)** Pearlite with HRC 38.

12-32 **(a)** Pearlite with HRC 36. **(c)** Pearlite and martensite with HRC 46.

12-34 **(a)** 1.3 in. **(c)** 1.9 in. **(e)** greater than 2.5 in.

12-39   0.25 h.

12-43   0.05 mm: pearlite and martensite with HRC 53. 0.15 mm: medium pearlite with HRC 38.

12-48 δ-ferrite; nonequilibrium freezing; quench anneal.

12-52   2.4% Si.

12-53   Ductile iron is most hardenable; steel is least hardenable.

CHAPTER 13

13-3   Eutectic microconstituent contains 97.6% β.

13-9   27% β versus 2.2% β.

13-11   Al $-$ 10% Mg.

13-13   (a) 0.113 in., 0.0151 lb, $0.021. (b) 0.113 in., 0.0233 lb, $0.014.

13-17   Al: 440%. Mg: 130%. Cu:1100%.

13-19   Lead may melt during hot working.

13-21 γ' more at low temperature.

13-23   Large formed first at higher temperature; solubility decreases as temperature decreases.

13-25   Ti-15% V: 100% β transforms to 100% α', which then transforms to 24% β precipitate in an α matrix. Ti-35% V: 100% β transforms to 100% β', which then transforms to 27% α precipitate in a β matrix.

13-28   Al: 7.5×10^5 in. Cu: 5.5×10^5 in. Ni: 3.4×10^5 in.

13-31   Spalls off; cracks.

CHAPTER 14

14-43 $B = 2.4$; true $= 22.58\%$; fraction $= 0.044$.

14-55   1.257 kg BaO; 0.245 kg Li_2O.

14-57 PbO_2 is modifier; PbO is intermediate.

CHAPTER 15

15-8   (a) 1.598 kg. (b) 1.649 kg. (c) 4.948 kg.

15-10   (a) H_2O. (b) 26.77 kg, 5.81 kg, 30.96 kg.

15-13   (a) 2500. (b) 2.4×10^{18}.

15-15   (a) 4798. (b) 9597.

15-16   186.69 g/mol.

15-17   (a) 211. (b) 175.

15-22   Polybutadiene and silicone.

15-24   Polyethylene and polypropylene.

15-26   4 repeat units; 8 C atoms, 12 H atoms, 4 Cl atoms.

15-28   74.2%.

15-33   1246 psi.

15-35   (a) PE. (b) LDPE. (c) PTFE.

15-39   At $\varepsilon = 1$, $E = 833$ psi; at $\varepsilon = 4$, $E = 2023$ psi.

15-41   0.0105.

15-46   6.383 kg; 3.83 kg.

15-51   After.

CHAPTER 16

16-7 7.65×10^{13} per cm^3.

16-9   2.47%.

16-13   9.408 g/cm^3.

16-15   (a) 0.507. (b) 0.507. (c) 7.775 g/cm^3.

16-17   11.18 to 22.2 kg.

16-24   (a) 2.53 g/cm^3. (b) 29×10^6 psi. (c) 15.3×10^6 psi.

16-26   0.964.

16-28   188 MPa.

16-30   For $d = 20$ μm, $l_c = 0.30$ cm, $l_c/d = 150$.

16-36   Sizing improves strength.

16-41   Pyrolize at 2500°C; 250,000 psi.

16-44 $E_{parallel} = 10.03 \times 10^6$ psi; $E_{perpendicular} = 2.96 \times 10^6$ psi.

16-46 $E_{parallel} = 11.86 \times 10^6$ psi; $E_{perpendicular} = 10 \times 10^6$ psi.

16-48   0.417 g/cm^3; 20.0 kg versus 129.6 kg.

CHAPTER 17

17-1   (a) 0.26 gal. (b) 0.77 g/cm^3.

17-3   Expands 83.9 in perpendicular to boards.

17-5   (a) 640 sacks, 53.7 tons sand, 110.9 tons aggregate, 2100 gal. (b) 4070 lb/yd^3. (c) 1 : 1.79 : 3.69.

CHAPTER 18

18-1   (a) 3380 W. (b) 2.546×10^{14} W. (c) 1.273×10^{10} to 1.273×10^{11} W.

18-5 $d = 0.0865$ cm; 1174 V.

18-7   0.968.

18-9   0.234 km/h; 0.0146 miles/h.

18-12 3.03×10^5 $ohm^{-1} \cdot cm^{-1}$ at $-50°C$; 0.34×10^5 $ohm^{-1} \cdot cm^{-1}$ at 500°C.

18-14 $-70.8°C$.

18-16   At 400°C, $\rho = 41.5 \times 10^{-6}$ ohm $\cdot$ cm; $\rho_d = 8.5 \times 10^{-6}$ ohm $\cdot$ cm; $b = 178.9 \times 10^{-6}$ ohm $\cdot$ cm. At 200°C, $\rho = 37.6 \times 10^{-6}$ ohm $\cdot$ cm.

18-19   330,160 Oe.

18-21   39.3 A.

18-23 $\mu_{500} = 7.3 \times 10^{-30}$ $cm^2/V \cdot s$; $\sigma_{500} = 1.66 \times 10^{-25}$ $ohm^{-1} \cdot cm^{-1}$.

18-33   (a) $n(Ge) = 1.767 \times 10^{23}$ per cm^3. (b) $f(Ge) = 1.259 \times 10^{-10}$. (c) $n_0(Ge) = 1.017 \times 10^{19}$ per cm^3.

18-35   (a) 8.32×10^{-8} s. (b) 5.75×10^{-7} s.

18-37   Sb: 15.2 $ohm^{-1} \cdot cm^{-1}$. In: 3.99 $ohm^{-1} \cdot cm^{-1}$.

18-39   37.54 $ohm^{-1} \cdot cm^{-1}$.

18-41   (a) 1.485×10^{20} per cm^3. (b) 4280 $ohm^{-1} \cdot cm^{-1}$.

18-43 2600%.

18-50 (a) 4.85×10^{-19} m. (c) 1.12×10^{-17} m.

18-52 9.4 V.

18-59 0.001238 μF.

18-63 0.0155 in.

18-65 0.0003 cm/cm.

CHAPTER 19

19-6 Fe: 39,600 Oe. Co: 31,560 Oe.

19-12 6400 Oe.

19-14 8000 G; 1.49 mA.

19-29 (a) 13,000 G. (b) 14,000 G. (c) 800 Oe. (d) 5.8 G/Oe. (e) 15.6 G/Oe. (f) 6.8×10^6 G Oe.

19-36 15×10^5 G $\cdot$ Oe.

19-37 High saturation inductance.

19-40 $0.468 \text{A} \cdot \text{m}^2/\text{cm}^3$.

CHAPTER 20

20-10 Ice, water, Teflon.

20-12 4%; 0.36%.

20-14 (a) 6.60°. (b) 6.69°. (c) 10°. (d) 0.178 cm.

20-16 0.0117 cm.

20-25 1.84×10^{-5} s.

20-30 2.333 eV; green.

20-32 (a) 13.11×10^{-5} cm. (b) 77.58×10^{-5} cm.

20-34 (a) 1.853×10^{-4} cm. (b) 1.034×10^{-2} cm.

20-37 13,790 V.

20-39 7477 V.

20-41 (a) 24,825 V. (b) Cu, Mn, Si.

20-43 At 0.711 Å, $I = 8.7$; at 1.436 Å, $I = 45 \times 10^{-6}$.

CHAPTER 21

21-3 (a) 1900 cal, 7950 J. (c) 8500 cal, 35.564 J.

21-5 0.0375 cm.

21-7 1.975 m.

21-9 15,182 in.

21-11 Brass: 10.0246 in., Invar; 10.0020 in.

21-18 256,200 psi; tensile.

21-20 No; 24,510 psi; 0.0098 in. decrease in length.

21-22 (a) 78.5×10^6 cal/day. (b) 47.09×10^6 cal/day.

21-24 19.6 min.

21-27 Interconnected graphite flakes in gray iron.

CHAPTER 22

22-1 Graphitic corrosion.

22-3 -0.172 V.

22-5 0.000034 g/1000 ml.

22-7 55 A.

22-9 34 years.

22-11 187.5 g Fe lost/h.

22-13 1100 alloy is anode and continues to protect 2024; 1100 alloy is cathode and the 3003 will corrode.

22-15 Al, Zn, Cd.

22-17 Stress corrosion cracking.

22-19 Ti is the anode, carbon will be the cathode.

22-23 Cold worked will corrode most rapidly, annealed most slowly.

22-27 (a) 12.2 g/h. (b) 1.22 g/h.

22-29 Most ceramics are already oxides.

22-31 698°C.

Index

License Agreement for Wadsworth Group, a division of Thomson Learning, Inc.

You the customer, and Wadsworth Group incur certain benefits, rights, and obligations to each other when you open this package and use the materials it contains. BE SURE TO READ THE LICENSE AGREEMENT CAREFULLY, SINCE BY USING THE SOFTWARE YOU INDICATE YOU HAVE READ, UNDERSTOOD, AND ACCEPTED THE TERMS OF THIS AGREEMENT.

Your rights:

1. You enjoy a non-exclusive license to use the enclosed materials on a single computer that is not part of a network or multi-machine system in consideration of the payment of the required license fee, (which may be included in the purchase price of an accompanying print component), and your acceptance of the terms and conditions of this agreement.
2. You own the disk on which the program/data is recorded, but you acknowledge that you do not own the program/data recorded on the disk. You also acknowledge that the program/data is furnished "AS IS," and contains copyrighted and/or proprietary and confidential information of Wadsworth Group, a division of Thomson Learning, Inc.
3. If you do not accept the terms of this license agreement you must not install the disk and you must return the disk within 30 days of receipt with proof of payment to Wadsworth Group for full credit or refund.

There are limitations on your rights:

1. You may not copy or print the program/data for any reason whatsoever, except to install it on a hard drive on a single computer, unless copying or printing is expressly permitted in writing or statements recorded on the disk.
2. You may not revise, translate, convert, disassemble, or otherwise reverse engineer the program/data.
3. You may not sell, license, rent, loan, or otherwise distribute or network the program/data.
4. You may not export or re-export the disk, or any component thereof, without the appropriate U.S. or foreign government licenses. Should you fail to abide by the terms of this license or otherwise violate Wadsworth Group's rights, your license to use it will become invalid. You agree to destroy the disk immediately after receiving notice of Wadsworth Group's termination of this agreement for violation of its provisions.

U.S. Government Restricted Rights

The enclosed multimedia, software, and associated documentation are provided with RESTRICTED RIGHTS. Use, duplication, or disclosure by the Government is subject to restrictions as set forth in subdivision (c)(1)(ii) of the Rights in Technical Data and Computer Software clause at DFARS 252.277.7013 for DoD contracts, paragraphs (c) (1) and (2) of the Commercial Computer Software-Restricted Rights clause in the FAR (48 CFR 52.227-19) for civilian agencies, or in other comparable agency clauses. The proprietor of the enclosed multimedia, software, and associated documentation is Wadsworth Group, 10 Davis Drive, Belmont, California 94002.

Limited Warranty

Wadsworth Group also warrants that the optical media on which the Product is distributed is free from defects in materials and workmanship under normal use. Wadsworth Group will replace defective media at no charge, provided you return the Product to Wadsworth Group within 90 days of delivery to you as evidenced by a copy of your invoice. If failure of disc(s) has resulted from accident, abuse, or misapplication, Wadsworth Group shall have no responsibility to replace the disc(s). THESE ARE YOUR SOLE REMEDIES FOR ANY BREACH OF WARRANTY.

EXCEPT AS SPECIFICALLY PROVIDED ABOVE, WADSWORTH GROUP, A DIVISION OF THOMSON LEARNING, INC. AND THE THIRD PARTY SUPPLIERS MAKE NO WARRANTY OR REPRESENTATION, EITHER EXPRESSED OR IMPLIED, WITH RESPECT TO THE PRODUCT, INCLUDING ITS QUALITY, PERFORMANCE, MERCHANTABILITY, OR FITNESS FOR A PARTICULAR PURPOSE. The product is not a substitute for human judgment. Because the software is inherently complex and may not be completely free of errors, you are advised to validate your work. IN NO EVENT WILL WADSWORTH GROUP OR ANY THIRD PARTY SUPPLIERS BE LIABLE FOR DIRECT, INDIRECT, SPECIAL, INCIDENTAL, OR CONSEQUENTIAL DAMAGES ARISING OUT OF THE USE OR INABILITY TO USE THE PRODUCT OR DOCUMENTATION, even if advised of the possibility of such damages. Specifically, Wadsworth Group is not responsible for any costs including, but not limited to, those incurred as a result of lost profits or revenue, loss of use of the computer program, loss of data, the costs of recovering such programs or data, the cost of any substitute program, claims by third parties, or for other similar costs. In no case shall Wadsworth Group's liability exceed the amount of the license fee paid. THE WARRANTY AND REMEDIES SET FORTH ABOVE ARE EXCLUSIVE AND IN LIEU OF ALL OTHERS, ORAL OR WRITTEN, EXPRESS OR IMPLIED. Some states do not allow the exclusion or limitation of implied warranties or limitation of liability for incidental or consequential damage, so that the above limitations or exclusion may not apply to you.

This license is the entire agreement between you and Wadsworth Group and it shall be interpreted and enforced under California law. Should you have any questions concerning this License Agreement, write to Technology Department, Wadsworth Group, 10 Davis Drive, Belmont, California 94002.